Residuals and Biosolids 2017 Conference

Seattle, Washington, USA
8 - 11 April 2017

Volume 1 of 2

ISBN: 978-1-5108-6297-5

Printed from e-media with permission by:

Curran Associates, Inc.
57 Morehouse Lane
Red Hook, NY 12571

CURRAN ASSOCIATES INC. proceedings .com

Some format issues inherent in the e-media version may also appear in this print version.

Printed by Curran Associates, Inc. (2018)

Water Environment Federation (WEF)
601 Wythe Street
Alexandria, Virginia 22314
USA

Phone: 1-800-666-0206
Fax: 1-703-684-1545

csc@wef.org

Additional copies of this publication are available from:

Curran Associates, Inc.
57 Morehouse Lane
Red Hook, NY 12571 USA
Phone: 845-758-0400
Fax: 845-758-2633
Email: curran@proceedings.com
Web: www.proceedings.com

TABLE OF CONTENTS

VOLUME 1

SESSION 1

SESSION 2

SESSION 3

SESSION 4

SESSION 5

SESSION 6

SESSION 7

SESSION 8

SESSION 9

SESSION 10

VOLUME 2

SESSION 11

SESSION 12

SESSION 13

SESSION 14

SESSION 15

SESSION 17

SESSION 18

SESSION 20

ADDITIONAL PAPERS

Understanding the Role of Mixing and Viscosity in Rapid Volume Expansion Due to Gas Holdup in Anaerobic Digesters

Nicholas Bartek[1], Matthew J. Higgins[1], Sudhir N. Murthy[2], Steven Beightol[1], Ahmed Al-Omari[2]
[1]Bucknell University, Lewisburg, PA; [2]DC Water, Washington, DC

ABSTRACT

The objective of this research was to better understand the role of mixing and viscosity in rapid volume expansion (RVE) due to gas holdup in anaerobic digesters. Most RVE research has focused on the absence of mixing, which is the worst-case scenario for anaerobic digestion facilities; however, comparatively little work has investigated the effects of different mixing intensities on RVE. To prevent the retention of biogas bubbles in suspension, commonly known as gas holdup, bubbles must rise through digester contents and evolve into the headspace of the digester at a rate equivalent to the nucleation and growth of biogas bubbles. As bubble rise velocity slows, the rate at which bubbles are nucleating and growing will exceed the rate at which they are evolving into the headspace. This will lead to a larger overall volume of biogas retained in suspension resulting in gas holdup and RVE. Due to the shear thinning and non-Newtonian behavior of digester contents, the extent of RVE should decrease as mixing increases because, assuming bubble rise velocity can be modeled using Stoke's Law, an increase in mixing will lead to a decrease in viscosity and increase in bubble rise velocity. To assess this hypothesis, this study considered two different digester samples with RVE tests conducted at different mixing intensities. As expected, volume expansion decreased as mixing intensity increased with a good linear correlation between viscosity and volume expansion. This work can be applied to identify a 'critical' shear rate at which volume expansion begins to occur for designing systems that effectively prevent RVE and ensure proper, adequate mixing.

KEYWORDS: Volume expansion, gas holdup, anaerobic digestion, mixing, viscosity, shear thinning, non-Newtonian

PROJECT OVERVIEW AND OBJECTIVES

Volume expansion in anaerobic digesters is a common and well-documented problem that has complicated operations and contributed to reduced production and even digester damage (Ganidi et al., 2009). Previously, volume expansion was commonly attributed to "foaming," a frothy layer that can form on top of digester contents; however, Chapman and Krugel (2011) proposed an alternative explanation suggesting that for some volume expansion events, the root cause was not "foaming" but rather "rapid volume expansion" (RVE) due to gas holdup.

During volume expansion, biogas bubbles will nucleate and remain in suspension instead of rising to the surface, resulting in gas holdup. As bubbles continue to grow, their growth contributes to an increase in digestate volume, causing RVE (Chapman and Krugel 2011). Since the introduction of RVE, a more comprehensive understanding has developed and been successfully applied to mitigate RVE events in various facilities (Higgins et al. 2014, Bartek et al. 2015).

As demonstrated by Higgins et al. (2014), gas holdup occurs predominantly under low or no mixing conditions due to the non-Newtonian behavior of digester solids. When mixing was stopped in laboratory experiments, reduced gas production rates and increased digester volume were recorded, suggesting bubbles were being retained in suspension. As expected, when mixing was restored, there was a sudden release of gas and decrease in digester volume. This observation can be explained by observing the rheological properties of digester contents, specifically, viscosity and yield stress.

Considering the results from Higgins et al. (2014), Bartek et al. (2015) theorized that gas holdup and volume expansion should be most severe in the absence of mixing as digester solids behave like a solid instead of a fluid. As a solid, digest contents exhibit a yield stress that must be overcome before movement occurs and bubbles can begin to rise. According to Bartek et al. (2015), bubbles must generate a buoyant force that exceeds the yield stress of the digester contents in order to begin rising. Once the yield stress is overcome, bubbles will begin to rise, releasing the held-up volume of gas, causing rapid volume expansion to collapse.

Using this foundational concept, Bartek et al. (2016) developed a simple mechanistic model to predict the extent and duration of a RVE event based on the yield stress and gas production rates of a digester in the absence of mixing. The model was fit to 11 different data sets and was proven effective in predicting the progression of, maximum extent of, and time until maximum RVE in the absence of mixing.

The bulk of RVE research up to this point has been focused on the absence of mixing, which is the worst-case scenario for anaerobic digestion facilities. While this provides a baseline for the maximum extent of expansion and can guide the development and design of emergency overflow systems for full-scale facilities, this does not address volume expansion that can occur due to inadequate or improper mixing. Specifically, Higgins et al. (2014) demonstrated that RVE can occur can occur even under low mixing conditions, potentially from poorly designed mixing systems and the non-Newtonian behavior of digester solids.

The goals of this research were to better understand the role of mixing in RVE due to gas holdup in anaerobic digesters. This knowledge can then be utilized to identify operating conditions for utilities that will minimize the risks of RVE and prevent either overmixing or undermixing a digester.

RHEOLOGY HYPOTHESIS

As microbes perform their metabolic processes, they release biogas that nucleates in the form of bubbles. These bubbles then rise through digester contents before breaking the liquid surface and evolving into the headspace of the digester. To prevent the holdup of biogas and consequently RVE, the rate at which bubbles are rising and being released into the headspace of the digester must be equal to the rate at which biogas bubbles are nucleating and growing in the reactor; otherwise, biogas bubbles will remain in suspension and continue to grow. This will lead to a larger overall volume of biogas being retained in suspension and cause the digester volume to expand.

As explained by Higgins et al (2014), the shear thinning behavior of digester contents provides insight regarding bubble rise velocity and the extent of RVE. Unlike Newtonian fluids that exhibit a constant shear stress regardless of shear rate, digester solids exhibit a shear thinning behavior such that when shear rate increases, viscosity decreases and vice versa, as shown in Figure 1. Since bubble rise velocity is a function of the solutions viscosity (such as with Stoke's Law-shown in Figure 1), bubble rise velocity will decrease when viscosity increases, leading to gas holdup and RVE. In other words, as mixing decreases, viscosity increases, and the time for a bubble to rise to the surface increases because its rise velocity decreases. This means that the bubbles will remain in suspension longer leading to more gas holdup and greater volume expansion. As such, the extent of volume expansion should increase as mixing intensity decreases.

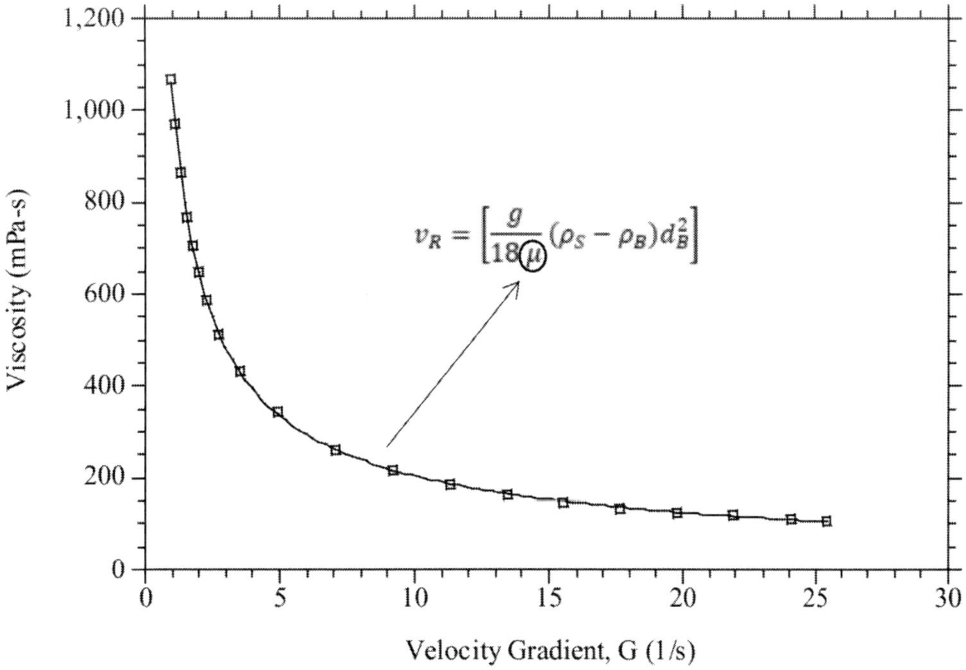

Figure 1. Apparent viscosity of digester solids (mPa-s) as a function of shear rate (1/sec) showing the shear thinning behavior of digester contents

Due to the non-Newtonian and shear-thinning behavior of digester contents, we hypothesized that as mixing increased and viscosity decreased, there would be a decrease in the degree of maximum volume expansion. Alternatively, as mixing decreased and viscosity increased, there would be a greater degree of volume expansion.

METHODS AND MATERIALS

The methods and materials in this study are identical to those used by Higgins et al. (2014), Bartek et al. (2015), and Bartek et al. (2016). For this study, only the 10-liter lab scale digesters made of see-through acrylic were used.

The 10 L reactors were part of other, ongoing studies, such that they were maintained and monitored daily. After the addition of inoculum, the reactors were sealed and the headspace was a connected to a respirometer to measure gas production data. Batch feeding and wasting were conducted daily through an inlet tube at the top of the digester and an outlet tube near the bottom of the reactor respectively. Mixing was provided to the reactor via a paddle and a mountable motor that was adjustable such that the mixing intensity could be changed. During testing, the reactor was fed and allowed to mix at its operating speed, 150 rotations per minute (rpm), for approximately 30 minutes to allow the feed to be dispersed throughout the reactor volume. Then, the mixing was decreased to the testing speed, typically in the range of 2-40 rpm. Different mixing speeds were tested on separate occasions, not in a continuous manner. Immediately after adjusting the mixing, the initial height of digester contents was marked on the reactor and the change in height was measured over time until contents reached a maximum height. During this time, gas production was also continuously monitored to observe the relationship between gas production and RVE. A picture of this experimental setup is shown in Figure 2.

Figure 2. Experimental setup for rapid volume experiments with various mixing speeds, provided with a variable speed motor. ***Not pictured: feed tube and hookup to respirometer***

As mentioned, the height of the digester contents over time was measured and RVE was then expressed as a percent of the original volume. The length of each test varied from sample to sample as the tests were conducted until the sample reached its peak volume expansion maintained a constant, steady-state extent of volume expansion. Eventually, the volume expansion would be collapsed or mixing would be restored and the volume expansion would be alleviated.

After each test, a sample of wasted digester contents was collected to assess various rheological properties using a viscometer (Brookfield DV-II +Pro). The data from the viscometer was collected on a computer using Rheocalc 32, a program that continuously records information regarding the viscosity, shear stress, and temperature of the sample at incremental shear rates from 0 to 25 /sec.

The results from the viscometer were then input to Kaleidagraph to fit the Ostwald–de Waele power relationship to the results to generate an equation to represent the viscosity of each sample. This relationship is represented by the following equation:

$$\mu = KG^{(n-1)} \tag{1}$$

where: μ is dynamic viscosity (Pa-s), K is flow consistency index (mPa-s^n), G is the mixing intensity (1/sec), and n is the flow behavior index (unitless).

To determine the resultant mixing intensity under a particular mixing regime, Equation 1 was rearranged to solve directly for G. Then Equation 1 was equated to Equation 2 and different values for the viscosity of the sample were iterated, using the Solver function in Excel, to determine a value that satisfied both Equation 1 and Equation 2. This value was then plugged into Equation 2 to determine G. Equation 3 was used to calculate the power imparted on the fluid, an input for Equation 2.

$$G = \left(\frac{P}{\mu V_D}\right)^{\frac{1}{2}} \tag{2}$$

$$P = K_L \mu n^2 D_I^3 \tag{3}$$

where: P is the power imparted on the fluid (watts), K_L is a impeller constant, μ is dynamic viscosity (Pa-s), n is revolutions per second (rev/s), and D_I is the diameter of the impeller (m)

RESULTS AND DISCUSSIONS

To assess the relationship between mixing, viscosity, and RVE, experiments were conducted on two different samples at various mixing speeds ranging from 5 to 50 RPM. For Sample 1, experiments were conducted at 2, 5, 10, and 25 RPM. Meanwhile, for Sample 2, experiments were conducted at 10, 20, 30, and 40 RPM. The viscosity profiles for each sample are shown in Figure 3. An overview of the parameters of each sample is provided in Table 1.

Table 1. Overview of characteristics of each sample considered in this study

Sample	Type of Digestion	Total Solids (%)	Volatile Solids (%)	Yield Stress (Pa)	Avg. Gas Production (mL/d)
Sample 1	MAD	3.57	2.38	1.46	8340
Sample 2	THP-MAD	5.74	3.86	3.08	16690

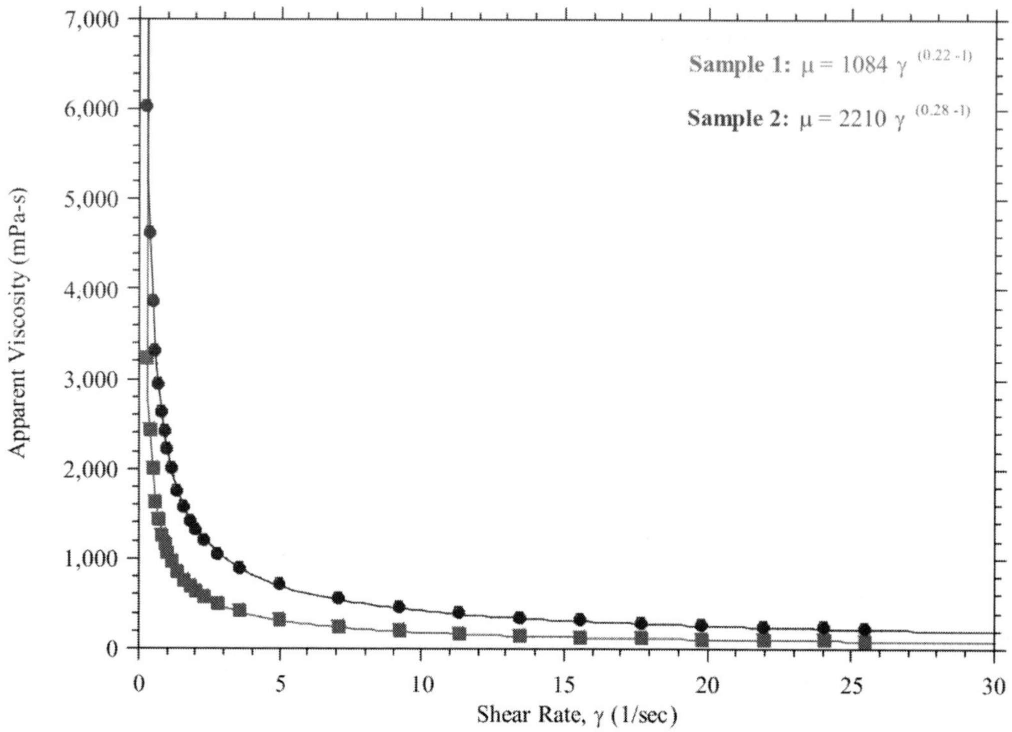

Figure 3. Apparent viscosity of Samples 1 and 2 (mPa-s) as a function of shear rate (1/sec) with the equation for modeling viscosity

As mentioned, using the equation generated by Kaleidagraph for the viscosity of the digester sample in addition to the formula to calculate G using the power imparted on a fluid, we were able to solve for G, velocity gradient or mixing intensity (1/s). Since the viscosity of digester contents changes with mixing intensity, this approach is necessary as calculating the power imparted on a fluid isn't a straightforward calculation as is the case when designing systems for water treatment. Using this data, we were then able to express the motor speed (RPM) of each test in terms of G and determine the viscosity of the digester contents during each trial. Based on Stoke's law, the amount of volume expansion measured should increase as mixing decreases and viscosity increases. The amount of volume expansion measured is compared to the viscosity and G for Samples 1 and 2 in Figures 4 and 5 respectively. Meanwhile, a strong linear correlation between viscosity and volume expansion is shown in Figures 6 and 7 for Samples 1 and 2.

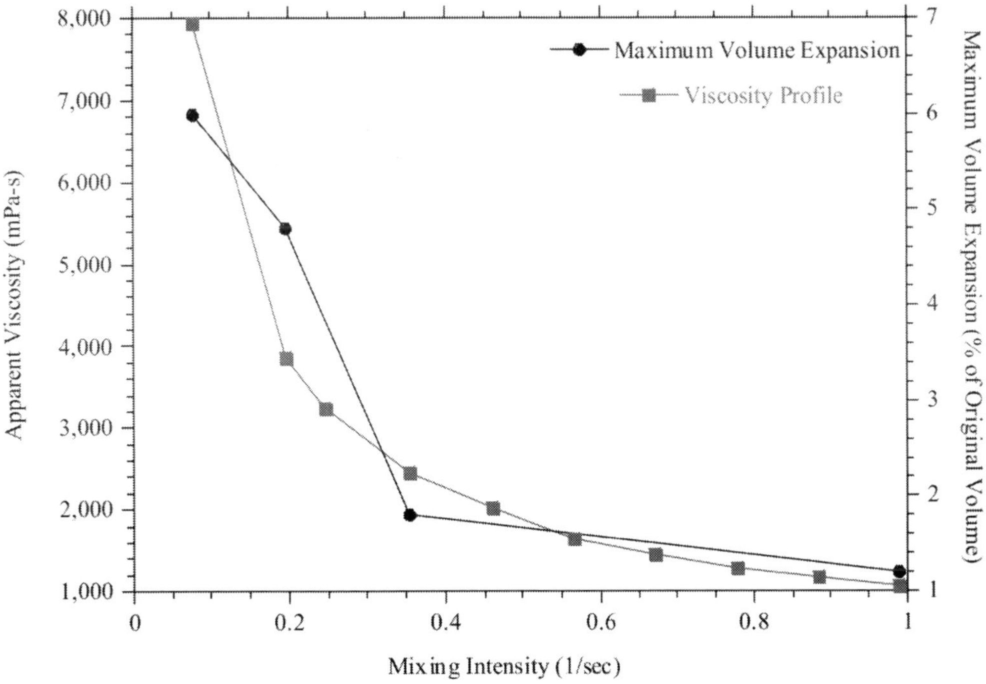

Figure
repres

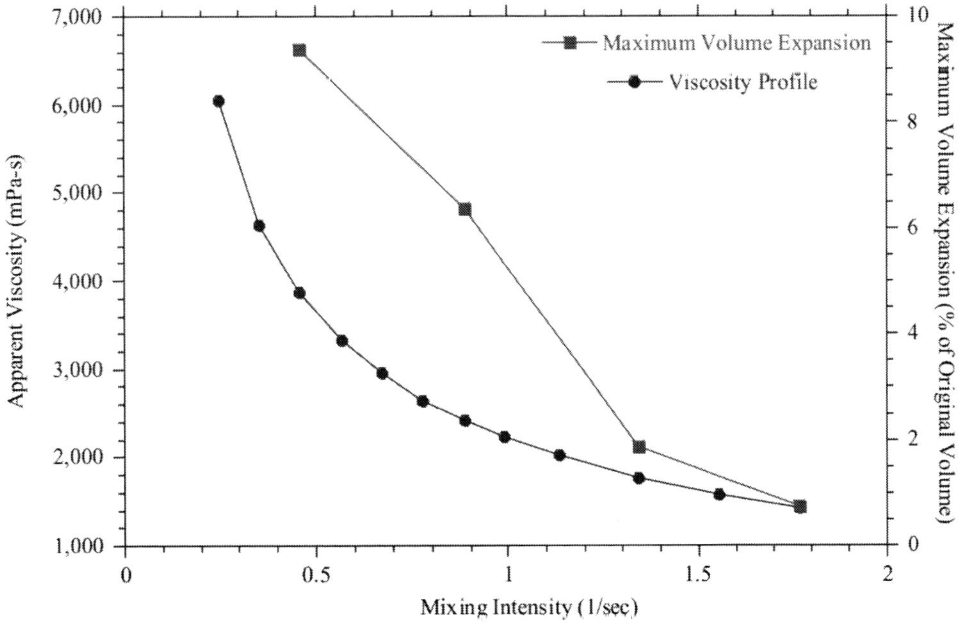

Figure 5. Viscosity profile and volume expansion for different mixing intensities, as represented by velocity gradient, G, for Sample 2.

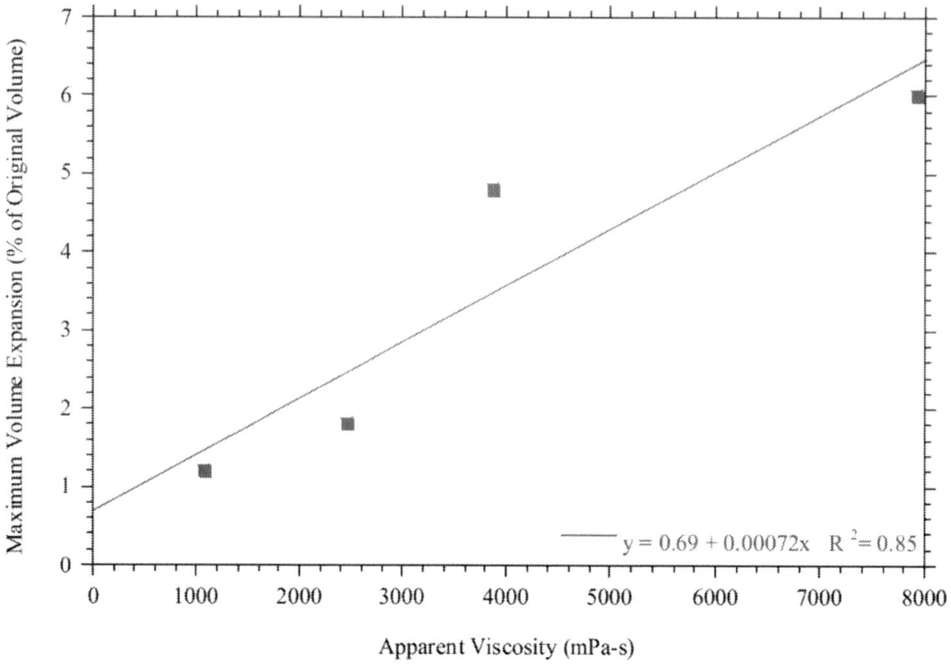

Figure 6. Volume expansion (% of original volume) expressed as a function of viscosity (mPa-s) for Sample 1.

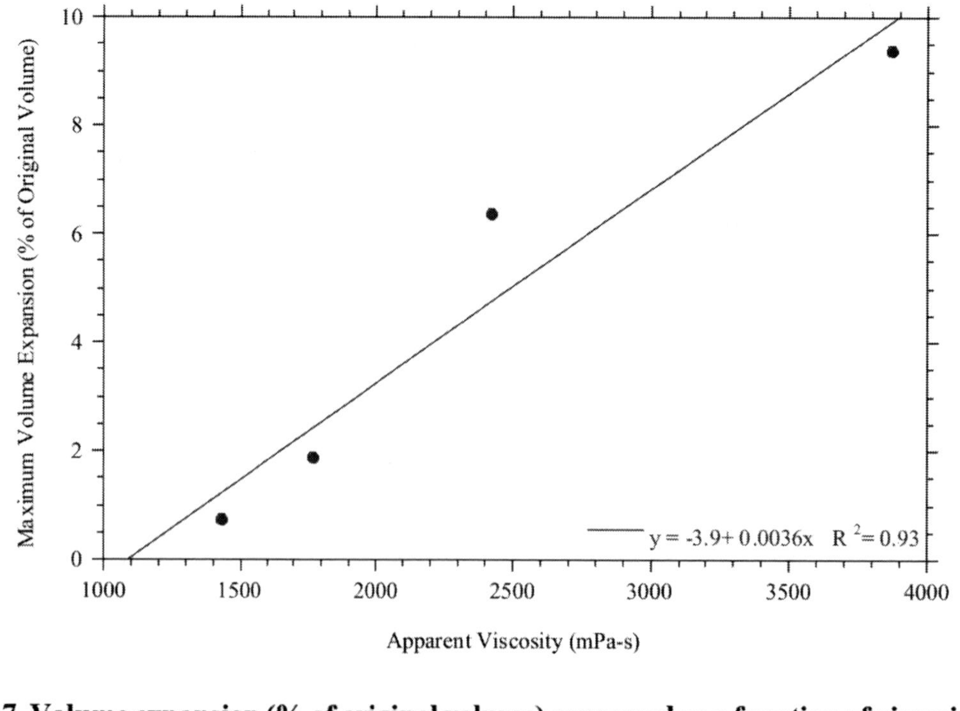

Figure 7. Volume expansion (% of original volume) expressed as a function of viscosity (mPa-s) for Sample 2.

As hypothesized, there is a clear, linear relationship between viscosity and the extent of rapid volume expansion. For both Samples 1 and 2, as mixing decreases and viscosity increases, the maximum amount of volume expansion also increases. The progression of each rapid volume expansion event, for which the maximum degrees of expansion are represented in the above Figures, can be found below in Figures 8 and 9 for Samples 1 and 2 respectively. The progression of rapid volume expansion in the absence of mixing has been provided for reference.

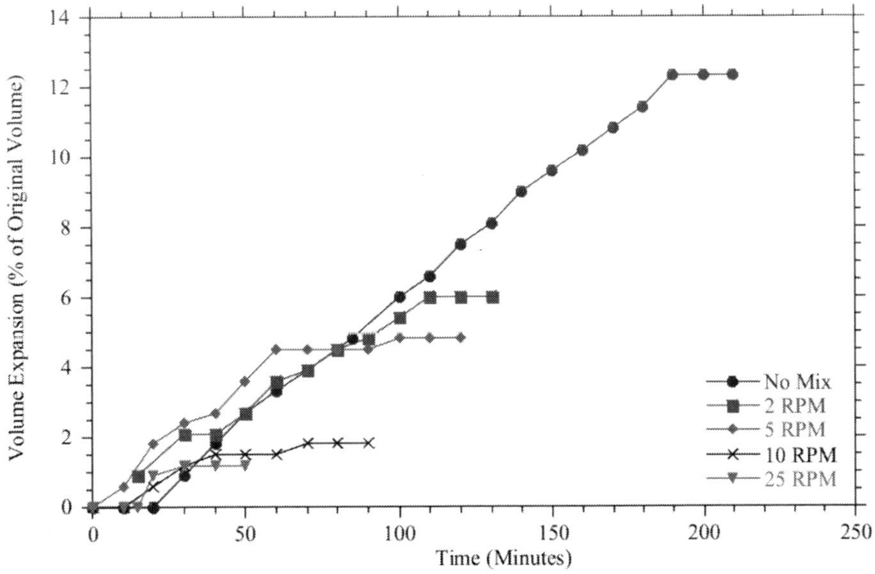

Figure 8. Progression of rapid volume expansion under different mixing regimes, including no mixing, for Sample 1.

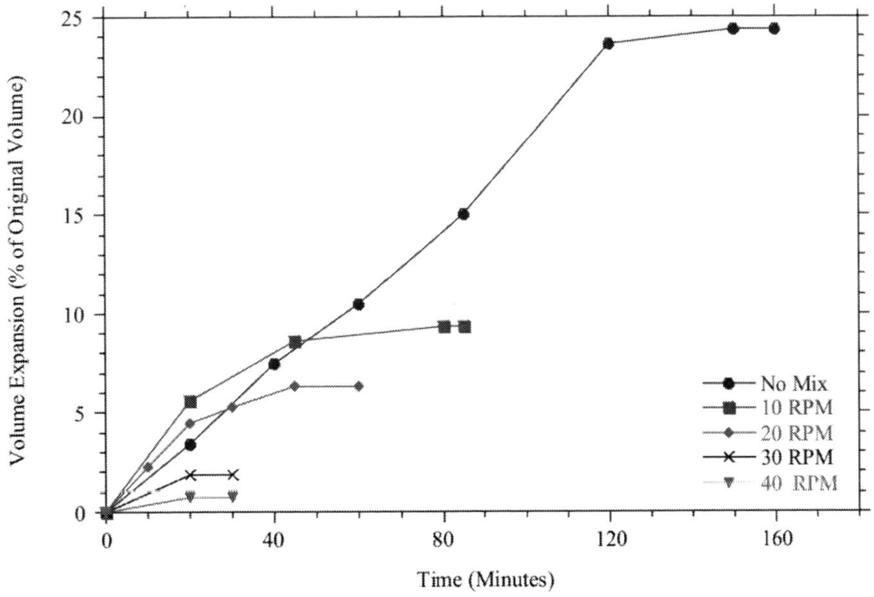

Figure 8. Progression of rapid volume expansion under different mixing regimes, including no mixing, for Sample 2.

As shown in Figures 8, even an incremental amount of mixing significantly reduces the amount of rapid volume expansion that occurs in the absence of mixing.

CONCLUSIONS

The results from this work show that RVE due to gas holdup is directly associated with the mixing intensity of a digester as mixing affects the rheological properties of digester contents. Specifically, assuming bubble rise velocity can be modeled using Stoke's Law, as mixing increases in a digester, viscosity decreases, and the bubble rise velocity increases. This leads to less gas holdup and less rapid volume expansion. As hypothesized, volume expansion decreased as mixing intensity increased. A good linear relationship between viscosity and RVE was proven and the relationship between the extent of RVE and mixing intensity appears to resemble that of mixing intensity and viscosity. Further work will include utilizing this information to design mixing systems for digesters that will minimize the risk of gas holdup and RVE by identifying a critical shear rate at which RVE will begin to occur. Ensuring mixing systems provide ample energy to achieve this minimum shear rate will prevent gas holdup and RVE. Further, this information will hopefully be integrated into the model developed by Bartek et al. (2016) to better understand the relationship between mixing, viscosity, and RVE .

ACKNOWLEDGEMENTS

This research has been funded in part by DC Water and WERF. The authors would like to thank the funding agencies for their support.

RERENCES

Bartek,N., Higgins, M.J., Murthy, S.N., Beightol, S., Peaslee, T. (2015). Causes and Cures of Rapid Volume Expansion in Anaerobic Digesters Due to Gas Holdup. *Proc. WEF Residuals and Biosolids Conference*

Bartek,N., Higgins, M.J, Murthy, S.N, Beightol, S., Al-Omari, A. (2016). Development of a Mechanistic Model to Understand Volume Expansion due to Gas Holdup in Anaerobic Digesters. *Proc. WEF Residuals and Biosolids Conference.*

Chapman, T. and S. Krugel. (2011). Rapid Volume Expansion – An Investigation into Digester Overflows and Safety. *Proc. WEF Residuals and Biosolids Conference*

Ganidi, N., Tyrrel, S., & Cartmell, E. (2009). Anaerobic digestion foaming causes – A review. *Bioresource Technology, 100*(23), 5546-5554. Retrieved December 19, 2014, from http://www.sciencedirect.com/science/article/pii/S0960852409006853

Higgins, M.J., S. Beightol, P.J. Nielsen, S.N. Murthy (2014). Insights into Rapid Volume Expansion in Anaerobic Digesters. *Proc. 2014 Water Env. Federation Annual Conference,* Austin, Texas.

Quantification of struvite content of biosolids is necessary to avoid bias in the assessment of digester and dewaterability performance

Alexandria Gagnon[1*], Christopher Wilson[1], Wendell Khujar[2], Ron Lattimer[2], Jeff Nicholson[1], Charles Bott[1]

[1] Hampton Roads Sanitation District, Virginia Beach, Virginia
[2] Hazen and Sawyer, Fairfax, VA
*Email: agagnon@hrsd.com

ABSTRACT

Spontaneous struvite ($MgNH_4PO_4 \cdot 6H_2O$) precipitation occurring within municipal anaerobic digesters can create a costly operational challenge if the deposited mineral impacts the mechanical operation of the digester and associated equipment. Additionally, for resource recovery facilities that recover struvite as a sellable product, the loss of struvite to nuisance precipitation within the digester can represent significant lost revenue. For these reasons, the application of technologies and approaches to prevent nuisance struvite precipitation within anaerobic digesters has been a recent industry focus (WERF NTRY1R12). Claims have been made regarding the use of these struvite mitigating technologies as a means for increasing cake solids dryness content, reducing polymer demand and improving digester volatile solids reduction performance at facilities practicing biological phosphorus removal. If true, these benefits to dewaterability and solids stabilization can result in significant costs saving for facilities.

This paper reviews the implications of struvite accumulation within digested sludge and accumulated digester grit with a view to providing insight into the impact that struvite can have on performance assessments. We will also discuss methods for assessing digested sludge and digester grit struvite content so that the impact of mineral deposition on the measurement of volatile solids destruction and sludge solids content can be routinely and adequately assessed.

KEYWORDS: struvite quantification, anaerobic digestion digester grit, sludge dewatering

INTRODUCTION

The conditions for the crystallization of struvite during anaerobic digestion has been documented elsewhere (e.g. Marti *et al.*, 2008). In general, struvite forms preferentially at higher pH of equimolar ratios of ammonium, magnesium, and phosphate, with magnesium and phosphate being the likely limiting constituents depending on upstream biological treatment process (e.g. EBPR) or geography (e.g. sea water influence). Struvite is well known for nuisance depositions within treatment plants, but significant quantities can precipitate as orthorhombic pyramidal crystals within the liquid column of an anaerobic digester (1) leaving with the biosolids or accumulating as grit within the digester over time (Figure 2).

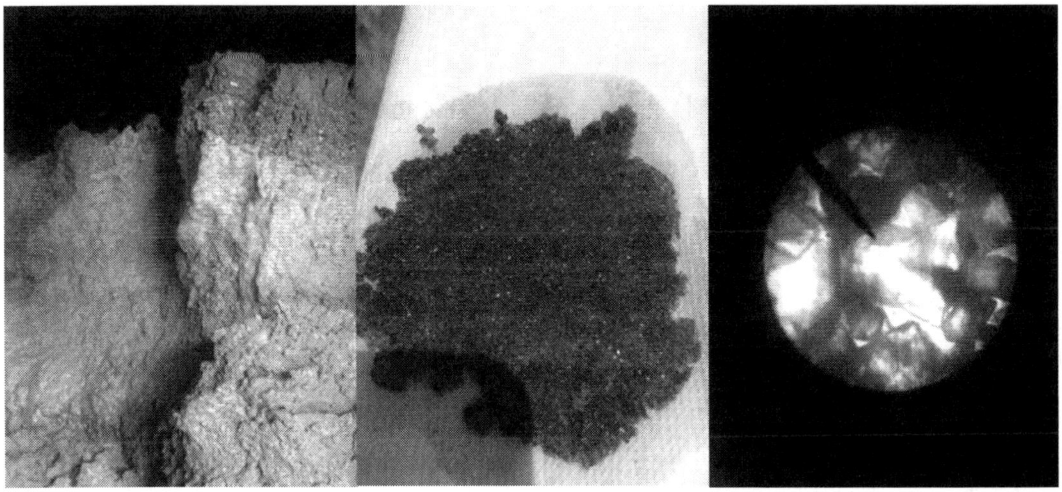

Figure 1. (Left)Deposited grit found during digester cleaning at HRSD's Nansemond Treatment, (Center) Air drying grit obtained from anaerobic digester during cleaning, (Right) Micrograph of washed grit recovered from liquid biosolids at HRSD's Nansemond Treatment Plant showing classic orthombic pyramidal crystal structure of struvite.

During conventional TS/VS analysis, struvite decomposes due to ammonia volatilization and evaporation of the waters of hydration from the high temperatures (104°C for total solids and 550°C for fixed solids) found in solids drying ovens. Struvite decomposes at temperatures as low as 55°C according to the reaction below, with full ammonia and water loss occurring around 250°C (Bhuiyan *et al.*, 2008):

$$MgNH_4PO_4 \cdot 6H_2O \rightarrow MgHPO_4 + NH_3 \uparrow + 6H_2O \qquad Equation\ 1$$

The loss of ammonium and water corresponds to a 51% reduction of mass of a mineral that should be assessed as inert fixed particulate solids within standard solids analyses. The loss of fixed solids during solids drying (functionally measuring mineral solids as water) creates a positive bias on residual VS/TS (i.e. percent volatiles) and a resulting negative bias on the measurement of volatile solids reduction (i.e. measurement of overestimated digested VS/TS relative to true raw VS/TS). In a conventional anaerobic digester treating EBPR sludge, the negative bias on volatile solids reduction measurement can be as much as 5% VSR, depending on the quantity of struvite in the digested sludge. This bias can be further compounded if dewaterability assessments (e.g., cake TS content (%TS) and polymer demand (lb active polymer/dry ton TS)) are used to benchmark performance. To understand, and properly correct for this bias, we must have reliable and easy to apply assessments of biosolids struvite content.

Struvite samples assessed during this study were taken from HRSD's Nansemond Wastewater Treatment Plant (NWWTP) is Suffolk, VA. As a15 MGD average 5- stage treatment plant with conventional anaerobic digestion under the influence of industry, the plant has higher than average influent phosphorus and nitrogen concentrations in the raw influent and throughout the plant. Due to elevated phosphorus concentrations in plant streams, NWWTP historically had issues struvite accumulation around the plant. In 2011 a struvite recovery facility (SRF) began removing phosphorus and nitrogen from the dewatering liquor from anaerobically digested sludge, where most of the struvite accumulated. Phosphorus removal as struvite occurred wholly post-anaerobic digestion and post-dewatering (i.e. on the dewatering centrate stream) after HRSD's NWWTP anaerobic digesters reducing recirculating phosphorus in the plant. It did not reduce struvite accumulation within the digester itself due to the high phosphorus concentration of waste activated sludge entering the digesters. Quantification of the struvite within the digester, both contained in the sludge and accumulated in the grit, had not been assessed before. NWWTP Digesters were cleaned in 2007 and 2008 and started undergoing cleaning again in May of 2016.

METHODOLOGY

The following are analytical methods were used to assess the quantity of struvite existing in digested sludge, sludge cake and digester grit samples from the anaerobic digester. Differential air drying was conducted using struvite produced in the SRF at NWWTP to show the relative mass loss associated with water and ammonia from struvite. Results from the following analytical methods will be compared against numerical methods for estimating the struvite content of various solids streams, including a calibrated EnviroSim BioWin and Visual MINTEQ.

Elutriation in an Inverted Cone

A method has been previously proposed to separate struvite and other dense materials from liquid sludges and liquors for observation and further analysis **(Rob Baur, Clean Water Services).** In this method, a liquid digester sample is placed in an inverted glass cone and water is added to the bottom of the cone allowing light particles to overflow. The obvious limitation of this method is that non-struvite grit will be collected in the same manner and very light fines may be removed in the overflow from the cone in an effort to remove organic particulates; however, separation of settleable material from liquid biosolids may produce a sufficiently high solids sample to allow the above differential drying methods to be applied to liquid samples, as described below.

A modification to the original method elutriation method was performed as follows:

A measured mass of digester cake or volume of digester sludge was combined with 500 mL of tap water (raised to a pH of 9 to prevent solubilization of struvite) and blended for 30-45 second on low (Waring Commercial Laboratory Blender.). The blended samples were poured into the 2 liter glass beaker and then filled with additional pH 9 tap water until a total volume of 2 liters was contained in the beaker. Gentle mixing was applied for 10 seconds using a polypropylene paddle followed by a 2 minute settling period. Samples were then gently decanted until less than 500 mL of solution remained and then refilled with tap water raised to a pH of 9. This was repeated until the sample liquid ran clear. Solids that accumulated in the bottom of the beaker were than separated from the remaining liquid mass using a 1.5 μm filter.

This method yielded visible struvite and miscellaneous material (ex. silica) from both the digester cake and digester sludge samples as seen in Figure 2. Due to the additional material collected during this experiment, weighing the removed solids could not be used as an accurate representation of the struvite content of the samples. In order to estimate struvite content samples underwent mineral acidification as detailed below.

Figure 2. Settled Material Collected During the Modified Rob Baur Method from Digester Cake (Left) and Digester Sludge (Right)

Differential drying temperature- standard temperature, low temperature and freeze drying
The premise of differential drying temperature is that at a sufficiently low temperature (approximately 55°C), the decomposition of struvite and loss of ammonium and water mass will not occur (Bhuiyan *et al.*, 2008). As a result, the comparative weight loss between low (Figure 3) and standard drying temperatures (assuming that at low temperatures the only mass lost is free water) can provide an estimate of ammonia and water lost from struvite.

An extension of the differential hot air drying temperature, lyophilization can be used as a more rapid alternative to low-temperature hot air drying. In principle, ammonia loss and water loss will be similar to low temperature air drying. The same limitation in differentiating between sources of ammonia and water mass loss exists for this method, but is a potentially more operationally convenient and repeatable (by achieving full water removal) method of analysis.

Struvite recovered from the struvite recovery facility was ground to fine powder and samples underwent one of the following: low temperature drying (55 C), high temperature drying (105C), or freeze drying. Initial final weights were taken and compared to estimated values of water and ammonia loss. Results from this analysis are not included in the manuscript, but will be discussed in the presentation.

Mineral Acidification
At low pH, struvite will dissolve to its ionic components; therefore, measuring the ratio and mass of released ions provides a measurement of struvite in a sludge sample. By measuring additional metal ions (e.g. calcium, ferric iron) it is possible to differentiate between the release of ammonium and phosphorus related to alternative mineral precipitates (e.g. hydroxyapatite and vivianite), though an assumption of few dominant minerals is likely practical to reduce the complexity of correlating mineral dissolution with soluble ion concentrations. Soluble ammonia, magnesium and phosphate are measured before and after acidification to pH 4.5 with sulfuric acid. Digester grit, cake, and sludge were added to 500 mL of D.I. water with sufficient mixing and the pH lowered to below 4.5 using 2N sulfuric acid. Once the pH was stable (no change in pH over a 15 min period), samples were filtered through 0.45 µm filters and sent to a NELAP accredited laboratory to be analyzed for ammonia, phosphate, magnesium and other metals including calcium, iron and sodium. The remaining solids collected after acidification can be seen in Figure 3.

Figure 3. Post-Acidification Settled Solids Collected from the Modified Rob Baur Method from Digester Cake (left) and Digester Sludge (right)

RESULTS AND CONCLUSIONS

Mineral Acidification Method

Solids recovered from the elutriation of digester material revealed a struvite content within the digester that ranged from 0.6 to 10.2 percent by weight as seen in Table 1. Digester cake was analyzed during two different time periods (May and December) and using two different methods. Samples acidified in May were not separated from the digester cake initially using the method above, but were diluted using D.I. water and acidified to a pH below 4.5 before filtering and analysis. In order to correct for the presence of phosphorus in the liquid, liquid removed as part of dewatering sludge through use of centrifuges .was assessed for phosphate, ammonia, and magnesium and metals content. After utilizing methods outlined above, struvite content of the digester cake ranged from 0.6 to 3.2 percent on average. On a molar basis, the weight by weight content standard deviation of the phosphorus, nitrogen, and ammonia between samples was less than 0.1 percent for digester cake. This proves it is unlikely that any other solids (hydroxyapatite, inert solids, etc.) could be contributing to analysis for estimation of struvite content.

Digester sludge analysis revealed a higher estimated struvite content ranging from 1.5 to 6.2% on average as seen in Table 1. It was expected that the struvite content in the sludge would be similar to that found in the cake, but was slightly higher than the estimated struvite content of the cake. It was determined that the struvite content of the digester sludge may be slightly high from precipitation induced by the addition of pH 9 solution and force of the blender during struvite separation. The difference in values could also be attributed to difficulty separating struvite from digester cake after centrifugation causing an underestimation of content. Standard deviation of the digester solids molar content on a weight by weight basis was also low showing a consistent method for separation of struvite from solids. The consistent difference between estimated struvite content of digester sludge and cake needs to be investigated further.

Digester grit showed an estimated struvite content that ranged from 35 to 75 percent as seen in Table 1. This struvite is accumulated in the digester over time and can contribute to decreased treatment volume and time between cleanings. During periods when the pH drops in the digester, this struvite is at risk to solubilize increasing the phosphorus, ammonia and magnesium in sidestream causing operational issues. Accurate estimation of digester struvite content within the grit and contained in the solids allows for estimation of cost-effectiveness of struvite mitigating technologies and assessment of digester performance with respect to volatile solids reduction..

Table 1. Mass of Struvite Component Ions and Estimated Struvite Content by Weight of Samples

Sample Type	Date	Mass of Ion Released			Estimated Struvite Content (% w/w)			AVG	STDEV
		g NH_4^+-N	g PO_4^{3-}-P	g Mg^{2+}	N-basis	P-basis	Mg^{2+}-Basis	%	± %
Dewatered Digester Cake	5/11/16	0.19	0.99	0.54	5.3%	12.4%	**8.6%**	8.8%	± 3.6%
	5/17/16	0.16	0.55	0.22	10.4%	16.1%	**8.4%**	11.7%	± 4.0%
	5/26/16	0.13	0.42	0.19	11.9%	17.7%	**10.1%**	13.2%	± 4.0%
	11/15/2016*			0.03			**2.3%**		
	11/29/16	0.01	0.03	0.03	3.1%	3.0%	**3.2%**	3.1%	± 0.1%
	12/6/16	0.01	0.03	0.03	2.7%	2.8%	**2.8%**	2.8%	± 0.0%
	12/12/16	0.00	0.01	0.01	1.3%	1.2%	**1.4%**	1.3%	± 0.1%
	12/21/16	0.00	0.01	0.00	0.6%	0.7%	**0.6%**	0.6%	± 0.1%
Digester Grit	5/11/16	4.17	9.33	7.51	36.2%	36.6%	**37.5%**	36.8%	± 0.7%
	5/17/16	3.24	7.61	5.34	48.7%	51.7%	**46.3%**	48.9%	± 2.7%
	5/26/16	0.59	1.35	1.05	53.3%	55.0%	**54.2%**	54.2%	± 0.8%
	11/15/2016*			0.22			**58.7%**		
	11/29/16	0.13	0.30	0.24	66.6%	67.8%	**69.4%**	67.9%	± 1.4%
	12/6/16	0.06	0.12	0.10	78.4%	70.6%	**74.2%**	74.4%	± 3.9%
	12/12/16	0.02	0.04	0.03	36.3%	37.8%	**35.4%**	36.5%	± 1.2%
	12/21/16	0.06	0.14	0.11	48.9%	55.2%	**53.2%**	52.4%	± 3.2%
Digester Sludge	11/15/2016*			0.03			**6.2%**		
	11/29/16	0.01	0.04	0.03	4.4%	4.8%	**4.7%**	4.7%	± 0.2%
	12/6/16	0.02	0.03	0.03	5.1%	4.4%	**4.6%**	4.7%	± 0.4%
	12/12/16	0.02	0.04	0.03	5.3%	5.4%	**5.2%**	5.3%	± 0.1%
	12/21/16	0.00	0.01	0.01	1.4%	1.5%	**1.5%**	1.4%	± 0.0%

*Phosphate and ammonia analysis was not conducted on November 15th samples.

REFERENCES

S. Beall, D. Jenkins, S. Vidanage. A systematic analytical artifact that significantly influences anaerobic digestion efficiency measurement, Water Environment Research, Volume 70, Number 5, July/August 1998, pp. 1019-1024.

M. Iqbal H. Bhuiyan, D.S. Mavinic, F.A. Koch, Thermal decomposition of struvite and its phase transition, Chemosphere, Volume 70, Issue 8, February 2008, Pages 1347-1356.

N. Marti, A. Bouzas, A. Seco, J. Ferrer, Struvite precipitation assessment in anaerobic digestion processes, Chemical Engineering Journal, Volume 141, Issues 1–3, 15 July 2008, Pages 67-74.

Driving Operational Improvement through the Use of Dashboards: A Regional Biosolids Pelletizing Facility Perspective

Amber Batson[1*], Raymond Schauer[2], Scott Trainor[2]

[1]CDM Smith, Houston, Texas.
[2]Solid Waste Authority of Palm Beach County, West Palm Beach, Florida.
[*]Email: batsonam@cdmsmith.com

ABSTRACT

In August 2009, the Solid Waste Authority of Palm Beach County (SWA) commissioned a regional Biosolids Pelletization Facility (BPF). The 600-wet ton per day (wtpd) drying facility utilizes landfill gas as its primary fuel source to create a beneficial end product sold as a fertilizer amendment. This unique facility processes a wide array of solids from wastewater treatment facilities—including undigested, aerobically digested, and anaerobically digested with solids content ranging from 13 to 20 percent total solids (TS). The combination of a varying feed stocks, as well as fluctuations in landfill gas quality and quantity, have presented interesting challenges for the contract operator. These challenges can result in high natural gas usage, ultimately increasing the operational costs for SWA and its utility partners.

Recognizing that there was a need to decrease these costs and improve other operational issues, SWA developed a dashboard benchmarking system with key performance indicators for the utility partners, the contract operator, as well as internal SWA staff that manage the landfill gas system. The dashboard system has proven to be an effective management tool for SWA and its utility partners. Results are presented.

KEYWORDS: Dashboard, benchmarking, pelletization, biosolids processing.

INTRODUCTION AND BACKGROUND

Prior to the construction of the regional biosolids pelletization facility, biosolids generated by wastewater treatment facilities located in south Florida were predominantly being disposed of by land application, which is permitted by the Florida Department of Environmental Protection (FDEP). The wastewater sludge was applied at permitted sites within the Lake Okeechobee, Everglades, and Indian River Lagoon Basins. Public concern, political action, and the continued efforts to restore the Everglades have resulted in regulatory changes that affect the current land application processes and limit potential land application sites due to nutrient loading effects. Therefore, many utilities across Florida, including those in Palm Beach County, were seeking a long-term, all weather solution for biosolids disposal.

Located centrally in Palm Beach County along the Florida Turnpike, the SWA Renewable Energy Park includes: Class I and Class III landfills, waste-to-energy facilities, a woody waste processing facility, and a recycling center as seen in Figure 1. As a condition of the land purchase agreement for the Renewable Energy Park site, SWA agreed to provide for the disposal of biosolids from the Seacoast Utility Authority for as long as the Authority operated disposal facilities on the site. In an ongoing effort to conserve landfill space and minimize odor nuisances early in the 1990's SWA constructed a facility to compost the Seacoast Utility Authority biosolids and yard waste on the Renewable Energy Park site. SWA also executed a 20-year agreement with the East Central Regional Water Reclamation to compost biosolids at the facility. Unfortunately the commercial markets for this compost material did not develop as projected and the compost operation did not prove to be financially viable for SWA. As a result, SWA sought alternate biosolids treatment and disposal options.

Given the potential environmental impacts of land application and the odor issues associated with landfill disposal SWA recognized the need to provide a safe, cost effective and sustainable treatment and disposal solution on a countywide basis. In addition, the landfill gas from the Class I and Class III landfills was being collected and flared onsite, wasting a valuable potential fuel source. Recognizing both the need for a better biosolids treatment option and the availability of a low-cost fuel source, SWA decided to develop a biosolids processing facility, in partnership with five of the municipal waste water treatment plant operators in the County, on their centrally-located Renewable Energy Park site to meet the long-term needs of Palm Beach County residents.

SWA procured the design and construction plus the 20-year operations and maintenance of the facility through a Design/Build/Operate contract. In August 2009, the regional BPF was commissioned.

FACILITY OVERVIEW

The BPF is a 600-wet ton per day (wtpd) rotary drum, direct drying facility. Multiple wastewater treatment facilities deliver wet sludge cake to this regional facility for processing. The incoming feedstock varies from 12 to 20+ percent total solids and can be undigested, aerobically digested, or anaerobically digested. The dewatering methodology also varies by wastewater treatment facility (typically belt filter presses or centrifuges). Incoming wet cake is stored at the facility then conveyed for processing.

Figure 2 below presents a process flow diagram for the solids processing and handling portion of the system. After being conveyed from the storage bunkers to the processing area, wet cake is mixed with previously dried solids and loaded into a rotary drum dryer. Inside the dryer, the moisture is driven off through direct contact with hot air, rendering the biosolids pathogen free. The dried solids exit the dryer and are transported to a cyclone separator on the roof of the facility. The separator collects the dried solids and discharges them through an airlock into a screener. Once in the screener, the granules of dried solids are divided into four categories: trash, oversize, product, and fines. The trash, which includes plastics or any other undesirable solids, is collected into bins for disposal. Oversized solids drop into a crusher then discharge into the recycle bin. Fines also go directly into the recycle bin. The recycle bin stores the dry material to be mixed with the wet cake before entering the dryer. The remaining properly-sized pellets are cooled and transported to storage silos. The pellets are sold as a fertilizer product.

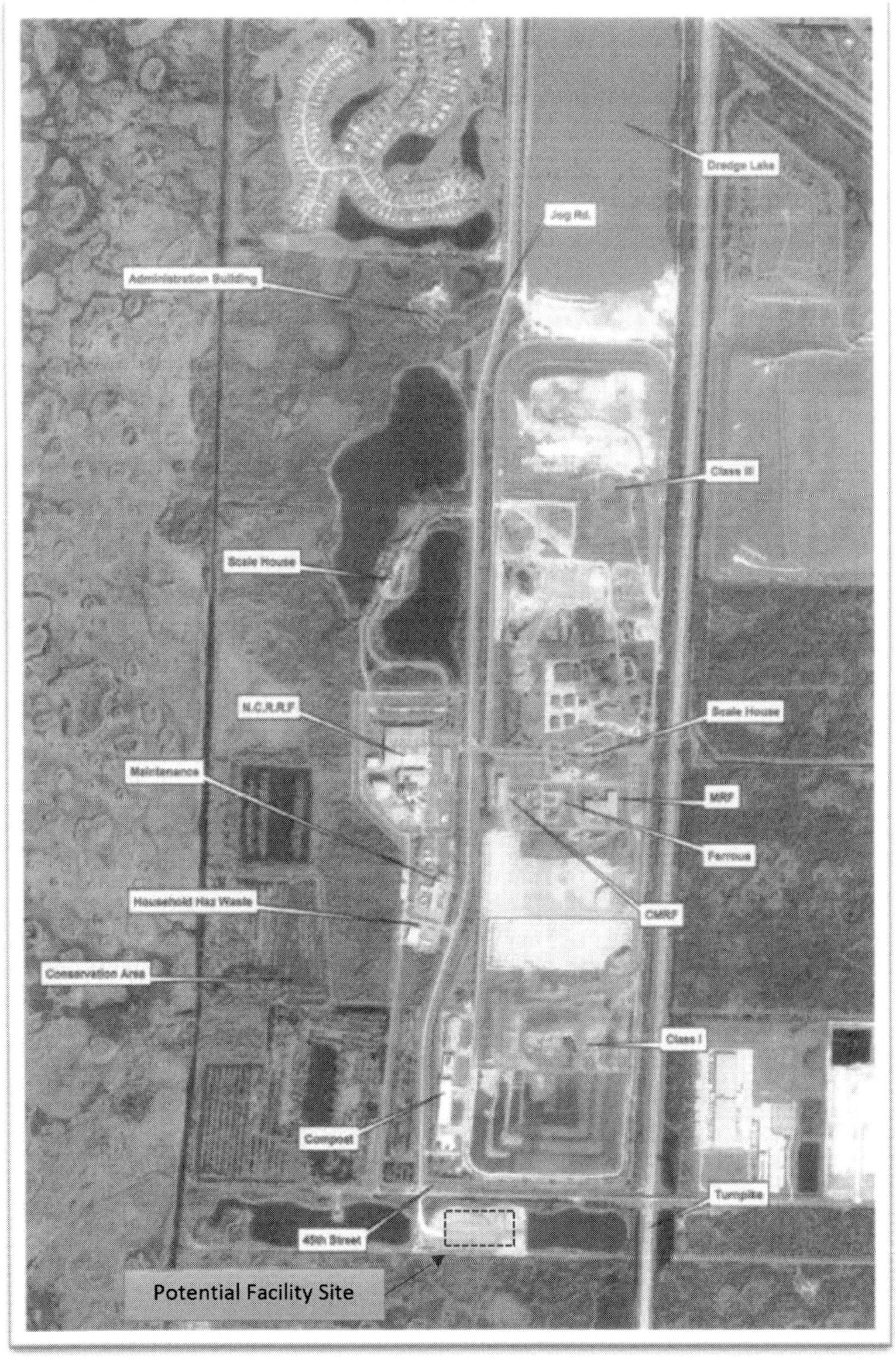

Figure 1. SWA Renewable Energy Park Campus (prior to construction)

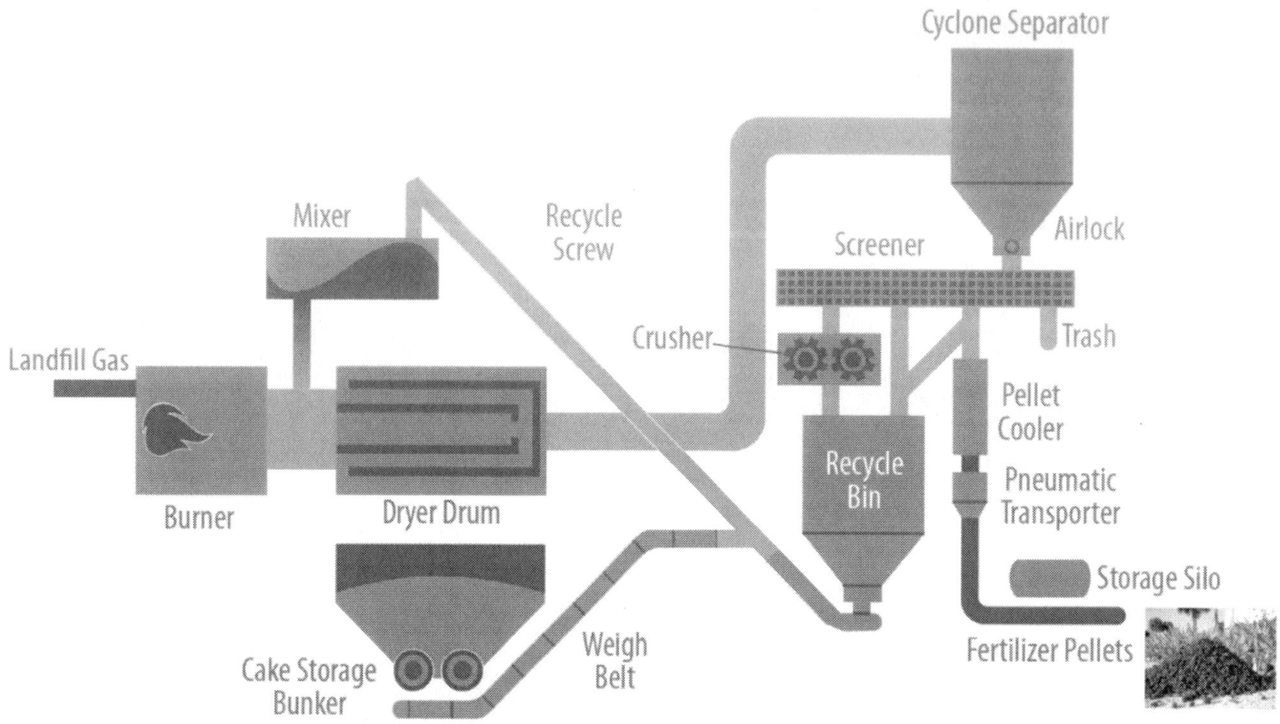

Figure 2: BPF Dryer and Solids Handling Process (Graphic courtesy of SWA and NEFCO)

The BPF air system has been presented in Figure 3 below. The BPF has a fully integrated air handling and emission control system. To get the heat necessary to evaporate the moisture from the wet cake, fuel is burned in air. The fuel source can be landfill gas, natural gas, or a combination of the two. Exhaust gases from the drying process are separated from the dried solids in the cyclone separator. The majority of the exhaust gases are recycled back to the inlet of the dryer system. Before being recycled, the gas must first be treated in a condensing scrubber to remove particulates and water vapor. Any excess gas not recycled is drawn through a venturi scrubber to polish any remaining particulates in the waste stream. It then is sent to a regenerative thermal oxidizer (RTO) that will destroy any odor-causing vapors before being discharged through the facility stack. By combining the wet scrubbing, condensing, exhaust recirculation, and thermal oxidation, the facility is able to have excellent control of odors from the drying process, minimal emissions, and economical operations.

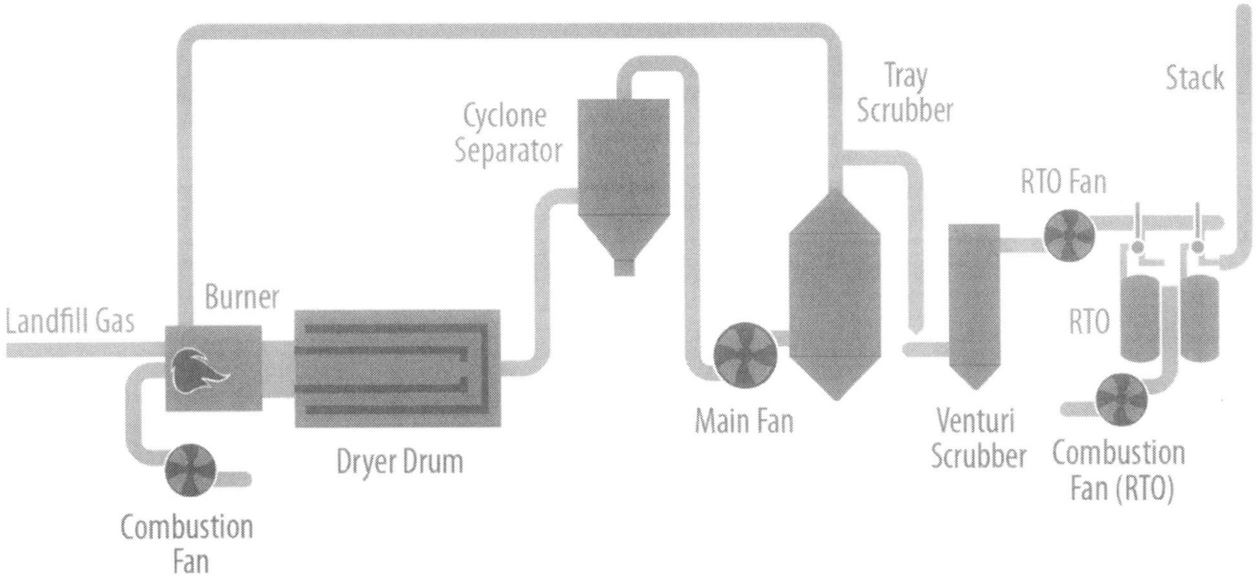

Figure 3: BPF Air Handling and Emission Control Process (courtesy of SWA and NEFCO)

LANDFILL GAS DELIVERY

A unique component for this facility is the use of landfill gas (LFG) as a primary fuel source. Landfill gas is collected from the nearby SWA municipal solid waste landfill. The landfill gas has varying quality (<40 to 60 percent methane) and quantity (0 to >3000 standard cubic feet per minute). Landfill gas is collected through a vacuum system with a series of extraction wells that deliver moist gas to a compression system. That system pressurizes the gas, eliminates particulates, cools the gas to significantly reduce moisture, then reheats the gas prior to delivery to the BPF via a mile-long pipeline at a pressure of roughly 25 pounds per square inch (psi).

OVERVIEW OF OPERATIONS POST-COMMISSIONING

From the time of facility startup, several challenges have presented themselves to both SWA and its contract operator. First, there has been a dramatic impact of the varying feedstock on the pelletization process. Ultimately this has resulted in the need for additional capital improvements to ensure the system can safely operate at design capacity, more diligent monitoring of critical operating parameters, and a more proactive maintenance strategy.

Second, the landfill gas system has a lot more variability than originally envisioned, which has impacted overall system performance. The availability of suitable quantity and quality of landfill gas for the BPF has been impacted by several things: (1) how the landfill is being managed – including landfill gas extraction well calibration, type and age of waste in place, and leachate water levels following rain events; (2) landfill gas compression system equipment uptime; and

(3) changing levels of impurities in the landfill gas (including siloxanes and hydrogen sulfide) that can damage the compression system equipment.

These two unique project considerations have ultimately led to higher than desired use of natural gas, ultimately increasing costs to SWA and the utility partners for operation of the system as presented in Table 1 below.

Table 1: BPF O&M Costs per Ton Prior to Implementation of the Benchmarking Dashboard

	FY09/10	FY10/11	FY11/12	FY12/13	FY13/14
Actual					
Base Processing Fee					
(Tiered Tipping Fee that reduces with larger quantities processed)					
Average (all Tiers)	$ 27.48	$ 28.00	$ 28.90	$ 29.14	$ 28.46
Cost Substantiation/Retention					
(Additional Approved Items not in Part of Base Processing Fee or Pass-Through Costs)					
Total Combined	$ -	$ -	$ 0.11	$ 0.06	$ -
Pass-Through Costs					
(Specific Costs Defined in Contract with Maximum Guarantees)					
Electricity	$ 5.20	$ 5.41	$ 4.84	$ 5.40	$ 5.45
Potable Water	$ 0.18	$ 0.45	$ 0.29	$ 0.37	$ 0.17
Industrial Water	$ 0.63	$ 0.89	$ 0.85	$ 1.00	$ 1.08
Wastewater	$ 0.72	$ 1.35	$ 1.29	$ 1.37	$ 1.32
Sodium Hypochlorite	$ 0.20	$ 0.24	$ 0.14	$ 0.25	$ 0.14
Caustic	$ 0.07	$ 0.08	$ 0.04	$ 0.06	$ 0.06
Natural Gas	$ 7.69	$ 8.76	$ 5.11	$ 10.89	$ 11.91
SWA LFG O&M	$ 2.00	$ 0.77	$ 2.01	$ 0.57	$ 2.47
SWA Admin, Consulting, Legal	$ 0.28	$ 1.46	$ 2.20	$ 2.70	$ 3.58
Actual O&M Cost/Ton - Total	$ 44.45	$ 47.41	$ 45.78	$ 51.81	$ 54.64

SALE OF EXCESS CAPACITY

Upon commissioning of the 600-wtpd facility, SWA and the original five utility partners determined that there was excess capacity available. Another local utility approached SWA and the original partners to inquire about purchasing the excess capacity to join as a sixth partner. This would allow each of the original partners to receive a refund on their capital investment for unused capacity; however, could impact long-term operating costs for the facility.

Extensive analyses were completed to determine if there would be sufficient capacity available for the new partner both on an annual average and on a peak week basis, given the limited existing storage capacity of the facility and the contractual daily processing limit. A landfill gas model was developed to project long-term gas availability. Both the projected tonnage and the declining landfill gas availability were evaluated to determine long-term operating costs with or without the utility partner.

Ultimately, the analyses showed that adding the new partner would not increase the unit operating costs for the facility so long as the partners (1) worked to even out deliveries to the facility on a daily basis, (2) an automated programmable logic controller (PLC)-based gas blending system was installed to blend both natural and landfill gas to maximize landfill gas utilization, and (3) the contract operator completed the necessary facility improvements to restore the original design capacity to the BPF.

DEVELOPMENT OF BENCHMARKING CRITERIA

Recognizing that the unit operational costs were increasing and that there was a need for improved communications and a focused effort to improve various issues while bringing on a new utility partner, SWA embarked upon a benchmarking effort. Key steps in the benchmarking process were as follows:

1. Gain an understanding of the overall goals and objectives for operation of the BPF. The following five goals were previously established:
    - Manage BPF Partner's incoming sludge deliveries within the daily and annual facility capacity limits to optimize process efficiency and limit disposal of sludge at the SWA landfill.
    - Ensure continuity of operations at the BPF by minimizing unscheduled downtime
    - Promote efficient BPF operations by minimizing residue generation
    - Continue safe operations at BPF by minimizing number of OSHA reportable incidents
    - Strive for continuous environmental compliance
    - Meet budgetary expectations for total O&M cost/ton for the fiscal year (FY)

2. Establish what metrics should be monitored. These could be related to strategic goals and objectives, environmental compliance, community stewardship, safety, operational efficiencies, costs, or industry standards. For FY2015, key performance indicators (KPIs) were established for the following:
    - The quality and quantity of solids being delivered to the BPF by each utility partner (to document contractual compliance and monitor loads being diverted from BPF due to daily contractual processing capacity limits)
    - Facility operations such as loads diverted, throughput, product quality, and residue generation (to document contract operator performance and operations efficiency)
    - Facility maintenance (to document effectiveness of contract operator's maintenance activities)
    - Safety (to document performance against industry standard)
    - Utility and chemical usage (to document operational cost components)
    - Landfill gas availability and natural gas usage (to document operational cost components)
    - Overall financial performance related to SWA's overall budget (to understand if achieving financial objectives)

- Environmental compliance
- Completion of small and large projects on schedule/budget (to gain an understanding of effectiveness of the risk-based prioritization of the initiative list system)
- Community stewardship such as odors, noise, and outreach efforts
- Knowledge retention including standard operating procedure (SOP) development and document updates

3. Establish what target levels warrant success (green level), caution (yellow level), or improvement needed (red level) for each KPI.

4. Determine data availability and ease of access for each KPI.

5. Develop an easy to read dashboard where a stakeholder could at a glance get an understanding of performance.

6. Get buy-in from the team on the KPIs, target action levels, and format to present the data to ensure continued success of benchmarking effort.

7. Populate the data, communicate results to stakeholders on routine basis, and analyze trends over time.

8. Determine root causes of issues, establish corrective action plans, implement the plans, and monitor success or failure of the corrective actions.

9. Refine the benchmark target levels as needed to support continued improvement.

A copy of the benchmarking dashboard through the end of FY2015 has been included as Attachment 1. The benchmarking metrics were developed in January 2015. As can be seen in the dashboard, landfill gas utilization is often in need of improvement (i.e. natural gas supplement usage needed to decrease). What is evident is that following implementation of the dashboard in January and after a few months of intense coordination efforts, the use of natural gas at the BPF decreased, lowering overall O&M costs.

What also can be seen is that the landfill gas skid costs exceeded monthly budget each month. This led to many discussions on gas impurities and deficiencies of the existing system. Ultimately, through good communication and timely reaction, the team was able to develop a possible solution to the impurities that were causing rapid deterioration of the landfill gas compression skid equipment. This reduced the O&M cost per ton each successive month of FY2015.

The dashboard system has proven to be an effective management tool for SWA and its utility partners. There have been several other management techniques that were completed concurrently with the dashboard system to drive operational improvement. They include prioritization via risk matrix of projects; open communication on a frequent basis with the utility partners, contract operator, and SWA staff; and the development of a new computerized maintenance management system (CMMS) that can be integrated with the plant's supervisory control and data acquisition (SCADA) system to provide predictive maintenance in addition to the traditional preventative and corrective maintenance.

RESULTS POST-IMPLEMENTATION

Since implementation of the dashboard system, several operational performance issues have come to light. Some of these issues have been related to performance of the utility partners who deliver the solids to the facility for drying. Other issues have been related to the utilization of landfill gas at the BPF, which is driven by various factors including how much landfill gas is available, the uptime of the landfill gas compression equipment, the performance of the landfill gas/natural gas blending system equipment, and operational decisions at the BPF. By identifying the issues, the dashboard has fostered communication between the different entities involved that ultimately impact operational costs at the facility. This has led to capital improvements, identification and procurement of critical spare parts to minimize downtime, new SOPs, and enhanced training. Ultimately, the combination of these tactics have led to decreased O&M cost per ton at the facility. The Fiscal Year (FY) 2014 cost per wet ton averaged $54.64 prior to implementation of the dashboard. The subsequent FY 2015 cost per wet ton decreased to $47.73 following implementation of the dashboard, even with bringing on the new utility partner.

Additionally, as can be seen in Table 2 on the following page, the team was able to meet each of the strategic performance goals for the BPF.

While the dashboard efforts have proven to be effective in helping to reduce O&M costs for the facility, there were some data gaps identified that would help better assess performance against overall program objectives. It was determined that there is not adequate information to monitor what is the root cause of the use of natural gas at the facility. Additional data and analysis is needed to further determine the root cause of excess natural gas utilization – supply limitations, LFG compression skid performance issues, BPF equipment-related issues and the repair activities, or BPF operational decisions. This likely requires the installation of additional instrumentation and the development of more SCADA reports to tie the data together to easily utilize it in the benchmarking dashboard.

Additionally, the contract operator has installed a new CMMS. Once the system is fully populated, additional KPIs are needed to monitor effectiveness of maintenance activities under the new CMMS. Suggested KPIs have included number of outstanding work orders, percentage of preventative work orders, and percentage of predictive work orders. These will be refined in the near future and added to the benchmarking dashboard.

Table 2: Performance against Strategic Goals and Objectives Post-Implementation of the Benchmarking Dashboard

Goal Number	Goal Description	Metric/KPI	Target	Actual for FY2015
BPF-1	Manage BPF Partner's incoming sludge deliveries to limit loads that go to the SWA landfill	Percentage of all incoming loads delivered to SWA landfill due to BPF being over capacity	< 1%	0%
BPF-2	Ensure continuity of operations at the BPF by minimizing unscheduled downtime	Percent of incoming loads diverted by contract operator to alternate site due to unscheduled facility downtime	< 1%	0.8%
BPF-3	Promote efficient BPF operations by minimizing residue generation	Percent of incoming sludge that leaves as residue	< 1%	0.17%
BPF-4	Continue safe operations at BPF by minimizing number of OSHA reportable incidents	Number of OSHA reportable incidents/yr	< 3	0
BPF-5	Strive for continuous environmental compliance	Number of reportable environmental incidents	< 12	6
BPF-6	Meet budgetary expectations for total O&M cost/ton for the FY	Percent of budget for predicted O&M cost/ton at the end of the FY	<97.5%	84.7%

SUMMARY AND CONCLUSIONS

SWA and its utility partners have found the development of the dashboard and benchmarking metrics to help improve communications and overall performance of the entire team, especially when combined with other organizational management techniques. Ultimately this has led to decreased operational costs at the facility. It continues to be used to monitor progress and identify potential issues early so that the team can collectively determine the best remedial action.

Lessons learned through the benchmarking dashboard process:

- Benchmarking dashboards provide an effective way for various stakeholders to know at a glance how the facility is performing, without needing to comb through multiple lengthy reports and data.
- They allow management to present not only areas where improvement are needed, but also to highlight areas of success in the organization.
- After implementing the dashboard, it likely will be determined that additional data is needed to assess performance and ultimately show root cause of issues. For example, it may show that additional meters are needed within the facility to monitor specific unit processes.
- Data availability is key and the ease of access to it will make the benchmarking dashboard process have long-term sustainability.
- Dashboards can be customized based upon the audience and specific priorities of individual stakeholders. Management could consider separate dashboards for the executive or political level versus what data is presented to operations and maintenance staff.
- Dashboards should be considered as an evolving tool, where KPIs are evaluated routinely and adjusted as needed to achieve overall goals and objectives within the organization.

ACKNOWLEDGEMENTS

The authors of this paper would like to acknowledge the contract operator, New England Fertilizer Company (NEFCO) for their vital role in the implementation of the dashboard and continued efforts to improve facility operations and maintenance; the BPF utility partners and SWA staff from various departments who have played an important role in making the dashboard effort a success; and lastly, Dr. Albrey Arrington of Loxahatchee River District, a BPF utility partner, who was instrumental in the development of the initial dashboard and metrics.

ATTACHMENT 1:
BENCHMARKING METRICS THROUGH FISCAL YEAR 2015

SWA

Biosolids Processing Facility
Monthly Facility Benchmarking

Prepared by: Amber Batson
Month: September 2015

Focus Area		Partner Deliveries												BPF Operations			
		Rejected Loads						Exceedance of BPF Daily Limits Resulting in Load(s) to Landfill						Avg Tonnage Received/ Day	Max Tonnage Received/ Day	Total Tonnage Received at BPF	Tonnage Diverted
		BR	ECR	LRD	PBC	SCR	SUA	BR	ECR	LRD	PBC	SCR	SUA				
Benchmark		# Loads	# Loads	# Loads	# Loads	# Loads	# Loads	# Loads	# Loads	# Loads	# Loads	# Loads	# Loads	wtpd	wtpd	wet tons	% of Incoming
Metric Number		1	2	3	4	5	6	7	8	9	10	11	12	13	14	15	16
Good Performance		0	0	0	0	0	0	0	0	0	0	0	0	Report Only	Report Only	Report Only	< 1%
Caution		1	1	1	1	1	1	1	1	1	1	1	1	-	-	-	< 5%
Improvement Needed		≥ 2	≥ 2	≥ 2	≥ 2	≥ 2	≥ 2	≥ 2	≥ 2	≥ 2	≥ 2	≥ 2	≥ 2	-	-	-	≥ 5%
FY2012 Baseline		0	0	0	0	0	0	N/A	N/A	N/A	N/A	N/A	N/A	246	445	90,899	25%
FY2013 Baseline		0	0	0	0	0	0	N/A	N/A	N/A	N/A	N/A	N/A	296	519	92,153	0.0%
FY2014 Baseline		0	0	0	0	0	0	N/A	N/A	N/A	N/A	N/A	N/A	321	516	101,262	0.4%
2014	Sep	0	0	0	0	0	0	N/A	N/A	N/A	N/A	N/A	N/A	369	516	11,057	0.9%
	Oct	0	0	0	0	0	0	0	0	0	0	0	0	446	593	12,082	0.0%
	Nov	0	0	0	0	0	0	0	0	0	0	0	0	378	537	10,194	6.9%
	Dec	0	0	0	0	0	0	0	0	0	0	0	0	494	576	12,856	0.0%
2015	Jan	0	0	0	0	0	0	0	0	0	0	0	0	502	583	14,002	0.4%
	Feb	0	0	0	0	0	0	0	0	0	0	0	0	495	584	11,890	0.0%
	Mar	0	0	0	0	0	0	0	0	0	0	0	0	471	598	13,672	0.9%
	Apr	0	0	0	0	0	0	0	0	0	0	0	0	471	543	12,729	0.4%
	May	0	0	0	0	0	0	0	0	0	0	0	0	430	573	11,607	0.4%
	Jun	0	0	0	0	0	0	0	0	0	0	0	0	373	544	10,776	2.2%
	Jul	0	0	0	0	0	0	0	0	0	0	0	0	359	553	10,768	0.0%
	Aug	0	0	0	0	0	0	0	0	0	0	0	0	387	525	11,602	0.0%
	Sep	0	0	0	0	0	0	0	0	0	0	0	0	395	514	11,052	0.0%
Consecutive Months at Green		74	74	74	74	74	74	12	12	12	12	12	12	N/A	N/A	N/A	3

N/A = not applicable
** = historical data not available

Revision Date: 10/22/2015

SWA

Biosolids Processing Facility
Monthly Facility Benchmarking

Prepared by: Amber Batson
Month: September 2015

Focus Area		BPF Operations (cont.)			Safety	Maintenance		Passthrough Usage						
		Average Throughput per Train	Final Product Meets Spec	Residue Generation	# of OSHA Recordable Injuries	Engineering Inspection Deficiencies	*TBD after new CMMS commissioned*	Electricity	Gas	Potable Water	Industrial Water	Waste water	Hypo	Caustic
Benchmark		Tons/Hour	%	%	# Injuries	# New Deficiencies	TBD	kWh/wt	mmBTU/ wt	gal/wt	gal/wt	gal/wt	gal/wt	gal/wt
Metric Number		17	18	19	20	21	22	23	24	25	26	27	28	29
Good Performance		≥ 12 and < 14.1	100%	≤ 1%	0	< 5	TBD	≤ 67.1	≤ 2.9	≤ 187	≤ 416	≤ 308	≤ 0.69	≤ 0.12
Caution		≥ 10 and < 12	-	-	-	< 8	TBD	> 67.1	> 2.9	> 187	> 416	> 308	> 0.69	> 0.12
Improvement Needed		< 10 or ≥ 14.1	< 100%	> 1%	≥ 1	≥ 8	TBD	-	-	-	-	-	-	-
FY2012 Baseline		10.3	100%	0.56%	5	**	**	52.2	1.9	29	539	422	0.14	0.03
FY2013 Baseline		9.9	100%	0.10%	0	**	**	62.8	2.7	58	696	635	0.28	0.04
FY2014 Baseline		10.8	100%	0.11%	0	**	**	61.3	2.7	13	848	854	0.17	0.07
2014	Sep	11.7	100%	0.07%	0	3	**	58.4	2.4	7	1,047	954	0.05	0.05
	Oct	12.2	100%	0.07%	0	14	**	55.1	2.4	2	899	846	0.19	-
	Nov	11.9	100%	0.05%	0	N/A	**	55.2	2.5	11	768	740	0.16	0.01
	Dec	12.0	100%	0.04%	0	N/A	**	54.0	2.5	46	618	634	0.11	0.04
2015	Jan	12.1	100%	0.08%	0	16	**	54.2	2.6	23	698	688	0.16	0.03
	Feb	11.9	100%	0.03%	0	2	**	56.3	2.6	20	703	674	0.16	0.04
	Mar	12.4	100%	0.36%	0	15	**	54.5	2.6	21	794	797	0.19	0.04
	Apr	11.9	100%	0.07%	0	4	**	58.0	2.5	20	887	828	0.20	0.04
	May	11.7	100%	0.13%	0	2	**	59.4	2.5	16	976	914	0.20	0.04
	Jun	12.0	100%	0.10%	0	5	**	59.3	2.3	20	1,107	1,002	0.16	0.06
	Jul	10.9	100%	0.22%	0	1	**	60.1	2.4	10	934	718	0.26	0.03
	Aug	11.4	100%	0.16%	0	1	**	58.2	2.4	12	563	489	0.22	0.03
	Sep	11.9	100%	0.81%	0	1	**	57.9	2.2	27	562	511	0.19	0.05
Consecutive Months at Green		0	74	74	36	3	**	24	74	74	0	0	74	21

Revision Date: 10/22/2015

Biosolids Processing Facility
Monthly Facility Benchmarking

Prepared by: Amber Batson
Month: September 2015

Focus Area	LFG Availability & NG Usage			Overall Financial Performance			Environmental Compliance			Initiative Completion	
	Average LFG Available	Skid Downtime	NG Supplement at BPF	Predicted LFG Skid Cost/Ton at End of FY	Actual O&M Cost/Ton, incl SWA costs	Predicted O&M Cost/Ton at End of FY	Reportable Incidents	Permit Non-Compliance	Env. Inspection Deficiencies	Open Items % on Schedule	Open Items % on Budget (SWA only)
Benchmark	mmBTU /hr	hours	%	$ / ton	$ / ton	% of Budget	# Incidents	# NOVs	# Deficiencies	%	%
Metric Number	30	31	32	33	34	35	36	37	38	39	40
Good Performance	Report Only	≤ 10% of total time	≤ 30%	≤ 95% of budget	≤ 95% of budget	≤ 97.5%	< 2	0	≤ 3	≥ 90%	≥ 90%
Caution	-	≤ 15% of total time	< 35%	≤ 100% of budget	≤ 100% of budget	> 97.5%	< 3	1	≤ 5	≥ 80%	≥ 80%
Improvement Needed	-	> 15% of total time	≥ 35%	> 100% of budget	> 100% of budget	> 99.0%	≥ 3	> 1	> 5	< 80%	< 80%
FY2012 Baseline	38.4	**	28%	$ 2.01	$ 45.78	86.7%	8	2	**	**	**
FY2013 Baseline	37.2	**	39%	$ 0.57	$ 51.81	96.5%	4	4	**	**	**
FY2014 Baseline	49.0	**	39%	$ 2.47	$ 54.64	96.6%	12	0	**	**	**
2014 Sep	51.8	44	33%	$ 2.46	$ 40.19	96.1%	1	0	1	100%	100%
Oct	55.5	45	32%	$ 1.62	$ 50.73	87.2%	1	0	2	97%	100%
Nov	48.7	20	54%	$ 1.75	$ 57.50	93.1%	1	0	1	97%	N/A
Dec	43.4	3	46%	$ 1.84	$ 54.26	93.3%	0	0	1	97%	N/A
2015 Jan	39.5	10	59%	$ 3.06	$ 59.98	92.9%	0	0	2	97%	N/A
Feb	37.2	23	44%	$ 3.08	$ 55.12	92.9%	1	0	0	79%	N/A
Mar	42.8	62	67%	$ 3.42	$ 62.27	93.4%	0	0	0	83%	N/A
Apr	51.5	105	33%	$ 3.63	$ 51.95	92.8%	0	0	3	86%	N/A
May	70.4	3	48%	$ 3.59	$ 45.35	92.9%	0	0	0	87%	N/A
Jun	71.5	2	27%	$ 3.74	$ 34.95	91.8%	0	0	0	86%	N/A
Jul	71.9	50	43%	$ 3.74	$ 35.48	91.4%	1	0	0	86%	N/A
Aug	74.4	13	46%	$ 2.92	$ 34.82	88.8%	1	1	0	86%	N/A
Sep	73.4	3	18%	$ 2.47	$ 29.49	84.9%	1	1	3	100%	N/A
Consecutive Months at Green	N/A	5	1	0	6	18	15	0	16	1	N/A

Comments:
33 - LFG Skid costs continue to be high due to increased H2S levels in LFG.
34 and 35 - Sept 2015 numbers are preliminary and subject to final reconciliation.
37 - NEFCO had NOVs in August and September for pH exceedances in their process wastewater discharge to WPB

Revision Date: 10/22/2015

Biosolids Processing Facility
Monthly Facility Benchmarking

Prepared by: Amber Batson
Month: September 2015

Focus Area	Community Stewardship			Knowledge Retention	
	Public Odor Complaints	Public Noise Complaints	Outreach Efforts (tours, papers, presentations)	SOP Development/ Document Updates	
				NEFCO	SWA
Benchmark	# Complaints	# Complaints	#	# of Docs Issued	# of Docs Issued
Metric Number	41	42	43	44	45
Good Performance	0	0	Report only	Report only	Report only
Caution	1	1	-	-	-
Improvement Needed	> 1	> 1	-	-	-
FY2012 Baseline	**	**	**	**	**
FY2013 Baseline	**	**	**	**	**
FY2014 Baseline	**	**	**	**	**
2014 Sep	**	**	**	**	**
Oct	0	0	0	0	1
Nov	0	0	0	0	0
Dec	0	0	0	0	0
2015 Jan	0	0	0	0	1
Feb	0	0	0	0	0
Mar	0	0	2	1	0
Apr	0	0	2	0	0
May	0	0	2	0	0
Jun	0	0	1	0	3
Jul	0	0	1	0	0
Aug	0	0	0	2	0
Sep	0	0	0	0	0
Consecutive Months at Green	12	12	N/A	N/A	N/A

Revision Date: 10/22/2015

The Trials and Tribulations of Commissioning a Large Biosolids Facility; An Operational Perspective

Kenneth Schnaars, PE[1] , Terry Highsmith[2]
[1] Brown and Caldwell, Nashville, Tennessee
[2] Metro Water Services, Nashville, Tennessee
Email.kschnaars@brwncald.com

ABSTRACT

The commissioning of a large biosolids facility can be challenging. It is even more challenging being called in after part of the system has been partially started, when one of the anaerobic digesters is experiencing a rapid rise event and a local news station is flying a helicopter overhead recording the problem. The Nashville Biosolids Facility is a large biosolids facility that processes 100 dry tons of solids per day (90 720 kg/d). The biosolids facility was constructed as a Design/Build/Operate type project. The facility consists of four 65 ft. (20 m) diameter dissolved air flotation thickeners, five 108 ft. (33 m) diameter anaerobic digesters, five solid bowl centrifuges and two 80 ton/day thermal direct drum dryer systems. The Nashville Biosolids Facility receives sludge and scum from two wastewater treatment plants; the 125 mgd [473 ML/d] (ADF) Central WRRF and the 37.5 mgd [142 ML/d] (ADF) Whites Creek WRRF.

During the initial commissioning phase of the facility there were process issues that had to be corrected by the design team that help minimize some of the initial rapid rise events and sludge mixing pump gas locking problems. A major digester design issue that was identified during the commissioning phase was that no liquid is removed from the surface of the digester thus trapping foam. Digester feeding, mixing and sludge removal all occurred below the liquid surface. A couple of the initial improvements that occurred during the commissioning phase was the addition of a foam suppression nozzle above the liquid surface and the extension of the sludge removal pipe to the liquid surface. After a 1.5-year commissioning/operating phase the biosolids facility passed a strict Reliability Testing Phase and was then turned over to the client. However, three months into the fully operational phase of the facility the City of Nashville suffered a major flood that shutdown the Biosolids Facility. During the facility commissioning phase, after the flood and during normal operation there were several membrane cover failures limiting the Owner's ability to store digester gas.

Through all of the trials and tribulations the Biosolids Facility has been successfully producing Class A solids that are pelletized and sold to produce fertilizer. Presently, the Biosolids Facility is going through several design improvements to reduce rapid rise events, improve reliability, increase production, and optimize process performance.

This presentation will provide an overview of the:

- Initial commissioning complexities
- Design modifications during commissioning
- Facility commissioning after a major flood
- Continuous digester rapid rise events
- Membrane gas holding cover failures
- Operator training programs
- Project improvements currently in design and/or near term design to minimize rapid rise events and optimize process operation

Biosolids facilities are becoming more sophisticated and complicated. Operators and maintenance personnel must understand that it not just a simple matter of opening up a valve. This presentation will provide a practical understanding to operators and maintenance personnel about what they should be aware of during the commissioning of a complex facility. It will also stress to plant personnel the importance of their involvement during project design phase so they can understand the process equipment that will be installed at their facility.

KEYWORDS: Anaerobic digestion, rapid rise, foam, centrifuge.

INTRODUCTION

The City of Nashville owns and operates three Water Resource Recovery Facilities (WRRF) referred to as the Central WRRF, Whites Creek WRRF and Dry Creek WRRF with average dry weather flow capacities of 125 mgd [473 ML/d], 37.5 mgd [142 ML/d] and 24 mgd [91 ML/d], respectively. The city of Nashville has struggled for years in determining the best way to dispose of the solids from their facilities. Up until the late 1990's or early 2000's, the biosolids from the Central and Whites Creek WRRF were incinerated in multiple hearth incinerators. After the incinerators were shutdown, the biosolids from the Whites Creek and Central WRRF were dewatered on belt filter presses and transported to a landfill. The biosolids at the Dry Creek WRRF are currently anaerobically digested, dewatered on belt filter presses and transported to a landfill. In 2004, Metro Water Services embarked on the design of a new $118 million dollar biosolids facility to process the biosolids from the Central and Whites Creek WRRFs. The project was designed and constructed using a design-build-operate procurement type format. The facility commissioning phase began in January 2008 after the construction phase testing was completed in December 2007.

Starting of a large and complex biosolids facility requires careful planning, good communication, coordination and cooperation between all parties, patience and numerous other tasks. The commissioning of a large facility does not mean you just press the start button and walk away. Problems normally occur when commissioning any system at a wastewater facility. However, knowing what to do and how to correct a problem safely and minimizing risk is important. This paper describes some of the problems that occurred at this particular biosolids facility and how these problems were managed or corrected.

BIOSOLIDS FACILITY

The Nashville Biosolids Facility serves the Central and Whites Creek WRRFs in Nashville, Tennessee. The combined primary sludge, waste activated sludge and primary and secondary scum from the Whites Creek WRRF are pumped approximately 6 miles to the Sludge Transfer Pumping Station located at the Central WRRF. At the Sludge Transfer Station, the Whites Creek solids are combined with the Central solids. The Central solids consist of primary sludge, waste activated sludge and primary and secondary scum. The solids from the both plants are mixed and pumped the biosolids facility located at the Central WRRF. The Nashville Biosolids Facility receives over 100 dry tons of solids per day (90 720 kg/d).

At the biosolids facility the solids are first thickened in four 65 ft. (20 m) diameter dissolved air flotation thickeners (DAFTs). The subnatant from the DAFTs is returned to the Central WRRF for re-processing. The thickened sludge from the DAFTs is pumped to four 108 ft. (33m) diameter primary anaerobic digesters each with a 2.5-million-gallon (9 463 m^3) capacity, as shown in Figure 1. Primary Digesters 1-3 have fixed covers and Digester 4 which can serve as primary or secondary digester was originally designed with a membrane type gas storage cover. The four primary digesters are operated as mesophilic digesters at a constant temperature of 95°F (35°C). The digested sludge from the primary digesters is transferred to a 108 ft. (33m) Digested Sludge Storage Tank (DSST) prior to being pumped to five solid bowl centrifuges. The Digested Sludge Storage Tank was designed with a membrane type gas storage cover.

Figure 1 --- Primary Anaerobic Digester

The sludge is pumped from DSST to five solids bowl centrifuges each designed for a solids loading rate of 2,600 lbs./hr (1 180 kg/h). The dewatered solids from the centrifuges can be transferred to a truck loading station where it can be hauled to an off-site landfill at a total solids concentration of 30 to 35% solids. The normal mode of operation is to transfer the dewatered solids to two thermal direct drum dryer systems as shown in Figure 2.

Figure 2 – Thermal Dryer Floor

The two thermal drum dryer systems reduce dewatered solids to pellets by maintaining the biosolids at a preset temperature for over 30 minutes. The temperature at the outlet end of the drum is maintained at 200°F (93°C). The final Class A solids are then trucked to a facility where it is converted to a marketable fertilizer product. The dewatered solids from the centrifuges to the dryers are maintained at a total solids concentration between 19 to 21%. The final pellet product is maintained at a pellet size of 4 mm and at a dry solids concentration of approximately 94%.

COMMISSIONING PHASE

The Nashville Biosolids Facility started its commissioning phase in January of 2008. By August of 2008 two primary digesters experienced rapid rise events causing the digester covers lift up and pull out their anchor bolts from the concrete. There were other rapid rise incidences that resulted in sludge spilling over the top of the digester walls but the cover hold down straps remained intact. It was determined that the digesters did not have a method of removing sludge from the digester surface. Digester feeding, mixing and solids removal all occurred below the liquid surface. To initially combat this problem, the digester operating levels were lowered until the design build team completed several modifications to the digesters.

- One of the modifications included extending one of the sludge mixing nozzles from the bottom of the digester to the top of the digester above the liquid level to act as spray nozzle to break down the foam.
- Another modification included extending the digested sludge take-off from the mid-point of the digester to the operating surface of the digester so sludge and trapped foam could be removed from the digester surface.
- The last modification to each of the fixed covered primary digesters included the installation of a weighed relief hatch on the digester cover. Even with these initial improvements, the commissioning team and plant operators climbed the top of the digesters approximately six times a day to monitor the foam level in the primary digesters.

The initial digester modifications helped the operation by removing solids from the liquid surface with the new withdrawal pipe and by spraying down the foam using the foam suppressing nozzle but did not completely solve the rapid rise issues. To this day, these digesters experience rapid rise events that an occur due to several reasons, but the main reason is when a digester mixing pump becomes gas locked. To help minimize the pump gas locking issue the Owner installed visual motor amperage readings onto the pump SCADA screen for each sludge mixing pump. This has allowed the operators to monitor the pumps and be proactive when they see the motor amperage change indicating the pump is about to gas lock. Once the operators notice that the pump amperage is changing then the operators can bleed the pump volutes before they stop completely due to gas locking.

Later in this paper, there is a discussion of the current digester improvements under construction to help minimize rapid rise events and their devastating effects.

During the commissioning phase the membrane cover on Digester 4 suffered a leak due to the over pressurization of the gas system during a rapid rise event. The Digester 4 cover was replaced before the design build contractor left the site, however this membrane cover suffered other failures after the facility was turned over to the Owner. These other membrane cover instances are described later in the paper.

When the centrifuges were first started they would operate for approximately four hours and then shut down due to blockages in the centrifuge and/or dewatered sludge conveyors. These blockages led to long hours of sludge clean up and dismantling of equipment to remove sludge blockages. It was determined that the centrifuges were overloaded causing them to shut down. This problem occurred due to operator inexperience in operating this new piece of equipment. After some additional operator training the centrifuges performed well. However, there was another issue with the centrifuges that hindered the commissioning team from efficiently operating and optimizing the centrifuges and that was because the five centrifuges were shipped without the piping to capture the centrate. The commissioning team could not effectively collect centrate samples to determine if the units were operating efficiently. Not knowing the centrate quality restricted the commissioning team from determining the capture rate of the individual centrifuges. The Owner installed the proper centrate piping after the facility was officially turned over to the City.

Thermal direct drum drying systems have numerous components that can overwhelm operators that have no experience with this type of process. The threat of fires, the numerous monitoring points, and the complexity of the process requires good training and actual operating time with the system to make this process successful. During the initial commissioning phase, it was determined that each dryer process train operated slightly different making operators sometimes favor operating one train over the other. Another issue that was occurring was the faulting and shutting down of the digester gas dryers that were supplying the digester gas to the thermal dryer burners. When the digester gas dryers would fault out, it would sometimes affect the thermal dryer and cause process swings that the operators would have to correct and get under control. The digester gas dryer manufacturer could not find the problem with their equipment. The Owner recently found a mechanical company that corrected the digester gas dryer faulting issue which has resulted in the facility using more digester gas to operate the thermal drum dryers.

There are other examples of process situations that occurred during commissioning, but they will not be provided now with this paper. As each problem arose, there was an organized and detailed corrective action plan that was taken to fix the situation.

NASHVILLE FLOOD

Early in 2010 the biosolids facility successfully completed a strict reliability test thus completing the commissioning phase of the project. Upon completion of the process reliability test in March of 2010 the facility was turned over to the Metro Water Services (Owner) ending the project. In May of 2010, Nashville was devastated by a major flood that resulted in submersing the biosolids facility in 4 feet (1.2m) of water. The flood shut down the facility for months. When the biosolids facility shut down two aeration tanks at the Central WRRF were converted into aerobic digesters so the Central and Whites Creek WRRFs could remain in operation until the biosolids facility was back on-line. During this time trailer mounted belt filter presses were brought in to dewater the aerobically digested sludge so it could be transported to a landfill.

While the biosolids facility was out of service the membrane covers on Digester 4 and the Digester Sludge Storage Tank (DSST) deflated and were damaged due to stretching. When the biosolids facility was going through flood recovery repairs new membrane covers were installed on Digester 4 and the DSST. After the covers were replaced on Digester 4 and the DSST they failed again. It was determined that the covers failed at the seams due to a change in the in the manufacturer's production procedures, so another set of membrane covers were installed on these two tanks. The new membrane for Digester 4 then failed again at the seams so another membrane was placed on the digester. The third membrane cover for Digester 4 during the flood repairs did not fail. During the replacement of Digester 4 membrane cover it was decided that original side gas outlet take-off needs to be modified to a top gas take-off design because of all the digester foam entering the gas piping through the side outlet.

Lingering and persisted problems followed after the flood, such as electrical issues, so additional repairs needed to be made. A number of electrical connects and cables were corroded, so the Owner contracted with an electrical firm to replace electrical cable and components affected by the flood water.

ON-GOING ISSUES and FACILITY IMPROVEMENTS

Since the commissioning phase, the Owner has made several improvements to the Biosolids Facility that have enhanced process performance, improved maintenance, and increased digester gas utilization. Along the way, the facility is still experiencing rapid rise events in their digesters that have lifted their fixed covers and pulled out anchor bolts from the concrete, as shown in Figure 3. Many of the rapid rise events occur when a sludge mixing pump becomes gas locked. When a rapid rise event occurs, it is very quick and expansive that the single relief hatch on the cover cannot effectively relief all of the sludge so the pressure under the cover becomes so great that it lifts the hold down straps.

Figure 3 -- Damaged Digester Hold Down Straps

In January of 2014 ice formed in the digester gas take off pipe on Digester 4 that plugged the pipe with ice. The plugged pipe restricted digester gas from leaving the digester thus resulting in a major failure of the membrane cover, as shown in Figure 4. The gas take-off pipe was plugged upstream of the digester pressure relief valves so these pressure relief valves never sensed a problem. The cover on Digester 4 is now being replaced with a fixed cover.

Figure 4 – Membrane Digester Gas Holder Failure

In December of 2016, construction started on an improvement project that will reduce rapid rise events and protect the fixed covers on the primary digesters. These improvements include the removal of the existing pumped mixing systems in the primary digesters (Digesters 1-4) and replacing them with internal draft tube mixers. Internal draft tube mixers were selected because the digesters are prestressed type tanks so cutting holes in the tanks is not possible. In addition, external standpipes are being installed to remove sludge during normal operation and safely remove large volumes of sludge during rapid rise events to protect the digester covers. New sludge transfer pumps are being installed that will remove the sludge from the standpipes during normal flow conditions and pump it to the DSST. During a rapid rise event, the transfer pumps will be operation but if they cannot keep up with the sludge from the digester then the sludge will be passively routed to a larger pump transfer station located on the biosolids complex. As stated earlier, this project also includes the replacement of the damage Digester 4 membrane cover with a fixed cover. There are other aspects to this project such as raising outside transformer pads so the transformers are above the flood of record. This digester improvement project is being performed as a design-build type project and should be completed in 2018.

OPERATOR TRAINING

Due to the complexity of the biosolids facility, the operators were provided extensive training. The operators were provided equipment manufacturer training during the construction phase of the project. Most of the manufacturer training occurred prior to the operation of the equipment. The field training during this period was noisy so many of operators could not hear the trainer. In addition, all this equipment was new and never seen before by most of the operators, so they did not know what questions to ask and they were confused when the equipment was initially placed on line. This was very evident during the commissioning phase of the project so additional training was required.

The operator training that occurred during the commissioning phase was performed by the Commissioning Engineer and it covered process theory, equipment operation, troubleshooting, process mathematics and other important process information. At this point of the project, the operators had some operating experience with the equipment and processes so this additional knowledge helped them to optimize the different processes. During these training sessions the operators were provided with a copy of the process Powerpoints that highlight design and operational consideration of each process. Other handouts included equipment and process troubleshooting information, process flow diagrams and other important information.

The Owner is very committed when it comes to operator training, and they see the importance of keeping their operators up to date with process knowledge. The Owner, like other municipalities, have plant staff leave the utility or they are reassigned within the utility, so new operators and maintenance personnel are replacing the personnel that have left. The Owner conducts refresher training courses that have involved the original Commissioning Engineer and/or manufacturers so the new operators and maintenance personnel can improve their knowledge of the various equipment and processes. The Owner periodically conducts specific equipment maintenance training, instrumentation training, safety, and other knowledge based training for their staff.

SUMMARY

Commissioning large a biosolids facility is time consuming and requires engineers and operators with experience and excellent troubleshooting skills. Commissioning large biosolids facilities requires the cooperation and understanding between all parties involved and those parties includes the manufacturers, engineers, contractor, and Owner. Never go into a start-up and not expect a problem to occur.

Before starting a biosolids facility prepare an operating plan, a risk management plan, safety plan and other documentation to help make the project go smoothly. Having good operating records will help monitor the processes and provide information to determine if the process is operating properly or if operational adjustments must be made to avoid a problem.

ACKNOWLEDGEMENT

The author would like to acknowledge the Nashville Metro Water Services operations and maintenance staff for their support and hard work through the commisioning phase and the continued operation of the Biosolids Facility.

Optimizing THP – The intermediate thermal hydrolysis process

Rus, E[*]; Perrault, A.[*]; Mills, N.[*]; Shana, A[*]; Molokwu, O.[*]; and Nilsen, P.J.[**]

[*] Thames Water, Wastewater Innovation, Reading, UK.
[**]CAMBI, R&D Department, As, Norway.
Ester.rusperez@thameswater.co.uk

Abstract

Thermal hydrolysis has proven to be an efficient pre-treatment process for sludge before anaerobic digestion (AD), by thermally enhancing organic matter hydrolysis, the rate limiting step in AD (Shana *et al.*, 2013). Recent research (Wilson and Novak, 2009, Shana *et al.*, 2012, Mills *et al.*, 2014) has shown that a new configuration with the existing technology can further enhance the efficiency of the system. The intermediate THP gave good results at laboratory scale, with volatile solids destruction of 64-67% and an average overall specific gas production of 500 m^3 per tonne of dry solids (Shana *et al.*, 2012).

For the past year, intermediate thermal hydrolysis has been tested in the Sludge and Energy Innovation Centre pilot plant located at Basingstoke sewage treatment works.

The pilot facility has allowed operational considerations to be explored and understood to inform the design and construction of full scale intermediate THP. Data and operational experience from the pilot plant is discussed and presented in this paper.

Key words

Intermediate THP; optimization; biogas; anaerobic digestion.

Introduction

The thermal hydrolysis process (THP) is a 20 year old technology that has proven to be successful at improving the efficiency of gas production during anaerobic digestion (AD) and reducing the organic matter and total solids mass present in digested sludge compared to that of conventional AD (Pickworth *et al.*, 2006). Furthermore, digester volume is reduced by about 50%. Dewaterability of digested sludge also improves, resulting in a reduction of transport costs, and the enhanced quality of the dewatered digested sludge makes this product more appealing for agricultural reuse. Its presence in the water industry is rapidly growing with 23 full scale THP worldwide currently operative or in construction (Mills *et al.*, 2014).

It is believed that the combined actions of high pressure, heat and sudden de-pressurisation further break down the organic matter, making it more available for bacteria to digest during the second stage digestion.

However, there are some limitations to the current THP setting. The process itself has a high energy demand in the form of steam being injected in the reactor to reach 6.5Bar for 30 minutes, resulting in additional support fuel required (Mills, 2011). Some research has started recently to optimise the process both in terms of increasing the gas yield and cutting down in energy demand.

The configuration that is discussed in this paper (intermediate THP) consists of a first stage of digestion, followed by THP and a second stage of digestion. At lab scale it showed an increase of 18% in biogas yield compared to that of THP-AD, with volatile solids destruction (VSD) of 65-67% (Shana *et al.*, 2012). The steam consumption is also reduced due to a smaller throughput after the 1st digester and additional heat is also available for steam production due to higher biogas yields. These results led to the construction of an intermediate THP (ITHP) pilot plant at Basingstoke STW. This paper shows the performance of the pilot plant during its optimisation. The objective of this project is to evaluate whether the process is operable at realistic scale as well as replicating the gas yield and improved dewaterability observed by Shana *et al.*, (2012).

Materials and methods

1. Plant design and configuration

Figure 1 shows the pilot plant at the Sludge and Energy Innovation Centre in Basingstoke STW.

Figure 1: Basingstoke ITHP Slugde and Energy Innovation Centre.

The layout of the ITHP plant built at the Sludge & Energy Innovation Centre is as follows (Figure 2).

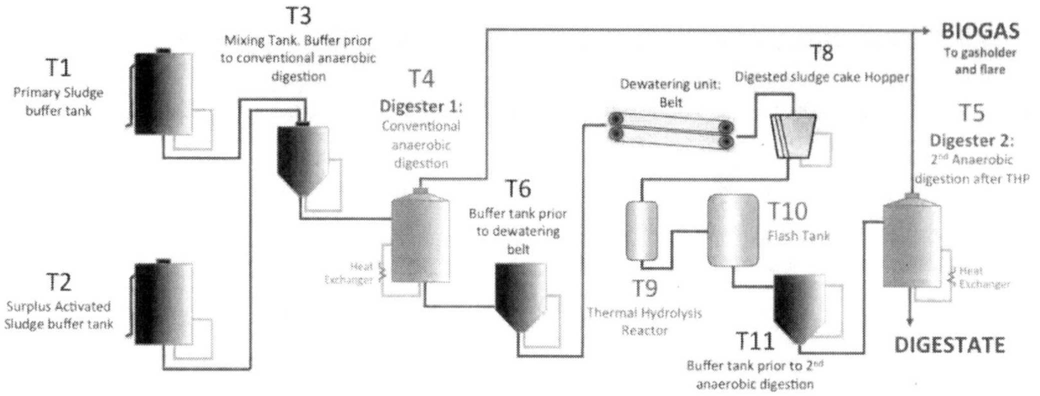

Figure 2: Plant layout.

The plant is not connected to Basingstoke sewage treatment works (STW). The two 20m^3 imports tanks (T1 & T2) at the front end can receive sludge from different sites by truck delivery, which allows for more flexibility in terms of sludge origin, type, dry solids (DS) content, age, etc. Typically, T1 is used for primary sludge (PS) and T2 for waste activated sludge (WAS). In order to keep the sludge well mixed, each tank is fitted with recirculation (40min/h) and air mixing (5min/h). Each tank would receive a weekly delivery of PS and WAS respectively.

The two types of sludge can be mixed at a desired ratio into the blending tank (T3) on a daily basis based on their respective DS. Tanks 1, 2, and 3 are fitted with recirculation line and air mixing in order to keep a well-mixed sludge and avoid stratification. A macerator was been added to tank 3 to break up potential rags or other screenings.

On an hourly basis, a controlled volume of sludge was added to the first digester (T4), based on the agreed organic loading rate (OLR) of 3 kgVS/m^3/d and measured % DS and volatile solids (VS) of the feed sludge. Both digesters were kept at mesophilic conditions between 38 and 39°C, by the use of automatic valves connected to two separate heat exchangers, heated by a water boiler. The gas produced was measured with flowmeters and continuously transferred into a gas holder.

Twenty minutes after each feed, the same volume of sludge was pumped out of digester 1 and transferred into a holding or buffer tank (T6). The volume of each digester was kept constant by checking the flows of each volume in and out. They were mixed by continuous recirculation of sludge.

Three times a week, conventionally digested sludge was dewatered using a belt-press. The liquid polymer is first diluted with high pressure final effluent and mixed with the sludge from T6 into the pipe delivering the flocculated sludge to the belt press. The cake obtained in the hopper after the dewatering stage averaged 23% DS. The cake was then transferred into a hopper (T8) to be diluted with hot water to about 14.5% DS. The dilution in T8 was controlled by a pressure gage located in the recirculation loop. Controlling the Klampress to produce a cake at 14.5% was found to be complicated and this pre-dilution system allowed for a consistent DS prior to the THP. The diluted cake was then recirculated for about 30 min and the pressure during recirculation of sludge was used as a reference for DS.

The pre-heated cake was transferred into the reactor tank (T9), where it was thermally hydrolysed. High pressure steam from the steam boiler was injected in the reactor bringing the mix to about 165°C and 6.5 bar. The sludge was kept under those conditions for 30 mins (set point can be changed if required) before being "flashed" in a matter of seconds into the flash tank (T10). Once hydrolysed, the sludge was transferred into another holding tank (T11). The tank was insulated in order to keep as much heat as possible and also fitted with a dilution to bring the %DS down to around 7%. It was observed that 7% DS was the maximum this particular site could cope with due to design limitations. This is not expected at full scale, with the option of going much higher in that case.

The operating regime of the second digester was similar to the first digester, where a set volume of hydrolysed sludge was transferred to the second digester (T5) on an hourly basis. Once again, this volume of sludge going in the second stage digester was based on an OLR of 5kgVS/m^3/d and measured %DS and %VS of the feed sludge from T11. The same volume was also pumped out of the digester (20 min after the feed) and directed to the drain. The biogas produced is measured and transferred into the same gas holder.

The gas holder allowed a constant pressure in the digesters by the use of a floating roof. When the roof is at a specified high level, the biogas is diverted to a flare stack where it is burned. An additional gas flowmeter was installed between the gas holder and the flare stack to measure total biogas generation. This point in the gas line proved to give the best most stable gas readings.

2. Analysis

All the tanks listed in Table 1 were sampled three times a week on site for DS, VS, volatile fatty acids (VFA), alkalinity, pH, ammonia. If any of the parameters looked into showed a problem in the process, sampling would be done on a daily basis until the digesters showed signs of stability. This was also done during ramp up periods. Table 1 shows the basic sampling schedule in more detail:

Table 1: Laboratory analysis done on Mondays, Wednesdays and Fridays.

Sludge Type	Analysis
Primary sludge (T1)	DS, VS*
Secondary Sludge (T2)	DS, VS*
Combined Sludge (T3)	DS, VS
First Stage Digested Sludge (T4)	**DS, VS, VFA, Alkalinity, pH, Ammonia**
Dewatered Cake (Belt Press)	DS, VS
Diluted Cake (T8)	DS, VS
Hydrolysed Sludge (T11)	DS, VS
Second Stage Digested Sludge	**DS, VS, VFA, Alkalinity, pH, Ammonia**

*Once a week and on delivery date.

Additionally, the biogas composition from each digester was measured with a gas analyser also at least three times a week.

Once a week, samples of each type of sludge are sent to Thames Water (TW) laboratories for comparison and ensuring data reliability.

Two different methods were used to record VFA:

(1) Titration: carried out three times a week, analysis of the four sludges previously specified is done onsite and on the same day of collecting samples. The methodology involves 2hr pre-centrifugation at 3000rpm. The analysis is done on 5ml of supernatant diluted in 50ml of DI water. Duplicates were done for the feed sludge and triplicates for digested sludge samples.

(2) Gas chromatography with Flame Ionization Detector: Done twice a week at Thames Water laboratory. This method allows for speciation of VFA and involves a pre-centrifugation of the samples.

Results and discussion

1. VFA, alkalinity, pH and ammonia

Figure 3 shows the OLR for (a) digester 1 and (b) digester 2 with their respective VFA concentrations for the entire period. Digesters needed to be ramped up in August 2014, December 2014, and May 2015 after periods of maintenance.

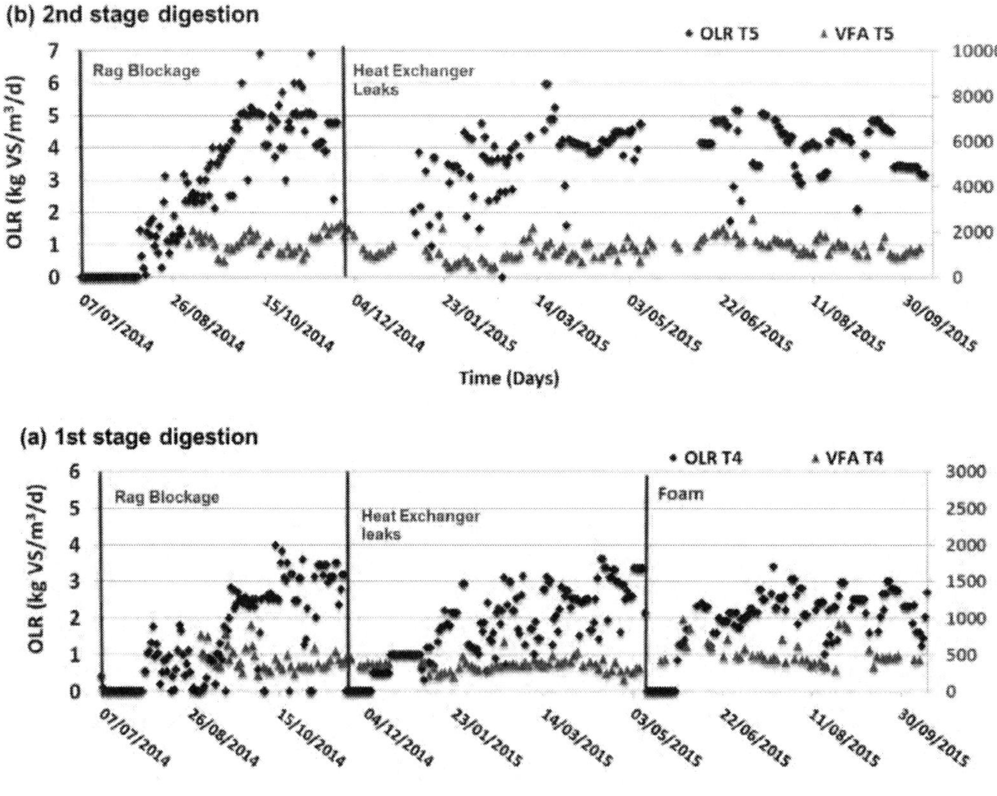

Figure 3: OLR and VFA of (a) digester 1 and (b) digester 2 for the period running from May to March 2015.

VFA remained fairly stable for both digesters at an average of 370±70 mg/L in digester 1 and 1448±478 mg/L in digester 2 with slight rises in both digesters during ramp up periods. Alkalinity also remained very stable for both digesters with average values of 3461±257mg/L and 4878±620mg/L in digesters 1 and 2 respectively. VFA/Alk ratios were within acceptable levels for digester 1 at 0.11±0.02. Digester 2 showed a VFA/Alk ratio of 0.32±0.13. Ammonium stayed below 3000mg/L in both digesters, with an average of 854±96 mg/L in digester 1 and 1678±246 mg/L in digester 2. The pH remained at 7.4±0.2 for digester 1 and slightly higher, 7.8±0.3, in digester 2.

2. Volatile solids destruction and specific gas production

During the initial period, depicted in Figure 4, the plant was still being commissioned. Many changes have taken place in order to optimise the pilot.

Rag blockages have been a big issue in the receiving tanks (T1, T2 and T3), blocking the feed pumps with the consequent stops in feed. The PS:WAS ratio in the feed was also difficult to control. Heat exchanger leaks also had an impact on the digesters, with a considerable drop in temperature. During this period, the average total gas yield was 330±23m^3/TDS and a VSD of 45±3%, or circa 30% less biogas and VSD than expected from a fully functioning ITHP set up based on the lab results from Shana *et al.*, (2012). Since October 2014, low temperatures (70°C) at 6.5 bar in the THP due to poor steam-sludge mixing in the reactor during batches was observed, which resulted in incomplete hydrolysis. Additionally, poor mixing in digester 2 was identified as a possible cause of low VSD in the second phase. A lithium test in this digester indicated a 20% reduction in the active working volume (WV).

After November 2014, DS in the feed sludge to the THP was dropped a few % points which led to an increase in temperature of the hydrolysis reactor to 158°C. A venting system was added to bleed off the inert gases from the reactor which rose the mean temperature to 161°C during batches. Figure 4 depicts the temperature and pressure in the THP reactor during a given batch. It shows the improvement in temperature of the reactor with the changes made:

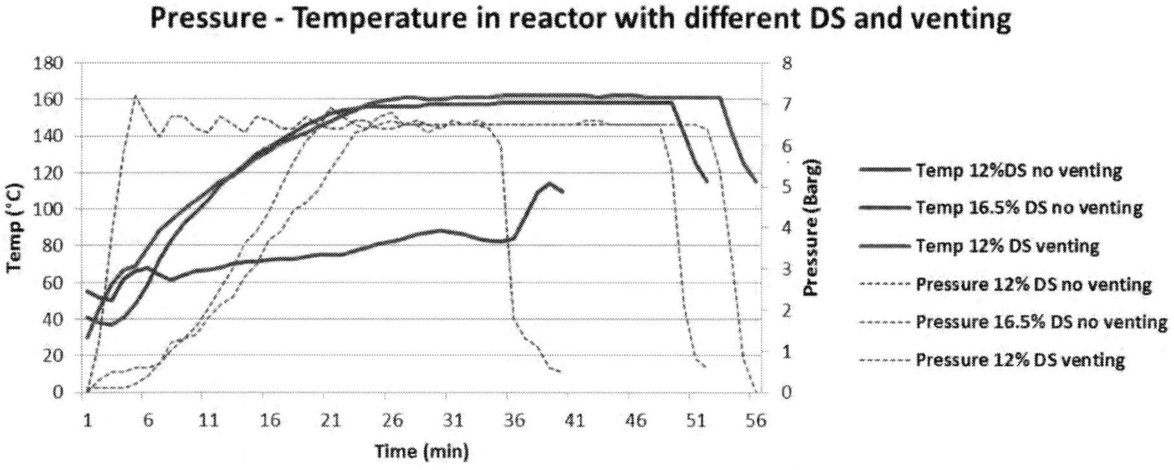

Figure 4: Pressure-Temperature relationship in T9 with DS and venting. THP batch profiles.

The feed in second stage digester was dropped from 9% to 7% DS and WV increased from 5.5 to 6 m^3. All adjustments had to be done to correct for the artefacts associated to this particular pilot plant so full scale expected results could be obtained. Furthermore, a 50:50 PS to WAS ratio was changed to 60:40 to mimic the lab scale work. These changes had an impact on the SGP and VSD as seen in Figure 4:

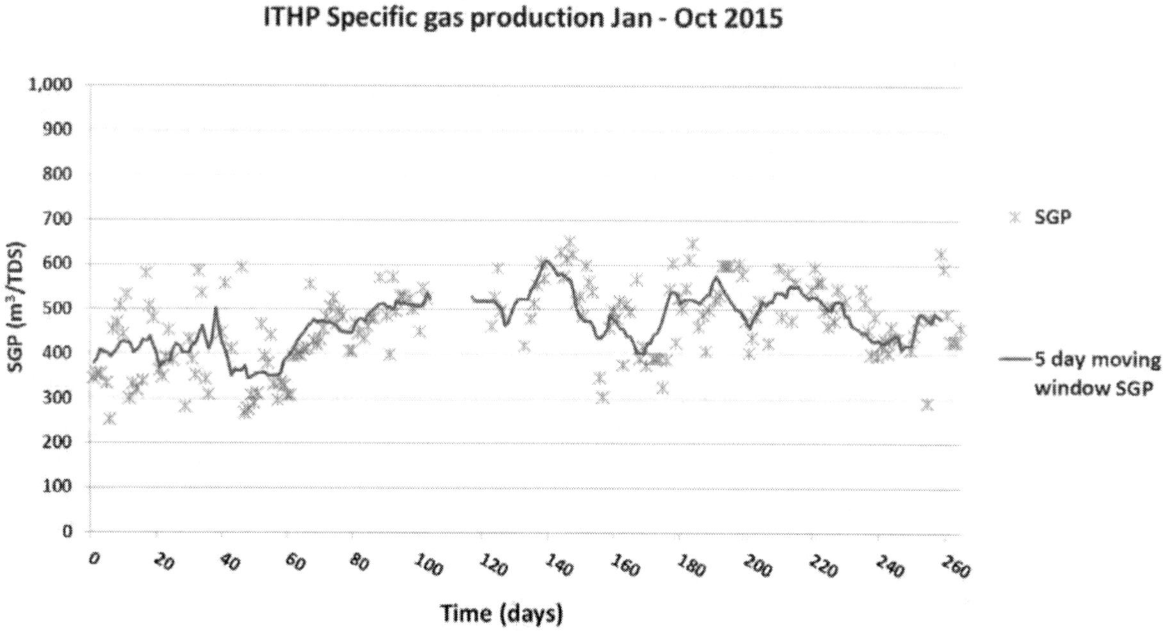

Figure 4: Specific gas production in m^3/TDS for the stable period (January – October 2015).

From January 2015, the site saw a clear improvement in overall gas production, averaging 505±81m^3/TDS for both digesters combined from March to October 2015. The lower points registered around day 160 coincide with a low VS% feed that the site received during those days. The efficiency of gas production dropped and quickly picked up after day 179 when a higher VS% feed was received. Some higher values of SGP around day 150 coincide with a higher PS % fed into digester 1 due to issues with WAS feeding.

VSD was calculated with a Mass Balance method and clearly correlates with the SGP. The period of March-October 2015 saw an average of 64±9% VSD across both digesters.

Both biogas yield and VSD found at pilot scale matched what was found previously at lab scale by Shana *et al.,* (2012): 500m^3/tDS and 64.5% VSD respectively.

3. Dewatering of final digested sludge:

The dewatering trials were carried out on two different dewatering units:

(1) Klampress belt press – The unit available onsite, used to dewater the 1st stage digested sludge, was used. The results showed DS of 34%.

(2) A pilot Bucher Press was trialled for 3 weeks, achieving 42%DS with different SNF's polymers and a maximum of 44% DS with Floplam HIB.

In both cases, ahigh energy mixer from the supplier SNF was used to allow for dual polymer injection. The sludge from digester 2 was held in a buffer tank prior to dewatering and hence, the temperature of the sludge at the time of dewatering had drop to 12°C. Lab scale work has shown that the same sludge dewatered at lower temperatures give lower DS in the cake. Therefore, the results obtained in both presses could see an improvement at temperatures closer to 35°C.

The Bucher results not only showed in the final cake (14% higher DS than what we commonly see onsite for conventional THP+AD digested sludges), but also a high rate of water release. This could lead to higher throughput in the full scale dewatering units.

Conclusions

The current results show that operating THP on digested sludge is achievable and that the second stage digester is very stable, being resilient to shock loads and sudden changes in temperature. The average process performance is similar to that seen in the lab work, with an average of 505±81m^3/TDS of biogas yield and 64±9% VSD over a stable period, making ITHP 17% more efficient in terms of gas generation than conventional THP. The final cake product achieved 44%DS with a Bucher press. The high rate of water release would also allow for high throughputs on the dewatering units. ITHP is therefore viable as a full scale option for Thames Water which currently holds a total of 9 sludge centres with THP. A regional ITHP project is currently being proposed for some of these sites which will allow for substantial energy savings.

References

Mills, N. (2011) The influence of heat balance on the economics of advanced anaerobic digestion processes. In: *16th Biosolids Conference: Organic Resources*. Exhibition. Aquaenviro. Leeds.

Mills, N.; Pearce, P.; Farrow, J.; Thorpe, R.B.; and Kirkby, N.F. (2014) Environmental and economic life cycle assessment of current and future sewage sludge to energy technologies. *Waste management*. **34**, 185 – 195.

Pickworth, B.; Adams, J.; Panter, K.; and Solheim, O.E. (2006) Maximising biogas in anaerobic digestion by using engine waste heat for thermal hydrolysis pre-treatment of sludge. *Water science and technology*. **54** (5), 101-108.

Shana, A., (2012) A study on the impact of an innovative Intermediate thermal hydrolysis process on the performance of anaerobic sewage sludge digestion process. *IWA World Congress on Water, Climate and Energy* (Dublin, Ireland).

Wilson, C.A., and Novak, J.T., (2009) Hydrolysis of macromolecular components of primary and secondary wastewater sludge by thermal hydrolytic pretreatment. *Water research*. **43**(18), 4489-4498.

Solubilization of Organics Due to Thermal Hydrolysis Pretreatment Leading to Stable Mesophilic Anaerobic Digestion

Manisha S. Berde[1], Domenec Jolis[1]
[1] San Francisco Public Utilities Commission, 750 Phelps Street, San Francisco, CA-94124

ABSTRACT

A study was conducted to determine the effect of solubilzation due to thermal hydrolysis pre-treatment on the microbial population responsible for anaerobic digestion at mesophilic temperature. Metagenomic and microbial population analysis was done on digester receiving thermal hydrolysis pretreatment as well as on conventional digester receiving no thermal hydrolysis pretreatment. The digester receiving thermal hydrolysis pretreatment was stable at lower detention times and showed shift in microbial population with predominance of methane producing population in comparison to conventional digesters receiving no thermal hydrolysis pretreatment. Among methanogens, *Methanosarcina* showed dominance in the THP digester. Conventional digesters had mixed culture of acid producing and methane producing population, with almost equal population of methane producing archaeal population - *Methanobacterium* and *Methanosarcina*. Kinetics showed high growth and substrate utilization rates in pilot digester receiving THP.

KEYWORDS: Thermal Hydrolysis Pretreatment, *Methanosarcina*, *Methanobacterium*, Mesophilic Temperature, Volatile Fatty Acids, Anaerobic Digestion

INTRODUCTION

Thermal hydrolysis pre-treatment (THP) is being widely used on primary and secondary sludge generated at sewage treatment plants. THP causes solubilisation of the organic matter which makes it more amenable to digestion (Jolis, 2008). It leads to formation of higher soluble COD which is easily taken up by the biomass, thus a stable performance with higher organic destruction and more biogas generation is observed than with the traditional anaerobic digestion at mesophilic temperature. Solubilisation of COD in sludge enhances biogas formation during anaerobic digestion (Bougrier, et al., 2008). In conventional digesters, carbohydrates, proteins, and lipids are hydrolyzed and further fermented to acids to form methane. On the contrary, these macromolecules are broken into smaller molecules during THP, which are easily transported through cell membranes during the anaerobic digestion process, thereby speeding up the digestion process. Pilot studies with THP followed by anaerobic digestion at mesophilic temperature revealed higher organic destruction as well as almost twice the biogas production, at detention time lower than 10 days (Jolis, 2008). Full-scale experience shows that anaerobic digesters can be loaded at a higher organic loading without compromising the digester stability at a minimum detention time of 8 days (Pook, 2013). Hence, determining the solubility of organic matter is important to predict the performance of digesters. Research suggests that higher soluble COD favors the growth of methanogen bacteria in the digester (Jolis, 2008), and experiments found roughly 30% of the COD in the THP sludge is in the dissolved form, with high concentrations of VFAs (Kepp et al., 2001). It is hypothesized that due to presence of solubilized organic matter, there is shift in microbial consortium which causes digesters to perform well at higher organic loading and lower detention time. Previous research has found that predominance of acetic acid in digester feed favors growth of *Methanosarccina* population - a robust type of methanogen that increases digester stability at lower detention times (Conklin et al., 2006).

OBJECTIVES

This study was conducted to validate the hypothesis of change in microbial consortium in digester receiving THP due to presence of soluble organics. The goal of this study was also to find formation of high concentrations of volatile fatty acids (VFA) that favor the growth of any specific methanogen population which are responsible for stable performance of digesters at high organic loadings and at lower detention times. When comparing microbial population, the results obtained from this study are compared with full scale digesters operated at mesophilic temperature and above 15 days detention time.

METHODS

Pilot Setup and Operation

Pilot set up comprised of anaerobic digester, pre-dewatering unit, and thermal hydrolysis pretreatment unit. A 113.5 Liters steel egg-shaped anaerobic digester (113.5 Liters) was set up to receive thermally hydrolyzed sludge. Pilot digester comprised of following appurtenances: a) positive displacement pump for recirculation and heating of the sludge, b) water bath with jacket heaters, and c) peristaltic pump to feed hydrolyzed sludge to the digester tank. The digester tank had four recirculation openings, two along the side and two at the top of conical portion of the tank, one feed port to feed the sludge, one sludge withdrawal port at the bottom conical portion of the tank for recirculation and heating of the sludge and one drain port at the side of the bottom conical portion of the tank to drain the tank if required. The interconnecting piping was 1 inch, whereas port sizes were ¾ inch. The tank was equipped with a sampling port to grab samples for lab analysis and a sludge overflow port. The sludge overflow port was connected to a plexiglass weir which in turn was connected to the drain system to discharge the wasted/digested sludge.

Digester was fed using peristaltic pump (Cole Parmer Instrument, Company, Vernon Hills, Illinois) with Hydrolyzed Combined Primary and Secondary Sludge (HCPAS). The digested sludge was wasted through the sludge overflow port which was connected to the drain.

The temperature in the digester tank and water bath was monitored using Resistance Temperature Detectors (RTD) (Pyro Mation Inc., Fort Wayne, Indiana), each installed in water bath and digester tank. The amount of gas formed during the anaerobic process was determined using thermal dispersion type mass flow meters (Sierra Instruments, Monterey, California). The gas meters were mounted at the top of the digesters. The gas leaving digester was passed through heater (Watlow, St. Louis, Missouri) to avoid any damage to the gas meter caused by condensation of water vapors. Occasional foaming of the digesters was anticipated; therefore, a foam trap was installed before the heaters to avoid foam getting into the heater and the gas meters.

The digester had cylindrical body with conical sections at the top and bottom of the body. The digesters were maintained under completely mixed conditions at mesophilic range (35 deg C).

Mixing and heating of the sludge at mesophilic temperature was achieved by progressive cavity pump (Moyno, Dayton, Ohio) that would withdraw the sludge from the bottom of the tank, circulate it through heat exchanger coil immersed in hot water bath, and discharge it back into the digester tank through ports located at the top conical portion of the tank. A (208 liters) drum was used as a water bath. The temperature in the water bath was achieved by means of jacket heaters (Cole Parmer Instrument Company, Vernon Hills, Illinois) wrapped around the drum. The jacket heater was furnished with thermostat to achieve the desired temperature. The digester tanks, interconnecting piping, and water bath were wrapped in fiberglass insulation to reduce heat loss.

The temperatures of the digester tanks and gas formation were logged on a continuous basis. Pilot set up is shown in Figure 1.

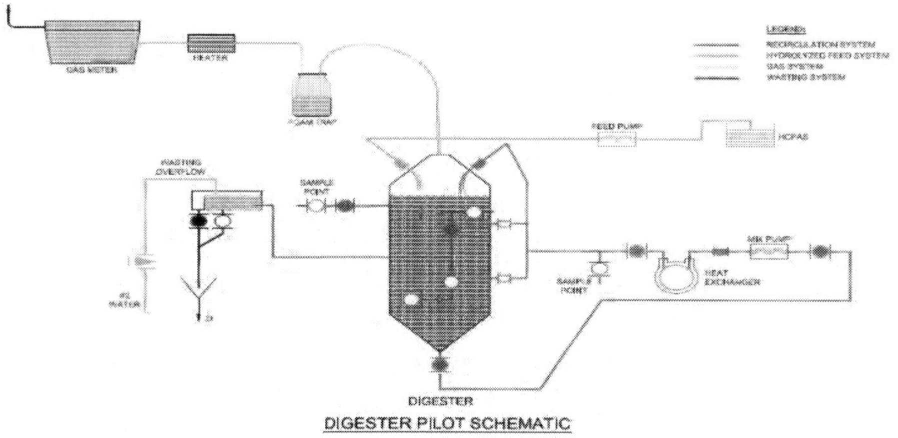

Figure 1: Digester Pilot Schematic

Pre-dewatering of the Combined Primary and Activated Sludge (CPAS)

Prior to thermal hydrolysis pretreatment of CPAS, sludge is pre-dewatered to obtain 13% -14% solids concentration. Pre-thickening is achieved using High Consistency Rotary Screen Thickener (FKC, Port Angeles, Washington). CPAS and polymer were fed to the HC-RST to obtain the desired solids consistency. Typical plant CPAS solids concentrations are observed in the range of 4% - 5%. HC-RST was operated once a week to produce required pre-dewatered sludge to desired consistency. The product was stored in the refrigerator until ready to run through the THP unit (Cambi Test rig, Asker, Norway).

Thermal Hydrolysis Pretreatment

Pre-dewatered sludge at 13% - 14% solids concentration was fed in to the thermal hydrolysis unit where the sludge was subjected at a temperature of 150°C and pressure of 5.5 bar at 30 minutes detention time followed by release of pressure. Cell lysis and dissolution of organic matter takes place at high temperature and pressure leading to product that is easily biodegradable. Thermally hydrolyzed product contained solids at approximately 10% concentration.

Analytical Methods

Samples were grabbed twice a week from the digester to measure soluble COD (sCOD), TS, and TVS using *Standard Methods* (APHA et al., 1992). Soluble COD (sCOD) was measured based on Method 5220D (APHA et al., 1992) by analyzing the supernatant after centrifuging the sludge samples. The microbial population was determined by metagenomic testing and quantified using Quantitative Polymerase Chain Reaction (qPCR) technique.

During metagenomics analysis, DNA was extracted from the samples using PowerSoil DNA isolation kit (MO BIO, Carlsbad, California). The extracted DNA was then amplified to determine 16S rRNA population using polymerase chain reaction (PCR). The amplicons obtained from purified PCR product were then sequenced using MiSeq reagent kit V3 chemistry on the MiSeq sequencer (Illumina, San Diego, California). A library of the sequenced amplicons was generated by KAPA DNA library preparation kit (Kapa Biosystems, Wilmington, Massachusetts). The copy number of 16S rRNA was quantified using real time PCR. Real time PCR was conducted by ABI 7500 (Applied Biosystems, Carlsbad, California). Gene copy number was calculated based on E.coli genomic DNA. The raw sequence data obtained was analyzed using FastQC for quality control check. The metagenomics analysis was performed at IEH laboratories & Consulting Group, Washington.

Quantification of *Methanosarcina* and *Methanobacterium* population in pilot digester as well as in full scale digester was conducted using qPCR. Taqman was used with a *Methanosarcina* and *Methanobacter* specific primer. A forward and a reverse primer were selected to amplify the two mentioned species. *Methanosracina* specific primer sets S-G-MSar-0450-S-a-19 (5'-TAGCAAGGGCCGGGCAAGA-3') and S-G-Msar-0589-S-a-20 (5'-ATCCCGGAGGACTGACCAAA - 3') (Sawayama, et al., 2006) were used. To amplify *Methanobacter* specific primer sets S-F-Mbac-0398-S-a-20 (5'-CCCAAGTGCCACTCTTAACG-3') and S-G-Mbac-0578-A-a-22 (5'-AGACTTATCAARCCGGCTACGA - 3') (Sawayama, et al., 2006) were used.

EPA method 8310 (EPA – Test Methods SW-846) was used as guidance to determine volatile fatty Acids (VFAs). The qualitative and quantitative analysis was performed using High Performance Liquid Chromatography. Analytical column used was; Restek Ultra Aqueous C18 reversed phase with no end cap with following specifications: 150-mm x 4.6-mm ID, 5-µm particle size, 100A pore size, 2.5-8 pH range, 80°C temperature limit, 15% carbon load. Mobile phase was 0.05-M Phosphate buffer. Detection was achieved using Agilent G1314A variable wavelength ultraviolet detector with following components: Standard flow cell, 10-mm path length, 14-pL volume, Deuterium lamp, 190-600-nm ultraviolet (UV) wavelength range. VFA analysis was performed at Eurofins lab – Monrovia, California.

Feed and Digested Sludge Characteristics

A 113.5 liters pilot digester was set-up to receive HCPAS at 12 days detention time at mesophilic temperature (35 °C). The feed rate to digester was 9.5 L/d. The digester was fed twice a day @ 4.7 L/feed using peristaltic pump. The experiment was run for 5 months (~150 days).

The digester received HCPAS at TS 10±1.3%, VS 80±6%, and sCOD 46±6.2 g/L, whereas digestate had concentrations of TS 3.6±0.6%, VS 66±5%, and sCOD 5.9±1.2 g/L. The pH in the digester was stable and was observed at 7.4±0.1.

Full scale digesters operated at mesophilic temperature and 15 days detention time received CPAS at TS 5%±0.7%, VS 83±3%, and sCOD 15±2 g/L, whereas digestate had TS 2.3%±0.2%, VS 69±4%, and sCOD 2±0.2 g/L

The performance of digester was determined based on the microbial population, and kinetics.

RESULTS

Microbial population

The microbial population in the pilot digester as well as in full scale digesters was determined using Metagenomic and qPCR analysis using methods described previously, in full scale digester with no THP and pilot digester receiving THP is shown in Table 1.

Table 1 - Comparison of microbial population in pilot digester receiving THP and full scale digesters at SEP and OSP with no THP

	Pilot Digester with THP	Full Scale Digester with no THP
Total Bacteria, 16SrRNA genes (per mL), Avg.	3.20E+09 ± 7.64E+08	3.60E+09 ± 1.73E+09
Acid Producing bacteria, % of Total Bacteria, Avg.	25 ± 3	24± 4
Total Archaea, 16SrRNA genes (per mL), Avg.	5.72E+08 ± 9.31E+07	4.45E+08 ± 1.84E+08
Methanogens, % of Total Archaea, Avg.	97 ± 2	94 ± 0

±Standard Deviation

Results revealed that pilot digester receiving hydrolyzed CPAS had a predominant population of methanogens (Archaea) compared to acid producing bacteria. Among methanogens, the acetate consuming archaea showed dominance and its population was found to be higher than the full scale digester, whereas the full scale digester had mix of acid producing and methanogen population. Overall, the total bacterial population in full scale digester was higher by almost 12% than the pilot digester receiving THP, but the archaeal population was higher in pilot digester compared to full scale digester. Hence, the result confirms the hypothesis of shift in microbial population due to solubilization during THP, governing the growth of methane producing archaeal population.

The metagenomics analysis and qPCR suggest that solubilization of organics during THP leads to formation of fermentation products which can be utilized for methane formation. The microbial consortium present in the pilot digester indicates that the archaeal population is about 30% higher than the methanogen population found in full scale mesophilic digesters, with *Methanosarcina* comprising about 65% population among methanogens. Hence, from the data obtained, the pilot digester mainly presented methanogenic activity taking place. Further investigation was carried out to determine the methanogen species such as *Methanobacterium* and *Methanosarcina* in pilot digester and full scale digester using qPCR method described previously. The methanogen species present in pilot digester receiving THP, full scale digester receiving no THP are shown in Table 2.

Table 2 – qPCR results for Methanobacterium and Methanosarcina in pilot digester and full scale digesters

Methanobacterium	**Target Sequence Copies (TSC/mL)**
Full Scale Digester- 1	1.14E+07 ± 1.13E+07
Full Scale Digester - 2	3.32E+07 ± 1.99E+07
Pilot Digester	< 1.0E+01
Methanosarcina	
Full Scale Digester- 1	1.26E+06 ± 1.02E+06
Full Scale Digester- 2	2.75E+06 ± 1.91E+06
Pilot Digester	2.20E+09 ± 5.52E+08

It can be observed from the qPCR results that *Methanosarcina* dominates in pilot digester. The concentration of solubilized organics due to THP is a deriving factor to predict the growth of specific microbial population. *Methanosarcina* is believed to be the dominant species in presence of high acetate concentration (Conklin et al., 2006). The qPCR results strongly suggest presence of acetate concentrations in the pilot digester receiving THP to the levels that promote the growth of *Methanosarcina*. Hence, it is assumed that the formation of acetate at higher concentrations takes place during thermal hydrolysis process, which is consumed rapidly to produce methane gas. Acetic acid, propionic acid and butyric acid are the dominant acids formed that are responsible in methane formation process (McCarty et al., 1963). Feed sludge and digested sludge samples from pilot digester were analyzed to determine the presence of VFAs, and their concentration in the feed and digested sludge is given in Table 3.

Table 3 – VFAs in feed and digested sludge of pilot digester receiving THP

VFA	Feed Sludge			Digested Sludge		
	mg/kg	mg/L as acetic acid	mg/L COD_t	mg/kg	mg/L as acetic acid	mg/L COD_t
Acetic Acid	1045 ± 78	1045 ± 78	1120±83	73 ± 28	73 ± 28	78±30
Butyric Acid	4950 ± 2475	3372±1685	9000±4500	140	95	173
Pyruvic Acid	160 ± 213	109±145	146±194	6	4	5

±Standard Deviation, COD_t – Theoretical COD

The VFA results indicate presence of high concentrations of acetic acid and butyric acid. This validates the assumption of *Methanosarcina* population outcompeting *Methanobacterium* which thrive on hydrogen to produce methane. Dominance of *Methanosarcina* indicates the conversion of butyric acid to acetic acid. In comparison, in full scale digesters, a mixed population of *Methanobacterium* and *Methanosarcina* was observed with *Methanobacterium* being the predominant population. Growth studies have shown that compared to other methanogen species, *Methanosarcina* is believed to be more resistant to high ammonia concentrations (Sprott et al., 1986). Formation of high concentrations of ammonia takes place in anaerobic digester receiving THP, with typical concentrations of ammonia found in pilot digester in the range of 2 g/L – 2.5 g/L. Hence, combination of high acetate and ammonia concentrations in the pilot digester receiving THP play an important role for the dominance of *Methanosarcina* population.

A theoretical model (Lawrence et al., 1969) was used to determine the distribution of microbial population based on the volatile acids present in the digesters using the COD converted to methane gas (COD_R) given in Table 4. The rate of utilization as acetic acid (mg/volume) is represented by (dF/dt). The equation for determining the substrate-specific fraction of microbial mass (ssf_x):

$$ssfx = \left(\frac{dF}{dt}\right)x\ (COD_R)x/\Sigma\left(\frac{dF}{dt}\right)(COD_R) \qquad (1)$$

Table 4 – Theoretical values for methane formation from COD conversion (Lawrence et al., 1969)

Substrate	Grams Methane COD/Mole of Substrate Fermented, COD_R	Grams Methane COD/g (as acetic acid) of Substrate Fermented, COD_R
Acetic Acid	64	1.07
Butyric Acid	32	0.53

Using the equation (1) and the values in Tables 3 and 4, the ssf_x for butyric acid is 0.35, whereas for acetic acid is 0.65. This means 35% of the microbial population is butyric acid fermenters and 65% population is acetic acid fermenters. The theoretical population evaluation of acetic fermenters correlates with the metagenomics analysis. Hence, it is strongly suggested by the results that the dominance of one specific type of microbial population can be achieved for stable digester performance due to THP.

Kinetics

Based on the data collected, following kinetic parameters were obtained as shown in Table 6. Specific substrate utilization rate (sCOD g/VS g.d) is obtained by finding the ratio of soluble COD consumed to the volatile solids in the reactor. Delta S (ΔS) for digester receiving THP was observed to be 0.385 kg/d, whereas for digester with no THP was found to be 0.098 kg/d. The yield coefficient was determined based on the methane gas formation and soluble COD destruction at 12 days detention time in digester receiving THP and at 15 days detention time in digesters with no THP. The methane content of the gas in the digester receiving THP was found to be around 65%, whereas for digester receiving no THP, the methane gas content was found to be 60%. The decay coefficient value is taken from the literature (MetCalf & Eddy, 2003), which is the overall combined for fermentation and methanogenesis. The cell tissue produced and yield coefficients in the pilot digester receiving THP as well in digester receiving no THP are calculated using the equation (2) and (3) (MetCalf and Eddy, 2003) for complete-mix reactors without recycle as follows:

$$V_{CH4} = 0.4[\Delta SQ - 1.42Px] \text{.........} (2)$$

$$Px = YQ\Delta S/(1 + k_d SRT) \text{..........} (3)$$

The data in Table 5 is overall growth rate in the digester and not specific to VFA utilization. The maximum specific growth rate (μ_{max}) was determined from Yield coefficient (Y) and Substrate utilization rate (Y x Substrate utilization).

Table 5 – Kinetic values for the digester receiving THP and with no THP

	Digester Receiving THP	Digester Receiving no THP
Substrate utilization sCOD g/VS g.d	1.8	1.2
Methane Production (m^3/d)	0.114	0.055
Methane Volume (m^3CH_4/m^3 reactor.d)	1.01	0.5
Decay Coefficient k_d (day-1)	0.03	0.03
Yield Coefficient , Y (g VSS/g bCOD)	0.25	0.18
Maximum Specific Growth Rate μ_{max} (day-1)	0.45	0.22

The kinetic values indicate that the digester with THP has higher growth rate, yield, and substrate utilization rates, which explains superior performance of the digester receiving THP and point to the possibility of stable process at reduced HRT.

DISCUSSION

Anaerobic digesters receiving THP can be loaded 2 – 3 times higher than the conventional digesters without affecting its stability. The key to superior performance appears to be the change in the nature of the organics present after THP. The results clearly indicate that presence of soluble organics promotes the growth of methanogenic activity. Hence, a change in microbial consortium is observed in digester receiving THP when compared to microbial population in full scale mesophilic digester without THP. The proportion of acid producing bacterial population in digester receiving THP is less than the methanogen population, whereas the opposite is true for anaerobic digesters without pretreatment. A conventional mesophilic digester with no pre-treatment requires a balance of acid producing and methane producing bacteria for stabilization of organic waste through the three steps process of: hydrolysis, acid production, and methane formation. On the contrary, hydrolysis process takes place during THP, so that the digester with THP can be operated mainly as a methanogenic reactor.

The study also observed the dominance of M*ethanosarcina* – methane forming strain in the digester receiving THP. The formation of methane gas takes place predominantly from conversion of acetate. Previous research (Zinder et al., 1979, and Conklin et al., 2005) observed dominance of M*ethanosarcina* in presence of high acetate concentrations. The VFA results during the study also verify presence of high butyric and acetic acid, hence promoting the growth of M*ethanosarcina* in the digester receiving THP. The doubling time of M*ethanosarcina* is about 12 h (Zinder et al., 1979), which explains higher growth rates and higher substrate utilization rates for digesters with M*ethanosarcina* predominance. Further methanogen washout study needs to be conducted to determine the minimum detention times at which the digester can be operated, thereby increasing the treatment capacity. Since, the results showed dominance of M*ethanosarcina,* this information can be utilized in design of full scale digesters receiving THP.

CONCLUSION

The hypothesis of change in microbial consortium in digester receiving THP due to presence of soluble organics is very well validated from the study as well as from comparison made with full scale mesophilic digesters that due to thermal hydrolysis there is change in the chemistry which causes a shift in microbial population. In digester receiving THP, higher population of methanogens were observed with predominance of M*ethanosarcina* which are believed to be more robust and can utilize substrate at higher rates. One of the reasons for this shift is the presence of higher acetate concentration that is readily available for microbes for stabilization. Based on the kinetic parameters the growth rate and substrate utilization rates are higher. Hence, digester could be operated at lower detention times.

ACKNOWLEDGEMENTS

The authors would like to thank operations, maintenance, and lab personnel for their assistance during the project. Smooth operation of the pilot was possible due to their tremendous support.

REFERENCES

Bougrier, C.; Delgenes, J. P.; & Carrère, H. (2008) Effects of thermal treatments on five different waste activated sludge samples solubilisation, physical properties and anaerobic digestion. *Chemical Engineering Journal*, ***139***(2), 236-244.

Conklin A.; Stensel D.H.; Ferguson J. (2006) Growth Kinetics and Competition between *Methanosarcina* and *Methanosaeta* in Mesophilic Anaerobic Digestion, *Water Environ. Res.*, **75**(5), 486-496

Jolis, D. (2008) High-solids anaerobic digestion of municipal sludge pretreated by thermal hydrolysis. *Water Environment Research*, ***80***(7), 654-662.

Kepp, U.; Panter, K.; & Solheim, O. E. (2001) High dry solids digestion. In *CIWEM/Aqua Enviro 6th European Biosolids and Organic Residiuals Conference November*.

Lawrence W. A.; McCarty L. P. (1969) Kinetics of Methane Fermentation in Anaerobic Treatment, *Water Environ. Feder.*, **41**(2), R1-R17

McCarty L. P.; Jeris S. J.; Murdoch W. (1963) Individual Volatile Acids in Anaerobic Treatment, *Water Environ. Feder.*, 35-12, 1501-1516

MetCalf and Eddy (2003) *Wastewater Treatment and Reuse*, 4th Edition, McGraw Hill; New York, NY

Pook M.; Mills N.; Heitmann M.; Panter K.; Walley P. (2013) Exploring the Upper Limits of Thermal Hydrolysis at Chertsey STW, *Proceedings of 18th European and Biosolids & Organic Resources Conference and Exhibition, November*

Sawayama, S.; Tsukahara, K.; and Yagishita, T. (2006) Phylogenetic description of immobilized methanogenic community using real-time PCR in a fixed-bed anaerobic digester, *Bioresour Technol* **97**, 69–76.

Sprott D. G.; Patel B. G. (1986) Ammonia Toxicity in Pure Cultures of Methanogenic Bacteria, *System Appl. Microbiol.*, **7**, 358-363

Zinder H. S.; Mah A. R. (1979) Isolation and Characterization of a Thermophilic Strain of *Methanosarcina* Unable to Use H_2-CO_2 for Methanogenesis, *Appl. And Environ. Microbiol.*, **38**(5), 996-1008

Does operation at increased ammonia concentration impact hydrolysis rates

Baoqiang Li[1,2], Adrian Romero[1,2], Elizabeth Manning[1,2], Matthew Higgins[3], Ahmed Al-Omari[2], Sudhir Murthy[2], Rumana Riffat[1], Haydée De Clippeleir[2].

[1] The George Washington University, Washington, DC, USA. (Emails: *baoqiangli@gwu.edu; aromerof@gwu.edu, riffat@gwu.edu*)
[2] DC Water, Washington, DC, USA. (Emails: *Ahmed.Al-Omari@dcwater.com; Haydee.DeClippeleir@dcwater.com; Sudhir.Murthy@dcwater.com*)
[3] Bucknell University, Lewisburg, Pennsylvania, USA. (Email: *mhiggins@bucknell.edu*)

Keywords: Thermal hydrolysis, Intensified anaerobic digestion, Recuperative thickening, Ammonia, ammonium, Hydrolysis rate;

ABSTRACT
This study investigated the influence of ammonia (or ammonium) on the hydrolysis rates of mesophilic Anaerobic Digestion (AD). Three lab scale mesophilic digesters fed with THP sludge operated at different SRT together with batch respirometry tests to obtain the experimental data. Results showed different inhibition threshold for long term digester run and batch respirometry tests. Recuperative thickening (RT) was applied to 2 additional digesters to evaluate the impact of biomass and ammonia nitrogen. Results suggested that although RT extending the SRT from 15 days to 30 days had lower ammonia levels, it did not have significant impact on overall performance.

INTRODUCTION
AD is often used for municipal wastewater sludge stabilization and volume reduction, while at the same time it allows for energy recovery from organics. Four steps are involved in AD process: hydrolysis (breakdown of particles and larger molecules), acidogenesis, acetogenesis, and methanogenesis. Hydrolysis is typically the rate-limiting step which means hydrolysis limits the overall degradation rate and subsequent biogas production (Pavlosthathis and Giraldo-Gomez, 1991; Vavilin et al., 1997, 1999). It is important to understand the limiting factors of the hydrolysis step to develop strategies to increases the extent of hydrolysis, thus increasing biogas production, reducing solids and providing significant economic and environmental benefits.

THP has been demonstrated to improve waste activated sludge (WAS) digestibility from wastewater treatment plants (Nielsen et al., 2011). THP partially hydrolyzes solids prior to digestion and increases the readily degradable substrate, and smaller particle sizes might be one of the reasons for overall increased performance. With the decreased viscosity after THP, THP digesters can be loaded at twice the loading rate and thus operate at increased ammonium and ammonia concentrations. This has led to the selection for acetate oxidizing bacteria (ACOX) and hydrogenotrophic organism over acetoclastic methanogens as the latter more sensitive for free ammonia (I.W. Koster, G. Lettinga, 1984; A. Schnurrer, A. Nordberg, 2008). Although much research has evaluated the impact of elevated ammonia concentrations on overall anaerobic digestion process, and specifically methanogenesis, much less research has specifically focused on the impact of ammonia concentration on the hydrolysis step. For instance, Lu et al. (2007) showed decreased hydrolysis of high protein substrate (fish residue) at initial ammonia concentrations of 2 g-N/L, but with methanogenesis still occurring at concentrations as high as 16 g-N/L. Inhibition of hydrolysis of carbohydrates and lipids under high ammonia concentrations has been studied with no decrease in the rates at ammonia levels of 957 mg NH_3-N/L for adapted cultures (Fernandes et al.,

2012). Paige (2012) showed that acclimated biomass can maintain hydrolysis rates of solids waste (food scrap and manure) at total ammonia nitrogen concentrations of 5 g-N/L.

To date, the knowledge of the impact of ammonia/ammonium on hydrolysis rates of sewage sludge is limited, thus limiting the understanding of the potential impacts of technologies and approaches aiming for an overall increase in biogas production and/or sludge treatment. In this paper we employed bench scale digesters and batch test to evaluate the impact of ammonia on hydrolysis rates under different Solids Retention Times (SRT) with and without Recuperative Thickening (RT).

MATERIALS AND METHODS

Continuously stirred, lab scale mesophilic anaerobic digesters (10L) were fed THP sludge at different feed concentrations but the same volatile solids (VS) loading rates. This resulted in different SRTs, but also varied the ammonia concentrations in the digester such that the effect of ammonia on hydrolysis rates could be evaluated. Analyses included TS, VS, pH, TCOD, sCOD, NH_4^+, alkalinity, and VFA. Real time gas production profiles were collected by a pulse-flow based PF-8000 Respirometer System (Respirometer Systems and Applications, LLC, AR, USA) attached to all digesters.

Recuperative thickening was applied to three additional digesters to separate the impact of biomass and ammonia nitrogen. Fix volume of sample was collected each day at the same time with feeding for recuperative thickening. Polymer solution was added to the collected sample, mixing with jar tester paddles at 300 rpm for 20 seconds. After 30 seconds awaiting time, mixed sample filtered through belt filtration membrane by gravity filtration, target volume of the supernatant was removed and concentrated sample was stored in 4°C for next day used. Diagram shown in **Figure 1**.

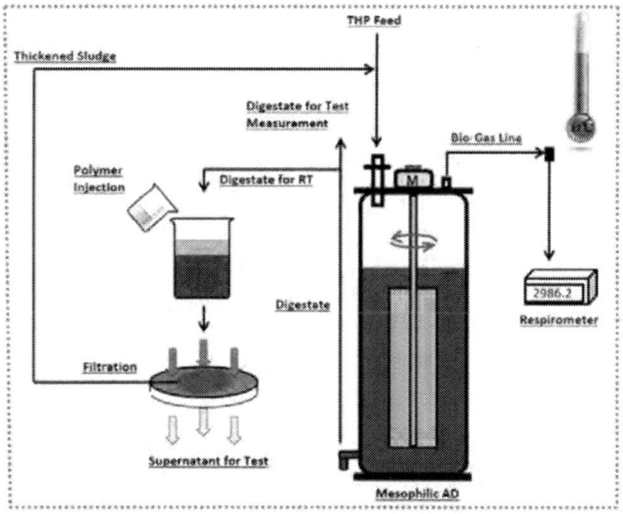

Figure 1. General overview of digester operation including recuperative thickening step.

Recuperative thickening allows for uncoupling of HRT and SRT and thus allows to operate at longer SRT compared to CSTR operation. The impact of operation at longer SRT was tested in Phase I by operating RT reactors at an SRT of 30 days while two of the conventional systems were maintained at SRT of 15 days. The HRT in both systems was kept constant and thus also maintaining the same organic loading rate. It was hypothesized that if hydrolysis was biomass limited, increased VSR should be achieved at increased SRT given a constant hydrolysis rate. In a second phase, the impact of operating at shorter HRT given a fixed SRT of 15 days was tested. To allow for a valid comparison, organic loading rates were kept constant by diluting the feed for the RT reactors. In this phase the impact of diluted concentrations of ammonia and organic substances was evaluated.

Batch respirometry tests were also used to estimate the impact of ammonia concentrations on hydrolysis rates. Seed sludge was collected from DC Water Blue Plains Advance Wastewater Treatment Plant and incubated 24 hours in advance of spiking Ammonium bicarbonate. Observed volumetric hydrolysis rate obtained from real time respirometer data was used to investigate the impact of NH_3 and SRT on hydrolysis step in anaerobic digestion, specifically with THP and RT process.

RESULTS AND DISCUSSION

Generally, a representative 24h biogas production rate profile where the reactors are fed at time equals zero, the gas production rate increases to a peak level within the first two hours and then begins to decline reaching a fairly constant level at 17 hours until the end of the cycle. The last few hours of the 24 hour cycle represent the hydrolysis limited phase. These profiles suggest that the overall hydrolysis rates were very similar for all SRT. However, when normalizing for the VS concentration in the system, one observes a decreased specific hydrolysis rates with increased SRT and increased ammonium and free ammonia concentration.

A batch experiment was performed using sludge from the full-scale THP digesters at DC Water and the direct impact of ammonium and ammonia was evaluated. The results showed that the hydrolysis rate was impacted at ammonia above 250 mg N/L. This was different from the observation in the long term reactor runs, where a decrease in hydrolysis rates was observed at lower NH_3 concentrations than 250 mg N/L. Potential adaptation to higher levels of total ammonia nitrogen by the biomass in full scale digesters could explain the lower inhibition in the batch tests (Paige, 2012).

On the other hand, due to the increased SRT targeted in the RT reactors (30 days SRT), increased TS concentration were obtained (Figure 3.1.7). The reactors with RT stabilized at 7.94±0.26% TS compared to 5.21±0.26% TS in the conventional system. Despite the operation at longer SRT and higher TS, ammonium levels were only slightly lower in the RT reactors. In addition, overall biogas production and volatile suspended solids reduction (VSR) was not significantly different between the two systems. The 24 hours biogas production curve of the 5 systems looked completely identical.

By maintain the same organic loading rates to the digesters but lower TS concentration in the feed to the RT reactors operating at 15 days SRT (7.5 days HRT) in the second phase, the TS in all digesters (conventional and RT) was about 5%. In addition, ammonium concentration as well as COD concentrations were significantly lower in the RT reactors. Average ammonium concentrations in the RT reactors were 56% of the conventional reactors and thus close to the 50% as expected by dilution. In case of soluble COD, about 66% of the sCOD in the conventional reactors was observed. This indicated that a larger accumulation of sCOD was observed in the RT reactors than the conventional reactors which could potentially indicate increased hydrolysis rates. However, when considering the overall biogas rates and VSR results, no significant changes in overall digestion performance was observed. At this point we concluded that, operation at similar organic loading rates but decreased HRT did not improve overall digester performance significantly. There were indications of increased sCOD (and minor ammonium) accumulation in the RT system as well as potential difference in reactor kinetics. However, this did not result in a significant improvement of overall anaerobic digestion.

CONCLUSION

Digesters fed with THP and operated under RT could potentially enhance the hydrolysis process in anaerobic digestion, which is typically the limiting step. However, both approaches are associated with an increase in ammonia/ammonium concentration. The results from this paper suggest that operating at higher total ammonia concentration may impact hydrolysis rates and thus decreasing the potential for higher biogas production. In addition, it was observed that applying RT may lower ammonia concentration in digester by eliminating supernatant.

REFERENCES

1. APHA (2012): Standard Methods for the examination of water and wastewater. Washington, D. C, American Public Health Association, Water Pollution Control Federation.
2. A. Schnurrer, A. Nordberg, 2008. Ammonia, a selective agent for methane production by syntrophic acetate oxidation at mesophilic temperatureWater Sci Technol, 57 (2008), pp. 735–740
3. Bader FG. 1978. Analysis of Double-Substrate Limited Growth. Biotechnology and Bioengineering, 20:183-202.
4. D.E. Contois. 1959. Kinetics of bacterial growth: relationship between population density and specific growth rate of continuous cultures. J. Gen. Microbiol., 21 (1959), pp. 40-50.
5. Fernandes T.V., Keesman K.J., Zeeman G., Lier J. van. 2012. Effect of ammonia on the anaerobic hydrolysis of cellulose and tributyrin. Biomass and bioenergy, 2012.
6. I.W. Koster, G. Lettinga, 1984. The influence of ammonium–nitrogen on the specific activity of pelletized methanogenic sludge Agric Wastes, 9 (1984), pp. 205–216
7. Lu F., He P.J., Shao L.M., Lee D.J., 2007. Effects of ammonia on hydrolysis of proteins and lipids from fish residues. Environmental biotechnology, 2007.
8. Nielsen, H.B., Thygesen, A., Thomsen, A.B., Schmidt, J.E., 2011. Anaerobic digestion of waste activated sludge-comparison of thermal pretreatments with thermal inter-stage treatments. J. Chem. Technol. Biotechnol. 86, 238-245.
9. Paige L. 2012. Anaerobic digestion of organic wastes: the impact of operating conditions on hydrolysis efficiency and microbial community compostition. Master thesis. Colorado State University, 2012.
10. Pavlosthathis, S.G., Giraldo-Gomez, E., 1991. Kinetics of anaerobic treatment. Water Sci. Technol. 25 (8), 35±61.
11. Stewart H., Al-Omari A., Bott B., De Clippeleir H., Massoudieh A., Murthy S., Su C., Takacs I., Wadhawan T., Wett B., 2016. Impact of dual substrate limitation model choice on nitrogen removaldesign. WEF/IWA 2016 Nutrient removal and recovery conference, Denver, USA.
12. Vavilin, V.A., Rytov, S.V., Lokshina, L.Y., 1997. A balance between hydrolysis and methanogenesis during the anaerobic-digestion of organic-matter. Microbiol. 66 (6), 712±717.
13. Vavilin, V.A., Rytov, S.V., Lokshina, L.Ya., Rintala, J.A., 1999. Description of hydrolysis and acetoclastic methanogenensis as the rate-limiting steps during anaerobic conversion of solid waste into methane. In: Mata-Alvarez, J., Tilche, A., Cecchi, F. (Eds.), Proceedings of the Second International Symposium on Anaerobic Digestion of Solid Wastes, Barcelona,1999.

Are Thermal Hydrolysis Digesters Robust Because of their Unique Microbial Communities?

Joshua Mah, P.E. (Virginia Tech and Haarslev)
John Novak, Ph.D., P.E (Virginia Tech)
Kaisen Lin (Virginia Tech)

ABSTRACT

The goal of this project was to provide an understanding of the differences between microbial community structures in anaerobic digesters fed with thermally pretreated ("hydrolyzed") municipal sludge and conventional digesters fed with non-pretreated sludge. This project aims to provide a better understanding of the microbial communities of THP digesters and the opportunities and/or vulnerabilities they may present.

KEY WORDS

Thermal hydrolysis, anaerobic digestion, microbial community, biosolids

INTRODUCTION

Anaerobic digestion is an important technology for sustainable production of renewable energy and other renewable resources from waste materials. Thermal hydrolysis pretreatment (THP) of feed sludge before digestion is a relatively newly adopted enhancement of the digestion process that results in additional benefits as demonstrated at more than 50 full-scale installations worldwide. THP can reduce the required digester volume and the associated capital costs, improve dewaterability and reduce the related sludge disposal costs, and facilitate the production of exceptional quality biosolids, which has public health benefits (Wilson et al., 2011).

Each of these aspects generally increases the favorability of digester installations. However, in the enthusiasm to expand the use of THP, little attention has been given to the effects it may have on the biological engine of the digester, the microbial community. The entire principle of creating a sterile feed sludge and a closed microbial community has been almost completely ignored because it is not the primary or intended objective of the process (although the process has been credited for the destruction of pathogens). The sterilization of feed sludge could have significant positive and/or negative impacts on the overall microbial community structure (beyond pathogen control) and digester stability that have been overlooked because the other benefits are so much easier to see.

Microbial communities in anaerobic digesters can be defined as complex metabolic relationships between populations (i.e. individual species) of microorganisms, each of which may play a highly specialized and/or symbiotic role in the overall catabolism of sludge to biogas. The observed performance of a given digester is directly attributable to the structure of the microbial community (or dysfunction thereof) and its collective fit to the environmental conditions of the digester such as the substrate type, loading rate, temperature and pH. The structure of the community and its relationships is in turn governed by both the metabolic characteristics of the individual populations and the manner in which the anaerobic digestion process is operated. In a properly operating digester, the microbial community and the digester conditions adapt and adjust to each other to reach an equilibrium.

Thermal hydrolysis pretreatment (THP) of digester feed sludge may play an additional key role in the structure of digester microbial communities because THP heats the sludge to at least 160°C for at least 20 minutes, which completely sterilizes the feed sludge and therefore creates a closed microbial community (in the most common THP process configuration). In a properly operating THP digester (no bypasses, leaks or contamination), no new microbes will be introduced into the digester after seeding unless a deliberate supplemental augmentation is performed. Without continuous inoculation of microbial diversity, the roles of competition, selection and extinction may have significantly stronger influences in the structures of microbial communities in THP digesters as compared to conventional systems. Furthermore, short of intentional or unintentional augmentation, the microbial community of the THP digester should be limited to that of the seed sludge, or less if some extinction or washout occurs. Questions begin to arise as to whether a certain type of seed sludge (startup inoculum) is necessary for THP digesters and whether the most favorable microbes for intense and/or high ammonia digestion conditions will survive an initally dilute startup procedure. Is there a way to evaluate and select seed sludges? How diminished does a population have to be to become "extinct"? In general, is there a "magic THP bug" as some proponents of THP claim?

Many more questions arise as to whether the sterilization effect of THP presents an advantage or disadvantage for the digestion process. One could argue that the "natural selection" process occuring within the closed ecosystem results in a highly refined microbial community that consists of only the top-performing microorganism for each degradable substrate or intermediate compound that is actually present. Assuming that the composition of the feed sludge is consistent over time, the highly selected community may equate to a well-trained army of substrate degraders ready to tackle varying loading rates thrown at it. Without competition from interloping microbes, the highly refined community can reproduce quickly to adjust to swinging substrate loads without the normally requisite condition of substrate pressure. In other words, when multiple competitive populations are present, a population shift will normally be driven by a high substrate pressure, which is generally indicitive of an undesireable process condition in digesters. With only a single population per substrate, the population shift doesn't occur. On the other hand, one could argue that the highly refined community might be poorly equiped to handle abrupt changes in substrate composition such as the introduction of a slug load of a new kind of waste (eg. codigestion). If the THP digester had been operated sufficiently long on a consistent substrate composition, there may be few if any microbes remaining that have specialized metabolism capabilities for substrates not in the consistent mixture. By definition, they would have been outcompeted and/or washed out during the process of developing a highly specialized or refined community. Also, for the previously reference single population per substrate postulate, it is possible that the single population is at any given time not performing under its ideal substrate growth conditions.

Although the detailed experimental tests needed to completely answer these questions were truncated from the overall scope of this research project, the proceeding results provide new insight into the microbial population dynamics in THP digesters and begin to answer these questions.

METHODS

The four THP digester samples were collected from the four Blue Plains digesters in August 2015, which represented approximately eight months of full-load operation for Digesters 1 and 2 and approximately five months of full-load operation for Digesters 3 and 4. Each of the four samples was sequenced in triplicate. Four more THP digester samples collected during the startup period (November 2014) were also included only in Figure 4. They were also each sequenced in triplicate. The six conventional mesophilic digester samples were collected during the summer of 2015. One of those six digester samples was obtained from the gas phase digester of an acid-gas phased digestion process plant (labeled in the Figures as "54MGD G/P"). Due to limited sequencing availability, some of the samples were sequenced in duplicate and one (9MGD) was sequenced in quintuplicate for a detailed statistical analysis. Three of the four digester feed (raw) samples shown in Figure 4 were obtained from the conventional plants as shown and one was obtained from a pre-THP sampling point at Blue Plains (384MGD). The seed samples were obtained from the Alexandria, VA plant in the middle of seeding at Blue Plains (10/10/2014) and after seeding (11/7/2014 with a field duplicate sample for quality control).

Samples were frozen at -20°C until DNA extraction. Genomic DNA was extracted using a phenol-chloroform-beadbeating procedure and it was frozen after cleanup at -80°C until the extracts were amplified by PCR for 16S V4. The extracts were indexed with unique barcodes and pooled at up to 108 extracts per library for multiplexed sequencing. The pools were sequenced using MiSeq v2 chemistry with 250 base pair paired-end reads.

The resulting data set of reads were quality filtered including chimera checking using both QIIME and mothur. The data in Figures 1, 2 and 3 in this paper came from QIIME (Caporoso et al., 2010) and Figure 4 was created with data generated in the mothur pipeline .

RESULTS

Microbial diversity in the THP digesters was lower than comparable conventional digesters

To begin with the microbial community data, Figure 1 shows the distribution of microbes in each of the six samples from conventional digesters across all 78 possible phylum-level bins in the Greengenes taxonomy (QIIME data). The overall distribution across the taxa is relatively consistent between the six conventional samples and generally typical for mesophilic municipal anaerobic digesters. Clearly, many of the bins are virtually devoid of abundance and attention can therefore be focused on a smaller subset of the bins.

Figure 1: Relative Abundances across all 78 Phylum-Level Taxa

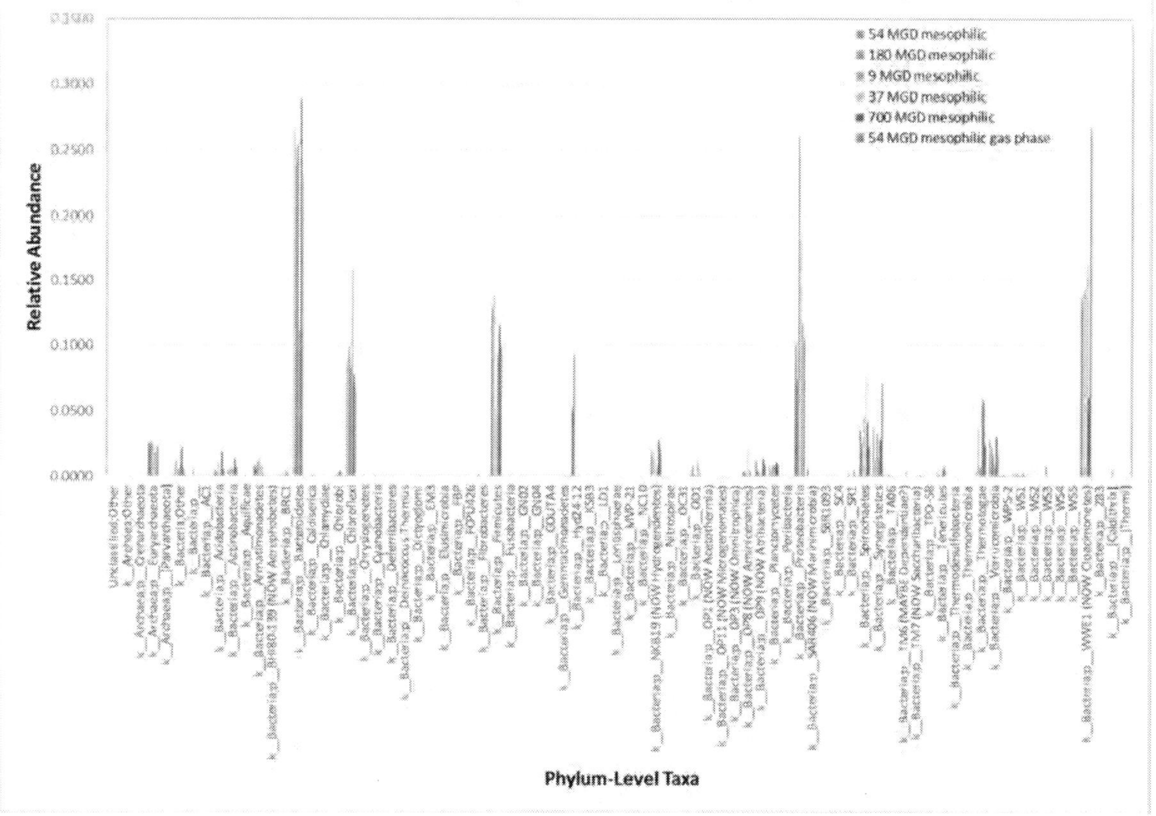

Figure 2 shows a subset of the 11 most dominant phyla that constitute the left portion of Figure 1. The phyla are sorted from left to right in order of the combined abundance of the six conventional samples. In addition, the four THP digester samples have also been included. They are deliberately shown in bright and similar colors to emphasize the consistently significant difference in distribution across the phyla. About half of the abundance in each of the four THP digesters is concentrated into a single phyum, Firmicutes. Although the figure does not present a formal measure of diversity, it is obvious that the THP digester communities are much less even (abundance distributed unequally between phyla). The figure also illustrates a reversal in the Firmicutes to Bacteriodetes ratio, but the shift is due to an almost five-fold increase in Firmicutes, not a reduction in Bacteriodetes.

Figure 2: Relative Abundances Across The 12 Most Dominant Phylum-Level Taxa

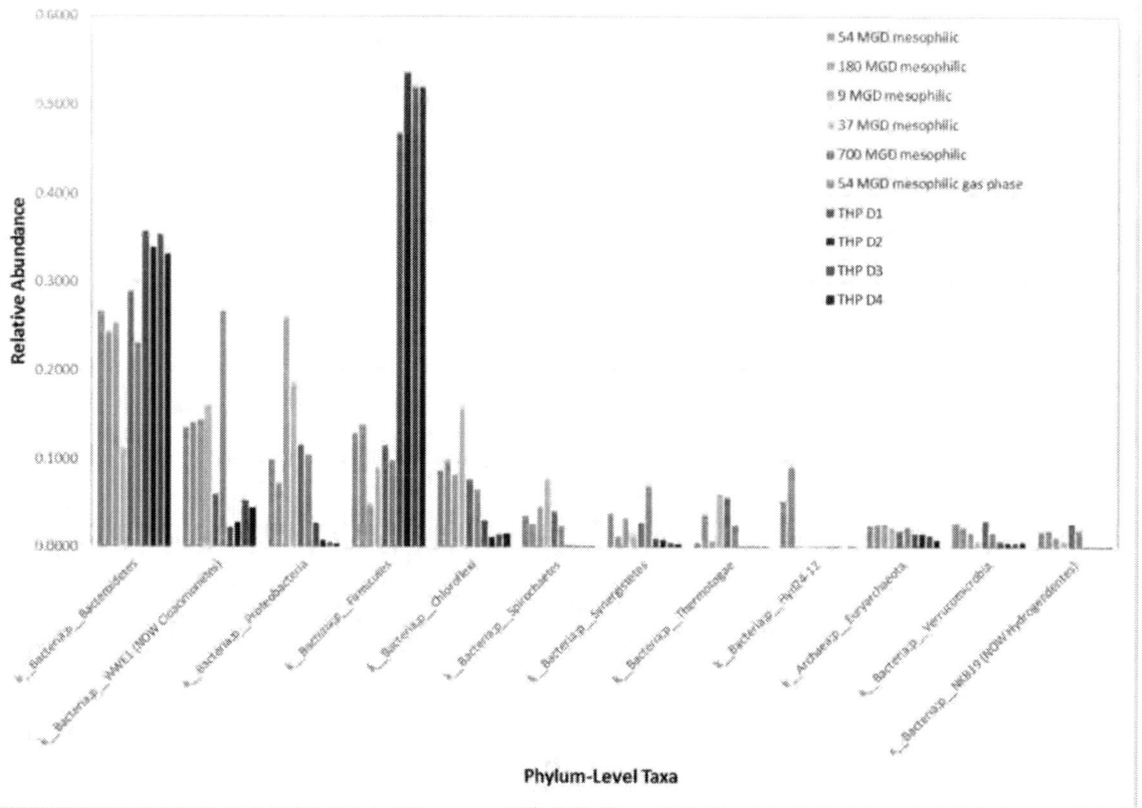

The term "diversity" as used in microbial ecology is highly ambigous and must be used with caution. As an analogy, many universities tout the geographical diversity of their student populations with statements such as "Our university has students from all 50 states". However, one must ask the question "Is the population really diverse if there is only one student from each of 49 states and 9,951 students from the home state?" On the other hand, would a population of students that represent only 25 states but each of the states equally be less "diverse"? The term "diversity" is ambiguous in microbial ecology and it is preferable to use more specific terminology.

There are a number of microbial ecology metrics known as alpha diversity factors that can be used to further examine the data and formally calculate richness and evenness measures for intersample comparisons. The term "richness" is used to quantify the number of unique types of members of the population, analgous to the number of states in the preceding analogy.

There are several measures of richness (number of actual species observed, rarefied species observed, and extrapolative measures such as ACE and Chao1). The term "evenness" is used to quantify the variation in distribution of abundance across the types of members in the population and there are also several measures of evenness. There are also some metrics that attempt to combine these two orthogonal measures into a single value (ie. Simpson or Shannon indices), but the problem is that there are infinte combinations of richness and evenness values that can lead to any single value of the combined measures.

A better tool for answering some of the "diversity" questions raised by this paper and for directly demonstrating the stark difference in community evenness between the conventional and THP digesters is the rank abundance plot. The rank abundance plot illustrates in an orderly manner the variation in abundance distribution among the represented taxa. To create this plot, the taxa represented within a single sample are ordered or "ranked" from that with the highest relative abundance to that with the lowest relative abundance without any regard to the identity of the taxa. The value of the abundance for the top ranking taxa is plotted in position 1, then so on with the next taxa and finally a line is drawn between the points to create either a flat or downward sloping line. If all of the OTUs in a given sample are equally abundant, the plot results in a flat line. For example, if a sample had ten taxa each having 10% of the abundance, the plot would be a flat line at 0.1. Steeper lines indicate less even distribution of abundance between OTUs. When the plots of multiple samples are overlaid, it is important to note that the identities of each samples' #1 ranked taxa might be different. Again, the point of this plot is to illustrate the evenness of abundance distribution among taxa, not to single out the identity of the taxa.

The rank abundance plot of Figure 3 shows a clear difference between the evenness of THP and conventional digester communities. Note that this plot contains essentially the same data shown in Figure 2, except the bars of each sample are ordered left to right without being confined to the taxa groups shown on the horizontal axis of Figure 2. The steep curve displayed by the THP communities indicates that their abundance is much less evenly distributed as they are dominated by only a few taxa. In fact, the THP digesters have about half of their relative abundance in a single taxa (rank #1) and another 35% in the next most dominant taxa (rank #2). Another 5% is in the third-ranked taxa and all of the remaining 10% relative abundance has to be split across all of the remaining minority taxa. On the other hand, the community abundance is spread much more evenly across the taxa in the digester, although the distribution is still far from completely even.

The reduced evenness and domination by only a few taxa for the THP digesters shown in Figure 3 implies a lower "diversity" of the THP digester microbial communites. The reduced evenness is itself one component of diversity and the lack of residual abundance to be spread across minority taxa is functionally equivalent to a reduced richness component. The implied lower richness is confirmed by traditional measures of actual observed richness and richness at the lowest common rarefaction value (data not shown). The values for both measures are significantly lower for the THP digester samples, in general half of the values common to the conventional digesters.

Figure 3: Phylum Level Rank Abundance Plot

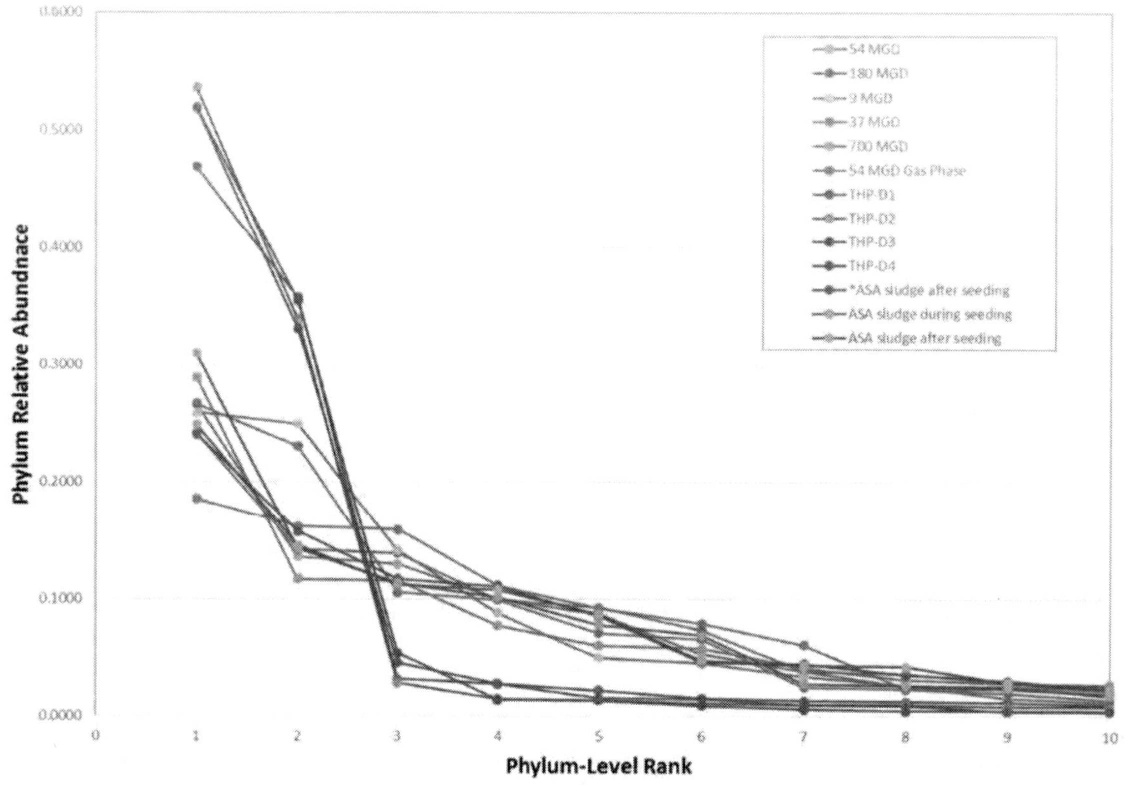

<u>The microbial communities of the THP digester samples were unique</u>

As previously mentioned, “diversity” is an ambigous term in microbial ecology. The previous section addressed alpha diversity, which is a method of comparing samples based on their internal diversity parameters (richness and evenness) as opposed to comparing how two or more samples share similar community members. The shared (or not shared) aspect of microbial communities is known as beta diversity in microbial ecology and it can be used to rigorously determine whether the communities of two or more samples are similar or unique.

There are numerous methodologies for comparing beta diversity in microbial communities, but one of the most efficient and effective visualization tools is the multidimensional scaling plot. Several forms of multidimensional scaling plots such as principle components analysis (PCA) or principle coordinates analysis (PCoA) are commonly used in engineering manuscripts and presentations. In essence, both PCA and PCoA seek to parameterize a large data set of of samples with a single dependent variable and numerous independent variables and reduce it into a numerical model of only a few ("principle") dimensions. While it can yield useful insight to certain types of data sets, the widely disparate relative abundance values in microbial communities present a signficant challenge.

A conceptually similar methodology is the nonmetric multidimensional scaling (NMDS) plot. In essence, it uses the ranks of the microbial taxa abundances instead of the actual values (hence the "nonmetric" or "nonparametric" term). The ranks fundamentally represent the importance of a given taxa to a sample (e.g. the #1 most important microbe), whereas the abundance values are not necessarily fundamentally indicative of a certain level of importance. While this may sound complicated, the procedure for processing the data is very simple in statistical programs such as R with the Vegan package. The input to the software is a simple table of OTUs (species) by samples with the OTU abundance counts in the cells. The software tool then calculates the Bray-Curtis dissimilarity metric between every possible pair of samples. Speaking in essence, the metric is equal to zero if the pair of samples share every OTU and it is equal to one if no OTUs are shared, with a range of values in between for partially shared communities (think of a two-circle Venn diagram). In theory, the resultant matrix of values that shows the "distance" between each possible pair of samples could be plotted in a multidimensional space, with the maximum number of necessary dimensions equaling the number of samples minus one. In other words, with a set of 10 samples in which no OTUs were shared by any of the samples (all distances between sample pairs equal to one), one would need a 9-dimensional space to allow all 10 samples to be exactly one unit away from each other. This is of course difficult for most people to visualize.

In order to better display and comprehend the relationships between samples, the NMDS algorithm seeks to squash the multidimensional space into a simpler form (a two dimensional plot in this case) while preserving the calculated "distances" between sample as best as possible It is analgous to streching a 3-D world globe onto a 2-D world map. In this sense, the resultant NMDS plot can be read sort of like a map with Euclidean distance being roughly proportional to differences between microbial communities in samples. Samples that are clustered together are "similar". Furthermore, the Vegan package calculates the same metric for OTUs using the samples that they do (or don't) share such that they can be plotted closer to (or further) from the samples that they are (or are not) strongly related to. The locations of the OTUs are also plotted to show their relationship to the samples. In an ideally fitted NMDS plot, a line drawn between any two samples would show the importance of intersected OTUs as a proportion of the distance along the line. In other words, an OTU halfway between two samples would be equally important to each of those samples. However, the NMDS plot must be interpreted with caution as some streching occurs in the reduction of many dimensions to a few.

Figure 4 is a Bray-Curtis NMDS plot that includes samples from THP digesters, conventional digesters, the THP digester startup seed, and raw (digester feed) sludges.

Figure 4: NMDS Plot of Bray-Curtis Dissimilarity Metrics

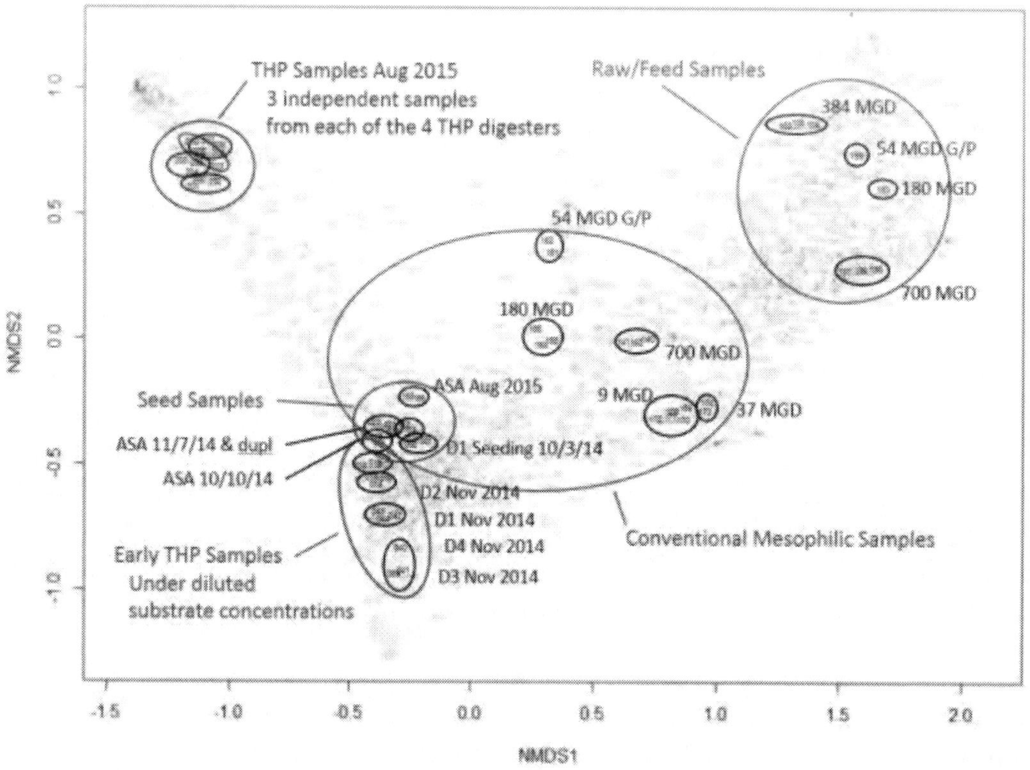

In Figure 4, individually sequenced sample replicates are shown as three digit black numbers and OTUs are shown as faint red numbers. The clustering of sample types is quite remarkable, with clear differences between THP, conventional and raw samples. Even within sample replicates and the one true field sample duplicate, the individual sample clusters are tight.

Many conclusions can be inferred directly from the NMDS plot.

The bugs may or may not be magic, but there are "THP Bugs": The NMDS plot confirms that the THP digesters have some very unique OTUs. In fact, the OTUs shown to the upper left of the THP digester cluster are so unique to THP digesters that they were "pushed" beyond the THP cluster in order to create as much distance as possible from the OTUs in between the THP and Conventional clusters (partially-shared OTUs). Ideally, the "pushed" OTUs would be directly under the blue circle, but the reduction of dimensions lead to some streching. In general, most of these OTUs are low in abundance, but OTUs 1, 10, 4 and 14 are dominant solely in the THP digesters. They will be confirmed and discussed in subsequent sections.

While not rigorously confirmed, the NMDS plot is likely to be useful for evaluating whether a given potential seed sludge might be a good fit for the startup of a THP digester. The addition of many new samples to an existing NMDS plot will likely warp the shape of the plot. However, the introduction of one or a few new samples will likely show where it falls within the range of existing sample types contained in the plot. Furthermore, while additional longer-term sampling is needed to confirm, the consistency of the microbial community of the Blue Plains seed source over time (ie. the October and November 2014 samples versus the August 2015 samples) shows that the inevitable delay between sampling a potential seed and receiving the sequencing results (up to 1-2 months) may not be a significant issue for the evaluation of potential seed sources.

DISCUSSION

This study clearly shows that the microbial communities of the sampled THP digesters had a lower richness and evenness than conventional mesophilic digesters. In general, it is accurate to say that they were less "diverse". While this study could not rigorously test whether the lower diversity was favorable or not, more than 20 years of THP operating experience suggests that the performance of THP digesters and there microbial communities is equivalent to or exceeds that of conventional mesophilic digesters. Although the results are not yet published, the performance testing requirements of the design-build contract for the THP digester system at Blue Plains demonstrated that the THP-fed digesters could sustain a 5% per day increase in solids loading for a period of weeks.

This study also showed that the microbial communities of the sampled THP digesters are unique and distinct from the communities found in conventional mesophilic digesters.

ACKNOWLEDGEMENTS

The authors thank DC Water for funding this research project.

REFERENCES

Caporaso, J Gregory , Justin Kuczynski , Jesse Stombaugh , Kyle Bittinger , Frederic D Bushman , Elizabeth K Costello , Noah Fierer , Antonio Gonzalez Peña , Julia K Goodrich , Jeffrey I Gordon , Gavin A Huttley , Scott T Kelley , Dan Knights , Jeremy E Koenig , Ruth E Ley , Catherine A Lozupone , Daniel McDonald , Brian D Muegge , Meg Pirrung , Jens Reeder , Joel R Sevinsky , Peter J Turnbaugh , William A Walters , Jeremy Widmann , Tanya Yatsunenko , Jesse Zaneveld & Rob Knight (2010). QIIME allows analysis of high-throughput community sequencing data. Nature Methods, Volume 7, pp. 335-336.

Wilson, C.A, Tanneru, C.T., Banjade, S., Murthy, S.N., and Novak, J.T. (2011). Anaerobic Digestion of Raw and Thermally Hydrolyzed Wastewater Solids Under Various Operational Conditions. Water Environment Research, Volume 83, Number 9, pp. 815-825(11).

Schloss, Patrick D., Sarah L. Westcott, Thomas Ryabin, Justine R. Hall, Martin Hartmann, Emily B. Hollister, Ryan A. Lesniewski, Brian B. Oakley, Donovan H. Parks, Courtney J. Robinson, Jason W. Sahl, Blaz Stres, Gerhard G. Thallinger, David J. Van Horn and Carolyn F. Weber (2009). Introducing mothur: Open-Source, Platform-Independent, Community-Supported Software for Describing and Comparing Microbial Communities. Applied and Environmental Microbiology, Volume 75, Number 23, pp. 7537-7541.

Free Nitrous Acid Pretreatment of Thickened Waste Activated Sludge Improves Anaerobic Degradability

Elsayed Elbeshbishy[1], Frances Okoye[1], Dang Ho[2], Siva Sarathy[2], Wayne Parker[3]
[1]Ryerson University, 350 Victoria Street, Toronto, ON, Canada, M5B 2K3
[2]Trojan Technologies, 3020 Gore Rd, London, ON, Canada, N5V 4T7
[3]University of Waterloo, 200 University Ave W, Waterloo, ON, Canada, N2L 3G1

Executive Summary: Anaerobic digestion of waste for biological methane production can be a sustainable source of renewable energy as well as an efficient waste management technique. Significant research efforts on digester design, process enhancement, and odors control are vital. Generally, the rate limiting step of anaerobic digestion of solid waste is hydrolysis where the cell wall is broken down allowing the organic matter inside the cell to be available for biological degradation. Therefore, many studies have been conducted to enhance the hydrolysis using feed pretreatment. In this study, the effect of free nitrous acid (FNA) on the biodegradability of thickened waste activated sludge (TWAS) was evaluated. The effect of FNA chemical dose ranging from 0.4 to 2.8 mg N/L on the degree of disintegration as well as the anaerobic digestability of TWAS was investigated. All of the pretreatment tests were conducted at a pH of 5.5. The results of this study showed that the soluble chemical oxygen demand (SCOD) increased by 140%. The biochemical methane potential tests showed an increase in methane production by 20% when the FNA dose was 1.4 mg N/L.

Keywords: Anaerobic Digestion, Waste Activated Sludge, Pretreatment, Biodegradability, Solubilization

1.Introduction

Background: Waste activated sludge (WAS) is produced from any facility that includes a biological process in its treatment of wastewater. In cities and areas that are heavily populated, the amount of WAS produced can be very significant. It is composed of 59–88% (w/v) of organic matter including a high fraction of volatile solids (VS) and normally dewatered to form thickened waste activated sludge (TWAS) by means of thickening techniques [1]. Proper management of WAS contributes to about 60% of the total costs of a municipal wastewater treatment plant (WWTP). Coupled with a steady increase in the amount of WAS produced due to population growth and more stringent environmental regulations in terms of wastewater collection and treatment, WAS management presents an opportunity for research that can provide innovative and cost effective solutions. Among several methods including incineration and land application for handling WAS, anaerobic digestion (AD) is the most commonly used due to the potential to reduce the sludge volume and recover bioenergy during its process [2 & 3]. AD is composed of four stages; hydrolysis, acidogenesis, acetogenesis, and methanogenesis. However, the digestion of WAS often requires long retention times of 20 - 30 days to obtain only 30 – 50% digestion efficiency. This is because WAS constituents are held together by a polymeric network formed by microbial extracellular polymeric substances (EPS) and cations which are resistant to the hydrolysis stage. Destruction of EPS network could promote hydrolysis and WAS biodegradability as well as enhance the dewaterability during and after anaerobic digestion. Through the use of pretreatment methods on WAS prior to AD, organic matter can be released by the destruction of the EPS network. A total COD (TCOD) solubilization of 40 to 60% can be achieved using pretreatment methods [1, 4-7]. Application of pretreatments have the ability to increase biogas generation, reduce capital and operational costs through increased digestion throughput or reduced retention time and or digester footprint as well as reduce adverse environmental impacts and greenhouse gas emissions.

Significant research efforts have gone into sludge pretreatment techniques. They can be broadly classified into Biological, Thermal, Mechanical and Chemical methods. Hybrid techniques also exist, where a combination of the aforementioned methods are used to improve efficiency or reduce drawbacks. Many of these methods are characterized by intensive costs and/or negative environmental impacts. In comparison with other chemical pretreatment methods, FNA pretreatment has two major advantages: it can be produced in a sludge reject water treatment system making it an inexpensive material and it is environmentally friendly since it will be transformed to nitrogen gas in subsequent chemical reactions [8 & 9]. However, the effect of FNA pretreatment on exploiting internal carbon source contained in WAS has not been thoroughly studied. FNA is believed to accelerate the disruption of both extracellular polymeric substances and cell envelopes to form more short-chain fatty acids (SCFA) [10].

Free Nitrous Acid (FNA) is known to oxidize cell walls which releases organic matter. Studies have shown that this biocidal agent can increase sludge internal carbon bioavailability by 50%, improve biochemical methane production and reduce sludge volume by 87% after 24 – 48 hours of pretreatment [10]. However, more research is required to understand the full potential and impacts of FNA as a pretreatment for AD of Thickened WAS (TWAS).

Objectives: The main goal of this study was to evaluate the effect of FNA pretreatment on the biodegradability of TWAS. It is intended to observe the impact of this pretreatment on the hydrolysis and methanogenesis stages in AD using solubility, VFA and biochemical methane potential (BMP) assays.

2.Methodology

TWAS was collected from Adelaide WWTP London, Ontario and the pretreatment procedure was carried out on the same day. FNA concentrations were determined using its equilibrium relationship with nitrite concentrations, pH and temperature. Nitrite was prepared by dissolving sodium nitrite in deionized water. Based on the literature, the methane production increases with increasing the FNA dose until 2.13 mg N/L, after this dose, there was no significant improvement in the process performance. For the first set of experiments, FNA concentrations ranging between 0.4 – 2.8 mg N/L were investigated at a constant pH of 5.5 and contact time of 24 hours. The experiment bottles were maintained at a temperature of 25°C and constant mixing of 100 rpm. The pH was maintained and monitored over the first 12 hours using HCl. To account for pH effect, two bottles were set up without FNA; one of which was maintained at pH 5.5, the other was not adjusted. Table 1 shows the pretreatment condition for this set in which nitrite concentrations ranged from 50 mg N/L to 400 mg N/L.

Table 1. The first set of pretreatment conditions for the TWAS samples

Exp.	Nitrite conc.	pH	Temp	FNA*
#	mg N/L		°C	mg N/L
1	0		25	0.0
2	0	5.5	25	0.0
3	50	5.5	25	0.4
4	100	5.5	25	0.7

5	150	5.5	25	1.1
6	200	5.5	25	1.4
7	250	5.5	25	1.8
8	300	5.5	25	2.1
9	350	5.5	25	2.5
10	400	5.5	25	2.8

FNA = $S_{NO-2\_N} / (K_a \times 10^{pH})$, $K_a = e^{-2300/(273+T)}$, S_{NO-2-N} = Nitrite concentration, T = temp. in ^{0}C.

After 24 hours, BMP tests were carried out. The BMP test was carried out using seed from St Mary's WWTP, Ontario using 0.4 g COD/g VSS food to microorganism ratio. Bottles were maintained at a temperature of 37°C and mixed at 150 rpm. The biogas was passed through sodium hydroxide salt to capture the carbon dioxide and the volume of methane was measured by releasing the headspace pressure using a manometer. At the end of the BMP tests, wet chemistry analysis was carried out. A follow up set of experiments was conducted with varying contact times (6, 12, 24 hours), pH (4.5, 5, 5.5) and nitrite concentrations (50, 100, 200 mg N/L). These conditions are presented in Table 2.

Table 2. The second set of pretreatment conditions for the TWAS

Exp. #	Time hr	Nitrite conc. mg N/L	pH	Temp °C	FNA mg N/L
1	12	50	4.5 ± 0.1	25	3.6
2	12	50	5.5 ± 0.1	25	0.4
3	12	200	4.5 ± 0.1	25	14.2
4	12	200	5.5 ± 0.1	25	1.4
5	6	100	4.5 ± 0.1	25	7.1
6	6	100	5.5 ± 0.1	25	0.7
7	24	100	4.5 ± 0.1	25	7.1
8	24	100	5.5 ± 0.1	25	0.7
9	6	50	5.0 ± 0.1	25	1.1

10	6	200	5.0 ± 0.1	25	4.5
11	24	50	5.0 ± 0.1	25	1.1
12	24	200	5.0 ± 0.1	25	4.5
13	12	100	5.0 ± 0.1	25	2.2
14	6	Control			

Portions of each sample were collected for analysis before and after the pretreatment procedure was conducted. Samples were filtered through 0.45 µm filter paper for the soluble parameters analysis. Liquid samples were analyzed for total solid (TS), volatile solid (VS), total suspended solid (TSS), and volatile suspended solid (VSS), using standard methods (APHA, 1998). Total, semi-soluble, and soluble chemical oxygen demand (TCOD, ssCOD, SCOD), and ammonium nitrogen (NH4-N) were measured using HACH test kits (HACH Odyssey DR/2500). Soluble parameters were quantified after filtration with 0.45 µm filter paper (RK-02915-14, Cole-Parmer, USA). The ssCOD was measured after filtration with 1.2 µm filter paper. Volatile fatty acids (VFAs) including acetate, propionate, n-butyrate, n-valerate, iso-butyrate, and iso-valerate were quantified using a gas chromatography (GC) (HP 5890 Series II, country) equipped with a Nukol fused-silica capillary column and flame ionization detector (FID). All liquid analyses were conducted in triplicates.

3.Results and Discussion

The results were normalized to per mass of VS for comparison. In the first set of experiments where pH was 5.5 and contact time was 24 hours, there was no significant change in TCOD before and after pretreatment as expected. However, the concentrations of SCOD were affected by FNA concentrations. Figure 1 shows the SCOD concentration trend for the first set of experiments. The concentration of SCOD increases from 0.08 to 0.2 mg COD/mg VS with increasing FNA concentrations until 1.4 mg N/L after which there is no significant change. This indicates that FNA enhances solubilization as compared with the untreated sample. No real difference was observed in the VSS concentrations after pretreatment.

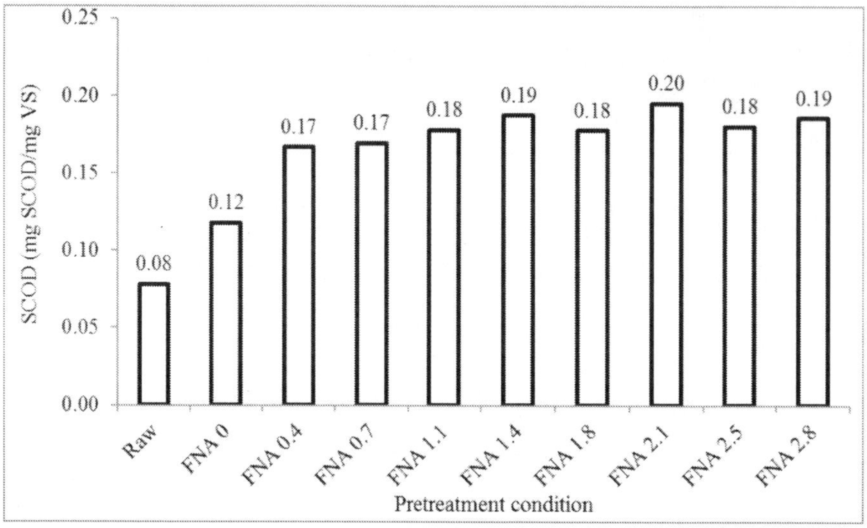

Figure 1. Concentrations of SCOD per mass VS after FNA pretreatment for first set of experiments

However, the concentration of total VFAs increased as shown in Figure 2 with the maximum of 0.17 mg COD/mg VS occurring at FNA dose of 2.1 mg N/L. This supports the notion that FNA treatment increases the bioavailability of organic material to microbes.

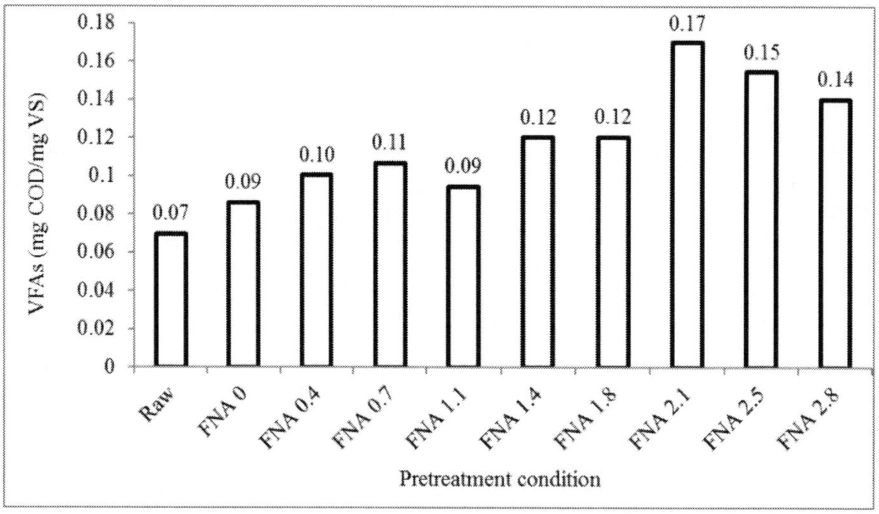

Figure 2. Concentrations of VFA per mass VS after FNA pretreatment for the first set of experiments

The concentrations of ammonia were observed to decrease slightly after pretreatment. The methane production yields were normalized to per volume of TWAS added, per mass of VSS and TCOD shown in Table 3. The methane yield per volume of TWAS was maximum as FNA 1.4 mg N/L at 6.1 L/L TWAS compared with the untreated sample of 4.7 L/L TWAS. The methane yield per mass VS and per mass of TCOD were highest at FNA 2.1 mg/L of 209 mL/g VS and 131 mL/g COD respectively. The maximum methane production rate was observed at FNA 1.8 mg N/L of 31 mL/day.

Table 3. Summary of methane production for first set of experiments

Pretreatment conditions				**Methane**				
Exp.	**pH**	**Temp**	**FNA**	**Volume**	**Yields**			**Maximum production rate**
#		**°C**	**mg N/L**	**mL**	**L/L_{TWAS}**	**mL/g VS_{added}**	**mL/g $TCOD_{added}$**	**mL/d**
1	6.7	25	0.0	117	4.7	158	97	19
2	5.5	25	0.0	129	5.2	172	108	20
3	5.5	25	0.4	137	5.5	185	114	24
4	5.5	25	0.7	140	5.6	179	120	28
5	5.5	25	1.1	145	5.8	193	120	30
6	5.5	25	1.4	152	6.1	203	127	30
7	5.5	25	1.8	149	6.0	199	130	31
8	5.5	25	2.1	151	6.0	209	131	30
9	5.5	25	2.5	147	5.9	196	123	30
10	5.5	25	2.8	147	5.9	196	126	30

The second set of experiments featured maximum SCOD concentrations at pH 5.5 with the combination 100 mg N/L nitrite and 24 hours contact time yielding the highest SCOD followed by 200 mg N/L nitrite and 12 hours. The maximum total VFA occurred at pH 5 with the combination of 200 mg N/L nitrite and 24 hours contact time followed by 200 mg N/L and 6 hours. Except for 2 samples, all the pretreated samples yielded higher methane volumes than the untreated sample with the highest yield occurring in the sample with treatment combination 50 mg N/L nitrite, 6 hours and pH 5 as seen in Table 4. It is worthy to note that at higher nitrite concentrations, higher methane yields were observed at pH of 5.5 with same pretreatment time (see exp. # 2 and 4), however at lower pH (either 5 or 4.5), this observation was opposite i.e. at pHs of 5 or 4.5, the lower the nitrite, the higher the methane yields (compare exp. # 9 with 10 and compare exp. # 1 with 3). This observation needs further investigation.

Table 4. Summary of the methane production for the second set of experiments

Pretreatment conditions						**Methane**			
Exp.	**Time**	**Nitrite conc.**	**pH**	**Temp**	**FNA**	**Volume**	**Yield**		
#	**hr**	**mg N/L**		**oC**	**mg N/L**	**mL**	**L/LTWAS**	**mL/g VS_{added}**	**mL/g $TCOD_{added}$**
1	12	50	4.5	25	3.6	281	5.6	255	103
2	12	50	5.5	25	0.4	286	5.7	269	102
3	12	200	4.5	25	14.2	263	5.3	245	94
4	12	200	5.5	25	1.4	315	6.3	268	123
5	6	100	4.5	25	7.1	289	5.8	255	109
6	6	100	5.5	25	0.7	277	5.5	268	94
7	24	100	4.5	25	7.1	275	5.5	248	100
8	24	100	5.5	25	0.7	295	5.9	259	103
9	6	50	5.0	25	1.1	319	6.4	286	114
10	6	200	5.0	25	4.5	240	4.8	245	83
11	24	50	5.0	25	1.1	294	5.9	271	112
12	24	200	5.0	25	4.5	238	4.8	242	88
13	12	100	5.0	25	2.2	276	5.5	252	108
14	6	Control				251	5.0	238	90

4. Conclusion

FNA pretreatment of TWAS enhances solubilization as observed in the increase in SCOD by over 100%. Biodegradability is also improved after treatment with FNA as compared to untreated TWAS by at least 20%. Further studies are required to determine the most economic pretreatment conditions for FNA pretreatment method and to study the phenomenon that was observed which led to higher methane concentration with lower nitrite addition at pH below 5.5. The effect of FNA pretreatment on the anaerobic digestibility of TWAS in a continuous stirred tank reactor (CSTR) and on the dewaterability of digested TWAS is ongoing.

References

[1] Tyagi, V. K., Lo, S. L. (2011) Application of physico-chemical pretreatment methods to enhance the sludge disintegration and subsequent anaerobic digestion: an up to date review, Rev Environ Sci Biotechnol, 10:215-242.

[2] Zhang, S., Zhang, P., Zhang, G., Fan, J., Zhang, Y. (2012) Enhancement of anaerobic sludge digestion by high-pressure homogenization, Bioresource Technology, 118:496–501.

[3] Xu, J. et al. (2014) Evaluation of thermal, thermal-alkaline, alkaline and electrochemical pretreatments on sludge to enhance anaerobic biogas production, Journal of the Taiwan Institute of Chemical Engineers, 45:2531-2536.

[4] Wahidunnabi, A. K., Eskicioglu, C. (2014) High pressure homogenization and two-phased anaerobic digestion for enhanced biogas conversion from municipal waste sludge, Water Research, 6:430-446.

[5] Wang, Z., et al. (2009) Digestion of thermally hydrolyzed sewage sludge by anaerobic sequencing batch reactor, Journal of Hazardous Materials, 162:799–803.

[6] Devlin, D.C., et al. (2011) The effect of acid pretreatment on the anaerobic digestion and dewatering of waste activated sludge, Bioresource Technology, 102:4076-4082. [20] Yiying, J. et al., (2009) Combined alkaline and ultrasonic pretreatment of sludge before aerobic digestion, Journal of Environmental Sciences, 21:279–284.

[7] Shahriari, H. (2011) Enhancement of Anaerobic Digestion of Organic Fraction of Municipal Solid Waste by Microwave Pretreatment, PhD Thesis, University of Ottawa.

[8] Pijuan, M., et al. (2012) Improving secondary sludge biodegradability using free nitrous acid treatment, Bioresource Technology, 116:92-98.

[9] Law, Y., Ye, L., Wang, Q., Hu, S., Pijuan, M., Yuan, Z. (2015) Producing free nitrous acid - A green and renewable biocidal agent – From anaerobic digester liquor, Chemical Engineering Journal 259:62–69.

[10] Zhao, J., et al. (2015) Free nitrous acid serving as a pretreatment method for alkaline fermentation to enhance short-chain fatty acid production from waste activated sludge, Water Research, 78:111-120.

Thermal Hydrolysis Reduces Carbon Footprint of Biosolids treatment

Dr WPF Barber

Cambi Inc, 279 Great Valley Parkway, Malvern, PA, 19355
Email: bill.barber@cambi.com

ABSTRACT

Globally, wastewater and sludge treatment plays a fundamentally important role in control of anthropogenic carbon impacts. Unfortunately, most of the infrastructure we see today was designed in the late 19th (e.g. thickening, dewatering, primary treatment, anaerobic digestion; thermal drying; nutrient extraction; Biosolids recycling as fertilizer) or early 20th centuries (e.g. activated sludge treatment) and therefore was not designed to be either carbon, or energy efficient. In absence of demolishing entire wastewater treatment plants due to their inefficiency, it is necessary to adapt existing works to become more relevant with respect to meeting modern demands. One successful way of doing this with respect to Biosolids treatment, is the application of thermal hydrolysis to increase renewable energy generation and reduce Biosolids quantity and therefore reduce downstream processing requirements. This paper presents results of a detailed model which calculates the carbon impact of numerous processes, and highlights the important role which thermal hydrolysis plays with respect to reducing carbon impact.

KEYWORDS: Biosolids treatment, carbon footprint, anaerobic digestion, thermal hydrolysis

INTRODUCTION

It has eventually and sometimes reluctantly, become understood that climate change is directly influenced by the anthropogenic release of what are known as "Green House Gases – GHGs". Their release into the earth's atmosphere has been correlated with an increase in global temperature which in turn, has been linked to weather change with potentially catastrophic impacts (IPCC, 2001). With such significant global impacts expected, it is therefore no surprise that a great deal of interest is being shown by industry and the media alike on highlighting the issues and also proposing methods of reducing the generation of GHGs.

GHGs have been classified by the Intergovernmental Panel on Climate Change (IPCC), and include: carbon dioxide; methane; nitrous oxide; tetraflouromethane; sulfur hexafluoride; HCFs; CFCs, and others. Determination of carbon footprint involves measuring the amount of these compounds emitted during a process and converting the data to a carbon dioxide equivalent. The carbon dioxide equivalent, (also expressed as CO_2eq or CO_2e), is an internationally accepted factor that expresses the global warming potential a GHG has in terms of the quantity of carbon dioxide that would have an identical impact. For example, the CO_2e for methane is 21 (IPCC, 2001) implying that 1 unit of methane has the same global warming potential as 21 units of carbon dioxide.

There is an implicit link between power consumption and carbon footprint, consequentially the power sector is responsible for 37% of all anthropogenic CO_2 (Ashton, 2005). In the UK, the Water Industry accounts for 3% of all power consumed (Water UK, 2007). Most of this power is used to run processes and pump water. The power demand of the Water Industry is partially offset by the generation of a methane-enriched biogas – produced during the anaerobic digestion of sewage biosolids – to electricity and heat via combined heat and power plants.

Biosolids Treatment
Power is required to treat biosolids to an acceptable level before use. As mentioned in the previous section, anaerobic digestion can be used as a process step and this may reduce overall power requirements. By law, biosolids must be treated to reduce the risk from pathogens and vectors prior to their application to or placement on the land for beneficial use or disposal (US EPA, 1999). Treatment types fall into two categories: Processes to Further Reduce Pathogens (PFRP), known as Class A; and, Processes to Significantly Reduce Pathogens, labeled Class B. The attainment of Class A quality is the more stringent of the standards and requires:

- a density of fecal coliforms in the sludge of less than 1000 most probable number (MPN) per gram of total dry solids, **or**
- a density of Salmonella sp. bacteria lower than 3 MPN/4 g of total dry solids, **and**
- Time/temperature relationship according to various equations

These standards apply to the point at which the biosolids are used. Class A can be attained via numerous treatment systems such as: composting, heat drying, heat treatment, thermophillic anaerobic digestion, β and γ radiation, and pasteurization (70°C for 30 minutes). Time temperature relationships required for the pasteurization of various micro-organisms have been well documented in the work of Feacham *et al.* (1983).

The requirements for Class B, or Processes to Significantly Reduce Pathogens (PSRP), are lower and therefore Class B treated biosolids have more restrictions on application. Here, biosolids quality must have a geometric mean fecal coliform density of < 2,000,000 colony forming units (cfu) or most probable number (MPN) per gram of solids dry weight. Class B systems include: aerobic digestion; mesophilic anaerobic digestion (MAD); composting; lime stabilization and air drying.

Thermal Hydrolysis
Thermal hydrolysis of sewage sludge, which involves the application of heat at above autoclave temperature for a defined time period prior to anaerobic digestion, is an established commercially available technology since the first full-scale plant in HIAS in 1995. The heat is typically provided by live steam injection at design temperature and concomitant pressure which is then rapidly released (exploded), although some configurations use standard heat exchange. At the time of writing, there are 75 facilities of which 39 are operating and the remaining are in various stages of design. In total, 1.65 million metric dry tonnes of sludge per year are, or will be, processed with thermal hydrolysis. Although facilities cover most major continents, the vast majority are in Europe (57) within which the UK boasts 19 operational plants. Drivers for installation are geographically market driven but typically include: increased loading rates (to minimize size of new digestion plants, or maximize use of existing facilities); improved sludge

cake dewaterability which reduces downstream transport and processing costs; increased production of renewable energy, and sterilization of sludge.

The first thermal hydrolysis plant in North America was installed at DC Water's Blue Plains Wastewater Treatment Works. The plant processes approximately 300 dry tons/d of sludge and has resulted in the generation of 9 MW of renewable energy, whilst reducing Biosolids volume from 1200 wet tons/d to 450 wet tons/d by eliminating the need for liming, reducing volume due to digestion, and improving dewatering. A study by Willis *et al.*, (2010) found that thermal hydrolysis and advanced nutrient removal which didn't require methanol resulted in a large reduction in carbon footprint for DC Water. The findings were similar to a previous study (Barber, 2009) which demonstrated installation of thermal hydrolysis as reducing carbon footprint of United Utilities by 32,000 tonnes/year, equivalent to 8% of the entire business' carbon impact.

In fact a large number of independent studies looking at carbon impacts of Biosolids process options have concluded that the installation of thermal hydrolysis results in lowest potential carbon footprint regardless of the end point or use for the sludge (Auty and Blake, 2009, Asian Development Bank, 2012). This appears to be due to a number of reasons as shown:

- Increased production of renewable energy
- Better volatile solids destruction resulting in less biosolids downstream for transport and further processing
- Better dewatering which further reduces biosolids for downstream processing. Also significant reduction in fossil fuel requirements for downstream drying
- Higher dewaterability increases energy content in cake which provides greater energy recovery benefit in downstream incineration, whilst improved volatile solids destruction reduces the quantity of material which needs to be incinerated
- Higher grade of biosolids means more landbank is opened up which reduces transport of biosolids
- Higher loading rates in digestion so less material used in construction which reduces embodied carbon impact

Calculating carbon footprints

Generally, the vast majority of the work involved in determining carbon footprints entails calculating the inputs required for carbon modeling. The inputs (e.g. power; transport; chemical consumption; quantity and type of building material used) are then multiplied by emission or conversion factors. The most well-known attempts to determine a Water Industry carbon footprint started in the UK and were made by UKWIR, who have developed a model entitled "Workbook for Quantifying Greenhouse Gas Emissions" (2005). The model includes Water Industry-specific data involving emissions of GHGs from wastewater treatment and application of sewage biosolids to land. The model has been the foundation for further work on carbon calculating (Entec UK, 2006). Since then, UKWIR is in the process of updating the workbook with revisions and additions, such as inclusion of embodied carbon calculating. This work is currently ongoing.

Subsequently, several other models have been developed looking at various aspects of wastewater and sludge treatment. One of the better ones is BEAM (Biosolids Emissions Assessment Model) developed in 2009 for the Canadian Council of Ministers of the Environment. However, none of the previous modelling attempts have gone into sufficient process detail where differences can be made between technology suppliers. The purpose of this paper is to explain the development of a model which is a further development of previous work (Barber, 2009) which undertakes very detailed mass and energy balances for both wastewater and sludge processes and can differentiate between different configurations of digestion. It also contains detailed energy balances for raw sludge options as well as drying and incineration.

METHODOLOGY

There fundamental differences between the current model presented here and previous attempts described above at calculating biosolids carbon footprints. The first difference is that it adds a layer of calculation which performs all the necessary mass and energy balances – previous models required the outputs of a mass and energy balance as an input, i.e. input kilowatt hours consumed. This allows inputs to be simpler, for example entering the throughput, BOD and ammonia concentrations and expected dry solids output from a dewatering device, the model will automatically calculate the power required for dewatering and liquor treatment and the chemicals, (both polymer and alkalinity for total nitrogen removal) and then convert them into a carbon footprint. The second difference is that the model has been designed to easily benchmark whole solutions, for instance all biosolids processing at the sewage works, transport of biosolids from the works and application of the biosolids to land, rather than individual process unit items. It calculates different biosolids solutions based on the same basic inputs which allows a quick comparison on which outlet to pursue for a particular project. Other more detailed differences are described elsewhere (Barber, 2008).

Central to the model is the impact of anaerobic digestion on down-stream processing. The model calculates a performance level dependent on the sludge type and pre-treatment provided. Based on external conditions and temperature, it calculates additional heat requirements – if required – for TAD, liquid pasteurization and ATAD. The additional heat required for ATAD is calculated after exothermic heat release has been determined based on the strength of the biosolids and hydraulic retention time. For thermal hydrolysis, the model determines the steam required for the pre-treatment plant and compares it to the steam available from the CHP plant. It then redirects biogas from the CHP plant towards a boiler to make up the deficit in energy requirement. Finally, for MAD combined with drying, the model can be run to redirect biogas into the dryer or to use biogas for energy generation and use natural gas for drying. For incineration options, the model determines the energy value of the sludge and therefore the minimum dry solids required to meet autothermic operation. If insufficient energy is available, it determines an auxiliary fuel demand.

Much in the same way that biogas can be used to displace fossil-fuel derived energy, a fertilizer benefit was calculated based on the fertilizer value of the biosolids and the energy required to manufacture an equivalent quantity of fertilizer (Kroiss and Zessner, 2007). In this study, it is assumed that only half of the displacement benefit is counted. The model is also set up to account for differences in crop yields between fertilizer and biosolids to include for sequestering effects. However, in this paper, it is assumed that crop yields are similar for both. The model also determines sequestration effects depending on inputs of sludge type, calculations of digestion performance, and fraction of recalcitrant and labile carbon in the sludge.

The model works by calculating an operating carbon footprint by subtracting the carbon footprint benefits (of renewable energy generation from biogas; energy generation from biosolids incineration; displacement of fertilizer, soil carbon sequestration) from the factors which contribute to carbon (such as power consumption; chemical usage and biosolids transport). The results presented in the remainder of this manuscript are based on that method of calculation, and other features of the model are described later in relevant sections.

Operating carbon footprints were determined for a number of processes with and without anaerobic digestion. Two digestion options were determined, one with thermal hydrolysis as a pre-treatment stage, and another without. A detailed list is given below:

1. Mesophilic Digestion (although model can also determine the following variants: acid phase; temperature phase; liquid pasteurization; thermophilic digestion; autothermal Thermophilic Aerobic digestion)
2. Thermal hydrolysis and mesophilic digestion
3. Raw Liming
4. Liming with supplemental heat
5. Raw drying
6. Mesophilic digestion and drying
7. Thermal hydrolysis with mesophilic digestion and drying
8. Raw Incineration (the model can work with and without thermal drying upstream, but for this paper, sludge is only dried to meet autothermic requirements – also calculated by model)
9. Mesophilic digestion and incineration
10. Thermal hydrolysis with mesophilic digestion and incineration
11. Raw (EnVessel) composting (model determines requirement of green waste dependent on carbon:nitrogen ratio and dry solids)
12. Raw landfill (model also can determine digestion with landfill)

The model also distinguishes carbon impact by Scope, which is defined by the Greenhouse Gas Protocol as follows:

- Scope 1: All direct greenhouse gas emissions. Examples of scope 1 emissions during biosolids treatment include fugitive emissions, methane from lagoons, storage or anaerobic digestion.
- Scope 2: Indirect greenhouse gas emissions from consumption of purchased electricity, heat or steam. They occur principally at electricity generators as a result of electricity consumption at another facility. Examples of scope 2 emissions during sludge treatment include power consumed by dewatering equipment.
- Scope 3: All indirect emissions other than those categorised as Scope 2. Examples include emissions associated with the extraction, manufacture and production of products purchased by a company, waste-related emissions and any business travel. Biosolids related examples include emissions from the production of polymer used for thickening or dewatering, the

production of lime used for pathogen deactivation, and the application of biosolids on agricultural and forestry land.

With respect to transport options, these can be by road (using either gasoline or diesel); train, sea or air freight. For this paper, options were looked at with gasoline road transport.

The required inputs to the model are given in Appendix A, whilst a list of the outputs is given in Appendix B.

As the model requires a full mass and energy balance, a financial model was created to convert the outputs of the model into a comparative whole life cost model.

Note: this paper only shows results of operating carbon footprints and not those associated with the carbon release during manufacture and maintenance of the plant – known as embodied carbon. Detailed work on a number of solutions for embodied energy was performed. It was found that over a period of 30 years operation, the construction and maintenance carbon footprints of the solutions (for large construction projects – e.g. a raw biosolids incineration plant) contributed typically less than 2% of the whole life carbon, implying that yearly operation of the plant was far more significant with respect to carbon footprint than its construction.

RESULTS

The contributory factors for the processes under evaluation are shown in Figure 1.

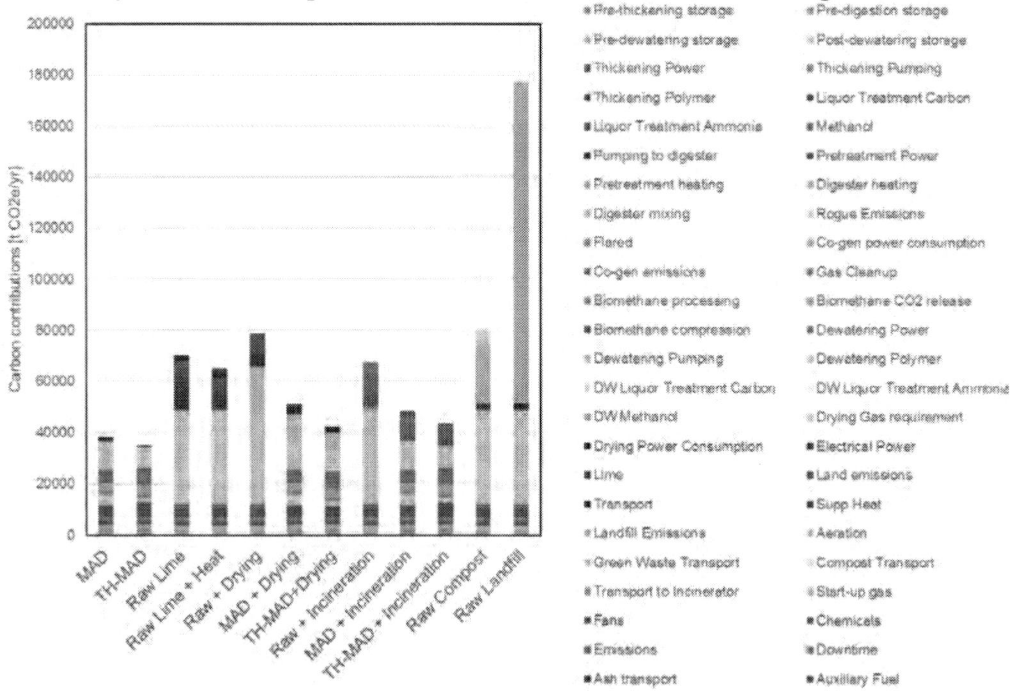

Figure 1. Contributory factors to carbon footprint – Note in subsequent graphs data for Raw landfill is not shown due to its size to make graphs easier to see.

Closer observation of Figure 1 reveals that the lowest 6 options are those which are preceded by anaerobic digestion, and the digestion variant with thermal hydrolysis is lower. Although more processing is required for thermal hydrolysis, e.g. the need for centrifuge thickening and also gas for steam generation, it has lower contributory factors than digestion options where it is absent. This is due to the large reduction in transport emissions which offsets the additional processing. Also, as anaerobic digestion plant is smaller with thermal hydrolysis and retention time lower, the mixing energy required in the digestion is also much lower. A breakdown of the individual components which make up the carbon footprints are given in Appendices C and D. Options with digestion are lower than those without due to destruction of sludge resulting in less sludge for downstream processing, whether it is transport, drying, incineration etc. In addition, Scope 3 emissions are much higher for options where digestion is absent – as shown in Figure 2 – due to due to the need of greater quantities of chemicals such as polymer, lime, and incineration chemicals.

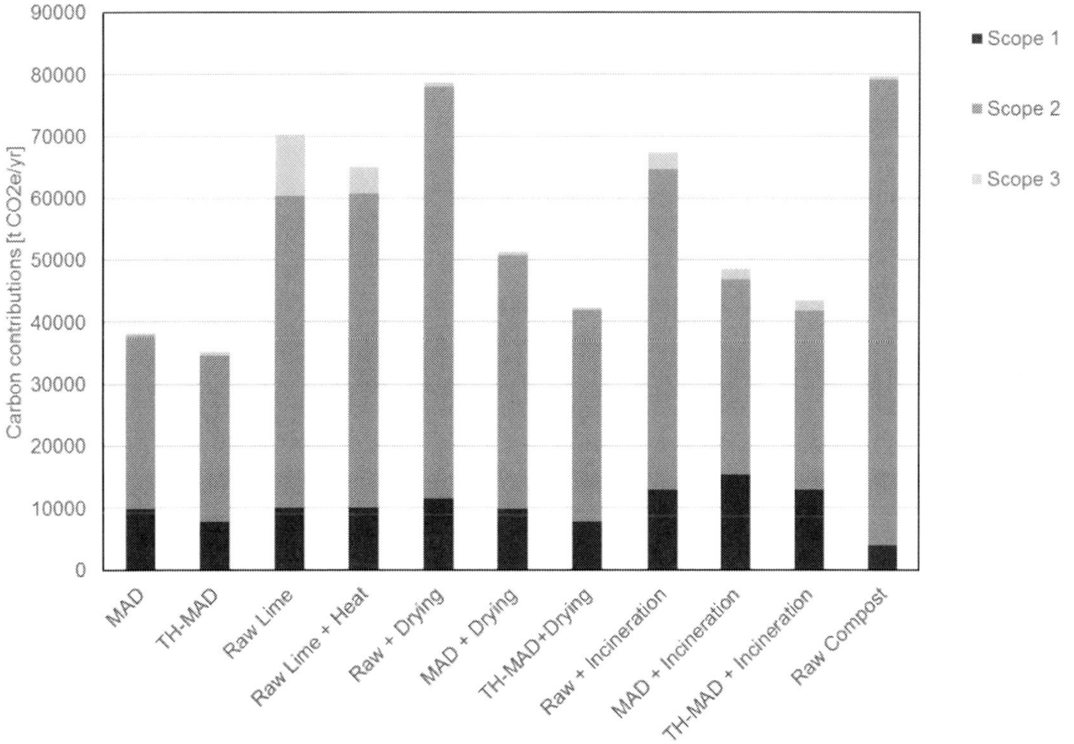

Figure 2. Contributory factors to carbon footprint by Scope

The carbon benefits are shown in Figure 3.

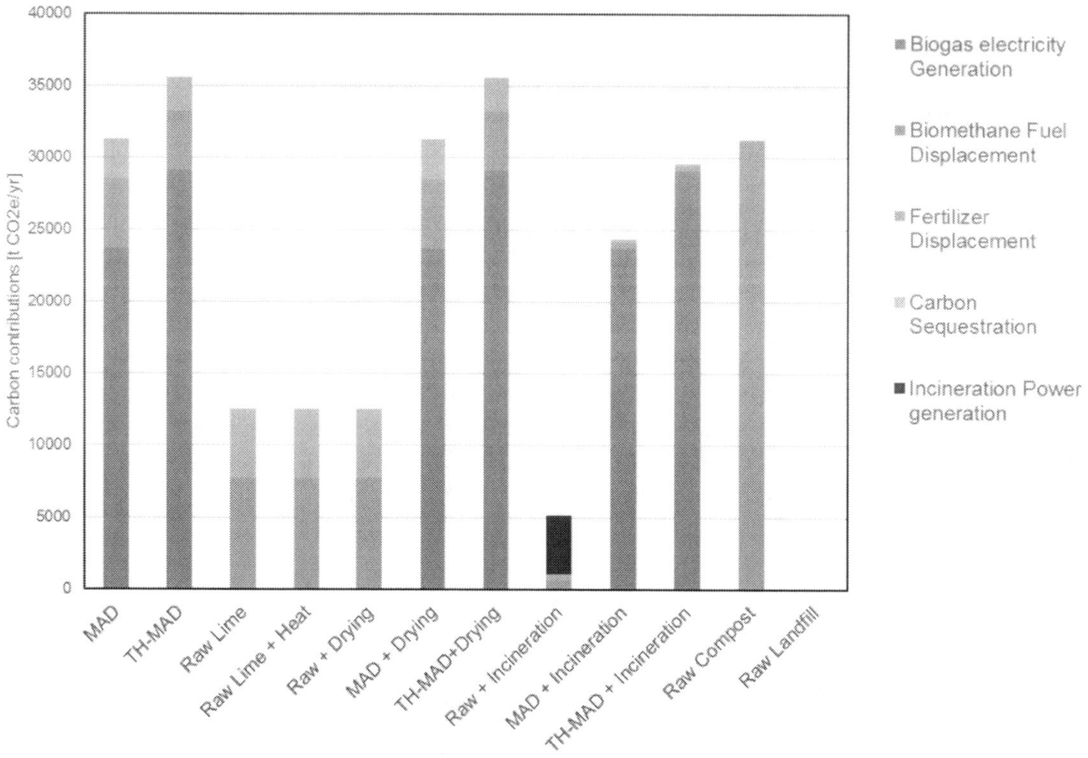

Figure 3. Contributory factors to carbon footprint by Scope

As before, where options involving anaerobic digestion showed the lowest contributions, the digestion options show the highest benefits due to the large renewable energy benefit from biogas. However, as less material is applied to land, the fertilizer displacement and sequestration benefits are lower than options where anaerobic digestion is absent. Composting also shows a high contribution due to a large fertilizer displacement, but, as we will see later, the large contributions shown in Figure 1 counteract the majority of this benefit. Regarding incineration, digestion and thermal hydrolysis with digestion result in no further power generation benefit, unlike the raw option. This may be explained by looking at the following table:

Table 1. Energy balance for Incineration

	Raw	MAD	Thermal Hydrolysis + MAD
Dry solids required for autothermic burning [%]	32.0%	39.0%	42.8%
Potential Generation [MW]	3.51	1.87	1.49
Energy required for pre-drying [MW]	2.99	3.11	1.72
Power generation [MW]	0.52	0	0
Auxiliary fuel requirement [MW]	0	1.25	0.24

As expected, removing volatile solids from the sludge by anaerobic digestion means that a higher dry solids is required to meet autothermic conditions and the potential energy generation also reduces. For raw sludge a minimum of 32% dry solids is required compared to 39% for digested and nearly 43% for thermal hydrolysis. However, similar dewaterability combined with lower calorific value means more energy is required for pre-drying for digested sludge compared to raw sludge. However, in spite of a lower calorific value on a dry basis, because the thermally hydrolyzed sludge dewaters so well, its calorific value is actually similar to raw sludge (Barber, 2016), and due to a much lower volume, less energy is required to divert to a pre-drying stage. In this example the energy required for pre-drying is only a fifth of that for the digestion option. In fact, the model was run at higher dry solids for thermal hydrolysis as observed on some full-scale plants, and in those simulations there was even a slight surplus of energy and hence low energy generation potential.

Combining Figures 1 and 2 present the operating carbon footprint of the options. This is shown in Figure 4.

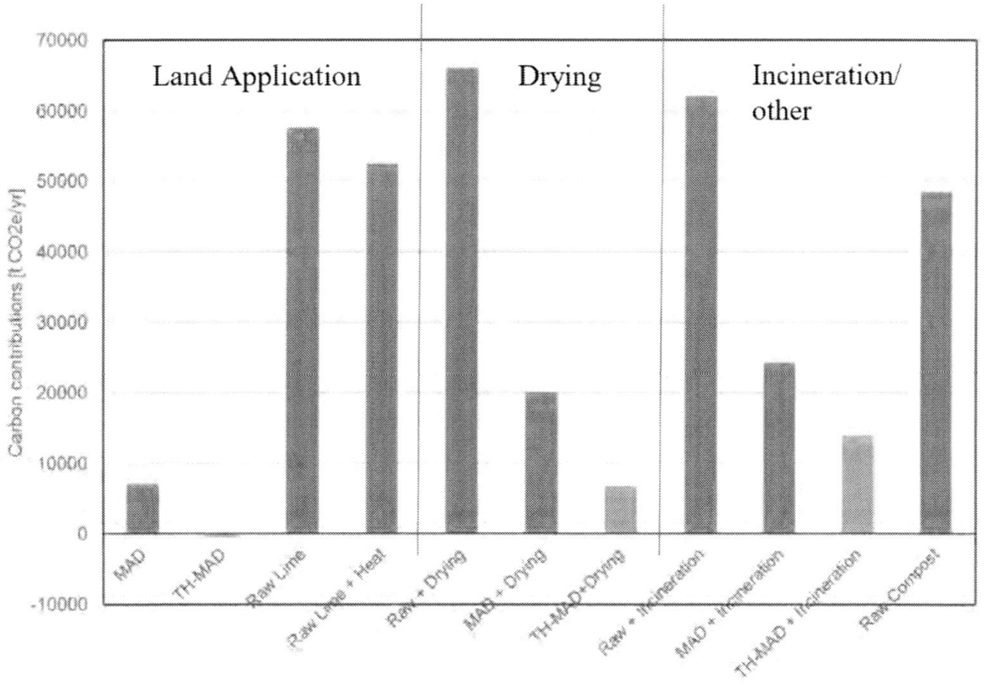

Figure 4. Operating carbon footprint of options tested

The Figure confirms previous studies in the field, i.e. the use of digestion, but also thermal hydrolysis with digestion can result in the lowest operating carbon footprints regardless whether the sludge is land applied, dried or burnt. This is due to a combination of the lowest quantity of contributory factors and the highest quantity of beneficial factors.

REFERENCES

Ainger, C., Baker, K. and Crishna, N., 2008. Carbon Accounting in the UK Water Industry: Guidelines for Dealing with 'Embodied Carbon'and Whole Life Carbon Accounting. UK Water Industry Research Limited, London, UK.

Ashton, J. (2005), "Europe Feels The Heat: The Power Sector and Extreme Weather", WWF-World Wide Fund For Nature (Formally World Wildlife Fund), PowerSwitch! Campaign. [online] http://www.wwf.org.uk/filelibrary/pdf/europefeelstheheatfinal040805.pdf

Asian Development Bank (2012) Promoting Beneficial Sewage Sludge Utilization in the People's Republic of China.

Auty, D. and Blake, J. (2009) Strategic carbon technology – how to use sludge to your advantage 13th European Biosolids & Organic Resources Conference & Workshop.

Barber, W. P. F (2009) Calculation of the carbon footprint of united utilities' Sludge Balanced Asset Programme, 13th European Biosolids & Organic Resources Conference & Workshop.

Barber, W. P. F. (2008) The influence of anaerobic digestion on the carbon footprint of various sewage sludge treatment options; *CIWEM Journal*.

Barber, W. P. F. (2016). Thermal hydrolysis for sewage treatment: A critical review. Water Research, 104, 53-71.

Barber, W.P., 2009. Influence of anaerobic digestion on the carbon footprint of various sewage sludge treatment options. Water and Environment Journal,23(3), pp.170-179.

BEAM Model (2009) The Biosolids Emissions Assessment Model (BEAM): A Method for Determining Greenhouse Gas Emissions from Canadian Biosolids Management Practices, Report prepared for Canadian Council of Ministers for the Environment, by Sylvis.

Department for Environment Food and Rural Affairs (DEFRA) (2007) *Guidelines to Defra's GHG conversion factors for company reporting*, [Online] http://www.defra.gov.uk/environment/business/envrp/conversion-factors.htm

Department for Environment Food and Rural Affairs (DEFRA) (2007) Guidelines to Defra's GHG conversion factors for company reporting, [Online] http://www.defra.gov.uk/environment/business/envrp/conversion-factors.htm

Entec UK Ltd., (2006) *Carbon Management for United Utilities*, Internal Report for Untied Utilities.

European Commission (2009) Reducing the carbon footprint of sewage sludge, Science for Environmental Policy, Journal of the European Commission.

Evans, T. D. (2006) "Sludge dewatering performance – a survey of real operational performance", Floculants & Coagulants Conference, Intertech Pira International, Budapest, 29–30 November.

Feacham, R. G., Bradley, D. J., Garelick, H., and Mara, D. D. (1983) "Health aspects of excreta and wastewater management", John-Wiley, Chichester.

Fjaergard, T., Sorensen, G., Solheim, O. E., and Seyffarth, T. (2006) "Cambi Thermal Hydrolysis combined with Thermophilic/Mesophilic Digestion – From pilot to full-scale", Proceedings of the 11th European Biosolids and Organic Resources Conference, Workshop And Exhibition, Wakefield, UK.

Harrison, K. G. (1993) Evaluation of two methods for lime stabilization of biosolids, J. R. Wauford & Company, Consulting Engineers, Inc.Nashville, Tennessee. [Online]

http://www.rdptech.com/tch8eval.htm
Hoyland, G., Newton, J., Riches, S., and Bland, R. (2006) Anglian Water's New Scheme Will Treat 80 000 tDS/year to Produce Enhanced Treated Sludge for Agricultural Use and Generate 5 MW of Renewable Electricity, 11th European Biosolids & Biowastes Conference, Wakefield, UK.
IPCC (2001) *Climate Change 2001 – The Scientific Basis*, Cambridge University Press, UK.
Kroiss, H., and Zessner, M. (2007) *Ecological and Economical Relevance of Sludge Treatment and Disposal Options*, Proceedings of the IWA Moving Forward, Wastewater Biosolids Sustainability: Technical, Managerial, and Public Synergy Conference, Moncton, New Brunswick, Canada, 47–54.
Le, S., Briddon, T., Harrison, D., and Werker, A. (2006) *Enzymic Hydrolysis Technology Demonstration – Production of Enhanced Treated Biosolids for Agricultural Recycling*, 11th European Biosolids & Biowastes Conference, Wakefield, UK.
Le, S., Harrison, D., and Werker, A. (2007) "Modelling the economic and environmental impact of enzymic hydrolysis in sludge treatment", 12th European Biosolids & Biowastes Conference, Wakefield, UK.
Marsh, A., (2002) *Tightening WwTW Emission Standards: A Review of the Treatment Technologies and their Impact on Climate Change*, UKWIR Report No. 02/WW/04/1.
Metcalf and Eddy, (2004) *Wastewater engineering. Treatment, disposal, reuse*, International Edition, 4th Edition, McGraw-Hill Publishing.
Monteleone, M.C., Furness, D., Jefferson, B.,and Cartmell, E., (2004) Fate of E. coli Across Mechanical Dewatering Processes. *Environmental Technology*, **25**, 825–831.
Outwater, A. B. (1994) *Reuse of Sludge and Minor Wastewater Residuals,* Lewis Publishers, CRC Press.
Panter, K., and Kleiven, H., (2005) "Ten years experience of full scale thermal hydrolysis projects", 10th European Biosolids & Biowastes Conference, Wakefield, UK.
Spooren, T. (2007) *Evaluation of the Effect of Davyhulme Sludge on the Shell Green EfW Plant*, MWH Internal Report for United Utilities, Report No. 47501215.001.R01.
UKWIR (2005) *Workbook for Quantifying Greenhouse Gas Emissions.*, Report (05/CL/01/3)
UKWIR (2005) Workbook for Quantifying Greenhouse Gas Emissions., Report (05/CL/01/3)
US EPA (1999) "*Control of pathogens and vector attraction in sewage sludge* – Including domestic sepatage under 40 CFR Part 503".
US EPA (2014) Greenhouse Gas Emissions from a typical passenger vehicle. Office of Transportation and Air Quality EPA-420-F-14-040a
US EPA GHG Equivalencies Calculator. Online, accessed 2016
Waste Incineration Directive (2000) Directive 2000/76/EC of the European Parliament and of the Council of 4 December 2000 on the incineration of waste, *Official Journal of the European Communities* L 332/91.
Water UK (2007) "Briefing Paper – Climate Change", Press Release. [online] http://www.water.org.uk/home/policy/climate-change/briefing-paper
WEF Manual of Practice No. 8 (1992) Design Of Municipal Wastewater Treatment Plants. Volume 2: Chapters 13 – 20, (1992). *Water Environment Federation*, VA, US.
Werker, A., G., Carlsson, M., Morgan-Sagastume, F., Le, M. S., Harrison, D. (2007) *Full Scale Demonstration And Assessment Of Enzymic Hydrolysis Pre-Treatment For Mesophilic Anaerobic Digestion Of Municipal Wastewater Treatment Sludge*, [Online] http://enzymichydrolysis.com/images/stories/Papers/werker_etal.pdf

Willis, J., Bailey, W., Jolly, E., Murthy, S., Peot, C., & Trueblood, D. (2010). DCWASA'S Certifiable GHG Inventory and Projected GHG Reductions from Cambi/Anaerobic Digestion Upgrades. Proceedings of the Water Environment Federation, 2010(2), 141-147.

Wilson, T. E., Iranpour, R., and Windau, T. D. (2004) Thermophilic Anaerobic Digestion in the US: Selected case histories, 9th European Biosolids & Biowastes Conference, Wakefield, UK.

Appendix A: Model Inputs

Sludge Input Data		
Sludge input	100	TDS.d
Dry Solids	2%	DS
Volatile Solids	75%	VS
Primary	60%	
Activated	40%	
Thermal Hydrolysis	All sludge	
N content of Primary VS	3%	
N content of WAS VS	9%	
Available N	80%	
COD/VS ratio	1.6	
Carbon content of VS	55%	
Calorific Value VS	24000	kJ/kg

Storage	Raw	Digestion1	Digestion2	
Pre-processing	0%	0%	0%	of sludge stored
Pre-digestion or raw pre-dewatering	100%	100%	100%	of sludge stored
Pre-dewatering or raw post-dewatering	100%	0%	0%	of sludge stored
Post-digestion		100%	100%	of sludge stored

Thickening	Raw	Digestion1	Digestion2	
Pumping head	200	200	200	m
Pumping efficiency	60%	60%	0.6	%
Thickening Type	Centrifuge	Centrifuge	Centrifuge	
Capture rate	95%	95%	98%	%
Operating hours	16	16	16	hrs/d
Poly consumption	5	5	7	kg/TDS
Dry Solids Output	5.00%	5.00%	16.17%	%DS
Quantity of liquors treated	100%	100%	100%	of flow
Quantity of methanol used for TotalN	100%	100%	5%	of flow
BOD concentration	3000	3000	3000	mg/l
Ammonia Concentration	100	100	100	mg/l
Type of liquor treatment	Nitrification	Nitrification	Nitrification	

Digestion	Digestion1	Digestion2	
Dry solids entering	5%	10%	%DS
Ambient temperature	22	22	C
Pumping Head	200	200	m
Pumping efficiency	60%	60%	efficiency
Black Box type	MAD	TH	
Black Box power demand	0	10	kWhr/TDS
Black Box operating temperature	22	22	C
Black Box HRT	2.00	2.00	days
Black Box convection losses	10%	10%	
Black Box Heating efficiency	80%	80%	
Anaerobic Digestion Operating Temperature	35	40	C
Anaerobic Digestion HRT	20.0	14.0	days
Anaaerobic Digestion Mixing intensity	0.006	0.006	kW/m3
Anaerobic Digestion - Unit volume	12500	7500	m3
Anaerobic Digestion - Convective heat loss	10%	10%	
Anaerobic Digestion - Heat efficiency	80%	80%	

Anaerobic Digestion main outlet	Digestion1	Digestion2	
Transport distance	60	40	miles
Transport type	Road - Gasoline	Road - Gasoline	
Outlet type	Land application	Land application	
Fertilizer Displacement quantity	50%	50%	% of benefit counted
Carbon sequestration sludge type	Digestion1	Digestion2	

Biogas Options	Digestion1	Digestion2	
Quantity of gas cleaned	100%	0%	
Power requirement for gas cleanup	0.26	0.26	kWhr/m3 biogas
Rogue Emissions	3.00%	0.00%	
Use or non-use of biogas for pre-treatment	Don't use	Use	
Quantity of gas diverted to boiler	0.00%	0.00%	% of total biogas
Boiler efficiency	80%	80%	
Quantity of gas flared	8.00%	8.00%	% of total biogas
Quantity of gas converted to biomethane	0.00%	0.00%	% of total biogas
Power consumption of co-gen	5%	5%	% of power generation
Co-gen electricity conversion efficiency	38%	38%	% of gas calorific value
Co-gen high grade heat recovery	25%	25%	% of gas calorific value
Co-gen low grade heat recovery	20%	20%	% of gas calorific value
Type of biomethane processing plant	Amine Scrubbing	Water Scrubbing	
Required biomethane methane content	98%	98%	% methane
Quantity of biomethane compressed	100%	100%	% of biomethane produced
Energy required for compression	0.49	0.49	kWhr/m3 biomethane
Displacement fuel type	Diesel	Diesel	

Dewatering	Raw	Digestion1	Digestion2	
Pumping head	200	200	200	m
Pumping efficiency	60%	60%	60%	
Dewatering Type	Centrifuge	Centrifuge	Centrifuge	
Capture rate	98%	98%	98%	
Operating hours	16	16	16	hrs/d
Poly consumption	8	8	8	kg/TDS
Dry Solids Output - Raw	25.00%			%DS
Quantity of liquors treated	100%	50%	50%	% of liquors
Quantity of methanol used for TotalN	5%	0%	5%	% of methanol required
Type of liquor treatment	Nitrification	Nitrification	Deammonification	

Drying	Raw	Digestion1	Digestion2	
Output Dry Solids required	92%	92%	92%	%DS
Thermal energy requirement	3500	3500	3500	MJ/t we
Use or non-use of biogas for drying		Don't use	Don't use	
Electrical energy requirement	60	60	60	kWhr/t we

Incineration	Raw	Digestion1	Digestion2	
Generation efficiency	20%	20%	20%	
Induced Draft energy consumption	2.77	2.77	2.77	kWhr//TDS
Forced Draft energy consumption	42.93	42.93	42.93	kWhr//TDS
Transport distance to incinerator	50	50	50	
Transport type	Road - Diesel	Road - Diesel	Road - Diesel	
Chemical consumption	0.0458	0.0458	0.0458	t chemical/tDS burnt
Start-up gas	0.697	0.697	0.697	m3/t DS
Downtime	1	1	1	months
Ash transport distance	50	50	50	
Ash transport type	Road - Gasoline	Road - Gasoline	Road - Gasoline	

Raw Liming	Raw	
Electrical Power consumption	30	kWhr/ t w
Lime added	50%	% dry basis
Storage cake shrinkage	2%	% DS points
Limed Cake transport distance	70	miles
Limed Cake transport type	Road - Gasoline	
Fertilizer displacement count	50%	% of benefit counted
Sequester sludge type	Raw	

Lime plus electrical heating	Raw	
Electrical Power consumption	20	kWhr/ t w
Lime added	28%	% dry basis
Storage cake shrinkage	2%	% DS points
Ambient cake temperature	15	C
Required pasteurization temperature	55	C
Heating efficiency	80%	%
Limed Cake transport distance	70	miles
Limed Cake transport type	Road - Gasoline	
Fertilizer displacement count	50%	% of benefit counted
Sequester sludge type	Raw	

Landfill	Raw	
Sludge type	Raw	
Transport distance	100	miles
Transport type	Road - Gasoline	

En-Vessel Composting	Raw	
Electrical Power demand	20	kWhr /t DS composted
Green waste dry solids	60%	%DS
Green waste C:N ratio	40	
Green waste voidage	70%	
Green waste transport distance	50	miles
Green waste transport type	Road - Gasoline	
Aeration requirement	2	kg O2/kg VS
Aeration efficiency	2	kg O2/kWHr
Compost desired dry solids	50%	%DS
Compost volume reduction	60%	% reduced compared to raw
Compost nitrogen release	60%	% of total nitrogen
Compost transport distance	50	miles
Compost transport type	Road - Gasoline	
Compost outlet type	Land application	
Compost fertilzer displacement	50%	% of benefit counted
Compost sequestration effect	0	t CO2e/T DS applied

Strategic Carbon	Raw	Digestion1	Digestion2
Quantity incinerated	50%	20%	0%
Quantity dried	13%	0%	0%
Quantity land applied	20%	15%	100%
Quantity Composted	12%	15%	0%
Quantity landfilled	5%	50%	0%
Transport distance to incinerator	30	30	30
Transport distance to drying	15	15	15
Transport to land application	60	60	60
Transport to composting plant	33	33	33
Transport to landfill	31	31	31
Transport type to incinerator	Road - Gasoline	Road - Gasoline	Road - Gasoline
Transport type to drying	Road - Diesel	Road - Diesel	Road - Diesel
Transport type to land application	Road - Gasoline	Road - Gasoline	Road - Gasoline
Transport type to composting plant	Rail	Rail	Rail
Transport type to landfill	Marine	Marine	Marine
Quantity of sludge which is raw	20%		
Quantity of sludge which is digestion1		30%	
Quantity of sludge which is digestion2			50%

Appendix B: Carbon Outputs from model

The following outputs are determined from the model

Contributory Factors

Storage	Pre-thickening storage
	Pre-digestion storage
	Pre-dewatering storage
	Post-dewatering storage
Thickening	Thickening Power
	Thickening Pumping
	Thickening Polymer
	Liquor Treatment Carbon
	Liquor Treatment Ammonia
	Methanol
Digestion	Pumping to digester
	Pretreatment Power
	Pretreatment heating
	Digester heating
	Digester mixing
Biogas	Rogue Emissions
	Flare
	Co-gen power consumption
	Co-gen emissions
	Gas Cleanup
	Biomethane processing
	Biomethane CO_2 release
	Biomethane compression
Dewatering	Dewatering Power
	Dewatering Pumping
	Dewatering Polymer
	DW Liquor Treatment Carbon
	DW Liquor Treatment Ammonia
	DW Methanol
Drying	Drying Gas requirement
	Drying Power Consumption
Raw Options	Electrical Power
	Lime
	Land emissions
	Transport
	Supplemental Heat
	Landfill Emissions
	Aeration (for composting)
	Green Waste Transport
	Compost Transport

Incineration	Transport to Incinerator Autothermicity required for burning Start-up gas Power for forced and induced draft fans Chemicals Emissions Downtime Ash transport Auxiliary Fuel

Beneficial Factors

- Biogas electricity Generation
- Biomethane Fuel Displacement
- Fertilizer Displacement
- Carbon Sequestration
- Incineration Power generation

Appendix C. Breakdown of contributory carbon footprint for various anaerobic digestion components

	Standard Digestion	With thermal hydrolysis
Thickening	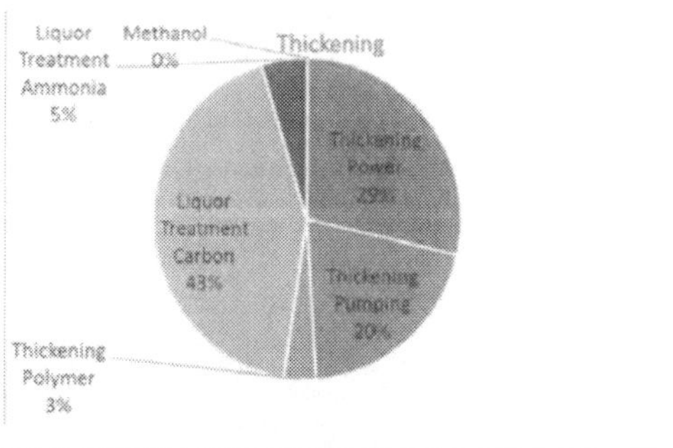	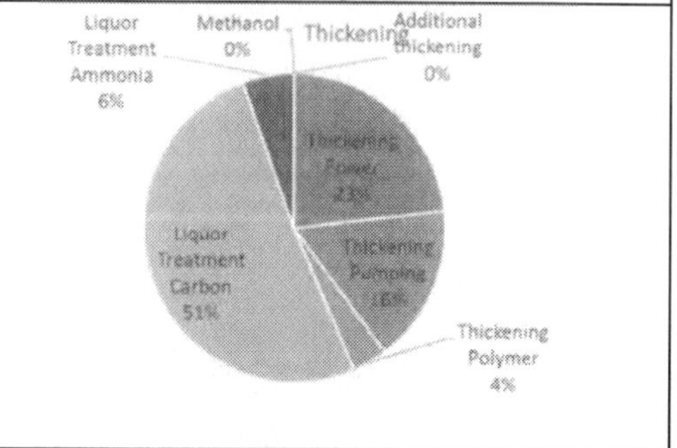
Digestion	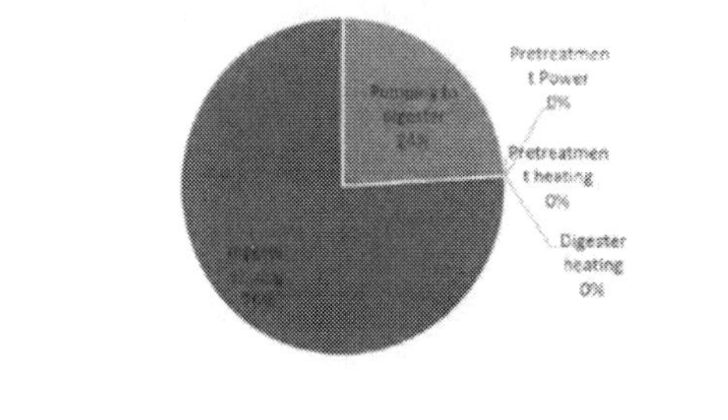	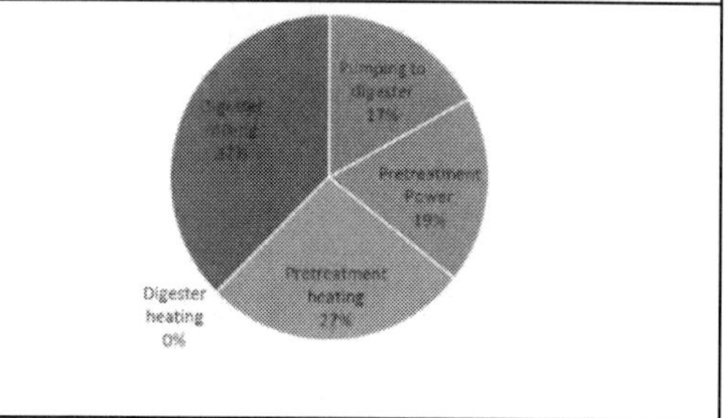
Biogas Processing	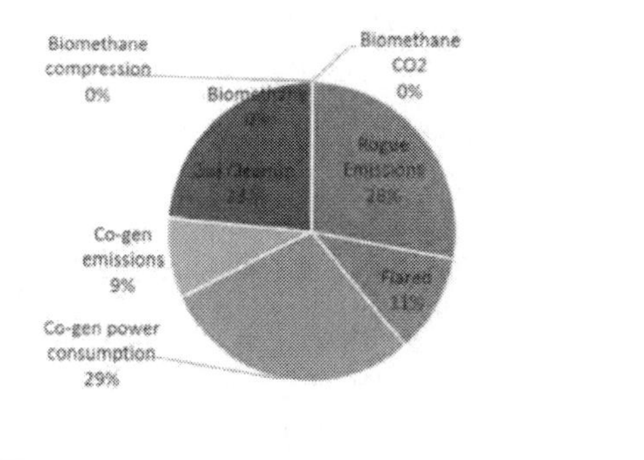	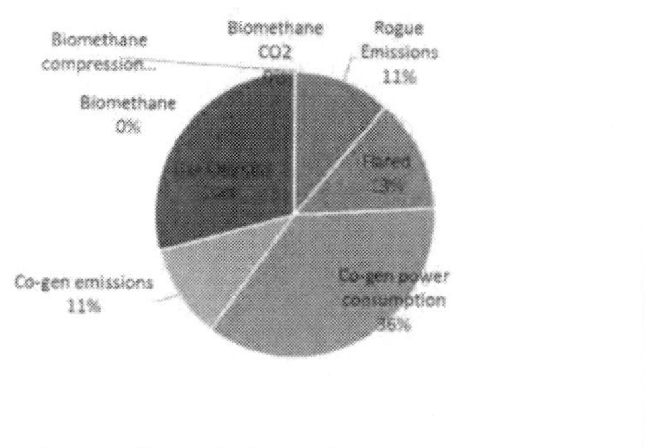
Dewatering	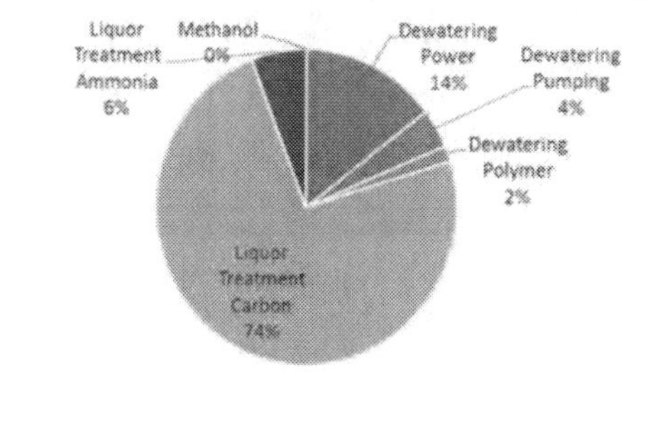	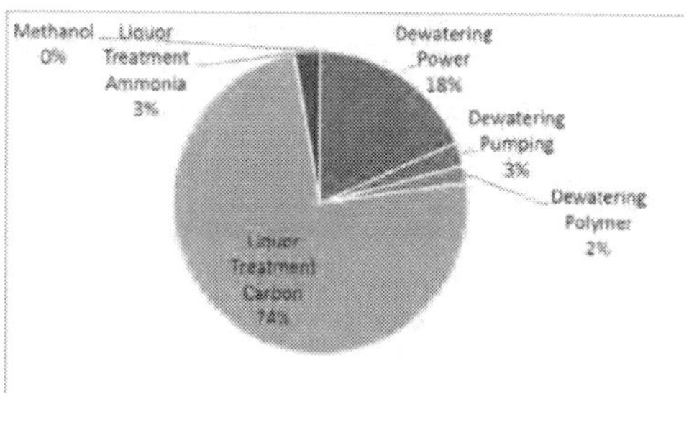

<table>
<tr><th></th><th>Standard Digestion</th><th>With thermal hydrolysis</th></tr>
<tr><td>Drying</td><td>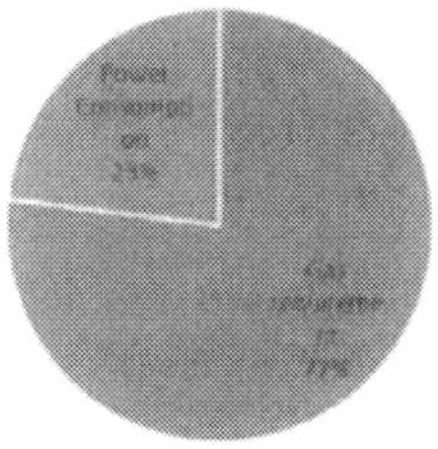
</td><td>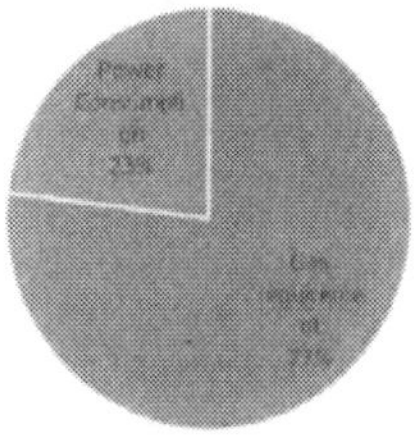
</td></tr>
<tr><td>Incineration</td><td>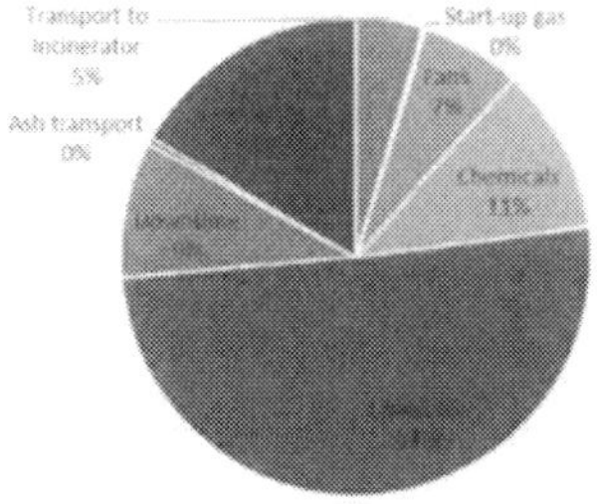
</td><td>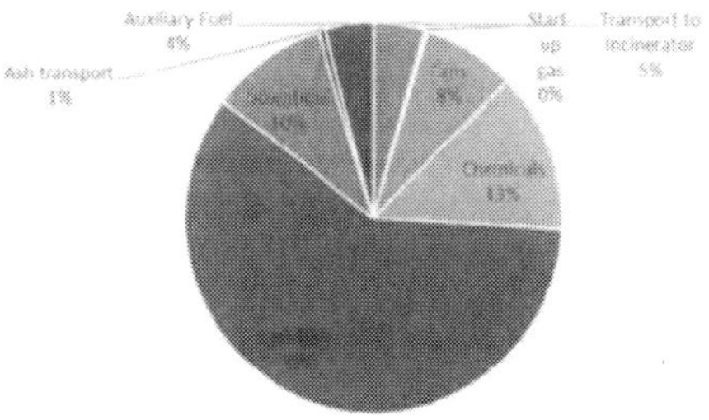
</td></tr>
</table>

Appendix D. Breakdown of contributory carbon footprints for raw sludge options

Drying	
Incineration	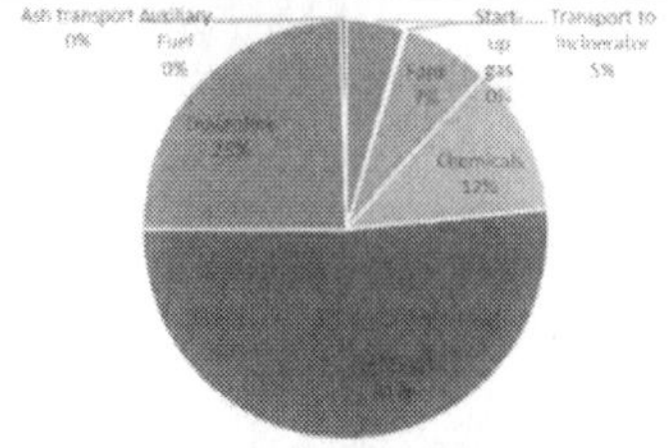
Liming	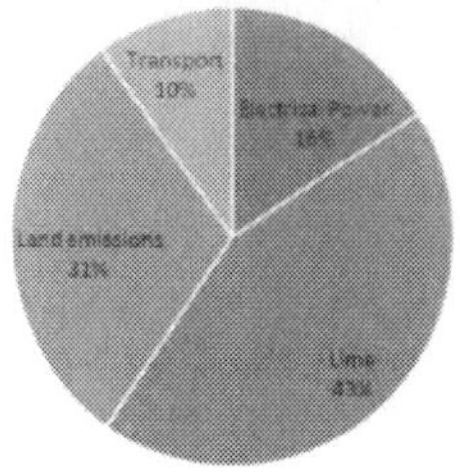
Lime with supplemental heat	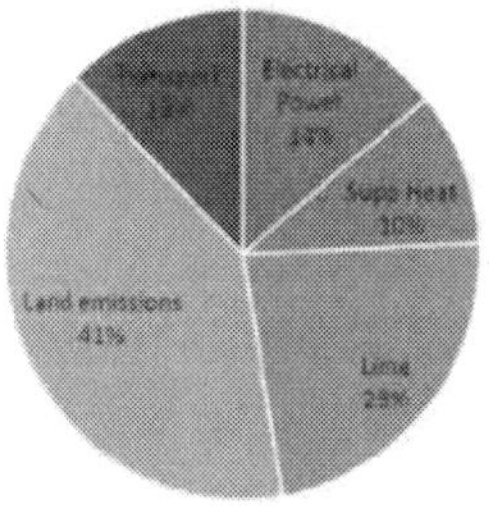
EnVessel Composting	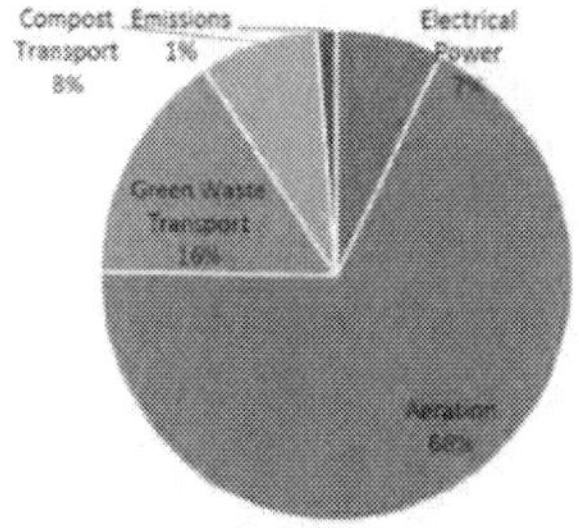
Landfill	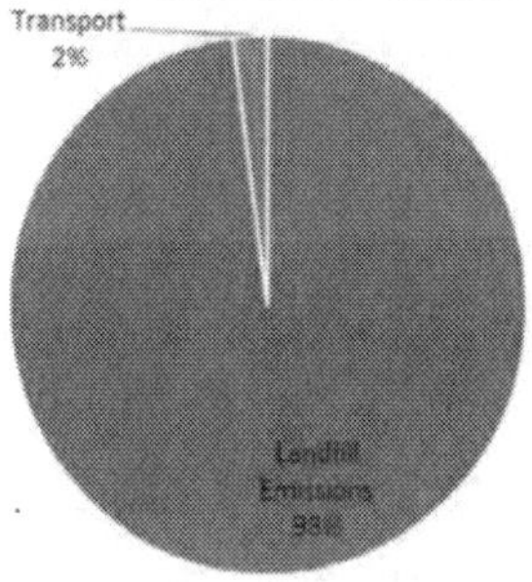

This page intentionally left blank.

Batch vs. Continuous Thermal Hydrolysis - Which is right for you?

Chris Thomson[1]; Rich Dimassimo[1]; Robert Clay[1]; Sudhakar Viswanathan[1] *; Dr. Jongmin Kim[2]; Dr. Nicholas Landes[3]

[1] Veolia Water Technologies, 4001 Weston Parkway, Cary, NC 27513; [2] University of Texas Rio Grande Valley, 1201 W. University Dr. Edinburg, TX 78539; [3] Freese and Nichols, Inc., 4055 International Plaza, Suite 200, Fort Worth, Texas 76109

*Corresponding author, Email: sudhakar.viswanathan@veolia.com

Abstract: A circular economy should benefit the consumer without producing waste or pollutants. In the wastewater industry, it is imperative we identify opportunities and develop solutions to harvest materials that appear to have little value and turn them into a valuable resource in order to realize the benefits of a circular economy. Thermal Hydrolysis is a proven cell lysing application for resource recovery in a sustainable and profitable manner from wastewater treatment plant. Two distinct thermal hydrolysis solutions have evolved, one a batch process and the other a continuous process. The two solutions are applicable in various configurations and each with unique features that will readily improve performance of existing anaerobic digesters or add considerably value to new digestion facilities. Often, the right solution for a wastewater treatment facility is not obvious. With energy cost on the rise and increase in population stressing the available natural resources, it is imperative that thermal hydrolysis efficiencies are improved to lower environmental impact. Numerous facilities across the globe have benefited from thermal hydrolysis, information from these facilities can be used to determine the ideal solution for most wastewater treatment plants. The design evolution, performance improvements, and lessons learnt over the past decade can also be used to assist wastewater treatment plants to fully realize the benefits of the circular economy.

Keywords: Biosolids; Thermal Hydrolysis, Sustainability, Anaerobic Digestion

Introduction

It is well known that anaerobic digestion (AD), with the right design and additional organic feedstock, integrated with a combined heat and power (CHP) can provide a vehicle to deliver cost-effective and energy net neutral solutions for wastewater treatment plants. The feasibility and sustainability of AD/CHP can be improved considerably with pre-treatment, such as cell lysing, which will enhance digester performance, improve energy recovery in the form of increased biogas production and achieve Class A quality biosolids.

Thermal Hydrolysis Process (THP) was first applied to improve sludge dewaterability (Haug et al.,1978), but has since grown into a pre-treatment for anaerobic digestion. THP is the engineered application of temperature (320 to 330°F) and pressure (87 to 130 PSI) to more effectively break down sludge. The lysing process decreases the viscosity of the material and increases available volatile solids for anaerobic digestion. Decreased viscosity alters the rheological properties to improve the digested sludge's dewaterability. Increased volatile content entering the digester increases methane production and reduces the amount of biosolids for disposal. The secondary benefits of lysis include lowering downstream process equipment requirements, reducing process chemical costs and production of Class A biosolids for ease of disposal and beneficial use as bio-fertilizer.

In short, thermal hydrolysis solves numerous challenges at a wastewater facility including increasing treatment capacity, reducing biosolids handling, increasing biogas production to reduce energy utilization, reduce operation cost for biosolids processing, produce renewable bio-fertilizer.

The typical process flow for THP includes a pre-dewatering step upstream (**Error! Reference source not found.**) of the TH process. Unlike conventional process flow schemes that include thickening upstream of the AD, pre-dewatering for THP achieves solids content as high as 20% prior to digestion. This allows for processing a larger volume of solids within a small THP footprint. Post THP sludge may require dilution to maintain homogenous solids dispersion, uniform heating and sub-toxic ammonia levels.

The typical, or standard, process flow for THP has been successful with close to 70 installations worldwide. When compared to traditional mesophilic digestion without THP pre-treatment, the typical process flow generally achieves the following improvements:

- 10 to 25% higher volatile solids reduction in the digesters
- 10 to 20% higher methane generation
- 5 to 15% greater cake solids from dewatering.

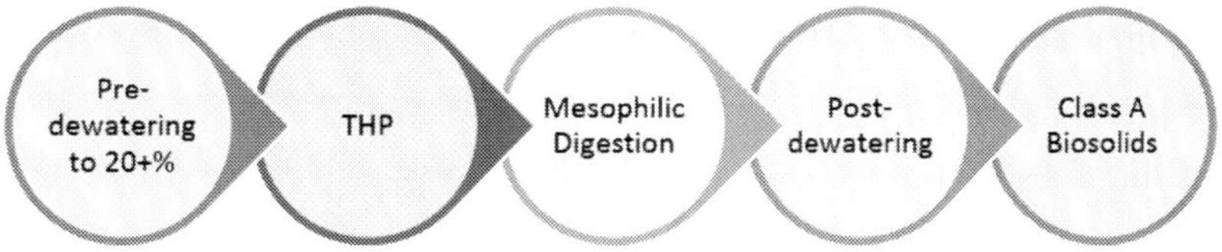

Figure 1: Typical Process Flow for the Thermal Hydrolysis Process

The success of THP, as evidenced by the increasing rate of installations and the treatment capacity in recent years (Figure), has been attributed to two methods for achieving thermal hydrolysis: one a batch process and the other a continuous process. The two solutions are applicable in various configurations and each with unique features that will readily improve performance of existing anaerobic digesters or add value to new digestion facilities.

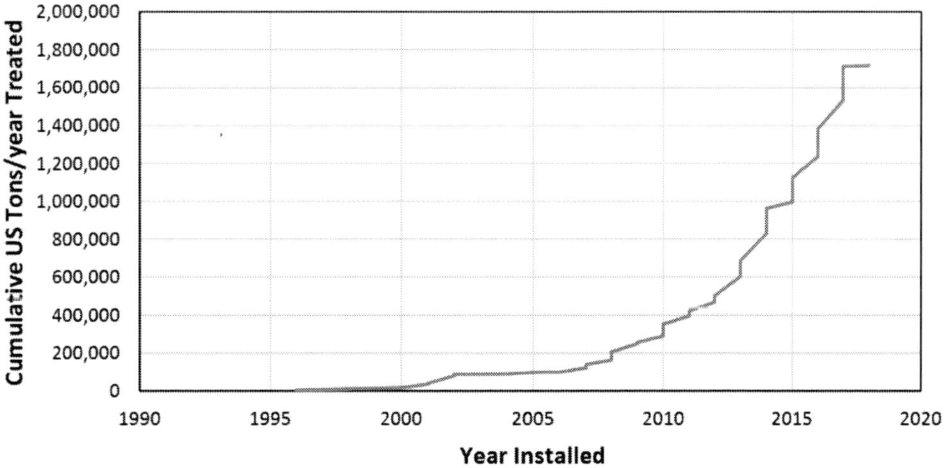

Figure 2: Growing Treatment Capacity of THP Installations Worldwide

Various factors such as flow rate, strength of waste, existing infrastructure, disposal method, operation experience and federal, state and local funds available for the facility dictate the direction of the solution. The process has also evolved over the years, adding new features and development that has resulted in compact, efficient process solutions for converting sewage sludge to a source of renewable energy and a usable biosolids end product.

Objectives

Although batch and continuous thermal hydrolysis processes have been successful, advances in both approaches may further enhance the established benefits of THP. In this paper, various types of thermal hydrolysis solutions along with their specific design features, system configurations, and operational variations are discussed, which aims to achieve the following:

1. Introducing adaptive application of THP technologies;
2. Providing the most feasible and economical THP configuration per specific sludge characteristics and unique field conditions;
3. Finally providing experts' insights about THP design and operation as a reference for municipalities and/or industries considering adopting advanced pre-treatment technologies for their organic wastes

Comparison

Thermal hydrolysis evolution over the last three decades is summarized in the schematic below. Two distinct thermal hydrolysis solutions have evolved, one a batch process and the other a continuous process. The former is well established technology but it requires minimum thickened sludge volume due to its steam delivery method using lances and its application is limited to the Lysis - Digestion (LD) using THP followed by AD mode only. The latter is relatively new and its applications are versatile due to the steam delivery method using dynamic mixers and can be configured in the LD mode, the Digestion - Lysis - Digestion (DLD) mode, and the Digestion - Lysis (DL) mode. Industry consensus is that the continuous THP system is relatively more adaptive to unique field conditions (Abu-Orf, et al, 2012).

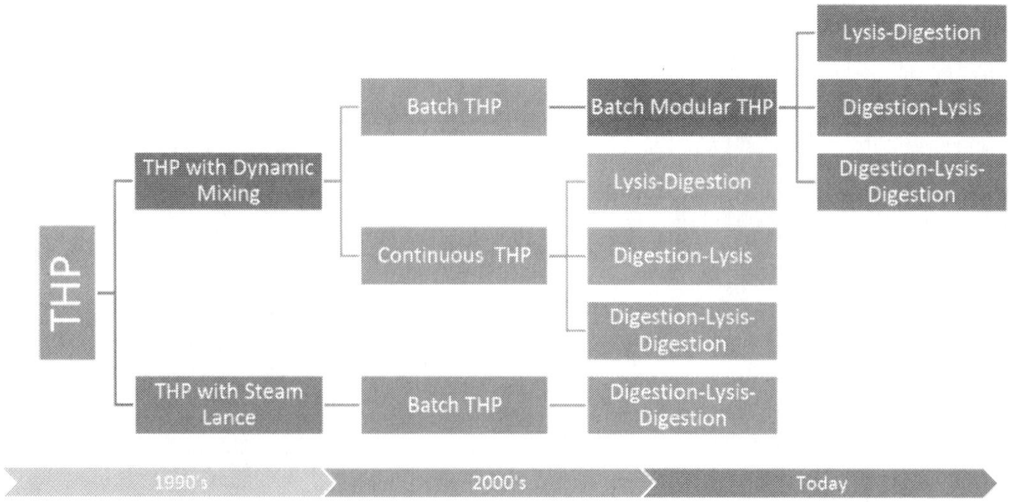

Figure 3: Evolution of THP

The ability to apply high pressure steam directly on to dewater sludge is a measurable efficiency of the TH process. The efficiency of thermal hydrolysis process to lyse the cells, which in effect is responsible for both the change of fluid rheology and the subsequent benefits described earlier is directly proportional to the efficiency of heat transfer from high pressure steam to directly on to the sludge.

The principal difference between thermal hydrolysis designs is the mode of operation. While the majority of the installations worldwide are operating in a batch mode, the continuous mode process is fast becoming an acceptable alternate to the design with various added value and advantages over the batch system.

The batch process has been in use for over thirty years, however the process has evolved over time. The process represented below is the next generation thermal hydrolysis. This new design of the process operates at a temperature range of 310-330 °F and at 100 psi. The process requires energy, but the majority of this energy is created by the process in the form of biogas combustion in combined heat and power systems and the subsequent heat captured from the exhaust of the CHP. Furthermore, the system is insulated and cladded resulting in minimal heat loss and also allows the flexibility to place the system outdoors.

In the process shown in Figure 4, the actual pressure heating of the sludge cake is taking place at 310 – 330 °F and 100 psi pressure for 20 minutes or more inside one of the four reactors.

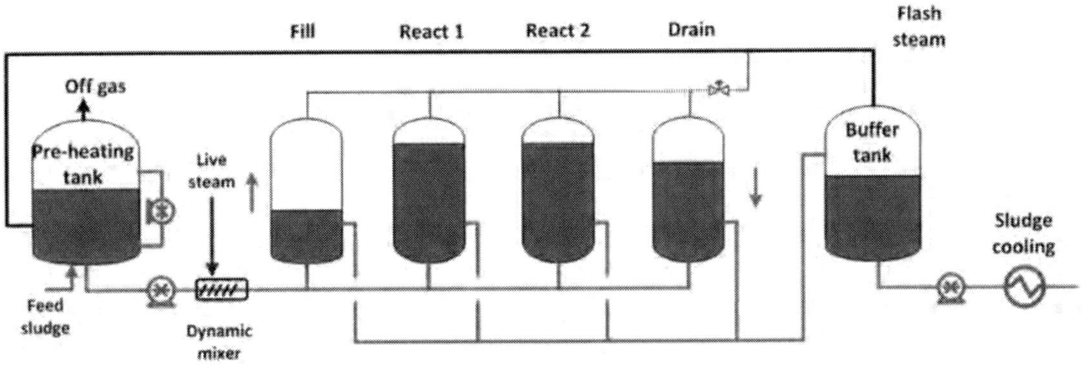

Figure 4: Batch System – Typical Scheme

The process, while by definition considered a batch operation, is a continuous flow batch THP system that utilizes three separate stages. They are typically (i) a Pre-Heating Tank, (ii) n number of batch TH Reactors (usually four, as shown in figure 4), and (iii) a Buffer Tank.

The sludge cake is pre-heated to 80 to 100°C in the Pre-Heat Tank. Here it is continuously receiving ambient temperature sludge cake and flash steam (from the Buffer Tank). This pre-heated sludge is then conveyed via progressive cavity pumps through a dynamic mixer where live steam is introduced and mixing and full condensation takes place.

The sludge structure and rheology is modified by two effects:

- Mechanical shear

- Temperature and pressure impacts

As a result, the viscosity of the sludge is significantly decreased. The dynamic mixers are a commonly used technology in the food, cosmetics, paint and dairy industries.

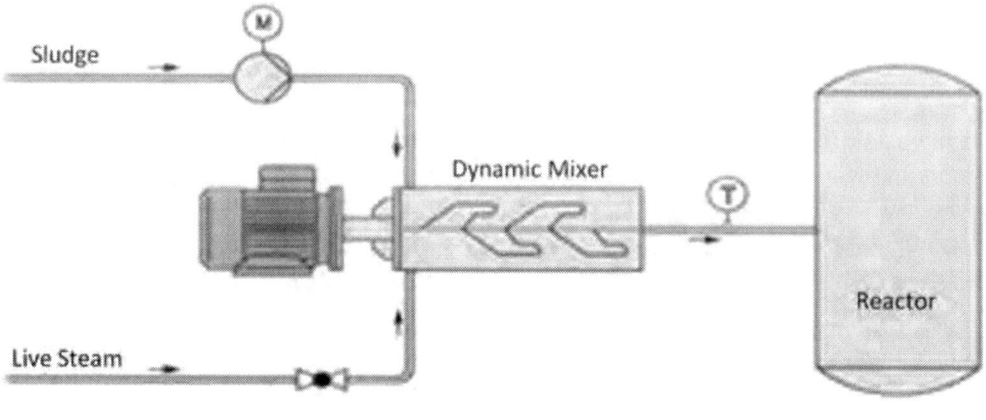

Figure 5: Dynamic Mixer used to condense live steam on incoming sludge

In the newer designs, live steam is directly injected in line upstream of the TH Reactors, allowing further optimization of the equipment designs such as tanks sizes, plant layout, and plant height. No heat transfer through steam lances is performed inside the TH Reactors; this reduces mechanical maintenance requirements and reduces system downtime. The process reactors operate in a sequence of Fill, React and Drain.

In the process the TH reactor vessels operate in a sequenced Fill, React, Drain configuration. The sludge feed is continuous, with each reactor fed in turn while the other reactor(s) are in React phase or being drained/transferred to the Buffer Tank. The depressurisation associated with the sludge transfer to the Buffer Tank produces flash steam which is recovered in the Pre-Heat Tank.

This sequence ensures that maximum benefit is realized from the energy value of the steam while reducing capital cost for tanks, structures, and equipment access platforms, and decreasing the overall foot print of the system.

In the older designs, live steam was injected into the TH reactors via steam lances. This limits the operational range of the TH in terms of solids concentration to about 17% dry solids. This requires water to be added to dewatered sludge to dilute the concentration down which creates an inefficiency within the process train as most modern dewatering units are capable of dewatering sludge to a much greater concentration with minimal effort and cost. This compels results in increased chemical and operating costs for the older THP design.

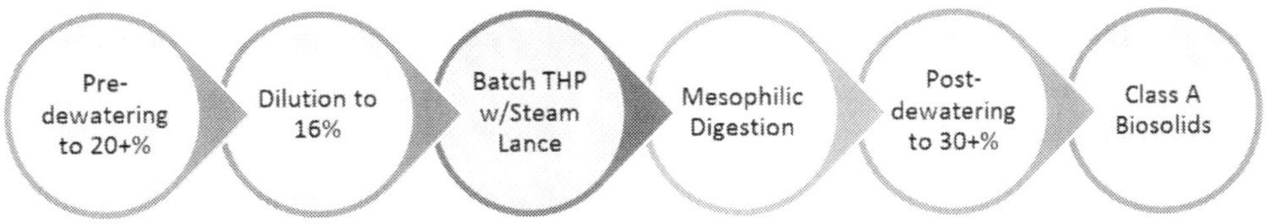

Figure 6: Old, process is limited by solids loading, requiring dilution of dewatered sludge

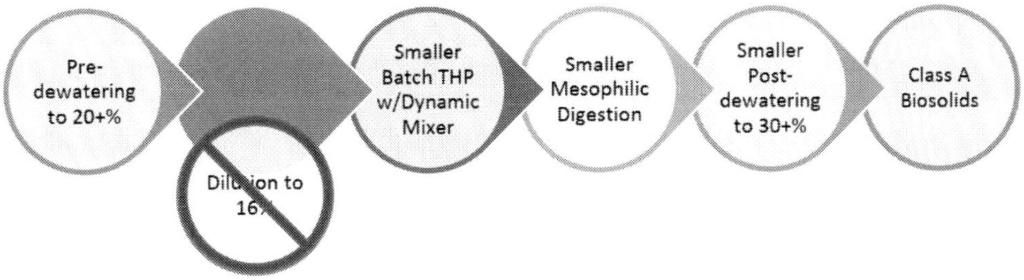

Figure 7: New, more efficient design allows for smaller footprint and higher solids load

Batch to continuous mode of operation is a normal evolution of a process giving a smaller footprint, simpler control and easier service and maintenance (Stedman, 2010). The continuous thermal hydrolysis process operates at a temperature range of 285°F - 330°F at pressures between 130 – 220 psi. Similar to the batch process, pre-dewatering is required.

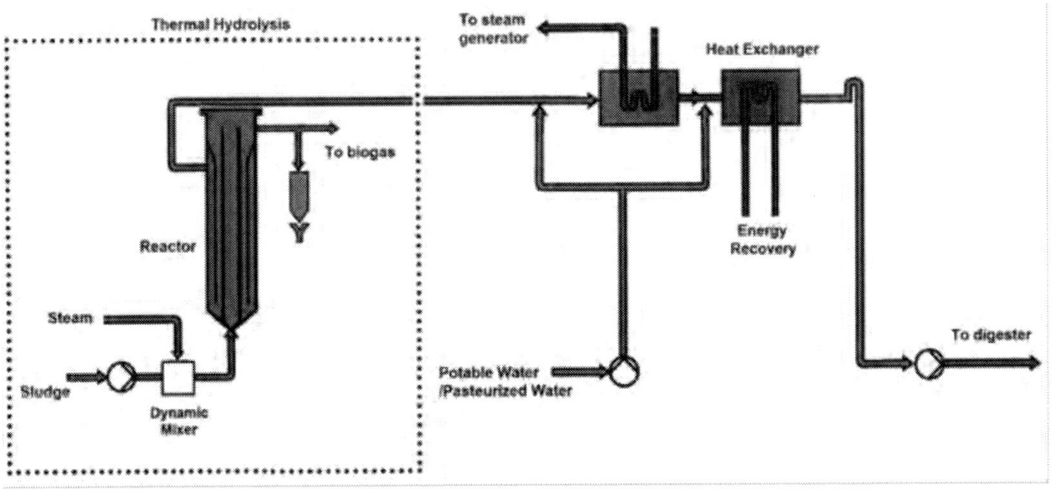

Figure 8: Continuous System – Typical Scheme

Dewatered sludge from a storage silo or hopper is continuously fed to the process via a - positive displacement pump. As the sludge enters the system, steam is continuously fed and mixed with the incoming sludge using a dynamic mixer, at the bottom of the reactor. This instantaneously blends live steam on to the sludge starting the condensation process.

As this mixture moves up and in to the reactor, the steam continues to condense onto the sludge, thereby transferring heat energy to the sludge and raising the temperature to that required for hydrolysis to occur. The sludge flow in the process combined condenser and reactor sections ensure a homogenous distribution of heat through the sludge and also capture any steam that has not condensed earlier. This ensures that all of the energy available from the steam injected is utilized effectively in the reactor, maximizing the energy efficiency of the process.

The reactor is typically a three pass tube in tube design, sludge at the required hydrolysis temperature and pressure at the top of the first concentric pipe flows to the middle pipe or first reactor section. In this section the sludge flows at a very low velocity, leading to plug flow conditions. This plug-type flow ensures there is no short-circuiting and all the sludge is under thermal hydrolysis conditions for the required amount of time. Subsequently, when sludge reaches the bottom of the first reactor section, it flows to the second reactor section or outer section. In the second reactor section, the sludge moves up completing the hydrolysis process. Once the sludge has been held for the required retention time of 20 minutes or more, it leaves at the top of the reactor.

Upon exiting the reactor, the sludge enters a heat exchanger system to cool the sludge to a suitable temperature for the downstream mesophilic digestion process without requiring supplemental heating in the digesters. The heat exchangers allow for easy cleaning and maintenance. The cooling water from the heat exchanger system can be used for pre-heating incoming cake, boiler water, building heat, or for other low temperature heat sources at the plant.

After the heat exchanger system, there are provisions to inject potable water into the sludge. The water injection can be used to dilute and cool the sludge as required by the downstream digestion process.

The final component of the continuous thermal hydrolysis system is the pressure holding pump. This pump is controlled to maintain a pressure set point within the system. In most cases, the pressure holding pump can also be used to feed the downstream digestion process.

The continuous thermal hydrolysis process also allowed for new configurations that were previous not in use.

The three commonly used configuration are:

- LD – Thermal lysis (L) followed by digestion (D)
- DLD – Digestion (D) followed by thermal lysis (L), which is followed by digestion (D) is a patented process that was first demonstrated in 2009 and validated in Denmark (Gurieff et al., 2012)
- DL – digestion (D) followed by Thermal lysis (L)

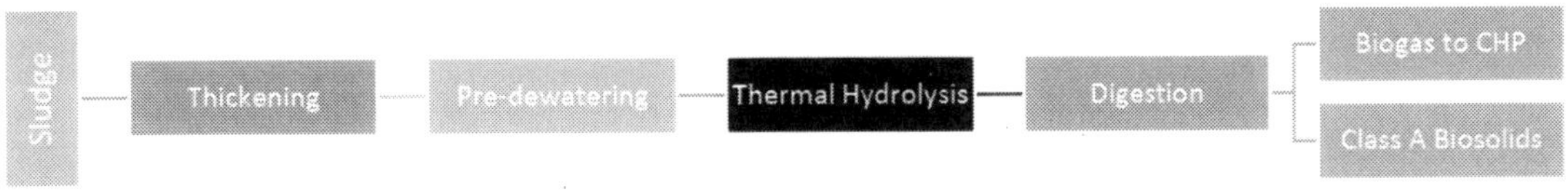

Figure 9: Sludge Lysing followed by Digestion Configuration

Lysis followed by digestion (LD) is the most traditional way to configure a thermal hydrolysis process. It has proven to be very effective in enhancing digestion performance while at the same time being cost effective. This configuration is typically used for larger facility with limitation of digestion capacity. Most batch thermal hydrolysis processes are utilized in this configuration. The most popular reference for this configuration is at a facility in Washington D.C, USA, using the CambiTHP™ process.

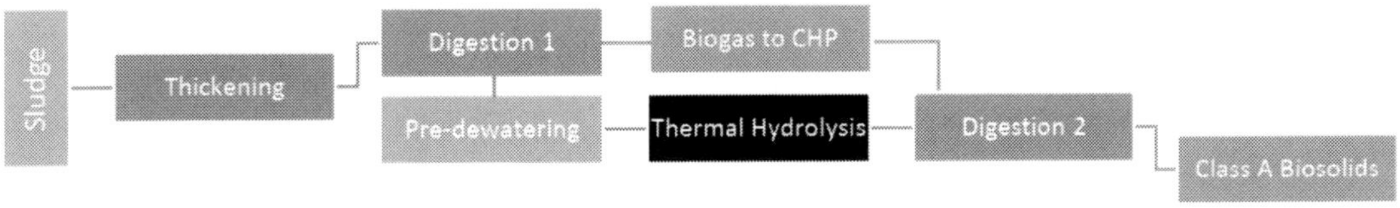

Figure 10: Digestion followed by Sludge Lysing and second stage Digestion Configuration

Digestion followed by lysis, followed by digestion (DLD) is a patented configuration and currently exclusively used with the continuous thermal hydrolysis process. It allows for the digestion process to degrade the simple, "easy to digest" organic compounds, followed by a small thermal hydrolysis process to break down the undigested organics, which is then sent back to a digester for degradation. This configuration works for both small and larger facility with excess digestion capacity. This configuration is especially advantageous if a continuous TH process is applied between two existing digesters. A well know reference for this configuration is a facility in Billund, Denmark using the Exelys™ process.

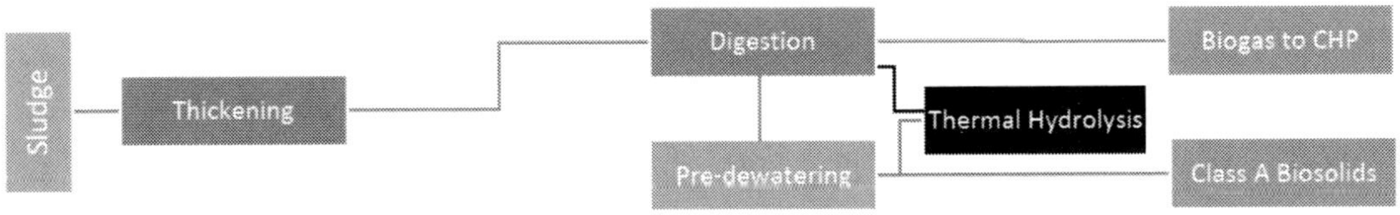

Figure 11: Digestion followed by Sludge Lysing Configuration

Digestion followed by lysis (DL) is a relatively new configuration and combines the benefits of the aforementioned configurations. It is a cost effective manner of achieving Class A biosolids and improve the energy profile of the treatment facility. This configuration is used in both small and larger facility with limitation of digestion capacity and limitations of room for expansion. The first known reference for this configuration is at a facility in Versailles, France, also using the Exelys^{TM} process.

A 2013 study that compared the various configurations of THP concluded that utilization of THP in any of the aforementioned configuration would result in smaller foot print of the anaerobic digestion facilities or increase of existing digestion capacity, high energy potential from biogas production and significant solids reduction across the process (Chauzy et al, 2013).

The central differentiating module that separates the older and newer design is in the mechanism of steam delivery. Older design relies on steam lances while the newer design utilizes dynamic mixers. This evolution extends the operational range to match the efficiencies of the dewatering unit, eliminating the need for dilution of dewatered sludge down to 16-17% DS. The new design is capable of operating at 28% DS, although typically designed at 23%, reducing footprint while improving energy utilization, production and management. The table 1 summarizes the major differences between the old and new design.

Table 1: Summary of Old vs. New TH Design

	Old	*New*
Mode	Batch	Continuous or Batch
Configuration	LD	LD, DLD, DL
DS%	16.5	Up to 28%
Operator Intensive	Yes	No
Dilution Required	Yes	No
Steam Injection	Lance	Dynamic Mixer
Lysing Efficiency	Medium	High

The graph in Figure 12is a plot of theoretical calculated biogas production over a range of input solids loading. The graph also shows the steam utilization on the secondary axis. In the new TH design, biogas potential is not impacted by concertation of solid, however there is a significant decline in biogas potential due to inefficient lysing of high concertation solids using steam lance. The steam requirement decreases with higher concertation solids input to the TH reactors. This is possible only if the steam delivery method is capable of handling high concentration waste. While dilution might be required post THP due to ammonia loading, higher solids TH allows the ability to reduce foot print, reduce energy usage in the form of steam, maintain the efficiency of upstream dewatering (pre-THP) process and improve overall efficiencies across the TH process.

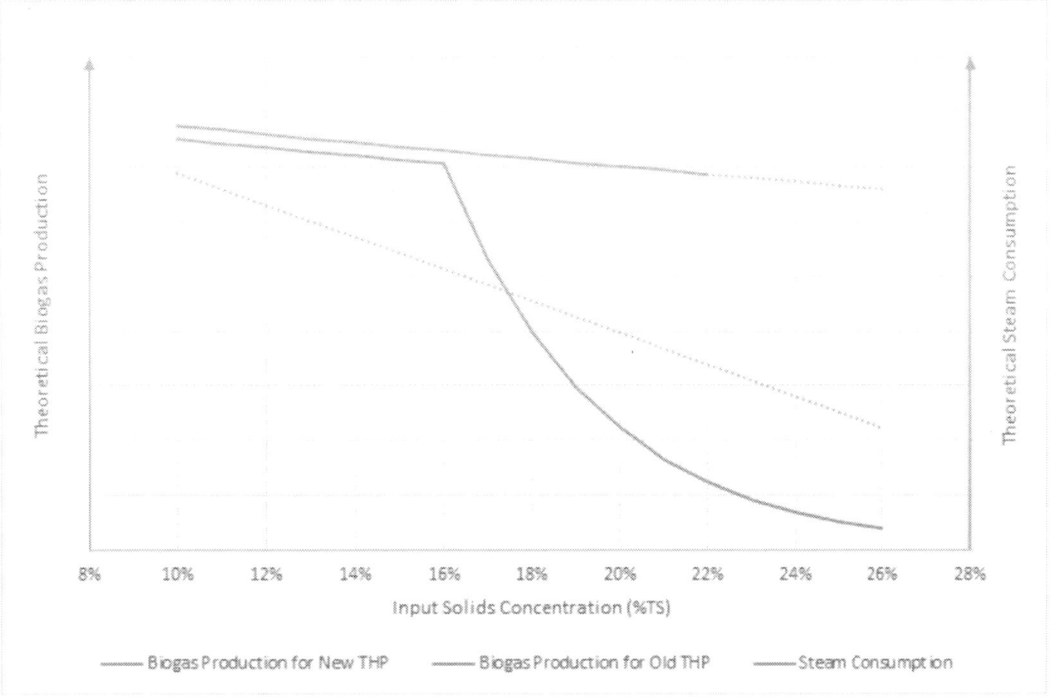

Figure 12: Improvements in design extends the operable solids loading concertation

Choosing the Right Technology

Careful consideration is required while selecting a thermal hydrolysis process. With information currently available and taking into account the evolution over the last three decades, it is now possible to evaluate feasibility and design application for any wastewater facility.

A high level evaluation of needs for any wastewater facilities should include study of:

- Current and future capacity of the plant
- Current and future energy needs and sources
- Current and future process efficiencies and available treatment options

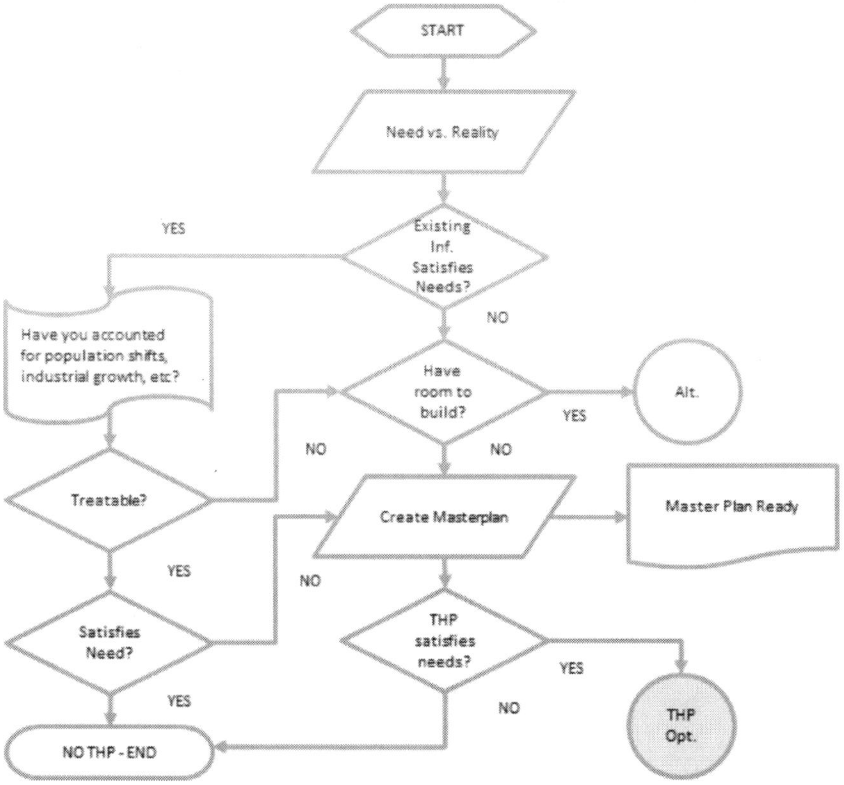

Figure 13: Example of Needs Analysis for THP

Other factors to consider during the needs analysis are sources and costs of energy, availability, conditions of existing infrastructure and capacities of:

- Mesophilic anaerobic digesters (MAD) - necessary for all THP
- Dewatering units - both pre-THP and post MAD
- Combined heat and power systems - to benefit from increased biogas production
- Biogas handling - especially to accommodate the increase in production
- Steam boilers - required for THP
- Heat exchanges - required to cool TH sludge

If it were determined that a THP is a feasible and economical solution for a wastewater facility, the analysis is shifted to determine the type of process to select. The following is an example of one such analysis.

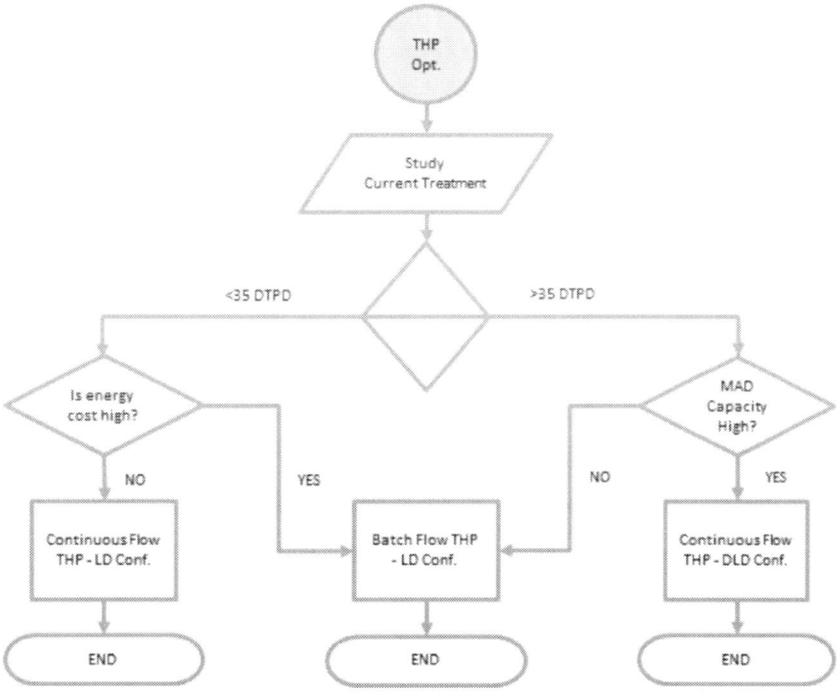

Figure 14: Example of Analysis to Determine THP Solution

An amalgamation of publically available data and proprietary information was used to determine the loading capacity that would benefit from a batch process as opposed to a continuous process. This study took into consideration

In the analysis to determine the right TH solution, it is evident that there is emphasis on economics based on loading capacity, this factor is not absolute and was determined based on publically bid THP projects worldwide and data correction to inflation, scope and currency. It is clear that there is a need and specific advantages to choosing the batch or the continuous process depending on the facility needs, status of current and future infrastructure and drivers such as capacity and regulation along with economic drivers like funding and available rebates.

Conclusions

With growing strain on resources due to population growth, thermal hydrolysis process is fast becoming the most versatile tool in biosolids management. The design evolution, performance improvements, and lessons leant over the past decade can also be used to determine the need and logically identify the solutions.

Thermal hydrolysis process solves not one but numerous challenges faced by wastewater treatment facilities including improvement in digester loading capacities by lowering viscosity, better destruction of organic waste by lysing, improve dewatering by eliminating extracellular polymeric substances, reduce volume of biosolids and reduce downstream equipment needs and chemical costs. The aim of this effort was to offer information and a tool demonstrating the differences, advantages and disadvantages of batch and continuous process and to determine the right THP solution for a given facility. Identifying and using the right solution will help wastewater treatment facility to fully realize the benefits of the circular economy.

References

Haug, R.T.; Stuckey, D.C.; Gossett, J.M.; Mac Carty, P.L. (1978) Effect of thermal pretreatment on digestibility and dewaterability of organic sludges. J. Water Pol. Control Fed. January, 73–85.

Abu-Orf M.; Goss T. (2012) Comparing Thermal Hydrolysis Processes (CAMBI™ and EXELYS™) for solids pretreatment prior to anaerobic digestion, Proceedings of the Residuals and Biosolids Conference, WEF RBC 2012, Raleigh, NC

Stedman, L. (2010) Green, efficient and low-impact: advances in sludge treatment technology. Water 21, Dec 2010, 35-36.

Gurieff, N. (2011) Maximizing Energy Efficiency and Biogas Production: EXELYS™ - Continuous Thermal Hydrolysis. Proceedings of the Water Environment Federation, WEFTEC 2011, Los Angeles, CA.

Chauzy J; Dimassimo R; Kline M; Howell G (2013) What is the Best Arrangement for Implementing THP (Thermal Hydrolysis Process) on my Sludge Treatment Facilities?, Proceedings of the Water Environment Federation, WEFTEC 2014, New Orleans, LA

Pilot-scale Evaluation of AirPrex® for Digestate Treatment

Blair Wisdom[1]*, Brad Van Anderson[1], Isaac Avila[1], Troy Gottschalk[1], Kurt Carson[1], Liam Cavanaugh[1]

[1] Metro Wastewater Reclamation District, Denver, Colorado
* Email: BWisdom@mwrd.dst.co.us

ABSTRACT
Waste sludge from biological phosphorus removal systems has been associated with negative impacts to anaerobic solids digestion and dewatering processes. Ahead of instituting a large-scale full plant biological phosphorus removal process, Metro Wastewater Reclamation District staff conducted an in-depth pilot-scale investigation of AirPrex® (CNP-Technology Water and Biosolids Corp., Kenosha, WI) as a possible approach to mitigate undesirable effects of Bio-P on the facility's solids treatment and handling processes. Results from the evaluation concluded that pretreatment of digestate by an AirPrex® type technology to promote in-vessel struvite formation was effective at sequestering phosphorus from the dewatering recycle stream and reducing struvite formation potential in downstream equipment. In addition, this pretreatment resulted in a 15 to 20 percent reduction in polymer required for dewatering and a notable increase in dry cake solids leading to an overall 7 to 10 percent reduction in biosolids hauling requirements depending on the degree to which struvite is recovered as a separate fertilizer stream or commingled with the biosolids.

KEYWORDS: Phosphorus, struvite, dewaterability, resource recovery

INTRODUCTION
The Metro Wastewater Reclamation District's (District) 220 MGD (833 MLD) Robert W. Hite Treatment Facility (RWHTF) currently serves 1.8 million population. This large-scale facility with anaerobic digestion is transitioning to enhanced biological phosphorus removal (Bio-P) to address nutrient over-enrichment of the South Platte River watershed. Waste Bio-P sludge has been associated with negative impacts on anaerobic solids digestion and dewatering processes (Shimp, Barnard, & Bott, 2013). It is also well known that Bio-P can exacerbate struvite formation in anaerobic digesters and downstream processes. The District conducted an in-depth pilot-scale investigation of AirPrex® (CNP-Technology Water and Biosolids Corp., Kenosha, WI) as a possible approach to mitigate undesirable effects of Bio-P on the RWHTF's solids treatment and handling processes.

AirPrex® (and other similar digestate treatment processes) imparts a slight aeration velocity to the digestate inside a dedicated reactor to strip some of the carbon dioxide (CO_2) and elevate the pH. Under elevated pH conditions magnesium (Mg) is added to the reactor to precipitate with phosphate (PO_4^{3-}) in order to predominately form struvite ($NH_4MgPO_4 \cdot 6H_2O$). The resulting change in chemistry of the digestate has been associated with reduced polymer demand and increased biosolids cake dryness. Several theoretical explanations are proposed to help describe these observations (Higgins & Novak, 1997) (Bergmans, Veltman, van Loosdrecht, van Lier, & Rietveld, 2014).

The District used five drivers to help define an effective overall approach for management of phosphorus once removed from wastewater. These drivers are to (i) break the phosphorus recycle loop that otherwise occurs during dewatering of anaerobically digested sludge (ii) to counter the impact that Bio-P can have on dewatering performance and biosolids hauling costs, (iii) reduce or eliminate problematic struvite scaling in piping and equipment and accumulation inside the digesters (Cavanaugh, Khunjar, Atwater, Carson, & McQuarrie, 2016), (iv) reduce the phosphorus relative to nitrogen in the biosolids product to create a more balanced product for nitrogen-intensive wheat and corn crops and (v) recover phosphorus in a form that is valued as a fertilizer product.

MATERIALS AND METHODS

The AirPrex® pilot system is a stand-alone mobile treatment process. The main setup for the pilot-scale evaluation included a 40 foot (12.2 meters) tall AirPrex® reactor with struvite recovery system, magnesium chloride ($MgCl_2$) chemical feed, and a 10-inch (25.4 centimeter) diameter dewatering centrifuge (Centrysis CS10-4) as shown in Figure 1.

Figure 1: AirPrex® Reactor (left) and Centrysis Centrifuge Dewatering (right)

A turbo blower and air compressor each supply air to two sparging rings installed at different heights within the reactor. Air sparging circulates and mixes the reactor contents and strips some CO_2 from the digestate increasing the pH to around 7.9. As magnesium concentration is typically limiting the struvite precipitation reaction, a liquid solution of 30 percent $MgCl_2$ was fed to create the desired molar ratio of Mg^{2+} to phosphorus (Mg:P) entering the reactor with the digestate. Once precipitated, the struvite crystals (specific gravity $\approx$ 1.7) increase in size with a fraction settling in the conical section of the reactor. The remaining struvite crystal fraction that did not settle remained within the biosolids matrix and discharged with the reactor effluent. The settled struvite was harvested from the reactor by periodically opening a discharge valve at the bottom of the cone with the flow routed to a grit classifier. The classifier used plant water to clean the struvite. The harvested struvite was discharged from the classifier into a collection container.

Hach test kits were used to field monitor the incoming PO_4^{3-} concentration of the digestate and the Mg dose was adjusted accordingly. During the final three weeks of the evaluation the Mg:P ratio was varied to examine the effect that changing this parameter would have on performance. Both the Untreated and Treated digestate feeds were sampled and analyzed each week by a certified laboratory as summarized in Table 1.

Table 1 - Sampling Schedule and Analyses[1,2]

Parameter	Reactor			Centrifuge		
	Influent	Underflow	Effluent	Influent	Centrate	Cake
pH	•		•	•	•	
Temperature	•		•	•	•	
Total Solids (TS)	•	•	•	•		•[3]
Volatile Solids (TVS)	•		•	•		•[3]
Total Ca, Mg, Fe, K, Na	•	•	•	•	•	
Dissolved Ca, Mg, Fe, K, Na	•		•	•	•	
Orthophosphate	•		•	•	•	
Total Phosphorus		•				
Water Extractable Phosphorus		•[4]				•[4]
Acid-cake Struvite Test						•[5]
Ammonium	•		•	•	•	
Conductivity	•		•	•	•	
Total Alkalinity	•		•	•	•	
COD_{EPS}	•		•	•		

[1] All analysis conducted once per week unless otherwise noted.
[2] All analysis conducted using grab samples.
[3] Daily.
[4] One time during study.
[5] Untreated cake samples collected on 1/21/2016 and 8/29/2016 and AirPrex treated cake samples collected on 7/29/2016 and 8/3/2016.

Throughout the evaluation, both polymer consumption and dewatered cake solids measurements were collected weekly on centrifuge operation for both the Untreated Digestate feed as well as the AirPrex® Treated Digestate. For each feed type the rate of polymer addition progressed through a prescribed dosage range. Machine performance data was collected for each setting.

Biosolids dewatering performance was assessed using two criteria; (i) dewatering polymer throughput use and (ii) the ultimate wet tons of biosolids generated from dewatering. Both assessment criteria require that special consideration be given to the influence that struvite mass has on the results as well as the analytical corrections required to fully address struvite mass in a given sample.

Correction to total solids results for struvite content

The ammonium (NH_4^+) and six water molecules that contribute to struvite have been observed to decompose under thermal heating. Mass loss begins at a temperature of 55 °C and is complete at a temperature of 250 °C (Bhuiyan, 2007). A correction was applied to the laboratory measured dewatered biosolids TS result to account for mass loss when the sample was dried at 104 °C. This study assumed that all of the water and NH_4 evaporated during heated drying resulting in a 51.4 percent loss (126 g/mol out of 245 g/mol total) of struvite mass. The TS cake solids laboratory measurements were thus corrected using Equation 1.

An intention of the AirPrex® reactor is to generate struvite out of the soluble PO_4^{3-} entering the reactor with the digested solids. Some of this struvite settles in the reactor and leaves with the underflow. However, a majority of the additional struvite generated remains in the biosolids matrix. The struvite retained in the biosolids increases the dry mass of dewatered biosolids produced at the facility. In order to estimate the additional struvite, acid extraction tests (acid-cake) were performed to quantify Mg^{2+} and PO_4^{3-} released from the cake sample and generate an estimate of the percentage of the overall dewatered biosolids mass that was attributed to struvite.

Equation 1 below was used to correct the %TS of the cake solids for untreated biosolids to account for the mass loss due to struvite decomposition during sample drying.

Equation 1

$$\%TS_{Corrected\ Untreated} = \%TS_{Measured} * (1 + (\%_{Struvite\ \mathrm{Content}} * 0.514))$$

Whereas,

$\%TS_{Corrected\ Untreated}$ = Lab result %TS corrected for struvite content in untreated biosolids,
$\%TS_{Measured}$ = Laboratory result %TS, and
$\%_{Struvite\ Content}$ = Estimated mass fraction of struvite in untreated biosolids.

Similarly, Equation 2 provides the corrected dry ton production for untreated biosolids to account for struvite mass loss. As described in Discussion of Results, full-scale dry ton data from the RWHTF was used in conjunction with the pilot-scale performance data to project and communicate comparative dewatering performance between the Untreated and AirPrex® Treated feed streams.

Equation 2

$$Total\ DT_{Corrected\ Untreated} = DT_{Measured} * (1 + (\%_{Struvite\ \mathrm{Content}} * 0.514))$$

Whereas,

$DT_{Corrected\ Untreated}$ = Measured dry ton result corrected for its struvite content,
$DT_{Measured}$ = From full-scale centrifuge throughput and performance data, and
$\%_{Struvite\ Content}$ = Estimated mass fraction of struvite in untreated biosolids.

Similar to Equation 1, Equation 3 corrects the measured %TS results for struvite mass loss during drying.

Equation 3

$$\%TS_{Corrected\ Treated} = \%TS_{Measured} * (1 + (\%_{Struvite\ Content} * 0.514))$$

Whereas,

$\%TS_{Corrected\ Treated}$ = Lab result %TS corrected for struvite in AirPrex® treated biosolids,
$\%TS_{Measured}$ = Laboratory result %TS, and
$\%_{Struvite\ Content}$ = Estimated mass fraction of struvite in AirPrex® treated biosolids.

For the AirPrex® Treated feed stream to the centrifuge, it was also necessary to estimate the additional struvite mass added to the biosolids during treatment. The amount of struvite additionally produced was estimated based on observed PO_4^{3-} conversion across the reactor. This value was then adjusted based on expected struvite recovery determined from the acid-cake test results. The total dry tons of dewatered biosolids, including struvite, was calculated by adding the unrecovered AirPrex® generated struvite to the struvite-corrected facility dewatered biosolids dry tons (See Equation 4).

Equation 4:

$$Total\ DT_{Corrected\ Treated} = DT_{AirPrex\ Struvite} + DT_{Corrected\ Untreated}$$

Whereas,

$Total\ DT_{Corrected\ Treated}$ = Corrected Untreated Total DT plus added struvite,
$DT_{Airprex\ Struvite}$ = Estimated struvite added based on PO_4^{3-} conversion and recovery
$DT_{Corrected\ Untreated}$ = From Equation 2.

Finally overall facility wet hauling requirements resulting from either type of biosolids stream were calculated using Equation 5

Equation 5:

$$Hauled\ Wet\ Tons = \frac{Total\ DT_{Corrected}}{\%TS_{Corrected}}$$

Whereas,

$Total\ DT_{Corrected}$ = From Equation 2 or Equation 4, and
$\%TS_{Corrected}$ = From Equation 1 or Equation 3.

Assessment of Struvite Formation Potential
Two different models were used to predict struvite formation potential. Struvite Tool (Office of Water Programs, Sacramento, California) was developed to predict struvite formation potential in a fluidized bed precipitation reactor. The model produced comparable results of struvite formation potential when compared with Visual Minteq (Ohlinger & Mahmood, 2003). The AirPrex® reactor, while not technically a fluidized bed reactor, operates similarly. Struvite Tool requires inputs for Mg, NH4, OP, conductivity, pH, temperature, and flow, and outputs struvite production, residual Mg, residual NH4, residual OP, and saturation index. Struvite Tool can generate struvite production and formation potential estimates with simplified inputs and does not incorporate other precipitation reactions, making the tool simple to use. This was the primary tool used to assess struvite formation potential

The second tool, Visual Minteq (KTH, Sweden) is a chemical equilibrium model program with a vast database of thermodynamic constants used for modeling precipitation in natural waters. It has been used to model struvite precipitation in complex sludge matrices in various studies (Ali, Schneider, & Hudson, 2003; Celen, Buchanan, Burns, Robinson, & Raman, 2007). A working knowledge of equilibrium chemistry is required to apply the model appropriately for systems with complex chemical and biological dynamics. The procedure outlined by Celen et al, 2007, for using Visual Minteq to model mineral precipitation in anaerobic digester was followed for this study

While the above tools are useful in determining struvite formation potential in the Untreated and Treated digestate streams, a calibrated and validated full plant BioWin model (EnviroSim Associates, Hamilton, Ontario) was needed to assess the impacts that breaking the phosphorus recycle loop could have on struvite formation in the digesters.

DISCUSSION OF RESULTS

The pilot system began operation on June 6, 2016 and ended August 3, 2016. With the exception of a few operational interruptions due to scheduled plant maintenance, the AirPrex® reactor operated continuously. Digestate from the full-scale digested sludge holding tank was pumped to the reactor at a constant flow of 11 gallons per minute (gpm) (41.6 lpm). The pilot centrifuge was configured such that it could be fed Untreated Digestate directly from the full-scale holding tank or AirPrex® Treated Digestate from the reactor. The centrifuge was operated for 6 to 8 hours per day during weekdays to generate dewatering performance curves across a set range of polymer doses for each of the two feed types; Untreated Digestate and AirPrex® Treated Digestate. Initial testing was conducted using Praestol K260-FL and K290-FL polymers, however following the polymer optimization, the majority of testing utilized BASF Zetag 8849.

For most of study, the Mg:P ratio for the AirPrex reactor was held between 1.1:1 and 1.5:1 with a target PO_4^{3-} conversion of 90 percent. For each centrifuge machine run the polymer dose was adjusted in 5 pound per dry ton increments with accompanying total solids (TS) and volatile solids (VS) measurements. Machine performance data was collected using polymer doses between 1.60 and 6.20 active pounds per hour which equates to a range between 24.8 and 73.7 active pounds per dry ton (12.4 to 36.9 kilograms per metric ton) based on feed solids TS measurements determined using standard drying procedures at 104 °C. The volumetric sludge throughput of the centrifuge was varied between 8 and 20 gpm (30.3 to 75.7 lpm) though the vast majority of testing was conducted at machine feed rates of either 10 or 15 gpm (37.9 to 56.8 lpm). The hydraulic pressure in the centrifuge was adjusted by the Centrisys operator to find ideal dewatering operating conditions for each feed type. The centrifuge hydraulic pressure ranged from 41 to 131 bar with an average of 80 bar on the Untreated Digestate runs, and ranged from 39 to 210 bar with an average of 102 bar on the AirPrex® Treated runs.

Breaking of the Phosphorus Recycle Loop

Figure 2 summarizes the centrate phosphorus results collected during the study. The average Total Phosphorus (TP) concentration of the untreated recycle centrate was 259 mg-P/L. The PO_4^{3-} concentration was observed to decrease in the centrate as the Mg:P molar dosing ratio increased. Digestate pretreatment in the AirPrex® system using a Mg:P molar dosing ratio of 1.4:1 resulted in greater than 90 percent conversion of PO_4^{3-} and corresponding removal of TP in the centrate return stream. At a still higher 1.7:1 dose ratio, the PO_4^{3-} concentration further decreased to an average concentration of 8.5 mg/L. However, TP began to increase, possibly indicating a high formation of smaller struvite crystals or fines that are not removed with the biosolids matrix. Higher concentrations of Mg have been shown to affect the size distribution of struvite crystals, as well affecting the purity due to an increase of additional Mg precipitates (Korchef, Saidou, & Amor, 2011). A magnesium chloride feed dose corresponding to a Mg:P molar ratio of 1.4:1 was observed to be the optimal ratio for limiting the TP in the recycle stream.

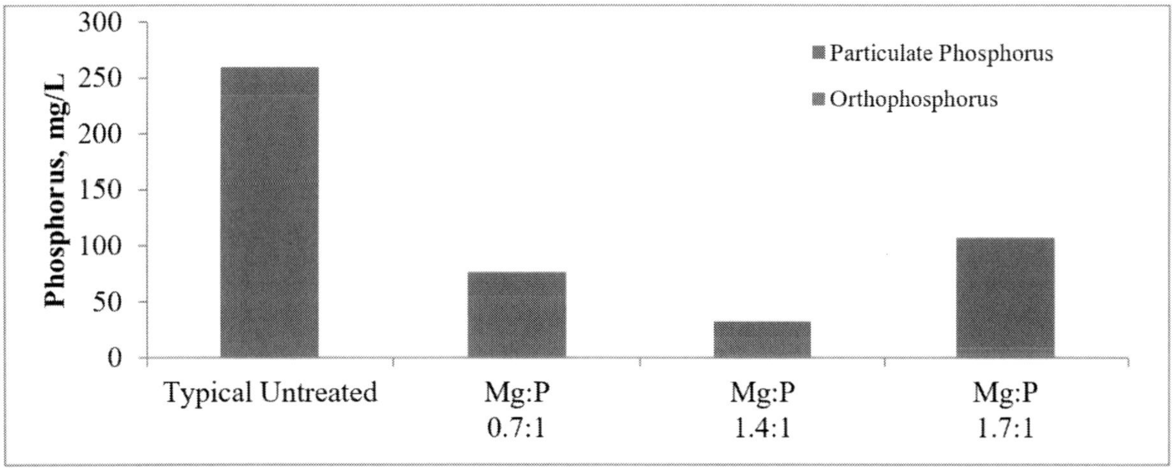

Figure 2: Phosphorus Recycle Load Control Untreated Digestate Centrate (left) and AirPrex® Treated Centrate at Three Different Mg Doses (right)

Figure 3 illustrates steady-state BioWin model results which help to show how lowering the recycle phosphorus load helps to reduce the secondary effluent from 0.26 mg/L P to 0.1 mg/L P.

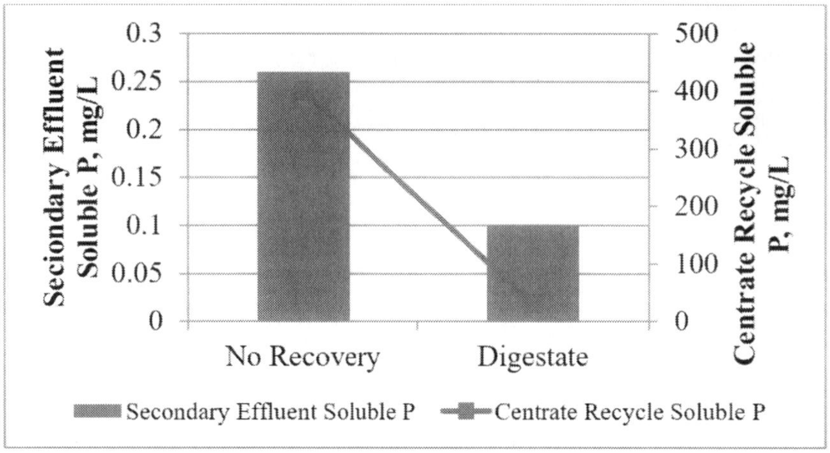

Figure 3: Phosphorus Recycle Load Control and Resulting Secondary Effluent PO_4^{3-} Concentrations

Biosolids Dewatering

Equations 1, 2 and 3 defined in the Materials and Methods require an estimation of the struvite content in the dewatered biosolids. Table 2 summarizes the results from the acid-cake tests. On average about 5.3 percent of the untreated biosolids mass was in the form of struvite. In comparison, about 11.7 percent of the treated biosolids mass was in the form of struvite. Given the recovery of the pilot AirPrex® treatment system, the total dry solids production increased by approximately 6 percent. Based on the results of the acid-cake testing on the AirPrex® treated digestate and the calculated struvite production (based on converted PO_4^{3-} in the reactor), it was estimated that the struvite product capture averaged 20 percent during the pilot, resulting in 80 percent of the produced struvite remaining in the biosolids.

Table 2: Measured Struvite Concentration in the Dewatered biosolids

Sample Date	Sample Stream	Average Concentration of Struvite in Dewatered Biosolids (% mass struvite per total mass)
1/21/2016	Untreated	5.6%
7/29/2016	AirPrex Treated	11.1%
8/3/2016	AirPrex Treated	12.3%
8/29/2016	Untreated	5.0%

Table 3 summarizes the results of the dewatering improvements assessment. The TS measurements were corrected as described in the Materials and Methods using Equations 1 and 3. Pilot-scale performance results were projected to full-scale using the average facility dewatered biosolids dry mass generated during the study period. During June and July 2016; the RWHTF generated 84.3 dry tons of biosolids per day. Using Equation 2 to adjust for struvite mass loss during sample drying, the corrected value for Untreated is 86.6 dry tons of biosolids per day.

The amount of struvite produced in the AirPrex® reactor was estimated to be 7.4 tons per day if applied at full scale. This is based on 90 percent conversion of the PO_4^{3-} passing through the reactor with the digestate. Accounting for the observed 20 percent product recovery, the resulting increase in dewatered biosolids mass from the AirPrex® process would be 5.9 tons per day. From Equation 4, the total dry tons of dewatered biosolids, including struvite, leaving the facility post-AirPrex® would be 92.5 dry tons per day.

Dewatering Polymer Consumption

Total polymer consumption was based on a mass flow basis (active pounds of polymer per hour) for the purposes of this comparison. The doses in pounds of active polymer per hour were averaged for each dataset for the digestate and the AirPrex® treated digestate centrifuge runs, and the variation in total polymer use between them was compared on a percent change basis.

Four datasets were generated from the AirPrex® study period for the examination of impacts on biosolids dewaterability. Variables in the centrifuge operation were systematically removed as the datasets progressed in order to perform more refined calculations on the anticipated full-scale impacts of an AirPrex® system at the RWHTF. The four datasets that were refined for further examination are:

- Dataset 1 (n=297) - All valid study period centrifuge dewatering data was included in this dataset. Only data points with obvious data errors (such as zero polymer flow) were eliminated to develop this dataset.
- Dataset 2 (n=197) - The data points in Dataset 1 were sorted to remove variability in centrifuge sludge throughput rates and polymer types. Only data points with 10 gpm and 15 gpm centrifuge sludge feed throughputs were accepted due to the fact that the majority of centrifuge runs were conducted at those two throughputs to simulate anticipated scale-up throughputs. Additionally, only data points where the BASF Zetag 8849 polymer was used for the centrifuge dewatering were accepted due to that polymer being solely used during the majority of the study.
- Dataset 3 (n=61) - The data points in Dataset 2 were evaluated to further remove variability in centrifuge sludge throughput rates, and to eliminate the variability in AirPrex® Mg dosing rates for different OP removal targets. Only data points with a 10 gpm centrifuge sludge feed throughput were accepted based on the 10 gpm having the largest number of data points to simulate the design point for future full-scale centrifuge dewatering facilities. Additionally, only data points with a 90% AirPrex® OP removal target were accepted (correlating to an approximate 1.4:1 Mg:P dosing ratio) in order to match the anticipated design point for a full-scale system that would remove the maximum amount of TP (as described in the analysis of the effectiveness of AirPrex® at breaking the recycle phosphorus loop).
- Dataset 4 (n=20) - The data points in Dataset 3 were assessed to further remove variability in the centrifuge operational settings, and to remove variability in digestate characteristics. Data pairs of one digestate data point and one AirPrex® treated digestate data point were pulled from Dataset 3. The pairs had to meet the criteria of both being within three days of each other in order to ensure consistent digestate OP concentrations, and to have centrifuge hydraulic pressures within 20 bar of each other in order to eliminate the biasing effect of the centrifuge to achieve better dewatering results at higher hydraulic pressure settings. Once a data point was pulled into a pair, it was not used again in another pair. Ultimately, Dataset 4 contained 20 data points (10 pairs) that met the evaluation criteria. Two pairs were three days apart, four pairs were one day apart, and four pairs were on the same day. The average hydraulic pressure of the data points was

93.5 bar, and the average difference in the hydraulic pressures of the data pairs was 7.1 bar, with a maximum of 16 bar and a minimum of 1 bar. Four data pairs had a higher hydraulic pressure for the AirPrex® data point, one pair had a higher hydraulic pressure for the digestate data point, and five data pairs had a hydraulic pressure difference of 5 bar or less. The data pairs in Dataset 4 were evaluated both by averaging all digestate and AirPrex® data points when each were lumped together, and by generating the dewatered biosolids TS and polymer dose differences for each pair and then averaging those individual results.

Table 3 provides the results of the dewatering analysis for the four data sets including raw measurements and corrections as described in the Materials and Methods section. Curves showing the polymer consumption and uncorrected dewatered biosolids TS concentrations are displayed in Figure 4a, b, c and d for each of the data sets.

Table 3 - Summary of AirPrex® Biosolids Dewaterability Impacts

Column	1	2	3	4	5	6	7	8	9	10	11	12
Data Set #	Untreated Dewatered biosolids Mass Dry Tons per Day[1]	Untreated Dewatered biosolids %TS[2]	Untreated Dewatered biosolids %TS (Corrected)[3]	Untreated Corrected Dewatered Biosolids Wet Tons per Day[4]	Untreated Polymer Dose Active lbs/hr[5]	AirPrex Dewatered biosolids Mass Dry Tons per Day[6]	AirPrex Dewatered biosolids %TS[7]	AirPrex Dewatered biosolids %TS (Corrected)[8]	AirPrex Treated Polymer Dose Active lbs/hr[9]	AirPrex Corrected Dewatered biosolids Wet Tons per Day[10]	AirPrex Wet Tons Reduction[11]	AirPrex Polymer Reduction[12]
1	86.6	20.68%	21.23%	407.9	3.50	92.5	23.52%	24.90%	3.53	371.5	8.9%	-0.86%
2	86.6	21.16%	21.72%	398.7	3.46	92.5	23.76%	25.15%	3.51	367.8	7.8%	-1.45%
3	86.6	21.17%	21.73%	398.5	2.99	92.5	25.05%	26.52%	2.97	348.8	12.5%	0.67%
4	86.6	21.31%	21.89%	395.6	3.18	92.5	24.17%	25.62%	2.62	361.0	8.7%	17.61%

Footnotes

1. Measured from full-scale dewatered biosolids production during the pilot study period.
2. Average measured total solids of untreated digestate dewatered biosolids from the pilot centrifuge.
3. Corrected total solids concentrations of untreated digestate dewatered biosolids from Equation 1 in Materials and Methods.
4. Average wet tons production of untreated digestate from Equation 5 in Materials and Methods.
5. Polymer consumption for dewatering untreated digestate expressed on a mass flow basis.
6. Estimated dewatered biosolids production of AirPrex treated digestate based on observed AirPrex pilot OP conversion, 20% product recovery, and full-scale biosolids flows and loads.
7. Average measured total solids of AirPrex treated digestate dewatered biosolids from the pilot centrifuge.
8. Corrected total solids concentrations of Airprex treated digesatate dewatered biosolids from Equation 2 in Materials and Methods.
9. Polymer consumption for dewatering AirPrex treated digestate expressed on a mass flow basis.
10. Average wet tons production of AirPrex treated digestate from Equation 5 in Materials and Methods.
11. Percent difference between the wet tons of untreated and AirPrex treated dewatered biosolids.
12. Percent difference between the untreated and AirPrex treated polymer dose.

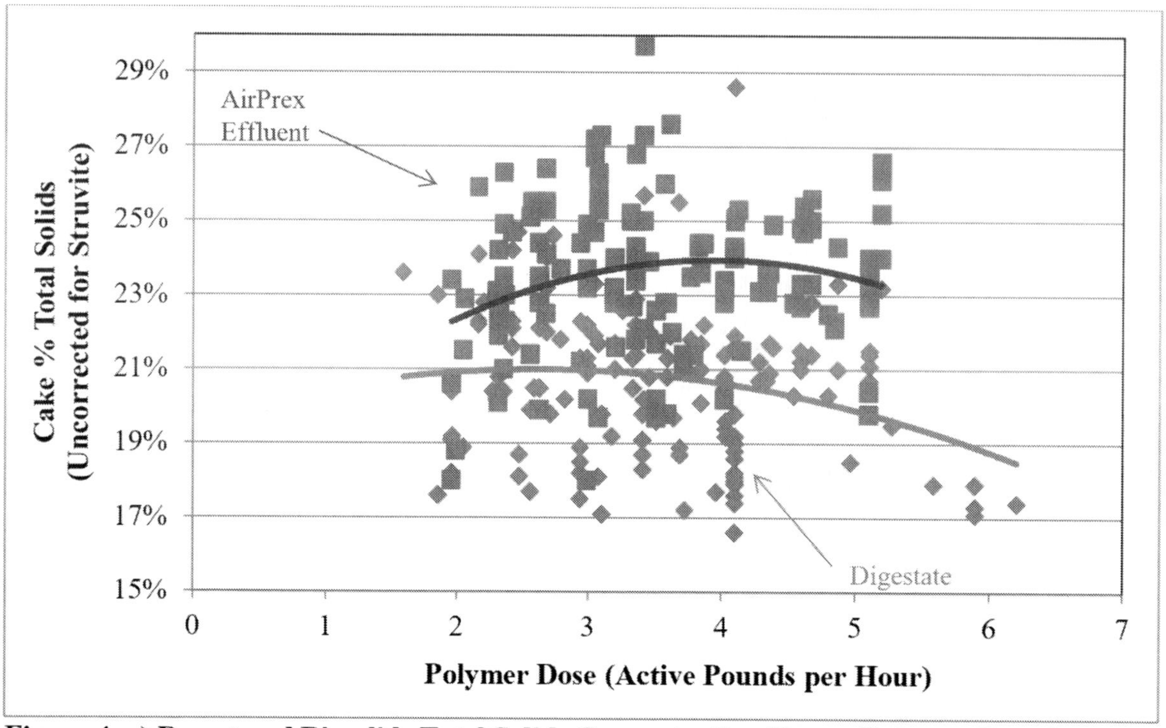

Figure 4. a) Dewatered Biosolids Total Solids Concentration Versus Polymer Dose for Dataset 1

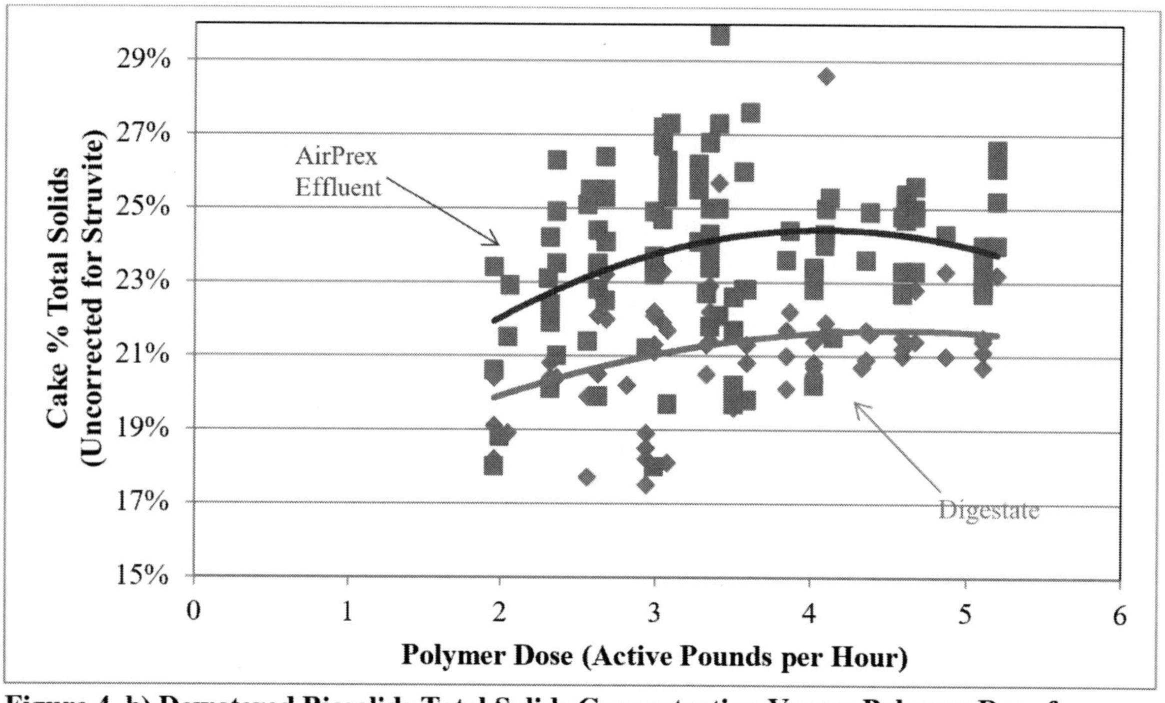

Figure 4. b) Dewatered Biosolids Total Solids Concentration Versus Polymer Dose for Dataset 2

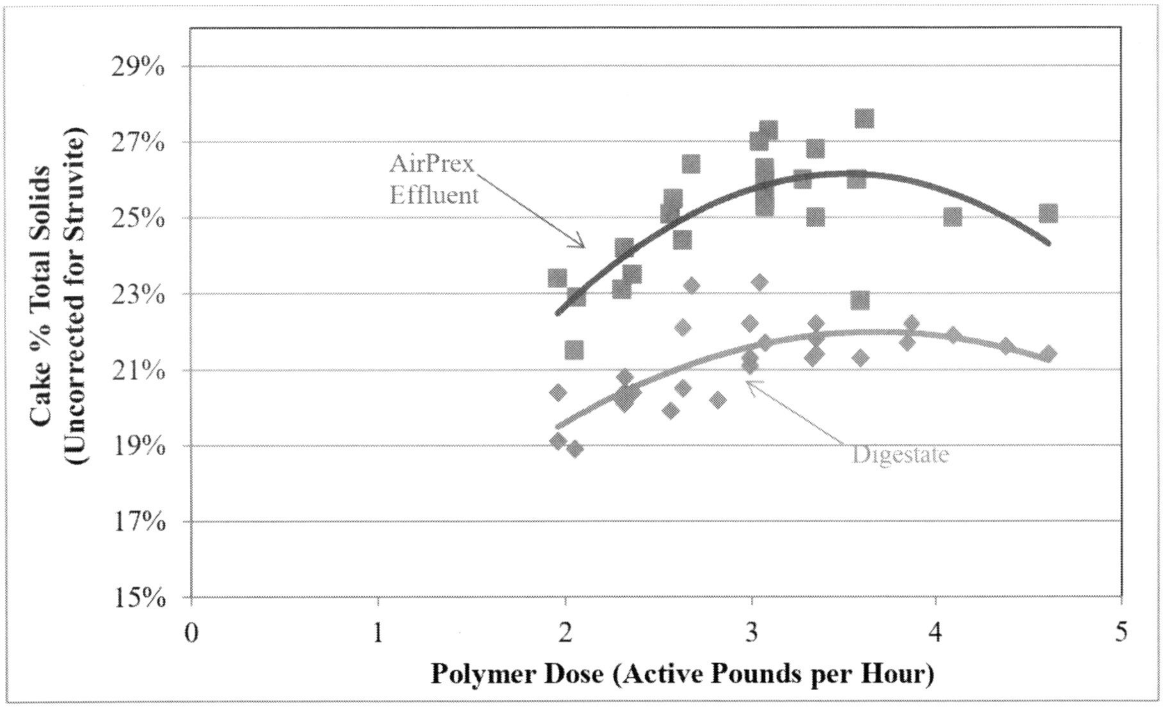

Figure 4. c) Dewatered Biosolids Total Solids Concentration Versus Polymer Dose for Dataset 3

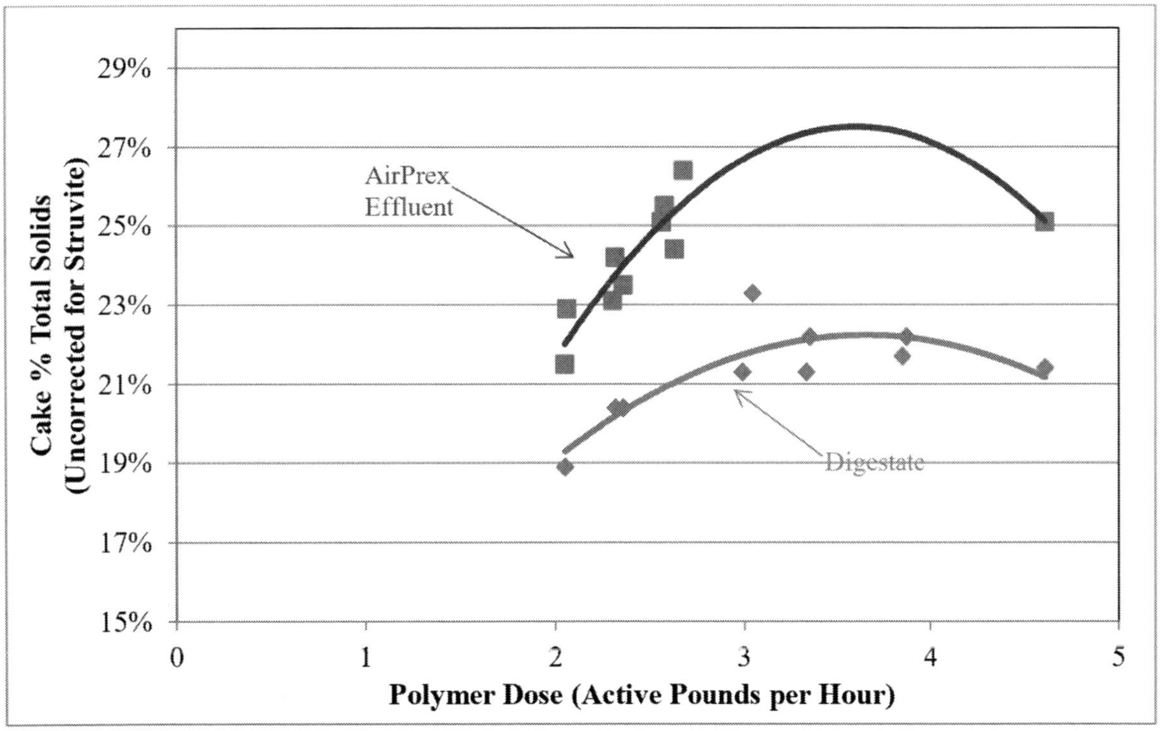

Figure 4. d) Dewatered Biosolids Total Solids Concentration Versus Polymer Dose for Dataset 4

There is similarity in polymer doses when averaging the large amount of data in Datasets 1-3 because both the untreated digestate and AirPrex® treated digestate centrifuge tests were prescriptively run through the same range of polymer doses to generate the necessary detailed dewatering information. When specific pairs of dewatering points were identified for Dataset 4 by eliminating all other factors, there was a decrease of 17.6 percent polymer consumption for the AirPrex® effluent centrifuge runs over the untreated digestate. Additionally, when polymer differences were calculated for each data pair, and then averaged, the calculated decrease in polymer consumption for Dataset 4 was 16.5 percent for the AirPrex® effluent. There were no instances in the data pairs where the untreated digestate polymer dose was lower than the AirPrex® treated digestate polymer dose.

Additional analyses of the decrease in total wet tons from AirPrex® treatment were performed on Dataset 4 to evaluate the sensitivity of the reduction in total wet tons of either dewatered biosolids produced when the assumed struvite recovery in the AirPrex® reactor varied, or when the struvite mass evaporated was modified. When the struvite recovery amount in the pilot reactor was varied from 0 percent to 35 percent of the AirPrex® generated struvite, the wet tons reduction ranged from 7.26 percent to 9.82 percent. Table 4 presents the results of the struvite recovery variations. Full-scale recovery amounts would not necessarily significantly change these wet tons projections, because the calculations are based on the pilot-scale study data for the dewaterability improvements.

Table 4 - Sensitivity of Total Biosolids Mass Reduction to Pilot-Scale Struvite Recovery Assumption

AirPrex Struvite Recovery Efficiency	0%	10%	20%	35%
AirPrex Wet Tons Reduction	7.3%	8.0%	8.7%	9.8%

To check the assumptions made about struvite evaporation occurring during the TS test, the assumed mass of struvite evaporation was modified from 51.4 percent (which assumed that the NH_4 and six water molecules evaporate during the test) to 29.4 percent (which equates to the evaporation of NH_4 and three water molecules). When examining the impact on the total wet tons reduction from AirPrex® treatment using Dataset 4, the total wet tons reduction changed from 8.7 percent at 51.4 percent struvite evaporation to a reduction of 7.4 percent at 29.4 percent struvite evaporation. This difference in the assumption of struvite evaporation is significant, because it equates to a difference of approximately 1,930 wet tons of biosolids being hauled annually.

The results from the analysis of all four datasets show two significant benefits from the AirPrex® process. First, throughout all four datasets, the total wet tons hauled showed a decrease of at least 7.7 percent with the potential for further improvement. When the increased hydraulic pressure bias for the AirPrex® treated digestate centrifuge runs was eliminated from the data, the wet tons decrease was still 8.7 percent, which reduces any concern that centrifuge operating set points were the driver behind the improved dewatering performance. Second, when the data is paired to ensure consistent digestate composition and centrifuge operations between the untreated digestate and AirPrex® treated digestate, the data showed at least a 16.5 percent reduction in total polymer use on a strictly active pounds consumed basis.

Struvite Reduction

Table 5 summarizes the average analyte concentrations and water quality parameters as determined through the weekly pilot sampling and analyses; these parameters were used as inputs in the models. Struvite formation potential was modeled downstream of the AirPrex® reactor to reflect potential nuisance struvite formation in the dewatering process. Digester effluent acted as the control stream. Struvite formation potential was estimated to determine the impact that the AirPrex® system would have on the nuisance struvite production in the dewatering complex and centrate piping and holding tanks. The AirPrex® reactor contents were determined via mass balance of the inputs, digester effluent, and $MgCl_2$ dose. Centrate was modeled for both control streams and test streams to determine struvite formation potential occurring in pipes and equipment downstream of dewatering.

Table 5 - Average Composition of Modeled Streams

Parameter	Digester Effluent	AirPrex® Reactor	Untreated Centrate	Treated Centrate
Dissolved Calcium, mg/L	39	46.2	40.1	57.9
Dissolved Iron, mg/L	1	1.1	0.2	0.6
Dissolved Potassium, mg/L	262	323	261	279
Dissolved Magnesium, mg/L	10	336	9.7	66.7
Dissolved Sodium, mg/L	112	144	129	141
Ammonium, mg/L	1,593	1,598	1,243	1,165
Orthophosphate, mg/L	253	295	234	6.7
pH, S.U.	7.8	7.8	7.9	7.9
Temperature, C	35	35	35	35
Conductivity, *uhmo/cm*	8,620	--	9,660	10,250
Alkalinity, *mg* $CaCO_3/L$	4,723	4,723	3,943	3,140

Figure 8 shows a comparison between the model outputs of both Visual Minteq and Struvite Tool looking at three different streams; within the AirPrex reactor (Top), treated centrate (bottom left), and untreated centrate (bottom right). Looking at the treated centrate stream, there is some degree of variability in the outputs between the two models. For example, Visual Minteq predicts the precipitation of significant quantities of magnesite alongside the struvite precipitation, hence the discrepancy between the two. However, when looking at untreated streams, the two models are closer in agreement with each other. This is due to the fact that the propensity for struvite to precipitate is greater in the untreated streams, and Visual Minteq predicts that struvite precipitation will dominate over the other minerals. The rest of the thermodynamic modeling results presented are from Struvite Tool.

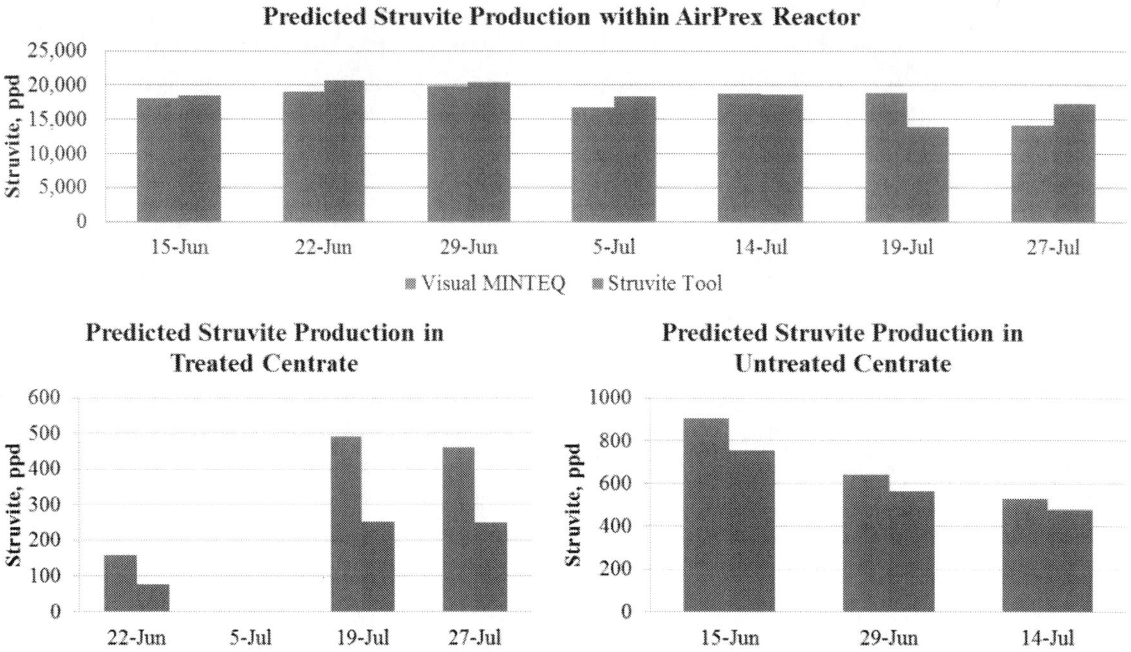

Figure 8: Comparison of Struvite Formation Potential Between Struvite Tool with Visual Minteq.

This reduced return of phosphorus to the secondary treatment process would reduce phosphorus uptake in the biological sludge, reducing phosphorus and Mg loading to the anaerobic digestion process, thus mitigating the impacts of nuisance struvite formation. A full plant BioWin model was used to determine the effect that reduced phosphorus recycle would have on struvite formation in the digesters. Figure 9 illustrates how the operation of an AirPrex® system would be anticipated to reduce the struvite formation in the digesters by 28 percent.

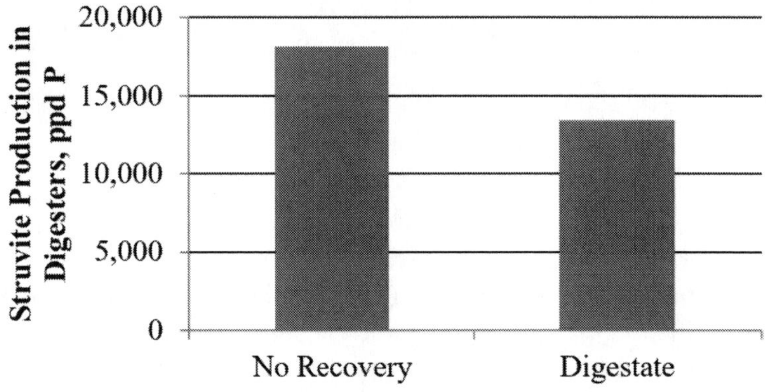

Figure 9: Struvite Formation in Gas Phase Digesters

Figure 10 summarizes the model estimates of struvite saturation index for the untreated digestate, AirPrex® treated digestate, and the treated centrate. Saturation index is an indicator for the propensity of struvite to precipitate, the higher the saturation index, the greater the potential. The results from this analysis show that AirPrex® can significantly reduce the propensity of struvite to form in downstream equipment.

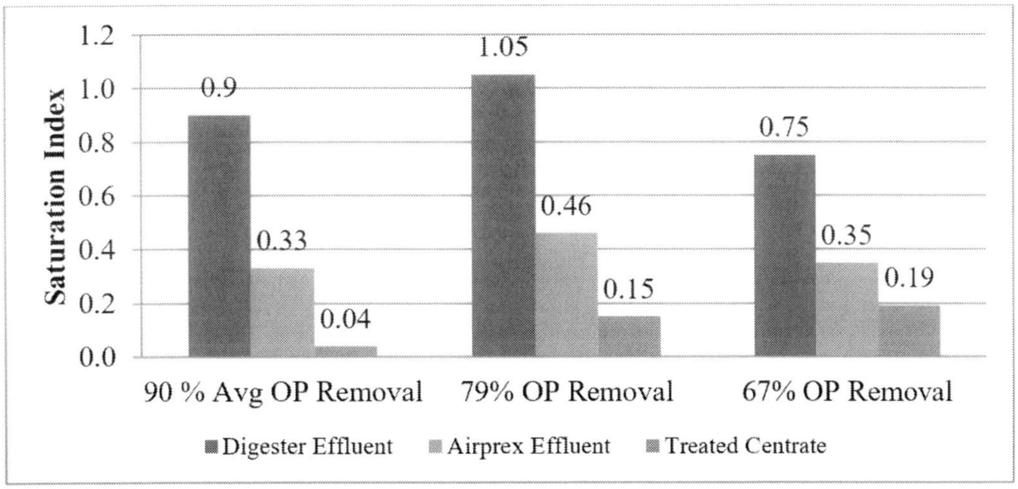

Figure 10: Struvite Formation Potential in Digesters and Downstream Equipment at Different OP Removals in AirPrex®

Figure 11 summarizes the model estimates of the magnitude of struvite formation in the digester effluent (left axis), treated and untreated centrate (left axis), and the AirPrex® reactor (right axis) throughout the course of the pilot. The first data points collected were during startup, and the optimum recovery had yet to be optimized; hence, the predicted struvite production of the AirPrex treated digestate on June 15 did not differ much from the untreated digestate. The next four data points were collected during targeted 90 percent OP conversion. At the targeted 90 percent OP conversion, the struvite estimates illustrate the benefit of intentional struvite precipitation to help with downstream nuisance struvite issues. Starting on July 19, lower Mg:P dosing ratios were fed to the reactor, resulting in increased struvite production estimates for locations downstream of the system.

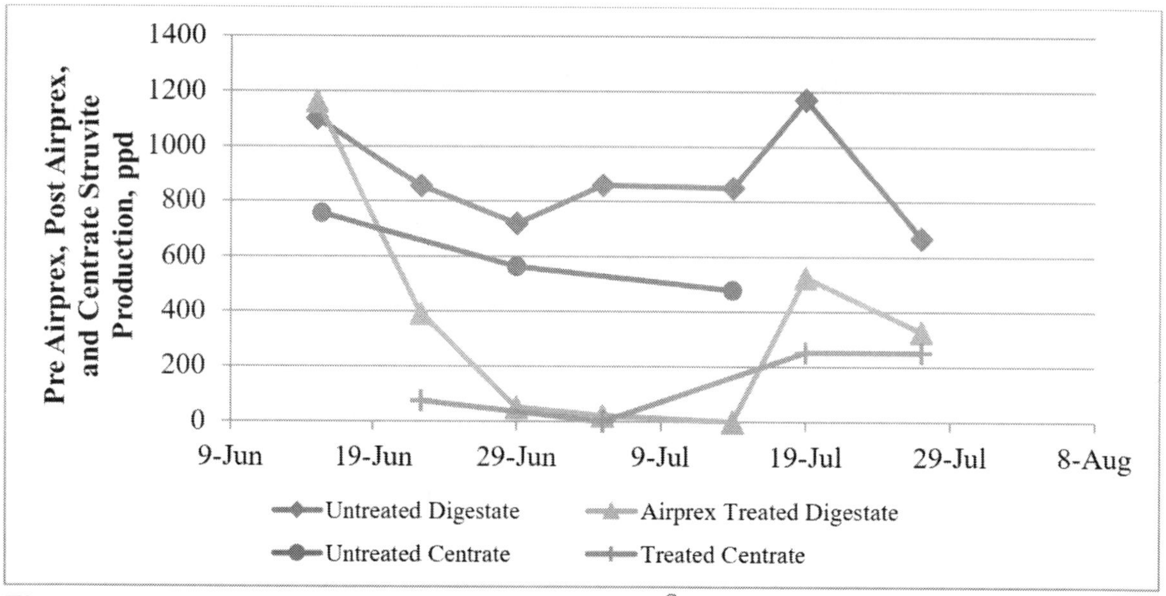

Figure 11: Struvite Production in Digesters, AirPrex® Reactor, and Recycle Streams throughout the Entirety of Pilot Study

Phosphorus Index

Phosphorus concentrations in the biosolids directly affect the loading rates of land application sites and therefore hauling and dispersal operating costs. Accumulating phosphorus in the biosolids due to tightening treated effluent limits can be detrimental to land application unless recovery of phosphorus as a separate "biosolids" stream with a separate management scheme is implemented. The mass of phosphorus recovered from the reactor was difficult to determine during the pilot due to periodic removal of the reactor struvite product but as presented in the dewatering section, recovery was estimated at around 20 percent with the potential to be higher for full-scale operation depending on reactor configurations and operation. This recovery results in an average of 80 percent of the produced struvite remaining within the biosolids. Recovery will directly affect TP in the biosolids (and thus land application of dewatered biosolids). From analyses on the dewatered biosolids, the untreated dewatered biosolids had a TP concentration of 2.7 on average while the AirPrex® treated dewatered biosolids had an average TP concentration of 3.4 percent. This concentration was consistent with TP concentrations obtained from BioWin modeling of the treatment process. As a comparison, ferric sequestration of phosphorus in the biosolids would result in a TP concentration of 3.8 percent.

Phosphorus Recovery

Metals analysis and pathogen counts were performed on the recovered struvite product. The results of the metals analysis are summarized in Table 6. Metals concentrations were all within the High Quality pollution limits set by the EPA. Fecal coliform testing was run for samples collected on three occasions to determine whether the recovered product was anticipated to require additional treatment in order to be classified as a Class A material per EPA 503 regulations. The fecal coliform counts are summarized in Table 7. The initial test resulted in a concentration above the cut-off level for Class A designation while the subsequent tests were well within the limits. While preliminary results indicate that it may be possible to meet Class A criteria without further treatment, supplemental treatment technologies such as solar drying may be required to further deactivate pathogens as required.

Table 6: Struvite Product Metals Analysis

Pollutant	"High Quality" Pollutant Concentration Limits** (mg/kg)	Pilot Struvite Results	Lab Analysis from Previous Pilot
Arsenic	41	BDL[2]	
Cadmium	39	BDL	BDL
Copper	1500	71	19
Lead	300	18.9	
Mercury	17	1.7	
Molybdenum[1]	< 75	0.8	
Nickel	420	BDL	BDL
Selenium	100	BDL	
Zinc	2800	33	26

1 There is currently no limit for Molybdenum for High Quality while rules are awaiting EPA investigation. Ceiling limit is 75 ppm.
2 Levels were below detection limit

Table 7: Struvite Product Pathogen Analysis

Date	Fecal Coliforms
	MPN/g-TS
7/19/2016	1000
8/2/2016	31
9/12/2016	<0.31

CONCLUDING REMARKS

The pilot data allowed for a quantitative analysis of the AirPrex® treatment system against each of the five established drivers and will ultimately be used to compare the system against other potential technology solutions. This data will feed into economic models, imparting confidence in the operational costs and cost-benefit analysis for this system. Although extensive piloting of technologies is time consuming and expensive, the process can provide system specific information that is not readily available for technologies with limited full-scale installation experience. Findings from this study showed that AirPrex® would be an effective technology for reducing the impacts of centrate phosphorus recycle on effluent quality and struvite generation in the digesters. Findings from the study also indicated that the technology would reduce dewatering polymer consumption by 15 to 20 percent along with a 7 to 10 percent reduction in wet ton hauling requirements. The initial quality of the recovered and cleaned struvite also made it conceivable that the material could be managed such that it could meet 503 Class A requirements and be distributed as a fertilizer.

REFERENCES

Ali, M. I., Schneider, P. A., & Hudson, N. (2003). Assessing nutrient recovery from piggery effluents. *International Congress on Modelling and Simulation (MODSIM03)*. New Zealand.

Bergmans, B., Veltman, A., van Loosdrecht, M., van Lier, J., & Rietveld, L. (2014). Struvite formation for enhanced dewaterability of digested wastewater sludge. *Environmental Technology*, 549-555.

Bhuiyan, I. H. (2007, October). Investigation into Struvite Solubility, Growth, and Dissolution Kinetics in the Context of Phosphorus Recovery from Wastewater. The University of British Columbia.

Cavanaugh, L., Khunjar, W., Atwater, A., Carson, K., & McQuarrie, J. (2016). Making the Case for Phosphorus Recovery: Theoretical and Full-Scale Business Case Evaluations at an 830-MLD Wastewater Treatment Facility. *WEF/IWA Nutrient Removal and Recovery*. Denver.

Celen, I., Buchanan, J. R., Burns, R. T., Robinson, R. B., & Raman, D. R. (2007). Using a chemical equilibrium model to predict amendments required to precipitate phosphorus as struvtie in liquid swine manure. *Water Research, 41*(8), 1689-1696.

Higgins, M., & Novak, J. T. (1997). The effect of cations ont he settling and dewatering of activated sludges. *Water Environment Research*, 215-224.

Korchef, A. Saidou, H. & Amor, M.B. (2011). Phosphate recovery through struvite precipitation by CO2 removal: Effect of magnesium, phosphate, and ammonium concentrations. *Journal of Hazardous Materials*, 602-613

Ohlinger, K. N., & Mahmood, R. J. (2003). Struvite scale potential determination using a computer model. *World Water & Environment Resources Congress.*

Shimp, G. F., Barnard, J. L., & Bott, C. B. (2013). Seeking to Understand and Address the Impacts of Biological Phosphorus Removal on Biosolids Dewatering. *WEFTEC 2013*, (pp. 5668-5685).

Enhanced Dewatering with Struvite Recovery: Pilot Testing of AirPrex® Technology at Miami's South District WWTP

Brian, Stitt[1]; Terry Goss[1]; Manual Moncholi[2]; Mohammad Abu-Orf[3]; Ismael Diaz[4]
[1]AECOM Water, [2]Miami-Dade Water and Sewer Department, [3]Formally AECOM Water, [4]Gannett Fleming

ABSTRACT

Magnesium ammonium phosphate (MAP) also known as struvite is a common problem in wastewater treatment plants which can foul piping and equipment. Struvite accumulation and fouling has historically been one of the major maintenance issues for the South District Wastewater Treatment Plant (SDWWTP) operations, consuming resources for continuous pipe cleaning to maintain steady, un-interrupted operation of the existing digestion and dewatering processes. Struvite formation results from high soluble orthophosphate (PO_4-P) concentration in the digested biosolids and the potential is expected to increase in the future with improved thickening prior to digestion. Moreover, biosolids with high soluble PO_4-P appears to result in poor dewatering performance in terms of lower cake solids and higher polymer dosing requirements. One method to deal with high orthophosphate concentrations is to precipitate out the struvite. In the AirPrex® process, marketed in North America by CNP Corporation, struvite is crystallized directly from the biosolids stream from an anaerobic digester prior to dewatering. AirPrex piloting was conducted at Miami Dade's SDWWTP in April 2016 and the AirPrex treated biosolids was dewatered using a pilot dewatering centrifuge. The feed to the pilot centrifuge was set up to allow the pilot to test both the AirPrex and non AirPrex treated digested biosolids in the same day. The results showed that the AirPrex process consistently achieved 91% removal of soluble orthophosphate and 14 to 16% removal of ammonia (NH_3-N). From a dewatering perspective, the AirPrex pretreatment provided a 2-4 points increase in the cake solids from the pilot centrifuge, in comparison to non AirPrex treated sludges with similar polymer dosages.

KEYWORDS

Biosolids, Struvite, Dewatering, Phosphorus, Anaerobic Digestion, Resource Recovery

INTRODUCTION

Struvite accumulation and fouling has historically been one of the major maintenance issues for the South District Wastewater Treatment Plant (SDWWTP) operations, consuming resources for continuous pipe cleaning to maintain steady, un-interrupted operation of the existing digestion and dewatering process. Struvite is magnesium ammonium phosphate ($MgNH_4PO_{4(s)}$) and results from high soluble orthophosphate (PO_4-P) concentrations in the digested biosolids when adequate ammonia and magnesium are present. The potential for struvite formation is expected to increase in the future with improved thickening prior to anaerobic digestion. Moreover, increasing the PO_4-P concentration in the biosolids has been reported to reduce dewatering performance in terms of lower cake solids and higher polymer dosing requirements (Kopp et al., 2016).

Goss et al (2017) presented the results from centrifuge thickening piloting, digestion high rate piloting and centrifuge dewatering piloting at the SDWWTP. SDWWTP (Digester 9) was isolated to receive mechanically thickened sludge (from a pilot thickening centrifuge) to simulate future high rate anaerobic digestion. The digester was operated in this manner to allow the digester to reach a steady state with mechanically thickened sludge. The solids content in Digester 9 was increased from 2 to 2.5% total solids (TS) to approximately 3.4 to 3.5% TS. Once three digester solids retention time turnovers were achieved in Digester 9, pilot centrifuge dewatering testing was conducted, but the results showed that only 18% TS could be achieved with active dry polymer dosages of 20 to 30 pounds per dry ton (lb/DT) or 10 to 15 grams per dry kilogram (g/kg). The initial goal was to achieve greater than 20%TS with greater than 95% solids recovery using an active dry polymer dose of less than 25 lb/DT (12.5 g/kg).

Since it was desired to improve dewatering and mitigate the maintenance issues associated with struvite fouling, SDWWTP staff evaluated methods for struvite control which could also enhance dewatering. In the AirPrex® process, marketed in North America by CNP Corporation, struvite is crystallized directly from the biosolids stream from an anaerobic digester prior to dewatering. The precipitation of struvite prior to dewatering is one potential method to achieve both improved dewatering and reduced maintenance costs. The objective of the pilot study presented herein was to demonstrate the technology at SDWWTP and document the struvite precipitation and centrifuge dewatering performance results. The AirPrex® pilot testing was conducted after a series of thickening, digestion and dewatering pilot testing were completed at SDWWTP (Goss et al., 2017).

BACKGROUND ON STRUVITE FORMATION

Magnesium ammonium phosphate (which is sometimes referred to as MAP) precipitation is a common problem in wastewater treatment plants which can foul piping and equipment. Struvite typically forms in plants that contain anaerobic digesters with upstream biological phosphorous removal. Struvite precipitation occurs when the release of orthophosphate and ammonia from cell hydrolysis during anaerobic digestion reacts with magnesium ions at pH conditions conducive for struvite formation (pH of 7.5 to 10). Struvite accumulation tends to occur at locations where pressure is low and CO_2 is released from solution thus increasing the pH. Unwanted struvite fouling has traditionally been solved by manual cleaning, dilution, dosing an iron salt to precipitate the phosphorous or using an anti-scaling agent to lower the pH.

The following chemical equation dictates struvite formation (Snoeyink, et al).

$$MgNH_4PO_4{*}6H_2O \leftrightarrow Mg_2^{+} + NH_4^{+} + PO_4^{3-} + 6H_2O \qquad pK_{SO} = 12.6$$

Under these conditions, the activities $\{Mg^{2+}\}\{NH_4^{+}\}\{PO_4^{3-}\}$ can increase above the solubility product or solubility equilibrium, defined at K_{SO}, causing struvite precipitation. The common places for struvite accumulation are locations where pressure is low and CO_2 is released from the solution thus increasing the pH (Snoeyink, et al).

For every kilogram of phosphorus recovered, 7.9 kilograms of dry struvite is produced. Typically, magnesium concentration in the wastewater or in the anaerobic digester is at a lower molar ratio than the phosphorous so magnesium is generally the limiting reagent for unintended struvite formation. Therefore, the addition of a magnesium salt is required and a common feature of most controlled struvite precipitation and removal processes.

AIRPREX PROCESS

The AirPrex process, was developed and patented by Berliner Wasserbetriebe (Germany) in collaboration with the Berlin Institute of Technology and is marketed in North America by CNP Corporation. In this process, struvite is crystallized directly from the biosolids stream out of an anaerobic digester, rather than from centrate, as is the case with more developed struvite crystallization processes such as the Ostara Pearl process. A general process flow diagram for the AirPrex process is provided in **Figure 1**. **Figure 2** provides a more detailed process diagram of the AirPrex reactor.

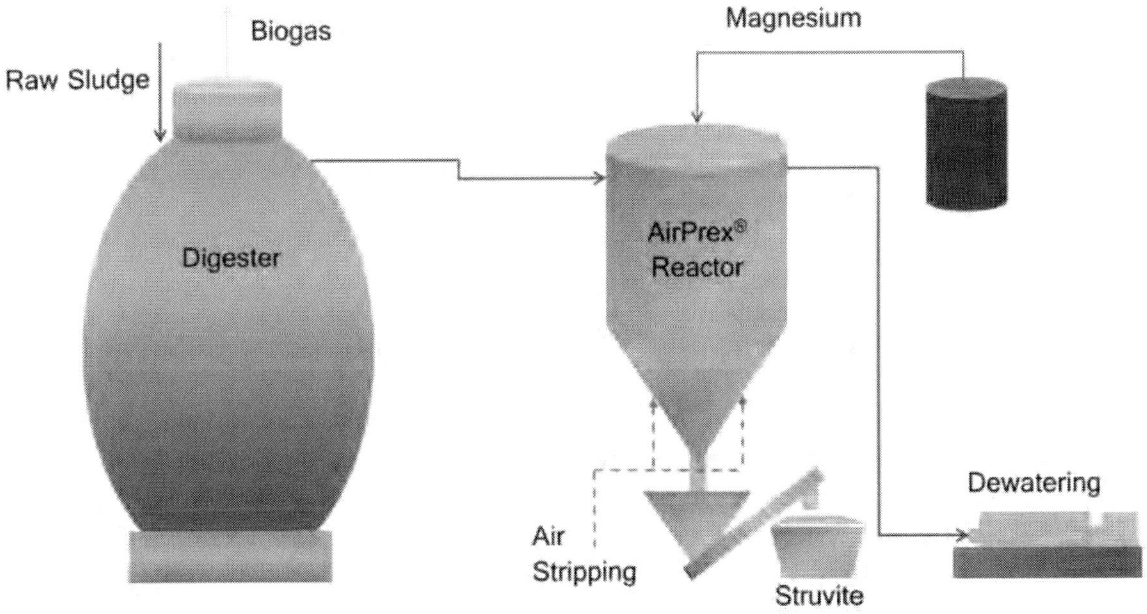

Figure 1: Typical AirPrex Process Flow Diagram (Courtesy of CNP)

AIRPREX PILOTING AT SDWWTP

The AirPrex piloting was conducted in April 2016 and the AirPrex treated biosolids was dewatered using a pilot dewatering centrifuge. The reactor was equipped with aerators that strip out carbon dioxide to increase the pH to between 7.9 and 8.2. The aeration also provides circulation of the struvite crystals inside the reactor (**Figure 2**), which grow until they reach a sedimentation point and settle to the bottom of the cone shaped reactor. Magnesium chloride was also dosed to the reactor as a 30% liquid solution and the dosing was set to be proportional to the orthophosphate concentrations and biosolids flow. For the pilot operation, the magnesium chloride dosing rate was set at 1.8 liters per cubic meter of digested biosolids (or 1.8 gallons per 1,000 gallons of digested biosolids).

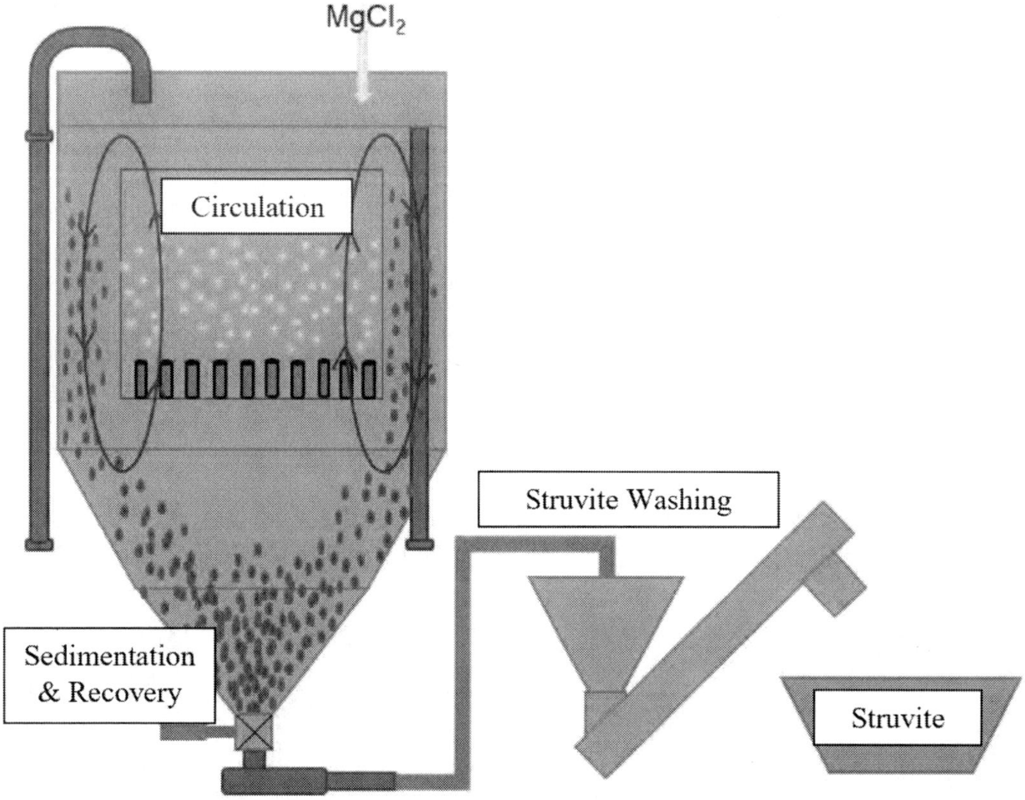

Figure 2: Detailed AirPrex Process Diagram (Courtesy of CNP)

During the pilot period, the AirPrex unit operated continuously with a digested biosolids flow that ranged from 30 to 45 liters per min (L/min) or 8 to 12 gallons per minute (gpm) and the treated sludge was stored in a mobile frac tank equipped with mechanical mixers that provided a buffer and storage between the AirPrex system and pilot centrifuge. The frac tank was filled continuously during operation but since the dewatering pilot throughput was up to four to ten times the flowrate of the AirPrex pilot, the pilot dewatering unit needed about 2 to 4 hours of operation time to empty the frac tank. Photos of the pilot unit reactor and frac tank are provided in **Figure 3.**

The feed to the dewatering pilot was set up to allow testing of both the AirPrex and non AirPrex treated digested biosolids in the same day, which allowed for consecutive testing to be conducted to determine impact of the technology on the dewaterability of the digested biosolids. The objective of the demonstration was to verify the performance of the technology in terms of:

- Percent of orthophosphate (PO_4-P) removal from the digested biosolids,
- Change in dry cake solids against the baseline,
- Change in polymer consumption compared to the baseline, and

- Ability to generate MAP (Struvite) that can be recovered.

Specified dewatering requirements were to achieve greater than 20% TS with greater than 95% solids recovery at an active polymer dose of 25 lb/DT (12.5 g/kg) or less (MWH, January 2016).

Figure 3: Picture of AirPrex Pilot and Frac Tank

The AirPrex reactor was first fed digested biosolids from Digester 9 on April 4, 2016 and the dewatering centrifuge first started processing AirPrex treated biosolids on April 5, 2016. Digester 9 was chosen since it was being operated as a pilot digester which was receiving mechanically thickened sludge for four months prior to the start of testing. The first week of dewatering operation was performed to optimize the dewatering centrifuge for the AirPrex treated biosolids. After a few days of operation, however, it was noted that the feed solids to the centrifuge from Digester 9 were decreasing rapidly. It was found that a flush valve was left open for several days that allowed water to fill Digester 9, diluting the digester. Based on the trend shown in **Figure 4**, it appears that dilution started at the end of March 2016, reducing the concentration in Digester 9 from 3.3 to 3.5% TS down to 2% TS. Because of the dilution, the feed to the AirPrex reactor was switched from Digester 9 to Digester 10 on April 12, 2016. Digester 10 was acting as a secondary digester that was receiving only mechanically thickened digested sludge from Digester 9 and the concentration in the digester was steady at approximately 2.5% TS.

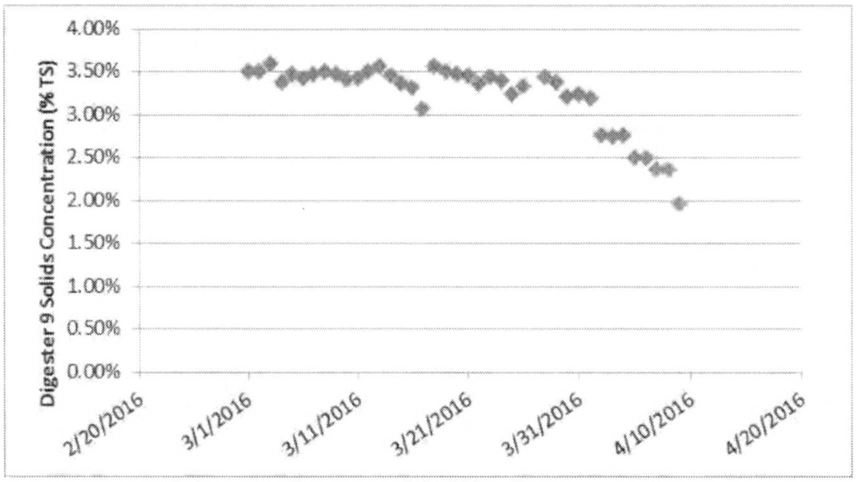

Figure 4: Digester 9 Solids Concentration

AIRPREX PERFORMANCE DATA

Daily sampling was conducted from the AirPrex reactors to monitor the orthophosphate and ammonia concentrations, as well as the pH of the inflow feeding the AirPrex unit and the outflow, which fed the frac tank (and was the feed for AirPrex treated biosolids dewatering testing). Centrate samples from the dewatering unit were also collected for a period to monitor both ammonia and orthophosphate concentrations, as well as pH. **Figure 5** summarizes the data monitored during the AirPrex testing and a vertical red line was added to the figures to depict the point where the feed to the AirPrex unit was switched from Digester 9 to Digester 10. The data did not show a large difference in the orthophosphate concentrations when switched from Digester 9 to Digester 10. However, the ammonia concentration in Digester 9 decreased over time as the digester was diluted. When switching to Digester 10, both the ammonia concentration and the pH of the inflow biosolids increased. When operating with feed biosolids from Digester 9, the pH averaged 7.6 but when switched to Digester 10, the pH averaged 7.8. The results from **Figure 5** also show that the centrate orthophosphate and ammonia concentrations, as well as pH, measured similar values and followed the trends of the AirPrex outflow biosolids.

After the first day of AirPrex operation, the process was optimized and approximately 91% orthophosphate reduction was maintained throughout the pilot reactor (ranging from 89 to 93%), reducing the concentration from approximately 200 milligrams per liter (mg/L) down to less than 20 mg/L. In addition, the AirPrex process also provided a 14 to 16% reduction in ammonia concentration. **Table 1** provides the average orthophosphate and ammonia concentrations in and out of the AirPrex pilot reactor throughout the test. The data broken down for the period when testing was conducted from Digester 9 and Digester 10 is also presented in **Table 1**.

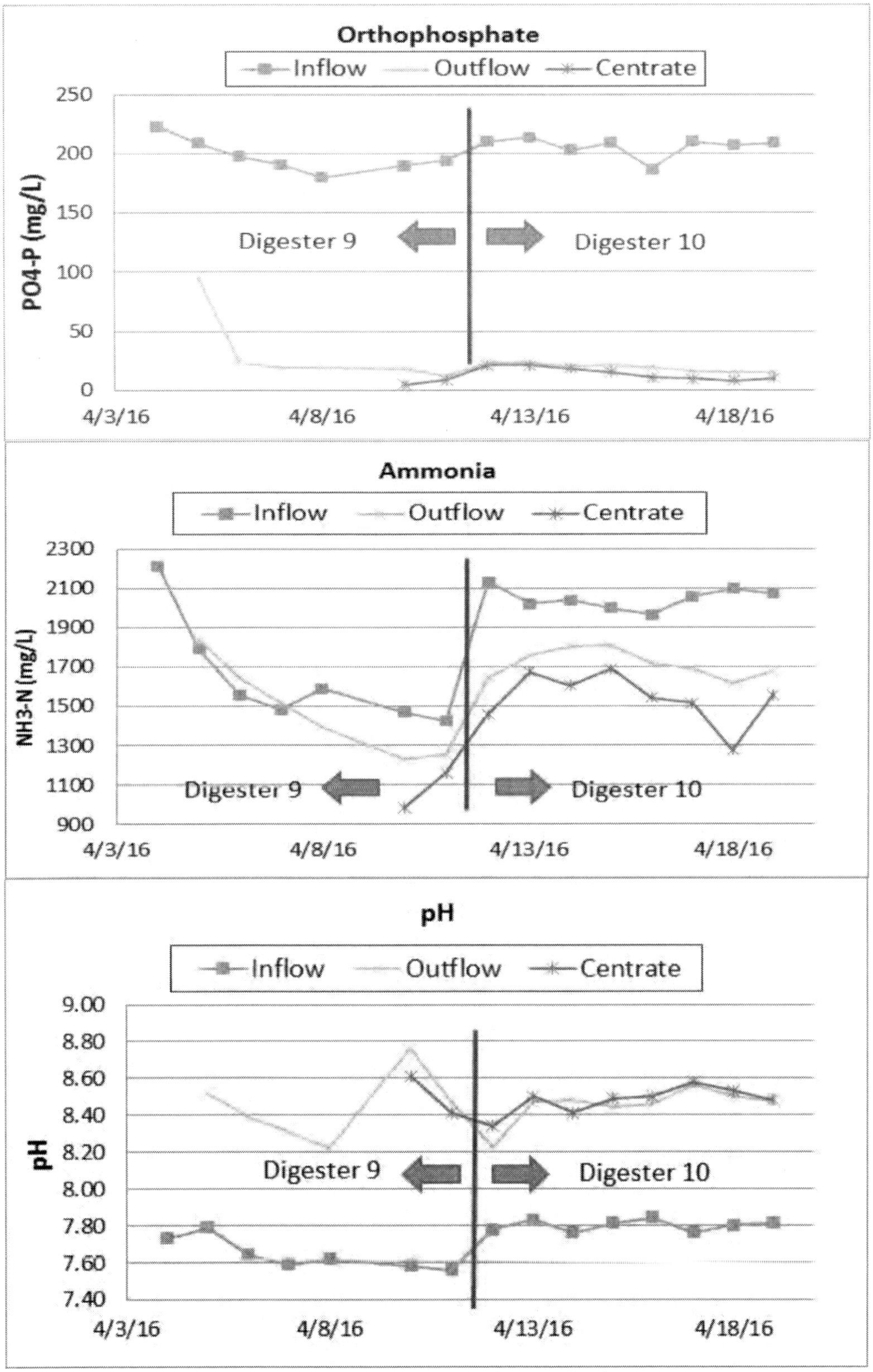

Figure 5: Summary of AirPrex Monitoring

Table 1: Summary of AirPrex Performance Data

	Inflow (mg/L)	Outflow (mg/L)	Reduction
Orthophosphate			
Average for entire test	202.3	18.7	90.7%
Digester 9	197.9	18.2	90.8%
Digester 10	206.1	19.0	90.8%
Ammonia			
Average for entire test	1,862	1,597	14.2%
Digester 9	1,648	1,408	14.6%
Digester 10	2,049	1,716	16.3%

From the reactor, the struvite collected in the cone was sent to a grit washer but limitations in the pilot set up showed that the struvite recovered was in a crude form and present with sludge and other debris. Pictures of the struvite product are provided in **Figure 6**. Furthermore, the grit washer was located in the test trailer beside the reactor and the struvite had a tendency to accumulate in the interconnecting hose. The grit washer was also oversized for the application so some seed sand had to be added to provide enough pressure for the auger's pressure sensor to be activated. For a full-scale application, the grit washer would be placed directly underneath the reactor and more time would be available to seed the system with struvite.

Figure 6: Pictures of Struvite Product

DEWATERING RESULTS

The majority of the dewatering testing with Airprex was conducted with the same dry polymer used in the plants existing dewatering centrifuges and previously used in the dewatering tests (Polydyne C-3283) at a targeted concentration of 0.4% TS. However, some additional testing was also conducted with Polydyne emulsion polymer, based on recommendations from onsite jar testing.

The first week of dewatering operation, using sludge from Digester 9, was performed to optimize the machine for the AirPrex treated biosolids but because of the dilution issue, the results could not be directly compared to the previous pilot dewatering testing conducted without AirPrex pretreatment. Optimization included adjusting pool depths, bowl speeds, differential speed, and polymer dosing. The initial dewatering results were, however, promising and results of greater than 21% TS were being achieved compared to 18% TS prior to starting the AirPrex Pilot.

Because of the dilution in Digester 9, it was decided to conduct sequential testing with and without AirPrex treated digested biosolids to gauge the impact of the technology on dewaterability of SDWWTP's digested biosolids. Since the concentration in Digester 9 was diluted, the feed to the AirPrex unit and to the pilot centrifuge was switched from Digester 9 to Digester 10 on April 12, 2016 and the remainder of the AirPrex and Centrifuge pilot dewatering tests were conducted using the digested biosolids from Digester 10.

Dry Polymer Curve Testing

In order to better gauge the impact that the AirPrex treatment had on the dewaterability of the digested biosolids from Digester 10, several dry polymer curve tests were conducted at 45, 60 and 80 gpm (170, 227, and 303 L/min) with and without AirPrex pretreatment. The flowrate, feed concentrations, dry polymer concentration and dates for these tests is summarized in **Table 2**. For all of the tests the bowl speed was maintained at 93% which is equal to 3,100 revolutions per minute (RPM).

Table 2: Summary of Dry Polymer Curve Tests from Digester 10

Flow (gpm)	Test with AirPrex			Test without AirPrex		
	Feed (% TS)	Polymer (% TS)	Date	Feed (% TS)	Polymer (% TS)	Date
45	2.40% TS	0.44% TS	4/16/16	2.45% TS	0.46% TS	4/15/16
60	2.48% TS	0.43% TS	4/13/16	2.40% TS	0.43% TS	4/13/16
80	2.54% TS	0.43% TS	4/13/16	2.42% TS	0.42% TS	4/14/16

All of the dry polymer curves conducted with and without AirPrex on the digested biosolids from Digester 10 is summarized in **Figure 7**. The data for all three polymer curves show that with AirPrex pretreatment, the sludge dewatering was improved, allowing close to a 3% point increase in the dry solids content at the same polymer dosing rate. The data also shows that the driest cake achievable without AirPrex pretreatment can be achieved with AirPrex pretreatment at a lower polymer dose. When comparing the trends with and without AirPrex, however, it can be seen that without AirPrex, the optimal polymer dose, meaning the point where additional polymer dose does not improve cake solids, is lower without AirPrex than with AirPrex. **Table 3** summarizes the optimal point based on the polymer curves with and without AirPrex treatment.

Table 3: Optimal Settings for Flow based on the Polymer Curves from Digester 10

Flow (gpm)	AirPrex				No AirPrex			
	Cake (% TS)	**Solids Recovery (%)**	**Polymer (lb/DT)**	**Differential Speed (RPM)**	**Cake (% TS)**	**Solids Recovery (%)**	**Polymer (lb/DT)**	**Differential Speed (RPM)**
45	23.0%	93.9%	31.1	1.4	20.4%	95.0%	29.8	2
60	21.5%	93.8%	32.0	2.2	18.7%	96.9%	29.7	3.3
80	20.8%	95.6%	30.9	3.1	18.4%	96.5%	28.7	4.1

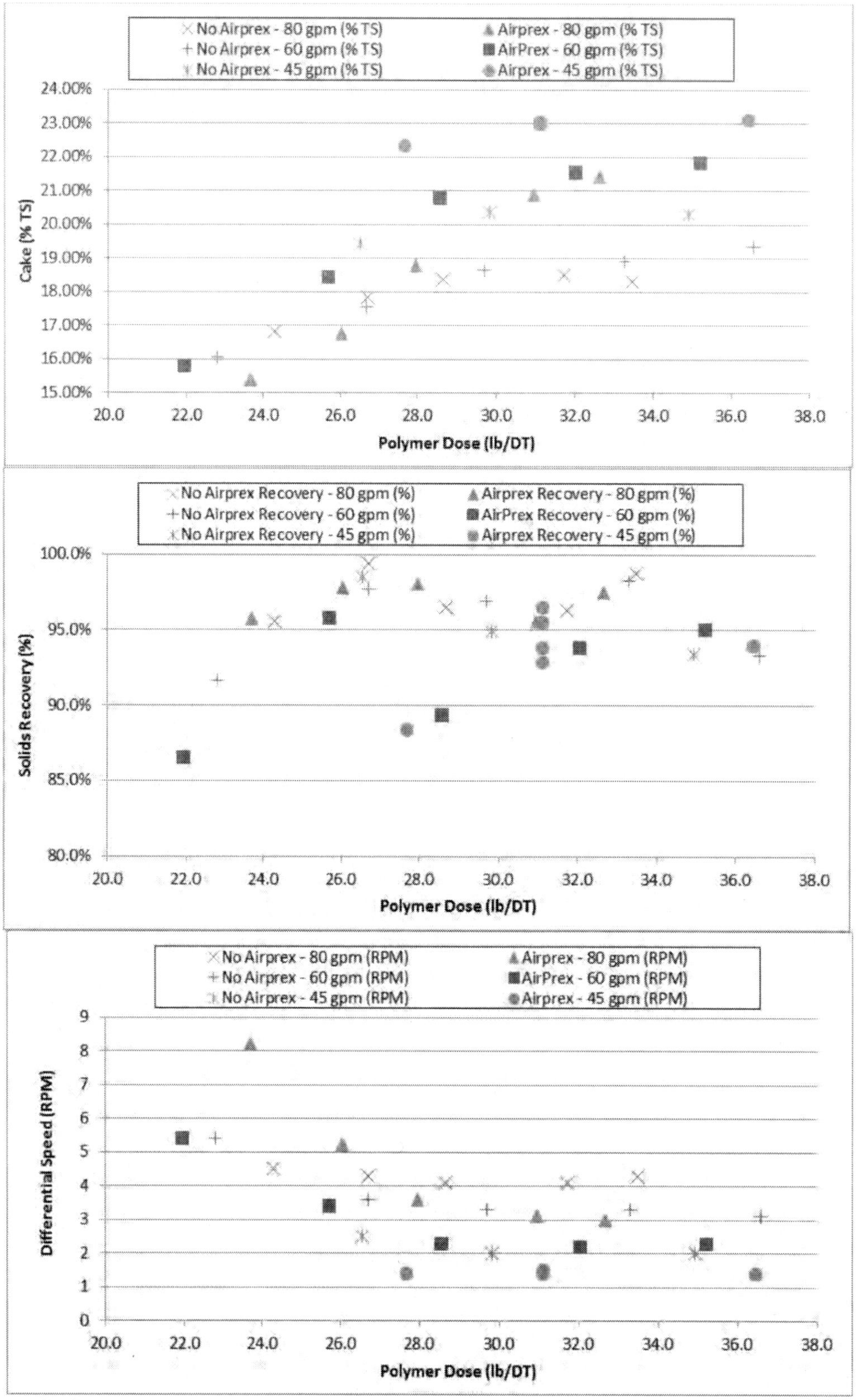

Figure 7: Summary of All Digester 10 Dry Polymer Curves with and without AirPrex

The data shows that with lower throughputs, lower differential speeds can be maintained and dryer cake can be produced with AirPrex treatment compared to operation without AirPrex treatment. The results also indicated that with AirPrex treatment, up to 21 to 23% TS cake could be achieved with recoveries at or above 93% when a 30 to 32 lb/DT (15 to 16 g/kg) active polymer dosage rate was used. This is compared to operation without AirPrex that shows 18 to 20% TS cake can be achieved with recoveries at or above 95% when a 29 to 30 lb/DT (14.5 to 15 g/kg) active polymer dosage rate was used.

Extended Operation Testing

In order to test the stability of dewatering operation for the AirPrex treated digested biosolids, several extended operation tests were conducted at 45, 60 and 80 gpm (170, 227, and 303 L/min) using digested biosolids from Digester 10. For all of the tests, the bowl speed was maintained at 93% (3,100 RPM) and polymer concentrations were approximately 0.4% TS. With the tests conducted at 60 and 80 gpm (227 and 303 L/min), the testing started with an extended run on non AirPrex treated biosolids based on the optimal settings and then switched to AirPrex treated biosolids to see the impact over the course of the run.

On April 14, 2016, an extended operation test was conducted at 60 gpm (227 L/min), targeting the optimal setting from **Table 3**, and the results are summarized in **Figure 8**. When running on the non AirPrex treated biosolids with a differential speed of 3.3 RPM and a polymer dose of 27.6 lb/DT (13.8 g/kg) active, the centrifuge dewatered the biosolids from approximately 2.4% TS to 18.9% TS with a solids recovery of 97.7%. When the centrifuge feed was switched to AirPrex treated biosolids at the same polymer dose (27.6 lb/DT or 13.8 g/kg active) but with a lower differential speed of 2.5 RPM, the dewatered cake solids increased to 20.1% TS (starting with a 2.5% TS feed) and recoveries were maintained at 98.2%. Further increasing the polymer dose to 31 lb/DT (15.5 g/kg) active improved the dewatered cake solids to 21.3% TS and recoveries were near 100%. Because of the high recovery, the differential was further reduced to 2.3 RPM but this did not show improvement in dewatering and recoveries were still at 98.5%. During the run on April 14, 2016, the non AirPrex treated feed averaged 2.4% TS and the AirPrex treated feed averaged 2.5% TS. The polymer concentration averaged 0.42% TS. The results, based on the optimal setting, matched the results indicated by the previously conducted polymer curve shown in **Figure 7**.

On April 15, 2016, an extended operation test was conducted at 80 gpm (303 L/min) and the results are summarized in **Figure 9**. The test again targeted the optimized setting outlined in **Table 3** but the testing was further expanded to gauge the impacts on the differential speed and polymer dose on dewaterability. The test started with non AirPrex treated biosolids using the optimized differential speed settings (3.1 RPM) and polymer dose settings (31.4 lb/DT or 15.7 g/kg) for the AirPrex treated biosolids. At these settings, up to 19.3% TS cake was produced but recovery was only 83.6%. When the differential was increased to 4.2 RPM and the polymer dose was reduced to 28.4 lb/DT (14.2 g/kg), the dewatering was reduced to 18.3% TS but recoveries improved to 93%. When switching to the AirPrex treated biosolids without adjusting any of the centrifuge parameters, the dewatered cake solids improved to 19.9% TS and recoveries improved to 98.1%. Lowering the differential to 3.2 RPM with the same polymer dose increased the dewatered cake solids to 20.7% TS and recoveries were still high at 97.1%.

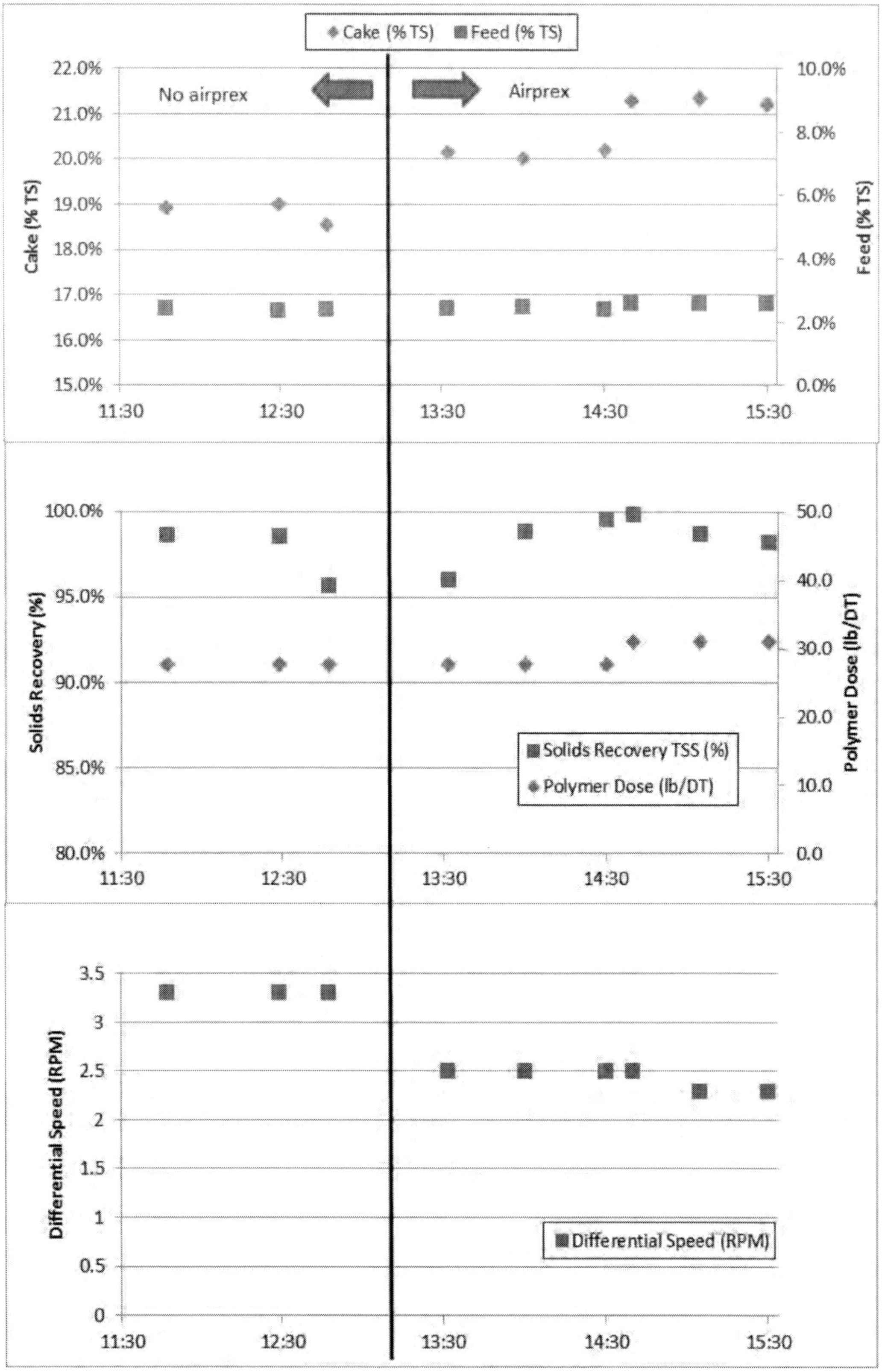

Figure 8: 60 gpm Operation – Dry Polymer (4/14/16)

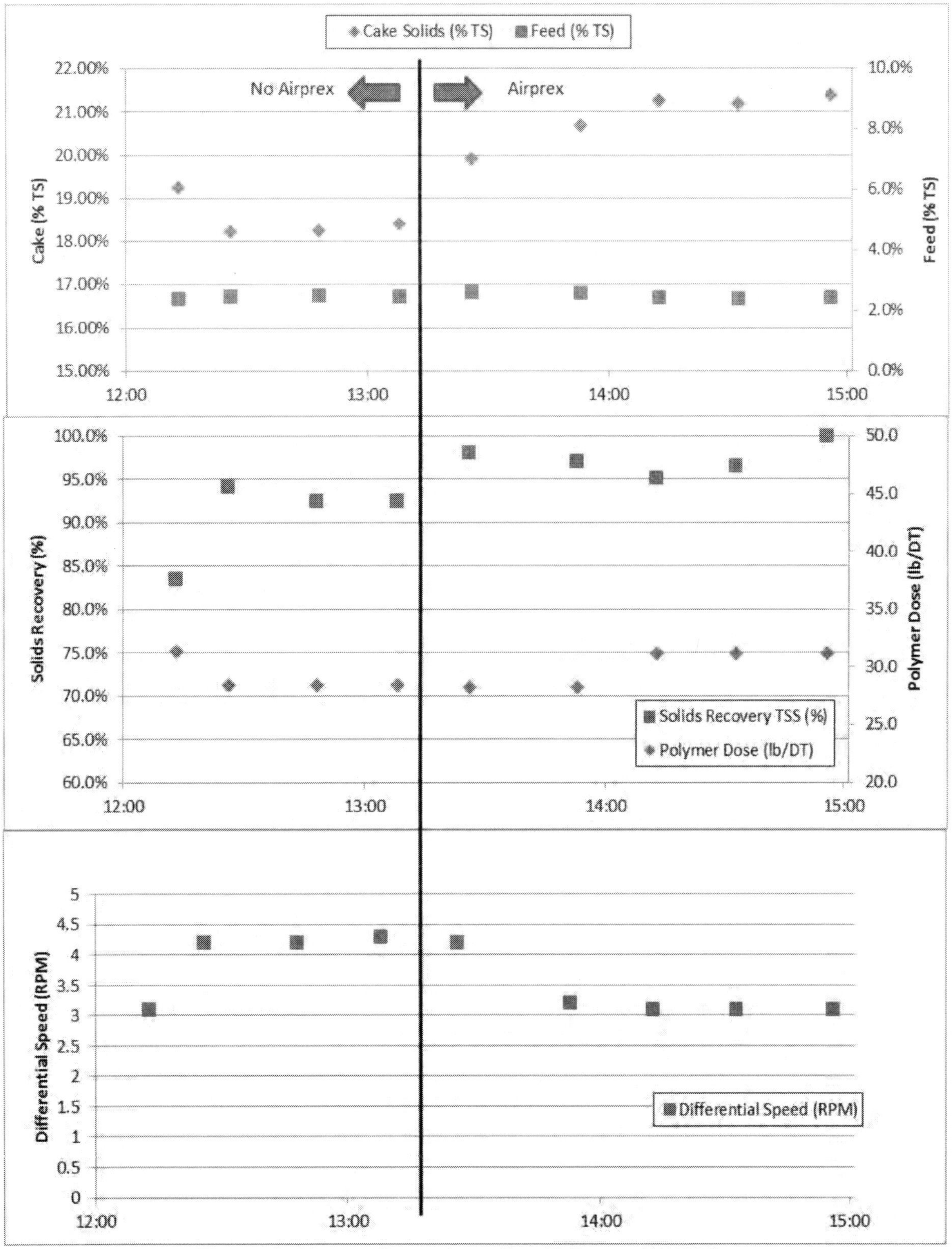

Figure 9: 80 gpm Operation – Dry Polymer (4/15/16)

Finally, when adjusting the differential speed and polymer to the optimized AirPrex setting outlined in **Table 3** (3.1 RPM and 31.2 lb/DT [15.6 g/kg] active), the dewatered cake solids improved to 21.3% TS with recoveries of 97.3% showing slightly better results than indicated by the previously conducted polymer curve (**Figure 7**). Throughout the testing, the feed biosolids concentration (both with and without AirPrex treatment) averaged 2.5% TS and the polymer concentration averaged 0.43% TS.

On April 16, 2016 an extended operation test was conducted at 45 gpm (170 L/min) right after conducting the 45 gpm (170 L/min) polymer dose test with AirPrex treated sludge. The testing showed that when operating at a 1.5 RPM differential speed and an active polymer dose of 31.1 lb/DT (15.6 g/kg) that dewatering up to 23.0% TS with recoveries at 95% were possible. The marginal increase in differential speed allowed the recoveries to improve to 95% compared to operation at 1.4 RPM differential speed. The feed solids concentration during this run averaged 2.4% TS and the polymer concentration averaged 0.44% TS.

EMULSION POLYMER CURVE TESTING

To provide a comparison to the dry polymer, testing was also conducted using emulsion polymer to determine the impacts on dewatering biosolids from Digester 10. Based on jar testing conducted by Polydyne, three different high molecular weight cationic emulsion polymers were tested on April 18, 2016 at 60 gpm or 227 L/min (C-6272, C-06292, and C-6275). Both the C-6272 and C-6275 polymers had average activities of 52.5% (ranging from 49 to 56%) and the C-6292 polymer had average activities of 51.5% (ranging from 48 to 55%). The results did not show a significant difference in performance between the three polymers but the C-6272 polymer shows slightly better performance in terms of solid recovery when compared to the other polymers. With AirPrex pretreatment and emulsion polymer, over 26% TS was possible but polymer dosage requirements were over 50 lb/DT (25 g/kg) active which is more than 40% more polymer than what was required when using dry polymer.

The data collected on April 18, 2016 allowed a C-6272 emulsion polymer curve to be generated at 60 gpm (227 L/min) and additional testing was conducted on April 19 and April 20, 2016 to generate a C-6272 emulsion polymer curve at 45 gpm (170 L/min). The feed concentration of the AirPrex treated sludge averaged 2.1 to 2.2% TS and the feed concentration of the non AirPrex treated sludge averaged 2.3 to 2.5% TS.

The data for both polymer curves show that with AirPrex pretreatment, the biosolids dewatering was improved allowing for a 2 to 4% point increase in the dry solids content at the same polymer dose. The data also shows that the best performance seen without AirPrex pretreatment can be achieved with AirPrex treatment at a lower polymer dose. The dewatered cake solids continued to increase at higher polymer dosages in the range tested and up to 26 to 27% TS was possible with recoveries near 95% with polymer dosages above 52 lb/DT (26 g/kg) active. This is compared to operation without AirPrex that shows 22 to 23% TS cake can be achieved but recoveries were below 93% when the polymer dosage rate is approximately 50 lb/DT (25 g/kg) active. All of the emulsion polymer curves conducted with and without AirPrex on the biosolids in Digester 10 is summarized in **Figure 10**.

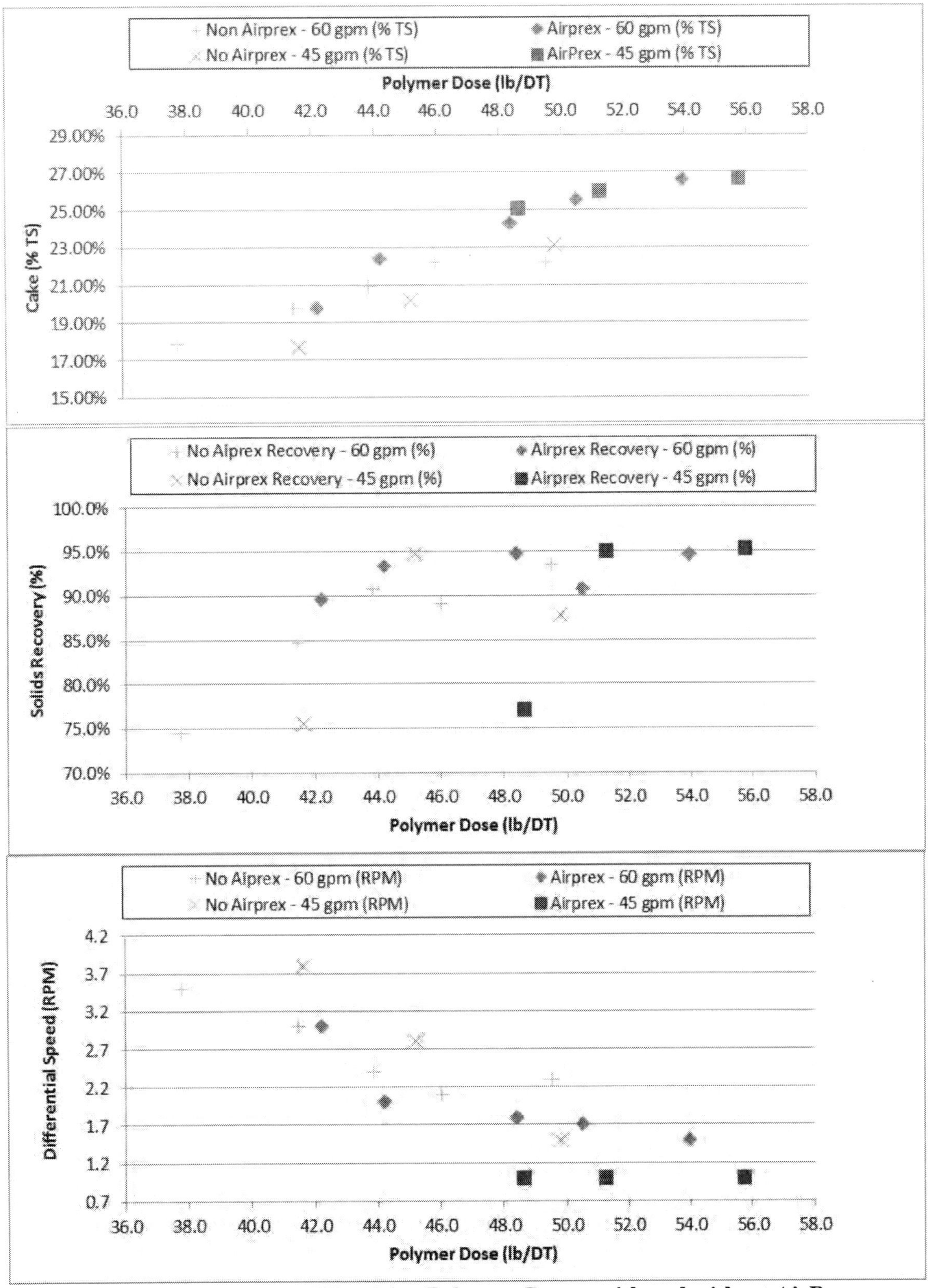

Figure 10: Summary of Emulsion Polymer Curves with and without AirPrex

CONCLUSIONS

The Results showed that the AirPrex process consistently achieved 91% removal of orthophosphate and 14 to 16% removal of ammonia (NH_3-N) from the digestion process and recovered struvite from the cone of the reactor. From a dewatering perspective, the AirPrex pretreatment provided a 2 to 4 points increase in the cake solids from the pilot centrifuge compared to operation without AirPrex pretreatment and similar active polymer dosages. With dry polymer, up to up to 22% cake solids was achieved with AirPrex pretreatment as compared to 19% without pretreatment using 25 to 35 lb/DT (12.5 to 17.5 g/kg) active polymer dosages. In the case of the emulsion polymer, up to 27% cake solids could be achieved with the AirPrex pretreatment as compared to the 23% without pretreatment but required polymer dosages in excess of 50 lb/DT (25 g/kg).

ACKNOWLEDGEMENTS

The team would like to thank all of the plant staff at SDWWTP, especially Leo Pou and Mike Garcia for the outstanding support during this pilot testing. The team would also like to thank the staff from Centrisys and CNP for their commitment to making the testing a success. We also thank the interns Andres McEwen, Yareliz Negron, and Kate Ireland who helped make this project possible by monitoring pilot operations, collecting samples and running numerous laboratory tests.

REFERENCES

Goss C., Stitt B., Moncholi M., Abu-Orf, M., Diaz I., (2017) "Piloting to Establish Performance Conditions at Miami's South District WWTP" *Proceedings of the WEF Residuals and Biosolids Management Conference 2017*, Water Environment Federation, Alexandria, VA

Kopp, J., Yoshida, H., Forstner, G. (2016) "Impact of Hydrolysis and Bio-P Removal Processes on Biosolids Dewaterability and Polymer Consumption in the Dewatering Process" *Proceedings of the WEF Technical Exhibition and Conference 2016*, Water Environment Federation, Alexandria, VA

S-884 - SDWWTP SLUDGE THICKENING & DEWATERING BLDG CD 1.06, 1.08, Specification Section 46 71 36 – CENTRIFUGES, MWH, January 20, 2016.

Snoeyink, V. and Jenkins, D. (1980) *Water Chemistry*. John Wiley and Sons, Inc., New York, NY.

Construction and Startup of the Rocky Mountain Region's First Anammox Sidestream Centrate Treatment System

Tom Dingemen[1], Matt Gough[2], Gina Rust[2], Adam Parmenter[2]

[1]City of Greeley WWT Division Manager, Greeley, Colorado
[2]HDR Engineering, Denver, Colorado

ABSTRACT
The City of Greeley recently installed a World Water Works (WWW) DEMON® sidestream centrate treatment system to reduce the ammonia concentration returning to the influent of the biological nutrient removal system. Construction of the system was completed in the fall of 2015 with startup in October. The system has now been operational for more than a year and while the removal rate is getting close to the specified requirement of 75 percent, there have been numerous issues which have slowed startup and implementation. This paper describes the challenges encountered and the lessons learned.

Key Words: Deammonification, anammox, centrate treatment, sidestream treatment, granules, startup.

INTRODUCTION
The City of Greeley owns and operates a secondary treatment wastewater treatment facility. Master planning in 2012 identified sidestream centrate treatment as a way to reduce recycle ammonia load on the secondary treatment system and help the facility to meet upcoming total inorganic nitrogen (TIN) requirements without construction of additional bioreactor volume. HDR worked with the City of Greeley to select, design, construct, and assist in startup and operation of the Rocky Mountain Region's first anammox treatment system. The City's experience in the technology selection, design and construction, and startup and operations of the facility are presented in the following sections. Additionally, a summary is provided of the startup operations to date along with recommendations for future facilities.

PROJECT DRIVERS
The City of Greeley, Colorado is located in Weld County approximately 60 miles northeast of Denver. The Greeley Wastewater Pollution Control Facility (WPCF) has a current average daily flow are approximately 8 MGD and permitted hydraulic capacity of 14.7 MGD. The treated effluent is discharged to the Cache la Poudre River. The current discharge permit does not include a limit on nitrate or TIN. However, a TIN limit of 15 mg/L will be included in the next permit cycle within the next year along with an effluent total phosphorus limit of 1 mg/L on an annual median basis. The WPCF liquid treatment system utilizes preliminary screening and grit removal, primary sedimentation, anoxic/aerobic biological nutrient removal bioreactor system, secondary clarification, and ultra violet light disinfection. Solids handling is accomplished through anaerobic digestion, centrifuge dewatering, and deammonification treatment of the centrate. Figure 1 provides a process treatment schematic of the system.

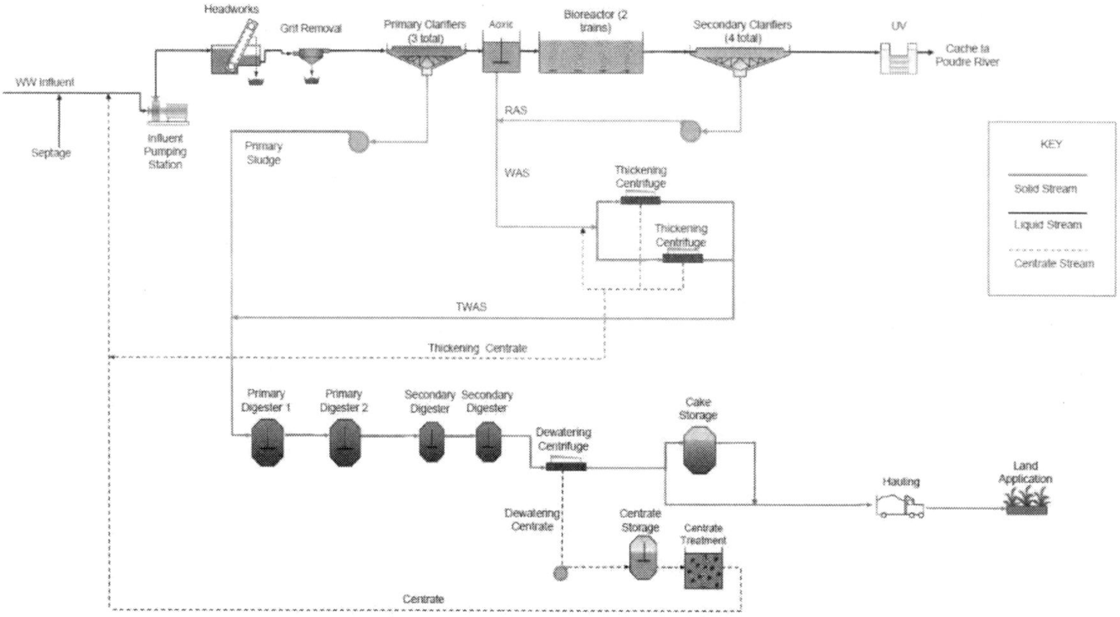

Figure 1 – Greeley WPCF Process Treatment Schematic

HDR completed a Solids Handling Master Plan for the City in 2012 which determined that under future conditions, without additional bioreactor volume, the secondary process would require a high MLSS and a short HRT to meet the upcoming TIN limit of 15 mg/L. Sidestream treatment of centrate was identified as a way to significantly reduce the recycle ammonia load from the anaerobic digesters and maximize available bioreactor volume. Additionally, the Colorado Department of Health and Environment (CDPHE) was providing up to $1 million in grant funding for projects which improved a facility's ability to remove nutrients. HDR and the City developed an application and the City was awarded the maximum amount toward the deammonification project.

SELECTION OF CENTRATE TREATMENT EQUIPMENT

Several centrate treatment alternatives were evaluated including the following innovative and emerging technologies- conventional nitrification/denitrification, sidestream nitrification/denitrification, and sidestream anammox based processes (deammonification). The key functions of the centrate treatment system were to reduce recycled ammonia load on the bioreactors, provide a small footprint, and minimize carbon/energy/alkalinity required to treat the ammonia.

In addition to a new centrate treatment system, centrate storage was also provided to enhance operational flexibility and help with equalizing energy demands by avoiding energy use spikes that result from increased oxygen demands resulting from sidestream during peak day raw wastewater loadings. A good example of the deleterious effect that a highly nutrient-rich sidestream can have on mainstream process is illustrated in Figure 2 which depicts the effect of the typical return of dewatering sidestream containing high ammonia concentration during peak load time of the day. The blue dashed line represents a non-equalized sidestream return and the amount of ammonia that must be oxidized during midday. The red solid line represents a sidestream that is equalized. One can see that equalization provides operators the ability run a more stable and efficient operation by controlling the ammonia loading to the mainstream process.

Based on a projected 2032 maximum month dewatering centrate flow of 230 gpm (generated from 5 days a week for 8 hours a day dewatering schedule), the resulting centrate volume generated is approximately 552,000 gallons per week. The required volume for dewatering recycle storage and equalization depends on the desired operation strategy. For full equalization, the storage tank has to hold enough recycle to return the average daily recycle flow each day between the two dewatering days, whichever are furthest apart (i.e. Friday of one week and Monday of the following week). To accommodate a three day weekend and assuming a consistent centrate treatment throughput of 55 gpm, a storage and equalization volume with approximately 310,000 gallons is required. To provide for redundancy, the tank was split into two 155,000 gallon sections. This provided a simple, cost effective solution to centrate storage.

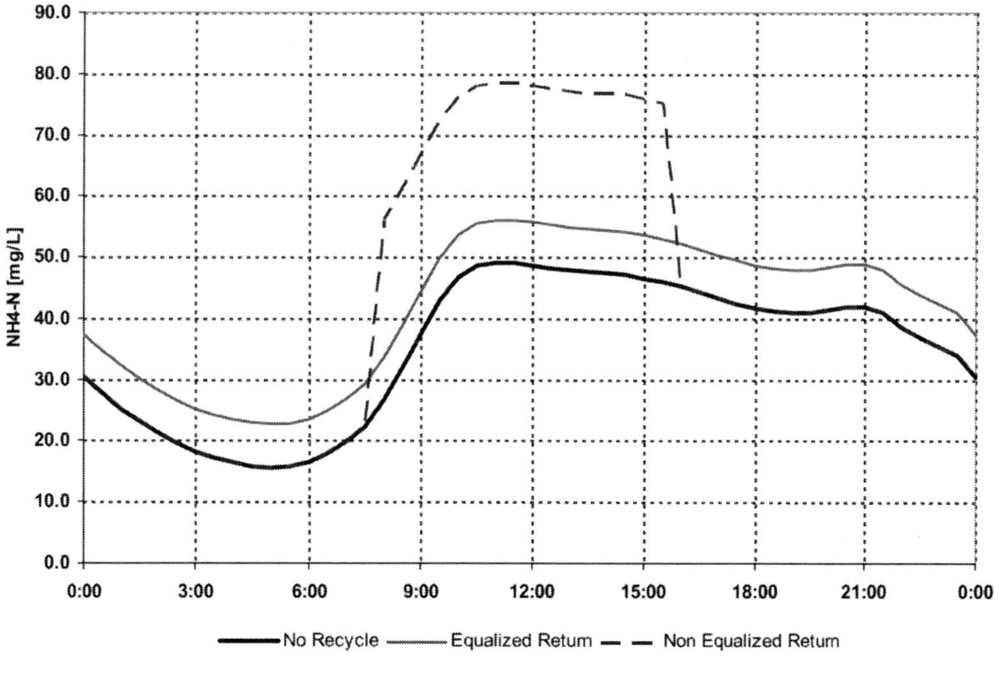

Figure 2 - Impact of Ammonia NH4 Load during Raw Influent Peak Loading

Alternative Selected – Deammonification

Deammonification has emerged as a cost effective, efficient, and reliable option to treat high strength ammonia wastewater treatment streams, in particular to treat in-plant return side streams from dewatering of anaerobic digested sludge. The technology has been applied to more than 50 full scale facilities worldwide. These installations operate well and require a modest level of operator attention. However, there are currently only a handful of full scale deammonification plants in operation in the U.S.

At the time of process selection, there were only two technologies available in the U.S. - the World Water Works DEMON® system and the Anita™ Mox system provided by Kruger. The WWW system is based on a sequencing batch reactor (SBR) process with the use of a hydrocylone to maintain the anammox bacteria population. SBR technology has been in use for more than 20 years. Due to the simple operating system, SBRs were typically used for small systems. The SBR system operates on a fill/react/settle/decant cycle. Due to the fill and draw nature of the process, the pumps, mixers, and blowers must cycle on and off multiple times per hour. The WWW system includes instrumentation to measure DO, pH, and temperature. These sensors are similar to existing units used by the City and are fairly simple to maintain.

The Kruger system is based on using moving bed biofilm reactor (MBBR) media and a fixed film process. The system is self-regulating and a continuous flow through design. MBBR technology has been in use for secondary treatment for more than 10 years with multiple installations in Colorado. The system is easy to operate and robust. Due to the flow through design of the system, the number of blower and pump start/stops per hour is kept to a minimum. The Kruger system typically includes instrumentation for DO, pH, ammonia, and nitrate. The ammonia and nitrate instruments require additional maintenance.

Based on information gathered from reference sites, due to the flow through nature of the Kruger MBBR system, it tends to require less attention to operate. However, the use of nitrate and ammonia instrumentation requires additional maintenance on the sensors. The batch nature of the WWW system results in more operator attention but the instrumentation is less complex. Overall, both systems are similar with regard to required operation and maintenance time. Both systems can provide up to 75 percent ammonia and 65 percent TIN removal efficiency depending on the influent alkalinity concentration and temperature of the reactor.

Centrate Treatment Process Selection

Due to the differences between the two deammonification treatment systems, selection of a technology was required prior to finalizing the construction documents. Table 1 provides the specified required minimum performance criteria, Table 2 summarizes the treatment system influent characteristics, and Table 3 provides a summary of the anammox centrate treatment system design criteria.

Table 1 – Minimum Required Performance Criteria

Parameter	Units	Value
Ammonia Removal Efficiency	%	≥ 75%
Total Inorganic Nitrogen Removal Efficiency	%	≥ 65%

Table 2 – Deammonification System Operational Criteria

Parameter	Units	Value
Flow, average	cfs	0.11
Flow, peak	cfs	0.13
Centrate TSS, average	mg/L	350
Centrate TSS, maximum	mg/L	950
Centrate BOD, average	mg/L	105
Centrate BOD, maximum	mg/L	200
Centrate COD, average	mg/L	600
Centrate COD, maximum	mg/L	1,000
Ammonia, average	mg/L	1,300
Ammonia, maximum	mg/L	1,500
Alkalinity, average	mg/L	4,000
Temperature, average	degC	28
pH, average	s.u.	8

Table 3 – Deammonification Treatment System Design Criteria

Design Element	Criteria
Ammonia Load	353.8 kg/d
Reactor Volume	121,176 gal
SBR Cycle Time	8 hours
Operation	24/7
Centrate Equalization	(2) 155,000 gal tanks
Struvite Prevention	PVDF piping

Proposals were received on December 4, 2013 from Kruger, Inc. and World Water Work, Inc. in response to the Request for Proposals dated November 8, 2013. The systems were evaluated based on both economic and non-economic criteria based on experience, operations and control, ease of use, quality, layout and footprint, energy efficiency, performance, cost and reference projects. Table 4 provides a qualitative summary of the vendor scoring to the evaluation criteria.

Table 4 – Deammonification System Evaluation

Equipment	Anita™ Mox (MBBR)	DEMON® (SBR)
Price	$994,000	$636,360
Net Present Value 20 year	$1,600,000	$1,100,000
Responsiveness	Met requirements	Met requirements
Key Personnel	No staff local to Colorado	Startup staff local to Colorado
Ease of Operations	Less operator attention	More operator attention
Ease of Maintenance	More instrumentation	Less instrumentation
Construction	Met requirements	Met requirements
Layout/Footprint	733 square feet	900 square feet
Energy Efficiency	Higher total required power	Lower total required power
Negotiated Performance Removal Guarantee Ammonia TIN	 80% 70%	 90% 80%
References	Positive references	Positive references
Schedule- Delivery of Equipment Start-up Duration	 10-16 weeks 3 months	 8-12 weeks 3 months
Instrumentation	DO, pH, NH3, NO3	DO, pH, Temp

Based on the available information in the proposals, the 20 year net present value, references, and a technical review of each system, the World Water Works DEMON system was selected for the Greeley Solids Handling and Treatment Improvements Project.

PERMITTING

Due to the relatively new nature of the deammonification process, the technology required approval through the CDPHE Alternative Treatment Technology review process. WWW worked with the regulatory agency and submitted the required information including sizing, equipment, and operational criteria. Due to a lack of adequate performance data, CDPHE required the installation of 60 horsepower blowers compared to the 40 horsepower blowers recommended by WWW to meet the anticipated oxygen demand. At the time, this requirement appeared to be minor with minimal impacts to the system. However, as noted in upcoming sections, the larger blowers have adversely impacted startup of the system.

DESIGN COMPLETION

Once the equipment system was selected, final design of the centrate storage and treatment process was completed. A number of features were included to reduce maintenance and improve operability including the following:

- Redundant centrate storage tanks
- Mixing in centrate storage

- Use of PVDF piping to minimize struvite formation
- Cleanouts at all bends
- Sample ports
- Chemical addition
- Provisions for bypass of system

CONSTRUCTION, STARTUP AND OPERATIONS

Construction of the new sidestream treatment system began July 2014 and was completed in December 2015. Figure 3 shows the design model of the centrate building compared to the completed structure.

Figure 3 –Centrate Treatment System

After equipment testing and validation, the centrate treatment system was seeded with anammox bacteria in October 2015 (12 totes equating to approximately 3,168 gallons of seed sludge). After seeding, WWW was onsite to assist in system startup.

LESSONS LEARNED

Since this technology is relatively new, there were a number of issues which had to be resolved during construction and startup. Below is a summary of key items-

- Regulatory Review Changed Design Resulting in Oversize at Startup-

  - Because this system was the first of its kind in Colorado, the regulators didn't fully understand the design and required upsizing of components including the blowers.
  - Blowers were upsized from 40 hp (375 scfm) to 60 hp (635 scfm).
  - This has resulted in significant challenges during startup as the blowers cannot be throttled down causing over-aeration.
  - The City has used a blowoff valve to help modulate resulting in poor control.

Figure 4 – Oversized Aeration Blowers

- Centrate Tank Heating-
  - A construction delay resulted in sludge seeding occurring in the fall to early winter. Successful startup of an anammox system requires a minimum reactor temperature of 25 degC.
  - The cooler temperatures and a reduced flow of centrate resulted in a treatment tank temperature below 20 degC.
  - After two weeks and minimal activity of the anammox bacteria, a temporary heater system was installed.
  - By the beginning of November, the anammox bacteria began to show activity as shown by ammonia and total inorganic nitrogen removal. A permanent heating system would have provided the City greater flexibility and is being evaluated as a retrofit.

Figure 5 – Tank Heating System

- Slow Startup-
  - By January 2016, the system still was not providing the required removal rates resulting in WWW deciding to provide the City with an additional 6 totes of anammox bacteria (1,584 gallons). The purpose was to increase the anammox bacteria concentration and speed up acclimation. The additional bacteria were harvested from a facility in Europe and delivered in March 2016.
  - Additionally, research into the activity of the Greeley DEMON® reactor contents showed the system performance could be improved through micro-nutrients addition. The Greeley system was designed to allow for temporary metering of chemicals. Greeley began adding micro-nutrients to the system in the spring of 2016. Unfortunately, nutrient addition has not resulted in improved performance of the system to date.

Figure 6 – Additional Totes of Anammox Bacteria

- Centrate Quality-
  - Due to issues with the dewatering centrifuges, centrate quality varied significantly during the startup of the anammox system. This resulted in the need to install a centrate diversion valve to maintain centrate quality to the treatment system.
  - Additionally, excess polymer due to dewatering issues, resulted in the anammox bacteria being coating and floating to the surface which caused a loss of bacteria through the overflow decanter. The polymer dose has since been resolved and is no longer causing issues, however, the bacteria sludge in the reactor still has a tendency to float quicker than other facilities. To resolve this issue, the City currently decants all the reactor contents through the hydrocyclone and is currently not using the decanter. However, at buildout flow rates, the hydrocyclones and feed pumps are not adequately sized to process the design flow of 60 gallons per minute. As such, these units will be replaced with larger units.

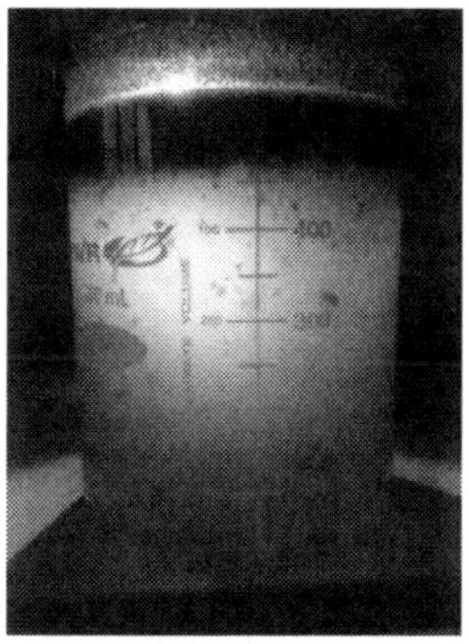

Figure 7 – Floating Sludge from Reactor

- Equipment Issues-
  - The key to maintaining the anammox bacteria is proper setup of the hydrocyclone. Due to poor installation, the cyclone was not retaining the bacteria resulting in the loss of the system within two months of seeding.
  - Based on information provided by WWW, the hydrocyclone may only provide 60 to 70 percent capture of the anammox bacteria. All bacteria not captured are wasted out of the system. The capture rate of the hydrocyclone has been proven successful for other fully operational systems but has been one of the causes for the slow startup of the Greeley system due to the quantity of anammox wasted in the overflow. It is unknown if this is an issue unique to the City of Greeley or a larger issue associated with the design of the hydrocyclones.
  - An additional six totes of anammox bacteria were delivered to the site in the summer of 2016.
  - To better capture the anammox bacteria, an additional screen was recommended for installation on the hydrocyclone overflow. The screen has an opening of 30 micro-meters and is able to capture up to 98 percent of the anammox bacteria.

Figure 8 – Additional Screen System on Hydrocyclone

- Centrate Storage Tank Draining and Cleaning-
  - As noted previously, early in the startup process excess solids from the dewatering centrifuge were sent to the centrate storage tanks. The tanks are designed to be drained by gravity or through the use of the treatment tank feed pumps. Due to the solids concentration, neither draining mechanisms were able to empty the contents resulting in significant operational attention.

    - To resolve this issue, the centrate feed pumps, which are currently end suction centrifugal, will be replaced with a pump that has a greater solids handling capability and lower net pump suction head required. This will also allow for transfer of contents from one equalization tank to the other providing greater operational flexibility.
- Daily Inspections and Operator Attention-
    - While the DEMON® system provides significant automation through the controls package, it is necessary for operations staff to perform daily visual inspections of the reactor contents, hydrocyclone overflow and underflow. Additionally, lab testing of the operating parameters should be taken to track the progress of the system. These include ammonia, nitrate, nitrite, DO, temperature, alkalinity, and pH.
    - During startup it is even more critical to monitor the performance of the hydrocyclone. This can be done by taking a sample of the overflow and underflow then running the sample through a centrifuge to separate out the anammox granules from the lighter floc. The samples can then be compared over time to determine the concentration of anammox granules in the system. Figure 9 provides progressive photos of the centrifuged samples from the Greeley system from October 2016 through December 2016. As shown, the anammox population has gradually increased over the three month period. Figure 10 provides photos of the reactor contents after each stage of the operational cycle.

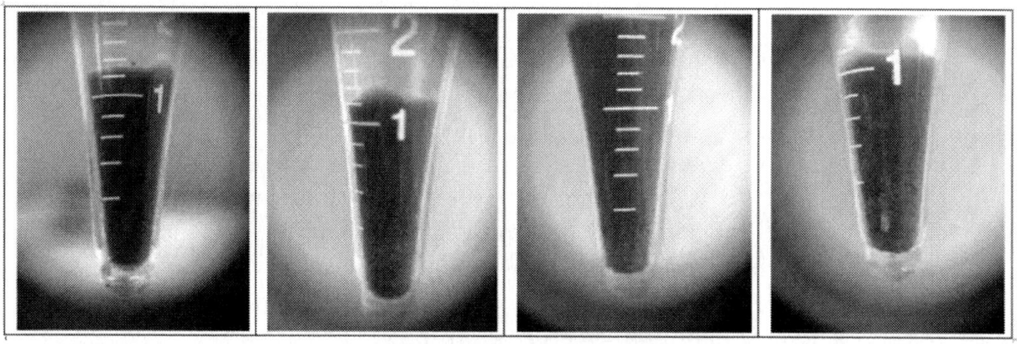

Figure 9 – Progression of Greeley Reactor Contents Centrifuged from October 2016 through December 2016

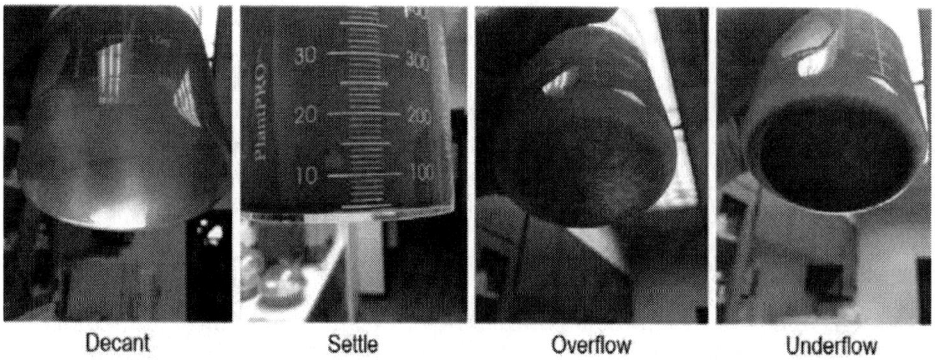

Figure 10 – Photos of Reactor Contents from Each Phase of the Process

SUMMARY AND CONCLUSIONS

Sidestream treatment of ammonia rich dewatering centrate provides a great opportunity to reduce the load on a secondary treatment system at significantly reduced energy, carbon, and alkalinity compared to nitrification and denitrification. The Greeley system was designed to provide a minimum ammonia removal of 75 percent. To date the system is averaging approximately 70 percent. However, visual inspection of the system shows the anammox population to be a lower concentration compared to other facilities. Recent DNA testing has confirmed this. More than a year after startup and the system is still having difficulty meeting the designed operating parameters. However, a majority of the operational problems could have been resolved during startup by implementing these key recommendations:

- Plan for additional methods for capture of the anammox bacteria. Based on our experience, utilizing hydrocyclones singularly results in the loss of excessive amounts of anammox bacteria adversely impacting operation of the system. To resolve this issue, a secondary screen on the hydrocyclone may be required.
- Ensure the regulatory review agency understands the process and is on board with the design parameters early in the project. Discuss the importance of sizing equipment correctly for the startup conditions. Consider blower style with a greater turndown to meet the varying load.
- Once the system is operational, the heat load from anaerobically digested sludge centrate is typically warm enough to keep the system operational above 28 degC. However, during startup, the load may be only 5 percent of normal operation. If startup occurs during the winter, supplementary heat will be needed in colder climates.
- Plan for multiple chemical injection points into the reactor. Depending on the site constraints and the makeup of the centrate, supplemental micronutrients and/or alkalinity may be needed to optimize the system.
- While both WWW and Kruger indicated the system could handle suspended solids as high as 950 mg/L, the greater the carbon concentration in the system the higher the bacteria population that will compete with the anammox. Ensure the centrate discharge pipe has a diverter valve to direct flow during centrifuge startup and shutdown to the plant drain instead of the centrate treatment system.
- Plan for and provide flexibility with regard to draining tanks with high solids. Gravity drain lines will plug quickly and pumps should be designed to handle the solids.
- Provide hands on training for the operating staff that includes an understanding of the daily inspection requirements. Each parameter has a “normal” operating range and the system should be adjusted if values are trending in the wrong direction. Understanding of these values comes from operating the system, looking at the sludge, and testing the reactor contents.

Roadmap for setting up an optimal treatment train configuration for nutrient recovery from (digested) residuals

Vaneeckhaute C[1,4], Belia E[2], Meers E[3], Tack FMG[3], Vanrolleghem PA[4]
[1]BioEngine – Research team on green process engineering and biorefineries, Chemical Engineering Department, Université Laval, 1065 ave. de la Médecine, Québec, QC, Canada, G1V 0A6, celine.vaneeckhaute@gch.ulaval.ca
[2]Primodal Inc., 145 Rue Aberdeen, Québec, QC, Canada, G1R 2C9
[3]EcoChem, Laboratory of Applied Ecochemistry, Ghent University, Coupure Links 653, Ghent, Belgium
[4]model*EAU*, Université Laval, 1065 ave. de la Médecine, Québec, QC, Canada, G1V 0A6

ABSTRACT

This paper advances the development of a generic roadmap for setting up strategies for nutrient recovery from digested waste. First, guidelines were presented to set up an optimal bio-based fertilization strategy as function of local/region-specific fertilizer legislations. Next, instructions were provided to evaluate the feasibility of bio-based fertilizer production as a function of input waste characteristics. Finally, an algorithm was developed aiming at the configuration and optimization of nutrient recovery treatment trains. Important input waste characteristics to measure, and essential factors for monitoring and control were identified. As such, this paper should provide a useful and comprehensive guide for wastewater and residuals processing utilities aiming to implement nutrient recovery strategies. This, in turn, may stimulate and hasten the global transition from wastewater treatment plants to water resource recovery facilities.

KEYWORDS

anaerobic digestion, mathematical modelling, process design, renewable fertilizers, residuals management, resource recovery, waste valorization

INTRODUCTION

A recent review of nutrient recovery technologies for digestate treatment (Vaneeckhaute et al., 2017) has highlighted the potential for nitrogen (N) recovery as ammonium sulfate (AmS) fertilizer (Bonmati and Flotats, 2003), as well as for phosphorus (P) recovery as struvite, $MgNH_4PO_4{:}6H_2O$ (and/or calcium (Ca) / magnesium (Mg)-P precipitates) (Rahman et al., 2014). Through field trials (Vaneeckhaute et al., 2014) and greenhouse experiments (Vaneeckhaute et al., 2016), the agronomic potential of these fertilizers has been demonstrated. The economic and ecological benefits of bio-based fertilization scenarios using these products have also been confirmed (Vaneeckhaute et al., 2013). Nevertheless, implementation of nutrient recovery strategies is still limited due to regulatory constraints, (operational) problems associated with the (variability of the) quality and quantity of the fertilizers produced, as well as the persisting uncertainty of fertilizer sales and inconsistency of marketing prices in regions where commercialization is possible (Seymour, 2009). Indeed, the fact that water resource recovery facilities

(WRRFs) must aim at delivering high-value products that can partially replace those produced by other means (e.g., chemical mineral nitrogen (N) production through the Haber-Bosch process) leads to a paradigm shift in specifications of the outputs of the facility: no longer treated wastewater and biosolids, but products that have to compete with what is already on the market. Finding the appropriate combination and sequence of technologies to treat a particular waste flow and the optimal operating conditions for the overall treatment train are key concerns (Carey et al., 2016; Guest, 2015).

This paper aims to provide a roadmap for **setting up optimal nutrient recovery treatment train configurations as function of waste stream (digestate) properties, as well as local/regional fertilizer legislations and markets.** The scope of the study includes anaerobic digestion and the selected best available technologies (and resulting bio-based products) applied at full-scale for the recovery of nutrients as marketable fertilizer commodities (Vaneeckhaute et al., 2017), i.e. P precipitation/crystallization (struvite, Ca/Mg-P precipitates), NH_3 stripping/absorption (AmS fertilizer), and acidic air scrubbing (AmS fertilizer). The selection of these technologies (and products) was made on the basis of the stage of implementation, the technical performance, and financial aspects, next to the fertilizer marketing potential. Besides the information acquired in Vaneeckhaute (2015), additional data were obtained through contact with technology providers. Hence, the roadmap is (partially) based on full-scale operational experience. As such, this paper may provide a helpful tool for residuals and wastewater processing utilities considering the implementation of anaerobic digestion and subsequent recovery and recycling of nutrients as marketable agricultural commodities.

METHODS

First, a generic nutrient recovery model (NRM) library was developed and validated at steady state (Vaneeckhaute, 2015), including mathematical process models for the best nutrient recovery systems currently available (as identified in Vaneeckhaute et al., 2017). Dynamic physicochemical three-phase process models for precipitation/crystallization (target product: struvite), stripping and acidic air scrubbing (target product: ammonium sulfate) as key unit processes were developed. In addition, a compatible biological-physicochemical anaerobic digester model was built. The latter includes sulfurgenesis, biological nitrogen, phosphorus, potassium, and sulfur transformations, interactions with organics, among other relevant processes, such as precipitation, ion pairing, and liquid-gas transfer. The models are based on detailed solution speciation and reaction kinetics. This NRM library is a compilation of the large body of knowledge on nutrient recovery processes that is currently available from research studies and operational experience. In addition to the development of a generic physicochemical modelling framework, a critical and challenging step when combining (stiff, i.e. numerical unstable) biological and physicochemical differential equations is their numerical solution (Lizarralde et al., 2015). Hence, a generic methodology to allow for accurate chemical speciation at minimal computational effort was also proposed.

Second, the NRM library was subjected to a global sensitivity analysis (GSA) so as to find the main factors that impact a wide range of 25 performance indicators of a nutrient recovery treatment train, including methane and biogas production, digestate composition and pH, ammonium sulfate recovery, struvite production, product physicochemical purity (e.g., co-precipitation), particle size and density, air and chemical requirements (acid, base), scaling potential, etc. Model simulation outputs were found very sensitive to input waste stream characteristics through their direct effect on pH, which is adequately determined by means of the integrated chemical speciation calculation. Moreover, new data needs appeared, especially related to the physicochemical kinetic precipitation/dissolution and liquid-gas/gas-liquid transfer coefficients.

Based on the obtained generalizable insights, an optimal treatment train consisting of several unit processes was developed and assessed in terms of nutrient recovery performance and operating costs. Using model-based optimization the operating conditions and certain design variables have been optimized in view of maximized revenue at minimal environmental impact. It was revealed that the optimal configuration and associated operating conditions also depend on local legislations and fertilizer markets, next to the high influence of the input waste flow characteristics.

Based on all acquired knowledge, a generic roadmap for setting up nutrient recovery strategies as function of fertilizer markets, legislations and input flow characteristics was developed. It includes:

1. An overview of bio-based fertilization recommendations as function of fertilizer legislations;
2. Guidelines for determining the feasibility of nutrient recovery based on operational experience;
3. An algorithm for configuration and optimization of nutrient recovery treatment trains as function of input waste characterization and fertilizer markets.

RESULTS AND DISCUSSION

The global sensitivity analyses (GSA) provided important generic insights in the interactions between process inputs and outputs for the three different waste streams studied so far, i.e. sewage sludge, manure and a co-digestion mixture. For all unit processes, variations related to the input waste composition resulted in major effects on the output variation through their direct effect on the operational pH and ionic strength. Major findings involve, among others, the impact of chloride (Cl) inhibition on ammonia removal in the stripping unit (so, MgO or $Mg(OH)_2$ is to be preferred over $MgCl_2{:}6H_2O$ when applying preceding phosphorus (P) precipitation), the impact of calcium (Ca), iron (Fe) and aluminium (Al) inhibition on P recovery in the precipitation unit (suggesting the inclusion of a Ca/Fe/Al precipitate separator after the anaerobic digester), and the interaction between Fe/Al, sulfur (S) and methane (CH_4) production in the anaerobic digester. By using $MgO/Mg(OH)_2$ in the struvite precipitation unit, pH is increased which is also beneficial for a subsequent ammonia stripper and thus reduces the need for base addition. Finally, if struvite is to be recovered, the implementation of the precipitation unit after digestion is also beneficial as the GSA showed that higher temperatures increase struvite purity.

Based on the results, it was possible to propose an optimal generic treatment train configuration for nutrient recovery aiming at the production of high-quality fertilizers (including struvite and AmS) at minimal cost (Figure 1).

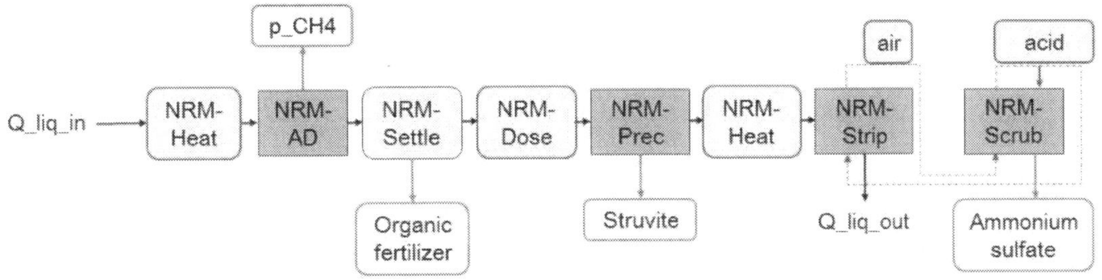

Figure 1 Proposed treatment train configuration targeting struvite and ammonium sulfate fertilizer; red = consumable (= cost); green = recovered resource (= revenue). AD = anaerobic digestion; Dose = chemical dosing; Heat = heat exchanger; Prec = precipitation/crystallization; p = partial pressure in the biogas; Q_liq = liquid flow rate; Scrub = scrubber; Strip = stripper.

Two important factors determining the optimal treatment train configuration for nutrient recovery are i) (local/regional) fertilizer legislations and markets, and ii) input characteristics of the waste flow (digestate in this case) to be treated (see above). A three-step roadmap for setting up nutrient recovery strategies as function of these determining factors is presented and discussed below.

Step 1: Set-up bio-based fertilization strategy as function of fertilizer legislations

If one wants to install a nutrient recovery treatment train, first contact should be sought with local/regional agronomic agencies and/or consultants in order to obtain insights in fertilizer related legislation and the corresponding market demand. If no local market would exist for recovered nutrient products, interest can be sought abroad. Depending on the targeted region, N or P can be the limiting factor for manure and digestate application as organic/organo-mineral base fertilizer. The latter determines for which fertilizers the market demand is the highest in the particular region. Overall, based on the analysis performed in this study, in P-poor regions, among the considered best available bio-based products to date, the agricultural demand for digestate (base fertilizer), recovered AmS and/or struvite, and Ca/Mg-P fertilizers is expected to be high. In P-saturated regions, among the considered recycled products, the most interesting fertilizers for agricultural purposes are the liquid fraction of digestate (as base fertilizer, whether or not mixed with raw digestate) and AmS.

Step 2: Evaluate feasibility of bio-based fertilizer production as function of input waste characterization

An important point to consider when designing nutrient recovery is the physicochemical characterization of the input waste stream to be treated. Obviously, first the macronutrients, especially N and P, of the waste flow have to be measured in order to check whether there is effectively an interest for N and P recovery. As such, technology providers confirmed that P recovery is only of interest if the P-load is higher than 80 kg d^{-1}, whereas N recovery using air stripping and scrubbing only becomes economically feasible at concentrations in the range of 400-500 mg N L^{-1}. Moreover, struvite production is only of interest if the waste flow has an N:P molar ratio above 1. The optimal N:P-ratio to maximize struvite recovery and purity would be higher than 6. Below that ratio, significant co-precipitation (e.g., with Ca-P) is to be expected. If, based on the above measurements, the recovery of N and P seems feasible, additional physicochemical analyses will have to be conducted in order to set up an optimal nutrient recovery treatment train configuration (see Step 3).

Step 3: Use of conceptual algorithm for treatment train configuration and optimization

A conceptual algorithm was developed on the basis of the findings in this research and contact with technology providers. It gives an overview of guidelines for configuration of nutrient recovery treatment trains, taking in account input waste characteristics and fertilizer market demands. The following digestate treatment train configurations depending on the feasibility scenario were identified:

1. *N and P recovery not feasible:* No action should be taken;
2. *P recovery not feasible, N recovery feasible:* The recommended treatment train configuration concerns NH_3 stripping and acidic air scrubbing (after anaerobic digestion), with optional pre- and post-treatments, depending on the nature of the waste material;
3. *P recovery feasible, N recovery feasible or not feasible + market for struvite available at acceptable transportation costs:* The recommended treatment train configuration concerns struvite precipitation, followed by AmS recovery (depending on the N content);
4. *P recovery feasible, N recovery feasible or not feasible + N:P-ratio and/or fertilizer markets not favourable for struvite recovery:*
    a. Digestate produced via thermophilic digestion: The recommended treatment train configuration concerns AmS recovery (depending on the N content), followed by Ca/Mg-P precipitation;
    b. Digestate produced via mesophilic or psychrophilic digestion: The recommended treatment train configuration concerns Ca/Mg-P precipitation, followed by AmS recovery (depending on the N content).

CONCLUSIONS

A generic three-step roadmap for setting up strategies for nutrient recovery from digestate was presented. It includes:

1. An overview of bio-based fertilization recommendations as function of fertilizer legislations;

2. Guidelines for determining the feasibility of nutrient recovery based on operational experience;

3. An algorithm for configuration and optimization of nutrient recovery treatment trains as function of input waste characterization and fertilizer markets.

As such, this paper provides useful guidance for wastewater and residuals processing utilities considering the implementation of nutrient recovery practices.

ACKNOWLEDGEMENTS

The first author has been funded by the Natural Science and Engineering Research Council of Canada (NSERC), the Fonds de Recherche de Québec sur la Nature et les Technologies (FRQNT) and Primodal Inc. Peter Vanrolleghem holds the Canada Research Chair in Water Quality Modelling.

REFERENCES

Bonmati, A., Flotats, X., 2003. Air stripping of ammonia from pig slurry: Characterisation and feasibility as a pre- or post-treatment to mesophilic anaerobic digestion. Waste Manage. 23(3), 261-272.

Carey, D.E., Yang, Y., McNamara, P.J., Mayer, B.K., 2016. Recovery of agricultural nutrients from biorefineries. Bioresource Technol. 215, 186-198.

Guest, J., 2015. Tailoring innovation to achieve financially viable energy and nutrient recovery from wastewater. Proceedings, Government Affairs Conference IWEA, University of Illinois, IL, USA.

Lizarralde, I., Fernández-Arévalo, T., Brouckaert, C., Vanrolleghem, P.A., Ikumi, D.S., Ekama, G.A., Ayesa, E., Grau, P., 2015. A new general methodology for incorporating physico-chemical transformations into multi-phase wastewater treatment process models. Water Res. 74, 239-256.

Rahman, M.M., Salleh, M.A.M., Rashid, U., Ahsan, A., Hossain, M.M., Ra, C.S., 2014. Production of slow release crystal fertilizer from wastewaters through struvite crystallization: A review. Arab. J. Chem. 7(1), 139-155.

Seymour, D., 2009. Can nutrient recovery be a financially sustainable development objective? Presentation, Annual Conference Pacific North West Clean Water Association, Boise, Idaho, USA, September 13-16 2009.

Vaneeckhaute, C., 2015. Nutrient recovery from bio-digestion waste: From field experimentation to model-based optimization. PhD Thesis, Ghent University, Belgium, and Université Laval, Québec, Canada.

Vaneeckhaute, C., Ghekiere, G., Michels, E., Vanrolleghem, P.A., Tack, F.M.G., Meers, E., 2014. Assessing nutrient use efficiency and environmental pressure of macro-nutrients in bio-based mineral fertilizers: A review of recent advances and best practices at field scale. Adv. Agron. 128, 137-180.

Vaneeckhaute, C., Janda, J., Vanrolleghem, P.A., Tack, F.M.G., Meers, E., 2016. Phosphorus use efficiency in bio-based fertilizers: Bio-availability and fractionation. Pedosphere 26(3), 310-325.

Vaneeckhaute, C., Lebuf, V., Michels, E., Belia, E., Vanrolleghem, P.A., Tack, F.M.G., Meers, E., 2017. Nutrient recovery from digestate: Systematic technology review and product classification. Waste Biomass Valor. 8(1), 21-40.

Vaneeckhaute, C., Meers, E., Michels, E., Buysse, J., Tack, F.M.G., 2013. Ecological and economic benefits of the application of biobased mineral fertilizers in modern agriculture. Biomass Bioenerg. 49, 239-248.

Utilization of Municipal Waste and Biosolids as Sources of Phosphorus for Land Restoration Projects

Principal Mdolo[1,2*] and Magnús H. Jóhannsson[2,3]

[1]Department of Natural Resources,Lilongwe University of Agriculture and Natural Resources,
[2]United Nations University Land Restoration Training Programme (UNU-LRT)
[3]Soil Conservation Service of Iceland (Landgræðsla ríkisins)
[*]Email: pmdolo@nrc.luanar.ac.mw

ABSTRACT

Waste disposal is one of the environmental challenges facing city authorities. However, waste contains considerable amounts of nutrients and organic matter required for plant growth. It is important to quantify the amount of waste generated and the quantity of nutrients that can be recovered. Further, the cost of applying the waste on land should be evaluated in order to make wise investment decisions. In Iceland, most of the waste generated is disposed without resource recovery. Therefore, the objectives of this study were to 1) quantify the amount waste generated in the Reykjavík capital area, 2) determine the quantity of phosphorus contained in waste generated 3) evaluate the cost of transporting the waste to nearby restoration sites and spreading it on the field. Operational data for 2015 was collected from wastewater and municipal waste operators in Reykjavík capital area. Phosphorus content in sewage sludge and municipal waste was calculated based on their dry matte and their typical phosphorus content reported in literature. The amount of waste generated, average distance to nearest restoration sites and recommended application rates formed the basis for calculating the cost of utilizing the waste in restoration. The cost of spreading the waste in the field was based on figures used by the Soil Conservation Service of Iceland. We found that phosphorus associated with wastewater treatment (effluent discharge and sludge) was 68%. Municipal waste accounted for 32% of the phosphorus generated. Phosphorus contained in all the waste was 400% higher than that used for land restoration. This phosphorus was, however, either landfilled or discharged into the ocean in waste. If this phosphorus was recovered, it would meet the phosphorus requirement for land restoration. The cost of utilizing the waste in restoration was estimated to be 140,000 ISK/ha (approximately $1,200/ha) with VAT included at 24%. This cost could be higher than using inorganic fertilizers. However, land application of waste is more economical than other waste treatment options. Furthermore, utilizing wastes reduces eutrophication in surface waters, and maybe most importantly, conserving the non-renewable phosphorus reserves. The study recommends that phosphorus recovery systems should be installed in wastewater treatment to improve phosphorus recovery. Legislation should be put in place to encourage waste utilization in land restoration.

Key word: Biosolids, compost, land restoration, municipal waste, phosphorus

INTRODUCTION

Waste is a man-made resource that is considered of no beneficial use to the owner or is not able to perform with respect to its desired function (Pongrácz and Pohjola 2004). This definition implies that it depends on the functionality of the material whether it is waste or not. Waste can be categorized as agricultural, industrial or municipal. Agricultural waste includes waste originating from agricultural activities such as crop and livestock production and agro-processing (Jana and De 2015). Municipal waste is highly organic (Hargreaves et al. 2008; Cesaro et al. 2015) due to the nature of activities that lead to its generation. Waste management is one of the major environmental challenges facing municipalities (Odlare et al. 2011). However, waste management practices vary with the economy and environmental regulations of a country (Desmidt et al. 2015). For example, in developed countries most households are connected to the main sewer and dispose of municipal waste through landfilling, incineration or composting (Gentil et al. 2009). On the other hand, most countries in Sub Saharan Africa do not have adequate waste management infrastructure (Couth and Trois 2010, 2011) such as wastewater treatment systems and solid waste management facilities. Malawi, for example still disposes of municipal solid waste in uncontrolled open dumps and discharge partially treated wastewater effluent into urban streams (Sajidu et al. 2007; Kumwenda et al. 2012; Wanda et al. 2015). In Iceland, partially treated wastewater is discharged in the ocean while municipal waste is landfilled. These practices cause environmental pollution such as eutrophication (Wang et al. 2014) and soil contamination. If utilized in land restoration projects, the waste can add organic matter to the soil which improves its biological and physicochemical characteristics such as improving aggregate stability and decreasing bulk density (Diacono and Montemurro 2010). Organic matter increases soil organic carbon and nitrogen (Diacono and Montemurro 2010) and can reduce plant parasitic nematodes (Treonis et al. 2010). Higher yields can be realized for crops grown on soils applied with waste (Antolín et al. 2005). However, crops can also accumulate toxic metals if the waste is applied excessively, given that the waste is rich in toxic metals (Antolín et al. 2005). Therefore, use of sewage sludge and municipal waste on non-food plants is a potential alternative.

If land application of waste is adopted in Iceland it will reduce the amount of waste landfilled, discharged into the ocean and consequently reduce the use of inorganic fertilizer for restoration. Since Icelandic soils are predominantly basic (Arnalds 2015), application of sewage sludge and municipal waste is expected to have minimal impact on heavy metal contamination. The pH value plays a major role in the sorption of heavy metals since there is low heavy metal uptake by plants at pH values above 7 (Smith 2009). At such pH values, the soluble form of heavy metals is converted into insoluble metal hydroxides or oxides which cannot be taken up by plants (Brady and Weil 2008). However, in Iceland, as vegetation establishes in restored areas, the pH values tend to decrease (Arnalds 2015) which could lead to sorption of heavy metals making them available to plants and hence grazing animals. Therefore, continued application of sewage sludge and municipal waste should be followed by periodic monitoring of the soils and plants (Singh and Agrawal 2008) for heavy metal toxicity.

Regulations on land application of waste on non-food plants are less restrictive than those regulating application on food crops. For example, land application of sewage sludge in the EU is regulated by the EU sewage sludge directive 86/278/EEC (Council of the European Communities 1986). Within this regulation, grazing animals can be allowed to graze in sewage sludge amended fields after three weeks of application (Council of the European Communities 1986). In Iceland, regulation 799/1999 regulates land application of sewage sludge (Ministry of the environment 1999). Within this regulation, untreated sewage can be applied to restoration fields if it is ploughed into the soil (Ministry of the Environment 1999). Further, the ministry allows a one-year period to elapse after application of sewage sludge before the land can be put to beneficial use (Ministry of the Environment 1999). Therefore, for land restoration purposes, application of sewage sludge and municipal waste can be ideal and the restored fields can be used for grazing in future. In Malawi, national standards regulating land application of sewage sludge and municipal waste do not exist. Consequently, untreated sewage sludge is often applied on home gardens and lawns. The public often manually excavate the sludge from wastewater treatment ponds and transport it to their homes or gardens for application. Environmental monitoring is usually not undertaken in areas where the sludge has been applied to assess the environmental risk to public health and the receiving soils. Physico-chemical analysis of sewage sludge is most often not performed by the operators.

Data on beneficial use of sewage sludge and other organic waste for Iceland are lacking, but small trials are under way (MH Jóhannsson, 12 May 2016, Soil Conservation Service of Iceland, Gunnarsholt, Iceland, personal communication) in which different types of waste are applied in restoration fields and vegetative growth monitored. In Sub Saharan Africa, especially Malawi, data on waste generation and land application of composts are also lacking. If these data were available, they would form a basis for lobbying governments to invest in composting facilities and land application of composted waste. If waste is collected and properly treated and applied on land, the demand for inorganic fertilizers would considerably decrease and hence conserve the declining non-renewable phosphorus resource.

The objectives of this research were:

i) To quantify the amount of waste generated in the Reykjavík capital area in 2015, including: wastewater, sewage sludge and municipal waste
ii) To determine the quantity of phosphorus in the amount of waste generated in the capital area for 2015
iii) To evaluate the cost of transporting the waste to nearby restoration sites and spreading on the field.

METHODOLOGY

Description of the Study Area

This study was conducted in the Reykjavík capital region in Iceland, which comprises seven municipalities. The population of the capital region in 2015 was 211,282 (Statistics Iceland 2015) which was 64% of the national population. The names and distribution of population in the seven municipalities making up the capital region are shown in Table 1.

Table 1. Population in the Reykjavík capital region (Statistics Iceland 2015)

Municipality	Population
Reykjavík	121,822
Kópavogur	33,205
Hafnarfjörður	27,875
Garðabær	14,453
Mosfellsbær	9,300
Seltjarnarnes	4,411

Kjósarhreppur	216
Total	**211,282**

Waste management regulations in Iceland are based on the EU regulations and each municipality develops their own waste management plans. The Environmental Agency of Iceland (Umhverfisstofnun) is responsible for developing the national waste management plan. Local authorities are supposed to develop their own waste management plans based on the national plan. In the Reykjavík capital region, solid waste management is coordinated and handled by SORPA, a company jointly owned by municipalities in the capital area. Wastewater treatment and sewage sludge management is operated by Reykjavík Energy.

Data Collection and Phosphorus Quantification

The study focused on waste which can readily be composted and utilized in land restoration. Hazardous waste, textiles, metals and plastics were not considered. Timber, wood, packaging waste, paper and cardboard were also not included in the analysis due to their low phosphorus content. Furthermore, these waste fractions are already exported to Sweden for recycling (Saltiola 2014). To quantify the amount of phosphorus available in the three major waste streams (wastewater, sewage sludge and municipal solid waste), operational data from wastewater and municipal solid waste operators in Reykjavík were used. Data collected was on waste generation rates for the year 2015. The operator sampled the effluent wastewater in March, June, September and December 2015 for monitoring nutrient content of the effluent. Phosphorus content in sewage sludge and municipal solid waste was calculated based on phosphorus content values (Table 2) and their moisture content (Table 3) reported in literature.

Table 2. Typical phosphorus content in different waste fractions.

Item	P content (g P/kg DS)	Source
Organic waste[a]	4	Kalmykova et al. 2012
Sewage sludge (% P_2O_5, TS)	2.5	Metcalf and Eddy 2003
Timber/wood	0.31	Sokka et al. 2004
Paper and cardboard	0.24	Sokka et al. 2004

TS= Total solids, DS= Dry solids, P= phosphorus, *a* (organic waste include kitchen waste, garden waste and other organic waste)

Table 3. Moisture content in different waste fractions (Kalmykova et al. 2012)

Waste fraction	Moisture content (% wet weight)
Paper and cardboard	0
Timber/wood	19
Kitchen waste	65
Garden waste	65
Other organic waste	65

Phosphorus content in wastewater was calculated from wastewater flow rate (*Q, L/sec*) and phosphorus concentration (mg/L) data acquired from the operator (Reykjavík Energy), using the following equation;

$$C = \frac{m}{Q}$$

Where C = concentration of phosphorus (mg/L), m = mass of phosphorus (mg), Q = wastewater flow rate (L/sec).

The amount of waste generated, average distance to nearest restoration sites and recommended application rates formed the basis for calculating the cost of utilizing the waste in restoration. The cost of transporting and spreading the waste on the field were acquired from the Soil Conservation Service of Iceland. Figure 1 shows map of the study area with possible restoration sites, within a radius of 30 km to 50 km from Reykjavík, with erosion scale of 3-5. Erosion severity is based on a scale of 0-5 with 0 representing no erosion and 5 indicating most severe erosion (Table 4) (Arnalds et al. 2001).

Table 4. Erosion severity scale, colour codes and suggested management actions (Source: Modified from Arnalds et al. 2001)

Erosion scale	Description	Colour code	Suggested management action
0	No erosion	Green	No suggestion
1	Little erosion	Green	No suggestion
2	Slight erosion	Green	Care needed when grazing
3	Considerable erosion	Yellow	Reduce and manage grazing
4	Severe erosion	Orange	Protected (no grazing)
5	Extremely severe erosion	Red	Protected (no grazing)

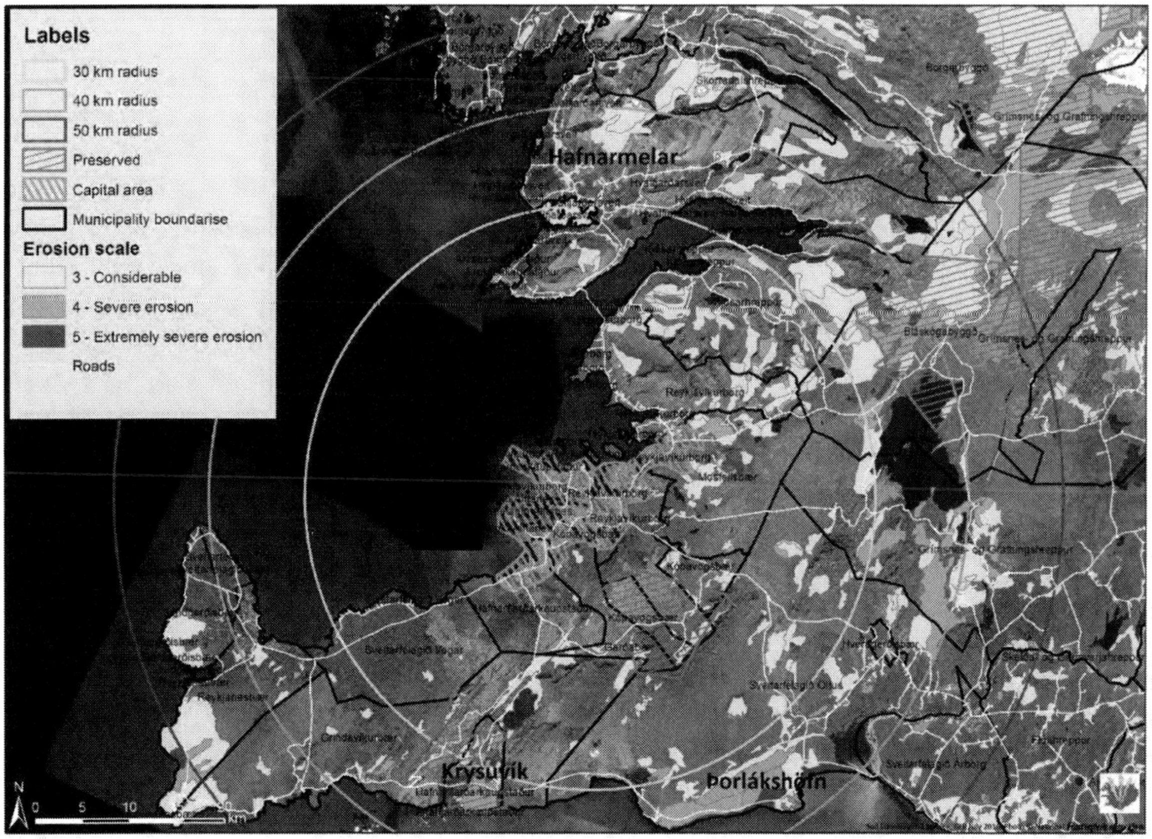

Figure 1. Map showing possible restoration sites within a radius of 30-50 km from Reykjavík with erosion scale of 3-5 (Source: Guðný H. Indriðadóttir, Soil Conservation Service of Iceland 2016).

RESULTS

Waste Generation Rates and Phosphorus Content in Sewage Sludge, Municipal Waste and Wastewater

Results for waste generation rates and associated phosphorus content in each waste fraction of sewage sludge, paper and cardboard, timber/wood, kitchen waste, garden waste, organic waste and wastewater are presented in Table 5. Phosphorus concentration in wastewater was based on analyses on samples collected from effluent four times in 2015. The average wastewater flow rate and phosphorus concentration was 2,601 L/sec and 2.4 mg/L, respectively. The percent phosphorus contribution of each waste fraction is further presented in Figure 2.

Table 5. Amount and different types of waste generated annually in the Reykjavík capital region and their phosphorus content.

Waste fraction	Units/year	Amount	Dry matter (%)	Dry matter (tons)	Total P (tons/year)
Sewage sludge	tons	722	-	-	8
Paper and cardboard	tons	18,852	100	18,852	5
Timber/wood	tons	11,275	81	9,132	3
Kitchen waste	tons	65,721	35	23,000	92
Garden waste	tons	3,015	35	1,055	4
Other organic waste	tons	27	35	9	0.04
Wastewater	m^3	82,249,862	-	-	195
Total					**306**

Wastewater contained the highest phosphorus (Fig. 2) of all waste fractions studied. Kitchen waste was the highest solid waste generated of all the municipal waste (Table 5). Despite being about 25 times smaller in the quantity generated to paper and cardboard, sewage sludge contributes 3% phosphorus while paper and cardboard contributed 1%. Timber/wood and garden waste both contribute 1% phosphorus.

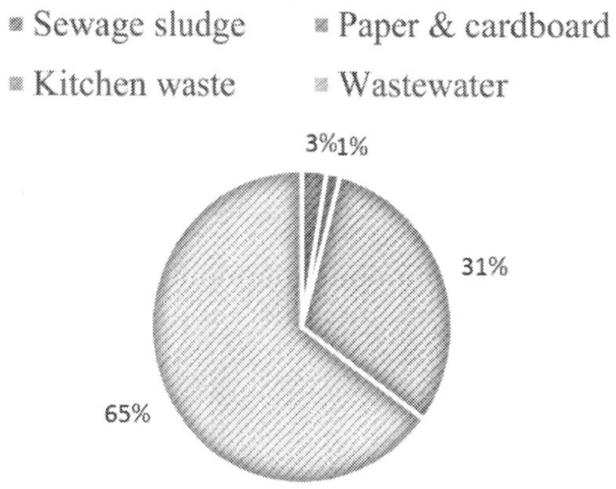

Figure 2. Percent (%) contribution of phosphorus from each waste fraction.

Cost of Utilizing Sewage Sludge and Municipal Waste in Land Restoration
The cost of loading, transporting and spreading the waste in land restoration field is presented in Table 6. Calculations were based on loading the waste in a trailer, transporting it to nearest restoration field (50 km) and spreading it on the fields at an application rate of 30 tons/ha. One round trip to the nearest restoration field using a trailer which can take 30 tons of waste would be 100 km.

Table 6. Breakdown of the cost associated with applying 30 tons of waste on a 1 ha of restoration field.

Item	Unit	Quantity	Cost (ISK/unit)	Total Cost (ISK)
Loading	hour	2	5,000	10,000
Transport	km	100	480	48,000
Spreading	hour	2	27,000	54,000
Total				**112,000**

The cost of utilizing the waste on restoration field is approximately 112,000 ISK/ha. With value added tax (VAT) included at 24%, the cost will be around 140,000 ISK/ha. This is, however, likely to be an upper estimate. The cost of producing 30 tons of compost would be around 60,000 ISK (Wei et al. 2001; Song and Lee 2010).

DISCUSSION

Waste Generation and Phosphorus Content
The contribution of timber, paper and cardboard to phosphorus was almost negligible. Therefore, in cases where the waste management objective is to recover phosphorus from waste, it is important to consider the phosphorus content of the waste fraction than the amount of waste generated. It has been shown in this research that sewage sludge contributed more to the amount of phosphorus than paper and cardboard. In Reykjavík, sewage sludge and kitchen waste should be preferred for phosphorus recycling to timber, paper and cardboard.

Over 82 million m^3/year of wastewater was generated in Reykjavík in 2015. This wastewater contained 195 tons/year of phosphorus. In comparison, this amount of phosphorus in effluent was four times higher than the amount in wastewater treatment plants in Gothenburg city in Sweden (Kalmykova et al. 2012), a city with a population almost double that of the capital region in Iceland. This observation can be explained by the different treatment systems employed in these two cities. In Gothenburg, advanced wastewater treatment systems are employed to remove phosphorus from effluent before final disposal (Kalmykova et al. 2012; Wu et al. 2016). In Reykjavík, sewage sludge and wastewater, combined, contributed 67% of phosphorus. Therefore, in urban areas, most phosphorus is contained in wastewater and sewage sludge (Chowdhury et al. 2014; Wu et al. 2016). However, most of this phosphorus is discharged as unproductive outflow into water bodies (Chowdhury et al. 2014) or landfilled (Villarroel et al. 2014).

Economic Evaluation of Transporting and Spreading the Waste in Restoration Field
Sites within a 30-50 km radius from the capital with erosion levels of 3-5 are many. However, not all of them are suitable for applying waste. Some of them are grazed and only a few are fenced off

from grazing. In addition, there are restrictions such as water protection areas and issues of land ownership that need to be sorted out before restoration activities are undertaken. Also, these sites may be inaccessible due to slope, height and rocks on the surface. Based on the map (Fig. 1), potential sites for waste application might be Krýsuvík in Hafnarfjörður, or Hafnarmelar in Borgarfjörður, all of which are active reclamation sites.

In 2015, about 2,500 tons of chemical fertilizer was used for land restoration with an average phosphorus content (P_2O_5) of 4.5% (MH Jóhannsson, 26 July 2016, Soil Conservation Service of Iceland, Gunnarsholt, Iceland, personal communication). The amount of phosphorus used as P_2O_5 was 113 tons. Since P_2O_5 contains 43.6% P, the amount of phosphorus as P used in 2015 was 49 tons. If sewage sludge generated in Reykjavík was properly treated, the phosphorus it contained was enough to provide 16% of the phosphorus used for restoration by the Soil Conservation Servise of Iceland. If all kitchen waste was composted and applied in land restoration, it would exceed the phosphorus budget for land restoration in 2015. Phosphorus from kitchen waste alone was almost twice (95 tons) the amount of phosphorus used (49 tons) for land restoration in 2015. A potential of 98 tons of phosphorus would be recovered in the effluent generated. Combining all these phosphorus values, the amount of phosphorus wasted through waste disposal in Reykjavík exceeded the phosphorus used for land restoration by 4 times. At a national scale, the phosphorus contained in waste would have met 18% of the total amount of phosphorus imported into Iceland in 2015. With the ongoing efforts in land restoration in Iceland, application of sewage sludge and municipal waste in land restoration could be a sustainable waste management option. As many other countries are strugling to meet stringent regulations on agricultural use of waste, Iceland could take this as an opportunity since the waste would be applied in restoration fields where nonfood plants grow.

Utilizing waste on nearest restoration field from Reykjavík would cost around 140,000 ISK/ha, VAT inclusive at 24%, using an application rate of 30 tons/ha. This application rate is recomended by the United States Environmental Protection Agency (USEPA) on highly degraded land (USEPA 2007), like the ones in the nearest restoration fields from Reykjavík. The cost of utilizing waste could be higher than using inorganic fertilizer considering that recovery and reuse of phosphorus from waste is currently expensive (Cornel and Schaum 2009). This cost could be lowered if the service of utilizing waste in restoration was outsourced through an open tender. Further, the cost of electricity in Iceland is comparatively lower than other European countries which could further reduce the cost of producing the compost. It is also worth noting that land application of waste not only provides nutrients, but also other benefits such as organic matter and elements required for plant growth (Mayer et al. 2016). On the other hand, other technologies of managing waste such as scrubbing of product gas after incineration, pyrolysis, wet oxidation and gasification are high (Fytili and Zabaniotou 2008) and they emit green house gases, dioxins, fumes and heavy metals. In comparison to such technologies, land application of waste is economical (Lundin et al. 2004). Furthermore, land application of waste is a way of conserving resources, especially phosphorus, which is nonrenewable. In addition, recovery of nutrients from waste reduces the need for importing inorganic fertilizers. Consequently, foreign currency is conserved as a result of reduced importation costs and the country's carbon footprint is reduced.

CONCLUSION

Results of the study suggest that municipal waste contains enough phosphorus that can considerably meet the phosphorus requirement for restoration works in Iceland. If proper phosphorus recovery treatment systems were employed, the amount recovered can surpass that used for operations like land restoration. Costs associated with composting, transporting and spreading the waste are higher than those associated with inorganic fertilizer use. However, these costs can be reduced if the service were to be outsourced through an open tender. There are other benefits of waste utilization that must be considered, such as pollution control, improvement of soil quality among others, when evaluating the total value of utilizing it in land restoration.

Proper care should be taken when applying waste to land to minimize the harmful effects they can cause to soil and plants, especially that the restored land could be used for grazing in future. For example, a detailed physico-chemical analysis of the waste (sewage sludge and municipal waste) and soil where the waste will be applied should be undertaken. Plants and their heavy metal accumulations tendencies should be studied before growing them on waste amended soils. Therefore, studies are required to understand the waste-soil-plant metal interaction and nutrient mineralization especially in Iceland with its unique soils and vegetation.

ACKNOWLEDGEMENT

We acknowledge financial support from the United Nations Land Restoration Training (UNU-LRT) programme which led to this research project. We are also grateful to the Soil Conservation Service of Iceland, SORPA, Reykjavík Energy and the local authority in Reykjavík for their support during data collection.

REFERENCES

Antolín, M.C.; Pascual, I.; García, C; Polo, A; Sánchez-Díaz, M. (2005) Growth, Yield and Solute Content of Barley in Soils Treated With Sewage Sludge under Semiarid Mediterranean Conditions. *Field Crops Research,* **94** (2/3), 224–23

Arnalds, O.; Arnason, A.; Thorarinsdottir, E.F.; Metusalemsson, S.; Jonsson, A.; Gretarsson, E. (2001) *Soil Erosion in Iceland*. Soil Conservation Service of Iceland and Agricultural Research Institute, Reykjavík

Arnalds, O. (2015) *The soils of Iceland*. Springer, New York

Brady, N.C.; Weil, R.R. (2008) *The nature and properties of soils*. 14th ed.; Pearson Education Inc. New Jersey

Council of the European Communities (1986) *Council directive of 12 June 1986 on the protection of the environment, and in particular of the soil, when sewage sludge is used in agriculture (86/278/EEC)*. Official Journal of the European Communities. http://eur-lex.europa.eu/legal-content/EN/TXT/PDF/?uri=CELEX:31986L0278andfrom=EN (accessed 8 August 2016)

Cesaro, A.; Belgiorno, V.; Guida, M. (2015) Compost from organic solid waste: Quality assessment and European regulations for its sustainable use. *Resources, Conservation and Recycling,* **94,** 72–79

Chowdhury, R.B.; Moore, G.A; Weatherley, A.J.; Arora, M. (2014) A review of recent substance flow analyses of phosphorus to identify priority management areas at different geographical scales. *Resources, Conservation and Recycling,* **83**, 213–228

Cornel, P.; Schaum, C. (2009) Phosphorus recovery from wastewater: Needs, technologies and costs. *Water Science and Technology,* **59**, 1069–1076

Couth, R.; Trois, C. (2010) Carbon emissions reduction strategies in Africa from improved waste management: A review. *Waste Management,* **30**, 2336–2346

Couth, R.; Trois, C. (2011) Waste management activities and carbon emissions in Africa. *Waste Management,* **31**, 131–137

Desmidt, E.; Ghyselbrecht, K.; Zhang, Y.; Pinoy, L.; Van der Bruggen, B.; Verstraete, W.; Rabaey, K.; Meesschaert, B. (2015) Global phosphorus scarcity and full-scale P-recovery techniques: A review. *Critical Reviews in Environmental Science and Technology,* **45,** 336–384

Diacono, M.; Montemurro, F. (2010) Long-term effects of organic amendments on soil fertility. A review. *Agronomy for Sustainable Development*, **30**, 401–422

Fytili, D.; Zabaniotou, A. (2008) Utilization of sewage sludge in EU application of old and new methods-A review. *Renewable and Sustainable Energy Reviews*, **12**, 116–140

Gentil, E.; Clavreul, J.; Christensen, T.H. (2009) Global warming factor of municipal solid waste management in Europe. *Waste Management and Research* **27**, 850–860

Hargreaves, J.C.; Adl, M.S.; Warman, P.R. (2008) A review of the use of composted municipal solid waste in agriculture. *Agriculture, Ecosystems and Environment,* **123**, 1–14

Jana, K.; De, S. (2015) Polygeneration using agricultural waste: Thermodynamic and economic feasibility study. *Renewable Energy*, **74**, 648–660

Kalmykova, Y.; Harder, R.; Borgestedt, H.; Svanang, I. (2012) Pathways and management of phosphorus in urban areas. *Journal of Industrial Ecology*, **16**, 928–939

Kumwenda, S.; Tsakama, M.; Kalulu, K.; Kambala, C. (2012) Determination of biological, physical and chemical pollutants in Mudi river. *Journal of Basic and Applied Scientific Research,* **2**, 6833–6839

Lundin, M.; Olofsson, M.; Pettersson, G.J.; Zetterlund, H. (2004) Environmental and economic assessment of sewage sludge handling options. *Resources, Conservation and Recycling*, **41**, 255–278

Mayer, B.K.; Baker, L.A.; Boyer, T.H.; Drechsel, P.; Gifford, M.; Hanjra, M.A.; Parameswaran, P.; Stoltzfus, J.; Westerhoff, P.; Rittmann, B.E. (2016) Total value of phosphorus recovery. *Environmental Science and Technology*, **50**, 6606-6620

Metcalf, E.; Eddy, H. (2003) *Wastewater engineering: treatment and reuse. Wastewater engineering, treatment, disposal and reuse.* 4th ed.; McGraw Hill, New York

Ministry of the Environment. (1999) *Reglugerð um meðhöndlun seyru [Regulations regarding handling of sewage sludge]*. Ministry of the Environment, Reykjavík (in Icelandic) http://www.reglugerd.is/reglugerdir/eftir-raduneytum/umhverfisraduneyti/nr/4292 (Accessed 24 August 2016)

Odlare. M.; Arthurson, V.; Pell, M.; Svensson, K.; Nehrenheim, E.; Abubaker, J. (2011) Land application of organic waste - Effects on the soil ecosystem. *Applied Energy*, **88**, 2210–2218

Pongrácz, E.; Pohjola, V.J. (2004) Re-defining waste, the concept of ownership and the role of waste management. *Resources, Conservation and Recycling*, **40**, 141–153

Sajidu, S.M.I.; Masamba, W.R.L, Henry, E.M.T.; Kuyeli, S.M. (2007) Water quality assessment in streams and wastewater treatment plants of Blantyre, Malawi. *Physics and Chemistry of the Earth* **32**, 1391–1398

Saltiola, S. (2014) *Evaluation of new alternatives to improve landfil waste diversion in Reykjavík, Iceland*. Masters thesis, University Jyväskylä, Finland

Smith, S.R. (2009) A critical review of the bioavailability and impacts of heavy metals in municipal solid waste composts compared to sewage sludge. *Environment International*, **35**, 142–156

Sokka, L.; Antikainen, R.; Kauppi, P. (2004) Flow of phosphorus in municipal waste: A substance flow analysis in Finland. *Progress in Industrial Ecology*, **1**, 165-184

Statistics Iceland (2015) *Statistical year book*. Statistics Iceland, Reykjavík http://www.statice.is/media/49296/landshagir2015.pdf (accessed 7 July 2016)

Treonis, A.M.; Austin, E.E.; Buyer J.S.; Maul, J.E.; Spicer, L.; Zasada, I.A. (2010) Effects of organic amendment and tillage on soil microorganisms and microfauna. *Applied Soil Ecology*, **46**, 103–110

United States Environmental Protection Agency (2007) *The use of soil amendments for remediation, revitalization and reuse*. Environmental Protection Agency

Villarroel, W.R.; Beck, M.B.; Hall, J.W.; Dawson, R.J.; Heidrich, O. (2014) The energy-water-food nexus: Strategic analysis of technologies for transforming the urban metabolism. *Journal of Environmental Management,* **141**, 104–115

Wanda, E.M.M.; Mamba, B.B.; Msagati, T.A.M. (2015) Determination of the water quality index ratings of water in the Mpumalanga and north west provinces, South Africa. *Physics and Chemistry of the Earth*:1–9

Wang, H.; Wang, T.; Zhang, B.; Li, F.; Toure, B.; Omosa, I.B.; Chiramba, T.; Abdel-Monem, M.; Pradhan, M. (2014) Water and wastewater treatment in Africa - current practices and challenges. *Clean - Soil, Air, Water,* **42**, 1029–1035

Wu, J.; Franzén, D.; Malmström, M.E. (2016) Anthropogenic phosphorus flows under different scenarios for the city of Stockholm, Sweden. *Science of the Total Environment,* **542**, 1094–1105

Extractive Nutrient Recovery is a Viable Nutrient Control Alternative for Water Resource Reclamation Facilities

Wendell Khunjar[1*], Ronald Latimer[1], Samuel Jeyanayagam[2], Chirag Mehta[3], Damien Batstone[3]

[1]Hazen and Sawyer, 4035 Ridge Top Road, Suite 400, Fairfax VA 22033 USA.
[2]CH2MHILL
[3]University of Queensland, Australia

*Email: wkhunjar@hazenandsawyer.com

ABSTRACT
Resource management strategies that support the water-energy-nutrient nexus are pivotal components of the Basins of the Future initiative. In this paper, we present information generated as part of a Water Environment Research Foundation (WERF) study that focuses on closing the gap between innovation and practice thereby accelerating wider adoption of extractive nutrient recovery. Extractive nutrient recovery, defined as the production of chemical nutrient products devoid of significant organic matter, represents a disruptive technology. It often entails a three-step framework to include accumulation, release, and extraction of a chemical nutrient product that is marketable and could potentially off-set operating costs and synthetic fertilizer use.

KEYWORDS: Phosphorus recovery, struvite, nutrients

INTRODUCTION
Nitrogen (N) and phosphorus (P) are essential elements of all life forms and are extensively used in synthetic fertilizer production. These nutrients once incorporated into crops are ingested and reintroduced to the environment via liquid and solid wastestreams. To avoid the accumulation of these nutrients in the environment, N and P is typically removed prior to discharge to a water body.

Recovery and reuse of nutrients provides an opportunity to reduce reliance on nutrient cycling and fundamentally change the way that we wastestreams are viewed. It should be acknowledged that nutrient recovery and reuse is not a new concept. It has been applied in different forms in the past e.g., land application of biosolids and reuse of secondary effluent for irrigation; however, extraction of a chemical nutrient product with low organic matter content, defined here as extractive nutrient recovery, has not been widely applied within the wastewater treatment industry.

This paper reviews the current state of knowledge on extractive nutrient recovery. The information presented herein was generated as part of the ongoing Water Environment Research Foundation (WERF) project titled: Towards a Renewable Future: Assessing Resource Recovery as a Viable Treatment Alternative (Latimer et al., 2012).

FRAMEWORK FOR RECOVERY

The nutrient concentration in the influent to municipal wastewater resource recovery facilities (WRRFs) typically ranges from 10 to 50 mg N/L for N and from 1 to 10 mg P/L for P. Since the efficiency of extractive nutrient recovery technologies is lower at these concentrations, it is proposed that that extractive nutrient recovery will most likely be viable if performed within a three step framework (Figure 1):

1. Accumulation of nutrients to high concentrations,
2. Release of nutrients to a small liquid flow with low organic matter and solids content,
3. Extraction and recovery of nutrients as a chemical nutrient product.

In this approach, biological, physical, thermal, and chemical mechanisms can be used in each step to transform the nutrients present in domestic wastewater to a chemical nutrient product that has a secondary market value. One of the advantages to using this approach is that multiple options for each stage of treatment can be developed and optimized separately; thereby allowing utilities to select the most appropriate solution for their needs. It is also possible that some utilities may not need capital investment for all three processes since existing infrastructure can be reused. A complete review of state of the art options available for the accumulation, release and extraction framework is provided as part of the WERF Nutrient Recovery Challenge (Latimer et al., 2015). This manuscript provides a brief description of the options.

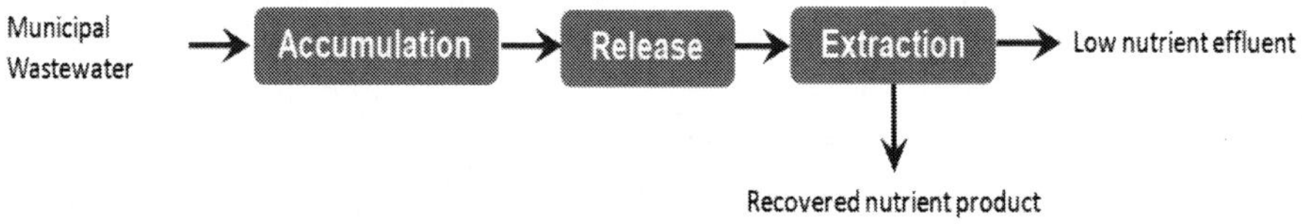

Figure 1. Components of Extractive Nutrient Recovery (Latimer et al., 2015)

Nutrient Accumulation Options

Nutrient accumulation technologies focus on concentrating the low nutrient content of municipal wastewater (Table 1). This can be accomplished using biological (N and P), physical (N and P) and chemical (mainly P) techniques. Biological accumulation techniques center around microbial accumulation in which specially adapted microorganisms (e.g. microalgae, polyphosphate-accumulating bacteria (PAOs), purple non-sulfur bacteria (PNSB), cyanobacteria) are able to uptake (N and P) and store nutrients (P). Plants such as duckweed can also be used as part of passive nutrient treatment/accumulation strategies. Research has shown that biological systems can remove between 70 to 90 % of N and P from wastestreams and are effective for treating a wide range of nutrient concentrations including the dilute content of nutrients typically associated with municipal WRRFs.

Biological processes have already been extensively applied for wastewater treatment, with over 84% of WRRFs in the USA employing some form of biological process (CWNS, 2004). Since these processes are expected to be further employed as effluent nutrient regulations become more stringent, this represents a fortuitous opportunity for the extractive nutrient recovery field, as it becomes possible to stage implementation of the extractive nutrient recovery over multiple years, with the first step initially being use of biological nutrient accumulation processes. Key requirements for using biological accumulation processes are an effective solid-liquid separation process like clarification or membranes to allow recovery of the nutrient rich biomass, as well as an appropriate release technology for subsequent processing.

Chemical accumulation using metal salt addition is another option that can be used to help accumulate nutrients (mostly P). In this process, the metal salt reacts with soluble P to form an insoluble phosphate complex, which is solid and can then be physically separated from the wastestream. Aluminum and iron solutions are often used for this purpose and can achieve greater than 85% P removal from the dilute stream, with the chemical solids being separated during clarification or filtration. One of the key challenges with using chemical accumulation techniques is that the chemically accumulated P is less useful because of the high metal salt content of the final product. This restricts its use in agricultural applications. Chemical accumulation is widely applied at domestic WRRFs. Therefore, it may be possible to accomplish extractive nutrient recovery at existing WRRFs by implementing suitable release and extraction processes to process the solids generated.

Another strategy that can be used to accumulate nutrients from the mainstream flow is adsorption and/or ion exchange. These processes can be used to remove N and P from dilute wastestreams, with removal efficiencies ranging between 50-90%. In this approach, a sorbent or ion exchange material is packed into a column. As the wastewater flows through the column, N or P (depending on the material) is either sorbed or chemically attracted to specific sites on the material. This approach has been used at pilot and full-scale for in tertiary filtration applications to help remove phosphorus. One of the biggest challenges with using adsorption and/or ion exchange for nutrient accumulation is the regeneration step which requires use of costly chemical brines and the need for replacement of spent adsorption media. Therefore, it may not currently be economically feasible to implement adsorption and/or ion exchange at larger plants. Embryonic research at the lab-scale is investigating the use of bio-regeneration as a method to help reduce costs associated with regeneration and replacement of sorbent material. It is expected that these processes will continue to become more important as WRRFs are increasingly asked to achieve effluent TP limits below 0.1 mg/L. In this scenario where adsorption and/or ion exchange is used in a tertiary filtration step, it may be possible to harvest the nutrient from this material for beneficial reuse.

Table 1. Summary of Technologies Suitable for Nutrient Accumulation at Full-Scale WRRFs (From Latimer et al., 2015)

		Operating conditions		Pre-treatment required	Chemical Input	Commercial process	Nutrient(s) Accumulated
		Temp. (°C)	pH				
Embryonic	Microalgae	15 - 30	7.5 – 8.5	-		Lemna Technologies	N and P
	Cyanobacteria	5 - 40	6.5 - 8	-	Carbon source	-	N and P
Innovative	Adsorption/Ion exchange	NA	<8.0	Solid-liquid separation	Adsorbent, Regeneration solution	P-ROC, RECYPHOS, PHOSIEDI, RIM NUT, BIOCON	N and P
Established	EBPR	5 - 40	6.5 - 8	-	Carbon	-	P only
	Chemical	25 - 40	6.5 - 10	-	Metal salts (Al or Fe)	-	P only

embryonic – technologies that are in the developmental stage (bench/pilot scale)
innovative – developed technologies with limited full-scale application
established - commercially viable technologies with a proven history of success

Nutrient Release Options

Release technologies allow us to recover the nutrients into a low flow high nutrient content stream with minimal solids content, which can be used for extraction processes (Table 2). Release technologies typically employ some combination of biological, thermal, chemical or physical processes. Biological release is the most commonly used process that has been implemented at WRRFs. In this process, the biomass is broken down and the organic carbon, nitrogen and phosphorus are converted to carbon dioxide and methane, ammonia and soluble phosphorus, respectively. Biological release can occur under anaerobic conditions (e.g., anaerobic digestion) or under aerobic conditions (e.g., aerobic digestion), and the extent of nutrient release is dependent on the conditions employed during digestion. After biological release, the effluent streams can contain greater than 100 mg P/L and 1000 mg N/L, as well as particulate matter that must be removed. One of the biggest advantages of using an anaerobic biological release process is the opportunity to not only recover nutrients, but also the biogas. Indeed, anaerobic digestion has been extensively applied as a cost effective option for reducing the solids content of primary and waste activated sludges. Recent work performed as part of the WEF funded a national survey of anaerobic digestion and biogas use indicates that close to 25% of all WRRFs greater than 1 mgd current employ anaerobic digestion (Beecher and Qi, 2012). Also, as the industry increasingly aims to achieve energy neutrality, it is expected that anaerobic digestion processes will be increasingly implemented. In a WRRF that already employs biological and/or chemical accumulation followed by biological release using anaerobic digestion, implementation of extractive nutrient recovery would simply require the installation of the extraction step. This latter upgrade has been done at several full-scale facilities in the US and Europe.

Another option for biological release is enhanced P release processes. In this approach, phosphorus that has been biologically accumulated by PAOs is selectively released from these microorganisms (in WAS). One variation of this process is called WASStrip™ and has been patented by Clean Water Services. Enhanced P release processes can be used in combination with anaerobic digestion to help minimize the O&M requirements associated with nuisance struvite/vivianite formation that is associated with the operation of biological accumulation processes.

Another commonly used practice for releasing nutrient bound in biomass and chemical sludge is thermochemical processes coupled with chemical release. Thermochemical options can include wet oxidation, incineration, gasification or pyrolysis. In these processes, high temperature is used to destroy organic material and produce a solid product containing P, which can then be chemically released. It is important to note that N is typically lost through gaseous emissions during these processes. As a result, thermochemical processes are most suitable for extractive nutrient recovery of P. Great advances in thermochemical processes have been achieved over the past 5-10 years and they have emerged as innovative alternatives to using anaerobic digestion for managing solids at municipal WRRFs. Chemical release of nutrients can from the char, ash, biosolids (digested, dewatered activated sludge) or undigested sludge can then accomplished using concentrated acids or bases at temperatures between 100 to 200^0C. The liquid stream is then subjected to extraction technologies to recover the nutrients. Coupling these processes to the extractive nutrient recovery framework could allow facilities who have limited expansion capacity or are limited in disposal options for biosolids to still become resource recovery treatment plants. One of the biggest challenges associated with this release option is presence of heavy metals in the liquid stream that is generated. Post-treatment will be required to limit the heavy metal content of the chemical product. This additional treatment steps can make this option economically challenging to implement at the current market value of the chemical nutrient products that are typically recovered from these processes.

Table 2. Summary of Technologies Suitable for Nutrient Release at Full-Scale WRRFs (From Latimer et al., 2015)

		Operating conditions		Pre-treatment required	Chemical Input	Commercial process	Nutrient(s) Released
		Temp. (°C)	pH				
Innovative	Chemical extraction	25-200	1 - 3	-	Leaching solution (sulfuric acid, hydrochloric acid, nitric acid, citric acid, oxalic acid, EDTA)	SEABORNE, STUTTGARTER VERFAHREN, LOPROX/PHOXAN, CAMBI, KREPCO, BIOCON, SEPHOS, AQUARECI, SESAL-PHOS, PASCH	N and P
	Thermochemical	150 – 1100	all	Temperature adjustment	-	MEPHREC, ASHDEC, THERMPHOS	P only
	Enhanced P Release from WAS	5 - 40	6.5 - 8	-	Carbon (volatile fatty acids)	WASStrip, PRISA	P only
Established	Anaerobic digestion	35 – 60	6.5 – 7.5	-	-	-	N and P

Nutrient Extraction Options

The next step of the extractive nutrient recovery process is the extraction and recovery of chemical nutrient products from concentrated liquid streams (Table 3). These extraction processes can be inserted downstream of accumulation or release technologies. At present, each extraction technology requires pre-treatment to reduce the solids content and/or change the temperature or pH of the liquid stream to a suitable condition for the extraction technology. One example of a commonly applied extraction technology is chemical crystallization. In this process, the soluble nutrient is precipitated and recovered as crystalline products. Products that can be generated by this process include are struvite (magnesium ammonium phosphate) and calcium phosphate (hydroxyapetite, brushite P only). In the case of struvite formation, the pH and concentration of magnesium, phosphate and ammonium is controlled to allow the precipitation of the chemical nutrient product which is then separated from the liquid stream via gravity or mechanical separation. Further drying and processing of the product is also commonly performed.

There are multiple variations of this chemical crystallization process that have been commercialized. In each of these systems, soluble P removal efficiencies up to 90% and ammonia removal efficiencies up to 30% can be expected if struvite is the product of choice. Addition of magnesium chloride or hydroxide as well as caustic (NaOH) is typically needed for the process to proceed. Precipitation of calcium phosphate is also possible with the addition of calcium instead of magnesium. There are over 49 full-scale installations of these processes throughout the world. For municipal WRRFs, these processes are commonly installed downstream of biological accumulation (e.g., EBPR) and biological release technologies (anaerobic digestion).

In order to recovery N only products, liquid-gas stripping of ammonia can be used. In order to extract the ammonia from the nutrient rich liquid stream, it is necessary to raise the pH above 9.3 and increase the temperature above 80^{0}C. Air can then be bubbled through the mixture. This forces the soluble N into the gas phase. This gas phase ammonia is then recovered by bubbling the nutrient rich gas into sulfuric or nitric acid. This process produces ammonium sulfate or ammonium nitrate, respectively. In this process, ammonia removal efficiencies up to 98% is possible; however, the relatively high cost of this method makes this option challenging for implementation in wastestreams with N content less than 2000 mg/L. As thermal hydrolysis processes like CAMBI™ and Exelys™ are increasingly implemented at municipal WRRFs, liquid-gas extraction of ammonia will become more technically feasible; however, the ultimate implementation of this process will be dependent on the cost of the products that will be recovered. While this process is established in industrial applications, it has not been extensively applied for recovery of N from municipal WRRFs.

Electrodialysis represents an embryonic extraction technology that allows for the recovery of all ions from nutrient streams at nutrient concentrations below 2000 mg/L. It represents a highly promising technology to the extractive nutrient recovery field. In this process, an electrical current is used to separate anion and cations across an ion exchange membrane.

Table 3. Summary of Technologies Suitable for Nutrient Extraction at Full-Scale WRRFs (From Latimer et al., 2015)

		Operating conditions		Pre-treatment required	Chemical Input	Commercial process	Nutrient(s) Extracted
		Temp. (°C)	pH				
Embryonic	Electrodialysis	10 - 40	< 8.0	Solid-liquid separation	Electricity	GE Water	N and P
Innovative	Liquid-gas stripping	>80°C	> 9.5	pH and temperature adjustment	Caustic	ThermoEnergy Castion™	N only
Established	Struvite crystallization	25 - 40	8 – 9	Solid-liquid separation	Caustic, Magnesium or Calcium	PHOSTRIP, PRISA, DHV CRYSTALACTOR, CSIR, KURITA, PHONIX, OSTARA, BERLINER VERFAHEN, FIX-PHOS	N and P

CONSIDERATIONS FOR CHEMICAL NUTRIENT PRODUCTS

At present, commercial technologies for extractive nutrient recovery primarily produce chemical nutrient products that are used in agricultural applications. This is because 85% of all nutrient products are associated with agronomy. Since food demand is expected to rise with an increasing global population, it is expected that demand for chemical nutrient products will also increase. This represents an opportunity for the wastewater treatment market to develop niche products that can be used in this field.

Biosolids are the primary product used to recycle nutrients from wastewater. One of the biggest challenges with biosolids is the expense associated with transporting a product with a high moisture content (~80% - 90%). Since the current value of nutrients in biosolids (~ $US 8 per tonne) is a fraction of the transport costs ($US30 per tonne to transport 50 km in USA or Australia with higher costs in Europe), nutrient recovery via biosolids can be an expensive undertaking. Even in scenarios where thermal processes are used to reduce the moisture content, the energy required (~800 kWh of energy (as gas) required to evaporate one tonne of water) is significant.

Consequently, recovery of nutrients into chemical nutrient products like struvite is the primary focus of several commercial extractive nutrient recovery technologies (Error! Reference source not found.). In addition to struvite, other products like calcium phosphate, (hydroxyapetite), iron phosphate (vivianite), phosphoric acid, ammonium sulfate, ammonium nitrate can also be recovered depending on the nature of the wastewater as well as the secondary market being targeted. An additional advantage of recovering chemical nutrient products is the fact that some of these products have use in alternative industries (Table 4).

Table 4. Summary of Chemical Nutrient Products resulting from Extractive Nutrient Recovery Processes (From Latimer et al., 2015)

Common Name	Chemical Name	Product Form	Uses
Struvite	Magnesium ammonium phosphate	Solid	Agricultural and ornamental crop fertilizer
Hydroxyapatite	Calcium phosphate	Solid	Agricultural and ornamental crop fertilizer. Sorbent for heavy metal contained in flue gas.
Vivianite	Iron Phosphate	Solid	Ornamental crop fertilizer. Inexpensive blue pigment for arts and crafts.
Phosphoric acid	Phosphoric acid	Liquid	Agricultural and ornamental crop fertilizer. Removal of rust, de-scaling of boilers and heat exchange tubes.
Ammonium nitrate	Ammonium nitrate	Liquid or Solid	Agricultural and ornamental crop fertilizer. Oxidizing agent in explosives.
Ammonium sulfate	Ammonium sulfate	Liquid or Solid	Agricultural and ornamental crop fertilizer. Used in flame retardant materials.

In order for chemical nutrient products to be used for agricultural purposes, they must meet some minimum requirements. For instance, all products must have consistent nutrient content and possess no/minimal odors. Solid products must have uniform size, comprise no less than 95% total solids, have less than 1% dust content and have a minimum bulk density of at least 45 pounds per cubic foot. Due to the limited mass production rate of the wastewater treatment sector, it will be challenging to compete with existing supply chains. Instead, recovered products from WRRFs should be marketed within niche markets to maximize resale. An example of this is the case of the Ostara Crystal Green™ product, which is used as a soil amendment product.

CONSIDERATIONS FOR IMPLEMENTING EXTRACTIVE NUTRIENT RECOVERY AT FULL-SCALE FACILITIES

Although there appears to be a general consensus that nutrient recovery can benefit the industry, there remain technical, social and economic challenges towards an industry-wide adoption of this approach (Figure 2; Guest et al. 2009). Many of these barriers largely revolve around a lack of technical and economic knowledge.

Site specific economics were identified as the primary barrier that will continue to constrain implementation of recovery technologies like struvite crystallization. To circumvent this challenge, it is important to properly quantify benefits associated with performing nutrient recovery.

Quantifiable benefits include but are not limited to:

- Mitigating the impact of the sidestream nutrient content on the mainstream nutrient removal process and providing a factor of safety for mainstream nutrient removal processes,
- Minimizing nuisance struvite formation and re-gaining infrastructure capacity,
- Reducing chemical and energy costs,
- Reducing chemical sludge production (if metal salt addition is practiced),
- Manipulating the nutrient (P and nitrogen) content of biosolids, and
- Improving sludge dewaterability.

By considering these benefits in concert with revenue potential from re-sale of recovered nutrient product, it has been shown that payback of investment can occur in less than 10 years (Latimer et al., 2015).

CONCLUSION

Extractive nutrient recovery entails the extraction of N and P contained in municipal wastewater as a nutrient product that has value in a secondary market. Hence it can represent a revenue stream for the utility. In addition, adoption of this strategy represents a move toward more sustainable WRRF operation and can complement traditional methods of achieving N and P removal. On a global scale, the nutrients that can be recovered from human waste approximates 17 percent of N and 25 to 30 percent of P produced as fertilizer (Rogalla, 2012). It should be noted that this in itself will not offset a significant portion of the global nutrient demand and must be combined with other strategies to make a difference such as nutrient recovery from animal wastes (animal wastes represent a much larger nutrient pool than human waste) and improved agricultural practices that reduce nutrient loss.

ACKNOWLEDGEMENTS
This work was supported in part by the Water Environment Research Foundation (WERF). We thank Joseph Rohrbacher and Vivi Nguyen with Hazen and Sawyer P.C. and Todd Williams (CH2M HILL),and Ron Alexander (R. Alexander Associates Inc) for their valuable input to this project.

REFERENCES

Beecher, Ned, Qi, Y. (2012) *DRAFT Biogas Study Report – Use of Biogas at Wastewater Treatment Plants in the United States*. Water Environment Federation (WEF), 2012.

Guest, J.S., Skerlos, S.J., Barnard, J.L., Beck, M.B., Daigger, G.T., Hilger, H., Jackson, S.J., Karvazy, K., Kelly, L., Macpherson, L., Mihelcic, J.R., Pramanik, A., Raskin, L., Van Loosdrecht, M.C., Yeh, D. and Love, N.G. (2009) A new planning and design paradigm to achieve sustainable resource recovery from wastewater. Environ Sci Technol 43(16), 6126-6130.

Latimer, R., Khunjar, W.O., Jeyanayagam, S., Mehta, C., Batstone, D., Alexander, R., (2015) Towards a Renewable Future: Assessing Resource Recovery as a Viable Treatment Alternative (NTRY1R12). Water Environment Research Foundation.

Rogalla, F. (2012). Unlocking the Energy and Nutrient Potential of Wastewater. 2012 IWA Yearbook.

SELECTION OF A THERMAL HYDROLYSIS PROCESS FOR THE CITY OF CALGARY'S BONNYBROOK WASTEWATER TREATMENT PLANT

Jennifer Peters, CH2M, Kari MacDonald, City of Calgary, Kim Fries, CH2M, Ryan Roberts, Stantec Consulting.

ABSTRACT

As part of the Plant D Expansion at the City of Calgary's Bonnybrook Wastewater Treatment Plant, increased sludge handling capabilities are required due to the greater mass of waste solids generated by the expanded plant. Various alternatives for solids stream treatment were considered during Conceptual Design and evaluated using a triple-bottom-line assessment for their component infrastructure requirements along with technical considerations such as reliability, robustness, resiliency, future proofing, operations and maintenance considerations and impacts on community and resources.

Three solids handling alternatives were short-listed. These solids management processes were:

1. Conventional Mesophilic Anaerobic Digestion
2. Acid Gas Digestion
3. Thermal Hydrolysis of Secondary Sludge

The solids options were highly sensitive to changes in economic and non-economic factors as well as whether planning for a Plant D expanded capacity or a larger expansion. Of the three short-listed options, a thermal hydrolysis process (THP) of secondary sludge demonstrated the most significant advantages for the plant. One of the key advantages was eliminating the need for future digesters when considering expansions in the short or longer term. Other advantages include: reduced sludge viscosity; increased volatile solids reduction (VSr); increased biogas production; increased dewaterability of digested sludge; and elimination of digester foaming.

This manuscript will discuss how a thermal hydrolysis process (THP) of secondary sludge was selected to achieve the increase in solids treatment capacity required as part of the Plant D Expansion. It will then discuss the key advantages of THP and how they relate to plant operations and future expansions at Bonnybrook as well as providing a brief description of the design basis for the facility at the Bonnybrook Wastewater Treatment Plant (WWTP).

Key Words: Thermal Hydrolysis, Digestion, Biosolids Management, Biosolids Planning

INTRODUCTION

The Bonnybrook WWTP was originally built as a primary wastewater treatment plant in 1932 and has been expanded over the years to include three secondary treatment plants and complimentary solids handling processes. Figure 1 shows an aerial photograph of the existing plant site.

Figure 1: Aerial Photo of Bonnybrook WWTP

The current treatment capacity of the Bonnybrook WWTP is approximately 1,000,000 equivalent population (EP). The population of Calgary has been steadily increasing over the past few decades and to address this growth, The City of Calgary needs to expand its wastewater treatment capacity. Part of the The City's expansion plans include the Plant D Expansion project at Calgary's Bonnybrook WWTP which will increase the plant capacity by more than 30 percent to accommodate an additional 325,000 EP.

Biosolids management is one of The City's key considerations when addressing plant capacity needs. As part of the Plant D Expansion project, a biosolids technology evaluation was performed to select the optimal solution to meet Bonnybrook's expanded capacity. This manuscript will describe the technology evaluation process which led to the selection of thermal hydrolysis of secondary sludge and discuss how the vendor was pre-selected based on overall life cycle costs, supplier experience and qualifications, as well as other criteria. The design basis for the Bonnybrook THP facility will also be described.

THE CITY OF CALGARY'S BIOSOLIDS MANAGEMENT PROGRAM

The City of Calgary's current biosolids management program includes mesophilic anaerobic digestion (MAD) to produce stabilized biosolids and biogas at all three WWTPs. The digested biosolids are transferred and stored in the Shepard Sludge Lagoons where they settle by gravity and are used in an agricultural land application program (Calgro™) in the summer months. The current biosolids generation rate is between 18,000 to 23,000 tonnes dry solids per year (tDS)/yr and is projected to increase to about 35,000 tDS/yr by 2030 to 2035. The surrounding agricultural lands can accept Class B biosolids at a rate of 11,500 to 22,000 tDS/yr. Because of the continued shortfall between biosolids generation and land application, The City has had to implement a series of alternative beneficial reuse programs through private service providers over the last five years.

Currently the biosolids generated at Bonnybrook WWTP are processed through nine mesophilic anaerobic digesters. Prior to digestion, primary solids are thickened and/or fermented using a system of both complete mix and static fermenters. Secondary solids are thickened using a dissolved air flotation system and then mixed with primary sludge in blend tanks prior to being pumped to the digesters. Following digestion, sludge is sent to a digested sludge storage tank (DSST) and pumped to the Shepard Lagoons for dewatering and land application.

To address the shortfall between biosolids generation and land application in Calgary in the future, The City is currently constructing a dewatering facility at the Bonnybrook WWTP and an off-site composting facility that will be able to accept approximately10,000 tDS/yr of biosolids under the first phase of the project. Future options for handling biosolids generation from a growing population include expansion of the composting facility or implementation of thermal drying, lime stabilization, or other beneficial reuse methods.

SELECTION OF A THERMAL HYDROLYSIS PROCESS

As part of the Plant D Expansion at Bonnybrook, increased sludge handling capabilities are required. Various alternatives for solids stream treatment were considered and short-listed technologies were evaluated using a triple-bottom-line (TBL) assessment. The selection process is described in further detail in the following sub-sections.

Design Considerations

The population in the Bonnybrook WWTP catchment has been growing rapidly over the past several years, and current projections indicate that the catchment population could grow to 1.8 million people by 2076. Although the original technology selection was based on an expanded population of 250,000 EP, subsequent projections lead to an increase in the design basis for the Plant D Expansion to 325,000 EP. It is also important to consider how the selected solids handling technology could accommodate additional population growth beyond the Plant D design horizon.

Historical data was used to develop estimated per capita solid loads in the Waste Organic Materials and Biosolids Master Plan (CH2M, 2011). These sludge loads, shown in Table 1, can be used in conjunction with population projections used to determine potential future solids loads at the Bonnybrook WWTP.

Table 1. Bonnybrook WWTP - Predicted Loads for Technology Evaluation

Sludge Source	Design Load
Primary Sludge	52 g/c/d
Fermented Primary Sludge	49 g/c/d
Waste Activated Sludge	53.5 g/c/d
Thickened WAS	52 g/c/d

Sludge Source	Design Load
Blended Sludge	101 g/c/d
Digested Sludge	58.5 g/c/d

Development at Bonnybrook WWTP is constrained by the available land on the existing plant site. Since 1932 the plant has grown significantly and the majority of land on the site is occupied by WWTP processes. Therefore technology footprint is also critical consideration in the evaluation and selection of a solids handling process that can meet the needs of the plant into the future.

Based on the considerations discussed above, three solids handling alternatives were short-listed for evaluation including: Conventional Mesophilic Anaerobic Digestion, Acid Gas Digestion, and Thermal Hydrolysis Process for thickened waste activated sludge (TWAS).

These three processes were evaluated using a triple-bottom-line assessment. The infrastructure requirements for each technology and the TBL assessment methodology are described in more detail below.

Triple Bottom Line Methodology

The methodology used in the triple bottom line assessment is MODA – Multi-Objective Decision Analysis. MODA is a tool which has the ability to consider various economic, social and environmental inputs to assist in reaching a defined decision goal. The steps involved in this process are as follows:

1. Establish the decision goal
2. Identify and specify evaluation criteria
3. Develop performance measures (measurement scales) to measure how well options meet each evaluation criteria
4. Assign scores for each evaluation criteria under each option
5. Assign weights to the evaluation criteria
6. Calculate total value scores and conduct a sensitivity analysis

The team defined the decision goal for this project as: "Select the best solids stream technology for the Bonnybrook WWTP Plant D Expansion project". Economic factors considered in the evaluation include total capital cost, annual operations and maintenance costs, and Net Present Value. The non-economic evaluation criteria used in the evaluation are broken down into the following categories: reliability, robustness, resiliency, future proofing, operations and maintenance (O&M) considerations and impacts on community and resources. Future proofing looked at two sub-criteria: the ability to produce a Class A biosolids immediately or after conversion; and the ability to meet the long term expansion to 1.8 million EP. Class A biosolids is not currently a requirement for The City of Calgary and is not expected to become a requirement in the near future, however the ability to convert the selected process to meet Class A biosolids in the future provides process flexibility and is advantageous in the event that regulations or requirements for beneficial reuse change.

Triple Bottom Line Analysis

The following sections describe the component infrastructure requirements for each short-listed technology and provide a summary how each option aligns with the non-economic criteria established for the project.

Conventional Mesophilic Anaerobic Digestion

This option involves the expansion of the existing MAD system at the Bonnybrook WWTP. The new infrastructure considered in this technology evaluation includes two new digesters with a volume of 8,000m^3, a new digester pump house and the expansion of the existing WAS thickening system through the addition of two new rotary drum thickeners. Table 2 below provides a summary of the non-economic considerations for this technology.

Table 2. Non-economic considerations for Mesophilic Anaerobic Digestion

Reliability	There are many full scale operations using this technology including the Bonnybrook WWTP. For large plants in Western Canada and North America, it is the most prevalent process and has proven to successfully stabilize wastewater sludge from diverse liquid stream configurations. However, there are some issues with foaming in the digesters at Bonnybrook WWTP which results in a 'loss' in volume of about 12 percent.
Robustness	Conventional anaerobic digestion is conservatively sized to accommodate substantial variations in feed sludge volume and characteristics. Toxic shocks can occur in some facilities when inhibitory materials are discharged to the plant with industrial wastewater; however, given the size of the Bonnybrook catchment and the proportion of residential wastes, this type of occurrence is unlikely.
Resiliency	If anaerobic activity were lost for some reason (inhibitory substances in influent; a substantial upset in digester operation (reactors go 'acid'), a long period when feeding was stopped; or similar), the process would need to be re-started. However, it would be almost impossible to 'lose' the existing sludge inventory at a plant the size of Bonnybrook and after a disruption in service, anaerobic performance would be expected to be re-established relatively quickly although there may be a period of time in which the feed rates had to be lowered as the biological community re-established a stable population and proportions.
Future Proofing – Class A Biosolids	Inherent Ability to Produce Class A Biosolids: MAD would not achieve high quality sludge (Class A) without a change to the process to thermophilic digestion or pre-pasteurization. Ability to be Reconfigured to Produce Class A Biosolids: The existing digesters were not designed to withstand higher thermal stresses associated with thermophilic operation. Alternatively, thermal hydrolysis, pre-pasteurization, or some other similar system could be employed to thermally pre-condition the sludge so that it met Class A requirements after mesophilic digestion (pathogens would be deactivated in the pretreatment stage).
Future Proofing - Long Term Population	With the completion of the Plant D Expansion (assuming a 250,000 EP expansion), the installed and firm capacities for the plant will be approximately 1,291,000 EP and 1,162,000EP respectively. The addition of two anaerobic digesters, as is described for this alternative, would provide slightly more capacity than needed for that population (say 2.5 percent). Space for further

Projection (1.8M EP)	anaerobic digesters is very limited; hence, a change in technology from mesophilic anaerobic digestion would be required to accommodate a long term (2076) population of approximately 1.8 million residents on the existing Bonnybrook WWTP site. New technologies which could be considered include conversion to Acid-Gas digestion, retrofit to Thermal Hydrolysis Process (THP) for TWAS, retrofit to Thermal Hydrolysis Process (THP) for all sludge, or building new solids handling facilities on adjacent property.
O&M - Complexity	Although mesophilic anaerobic digestion is relatively complex, City staff have a well-founded appreciation of the interactions between liquid stream and digestion performance and are able to be proactive in the operation of the digestion systems.
O&M - Familiarity	City of Calgary staff are very familiar with this process through experience at all of Calgary's WWTPs.
O&M - Staffing Requirement	No new additional full time employees are likely to be required to operate and maintain the additional infrastructure associated with this option.
O&M - Safety	Status quo. Although anaerobic digestion is inherently a hazardous process due to the presence of explosive and hazardous gases, the designs for anaerobic facilities are fashioned to mitigate most unexpected risks and City staff are well aware of the safety issues and are prepared to deal with those concerns proactively.
O&M - Flexibility	Given the number of digesters and the redundancy in the system, anaerobic digestion is sufficiently flexible to handle most variations in sludge quantity and quality. Multiple complementary components and vendors are present in the marketplace, with very few proprietary issues.
Community - Odour	Status quo. Risk relative to current plant process is unchanged.
Community - Noise	
Other Community Issues	
Resources - Beneficial Use of Energy	Biogas generation will increase at a rate proportional to the serviced population. Continued use of the biogas in cogeneration will contribute to the plant's self-generated energy.
Resources - Beneficial Use of Nutrients	Anaerobic digestion will continue to release nutrients from the sludge to the liquid stream, where they may be recovered for beneficial use. There will be no change to the proportional quantities.
Resources - Dewaterability/ Polymer Use	The continued use of anaerobic digestion will not change the dewaterability nor polymer demands as compared to the current site.

Acid Gas Digestion

One of the key benefits of an acid-gas digestion system is the potential for elimination of foam in the digesters. However, the elimination of foam is not a guarantee with the implementation of this technology. Therefore, the acid gas digestion option was broken down into two sub-options which considered whether the technology would be successful in eliminating digester foaming issues. The infrastructure capacity requirements vary significantly depending on whether foam is considered, as it can occupy a significant amount of volume in the digesters. If it is assumed that foam control is successful, the required infrastructure would include three new acid gas reactors, a new acid gas reactor pumphouse, and an expansion of the waste activated sludge (WAS) thickening system. No new digesters would be required. If it is assumed that foam control is not successful, the required infrastructure would include three new acid gas reactors, a new acid gas reactor pump house, one new digester, a new digester pumphouse and an expansion of the WAS thickening system. Table 3 below provides a summary of the non-economic considerations for this technology.

Table 3. Non-economic considerations for Acid Gas Digestion

Reliability	There are many full scale operations in place in the U.S. and other countries that have demonstrated reduction in solids, improvements in dewaterability as well as an increase in gas production. There are no plants of this configuration in Western Canada, but there are acid gas digestion systems in nearby states of the US.
Robustness	The acid phase must be maintained at high volatile solids loading rates (VSLRs) to function as an acid phase digester. When the high VSLR is reduced, such as during a power outage or a changeover of a reactor, the volatile fatty acids (VFAs) will be consumed rapidly causing a rise in the pH and a transient surge in gas production. The default operation then becomes that of a conventional high rate digester, so long as some feeding is maintained. The effect is reversible and no special seeding is needed to reverse the transition. As a result the process has no difficulty responding to the typical daily transients in VSLR. The gas reactor is similar in its ability to handle fluctuations as a conventional anaerobic digester. Of course, the feed rate and composition is somewhat buffered by the upstream acid reactors, so this phase is slightly more robust than the conventional system. Acid-gas digestion has some inherent potential to reduce or eliminate foaming; however, it is not proven for all situations.
Resiliency	Experience has shown that the process can be restored in a matter of days to full loading after a shutdown or major reduction in VSLR. However, if the shutdown is long term, then it will take a significant time to re-establish operation as described for the conventional digestion option.
Future Proofing – Class A Biosolids	Inherent Ability to Produce Class A Biosolids: There is a migration path to achieving Class A sludge with thermophilic digestion in the acid phase, the gas phase or both. Batch feeding would likely be necessary to eliminate pathogen contamination by backmixing. Experience has shown that conversion to thermophilic gas phase digestion at a reference plants can result in other effects, including gas binding of the original mixing and recirculation pumps, and additional struvite formation.

	Ability to be Reconfigured to Produce Class A Biosolids: May not require reconfiguration if acid phase reactors constructed to meet Class A thermophilic digestion requirement. If future conversion to THP, the acid phase reactors would no longer be required.
Future Proofing - Long Term Population Projection (1.8M EP)	To accommodate a long-term (2076) population of approximately 1.8 million residents on the existing Bonnybrook WWTP site, the following upgrades or expansion would be required: add more Gas Phase Digesters, retrofit to Thermal Hydrolysis Process (THP) for TWAS and abandon acid reactors, retrofit to Thermal Hydrolysis Process (THP) for all sludge, or build new solids handling facilities on adjacent property.
O&M - Complexity	Process is similar to the conventional digestion process already operated by City staff. The major difference is the need to operate and monitor the acid reactors so that they generally achieve a reasonable level of acidification without commencing gasification.
O&M - Familiarity	The City operations staff do not have experience with operating acid phase digestion; however, given that this process is similar to conventional digestion, it is not anticipated that this conversion would create any significant issues.
O&M - Staffing Requirement	No new staff required
O&M - Safety	When compared to conventional anaerobic digestion, there is no significant difference in the hazards associated with the process. The facility can be designed to eliminate most risks.
O&M - Flexibility	Given the number of digesters and the redundancy in the system, acid gas digestion is sufficiently flexible to handle most variations in sludge quantity and quality. Multiple complementary components and vendors are present in the marketplace, with very few proprietary issues.
Community - Odour	Ammonia odours in dewatered biosolids are stronger than that from a conventional high rate digestion facility, and must be managed accordingly. Engineering practices needed to manage odours are common knowledge so risk is unchanged.
Community - Noise	Unchanged risk
Other Community Issues	Unchanged risk
Resources - Beneficial Use of Energy	Biogas generation will increase due to the increase in gas generation with the conversion to acid gas digestion. The immediate change would be in the range of 5 to 10 percent, although further increases would occur at a rate proportional to the serviced population. Continued use of the biogas in cogeneration will contribute to the plant's self-generated energy.

Resources - Beneficial Use of Nutrients	Anaerobic digestion will continue to release nutrients from the sludge to the liquid stream, where they may be recovered for beneficial use. There will be no change to the proportional quantities.
Resources - Dewaterability/ Polymer Use	The experience at most installations has been that the reduction in digested biosolids resulting from even a small increase in the amount of volatile solids destroyed will offset a mildly greater specific polymer demand. The result is that the mass consumption of polymer will remain about the same.

Thermal Hydrolysis of TWAS

The infrastructure considered for the THP of TWAS includes three new centrifuges, sludge storage and conveyance, a new THP system, sludge blending with primary sludge, new digester feed pumps, as well as a new foul air treatment system and an expansion of the WAS thickening processes at the plant. The design population used for the base analysis is 1,291,000.

One of the key reasons that THP of secondary sludge was short-listed as opposed to THP of both primary and secondary sludge is that a Class A biosolids is not currently required in Calgary and a WAS-only THP facility is much smaller and less costly while still realizing most of the benefits of thermal hydrolysis of all the plant's sludge. This is discussed in further detail below.

Table 4 below provides a summary of the non-economic considerations for this technology.

Table 4. Non-economic considerations for THP of TWAS

Reliability	There are many THP systems in operation in the world, including 40 systems in Europe and some in Asia and Australia. However, THP is limited in large scale applications. No references are available for Canada or comparable cold climate conditions. In most installations, one of the desired benefits is pathogen free sludge, which necessitates that all of the sludge flow is directed to THP.
Robustness	A THP system is relatively robust – biosolids are extremely biodegradable after THP pretreatment and digestion is stable. Significant changes in the influent sludge flows and loads are managed without upset, within reason. Generally, the digestion phase is conservatively sized (12 to 15 day solids retention time (SRT)) for this process, so there is minimal chance of upset. THP has highest likelihood to resolve foaming of the short-listed technologies.
Resiliency	When an entire THP system is removed from service for a period, sludge needs to be diverted to another location until the system is returned to service. In Calgary's case, it might be possible to use the additional digesters to extend the holding capacity until the THP units were returned to service. Following loss of THP, digestion would need to re-acclimate to the specific characteristics of the THP discharge stream once normal operations are re-established. Although not considered onerous, this process would take more time and supervision as compared to other short-listed options. Most THP systems are designed such that only one reactor needs to be taken out of service at a time, which allows for continued operation of the THP system. however, certain vessels do not have redundancy unless multiple trains are provided and when these vessels are out of service, the sludge would have to be stored upstream or offline or bypassed directly to the digesters. These shutdowns can be managed to reduce the period a complete system is out of service.

Future Proofing – Class A Biosolids	Inherent Ability to Produce Class A Biosolids: THP of secondary sludge only does not produce Class A biosolids. Ability to be Reconfigured to Produce Class A Biosolids: Readily reconfigurable to produce Class A biosolids via diversion of primary sludge to the THP process. Digestion capacity is already in place to accommodate this. There may be a further benefit of THP in the future treatment of contaminants of emerging concern.
Future Proofing - Long Term Population Projection (1.8M EP)	To accommodate a long-term (2076) population of approximately 1.8 million residents on the existing Bonnybrook WWTP site, one of the following upgrades or expansion would be required: • Return of two retired digesters to service and expand THP system to handle additional loads • Expansion and conversion of the THP system to handle primary and secondary sludge (as noted previously). There is sufficient space available on site for additional thermal hydrolysis processing capacity. • Greenfield expansion on adjacent property
O&M - Complexity	This system is more complex than the conventional mesophilic and acid gas digestion options. There is an additional dewatering step and a steam generation system is required which needs specialty operations.
O&M - Familiarity	City staff do not have experience with THP.
O&M - Staffing Requirement	Additional certification is required for high pressure systems. The dewatering and THP system would mandate the addition of at least 7 O&M staff (full-time equivalent). This additional staffing requirement would add at least one new operator to each shift and would add the equivalent of two mechanical maintenance staff and one electrical, instrumentation and controls maintenance staff member.
O&M - Safety	Process would be designed to eliminate most risks; however, high pressure and high temperature operation add some inherent risk to this option above those risks associated with conventional anaerobic digestion.
O&M - Flexibility	A THP system is designed with sufficient redundant elements to handle most operating and maintenance variables. The number of extra trains and reactors allow it to operate with units out of service, with high or low sludge loads, and with some level of deterioration of upstream processes. There are various vendors that have developed this technology; however, there is only one vendor that has a significantly larger portion of the systems installed in the world than the rest (Cambi). Furthermore, once a supplier is selected there is tendency to be tied to that system.
Community - Odour	The processes generate a noxious intermediate product, from which the off-gases have to be handled appropriately. This would be mitigated through design.
Community - Noise	No change

Other Community Issues	Decrease in truck movements – reduction of sludge transportation and disposal cost
Resources - Beneficial Use of Energy	Net zero difference to conventional digestion. For the two digesters which can be retired, there is a reduction of 75kW per digester. Increased biogas production adds to the amount of electricity that can be produced through cogeneration. An increase in gas production able to sustain approximately 0.5 MW of additional electrical energy production is anticipated. Additional thermal energy would be generated as well; however, this energy would be consumed for the most part by the THP system.
Resources - Beneficial Use of Nutrients	Due to the higher volatile solids reduction (VSr) achieved through this process, more nutrients would be solubilized through digestion, making them available for recovery as nutrient sources.
Resources - Dewaterability/ Polymer Use	Due to the high VSr, the digested sludge has a lower VS:TS ratio, improving its dewaterability substantially. This improvement would result in lower polymer consumption, higher dewatered sludge concentrations, higher throughput through the centrifuges, or some combination of the above.

Economic Analysis

Cost estimates were developed for each technology based on the component infrastructure requirements described in the technology summaries above. These estimates are shown in Table 5 below.

Table 5. Conceptual Cost Estimates for Short-listed Technologies

Technology	Capital Cost	Annual O&M Cost	NPV
Conventional MAD	$ 95,300,000	$ 1,110,000	$ 114,600,000
Acid Gas (No Foam Allowance)	$ 92,000,000	$ 21,000	$ 92,300,000
Acid Gas (Foam Allowance)	$ 129,400,000	$ 160,000	$ 132,200,000
THP of TWAS	$ 89,700,000	$ 1,830,000	$ 121,500,000

Triple Bottom Line Analysis Results and Discussion

The results of the TBL analysis indicated that the selected solids stream technology varied considerably depending on the capacity considered in the evaluation, and whether planning was for a 30 percent Plant D expanded capacity or a larger future expansion. When considering the capacity of Plant D only, acid-gas digestion assuming no foaming issues scores the highest. However, since foam elimination is not a guarantee with this technology, it is important to consider how the additional capacity required to accommodate digester foaming would impact the results. If foam capacity is considered, acid-gas digestion scores between low and average for sensitivity analyses considering both economic and non-economic factors. Conventional MAD scores high when considering Plant D capacity only, but scores lower when considering future capacity requirements. Of the three short-listed options, a thermal hydrolysis process (THP) of secondary sludge demonstrated the most significant advantages for the plant at the lowest overall cost when considering future plant expansion requirements. Figure 2 below provides a graphical representation of the value vs. cost for each short-listed option when considering future capacity requirements.

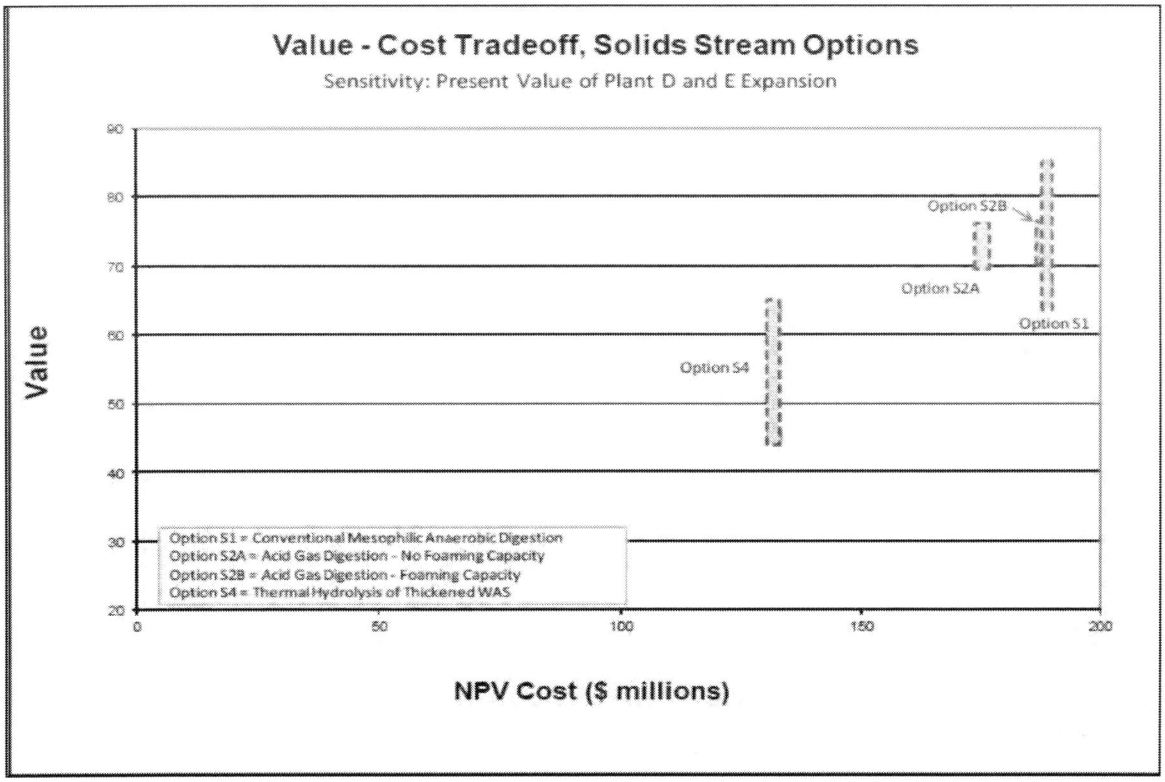

Figure 2: Value vs. Cost Tradeoff for Plant D and Future Expansions

A THP facility receiving only secondary sludge was ultimately selected for Bonnybrook. One of the key reasons in favour of a secondary sludge THP system (as opposed to complete THP for both primary and secondary sludge) is that a Class A biosolids is not required for The City's successful land application program. It currently operates using Class B biosolids and there is no indication that more stringent regulations or other factors will drive the need for Class A biosolids in the foreseeable future. Furthermore, the capacity requirements at Bonnybrook are such that hydrolysis of secondary waste activated sludge (WAS) provides sufficient capacity beyond the next expansion without the need to expand the anaerobic digestion capacity. The selection of THP would allow for the retirement of two small digesters and one large digester at the Bonnybrook WWTP. These tanks could then be used to accommodate additional solids loads associated with future plant expansions or, as is the case for the larger digester, act as a standby digested sludge storage tank until such time as additional digestion capacity is required.

The key benefits of a WAS only THP facility include:

- A smaller THP facility
- Lower overall life cycle costs
- A more favorable overall energy balance

Approximately fifty percent less steam and heat energy is required with WAS only THP and little or no supplemental cooling or heating will be required since the blended sludge (primary plus hydrolysed secondary sludge) will have a resulting temperature in the range of that required for mesophilic digestion.

A WAS only THP system has the same benefits of a complete THP system including:

- Reduced sludge viscosity
- Increased volatile solids reduction (VSr)
- Increased biogas production
- Increased dewaterability of digested biosolids
- Minimal potential for anaerobic digester foaming

The WAS only THP system will result in an increase in the anaerobic digester throughput without increasing digester tankage and will produce sufficient biogas to meet the THP steam requirements.

As with any technology, there are some disadvantages to a THP system as well. THP is new to The City of Calgary and is a complex system that requires a pre-dewatering step as well as a source of steam. The main process related disadvantage of THP is an increase in the ammonia and total phosphorus loading in sludge and the return streams from the dewatering processes, which can increase nutrient loadings to the plant's biological nutrient removal processes. Increased ammonia concentrations can also cause upset conditions in the digesters, which are typically mitigated by diluting the THP sludge with effluent water. However, the addition of effluent water results in a lower TS concentration in the sludge which can increase the volume requirements in the digesters. Preliminary calculations predict an ammonia concentration in the blended sludge of about 2500 mg/L. Industry experience has shown that THP digesters can be operated at up to 3000 mg/L without experiencing any upsets. Therefore, it is unlikely that dilution water will be required. However, provisions have been made to add effluent water to the suction of the HWAS pumps or directly to the blend tanks in the event that dilution is required. Dilution with effluent water also allows for cooling if the temperature of the digesters starts to rise above optimal levels for anaerobic digestion (35 to 40°C).

High total phosphorus (TP) concentrations can be experienced in the return streams from sludge dewatering processes. A nutrient removal system is currently being considered for the Bonnybrook WWTP which would substantially reduce the TP in the return streams. The THP facility has been designed such that the centrate from the TWAS dewatering process can be sent to this potential future nutrient removal system to remove excess phosphorus.

EQUIPMENT PRE-SELECTION

An equipment pre-selection and evaluation process was undertaken through a request for proposal (RFP) to select the THP system that would be most advantageous to The City. Since sludge feed to the THP system is such a critical component, the dewatered waste activated sludge (DWAS) storage silos and conveyors (including the classifying conveyors downstream of the centrifuges, the distribution conveyors and the extraction conveyors, as well as the THP feed pumps) were included in the THP vendor's scope of supply to ensure unit responsibility for the DWAS feed and THP systems.

The RFP also included an optional service agreement and required that proponents submit pricing for services to assist plant staff with operations and system optimization for approximately one year after commissioning. The City can choose to enter into this Service Agreement with the successful vendor at any time.

Proven performance on a larger scale is critical to The City; thus it was mandatory that proponents provide at least one reference installation that is operating at a peak capacity of 40 tDS/d or greater and has been in operation for at least 12 months. Vendors that could not meet this and other mandatory criteria did not move on to Stage II of the evaluation.

During Stage II of the evaluation, references were contacted to elicit feedback concerning equipment reliability and vendor responsiveness. Each vendor had to provide three references and could receive a maximum of 20 points per reference. If the reference facility reported any defects in manufacturing quality, poor service, unavailability of replacement parts or, operational concerns, points were deducted. Scoring for supplier experience and qualification for THP equipment represented almost 30 percent of the total score, which includes the rated criteria and life cycle costs. Supplier experience for silos and conveyors represented another 5 percent of the total score. The ability to meet technical specifications represented about 14 percent of the score and the remaining criteria (listed in Table 6 below) accounted for 12 percent of the score. Overall, the scoring from Stage II of the evaluation represented over 60 percent of the total score.

Stage III of the evaluation involved a life cycle cost analysis which accounted for about 40 percent of the total score. In addition to the capital cost and cost of site services, commissioning, guaranteed performance testing and training, the life cycle costs included the cost of the building structure, building mechanical, electrical, instrumentation and controls, as well as the cost to re-assemble the equipment on site and annual operations and maintenance costs. Vendors were asked to provide guarantees for the thermal efficiency (defined as the total amount of heat energy (kJ) required by the THP system to process a tonne of dry material in the feed DWAS), peak instantaneous peak steam consumption (kg steam per hour), optimal steam consumption (tonne of steam per tonne of dry solids), and peak system power consumption (kW), in addition to a list of other parameters, which were used in the life cycle cost analysis. As throughput and thermal efficiency have a significant impact on the life cycle costs, special damages were included in the RFP, with severe penalties if the guarantees are not met.

The opportunity for a commercially confidential meeting was offered to each vendor who was interested in bidding on the project, and individual meetings were set up upon vendor request. , These meetings were held during the proposal period in the presence of a third party fairness moderator to answer questions from the vendors regarding the details of the RFP. If the answers to the questions or clarifications resulting from each meeting were not considered confidential, the responses were issued formally to all proponents. Because these meetings were held during the proposal period, it was not necessary to go to Stage IV of the evaluation and call each vendor in for an interview.

The scoring criteria for the THP proposal evaluation are listed below in Table 6.

Table 6: THP Proposal Evaluation Scoring Criteria

Category: Rated Criteria	**Maximum Potential Score**
Supplier Experience and Qualifications - THP Equipment	60
Supplier Experience and Qualifications - Silos and Conveyors	12
Maintenance and Support Personnel Response Time	10
Time to Complete and Submit Final Shop Drawings, Training Plans and Training Materials	5
Ability to Meet Technical Specifications	30
Compliance with Contract Terms and Conditions	10
Total Maximum Rated Criteria Score	***127***
Category: Price	**Maximum Potential Score**
Life Cycle Cost	*82*
TOTAL MAXIMUM EVALUATION SCORE	**209**

The evaluation of the vendors was extensive to ensure a fair comparison and the results were very close. Ultimately, Cambi was selected as the successful THP supplier for the Bonnybrook Plant D Expansion.

PROCESS DESCRIPTION

Two of Cambi's B6x4 trains will be installed, operating as independent trains, each with the capacity to handle the plant's projected maximum week DWAS loading of 73,000 kgDS/d. Each B6x4 train consists of one pulper tank, four reactors, one flash tank, a process gas handling unit, internal transfer/recirculation pumps, and the discharge pumps. A schematic of Cambi's B6x4 system is provided below in Figure 3 and a 3-dimensional representation is provided in Figure 4.

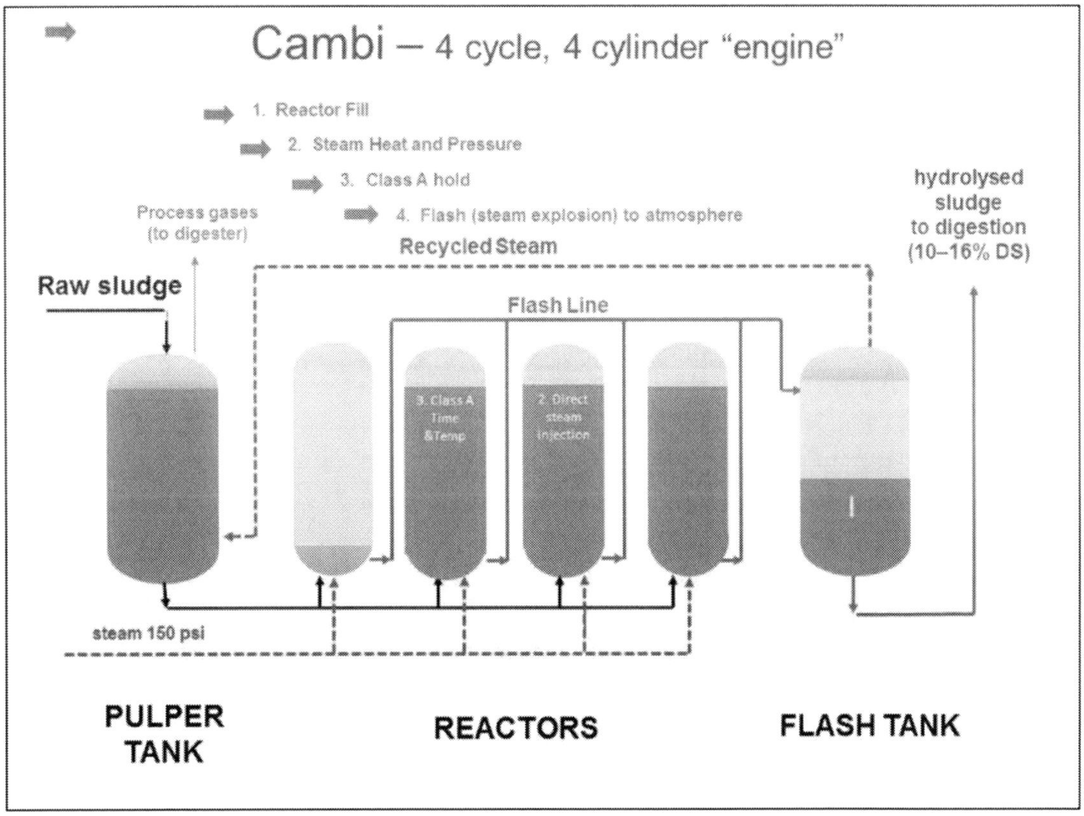

Figure 3: Schematic of Cambi's B6x4 System (courtesy of Cambi)

Figure 4: 3D Representation of a Cambi B6x4 Train (courtesy of Cambi)

The overall THP facility will include three dewatering centrifuges, centrifuge feed and polymer systems, dewatered sludge classification, storage and conveyance, THP, and blend tanks to mix hydrolyzed sludge with thickened fermented/thickened primary sludge before all sludge is pumped to the existing anaerobic digesters. Steam generation is part of separate upgrades to the plant's cogeneration system which include a new biogas powered steam generating turbine. Refer to the THP Facility Process Flow Diagram below, in Figure 5 on the next page.

TWAS is sent from the dissolved air flotation (DAF) facility to the centrifuge feed tanks in the THP facility where it is thoroughly mixed using hydraulic mixing to ensure consistency. The DAF facility has been recently upgraded to improve thickening capacity. There are plans to install new rotary drum thickeners in the future when additional thickening capacity is required, as discussed in the technology evaluation sections above. From the DAF facility, TWAS is pumped to the centrifuges for dewatering upstream of the THP system. The dewatered sludge (DWAS) produced by the centrifuges passes through classifying conveyors and horizontal conveyors which evenly distribute the sludge between two storage silos. Horizontal and vertical extraction conveyors located at the base of the silos transfer sludge to the THP feed pumps which feed the THP equipment.

The THP feed pumps feed the pulper tanks (one per Cambi B6x4 train) where the sludge is pre-heated using recycled flash steam. Live steam is injected into the reactors, raising the temperature to between 160 and 165°C and the pressure to between 6 and 8 barg to hydrolyze the sludge. The contents are allowed to react for 20 to 30 min to provide complete thermal hydrolysis before a valve is opened allowing the contents to flash to the flash tank. Process gases generated through thermal hydrolysis are collected in a process gas unit, cooled, condensed and injected into the blend tanks below the liquid level. The blend tanks' headspaces are connected to the biogas storage and treatment system which also collects biogas from the digesters and feeds the cogeneration system to produce steam for the THP as well as heat and power for the Bonnybrook WWTP.

Hydrolyzed sludge (HWAS) from the THP system is pumped to the blend tanks where it is mixed with primary sludge (thickened fermented/waste fermented sludge (TFS/WFS) and thickened primary sludge (TPS)). Primary scum and fats, oils and grease (FOG) can also be sent to the blend tanks, as an alternative to going directly to the digesters. Digester feed pumps transfer the blended sludge to the digesters (one dedicated pump and pipeline per digester). The digester mixing system is currently being upgraded to a hydraulic system to improve mixing within the digesters.

Centrate from the centrifuges flows by gravity into the DAF subnatant channel where it then returns to the front end of the plant. In the future, if nutrient recovery is implemented at the plant, the centrate may be transferred to the nutrient recovery facility. Waste sludge from the centrifuges (slop) is flushed to a slop tank located on the lower level of the facility and then pumped to the centrifuge feed tanks since it cannot flow by gravity into the feed tanks. Polymer is injected into the centrifuge feed lines to each centrifuge, at three separate locations.

The sludge dewatering and storage equipment is covered and foul air is collected and sent to the bio-trickling filters via positive discharge odour control fans. The bio-trickling filters discharge through a large dispersion stack which also receives exhaust building air. Process gas generated in the THP system is handled separately, through vendor supplied process gas units (one per train). The process gas units are located on the THP facility roof in a separate enclosure which is ventilated to the odour control system. Each vessel in the THP train is equipped with a pressure relief valve to relieve on over-pressure conditions. The relief valves relieve to a caisson on each THP train which is piped to a pressure relief stack located on the roof.

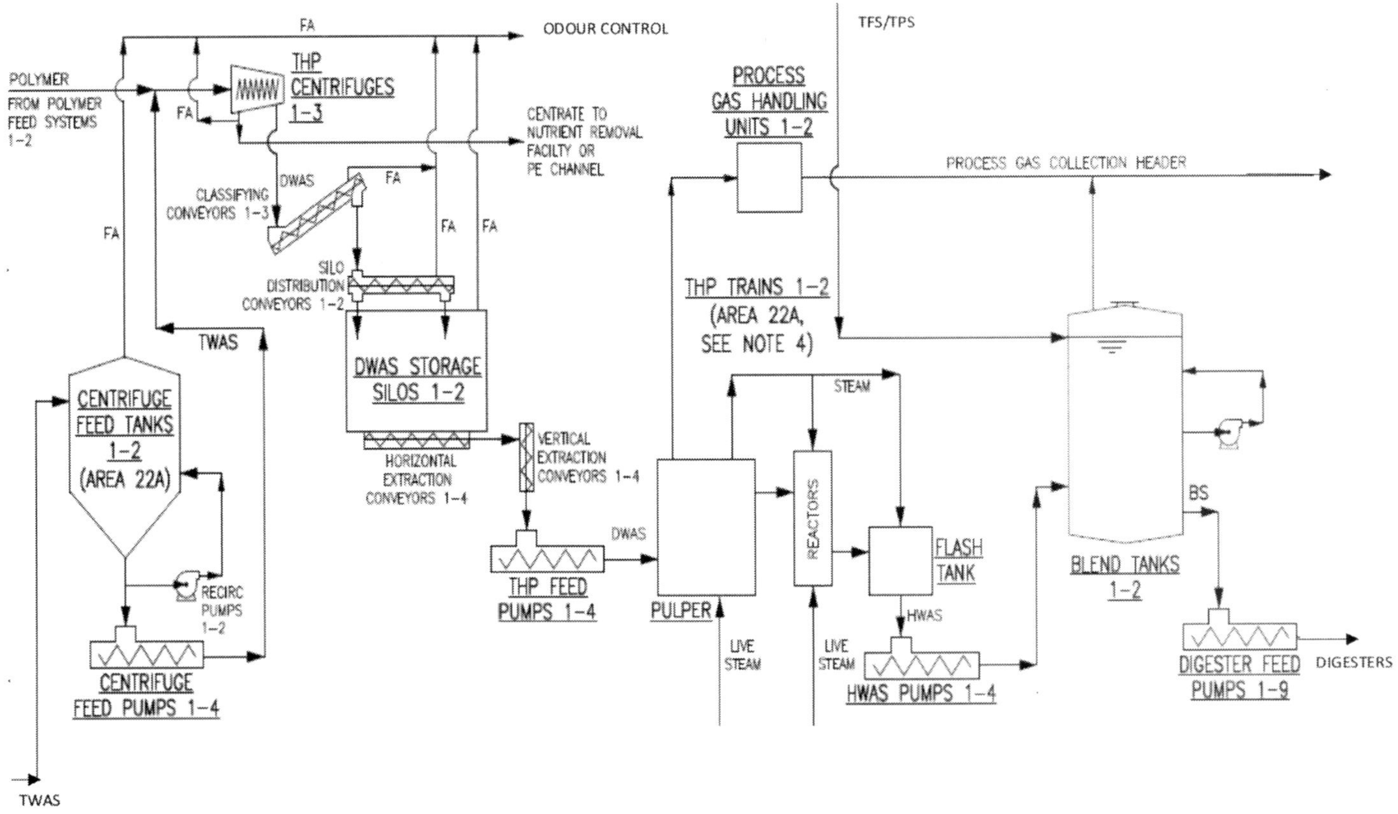

Figure 5: THP Facility Process Flow Diagram

DESIGN BASIS

A THP plant must be sized to handle shorter term peak flows than a digestion system. The THP facility at Bonnybrook WWTP t has been designed to handle the predicted 2037 maximum week hydraulic and solids loads based on the combined TWAS loads for the Bonnybrook WWTP, including the Plant D Expansion.

The TWAS is typically thickened to between 3.75 and 4.5 percent total solids (TS) in the recently upgraded DAF facility. The TWAS flows have been calculated assuming an average TS concentration of about 4.2 percent. However, the centrifuges and sludge pumps have been sized based on the minimum solids concentration of 3.2 percent and maximum flow.

The TWAS needs to be dewatered to generate DWAS at 16.5 to 18 percent TS to meet THP requirements. Cambi prefers to receive the sludge at 16 percent at the pulper to prevent bridging and short circuiting of the steam in the reactors. Based on preliminary dewaterability testing, it is feasible to dewater the secondary sludge using centrifuges to a concentration of 18 percent without the need to use significantly more polymer. Therefore, the TWAS centrifuges have been sized to dewater the TWAS to a dry solids concentration between 16 and 18 percent. An average of 16 percent TS was used to size the THP pumps. Dilution water can be added after the centrifuges if required.

Boundary layer injection rings will be provided on the discharge piping from THP feed pumps to reduce headlosses through the pipelines. Effluent water may be added upstream of the THP feed pumps if needed to bring the solids concentration down to 16 percent at the pulper tank. The temperature of the feed sludge impacts the steam requirements for the THP process. Typically the sludge is between 10°C and 18°C, depending on ambient conditions.

The hydrolysed sludge exits the flash tank at around 100 to 110°C and should be cooled to protect the HWAS pump stators and downstream equipment as well as prevent the digesters from over-heating. Cambi typically achieves this by injecting plant effluent water upstream of the HWAS pumps. However, in Bonnybrook's case, where only the secondary sludge is being hydrolysed and then blended with cool primary sludge, the resulting heat of the blended sludge will typically be within the normal operating range required for anaerobic digestion. Therefore, rather than cooling the HWAS with effluent water and risk bringing the blended sludge temperature down below that which is optimal for digestion, a side stream of blended sludge will be recycled to the suction of the HWAS pumps to provide sufficient cooling for equipment protection without losing all the heat generated through THP. The option to use effluent water for dilution and cooling is included as well and may be needed during the summer months or in the event that ammonia levels start to exceed acceptable concentrations in the digesters.

The design basis for the THP system is provided in Table 7.

Table 7: THP System Design Basis

	Average Month	Maximum Month	Maximum Week (Design Basis)
TWAS Mass Loading	50,485 kg/d	61,556 kg/d	75,903 kg/d
TWAS Flow	1,122 m^3/d	1.641 m^3/d	2,024 m^3/d
DWAS Mass Loading	48,380 kg/d	59,200 kg/d	73,000 kg/d
DWAS Flow	300 m^3/d	370 m^3/d	456 m^3/d
HWAS Mass Loading[1]	40,800 kg/d	50,000 kg/d	61,400 kg/d
HWAS Flow[1]	355 m^3/d	435 m^3/d	535 m^3/d

Note 1: HWAS flows and loads exclude blended sludge recycle.

Major equipment sizing is provided in Table 8.

Table 8: THP Major Equipment Sizing

Parameter	Value
Dewatering System	
Centrifuges	
Number of Units	3 (2 duty/1 standby)
Design Solids Loading per unit	1594 kg/h
Maximum Solids Loading per unit	2000 kg/h
Predicted Polymer Consumption	8 to 10 kg active polymer per tonne dry solids
TWAS Solids Concentration	3.75% to 4.5%, average 4.2% TS
DWAS Solids Concentration	16% to 18% TS
DWAS Feed Temperature	10˚C to 18˚C
HWAS Concentration Post Dilution	10-13.5% TS (concentration depends on dilution or blended sludge recycle)
HWAS Temperature Range – before cooling	100–110°C
HWAS Temperature Range – after cooling	80–92°C
Steam Consumption Range	2.0–3.1 t/h
Maximum Instantaneous Steam Demand	3,800 kg/h
Minimum Steam Pressure at THP Interface	11.5 bar(g)
Minimum Steam Temperature at THP Interface	190 °C

Future Expansion

The THP facility can be expanded to handle roughly 25 percent higher solids loading with the addition of a fourth centrifuge (space has been allocated for this purpose). The THP facility can also be expanded by approximately 50 percent by replacing the centrifuges with the next model size up. The buffer tanks (centrifuge feed tanks, dewatered sludge storage silos, and blend tanks) have been sized for the increased flow resulting from a 50 percent plant expansion. The THP trains provided by Cambi can actually treat about 85,000 kgDS/d each and therefore provide sufficient capacity for up to approximately year 2039 while still maintaining The City's requirements for redundancy. Beyond that point, The City would require a new Cambi B6x4 train unless interconnections between trains could be provided. It is more likely that a new THP facility will be required to provide additional THP trains as well as upstream dewatering storage and conveyance equipment and downstream blending and digester feed equipment. Space for this new facility is available to the east and north east of the first THP facility.

Detailed design commenced in August 2016 and is expected to be complete in early summer of 2017.

REFERENCES

Waste Organic Materials and Biosolids Master Plan (CH2M, 2011)

Bonnybrook WWTP Plant D Expansion – Conceptual Design Report (Stantec, 2014)

Reinventing Your Biosolids Management Program and Unlocking Resource Recovery with Thermal Hydrolysis – The City of Raleigh's Journey

C. Michael Bullard[1], Amy Hanna[1], Greg Knight[2], Kent Lackey[2], Brenan Buckley[3], Perry Schaefer[3], T. J. Lynch[4], and Aaron Brower[4]

[1] Hazen and Sawyer
[2] Black and Veatch
[3]Brown and Caldwell
[4] City of Raleigh

Corresponding Author: C. Michael Bullard, mbullard@hazenandsawyer.com

ABSTRACT

The City of Raleigh operates the Neuse River Resource Recovery Facility (NRRRF), a 75 MGD (11.83 ML/hr) capacity advanced nutrient removal treatment facility. The treatment facility includes primary and advanced secondary treatment coupled with tertiary filtration to produce a high-quality effluent with typical effluent total nitrogen concentrations less than 3.0 mg TN/L. Residuals generated during the treatment process are currently managed via a range of product streams with different characteristics and different levels of stabilization.

During its most recent residuals masterplanning cycle the City adopted a course of action which will consolidate a variety of residuals stabilization processes in to a single stabilization process utilizing thermal hydrolysis pre-treatment (THP) coupled with mesophilic anaerobic digestion. Following the masterplanning process, the City of Raleigh retained an engineering team to develop the recommended masterplan concept into a Preliminary Engineering Report (PER) and is currently underway with detailed design activities with project delivery using a Construction Manager at Risk (CMAR) delivery method.

This paper presents the approach to, and results of, the concept development advanced through the preliminary engineering report phase of the work. Major facilities evaluated in the PER and currently under detailed design include:

- Fats, Oils and Grease (FOG) Receiving and Handling
- Secondary Sludge Thickening Improvement
- Blended Sludge Screening and Pre-Dewatering
- Thermal Hydrolysis Pretreatment and Mesophilic Anaerobic Digestion
- Post-Dewatering System Improvements
- Sidestream Nitrogen Removal via ANNAMOX bacteria

This paper shares the City of Raleigh's journey to reinvent its biosolids management program by applying a THP and anaerobic digestion process to produce on-site a Class A biosolids product from all of its residuals.

Background

The City of Raleigh operates the Neuse River Resource Recovery Facility (NRRRF), a 75 MGD (11.83 ML/hr) capacity advanced nutrient removal treatment facility, which serves most of the municipalities located in the eastern half of Wake County, North Carolina in the rapidly growing Research Triangle Park metropolitan area. The treatment facility includes primary and advanced secondary treatment coupled with tertiary filtration to produce a high-quality effluent with typical total nitrogen concentrations less than 3.0 mg TN/L suitable for direct discharge to the Neuse River or for utilization as a high quality reuse water through a growing reuse program. Residuals generated during the treatment process are currently managed by the City in a variety of outlet streams, including:

- Alkaline Stabilization to Class A and marketing and distribution as "RaleighPlus"
- Off-Site Contract Composting of dewatered unstabilized cake
- Off-Site Class B Liquid Land Application of aerobically-stabilized residuals
- On-Site Class B Liquid Land Application of aerobically-stabilized residuals

The City has had a long term interest in environmental stewardship and resource recovery and has been a long-term participant in the National Biosolids Partnership's Environmental Management Program as a Platinum Certified Member since December 2006. As a part of their NBP-EMS program the City had conducted a residual management masterplan study in 2008 with a subsequent update in 2013. The 2008 masterplan evaluated a number of management technologies and approaches and recommended upgrading the management program to include mesophilic anaerobic digestion and thermal drying to produce a Class A product stream. However, economic considerations immediately following completion of the 2008 plan resulted in a delay in implementation of the concept and resulted in a desire to update the masterplan in 2013 to affirm, or change, decisions developed and recommended in the 2008 plan.

The 2008 masterplan evaluated the thermal hydrolysis pretreatment (THP) process and classified it as an emerging technology for biosolids management. The 2013 masterplan update; however, acknowledged the continued development and commercialization of the THP process over the intervening five-years and recommended that the anaerobic digestion process be coupled with THP to produce a Class A product. Drying was considered as a "follow-on" process in the 2013 masterplan for additional mass reduction and for the production of a differentiated product stream for marketing and distribution. Following the 2013 masterplan update the City of Raleigh retained an engineering team to develop the recommended masterplan concept into a Preliminary Engineering Report (PER) with a plan to move the project into detailed design, construction, and ultimately into commissioning.

The following sections of the paper will present the approach to, and results of, the concept development as it has advanced through the preliminary engineering report phases and initial detailed design phases of the work. The facilities envisioned as part of the City of Raleigh's BioEnergy Recovery Program (BERP) are planned to be delivered in at least two phases. The first phase of improvements will provide infrastructure to support current plant design loadings (75 MGD / 11.83ML/hr) and a second phase will ultimately expand capacity to meet design loading rates associated with 90-MGD (14.20 ML/hr), the plant's 2035 expected volumetric permitted flow capacity.

BioEnergy Recovery Program Facility Description

The major facilities recommended in the PER to be constructed as part of the BERP include:

- Fats, Oils and Grease (FOG) Receiving and Handling
- Secondary Sludge Thickening Improvements
- Blended Sludge Screening and Pre-Dewatering
- Thermal Hydrolysis Pretreatment and Mesophilic Anaerobic Digestion
- Post-Dewatering System Improvements
- Sidestream Nitrogen Removal via ANNAMOX bacteria
- Digester Gas Treatment and Utilization

The general process flow arrangement for the major facilities are shown in Figure 1. The major unit process components envisioned are described in greater detail below.

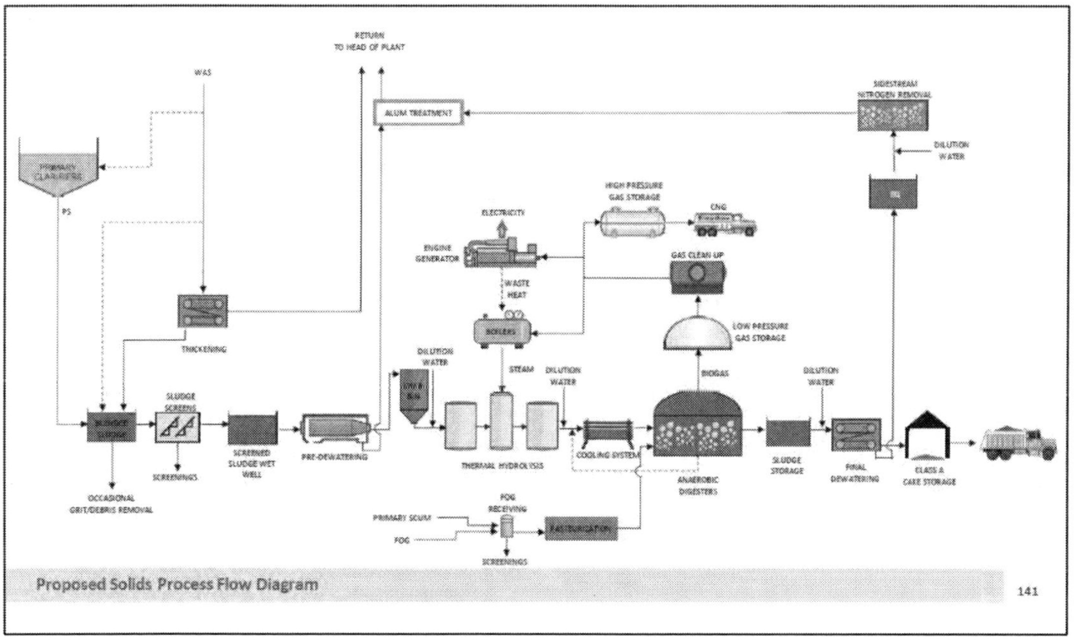

Figure 1 – BERP Process Flow Schematic

The general facility arrangement and conceptual layout rendering on the NRRRF site is shown in Figure 2.

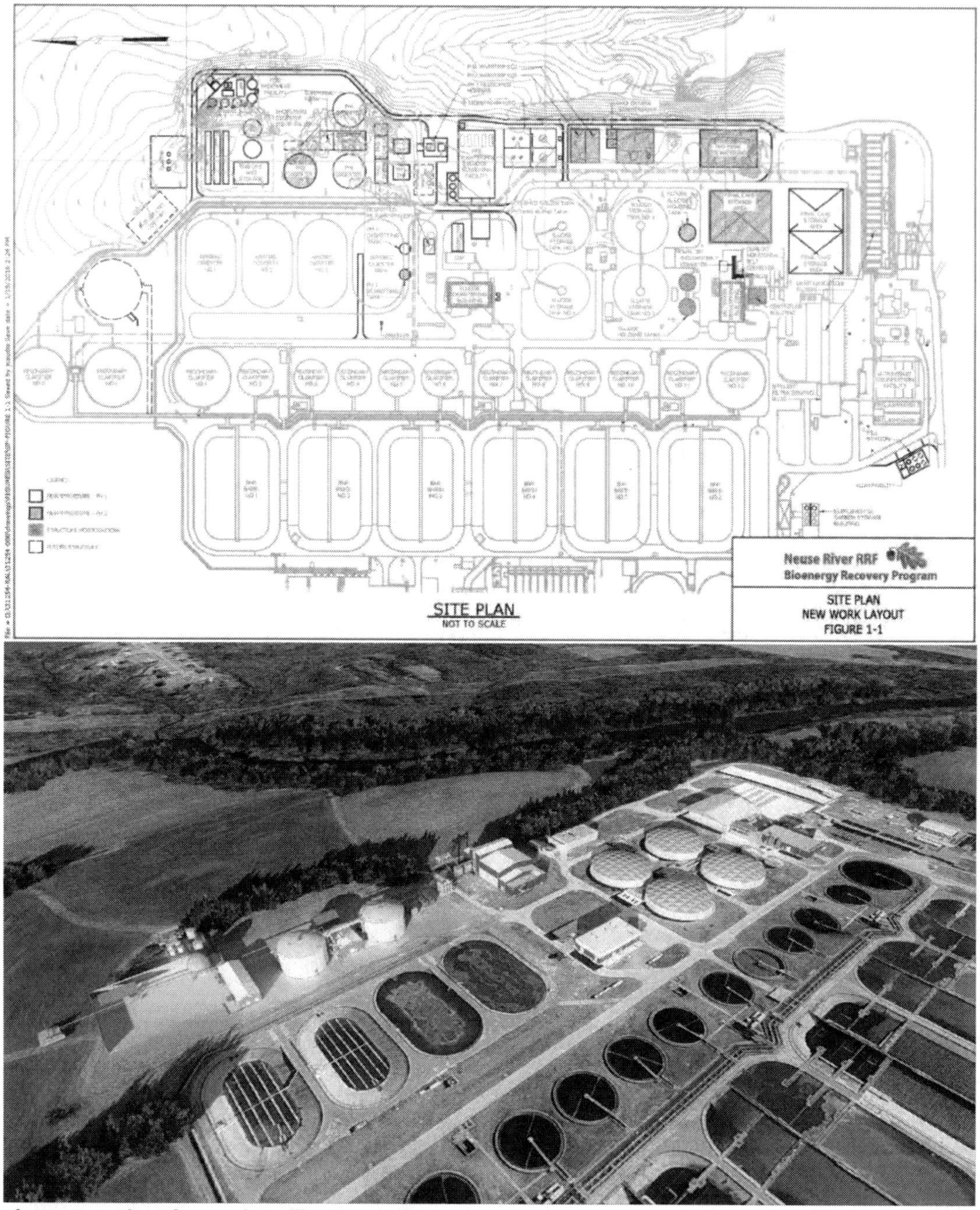

Figure 2. – Site Plan and Facility Rendering

Fats, Oils and Grease (FOG) Receiving and Handling

A market analysis study was conducted to estimate the quantity of grease interceptor waste (GIW), commonly referred to as "FOG", in the plant's Wake County service area and typical tipping and receiving costs that haulers in the region were paying for handling and disposal. Estimates were developed using several methods including: (1) service area population and (2) service area food service. Based on the market analysis a decision was made to include provision for receiving up to 420,000 gallons per week spread across six (6) delivery days with continuous seven (7) day per week addition to the digestion process.

The GIW/FOG receiving station will include two unloading connections with material routed through rock traps and screening prior to delivery to a GIW/FOG receiving wet well. The wet well included chopper pumps for both mixing and transfer service. The GIW/FOG will not be routed directly to the THP unit process but will enter the digestion process downstream as a pasteurized material using the 40 CFR 503 time and temperature provisions to achieve pasteurization. The pasteurization process will use multiple batch reactors operating in a fill-heat-hold-drain configuration. Pasteurized GIW/FOG will be routed to storage tankage for continuous feed to the anaerobic digester unit process.

A preliminary general arrangement rendering for the full facility is shown in Figure 3. The receiving station will include two screening and handling units discharging to a receiving wet well equipped with dual chopper pumps for both mixing service and for transfer of received materials to the downstream pasteurization tanks. Initially four pasteurization reactor vessels will be provided with the ability to expand the facility to include two (total of six) reactors. Following batch pasteurization the material will be routed to two large volume storage tanks from which the pasteurized material will be feed to the anaerobic digester tanks. Full system containment will be provided for the receiving, pasteurization, and storage tank areas.

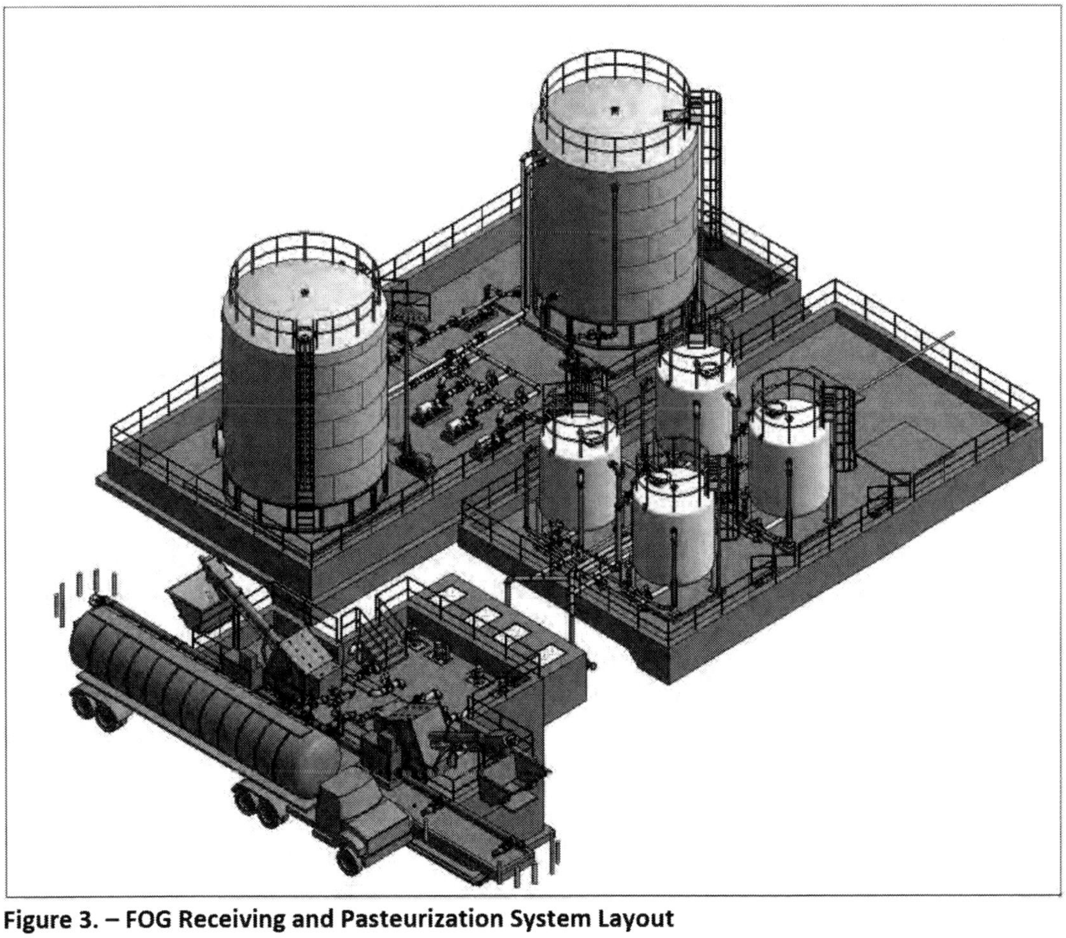

Figure 3. – FOG Receiving and Pasteurization System Layout

Enlarged renderings of the pasteurization system and the storage tankage are shown respectively in Figure 4 and Figure 5.

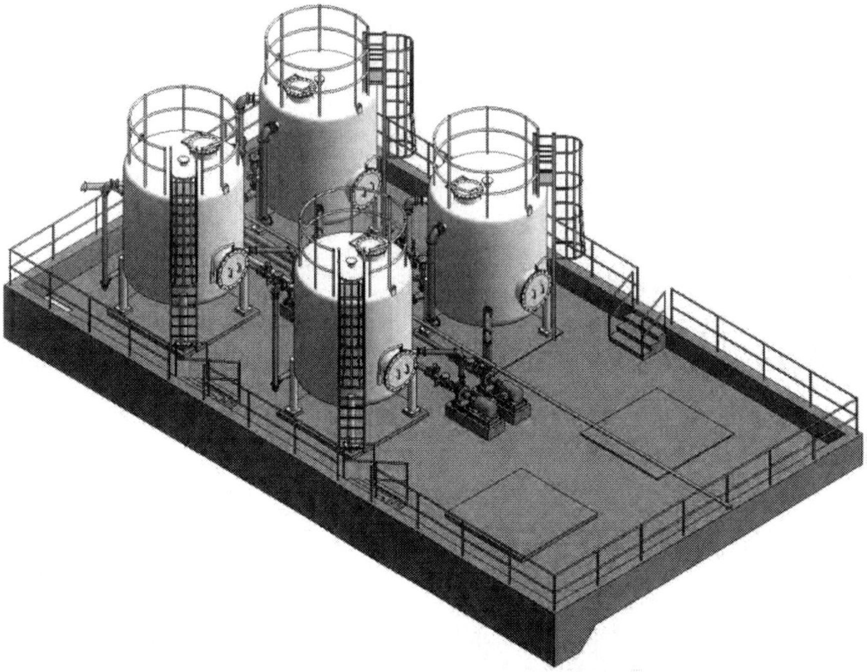

Figure 4. – FOG Receiving and Pasteurization – Pasteurization Tanks

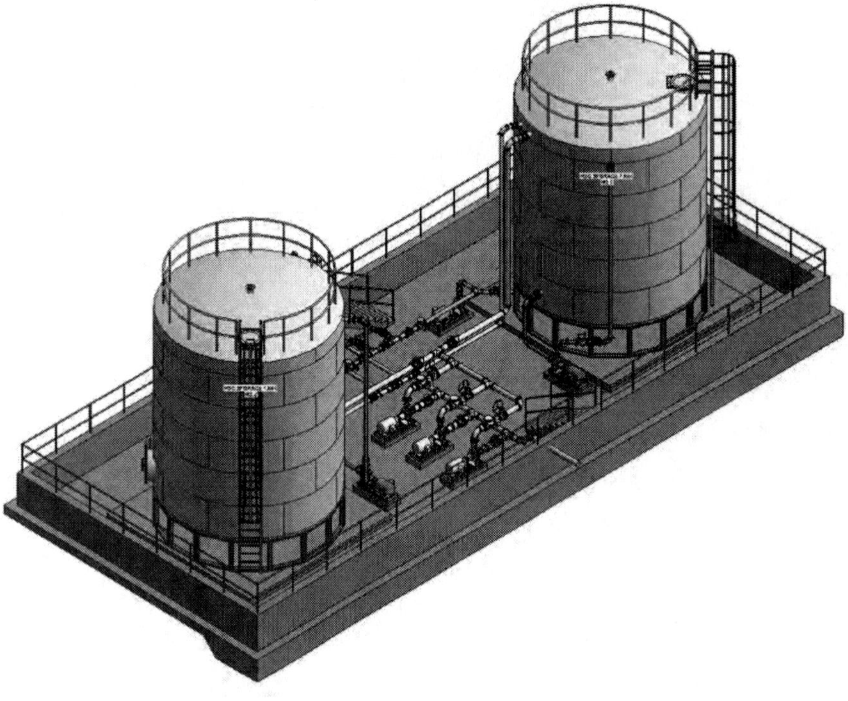

Figure 5. – FOG Receiving and Pasteurization – Storage Tankage

Secondary Sludge Thickening Improvements

Waste activated sludge (WAS) will be thickened upstream of the screening and pre-dewatering unit process using gravity belt thickening (GBT) unit process equipment. The existing GBT building includes three (3) two meter (2.0-m) GBT thickening units. The envisioned upgrades will include installation of a fourth GBT in an open slot in the thickening room. The upgrade to the thickening unit process will also include replacement of GBT feed pumps, upgrade to the polymer preparation and feed equipment and miscellaneous improvements for the piping and valving to provide for improved process control.

Blended Sludge Screening and Pre-Dewatering

Thickened waste activated sludge (TWAS) will be blended with primary sludge (PS) in a new Screening and Pre-Dewatering Building to be constructed on the plant site. This process building will include the following:

- Blended Sludge Storage – Three storage reactors with approximately 60,000 gallons working volume to be used for receiving and blending TWAS and PS upstream of screening and dewatering. These storage tanks will be fully mixed using vertical mixing equipment. This storage volume will be initially sized to meet future Phase II throughput demands.
- Screening – The Phase I improvements will include four (4) sludge screens with provision for adding two (2) more units with the Phase II expansion. These screening units will removed debris and materials greater than 5-mm in size to protect the downstream THP system equipment following dewatering. Screenings will be collected in containers and hauled to a local landfill for disposal.
- Screened Sludge Storage – Three (3) storage reactors similar in size and configuration to the blended sludge storage reactors will be provided to decouple the screening and the dewatering processes and provide additional dampening in the system upstream of the THP treatment process.
- Pre-Dewatering – The Phase I improvements will include three (3) horizontal decanter centrifuges each with a dewatering capacity of 4,000 dry pounds per hour. Provision will be made to expand the pre-dewatering capacity under Phase II by adding two (2) additional horizontal decanter centrifuges. An overhead bridge crane will be provided for equipment maintenance.
- Dewatered cake storage will be provided by relocation of two existing dewatered cake sludge storage hoppers. The THP feed pumps will be located integral with the bulk dewatered cake storage facility and provisions are also made for truck loadout from the pre-dewatering building to meet plant sludge disposal needs when the THP system would be out of service for its annual maintenance inspection.
- Polymer Storage and Conditioning – The Phase I improvements will include two (2) bulk neat polymer storage tanks and associated mixing and neat polymer make-up equipment. Provision will be made in the facility layout for up to two (2) additional bulk neat polymer storage tanks.
- Building Support Systems – This facility will also include spaces for men's and women's bathrooms and locker facilities, a control room, a wet lab space, and mechanical and electrical rooms.

Renderings of the intended facilities are shown as follows:

- Figure 6. Screening and Dewatering Building 3-D Rendering
- Figure 7. Screening and Dewatering Building – Upper Floor
- Figure 8. Screening and Dewatering Building – Lower Floor

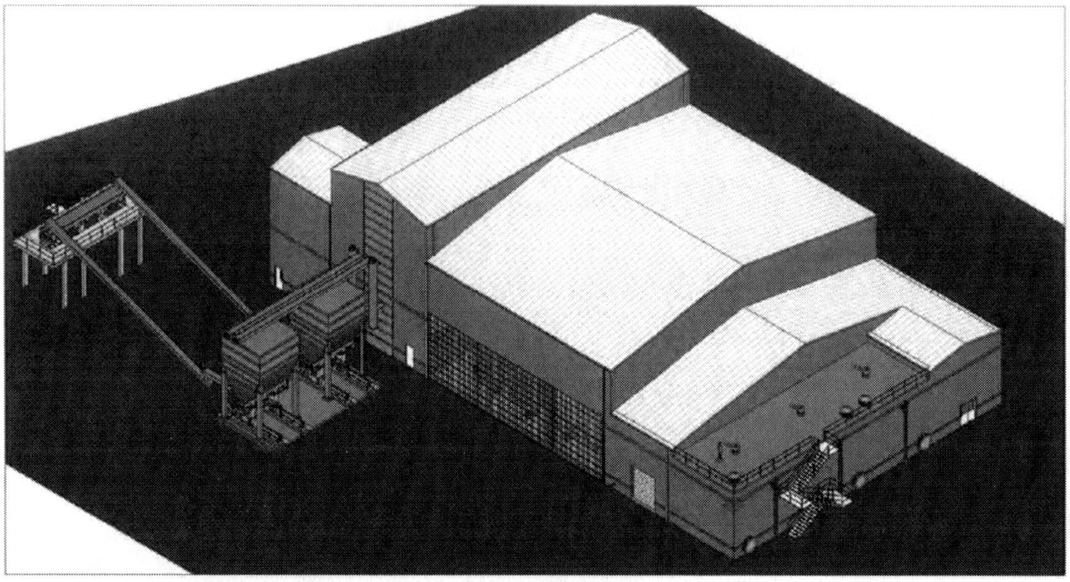

Figure 6. – Screening and Dewatering Building 3-D Rendering

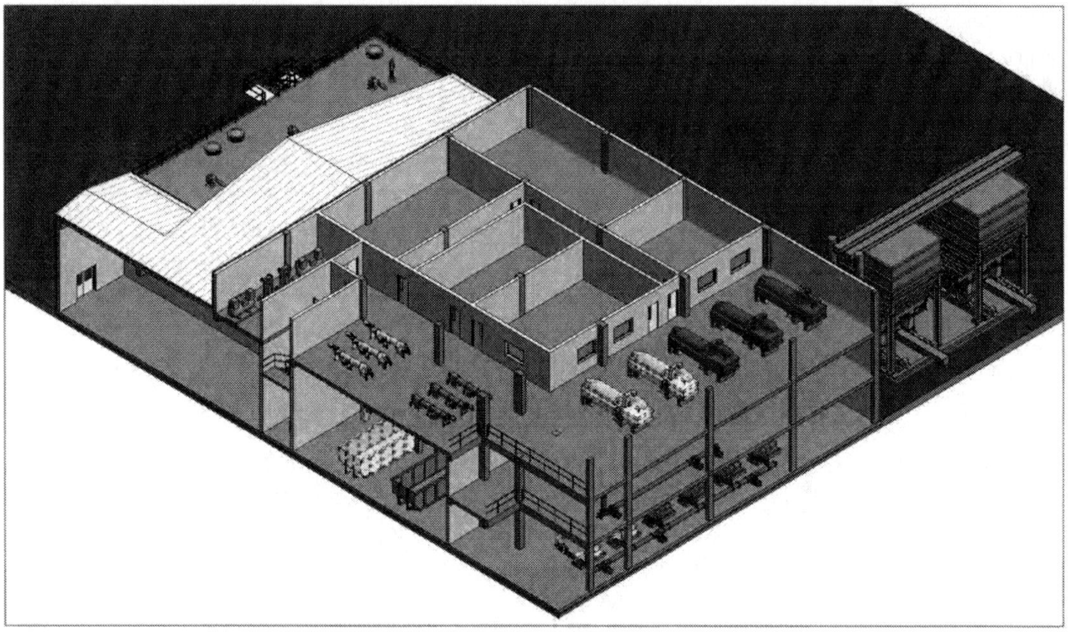

Figure 7. – Screening and Dewatering Building – Upper Floor

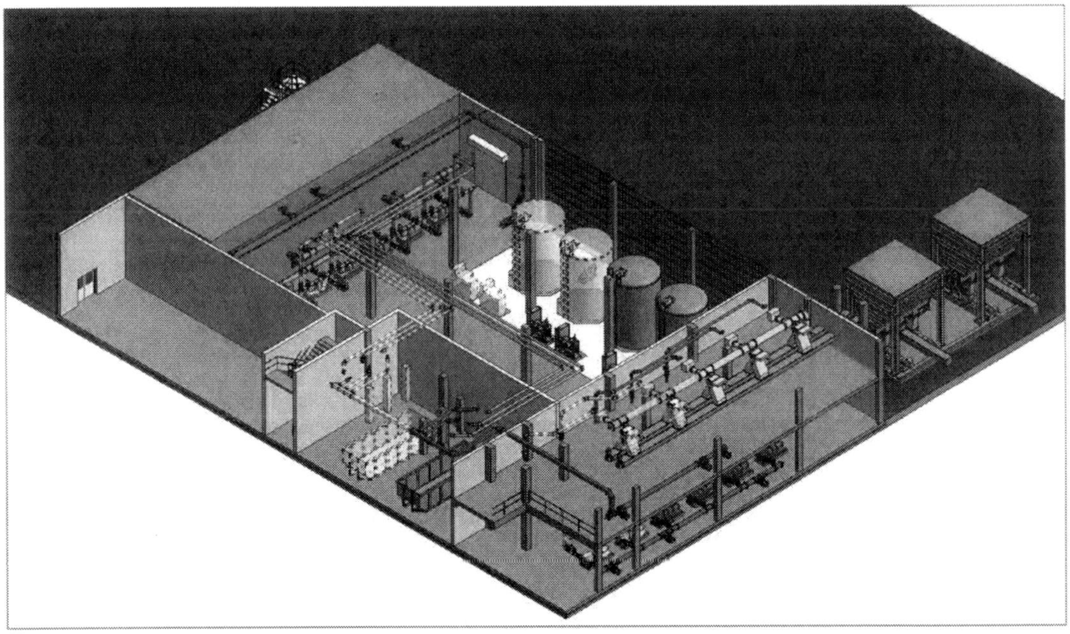

Figure 8. – Screening and Dewatering Building – Lower Floor

Thermal Hydrolysis Pretreatment and Mesophilic Anaerobic Digestion (THP+MAD)

The Phase I thermal hydrolysis pretreatment (THP) process initially have a design throughput capacity of 184,000 dry pounds per day (83,600 kg per day) at 16.5% feed solids content. Provision will be made in the design and general arrangement for the ultimate addition of a second THP treatment train in the Phase II facility expansion. Key elements of the THP+MAD system include:

- THP Reactor Process – During the PER process preliminary sizing and design was developed around the CAMBI B6-4 reactor system and the Kruger BIOTHELYS batch reactor treatment systems. Following completion of the PER and during the early detailed design development the City decided to move forward with the design based on the CAMBI B6-4 reactor system.
- Post THP Cooling – Cooling of the post-THP hydrolyzed sludge is accomplished initially by dilution of the post-THP sludge, which has a concentration of 13.5% to 14.0% based on condensed steam water mass, to a digester feed concentration in the 9.5% to 10.5% range required to manage digester ammonia concentrations. Following this initial dilution, which will cool the post-THP sludge from approximately 220°F (105°C) to about 170°F to 160°F (75°C to 70°C), supplemental cooling will be required to maintain anaerobic digester temperature in the 104°F (40°C) range. Final cooling will be provided by a tube-in-tube heat exchanger provided as part of a digested sludge recirculation loop to maintain velocity in the heat exchanger.
- Anaerobic Digestion – The Phase I anaerobic digestion system consists of two (2) anaerobic digesters with a nominal straight shell operating volume of 2.1 million gallons (17.5 ML). These digester tanks will be provided with fixed covers and pumped mixing systems and be designed to operate at a fixed volumetric overflow level with special provisions to accommodate storage of up to 10% volume inside the tank for rapid volume expansion in addition to being provided with an emergency overflow for volume expansion beyond this amount.

- Digester Control Building – A digester control building is provided with will house the mixing pumps, recirculation and the transfer pumps. Full redundancy is provided for the mixing pumps with an installed spare pump provided for each of the two digester tanks. Recirculation pumps are provided to support the digester cooling system recirculation loop. Transfer pumps are provided to move digested sludge from the anaerobic digester complex to the digested sludge storage tanks immediately upstream of final dewatering.

Renderings of the THP and anaerobic digestion system are shown in the following:

- Figure 9. Anaerobic Digester Typical Configuration
- Figure 10. Digester Complex Layout
- Figure 11. Digester Control Building General Arrangement

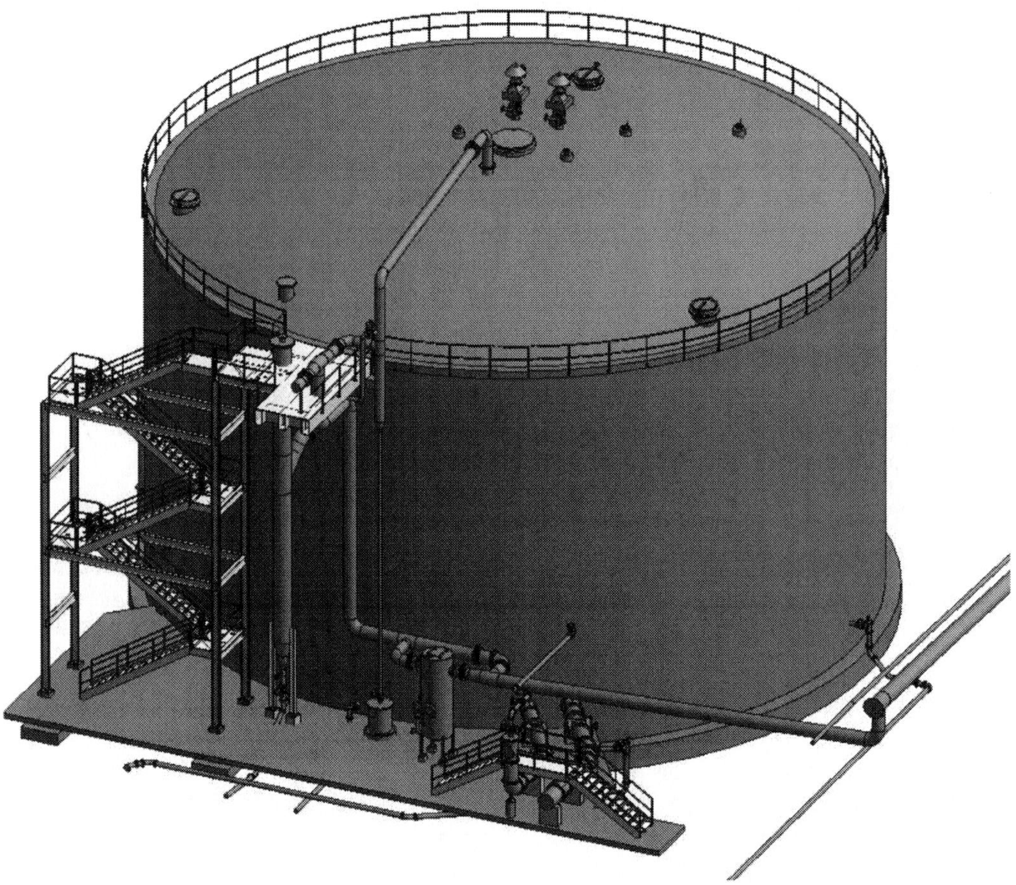

Figure 9. – Anaerobic Digester Typical Configuration

Figure 10. – Digester Complex Layout

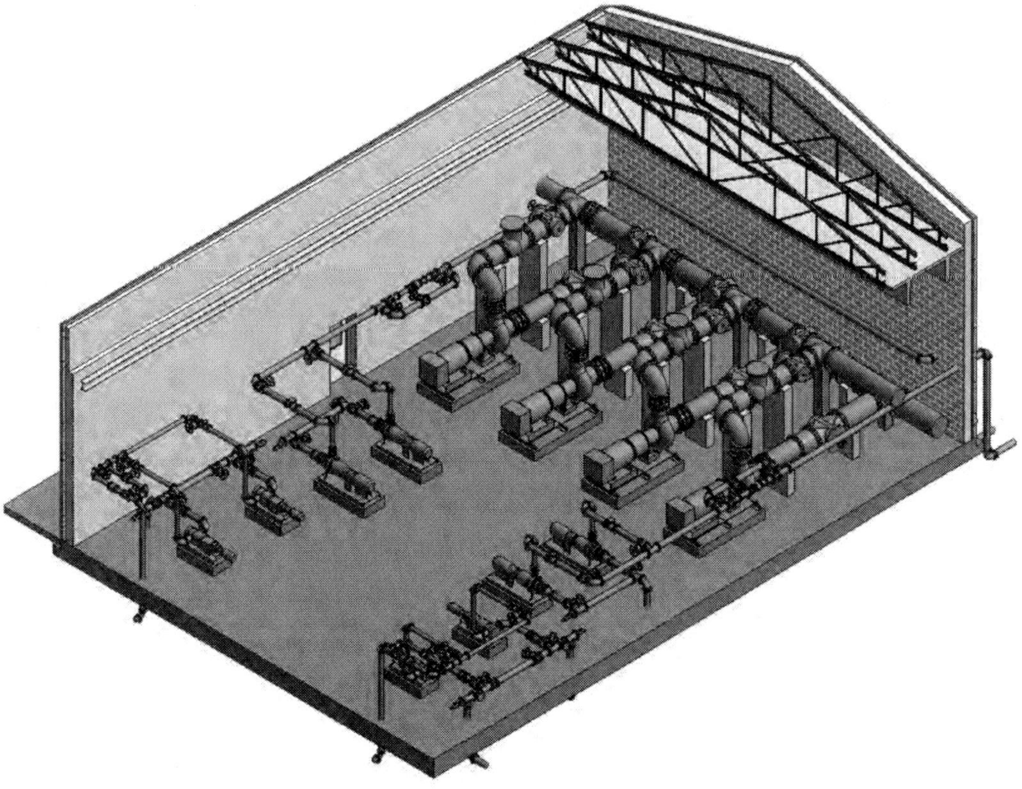

Figure 11. – Digester Control Building General Arrangement

Post-Dewatering System Improvements

The Neuse River RRF has three 2.0-meter belt filter presses (BFP) and one 650-mm nominal diameter dewatering centrifuge (DCEN) which prior to the upgrades were utilized to produce a dewatered cake for downstream on-site alkaline stabilization or for off-site composting by a 3rd party service provider. The improvements required to accommodate a transition to the post-THP Class A dewatering service include:

- Digested Sludge Storage Tanks – The two existing aerated and pumped mixed sludge storage tanks upstream of dewatering will be retained in service. The aeration system will be retained; however, it is not normally intended to operate on the anaerobically digested sludge and mixing will be provided by the pumped jet mixing system. Minor work will be completed on these tanks for rehabilitation of the geodesic dome covers.
- Dewatering Unit Feed Pumping – Feed pumping to the current BFP and DCEN units is provided by progressing cavity pumps sized to provide a feed rate to the downstream dewatering units of up to 300-gallons per minute at a typical feed solids concentration of 2.0%. However, the post-THP residuals concentration is expected to be in the 4.5% to 5.5% total solids range making the existing pumps oversized for feeding the dewatering equipment. These existing pumps will be replaced with smaller volumetric capacity pumps to provide for a mass feed rate to the BFP dewatering equipment of up to 2,500 dry pounds per hour (1,140 kg/hour) and to the DCEN equipment of up to 3,000 dry pounds per hour (1,360 kg/hour).
- Pre-Dewatering Dilution Water - Post-feed pumping dilution capacity to support a 1:1 dilution will be provided as part of the improvements. The dilution water will serve to reduce the feed concentration to the dewatering units to 2.25% to 2.75% for enhanced polymer dispersion during conditioning and also to provide dilution to meet sidestream treatment conditioning requirements.
- Belt Filter Press Dewatering – The existing belt filter press dewatering equipment is nearing the end of its useful life (> 20 years age) and will be replaced as part of this project. The existing BFP equipment has eight (8) pressure rolls; however the replacement dewatering equipment will be provided with twelve (12) pressure rolls for enhanced dewatering of the post-THP anaerobically digested sludge. Special provision will be made on the BFP dewatering units to separate filtrate from the washwater. Filtrate will be collected and routed to sidestream treatment and washwater will be routed back to the plant headworks through the plant drain system.
- Centrifuge Dewatering – The existing DCEN unit was installed less than 10-years ago and will continue in operational service. Minor piping modifications will be required to route the centrate to the sidestream treatment system. Furthermore, downstream screw conveyance equipment will be replaced with a belt conveyor to integrate with the other dewatered cake conveyance system modifications.
- Dewatered Cake Storage - The existing dewatered cake storage bins utilized to support the off-site composting operation will be relocated as part of this project and repurposed for wet cake receiving from the pre-dewatering process and used to support the THP feed pumping system.

These improvements leverage the plant's existing asset base by making use of the current dewatering building and significant portions of the conveyance and post-dewatering covered cake storage area.

Sidestream Treatment Improvements

Several sidestream treatment alternatives were evaluated during the PER phase of the project, including:

- Base Case – This option would have returned the nitrogen and phosphorus rich final dewatering filtrate to the plant headworks with minimum nutrient removal and/or recovery processing. This established the economic baseline upon which the other sidestream treatment alternatives would be assessed. Phosphorus removal for the base case would utilize chemical phosphorus removal to achieve effluent total phosphorus limits. Nitrogen removal for the based case would use a 4-stage BNR activated sludge process coupled with deep bed denitrification filters with supplemental carbon addition.
- Phosphorus Removal Only Option – In lieu of chemical phosphorus removal in the mainstream this alternative evaluated using a struvite recovery technology to create a marketable fertilizer product. The nitrogen removal under this alternative remained in the mainstream treatment process.
- Nitrogen Removal Only Option – In lieu of mainstream nitrification and denitrification this alternative evaluated utilizing deammonification for sidestream nitrogen removal. Several different vendor furnished technologies were considered as part of the evaluation. The phosphorus removal under this alternative remained chemical addition in the mainstream treatment process.
- Nitrogen and Phosphorus Removal Option – This option coupled the struvite recovery and deammonification processes into a combined sidestream treatment approach for both phosphorus and nitrogen removal.

The recommended option retains chemical phosphorus removal as the selected alternative; however, provisions have been made in the process flow and site layout to convert to struvite harvesting and recovery should economic and/or regulatory conditions support a treatment technology transition. The nitrogen removal approach recommended was to provide deammonification to the sidestream from the final dewatering process. Several technology approaches will be further developed during the detailed design phase to the 30% design level at which point a final technology selection will be made for implementation as part of the overall Phase I improvements project.

Digester Gas Treatment and Utilization

Multiple options for digester gas utilization were evaluated during the PER phase of the project. All options included low pressure digester gas storage using ground level dual membrane gas storage coupled with fixed cover anaerobic digestion and digester gas utilization for THP process steam generation. Excess digester gas from the stabilization process was evaluated with options ranging from waste gas flaring (no energy recovery), on-site electrical generation without heat recovery, full on-site combined heat and power, cleaning for municipal fleet vehicle fuel, and cleaning for pipeline injection as a renewable natural gas (rNG) resource. These alternatives, and the associated economic analysis, are more fully described in Knight, et. al. (WEFRBC, 2017). The final selected alternative includes cleaning gas to pipeline injection quality and utilizing the natural gas pipeline to transport the rNG resource to a fleet fueling station for use in municipal vehicles with surplus rNG to be marketed to others as a vehicle biofuel.

The preliminary general arrangement drawing, excluding the final digester gas treatment train, is shown in Figure 12. During the PER phase several gas cleaning train approaches were evaluated for hydrogen sulfide, moisture, particulate, siloxane and carbon dioxide removal. Final selection of the gas cleaning train system, not shown in Figure 12, will be completed during the final design phase based on site visits, additional technology assessment and improved costing and economic analysis of total lifecycle costs including capital and operational and maintenance costs.

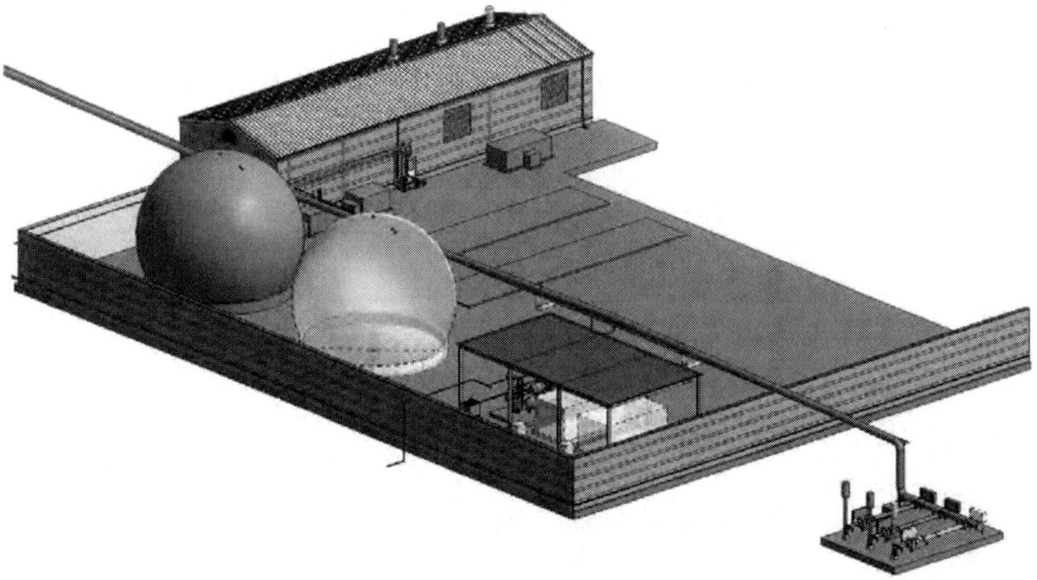

Figure 12. – Digester Gas Storage and Boiler Building General Arrangement

Current plans are to utilize pipeline natural gas for firing the steam boilers to support the THP system. Federal renewable fuel standard requirements (40 CFR Part 80, Subpart M – Renewable Fuel Standard) allow biogas generated from wastewater treatment to qualify as a cellulosic biofuel (Category 3 D-Code) which makes this fuel stream eligible to generate a marketable RIN (Renewable Identification Number) as part of its production. To the extent that a marketable RIN can be generated this provides and avenue for the City of Raleigh to further monetize the benefits associated with production of the rNG product. Therefore, the highest economic benefit for the city can be realized by purchasing pipeline NG for firing the steam boilers and maximizing production of rNG from the digester's biogas.

A preliminary general arrangement of the boiler building is shown in Figure 13. Two boilers, a duty and standby, will be installed as part of the Phase I improvements. As shown, provision has been made in the layout for the future addition of a third boiler to accommodate increased steam demands from the Phase II improvements which will add a second THP train to the plant's overall processing capacity. Additional systems in the boiler building include boiler water softening and treatment in addition to de-aeration and boiler feed water pumping. The boiler building also serves as an electrical distribution hub for the digester gas cleaning and associated equipment located in this quadrant of the overall plant facilities.

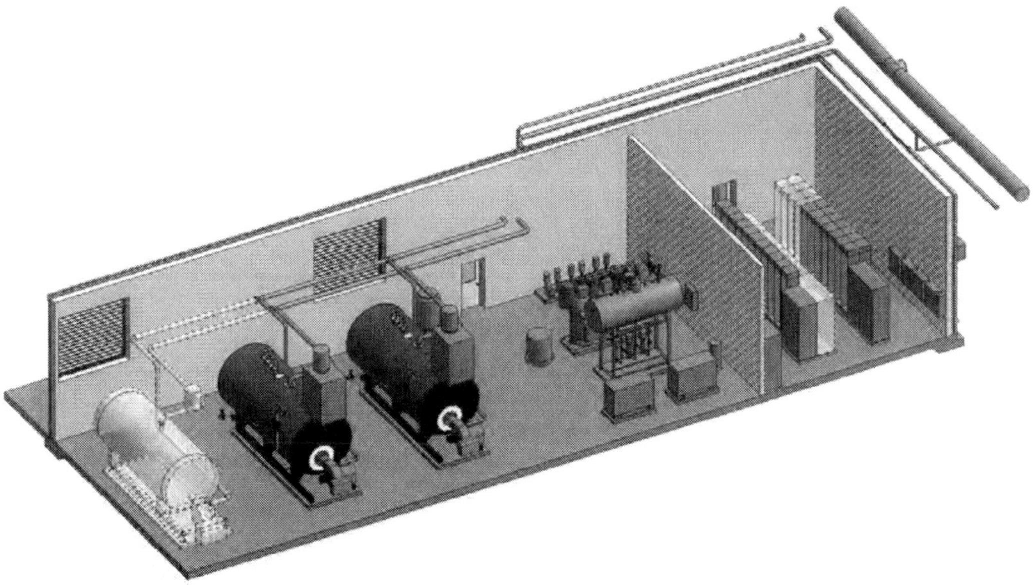

Figure 13. – Steam Boiler Building General Arrangement

Energy Management Considerations

The recommended facilities to be installed under the Phase I improvements will result in a net energy positive operation when operating at their design loading rates as summarized in Figure 14. The resultant high level energy balance is summarized as follows:

Digester Gas Energy Production & Consumption	
Total Digester Gas Energy Production	11.08 MW
Digester Gas Boiler Steam Input Energy for THP and FOG	-4.34 MW
Net Digester Gas Energy Production (after steam)	6.74 MW
Electrical Energy Utilized for Treatment	-4.62 MW
Net Energy Balance (Production – Consumption)	**2.12 MW**

The ability of the project to clearly demonstrate a net energy production allowed the City of Raleigh to secure a commitment from the North Carolina Division of Water Infrastructure's Clean Water Revolving Loan Fund (CWRLF) as a "Green Project" for partial funding of the anticipated $80MM to $100MM facility capital cost. The $50MM CWRLF commitment as a Green Project allows the City of partially finance the anticipated improvements with a 0% interest 20-year loan with a resultant net present worth cost savings of approximately $13MM over a conventional 30-year duration bond financing project.

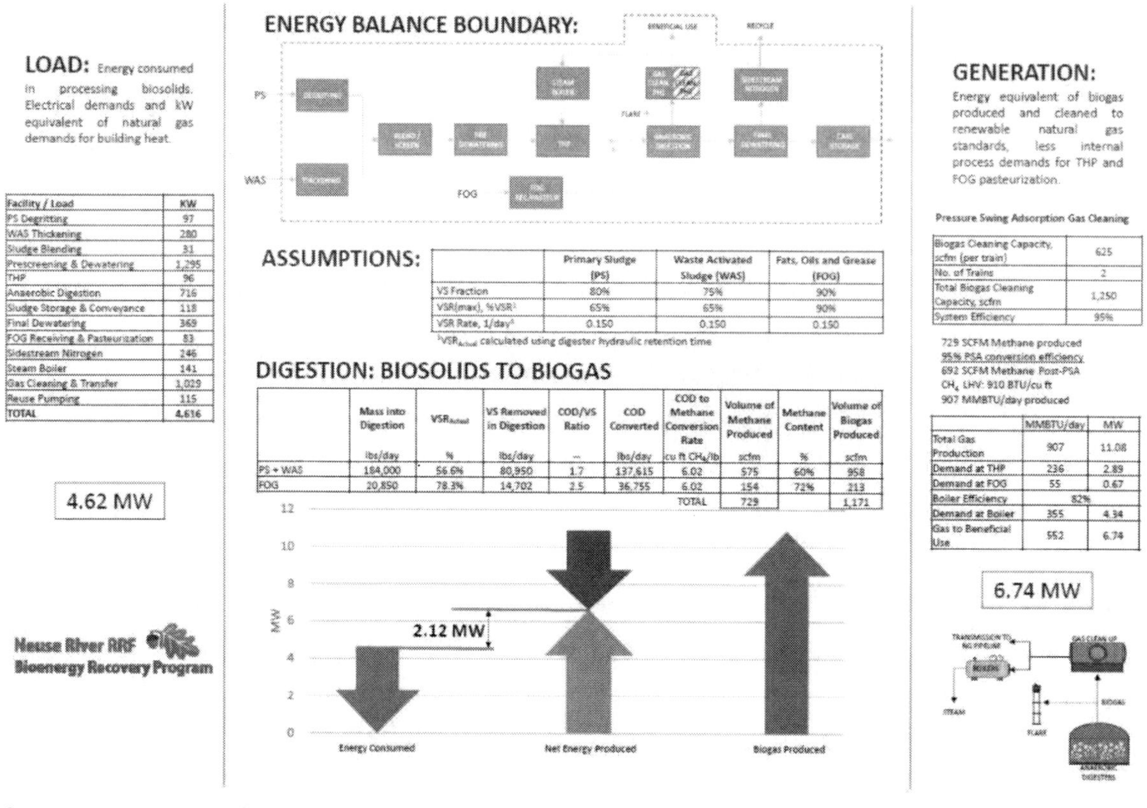

Figure 14. – Energy Balance Summary

Discussion and Conclusions

The City of Raleigh is currently in the detailed design phase for the improvement project described in this paper. Several significant milestones have been achieved since the completion of the PER, including:

- Project Delivery Method – Several delivery approaches were considered for the project including conventional design-bid-build, progressive design-build, and construction manager at risk (CMAR). Ultimately the City selected to deliver the project using the CMAR method and has completed the selection of a CMAR for the project who is now working with the detailed design team to finalize the construction documents.
- Thermal Hydrolysis Technology Supplier – Both the CAMBI and BIOTHLYS systems were carried through the PER development phase. Since completion of the PER phase the City has selected CAMBI as the system supplier for the THP system.
- Preliminary Design Document Preparation – The detailed design team has completed the 30% design documents for the project and submitted them for review and comment by both the City and the CMAR. The CMAR will be providing a preliminary opinion of probable construction cost based on the 30% documents for financial planning purposes and to guide decisions as the project documents move toward completion.
- HAZOP Study and Workshop – A HAZOP study was completed on the 30% design documents for each of the major treatment processes associated with the Phase I improvements and findings

from the HAZOP are being incorporated into the final design for enhanced process safety and control.

The current project schedule is to complete detailed design activities during 2017 with the project moving into construction during 2018.

References

Knight, G., Bullard, C. M., Lackey, K., Polo, C., Kemp, J., Brower, A., and Lynch, T. J., A Greener Future for the City of Raleigh: Turning Biosolids into Vehicle Fuel, Proceedings Water Environment Federation Residuals and Biosolids Conference, Seattle, WA, 2017.

A Commitment to Improving the Odors of DC Water Biosolids Leads to Improved Acceptance

Bill Brower[1]*, Chris Peot[1], Al Razik[2]

[1] District of Columbia Water and Sewer Authority, Washington, DC.
[2] Maryland Environmental Service, Millersville, MD.
*Email: bill.brower@dcwater.com

ABSTRACT

Odors were the largest issue facing the DC Water biosolids beneficial reuse program in the early 2000s, and there were serious concerns about the future of the program. DC Water took steps to address odor issues, including obtaining odor readings from trained inspectors at all beneficial reuse sites. The result is a dataset that is unique in the industry—odor readings from essentially every site receiving Blue Plains biosolids spanning more than a decade. Odor strength readings improved considerably over this period, and a reduction in incidents mirrored this trend. DC Water now has such a high-quality material that it is selling biosolids to customers using it as a soil amendment in such sensitive sites as landscaping for private homes and high profile, high foot-traffic locations. This dataset has affected plant operations, solids handling and, ultimately, greatly improved the strength of the DC Water biosolids program.

KEYWORDS: Odor, biosolids, land application, public perception, marketing.

INTRODUCTION

In the early 2000s, odors were the largest issue facing the District of Columbia Water and Sewer Authority (DC Water) 1100 wet-tonnes-per-day (1200 wet-tons-per-day) biosolids program because of the emotional response malodorous material elicited near land application sites. There were serious concerns about the future of the beneficial reuse program due to the volume of complaints. The Class B lime-stabilized product coming from the Blue Plains Advanced Wastewater Treatment Plant in Washington, D.C. was inconsistent, both in terms of odor and pathogen levels, and staff was reacting to complaints rather than proactively addressing the underlying causes.

In the late 1990s and early 2000s, staff estimate they received well over 100 complaints each year. In the subsequent years, DC Water took the following steps to address odor issues:

- Determine the source of odors at the plant (2000-2008)
- Make process and equipment changes for lime stabilization system (2000-2009)
- Use odors and product quality as significant factors in choosing new solids equipment (2000-2012)
- Increase presence at land application sites and communication with field staff (2003-present)
- Obtain subjective and objective odor readings by trained inspectors at all beneficial reuse sites (2004-present)
- Experiment with additional handling to reduce odors (2016-present)

Previously published research described how DC Water determined the source of odors at the plant, process changes that were made to minimize odors and evaluation of solids processing equipment to inform capital purchases. Murthy, et al., (2001) showed that more odorous material remained more odorous after stabilization and that improperly mixed polymer used in dewatering could lead to fishy odors, a major problem in the field. In Higgins, et al., (2008), samples were taken throughout the solids and liquids streams at twelve wastewater treatment plants to determine the dominant sources of odors. Based in part on these findings, DC Water developed an odor diagnostic tool specific to Blue Plains.

Murthy, et al., showed in 2002 that inadequate mixing of biosolids with lime in pug-mill mixers at Blue Plains was leading to odor issues. Based on this finding, new mixers with a greater mixing energy were installed and odors significantly improved. This work also found that in digested material, high rates of shear leads to more odorous material. This was a major factor in the selection of belt filter presses over centrifuges for final dewatering when new solids equipment was installed in 2014.

In the next several years DC Water performed significant additional research to determine the best replacement for aging lime stabilization equipment. Murthy, et al., (2009) looked at two factors—odors and fecal coliform regrowth—"to minimize the risk for a sustainable Class A biosolids program." Thermal hydrolysis, which was eventually chosen as a pretreatment step to anaerobic digestion, was found to produce lower odors than conventional mesophilic digestion alone. It is also possible to dewater thermally hydrolyzed solids to a high percent solids on a belt filter press, which as indicated above generally produces lower odors for digested material than centrifuges.

To show how these changes made a difference in the field, this paper will focus on a unique dataset—odor readings from essentially every land application site receiving Blue Plains biosolids spanning more than a decade—and how these data have affected plant operations, solids handling and, ultimately, the strength of the DC Water biosolids program.

METHODOLOGY

Starting in August 2004, DC Water began contracting with Maryland Environmental Service (MES) to inspect every land application site—including taking subjective and objective odor readings. Through 2016, this amounted to over 29,000 odor readings. MES staff were trained in the use of olfactometers (Nasal Ranger®; Figure 1), a handheld device which allows users to determine an objective value for the strength of an odor. Ambient air is mixed with filtered air in discrete volume ratios. The user holds the device tight to their face, breathes filtered air as a blank, then starts at a "dilution-to-threshold" (D/T; volume of filtered air:volume of ambient air) of 60, i.e. a very dilute sample. If the user is able to smell the odors, the ambient air is very odorous. If the user is unable to detect any odors they adjust the D/T to 30, and so on, until an odor is detected.

Figure 1 - MES inspector demonstrating the use of an olfactometer

MES also characterize odors by hedonic tone (the pleasantness or unpleasantness of an odor), level of odor (None, Slight, Moderate, and Strong) and standard odor descriptors (e.g. earthy, fishy, ammonia, musty).

MES inspectors routinely are tested for odor sensitivity using standardized n-butanol pens. The average score (9.1) correlates well with published values (7.6-8.5). Individual inspectors generally fall within the range of the "smell bell" curve expected for people with normal odor sensitivity.

MES sends customer satisfaction surveys to all farmers receiving Blue Plains biosolids, tracking the results in a database. Through 2016, MES had distributed nearly 1,700 surveys, with a 29% return rate. From the responses to a question about odors, DC Water can see how farmers' perceptions are changing over time.

Drawing on reports from the field MES also tracks any incidents that occur from the time the biosolids leave the plant—from vehicle accidents to odor complaints and neighbor complaints. (Note: MES began tracking incidents from another source, the Virginia Department of Environmental Quality Pollution Response and Preparedness database, in June of 2015. The four incidents from this source have not been included here for consistency throughout the period.)

RESULTS

In the late 1990s DC Water decided to replace the aging lime stabilization equipment at Blue Plains. Based on the work of Murthy, et al., (2002 and 2009) anaerobic digesters, thermal hydrolysis and belt filter presses were chosen to handle DC Water biosolids. Odors played a significant role in these choices. This was a discretionary project and DC Water made the decision to choose equipment with final product quality in mind throughout, even when it came at a premium.

In November 2014, DC Water began producing digested biosolids. In a gradual transition, the amount of digested material steadily increased, resulting in a corresponding decrease in the amount of limed material produced. In 2014, 1% of the biosolids leaving the plant were digested, increasing to 87% in 2015 and 100% in 2016. Largely due to the conversion of organic matter to methane and carbon dioxide in the digesters, higher solids concentration following dewatering and the elimination of lime addition, average daily production from Blue Plains reduced from 1100 wet-tonnes-per day to just above 400 wet-tonnes-per-day after the transition to the new equipment.

The reduced volume from Blue Plains means that more sites receiving Blue Plains material are also receiving biosolids from other wastewater treatment plants. The presence of other sources was not recorded in the MES database in prior years so the impact of these other sources cannot be precisely known; however, Blue Plains material typically accounts for most of the biosolids at a given site so results will largely reflect changes made by DC Water.

Dilution-to-threshold readings show the positive effect the increased attention to addressing odor issues has had in the field. Figure 2 shows improvement in the subjective strength of Blue Plains biosolids over the past decade. A focus on improving lime and polymer mixing and other process changes resulted in continuously improving odor readings from 2006-2014. The jump in 2015, when nearly all material was rated as having minimal odor, can be attributed to the switch to digested material. This is highlighted in Figure 3, where the benefits of focusing on product quality in the newly installed solids processing equipment is evident.

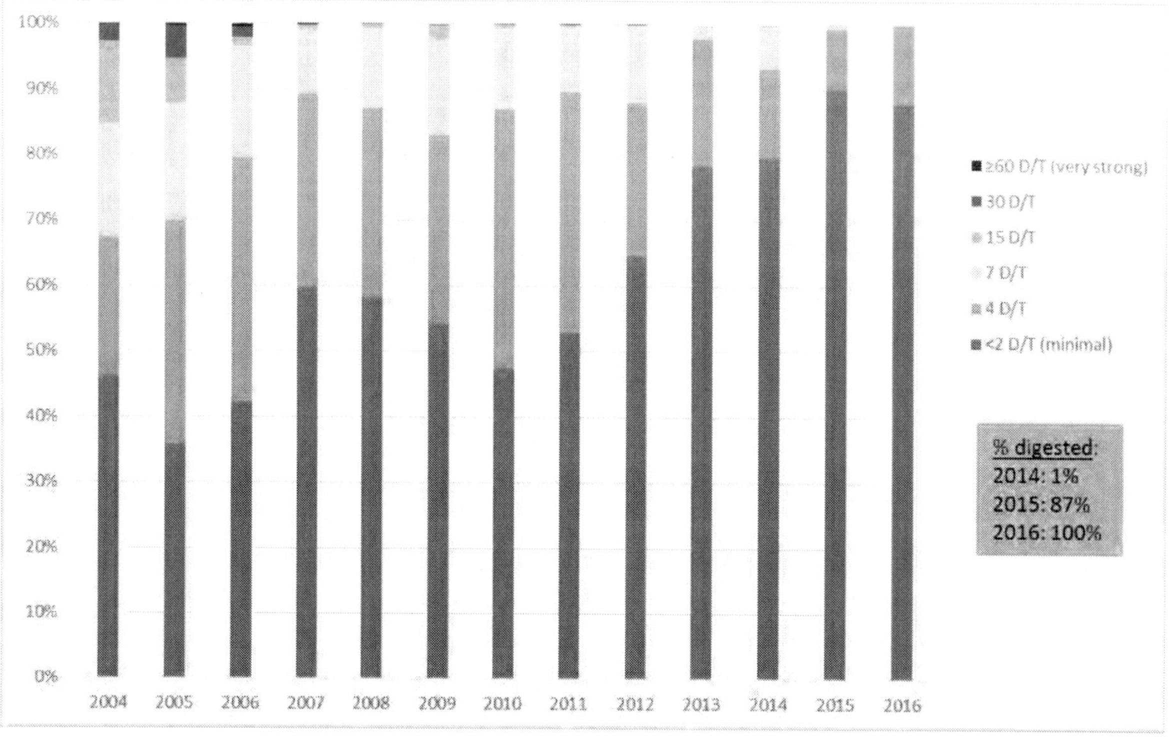

Figure 2 - Distribution of dilution-to-threshold readings by year

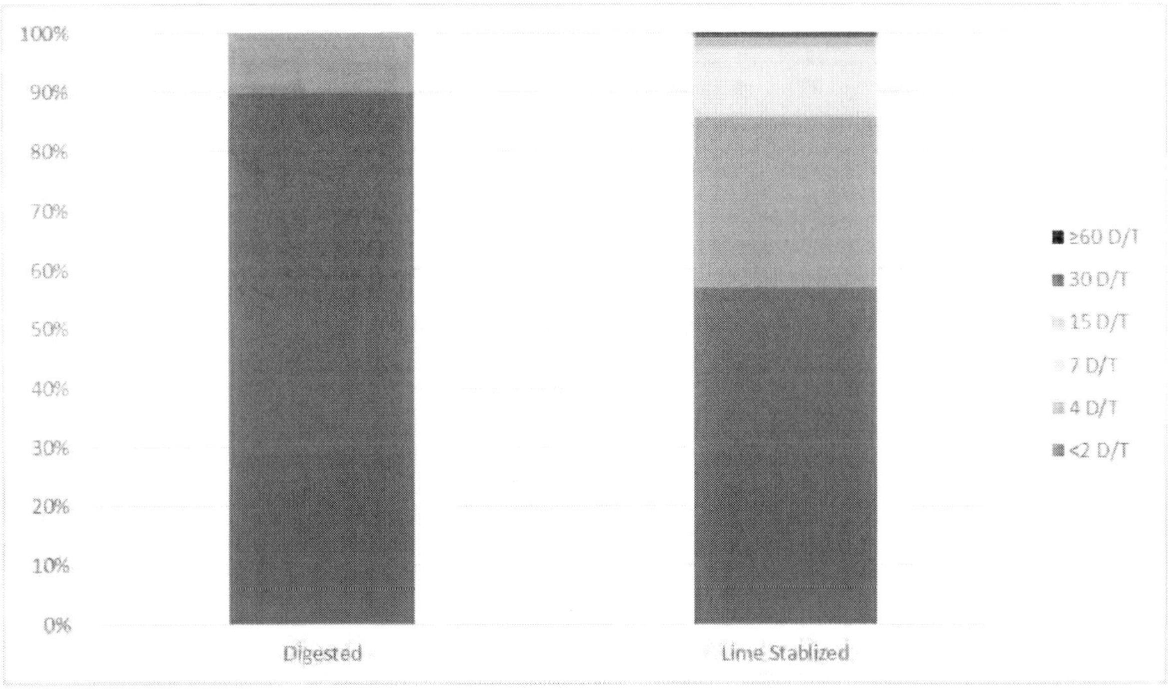

Figure 3 - Distribution of dilution-to-threshold readings by biosolids type

An analysis was performed to quantitatively verify if there were statistically significant differences between the olfactometer readings taken for DC Water lime-stabilized material and digested biosolids as appears to be apparent in Figure 3. A histogram of the data (not shown here) was generated and clearly showed that these data did not follow a normal data distribution. Because neither the lime stabilized nor digested material followed a normal distribution, a non-parametric test method, the Mann-Whitney U test, was used to evaluate the data. The Mann-Whitney U test is the non-parametric equivalent to the t-test, and is used to compare the means of two independent population sets to test whether they are equal or not.

There were a total of 29,060 olfactometer measurements (2,789 readings for digested material and 26,271 readings for lime-stabilized biosolids) taken during the period of August 2004 through December 2016. At a significance level $\alpha = 0.05$, the results of the Mann-Whitney U test show that there is a difference between the datasets. Hence, there is a statistically significant difference between the lime-stabilized and digested olfactometer readings.

Although the data are not normally distributed, a one-way Analysis of Variance (ANOVA) test was also run on the datasets. ANOVA is typically applied to normally distributed data, but because this dataset has so many data points it can be used in place of a non-parametric test. This test also concluded that there was a statistically significant difference between the two independent data sets. These analyses quantitatively show that there is a difference between the lime-stabilized and digested olfactometer readings—i.e., the odor readings from the lime-stabilized material are higher than those of the digested material.

Figure 4 shows a similar trend as dilution-to-threshold for hedonic tone, or the pleasantness (positive values) or unpleasantness (negative values). One inspector has even rated the digested material as having positive values for hedonic tone, explaining that he actually enjoys the odors.

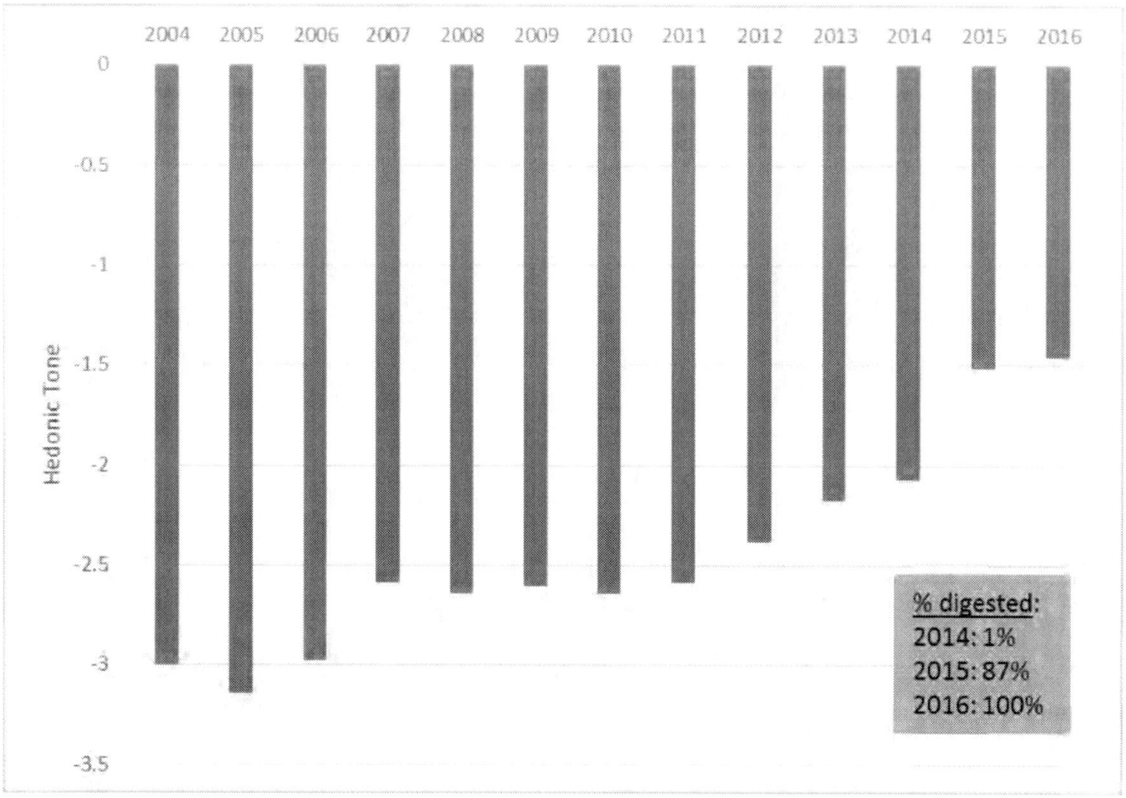

Figure 4 - Average hedonic tone per year (0=neutral)

An additional parameter that MES inspectors record is a subjective description of their perception of the "level" of the odor—strong, moderate, slight or none. Figure 5 shows a trend that seemingly contradicts the results above, with the number of inspectors indicating that the biosolids had no odor decreasing over time. However, MES staff indicate that this was likely due to changes in guidelines for this parameter since the program started. A trend from moderate to slight odors is nonetheless apparent over time, and in particular since the introduction of digested material.

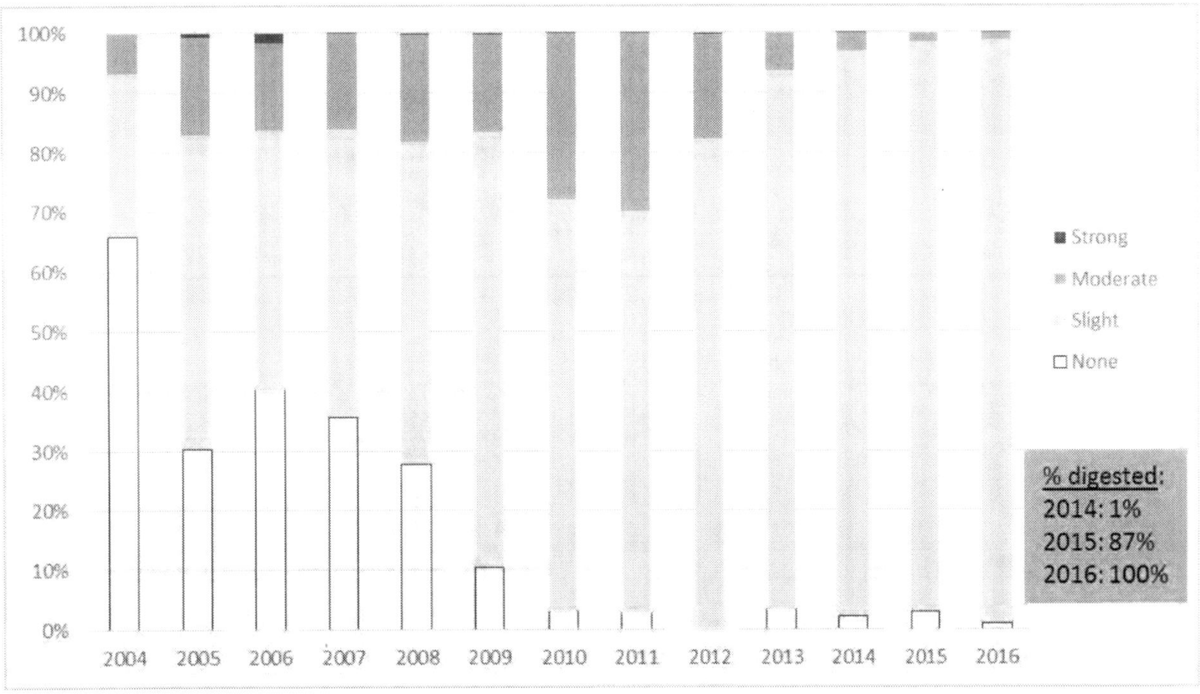

Figure 5 - Distribution of odor level descriptors by year

In addition to the objective measurements of dilutions-to-threshold and the subjective descriptors of hedonic tone and odor strength, MES inspectors are trained to provide a subjective descriptor of the odors in the field (e.g. "woody" or "Earthy"). Figure 6 shows the distribution of the descriptors used by year. (Note: Only one descriptor was used in 2004 so this year has been omitted.) Ammonia odors, characteristic of lime-stabilized biosolids, have fallen off significantly since the move to digested material, as have "musky" and "manure" odors. Roughly a third of the beneficial reuse sites in 2016 were given the fairly neutral descriptor of "Earthy" by inspectors. As was discussed early, Murthy, et al., (2001) showed that inadequately mixed dewatering polymer could lead to fishy odors and fishy odors were indeed noted in the field 93 times from 2005-2014.

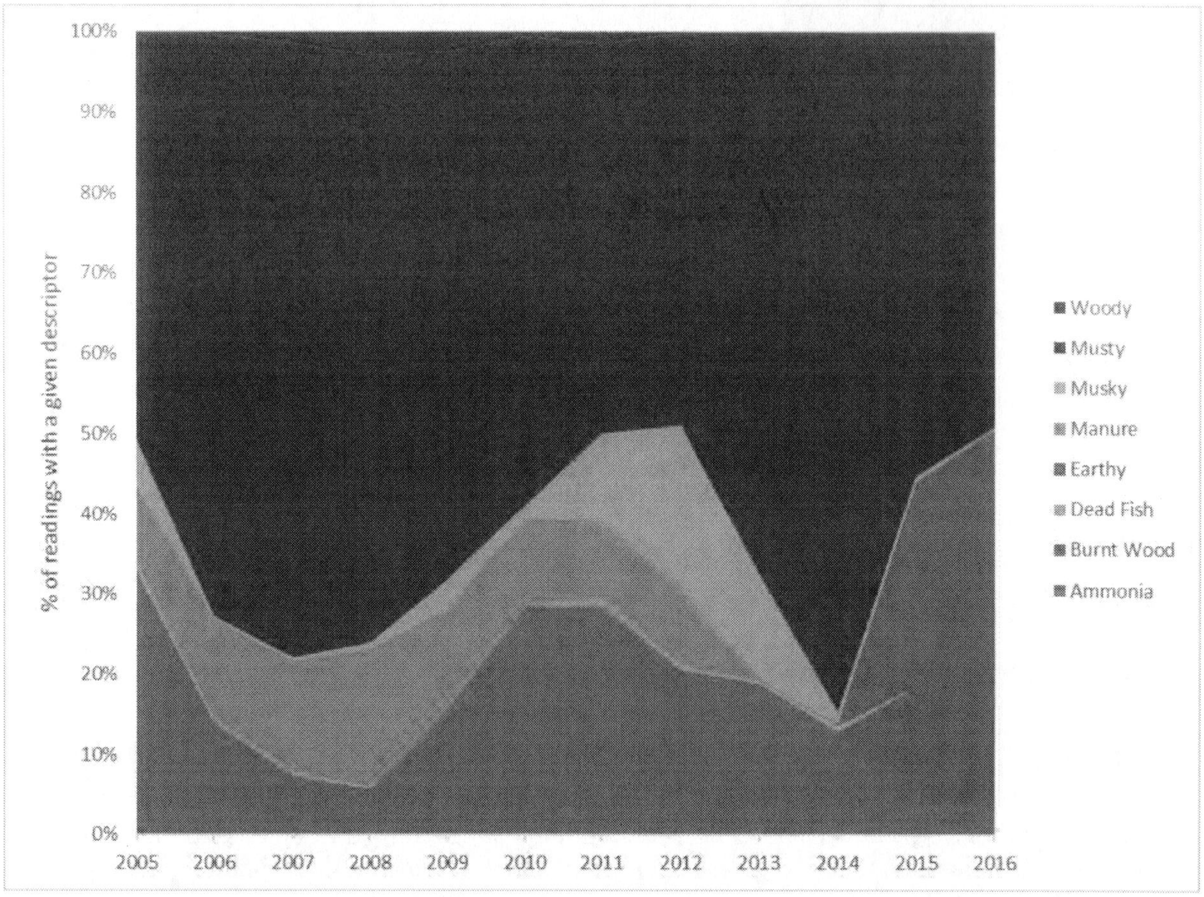

Figure 6 - Distribution of odor descriptors by year (Note: descriptors with fewer than 90 total uses were omitted from this graph.)

After the objective readings of the trained staff of MES, perhaps the group most qualified to rate the odors of biosolids are the farmers receiving the material. MES sends customer satisfaction surveys to all farmers receiving Blue Plains biosolids. Through 2016, MES had distributed over 1,700 surveys, with a 29% return rate. As shown in Figure 7, an increasing number of farmers have indicated that they did not notice or were indifferent to the odors associated with land application. There was a dip in 2016 for reasons that are not immediately clear.

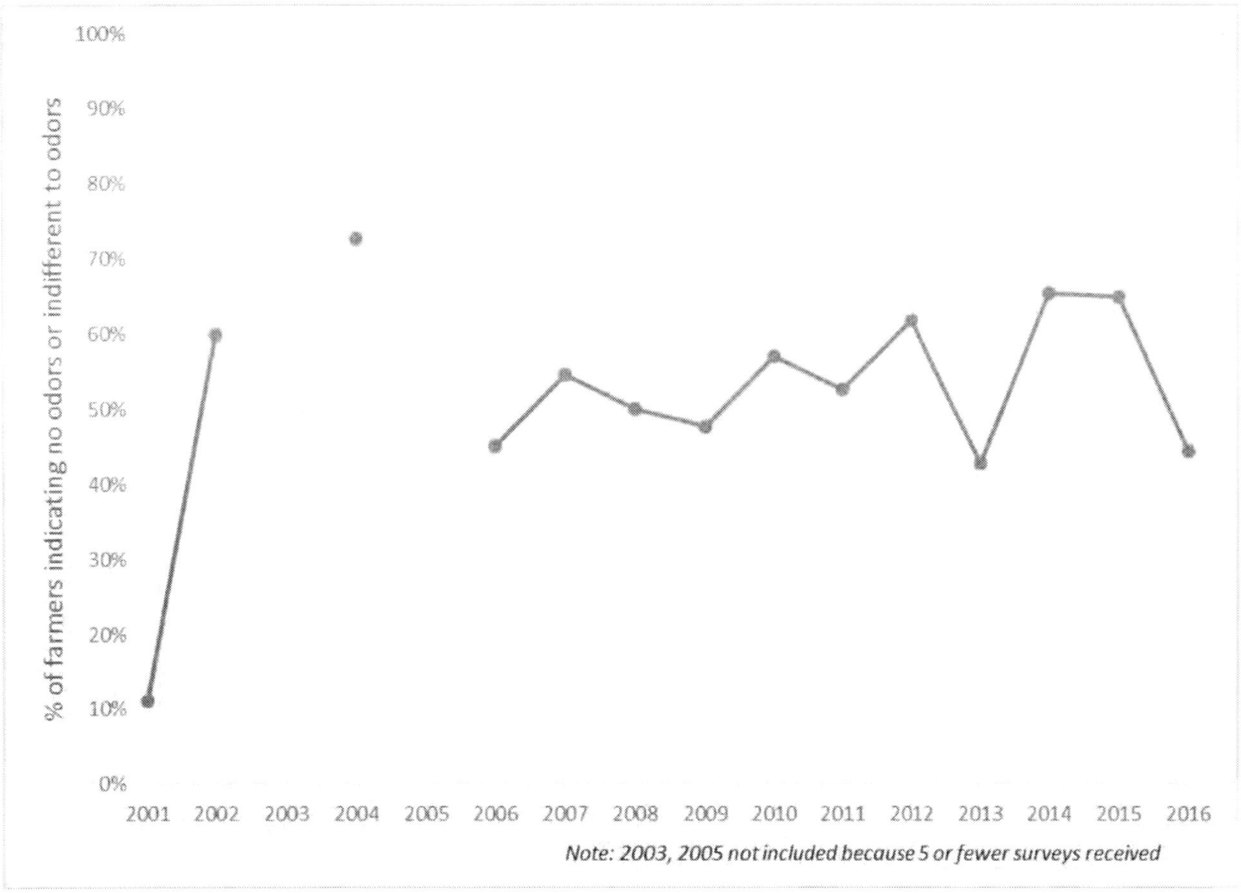

Figure 7 - % of farmers indicating noticing no odors or indifferent to odor associated with land application of Blue Plains biosolids in surveys

MES tracks all incidents that occur after biosolids leave the plant. Excluding issues that would be independent of biosolids quality (e.g. vehicle accidents), the total number of issues and those involving odor have been in decline over the past decade (see Figure 8). In the early 2000s, DC Water staff estimate that well over 100 complaints were received each year. Anecdotally, odorous material seems to attract incidents of all types. Odors are typically the first issue with opposition groups, who may complain about additional issues once odors have drawn their attention. In 2016, when nearly 10,000 loads were transported, there were only four total incidents, only one of which was an odor complaint. This low number of incidents is particularly impressive given the volume of biosolids produced at Blue Plains; typically DC Water has represented about half of the biosolids applied in the entire state of Virginia.

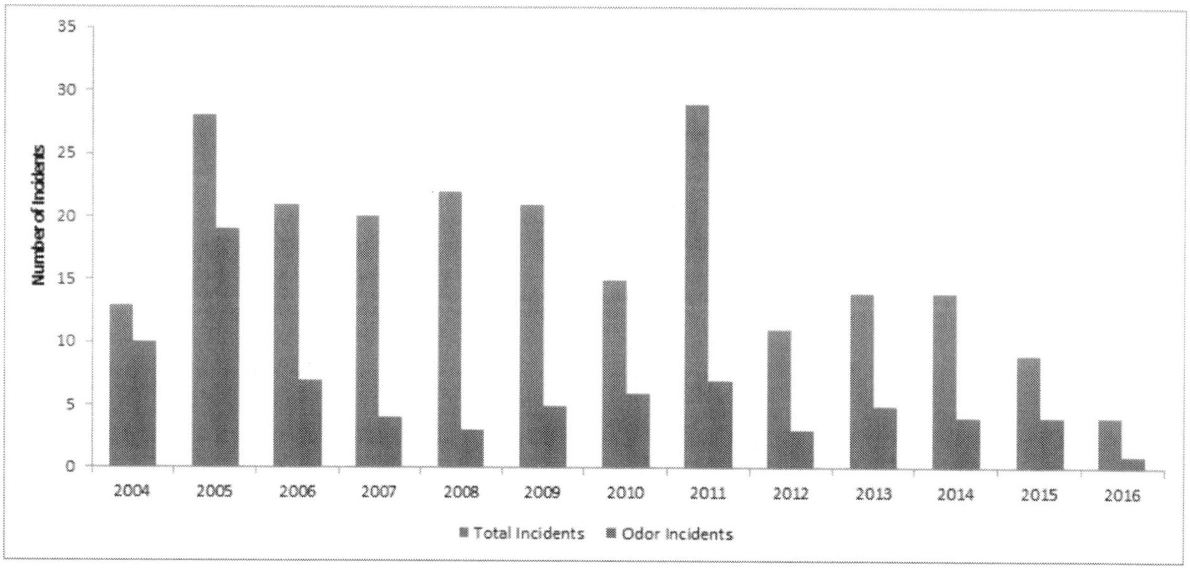

Figure 8 - Incidents recorded by MES by year (Note: Excludes vehicle accidents and other vehicle issues that are independent of biosolids quality.)

DISCUSSION & CONCLUSIONS

Following years of land applying an inconsistent and odorous Class B product and the resultant complaints that threatened the existence of the beneficial reuse program, DC Water got serious about product quality about a decade ago. They are not regulated but DC Water has found that taking odors as seriously as regulated parameters greatly improves the strength of the program. It has taken a significant investment of time and money to obtain the database of field data DC Water now has but this effort has helped to drastically improve acceptance.

Having inspectors in the field daily taking odor readings allowed DC Water to make necessary changes in near real time. For instance, if an odor was observed in the field, MES staff would communicate their odor readings to DC Water staff. Using a separate scale database, staff could determine where that particular material was loaded. If the problem persisted at the plant, staff could direct the remainder of the material in that storage area to a more remote land application site, or in more severe cases direct it to landfill. Then, based on previous work that determined the root cause of different odors in the biosolids, the odor description from MES could help diagnose steps to correct the problem. For instance, if inspectors detected "fishy" odors (as shown in Figure 6), staff would know the likely cause was stratified emulsion polymer. Circulating the polymer for thirty minutes prior to incorporating it into the solids before centrifuge dewatering consistently removed fishy odors.

The MES contract represents an approximately 5% premium on top of the DC Water biosolids reuse contract. Virginia, where the bulk of our material has been applied over this time period, has a program for inspection but does not have the resources to be at every site where biosolids are being applied. DC Water feels that it is worth the additional cost to have MES in every field for several reasons. In addition to providing a face in the field for landowners and neighbors, if inspectors notice an odor issue it is quickly communicated to DC Water staff who can attempt to minimize the impact of odorous material by fixing an issue at the plant and/or quarantining material for further processing or disposal in a landfill.

Perhaps the best indicator that the focus on improving odors has been successful is that DC Water has begun marketing its Class A digested cake. The fact that respected soil blenders, landscapers, arborists, and other companies, as well as community gardens and the public are willing to pay money for DC Water biosolids, branded as Bloom® (bloomsoil.com), is the ultimate indicator of acceptance.

The DC Water effort to market Bloom began in earnest in the fall of 2016. Through the end of 2016, DC Water marketed the equivalent of 900 tons of biosolids (or approximately two days' worth of production). The marketing goal for 2017 is 9,000 tons, and through mid-February (before spring peak demand) nearly 600 tons have already been sold at a competitive market rate.

To make the material more marketable to certain customers, DC Water has been experimenting with "curing" the biosolids cake. Similar to windrow composting, the curing process consists of placing dewatered cake into covered or uncovered long piles and turning a few times per week for two to three months. This process improves several product attributes (handle-ability, dryness, pH, ammonia levels, stability) which improve the marketability. Crucially, it also reduces odors beyond the already relatively low levels in uncured cake. Many customers are applying this material in particularly sensitive sites, including the landscaping at private homes and high-profile, heavily trafficked sites in the nation's capital, where odors must be at a minimum. DC Water is selling "Cured Bloom" at double the price of "Fresh Bloom" cake (which has received no further processing beyond dewatering on belt filter presses to approximately 31% solids).

DC Water will continue to closely track odors in final products and at application sites, and look for additional ways to improve odors. As the proportion of material being used in more densely populated areas increases the importance of producing a consistently low odor and inoffensive material will continue to grow.

ACKNOWLEDGEMENTS

The authors gratefully acknowledge the daily efforts of Maryland Environmental Service inspectors, which provide our industry with this unique dataset. We also acknowledge the National Biosolids Partnership for encouraging us to elevate odors to a primary concern through our Biosolids Management Program.

REFERENCES

Murthy, S. Sadick, T.; Bailey, W.; Peot, C.; Strawn, M. (2001) Odor Mitigation from Lime Stabilized Biosolids. *Proceedings of the Water Environment Federation, WEF/AWWA/CWEA Joint Residuals and Biosolids Management 2001*; Water Environment Federation: Alexandria, Virginia; 725-736(12).

Murthy, S.; Peot, C.; North, J.; Novak, J.; Glindemann, D.; Higgins, M. (2002) Characterization and Control of Reduced Sulfur Odors from Lime Stabilized and Digested Biosolids. *Proceedings of the Water Environment Federation, Residuals and Biosolids Management 2002*; Water Environment Federation: Alexandria, Virginia; 1105-1124(20).

Higgins, M.J.; Chen, Y.C.; Novak, J.T.; Glindemann, D.; Forbes, R.H.; Erdal, Z.; Witherspoon, J.; McEwen, D.; Murthy, S.; Hargreaves, J. R.; Adams, G. (2008) A Multi-Plant Study to Understand the Chemicals and Process Parameters Associated with Biosolids Odors. *Environmental Engineer: Applied Research and Practice*. 5, 1.

Murthy, S.; Higgins, M.; Chen, Y.; Novak, J.; Wilson, C.; Riffat, R.; Aynur, S. (2009) Impacts of Enhanced Digestion Processes on Biosolids Quality Parameters: Odors and Indicators. *Proceedings of the Water Environment Federation, WEFTEC 2009*; Water Environment Federation: Alexandria, Virginia; 3936-3943(8).

1/11/17

Developing a Large-scale CAMBI Biosolids Marketing and Distribution Program

AUTHOR: Ronald Alexander, R. Alexander Associates, Inc. and Chris Peot, District of Columbia Water & Sewer Authority

EXTENDED ABSTRACT:

DC Water & Sewer Authority (DC Water) is now operating the first CAMBI thermal hydrolysis biosolids treatment facility in the US; it is also the largest CAMBI facility in the world. The facility is producing a Class A, thermophilically treated biosolids cake with unique physical properties. The product is easier to handle and blend, than many other biosolids cakes, and it is low in odor. At full capacity, the DC facility will generate approximately 450 wet tons per day of CAMBI treated biosolids.

The long-term (10 year) goal of DC Water is to take its current biosolids distribution program, where it pays all associated costs for distribution and application, and turn it into a program which is 'cost neutral'. The primary long-term uses for the CAMBI treated biosolids, now branded as "Bloom", is planned to be as an ingredient in blended soils used in landscaping, turf and environmental applications, and after further stabilization (aging/curing), as a soil amendment used in similar applications.

Knowing that it would generate such a large volume of biosolids, DC Water invested into extensive planning and market/product research efforts. To start off the product development efforts, once a uniform dewatered biosolids cake began being produced, DC Water started an extensive laboratory testing program on the 'fresh' dewatered biosolids. Interestingly, the fresh Bloom product had some similar characteristics to well established biosolids composts produced from their region (see Table 1). However, it possessed a much higher respiration rate, which relates to a lack of stability of the carbon within the product. With this in mind, stronger consideration was given to marketing an unblended Bloom product as a soil amendment. This strategy could also be implemented faster and with less financial risk than marketing the product through an internal blending program alone.

Table 1 - BLOOM Comparison to Regional Biosolids Composts

Characteristics	Units of Measure	CAMBI Digested Cake (BLOOM)	Biosolids Compost 1	Biosolids Compost 2
Nutrients				
Total Nitrogen	%, wet wt basis	1.6	2	2
Ammonia Nitrogen	ppm, wet wt basis	1707	2100	1100
Organic Nitrogen	%, wet wt basis	1.4	1.8	1.9
Available Phosphorus (P_2O_5)	%, wet wt basis	2.01	3.2	2.4
Soluble Potash (K_2O)	%, wet wt basis	0.03	0.68	0.42
Calcium (Ca)	%, wet wt basis	0.78	2.1	2.1
Magnesium (Mg)	%, wet wt basis	0.09	0.28	0.21
Metals				
Arsenic (As)	ppm, dry wt. basis	1	11	13
Cadmium (Cd)	ppm, dry wt. basis	0.6	1.2	<1.0
Copper (Cu)	ppm, dry wt. basis	114.4	180	210
Lead (Pb)	ppm, dry wt. basis	16.7	20	18
Mercury (Hg)	ppm, dry wt. basis	0.31	<1.0	<1.0
Molybdenum (Mo)	ppm, dry wt. basis	5.4	7.8	3.2

Nickel (Ni)	ppm, dry wt. basis	7.1	16	11
Selenium (Se)	ppm, dry wt. basis	3.8	3.3	2
Zinc (Zn)	ppm, dry wt. basis	266.4	580	330
Physical / Chemical				
Moisture	%	68.8	30.6	28.9
pH Value	Units	7.9	8.28	8.02
Electrical Conductivity	mmhos/cm	3.61	9.4	3.7
Organic Matter	%, dry weight	56.1	53.4	64.3
Ash	%, dry weight	43.9	46.6	35.7
Bulk Density - as received	lbs/cu.ft.	44.3	37	29
C:N Ratio	Ratio	6.6	9.8	12
Biological				
Maturity/Stability				
Dilution ratio		1:1	1:2	1:1 / 1:3
Emergence	% of control	95	93.3	100/100
Seedling Vigor	% of control	100	14	0/0
Description of plants		NA	stunted	Stunted
CO_2 Evolution Rate				
mg CO_2-C/g organic matter/day		9.6	1.7	5
mg CO_2-C/g solids/day		5.4	0.9	3.2
Pathogens				
Fecal Coliform	<2.0 MPN/g	Pass	pass	Pass
Salmonella	<3.0 MPN/4g	Not tested	pass	Pass

With this in mind, a well-managed pilot project was initiated which pragmatically monitored Bloom characteristics over time. 50 to 100 ton windrows were established and both aging and turning was done; with piles placed under cover (open walled building) and out of doors. Data from the approximately 2-month long field trials are outlined in Table 2. The research found that the ammonia-N, moisture content, pH, electrical conductivity levels all reduced over time, along with the product's respiration rate. It is assumed that the reduction in pH and electrical conductivity is related to the drop in ammonia-N. Further, in the 2 to 3 week growth trials, seed emergence (germination) was excellent in fresh and aged Bloom, while seedling vigor (healthy growth) improved with the age of the material. The related results (Table 2) illustrate that improvements in physical, chemical and biological characteristics take place in the product over time. From this trial, it cannot be stated that aging the product under cover was any better than aging it out of doors. However, this could certainly be the case in high precipitation and under other conditions.

The product improvements were shown to be linear in nature. This could hopefully mean that if a well-run aging / curing operation is developed, then improvements in product quality could be predictable and optimal process techniques could be established.

Table 2 - BLOOM Aging/Curing Data

Characteristics	Date Received	10-Mar-16	18-Mar-16	31-Mar-16	14-Apr-16	29-Apr-16	19-May-16	10-Mar-16	18-Mar-16	31-Mar-16	14-Apr-16	29-Apr-16
	Sample ID	Bloom uncovered + 2 day post flip	Bloom uncovered + 10 day flip yesterday	Bloom uncovered + 23 day post flip	Bloom uncovered + 37 days flip + 1	Bloom Uncovered +52 d. flip + 0	Bloom uncovered +72 days, flip +2	Bloom windrow + 9 day post flip	Bloom windrow + 17 day flip yesterday	Bloom windrow + 30 days post flip	Bloom windrow + 44 days flip + 6	Bloom Windrow +59 d, flip + 0
NUTRIENTS												
Total Nitrogen	%, wet wt. basis	1.4	1.4	1.6	1.7	1.7	1.5	1.46	1.6	1.8	1.8	2
Ammonia (NH_4-N)	ppm, wet wt. basis	1,100	1,400	970	950	610	210	1,500	1,500	690	120	120
Nitrate (NO_3-N)	%, wet wt. basis	<0.3	<0.3	3.8	2.8	13	3.7	<0.3	4.7	2.9	14	29
Org. Nitrogen	%, wet wt. basis	1.4	1.3	1.6	1.6	1.6	1.5	1.4	1.5	1.7	1.8	2
Phosphorus (P_2O_5)	%, wet wt. basis	2.2	2.5	2.7	3.4	4	3.5	2.2	2.9	3.3	3.9	0.32
Potassium (K_2O)	%, wet wt. basis	0.038	0.042	0.042	0.059	0.067	0.056	0.035	0.044	0.048	0.069	0.42
Calcium (Ca)	%, wet wt. basis	0.78	0.9	0.99	1.2	1.4	1.3	0.84	1	1.2	1.4	1.2
Magnesium (Mg)	%, wet wt. basis	0.12	0.13	0.14	0.18	0.22	0.19	0.12	0.15	0.17	0.21	0.43
Sulfate (SO_4-S)	ppm, wet wt. basis	860	1,200	1000	1400	1100	1300	1,500	1,700	1200	1100	880
Sodium (Na)	%, wet wt. basis	0.014	0.017	0.018	0.023	0.052	0.024	0.023	0.031	0.04	0.047	0.13
Chloride (Cl)	%, wet wt. basis	0.018	0.017	0.014	0.021	0.02	0.021	0.028	0.034	0.029	0.039	0.037
Iron (Fe)	ppm, dry wt. basis	72,000	84,000	74,000	86,000	89,000	89,000	73,000	88,000	86,000	97,000	15,000
OTHER PARAMETERS												
Moisture	%, wet wt. basis	66.1	65.4	58.4	53.6	47.5	52.5	65.9	62.4	57.9	54.1	46.9
pH Value	Units	7.8	7.94	7.44	7.12	7.09	6.4	7.72	7.17	6.78	6.19	6.37
Bulk Density	lbs/cf, wet wt basis	52	54	42	45	50	51	48	49	46	46	53
Soluble Salts Electrical Conductivity	dS/m2 (mmhos/cm), dry wt. basis	6.4	8.1	5	6	3.8	4.2	8.7	9.2	5.4	4.6	3.3
Organic Matter	%, dry wt. basis	52.9	53.3	48.4	48.6	46.5	46	53.3	54.2	52.2	50.6	50.7
C:N Ratio	Ratio	7	7	7	7.3	7.7	7.6	6.4	6.7	6.9	7.1	7
Ag Index	Ratio	>10	>10	>10	>10	>10	>10	>10	>10	>10	>10	>10
Stability-Respirometery	mg CO2-C/g OM/day	8.8	5.1	4.6	1.9	1.5	1.7	9.5	3.1	2.8	2.7	1.6
Maturity – Emergence	average % of control	100	93	93	100	93	100	100	100	100	93	100
Maturity – Vigor	average % of control	84	72	77	69	99	92	55	65	111	102	96
Description of Plants	N/A	*Healthy*	*Stunted*	*Stunted*	*Stunted*	*Healthy*	*Healthy*	*Stunted*	*Stunted*	*Healthy*	*Healthy*	*Healthy*

Other related activities include:

- Product Development Research / Activities (using both university and private entities)
  - Creating a variety of potential soil products and growing media (completed 2015-2016),
  - Including the development of a trial sized on-site soil blending plant (currently under construction), and
  - Completing an initial and now on-going CAMBI biosolids stabilization trials.
- Market Research and Development
  - Identified opportunity to market the 'stand-alone' BLOOM product as a soil amendment, with easier access into the marketplace,

- Currently identifying end users and marketing both 'fresh' and 'cured' BLOOM. Activities and strategies within these overall tasks are extensive and will be discussed. They include:

  Market Research and Development
  - Gathering regional data regarding popular soil products produced and marketed in the MD/DC/VA marketplace,
    - Including the identification of large soil blenders and soil amendment suppliers, and pricing data
  - Completing end user demographic research, and
  - Beginning marketing activities (will provide update).

KEYWORDS: CAMBI, biosolids, market development

Business Case Concept Portfolio Supporting a 90 dtpd Biosolids Reuse Program

Alicia Gilley, PE, Director of Resource Recovery and Reuse
Steven Rogowski, PE, Manager of Operations and Maintenance
Metro Wastewater Reclamation District
6450 York Street
Denver, CO 80229
(303) 286-3000

ABSTRACT

Biosolids management program business cases have historically been developed to achieve financial and reliability goals from a wastewater treatment standpoint. Industry perspectives have evolved to emphasize biosolids as a resource and some viewpoints have recently been formalized in concepts such as Utility of the Future (UTOF) which have encouraged broader business case definitions. The Metro Wastewater Reclamation District (Metro District or District) has implemented diverse biosolids management programs and concepts over time, including the current 90 dry ton per day (dtpd) agricultural land application program that aligns with many UTOF concepts. While the program can be viewed from a singular business viewpoint it is more fully characterized by a portfolio of business cases that include strategic, economic, financial, commercial, and management considerations. Three interrelated business cases are presented to illustrate internal business process, regional clean water utility, and external customer viewpoints. Business case examples and data are presented on anaerobic digestion, biosolids dewatering, and private agricultural land application portfolio elements of the program.

KEYWORDS

Dewatering, combined heat and power, beneficial reuse, land application.

INTRODUCTION

The Metro Wastewater Reclamation District's (Metro District or District) is a clean water utility that serves about 1.8 million people in the metro Denver, Colorado region. The utility operates two plants – the 833,000 cubic meter per day (220 million gallons per day, maximum month) Robert W. Hite Treatment Facility (RWHTF) and the 90,000 cubic meter per day (24 million gallons per day, annual average) Northern Treatment Plant. The District current produces an average of 82 dry metric tons (90 dry tons) of Class B biosolids per day.

The RWHTF currently operates with primary clarification, parallel secondary processes in a Modified Ludzack-Ettinger configuration for nitrogen removal, and two-phase anaerobic digestion. On average 0.9 MGD of digested solids, at 2.0% Total Solids (% TS) or higher, is dewatered with high-speed centrifuges using cationic polymer and ferric chloride as a flocculant. Annually, the dewatering process produces 27,215 dry metric tons (30,000 dry tons) of Class B biosolids which are land applied on the District-owned farm and nearby private farms. The organizational responsibilities associated with these operations are covered in two departments –

the RWHTF Operations & Maintenance Department (O&M) manages liquid treatment, solids processing, including biosolids dewatering, while the Resource Recovery and & Reuse (RR&R) Department manages the transportation and land application of all of the produced biosolids.

Portfolio Priorities for Success

Based on the District's experience in developing, evolving, and maintaining its successful agricultural land application program, a number of portfolio priorities for continued success have been identified, including the following:

- Biosolids quality alignment with commercial fertilizer value: A key success factor for land application has been a implementing a customer service approach that allows biosolids to be on par or compete with commercial agricultural fertilizer. This includes favorable pricing and capability to meet fertilizer application rates that meet crop needs.
- Application timeliness consistent with commercial processes: In addition to meeting fertilizer application rates, it is critical to have the ability to do it in a timely matter that is not disruptive to agricultural farming practices. This is achieved through materials handling and managing multiple outlets to account for seasonal limitations in specific outlets.
- Managed transport volume and timeliness: To achieve the application timeliness priority requires alignment with the transport of biosolids, demanding that adequate transport capacity be both forecast and accomplished in the required amount of time.
- Consistent biosolids dewaterability and preheating: Dewatering is the traditional line of demarcation between "wastewater operations" and "biosolids operations." Managing anaerobic digester operations with goals that support the biosolids dewaterability assures both biosolids volume goals and fertilizer application rates can be met.
- Targeted and forecasted dewatered solids volumes: The ability to forecast and deliver total dewatered biosolids output volume is key to meeting fertilizer application rates and scheduling biosolids transport and land application operations.
- Energy recovery aligned with tariffs and regional energy initiatives: Although energy recovery may appear somewhat removed from solids processing and biosolids land application, it provides for reuse of digester gas to support broad District sustainability goals and provide a reliable energy source for biosolids preheating to achieve dewatering goals.

A business case can be defined as a clear and rational basis for decisions and investment. In order to support the decisions and investments related to this portfolio of priorities for the biosolids management program, the District routinely develops business concepts to define facility and business practice development needs and establish operational goals and metrics. Three cases that take into consideration different viewpoints are as follows:

Dewatering Case. An internal customer viewpoint is illustrated by the Dewatering Case that evaluates centrifuge dewatering and considers the needs of multiple operating departments related to centrifuge dewatering and land application operations to optimize costs and impacts.

Digestion Case. A regional clean water utility viewpoint is illustrated by the Digestion Case that includes anaerobic digestion and supporting the wastewater treatment needs through biosolids quality, nutrient content, and combined heat and power resource recovery.

Private Agriculture Case. An external customer viewpoint is illustrated through biosolids fertilizer value impacts to private agricultural producer operations.

The relationship between these individual business cases and the portfolio priorities are presented graphically on Figure 1.

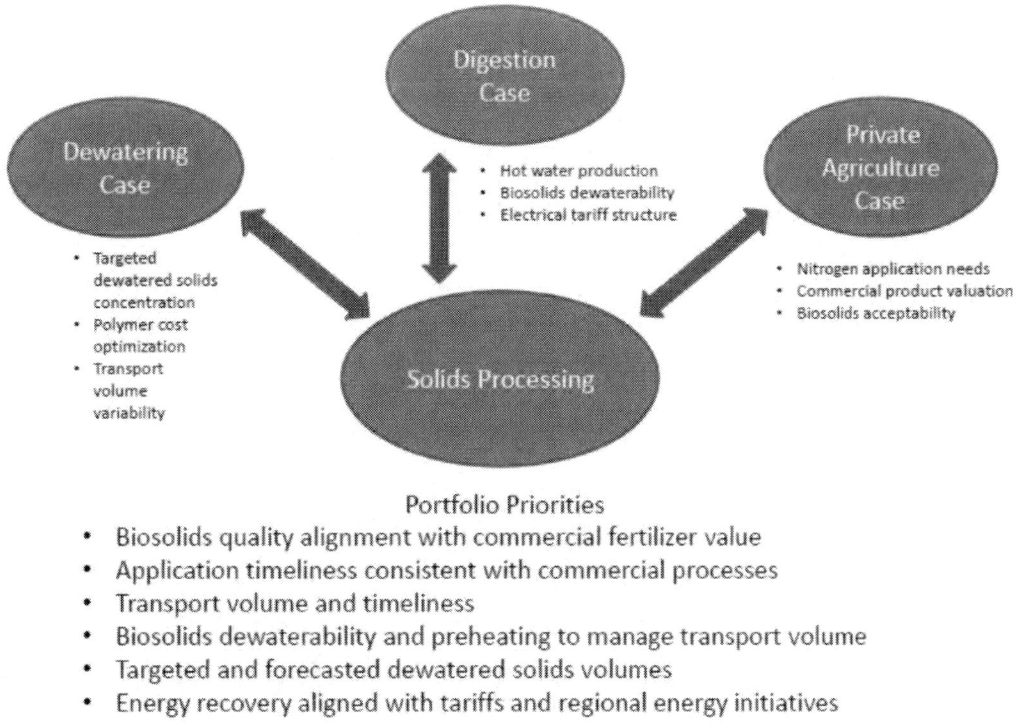

Figure 1 – Biosolids Business Case Concepts Individual and Portfolio Priorities

Business Case Considerations

As stated previously, a business case can be defined as outlining a clear and rationale basis for decisions and investment. In addition to potentially having different viewpoints as discussed above, business cases can be established to support a variety of business case considerations that include:

- Strategic –defines alignment with District goals or a compelling need for change
- Economic – provides good value for expenditures
- Commercial – coordinates with external business practices and activities
- Financial – can be supported by District budget processes and constraints
- Management- measures how well the concept can be achieved with reasonable resources and oversight.

DEWATERING CASE

A number of concepts related to Dewatering Case were developed by the District through a Whole Cost Dewatering Initiative. The objective was to evaluate centrifuge dewatering-related costs and impacts to both the O&M and RR&R Departments. The approach identified cost and performance variables and tradeoffs between District functions and yielded several results and priorities: (a) a moderated dewatered biosolids solids concentration target to reduce polymer expenditures, (b) dewatered cake solids performance metrics to reduce dewatering variability and support transport volume needs related to land application, and (c) a recognition of the need to manage biosolids dewaterability to improve the predictability of dewatering performance, cost, and dewatered biosolids volume impacts. The dewaterability element will be explored further in the Digestion Case.

For the Metro District the ideal target for the dewatered biosolids total solids percentage (% TS) was determined to be between 20% and 22% (Table 1). To develop this recommendation the variable O&M and RR&R expenses were examined to determine their economic impact to each department and the District as a whole of varying dewatered biosolids % TS. Several direct expenses for O&M and RR&R that vary with the biosolids % TS were evaluated to determine the % TS that provides the best overall value. These variable expenses include dewatering polymer quantity and type, as well as RR&R transportation and distribution operating costs, maintenance costs, labor costs, and equipment costs for additional tractors, trailers, and application equipment.

Table 1. Total District combined variable expense summary for dry polymer use as a function of biosolids % TS.

Total District Combined Annual O&M and RR&R Variable Expenses - Dry Polymer				
	Biosolids % TS			
	18	20	22	24
RR&R	$4,031,977	$3,549,763	$3,316,589	$3,226,443
O&M	$1,294,520	$1,690,764	$2,208,296	$2,884,241
Total	$5,326,497	$5,240,527	$5,524,885	$6,110,684

Dewatering polymer is the primary dewatering expense for the O&M Department and Table 1 shows that over the entire range of % TS values. To establish the relationship between O&M dewatering polymer costs and variable biosolids % TS (Figure 1), daily polymer doses and corresponding biosolids % TS data were plotted against each other and analyzed. The dataset used for this examination originated from bi-weekly standard centrifuge tests, where parameters such as torque, sludge and polymer feed rates, centrifuge identification, polymer solution strength, and polymer type were adjusted to monitor changes in biosolids % TS.

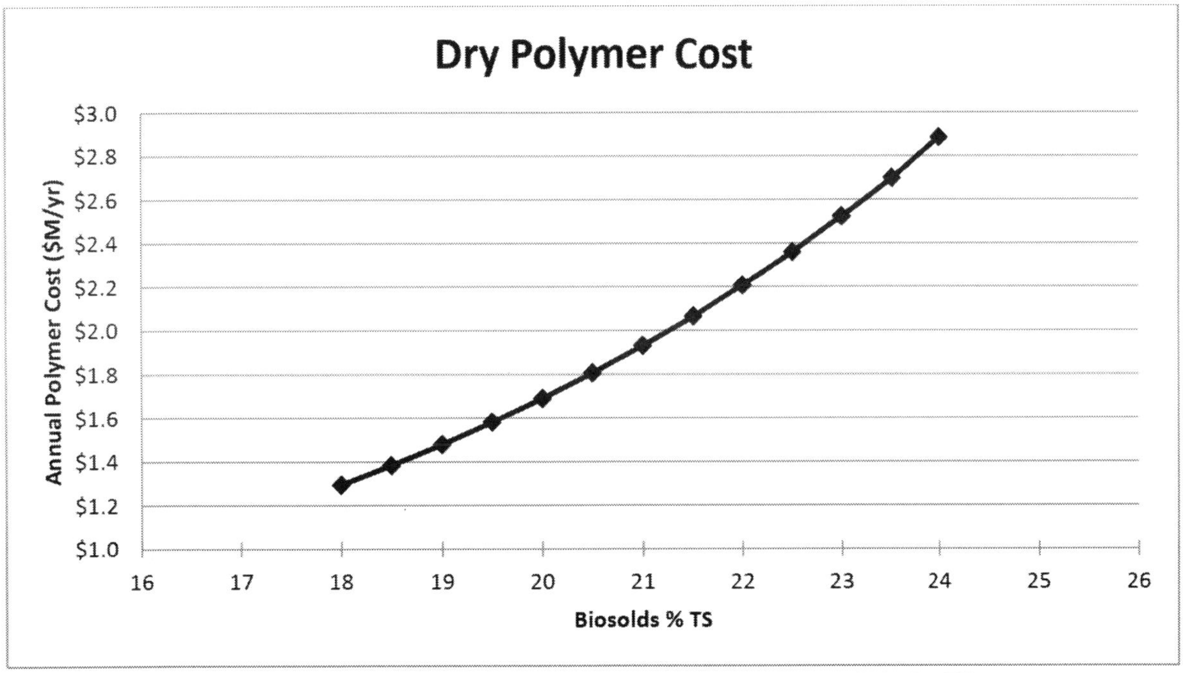

Figure 2. O&M dry dewatering polymer cost as a function of biosolids % TS.

The generation of the O&M annual dewatering polymer expenditures involved several important assumptions, namely:

- O&M variable costs for this analysis consider only polymer costs.
- Polymer dose data is limited by the standard centrifuge testing regime and a small amount of available dry polymer data, therefore the variability observed in daily operations is not included in the analysis.
- Logarithmic equations are the best fit for the data sets examined

Based on current RR&R biosolids recycling practices, the impact of total biosolids volume changes due to the variable cake solids concentration was considered. For RR&R direct variable cost, one additional operator will be needed for each 1% decrease in biosolids % TS, and that one additional mechanic will be needed for each 2% decrease in biosolids % TS. Each new employee is assumed to cost $100,000 per year. Additionally, one truck and trailer combination will be required for each 1% decrease in biosolids % TS. The truck and trailer combination is assumed to cost $250,000 with a ten-year service life, so that discrete expense is included in this analysis as an additional $25,000 per year for each 1% decrease in biosolids % TS. Although labor costs and equipment costs vary with changing biosolids % TS, the costs associated with biosolids % TS values greater than 21.5% are not effected in this analysis since the Metro District already had the equipment and employees needed. Table 2 (below) breaks out the variable costs for RR&R as a function of biosolids % TS.

Table 2. RR&R total annual variable expenses as a function of biosolids % TS with the linear and step function combined variability demonstrated.

Biosolids % TS	Wet Tons	Total RR&R O&M Variable Expense	Total Labor Expense	Additional Equipment Expense	Total RR&R Variable Expenses
unit	**1**	**$8.21**			
18	161,111	$1,322,721	$2,634,841	$75,000	$4,032,562
18.5	156,757	$1,286,975	$2,634,841	$75,000	$3,996,816
19	152,632	$1,253,109	$2,534,841	$50,000	$3,837,950
19.5	148,718	$1,220,975	$2,534,841	$50,000	$3,805,816
20	145,000	$1,190,450	$2,334,841	$25,000	$3,550,291
20.5	141,463	$1,161,411	$2,334,841	$25,000	$3,521,252
21	138,095	$1,133,760	$2,234,841	$0	$3,368,601
*21.5	134,884	$1,107,398	$2,234,841	$0	$3,342,239
22	131,818	$1,082,226	$2,234,841	$0	$3,317,067
22.5	128,889	$1,058,179	$2,234,841	$0	$3,293,020
23	126,087	$1,035,174	$2,234,841	$0	$3,270,015
23.5	123,404	$1,013,147	$2,234,841	$0	$3,247,988
24	120,833	$992,039	$2,234,841	$0	$3,226,880

***21.5 % TS corresponds to the actual RR&R budget.**

The combination of the O&M and RR&R variable cost as a function of biosolids % TS lead to the determination that the optimal target biosolids % TS was between 20% and 22%, as shown previously in Table 1. For this information to be beneficial as a management tool on an ongoing basis, it is necessary to revisit and update the information as conditions changes.

DIGESTION CASE

The Digestion Case provides a regional clean water utility viewpoint for establishing portfolio business case components. Anaerobic digestion is widely used in the industry and has evolved from primarily a wastewater treatment perspective with emphasis on solids reduction and stabilization to one that considers resource recovery to a greater degree. Anaerobic digestion provides the critical functions of ensuring Class B regulatory compliance and solids processing to support liquid treatment operations. Management of digestion operations has developed, however, to include resource recovery elements including dewaterability, product nutrient content, and methane reuse through combined heat and power facilities.

As indicated in the Dewatering Case, managing biosolids dewaterability has been recognized as central to reliably achieving centrifuge dewatering targets. Producing a digested solids with predictable and consistent dewatering attributes contributes to the portfolio priorities in a number of critical areas.

As will be further developed in the Private Agriculture Case, the ability to deliver the targeted plant nutrient application rate is critical for the program. Because biosolids are delivered and applied before final laboratory quality results are available, a number of parameters are forecast and actual results deviation from forecast results in under-or over-application relative to nutrient targets. For instance, producing cake solids of 18 percent TS rather than 21 percent TS will result in nutrient under application of plant available nitrogen by 14 percent - not meeting the agricultural producer expectations. Similar predictability cases can be evaluated for individual nutrients (nitrogen and phosphorus) and biosolids hauling volume.

To manage dewaterability and its impact on operations, the District has implemented metrics to measure dewatering variability in the form of cumulative distribution charts. While not a direct measurement of dewaterability, the approach provides a measurement of the variability in terms of process impacts. Figure 3 shows the cumulative distribution of cake solids for an individual month and annual basis. The metric provides a key visual in that a flat curve is indicative of limited variability and provides objective performance measurement to the plant operations staff.

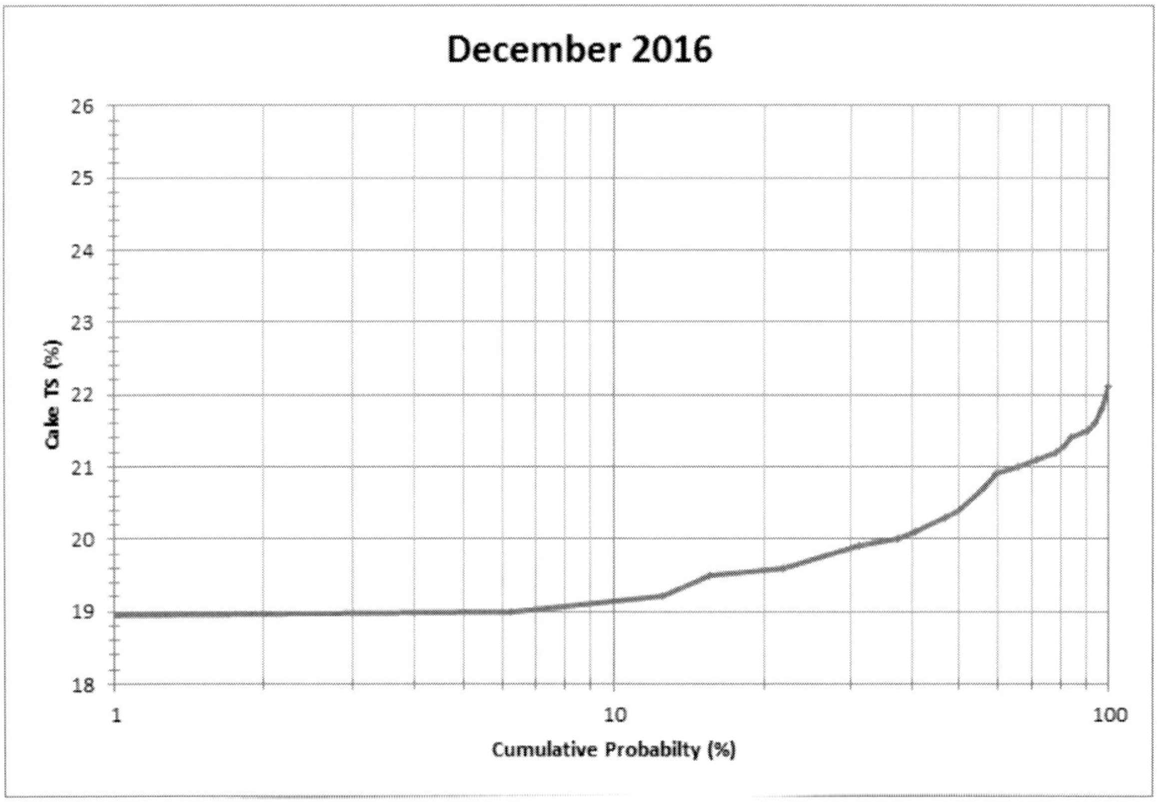

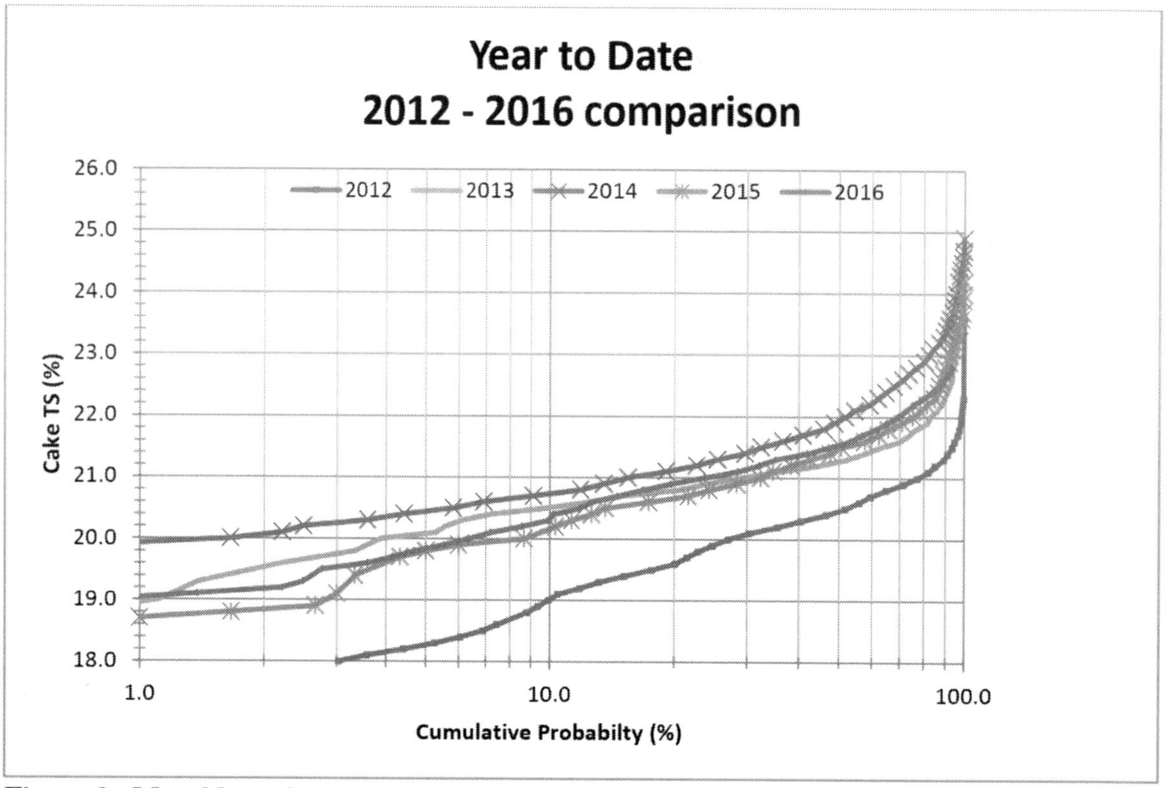

Figure 3. Monthly and Annual Cumulative Probability Dewatering Plots.

PRIVATE AGRICULTURE CASE

The Metro District began its beneficial reuse agricultural land application program in 1981 and currently applies to the District-owned METROGRO Farm (Farm) and permitted private agricultural land. The District recovers some operating costs by charging private farmers for the METROGRO® Cake product through a uniform price per pound rate structure. The acceptability and value of biosolids as a fertilizer and the willingness of farmers to pay for biosolids are based on the difference between a prospective farmer's evaluation of the benefits and costs of chemical fertilizers when compared with biosolids.

When biosolids are given away for free, the tendency is for the farmer to assume a low value of the biosolids and question the benefits. If the biosolids are sold and the price is too high, the farmers tend to perceive the benefits of biosolids as high, but may be unwilling to pay the asking price when a known product such as chemical fertilizers is available at a comparative price. Biosolids must be priced high enough so that the farmer identifies a value, yet low enough to be economically attractive, given unfamiliar sources, products, and application considerations.

Plant Available Nitrogen

Most commercial mineral nitrogen fertilizers are one hundred (100) percent plant available when applied to the soil. Fertilizers such as ammonium nitrate, anhydrous ammonia, urea, and ammonium sulfate are in a form that is readily available to a plant or convert rapidly to a form usable by the plant when soils have sufficient warmth and moisture for crop growth.

Biosolids, however, contain both organic nitrogen and mineral forms of nitrogen. The organic nitrogen is an integral component of cells in organic material in the biosolids. Plants cannot directly use organic nitrogen. Instead, plants rely on microbes in the soil to break down the organic nitrogen in a process called mineralization and release it into the soil in a mineral form plants can use.

The plant available nitrogen (PAN) in Metro District biosolids can be calculated and PAN becomes the common denominator to compare biosolids and chemical fertilizer. The average nitrogen content of the District's biosolids produced in 2016 is shown in Table 3 below.

Table 3. Nitrogen Content of Metro District Biosolids Produced in 2016

Plant nutrient	Range	Mean	Median
Organic N, %	5.25 – 6.08	5.68	5.69
NH4-N, %	0.88 – 1.09	1.0	1.0
$NO_2 + NO_3$, %	0.0	0.0	0.0
Total PAN, lbs/dry ton	**35.80 – 42.8**	**40.1**	**39.8**

Biosolids and Commercial Fertilizer Cost Comparison

The Metro District recovers some operating costs by charging private farmers for the METROGRO® Cake product. The fee structure for the private farmers receiving the METROGRO® Cake product ranges from $8 per acre for 75 pounds or less of nitrogen applied per acre to $12 per acre for 76 pounds or more of nitrogen applied per acre. The private farmers also pay an additional fuel surcharge, which is 2 percent of the District's current fuel cost per gallon times the total number of miles driven to a specific land application site. The pricing system used by the District is cost effective for the farmers; in effect, subsidizing their nutrient application activities. Table 4 shows current commercial fertilizer prices, including a breakdown of PAN costs per pound.

Table 4. Commercial Fertilizer Costs

Commodity, % (PAN-P-K)	Cost[1]	Cost/lb	Cost/lb PAN
28-0-0	$325.00/Ton	$0.1625	$0.58
46-0-0	$440.00/Ton	$0.2200	$0.48
82-0-0	$615.00/Ton	$0.3075	$0.38
32-0-0	$335.50/Ton	$0.1678	$0.52
Average			**$0.49**

[1] Prices effective September 2, 2015

On average, dryland winter wheat requires a maximum of 75 pounds per acre of PAN. Using the mean nitrogen content of 40.1 pounds of PAN per dry ton of biosolids, this equates to an application rate of 1.875 dry tons of biosolids per acre, or 3,750 pounds of biosolids per acre, required to provide the maximum agronomic nitrogen requirement. Under the Metro District's current rate structure, the cost of application to this crop would be $8.00 per acre plus a fuel surcharge. In comparison, the average commercial fertilizer cost associated with 75 pounds per acre of PAN equates to $36.75 per acre plus the application cost. The comparison for a 320-acre site is shown in Table 5 below.

Table 5. Cost Comparison of a 320-acre Winter Wheat Application Site

Biosolids Rate Structure 75 lb PAN/acre		**Commercial Fertilizer Rate Structure 75 lb PAN/acre**	
Cost/acre	$ 8.00	Cost/acre	$ 36.75
Total Biosolids cost	$2,560	Total Commercial Cost	$11,760
Fuel Surcharge[1,2]	$ 970	Application Cost	$ 1,600
Total Application Cost	**$3,530**	**Total Application Cost**	**$13,360**

[1] Fuel Surcharge = Miles driven to site * Current fuel price * 0.02
[2] Average distance traveled per load = 123 miles; Average fuel price for 2014 = $3.0325/gal

For an irrigated corn site, 225 pounds per acre of PAN is an average application rate of biosolids. Using the mean nitrogen content of 40.1 pounds of PAN per dry ton of biosolids, this equates to an application rate of 5.6 dry tons of biosolids per acre, or 11,222 pounds of biosolids per acre. Under the Metro District's current rate structure, the cost of application to this crop would be $12.00 per acre plus a fuel surcharge. In comparison, the average commercial fertilizer cost associated with 225 pounds per acre of PAN equates to $110.25 per acre plus the application cost. The comparison for a 320-acre irrigated corn site is shown in Table 6 below.

Table 6. Cost Comparison of a 320-acre Irrigated Corn Application Site

Biosolids Rate Structure 225 lb PAN/acre		**Commercial Fertilizer Rate Structure 225 lb PAN/acre**	
Cost/acre	$12.00	Cost/acre	$110.25
Total Biosolids cost	$3,840	Total Commercial Cost	$35,280
Fuel Surcharge[1]	$2,909	Application Cost	$ 1,600
Total Application Cost	**$6,749**	**Total Application Cost**	**$36,880**

[1] Fuel Surcharge = Miles driven to site * Current fuel price * 0.02
[2] Average distance traveled per load = 123 miles; Average fuel price for 2014 = $3.0325/gal

Based on the Metro District's current rate structure, the invoiced price of application of biosolids (at the maximum agronomic rate) to dryland winter wheat sites is approximately 25% of the cost of equivalent commercial fertilizer. On an average application rate to irrigated corn sites, the District's rate structure is approximately 20% of the cost of commercial fertilizer.

The value of biosolids over chemical fertilizers may be reduced due to a number of factors such as scheduling biosolids transport and application at times that are compatible with agricultural planting and harvesting. Other conditions such as rain or snowfall, regulations, and the characteristics of biosolids may require applications to occur over a period of days rather than commercial fertilizers which can be applied quickly when the farmer requests.

In addition to the nitrogen supplied by the biosolids, additional benefits to agricultural land include increased organic matter, phosphorus, potassium and micronutrients. It is difficult to include these benefits on cost-basis as there is no fertilizer recommendation for the micronutrients. Nevertheless, these constituents are present in biosolids, and provide benefits to the crops and to the soils at land application sites.

Fee Structure Alternatives

With the majority of biosolids generated by the Metro District being applied to private farm sites, it is advantageous to the District to recover a fair portion of the costs associated with the land application activities from the private farmers that benefit from the application of biosolids to their crop land. In addition to maintaining the current fee structure, providing a uniform price per pound rate structure and an alternate tiered rate structure are also potential alternatives.

A uniform price per pound of biosolids PAN would be one alternative to consider as an option to the current rate structure. Unit pricing in the suggested initial range of \$0.14 - \$0.16 per pound of applied PAN per acre would provide a reasonable return to the Metro District and still provide a value to the agricultural producer. Under this alternative, it is recommended the fuel surcharge be discontinued so the perception is the farmer is paying for the value of biosolids and not the value of fuel.

A tiered rate structure would provide for a set cost per acre for an increased number of levels of PAN applied compared to the District's existing rate structure. This tiered rate structure, shown in Table 7 below, will still provide a significant cost-savings benefit to the farmers and will provide a similar return to the District as the current rate structure. Under this alternative, it is recommended the fuel surcharge be discontinued so the perception is the farmer is paying for the value of biosolids and not the value of fuel.

Table 7. Tiered Rate Structure

Pounds of PAN/acre	Rate/acre
0-50	$8.00
51-100	$10.00
101-150	$15.00
151-200	$17.00
200-250	$22.00
251-300	$24.00

A cost comparison of the current rate structure, uniform rate structure, tiered rate structure, and commercial fertilizer for both the 320-acre dryland winter wheat site and 320-acre irrigated corn site are shown in Table 8 and 9 below.

Table 8. Cost Comparison of a 320-acre Winter Wheat Application Site (75 lb PAN/acre)

Rate Structure	Cost Basis	Total Biosolids Cost	Fuel Surcharge[1]	Total Application Cost
Current	\$8.00/acre	\$ 2,560	\$ 970	**\$ 3,530**
Uniform	\$0.15/lb PAN/acre	\$ 3,600		**\$ 3,600**
Tiered	\$10.00	\$ 3,200		**\$ 3,200**
Commercial	\$0.49/lb PAN/acre	\$11,760		**\$11,760**

[1] Fuel Surcharge = Miles driven to site * Current fuel price * 0.02
[2] Average distance traveled per load = 123 miles; Average fuel price for 2014 = \$3.0325/gal

Table 9. Cost Comparison of a 320-acre Irrigated Corn Application Site (225 lb PAN/acre)

Rate Structure	Cost Basis	Total Biosolids Cost	Fuel Surcharge[1]	Total Application Cost
Current	\$12.00/acre	\$ 3,840	\$2,909	**\$ 6,749**
Uniform	\$0.15/lb PAN/acre	\$10,800		**\$10,800**
Tiered	\$22.00/acre	\$ 7,040		**\$ 7,040**
Commercial	\$0.49/lb PAN/acre	\$35,280		**\$35,280**

[1] Fuel Surcharge = Miles driven to site * Current fuel price * 0.02
[2] Average distance traveled per load = 123 miles; Average fuel price for 2014 = \$3.0325/gal

A uniform price per pound of biosolids is the recommended alternative to consider as an option to the current rate structure. Unit pricing of \$0.15 per pound of applied PAN per acre would provide a reasonable return to the Metro District and still provide a value to the agricultural producer. This uniform rate will increase the cost to the farmer of an average dryland winter wheat site by only 2 percent and will increase the cost to the farmer of an average irrigated corn site by nearly 60 percent. At the same time, this uniform rate is only 31 percent of the commercial fertilizer cost at the winter wheat application site and 31 percent of the commercial fertilizer cost at the irrigated corn application site.

CONCLUSION

By considering the different business case viewpoints presented in this paper, the District is able to leverage results so that individual business cases contribute to the portfolio of priorities and collectively address key business case considerations.

The Dewatering Case considered the needs of multiple operating departments related to centrifuge dewatering and land application operations to optimize costs and impacts. This case optimized dewatering targets to minimize impacts to biosolids quality. This evaluation resulted in a savings of nearly \$800,000 by reducing the targeted total solids from 24 percent to 21.5%.

The Anaerobic Digestion case provided the critical functions of ensuring Class B regulatory compliance and solids processing to support liquid treatment operations. Management of digestion operations has developed, however, to include resource recovery elements including dewaterability, product nutrient content, and methane reuse through combined heat and power facilities.

The Private Agriculture Case evaluated the acceptability and value of biosolids as a fertilizer and the willingness of farmers to pay for biosolids are based on the difference between a prospective farmer's evaluation of the benefits and costs of chemical fertilizers when compared with biosolids. This evaluation is estimated to increase the revenue received from land application of biosolids by nearly 65 percent.

The net result of these three business cases provides the District with established operational goals and metrics which generate additional revenues, reduce costs, and contribute to the District's Mission: To protect the region's health and environment by cleaning water and recovering resources.

Biosolids on Trial – Recent Litigation Wins for Land Application

James B. Slaughter, Esq.[1]

[1] Partner, Beveridge & Diamond, P.C., Washington, DC.
Email: jslaughter@bdlaw.com.

ABSTRACT

Recent courtroom wins in California, Pennsylvania and Virginia for land application of biosolids have underscored the safety and acceptability of this critical recycling practice. The two week trial and recent judgment striking down Kern County, California's biosolids ban is believed to be the first ever full trial over the safety of land application. *City of Los Angeles v. Kern County*, 2016 WL 7175653 (Tulare Co. Super. Ct. Nov. 28, 2016). After hearing evidence from twenty witnesses, the court ruled unconstitutional Kern County's ban on land application of biosolids. The trial and 48 page decision provide valuable resources for biosolids managers for the defense of land application among regulators, the public, and in the courts.

The Pennsylvania Supreme Court handed land application another important victory in *Gilbert v. Synagro*, 131 A.3d 1 (Pa. 2015), ruling unanimously ruled that farming with biosolids is a normal agricultural operation entitled to protection under Pennsylvania's Right to Farm Act. The decision resulted in the dismissal of nuisance and negligence claims brought by plaintiffs residing near a land application site, capping eight years of litigation.

Finally, the Richmond, Virginia, Circuit Court in December 2016 upheld Virginia's regulations for land application against challenges from environmental groups. *Kelble v. Commonwealth*, 2016 WL 7491575 (Richmond, Va. Cir. Ct. Dec. 7, 2016). The Court rejected claims that the regulations were insufficiently protective of groundwater and that the regulations would allow excessive phosphorous loading in waterways from alleged biosolids runoff.

This paper summarizes and analyzes the latest developments in biosolids law to provide wastewater and residuals professionals current information to guide decisions on risk, liability, management and planning, community outreach, and government relations, including ordinances, legislation and lawsuits.

KEYWORDS: Biosolids, Bans, California, Class A, Class B, Health Effects, Land Application, Law, Local Ordinances, Nuisance, Odor, Pennsylvania, Preemption, Perfluorocarbons, Toxic Tort, Virginia.

INTRODUCTION

The success of land application of biosolids in recent decades and scientific research have built a strong record for the defense of this important recycling practice, whether in the media, with regulators and local governments, in the courts, and in the court of public opinion. The commitment of biosolids generators and their contractors, working in conjunction with their farm partners, to defending land application through sustained litigation has produced concrete wins in 2015 and 2016 that stakeholders and advocates should emphasize to deter opponents of the practice. These cases also provide valuable lessons on how the data and science developed around land application play out in a court room setting, underscoring the importance of robust data gathering, odor monitoring, and publishing and communicating the results.

Biosolids Ban Struck Down – *Los Angeles v. Kern County*

Kern County, California, one of the nation's largest agricultural counties, passed a complete ban on all land application of biosolids through a voter initiative in 2006 named "Measure E." The City of Los Angeles has land applied its biosolids at a 4,700 acre farm in Kern County since 1994, and the City purchased the farm in 2002. Faced with the ban that would force the shutdown of its farm, the City promptly organized a coalition of other wastewater agencies (Orange County Sanitation District and Los Angeles County Sanitation Districts), contractors, truckers and farmers to prosecute a lawsuit against the Measure E ban.

The Plaintiffs quickly won an injunction against Measure E that prevented the ban from going into effect. Since 2006, the voter initiative ("Measure E") has been under a series of injunctions pending trial. *City of Los Angeles et al. v. Kern County*, 462 F. Supp. 2d 1105 (C.D. Cal 2006); 509 F. Supp. 2d 865 (C.D. Cal. 2007); 214 Cal. App. 4th 394 (2013). Procedural issues were litigated for several years until extensive discovery occurred in 2015 and a bench trial was held in the spring of 2016.

The court heard over twenty witnesses (including leading biosolids experts, hydrogeologists, chemists, and risk assessment experts) and received over 700 exhibits. The evidence included the results of extensive soil and groundwater analysis conducted by Kern County's experts at Green Acres Farm, plus the detailed records accumulated by the City based on twenty-two years of land application. Witnesses for the Plaintiffs challenging the ban included Ian Pepper of the University of Arizona, Greg Kester of the California Association of Sanitation Agencies (CASA) (also a Plaintiff), hydrogeologist Tom Johnson, and risk assessment expert Dr. Robert Scofield.

The Plaintiffs' data, experts and fact witnesses demonstrated that after two decades of land application (and daily irrigation with treated wastewater effluent from the City of Bakersfield), the levels of metals in the soil at Green Acres Farm were low, as were the levels of other anthropogenic compounds not regulated by the Part 503 rules, such as perfluorinated compounds. Minute detections of trace compounds in soil and groundwater at depth posed no risk to off-site aquifers. A risk assessment showed that consumption by cattle of the feed crops grown at Green Acres Farm posed not risk for consumption of the dairy products. The consistent compliance of Los Angeles and its contractors with state and Kern County permits buttressed the evidence of minimal risk from land application.

Importantly, Kern County acknowledged at the start of the trial that it had no evidence of environmental or public health harm from land application. Instead, the County defended its biosolids ban on "the risk for potential harm" from trace compounds in biosolids, focusing on perfluorocarbons (PFCs), particularly PFOA and PFOS, and the alleged risk to groundwater. Kern's two main experts – chemists Gwynn Johnson of Portland State and Christopher Higgins of the Colorado School of Mines – could not offer a conclusion to a reasonable degree of scientific certain that land application of biosolids constituted a risk to human health. Kern's experts confirmed that Kern's sampling demonstrated that metals and trace compounds largely remained in the shallow soil zone and there was little migration to depth. No constituent found in soil or groundwater exceeded a regulatory standard or advisory screening levels. Kern was also unable to show any potential impacts on a drinking water source from land application or that degraded groundwater conditions were not caused by historic oil and gas drilling work in the area.

Based on the trial evidence and post-trial briefs, Judge Lloyd Hicks of the Tulare County Superior Court issued a 48 page opinion in November 2016 striking down Measure E on two independent grounds. The Court ruled that Kern's biosolids ban "is invalid and void for all purposes, for the dual reasons that it exceeds Kern's police power authority and is preempted by state law." *City of Los Angeles v. Kern County*, 2016 WL 7175653, 2016 Cal Super Lexis 9727 (Tulare Co. Super. Ct. Nov. 28, 2016). The Court's opinion also made numerous findings that land application of biosolids posed no risk, providing an important endorsement of the scientific and regulatory consensus regarding the safety of land application.

On the police power claim, the Court held that "the overwhelming weight of the evidence is that there is no basis in fact for any determination that land application of biosolids poses any risk to Kern County residents, let alone a real and substantial risk that would be alleviated by banning such land application.... Los Angeles has met its burden of producing evidence to the effect that there is no basis in fact for Measure E's public welfare claims....There is no evidence of risk to human health." The Court concluded that because the biosolids ban bore no substantial relation to the public welfare, it was an invalid exercise of the county's police power under the California Constitution.

The Court also scrutinized how banning land application would affect agencies and individuals outside of Kern County to determine whether the ballot measure exceeded the locality's police power. Judge Hicks found that Measure E did not accommodate regional interests in cost-efficient management of biosolids, and no countervailing interests in Kern were served by the ban.

The preemption claim contended that the state recycling law mandated promotion of biosolids recycling and that the state mandate had primacy over a local ban that was in opposition to the goals of state law. The Court ruled that the California Integrated Waste Management Act ("CIWMA"), which requires that all local governments in California promote and maximize recycling, was controlling over a local voter initiative and thus preempted Measure E. The Court held that "Measure E prohibits what CIWMA requires.... Measure E is in direct conflict with, and inimical to, CIWMA, and is therefore, for that reason, also void. . . . Banning a commonly used and cost efficient method of recycling and re-use is not consistent with and is destructive of the state's policies and requirements." Judge Hicks also found on the trial record that composting biosolids is not an alternative to land application because compost must be land applied.

The *Los Angeles v. Kern County* decision striking down Measure E on preemption grounds built on a trend of other state and federal courts that have found that local biosolids bans are contrary to state law and therefore are illegal. Pennsylvania, Virginia, North Carolina, Maryland and other states have all upheld the primacy of state over local regulation of biosolids. *See, e.g.*,

Washington Department of Ecology v. Wahkiakum County, 337 P.3d 364 (Wash. App. 2014); *Liverpool v. Stephens*, 900 A.2d 1030 (Pa. Commw. Ct. 2006); *Granville Farms v. County of Granville*, 612 S.E.2d 156 (N.C. Ct. App. 2005); *O'Brien v. Appomattox County*, 293 F.Supp.2d 660 (W.D. Va. 2003); *Synagro v. Rush Township*, 299 F. Supp. 2d 410 (M.D. Pa. 2003); *Soaring Vista Properties v. Board of County Commissioners of Queen Anne's County*, 741 A.2d 1110 (Md. 1999). The *Los Angeles v. Kern* decision is particularly important because it is based on a full trial record of the benefits and lack of risk of biosolids.

Pennsylvania Supreme Court Finds Land Application a Normal Agricultural Operation – *Gilbert v. Synagro*

In late 2015, a unanimous Pennsylvania held that land application of biosolids is an agricultural activity shielded from untimely litigation by Pennsylvania's Right to Farm Act (RTFA), and dismissed nuisance and negligence claims brought by neighbors of farms using biosolids. *Gilbert v. Synagro*, 131 A.3d 1 (Pa. 2015). Almost all states have right to farm laws and the *Gilbert* decision establishes important precedent that biosolids are an important bulk organic fertilizer and that land application is well embedded in American farming.

The *Gilbert* lawsuit arose from farming operations in York County, Pennsylvania. Neighboring landowners complained of odors from farmers' use of biosolids and filed a lawsuit asserting claims of nuisance, negligence and trespass. The trial court granted the defendants' motion for summary judgment on the ground that the plaintiff landowners filed their claims outside the Pennsylvania Right to Farm Act's (RTFA) one-year statute of repose. The statute of repose required that the Plaintiffs file their lawsuit within one year of smelling odors, which they did not. On appeal, the Superior Court of Pennsylvania reversed the trial court, holding that whether biosolids application qualified as a "normal agricultural operation" protected by the RTFA required a jury to evaluate the manner in which biosolids had been applied.

The Pennsylvania Supreme Court rejected the lower court's rationale in full. In language with implications for future tort cases, the justices emphasized that the RTFA, like any statute of repose, is jurisdictional and thus, its "applicability is a question for the trial court, not the jury." The Court further explained that whether a particular activity, such as biosolids application, falls within the definition of a term used in a statute of repose is to be treated as a question of law that only a judge can resolve. The opinion makes clear that the RTFA can only function properly if its bar against litigation is not contingent on a jury's determinations.

The *Gilbert* decision confirmed that biosolids application is a normal agricultural operation entitled to protection by the RTFA and likely will influence courts across the country in their review of legal issues regarding biosolids recycling to farmland. Citing numerous statistics and the friend of court briefs of a broad coalition of farm and municipal groups, the Court acknowledged the widespread use of biosolids in agriculture in Pennsylvania and across the country. *Gilbert* represents a clear statement by a state's highest court that biosolids use is an accepted and common practice in modern farming that is entitled to protection from untimely and burdensome litigation.

Virginia Biosolids Regulations Upheld Over Challenge by Environmental Groups – *Kelbe v. Commonwealth of Virginia*

In December 2016, a Virginia trial court upheld the State's comprehensive biosolids regulations over challenges from environmental groups that the regulations were not sufficiently protective against run-off from land application sites and that land application could endanger groundwater, particularly in areas with karst topography. *Kelbe of the Potomac Riverkeeper v. Virginia State Water Control Board*, 2016 WL 7491575 (Richmond, Va., Cir. Ct., Dec. 7, 2016). The Court affirmed the Virginia Department of Environmental Quality Biosolids Regulations against the Potomac/Shenandoah Riverkeeper's legal challenge. The Court found there was substantial evidence to support the rules.

The Court's eight page opinion quoted favorably the pro-land application conclusions of an expert panel assembled by the State to review allegations of health and environmental impacts from biosolids. The Court noted that the "Expert Panel concluded that land application is a viable reuse of biosolids that has been shown to be protective of the environment when applicable law and regulations are followed." Judge Taylor then quoted the expert panel's finding that it "uncovered no evidence or literature verifying a causal link between biosolids and illness . . . . As long as biosolids are applied in conformance with all state and federal law and regulations, there is no scientific evidence of any toxic effect . . . . It is important that biosolids be viewed as a resource . . . . Many Virginia farmers depend on biosolids to provide nutrients and organic matter that enhance soil and crop production . . . . It is sensible to take advantage of the benefits of a product that is ever present and must be managed."

CONCLUSION

Biosolids generators, transporters, appliers, farmers and other end-users need to be aware of the many legal tools available to challenge local attempts to restrict beneficial use and to defend against tort lawsuits. American law generally favors farming. Biosolids must be placed firmly in the context of modern, sustainable farming and the use of organic fertilizers to improve yields and soil. Biosolids must also be portrayed and defended as a leading recycling practice. . The great weight of science and experience supports beneficial use of biosolids. The future success of biosolids recycling depends both on educating the public and asserting the legal rights of biosolids programs by fighting restrictive ordinances and lawsuits. Recent courtroom wins for biosolids underscore the value and need for aggressive litigation strategies.

Creating Cost-Effective Centralized Solids Management

Perry Schafer[1], Natalie Sierra[2], Steve Wilson[2]

[1]Brown and Caldwell – 1600 Duke Street, Suite 310, Alexandria VA 22314. Email: PSchafer@BrwnCald.com
[2]Brown and Caldwell

ABSTRACT

There are many examples of centralized or regional solids processing systems for municipal wastewater treatment in North America and overseas. These have been implemented in response to basic economic drivers as well as many other factors that play into these decisions. Some WWTP sites are ill-suited for solids processing functions due to inadequate space, difficult access, neighborhood problems, or aesthetic considerations. Centralizing can sometimes create a major economy of scale that provides significant economic benefits for the owner. Solids are usually transferred to other sites either via pipelines and sewers, or trucked, either as a slurry in a tanker truck, or as cake material in dump trucks or similar bulk-material trailers. Occasionally, agencies use other modes of transport including boats, barges, or rail. Understanding the reasons and drivers for centralized solids processing is an important aspect of learning why and how these systems have been implemented in a wide variety of situations.

KEY WORDS

Solids processing, biosolids management, centralized, regional

INTRODUCTION

Centralized or regional wastewater solids processing usually involves transporting the solids to a central location where a combined-scale system can create economic or other benefits for the owners and operators of the system. Central processing is sometimes aimed at creating a high-quality product from the combined mixtures of the solids. However, more frequently, there are other reasons that push agencies to consider centralizing the solids processing. These reasons include overall economic considerations, staffing limitations, and, especially, space and environmental limitations at some sites. Understanding the reasons and drivers for centralized systems is an important aspect of learning why these systems have been implemented.

There are many examples of this type of regional or centralized solids management activity today at wastewater treatment plants (WWTPs) in North America and overseas. These involve several methods of solids consolidation for joint processing, but, in general, they occur within the framework of the following transportation categories:

- Pipeline transfers of solids slurries are quite common to provide larger scale systems (at a central location), or to move solids out of a sensitive plant location to a site better suited to processing. Examples include Chicago (O'Brien Plant solids slurry piped to Stickney);

Miami-Dade (North Plant solids pumped to Central); San Diego's Metro Biosolids Center (see case history below); Northeast Ohio Regional Sewer District (Easterly solids pumped to Southerly Plant); as well as systems in Houston, City of Los Angeles, and LA County Sanitation Districts.

- Trucking is often used to transport solids slurries from small WWTPs to larger facilities. This has been going on for decades largely because of high unit costs for a small plant to afford and operate a full suite of solids management processes. Over the years, vendors have created small-scale systems to meet the need for small plants, but the practice of trucking persists.

- Larger trucking transfers to central locations also exist. Examples include hauling solids material to a centralized composting facility (examples in California and other states), and transporting cake from multiple plants to an incineration or anaerobic digestion facility. Several solids processing centers operate in the UK, where unstabilized solids are trucked to a combined facility involving advanced anaerobic digestion, and biogas/energy production (Example: Davyhulme Plant – see the case history below).

- Rail/boat transfers are less frequently used, but New York City and other agencies have these types of transport in use today.

CASE HISTORIES

Four case histories are reviewed here, to help understand how these centralized systems have developed, and to review the drivers and objectives for them.

San Diego Centralization at Metro Biosolids Center, California

The Pt. Loma WWTP in San Diego is one of the most physically constrained treatment plant sites in the country, situated as it is on a narrow shelf of land adjacent to the Pacific Ocean. Several decades ago, a digested solids force main was constructed to transfer the anaerobically digested slurry from the plant to an air drying area (Fiesta Island) several miles north of the plant. This system of solids processing was operated successfully by the City of San Diego for many years, but eventually the air drying beds became an odor and aesthetic problem in an increasingly recreational portion of the City. In the 1990s, along with needed wastewater system and water recycling expansion, a centralized solids processing center was created – the Metro Biosolids Center, or MBC. MBC was sited adjacent to the Miramar Landfill, at a location with adequate footprint and good freeway access. The digested solids pipeline was extended to MBC - a total distance of 28 km (17 miles) from the Pt. Loma plant (San Diego, 2016).

MBC is a unique regional consolidation system for both biosolids processing and efficient energy production/management. MBC has thickening and anaerobic digestion facilities to process solids from the North City Water Reclamation Plant. Unstabilized solids from North City are pumped to MBC through another 7 km (4.5 mile) force main (see Figure 1). The digested solids from North City are combined with the pumped digested solids from Pt. Loma, and jointly dewatered with centrifuges at MBC in a large-scale Class B dewatering operation.

Space has been reserved at the MBC site for further processing of biosolids in the future, if that should prove necessary or advisable for the City.

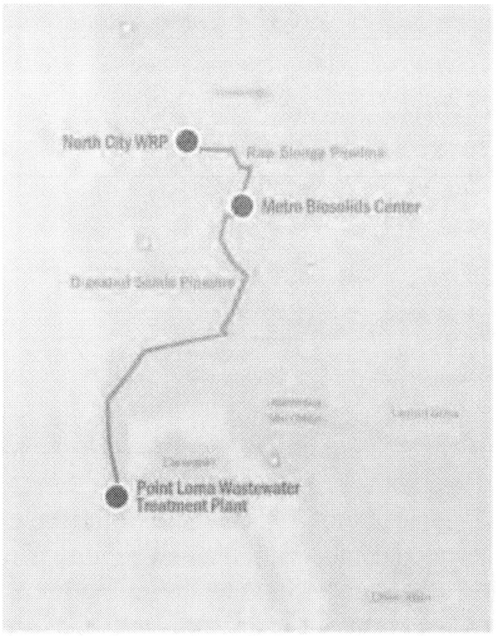

Pipelines bring digested and unstabilized solids to the Metro Biosolids Center (MBC) for processing and centralized Class B dewatering and loadout.

Figure 1. San Diego, California Centralized Solids Handling at MBC

Landfill gas is combined with on-site digester gas at MBC to fuel a privatized engine-based system which produces power for much of the MBC facilities, as well as heat for anaerobic digestion. This system will be expanded in the coming years and could involve electrical power export from the site.

The City's Class B biosolids are all beneficially used. Most of the biosolids production in recent years has gone to Alternative Daily Cover at landfills, but also for agricultural land application. About 100 dry metric tonnes/day (110 dry US tons/day) of Class B biosolids are produced at MBC and dewatered to about 28 percent solids for trucking and disposition.

St. Petersburg Biosolids/Energy Consolidation, Florida

The City of St. Petersburg is proceeding to implement a centralized wastewater solids processing system at its Southwest Water Reclamation Facility (SWWRF) (St. Petersburg, 2017; and Willis et al, 2012). Solids from two other wastewater plants (the Northwest WRF and the Northeast WRF) will be routed to the SWWRF via force mains and existing sewers. A fourth wastewater plant (the AWWRF) has already been removed from service, with its wastewater re-directed for treatment at the SWWRF (see Figure 2 for a system overview). To accommodate increased wastewater treatment capacity at the SWWRF, primary clarifiers, biological treatment upgrades and related facilities are being implemented. With all solids processing centralized at the SWWRF, additional thickening, anaerobic digestion, and dewatering capacity is being built at this plant. Since one of the primary objectives of the program is to develop Class A (pathogen-free) biosolids, the digestion process selected is a thermophilic/mesophilic system (Temperature

Phased Anaerobic Digestion, TPAD)) which includes thermophilic batch tanks to meet the time/temperature requirements for Class A designation.

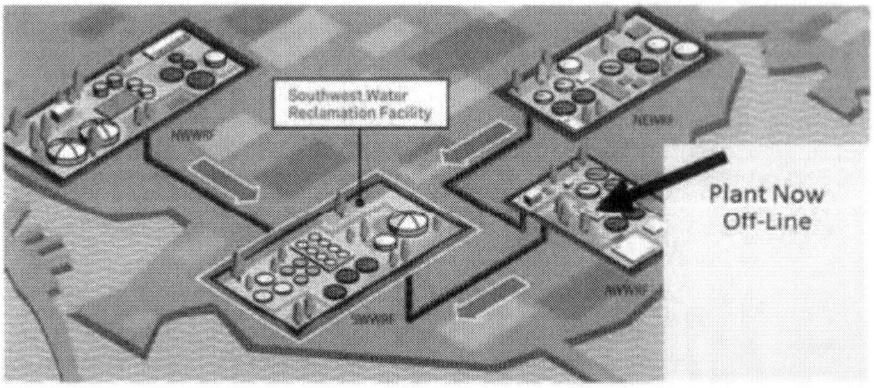

Figure 2. Centralized Solids Processing at the Southwest WRF in St. Petersburg

In Florida, there are significant regulatory and economic advantages with producing Class A biosolids over Class B material. With Class A material (actually called Class AA in Florida), the biosolids can be classified as "fertilizer", and can thus bypass some of the typical Class B limitations and controls. The Class A dewatered cake from the new facility is planned for agricultural land application in the region near Tampa/St. Petersburg, Florida.

The SWWRF was the most logical location for a consolidated solids processing system because of its capacity and existing facilities, and also because the SWWRF site had sufficient space to accommodate the new or modified tanks and equipment. The wastewater treatment system in St. Petersburg totals about 130 to 150 megaliters/day (35 to 40 mil gal/day) of average-flow capacity. The centralized solids system at SWWRF is planned for startup in 2018 and 2019.

The digester gas at the centralized SWWRF facility will be processed to remove carbon dioxide and various contaminants and then used for vehicle fuel within the City's solid waste trucking fleet. This allows the City to collect incentive payments available for renewable fuel production. Natural gas will be used to provide fuel for an engine-based Combined Heat and Power system at the SWWRF, with the engine waste heat used to heat the digesters to thermophilic temperatures.

United Utilities Regional Sludge Processing (United Kingdom)

United Utilities (UU) operates several WWTPs in the Manchester/Liverpool Region of Northwest England and has consolidated its wastewater solids processing system over the years. One of the major concepts instituted about 1980 was the construction of an 86-kilometer (53 mile) solids slurry pipeline which was originally used to transport solids from Mersey Valley WWTPs to the ocean. In the 1990s, ocean disposal was terminated and a major dewatering/incineration plant was implemented at Shell Green (see Figure 3 for regional location of facilities). The pipeline has been instrumental in transporting solids to the Shell Green site for this consolidated incineration program.

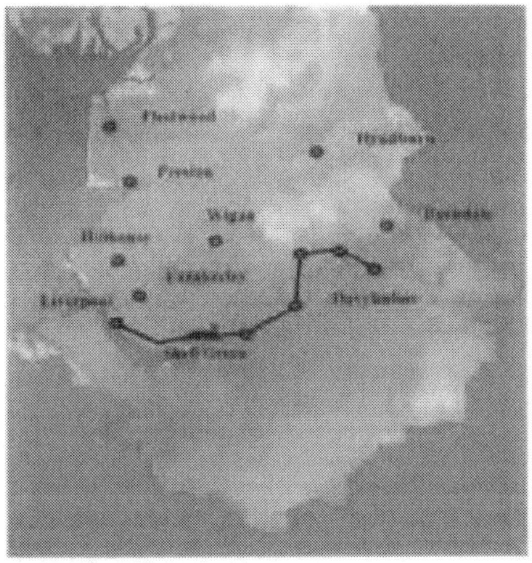

Dewatered raw cake is trucked from 7 plants (shown as red dots) to the centralized THP/digestion facility at the Davyhulme Plant. In addition, digested solids are piped from Liverpool and Davyhulme (Blue line) to centralized incineration at the Shell Green location.

Ref. – Edgington et al, 2014

Figure 3. United Utilities – Manchester, UK Centralized Solids Processing

For several reasons, UU has developed a centralized advanced digestion system (using Cambi THP pre-treatment) and an energy production facility at the large Davyhulme Wastewater Plant in Manchester. This project dovetailed with the already-built Mersey Valley pipeline. The primary drivers for this Davyhulme digestion/energy project are summarized here (Edgington et al, 2014):

- Reduce the quantity of biosolids material for final disposition.
- Need for an enhanced quality biosolids product to use in agricultural land application in the regional land bank available in the Manchester area.
- Maximize methane/energy recovery from anaerobic digestion of wastewater solids.
- Maximize economy of scale and better utilize assets available, including eight large existing digesters at the Davyhulme Plant site.
- Reduce the utility's carbon footprint.

The centralized THP/digestion/energy program includes trucking unstabilized dewatered solids from seven outlying or feeder WWTPs as shown on Figure 3. The quantity of solids trucked from the outlying plants is almost 60 percent of the total fed to the centralized Davyhulme facility, and involves 20 to 30 truck trips per day to bring the material to Davyhulme. The remaining 40 percent of the feed comes from the indigenous solids produced at the Davyhulme Plant. Two truck discharge stations are located at Davyhulme as well as large storage silos. Five Combined Heat and Power (CHP) engines (maximum capacity of 12 MW of power production) are located at the site. Also, there are three composite boilers that utilize both engine heat and direct biogas firing to produce the 12-bar (175 psig) steam required by the Cambi THP system.

Inland Empire Regional Composting Facility (IERCF) – California

About 10 years ago, the Inland Empire Utilities Agency (IEUA) and the Los Angeles County Sanitation Districts (LACSD) joined forces to implement a regional biosolids composting facility in Rancho Cucamonga, California. The two agencies formed a Joint Powers Authority to modify an existing, large warehouse and turn it into a very large indoor composting plant (10 acres of fully enclosed composting) that uses the aerated static pile composting method. Each agency trucks about 68,000 wet metric tonnes/year (75,000 wet US tons/year) of mostly Class B dewatered biosolids to the facility. The IERCF also utilizes green waste for the composting process and creates a Class A, exceptional quality product that is all marketed. A very large, high-performance biofilter is used to treat and control odors from the IERCF.

Trucking Class B anaerobically-digested and dewatered material from the two agencies' WWTPs involves many truck trips per day and has been a major part of each agency's efforts to improve their sustainability of wastewater solids management. Both agencies were looking for cost-effective solids solutions that would provide diversity in end use, as well as advance successful and long-term beneficial use of the biosolids they produced, and help minimize future risks as regulatory factors and public perception change over time. Figure 4 shows an aerial view of the IERCF which sits adjacent to one of the IEUA's water reclamation plants.

This aerial view shows the Inland Empire Regional Composting Facility (IERCF) and its immediate surroundings which include an IEUA water reclamation plant, a detention center, and a private power plant.

Figure 4. Inland Empire Regional Composting Facility

DRIVERS FOR CENTRALIZED SOLIDS MANAGEMENT

Table 1 presents a summary of several centralized solids management systems. Most of these are in the US, but some are located overseas. The drivers for these centralized systems are noted where possible to gain an understanding of why the agency/organization moved into this type of arrangement.

Table 1. Examples of Centralized Solids Processing Systems and Drivers		
Agency/System	**Summary Description**	**Drivers/Comments**
Pipeline Transfer Systems		
MWRD of Greater Chicago – Centralized at Stickney Plant	Unstabilized solids slurry pipeline from the O'Brien WRP to Stickney Plant for combined/joint digestion and advanced solids processing.	Limitations on sludge processing at the O'Brien Plant compared to Stickney. Economy of scale of very large system at Stickney.
San Diego – Metro Biosolids Center (MBC)	Digested solids pipeline from Pt. Loma to MBC. And, unstabilized solids pipeline from North City Plant to MBC. Digestion/energy and large combined dewatering at MBC.	Economies of scale and major site advantages at MBC. The Pt. Loma WWTP site has severe restrictions of space, and difficult access.
Northeast Ohio Regional Sewer District – Centralize at Southerly	Unstabilized solids are pumped from Easterly plant to Southerly plant for combined processing (dewatering and incineration).	Inadequate space at Easterly for processing. Economics of combined system and site availability at Southerly.
Combined processing at Nashville's Central WWTP	Unstabilized solids are pumped from White's Creek plant to the Central Plant for combined and centralized processing using anaerobic digestion and thermal drying.	Economy of scale with a larger combined system at the Central Plant.
City of Houston. Sludge transfers.	Wastewater solids are often transferred via the sewer system for centralized processing at larger Houston plants.	Many smaller WWTPs in Houston do not have solids processing – costly and difficult to implement at smaller WWTPs. Also, staffing issues at smaller WWTPs.
Truck Transfer Systems		
United Utilities, Manchester UK. Large, combined digestion/energy Program.	Cake truck transport from 7 outlying WWTPs to centralized Davyhulme plant for Cambi THP and	Overall economic and energy drivers, advance the utility's sustainability goals, GHG reductions,

Table 1. Examples of Centralized Solids Processing Systems and Drivers		
Agency/System	**Summary Description**	**Drivers/Comments**
	digestion/energy program. Also central incineration at the Shell Green site.	and maximize use of existing facilities.
West Central Wisconsin Biosolids Facility	Trucking solids slurry from about 20 small WWTPs to centralized dewatering and alkaline stabilization system with solids reuse program.	Economics provided impetus for larger combined system and biosolids reuse program. Ease of contracting and system operation, including staffing.
South Kern Composting Facility by Synagro, Southern California	County Sanitation Districts of LA County, and other Southern Calif municipal agencies truck Class B biosolids to this private composting operation in Kern County.	Synagro operates this as a merchant facility – i.e., for-profit business, accepting multiple agency material for joint composting and product marketing and sale.
Washington Suburban Sanitary Commission, MD – Centralized Program.	Dewatered solids from four WSSC WWTPs are planned to be trucked to central facility at the Piscataway Plant for joint THP/digestion/energy Program.	Overall economics and energy benefits are drivers for WSSC. Much improved biosolids product quality and also sustainability issues with long-term risk reduction. Program is now in the design stage.
Boat and Rail Transfer		
New York City. Biosolids transfers by Dept of Environmental Protection (DEP) boats.	Digested solids slurry is transferred via boat between some of the city's WWTPs for joint dewatering and truck loadout.	Overall economics drives this approach at NYC, along with limited footprint at some plants for dewatering/loadout facilities.
Passaic Valley Sewerage Commission – Rail and Boat transfer.	PVSC accepts rail, truck and boat/barge deliveries of wastewater solids and similar materials at its Newark, NJ plant for combined/joint processing.	PVSC accepts wide variety of materials – this is a business-oriented operation by PVSC due to its large/available solids processing capacity.

The private sector has also developed joint use facilities, sometimes called merchant facilities, whereby multiple agencies can contract to process their biosolids into a more marketable product. Large-scale composting is often conducted in such private facilities, but there have also been chemical treatment facilities which create fertilizer products. In Southern California, several of the drivers listed above have led to the development of regional composting facilities used by both large and small wastewater utilities.

Solids Transport Via Sewers and Pipelines

Wastewater sewers often present an easy alternative to transport wastewater solids to a different plant or location where centralized processing can occur. Sometimes a solids force main is used to get the slurry to an adequately sized sewer to ensure sufficient capacity and velocity to move the solids reliably downstream. There has been concern over the years that, for at least warmer climates, the introduction of more solids into the wastewater stream may exacerbate sulfide production and cause increased corrosion rates and more odor. When San Diego was considering how to move solids between plants/sites in the early 1990s, the staff made a conscious decision to implement separate force mains for solids slurries and eliminate the potential negative consequences of putting these solids in gravity-flow wastewater sewers.

Force mains are often used to transport solids. Design details for solids slurry pumping are outlined in the Federation's MOP 8 text (WEF, 2010) and in other documents on the subject. Most agencies pump relatively thin slurries of solids, in the range of one percent solids content, to minimize headloss and to reduce transit time in the pipe so that sulfide and odor production are minimized. Thickening of the solids at the downstream/receiving end of the pipeline is then a common approach.

SUMMARY OF CENTRALIZED SYSTEM DRIVERS

The drivers for centralized solids processing facilities in North America have typically included at least some of the following factors:

- Footprint Issues. WWTP sites sometimes have footprint limitations that prevent or greatly limit the development of a solids processing system. Therefore, developing the solids processing system at a different site which has adequate space creates a major advantage.

- Economics. Cost for solids processing and use have risen over time, making consolidated systems more attractive (economy of scale).

- Staffing. Providing adequate staffing for solids processing functions is often a key issue, especially for smaller WWTPs.

- Transportation. Centralizing biosolids processing near convenient transportation routes can ease operational concerns and help control transportation costs associated with biosolids products and other trucking needs.

- End use restrictions and biosolids quality. More restrictions on landfill use, tightening air rules (on incineration, for example), and improving biosolids product quality can push agencies to consider more advanced processing which either takes more space or makes sense at larger scale.

- Energy Issues. Energy production and carbon footprint considerations can tip the scale in terms of combining more feedstock into a larger, central system.

- Odor, Access, and Environmental Concerns. Odor from sludge processing systems is frequently a problem for wastewater agencies, and consolidating such processing at a better site for odor control is sometimes helpful. Also, there can be truck access or other issues at some sites which pushes owners to use pipelines to transfer the sludge to a better location for processing.

The various forces at play in North America seem to be providing creative energy for agencies and engineers to think more broadly about biosolids solutions than was the case previously. Additional centralized systems are likely to be implemented over time both in North America and overseas, since the drivers for these systems cover a broad range of issues that are likely to continue to affect wastewater utilities world-wide.

REFERENCES

City of San Diego, 2016. Information from the San Diego Wastewater Department website and reports.

City of St. Petersburg, 2017. Information from the St. Petersburg website concerning the city's wastewater system.

Edgington, Rachel et al, 2014. Thermal Hydrolysis at Davyhulme WwTW – One Year On. Presented at the 19th European Biosolids Organic Resources Conference, Manchester UK.

Fonda, Ken, Dwight Corriea, Jack Swerlein, 2002. Startup Experience of the San Diego Metro Biosolids Center Centrifuge Thickening Process. Presented at the Water Environment Federation annual conference (WEFTEC) in October 2002.

Inland Empire Utilities Agency (IEUA) 2017. IEUA website information on its biosolids facilities, and the composting program at the Inland Empire Regional Composting Facility.

Los Angeles County Sanitation Districts (LACSD), 2017. Information from the LACSD website and on LACSD's biosolids facilities and the biosolids use and disposition.

United Utilities, 2017. Information from the United Utilities website and information about its solids processing and biosolids management systems, Manchester UK.

Water Environment Federation, 2010. Design of Municipal Wastewater Treatment Plants, fifth edition. Volume 3 – Solids Processing and Management. WEF Manual of Practice No. 8.

Willis, John, Steve Marshall, George Cassady, and Todd Bosso, 2012. Biosolids and Yard-Waste-to-Energy Feasibility Study in St. Petersburg, FL. Presented at the Water Environment Federation annual conference (WEFTEC) in October 2012.

High-Tech Analysis of Low-Cost, Low-Tech Methods for Sustainable Class A Biosolids Production: Set up and Initial Pilot-Scale Data

Christa L. Meingast[1], Jennifer G. Becker[1], Eric A. Seagren[1]*

[1] Michigan Technological University
* Email: eseagren@mtu.edu

ABSTRACT

Many water resource recovery facilities (WRRFs) across the U.S. are seeking to upgrade their biosolids treatment systems and produce Class A biosolids, thereby opening additional avenues for land application and distribution of biosolids. However, many WRRFs lack the resources to evaluate and implement the systems that are currently accepted for achieving Class A status, including the Processes to Further Reduce Pathogens and other U.S. EPA-approved processes that tend to be energy-intensive, high-tech, and expensive. The overall goal of this multi-phase project is to develop a rational and universal approach for the design of a low-cost, low-tech (LCLT) Class A biosolids treatment process. To achieve this goal, a broad research effort has been initiated in which fundamental information on the impact of key process parameters on the kinetics of inactivation of pathogen and indicator organisms will be collected under a range of conditions in controlled laboratory studies, and validated in pilot-scale studies. This presentation is focused on the set up of the pilot scale studies and the data collected to date. Ultimately, the results of this study will contribute to greater implementation and approval of LCLT treatment processes, allowing more WRRFs to sustainably and economically produce Class A biosolids.

KEYWORDS: Biosolids, Class A, Sustainable, Low-Cost, Low-Tech.

INTRODUCTION

Water resource recovery facilities (WRRFs) are facing growing challenges, including increasing public scrutiny, which may limit the sustainable reuse of biosolids. Accordingly, many WRRFs across the U.S. are seeking to upgrade their biosolids treatment systems to produce Class A biosolids, thereby opening additional avenues for land application and distribution of biosolids. However, although some WRRFs are able to evaluate and the implement the systems currently accepted for producing Class A biosolids, such as the processes to further reduce pathogens (PFRPs), the vast majority of U.S. WRRFs serve small communities that do not have the resources to implement these mostly high-tech, high cost treatment systems. As a result, there is a growing need for low-cost, low-tech (LCLT) treatment options for Class A biosolids production.

Sustainable and cost-effective alternatives to PFRPs do exist and have been successfully implemented. For example, the Stickney Water Reclamation Plant (Metropolitan Water Reclamation District of Greater Chicago, MWRDGC) implemented a successful LCLT process of long-term storage, air drying, and cake storage. However, widespread adoption and regulatory approval of these LCLT methods has been hindered by a lack of information on the impacts of environmental factors and biosolids properties on pathogen and indicator organism (PIO) inactivation during LCLT treatment. Examples of environmental factors and biosolids properties that can affect PIO reduction include pH, temperature, volatile fatty acids (VFAs), ammonia concentration, volatile solids (VS), and moisture content. Fundamental information on how these key environmental parameters affect PIO inactivation is needed so that the time required to achieve Class A pathogen inactivation requirements in LCLT biosolids treatment systems can be accurately and reliably predicted *a priori*.

The overall goal of this multi-phase project is to develop a rational and universal approach for the design of a LCLT Class A biosolids treatment process. This design approach will be analogous to the time-temperature criteria used to establish most PFRPs. Application of such a design approach will help ensure that LCLT processes can be reliably and consistently used to achieve Class A requirements. To achieve this goal, a broad research effort has been initiated in which fundamental information on the impact of key process parameters on the kinetics of inactivation of PIOs will be collected under a range of conditions in controlled laboratory studies, and validated in pilot-scale studies. The pilot-scale studies, which are the focus of this paper, are being performed to achieve the following specific research objectives (**ROs**):

1. **RO1**: Design and set up pilot-scale LCLT Class A biosolids treatment systems at two small WRRFs.

2. **RO2**: Monitor key environmental conditions (ambient environmental conditions, biosolids storage temperature, moisture content (measured as total solids, TS), organic matter content (measured as VS), NH_3 content, VFAs, alkalinity, and pH) over the course of one or two year(s).

3. **RO3**: Apply culture- and molecular-based methods to quantify the inactivation of PIOs (fecal coliforms (FC), coliphage, enteric viruses, and helminth ova) over time in the pilot-scale systems.

4. **RO4**: Evaluate the PIO inactivation rates in the pilot-scale systems as a function of the environmental conditions.

This paper reports on the design and set up of the pilot-scale studies (**RO1**), and presents some of the data that have been collected to date as part of **RO2** and **RO3**.

MATERIALS AND METHODS

Design and Set Up of Pilot-Scale Systems

Pilot-Scale Locations and Overall Approach

The pilot-scale studies are being performed in collaboration with two WRRFs (Figure 1): the Geogebic Iron Wastewater Authority (GIWA) in Ironwood, Michigan, USA, and the Portage Lake Water and Sewage Authority (PLWSA) in Houghton, Michigan, USA. These two similarly sized activated sludge plants (design flow ~3 to 3.5 MGD) perform mesophilic anaerobic digestion (MAD) for vector attraction reduction, followed by dewatering via belt-filter press, and produce materials that are classified as Class B biosolids. The dewatered solids are being used to simulate the combined high-solids, long-term storage/air drying process used at the Metropolitan Water Reclamation District of Greater Chicago. Large uninsulated storage sheds are currently used for winter biosolids storage at PLWSA and GIWA. Whether biosolids are stored indoors or outdoors affects solar radiation, temperature, and moisture content. Therefore, two pilot-scale tests are being performed in triplicate at each site, with pilot test beds located inside a storage shed in Treatment 1, and outdoors in Treatment 2. For each location and treatment, the studies are being conducted in two stages. In Stage 1, digested and dewatered biosolids are placed in the test beds and stored for 1 to 2 years, without addition of new biosolids or disturbance, except for monitoring. In Stage 2, the biosolids will be removed from the test beds and formed into windrows for continued air drying. This presentation is focused on the initial data gathered during Stage 1.

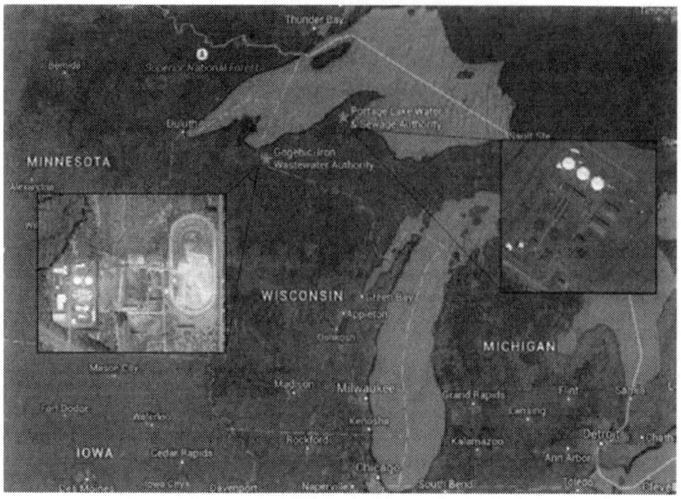

Figure 1: Locations of GIWA and PLWSA

Pilot-Scale Test Beds

The pilot-scale test beds were designed to: 1) maintain structural integrity while holding ~two tons of solids, plus precipitation, throughout the year, and undergo freeze/thaw cycles of the biosolids; 2) remain watertight for up to two years, not allowing leachate to leak from the structure; 3) allow drainage water from the biosolids to be collected and disposed of properly; and 4) withstand ambient temperatures (~20 °F to 95 °F). To simulate average dimensions of a biosolids pile, boxes were built with nominal interior dimensions of 4' high, 4' wide, and 8' long, using exterior-grade plywood, with 2-in x 4-in braces, and additional reinforcements (Figure 2). After the construction of the boxes, potable-grade vinyl liners were installed, along with a drainage system and temperature sensor stacks (described below). To initiate the study, the test

Figure 2: Completed Pilot-Scale Test Bed

beds were carefully filled with freshly pressed biosolids to a nominal depth of 3 ft.

Seeding Biosolids with Ascaris Ova and Human Poliovirus

To certify a biosolids treatment process as a "Process Equivalent to a PFRP" under Alternative 6 of the EPA 503 rule, the following mandatory minimum pathogen reduction requirements must be demonstrated: 1) ≥ 3 log reduction of total enteric viruses, 2) ≥ 2 log reduction of viable helminth ova, and 3) ≥ 1 log reduction of fecal coliforms. However, biosolids produced via MAD in the U.S. typically do not naturally have sufficiently levels of helminth ova and enteric viruses to validate the reduction requirements; therefore, these organisms were seeded into the pilot scale boxes, with target levels of 10^5 to 10^6 PFU enteric virus/g TS and ≈ 10^4 *Ascaris* ova/g TS. Vaccine-grade type 1 human poliovirus, which has been attenuated and therefore cannot cause polio infection, was used as a representative enteric virus.

To prevent the release of *Ascaris* ova and poliovirus into the environment during the study, sentinel chambers incorporating a membrane containment system were designed for spiking the biosolids with these pathogens. The membranes prevent exposure of personnel to the pathogens and the release of pathogens into the environment, while still allow exchange of gases and liquids with the bulk biosolids in the test beds. Therefore, the physical-chemical conditions inside the sentinel chambers are essentially the same as in the surrounding bulk biosolids. Because the sizes of the helminth ova (~ 35 to 50 µm diameter) and poliovirus (~28 nm diameter) are very different, two designs were used for the sentinel chambers.

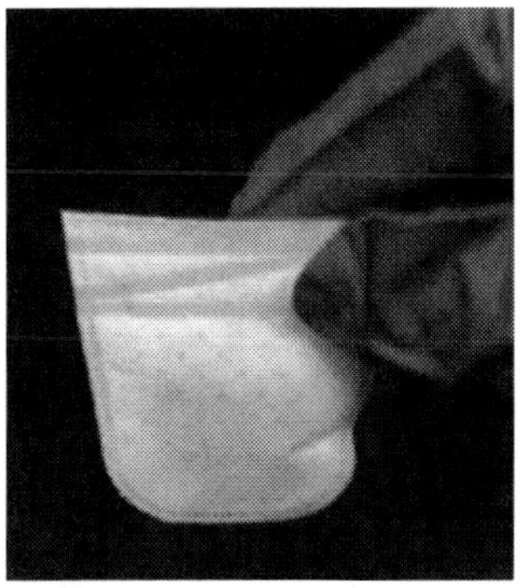

Figure 3: Helminth Ova Bag Design

The helminth ova sentinel chambers were designed following the approach of Gould et al. (1). Specifically, ANKOM F57 filter bags were used because of their pore size of 25 µm. These bags are also chemically resistant and can be heat sealed. A complete sentinel bag for the helminth ova is shown in Figure 3.

The sentinel chamber for the human poliovirus (Figure 4) was adapted from the design of Sidhu and Toze (2). This design incorporates a Teflon chamber that contains the biosolids and virus, and is secured on each end by a membrane (25 mm-diameter, 0.025-µm pore size, Product number: VSWP02500, Millipore), a marine-grade stainless steel mesh (0.5 mm), an O-ring that sits within a groove in the end of the chamber, and a brass pressure plate. Brass screws and nuts hold the assembly together.

Figure 4: Photograph of Poliovirus Sentinel Chamber

The same general approach was used for spiking each type of sentinel chamber: 1) fresh dewatered biosolids were collected, 2) the biosolids were added to the chamber, 3) a stock solution of the target organism was added to the chamber and mixed with the biosolids, and 4) the chamber was sealed. Once prepared, the sentinel chambers were installed in the pilot-scale test beds via stacks constructed using a pipe-within-a-pipe design. First, the sentinel chambers were secured within an inner pipe constructed from a 2.5-in square PVC pipe, which was sealed at each end. In the case of virus sentinel chambers, the chambers were attached to the square inner pipe by inserting the chambers through holes on each side of the pipe so the chamber extended across the pipe, and securing the chamber to the pipe on each end. For the *Ascaris* sentinel chambers, the chambers were placed within a 12 in × 18 in, nylon monofilament, drawstring bag with a 25-µm mesh size (filterbag.com), that was hung within the inner pipe. For each treatment, a nylon drawstring bag suspended in a "low" position in one test bed, in a "middle" position in a second test bed, and at a "high" position in a third test bed. Each inner pipe containing either virus or *Ascaris* sentinel chambers was subsequently placed inside a 4-in diameter, circular, PVC drainage pipe, which was capped on the bottom end, and biosolids packed into the space between the two pipes surrounding the chambers. At each sampling event, the square inner pipe is lifted out, a randomly chosen chamber is removed and processed for analysis, and the square pipe is replaced within the outer, circular pipe. In addition to the chambers installed in the pilot-test beds, control chambers containing pathogen-spiked biosolids were placed in self-sealing plastic bags (Ziploc) to prevent desiccation, stored at 4 °C, and analyzed at each sampling event.

Monitoring Methods

Environmental Conditions

The environmental conditions (ambient and *in situ* within the biosolids piles) are being monitored for the indoor and outdoor test beds over the course of the experiment. Ambient environmental conditions (precipitation, temperature, humidity, solar radiation, and UV radiation) are being monitored using Davis Wireless Vantage Pro2 Plus Weather Stations placed indoors and outdoors at GIWA and PLWSA. In situ biosolids temperatures are measured in each pilot test bed at depths of 0.5', 1.5' and 2.5' from the top (the total depth of the biosolids is 3'), using three stacks of Thermochron iButton sensors (Embedded Data Systems, Cat. No. DS1921G-F5#), for a total of 9 iButton sensors per box.

Physical/Chemical Parameters

Physical/chemical qualities of the biosolids that can influence the inactivation of PIOs include the moisture and organic content (measured as TS and VS, respectively), pH, VFAs, and ammonia. Therefore, the pilot test beds are sampled monthly (except when biosolids are frozen during the winter season) at each WRRF for analysis of these parameters. Sampling is performed following the guidance in the USEPA (1989) guidance for sampling dewatered sludge drying beds: (1) grab samples are removed from each quarter of each box by collecting a core across the full depth of the bed, (2) the cored material from each quadrant is composited, and (3) sub-samples are taken and preserved appropriately for the various analyses.

EPA method 1684 (3) is used to measure the TS and VS of each sample. Quality Control (QC) tests include demonstrating laboratory capability, analysis of blanks, analyses of on-going precision and recovery samples, and duplicate analyses. EPA Method 9045D (4) is used to determine the pH of biosolids samples, except that a 10-g (wet weight) sample is mixed with 20 mL of distilled water (15). A Thermo Scientific Orion Dual Star pH/ISE Benchtop meter with a Thermo Scientific Orion 9156SjWP Double Junction Low Maintenance pH electrode is used with a two point calibration and duplicate analyses for QC. The titration method of Anderson and Yang (5) is followed to determine VFAs, bicarbonate alkalinity, and total alkalinity. QC includes analysis of duplicates and titration of standard solutions. Finally, the ammonia-nitrogen content in biosolids samples is determined using the HACH company TNTplus 832 (range 2 to 47 mg/L NH_3-N) Ammonia Method 10205 colorimetry procedure, which is equivalent to the USEPA Method 350.1. Acidified samples are diluted and prepared for analysis by distillation following Standard Method 4500-NH_3 (6). The absorbance of the distillate is measured using a HACH DR6000 spectrophotometer, which has a built in calibration curve. QC tests are performed to demonstrate laboratory capability, analyses of on-going precision and recovery samples, duplicate analyses.

Microbial Assays

To demonstrate compliance with the Class A biosolids pathogen requirements, and equivalency to the PFRPs (14), fecal coliforms, total culturable (enteric) viruses and helminth (*Ascaris*) ova levels are being monitored throughout the study. Fecal coliforms are measured by following the most probable number (MPN) procedure found in EPA Method 1681 (7). QC includes on-going precision and recovery samples, spiked samples, and duplicate analyses. Quantification of the enteric virus is a two-step process. First, the human poliovirus is extracted from the spiked biosolids using the virus recovery procedure described in ASTM D4994-89 (9). Then, the virus is quantified with the use of one of two procedures: (1) the plaque assay found in Appendix H in *Control of Pathogens and Vector Attraction in Sewage Sludge* (10), which is used to count high quantities of poliovirus; and (2) the quantal assay described in Chapter 15 of the USEPA Manual of Methods for Virology (11), which is used to quantify smaller amounts of virus. Recovery and examination of the *Ascaris* ova from the spiked biosolids is performed following the method of Bowman et al. (12).

Coliphage (viruses that infect *Escherichia coli*) levels are not required for demonstrating PFRP equivalency, but are measured in this study as an additional PIO. Coliphages have similar size, morphology and structure to enteric viruses and, therefore, behave similarly to enteric viruses under analogous biosolids conditions. Importantly, compared to enteric viruses, coliphages are much more abundant in wastewater and can be analyzed using a relatively simple and rapid culture-based assay. Male-specific and somatic coliphages are measured using the single agar layer procedure, according to EPA Method 1602 (8).

RESULTS AND DISCUSSION

In July and August of 2016, the full pilot-scale studies were initiated at PLWSA and GIWA, respectively. Six pilot-scale test beds were constructed and are being monitored at each WRRF. Three test beds are located indoors (Treatment 1, boxes 4, 5, and 6), and three are located outdoors (Treatment 2, boxes 1, 2, and 3). Results that have been collected thus far include the full suite of environmental conditions, key physical/chemical properties (i.e., TS, VS, pH, VFAs, ammonia), and the PIOs (FCs, *Ascaris* ova, and coliphages). The data reported here were collected from the start date up to December 2016 at which point the biosolids froze. Data collection and analysis will continue for up to one year at PLWSA, and up to two years at GIWA, without the addition of new biosolids. Representative data from the two WRRFs are presented here to demonstrate the data being collected, and illustrate key trends.

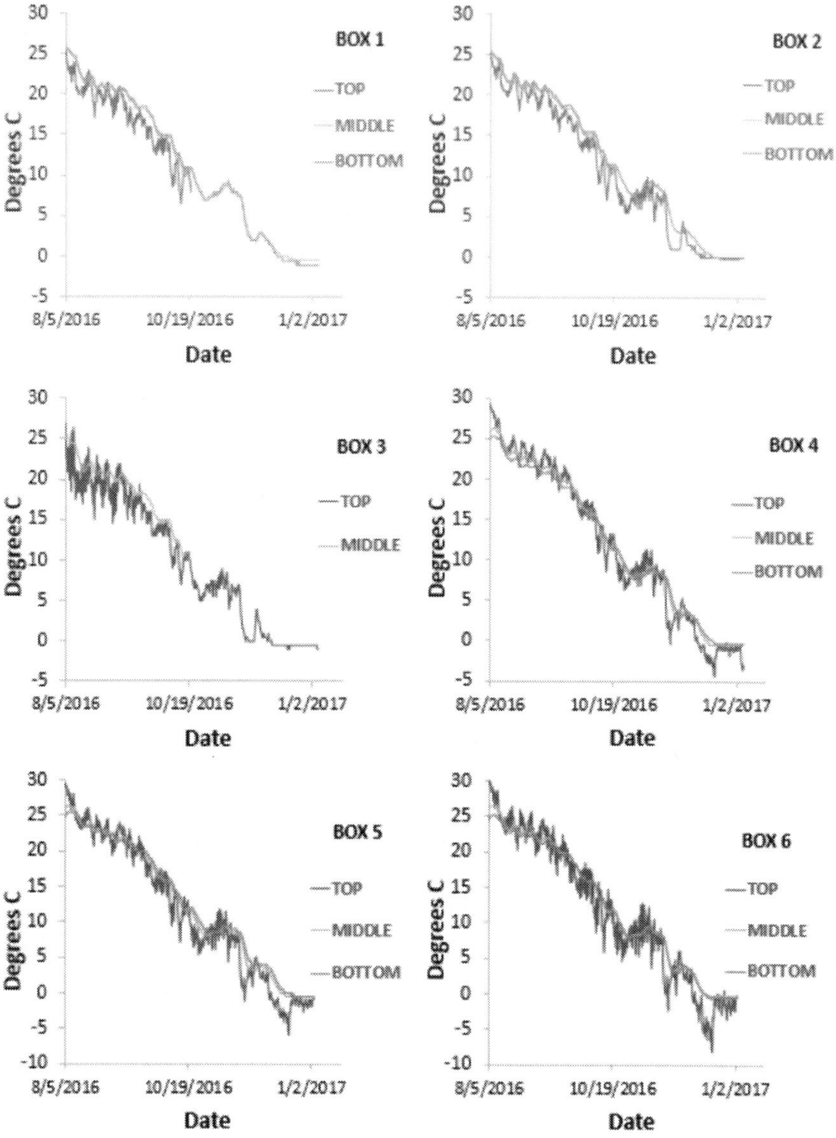

Figure 5: *In Situ* Biosolids Temperatures at PLWSA. Each data point represents the average of duplicate analyses.

Environmental Conditions

The *in situ* biosolids temperatures measured at depths of depths of 0.5 ft, 1.5 ft and 2.5 ft at PLWSA are presented in Figure 5. As expected based on ambient temperatures, the *in situ* biosolids temperature followed an overall decreasing trend from August through December in Treatments 1 and 2. Although similar temporal patterns were observed in the temperature profiles, the data demonstrate that as the depth in the biosolids increased, the amount of variation in temperature decreased.

Physical/Chemical Parameters

The TS and VS values determined to date at GIWA and PLWSA illustrate some differences between the plants and treatments (Figure 6). At PLWSA, the TS is relatively constant in both treatments, while at GIWA, the TS varies to a greater extent, with the TS trending higher in the inside test beds (4, 5, and 6) compared to the outside test beds (1, 2, and 3). Interestingly, the VS at PLWSA and GIWA are decreasing with time for both treatments, but at PLWSA, the VS experienced a large decrease (60% to 20%) between day 28 and 59, while at GIWA the VS has been dropping more gradually from around 50% to 30%.

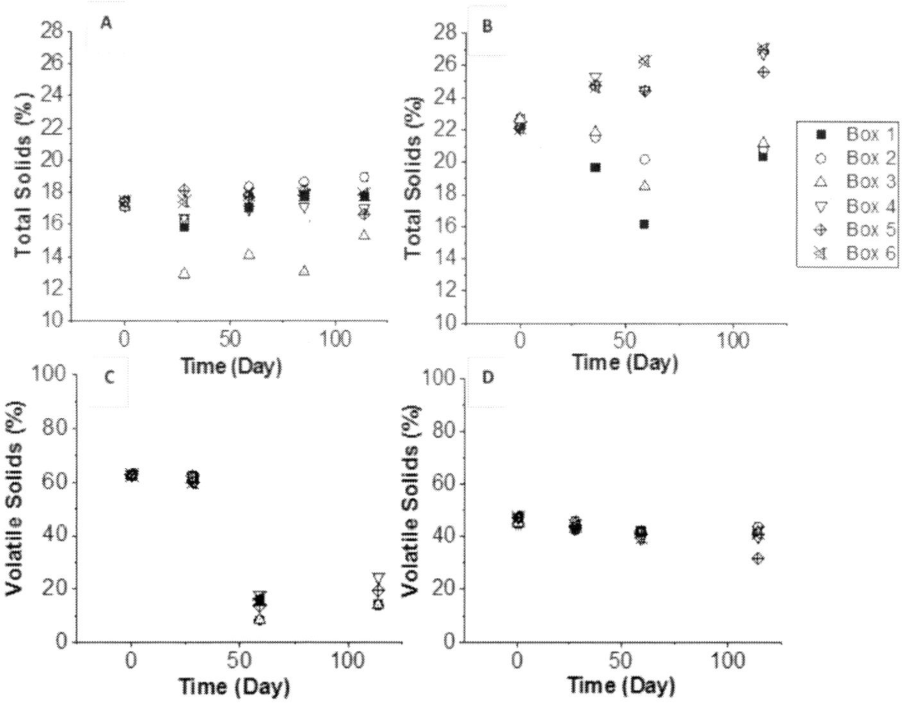

Figure 6: Total Solids at (A) PLWSA and (B) GIWA, and (C) Volatile Solids at PLWSA and (D) GIWA. Each data point represents the average of duplicate analyses.

pH levels observed to date in the test beds are shown in Figure 7. The initial pH levels observed at both plants were above 8. Subsequently, the pH at PLWSA dropped slightly, after which it has leveled off, with similar trends for both treatments. The pH at GIWA has also shown decreasing levels over time, and the inside (4, 5, 6) and outside (1, 2, 3) boxes may be starting to differentiate.

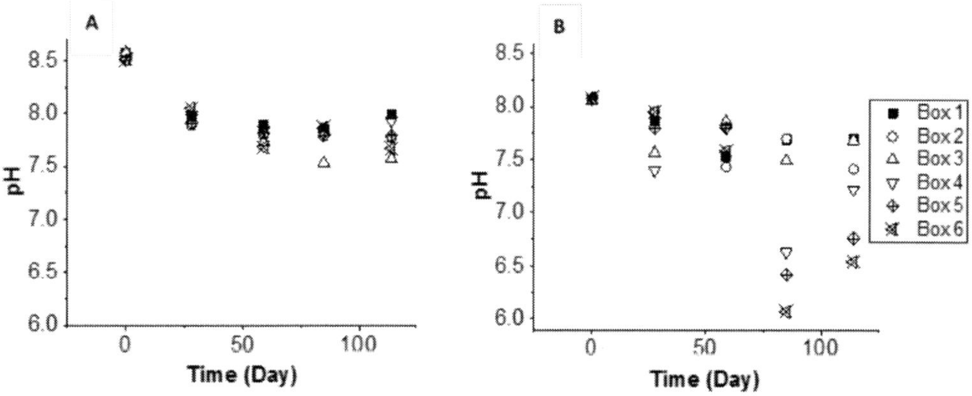

Figure 7: pH at (A) PLWSA and (B) GIWA. Each data point represents the average of duplicate analyses.

At this point, it is difficult to discern any distinct trends in the bicarbonate and VFA levels (Figure 8). At both WRRFs, the bicarbonate and VFA levels have generally remained below 40,000 mg/kg.

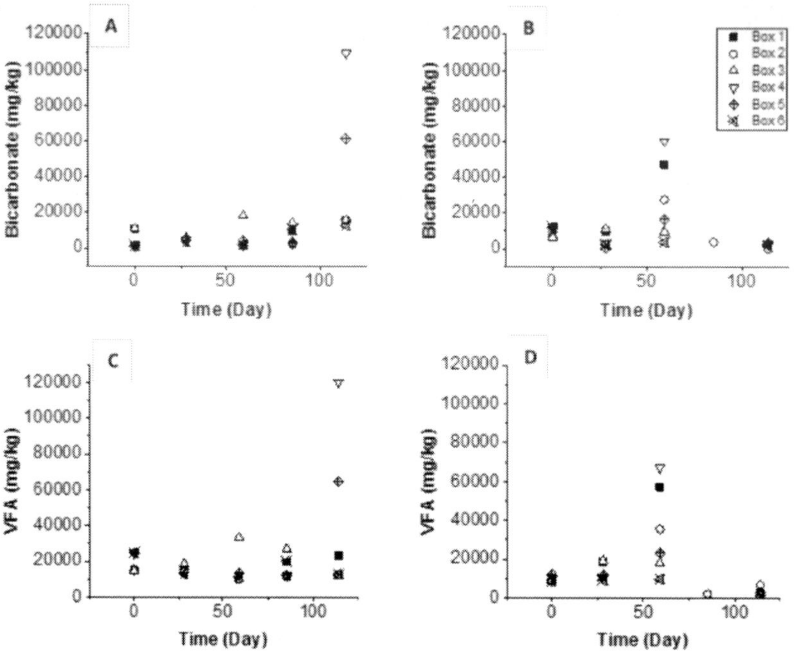

Figure 8: Bicarbonate at (A) PLWSA and (B) GIWA, and VFA at (C) PLWSA and (D) GIWA. Each data point represents the average of duplicate analyses.

The ammonia concentration data presented in Figure 9 reveal some differences between the test beds at the two plants. Specifically, while the ammonia concentrations are generally increasing with time at PLWSA, there is no clear trend at GIWA.

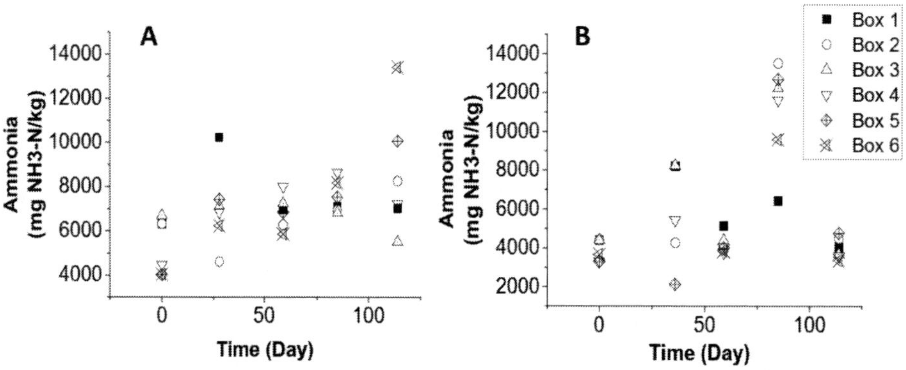

Figure 9: Ammonia at (A) PLWSA and (B) GIWA. Each data point represents an individual measurement.

Microbial Assays

The FC data obtained to date are summarized in Figure 10 and show that FC levels decreased steadily at both locations, although some regrowth occurred at both plants on day 59.

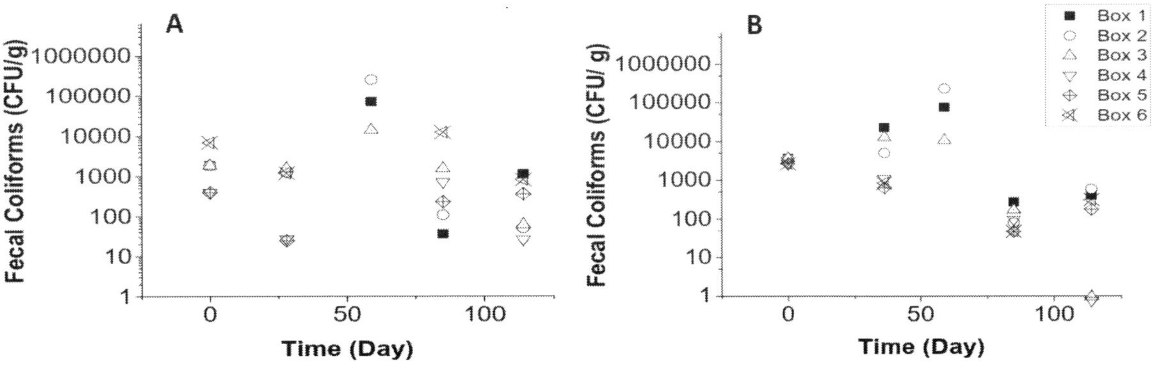

Figure 10: Fecal Coliforms at (A) PLWSA and (B) GIWA. Each data point represents an individual measurement.

Male specific and somatic coliphage concentration were measured on day 0 and 114 (Figure 11). At both GIWA and PLWSA, the coliphage concentration started out at non-detectable levels. By day 114, coliphage numbers had increased in almost all test beds at PLWSA. Male specific coliphage concentrations were generally higher than the levels of the somatic coliphage. The coliphage values also increase over time at GIWA, but there is more variability in the day

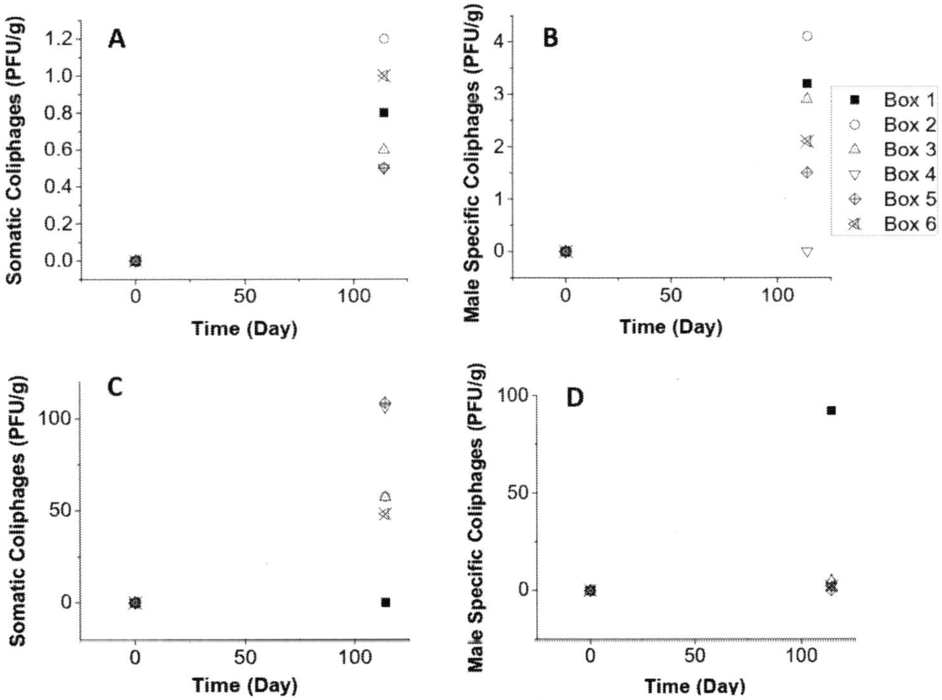

Figure 11: Somatic Coliphages at (A) PLWSA and (C) GIWA, and Male Specific Coilphages at (B) PLWSA and (D) GIWA. Each data point represents an individual measurement.

114 values at GIWA and, interestingly, the somatic coliphage levels are generally much higher than the male specific values.

CONCLUSIONS

This paper presents the set up and initial data from pilot-scale LCLT treatment systems at two WRRFs. The initial results from the pilot scale study demonstrate that changes in environmental conditions, physical-chemical parameters, and microbial populations have occurred over the course of the study to date (August 2016-January 2017). However, additional data on the overall changing conditions at PLWSA and GIWA during the course of a year are needed to determine the efficacy of a LCLT process. Nevertheless, FC have begun to drop since the beginning of the study. Based on our preliminary work (13), it is speculated that freeze-thaw cycles over the course of the winter will produce more drastic physical-chemical changes in the biosolids, and further reductions in the PIOs.

When completed, this research will improve our understanding of PIO inactivation in LCLT treatment systems under a broad range of conditions, enabling a rational design of LCLT treatment systems for Class A biosolids production. Ultimately, the results of this study should contribute to greater implementation and approval of LCLT treatment processes, thus allowing more WRRFs to sustainably and economically produce Class A biosolids.

ACKNOWLEDGEMENTS

Funding for this work was provided by the Water Environment & Reuse Foundation (Project NTRY11T15), the Michigan Department of Environmental Quality via a Stormwater, Asset Management, and Wastewater (SAW) grant (Project Number 1085-01), and Lystek, Inc. The authors also gratefully acknowledge the assistance provided by Mr. Mark Bowman at GIWA, and Mr. Zane Mackenzie at PLWSA, as well as the staff at each facility.

REFERENCES

1. **Gould JC, Rossano MG, Lawrence LM, Burk SV, Ennis RB, Lyons ET.** 2013. The effects of windrow composting on the viability of *Parascaris equorum* eggs. *Veter Parasitol* **191:**73-80.
2. **Sidhu JPS, Toze** S. 2012. Assessment of pathogen survival potential during managed aquifer recharge wtih diffusion chambers. J Appl Microbiol 113:693-700.
3. **USEPA.** 2001. Method 1684: Total, Fixed, and Volatile Solids in Water, Solids, and Biosolids. Washington, DC.
4. **USEPA**. 2004. SW-846 Method 9045D: Soil and Waste pH, part of Test Methods for Analyzing Solid Waste, Physical/Chemical Methods. Washington, DC.
5. **Anderson GK, Yang G**. 1992. Determination of bicarbonate and total volatile acid concentration in anaerobic digesters using a simple titration. Water Environ Res 64:53-59.

6. **APHA, AWWA, WEF**. 2006. Standard Methods for the Examination of Water and Wastewater, 22nd ed. American Public Health Association, American Water Works Association, and Water Environment Federation, Washington, D.C.
7. **USEPA**. 2010. Method 1680: Fecal Coliforms in Sewage Sludge (Biosolids) by Multiple-Tube Fermentation using Lauryl Tryptose Broth (LTB) and EC Medium Washington, DC.
8. **USEPA**. 2001. Method 1602: Male-specific (F+) and Somatic Coliphage in Water by Single Agar Layer (SAL) Procedure. U.S. Environmental Protection Agency, Office of Water, Washington, D.C.
9. **ASTM International**. 2014. D4994-89 Standard Practice for Recovery of Viruses from Wastewater Sludges, vol D4994-89. ASTM International, West Conshohocken, PA.
10. **U.S. EPA**. 2003. Control of Pathogens and Vector Attraction in Sewage Sludge. Office of Research and Development, Cincinnati, OH.
11. **USEPA**. 2001. USEPA Manual of Methods for Virology. Washington, D.C.
12. **Bowman DD, Little MD, Reimers RS**. 2003. Precision and accuracy of an assay for detecting Ascaris eggs in various biosolid matrices. Water Res 37:2063-2072.
13. **Meingast CL, Becker JG, Seagren EA**. Effects of biosolids moisture content on indicator organism reduction via freezing and thawing. Water Environment Federation/International Water Association, Biosolids and Residuals Conference 2015. Washington, DC, June 7-10, 2015.
14. **USEPA**. 2015. Biosolids Equivalency Criteria. https://www.epa.gov/biosolids/equivalency-criteria.
15. **North, J.M., J.G. Becker, E.A. Seagren, M. Ramirez, and C. Peot.** (2008). Methods for quantifying lime incorporation into dewatered sludge. I: Benchscale evaluation. Journal of Environmental Engineering (ASCE), 134(9), 750761.

Bay Area Biosolids to Energy (BAB2E) Coalition: The San Francisco Bay Area's Regional Approach to Biosolids Management

Sarah A. Deslauriers[1]*

[1] Carollo Engineers, Inc., Walnut Creek, CA (BAB2E Coalition Program Manager)
*sdeslauriers@carollo.com

ABSTRACT

The Bay Area Biosolids to Energy (BAB2E) Coalition originally formed in 2004 when a group of agencies came together to evaluate the feasibility of a regional biosolids management project to avoid the threat of a potential ban on land application of biosolids. Recognizing the need for long-term sustainable biosolids management, by 2008 the membership expanded and the group decided to officially brand itself the BAB2E Coalition to take advantage of opportunities anticipated to be developed under new state legislation (specifically, Assembly Bill 32 or AB 32). Assembly Bill 32 was adopted in 2006 and requires the state to reduce greenhouse gas (GHG) emissions to 1990 levels by 2020 (with further reduction by 2050). To achieve GHG reductions, the state created numerous programs incentivizing renewable energy and low carbon fuel production. This legislation served as a driver to prioritize the conversion of biosolids to energy over other Class A or B options. This poster presentation will provide an update on the challenges for biosolids management faced in the Bay Area and California associated with GHG legislation, as well as the progress made to date on innovative technologies and solutions to provide new options for biosolids and resource recovery in the future.

KEYWORDS

Biosolids, regional, sustainable, resource recovery, renewable energy

INTRODUCTION

The Coalition began by investigating a single regional merchant facility to be located centrally among Coalition members. With diligent review and evaluation, this approach was found to have practical and environmental limitations - including being a single management option resulting in GHG emissions from the truck transport required to haul member agency solids to the facility. The lessons learned from that process, along with the evolving legislation, resulted in member agencies deciding to initiate evaluations of and hosting subregional projects that look beyond converting biosolids to energy. The projects are seeking to understand all benefits from the use of biosolids and to responsibly recycle back to the environment value-added products of biosolids (including conversion to energy). At the same time, these projects are diversifying biosolids management options beyond land application (i.e., landfilling or land applying Class B biosolids in Solano County) to ensure Bay Area agencies have local, sustainable year-round options available.

This approach not only provides for year-round diversification of biosolids management options, it also provides the opportunity to demonstrate how well each option contributes toward achieving state goals, introduced by the Governor in 2015, to further reduce statewide GHG emissions by 2030 (going beyond AB 32's 2020 targets). These goals (also referred to as the Governor's "five pillars") are:

- Reducing today's petroleum use in cars and trucks by up to 50 percent
- Increasing from 33 to 50 percent our electricity derived from renewable sources
- Doubling the energy efficiency savings achieved at existing buildings and making heating fuels cleaner
- Reducing the release of methane, black carbon, and other short-lived climate pollutants
- Managing farm and rangelands, forests and wetlands so they can store carbon

These pillars recognize that several major areas of the California economy will need to reduce emissions. To enact these pillars, the Governor has recently signed legislation that either directly or indirectly impacts the disposal and use of POTW biosolids at landfills, as well as the diversion of other organic waste streams to POTWs. The good news is the regulatory agencies involved (i.e., Air Resources Board, CalRecycle, and State Water Resources Control Board) consider POTWs as part of the solution for diverting other organic waste streams from landfills (such as, food waste), in effort to reduce methane production/emissions at landfills. Diversion of organic waste could be accomplished through the use of existing or new anaerobic digestion capacity at POTWs for co-digestion with solids. The increased organic material sent to the anaerobic digesters will increase biogas generation for use in generating renewable energy onsite, production of pipeline grade biomethane, or production of a transportation fuel. All of these uses of biogas would further offset reliance on fossil fuels, reduce GHG emissions, and qualify for funding sources that have been or are being developed. However, legislation DOES NOT address how to handle the resulting biosolids/organic material or the increased nutrient loading sent to the liquids treatment train.

The Coalition's mission is to *develop a diverse and robust portfolio of beneficial biosolids resource recovery projects for the San Francisco Bay Area.* Since its formation, the Coalition has witnessed the availability of biosolids management options (disposal and beneficial use) continue to decrease. Land application is banned or limited in many counties throughout California. In Northern California, approximately 50 percent of the biosolids produced are either used as alternative daily cover (ADC) at a landfill or disposed of at a landfill during dry weather, and up to 90 percent of biosolids are sent to a landfill during wet weather. Because biosolids production largely depends on regional population, the amount of biosolids generated in the Bay Area is expected to increase as the region's population continues to grow.

The State of California has historically been a leader in the environmental movement with early adoption of recycling regulations and GHG emission reduction goals starting in the 1990s. The State continues to require increased waste diversion from landfills, with recent legislation being passed that requires 75 percent diversion of organic waste (including biosolids used as ADC) from landfills by 2025. These restrictions on traditional biosolids management options could mean additional costs to ratepayers and increased air pollutants as biosolids are transported greater distances for use in counties that do not have a ban on land application.

GOALS AND OBJECTIVES

With options for biosolids disposal and beneficial use shrinking, there is a need for creative and innovative solutions for biosolids management. In addition to potentially eliminating biosolids disposal and ADC options, the required diversion of organics from landfills also means that food wastes and other organics will now need to be managed in a different way. The state is looking to POTWs as part of the solution by co-digesting food waste and other organics with wastewater solids in anaerobic digesters. While these additions would add to the ultimate production of biosolids that must be managed, they also increase the feedstock for biogas production, thereby increasing the opportunity to produce renewable energy or a low carbon fuel.

Across the San Francisco Bay Area, agencies have implemented programs to capture methane gas through anaerobic digestion. However, energy value still remains in the biosolids after digestion. The Coalition seeks to create local sustainable solutions to biosolids management by utilizing the energy remaining in biosolids. The Coalition is currently exploring opportunities to diversify biosolids management options with the development of regional facilities that will use biosolids and possibly other renewable feedstocks to generate renewable energy and/or other value-added bio-products. While open to the use of any of the available technologies, the Coalition seeks to exceed air regulations and to maximize net renewable energy and bio-product production with state of art equipment. Locating regional facilities within the Bay Area has the added benefit of reducing vehicle miles, currently estimated to exceed one million miles per year, and the GHG emissions associated with hauling.

DISCUSSION

Producing energy from biosolids remains an emerging field with significant advancements in technology, research, and development currently taking place. The Coalition pursues a multi-pronged approach that includes:

- Advocacy for legislation, regulations, and funding to foster technology solutions and project implementation
- Solicitations to identify ready-to-implement commercial scale projects
- Technology incubation to foster technology research and development through facilitating partnerships and hosting demonstration projects.

California POTWs have the ability to contribute toward several of the Governor's pillars in a single project, and POTWs serve as the backbone of a community's resiliency in the face of changing conditions. Currently there are six subregional projects being investigated by five member agencies with the objectives of demonstrating viable, year-long (weather resilient) alternatives to land application that look beyond "biosolids to energy" and seek to responsibly recycle back value-added products of biosolids to the environment. Since its formation, key achievements to date of the Coalition include:

- Significantly advancing the industry state of knowledge on biosolids to energy technologies.

- Serving as a technology incubator. The Coalition evaluated and tested several promising technologies; some of these did not work out and others have advanced towards demonstration planning.
- Launching several Bay Area pilot, demonstration, and full-scale projects:
    - Dublin San Ramon Services District in partnership with Synagro and SCFI is currently evaluating the SCFI (AquaCritox®) technology to determine project viability. The proposed project site is at the District's treatment plant located in Dublin, CA. The technology will further break down volatile solids in the anaerobically digested sewage sludge (biosolids). The heat generated from the exothermic reactions can be used to power the process and also produce steam and/or hot water. This heat/energy can be used to heat the anaerobic digesters at the treatment plants offsetting the use of fossil fuel based energy and the associated GHG emissions. The product solids are rich in nutrients and can serve as a valuable soil amendment. For additional information on SCFI's AquaCritox® unit, visit www.scfi.eu/water/
    - West County Wastewater District is currently evaluating the SCFI (AquaCritox®) technology to determine project viability and needed support infrastructure. The project site is the District's treatment plant located in Richmond, CA. The project involves utilizing SCFI technology to treat secondary Waste Activated Sludge (WAS) in lieu of sending the material for digestion. This opens up 40 percent of the digester capacity to alternative bio-sources. The WAS is an ammonia source in the side stream treatment process that needs to removed, AquaCritox® allows us the opportunity to eliminate this side stream and reduce treatment needs. For additional information on SCFI's AquaCritox® unit, visit www.scfi.eu/water/
    - Silicon Valley Clean Water (SVCW) has partnered with BioForceTech (BFT) Corporation for a clean, green, sustainable method for handling biosolids derived from the wastewater treatment process. The BFT facility is located on SVCW's treatment plant site in Redwood City, CA, and is currently in construction. The anticipated startup is spring 2017 when approximately half of the biosolids produced from the wastewater treatment process will be delivered to BFT's facility. Once delivered, the biosolids will go through BFT's innovative BioDryer™ process within which the biosolids will be dried. This will be followed by pyrolysis to further reduce the biosolids mass. The byproducts of the process are syngas and biochar, both have high reuse value. For additional information on BFT's processing facility, go to www.bioforcetech.com.
    - Fairfield Suisun Sewer District (FSSD) in partnership with Lystek International has constructed a full-scale merchant project (fully funded with capital from the private sector) in Fairfield, CA, with a capacity to process 150,000 wet tons of biosolids per year into a licensed biofertilizer. FSSD is the first customer at approximately 13,000 wet tons per year. Lystek is currently marketing Bay Area wastewater agencies to send their biosolids to the facility as well to generate a licensed fertilizer. For additional information on Lystek's approach to biosolids management, visit www.lystek.com/.
    - Delta Diablo in partnership with Mt. Diablo Resource Recovery (MDRR) is currently evaluating PHG Energy's (PHGE) gasification technology to determine project size, phasing, optimal feedstock ratios, and needed support infrastructure. The District is also evaluating project site locations, either at the District's main treatment plant or at MDRR, as well as working with the Bay Area Air Quality Management District regarding pathways for obtaining an air permit for the gasification technology. The

initial project will convert 1 truck/day of green waste from MDRR and 25 percent of Delta Diablo biosolids to produce approximately 500 KW of electricity, reducing GHG emissions by 85 percent. Delta Diablo has also signed an MOA with MDRR to equally share the planning costs for this project. For additional information on PHGE's gasification system, visit www.phgenergy.com/gasifiers/new-gasification-technologyadvances-a-proven-method.
    - Delta Diablo in partnership with MDRR is also currently evaluating Anaergia's pyrolysis technology to determine sizing, phasing, and needed support infrastructure to convert 33 tons per day (tpd) of anaerobically digested biosolids and 80 tpd of food/organic waste into an ammonium sulfate fertilizer product, a high-nutrient bio-char soil amendment and 2.2 megawatts (MW) of carbon-negative renewable electricity. This project is estimated to result in ~25,000 metric tons of GHG emissions reductions annually, divert 22,880 tons of food and organic wastes from the landfill, and produce enough electricity to take Delta Diablo off the grid (1.3 MW) and export 0.9 MW to the grid. Delta Diablo has signed an MOA with MDRR to equally share planning costs. For additional information on Anaergia's system, visit: www.anaergia.com/services/municipal/waste-water
- Continuing to sustain land application in the Bay Area by seeking to create more capacity for biosolids in the Bay Area marketplace.
- Educating the public on biosolids management issues in California through public outreach efforts, including creation of a public website and obtaining media coverage.
- Growing momentum in advocacy for biosolids to energy, including sponsoring a bill in the State Legislature.

SUMMARY AND CONCLUSION

The Coalition's comprehensive, regional approach to develop biosolids to energy projects is the first in the nation and will be a model for major metropolitan areas across the United States that seek to diversify their biosolids use/disposal portfolio, create energy, lower GHGs, and advance solids processing technologies. The Coalition's plan for the next two years is to further diversify in the face of regulatory developments that require the eventual elimination of one of the Bay Area's most prevalent biosolids options - alternative daily cover (ADC) at landfills. In order to address this critical challenge and other unmet needs, over the next two years the Coalition plans to:

- Evaluate and select up to six all-weather (i.e., year-round) biosolids end use management options
- Provide ongoing support to active Coalition projects at member Wastewater Resource Recovery agencies
- Develop partnerships with academia to advance biosolids research
- Continue serving as a technology clearinghouse and incubator for biosolids projects
- Further develop a strategy for legislative advocacy
- Continue and expand community outreach and education on the value of biosolids products

Developing Biosolids Based Bioretention Soils for Green Stormwater Infrastructure

Julia G Jay[1] and Sally Brown[1]

[1]University of Washington

Keywords: Bioretention, stormwater, copper, nitrogen, phosphorus

Bioretention systems are increasingly used as alternatives to engineered systems for stormwater management in urban areas. These systems typically consist of sand and compost and are designed to allow for rapid infiltration and filtering of stormwater. In some cases, these systems drain directly into streams. In other cases, the water percolates through the subsoil into groundwater. The systems are also designed for a wide range of settings including curbside systems, systems in parking lots and systems alongside buildings. The contaminants of concern in the stormwater include nutrients (N and P), metals (primarily Cu, Pb and Zn), organics (primarily PAHs) sediment and pathogens. Recent work has shown that untreated stormwater is toxic to fish and that filtering stormwater through a sand, compost and bark mixture was sufficient to eliminate any negative impacts (McIntyre et al., 2016). The current regulations on bioretention systems for green stormwater infrastructure in WA State specify that the compost used not include municipal biosolids or animal manures. The rational for this guidance is that biosolids and manures are considered to have higher nutrient and metal concentrations than yard waste or food scraps. The current guidance does not take into account that composts can have widely different characteristics based on feedstock blends and different ratios of each component of the mixture. Previous work demonstrated that feedstocks were not a significant predictor of performance for bioretention systems (Brown et al., 2016). In contrast, the phosphorus saturation index (PSI), a test to estimate the potential for phosphorus leaching, was shown to be highly predictive of efficacy of the different mixtures for both nutrients and metals.

Methods

We conducted a replicated field trial to test a range of different bioretention soil mixtures (BSM) for their ability to remove metals and nutrients from stormwater. The study included 40% food yard waste compost and 60% sand by volume, the current recommended mixture in WA State and a 100% sand as control treatments. Other treatments included biosolids: yard waste compost produced by King County added at different rates with sand, and +/- Fe based water treatment residuals, and high Fe biosolids from DC Water, used both directly and following composting. A list of the treatments used in the study is shown below. The control treatments and the treatments that include biosolids from DC Water are shown in bold.

1. **Yard/food compost: sand 40:60**
2. Biosolids compost sand 40:60
3. Biosolids compost: WTR: sand 35:5:60- to same PSI as #2
4. Biosolids compost: sand 80:20 (2x compost volume from #2)
5. Biosolids compost: WTR: sand 70:10:20 – to same PSI as #4

6. Biosolids compost: sand 20:80 (0.5 compost volume from #2)
7. Biosolids compost: WTR: sand 17.5:2.5:80 – to same PSI as #6
8. **100% sand**
9. Biosolids compost: sawdust: sand 15:5:80
10. Biosolids compost: oyster shells: sand 15:5:80
11. **Sand: loamy clay: DC Water Biosolids 80:12:8**
12. **Sand: DC Water Biosolids 80:20**
13. **Sand: DC Biosolids compost 80:20**
14. **Sand: Sawdust: DC Water Biosolids:Mineral Fines 15:40:40:5**
15. **Sand: Sawdust: DC Water Biosolids 25:25:50**

A total of 12 leachings were conducted. The first 8 were done with synthetic stormwater at both high and low metal levels with the last 4 done with stormwater collected from highway runoff in Seattle, WA. The effluents from the columns were collected and analyzed for total and dissolved metals and P and total N and nitrate. Concentrations of nitrogen, phosphorus and copper in stormwater are shown in Table 1.

Table 1. Concentrations of total nitrogen, phosphorus and copper in stormwater used for the replicated greenhouse trial.

	Total Nitrogen	Copper	Phosphorus
	mg/L	ug/ L	
Synthetic high metal	0.92 ± 0.04	2792 ± 1153	1443 ± 125
Synthetic low metal	1.1 ± 0.04	525 ± 132	2009 ± 80
Actual stormwater	1.8 ± 1.0	39.8 ± 19.1	80.8 ± 32.8

Results- metals

In general, all treatments removed metals from the stormwater with effluent concentrations significantly lower than influent concentrations. The rates of removal were high for all metals measured (Cd, Cr, Cu, Pb, Zn). Of all of the treatments tested, the high Fe biosolids based BSM were the most effective at metal removal showing full removal for the synthetic stormwater treatments and very high removal rates from the actual stormwater. The percentage removal rate of metals for two high Fe biosolids treatments are shown in Table 2. These results are similar to those seen in an earlier trial where both biosolids: yard waste and food:yard waste composts showed high removal of Cu and Zn from synthetic stormwater (Brown et al., 2016). These results suggest that biosolids based bioretention soil mixtures can be used to remove metals from stormwater. It is important to note that the biosolids that were most effective had Fe added during wastewater treatment. The Fe can form highly adsorptive oxides and remove metals from solution. In a previous study high Fe compost from the same treatment plant was as effective as high rates of P addition for reducing the bioaccessibility of Pb in a smelter impacted soil (Brown et al., 2004). These results also suggest that properties other than feedstocks can be more effective at predicting BSM performance.

Table 2. Portion of total metal removed from stormwater by the bioretention soil mixtures containing high Fe biosolids or biosolids compost from DC Water. A removal rate of 1 means that all metal added in the influent was removed by the mixtures and metal concentrations in the effluent were below detection limits.

	Biosolids: sand		Biosolids compost: sand	
	Actual	Synthetic	Actual	Synthetic
Cd	1	1	1	1
Cr	0.92 ± .22	0.99 ± .01	0.92 ± .32	0.99 ± .01
Cu	0.86 ± .14	0.99 ± .02	0.83 ± .22	0.99 ± .01
Pb*	---	0.98 ± .02	---	0.99 ± .02
Zn	0.92 ± .07	0.99 ±.02	0.91 ± .10	0.99 ± .01

* Concentrations were below detection limits in stormwater so removal rates could not be calculated.

Results- nutrients

All mixtures were sources of N and P for the duration of the trial. Over time, concentrations of N decreased significantly. The decrease in total N for all treatments across the duration of the trial is shown in Figure 1. Unlike engineered systems, the performance of systems that rely on natural processes are likely to vary over time. A question that needs to be addressed with the use of these systems is what level of variability is acceptable and also what time frame is acceptable for the systems to come into equilibrium. For example the concentrations of N in the leachate from the first event and the final event for select treatments is shown in Figure 2. The purpose of this figure is two fold. Each of the treatments shown in the figure contain the high Fe biosolids, either as compost or as biosolids. Certain of these treatments had the highest initial N concentrations in the leachate for the first leaching events. By the end of the trial, the total N in these treatments was significantly reduced and was similar to many other treatments. Secondly, treatment #14 Sand: Sawdust: DC Water Biosolids:Mineral Fines 15:40:40:5 had consistently low N in the leachate. Here we mixed sawdust with the biosolids at a 1:1 dry weight ratio. This increased the C:N ratio of the mixture. The higher C:N ratio was a more significant factor in predicting N movement than the presence of biosolids in the BSM. The added C was sufficient to tie up the N and essentially eliminate N movement through the columns. Again, this is an example of variables other than feedstocks as being more important for predicting system performance. The trial was conducted without plants. Rye grass was seeded into the columns only after the 12 leachings had been completed. This was done to measure plant response. Previous work has shown that plants will significantly decrease nutrient concentrations in the effluent from these mixtures (Brown et al., 2016).

Figure 1. Total nitrogen in leachate across all treatments for the duration of the study. The decrease in total N over time is clear.

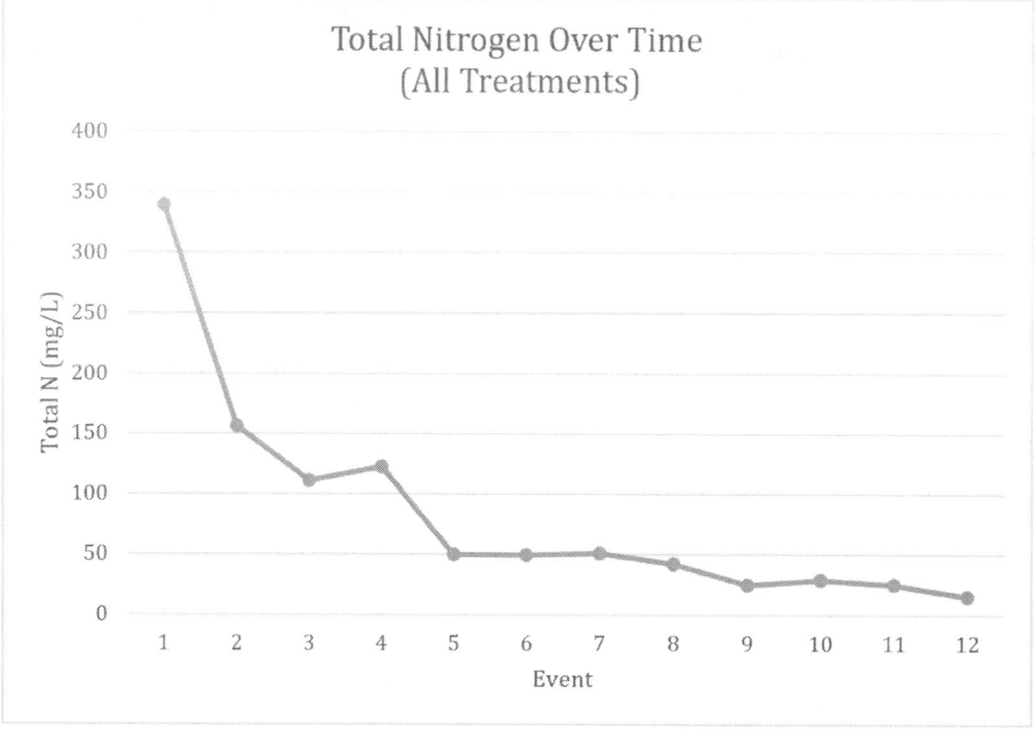

Figure 2. A comparison of total N in the DC Water biosolids treatments from the first leaching event and the last leaching event. A decrease in N in the leachate is clear for treatments 11,12, 13 and 15. Treatment 14 had high rates of sawdust added and showed minimal N movement throughout the trial.

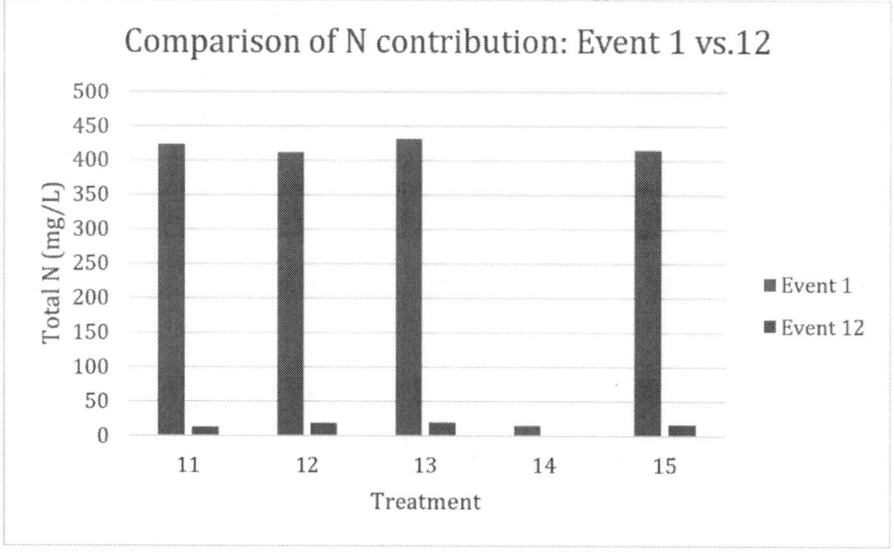

Conclusions

The results of this study indicate that biosolids products can be effective in bioretention soil mixtures. However, not all materials will be equally effective. It is critical to determine what predictive measures can be used to evaluate system performance so that different feedstocks can be reliably used for these systems. Developing standards that can reliably be tested and are effective for predicting performance of bioretention systems will allow use of boutique blends for these systems, but it will also allow for use of readily and locally available materials if they are able to perform. This will in turn allow for more reliable systems and more widespread use of green stormwater infrastructure.

References

Brown, S.L., W. Berti, R.L. Chaney J Halfrisch and J Ryan. 2004. In situ use of soil amendments to reduce the bioaccessibility and phytoavailibility of soil lead. J. Environ Qual. 33:522-531.

Brown, S.L., A.Corfman, K. Mendrey, K. Kurtz, and F. Grothkopp. 2016. Stormwater Bioretention systems- testing the phosphorus saturation index and compost feedstocks as predictive tools for system performance. J. Environ. Qual., 45:1:98-106

McIntyre, J.K., R.C. Edmunds, E. Mudrock, M. Brown, J.W. Davis, J.D. Stark, J.P. Incardona, and N.L. Scholz. 2016. Confirmation of stormwater bioretention treatment effectiveness using molecular indicators of cardiovascular toxicity in developing fish. Environ. Sci. and Tech.50:3:1561-1569

Biosolids Reuse: Continuous Flow-Through Column Testing of Biosolids-Derived Biochar to Sorb Micropollutants

L.K. Kimbell[1], Y. Tong[1], A. Avila[1], B. K. Mayer[1], P. J. McNamara[1]

[1]Marquette University

Email: lee.kimbell@marquette.edu

ABSTRACT

Trace organic compounds such as triclosan are commonly discharged to the environment in wastewater effluents and residual biosolids. Triclosan is of particular concern due to its biologically active nature, leading to its selection as a representative micropollutant in this study. Column adsorption experiments were conducted to determine the triclosan adsorption capacity of biosolids-derived biochar. The biosolids-derived biochar produced was successfully used to remove triclosan from synthetic and wastewater solutions. The biosolids-derived biochar adsorbed more triclosan than a wood-based biochar, but demonstrated less adsorptive capacity than commercial activated carbon.

KEYWORDS: Triclosan, Adsorption, Sustainability, Wastewater

INTRODUCTION

Trace organic compounds such as antibiotics and hormones are ubiquitous in urban surface waters and are commonly found in treated wastewater effluent[1]. These compounds are often referred to as micropollutants, and are of concern due to their ability to impact aquatic species' reproduction after being discharged from water resource recovery facilities (WRRFs) into receiving waters[2-4]. Triclosan (TCS) (a representative micropollutant selected for this study) is a biologically active antimicrobial compound used in personal care products due to its ability to specifically target *Bacteria*[5]. Elevated concentrations of TCS have been linked to antimicrobial cross-resistance as well as antibiotic resistance genes in engineered systems and the natural environment[6-7]. Conventional WRRFs are not designed to remove micropollutants such as TCS. As a result, treatment processes capable of removing micropollutants are increasingly important.

Carbonaceous adsorbents such as activated carbon and biochar have been used to remove a wide range of contaminants from aqueous solutions. Previous work demonstrated the ability of biosolids-derived biochar for TCS sorption in bench-top batch experiments[7], but the implementation of biosolids-derived biochar as a sorbent in a continuous flow-through column, which better reflects real-world applications, remains unknown. The primary objective of this project was to determine the ability of biosolids-derived biochar to remove TCS from water and wastewater solutions in a continuous flow-through column under a variety of conditions.

METHODS

The feedstock for the biochar in this study was Milorganite, which is a heat-dried blend of waste activated sludge and anaerobically-digested primary solids produced by the Milwaukee Metropolitan Sewerage District (MMSD) in Milwaukee, WI. The feedstock was pyrolyzed to produce biochar by purging Milorganite with inert gas, and heating it in a muffle furnace. The biochar was cooled in a desiccator prior to conditioning for adsorption tests. All biochar was washed with Milli-Q water to remove residual surface impurities and ash prior to acid washing. Acid pretreatment with HCl acid was necessary to activate the biochar and increase the adsorptive capacity[8].

Column sorption experiments were conducted to determine TCS removal from water and wastewater via biosolids-derived biochar under varying conditions, as shown in **Figure 1**. Specifically, the impacts of pH, flow rate, other micropollutants, and nutrients (ammonium and phosphate) on TCS removal were evaluated through rapid small-scale column tests. TCS was spiked into aqueous solutions at µg/L concentrations, and was quantified via liquid chromatography-mass spectrometry (LC-MS). Ammonium and phosphate were spiked into aqueous solutions at mg/L concentrations, and were quantified using appropriate Standard Methods for each compound. The TCS removal with biosolids-derived biochar was also compared to a commercial activated carbon and a wood-derived biochar.

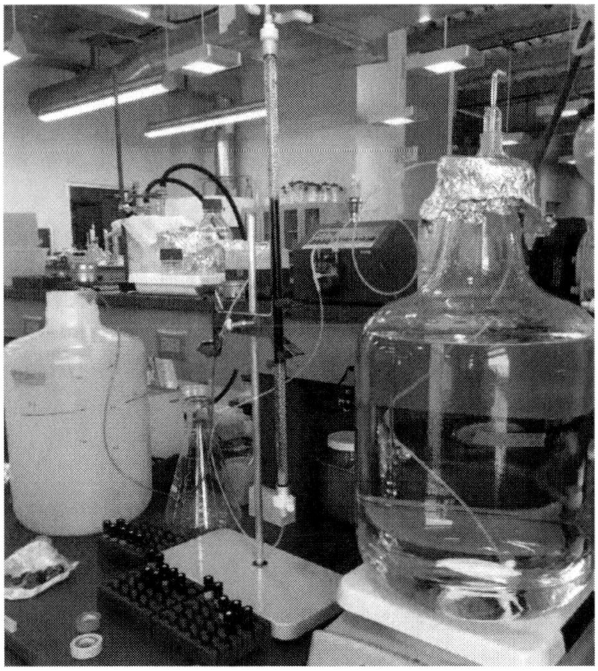

Figure 1. Column adsorption test for TCS removal with biosolids-derived biochar.

RESULTS

The solution pH had minimal impact on treatment efficiency; TCS removal efficiencies were similar for pH values of 7 and 8.5. The biosolids-derived biochar sorbent was able to remove over 80% of influent TCS over the course of the pH experiments. This demonstrates that neutral pH values are sufficient for TCS removal via biosolids-derived biochar.

Column adsorption tests were conducted with varying flow rates from 10 to 40 mL/min (2.58 to 10.32 gpm/ft^2). Generally, higher TCS removal was observed when columns were operated at lower flow rates and a longer empty bed contact time (EBCT). The reduction in TCS removal for higher flow rates was presumably due to a shorter EBCT. Breakthrough of TCS was modeled as the time at which the effluent TCS concentration equaled 5% of the influent TCS concentration.

Competition experiments were conducted with a mixture of nutrients and other micropollutants to investigate the feasibility of TCS sorption to biosolids-derived biochar in a complex matrix. The results demonstrated that ammonium and phosphate do not compete for sorption sites and had negligible effect on micropollutant sorption to biosolids-derived biochar. The removal of TCS was suppressed due to the co-presence of sulfamethoxazole (SMX) and 17β-estradiol (E2), and breakthrough was observed sooner than for the single solute adsorption tests.

The removal of TCS via biosolids-derived biochar was compared to two other carbonaceous adsorbents: a commercially available activated carbon and a wood-derived biochar. The biosolids-derived biochar effectively removed a higher amount of TCS from solution compared to the wood-derived biochar. The activated carbon removed the highest amount of TCS from solution amongst the three adsorbents used.

DISCUSSION/CONCLUSIONS

Overall, biosolids-derived biochar has demonstrated capability for removing hydrophobic compounds such as TCS from aqueous solutions in a continuous flow-through column. The reuse of dried biosolids as a sustainable feedstock for biochar production could provide multiple benefits including the reduction of biosolids that need to be managed and providing enhanced removal of trace organic compounds by WRRFs prior to discharge to surface waters. Biochar could be produced on-site and eliminate the need to purchase activated carbon. After sorption of micropollutants, the biochar could be regenerated via pyrolysis, which has been shown to remove micropollutants[9].

References

(1) Pal, A., Yiliang, H., Martin, J., Martin, R., Yew-Hoong Gin, K., 2014. Emerging contaminants of public health significance as water quality indicator compounds in the urban water cycle. Environment International, 71, pp.46–62.

(2) Fritsch, E. B., Connon, R. E., Werner, I., Davies, R. E. U. E., Beggel, S., Feng, W., & Pessah, I. N. (2013). Triclosan impairs swimming behavior and alters expression of Excitation Contraction coupling proteins in fathead minnow (Pimephales promelas). Environmental Science & Technology.

(3) McMurry, L. M., Oethinger, M., & Levy, S. B. (1998). Triclosan targets lipid synthesis. Nature, 394(6693), 531–532.

(4) Orvos, D. R., Versteeg, D. J., Inauen, J., Capdevielle, M., Rothenstein, A., & Cunningham, V. (2002). Aquatic toxicity of triclosan. Environmental Toxicology and Chemistry / SETAC, 21(7), 1338–1349.
(5) Carey, D.E. & McNamara, P.J., 2015. The impact of triclosan on the spread of antibiotic resistance in the environment. Frontiers in Microbiology, 5(January), pp.1–11.
(6) Carey, D.E. et al., 2016. Chronic exposure to triclosan sustains microbial community shifts and alters antibiotic resistance gene levels in anaerobic digesters. Environmental Science: Processes & Impacts, 18(8), pp.1060–1067.
(7) Carey, D.E. & McNamara, P.J., 2016. Altered antibiotic tolerance in anaerobic digesters acclimated to triclosan or triclocarban. Chemosphere, 163, pp.22–26.
(8) Tong, Y., Mayer, B.K. & McNamara, P., 2016. Triclosan adsorption using wastewater biosolids-derived biochar. Environ. Sci.: Water Res. Technol. Available at:
(9) Ross, J. J., Zitomer, D. H., Miller, T. R., Weirich, C. A., & McNamara, P. J. (2016). Emerging investigators series: pyrolysis removes common microconstituents triclocarban, triclosan, and nonylphenol from biosolids. Environ. Sci.: Water Res. Technol., 2, 282–289.

Trace Organic Contaminants in Biosolids: Recent Survey, Risk Analysis and Communication Tools

Kate Kurtz[1], Roberta King[1], Ashley Mihle[1], Sally Brown[2], Mark Cullington[3], Heather Brunelle[3], Laura Kennedy[3], Dana Devin-Clarke[3], Maile Lono-Batura[4]

[1]King County Wastewater Treatment Division
[2]University of Washington
[3]Kennedy/Jenks Consultants
[4]Northwest Biosolids
*Email: kate.kurtz@kingcounty.gov

Abstract
Scientific literature and individual sampling efforts inform us that wastewater contains trace amounts of organic chemical contaminants from pharmaceuticals and personal care products (PPCPs), however, biosolids treatment processes are generally not designed to remove these chemicals. Agencies and citizens want to know which contaminants are present and at what concentrations, in their own biosolids or typical biosolids products. Agencies, however, often do not have the ability to analyze for these contaminants in their routine monitoring. Moreover, they want to understand the risk of these compounds to human health. To help answer these questions, member agencies of Northwest Biosolids recently conducted three projects: 1) a voluntary survey of organic contaminants in biosolids products, 2) a risk analysis of those contaminants as they relate to human health effects, and 3) printed communications tools to explain the results, potential effects, and relative exposure to compared to a typical daily intake or therapeutic dose.

Survey of organic contaminants in biosolids products
To conduct the organic contaminants survey, eleven agencies submitted a total of 27 samples of biosolids, which included a variety of Class A and B products such as Class B cake, Class A cake blends or soil mixes, and Class A compost, to AXYS Analytical Services, Ltd. For comparative purposes, samples of yard and food waste compost and dairy manure compost were also analyzed. The laboratory analyzed all samples for a large suite of organic compounds including pharmaceuticals, antimicrobial agents, musks, flame retardants, and hormones. Data were then compiled and described by the University of Washington. Contaminant concentrations were found to be generally consistent with representative concentrations seen in the published scientific literature. Generally, composted biosolids had lower concentrations of organic contaminants than digested biosolids, mesophilic or thermophilic. Both the manure and the yard and food waste composts contained detectable concentrations of organic contaminants, including pharmaceuticals. While this was not surprising for the dairy manure, there were more detectable pharmaceuticals in the yard/food waste compost than anticipated. Of the 58 pharmaceuticals analyzed for, the biosolids compost had 14 detected compounds, the yard/food waste compost had 12 detected compounds, and the dairy manure compost had 12 detected compounds, while a representative anaerobically digested Class B biosolids had 21 detected compounds. Representative results depicting a Class B biosolids cake and a Class A biosolids compost are described in Table 1. For comparative purposes results from the yard/food waste compost, the manure compost, and the concentrations used for the risk analysis study are also described in Table 1.

While detectable concentrations of pharmaceuticals and other organic contaminants have been well documented in the scientific literature, and therefore not surprising to see consistent results in this survey, several pharmaceutical compounds detected in the yard/food waste compost and the manure compost were not anticipated. For example, ibuprofen was detected at a higher concentration in the yard/food waste compost than in both the biosolids cake and compost. Likewise, acetaminophen and naproxen, two other analgesic compounds, were also found at higher concentrations in the yard/food waste compost than both the biosolids cake and compost. This elicits the question if citizens are using their greenwaste bins to dispose of unwanted medication. While it was expected to find residual antibiotics in the manure compost, the detection of compounds such as fluoxetine (an antidepressant medication also sold under the brand names Prozac and Serafem), caffeine, triclosan (an antimicrobial agent commonly found in soap), and carbamazapine (an anti-seizure medication) was not widely anticipated.

Table 1. Representative results from a survey of biosolids products and composts analyzed for pharmaceuticals and trace organic compounds common in daily life.

Contaminant	Class B Anaerobically Digested Cake	Class A Biosolids Compost	Yard/Food Waste Compost	Dairy Manure Compost	Representative Concentration in Class B Biosolids (Used for Risk Analysis)	Representative Concentration in Class A Compost (Used for Risk Analysis)
	(mg/kg)	(mg/kg)	(mg/kg)	(mg/kg)	(mg/kg)	(mg/kg)
Acetaminophen	-	-	0.062	-	0.29	0.0015
Fluoxetine	0.266	0.049	-	0.005	0.087	0.036
Bisphenol A	0.945	0.547	0.725	-	9.0	9.0
Ibuprofen	0.190	0.030	0.278	-	0.35	0.35
BDE-209	0.71	0.31	0.06	0.06	1.7	0.24
Azithromycin	0.172	0.025	-	0.002	0.46	0.035
Ciprofloxacin	3.240	0.500	-	-	3.4	0.93
Erythromycin-H2O	0.004	-	-	-	0.02	0.006
Ofloxacin	2.020	0.487	-	0.066	1.8	0.66
Sulfamethoxazole	-	-	-	-	0.0056	0.001
Triclosan	6.840	2.050	-	0.233	17	1.2
Caffeine	-	-	41.000	0.036	*	*
Carbamazepine	0.033	0.015	-	0.002	*	*
Diphenhydramine	0.868	0.135	0.002	0.008	*	*
Diltiazem	0.007	-	-	-	*	*
Digoxin	0.007	-	-	-	*	*
Enrofloxacin	0.011	-	-	-	*	*
Flumequine	-	-	0.002	-	*	*
Miconazole	0.521	0.096	-	0.022	*	*
Oxolinic Acid	-	0.009	-		*	*
Sarafloxacin	-	-	-	0.020	*	*
Sulfanilamide	0.050	-	-	-	*	*
Thiabendazole	0.030	0.107	0.020	0.008	*	*
1,7-Dimethylxanthine	-	-	2.020	-	*	*
Furosemide	0.051	-	0.235	-	*	*
Gemfibrozil	0.029	0.030	0.005	0.002	*	*
Glyburide	0.006	-	-	-	*	*
Naproxen	0.007	-	0.026	-	*	*
Triclocarban	7.180	1.810	0.005	0.081	*	*

1) Analytes tested for, but not detected in any of the samples:
Carbadox, Cefotaxime, Clarithromycin, Clinafloxacin, Cloxacillin, Dehydronifedipine, Digoxigenin, Lincomycin, Lomefloxacin, Norfloxacin, Norgestimate, Ormetoprim, Oxacillin, Penicillin G, Penicillin V, Roxithromycin, Sulfachloropyridazine, Sulfadiazine, Sulfadimethoxine, Sulfamerazine, Sulfamethazine, Sulfamethizole, Sulfamethoxazole, Sulfathiazole, Trimethoprim, Tylosin, Virginiamycin M1, Glipizide, Hydrochlorothiazide, 2-Hydroxy-ibuprofen, Warfarin

2) * The first 11 compounds listed were also included in the risk analysis and therefore have an associated representative concentration used for the analysis. The other compounds tested as part of the survey were not included in the risk analysis.

3) - Indicates that the concentration was below the analytical detection limit.

Risk Analysis

Scientific literature and testing from a limited number of agencies suggest that organic contaminant compounds are found at low concentrations in biosolids. The scientific literature thus far, however, has a dearth of discussions related to likely exposure pathways and risk to human health from these organic contaminants. Biosolids managers, users, and the public alike want to know the risk of these compounds in a way that is meaningful and easily understood. In an attempt to answer these questions, King County, NW Biosolids, University of Washington, and Kennedy/Jenks Consultants recently performed a risk analysis to quantify the risk of exposure from biosolids.

The risk analysis method used in this study followed established U.S. EPA risk assessment methodology, which is a four-part process involving data evaluation, assessing exposure, assessing toxicity, and characterizing risk (risk = toxicity × exposure).
The data evaluation included a literature review focused on the occurrence of chemicals, primarily pharmaceutical and personal care products, in biosolids. Based on the results of the literature review, selected organic contaminant compounds for this analysis represent compounds that have been measured frequently in biosolids, are representative of a larger category of chemicals believed to have a potential human health risk and are recognizable by the public. The categories of chemicals included analgesics, antidepressants, birth control, antibiotics, antimicrobials, plasticizers and flame retardants. The compounds used for this analysis are listed in Table 2.

Table 2: Contaminants assessed in this study

Contaminant	Type of Substance	Common Usage
Acetaminophen	Over the Counter pharmaceutical	Analgesic
Fluoxetine	Prescription pharmaceutical	Antidepressant
Ethinyl Estradiol	Prescription pharmaceutical	Birth Control
Bisphenol A (BPA)	Organic compound used in common products	Various uses, primarily plastics manufacturing
Ibuprofen	Over the Counter pharmaceutical	Non-steroidal anti-inflammatory
Deca-BDE	Organic compound used in common products	Flame retardant
Azithromycin	Prescription pharmaceutical	Antibiotic
Ciprofloxacin	Prescription pharmaceutical	Antibiotic
Erythromycin-H_2O	Prescription pharmaceutical	Antibiotic
Ofloxacin	Prescription pharmaceutical	Antibiotic
Sulfamethoxazole	Prescription pharmaceutical	Sulfa-Antibiotic
Triclosan (TCS)	Personal care product	Antimicrobial

An exposure assessment quantified theoretical exposures to Class A compost and Class B biosolids cake. The representative concentrations of the selected compounds in biosolids, compost, water runnoff, and agricultural crops grown with biosolids used for this analysis were derived from published literature, as the voluntary survey administered by NW Biosolids had not been conducted at the time of this risk analysis. Risk analysis scenarios employ very conservative assumptions, assume worst case situations and, thus, are very protective of the populations assessed within these studies. The inherent, cautious approach of risk assessments maximizes the protection of the environment and human health.

The toxicity assessment utilized the findings of recent studies that established toxicological benchmarks for the compounds assessed in this study on the basis of extensive reviews of toxicological data. The toxicological benchmarks used in the risk assessment are referred to as acceptable daily intakes (ADIs). An ADI is the amount of a chemical someone can be exposed to daily over an extended period of time (usually a lifetime) without experiencing harmful effects (U.S. EPA 2014).

The risk characterization combined the findings of the exposure assessment and toxicity assessment to provide estimates of acceptable concentrations of the compounds assessed in Class A compost and Class B biosolids. The following individuals (scenarios) were evaluated for exposure to contaminants from dermal contact and incidental ingestion of biosolids:

- Child exposed while playing in a home garden or lawn fertilized with Class A biosolids compost. The child is assumed to be 33 pounds, and plays on lawns or in the home garden immediately after the application of compost, which is assumed to occur 52 times per year. It is assumed that the child ingests 200mg of compost and that the child's head, hands, forearms, lower legs, and feet are in direct contact with the compost all 52 times per year.

- Adult gardener exposed while working in a home garden fertilized with Class A biosolids compost. This scenario assumed a 176-pound adult who works in the garden immediately after compost application 52 days per year for 20 years. It is assumed that the gardener unintentionally ingests 100mg of compost and that the adult's head, hands, forearms, lower legs, and feet are in direct contact with the compost all 52 times per year.
- Occupational worker exposed while applying Class B biosolids to agricultural or forest land. This scenario assumed a 176-pound adult who applies biosolids to a forest application area or an agricultural field 220 days per year for 25 years. It is assumed that the adult worker unintentionally ingests 100mg of Class B biosolids and that the adult's arms and hands are in direct contact with the biosolids.
- Adult hiker exposed while hiking in a forested area fertilized with Class B biosolids. This scenario assumed a 176-pound adult who hikes in a forest application area immediately after the application of Class B biosolids 12 days per year for 20 years. It is assumed that the hiker unintentionally ingests 100mg of Class B biosolids and that the adult's arms and hands are in direct contact with the biosolids. It is further assumed that the hiker intentionally ingests one liter of surface water containing contaminants from runoff, however, to be conservative no dilution factor was assumed for the purposes of this risk analysis.

While it is highly unlikely that a child at play, adult gardener, occupational worker, or hiker would approach this level of exposure, it was assumed they would be so exceedingly exposed for the purposes of this analysis and to maintain a high level of conservancy.

Risk Analysis Results

The results of the risk analysis suggest that measured concentrations of organic contaminants in biosolids are much lower than the calculated acceptable concentrations for the evaluated, highly conservative, exposure scenarios. This indicates that exposure to Class B biosolids and Class A compost should not result in adverse health effects from the compounds assessed. In many cases it would take thousands or millions of years of exposure to exceed the acceptable daily intake. The representative concentration of the compound in biosolids (derived from the literature) versus the acceptable concentration of each compound in biosolids and the associated number of years of exposure to Class B biosolids or Class A compost without adverse health effects are presented in Tables 3 and 4 below. For all compounds analyzed it would take thousands or millions of years of exposure to reach potential adverse health effects. Moreover, the representative concentration of the compound is consistently orders of magnitude lower than then acceptable concentration. For illustrative purposes, the compound with the lowest and highest acceptable concentration will be discussed.

Ethinylestradiol (EE2), birth control, has the lowest acceptable concentration of all twelve compounds analyzed, which means it takes the least amount of exposure to biosolids to achieve or exceed the acceptable concentration. The acceptable concentration of EE2 for each exposed individual is shown below, as well as the years of exposure to biosolids without adverse health effects:

- Occupational Worker: Acceptable Concentration – 0.05 mg/kg; Years – 1,237
- Adult Hiker: Acceptable Concentration – 1.0 mg/kg; Years – 18,329
- Adult Gardener: Acceptable Concentration – 0.2 mg/kg; Years – 91,395
- Child Resident: Acceptable Concentration – 0.02 mg/kg; Years – 98

The representative concentration of EE2 in biosolids is 0.0011 mg/kg, which is an order of magnitude less than the lowest acceptable concentration for the child resident of 0.02 mg/kg. As shown above, the child would need to be exposed to weekly compost applications for 98 years, while the adults would need to be exposed for over a thousand years. Even in this most sensitive scenario, biosolids are not predicted to result in adverse health effects.

TCS, an anti-microbial agent found in many products, including hand soap, has the highest acceptable concentration of the compounds analyzed. This means it takes the greatest amount of exposure to biosolids to achieve the acceptable concentration. The acceptable concentration of TCS for each exposed individual is shown below as well as the years of exposure to biosolids without health effects:

- Occupational Worker: Acceptable Concentration – 93,925 mg/kg; Years – 136,519
- Adult Hiker: Acceptable Concentration – 1,740,383 mg/kg; Years – 2,023,701
- Adult Gardener: Acceptable Concentration – 269,120 mg/kg; Years – 4,935,335
- Child Resident: Acceptable Concentration – 31,114 mg/kg; Years – 155,568

Table 3: Years of Exposure to Class B Biosolids without Adverse Health Effects

Chemical of Potential Concern	Representative Concentration in Class B Biosolids (mg/kg)	Acceptable Level of Exposure of each compound (mg/kg)		Exposure to Class B Biosolids Without Adverse Health Effects (Years)	
		Adult Hiker	Occupational	Adult Hiker	Occupational
Analgesic					
Acetaminophen	0.29	62,847	3,392	4,334,286	292,391
Anti-depressant					
Fluoxetine	0.087	967	52	222,271	14,994
Birth Control					
17-a ethinylestradiol (EE2)	0.0011	1.0	0.05	18,329	1,237
Phthalate					
Bisphenol A	9.0	1,160,255	62,617	2,566,226	173,118
NSAID					
Ibuprofen	0.35	9,669	522	552,502	37,272
Flame Retardants - PBDE					
Deca-BDE	1.7	113,913	4,899	1,340,151	25,728
Antibiotics					
Azithromycin	0.46	24,172	1,305	1,050,956	70,897
Ciprofloxacin	3.4	9,669	522	57,638	3,888
Erythromycin - H2O	0.020	24,172	1,305	24,171,981	1,630,642
Ofloxacin	1.8	19,338	1,044	217,032	14,641
Sulfamethoxazole	0.0056	38,675	2,087	137,340,798	9,265,012
Anti-microbial					
Triclosan	17	1,740,383	93,925	2,023,701	136,519

<u>**Notes:**</u>

The purpose of this table is to show the number of years it would take to see a potential increase in adverse health effects from the Class B biosolids exposure scenarios in this risk analysis.

mg/kg = Milligrams per kilogram.

Representative Concentration in Class B biosolids based on average measured concentration listed in reviewed literature.

Acceptable Level of Exposure represents modeled minimum concentration below which adverse effects from exposure will not occur.

Table 4: Years of Exposure to Class A Compost without Adverse Health Effects

Chemical of Potential Concern	Representative Concentration in Class A Compost (mg/kg)	Acceptable Level of Exposure of each COPC (mg/kg)		Exposure to Class A Compost Without Adverse Health Effects (Years)	
		Adult Gardener	Child Resident	Adult Gardener	Child Resident
Analgesic					
Acetaminophen	0.0015	10,693	1,124	142,576,343	4,494,188
Anti-depressant					
Fluoxetine	0.036	165	17	91,395	2,881
Birth Control					
17-a ethinylestradiol (EE2)	0.0011	0.2	0.02	3,119	98
Phthalate					
Bisphenol A	9.0	197,413	20,742	436,635	13,763
NSAID					
Ibuprofen	0.35	1,645	173	94,006	2,963
Flame Retardants - PBDE					
Deca-BDE	0.24	17,646	6,374	1,470,486	159,343
Antibiotics					
Azithromycin	0.035	4,113	432	2,350,159	74,080
Ciprofloxacin	0.93	1,645	173	35,379	1,115
Erythromycin - H2O	0.0060	4,113	432	13,709,264	432,133
Ofloxacin	0.66	3,290	346	99,704	3,143
Sulfamethoxazole	0.001	6,580	691	131,608,932	4,148,482
Anti-microbial					
Triclosan	1.2	296,120	31,114	4,935,335	155,568

Notes:

The purpose of this table is to show the number of years it would take to see a potential increase in adverse health effects from the Class A compost exposure scenarios in this risk analysis.

mg/kg = Milligrams per kilogram.

Representative Concentration in Class A compost based on average measured concentration for each COPC listed in the reviewed literature.

Acceptable Level of Exposure represents modeled minimum concentration for each COPC below which adverse effects from exposure will not occur.

Relative Risk and Results

The risk analysis supports the conclusion that the representative concentrations of the analyzed organic contaminants in biosolids are below levels that result in adverse impacts to human health, however, the relative risk associated with these concentrations can be challenging to understand for the public and technical professionals alike. To provide context to the relative exposure of a contaminant in biosolids, we further calculated the number of years of exposure to reach a therapeutic dose of the pharmaceutical compounds, or an equivalent exposure to typical daily activity for non-pharmaceutical compounds (i.e., PBDEs, triclosan, and BPA). This comparison does not provide an indication of risk, but rather a point of comparison to relatable, typical exposure. The results of this relative comparison can be found in Table 3. For all pharmaceutical comparisons, it would take hundreds to millions of years of exposure to biosolids before achieving the exposure from a single dose of the examined compound. While the results were not as dramatic for the non-pharmaceutical compounds, Table 3 below shows it takes a minimum of one month of exposure to biosolids to accumulate the same dose or exposure of the compounds as a person might get from a single use during common activities. These typical daily intakes were determined from FDA and EPA exposure assessments.

Table 3: Exposure Required for Equivalent Therapeutic Dose or Typical Daily Intake

Chemical of Potential Concern	Therapeutic Dose or typical Daily intake (mg)	Description (a) (b)	Years of Exposure to Receive Equivalent Dose			
			Child Class A Compost	Adult Gardener Class A Compost	Adult Hiker Class B Biosolids	Occupational Class B Biosolids
Analgesic						
Acetaminophen	1,000	2 caplets of Extra Strength Tylenol	50,514,235	90,143,104	2,740,328	147,890
Anti-depressant						
Fluoxetine	20	Recommended starting daily dose of Prozac	42,095	75,119	182,689	9,859
Birth Control						
17-a ethinylestradiol (EE2)	0.01	Low dose birth control (Lo Loestrin)	718	1,282	7,533	407
Phthalate						
Bisphenol A	0.04		0.1	1	4	0.2
NSAID						
Ibuprofen	200	1 tablet of Advil	43,298	77,266	454,112	24,507
Flame Retardants - PBDE						
Deca-BDE	0.00014		0.04	0.07	0.06	0.003
Antibiotics						
Azithromycin	250	Lowest daily dose	541,224	965,819	431,900	23,309
Ciprofloxacin	250	Lowest daily dose	20,369	36,348	59,217	3,196
Erythromycin - H2O	1,000	Recommended daily dose	12,628,559	22,535,776	39,734,763	2,144,406
Ofloxacin	400	Lowest daily dose	45,922	81,948	178,383	9,627
Sulfamethoxazole	400	Lowest daily dose	30,308,541	54,085,862	56,441,424	3,046,031
Anti-microbial						
Triclosan	87	Intake from single use of anti-microbial soap at 0.3% triclosan	5,478	9,775	4,008	216

Notes:
The purpose of this table is to show the number of years of exposure it would take to receive the therapeutic dose of the various COPCs.

(a) Daily intake for Bisphenol A uses 90th percentile intake for adults (0.5 ug/kg-bw/day) from FDA's updated assessment, June 2014. Note that comparison exposure for child was modified based on ratio of child:adult body weight (15 kg:80 kg).

(b) Daily intake for Deca-BDE is the sum of exposure doses for BDE-209 from Table E-2 in EPA's exposure assessment of PBDEs, May 2010.

Risk Communication

After concluding the risk analysis and relative risk comparisons, staff from King County sought to create printed communications materials aimed at translating the technical results into a format that could be easily understood by the public. The full risk analysis contains powerful and enlightening information with great use for technical industry professionals, however, King County staff wanted a piece they could use to communicate the results with lay audiences. Generally, visual materials such as infographics, communicate information faster and clearer than written text. Understanding the power of this tactic, staff from King County's biosolids program created several draft brochures attempting to visually describe the results of the risk analysis. Next, they informally polled the reaction to these brochures from several adult readers. It was quickly determined that providing information on the "acceptable concentration" of a contaminant garnered strong negative reactions. Some readers reported that the only acceptable concentration to them was zero, and that it was offensive to indicate that having some pharmaceuticals in the biosolids was acceptable. Moreover, the definition of an acceptable concentration proved to be a bit too technical for a lay audience. As a result, this information was stripped from the final communications materials. Versions of the brochure with all 12 of the analyzed compounds described also garnered a negative reaction from readers. While most readers wanted to know what the risk of pharmaceuticals from biosolids was, they were surprised to see "such a long list" of detected compounds.

As a result, the final draft included three well-recognized representative compounds: an analgesic, an antibiotic, and the antimicrobial agent in hand soap. Infographics in which realistic pictures of pills were used like icons to depict the described compounds also polled negatively; the images subconsciously and incorrectly communicated to some individuals that large amounts of those compounds were found in biosolids. Compounds that took millions of years of exposure to reach either the acceptable concentration or a therapeutic dose were viewed with skepticism from several readers, as the time period seemed unfathomably long. As a result, we used compounds with a shorter risk timeline. Birth control and anti-depressants also polled negatively, particularly among male readers, and therefore those compounds were excluded from the final brochure. Ultimately, the team decided to focus on only three of the analyzed compounds: ibuprofen, azithromycin, and triclosan (an analgesic, an antibiotic, and an antimicrobial respectively). The brochure describes the compounds in terms of years of exposure in each of the four scenarios to reach the equivalent of a single dose, or one hand wash using anti-microbial soap. This approach proved to be relatable and easily understood by the polled adult readers. The brochure very succinctly discussed the definition of risk, the exposure scenarios, and general information about biosolids. The infographic and associated text is now published as two brochures, one from King County and the other from NW Biosolids, and also on King County's biosolids website LoopForYourSoil.com/The- science/loop-is-safe. While many of King County's printed communications materials aim to reach a fourth grade reading level, these materials are aimed at a high school reading level, given the complexity of the subject matter.

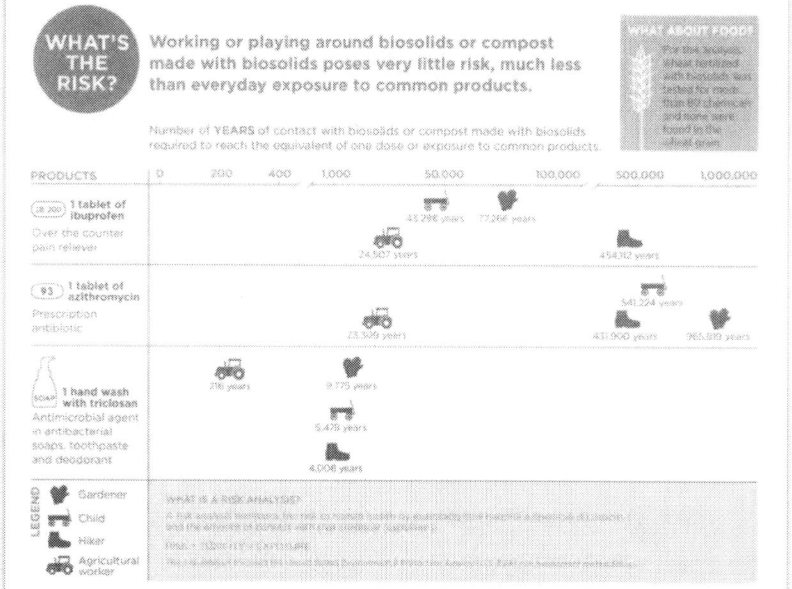

WHAT WERE THE SCENARIOS USED FOR THIS RISK ANALYSIS?

	Compost made with biosolids		Biosolids	
	Gardener (176 pounds)	**Child** (33 pounds)	**Hiker** (176 pounds)	**Agricultural worker** (176 pounds)
Parts of body in direct contact	Head, hands, forearms, lower legs, and feet	Head, hands, forearms, lower legs, and feet	Hands and arms	Hands and arms
Amount swallowed	100 mg (1/25 teaspoon)	200 mg (1/12 teaspoon)	100 mg (1/25 teaspoon) & 1 liter runoff water	100 mg (1/25 teaspoon)
Contact frequency	52 days/ year for 20 years	52 days/ year for 6 years	12 days/ year for 20 years	220 days/ year for 25 years

The average person working or playing around biosolids or compost made with biosolids would not regularly get it all over their body, accidentally eat some of it, or drink muddy runoff water. This risk analysis overestimated exposure to ensure confidence in the results.

Conclusions

Due to the completion of the survey, risk assessment, and communications materials, wastewater industry professionals now have improved resources to help answer questions about the risk of pharmaceuticals and other organic contaminants in biosolids. The risk analysis, including the relative risk portion, quantified the potential health risk to humans who are in contact with biosolids. While the risk assumptions were overly conservative for the sake of being protective, results nonetheless indicate that human health is not estimated to be adversely affected by exposure to the selected organic contaminants in biosolids. The voluntary survey of organic amendments from the Northwest region show the representative concentrations of organic contaminants used in the risk analysis are consistent with concentrations found in recent samples. Moreover, the survey results indicate organic amendments from animal manure and yard/food waste feedstocks are not necessarily less contaminated with organic compounds than are biosolids. The printed communications pieces aimed at explaining the results of the risk analysis for a lay audience use infographic techniques to visually communicate the results and relative risk to everyday activities and chemical exposure. Additional studies may be of interest to quantify other contaminants that are produced and used by the public as well as potential impacts to other environmental receptors such as wildlife.

PBDEs in Class A Biosolids Produced from Thermal Hydrolysis and Anaerobic Digestion Processes

Xuanzhao Wang[1*], Natasha Andrade[1], Alba Torrents[1], Mark Ramirez[2], Chris Peot[2]

[1] Department of Civil and Environmental Engineering, University of Maryland, College Park, Maryland
[2] DC Water, District of Columbia Water and Sewer Authority, Washington, D.C.
[*]Email: xzwang@umd.edu

ABSTRACT

Polybrominated diphenyl ethers (PBDEs) are ubiquitous in the environment and usually accumulate in the biosolids during wastewater treatment. In this study, we measure the impact of changing biosolids stabilization methods on PBDE concentration in biosolids. We investigated PBDEs concentrations in Class A biosolids produced from newly adopted thermal hydrolysis pretreatment and anaerobic digestion (THP-AD) in a Mid-Atlantic nutrient recovery facility. The total PBDE concentration in Class A biosolids was 720 ± 110 µg/kg d.w (n=21), lower than the total PBDE concentration found in previously produced Class B biosolids product from the same facility in 2010. Among the analyzed eight congeners (BDE-28, -47, -99, -100, -153, -154, -183, and -209), BDE-47, -99, and -209 were the most common congeners and contributed to about 87% of the total PBDEs concentration. From Class B to Class A biosolids production, the most prevalent congener was BDE-209, and its concentration was decreased from an average 1500 µg/kg d.w. to an average 240 µg/kg d.w. This congener also contributed less to the total PBDE concentration from 82% to 34%. In addition, despite the varied conditions of initializing a full scale anaerobic digestion system, no significant differences were observed between the startup period and the full-operation period of the THP-AD system.

KEYWORDS: Biosolids, PBDEs, Thermal Hydrolysis

INTRODUCTION

Polybrominated diphenyl ethers (PBDEs) are widely used as flame retardants in consumer products. Belonging to the group of brominated flame retardants, PBDEs can prevent ignition and slow down fires in the initial phase due to their physical and chemical properties (Harrad et al., 2008). In the 1970s, in order to replace some banned flame retardants, such as polychlorinated biphenyls (PCBs) and polybrominated biphenyls, PBDEs were widely added in polymers, such as electronic components, household appliances, furniture, textiles, etc. and became ubiquitous in our environment (Daso et al., 2010). Based on the IUPAC system, all PBDEs contain a diphenyl ether skeleton and different numbers of bromine atoms with 209 possible congeners (Fromme et al., 2009).

Three commercial formulations containing a variety of congeners have been used: penta-, octa-, and deca-BDE (Krol et al., 2012). According to the study of La Guardia et al., 2006, deca-BDE dominated 83.3% of the 2001 PBDE global market demand, followed by penta-BDE at 11.1% and octa-BDE at 5.6%. Therefore, eight congeners of PBDEs (BDE-28, -47, -99, -100, -153, -

154, -183, and -209), representing the major congeners found in commercial formulations, were chosen as the targets for our analysis.

Due to the lack of binding sites on polymers, PBDEs are simply integrated into materials but not chemically bound. PBDEs are semi-volatile organic compounds that can easily leach out of the polymers and enter air, dust, soil, etc. With similar structure and metabolites as PCBs, PBDEs may have carcinogenic and thyroid disturbing effects that bring great concern to human and ecosystem health (Knoth et al., 2007; Naert et al., 2007). Therefore, the production and uses of all formulations of PBDEs were banned within European Union in 2008 and they are currently being phased out in the U.S. and other parts of the world (Covavi et al., 2011).

However, the characteristics of persistence in the environment and the tendency of bioaccumulation means PBDEs are likely to be detected in the environment for many decades after their production ceases. As PBDEs are released into wastewater from consumer products, and they are hydrophobic in nature, they can be effectively removed with organic solids during wastewater treatment processes (Andrade et al., 2010). Consequently, biosolids produced from nutrient recovery facilities (NRFs, formerly known as wastewater treatment plants) generally have a relatively high concentration of PBDEs. In the U.S., more than 50% of biosolids are applied on agriculture land as a fertilizer. A few studies showed high concentrations of PBDEs exist in biosolids samples, and the application of large amounts of biosolids as fertilizer could put PBDEs into the food chain (Hale et al., 2012; Venkatesan and Halden, 2014). The study by Venkatesan and Halden, 2014, estimated about 24,000-36,000 kg/year (53,000 pound/year – 79000 pound/year) of PBDEs were released to the environment through land application of biosolids.

To ensure the safety of biosolids application in soil, the Environmental Protection Agency (EPA) has framed a classification system and considers biosolids which not only meet, but exceed the minimal requirements of pathogen reduction, metals content limits, and vector control to be of Class A "Exceptional Quality" (EQ) biosolids (U.S. EPA, 1993). In the 40 Code of Federal Regulations Part 503.32 rule (EPA, 1993), Class A biosolids must meet at least one of the following requirements: either the density of *Salmonella* sp. must be less than 3 MPN/4g dry weight or the density of fecal coliforms must be less than 1000 MPN/g dry weight. Class A classification is necessary if a user wants to apply the biosolids to residential lawns, home gardens, or other unrestricted public contact areas (U.S. EPA, 1994). In comparison, Class B biosolids has a looser requirement that the pathogen level could reach as high as 2 million MPN per gram dry weight of biosolids (U.S. EPA, 1994).

For either Class A or Class B biosolids, no regulation exists for PBDEs content in biosolids in the U.S. (Davis et al., 2015). Since November 2014, a large Mid-Atlantic NRF put online a new stabilization treatment process for biosolids that produces Class A product. The newly adopted process includes the Cambi^{TM} Thermal Hydrolysis Pretreatment (THP) and anaerobic digestion (AD) with the intent to enhance the microbial activity to increase biogas production and decrease pathogen levels in the biosolids (Oosterhuis, 2014). As a number of past studies showed that active microbial activity could effectively biodegrade organic pollutants in biosolids, the concentrations and distribution of PBDE congeners could be affected by the THP and AD processes (Stiborova et al., 2015b , Huang et al., 2014; Stiborova et al., 2015a; Tokarz et al., 2008).

According to a study by Stiborova et al. (2015a), biodegradation is an effective method to remove organic pollutants in biosolids. A number of studies showed the adoption of mesophilic anaerobic digestion in the wastewater treatment would debrominate PBDEs from high-brominated to low-brominated congeners (Huang et al., 2014; Stiborova et al., 2015b; Tokarz et al., 2008). Since lower-brominated PBDEs are considered to be more toxic than higher-brominated PBDEs (Liu et al., 2016), both the total concentration of PBDEs and the composition of PBDEs congeners in biosolids should be measured.

In addition, the switch from Class B biosolids production processes to THP-AD system Class A biosolids production processes at the target NRF was gradually introduced. Based on the biosolids analysis results on EPA criteria, from November, 2014 to the end of February, 2015, the period was named startup stage. In this stage, the Class A biosolids had just started and it was uncertain if the product was of steady process and quality. From March, 2015 on, the period was named full-operation stage and stable quality Class A biosolids were produced. Therefore, the PBDEs comparison between startup and full-operation stages may reveal the PBDEs treatment ability of THP-AD system.

The goals of this study were to: 1) determine the PBDEs concentrations and the distribution of different congeners in Class A biosolids products; 2) compare the PBDEs concentrations in Class A biosolids and the Class B biosolids previously produced at the same location to evaluate the overall impact of the newly adopted THP-AD treatment technology on PBDEs; and 3) compare the PBDEs concentrations' variation between the startup and the full-operation stages.

METHODS

Class A Biosolids samples were collected weekly from the final product belt conveyor at the target NRF from November 2014 until February 2015 and monthly from February 2015 until January 2016. Production of the biosolids using the two new processes started on November 26, 2014 and due to the possible high variability in the product from the treatment operation initiation, samples were collected on a weekly basis. Samples were collected using sterile instruments and containers. After sample collection, biosolids samples were immediately transferred into 250 mL amber jars (Kimble™, 16oz, USA) and were frozen at -20°C until processing.

The solids content of the samples was measured using EPA Method 1684. The sample was dried a minimum of 12 hours in an oven (Fisher Scientific, Isotemp, USA) at 103-105 °C for total solids (TS) calculation and a minimum of 2 hours ignition in a heat furnace ((Neycraft Vulcan, A-550, York, PA) at 550 °C for volatile solids (VS) calculation (U.S. EPA, 2001).

For PBDEs analysis, the sample preparation, chemical extraction and cleaning, and chemical analysis was based on EPA method 1614 (U.S.EPA, 2010). We utilized a modified method adopted from De la Cal et al. (2003) and Deng et al. (2015). Before processing, biosolids samples were thawed in a refrigerator at -4 °C overnight and then allowed to reach room temperature for chemical extraction. Approximated 1.5 g (±0.01 g) sample was mixed with 3.0 g of hydromatrix (Agilent Technologies, Hydromatrix, USA), and spiked with 40 ng of the surrogate polychlorinated biphenyl PCB-209 (Cambridge Isotope Laboratories, PCB-209-CS, Andover, MA) and 40 ng of the internal standard PCB-138 (Cambridge Isotope Laboratories, PCB-138-CS, Andover, MA). Samples were extracted in an Acceleration Solvent Extraction (ASE) system (Thermo Scientific, Dionex ASE 350 with Dionium Components Smartrun System & Solvent Saver System, Sunnyvale, CA). The extraction cycles were performed at a pressure of 1500 psi and temperature of 100 °C with n-hexane (Fisher Scientific, n-Hexane, USA) as solvent. After 5 minutes to heat up the oven, two static extractions of 10 min were developed at constant pressure and temperature, generating approximately 35 mL of extract.

Since the extract had high content of lipid and sulfur that could interfere the PBDEs analysis (Berton et al., 2016), the extract was passed through a packed glass chromatographic column (Kimble™ Kontes™ PTFE-Plugged Column, 200mm L*15mm ID, USA) for clean-up. From bottom to top, the column was packed with glass wool (Acros Organics, Glass wool, New Jersey), copper powder (Fisher, Copper C434-500 Powder, USA), 1 g activated silica gel (Alfa Aesar, 150 angstroms wide pore silica gel, USA), 2 g 3% basic silica gel with NaOH (Fisher, NaOH pellets, USA), 1 g activated silica gel, 4 g 40% acid silica gel with concentrated sulfuric acid (Fisher, A300-212 Sulfuric Acid, USA), 2 g activated silica gel and 1 g anhydrous sodium sulfate (Fisher, S429-212 Sodium Sulfate Anhydrous, USA). PBDEs were eluted with n-hexane, evaporated in evaporator (Zymark, TurboVap LV Evaporator, Hopkinton, MA) under a gentle nitrogen stream, and dissolved with n-hexane to 1 mL for chemical analysis.

Eight PBDEs congeners (BDE-28, -47, -100, -99, -154, -153, -183, and -209) were analyzed by Agilent gas chromatography-mass spectrometry (GC-MS) 6890 N/5975. Negative chemical ionization in selected ion monitoring mode and DB-5MS capillary column were used. The quantification of PBDEs was performed by monitoring the ion fragments with mass-to-charge ratio.

The recovery of PCB-209 was calculated for each sample set and the results were accepted with surrogate recovery from 90% to 110%. At least duplicate data for each biosolids sample was achieved for further analysis. For each sample batch (10 samples), a sand blank was analyzed for control. In addition, PBDE-spiked sand and PBDE-spiked biosolids sample (addition of 0.05 μg of BDE-28, -47, -99, -100, -153, -154, and -183, each and 0.5 μg of BDE-209) (Cambridge Isotope Laboratories, Andover, MA) were analyzed to calculate congener-specific recovery to ensure the reliability of the experimental method.

RESULTS AND DISCUSSION

In our study, a total of 21 samples of Class A biosolids samples from November 2014 to January 2016 were collected and analyzed. For each sample, TS were measured to calculate PBDEs concentration in a dry weight basis. The average TS of our Class A biosolids samples was 29.3 ± 2.35% (*n*=21).

Among the eight PBDEs congeners analyzed, the concentrations of BDE-47, -99, and -209 were the highest. The detection limit and quantitation limit of each chemical were measured and calculated. Due to low concentrations of BDE-28 (below detection limit), its results were not shown in this manuscript. The average concentrations and standard deviations of BDE-47, -99, -100, -153, -154, -183, -209, and total PBDEs were calculated (Table 1). BDE-209 concentrations were at similar levels to BDE-47 and BDE-99. Due to chemical partition and commercial production, BDE-209 is generally observed in significantly higher levels than other PBDE congeners in biosolids samples. In the studies by Venkatesan and Halden (2014), BDE-209 was present in U.S. biosolids as the majority congener with a mean concentration 5360 ± 5163 μg/kg d.w. BDE-99 and BDE-47 were the two other congeners with most contribution to the total with mean concentrations of 1044 ± 448 μg/kg d.w. and 789 ± 318 μg/kg d.w., respectively. However, results from the target NRF Class A biosolids indicated relatively low concentrations of BDE-209 in the Class A biosolids product (Table 1). This observed congener distribution was not commonly reported, and certainly did not reflect the congener distribution observed in the Class B biosolids that were previously produced at the facility (Andrade et al. 2015) (Table 1).

Table 1. Mean PBDEs concentrations and standard deviations in Class A biosolids samples and previously-produced Class B biosolids.

Congeners	Mean ± SD (μg/kg d.w.)	
	Class A	**Class B***
BDE47	190 ± 34	121 ± 36.1
BDE100	44 ± 6.9	20.5 ± 11.1
BDE99	200 ± 27	134 ± 42.1
BDE154	18 ± 2.6	7.97 ± 0
BDE153	22 ± 4.0	8.23 ± 0
BDE183	7.5 ± 1.6	6.64 ± 0
BDE209	240 ± 72	1490 ± 503
Total	720 ±110	1790 ± 528

*Data obtained from Andrade, et al. 2010

In a previous study that reported PBDE concentrations in the Class B product produced by the same facility, as shown in Table 1, BDE-47+BDE-99 concentrations hovered around 250 μg/kg d.w. (Andrade et al. 2015). The sum of the same two congeners in the Class A biosolids would be around 390 μg/kg d.w., indicating an increase in the concentration of these two congeners. Moreover, the concentration of BDE-209 in the previously-produced Class B material hovered around 1500 μg/kg d.w., which was higher than values observed in the Class A material at around 240 μg/kg d.w. (Andrade et al. 2015).

Furthermore, the comparison of the eight PBDEs congeners' contribution to the total concentration between Class A and Class B biosolids products was conducted (Figure 1). In the Class A biosolids, BDE-47, -99, and -209 contributed to about 87% of total PBDEs while in the Class B biosolids, these three congeners contributed to about 97% of total PBDEs. Most significantly, the percent contribution of BDE-209 was decreased from 82 ± 5.9% (*n*=62) to 34 ± 6.9% (*n*=21) from Class B to Class A biosolids (Andrade et al. 2015).

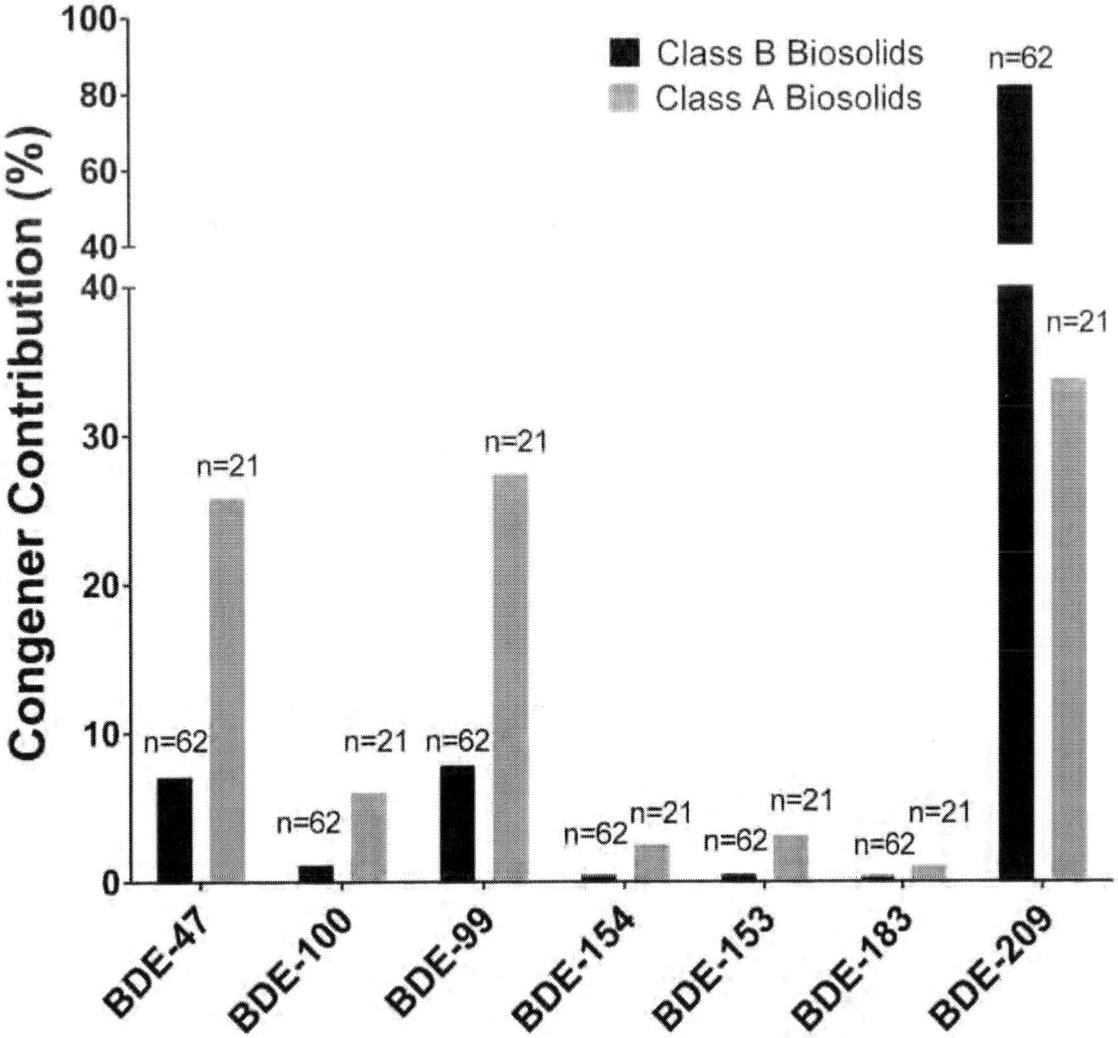

Figure 1. Average PBDEs congeners' contribution to the total PBDE concentration in Class A and Class B biosolids produced by the same NRF.

These results suggest that THP-AD could enhance removal of BDE-209, perhaps by the removal of bromines, efficiently debrominating the larger congener, generating more of the lower-brominated compounds. However, a more in-depth investigation of the PBDE congeners is required. This study provides information on the impact of THP-AD (with enhanced microbial activity) on PBDE concentrations, which can help researchers understand differences between biosolids stabilization processes on microconstituents. In addition, this study shows promise that this technology could be used for microconstituents removal and further research should be conducted to better understand the mechanisms in which these processes rely.

Biosolids samples were collected from the very first day of the Class A biosolids production. It is interesting then, to analyze the data in a temporal scale to observe the impact of the startup of a large-scale stabilization process in PBDEs' biosolids concentrations. The variation of PBDEs concentration from the startup stage to full-operation stage of the THP-AD system was investigated based on the concentrations of the three most common PBDE congeners: BDE-47, -99, and -209 (Figure 2). Due to the large size of the treatment system and beginning stage, the production of Class A biosolids at the target NRF, during Nov. 26th, 2014 to the end of Feb., 2015, the startup stage, the THP was operated but anaerobic digesters operation parameters were still being optimized. From March 2015 until the last sampling event for this study, we characterized as full-operation stage, when the treatment system reached stable conditions and qualified consistently as Class A biosolids. We hypothesized that PBDE concentrations could also be impacted by the differences between the two established stages due to constant changes to operating parameters.

In order to analyze significant differences between the startup stage and full-operation stage, about one year-long data were separated into startup stage (n=12) and full-operation stage (n=9). Concentrations between startup and full-operation stages were statistically not different for BDE-47 (p=0.60), BDE-99 (p=0.92) and BDE-209 (p=0.16) (unpaired t test). The results indicated the variation of THP-AD system in the early operation stage did not affect the debromination of PBDEs in the biosolids products. This could be an indication that the AD in this particular facility was efficient in PBDE removal even when the AD process was not operating at optimum conditions. It is also possible that the high temperatures and pressures of the THP process could enhance further debromination during the AD. However, more research is needed to investigate the reasons.

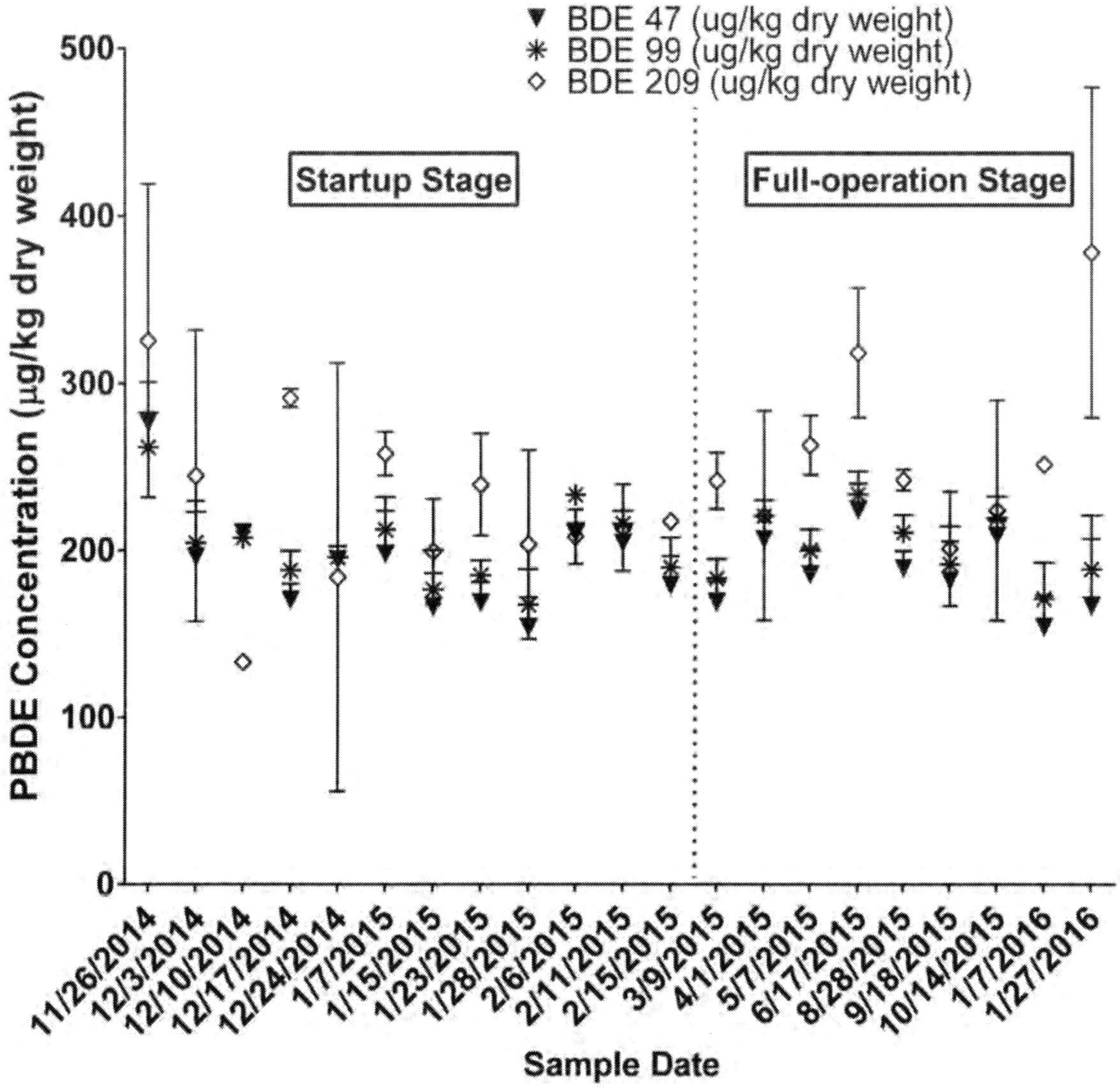

Figure 2. Concentrations of BDE-47, -99, and -209 in biosolids samples from Nov. 26th, 2014 to Jan. 27th, 2016.

CONCLUSION

Overall, this study determined eight common PBDEs congeners' concentrations in Class A biosolids product that was stabilized with THP and AD processes. The total PBDE concentration in the biosolids in the investigated period was 720 ± 110 µg/kg d.w (n=21). This reported concentration was lower than the total PBDE concentration found in previously produced Class B biosolids product from the same facility in 2010. Among the eight congeners, BDE-47, -99, and -209 were the most common congeners and contributed to about 87% of the total PBDEs concentration. The most prevalent PBDE congener in both Class B and Class A biosolids was BDE-209, however, from Class B to Class A biosolids production, the concentration was decreased from an average 1500 µg/kg d.w. to an average 240 µg/kg d.w. The contribution to the total concentration also decreased from 82% in Class B to 34% in Class A. It is unknown why the concentration of BDE-209 decreased in the Class A product compared to the Class B product. However, the study suggests that debromination may be enhanced due to extensive microbiological activity in the AD process and high temperatures and pressures to which the biosolids are exposed to in the THP process. In addition, the concentration of PBDEs were relatively constant during the startup stage to full-operation stage, which could indicate that the debromination processes may be efficient since the changing operational parameters of the AD in the startup stage seemed sufficient to reduce PBDE concentrations. Further research will investigate the mechanisms of debromination that occur in the THP and AD systems with samples analysis between treatment processes and more fundamental work.

ACKNOWLEDGEMENTS

We gratefully thanks Dana Armstrong and research assistant Taylor LaChance from University of Maryland, College Park, for the sampling the biosolids and the experiment assistances.

REFERENCES

Andrade, N. A.; Lozano, N.; McConnell, L. L.; Torrents, A.; Rice, C. P.; Ramirez, M. (2015) Long-term trends of PBDEs, triclosan, and triclocarban in biosolids from a wastewater treatment plant in the Mid-Atlantic region of the US. *Journal of Hazardous Materials*, **282**, 68–74.

Andrade, N. A.; McConnell, L. L.; Torrents, A.; Ramirez, M. (2010) Persistence of Polybrominated Diphenyl Ethers in Agricultural Soils after Biosolids Applications. *Journal of Agricultural and Food Chemistry*, **58** (5), 3077–3084.

Berton, P.; Lana, N. B.; Ríos, J. M.; García-Reyes, J. F.; Altamirano, J. C. (2016) State of the art of environmentally friendly sample preparation approaches for determination of PBDEs and metabolites in environmental and biological samples: A critical review. *Analytica Chimica Acta*, **905**, 24–41.

Cal, A. de la; Eljarrat, E.; Barceló, D. (2003) Determination of 39 polybrominated diphenyl ether congeners in sediment samples using fast selective pressurized liquid extraction and purification. *Journal of Chromatography A*, **1021** (1–2), 165–173.

Covaci, A.; Harrad, S.; Abdallah, M. A.-E.; Ali, N.; Law, R. J.; Herzke, D.; Wit, C. A. de. (2011) Novel brominated flame retardants: A review of their analysis, environmental fate and behaviour. *Environment International*, **37** (2), 532–556.

Daso, A. P.; Fatoki, O. S.; Odendaal, J. P.; Okonkwo, J. O. (2010) A review on sources of brominated flame retardants and routes of human exposure with emphasis on polybrominated diphenyl ethers. *Environmental Reviews*, **18** (NA), 239–254.

Davis, E. F.; Gunsch, C. K.; Stapleton, H. M. (2015) Fate of flame retardants and the antimicrobial agent triclosan in planted and unplanted biosolid-amended soils: Environmental Chemistry. *Environmental Toxicology and Chemistry*, **34** (5), 968–976.

Deng, D.; Chen, H.; Tam, N. F. Y. (2015) Temporal and spatial contamination of polybrominated diphenyl ethers (PBDEs) in wastewater treatment plants in Hong Kong. *Science of The Total Environment*, **502**, 133–142.

Hale, R. C.; La Guardia, M. J.; Harvey, E.; Chen, D.; Mainor, T. M.; Luellen, D. R.; Hundal, L. S. (2012) Polybrominated Diphenyl Ethers in U.S. Sewage Sludges and Biosolids:

Temporal and Geographical Trends and Uptake by Corn Following Land Application. *Environmental Science & Technology*, **46** (4), 2055–2063.

Fromme, H.; Körner, W.; Shahin, N.; Wanner, A.; Albrecht, M.; Boehmer, S.; Parlar, H.; et al. (2009) Human exposure to polybrominated diphenyl ethers (PBDE), as evidenced by data from a duplicate diet study, indoor air, house dust, and biomonitoring in Germany. *Environment International*, **35** (8), 1125–1135.

Harrad, S.; Ibarra, C.; Diamond, M.; Melymuk, L.; Robson, M.; Douwes, J.; Roosens, L.; et al. (2008) Polybrominated diphenyl ethers in domestic indoor dust from Canada, New Zealand, United Kingdom and United States. *Environment International*, **34** (2), 232–238.

Huang, H.-W.; Chang, B.-V.; Lee, C.-C. (2014) Reductive debromination of decabromodiphenyl ether by anaerobic microbes from river sediment. *International Biodeterioration & Biodegradation*, **87**, 60–65.

Knoth, W.; Mann, W.; Meyer, R.; Nebhuth, J. (2007) Polybrominated diphenyl ether in sewage sludge in Germany. *Chemosphere*, **67** (9), 1831–1837.

Król, S.; Zabiegała, B.; Namieśnik, J. (2012) PBDEs in environmental samples: Sampling and analysis. *Talanta*, **93**, 1–17.

La Guardia, M. J.; Hale, R. C.; Harvey, E. (2006) Detailed Polybrominated Diphenyl Ether (PBDE) Congener Composition of the Widely Used Penta-, Octa-, and Deca-PBDE Technical Flame-retardant Mixtures. *Environmental Science & Technology*, **40** (20), 6247–6254.

Liu, L.; Zhang, Y.; Liu, R.; Wang, Z.; Xu, F.; Chen, Y.; Lin, K. (2016) Aerobic debromination of BDE-209 by Rhodococcus sp. coupled with zerovalent iron/activated carbon. *Environmental Science and Pollution Research*, **23** (4), 3925–3933.

Naert, C.; Van Peteghem, C.; Kupper, J.; Jenni, L.; Naegeli, H. (2007) Distribution of polychlorinated biphenyls and polybrominated diphenyl ethers in birds of prey from Switzerland. *Chemosphere*, **68** (5), 977–987.

Oosterhuis, M.; Ringoot, D.; Hendriks, A.; Roeleveld, P. (2014) Thermal hydrolysis of waste activated sludge at Hengelo Wastewater Treatment Plant, The Netherlands. *Water Science & Technology*, **70** (1), 1.

Stiborova, H.; Vrkoslavova, J.; Pulkrabova, J.; Poustka, J.; Hajslova, J.; Demnerova, K. (2015a) Dynamics of brominated flame retardants removal in contaminated wastewater sewage sludge under anaerobic conditions. *Science of The Total Environment*, **533**, 439–445.

Stiborova, H.; Vrkoslavova, J.; Lovecka, P.; Pulkrabova, J.; Hradkova, P.; Hajslova, J.; Demnerova, K. (2015b) Aerobic biodegradation of selected polybrominated diphenyl ethers (PBDEs) in wastewater sewage sludge. *Chemosphere*, **118**, 315–321.

Tokarz, J. A.; Ahn, M.-Y.; Leng, J.; Filley, T. R.; Nies, L. (2008) Reductive Debromination of Polybrominated Diphenyl Ethers in Anaerobic Sediment and a Biomimetic System. *Environmental Science & Technology*, **42** (4), 1157–1164.

U.S. EPA. (1993) Standards for the Use and Disposal of Sewage Sludge. Code of Federal Regulations, Part 503, Title 40; U.S. Environmental Protection Agency: Washington, D.C.

U.S. EPA. (1994) A Plain English Guide to the EPA Part 503 Biosolids Rule, EPA-832/R-93-003; U.S. Environmental Protection Agency: Washington, D.C.

U.S. EPA. (Jan. 2001) Method 1684 Total, Fixed, and Volatile Solids in Water, Soils, and Biosolids, EPA-821-R-01-015; U.S. Environmental Protection Agency: Washington, D.C.

U.S.EPA. (May 2010) Brominated Diphenyl Ethers in Water, Soil, Sediment, and Tissue by HRGC/HRMS, EPA-821-R-10-005; U.S. Environmental Protection Agency: Washington D.C.

Venkatesan, A. K.; Halden, R. U. (2014) Brominated flame retardants in U.S. biosolids from the EPA national sewage sludge survey and chemical persistence in outdoor soil mesocosms. *Water Research*, **55**, 133–142.

Hydrothermal processes for simultaneous bioenergy recovery and destruction of bioactive microconstituents from biosolids

Young Hwan Shin[a], Lance Schideman[b*], Peng Zhang[a], John Scott[b], Michael Plewa[c], Yuanhui Zhang[a]

[a]Department of Agricultural and Biological Engineering, University of Illinois at Urbana – Champaign, 1304 W. Pennsylvania, Urbana, IL 61801, USA

[b]Illinois Sustainable Technology Center, a division of the Prairie Research Institute at University of Illinois at Urbana – Champaign, Hazelwood Drive, Champaign, IL 61820, USA

[c]Department of Crop Sciences, University of Illinois at Urbana – Champaign, IL 61801, USA

ABSTRACT

Biosolids from animal manure contain bioactive microconstituents that potentially contribute to ecological or health effects. For instance, estrone (E1), 17β-estradiol (E2), and estriol (E3) can have adverse effects on the reproductive biology of vertebrates at very low concentrations (Irwin et al., 2001; Schuh et al., 2011). Antibiotic resistant bacteria that cause life threatening infections can be developed from bacteria exposed to low levels of antibiotics excreted in urine and feces (Wise et al., 1998; Sarmah et al., 2006; Berge et al., 2006; Martinez, 2009). This study investigates the effects of Mixed algal-bacterial bioreactor (MABB) and hydrothermal processes such as hydrothermal liquefaction (HTL) and catalytic hydrothermal gasification (CHG) simultaneously produce biomass for bioenergy production and reduce the discharge of bioactive microconstituents. Thus, the integrated system improves the quality of water resources and increase energy security by removing organics/nutrients and producing bio crude oil/gas, respectively.

Fresh swine manure samples were filtered with 0.45 μm membrane to generate the liquid portion of swine manure (LPAM). GC/MS followed by Solid Phase Extraction (SPE) were used to analyze bioactive microconstituents and the concentration of E1, E2, E3, and EE2 in LPAM were ranged from 10,602 to 22,646 ng/L. cytotoxicity of the LPAM was investigated using a previously developed Chinese Hamster Ovary (CHO) cell assay (Hsie AW, 1975; Wagner et al., 1998), and the extracted LPAM organics had a cytotoxicity index of 2.38, which is slightly less than secondary municipal wastewater effluent, 2.64 (Pham et al., 2013). A fluctuation test was developed to measure antibiotic resistance in bacterial populations exposed to antibiotics in manure. E. coli 15597 was exposed to florfenicol (FF) at lethal (17 μg/mL) concentration that provides a selective advantage to individual bacterium that were acclimated to under non-lethal

FF selection (2 μg/mL), and this study confirmed that exposure to FF at a non-lethal concentration increased survival during an antibiotic challenge.

MABB with adsorbents was operated to extract bioactive microconstituents and other organics from the LPAM under various operating conditions (organic loading: 35.3 to 470 mg sCOD/L/d; light intensity 350 μmol photons/m2/s; temperature: 18 to 25 °C; Hydraulic retention time: 4 days; Solid retention time: 30 days), and biomass was harvested to use as a biofuel feedstock with maximum productivity of 819 mg/L/d.

HTL and CHG with 9 different conditions (Temperature: 200 to 550 °C; Time: 60 min; Catalyst: Ruthenium on alumina) were performed to determine the optimal conditions for renewable energy production and to study the effects on the fate of bioactive bioactive microconstituents. E1and E2 in the spiked biomass was removed up to 94.6 and 99.8% with HTL run at 300 °C/60 min which was the most optimal condition to produce bioenergy. The cytotoxicity of the HTL-WW under three conditions (sample I, II, and III: 250°C/60min, 300°C/60min, and 350°C/60min) was investigated CHO cell assay. Mean cytotoxicity index (CTI) values of sample I, II, and III were 3710.2, 2380.8, and 5373.3, respectively, which tells us that the descending cytotoxicity potency was Sample III > Sample I > Sample II and toxicity in HTL-WW is proportional to the temperature of HTL (250° and 300°C < 350°C).

In conclusion, the combination of MABB and HTL/CHG can simultaneously remove bioactive microconstituents from swine manure (E2, MABB: 92.3 ± 1.5%; HTL: 99.7 ± 0.1%, CHG: 99.9 ± 0.01%) and convert them into valuable bioenergy products under optimal condition (HTL: 300 °C/60min, CHG: 450 °C/60min/Ru). This will contribute to the development of cost-effective systems to reduce water pollution and increase bioenergy production as well as opportunities for reuse of the aqueous fraction.

KEYWORDS

Microconstituents, Mixed algal-bacterial bioreactor (MABB), Hydrothermal Liquefaction (HTL), Catalytic Hydrothermal Gasification (CHG)

INTRODUCTION

Natural and synthetic estrogenic hormones at trace levels in the environment have been reported in numerous studies and are of growing concern due to potential adverse effects on the reproductive biology of vertebrates at very low concentrations (10–100 ng/L) (Routledge et al., 1998; Schuh et al., 2011). Naturally occurring estrogens in animal waste can cause negative environmental impact through the disruption of the endocrine system in wild life, domesticated animals, and humans (Khanal et al., 2006). Excretion of steroidal estrogens from humans and farm animals is the major source of estrogenic compounds in the environment and can potentially contaminate surface and ground water (Finlay-Moore et al., 2000; Hanselman et al., 2004; Raman et al., 2004; Shore et al., 1993). The annual amount of estrogen excretion from livestock animals such as cattle, pigs, sheep, and chickens is estimated to be 39 tons in the European Union and 41 tons in the United States (Lange et al., 2002). High concentrations of estrogenic hormones and their partial breakdown products are often reported in wastewaters containing manure (Bradford et al., 2008; Hanselman et al., 2003; Hutchins et al., 2007; Kolodziej et al., 2004). For example, the concentration of E2 and E1 in surface water and well water near a cattle farm has been reported to be within 0.05–7.4 ng/L and 4.5 ng/L, respectively (Fine et al., 2003; Irwin et al., 2001). Moreover, concentrations of estrogens in wastewater originating from agricultural activities was found to be three to four times higher than municipal wastewaters (Shore et al., 1993). Excretion of estrogenic hormones from agricultural activities is a significant concern as they can contaminate water resources and cause adverse ecological effects at very low concentrations (Irwin et al., 2001; Jobling et al., 1998; Thorpe et al., 2003). Therefore, it is particularly important to better understand the fate, transport, and transformation of these bioactive compounds in livestock systems and to develop management practices that cost-effectively mitigate the associated risks.

To set up the advanced swine manure management systems, an integrated manure management system was suggested which includes a biological process for biomass production and hydrothermal biomass conversion for bioenergy. LPAM is the Liquid Portion of Animal Manure which pertains to the liquid portion of swine manure in this study. To produce LPAM from raw swine manure slurry, pretreatment methods for wastewater treatment such as sedimentation, screening, bag filtration, and microfiltration were used to separate the solids and liquids. Microalgae could be a key element for biofuel production and CO_2 fixation and, consequently, scientific efforts are focused on the conversion of algal biomass to bioenergy using hydrothermal liquefaction and gasification which could be promising technologies. However, the natural estrogenic hormones in the LPAM could not be removed completely by the adsorption to mixed algal – bacterial biomass, and the estrogenic hormones still have estrogenic properties that can adversely affect aquatic vertebrates (Ifelebuegu. 2011; Zhang et al., 2014). Manure contains bioactive microconstituents which potentially causes undesirable ecological or health effects that can contribute to high rates of feminization for aquatic species by reducing reproductive abilities and increasing deformities related to disruption of the endocrine system. Therefore, this study will investigate the fate of microconstituents from LPAM, referring to the Liquid Portion of Animal Manure which accounts for 80% - 95% of total manure quantities, which are subjected to hydrothermal bioenergy conversion processes. This study focuses on LPAM from swine manure sources and several common hormones from the swine farming systems such as estrone (E1), 17β-estradiol (E2), and estriol (E3).

The biomass or organic wastes that are used to produce biofuels should be analyzed the heavy metals that can affect the environment safety (Nzihou & Stanmore. 2013). Toxic heavy metals such as lead, arsenic, cadmium, and mercury can produce toxicological effects on human health and subsequently on the environmental safety (Brathwaite & Rabone. 1985; Zhou et al., 2015). Therefore, it is mandatory to investigate the chemical composition of biomass including heavy metals as well as heating value of the bioenergy feedstock to calculate the energy yield of bioenergy production process. If the heavy metals in inputs and outputs of bioreactors and hydrothermal bioenergy processes are characterized, the distribution of heavy metals in the integrated manure management system could be predicted and minimize the environmental impacts by removing residual toxic heavy metals.

The final goal is to develop technologies & tools for livestock manure management that can simultaneously provide significant environmental benefits and produce valuable bioenergy products. A novel manure treatment system that can synergistically capture microconstituents from animal manure and convert them into valuable bioenergy products instead of polluting the environment with microconstituents will be investigated. This provides a cost-effective way to reduce water pollution, increase bioenergy production, and increase opportunities for reuse of the aqueous portion of manure. The objectives of this study are 1) Demonstrate the ability to extract microconstituents and other organics from LPAM to create a feedstock for bioenergy production. 2) Investigate the effects of hydrothermal bioenergy production processes on the fate of bioactive microconstituents. 3) quantify and investigate the toxicity of organic mixture of LPAM, effluent of MABB, HTL-WW, and CHG-WW in the manure management system.

METHODOLOGY

1.1 Chemicals and reagents

Steroid hormones such as Estrone (E1, ≥99%), 17β – estradiol (E2, ≥98%), and Estriol (E3, ≥99%) were used as hormone standards which were purchased from Sigma-Aldrich Chemicals (Milwaukee, WI). Stock solutions of E1, E2, and E3 at 2 mg/ml were prepared in methanol. The Solid Phase Extraction (SPE) columns, Supelclean Envi-Carb (500mg/6ml, No. 57094) and Supelclean LC-Florisil (1 g/6ml, No. 57057), were purchased from Supelco (Bellefonte, PA, USA). Distilled water was obtained with a Milli-Q system (Millipore, Bedford, MA, USA). HPLC grade organic solvents such as n-hexane, acetone, acetonitrile, methanol, or dichloromethane were purchased from Fisher Scientific (Fair Lawn, NJ).

Filtrasorb®400 granular activated carbon (Calgon Carbon, Pittsburgh, PA) were screened using a mesh size No. 50 at 0.297 mm to achieve the desired average particle size of the GAC and washed several times using deionized water to remove dust and particles followed by drying at room temperature (AWWA B604, 1992). The cleaned GAC were packed into mesh bags and the GAC bags were vertically installed in the bioreactor to allow the airflow to provide contact between wastewater and GAC as the wastewater circulates. Table.1 shows the chemical and physical properties of natural estrogenic hormones that is commonly detectable in swine manure as well as nearby bodies of water.

Table 1
Properties of selected estrogenic hormones and antibiotic (Hanselman et al., 2003; Zhiquang et al., 2004, A.K. Sarmah et al., 2006).

Properties	**Estrone**	**17β-estradiol**	**Estriol**	**17α-estradiol**
Used abbreviation	E1	E2	E3	EE2
Class	Steroid	Steroid	Steroid	Steroid
Cas registry number	53-16-7	50-28-2	50-27-1	57-63-6
Molecular weight (g/mol)	270.4	272.3	288.4	296.4
Vapor pressure (Pa)	3×10^{-8}	3×10^{-8}	9×10^{-13}	6×10^{-9}
Water solubility (20°C, ppm)	0.8 - 12.4	3.9 - 13.3	3.2 - 13.3	4.8
pK_a	10.3 - 10.8	10.5 - 10.7	10.4	10.21
log K_{ow}	3.1 - 4.0	3.1 - 4.0	2.6 - 2.8	3.67, 4.15
Molecular formula	$C_{18}H_{22}O_2$	$C_{18}H_{24}O_2$	$C_{18}H_{22}O_3$	$C_{20}H_{24}O_2$
Structure	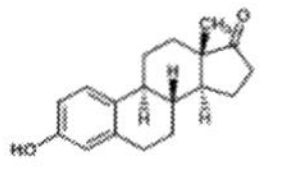	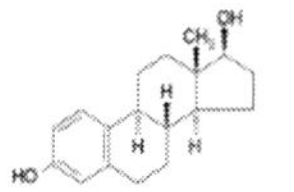	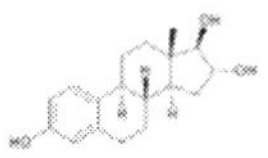	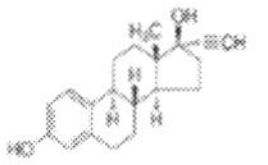

1.2 Analytical methods for estrogenic hormones

Figure.1 shows an integrated system that was suggested for swine manure management. The system is composed of a) mixed algal-bacterial bioreactor, b) hydrothermal liquefaction, and c) catalytic hydrothermal gasification.

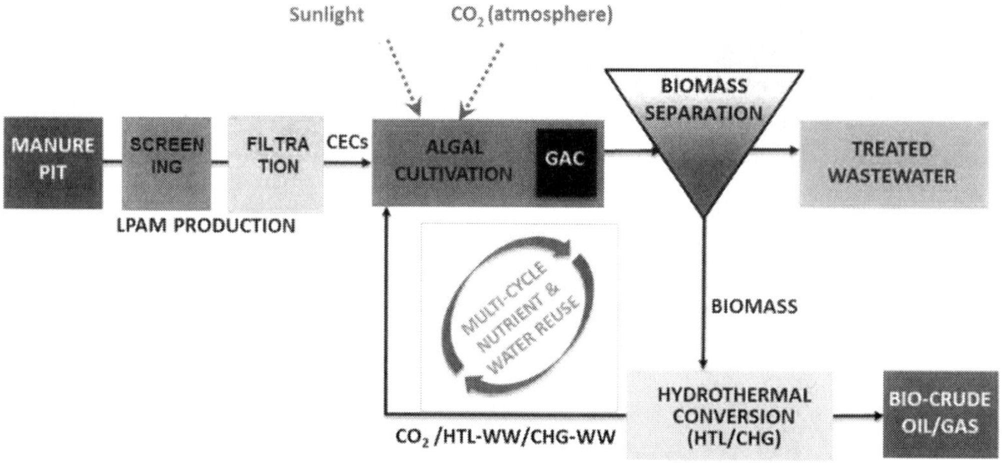

Figure. 1. Flow diagram of an integrated swine manure management system.

Samples preparation

To enhance the sensitivity of analysis by removing the matrix effects, the hormones were extracted and analyzed from both liquid and solid fractions in each step. The final data was obtained by averaging the results after at least 2 repetitions for both liquid and solid fractions. Fresh swine manure samples were collected from the farrowing barn, and the hormones were extracted from the feces and solid portion of manure slurry followed by centrifugation with methanol solvents. Solid Phase Extraction (SPE) was used to concentrate the target analytes from the extracts by serial methanol extraction, and matrix effects can be eliminated by removing background compounds (Hanselman et al., 2003; Singh et al., 2013; Zheng et al., 2008). The extracts were stored at -21˚C after evaporation with N^2 at 60˚C. Figure. 2 shows the schematic flow diagram for the sample preparation in liquid and solid fractions.

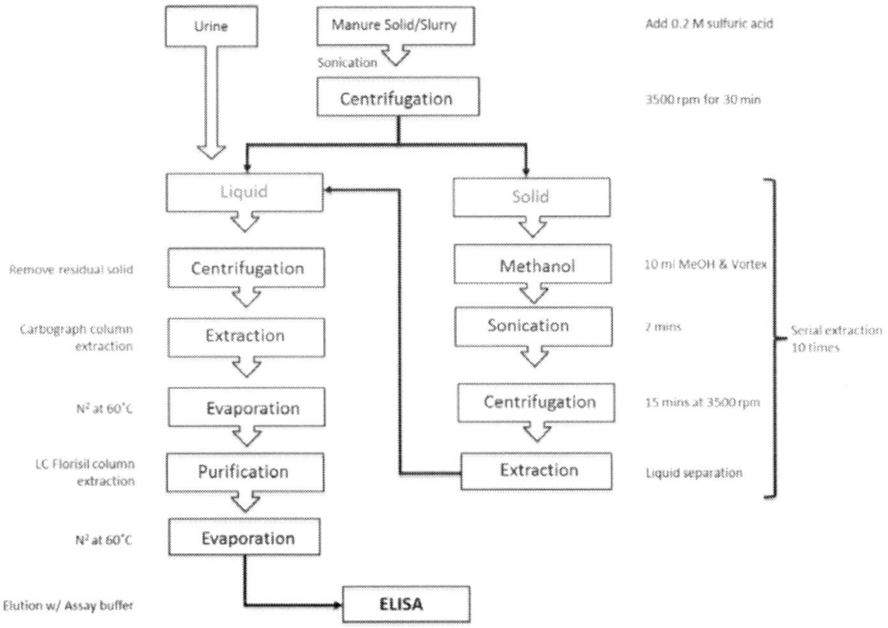

Figure. 2. Flow diagram of the sample preparation method for the analysis of E1, E2, E3, and EE2 by GC/MS.

A) SOLID FRACTION

According to Singh et al. (2013), estrogenic hormones more than 85% of the spiked hormones can be extracted from solid manure by serial extraction with methanol and sonication. In this study, the serial extraction method was used for feces and solid portions of manure slurry to extract natural estrogenic hormones. 5 mg of the solid manure sample was vortexed with 20 ml of methanol and centrifuged at 2,500 rpm for 15 minutes to separate the solid and liquid fractions followed by sonication for 4 minutes. Then, the liquid portion was collected in the glass container and stored at - 20 °C for liquid fraction analysis. The extraction was repeated to 10 times, and the solid fraction was stored at -20 °C for further analysis.

B) LIQUID FRACTION

The urine and liquid fractions of the samples were centrifuged at 2,500 rpm for 15 min to remove residual suspended solids to minimize Carbograph SPE column plugging. Before passing the liquid portion into the Carbograph column, all the columns were preconditioned by sequentially washing the Carbograph column with 10 ml dichloromethane/methanol (80:20), 5 ml of acidified methanol with formic acid, 5 ml of methanol, and 10 ml of DI water. The supernatant was passed through the washed Carbograph column using the vacuum and concentrated hormones which was eluted with 10 ml of dichloromethane/methanol (80:20). After evaporating the solvents with N^2 at 60 °C, the extracts were re-dissolved in the 2 ml of dichloromethane/hexane (50:50). The concentrates were passed through the Florisil SPE column to clean up the samples for removing matrix effects followed by the preconditioning of cartridge with 5 ml dichloromethane/methanol (80:20) and 5ml of dichloromethane/hexane (50:50).

After washing the column with 10 ml of dichloromethane/hexane (50:50), the hormones were eluted with 6 ml of dichloromethane/methanol (80:20) and then the samples were dried again with N^2 at 60 °C. According to the modified method by Zheng et al. (2008), 20 ml of hormones samples that were kept in the glass vial (Supelco, Belle-fonte, PA, USA) were extracted and purified using the Carbograph and Florisil cartridges. After the purification procedure for estrogenic hormones, the extracts were re-dissolved using dichloromethane/methanol (80:20) and evaporated again followed by separating eluents for each GC/MS analysis followed by derivatization of the extracts.

GC/MS analysis

A Hewlett-Packard (HP) 6890 GC in tandem with a Waters AutoSpec Ultima High Resolution mass spectrometer equipped with a Rtx-5MS column (30 m × 0.25 mm i.d. × 0.25-μm film thickness, Restek, Bellefonte, PA) was used for identification and quantitation of steroid hormones. The GC conditions were 1.0 mL/min carrier gas flow rate (He), 290°C inlet temperature, and 290°C interface temperature. The initial oven temperature was 80°C for 2 min; then the temperature was increased to 230 °C at 10 °C /min and held for 3 minutes, and then increased to 280°C at 10°C /min and held for 5 minutes.

Mass spectrometric analysis was carried out using both full scan and selected ion monitoring (SIM) modes. In the full scan mode, the electron impact (EI) mass spectra were generated using an electron energy of 70 eV and ions with m/z 200–900 were monitored. The full scan mode was used for the identification of analytes by fragmentation patterns and retention time as compared to known standards. For quantitation of hormones in all sample extracts, SIM mode was used. The following quantitation ions were used in the SIM: m/z = 664.1295 for 17α-estradiol and 17β-estradiol, m/z= 449.1352 for estriol, m/z = 466.1379 for estrone. For every hormones analysis, five-point calibration curves (10 - 10000 μg/L) were derived and then analyzed with the internal standard (Hexachlorobenzene). To gauge fluctuation, sample carry over, or possible internal contamination of the instrument, one solvent blank and one standard spike were analyzed between every 15 injections. Spiking 50μg of hormone solution in 1 L DI water determined the recovery. 71%-113% represents the recovery rates of the three steroid hormones in DI water. For most hormones, the relative standard deviation (RSD) was less than 5%.

1.3 Effects of the algal bioreactor on the removal of estrogenic hormones

Mixed algal-bacterial bioreactor

A mixed algal-bacterial bioreactor (MABB) and conventional activated sludge (CAS) reactor were operated in a sequencing batch mode to capture microconstituents and other organics into biomass for use as bioenergy feedstock. LPAM was fed into the bioreactors as a feedstock after diluting it to adjust the sCOD based on the specific organic loading rate. The total volume of each reactor was 50 gallons, and the settling and reaction time for each reactor were 1 and 47 hours, respectively. The aeration rate of each reactor was 6 L/min (0.03 vvm: L air/L reactor volume/min) for the MABB process and 11 L/min (0.06 vvm) for the CAS process to make the dissolved oxygen (DO) the same for both reactors.

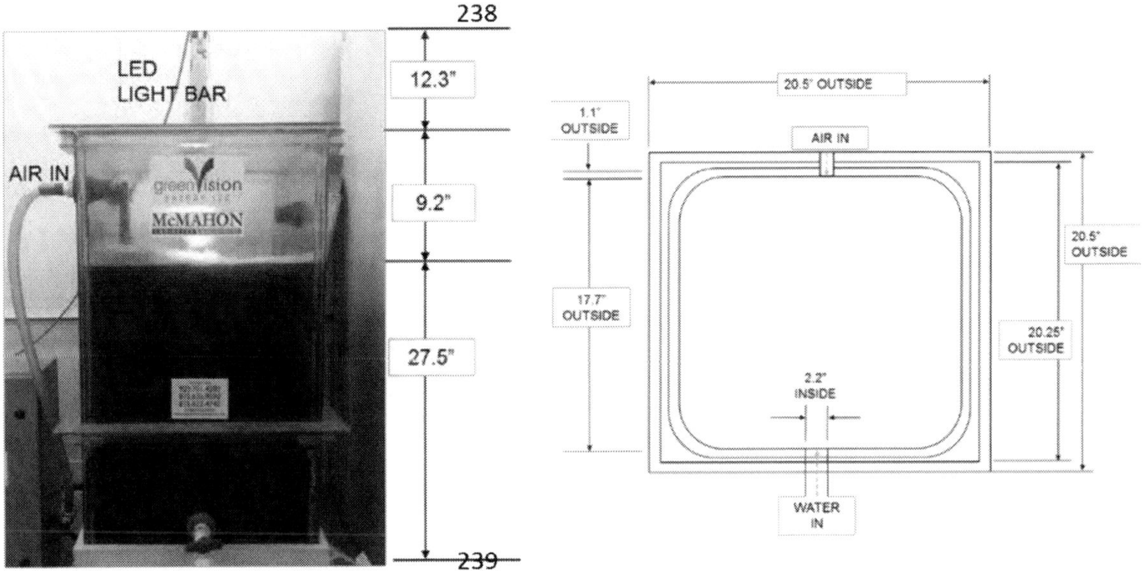

Figure. 3. Schematic diagram of mixed algal-bacterial bioreactor (MABB)

For the MABB operation, a light bar with red and blue LED lights was used to support photosynthetic algal growth, and the maximum light intensity at a 2" distance from the LED was 320 μmoles photons/m^2/s. The HRT and SRT of the reactor were 4 and 25-30 days. The uptake of dissolved organics and nutrients by microalgae and bacteria in the reactor was analyzed. The aeration flow rate was set to 6 L/min (0.03 *vvm*), which is lower than the aeration rate (1.2 *vvm*) for the maximum growth rate of *Spirulina* (Ronda et al., 2012). The other operating conditions of each reactor are listed in Table. 2.

In this study, the effluent water quality of each reactor was analyzed to compare the reactor performances of MABB and CAS, including the pollutant removal and biomass productivity. Also, the abilities to remove toxic compounds were compared by analyzing the cytotoxicity of the inputs and outputs from each reactor.

Table 2

Operating conditions for mixed algal-bacterial bioreactor

	Mixed Algal-Bacterial Bioreactor	**Conventional Activated Sludge reactor**
Reactor type	Sequencing Batch Reactor	Sequencing Batch Reactor
Total Volume (gal)	50	50
Light intensity (μ-photons/m^2/s)	350	-
Temperature (°C)	18	16
Aeration rate (L/min)	6 (0.03 vvm)	11 (0.058 vvm)
Organic Loading Rate (mg/L/d)	46.5 - 152	46.5 – 152
HRT (day)	1- 4	1- 4
SRT (day)	25 – 30	25 – 30
Fill volume ratio (V_F/V_T, %)	50	50

Granular activated carbon (GAC) addition

To investigate the effects of GAC on the ability to extract dissolved organics and estrogenic hormones from LPAM for bioenergy production, the removal of organics and estrogenic hormones in each bioreactor were monitored in one cycle of the sequencing batch operation. Filtrasorb®400 granular activated carbon (Calgon Carbon, Pittsburgh, PA) were screened using a mesh size No. 50 at 0.297 mm to achieve the desired average particle size of the GAC and washed several times using deionized water to remove dust and particles followed by drying at room temperature (AWWA. 2012). The cleaned GAC were packed into mesh bags and the GAC bags were vertically installed in the bioreactor to allow the airflow to provide contact between wastewater and GAC as the wastewater circulates. The treated LPAM samples from the bioreactors with and without GAC were collected to investigate the water quality, residual estrogenic hormones, and cytotoxicity.

Removal of estrogenic hormones with and without GAC addition

The LPAM was spiked with high and low concentrations of hormones (198.7 μg/L and 3 μg/L) from a stock solution including E1, E2, and E3, to investigate the consistency and eligibility of the data for developing a dynamic process model to describe the fate and transport of microconstituents in the bioreactors using the STELLA software platform. After spiking the reactors, the samples were taken regularly using glass containers for one cycle of operation time. All of the samples were centrifuged at 2,500 rpm for 15 min to minimize Carbograph SPE column plugging by removing the suspended solids, and the residual hormones in the liquid fraction were concentrated by Solid Phase Extraction based on the procedure of *2.2.3.1* to eliminate the matrix effects (Hanselman et al., 2003; Singh et al., 2013; Zheng et al., 2008). The concentrates were stored at -20°C for further analysis. The concentrations of E1, E2, E3, and EE2 in the liquid fraction were analyzed using GC/MS.

Effects of GAC on the fate of estrogenic hormones

According to Rossner et al. (2009), activated carbon is a more effective adsorbent for removing estrogenic hormone mixtures (≥ 98%) from lake water than zeolite and ion-exchange resin, due to its larger volume of pores. In addition, particular activated carbon (PAC) removed more than 90% of BPA, E2, and EE2 from drinking-water matrices (Yoon et al., 2003). Furthermore, previous studies have shown that sorption is the major mechanism for removing estrogenic hormones including E1 in an algal-based wastewater treatment system, and it is absorbed well by various strains of algae (Shi et al., 2010). Also, activated carbon (AC) was the most effective adsorbent for extracting hormones from the aqueous phase compared to adsorbents such as zeolite and ion-exchange resin (Rossner et al., 2009). In this test, GAC was used due to the ease of separating it from the bulk liquid. The enhancing effects of GAC on the extraction of hormones and organics from LPAM to create feedstock for bioenergy production were investigated in MABB and CAS. The hormones in the treated LPAM were analyzed based on the method in 2.3.1.4, and water quality parameters were monitored to investigate the effects of GAC on the performance of the bioreactors under steady-state operation.

Feedstock for Hydrothermal liquefaction and Catalytic hydrothermal gasification

To study the fate of hormones during hydrothermal liquefaction (HTL) and catalytic hydrothermal gasification (CHG) processes, artificially spiked biomass (E1, E2, and E3, 3 mg/L for each compounds) was used in this study. Wet mixed biomass was collected using a flat-sheet Kubota microfiltration membrane (0.4 μm, cartridge type 203) from mixed algal-bacterial bioreactor and stored at 4 °C in a cold room before processing. The solids content and volatile solids percentage of the homogenized biomass after a high shear mixer was 19.3 ± 0.2 % and 81.9 ± 0.04%, respectively. Detailed biomass preparation methods were explained in *3.2.2.3*.

Before the hydrothermal liquefaction and catalytic gasification, the concentrations of E1, E2, E3 and EE2 in the biomass and HTL-WW were analyzed using the analytical method for solid and liquid fraction explained in *2.2.3.2*. Before the hydrothermal liquefaction and catalytic gasification, the concentrations of E1, E2, E3 and EE2 in the biomass and HTL-WW were analyzed using the analytical method for solid and liquid fraction explained in *2.2.3.2*.

Table. 3 displays the residual concentrations of E1, E2, E3, and EE2 in the MABB biomass and HTL-WW, produced from HTL under 300°C/60min. Residual concentrations of natural estrogenic hormones in the biomass ranged from 19.1 to 377.8 μg/L with serial extraction with methanol based on the method in *2.2.3.1*. After HTL of biomass under 300°C/60min, the residual concentrations of hormones ranged from 0.3 to 21.4 μg/L in the HTL-WW. Total and volatile solids content of the biomass was 19.3 ± 0.2% and 81.9 ± 0.04%, respectively.

Table 3
Residual concentrations of hormones and solids content before and after HTL (300°C/60 min)

	E1	**E2**	**E3**	**EE2**	**Total**	**Total Solids (%)**	**Volatile solids (% of Total solids)**
Biomass (μg/L)	375.2 ± 2.5	351.5 ± 52.4	195.5 ± 17.9	19.1 ± 0.1	933.8 ± 18.6	19.3 ± 0.2	81.9 ± 0.04
HTL-WW (μg/L)	3.8 ± 0.1	10.3 ± 0.9	21.2 ± 0.3	0.3 ± 0.03	35.5 ± 1.0	2.5 – 3%	1.5 – 2.5%

1.4 Effects of the hydrothermal liquefaction (HTL) on the removal of estrogenic hormones

The HTL conversion of biomass cultivated from LPAM were conducted covering different combinations of temperature (200 - 400°C) and reaction time (60 min) without the addition of catalyst to define the optimal condition for bioenergy production and hormones removal (Table. 4), which were used in later modeling tasks. The major products of hydrothermal biomass conversion included bio crude oil, solid residues, gases, and aqueous products that was used to analyze the residual hormone concentration after the reaction. All HTL treatments were investigated using a small tubular batch reactor assembled with parts from Swagelok. The reactor was made of 3/4 inch OD tube (0.095-inch wall thickness) sealed with a cap and a needle valve with necessary fittings. Tubing was cut to length for 40 ml working volume; approximately 5 ml extra dead volume was attributed to fittings and valve. 10g (total weight) of biomass was loaded in the home-made Swagelok tubular reactor, then the reactor was purged with nitrogen for three times, and finally sealed at 120 psi. Then, the reactor was placed inside a preheated furnace at designated temperature (Azadi & Farnood. 2011; Bagnoud-Velasquez et al., 2014; Elliott. 2008; Pham et al., 2013; Schmieder et al., 2000). After the desired reaction time elapsed, the reactor was submerged into a bucket of water at room temperature to cool down for at least 3 minutes, and, after cooling, the excessive water outside the reactor was blown off using compressed air. Then the reactor was opened and the content was carefully collected after it reached room temperature.

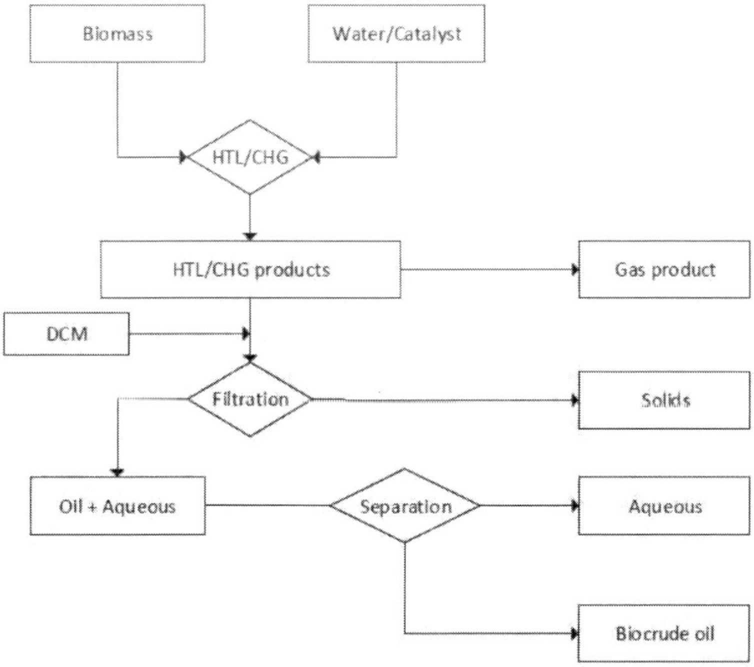

Figure. 4. product separation for hydrothermal liquefaction (HTL) and catalytic hydrothermal gasification (CHG)

As shown in Figure. 4, solid, aqueous, and oil products were mixed with 10ml dichloromethane (DCM) in the reactor body which then was placed on a shaker at 1000 rpm for 10 seconds after HTL reaction. Then the contents were transferred to vacuum filter (Whatman No. 4 Filter Paper) for separation. Reactor body was further rinsed with DCM in 10ml portion and shaken until the pour-out was clear. All content after rinsing was transferred onto the same filter. After filtration is done, the solids were left on the filter, aqueous and oil (in DCM) were collected in the flask. Aqueous and oil (in DCM) were separated by separation funnel, and they were transferred to sealed vials separately.

Table 4

Operating conditions for hydrothermal liquefaction (HTL) and catalytic hydrothermal gasification (CHG)

		Hydrothermal Liquefaction (HTL) & Catalytic Hydrothermal Gasification (CHG)						
Temperature (°C)		**250**	**300**	**350**	**400**	**450**	**500**	**550**
HTL	**60**	250/60	300/60	350/60	400/60	NA	NA	NA
CHG	**60**	NA	NA	350/60	400/60	450/60	500/60	550/60

To check the organics removal and break down of the hormones under the HTL tests, four different temperatures (250, 300, 350, and 400 °C) at 60 min were applied to HTL tests with spiked biomass. Table. 4 shows that the combinations of HTL operating conditions to test the effects of HTL on the fate of estrogenic hormones such as E1, E2, E3, and EE2 in the mixed biomass.

1.5 Effects of the catalytic hydrothermal gasification (CHG) on the removal of estrogenic hormones

To investigate the effects of different CHG operating conditions on the fate estrogenic hormones, CHG experiments of biomass were conducted under 16 different combinations of operating conditions including reaction temperature (350 - 550°C), reaction time. Based on our preliminary tests of oil yield under various operating conditions, 350°C and 60min condition was the optimal condition for highest oil yield and selected as a general operating condition to determine the different hormones removal rate with other variables.

Direct biomass gasification tests were conducted using the same 40 ml Swagelok batch reactor. MABB biomass and ruthenium on alumina (Ru) catalyst were premixed with the concentration of 50 mg Ru/g biomass before being added into the reactor. The reactor was loaded with the 10g of homogenized biomass and 0.5 mL of the stock solution (E1, E2, and E3; 1μg/ml in methanol). 10 mL of pure DI water instead of biomass without stock solution served as the blank control. After loading the spiked feedstock, the reactor was purged 3 times with pure nitrogen and pressurized at 120 psi to keep consistency with HTL runs. The furnace (Barnstead Thermolyme Co.) was preheated and maintained at the designated temperature for each reaction. Finally, the reactor was placed into the preheated furnace for the desired reaction temperature. Four reactors were simultaneously run under various operating conditions of the same temperature as shown in Table. 4, and the destruction of estrogenic hormones were recorded for each variation.

Afterwards, a water bath at room temperature was prepared to quickly cool down the reactors. The gas volume was measured by the pressure reading from the gauge attachment and a portion of the gas was kept in a gas sample vial for further analysis. All the samples collected from each operating condition of CHG were prepared for hormone analysis within 2 hours after being deposited according to the analytical methods in *2.8* to obtain accurate data of hormones before occurring the transformation.

1.6 Chemical and biological characterization of inputs and outputs

CHO cell mammalian cytotoxicity

Organic compounds in LPAM and municipal wastewater samples were extracted on XAD-2 and XAD-8 resins (Richardson 2011). The organic extracts were tested for cytotoxicity effect using the CHO chronic cytotoxicity assays. Details on extraction and assays are below. The XAD resins were prepared by consecutively washed with 0.1 N NaOH, distil water, and methanol for 30 min to extract the potential toxic agents from the LPAM. After another washing the resins with methanol, ethyl acetate, and methanol for 24 h, 100 mL of XAD-2 and XAD-8 resin were packed into a chromatography column (i.d. × length: 28 mm × 400 mm with a 1000 mL reservoir) in order. The column was continuously rinsed with 300 mL of 0.1 N NaOH, 100 mL of 0.1 N HCl, and ultrapure water. Prior to the extraction, LPAM samples were to a pH <1 using 12 N HCl and slowly passed through the packed resin beds and allowed to drain completely. Organic compounds that adsorbed to the resin were then eluted with 400 mL of ethyl acetate. The ethyl acetate was collected in a separator funnel, and the bottom layer was discarded. The remaining ethyl acetate was passed through organic free sodium sulfate to remove water and concentrated to 1mL with the aid of a RotoVap concentrator. The extracts were transferred to 1 mL conical vials and stored at –20°C before processing.

To analyze the chronic cytotoxicity, reduction in cell density on flat-bottom tissue culture 96-well microplates was measured as a function of a concentrated processed wastewater (CPW) sample over a period of approximately 3-4 cell divisions (72 h) (Plewa et al., 2009; Plewa et al., 2010). 4 replicate wells were prepared for each concentration of a specific organic extract for each CPW sample. The experiments were repeated at least 2 times, and a regression analysis was conducted for each concentration-response curve. The LC_{50} value is the concentration of the wastewater sample that induced a cell density of 50% as compared to the concurrent negative controls.

Heavy metals

Samples that were clear and contained no particulates were filtered through a 0.45 m filter prior to analysis. For solid samples and aqueous samples containing particulate, a sub-sample was microwave digested into solution by the United States Environmental Protection Agency (U.S. EPA) Method 3051A (U.S. EPA 2007a). Chromium, manganese, zinc, arsenic, rubidium, molybdenum, calcium, and lead measurements were performed with a VG Elemental PQ Excel ICP-MS per U.S. EPA Method 6020A (U.S. EPA, 1998). Sodium, calcium, potassium, and iron measurements were performed with a Varian Spectra AA 55B atomic absorption instrument per

U.S. EPA Method 7000B (U.S. EPA 2007b). Both instruments were calibrated using reference materials procured from SpexCertiprep and internal standards were used during the ICP-MS analysis. In addition, a certified reference material (Dogfish Liver Certified Material for Trace Metals, DOLT-3), a duplicate sample, and an analytical sample spike were performed during both assays as quality control parameters. The results of the reagent blank indicate no contamination was introduced during sample preparations.

Antibiotic resistance assay

To investigate the possible biological effects of antibiotics in manure, a fluctuation test was developed to measure antibiotic resistance in bacterial populations exposed to antibiotics. For the initial proof of concept, *E. coli* 15597 was exposed to florfenicol at a concentration that provides a selective advantage to individual bacterium in the population. *E. coli* 15597 was exposed to florfenicol at a concentration that provides a selective advantage to individual bacterium in the population. Because the antibiotic can kill or inactivate cells which lack the favorable genotype, cells with resistant genotypes have a reproductive advantage and gradually dominate the population. *E. coli* grown to the stationary phase in the Luria-Bertani (LB) medium with non-lethal antibiotics (2 mg/L) became resistant to normally lethal concentrations of antibiotics (17 mg/L). The "selective concentration" was a concentration of florfenicol that reduced the growth rate, measured as a change in optical density at a wavelength of 600 nm (OD_{600}) of an initial inoculation of *E. coli* in LB medium + florfenicol as compared to the same number of bacteria inoculated in LB only, and "lethal" concentration was a concentration of antibiotic in LB that eliminated growth. The results of this experiment showed that bacteria grown in 3 consecutive overnight cultures of LB + 2 mg/L florfenicol were more likely to produce growth in lethal (17 mg/L) concentrations of florfenicol. This test was applied to LPAM to define the baseline of antibiotic resistance. Subsequently, this test was applied to treated LPAM, processed by the MABB to define the effect of this bioprocess on the potential of LPAM to foster the development of antibiotic resistance. The positive and negative control of MABB effluent were compared to identify the antibiotic resistance generated by florfenicol.

RESULTS AND DISCUSSION

2.1 Effects of the algal bioreactor on the removal of estrogenic hormones

LPAM was fed into the MABB and CAS to remove the estrogenic hormones and produce biomass for bioenergy production. The concentrations of E1, E2, E3, and EE2 in the feedstock and the effluent of each bioreactor were analyzed to calculate the percent removal of each hormone before and after the bioreactor operations. The main removal mechanism of estrogenic hormones in the bioreactors was the adsorption onto biomass because hormones are hydrophobic compounds and there was some lag phase for the biological degradation by algae or bacteria in the bioreactor (i.e. 70 hours lag phase of E3-sulfate for hydrolysis by fecal coliform) (D'Ascenzo et al., 2003; Vos. 1996; Zhang & Henion. 1999).

To quantify the effects of GAC on the removal of estrogenic hormones, the percent removal of estrogenic hormones was investigated for each bioreactor under various operating conditions, including the addition of GAC as well as low and high spikes of hormones. The average percent removal of total hormones in the bioreactors with and without GAC were ranged from 85.5 to 95.0 %, and the total percent removal of each operating condition as follows:

MABB w/ GAC (95.0) > CAS w/ GAC (92.2) > MABB w/o GAC (85.8) > CAS w/o GAC (85.5)

These results state that more than 85.5% of estrogenic hormones in average can be removed in the bioreactors, and MABB is more effective at hormone removal than CAS regardless of GAC addition because the higher temperature and LED light in MABB could contribute to enhance the hormones adsorption to biomass and photochemical degradation of hormones by light energy (He et al., 2012; Lin & Reinhard. 2005; Puma et al., 2010; Whidbey et al., 2012).

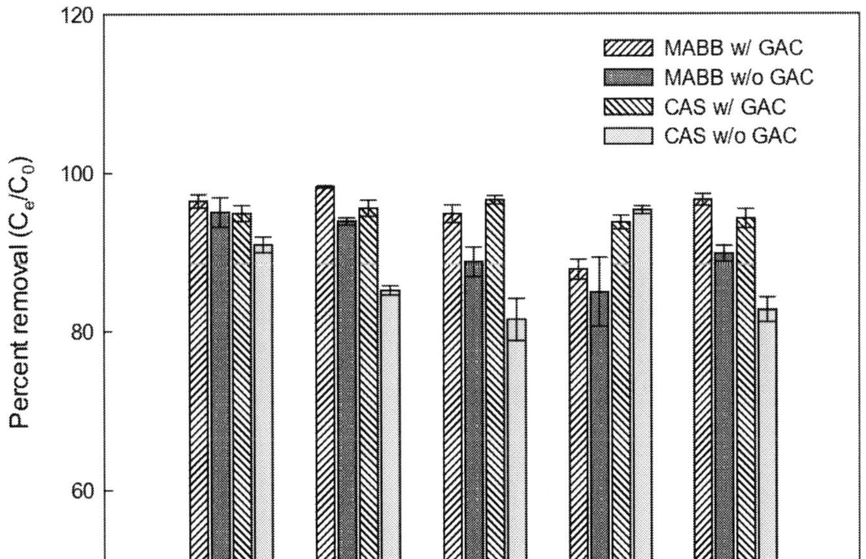

Figure. 5. Percent removal of estrogenic hormones in the MABB and CAS under the condition of w/ and w/o GAC

In addition, Figure. 5 showed that GAC increased the removal of total hormones up to 10.9 % and 8.1 % in MABB and CAS, respectively, due to additional adsorption of hormones to GAC (Westerhoff et al., 2005; Yoon et al., 2003). These results state that GAC addition can enhance the removal of estrogenic hormones in the bioreactors, and GAC is more synergistic with MABB than

CAS in removing hormones. In addition, removal of E2, E3, and EE2 were more sensitive to the addition of GAC than E1 in each reactor because the increased percent removal of E2 and E3 in the bioreactors were up to 1.9 and 2.0 times higher than that of E1.

2.2 Effects of the hydrothermal liquefaction (HTL) on the removal of estrogenic hormones

Figure. 6 showed that the results of preliminary experiments about the yield of biocrude oil and solid residue under HTL of biomass. The highest oil production was at 300°C and 60 min reaction time, where the bio crude oil yield reached at 39% (dry basis), and solid residue yield at 15% (dry basis). Little liquefaction happened under 200°C, as the oil yield was essentially the extract from the biomass itself. With the reaction temperature increased, the yield of bio crude oil increased steadily till 300°C, and then decreased when the temperature was at 350°C.

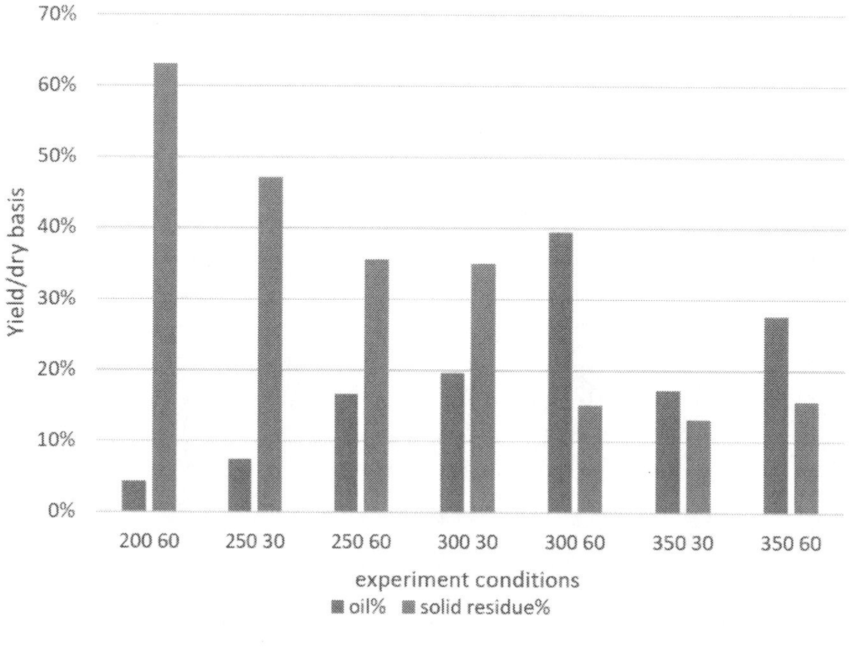

Figure. 6. Biocrude oil yield and solid residue of biomass under different reaction conditions

To investigate the effects of temperature on the removal and break down of the hormones, four different temperatures (250, 300, 350, and 400 °C) at a reaction time of 60 min were applied to HTL tests with spiked biomass. Table. 3 displays the residual concentrations (E1, E2, E3, and EE2) in the HTL-WW. Percent removal of E1, E2, E3, EE2, and total in the spiked biomass at 300 °C/60 min was 93.4 ± 0.8, 98.2 ± 0.7, 82.0 ± 1.8, 98.4 ± 4.1, and 96.5 ± 3.2%, which was higher than percent removal of total hormones at any other conditions. Therefore, total percent removal of hormones in the spiked biomass after HTL tests as follows: 300/60 > 350/60 > 400/60 > 250 °C

/60 min. Overall percent removal of E3 was up to 87.9%, which is lower than the percent removal of other compounds because the E2, EE2, and E1 hormones can be transformed to the E3 hormone in the reaction because E3 was one of main metabolites during estrogen degradation under anaerobic condition based on the previous study (Zheng et al., 2012).

In conclusion, the most optimal operating conditions for HTL processes to provide simultaneous bioenergy production and removal of hormones was 300°C for 60 min which results in up to 95.43% removal of total estrogenic hormones (Figure. 7). Percent removal of E3 was lower than the percent removal of any other compounds due to increased concentration by transformation from other hormones. Higher temperatures at 60 min for HTL tests provide more effective removal of E1 and E3. However, the optimal operating conditions of HTL for bioenergy production provided the highest removal of estrogenic hormones simultaneously.

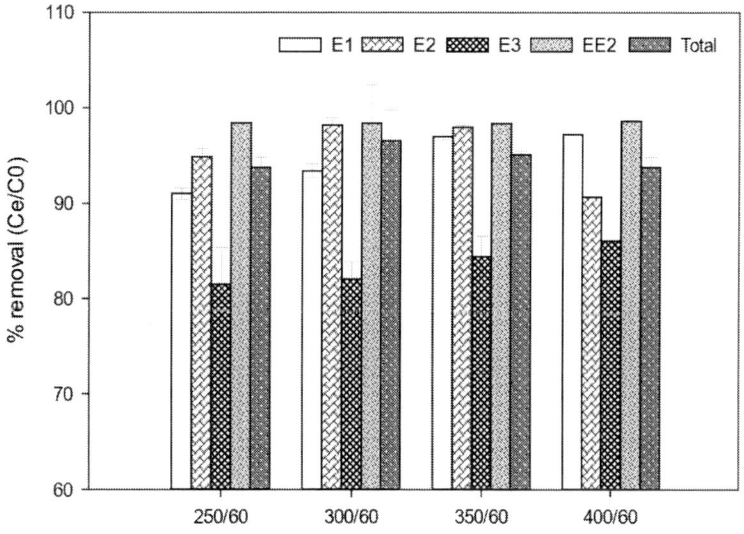

Figure. 7. Effects of different reaction temperature on the removal of E1, E2, E3 and EE2 in HTL tests (n≥1)

2.3 Effects of the catalytic hydrothermal gasification (CHG) on the removal of estrogenic hormones

Figure. 8 shows that percent removal of E1, E2, and E3 ranged from 99.87 to 99.99 under higher temperature than 450°C in CHG tests under 60min/Ru, which means that most of the main estrogenic hormones can be removed at the CHG of biomass.

Also, percent removal of EE2 and total hormones increased from 82.59 to 96.52% and 98.47 to 99.96% when temperatures were increased from 350 to 550°C in CHG under 60min/Ru. In conclusion, most of the E1, E2 and E3 hormones can be removed at temperatures higher than

450 °C, and removal of EE2 and total hormones are proportional to the temperature in the CHG of biomass under 60min/Ru. Removal of estrogenic hormones is sensitive to temperature changes in CHG.

If the temperature is above the critical point (temperature: 374°C, pressure: 22.1 MPa) in CHG, the physical properties of water changes significantly, and it can have good transport properties that can break down hydrocarbons to produce gases such as H_2, CO, CO_2, and CH_4 (Krishnan et al., 2016). This phyco-chemical reaction could contribute to the transformation and break down of estrogenic hormones. Therefore, temperatures above the critical point can increase the percent removal of estrogenic hormones such as E1, E2, E3, and EE2 in CHG tests.

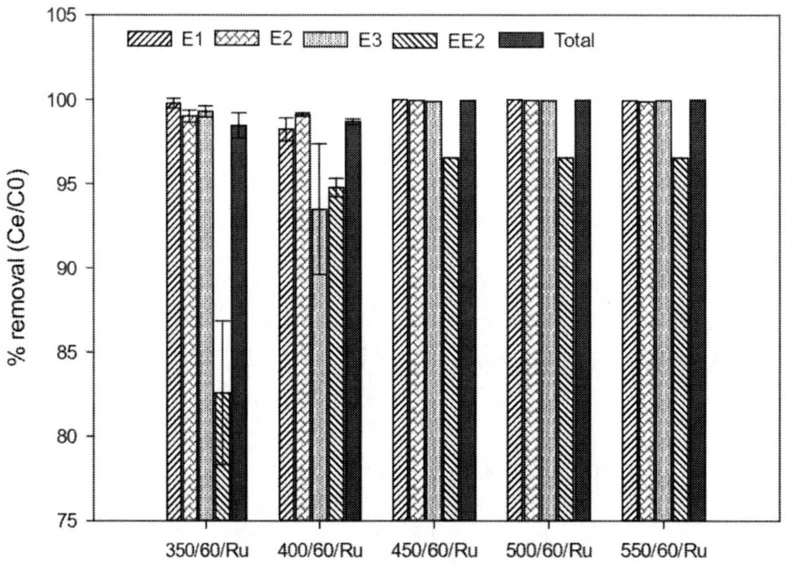

Figure. 8. Effects of reaction temperature on the fate of estrogenic hormones in CHG (n≥1)

2.4 Chemical and biological characterization of inputs and outputs

Cho cell mammalian cytotoxicity

The CHO cytotoxicity analyses were conducted to quantify the removal of toxicity after MABB and CAS with and without GAC addition. The soluble chemical oxygen demand (sCOD) of each sample was used to compare the relative cytotoxicity among the samples. Without spiking the microconstituents to reactors, cytotoxicity in the effluent from MABB without GAC was decreased when its compared to LPAM feedstock, which means toxic compounds in LPAM could be removed in the mixed algal-bacterial bioreactor. However, the cytotoxicity in the effluent from MABB/CAS was increased after spiking bioreactors with antibiotics. This results tells us that relatively increased concentration of nitrogenous organic compounds in the effluents from bioreactors could be the reason of cytotoxicity because most of the nitrifying bacteria were killed by antibiotics which resulted in the sharp increase of nitrogen concentrations. Thus, addition of GAC into bioreactors decreased cytotoxicity in the effluent from each MABB and CAS bioreactor, and the value of cytotoxicity as follows:

MABB with GAC < CAS with GAC < MABB without GAC < CAS without GAC

These results described that GAC was synergistic with MABB and CAS to remove the cytotoxicity of LPAM by the adsorption of toxic compounds or stabilized performance of each bioreactor. Figure. 9 have shown that MABB was more effective to remove toxic compounds in LPAM than CAS due to lower cytotoxicity in the effluent of MABB than CAS.

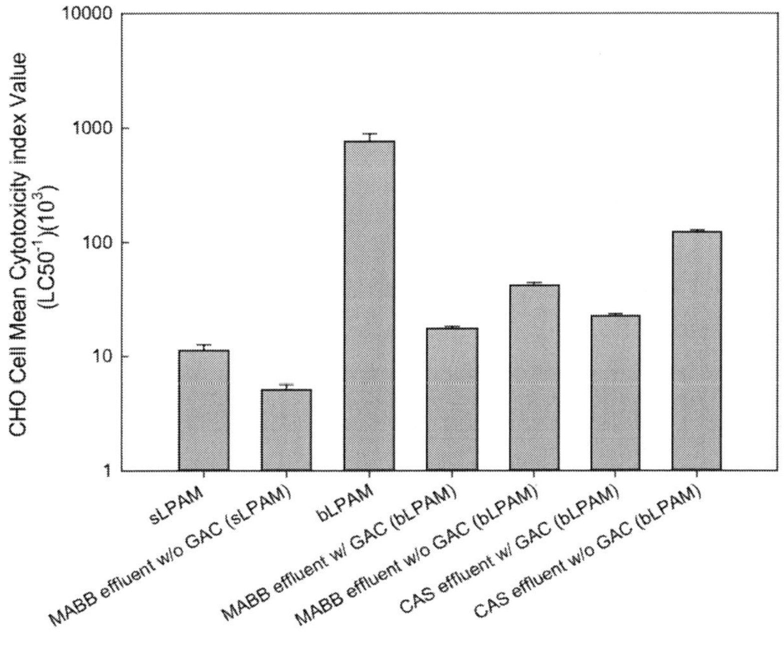

Figure. 9. CHO cell Cytotoxicity Index Values (CTI) of inputs and outputs of bioreactors

The cytotoxicity of the HTL-WW under three conditions (sample I, II, and III: 250°C/60min, 300°C/60min, and 350°C/60min) was investigated using a CHO cell assay. LC 50 and mean cytotoxicity index (CTI) values of sample I, II, and III were 0.275 and 3710.2, 0.413 and 2380.8, 0.186 and 5373.3, respectively. This analysis allows us to have conclusions as follows: 1. Each HTL-WW sample induced concentration dependent cytotoxicity in mammalian cells. 2. The descending cytotoxicity potency was Sample III > Sample I > Sample II. 3. Toxicity in HTL-WW is proportional to the temperature of HTL (250° and 300°C < 350°C). 4. Modifications in HTL may affect the resulting toxicity of the output product.

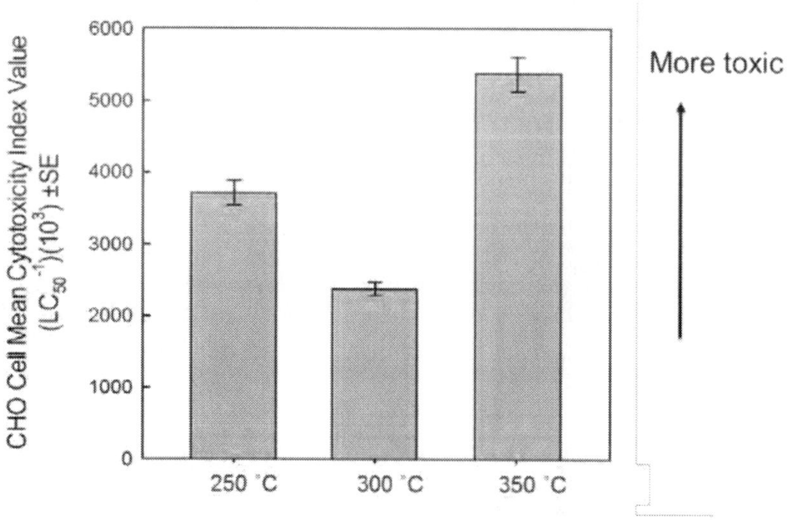

Figure. 10. CHO cell cytotoxicity index values for each HTL-WW sample.

The CHG-WW samples showed the lowest cytotoxic concentration that are the lowest concentration of the CPW sample in the concentration-response curve that induced a statistically significant reduction in cell density as compared to the concurrent negative controls. According to Fig 4.21, cytotoxicity in CHG-WW were positively related to increasing temperature from 300°C to 400°C, which was the highest cytotoxicity in CHG with biomass. However, cytotoxicity was sharply decreased at 450°C because types of soluble organics that can make toxicity could be changed with completed conversion of biomass to syngas under higher operation temperature.

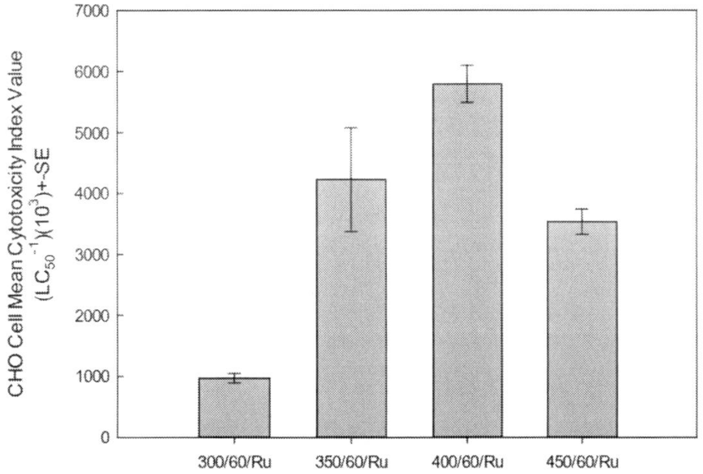

Figure. 11. CHO cell cytotoxicity index values for each CHG-WW sample.

Heavy metals

Excess amounts of metals are sometimes added to animal feeds to promote growth. For example, copper (Cu) and zinc (Zn) are typically used in swine diets to promote growth (Parker. 2007). Ayers et al. (1985) reported that the maximum recommended concentration of arsenic (As), Cu, and Zn in irrigation water is 0.1, 0.2, and 2.0 mg/L, respectively. Figure. 12 and 13 provided illustrative concentrations of heavy metals that were measured in the inputs and outputs of MABB with and without GAC addition. Effluents of MABB significantly lower than the recommended As, Cu, and Zn concentrations for irrigation water or water reuse, and all the samples from bioreactors with and without GAC addition contained from 94.3 to 97.8% of heavy metals in LPAM. Also, average 5.6 and 2.2 % of heavy metals in LPAM were concentrated into biomass during the mixed algal-bacterial bioreactor with and without GAC. This results meant that GAC enhanced the percent removal of heavy metals from LPAM by increasing biomass production along with the stabilized performance of bioreactor with the removal of recalcitrant compounds. Thus, Pb, Cd, Ni, Cr, and Zn could be removed additionally by the chemical adsorption of active carbon (Karnib et al., 2014). According to McBride. (1995), increased risk of phytotoxicity, and contamination of surface and ground water were possible adverse effects of high levels of these metals in the environment. Mixed algal – bacterial bioreactor with GAC addition was suggestive method to remove heavy metals from LPAM by concentrating them into biomass for bioenergy production.

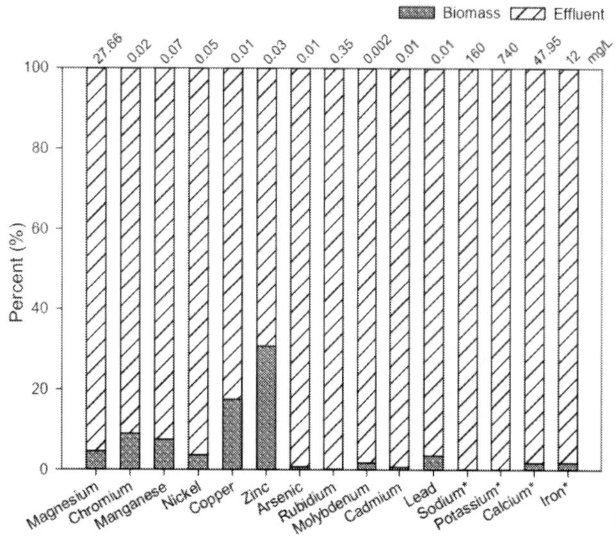

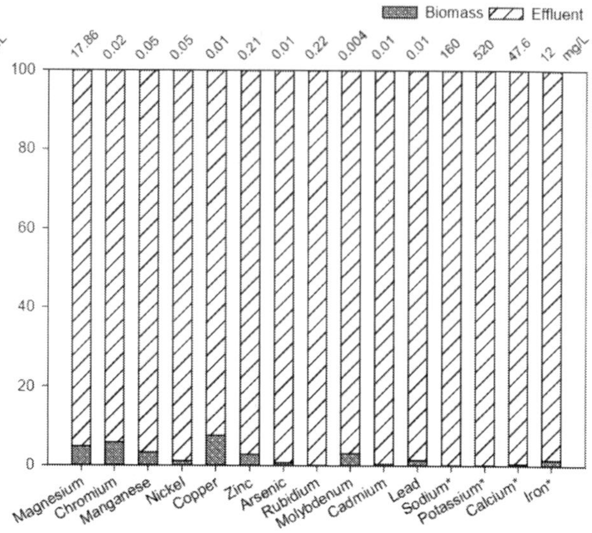

593

(Left) **Figure. 12.** Distribution of heavy metals in MABB with GAC (Right) **Figure. 13.** Distribution of heavy metals in MABB without GAC

Fig. 14 showed illustrative concentrations of heavy metals that were measured after the HTL of biomass feedstock. Based on the results of heavy metals in the biomass feedstock in Fig. 12 and Fig. 13, the percent distribution to the outputs of HTL such as aqueous phase, oil phase, and residue were calculated. Almost all the metal compounds are distributed to residue and HTL-WW up to 69.9% and 29.5%, respectively, which meant that there is still considerable amount of heavy metals left in the aqueous phase. Thus, more than 20% of Cu, As, Ru, Mo, Na, and K were left in the HTL-WW, but this could be a bottleneck to reuse the HTL-WW as irrigation or potential water source because As could produce toxicological effects on human health and environmental safety (Brathwaite & Rabone. 1985; Zhou et al., 2015). Also, because (Ayers et al., 1985). reported that the recommended maximum concentration of As for irrigation water was 0.1 mg/L, the fate and transport As in the integrated manure management system has to be characterized.

Figure. 15 provided the descriptive concentration of heavy metals in the aqueous, residue, and oil after the CHG of biomass feedstock. The weight of wet biomass feedstock was 5g with 80% of water content and the weight distribution of aqueous, oil, and residue after the CHG of biomass were 77.2, 20.6, and 2.2 %, respectively. The percent distributions of metal compounds in the outputs of CHG with biomass were calculated based on the original concentration in the biomass and weight distribution of aqueous, oil, and residue. The percent distribution of Cr, Cd, Pb, and As in the aqueous phase were ranged from 17.8 to 45.7% of heavy metals in the biomass feedstock. However, McBride. (1995) reported that hazardous metal compounds such as Pb and As could provide adverse and toxicological effects, and these metal compounds are needed to be monitored even though the concentration of Pb and As were lower than the suggestive maximum concentration for livestock drinking water, 0.2 and 0.1 mg/L (Ayers et al., 1985).

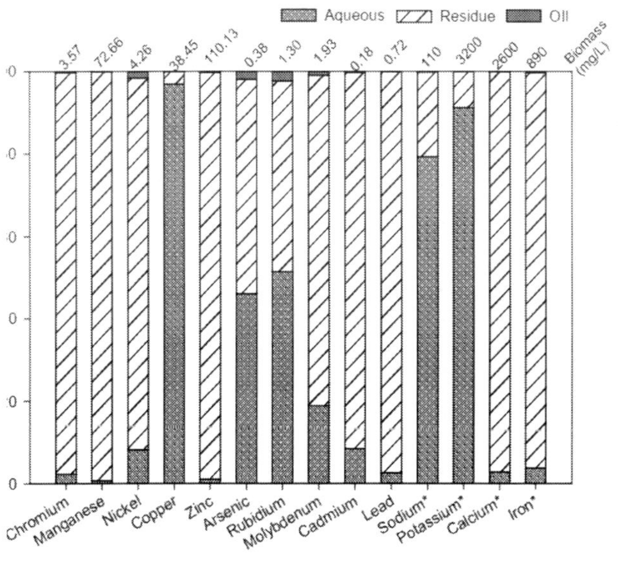

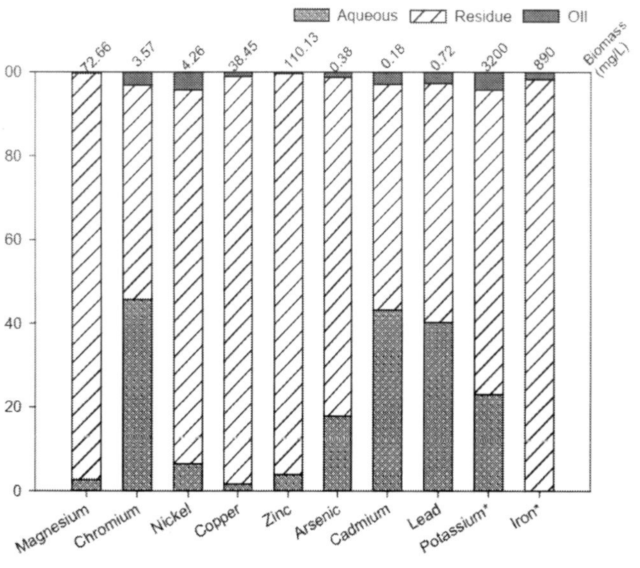

(Left) **Figure. 14.** Effects of different reaction time on the removal of E1, E2, E3 and EE2 in HTL tests (n≥2) (Right) **Figure. 15.** Effects of different reaction time on the removal of E1, E2, E3 and EE2 in HTL tests (n≥2)

Antibiotic resistance assay

A new analytical method was developed for the antibiotic resistance by florfenicol. As show in Figure. 16, exposure of *E.coli* to florfenicol at a non-lethal concentration (2 mg/L) contributed to increase the resistance to lethal concentration of florfenicol (17 mg/L), which is the antibiotic resistance to florfenicol under LPAM+LB medium. Antibiotic resistance assay was sensitive and could detect selective pressure induced by 2 mg/L florfenicol. Thus, when the negative and positive control of LB medium and MABB effluent were compared in Figure. 17, the strong resistance was observed in the negative control in LB medium and MABB effluent. These results described that three different types of medium such as LB, LPAM, and treated LPAM (MABB effluent), having different background matrix, were allowed to detect the response of antibiotic resistance generated by non-lethal concentration of florfenicol.

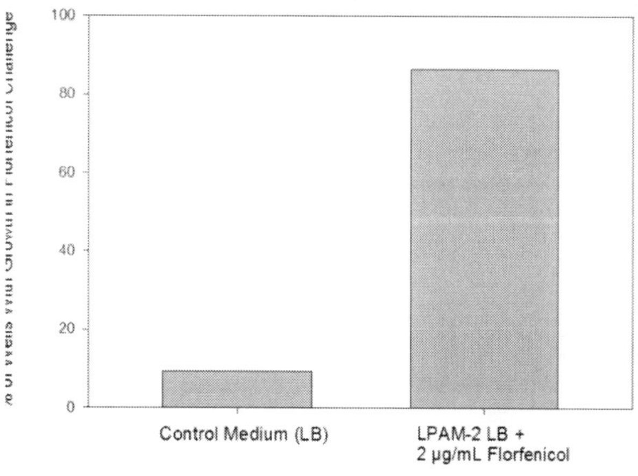

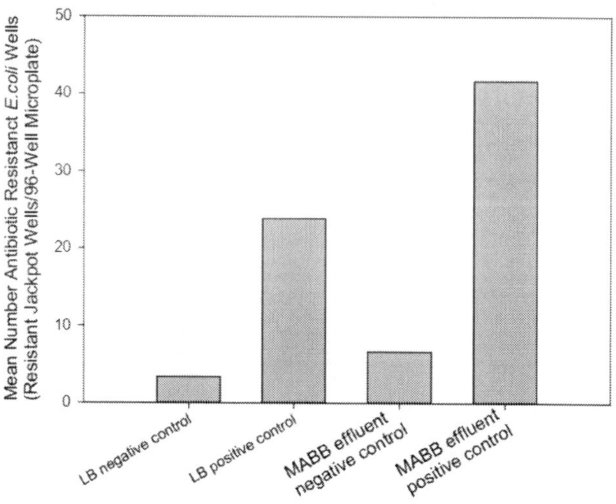

(Left) **Figure. 16.** Culturing *E.coli* in LPAM medium with 2 mg/L florfenicol to generate antibiotic resistance to lethal concentration (17 mg/L) (Right) **Figure. 17.** Antibiotic resistance assay under LB medium and MABB effluent with positive and negative control

To determine if HTL and CHG can reduce the capacity of bacterial cells to generate antibiotic resistance, antibiotic resistant bacteria (ARB) assay tests were conducted with HTL-WW and CHG-WW. Artificially spiked biomass and HTL-WW were processed by HTL and CHG, respectively, and then survival rates were measured to determine if the antibiotics in the selection phase conferred antibiotic resistance after culturing bacteria in each medium. Fig. 20 demonstrated that the ARB Assay is sensitive and can detect the selective pressure as illustrated in negative control (Sample 1) versus positive control data of LB (Sample 2). In conclusion, ARB assay is sensitive to measure antibiotic resistance, and the HTL and CHG processing could eliminate the capacity of Samples (Sample 3 to 8) to generate FF-resistant bacteria.

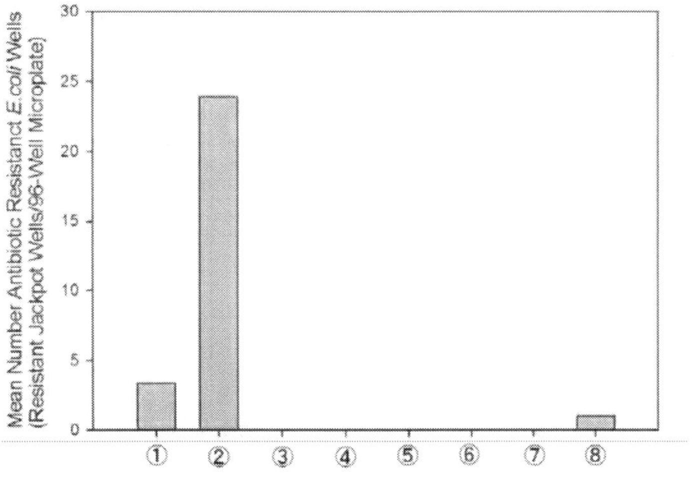

Figure. 18. The results of the ARB fluctuation assay of HTL/CHG samples generated low or no antibiotic resistant jackpot wells in the microplates.

1) LB negative control
2) LB positive control
3) HTL-WW with MABB biomass
4) CHG-WW with MABB biomass
5) Sequential HTL/CHG-WW with MABB biomass
6) HTL-WW with CAS biomass
7) CHG-WW with CAS biomass
8) Sequential HTL/CHG-WW with CAS biomass

CONCLUSIONS

MABB showed higher hormone removal (85.8%) due to light deformation and higher temperature compared to CAS (85.5%). Thus, GAC addition contributed to enhance the removal of hormones from 92.2 to 95% in MABB because of extra adsorption of hormones onto activated carbon. In the HTL tests, the reaction time and temperature were main parameters that can affect HTL performances and removal of hormones. 300°C/60min was the optimum condition that showed high removal of hormones, toxicity, and high oil yield. Hormones removal and oil yield were positively related to operating temperature in HTL. Also, hormone removal in CHG tests was correlated to operating temperature positively, and stabilized to 99.9% higher than 450 °C.

In the MABB, 94.3 - 97.8% of total heavy metals were left in the effluents and the others were accumulated in the mixed algal-bacterial biomass in MABB. In case of HTL tests, heavy metals were mainly distributed to residue (69.9%) and HTL-WW (29.5%), but As left in HTL-WW which is higher than the suggested concentration for irrigation water (0.1 mg/L) (Brathwaite & Rabone. 1985; Zhou et al., 2015). Thus, heavy metals after CHG tests were mainly distributed to residue (79.3) and CHG-WW (18.5%), but more than 20% of As and Pb were left in CHG-WW.

Cytotoxicity was decreased in the effluent from MABB and CAS compared to influent LPAM. Also, GAC contribute to decrease the cytotoxicity in the effluent from bioreactors because GAC addition improved water quality of bioreactors and enhanced removal of hormones and toxic compounds. Each HTL-WW sample induced concentration dependent cytotoxicity in mammalian cells and the optimum operating condition of HTL for the biocrude oil yield, 300°C/60min, showed the lowest cytotoxicity in HTL-WW due to complete conversion of biomass to oil (250° and 300°C < 350°C). In CHG tests, cytotoxicity was positively related to temperature, but sharply decreased higher than 450°C because the types of soluble organics that produce cytotoxicity could be changed with higher operating temperature. In conclusion, modifications in HTL and CHG may affect the resulting toxicity of the output product.

ARB assay was sensitive and could detect the selective pressure in negative control versus positive control data of LB that meant ARB assay was sensitive to measure antibiotic resistance. Thus, HTL and CHG processing could eliminate the capacity to generate FF-resistant bacteria.

ACKNOWLEDGMENTS

This project was supported by Agricultural and Food Research Initiative Competitive Grant no. 11332987 from the USDA National Institute of Food and Agriculture. The department of Agricultural and Biological Engineering at University of Illinois at Urbana-Champaign provided experimental facilities and support for setting up the reactors. Authors also thank John Scott for his analytical support.

REFERENCES

AWWA. 2012. AWWA B604 - 12 Granular Activated Carbon, American Water Works Association, pp. 56.

Ayers, R.S., Westcot, D.W., Food, Agriculture Organization of the United, N. 1985. *Water quality for agriculture*. Food and Agriculture Organization of the United Nations, Rome :.

Azadi, P., Farnood, R., 2011. Review of heterogeneous catalysts for sub- and supercritical water gasification of biomass and wastes. **36**(16), 9529-9541.

Bagnoud-Velasquez, M., Brandenberger, M., Vogel, F., Ludwig, C., 2014. Continuous catalytic hydrothermal gasification of algal biomass and case study on toxicity of aluminum as a step toward effluents recycling. **223**, 35-43.

Bradford, S.A., Segal, E., Zheng, W., Wang, Q., Hutchins, S.R., 2008. Reuse of concentrated animal feeding operation wastewater on agricultural lands. **37**(5), S97-S115.

Brathwaite, R.L., Rabone, S.D.C., 1985. Heavy metal sulphide deposits and geochemical surveys for heavy metals in New Zealand. **15**(4), 363-370.

D'Ascenzo, G., Di Corcia, A., Gentili, A., Mancini, R., Mastropasqua, R., Nazzari, M., Samperi, R., 2003. Fate of natural estrogen conjugates in municipal sewage transport and treatment facilities. **302**(1-3), 199-209.

Elliott, D.C., 2008. Catalytic hydrothermal gasification of biomass. **2**(3), 254-265.

Fine, D.D., Breidenbach, G.P., Price, T.L., Hutchins, S.R., 2003. Quantitation of estrogens in ground water and swine lagoon samples using solid-phase extraction, pentafluorobenzyl/trimethylsilyl derivatizations and gas chromatography-negative ion chemical ionization tandem mass spectrometry. **1017**(1-2), 167-185.

Finlay-Moore, O., Hartel, P.G., Cabrera, M.L., 2000. 17 beta-estradiol and testosterone in soil and runoff from grasslands amended with broiler litter. **29**(5), 1604-1611.

Hanselman, T.A., Graetz, D.A., Wilkie, A.C., 2003. Manure-borne estrogens as potential environmental contaminants: A review. Environ. Sci. Technol. **37**(24), 5471-5478.

Hanselman, T.A., Graetz, D.A., Wilkie, A.C., 2004. Comparison of three enzyme immunoassays for measuring 17 beta-estradiol in flushed dairy manure wastewater. **33**(5), 1919-1923.

He, Y.Q., Xu, G., Tang, L., Qian, D.Y., Ren, L.T., Hu, G.Y., Lei, J.Q., Wu, M.H. 2012. Research on Photolysis of Steroid Estrogens in Aquatic System. in: *Materials for Environmental Protection and Energy Application, Pts 1 and 2*, (Ed.) D. Wang, Vol. 343-344, pp. 241-245.

Hutchins, S.R., White, M.V., Hudson, F.M., Fine, D.D., 2007. Analysis of lagoon samples from different concentrated animal feeding operations for estrogens and estrogen conjugates. Environ. Sci. Technol. **41**(3), 738-744.

Ifelebuegu, A.O., 2011. The fate and behavior of selected endocrine disrupting chemicals in full scale wastewater and sludge treatment unit processes. Int. J. Environ. Sci. Tech. **8**(2), 245-254.

Irwin, L.K., Gray, S., Oberdorster, E., 2001. Vitellogenin induction in painted turtle, Chrysemys picta, as a biomarker of exposure to environmental levels of estradiol. **55**(1-2), 49-60.

Jobling, S., Nolan, M., Tyler, C.R., Brighty, G., Sumpter, J.P., 1998. Widespread sexual disruption in wild fish. Environ. Sci. Technol. **32**(17), 2498-2506.

Karnib, M., Kabbani, A., Holail, H., Olama, Z. 2014. Heavy Metals Removal Using Activated Carbon, Silica and Silica Activated Carbon Composite. in: *Technologies and Materials for Renewable Energy, Environment and Sustainability*, (Eds.) C. Salame, M. Aillerie, P. Papageorgas, Vol. 50, pp. 113-120.

Khanal, S.K., Xie, B., Thompson, M.L., Sung, S., Ong, S.-K., Van Leeuwen, J., 2006. Fate, transport, and biodegradation of natural estrogens in the environment and engineered systems. Environ. Sci. Technol. **40**(21), 6537-6546.

Kolodziej, E.P., Harter, T., Sedlak, D.L., 2004. Dairy wastewater, aquaculture, and spawning fish as sources of steroid hormones in the aquatic environment. **38**(23), 6377-6384.

Krishnan, V., Thachanan, S., Matsumura, Y., Uemura, Y., 2016. Supercritical Water Gasification on Three Types of Microalgae in the Presence and Absence of Catalyst and Salt. **148**, 594-599.

Lange, I.G., Daxenberger, A., Schiffer, B., Witters, H., Ibarreta, D., Meyer, H.H.D., 2002. Sex hormones originating from different livestock production systems: fate and potential disrupting activity in the environment. Anal. Chim. Acta. **473**(1-2), 27-37.

Lin, A.Y.C., Reinhard, M., 2005. Photodegradation of common environmental pharmaceuticals and estrogens in river water. **24**(6), 1303-1309.

McBride, M.B., 1995. TOXIC METAL ACCUMULATION FROM AGRICULTURAL USE OF SLUDGE - ARE USEPA REGULATIONS PROTECTIVE. **24**(1), 5-18.

Nzihou, A., Stanmore, B., 2013. The fate of heavy metals during combustion and gasification of contaminated biomass-A brief review. **256**, 56-66.

Parker, G., G. Cromwell, V. Hays, and J. McKean. 2007. *Feed additives for swine. Report Nutrition-6 (PIH-31), Pork Industry Handbook.*, Oklahoma State Univ., Stillwater.

Pham, M., Schideman, L., Scott, J., Rajagopalan, N., Plewa, M.J., 2013. Chemical and Biological Characterization of Wastewater Generated from Hydrothermal Liquefaction of Spirulina. **47**(4), 2131-2138.

Plewa, M.J., Wagner, E.D., Foundation, W.R. 2009. *Mammalian Cell Cytotoxicity and Genotoxicity of Disinfection By-products*. Water Research Foundation.

Plewa, M.J., Simmons, J.E., Richardson, S.D., Wagner, E.D., 2010. Mammalian Cell Cytotoxicity and Genotoxicity of the Haloacetic Acids, A Major Class of Drinking Water Disinfection By-Products. Environ. Mol. Mutagen. **51**(8-9), 871-878.

Puma, G.L., Puddu, V., Tsang, H.K., Gora, A., Toepfer, B., 2010. Photocatalytic oxidation of multicomponent mixtures of estrogens (estrone (El), 17 beta-estradiol (E2), 17 alpha-ethynylestradiol (EE2) and estriol (E3)) under UVA and UVC radiation: Photon absorption, quantum yields and rate constants independent of photon absorption. **99**(3-4), 388-397.

Raman, D.R., Williams, E.L., Layton, A.C., Burns, R.T., Easter, J.P., Daugherty, A.S., Mullen, M.D., Sayler, G.S., 2004. Estrogen content of dairy and swine wastes. Environ. Sci. Technol. **38**(13), 3567-3573.

Ronda, S.R., Bokka, C.S., Ketineni, C., Rijal, B., Allu, P.R., 2012. AERATION EFFECT ON SPIRULINA PLATENSIS GROWTH AND gamma-LINOLENIC ACID PRODUCTION. **43**(1), 12-20.

Rossner, A., Snyder, S.A., Knappe, D.R.U., 2009. Removal of emerging contaminants of concern by alternative adsorbents. **43**(15), 3787-3796.

Routledge, E.J., Sheahan, D., Desbrow, C., Brighty, G.C., Waldock, M., Sumpter, J.P., 1998. Identification of estrogenic chemicals in STW effluent. 2. In vivo responses in trout and roach. **32**(11), 1559-1565.

Schmieder, H., Abeln, J., Boukis, N., Dinjus, E., Kruse, A., Kluth, M., Petrich, G., Sadri, E., Schacht, M., 2000. Hydrothermal gasification of biomass and organic wastes. **17**(2), 145-153.

Schuh, M.C., Casey, F.X.M., Hakk, H., DeSutter, T.M., Richards, K.G., Khan, E., Oduor, P.G., 2011. Effects of field-manure applications on stratified 17β-estradiol concentrations. J. Hazard. Mater. **192**(2), 748-752.

Shi, W., Wang, L., Rousseau, D.P.L., Lens, P.N.L., 2010. Removal of estrone, 17 alpha-ethinylestradiol, and 17-estradiol in algae and duckweed-based wastewater treatment systems. Environ Sci Pollut Res. **17**(4), 824-833.

Shore, L.S., Gurevitz, M., Shemesh, M., 1993. ESTROGEN AS AN ENVIRONMENTAL-POLLUTANT. **51**(3), 361-366.

Singh, A.K., Gupta, S., Kumar, K., Gupta, S., Chander, Y., Gupta, A., Saxena, R., 2013. Quantitative analysis of conjugated and free estrogens in swine manure: Solutions to overcome analytical problems due to matrix effects. **1305**, 203-212.

Thorpe, K.L., Cummings, R.I., Hutchinson, T.H., Scholze, M., Brighty, G., Sumpter, J.P., Tyler, C.R., 2003. Relative potencies and combination effects of steroidal estrogens in fish. **37**(6), 1142-1149.

U.S.EPA. 1998. Inductively Coupled Plasma—Mass Spectrometry. Method 6020A (SW-846). Revision 1, (Ed.) U.S.E.P. Agency, U.S. EPA. Washington, DC.

U.S.EPA. 2007a. Microwave Assisted Acid Digestion of Sediments, Sludges, Soils, and Oils. Method 3051A., (Ed.) U.S.E.P. Agency, U.S. EPA. Washington, DC.

U.S.EPA. 2007b. Flame Atomic Absorption Spectrophotometery. Method 7000B., (Ed.) U.S.E.P. Agency, U.S. EPA. Washington, DC.

Vos, E.A., 1996. Direct ELISA for estrone measurement in the feces of sows: Prospects for rapid, sow-side pregnancy diagnosis. **46**(2), 211-231.

Westerhoff, P., Yoon, Y., Snyder, S., Wert, E., 2005. Fate of endocrine-disruptor, pharmaceutical, and personal care product chemicals during simulated drinking water treatment processes. Environ. Sci. Technol. **39**(17), 6649-6663.

Whidbey, C.M., Daumit, K.E., Nguyen, T.H., Ashworth, D.D., Davis, J.C.C., Latch, D.E., 2012. Photochemical induced changes of in vitro estrogenic activity of steroid hormones. **46**(16), 5287-5296.

Yoon, Y.M., Westerhoff, P., Snyder, S.A., Esparza, M., 2003. HPLC-fluorescence detection and adsorption of bisphenol A, 17 beta-estradiol, and 17 alpha-ethynyl estradiol on powdered activated carbon. **37**(14), 3530-3537.

Zhang, H.W., Henion, J., 1999. Quantitative and qualitative determination of estrogen sulfates in human urine by liquid chromatography/tandem mass spectrometry using 96-well technology. **71**(18), 3955-3964.

Zhang, Y., Habteselassie, M.Y., Resurreccion, E.P., Mantripragada, V., Peng, S., Bauer, S., Colosi, L.M., 2014. Evaluating Removal of Steroid Estrogens by a Model Alga as a Possible Sustainability Benefit of Hypothetical Integrated Algae Cultivation and Wastewater Treatment Systems. ACS Sustain Chem Eng. **2**(11), 2544-2553.

Zheng, W., Yates, S.R., Bradford, S.A., 2008. Analysis of steroid hormones in a typical dairy waste disposal system. Environ. Sci. Technol. **42**(2), 530-535.

Zheng, W., Li, X.L., Yates, S.R., Bradford, S.A., 2012. Anaerobic Transformation Kinetics and Mechanism of Steroid Estrogenic Hormones in Dairy Lagoon Water. Environ. Sci. Technol. **46**(10), 5471-5478.

Zhou, Y., Xu, Y.B., Xu, J.X., Zhang, X.H., Xu, S.H., Du, Q.P., 2015. Combined Toxic Effects of Heavy Metals and Antibiotics on a Pseudomonas fluorescens Strain ZY2 Isolated from Swine Wastewater. **16**(2), 2839-2850.

Can Autocatalytic Pyrolysis of Wastewater Biosolids be Energy Neutral and Generate Value-Added Products?

Zhongzhe Liu[1], Simcha Singer[2], Daniel Zitomer[1], and Patrick McNamara[1*]

[1]Department of Civil, Construction and Environmental Engineering, Marquette University, P.O. Box 1881, Milwaukee, Wisconsin 53201, USA
[2]Department of Mechanical Engineering, Marquette University, P.O. Box 1881, Milwaukee, Wisconsin 53201, USA
[*]Corresponding author email: patrick.mcnamara@marquette.edu

Abstract

Over eight million metric dry tons of wastewater biosolids are produced annually from water resource recovery facilities (WRRFs) in the United States. WRRFs are facing multiple challenges including energy generation, nutrient recovery, and pollutant removal requirements. Autocatalytic pyrolysis of wastewater biosolids is a promising process to improve energy generation and product recovery while removing pollutants. In particular, it can generate biochar (a valuable soil amendment) as well as high-yield pyrolysis gas, which is a renewable clean energy to complement the energy requirement at WRRFs. An energy analysis of autocatalytic pyrolysis was conducted with the assumption of 10% heat loss and using the highest catalyst loading. The results showed that the autocatalytic pyrolysis at 600°C was slightly exothermic but was endothermic at 700°C and 800°C. Therefore, some py-gas was used to supply the heat for 700°C and 800°C operation. However, the remaining fraction of py-gas still reduced the energy for biosolids drying by approximately 70%. At pyrolysis temperatures of 600°C, the remaining py-gas could cover 40% of the energy for biosolids drying. In summary, autocatalytic pyrolysis of wastewater biosolids can be energy neutral and generate the value-added product, biochar.

Keywords: Biochar; Energy content; Heat balance; Wastewater sludge; Catalyst; Bio-oil

Introduction

Over eight million metric dry tons of wastewater biosolids are produced annually from water resource recovery facilities (WRRFs) in the United States (Gude, 2015). WRRFs are facing multiple challenges including energy generation, nutrient recovery, and pollutant removal requirements (Eggen et al., 2014; Jhansi and Mishra, 2013; Mo and Zhang, 2013; Venkatesan et al., 2016). Pyrolysis could be a key technology to aid in energy generation and product recovery while removing pollutants.

Biosolids pyrolysis is a process that decomposes wastewater biosolids upon heating at elevated temperature under anoxic conditions. The process produces biochar, a valuable soil amendment, along with pyrolysis gas (py-gas), and bio-oil (pyrolysis condensate) (Bridle and Pritchard, 2004). Meanwhile, micropollutants such as triclosan and estrogens can be removed from biochar during pyrolysis (Hoffman et al., 2016; Ross et al., 2016). Moreover, pretreated biochar can be used as an adsorbent for nutrient and micropollutant removal from used water (Carey et al., 2015; Tong et al., 2016). A recent study concluded that biosolids pyrolysis could be a low waste solution to biosolids handling in Europe (Samolada and Zabaniotou, 2014). However, biosolids derived bio-oil normally accounts for approximately 40% (by mass) of the total products of biosolids pyrolysis (Inguanzo and Domınguez, 2002; McNamara et al., 2014), and the bio-oil requires costly upgrading before being used as a fuel due to its corrosive and unstable properties. Thus, autocatalytic biosolids pyrolysis (Figure 1) was developed by using biochar as the catalyst to convert the majority of bio-oil vapor to py-gas that needs no conditioning before use for energy recovery via combustion. Our previous work showed that using biochars such as biosolids derived biochar as a catalyst can significantly reduce bio-oil yield while increasing py-gas yield (Liu et al., 2016; Liu et al., 2016). Using our biosolids derived biochar catalyst, energy can shift from bio-oil to py-gas and the remaining condensate becomes cleaner with fewer high-molecular-weight hydrocarbons. The autocatalytic process has many potential benefits including maximizing energy generation, minimizing negative environmental impacts, and recovering nutrients.

To ensure that the autocatalytic process is sustainable, a complete energy balance analysis must be conducted. The main objective of this study was to evaluate the enthalpy of autocatalytic pyrolysis of dried biosolids. Also, the net energy in py-gas (i.e. the remaining energy after a portion of total energy in py-gas is used to heat the pyrolysis process) was compared to the heat required for biosolids drying in order to determine how much energy for drying can be saved by using the py-gas.

Methods

Dried biosolids (i.e., Milorganite®) were pyrolyzed with the introduction of biochar as a catalyst located downstream. The hot pyrolysis vapor exited the main pyrolysis reactor, passed through the hot catalyst, and was then separated by condensation. The biochar was obtained in the main reactor, and the bio-oil was collected in the condensers. The py-gas was collected in a Tedlar® bag for gas chromatography analysis. The pyrolysis temperature was varied from 600°C to 800°C. The catalyst temperature was maintained at the same value as the main pyrolysis reactor. The mass ratio of catalyst to feedstock ranged from 0.2 to 1. Control tests without catalyst addition were conducted with the downstream temperature fixed at 500°C. The energy content, higher heating value (HHV), of py-gas was calculated based on the gas composition and the corresponding component HHVs. The HHVs of biochar and bio-oil were determined by bomb calorimetry (Parr 1341, Parr Instrument Company, Moline, IL). The chemical energy (kJ) of the product per mass (kg) of biosolids pyrolyzed was calculated by multiplying the yield by the corresponding HHV.

The enthalpy of autocatalytic pyrolysis was calculated using a modified heat balance (Figure 2) based on our previous work (McNamara et al., 2016). The heat flow diagram shows that the input energy is the chemical energy of dried biosolids and the output energies are the chemical energies of biochar, bio-oil, py-gas, heat loss, and sensible and latent heat released when the products cooled down to the reference temperature of 25°C. 10% heat loss was assumed in this study. The sensible heat of py-gas was estimated based on the temperature-dependent heat capacity of each gas component. The sensible and latent heat of bio-oil was estimated according to the properties of petroleum with a similar specific gravity of 1. The enthalpy of autocatalytic pyrolysis was determined by the difference of output energy minus input energy. A negative difference meant exothermic and positive meant endothermic.

Results and Discussion

Autocatalytic biosolids pyrolysis increased the py-gas yield and decreased the bio-oil yield simultaneously (Liu et al., 2016). Meanwhile, the py-gas HHV increased and the bio-oil HHV decreased during autocatalysis compared to the control test without catalyst. In particular, when the catalyst to feedstock ratio increased to 1 at 800°C, the bio-oil energy content (HHV) decreased to its lowest value of 1,300 kJ/kg. Correspondingly, the energy contained in bio-oil decreased dramatically and more energy shifted from bio-oil to py-gas via autocatalysis. When 1 kg of biosolids was pyrolyzed at 800°C with the highest catalyst loading, the energy contained in the py-gas was 11,100 kJ while the energy in bio-oil was only 275 kJ.

At 600°C, 700°C, and 800°C, the total output energy for non-catalytic pyrolysis was lower than the input energy (biosolids input energy was 15,800 kJ), indicating these processes were slightly exothermic. However, for autocatalytic pyrolysis, at 700°C and 800°C with a catalyst to feedstock mass ratio of 1, the processes were endothermic because the total output energy was greater than the input energy.

Because autocatalytic pyrolysis is exothermic at 600°C, the energy in the py-gas can be totally used for biosolids drying. At 600°C with the highest catalyst loading, the reduction percentage of energy for biosolids drying was 40%. At 700°C and 800°C, even though a fraction of the py-gas was used to supply the heat for the endothermic autocatalytic pyrolysis, the remaining py-gas could reduce the energy for biosolids drying by approximately 70%.

Conclusions

Autocatalytic pyrolysis of wastewater biosolids can generate, biochar, a value-added product, as well as high-yield py-gas. Py-gas is a renewable clean energy source which can complement the energy requirement at WRRFs. Common micropollutants can be removed from biochar during pyrolysis. Biosolids derived biochar proved to be a cost-effective catalyst during autocatalytic pyrolysis to improve on-site energy recovery. An energy analysis was conducted with the assumption of 10% heat loss and at the highest catalyst loading. The results showed that the autocatalytic pyrolysis at 600°C was slightly exothermic but was endothermic at 700°C and 800°C. Therefore, py-gas was partially used to supply the heat for 700°C and 800°C operation. However, the remaining py-gas still reduced the energy for biosolids drying by approximately 70%. At 600°C, the remaining py-gas could cover 40% of the energy for biosolids drying. In summation, the autocatalytic pyrolysis of wastewater biosolids can be an energy neutral process that produces clean biochar.

Acknowledgments

This work was supported by the Milwaukee Metropolitan Sewerage District and Marquette University.

References

Bridle, T., Pritchard, D., 2004. Energy and nutrient recovery from sewage sludge via pyrolysis. Water Sci. Technol. 50, 169-175.

Carey, D.E., McNamara, P.J., Zitomer, D.H., 2015. Biochar from Pyrolysis of Biosolids for Nutrient Adsorption and Turfgrass Cultivation. Water Environ. Res. 87, 2098–106. doi:10.2175/106143015X14362865227391

Eggen, R.I.L., Hollender, J., Joss, A., Schärer, M., Stamm, C., 2014. Reducing the discharge of micropollutants in the aquatic environment: the benefits of upgrading wastewater treatment plants. Environ. Sci. Technol. 48, 7683–9. doi:10.1021/es500907n

Gude, V., 2015. Energy positive wastewater treatment and sludge management. Edorium J. Waste Management. 1, 10-15.

Hoffman, T.C., Zitomer, D.H., McNamara, P.J., 2016. Pyrolysis of Wastewater Biosolids Significantly Reduces Estrogenicity. J. Hazard. Mater. doi:10.1016/j.jhazmat.2016.05.088

Inguanzo, M., Domınguez, A., 2002. On the pyrolysis of sewage sludge: the influence of pyrolysis conditions on solid, liquid and gas fractions. J. Anal. Appl. Pyrol. 63, 209-222.

Jhansi, S.C., Mishra, S.K., 2013. Wastewater Treatment and Reuse: Sustainability Options. Cons. J. Sustain. Dev. 10, 1–15.

Liu, Z., McNamara, P., Zitomer, D., 2016. Biochar Production and Bio-oil Upgrading by Synergistic Catalytic Pyrolysis of Wastewater Biosolids and Industrial Wastes. Proc. Water Environ. Fed. 2016, 3182–3187. doi:10.2175/193864716819713682

McNamara, P., Koch, J., Zitomer, D., 2014. Pyrolysis of Wastewater Biosolids: Lab-Scale Experiments and Modeling. Proc. Water Environ. Fed. 2014, 1–14. doi:10.2175/193864714816196655

McNamara, P.J., Koch, J.D., Liu, Z., Zitomer, D.H., 2016. Pyrolysis of Dried Wastewater Biosolids Can Be Energy Positive. Water Environ. Res. 88, 804–810. doi:10.2175/106143016X14609975747441

Mo, W., Zhang, Q., 2013. Energy-nutrients-water nexus: integrated resource recovery in municipal wastewater treatment plants. J. Environ. Manage. 127, 255–67. doi:10.1016/j.jenvman.2013.05.007

Ross, J.J., Zitomer, D.H., Miller, T.R., Weirich, C.A., McNamara, P.J., 2016. Emerging investigators series: pyrolysis removes common microconstituents triclocarban, triclosan, and nonylphenol from biosolids. Environ. Sci. Water Res. Technol. 2, 282–289. doi:10.1039/C5EW00229J

Samolada, M.C., Zabaniotou, A.A., 2014. Comparative assessment of municipal sewage sludge incineration, gasification and pyrolysis for a sustainable sludge-to-energy management in Greece. Waste Manag. 34, 411–420. doi:10.1016/j.wasman.2013.11.003

Tong, Y., Mayer, B.K., McNamara, P.J., 2016. Triclosan adsorption using wastewater biosolids-derived biochar. Environ. Sci. Water Res. Technol. 2, 761–768. doi:10.1039/C6EW00127K

Venkatesan, A.K., Hamdan, A.-H.M., Chavez, V.M., Brown, J.D., Halden, R.U., 2016. Mass Balance Model for Sustainable Phosphorus Recovery in a US Wastewater Treatment Plant. J. Environ. Qual. 45, 84–9. doi:10.2134/jeq2014.11.0504

Zhongzhe Liu, Patrick McNamara, Daniel Zitomer, 2016. Product upgrading during biosolids pyrolysis by using a low-cost natural catalyst, in: Proceedings of the Water Environment Federation.

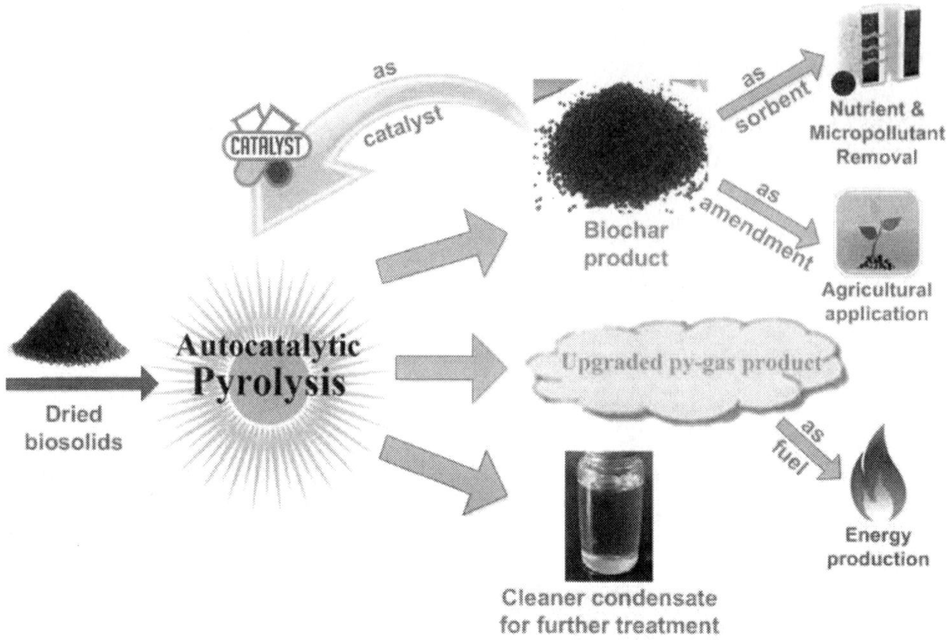

Figure 1. Autocatalytic pyrolysis process and its value-added products

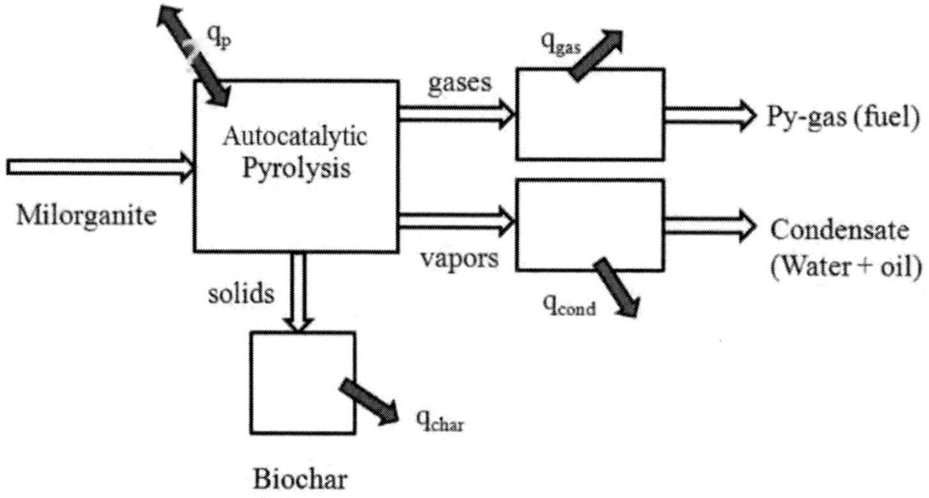

Figure 2. Relevant heat flows during autocatalytic pyrolysis of biosolids. Red arrow indicates uncertain heat required for pyrolysis and blue arrows indicate heat released from products (modified based on (McNamara et al., 2016)**)**

How the City of Lebanon TN Implemented Gasification for Biosolids Disposal and Power Generation

Introduction

The City of Lebanon, TN has completed construction of a waste-to-energy system sited at the wastewater treatment plant (WWTP). This project will not only create electricity, but will also utilize existing wood waste, WWTP sludge, and the county's discarded tires as fuel creating an ideal situation for reducing waste and producing renewable energy. The designed system is the integration of three well-established commercial technologies: 1) a biomass gasifier, 2) a thermal oxidizer/hot oil heat exchanger, and 3) Organic Rankine Cycle (ORC) power generator. The plant is designed for a 25-year operating life. The system design is fundamentally different from incineration since it uses gasification to reduce NOx, SOx, and fly ash. As such, the project design further supports the environmental goals of the Tennessee Department of Environment and Conservation since it first converts the biomass into a clean combustible gas and then combusts it in a thermal oxidizer yielding stack emissions similar to natural gas unlike traditional incineration-to-energy projects. Project revenues are derived from the avoided cost of purchased electricity and anaerobic digestion using ATAD and tip fees for processing wood waste and tires. As a result, the project is expected to provide the city a net savings of $300,000 annually, and $6 million during its operational lifetime through avoided costs and operational efficiencies while simultaneously meeting all selection criteria of the TDEC Energy Program and its broader goals.

The City of Lebanon's WWTP is supplied electricity by the Middle Tennessee Electric Cooperative (MTEC), a municipal distribution utility serving residential and commercial customers within the corporate limits of the City of Lebanon. MTEC is a separate entity and is not affiliated with the Lebanon WWTP. Lebanon WWTP is currently supplied electricity at an average cost of $.10/kW.

Equally important however, are the costs in managing and disposing of existing waste streams via traditional waste disposal methods. The City of Lebanon collects and surrounding industries contribute more than 27 tons of woody biomass per day that currently must be disposed of in a landfill. In addition, the city must also dispose of over 3 tons of biosolids per day from the wastewater treatment facility via land application. At current volumes, over 1840 trips a year are required to dispose of the material. As is explained below, the implementation of this project will consolidate these waste streams at the WWTP and eliminate the majority of disposal costs while adding substantial energy and environmental benefits that will contribute to the long term viability and continued growth of the City of Lebanon.

PROPOSED SOLUTION:

PHG Energy, a Tennessee based company, has proposed to design, engineer, install and commission a downdraft gasification-to-energy system that can convert approximately 32 tons per day of the wood, tires and sludge waste collected in the city into carbon neutral electricity for use at the WWTP.

Gasifiers, thermal oxidizers and ORC's are well-established commercial technologies individually. This project combines a state-of-the-art biomass gasification system with thermal oxidation equipment with ORC technologies to produce reliable electrical power with emission levels comparable to natural gas combustion. ORC power generators are often preferred over traditional steam cycle or internal-combustion engine (ICE) power generators because they can offer much lower operating and maintenance costs while operating without the need for dedicated attendance by an operator. Using gasification instead of incineration uses a fundamentally different thermochemical process to cleanly convert the city's waste streams into a combustible gas, called "producer gas," limiting the formation of NOx, SOx, and fly ash emissions inherent to an incineration process. Gasification offers the added benefit of feedstock flexibility; the same equipment can handle a wide variety of feedstock giving the city flexibility in handling its waste streams over the 25-year life of the plant. The PHG gasifier, from Antioch, TN-based PHG Energy, LLC, is an industrial-grade downdraft gasifier that converts biomass into a clean combustible producer gas. PHG gasifiers have over 45,000 hours of combined operation and run with relatively little need for operator interaction. The downdraft-style gasifier is well known to be the most efficient method of gasification and produces the cleanest gas from woody biomass.[1] The simplicity of the design enables the entire gasification process to occur with very few moving parts and relatively low parasitic loads.

Thermal oxidizers are the most common and most efficient technology employed today to destroy hydrocarbons in the form of Hazardous Air Pollutants (HAPs) and Volatile Organic Compounds (VOCs) and have been in use for more than 70 years. The principal behind thermal oxidation of HAPs and VOCs is that when combusted, the hydrocarbon pollutants will convert to CO_2 and H_2O in the outlet stream. The Thermal Oxidizer proposed for this project will include an attached hot water heater that will transfer the energy from the thermal oxidizer's exhaust flue stream into a closed loop circulating stream of hot water. The hot water stream will carry the heat energy into the ORC generator, where it will be used to drive the ORC's closed-loop power generation.

Hundreds of ORC's are operating in the field today from various manufacturers including General Electric and Pratt and Whitney. An ORC operates similarly to a steam-turbine generator, however, the ORC uses a working fluid inside an enclosed system that evaporates at a much lower temperature than water and is non-corroding to valves, tubing and turbine blades that drive the generator. Since the system is enclosed, there is no make-up fluid in the generator itself, and it does not require a boiler-certified operator to function. The ORC is simple to start and stop, operates very quietly and requires minimum operation and maintenance.[3]

Overall, the proposed system offers a combination of commercially proven technologies that are both robust in its design and flexible in their operation. The complete system will use landfill bound waste to provide a continuous stream of domestically produced energy that is clean, carbon-neutral, and renewable providing the city with both economic and long-term environmental benefits.

ENERGY EFFICIENCY

This project will produce measureable energy savings through onsite reductions in electrical energy purchases and offsite reductions in diesel fuel consumption through avoided trips to landfills. A generation meter and monitoring system will provide real time and accurate measurements of the amount of electrical energy specifically produced by the system. Since the system will be connected directly to the common electrical grid, it is assumed that the metered amount of electricity produced will result in a corresponding decrease in the amount of electricity purchased from the City of Lebanon's electrical utility supplier.

Currently, industrial and manufacturing waste within the city is hauled to a landfill located approximately 22 miles away from the source of the material, resulting in round trip distances of approximately 44 miles per trip. At current volumes, over 1680 round trips are needed annually to dispose of existing waste streams. By providing a central location within the corporate city limits for this material to be transported and converted into energy, thousands of truck miles will be saved. The effect of this project activity will be measurable through the identification of avoided individual tipping tickets at the currently utilized landfills.

Savings

Energy savings can be calculated and discussed in terms of both energy amount and dollar value. Assuming a conservative availability, over 748MW-hrs can be generated annually from the facility. This yields approximately $75,000/year in electrical energy savings. In addition, taking the ATAD offline will net an additional $175,000 annually in electricity savings.

Energy savings resulting from reduced transportation and combustion of diesel fuel needed to transport the waste streams to the gasification site is realized. Savings are calculated based on the transportation distance and the number of trips needed per year. Project implementation avoids approximately 1680 trips and more than 104,000 gallons of diesel fuel annually.

Finally, avoiding landfill tipping fees associated with the disposal of waste wood and sludge combined with the labor and fuel savings will save the approximately $248,000 per year. When combined with the electric savings, over $300,000 per year will be saved from all aspects of the project after expenses.

In addition to the saving realized above the City will be developing markets for the Biochar that will provide additional revenues for the project. Currently the City is able to give the Biochar away for free.

GENERAL PUBLIC BENEFIT

Immediate Results & Project Readiness

The site for the project is owned by the City of Lebanon and has been reviewed by the city council. Existing waste streams (fuel supplies) are already part of the City of Lebanon's existing services and infrastructure so there is no need to engage in lengthy and extensive fuel procurement negotiations. As the required technology is readily available and no long lead times are expected, the project can be fully operational within 9 to 12 months of grant approval. Once operational the project will produce immediate and meaningful beneficial results.

Long-Term Savings

The project will be designed, constructed and maintained for at least 25 years. The project can provide net savings to the citizens of Lebanon of more than $6 million over the life of the project. This estimate is considered conservative and is calculated as the sum of electrical savings and avoided costs associated with the transportation and disposal of existing waste.

Other Public Benefits

Since the wastewater treatment plant sludge will be gasified at the site of the Lebanon treatment facility, the need to transport the sludge or tires over highly traveled local highways will be eliminated. As a result, the risk of a spill will be greatly diminished ensuring that no environmental pollution due to an accident will occur. The same is true of the current wood waste and tire disposal methods.

Biochar is also an inert byproduct of the gasification system and can be used as a soil additive or mixed with compost to further enhance its economic value. Also independent companies seek and purchase biochar for reuse. This further eliminates any additional materials being sent to a landfill and opens opportunities for additional revenue streams.

Renewable Resources

Biomass comes in many forms and there is significant debate as to which forms should be considered renewable. Virgin timber and the use of materials from forestlands harvested unsustainably face considerable scrutiny when being evaluated as renewable. However, since the fuel to be used for the Lebanon project is derived from waste streams, it can be deemed renewable. As the city provides a service and maintains its streets and right of ways, the wood waste will continue to be produced without impacting carbon storage or contributing to deforestation. Wastewater sludge also may not normally be thought of as renewable, but as this stream will also to continue to be created everyday it too can be deemed as a renewable resource. The result is 100% renewable fueled power generation. The only fossil fuels used are to start up the thermal oxidizer and to maintain a pilot flame, which in totality are negligible and insignificant to the overall project implementation. No additional energy is needed for preparation of discarded tires as the nylon cords and steel belts do not have to be removed in order for the tire to be used as feedstock as they become part of the biochar.

PROTECTION OF ENVIRONMENT RESOURCES

Utilizing Existing Resources

The wood and wastewater sludge waste are expensive to dispose of and are not good for the air quality as presently disposed because it can convert to methane. The utilization of waste products for energy represents the highest value for otherwise low or no value materials. Innovation and utilization of existing resources has been a hallmark of Lebanon leadership and is consistent with their vision and future of their community. While it is true that many landfills currently utilize methane collection, capture and flare or energy generation technologies, the US EPA estimates that 60-90% of the methane is collected. This wasted energy represents a significant weakness in existing resource utilization strategies and presents a unique opportunity for forward-looking communities to control their own environmental and energy futures. Discarded tires are currently picked up and reused or recycled. Use of tires in the feedstock will further remove transport vehicles, labor costs and its emissions from the environment.

Conservation

Gasifying the wood waste, tires and sludge inside the city limits of Lebanon will result in a significant reduction in the use of coal and petroleum based products. Best practice energy and environmental strategies recommended by experts in the field consistently call for the use of efficiency and conservation strategies as the first line in forming a comprehensive plan for sustainable communities. The City of Lebanon hopes to become a leader and demonstrate in a concrete fashion to other communities in Tennessee and the nation how a conservation strategy aimed at utilizing existing resources can ultimately turn a liability into an asset.

Reduction in Carbon Intensity

Biomass, when gasified and used consistent with the proposed project design, will also emit CO_2. However, wood waste, tires and WWTP sludge are considered renewable and are accounted for as a short term biogenic carbon emissions. This proposal presents a significant opportunity to reduce the carbon intensity associated with the Lebanon WWTP. Reduction in the usage of diesel fuel used in the transportation of waste material to landfill and disposal sites also reduces the Lebanon WWTP's carbon intensity.

AIR QUALITY

Overall Emissions Reduction

The City of Lebanon is located within the Tennessee Valley Authority electric generation grid, specifically Sub region: SERC Tennessee Valley. This project will reduce the need for electricity to be provided from power plants located within the TVA service area through Lebanon Electric. Since TVA will not have to generate the 748,880 kWh provided locally by the Lebanon Waste Biomass Gasification to Energy Project, there will be a reduction of 9809 pounds of sulfur dioxide, 3719 pounds of NOx, and 2717 tons of carbon.

Non-Attainment Area Targeted

The most direct reduction in emissions from the offset of electric energy is assumed to occur in Lebanon/Wilson County. At present the wastewater sludge is disposed of by local land application. Wood waste is taken to a landfill in Rutherford County. Discarded tires are taken out of state. The elimination of transportation of these waste streams would further reduce GHG, Sulfur dioxide, VOC's, Nitrogen oxides and HAP's in Wilson and Rutherford counties.

ABILITY TO LEVERAGE (MATCH) FUNDING TO ENHANCE OVERALL PROJECT OBJECTIVES – The City of Lebanon has been granted a Qualified Energy Conservation Bond for this project. QEBC will pay 70% of the interest, a savings of $122,000 per year. Qualified Energy Conservation Bonds were first authorized by Congress in October 2008. At that time, Congress allowed a maximum of $800 million in QECB volume cap nationwide. In February 2009, through the American Recovery and Reinvestment Act, Congress increased the volume cap to $3.2 billion. These bonds may be issued by state, local and tribal governments to finance qualified energy conservation projects. A maximum of 30% of the aggregate bonds may be used to finance private activity projects.[9]

ESTIMATE OF REDUCTIONS AND/OR AIR EMISSIONS

This project is projected to result in over 2717 tons of CO_2, 3719 pounds of NOx, and 9809 pounds of SO_2 reduction since the fuel source is carbon neutral biomass using gasification and a thermal oxidizer to convert the biomass to heat. In addition, over 124 MW-hrs. of electricity from line losses will be saved since the power will be generated on-site.

INNOVATIVE TECHNOLOGY REDUCES MERCURY EMISSIONS TO KEEP OHIO'S SEWAGE SLUDGE INCINERATORS HOT AND COSTS COOL

The Road to SSI MACT Compliance for Cincinnati and Cleveland

Connor Smith (Black & Veatch), Scott Reed (Black & Veatch), Robin Rupe (Northeast Ohio Regional Sewer District), Sara Cramer (Metropolitan Sewer District of Greater Cincinnati)

ABSTRACT

Effective March 21, 2016, all Sewage Sludge Incinerators (SSIs) were required to be in compliance with the United States Environmental Protection Agency (USEPA) Maximum Achievable Control Technology (MACT) air pollutant emissions standards.

At the Northeast Ohio Regional Sewer District's (NEORSD) Southerly Wastewater Treatment Center (WWTC) and the Metropolitan Sewer District of Greater Cincinnati's (MSDGC) Mill Creek Wastewater Treatment Plant (WWTP), an analysis of historical feed-sludge mercury data from approximately 2012 to present, coupled with each facility's most recent incinerator emissions tests, confirmed that additional emissions controls were required for each of the two facility's combined six fluidized bed incinerators (FBIs) (three at each facility) to consistently bring the incinerators into compliance with the new regulations, more specifically with mercury emissions limits. Without the addition of enhanced emissions controls focused on mercury removal, each facility would likely not have been able to demonstrate consistent compliance with the new rule subsequent to the associated March 2016 regulatory compliance deadline, potentially resulting in the loss of permit to incinerate sludge. This loss of permit would have resulted in NEORSD and MSDGC needing to institute hauling programs to direct an average of 120+ dry tons per day (dtpd) of dewatered sludge cake from each facility to landfill. This operation was deemed by both institutions to not only be cost prohibitive, but also to be unsustainable.

Traditionally, fixed bed granular activated carbon (GAC) adsorbers have been used to reduce mercury emissions for incinerators operating existing wet scrubber emissions control systems like those at Southerly WWTC and Mill Creek WWTP. However, both NEORSD and MSDGC recognized the potential of a newer proprietary mercury removal technology, known as Sorbent Polymer Composite (SPC) absorption, which as part of several successful pilot studies was demonstrating significant mercury removal in emissions from municipal SSIs. When compared to large fixed bed GAC adsorbers, the SPC absorption vessels not only offered notable capital cost savings, but also required less ancillary equipment, leading to lower overall facility operations and maintenance (O&M) costs and greater ease of retrofitting the SPC vessel systems within the existing incinerator building space.

Following careful analysis, both NEORSD and MSDGC ultimately decided to install new SPC absorption vessel systems for each of their three respective FBIs. In early 2016, following installation of the new emissions controls systems, each of the SSIs underwent official SSI MACT compliance testing. This testing demonstrated that all six of the incinerators were in compliance with all regulated emissions limits, including mercury. Furthermore, all testing was complete prior to the March 21, 2016 regulatory compliance deadline, which was a significant accomplishment given the compressed time frame to analyze, design, install, and commission the

SPC vessel systems following publishing of the SPC technology pilot study results in 2014. Results and ongoing monitoring of the SPC technology's effectiveness are discussed herein.

KEYWORDS
Incineration, sewage sludge, fluidized bed, regulatory, air pollution control

INTRODUCTION

Regulatory Requirements

On March 21, 2011, the United States Environmental Protection Agency (USEPA) published *Title 40 Code of Federal Regulations (CFR) Part 60 – Standards of Performance for New Stationary Sources and Emissions Guidelines for Existing Sources: Sewage Sludge Incineration Units; Final Rule,* and thereby established more stringent air pollutant emissions standards for both pre-existing and planned future sewage sludge incinerators (SSIs) under a "model rule" (referred to herein as the SSI Maximum Achievable Control Technology (MACT) rule). In summary, the model rule stipulated that affected utilities were required to demonstrate compliance within a five year window (i.e. no later than March 21, 2016) or otherwise by an earlier date per an approved State Plan (refer to 40 CFR Part 60, §60.5245). In the State of Ohio, which did not submit a separate State Plan, March 21, 2016 was considered as the applicable final compliance deadline for all affected utilities operating SSIs.

Under the model rule, a distinction between SSIs that would be classified as "existing" versus "new" was included. This distinction was notable in that it represented a sharp difference in the level of compliance required for regulated air pollutants. In summary, an SSI classified as "new" would have stricter emissions limits when compared to an SSI classified as "existing," and would therefore likely require additional emissions control to maintain air pollutant emissions compliance. Requirements for classifying SSIs as either "existing" or "new" were defined as follows"

- **Existing SSIs:** Multiple hearth incinerators (MHIs) and fluidized bed incinerators (FBIs) where construction commenced on or before October 14, 2010.
- **Modified SSIs:** Existing MHIs and FBIs which undergo modifications later than September 21, 2011 and meet one or both of the following two criteria:
    - The cumulative cost of changes over the life of the SSI exceeds 50 percent (50%) of the original cost of building (construction the SSI) and installing the SSI (not including land cost) updated to current dollars. A summary of what items are within the boundary of the SSI and thus used to calculate these costs was initially provided under 40 CFR Part 60, §60.4775.
    - Any physical change in the SSI or change in the method of operating the SSI that increases the amount of any air pollutant emitted for which Section 129 of the Clean Air Act or Section 111 of the Clean Water Act has established standards.

  Modified SSIs under the above definition will be required to meet "new" MHI or "new" FBI air pollutant emissions limits.
- **New SSIs:** MHIs and FBIs for which construction commenced after October 14, 2010.

For reference, air pollutant emissions limits set by the model rule for "existing" and "new" MHIs and FBIs are provided in Table 1.

Table 1. SSI MACT Regulated Air Pollutants and Emissions Limits for

Pollutant[1]	Units[1]	EMISSIONS LIMITS			
		"Existing" MHIs[3]	"New" MHIs[4]	"Existing" FBIs[3]	"New" FBIs[4]
Particulate Matter (PM)	mg/dscm	80		18	9.6
Hydrogen Chloride (HCl)	ppmvd	1.2	1.2	0.51	0.24
Carbon Monoxide (CO)	ppmvd	3,800	52	64	27
PCDD/PCDF, TEQ[2]	ng/dscm	0.32	0.0022	0.10	0.0044
PCDD/PCDF, TMB[2]	ng/dscm	5.0	0.045	1.2	0.013
Mercury (Hg)	mg/dscm	0.28	0.15	0.037	0.0010
Oxides of Nitrogen (NOx)	ppmvd	220	210	150	30
Sulfur Dioxide (SO_2)	ppmvd	26	26	15	5.3
Cadmium (Cd)	mg/dscm	0.095	0.0024	0.0016	0.0011
Lead (Pb)	mg/dscm	0.30	0.0035	0.0074	0.00062

[1] All tested parameters corrected to 7% O_2
[2] Compliance with PCDD (Polychlorinated Dibenzodioxins) and PCDF (Polychlorinated Dibenzofurans) limits for TEQ (Toxic Equivalency Basis) or TMB (Toxic Mass Basis) is required, not both.
[3] 40 CFR Part 60 Subpart MMMM ("Existing")
[4] 40 CFR Part 60 Subpart LLLL ("New")

Air pollutant emissions limits set forth under the model rule, which took effect March 21, 2016 as previously noted, are enforceable by *Federal Implementation Plan (FIP) 40 CFR Part 62 Subpart LLL, RIN 2060-AR77.*

Utility Background

In the State of Ohio, several utilities operated MHIs and FBIs prior to the promulgation of the SSI MACT rule. Some of these utilities have since chosen, for a variety of reasons, to discontinue operation of their incineration facilities rather than invest in the upgrades needed to ensure compliance with the rule requirements. On the other hand, the Northeast Ohio Regional Sewer District (NEORSD) and the Metropolitan Sewer District of Greater Cincinnati (MSDGC) each chose to continue operation of their existing incineration facilities at the Southerly WWTC and the Mill Creek WWTP, respectively.

The decision to continue operation of existing incinerators was made by each utility, in part, due to the alternative requirement for each to haul an average of 120+ dry tons per day (dtpd) of dewatered sludge cake to landfill, which was deemed not only to be cost prohibitive, but also to be unsustainable. As a result, NEORSD and MSDGC separately initiated efforts to determine which upgrades would be required to ensure consistent compliance with the SSI MACT rule, and subsequently to implement the required upgrades. These efforts are discussed in detail herein.

A process schematic of a typical fluidized bed incinerator system, like the six total SSI systems installed at the Southerly WWTC and Mill Creek WWTP (three at each facility) is provided as Figure 1 for reference.

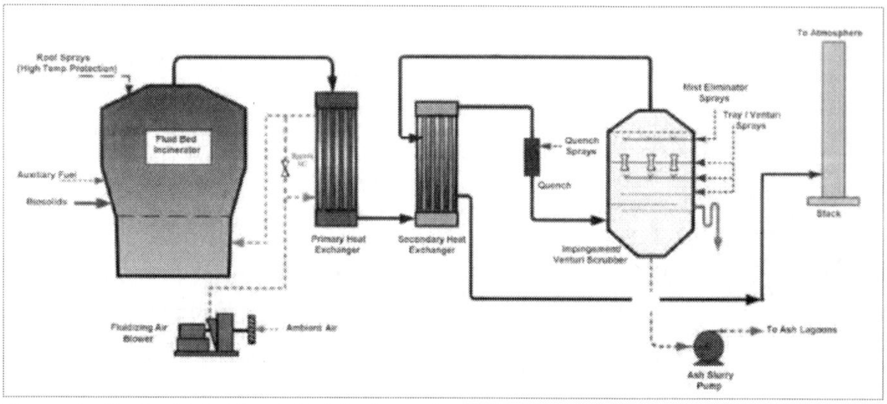

Figure 1. Typical Fluidized Bed Incineration Process Schematic

NEORSD Southerly WWTC

The Southerly WWTC serves approximately 600,000 customers within the Greater Cleveland area. The facility is the largest of NEORSD's three wastewater treatment plants and treats an average daily flow of 125 million gallons per day (MGD) while providing complete treatment of peak flows up to 400 MGD. The facility can also partially treat up to an additional 335 MGD of storm water through primary treatment. The solids stream of the facility begins in the primary settling tanks, and the first stage and second stage secondary settling tanks. Settled sludge from the primary settling tanks is conveyed to gravity thickeners to thicken the primary sludge to 4-7% solids. Excess activated sludge from the secondary settling tanks is conveyed to gravity belt thickeners likewise to thicken the activated sludge to 5-7% total solids.

From the gravity thickeners, thickened primary sludge (TPS) is directed into one of six sludge storage tanks (SSTs) which also receive thickened excess activated sludge (TEAS) from the gravity belt thickeners. The SSTs act as sludge feedwells for the downstream centrifuges, which were commissioned in 2012 with the construction of the Southerly WWTC Renewable Energy Facility (REF). The REF is generally comprised of three main building areas: the Incinerator Building, the Biosolids Handling Building, and the Truck Loading Facility. These buildings are further subdivided into various rooms and process areas.

The Dewatering Building houses nine dewatering centrifuges, which are fed TPS/TEAS from the SSTs via one of three centrifugal pumps. Once delivered to the dewatering centrifuges, sludge is dewatered and directed to one of three sliding frame collection bins (three centrifuges per collection bin). Six piston type cake pumps (two per collection bin) direct dewatered cake to one of three 100 dtpd FBIs located in the Incinerator Building.

Each of the Southerly WWTC FBIs commenced construction prior to October 2010, and thus are considered "existing" per the SSI MACT rule. Each of the incinerator trains are identical and consist of a fluid bed reactor, primary heat exchanger (PHE) for preheating combustion air, waste heat boiler (WHB) for steam production, secondary heat exchanger (SHE) for plume suppression, ash handling equipment, and auxiliary systems including power generation. In addition, each incinerator train was initially constructed with emissions controls sufficient to meet regulatory requirements for air pollutant emissions prior to the more stringent requirements under the SSI MACT rule. The existing emissions controls for each FBI consists of a quench and an impingement EnviroCare (ECI) VenturiPak wet scrubber with multiple fixed venturis.

MSDGC Mill Creek WWTP

The Mill Creek WWTP treats an average flow of 130 MGD and serves the City of Cincinnati and central Hamilton County, Ohio. It is the largest of MSDGC's three major wastewater treatment plants and has a firm capacity of 430 MGD through preliminary and primary treatment and 240 MGD through secondary treatment. The solids process stream for the facility begins in the primary settling basins and secondary clarifiers. Settled sludge from the primary settling basins is conveyed to gravity thickeners to thicken the primary sludge to 4-7% total solids. Waste activated sludge (WAS) from the secondary clarifiers is conveyed to thickening centrifuges, which thicken the activated sludge to 5-7% total solids.

From the gravity thickeners, thickened primary sludge (TPS) is directed into a common sludge feed header. This header also conveys the centrifuge thickened waste activated sludge (TWAS) to promote in-line blending of the combined thickened sludge (TPS/TWAS). TPS/TWAS may be transferred directly to the dewatering sludge feedwells, or first to one of twelve sludge storage tanks (former digester tanks), where TPS/TWAS is then pumped to the dewatering sludge feedwells in the Dewatering Building.

The dewatering sludge feedwells temporarily hold TPS/TWAS prior to it being pumped by one of seven progressing cavity pumps to one of six dewatering centrifuges. Centrifuge dewatered sludge cake is then directed to one of three live bottom collection bins. Six piston cake pumps, two per collection bin, direct dewatered cake to the three 96 dtpd FBIs located in the adjacent Incinerator Building.

Similar to the Southerly WWTC FBIs, the Mill Creek WWTP FBIs commenced construction prior to October 2010, and thus are considered "existing" per the SSI MACT rule. Each of the incinerator trains are also identical and consist of a fluid bed reactor, PHE, SHE, ash handling equipment and auxiliary systems. The Mill Creek WWTP incinerators are not used to generate power as are the Southerly WWTC incinerators, and thus are not equipped with WHBs. Similar to the Southerly WWTC incinerators; each of the Mill Creek WWTP incinerators was initially constructed with a quench and an impingement ECI VenturiPak wet scrubber.

Initial Compliance Evaluations

NEORSD and MSDGC each separately engaged Black & Veatch in late 2013 to assess the effectiveness of existing emissions controls (i.e. ECI VenturiPak wet scrubbers) equipment for the Southerly WWTC and Mill Creek WWTP FBIs to meet the more stringent air pollutant emissions standards under the SSI MACT rule. Black & Veatch's role included coordination, review, and interpretation of historical and newly collected data in preparation for the design and implementation of enhanced emissions controls improvements.

Initial compliance evaluations for both NEORSD and MSDGC followed similar paths, and each primarily relied on the results of previous and new emissions compliance tests to assess the effectiveness of existing emissions controls to meet the SSI MACT rule's requirements.

More specifically, in early 2014, NEORSD contracted a third party testing agency to assist in conducting Self-Evaluation Engineering Studies (i.e. unofficial emissions compliance tests) for each of the Southerly WWTC FBIs. These studies were conducted from March 25-26, 2014 and July 22, 2014, June 4-6, 2014, and June 20, 2014 for Southerly WWTC FBIs No. 3, No. 2, and No. 1, respectively. As previously noted, the goal of these studies was to compare air pollutant emissions levels for each of the incinerators against the respective SSI MACT rule's air pollutant emissions limits.

In general, the testing methods utilized during NEORSD's Self-Evaluation Engineering Studies and during similar efforts conducted by MSDGC (discussed later herein) were in conformance with those required per 40 CFR Part 60 Subpart MMMM, and as summarized in Table 2.

Table 2. Required SSI MACT Resting Analytes and Methods for Existing FBIs

Analyte[1]	Averaging Methods & Sampling Duration[3]	Sampling Method[3]
Particulate Matter (PM)	Average of three runs Minimum 1 dscm per run	EPA Method 5, 26A, or 29
Hydrogen Chloride (HCl)	Average of three runs Minimum 1 dscm per run	EPA Method 26A
Carbon Monoxide (CO)	Average of three runs Minimum one hour duration per run	EPA Method 10, 10A, or 10B
PCDD/PCDF, TEQ[2]	Average of three runs Minimum 1 dscm per run	EPA Method 23
PCDD/PCDF, TMB[2]	Average of three runs Minimum 1 dscm per run	EPA Method 23
Mercury (Hg)	Average of three runs Method 29: Minimum 1 dscm per run Method 30B: As specified	EPA Method 29, 30B, or ASTM D6784-02
Oxides of Nitrogen (NOx)	Average of three runs Minimum one hour duration per run	Method 7 or 7E
Sulfur Dioxide (SO_2)	Average of three runs Method 6: Minimum 60L per run Method 6C: Minimum one hour duration per run	Method 6, 6C, or ANSI/ASME PTC 19.10-1981
Cadmium (Cd)	Average of three runs Minimum 1 dscm per run	EPA Method 29
Lead (Pb)	Average of three runs Minimum 1 dscm per run	EPA Method 29
Fugitive Ash Emissions	Three 1-hour observation periods	EPA Method 22

[1] All tested parameters corrected to 7% O_2
[2] Compliance with PCDD and PCDF limit for TEQ or TMB is required, not both
[3] 40 CFR Part 60 Subpart MMMM

Similarly to NEORSD, MSDGC contracted a separate third party testing agency to perform unofficial emissions compliance tests for one FBI at the Mill Creek WWTP (FBI No. 3). This testing occurred February 4-7, 2013. Note that MSDGC opted only to test a single FBI at Mill Creek WWTP given that historical, official emissions compliance test results were also available for each of the Mill Creek WWTP FBIs from their commissioning in 2010.

The following provides a summary of the initial findings of the initial compliance evaluations for NEORSD and MSDGC.

NEORSD Southerly WWTC

Summary emissions testing results for each of the early-2014 Self-Evaluation Engineering Studies conducted by NEORSD are presented in Tables 3 through 6. Note that each table also indicates the level of compliance achieved for each pollutant per the SSI MACT rule for existing FBIs, indicated as "Percent of Allowable." Rows, which show **bolded** text indicate those pollutants which were found to either be non-compliant (above the associated MACT limit) or which were considered approaching non-compliant (>75% of the allowable limit).

Table 3. Southerly WWTC FBI No. 1 Emissions Compliance Test Results – June 2014

Pollutant	Measured Average[1,6]	Allowable[2]	Units[3]	Percent (%) of Allowable[6]
Particulate Matter (PM)	2.0	18	mg/dscm	11.1
Hydrogen Chloride (HCl)[5]	-	0.51	ppmvd	-
Carbon Monoxide (CO)	24.5	64	ppmvd	38.3
PCDD/PCDF, TEQ[4,5]	-	0.10	ng/dscm	-
PCDD/PCDF, TMB[4]	-	1.2	ng/dscm	-
Mercury (Hg)	**0.0352**	**0.037**	**mg/dscm**	**95.1**
Oxides of Nitrogen (NOx)	44.2	150	ppmvd	29.5
Sulfur Dioxide (SO_2)	**22.8**	**15**	**ppmvd**	**152.0**
Cadmium (Cd)	<0.00057	0.0016	mg/dscm	<35.6
Lead (Pb)	<0.00047	0.0074	mg/dscm	<6.4

[1] Based on average of three test runs conducted in June 2014 for Southerly WWTC FBI No. 1
[2] SSI MACT rule air pollutant emissions limits for "existing" FBIs
[3] All test parameters tested to 7% O_2
[4] Compliance with PCDD and PCDF limit for TEQ or TMB is required, not both
[5] HCL and PCDD/PCDF not tested for FBI No. 1 during June 2014 testing
[6] "<" symbol indicates that the compound was not present in quantities above the Minimum Detection Limit (MDL) for at least one sample

Table 4. Southerly WWTC FBI No. 2 Emissions Compliance Test Results – June 2014

Pollutant	Measured Average[1,6]	Allowable[2]	Units[3]	Percent (%) of Allowable[6]
Particulate Matter (PM)	1.87	18	mg/dscm	10.4
Hydrogen Chloride (HCl)[5]	<0.151	0.51	ppmvd	<29.6
Carbon Monoxide (CO)	7.74	64	ppmvd	12.1
PCDD/PCDF, TEQ[4,5]	0.00047	0.10	ng/dscm	0.5
PCDD/PCDF, TMB[4]	-	1.2	ng/dscm	-
Mercury (Hg)	**0.040**	**0.037**	**mg/dscm**	**108.1**
Oxides of Nitrogen (NOx)	103.1	150	ppmvd	68.7
Sulfur Dioxide (SO_2)	10.7	15	ppmvd	71.3
Cadmium (Cd)	**<0.0015**	**0.0016**	**mg/dscm**	**<93.8**
Lead (Pb)	<0.00047	0.0074	mg/dscm	<6.4

[1] Based on average of three test runs conducted in June 2014 for Southerly WWTC FBI No. 2
[2] SSI MACT rule air pollutant emissions limits for "existing" FBIs
[3] All test parameters tested to 7% O_2
[4] Compliance with PCDD and PCDF limit for TEQ or TMB is required, not both
[5] PCDD/PCDF TEQ tested for FBI No. 2 during June 2014 testing
[6] "<" symbol indicates that the compound was not present in quantities above the MDL for at least one sample

Table 5. Southerly WWTC FBI No. 3 Emissions Compliance Test Results – March 2014

Pollutant	Measured Average[1,6]	Allowable[2]	Units[3]	Percent (%) of Allowable[6]
Particulate Matter (PM)	2.77	18	mg/dscm	15.4
Hydrogen Chloride (HCl)[5]	0.158	0.51	ppmvd	31.0
Carbon Monoxide (CO)	9.28	64	ppmvd	14.5
PCDD/PCDF, TEQ[4,5]	0.0130	0.10	ng/dscm	13.0
PCDD/PCDF, TMB[4]	-	1.2	ng/dscm	-
Mercury (Hg)	**0.0297**	**0.037**	**mg/dscm**	**80.3**
Oxides of Nitrogen (NOx)	44.5	150	ppmvd	29.5
Sulfur Dioxide (SO_2)	**34.8**	**15**	**ppmvd**	**232.0**
Cadmium (Cd)	<0.00091	0.0016	mg/dscm	<56.9

Pollutant	Measured Average[1,6]	Allowable[2]	Units[3]	Percent (%) of Allowable[6]
Lead (Pb)	<0.00017	0.0074	mg/dscm	<2.3

[1] Based on average of three test runs conducted in March 2014 for Southerly WWTC FBI No. 3
[2] SSI MACT rule air pollutant emissions limits for "existing" FBIs
[3] All test parameters tested to 7% O_2
[4] Compliance with PCDD and PCDF limit for TEQ or TMB is required, not both
[5] PCDD/PCDF TEQ tested for FBI No. 3 during March 2014 testing
[6] "<" symbol indicates that the compound was not present in quantities above the MDL for at least one sample

Table 6. Southerly WWTC FBI No. 3 Emissions Compliance Test Results – July 2014

Pollutant	Measured Average[1,6]	Allowable[2]	Units[3]	Percent (%) of Allowable[6]
Particulate Matter (PM)	0.0018	18	mg/dscm	0.0
Hydrogen Chloride (HCl)[5]	-	0.51	ppmvd	-
Carbon Monoxide (CO)	26.2	64	ppmvd	40.9
PCDD/PCDF, TEQ[4,5]	-	0.10	ng/dscm	-
PCDD/PCDF, TMB[4]	-	1.2	ng/dscm	-
Mercury (Hg)	**<0.034**	**0.037**	**mg/dscm**	**91.9**
Oxides of Nitrogen (NOx)	41.2	150	ppmvd	27.5
Sulfur Dioxide (SO_2)	**<19.0**	**15**	**ppmvd**	**<126.7**
Cadmium (Cd)	<0.00033	0.0016	mg/dscm	<20.6
Lead (Pb)	<0.00142	0.0074	mg/dscm	19.2

[1] Based on average of three test runs conducted in July 2014 for Southerly WWTC FBI No. 3
[2] SSI MACT rule air pollutant emissions limits for "existing" FBIs
[3] All test parameters tested to 7% O_2
[4] Compliance with PCDD and PCDF limit for TEQ or TMB is required, not both
[5] HCL and PCDD/PCDF not tested for FBI No. 3 during July 2014 testing
[6] "<" symbol indicates that the compound was not present in quantities above the MDL for at least one sample

As shown in Tables 3 through 6, pollutants that were shown to exhibit emissions levels higher than their associated SSI MACT limits were mercury (Hg) and sulfur dioxide (SO_2). In addition, cadmium (Cd) was shown to test above the allowable limit for two of the nine test runs conducted (three per testing event), resulting in an average tested concentration of approximately 94% of the allowable limit for FBI No. 2. All other analytes measured well within the allowable limits per the SSI MACT rule, and were thus considered as being sufficiently controlled by existing air pollutant emissions controls systems.

MSDGC Mill Creek WWTP

Summary emissions testing results for the early-2013 emissions compliance testing conducted by MSDGC for Mill Creek WWTP FBI No. 3 are provided in Table 7, which displays the level of compliance achieved for each pollutant per the SSI MACT rule for existing FBIs, indicated as "Percent of Allowable."

Table 7. Mill Creek WWTP FBI No. 3 Emissions Compliance Test Results – February 2013

Pollutant	Measured Average[1,5]	Allowable[2]	Units[3]	Percent (%) of Allowable[5]
Particulate Matter (PM)	0.882	18	mg/dscm	4.9
Hydrogen Chloride (HCl)	0.0916	0.51	ppmvd	18.0
Carbon Monoxide (CO)	1.2	64	ppmvd	1.9
PCDD/PCDF, TEQ[4]	0.0058	0.10	ng/dscm	5.6
PCDD/PCDF, TMB[4]	0.0663	1.2	ng/dscm	5.5
Mercury (Hg)	**0.0350**	**0.037**	**mg/dscm**	**94.6**
Oxides of Nitrogen (NOx)	61.2	150	ppmvd	40.8
Sulfur Dioxide (SO_2)	2.09	15	ppmvd	13.9
Cadmium (Cd)	<0.000416	0.0016	mg/dscm	<26.0
Lead (Pb)	0.000896	0.0074	mg/dscm	12.1

[1] Based on average of three test runs conducted in February 2013 for Mill Creek WWTP FBI No. 3
[2] SSI MACT rule air pollutant emissions limits for "existing" FBIs
[3] All test parameters tested to 7% O_2
[4] Compliance with PCDD and PCDF limit for TEQ or TMB is required, not both
[5] "<" symbol indicates that the compound was not present in quantities above the MDL for at least one sample

As displayed upon Table 7, mercury was found to be the only tested parameter above or near respective limits per the SSI MACT rule. On average, mercury emissions during the February 2013 testing were 95% of allowable, with one test reporting mercury levels at 99.6% of allowable.

As previously indicated, in addition to the February 2013 emissions test results for Mill Creek WWTP FBI No. 3, official emissions test data from the commissioning of the Mill Creek WWTP incinerators were reviewed. These results are based on tests conducted in May, July, and September 2010 for Mill Creek WWTP incinerators No. 2, No. 3, and No. 1, respectively. Note that the 2010 emissions test results were conducted to establish compliance with a separate set of emissions standards, and as such do not include all regulated air pollutants under the SSI MACT rule; refer to Tables 8 through 10.

Table 8. Mill Creek WWTP FBI No. 1 Emissions Compliance Test Results – September 2010

Pollutant	Measured Average[1,6]	Allowable[2]	Units[3]	Percent (%) of Allowable[6]
Particulate Matter (PM)	2.94	18	mg/dscm	16.3
Hydrogen Chloride (HCl)[5]	-	0.51	ppmvd	-
Carbon Monoxide (CO)	2.63	64	ppmvd	4.1
PCDD/PCDF, TEQ[4,5]	-	0.10	ng/dscm	-
PCDD/PCDF, TMB[4]	-	1.2	ng/dscm	-
Mercury (Hg)	**0.070**	**0.037**	**mg/dscm**	**189.2**
Oxides of Nitrogen (NOx)	17.27	150	ppmvd	11.5
Sulfur Dioxide (SO_2)	**22.28**	**15**	**ppmvd**	**152.0**
Cadmium (Cd)	<0.000078	0.0016	mg/dscm	<4.9
Lead (Pb)	0.00023	0.0074	mg/dscm	3.1

[1] Based on average of three test runs conducted in September 2010 for Mill Creek WWTP FBI No. 1
[2] SSI MACT rule air pollutant emissions limits for "existing" FBIs
[3] All test parameters tested to 7% O_2
[4] Compliance with PCDD and PCDF limit for TEQ or TMB is required, not both
[5] HCL and PCDD/PCDF not tested for FBI No. 1 during September 2010 testing
[6] "<" symbol indicates that the compound was not present in quantities above the MDL for at least one sample

Table 9. Mill Creek WWTP FBI No. 2 Emissions Compliance Test Results – May 2010

Pollutant	Measured Average[1,6]	Allowable[2]	Units[3]	Percent (%) of Allowable[6]
Particulate Matter (PM)	1.07	18	mg/dscm	5.9
Hydrogen Chloride (HCl)[5]	-	0.51	ppmvd	-
Carbon Monoxide (CO)	0.71	64	ppmvd	1.1
PCDD/PCDF, TEQ[4,5]	-	0.10	ng/dscm	-
PCDD/PCDF, TMB[4]	-	1.2	ng/dscm	-
Mercury (Hg)	**0.066**	**0.037**	**mg/dscm**	**178.4**
Oxides of Nitrogen (NOx)	22.37	150	ppmvd	14.9
Sulfur Dioxide (SO_2)	0.12	15	ppmvd	0.8
Cadmium (Cd)	**<0.0022**	**0.0016**	**mg/dscm**	**<137.5**
Lead (Pb)	0.00070	0.0074	mg/dscm	9.5

[1] Based on average of three test runs conducted in May 2010 for Mill Creek WWTP FBI No. 2
[2] SSI MACT rule air pollutant emissions limits for "existing" FBIs
[3] All test parameters tested to 7% O_2
[4] Compliance with PCDD and PCDF limit for TEQ or TMB is required, not both
[5] HCL and PCDD/PCDF not tested for FBI No. 2 during May 2010 testing
[6] "<" symbol indicates that the compound was not present in quantities above the MDL for at least one sample

Table 10. Mill Creek WWTP FBI No. 3 Emissions Compliance Test Results – July 2010

Pollutant	Measured Average[1,6]	Allowable[2]	Units[3]	Percent (%) of Allowable[6]
Particulate Matter (PM)	2.48	18	mg/dscm	13.8
Hydrogen Chloride (HCl)[5]	-	0.51	ppmvd	-
Carbon Monoxide (CO)	15.54	64	ppmvd	24.3
PCDD/PCDF, TEQ[4,5]	-	0.10	ng/dscm	-
PCDD/PCDF, TMB[4]	-	1.2	ng/dscm	-
Mercury (Hg)	**0.057**	**0.037**	**mg/dscm**	**154.1**
Oxides of Nitrogen (NOx)	31.65	150	ppmvd	21.1
Sulfur Dioxide (SO_2)	**13.61**	**15**	**ppmvd**	**90.7**
Cadmium (Cd)	<0.0001354	0.0016	mg/dscm	<8.5
Lead (Pb)	0.00054	0.0074	mg/dscm	7.3

[1] Based on average of three test runs conducted in July 2010 for Mill Creek WWTP FBI No. 3
[2] SSI MACT rule air pollutant emissions limits for "existing" FBIs
[3] All test parameters tested to 7% O_2
[4] Compliance with PCDD and PCDF limit for TEQ or TMB is required, not both
[5] HCL and PCDD/PCDF not tested for FBI No. 3 during July 2010 testing
[6] "<" symbol indicates that the compound was not present in quantities above the MDL for at least one sample

As shown in Tables 7 through 10, pollutants that were shown to exhibit emissions levels higher than their associated SSI MACT limits were mercury, sulfur dioxide, and cadmium. All other analytes measured well within the allowable limits per the SSI MACT rule, and were thus considered as being sufficiently controlled by existing air pollutant emissions controls systems.

Note that elevated levels of sulfur dioxide and cadmium for various 2010 tests were at least in part attributed to process water restrictions to existing air emissions controls devices.

Mercury Emissions Controls

Elevated levels of sulfur dioxide and cadmium observed during various testing events for the Southerly WWTC and Mill Creek WWTP FBIs were addressed via separate improvements to the existing EnviroCare VenturiPak wet scrubbers already installed for each FBI at these facilities. These improvements are not the topic of this discussion, and are thus not included herein.

Aside from elevated emissions of sulfur dioxide and cadmium, data reviewed during the initial compliance evaluations strongly pointed towards the requirement for enhanced mercury emissions controls for FBIs at both Southerly WWTC and Mill Creek WWTP. As such, the following provides a brief review of the sources of mercury in SSI applications alongside a summary of technologies considered by the project teams to address mercury emissions at Southerly WWTC and Mill Creek WWTP.

The source of mercury in SSI exhaust gasses is the solids generated within the wastewater treatment facility. Unlike most metals which typically drop out of the incineration process with the ash, and which are removed in the wet scrubber or other particulate removal technology, mercury can exist in the following three forms following incineration:

- **Particulate Mercury:** Mercury entrained with the ash and particulate matter. This fraction of the total mercury is removed in the wet scrubber and sent to the wet ash lagoon with other metals and particulates. A very small portion of the total mercury is in this form.
- **Elemental Mercury:** Mercury is volatilized by combustion. As the gaseous elemental mercury is cooled through the remaining processes, it may react with other components of the flue gas to form oxidized gaseous mercury. This form of mercury is soluble in water and can be partially removed in the wet scrubber.
- **Oxidized Gaseous Mercury:** Elemental mercury which is oxidized by exhaust gas components such as: halogens (chlorine, fluorine, and bromine); oxides of sulfur such as sulfur dioxide (SO_2) and sulfur trioxide (SO_3); and nitrogen such as nitrogen dioxide (NO_2).

In most wastewater treatment plants, only a fraction of the elemental mercury is oxidized. Therefore, a fraction of the mercury discharged from the incineration process remains in the elemental form. Although some of the oxidized mercury can be captured by wet scrubber technologies (such as the EnviroCare VenturiPak wet scrubbers installed at Southerly WWTC and Mill Creek WWTP) this form of mercury removal is very inefficient. As such, proven technologies used for mercury removal include powder activated carbon (PAC) injection for plants with dry ash systems and fixed bed granular activated (GAC) scrubbers for wet ash systems (both the Southerly WWTC and Mill Creek WWTP FBIs use wet ash systems).

PAC Injection

Figure 2 illustrates a summary process schematic for a typical PAC injection system. Typical systems include PAC storage and handling components (not shown on Figure 2), PAC injection towers, fabric filters, and spent PAC truck loading equipment. For SSI applications, PAC injection has historically been installed upstream of the wet scrubber and used as the primary means for removing both ash and mercury. For such installations, dry ash handling is used. Because the Southerly WWTC and Mill Creek WWTP FBIs currently operate with a wet ash handling system to direct ash slurry to lagoons, use of a PAC injection system train would require the system to be installed downstream of the existing wet scrubber and after the SHE to minimize impact on current ash handling practices. As such, the existing SHE would be re-used to re-heat the exhaust gases above dew point temperatures before entering the fabric filter. The fabric filter bags are expected to form a PAC layer on the surface of each bag essential for gas-solid contact with the carbon for mercury adsorption. The PAC would need to be fed to the exhaust gases on a continuous basis to maintain this layer for proper mercury adsorption. Note that PAC injection is typically capable of achieving a mercury reduction of ~85%.

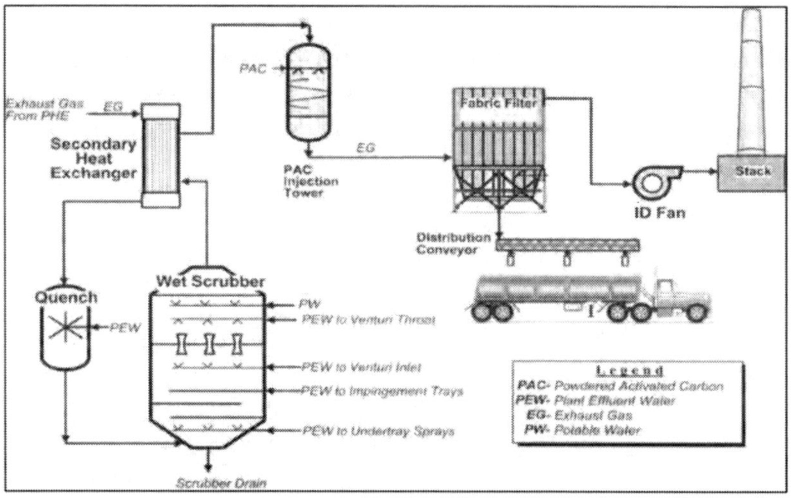

Figure 2. Typical PAC Injection System Schematic

Note that, in lieu of the dry ash load-out arrangement displayed on Figure 2, for FBIs which operate energy recovery facilities (and which are thus equipped with WHBs) it would theoretically be possible to direct ash collected from fabric filters to existing WHB ash slurry mix tanks. This would likely require modifications / partial re-design of existing ash slurry systems.

Given the notable considerations for PAC injection such as: 1) space requirements (each fabric filter is expected to have dimensions of approximately 45 ft. x 15 ft. x 50 ft. (L x W x H)); 2) typical applicability of a PAC injection system for SSIs operating dry ash systems; and 3) relative cost (including higher continuous power demand and chemical usage), this technology was not further considered by the project teams for Southerly WWTC and Mill Creek WWTP.

Fixed Bed GAC Media Filters

Figure 3 illustrates a summary process schematic for a typical fixed bed GAC media filter system. Typical systems are comprised of a demister and a multiple-celled, fixed-bed GAC media filter. The demister is intended to remove water droplets from the exhaust gases prior to entering the GAC media filter (scrubber). This droplet removal allows the mercury in the SSI exhaust gases to be effectively adsorbed to the GAC media. Once the adsorption capacity of the media is exhausted, it is removed and replaced. Each media filter is typically designed for one to two years of operation before media replacement is required. In addition, each scrubber system would be provided with a start-up skid including an electrical heater and blower (not shown on Figure 3) to warm-up the media filter prior to system start-up. Note that GAC media filter technology is typically capable of achieving mercury reduction greater than 95%.

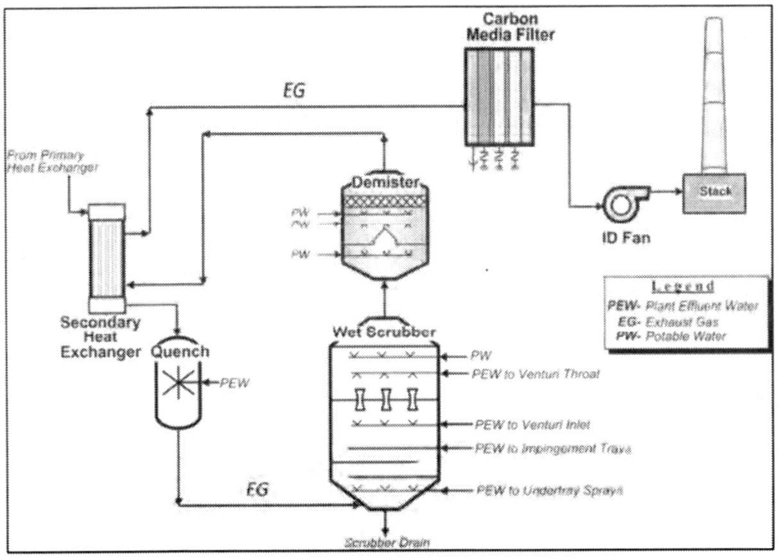

Figure 3. Typical GAC System Schematic

A benefit of the fixed bed GAC media technology is its ability to achieve superior mercury removal efficiencies when compared to other options discussed herein. A potential future change in regulations (requiring greater than 90% mercury removal) could be supported by the GAC system. As a result, GAC system manufacturers currently operating in the municipal SSI market (Carbon Process & Plant Engineering (CPPE) and Air Pollution Control Technologies, Inc. (APC)) are typically willing to provide performance guarantees designed to meet mercury limits for FBIs classified as "existing" and "new" per the SSI MACT rule

Downsides to the GAC media technology include space requirements (each tower is expected to have dimensions of approximately 12 ft. in diameter and up to 40 ft. high); these dimensions would likely require that such systems be installed in standalone building for existing installations). Another notable downside for this technology is its relative cost compared to other options, as discussed herein.

Sorbent Polymer Composite (SPC) Media Technology

Near to the beginning of evaluation efforts conducted by the project teams for Southerly WWTC and Mill Creek WWTP, an alternative to the PAC and GAC systems was identified. Previously unproven in the municipal SSI market, this alternative was a proprietary system offered by EnviroCare International (ECI) and W.L. Gore utilizing Sorbent Polymer Composite (SPC) media. At the onset of the evaluation efforts, this technology had undergone two pilot tests in SSI applications, and was being evaluated by a number of other utilities also operating SSIs.

The SPC media technology can be retrofitted as part of existing EnviroCare wet scrubbers or may be provided as a standalone unit. The system utilizes a proprietary media bed comprised of SPC media modules, which have the capability of capturing both the ionized and elemental forms of mercury from SSI exhaust gases. In contrast to GAC media scrubbers, the sorbent media modules are tolerant of water saturated gas streams and do not require a separate demister or a re-heating step upstream. It is important to note that the SPC's mercury removal efficiency is a function of media bed depth provided with each installation, or the number of sequential layers of SPC media modules provided. Similar to GAC media scrubbers, the sorbent media modules need to be replaced on an average bi-annual basis depending on inlet mercury loading and volume of the media bed. To measure performance of the sorbent bed over time, periodic media or gas grab samples are required; samples are analyzed to determine remaining life of the media bed.

Figure 4 illustrates a summary process schematic for an installation of the SPC media technology installed as a retrofit to an existing wet scrubber (with SPC media modules located downstream of the existing fixed venturi section).

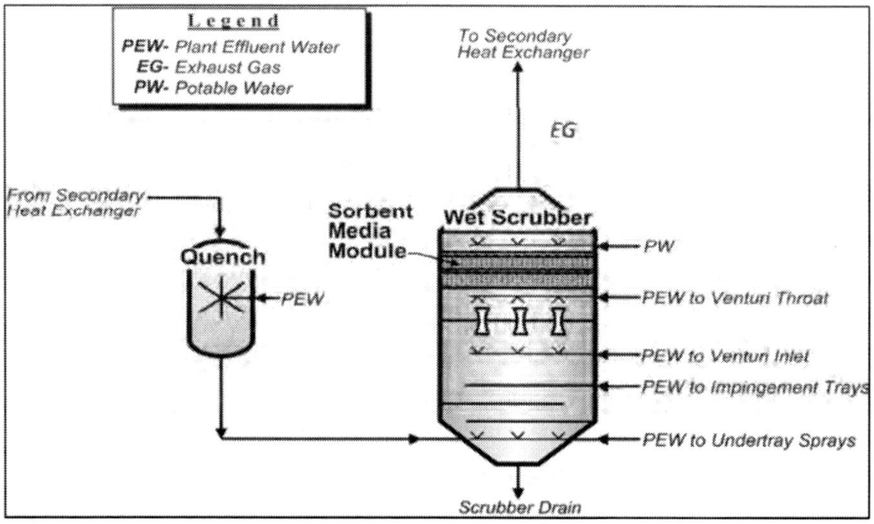

Figure 4. Typical SPC Integrated Into Wet Scrubber Schematic

An alternate approach to the integrated SPC media module arrangement (Figure 4) is to locate the SPC media modules within a standalone vessel downstream of the wet scrubber. This option allows for greater flexibility in terms of locating the system and designing for access requirements.

With the standalone configuration, the SPC media modules would be arranged within the standalone vessel in layers to prevent exhaust gas bypassing or "short circuiting," and to allow each layer of modules to be sampled or removed from the vessel without disturbing or removing other layers. In addition, each vessel would be equipped with a demister (mist eliminator) section and appropriate spray lances downstream (above) the SPC media modules for intermittent wash-down.

SPC TECHNOLOGY EVALUATION AND TECHNOLOGY COMPARISON

While it was recognized that the SPC media technology would result in notably less equipment and modifications when compared to GAC and PAC systems, thus resulting in notably lower capital cost, it was also understood that the technology had not been proven in a full scale SSI application prior to the evaluation efforts described herein. As a result, a detailed evaluation was performed by the project teams for Southerly WWTC and Mill Creek WWTP to determine the applicability of the technology for FBIs at these facilities. Note that the evaluations conducted for both facilities were similar in nature, and ultimately yielded similar conclusions. As a result, for the purposes of this discussion, only specific details associated with the NEORSD evaluation are provided.

In order to assess the level of compliance offered by the SPC technology in relation to Southerly WWTC's FBIs specifically, the project team conducted a review of the Self-Evaluation Engineering Study emissions testing results previously presented herein and of an extensive set of historical inlet biosolids mercury concentrations provided by NEORSD. This process resulted in an estimation of emissions compliance failure probability with the use of the SPC technology:

- 124 data points representing inlet-biosolids mercury concentrations (collected by NEORSD) were compiled. Of these, 23 were taken by NEORSD as background test data during test runs for each of the Self-Evaluation Engineering Studies for FBIs No. 1, No. 2, and No. 3 (refer to Tables 3 through 6).
- A comparison of the reported mercury emissions test results, biosolids feed rates, and stack discharge flow rates during each study, and of the inlet biosolids mercury concentrations provided by NEORSD during testing, was performed as a calibration for a mercury emissions model developed by the project team. It was determined that the existing Southerly WWTC ECI VenturiPak wet scrubbers are capable of capturing approximately 25% of the total mercury (i.e. the mercury is contained in the ash slurry effluent). As previously referenced, note that the wet scrubber efficiency in terms of mercury capture will vary depending on the form of mercury present in the exhaust gases, as only ionized forms can be effectively removed by water. Note also that mercury concentrations can vary seasonably and with the weather, or depending on the type and quantity of influent industrial discharge.
- Utilizing the calibrated mercury emissions model, the project team evaluated various mercury removal efficiencies across individual layers of SPC media modules to determine the total mercury removal efficiency required to ensure compliance with the SSI MACT rule.

SPC Compliance Analysis Results

Inlet biosolids mercury concentrations generated from the analysis of grab samples collected by NEORSD were compiled and sorted chronologically from January 2011 to June 2014 (latest data available during the evaluation). Note that this period begins after the point at which dental amalgam traps were required for dental facilities (starting in 2010), meaning that the mercury removal benefits resultant from this change are captured within the data set. Overall results are presented in Figure 5.

Note that the average inlet mercury concentration observed over the January 2011 to June 2014 period was ~0.57 mg Hg / kg biosolids, while the maximum was ~0.99 mg Hg / kg biosolids. Note also that the standard deviation of the data set was calculated as ~0.14 mg Hg / kg biosolids. This data indicates that Southerly WWTP FBI inlet mercury concentrations are highly variable and unpredictable, as is typical for municipal wastewater treatment facilities.

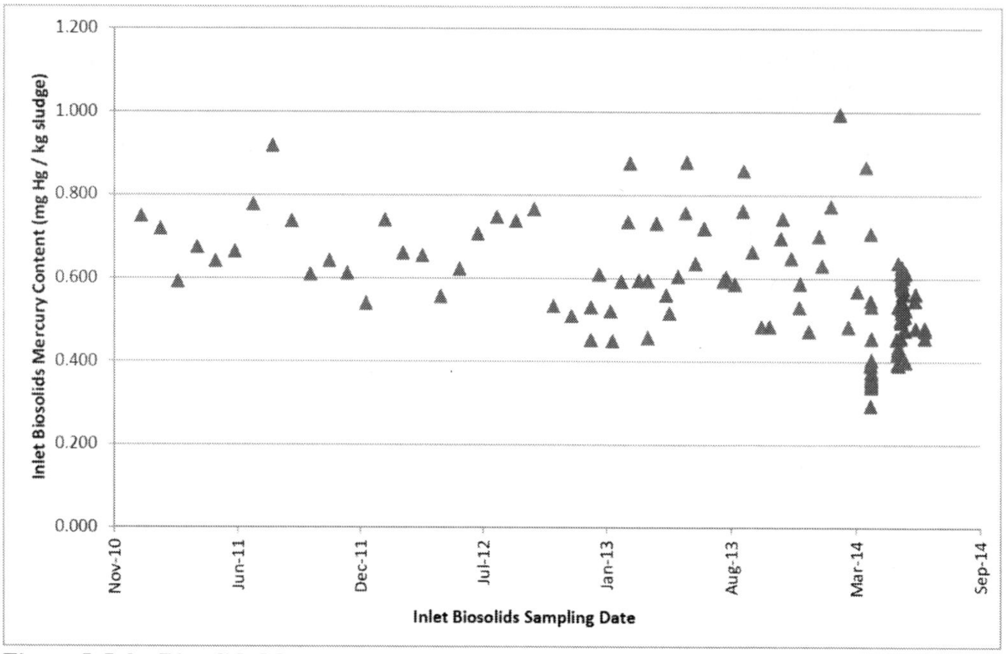

Figure 5. Inlet Biosolids Mercury Concentrations – January 2011 through June 2014

Utilizing the inlet biosolids mercury concentrations presented in Figure 5, in addition to the operations data collected by NEORSD during each of the Self-Evaluation Engineering Studies, a model calibration of predicted incinerator exhaust gas mercury concentrations (mg / dscm corrected to 7% O_2) was performed. This calibration, performed over 23 inlet-biosolids mercury concentrations taken during the studies, indicated that the existing Southerly WWTC ECI VenturiPak wet scrubbers are capable of capturing approximately 25% of the total mercury in the incinerator-exhaust-gas stream.

Assuming 25% total mercury reduction across the existing wet scrubbers, the calibrated model was then extended to the additional 101 inlet-biosolids mercury concentration data points provided to the project team by NEORSD. Assuming no additional mercury removal equipment is provided for Southerly WWTC's FBIs, exhaust gas mercury concentrations downstream of the wet scrubbers were predicted for each historical inlet concentration. The results of this analysis are presented in Figure 6.

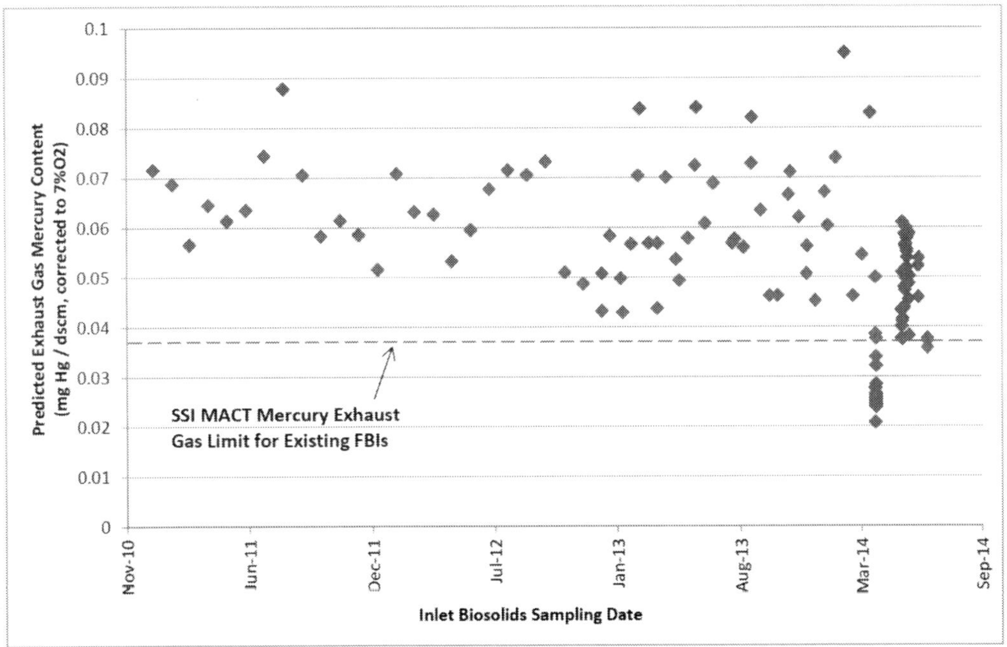

Figure 6. Estimated Mercury Exhaust Gas Content with Existing Emissions Controls

From Figures 5 and 6, note that the highest concentration of data collected over the January 2011 to June 2014 period was generated from late March 2014 to early June 2014. Also, note that inlet mercury loadings over this time were generally lower than those documented elsewhere within the historical data set. As a result, while the Self-Evaluation Engineering Study emissions test results documented exhaust gas mercury emissions below or only slightly higher than the applicable SSI MACT limit for existing FBIs, the mercury emissions model generally predicted higher mercury emissions values for the remaining data points. In short, this helped reinforce the need for enhanced mercury emissions controls at Southerly WWTC.

Based on discussions with the SPC technology provider (ECI) during the evaluation, a conservative assumption of guaranteed mercury removal efficiency across an individual layer of SPC media modules is approximately 25%. Note that the evaluation recognized that there was still some uncertainty regarding the effect of varying inlet mercury loadings on the removal efficiency achieved by the SPC media (i.e. whether higher or lower inlet mercury loadings increase or decrease removal efficiency) given that limited inlet versus outlet mercury emissions data is available. Of particular interest was the ability of the SPC media to handle inlet "spikes" in mercury loadings. However, for the purposes of the evaluation, the mercury emissions model assumes a 25% mercury-removal-efficiency guarantee across an individual layer of SPC media modules independent of inlet mercury loading. Using this approach, various predictive plots were generated assuming varying layers of SPC media modules and associated level of mercury removal. For a single layer of SPC media modules, approximately 65% of the data points were projected to be above the "existing" FBI MACT limit of 0.037 mg Hg / dscm exhaust gas (corrected to 7% O_2). For two layers of SPC media modules, approximately 20% of the data points were projected to be above the "existing" FBI MACT limit of 0.037 mg Hg / dscm exhaust gas (corrected to 7% O_2). For three layers of SPC media modules, only two data points (~1.6%) were projected to be above the "existing" FBI MACT limit of 0.037 mg Hg / dscm exhaust gas (corrected to 7% O_2). Note, however, that a number of other data points were also close to this limit. Lastly, a plot was generated for four layers of SPC media modules. Under this scenario, as shown on Figure 7, no data points were projected to be above the existing FBI MACT limit of 0.037 mg Hg / dscm exhaust gas (corrected to 7% O_2).

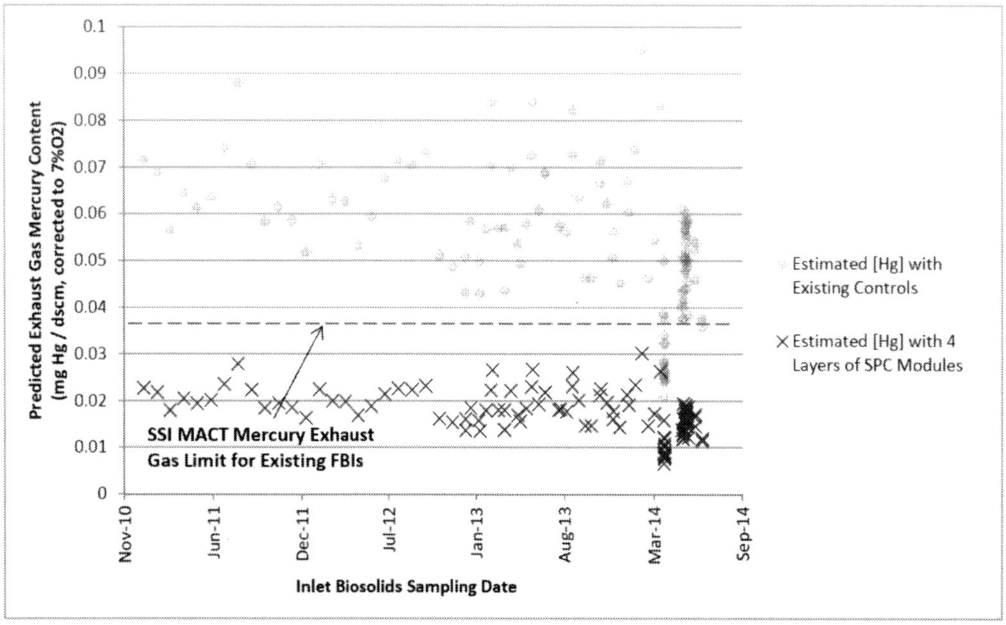

Figure 7. Estimated Mercury Exhaust Gas Content with Four SPC Media Layers

Considering the results of this evaluation, the project team concluded that the SPC media technology was a viable option for providing adequate control of mercury emissions in exhaust gas to meet the SSI MACT standard for existing FBIs at Southerly WWTC, as long as an appropriate number of layers of SPC media modules are provided and maintained. Note that a similar conclusion was generated for the FBIs at the Mill Creek WWTP.

Comparison of Mercury Removal Technologies

From the discussion above, two alternative mercury removal technologies (GAC media and SPC media) were identified as viable for providing the required enhanced mercury emissions controls to bring the Southerly WWTC FBIs and Mill Creek WWTP FBIs into compliance with the applicable SSI MACT standard for FBIs classified as "existing." Table 11 provides a summary comparison of these two technologies.

Table 11. Summary Comparison of Applicable Mercury Removal Technologies

GAC MEDIA FILTERS	SPC MEDIA TECHNOLOGY
Mercury Removal Efficiency	
Expected mercury-removal efficiencies of 95% or higher.	Sufficient mercury-removal efficiency for ECI to guarantee emissions of less than <0.037 mg/dscm for typical Southerly WWTC and Mill Creek WWTP FBI inlet biosolids mercury loadings.
Regulatory	
▪ Able to meet both "existing" and "new" SSI MACT mercury standards for FBIs. ▪ GAC system also capable of providing performance guarantees for PCDDs/PCDFs[1].	▪ Able to meet "existing" SSI MACT mercury standards for FBIs for typical inlet mercury loadings; GAC technology would need to be employed to meet "new" FBI MACT limits. ▪ Performance guarantees currently not provided for removal of PCDDs/PCDFs[1].
Equipment	
▪ Equipment located downstream of wet scrubber. ▪ Mercury captured on carbon media. ▪ Requires exhaust gas conditioning upstream of fixed bed GAC media filter to remove large water droplets and re-heat the gases to temperature above gas due point. Water droplet removal is achieved by a demister and re-heating through existing secondary heat exchanger. ▪ Requires upsized fluidizing air blowers or ID fans to accommodate additional pressure drop. ▪ Cannot be feasibly located within existing REF.	▪ Equipment integrated within existing wet scrubber or alternatively downstream of wet scrubber as a standalone vessel. ▪ No additional upstream gas conditioning is required. ▪ Downstream demister (mist eliminator) system required for periodic wash-down with standalone vessel configuration. ▪ Equipment can be located within existing REF.
Consumables	
GAC scrubbers would be sized for change out bi-annually with ~30 tons of media.	SPC media-module-life guarantees are for 18 months.
Cost	
▪ ~$11M total equipment cost to meet SSI MACT mercury and PCDD/PCDF standards for "new" FBIs (3 units, 1 per FBI)[2]. ▪ <$8M total equipment cost to meet SSI MACT mercury and PCDD/PCDF standards for "existing" FBIs (3 units, 1 per FBI)[3]. ▪ ~$175K per GAC media bed replacement.	▪ ~$1.5M equipment cost (3 units, 1 per FBI)[4]. ▪ ~$10K replacement cost per SPC media module (9 modules per layer; ~$90K total replacement cost per layer).

SPC Media Technology Recommendation

In total, both the GAC and SPC systems were found to be able to provide mercury removal efficiencies required to meet the SSI MACT limits for mercury emissions from FBIs classified as existing. Based on the information presented in Table 11 and summarized earlier in this document, it was ultimately recommended by the project teams for Southerly WWTC and Mill Creek WWTP that the SPC media technology be installed. The following presents additional information regarding this recommendation:

- **Reduced system complexity**. The GAC media technology system requires gas-conditioning steps, including upstream demister and exhaust gas reheat equipment, in addition to ancillary equipment, such as GAC media filter start-up heater skids. Furthermore, installation of new ID fans (in the case of Mill Creek WWTP) or upgrades to the existing ID fans (in the case of Southerly WWTC) were determined to be required to address additional pressure losses presented by the new system. In comparison, the SPC media technology only requires standalone SPC media module containment vessels located downstream of the existing wet scrubbers. This is due to low-pressure drop across each layer of SPC media modules (1/8 to 1/3 in. w.c.).
- **Reduced space requirements**. The SPC media technology can be housed entirely within the existing building space. Given the notably large footprints of the GAC system media filters, they cannot be feasibly installed within the existing building and would need to be housed within a new superstructure addition adjacent to the incinerator buildings for Southerly WWTC and Mill Creek WWTP.
- **Enhanced future mercury-removal efficiency**. While the SPC media technology does not provide mercury removal performance up to that provided by the GAC media technology (i.e. greater than 95% percent mercury removal, meeting existing and new SSI MACT requirements), the proposed layout allows for additional layers of SPC media to be installed to achieve greater mercury removal efficiency in the future. Note that with ECI's 2014 pilot at Asheville, NC, a five-layer arrangement of SPC media modules has resulted in greater than 90% mercury-removal efficiency. Based on these results, ECI preliminarily offered a 70% performance guarantee for mercury removal across a five-layer system.
- **Cost savings**. The SPC media technology appears to be about a quarter of the equipment cost required for the GAC media technology (not including installation, building modifications, new superstructure, ductwork, or ancillary equipment and controls).

RESULTS

SPC Media Technology Installation and Testing

Based on the information presented herein, and in considering the recommendation of the project teams, both NEORSD and MSDGC ultimately chose to install the SPC media technology for each of their respective three FBIs at Southerly WWTC and Mill Creek WWTP.

In-line with the findings of SPC technology evaluations conducted by the project teams, NEORSD chose to install a three layer standalone SPC media vessel for each FBI, with space to house an additional two layers of SPC media modules in the future, if required. MSDGC on the other hand chose to install a five-layer standalone SPC media vessel for each FBI, with space to house an additional two layers of modules. Note that MSDGC's decision to install an initial five layers of modules was based on elevated mercury loadings seen throughout the MSDGC service area, including at other wastewater treatment facilities, which may in the future direct dewatered sludge cake to Mill Creek WWTP for disposal.

Pictures of the final NEORSD and MSDGC SPC vessel installations are provided as Figure 8.

Figure 8. Typical SPC Vessel Installation (Southerly WWTC Left, Mill Creek WWTP Right)

Official Emissions Testing Results

Official SSI MACT rule air pollutant emissions compliance testing was conducted by both NEORSD and MSDGC in late 2015 through early 2016. Final results for testing are provided in Table 12 for Southerly WWTC FBIs No. 1, 2, and 3, and in Table 13 for Mill Creek WWTP FBIs No. 1, 2, and 3, respectively. Note that each table presents the results as a representation of compliance achieved for each pollutant per the SSI MACT rule for existing FBIs, indicated as "Percent of Allowable."

Table 12. Southerly WWTC FBI No. 1, 2, and 3 Official Compliance Testing – Dec 2015 through Feb 2016

Pollutant	Units	Allowable[2,3]	PERCENT (%) OF ALLOWABLE[1,3,5]		
			FBI No. 1	FBI No. 2	FBI No. 3
Particulate Matter (PM)	mg/dscm	18	6%	6%	5%
Hydrogen Chloride (HCl)	ppmvd	0.51	<5%	<35%	<18%
Carbon Monoxide (CO)	ppmvd	64	<2%	<4%	<2%
PCDD/PCDF, TEQ[4]	ng/dscm	0.10	4%	5%	18%
PCDD/PCDF, TMB[4]	ng/dscm	1.2	7%	<1%	<1%
Mercury (Hg)	mg/dscm	0.037	38%	8%	18%
Oxides of Nitrogen (NOx)	ppmvd	150	53%	68%	40%
Sulfur Dioxide (SO_2)	ppmvd	15	<54%	<39%	<57%
Cadmium (Cd)	mg/dscm	0.0016	<11%	<16%	<19%
Lead (Pb)	mg/dscm	0.0074	4%	10%	4%

Pollutant	Units	Allowable[2,3]	PERCENT (%) OF ALLOWABLE[1,3,5]		
			FBI No. 1	FBI No. 2	FBI No. 3

[1] Results based on average of three test runs conducted for respective FBIs
[2] SSI MACT rule air pollutant emissions limits for "existing" FBIs
[3] All test parameters tested to 7% O_2
[4] Compliance with PCDD and PCDF limit for TEQ or TMB is required, not both
[5] "<" symbol indicates that the compound was not present in quantities above the Minimum Detection Limit (MDL) of the analytical method for at least one sample fraction

Table 13. Mill Creek WWTP FBI No. 1, 2, and 3 Official Compliance Testing – Dec 2015 through Feb 2016

Pollutant	Units	Allowable[2,3]	PERCENT (%) OF ALLOWABLE[1,3,5]		
			FBI No. 1	FBI No. 2	FBI No. 3
Particulate Matter (PM)	mg/dscm	18	11%	7%	8%
Hydrogen Chloride (HCl)	ppmvd	0.51	<6%	<4%	<5%
Carbon Monoxide (CO)	ppmvd	64	4%	11%	<1%
PCDD/PCDF, TEQ[4]	ng/dscm	0.10	2%	9%	4%
PCDD/PCDF, TMB[4]	ng/dscm	1.2	2%	20%	7%
Mercury (Hg)	mg/dscm	0.037	3%	7%	33%
Oxides of Nitrogen (NOx)	ppmvd	150	24%	27%	12%
Sulfur Dioxide (SO_2)	ppmvd	15	<1%	1%	<1%
Cadmium (Cd)	mg/dscm	0.0016	<4%	<5%	<9%
Lead (Pb)	mg/dscm	0.0074	35%	23%	8%

[1] Results based on average of three test runs conducted for respective FBIs
[2] SSI MACT rule air pollutant emissions limits for "existing" FBIs
[3] All test parameters tested to 7% O_2
[4] Compliance with PCDD and PCDF limit for TEQ or TMB is required, not both
[5] "<" symbol indicates that the compound was not present in quantities above the Minimum Detection Limit (MDL) of the analytical method for at least one sample fraction

Notable in the results presented within Tables 12 and 13 is that one incinerator each for Southerly WWTC and Mill Creek WWTP exhibited reduced mercury capture (i.e. Southerly WWTC FBI No. 1 and Mill Creek WWTP FBI No. 3) when compared to testing results for the other incinerators at each facility, respectively. Additional data collected by MSDGC during official compliance testing offers some explanation as to this variation as discussed following Table 15. In summary, MSDGC opted to perform additional testing for the concentration of mercury and sulfur dioxide within gasses inlet to each SPC vessel. The primary purpose for this testing was to ensure performance guarantees provided by the SPC technology provider were met (i.e. evaluate air pollutant removal efficiencies across the SPC vessels). Sulfur dioxide was tested in addition to mercury given that pilot information provided by ECI indicated some level of sulfur dioxide removal across the SPC media modules.

Comparison of inlet versus outlet mercury and sulfur dioxide exhaust gas concentrations are provided in Table 14 for each of the Mill Creek WWTP FBIs.

Table 14. Mill Creek WWTP FBI SPC Vessel Performance Evaluation

Pollutant	Units	FBI No. 1[1,2,3]			FBI No. 2[1,2,3]			FBI No. 3[1,2,3]		
		Inlet	Outlet	Percent (%) Removal	Inlet	Outlet	Percent (%) Removal	Inlet	Outlet	Percent (%) Removal
Mercury (Hg)	mg/dscm	0.0340	0.0009	97.4	0.0499	0.0027	94.6	0.0312	0.012	61.5
Sulfur Dioxide (SO_2)	ppmvd	<0.91	<0.2	78.0	1.32	0.104	92.1	1.91	<0.1	>94.9

[1] Results based on average of three test runs conducted for respective FBIs
[2] All test parameters tested to 7% O_2
[3] "<" symbol indicates that the compound was not present in quantities above the Minimum Detection Limit (MDL) of the analytical method for at least one sample fraction

As shown in Table 14, inlet mercury levels to the SPC vessels for each Mill Creek WWTP FBI were consistent, ranging from 0.0312 to 0.0499 mg/dscm (corrected to 7% O_2). However, while resultant mercury removal efficiencies across the SPC vessels for FBI No. 1 and No. 2 were also consistent, the mercury removal percentage across the SPC vessel for FBI No. 3 was much lower. This can at least in part be attributed to a "break in" period required by the SPC technology manufacturer for the SPC media. In summary, the manufacturer states that a month (or so) is required for the SPC media to settle in on a long-term removal efficiency. As a result, test results for Mill Creek WWTP FBIs No. 1 and No. 2, which were collected for systems that had been in operation for less than two weeks, demonstrate higher mercury-removal efficiencies within the SPC media "break in" period. On the other hand, test results for Mill Creek WWTP FBI No. 3, which were collected after the system had been online for over a month, demonstrate lower mercury-removal efficiencies after the "break in" period. Figure 9, provided by ECI, provides pilot data that indicate anticipated long-term mercury removal across the SPC media. Note that towards the beginning of the SPC media life cycle, removal efficiencies hover around 95%, while after the "break in" period, efficiencies drop to a somewhat consistent ~80% level.

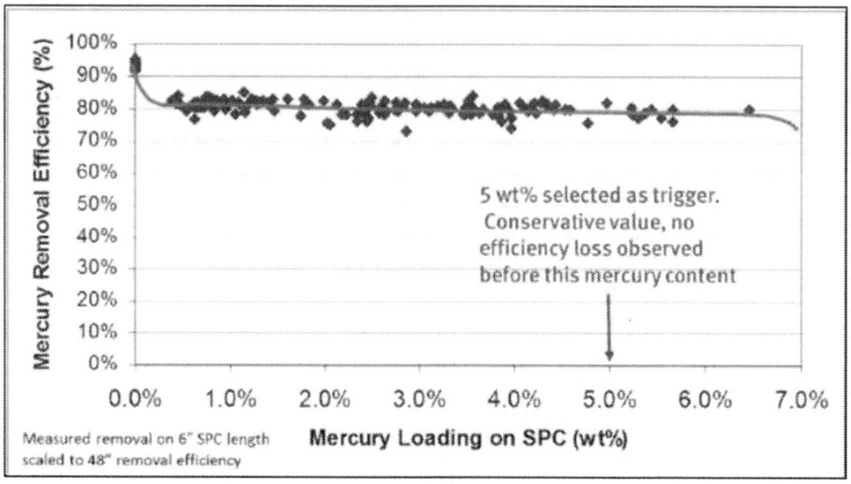

Figure 9. ECI Published Data for SPC Media Long Term Mercury Removal Efficiency

Sulfur Dioxide Removal

SPC vessel inlet versus outlet sulfur dioxide concentrations for each of the Mill Creek WWTP FBIs, as displayed on Table 14, point towards significant, but somewhat variable, sulfur dioxide removal across the SPC media. Based on discussions with ECI, this is due to the presence of a sulfur dioxide oxidation catalyst within the SPC media, which converts sulfur dioxide to sulfuric acid (H_2SO_4). In effect, the acid is collected on the surface of the SPC media, requiring periodic washdown of the SPC media. For installations at Southerly WWTC and Mill Creek WWTP, this was accomplished through use of an automatic spray wash-down system set to approximately 2 minutes of spray each hour.

It is anticipated that additional data for sulfur dioxide removal across the SPC vessels will be collected over time by ECI.

DISCUSSION

Long Term SPC Media Performance

As shown in Figure 9, ECI has conservatively recommended that individual SPC media modules be replaced after reaching a mercury concentration of approximately 5% by weight. Given that exhaust gas mercury loadings to the SPC vessel are not continuously monitored, it became apparent that periodic sampling and monitoring of the SPC media would be required in order to better determine proper media change-out frequency. This periodic monitoring was also required as part of Site Specific Monitoring Plans developed for Southerly WWTC and Mill Creek WWTP in conformance with the requirements of the SSI MACT rule.

Long-term SPC-media monitoring was thus structured in such a way as to allow sampling of SPC media modules on an approximate quarterly basis. This monitoring involves the following:

- Visual inspection of SPC vessel and media modules
- Rotation of SPC media modules per ECI's recommendation (to ensure equal mercury capture across modules)
- Location (physically within the SPC vessel) of each SPC media module and operational time of each system tracked
- Physical sample of SPC media to determine mercury concentration by weight

- Gas phase carbon trap (EPA Method 30B) samples taken to determine mercury removal across SPC media modules

Results of the most recent sampling/testing for Southerly WWTC and Mill Creek WWTP are provided in the following sections.

NEORSD Southerly WWTC

Table 15 provides pertinent information as to the operation of each Southerly WWTC FBI and respective SPC vessel since commissioning in 2015.

Table 15. Southerly WWTC SPC Vessel Operational Information

	FBI No. 1	FBI No. 2	FBI No. 3
SPC Vessel Installation Date	12/03/15	11/16/15	12/31/15
Startup Date	12/20/15	11/20/15	1/7/16
Dry Tons of Sludge Fired	14,901.3	7,868.4	8,497.5
Hour of Operation	5,293.6	3,000.5	2,819.9
Average Feed Rate (ton/hour)	2.81	2.62	3.01

As shown on Table 15, FBI No. 1 has been in operation for the longest of the three Southerly WWTC FBIs since commissioning of the SPC vessels in early 2016. As a result, it would be anticipated that the SPC media contained within this incinerator's SPC vessel would exhibit the highest levels of mercury concentration by weight. This is correlated via the information presented in Table 16 below.

Summary results for ECI's inspection and sampling of the Southerly WWTC SPC vessels, which occurred October 18-19, 2016, are provided below. Physical media sampling was not conducted for the SPC vessel for FBI No. 2 due to that incinerator being in-service during the inspection period.

Each of the Southerly WWTC SPC vessels was initially installed with three layers of SPC media modules. The first layer ("Layer 1") within each SPC media module is the layer through which inlet gasses pass first, and is located at the bottom of each SPC vessel. Table 16 indicates the remaining mercury adsorption capacity of each layer of SPC media modules, as determined by ECI during the October 18-19, 2016 sampling.

Each module was assumed capable of adsorbing up to 1.5 pounds prior (5% by weight) before requiring replacement. Through physical sampling of the SPC media and performance of testing to determine mercury content, remaining mercury adsorption capacity was determined.

Table 16. Southerly WWTC SPC Vessel Remaining SPC Media Mercury Adsorption Capacity

	FBI No. 1	FBI No. 2	FBI No. 3
Layer 1 Remaining Capacity	90%	Not Tested	92%
Layer 2 Remaining Capacity	93%	Not Tested	95%
Layer 3 Remaining Capacity	95%	Not Tested	95%

Based on the findings presented in Table 16, which represent ~10 months of operation for each of the Southerly WWTC SPC vessels, ECI has indicated that the SPC media modules appear to be operating as expected in terms of mercury capture. ECI has indicated that an estimate of remaining life of the SPC media modules will be made following the next quarterly sampling/testing event.

Unfortunately, during the October 18-19, 2016 sampling by ECI, it was discovered by ECI that EPA Method 30B (carbon trap sampling) to determine mercury removal across each layer of SPC media modules will not be effective for the SPC vessels at Southerly WWTC. This is due to a high negative pressure across the SPC vessels, which are installed immediately upstream of induced draft (ID) fans in the incinerator process train. As a result, the carbon contained within the standard Method 30B sample tubes was disrupted in most of the traps, and in some cases, the carbon was completed evacuated from the tube into the SPC vessel. For this reason, no data is currently available for mercury removal across each layer of SPC media modules for the October sampling event. ECI has indicated that modifications to the sampling methodology will be made, and sampling performed in coordination with NEORSD in the near term. Data will be appended to this document as it becomes available.

Mill Creek WWTP

Table 17 provides pertinent information as to the operation of each Mill Creek WWTP FBI and respective SPC vessel since commissioning in 2015.

Table 17. Mill Creek WWTP SPC Vessel Operational Information

	FBI No. 1	FBI No. 2	FBI No. 3
SPC Vessel Installation Date	02/23/16	01/19/16	12/22/15
Startup Date	On Standby	01/29/16	12/26/15
Dry Tons of Sludge Fired	0	7,524	8,510
Hour of Operation	0	3,473	4,207
Average Feed Rate (ton/hour)	0	2.17	2.02

As shown on Table 17, FBIs No. 2 and No. 3 has been in near continuous operation since commissioning of the SPC vessels in early 2016, while FBI No. 1 has been in standby for the same period. As a result, it would be anticipated that the SPC media contained within these incinerator's associated SPC vessels would exhibit similar mercury concentrations. This is correlated via the information presented in Table 19 below.

Summary results for ECI's inspection and sampling of the Mill Creek WWTP SPC vessels, which occurred July 26 through 28, 2016, 2016, are provided below. Physical media sampling was not conducted for the SPC vessel for FBI No. 1 due to that incinerator not having been in operation since commissioning in early 2016.

Each of the Mill Creek WWTP SPC vessels was initially installed with five layers of SPC media modules. The first layer ("Layer 1") within each SPC media module is the layer through which inlet gasses pass first, and is located at the bottom of each SPC vessel. Table 18 indicates the remaining mercury adsorption capacity of each layer of SPC media modules, as determined by ECI during the July 26 through 28, 2016 sampling.

Similarly to the approach for Southerly WWTC, each module was assumed capable of adsorbing up to 1.5 pounds prior (5% by weight) before requiring replacement. Through physical sampling of the SPC media and performance of testing to determine mercury content, remaining mercury adsorption capacity was determined.

Table 18. Mill Creek WWTP SPC Vessel Remaining SPC Media Mercury Adsorption Capacity

	FBI No. 1	FBI No. 2	FBI No. 3
Layer 1 Remaining Capacity	100%	96%	99%
Layer 2 Remaining Capacity	100%	98%	98%
Layer 3 Remaining Capacity	100%	98%	99%
Layer 4 Remaining Capacity	100%	99%	99%
Layer 5 Remaining Capacity	100%	99%	99%

Based on the findings presented in Table 18, which represent ~6-7 months of operation for the SPC vessels for Mill Creek WWTP FBIs No. 2 and No. 3, ECI has indicated that the SPC media modules appear to be operating as expected in terms of mercury capture. ECI has indicated that an estimate as to the remaining life of the SPC media modules will be made following the next quarterly sampling/testing event.

In addition to the physical media sampling performed for each of the Mill Creek WWTP SPC vessels, ECI was able to perform EPA Method 30B sampling upstream and downstream of each layer of SPC media modules for Mill Creek WWTP FBIs No. 2 and No. 3. Results for this sampling are provided upon Figure 10, provided by ECI to MSDGC, below.

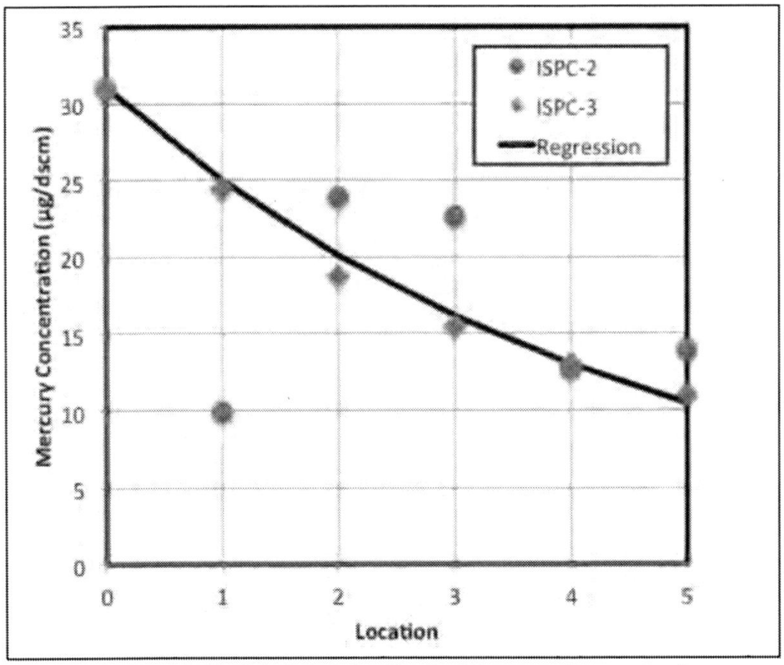

Figure 10. SPC Media Module Inlet/Outlet Mercury Concentrations

Per information provided by ECI, the solid black line shown on Figure 10 represents the anticipated mercury removal efficiency across five layers of SPC media modules, and generally depicts a 20% mercury reduction across each layer of modules with a total mercury mass reduction of 66% across all five layers.

From Figure 10, it is clear that the EPA Method 30B sampling for FBI No. 3's SPC vessel revealed mercury reduction across each layer of SPC media modules along the expected trend, with a total mercury reduction observed at approximately 65%. In reviewing data from official emissions compliance tests presented previously herein, note that Mill Creek FBI No. 3 previously tested at approximately 61% mercury removal across the SPC vessel. The findings of the subsequent EPA Method 30B sampling appear to confirm that, following the break-in period for the SPC media, approximately 60-70% mercury removal can be expected across five layers of SPC media modules.

While sampling for SPC vessel for FBI No. 3 aligned well with anticipated results, data for FBI No. 2's SPC vessel did not appear to correlate very well at all. Based on discussions with ECI, it is unclear why this is. Another iteration of EPA Method 30B is planned in the short term to further evaluate this. Data will be appended to this document as it becomes available.

CONCLUSION

The new SPC vessels at the Southerly WWTC and Mill Creek WWTP have now been in operation for about a year. The information presented herein has demonstrated that the decision by NEORSD and MSDGC to install this technology has, to-date, been a good one in terms of achieving compliance with the SSI MACT rule requirements. In addition, in comparison to the GAC technology, it is anticipated that the chosen approach resulted in at least \$15M - \$20M in capital savings for both NEORSD and MSDGC.

Since completion of the SPC vessel installations at Southerly WWTC and Mill Creek WWTP, several other facilities have installed the technology. Installations include facilities in New York, North Carolina, Virginia, and Washington State. In addition, a number of other utilities are currently in evaluation stages to determine the applicability of the technology for their MHI or FBI systems.

Long-term monitoring of the SPC technology's performance will be critical in determining the long-term effectiveness, longevity of SPC media modules and associated change out frequency, and any unforeseen operational issues that could arise over time. For now, all information appears to point toward a technology that is capable of providing the required SSI exhaust-gas mercury-removal efficiencies required to comply with the SSI MACT rule for FBIs classified as "existing" and MHIs classified as "existing" or "new."

REFERENCES

"New Mercury and SO_2 Control Technology by EnviroCare and W.L. Gore." Envirocare International | Engineered Solutions for Pollution Control. EnviroCare International, n.d. Web. 12 Jan. 2017.

Triple Bottom Line Analysis of Energy Recovery from Thermal Oxidation of Wastewater Solids Compared to Coal

Anna J. Munson[1*], Webster F. Hoener[1], Robert P. Dominak[2], James E. Welp[1]

[1]Black & Veatch.
[2]Friar Consulting.
[*]Email: MunsonA@bv.com

ABSTRACT

Many water resource recovery facilities (WRRFs) use thermal oxidation (incineration) to manage solids produced by the wastewater treatment process. While this management option has typically been used to achieve volume/mass reduction and sterilize the solids, energy recovery from thermal oxidation is of increased interest as a component of approaches to attain energy neutrality. Because of the increased interest in energy recovery, the Water Environment & Reuse Foundation (WE&RF) has undertaken a state of science review, with the objectives to:

- Evaluate the potential for energy and heat recovery from the thermal oxidation of wastewater solids based on the latest generation of thermal oxidation technology.
- Compare the value of energy recovered from wastewater solids by thermal oxidation with that from coal, based on a triple bottom line (TBL) approach, evaluating economic, environmental, and social criteria.
- Estimate the quantity of renewable energy available from thermal oxidation of wastewater solids and residuals from domestic wastewater and associated feedstocks, such as fats, oil, and grease (FOG), scum, and imported biomass.

This paper summarizes the findings from one of the major research activities; a TBL life cycle assessment comparing thermal oxidation of wastewater solids with coal. Documentation of the potential recoverable energy from wastewater solids and residuals and a state of technology summary reviewing the different systems available for recovering energy from incinerators were included in previous papers.

KEYWORDS: Thermal oxidation of wastewater solids, energy and heat recovery.

INTRODUCTION

Many water resource recovery facilities (WRRFs) use thermal oxidation (incineration) to manage solids produced by the wastewater treatment process. While this treatment option has typically been used to achieve volume/mass reduction and sterilize the solids, energy recovery from thermal oxidation is of increased interest as a component of approaches to attain energy neutrality. Trends in anaerobic digestion, such as co-digestion of FOG or imported organic wastes to increase energy production, raise the question if similar enhancement of energy recovery could be accomplished for thermal oxidation with the co-combustion of similar alternative feedstocks. Understanding the role energy recovery from solids through thermal oxidation could perform is an important factor in determining ways for publicly owned treatment works (POTWs) that practice incineration to reduce operating costs.

Because of the increased interest in energy recovery from thermal oxidation of wastewater solids, the Water Environment & Reuse Foundation (WE&RF) has undertaken a state of science review, supported by Target Collaborative Research funding, identified as project ENER13T14. The objectives of the research were:

- Evaluate the potential for energy and heat recovery from the thermal oxidation of wastewater solids based on the latest generation of thermal oxidation technology with combined heat and power (CHP) systems.
- Compare the value of energy recovered from wastewater solids by thermal oxidation with that from coal, based on a triple bottom line (TBL) approach, evaluating economic, environmental, and social criteria.
- Estimate the quantity of renewable energy available from thermal oxidation of wastewater solids and residuals from domestic wastewater and associated feedstocks, such as FOG, scum, and biomass at existing incineration facilities to assist WE&RF in informing the Department of Energy, regulators, and lawmakers.

The findings from two of the major research activities were presented previously. The evaluation of potential energy and heat recovery from thermal oxidation of wastewater solids with the latest generation of thermal oxidation technology was presented at WEF's 2016 Residuals & Biosolids Conference. A summary of the thermal oxidation technologies available and industry experience with the technologies was presented at WEF's 2016 WEFTEC Conference. This paper summarizes the findings from the remaining major research activity; a TBL life cycle assessment comparing the energy value from thermal oxidation of wastewater solids to the energy value from coal.

TRIPLE BOTTOM LINE (TBL) EVALUATION OF THE ENERGY VALUE FROM WASTEWATER SOLIDS COMPARED TO COAL

Sustainability is an important goal for the water resource recovery field. WRRF managers are advancing the sustainability of their operations through energy efficiency measures, renewable energy systems, and resource recovery projects. Sustainability is widely defined in business, industry, and the public sector as standing on three pillars: economic, environmental, and social. These are the three components of the TBL analysis; a relative simple concept to grasp, which is used widely by corporations and government entities worldwide.

As part of the research to achieve energy neutrality at WRRFs, WE&RF developed a TBL evaluation tool (Elenbaas, 2014) to help WRRF decision-makers assess solids management technology options and process configurations based on which can yield the most sustainable outcomes. The TBL evaluation tool was used to assess the sustainability of energy harnessed from thermal oxidation of wastewater solids compared with energy produced at a coal-fired power plant. Coal was selected for comparison because the combustion and energy recovery equipment used is similar to that for wastewater solids and is one of the most common energy sources for producing electrical power in the United States.

Energy Alternatives
The alternative for energy recovery from thermal oxidation of wastewater solids (Wastewater Solids Alternative) is based on new heat recovery and power production improvements to an existing fluidized bed incineration facility to produce electricity. The system configuration of an FBI co-fired with FOG and wastewater solids was selected as the basis for the TBL analysis as most representative of current technology for thermal oxidation systems in the United States using the most effective feedstocks for renewable energy production. The facility is based on a capacity of 90 dry metric tonnes (dmt) (100 dry tons (dt)) of wastewater solids per day and 378 megaliters (ML) (100 million gallons (MG)) of wastewater per day. The energy recovery system includes:

- waste heat boiler
- steam turbine generator
- steam system
- electricity transmission system
- building

Energy recovery from the system was estimated at 631 kWh per dry metric tonne of wastewater solids processed. For the purposes of evaluation, only capital and operating costs for the new energy recovery improvements were evaluated, as the solids thermal oxidation systems would be existing with a primary purpose of processing wastewater solids and operating costs for treatment would occur whether energy was recovered or not. The alternative is based on sufficient land being available at the facility for construction of new energy recovery systems.

The alternative for electrical power production from a coal-fired power plant (Coal Alternative) is based on a super critical (SC) pulverized coal (PC) plant with a single reheat cycle and a CO_2 capture system. A SC PC plant was selected to represent current technology for efficient power production from coal combustion. Use of a CO_2 capture system was selected for compliance with future 2030 Clean Power Plan emission requirements. The facility is based on power production capacity of 550 MW. The SC PC plant includes:

- coal storage and feed systems
- PC boiler
- generator with high-, intermediate-, and low-pressure turbine sections
- baghouse
- emission control system, including selective catalytic reduction, flue gas desulfurization, and carbon dioxide recovery systems
- steam system
- cooling water system
- electricity transmission system
- stack
- building

The Coal Alternative was based on coal supplied from existing mines, but mining activities were considered when evaluating social and environmental impacts. Similar to the energy recovery alternative, the cost for land acquisition was not included in evaluation of the Coal Alternative. Criteria regarding sizing, energy balance, emissions, and costs for the SC PC plant are based on Case 12 of the National Energy Technology Laboratory report *Cost and Performance Baseline for Fossil Energy Plants* (NELT, 2013).

TBL Categories and Scoring

The criteria selected for evaluating the Wastewater Solids and Coal Alternatives followed the general criteria established in the TBL evaluation tool (Elenbaas, 2014). Criteria and weightings from that tool were adapted for the scope and objectives of this research using the following considerations:

- Determine the value of each criterion to the decision-making process and eliminate criteria that result in similar scores among all options.
- Ensure that the criteria address all key elements of sustainability.
- Develop metrics for each criterion, using measurements that are appropriate and commonly understood by stakeholders.

Table 1 lists the criteria and weightings used in the TBL evaluation. Additional air quality criteria and workplace conditions were added to assess elements that were important to the stakeholders in this evaluation.

Each prime category of economic, environmental, and social impacts were given equal weight to reflect the importance of each of these categories. The research team adjusted the weightings for the sub-categories and criteria to reflect the study focus of energy resource recovery.

The economic category included lifecycle cost assessments and engineering/technical criteria of simplicity, flexibility, and state of technology. Lifecycle cost expresses the total cost of construction and operation of a given alternative. The net present value (NPV) of operations and maintenance (O&M) costs for 20 years was included in the lifecycle cost assessment for each alternative. The lifecycle cost was a major consideration for this evaluation and was weighted at 85% of the total economic category score.

Engineering and technical considerations comprised the remaining 15% of the economic category score. Within this subcategory simplicity, flexibility and the state of technology were weighted equally important. The simplicity criterion is a measure of how simple the system is to operate, maintain, and repair. Flexibility is a measure of how easily the alternative can be modified to support future process changes and capacity expansion. The state of technology is a measure of how well established an alternative is within the industry. Industry experience lowers the economic risk to maintain and optimize the well-established technology.

Table 1. TBL criteria and weighting for Wastewater Solids and Coal Alternatives.

TBL Category & Criteria	Category Weight	Sub-category Weight	Criteria Weight	% of TBL Score
Economic	33.3%			
Net Present Value		85%		28.3%
Engineering/Technical		15%		5%
Simplicity			33.3%	1.7%
Flexibility			33.3%	1.7%
State of Technology			33.3%	1.7%
Environmental	33.3%			
Conservation of Resources		50%		16.7%
Ash Use			15%	2.5%
Fixed Carbon - Energy Recovery			35%	5.8%
Fixed Carbon - Greenhouse Gases			35%	5.8%
Water Conservation			15%	2.5%
Air Quality Impact		25%		8.3%
NOx			17%	1.4%
SO2			17%	1.4%
Particulate Matter			17%	1.4%
Lead			17%	1.4%
Mercury			17%	1.4%
HCl			17%	1.4%
Regulatory		25%		8.3%
Social	33.3%			
Nuisance Issues		50%		16.7%
Dust			33%	5.5%
Traffic			33%	5.5%
Visual			33%	5.5%
Workplace Conditions		50%		16.7%
Workplace Environment			50%	8.4%
Illness/Injury			25%	4.2%
Fatality			25%	4.2%

The research team considered conservation/optimization of resources more important than the other sub-categories in this particular evaluation because the study focused on energy resource recovery. This sub-category was weighted as 50% of the environmental category score. The conservation/optimization of resources sub-category reflects the number of ways in which ash remaining from combustion of wastewater solids and coal can be reused and how efficiently each alternative uses carbon-based fuel and water.

Ash is a common byproduct of both alternatives and was considered a resource for this analysis. In some cases, ash is used for products such as cement add-mixtures, road base aggregate, and roofing shingle production. The ash criterion was assessed based on the volume of ash produced and its potential for reuse. The research team assigned ash production a weighting of 15% of the conservation sub-category score.

Similarly, conservation of water criterion was assigned a weighting of 15% of the conservation sub-category score. Each alternative was assessed based on the amount of water consumed per unit of electricity produced. The analysis considered only raw water used from external sources and not water recycled within the process itself.

Whether energy was from a renewable biogenic source or non-renewable fossil fuel source was evaluated in the Fixed Carbon - Energy Recovery sub-category and assigned a weighting of 35%.

The greenhouse gas criterion reflects the relative amounts of greenhouse gases produced by each alternative. The research team calculated greenhouse gas emissions for each alternative based on carbon dioxide equivalent emissions from existing FBIs and the emissions identified for coal power plants with carbon (CO_2) capture technology in *Cost and Performance Baseline for Fossil Energy Plants* (NETL, 2013). The research team assigned the greenhouse gas criterion a weighting of 35% of the conservation/optimization of resources sub-category score.

The sub-category of air quality was weighted as 25% of the environmental category score. Pollutant emissions from each alternative were evaluated based on the maximum amount of pollutant release allowed by regulatory requirements for each technology. For coal power plants, emissions must meet the Utility Maximum Achievable Control Technology (MACT) standards. Facilities using thermal oxidation of wastewater solids must meet the Sewage Sludge Incinerator (SSI) MACT standards. Both MACT standards contain emission limits for the following compounds:

- oxides of nitrogen (NOx)
- sulfur dioxide (SO2)
- particulate matter
- hydrogen chloride (HCl)
- mercury (Hg)
- lead (Pb)

The TBL air quality impact sub-category assessed the amount of each compound that could be released by the alternative per unit of energy produced.

The remaining 25% of the environmental category score was a measure of the alternative's flexibility to adjust to new environmental regulations. Both thermal oxidation of wastewater solids and coal power generation are subject to extensive regulations. For this TBL assessment it was assumed that future environmental regulations for combustion facilities will increasingly restrict exhaust gas emissions and byproduct use or disposal. This study assessed how each alternative could respond to more restrictive environmental regulations.

The social category represents the impact of each alternative on the community and was based on two sub-categories: nuisance issues and workplace conditions. The sub-categories were assigned equal weighting for the total social category score. Nuisance issues were evaluated based on criteria of dust, visual impact, and traffic. Workplace conditions were evaluated qualitatively on workplace environment, such as how comfortable the facilities are for employees, and quantitatively on workplace safety incidents. Workplace environment scores were assigned a workplace condition weight of 50%. The workplace safety analysis compared illnesses/injuries and fatalities reported to the U.S. Bureau of Labor Statistics from 2012-2014 for coal mining, coal power plants and wastewater treatment plant operations. The illness/injury scores and the fatality scores were equally weighted for the workplace conditions sub-category.

Economic Category Scores

The Coal Alternative received an economic category score of 4.8 whereas the Wastewater Solids Alternative scored 4.32. The criteria analyzed to determine the scores were the lifecycle cost of both alternatives and the engineering/technical considerations.

The lifecycle cost assessment included the cost of construction, operation and maintenance for each alternative. Both alternatives were to be constructed on property already owned by the facility. Construction cost included the equipment, building, instrumentation, electrical and all design and construction services. Operation and maintenance costs included electricity, chemicals, fuel, maintenance materials, water, waste disposal and labor.

The lifecycle cost for the Coal Alternative was much higher than for the Wastewater Solids Alternative due to the difference in size of each facility. The coal power plant was based on an electric production capacity of approximately 550,000 kW, whereas the wastewater solids energy recovery facility was based on an electric production capacity of 1,900 kW. Costs for both alternatives were normalized on a unit of energy basis ($/kWh). Normalized capital costs for the Coal Alternative were significantly lower than the Wastewater Solids Alternative, reflecting the economies of scale for the larger production facility. Normalized operating costs for the Wastewater Solids Alternative were significantly lower than the Coal Alternative, reflecting the relatively low incremental operating costs of operating energy recovery for an existing process facility. The costs are summarized in Table 2.

Table 2. Wastewater Solids and Coal Alternative costs.

Cost Parameter	Wastewater Solids	Coal
Capital Cost	$16,794,000	$1,867,567,000
Net Present Value (NPV) Operations and Maintenance (O&M) Cost	$3,088,000	$3,233,151,000
Total NPV Cost	$19,882,000	$5,100,718,000
Net Power Output (kW)	1900	549,970
Capital Cost ($/kW)	$8,839	$3,396
O&M Costs ($/kW)	$1,625	$5,879
Lifecycle Cost ($/kW)	$10,464	$9,275

The Coal Alternative received a score of 5.0 in this subcategory reflecting that it had the lowest lifecycle cost, while the Wastewater Solids Alternative received a score of 4.4.

The engineering/technical sub-category was scored qualitatively as shown in Table 3.

Table 3. Engineering/technical criteria scores.

Engineering/Technical Criteria	Criteria Weights	Wastewater Solids Score	Coal Score
Simplicity	33.3%	4	4
Flexibility	33.3%	3	3
State of Technology	33.3%	4	4
Weighted Average Score		3.67	3.67

The lifecycle cost and engineering/technical sub-category weights of 85% and 15%, respectively, were applied to the criteria scores to generate the economic category scores depicted in Figure 1. The Coal Alternative scored slightly higher than the Wastewater Solids Alternative because of the economies of scale associated with the large coal-fired power plant.

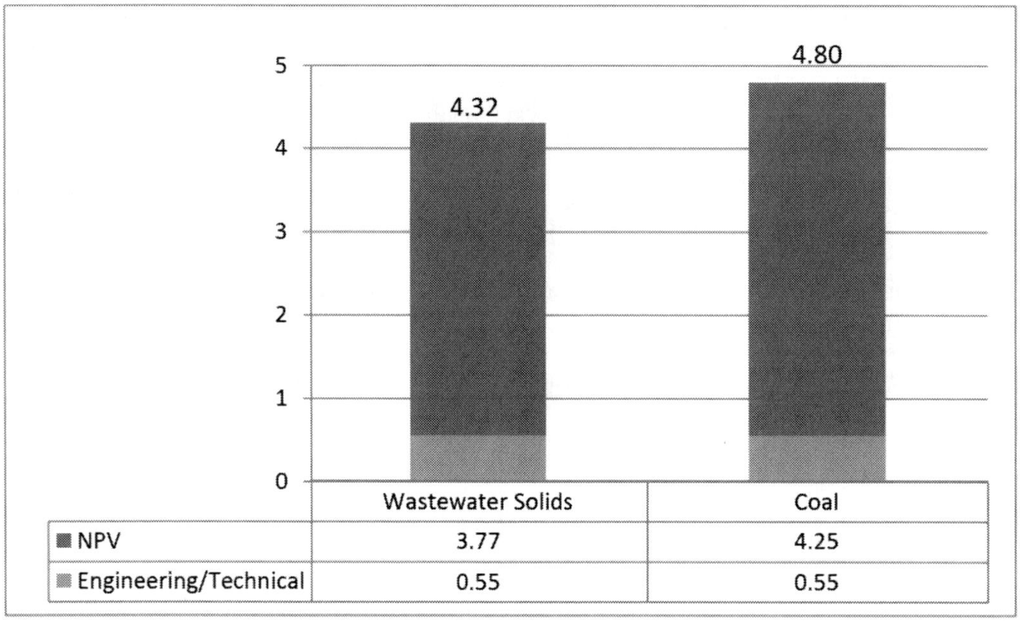

Figure 1. Category weighted economic results.

Environmental Category Scores

The Environmental Category evaluated each alternative for the conservation of resources and ability to adjust to regulatory changes as well as the air quality impacts of each alternative. The Wastewater Solids Alternative received an environmental score of 3.8 whereas the Coal Alternative received an environmental score of 2.1.

Conservation/optimization of Resources

The resources evaluated for this assessment were carbon, water and ash. The energy recovery portion of the wastewater solids thermal oxidation process does not consume or produce any of these resources. However, energy cannot be recovered without the operation of the incineration process, and the environmental impacts were evaluated for the entire thermal oxidation process to consider how this process compared overall with the energy produced from coal-fired power plants. The scores for this sub-category are provided in Table 4.

Table 4. Conservation/optimization of resources scores.

Conservation of Resources	Criteria Weight	Wastewater Solids Score	Coal Score
Ash Use	15%	3	4
Fixed Carbon—Energy Recovery	35%	5	0
Fixed Carbon—Greenhouse Gases	35%	3.6	5
Water Conservation	15%	5	0
Weighted Average Score		4.21	2.35

Air Quality Impact

The impact of each energy production alternative on air quality was considered in the environmental category assessment. The compounds evaluated were selected as representative of key pollutant impacts from the emissions regulated by the MACT standards for SSIs and utilities. The wastewater solids thermal oxidation system emissions were based on the emission limits of the MACT standards for existing fluid bed incinerators. The emissions from coal-fired power plants were based on Utility MACT standards emission limits and values were obtained from the 2013 NETL report. Emissions were normalized on an energy unit basis for comparison of alternatives. Table 5 shows the calculated emissions for each alternative.

Table 5. Air pollutant emissions.

Air Pollutant	Wastewater Solids, g/MWh	Coal, g/MWh
HCl	2.18	9.07
NO_x	862	454
SO_2	122	680
Hg	0.104	0.004
Pb	0.021	0.009
Particulate Matter	50.3	136

Each pollutant was weighted equally to determine the weighted average of the air emission criterion. The Wastewater Solids Alternative received a weighted score of 1.83, whereas the Coal Alternative received a weighted score of 1.67.

Regulatory Flexibility
The flexibility of each alternative to adapt to regulatory changes was assessed and scored qualitatively. The weighted score of each criterion were multiplied by the sub-category weight and summed to calculate the environmental category score for each alternative. The environmental category scores are depicted in Figure 2.

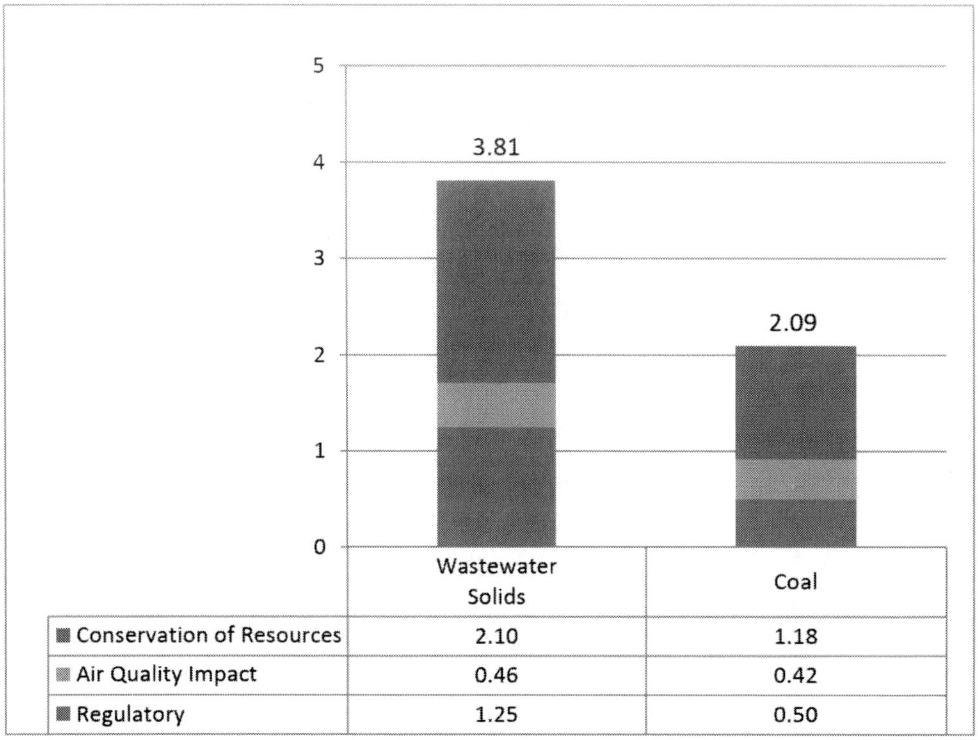

Figure 2. Category weighted environmental results.

Social Category Scores
The Social Category included nuisance issues that impact the general public and workplace conditions that impact plant personnel.

Nuisance issues considered were dust, visual impacts, and traffic. The Wastewater Solids Alternative would generate little dust outside of the facility. An energy recovery facility would look very similar to other buildings at the WRRF and would have minimal visual impact. Only minor increases in truck traffic would be expected with construction of an energy recovery facility.

The nuisance issues for coal power plants are more prominent than for the wastewater solids incineration facilities. Coal mining was included in this assessment because it is an integral activity of power production from coal. A coal power plant with coal mining operations produces more dust and traffic than a wastewater solids energy recovery facility. The visual impacts on the community are about the same for the respective alternative energy generation facilities, but the Coal Alternative scored lower in this category because a coal mine has negative visual impacts. The scores for the nuisance issues subcategory are shown in Table 6.

Table 6. Nuisance issues scores.

Nuisance Issues	Criteria Weight	Wastewater solids Score	Coal Score
Dust	33.3%	5	0
Visual	33.3%	4	2
Traffic	33.3%	4	1
Weighted Average Score		4.3	1

The workplace conditions sub-category included the criteria of workplace environment and workplace safety. The workplace environment assessment was qualitative.

Workplace safety scores were based on incident data available from the U.S. Bureau of Labor Statics for 2012 to 2014. Data from the categories for Mining/Coal Mining and Utilities/Fossil Fuel Electric Power Generation was used for the Coal Alternative. Data from the Utilities/Sewage Treatment Facilities was used for the Wastewater Solids Alternative. The inverse of average incidents was normalized to the percent of the maximum score and then scores were adjusted down so that the lowest score is 0.

Table 7. Workplace injuries and fatalities calculations and scoring.

Workplace Safety	Average Incidents Per Year	Inverse	Normalized Score	Adjusted Score
Illness & Injury				
Sewage Treatment Facilities	284,267	3.5E-06	5	3.64
Coal Mining and Power Generation	1,047,367	9.5E-07	1.357	0
Fatalities				
Sewage Treatment Facilities	2	0.5	5	4.48
Coal Mining and Power Generation	19.3	0.052	0.517	0

The workplace conditions score was calculated by applying a weighting factor of 50% to the workplace environment score, 25% to the illness/injury score and 25% to the fatalities score as shown in Table 8. The Wastewater Solids Alternative received a workplace conditions score of 4.03 and the Coal Alternative received a score of 1.0.

Table 8. Workplace conditions subcategory scores.

Workplace Conditions	Criteria Weight	Wastewater Solids Score	Coal Score
Workplace Environment	50%	4.00	2.00
Injuries	25%	3.64	0
Fatalities	25%	4.48	0
Weighted Average Score		4.03	1

The nuisance issues score and workplace conditions scores were each weighted at 50% of the Social Category score. Figure 3 depicts the weighted scores that were generated.

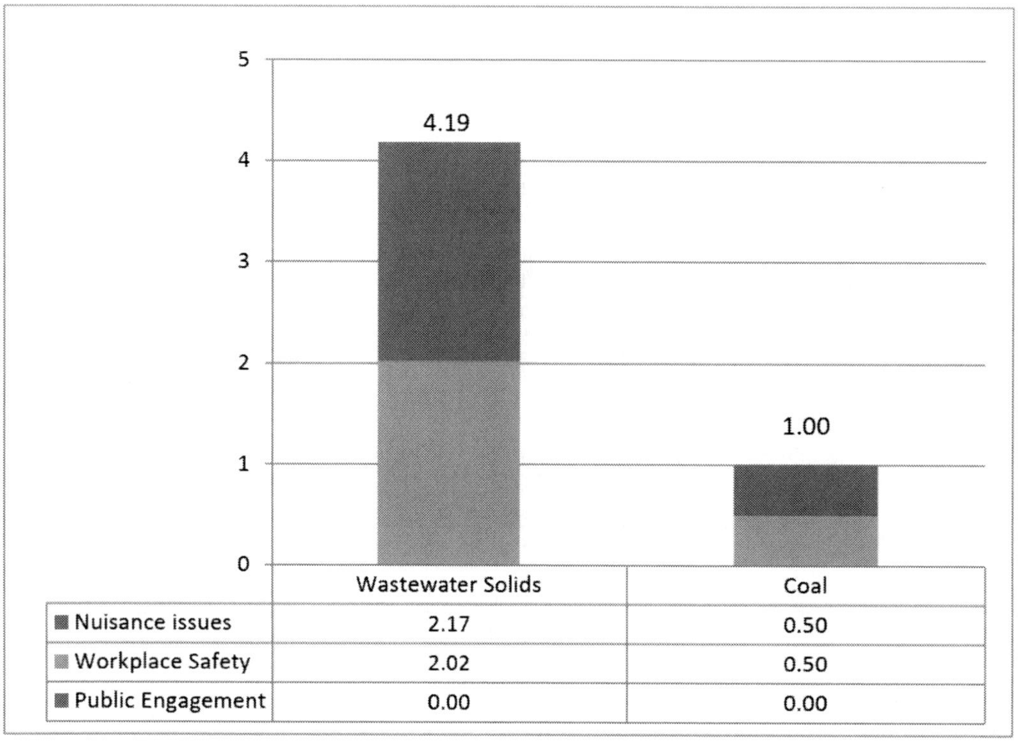

Figure 3. Social category results.

Overall TBL Evaluation
An overall weighting of the category scores resulted in a 4.1 rating for the Wastewater Solids Alternative and a 2.6 rating for the Coal Alternative. The overall TBL ratings demonstrate that based on overall sustainability considerations, energy recovery from wastewater solids has advantages for generation of electricity compared with coal-fired power plants. Figure 4 shows the Category Weighted TBL results.

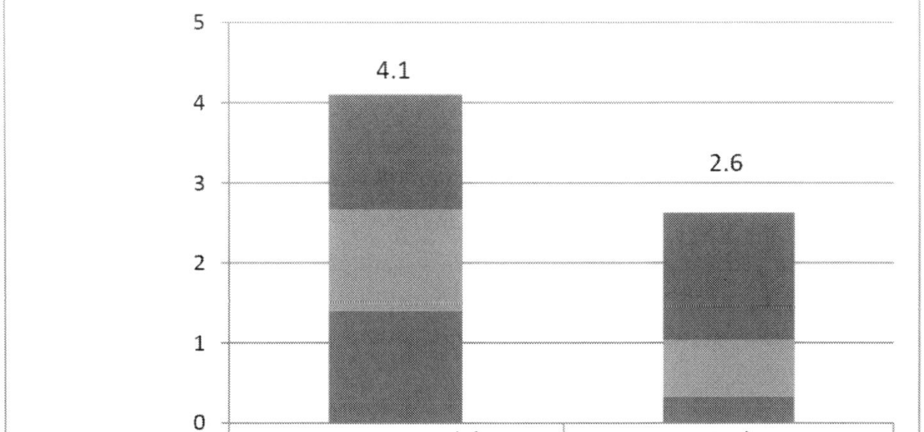

	Wastewater Solids	Coal
Economic	1.4	1.6
Environmental	1.3	0.7
Social	1.4	0.3

Figure 4. Category weighted TBL results for energy recovery from thermal oxidation of wastewater solids and coal-fired power production.

CONCLUSIONS FROM THE TBL ANALYSIS

The primary conclusion from the TBL analysis of power produced from energy recovered from thermal oxidation of wastewater solids compared to energy generated from coal is that the Wastewater Solids Alternative is more sustainable. This is mainly a result of the higher score that the Wastewater Solids Alternative achieves when environmental and social impacts are accounted for. Additional conclusions from the TBL analysis include:

- Lifecycle costs for the Wastewater Solids and Coal Alternatives are relatively similar, with lower costs for electricity produced from coal-fired power plants attributed to economies of scale.
- Conservation of resources and regulatory flexibility were significantly more favorable for the Wastewater Solids Alternative than the Coal Alternative, resulting in a higher environmental category score for the Wastewater Solids Alternative.
- Although a coal-fired power plant itself was not rated to be a major nuisance, the mining and transportation of the coal significantly decreased the social sustainability score for this alternative.

ACKNOWLEDGEMENTS

The authors thank WE&RF for its support of this research. In addition, the authors thank the many personnel at facilities that provided information and numerous hours of in-kind contributions and the ENER13T14 PSC team for their valuable guidance and oversight.

REFERENCES

Elenbaas, M., A. Patil, S. Tarallo, N. Beecher, A. Carpenter, and S. Brown. 2014. *Triple Bottom Line Evaluation of Biosolids Management Options*. Report No. ENER1C12a. Alexandria, VA: Water Environment Research Foundation /New York State Energy Research and Development Authority.

Dominak, R., W. Hoener, J. Welp, and G. Queiroz. 2016 WEFTEC Conference Presentation: Energy Recovery from Thermal Oxidation of Wastewater Solids: State of Science Review. Water Environment Research Foundation.

Hoener, W., R. Dominak, J. Welp, and G. Queiroz. 2016 Residuals & Biosolids Conference Presentation: Evaluation of Energy Potential and Heat Recovery from Thermal Oxidation of Wastewater Solids with the Latest Generation of Thermal Oxidation Technology. Water Environment Research Foundation.

National Energy Technology Laboratory. 2013. Cost and Performance Baseline for Fossil Energy Plants Volume 1: Bituminous Coal and Natural Gas to Electricity, Revision 2a; U.S. Department of Energy.

U.S. Department of Labor, Bureau of Labor Statistics, Census of Fatal Occupational Injuries (CFOI), Industry by private sector, government workers and self-employed workers. Summary Tables (2012, 2013, 2014). https://www.bls.gov/iif/oshcfoi1.htm (accessed Aug 2016).

U.S. Department of Labor, Bureau of Labor Statistics. Incidence rate and number of nonfatal occupational injuries by industry and ownership. Summary Tables (2012, 2013, 2014). https://www.bls.gov/iif/oshsum.htm (accessed Aug 2016).

Sewage Sludge Gasification as an Alternative to Incineration

Andrew Jones[1*]

[1] GV Jones & Associates Inc., Anchorage, Alaska
* Email: andrew@gvjones.com

ABSTRACT

While gasification had been used for many years as a means of converting carbonaceous fuels to useful heat and power, its use was largely phased out with the development and distribution of inexpensive and energy rich petroleum hydrocarbon fuels. However, there has been a recent renewed interest in gasification as a viable means to produce renewable energy while reducing the consumption of fossil fuels. Gasification can provide energy from renewable fuel sources while producing significantly less emissions than conventional combustion processes.

Sewage sludge gasification utilizes dewatered and dried sewage sludge as a feedstock. The gas produced in the gasifier, referred to as syngas, is comprised of hydrogen and carbon monoxide which can subsequently be combusted in boilers for heat or engines/turbines for electrical power generation.

Sewage sludge gasification as a process deployed at municipal wastewater treatment facilities is relatively new, with several pilot scale studies initiated between 2003 and 2007. Some of these initial efforts never advanced beyond a planning stage due to the absence of public support for implementation of technologies without an extensive proven operational track record (Stamford, CT). Others had started full scale operations, and subsequently shut down due to costs incurred to correct technical problems which could not be financially supported (Sanford, FL). However, there are two existing operational full scale sewage sludge gasification facilities in Europe (Balingen and Mannheim, Germany) and a third, the world's largest, in Asia (Tokyo, Japan).

The primary wastewater treatment facility in Anchorage, Alaska currently utilizes a multi-hearth incinerator to reduce the volume of wastewater solids which must be hauled to the local landfill. The city is considering sewage sludge gasification as a viable alternative to their incinerator to reduce their emissions, reduce sludge volume, and to operate their facility without being subject to the SSI Rule. In pursuit of that, the city is engaged in a full scale pilot project in which sludge from Anchorage is being dried on site, bagged, and shipped to a full scale operational gasifier for full scale testing. The results of this testing shall determine whether or not further steps towards the acquisition of a gasifier for the treatment plant in Anchorage will be taken.

KEYWORDS: Gasification, Sewage Sludge Incineration Rule, SSI Rule, Renewable Energy

INTRODUCTION

The John M. Asplund Wastewater Treatment Facility (AWWTF) located in Anchorage, Alaska handles the wastewater for the city of Anchorage, truck hauled sewage sludge from other nearby treatment plants and area septic tanks, and the wastes produced by nearby fish processers. The AWWTF currently processes mechanically dewatered primary sludge in a multi-hearth incinerator which is fired with natural gas. This volume reduction is necessary to reduce the amount of solid material which must be hauled to the landfill 20 miles away.

The incinerator has been in operation for over thirty years and the Anchorage Water and Wastewater Utility (AWWU) will soon be faced with the decision to either replace several worn components of the existing incinerator, or to replace the incinerator entirely. Under the Sewage Sludge Incineration Rule (SSI Rule) pursuing either of these strategies will trigger more stringent stack exhaust emissions standards for the incinerator. The manpower and emissions testing associated with maintaining compliance with the existing air emissions permit for the incinerator already represent a significant cost to AWWU and those costs are expected to increase if more stringent emissions standards must be met by the incinerator in the future.

Alternative sludge management strategies to incineration have been proposed, but Alaska's cold climate and relative remoteness make technologies like anaerobic digestion and composting less practical when compared to incineration. Neither technology offers a solids reduction which is comparable to that of incineration. Faced with these challenges, the AWWU Engineering Division has elected to consider another alternative to incineration; sewage sludge gasification.

Gasification processes appear similar, but are actually very different from incineration processes. Incineration processes are supplied with oxygen in excess of the stoichiometric requirement for full oxidative conversion of the carbon in sewage sludge to carbon dioxide, but in gasification processes oxygen is supplied at much less than this stoichiometric requirement. This leads to the generation of a synthesis gas (syngas) which consists primarily of carbon monoxide, hydrogen, and methane. This syngas can then be combusted to generate useful heat and electricity while producing much cleaner emissions than incinerators. In fact, the US EPA recently indicated that sewage sludge gasification does not fall under the scrutiny of the SSI rule. In a letter to the operator of a sewage sludge gasifier the EPA stated that "...no flame is applied or propagated in the gasifier and the gasifier prevents combustion by limiting the air-to-sludge ratio such that combustion cannot occur."

The fact that a gasifier operates in an oxygen limited environment results in emissions of reduced gas species rather than the oxidized gas species which are typical of an incinerator. Instead of oxidized nitrogen (NO_x), a gasifier produces ammonia gas which can then be eliminated from the gas stream in an acid scrubber. Instead of producing the oxidized sulfur compounds (SO_x) which are regulated in incinerator emissions, the sulfur in a gasifier is reduced to hydrogen sulfide. The addition of a small amount of limestone to a gasifier can then convert this gaseous sulfide to a solid material which is subsequently removed along with the ash from the gasifier.

$$CaCO_3 + H_2S \Leftrightarrow +CaS + H_2O + CO_2$$

The removal of sulfur in the gasifier can result in SO_x emissions which are orders of magnitude less than the SO_x emissions from an incinerator. (USDOE, 2000)

There are several different possible configurations for a gasifier, but the fluidized bed gasifier is shown here as an example to illustrate the overall process of a sewage sludge gasifier:

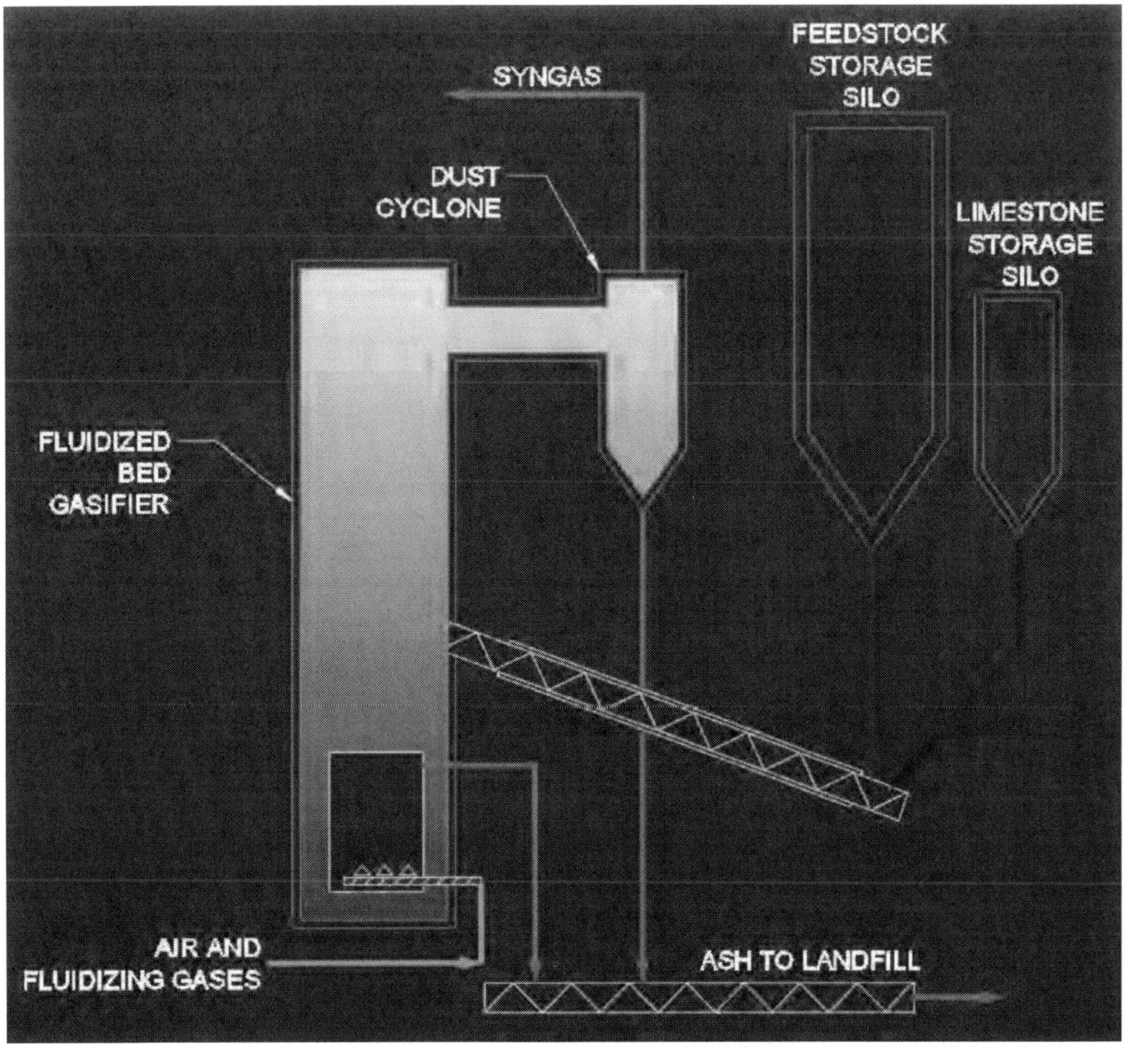

Figure 1 – Fluidized Bed Gasifier

The syngas produced by the gasifier can then be used to generate heat and electricity. The syngas can either be supplemented with natural gas and used to generate electricity in a gas turbine or the heat energy of the syngas can be directly extracted in a thermal oxidizer.

For the city of Anchorage this potentially means achieving heat and power generation for the treatment facility, reducing emissions of regulated contaminants, and not being subject to the scrutiny of the SSI Rule. To this end, the city of Anchorage is preparing to conduct a full scale test on an operational gasifier in Germany to measure the yield of syngas and heat energy per pound of Anchorage sewage sludge. This data can then be used as a basis for an economic and practical recommendation for the installation of a gasifier at AWWTF. This testing is to occur in Germany in the Spring of 2017. A conceptual energy balance example is shown in Figure 2:

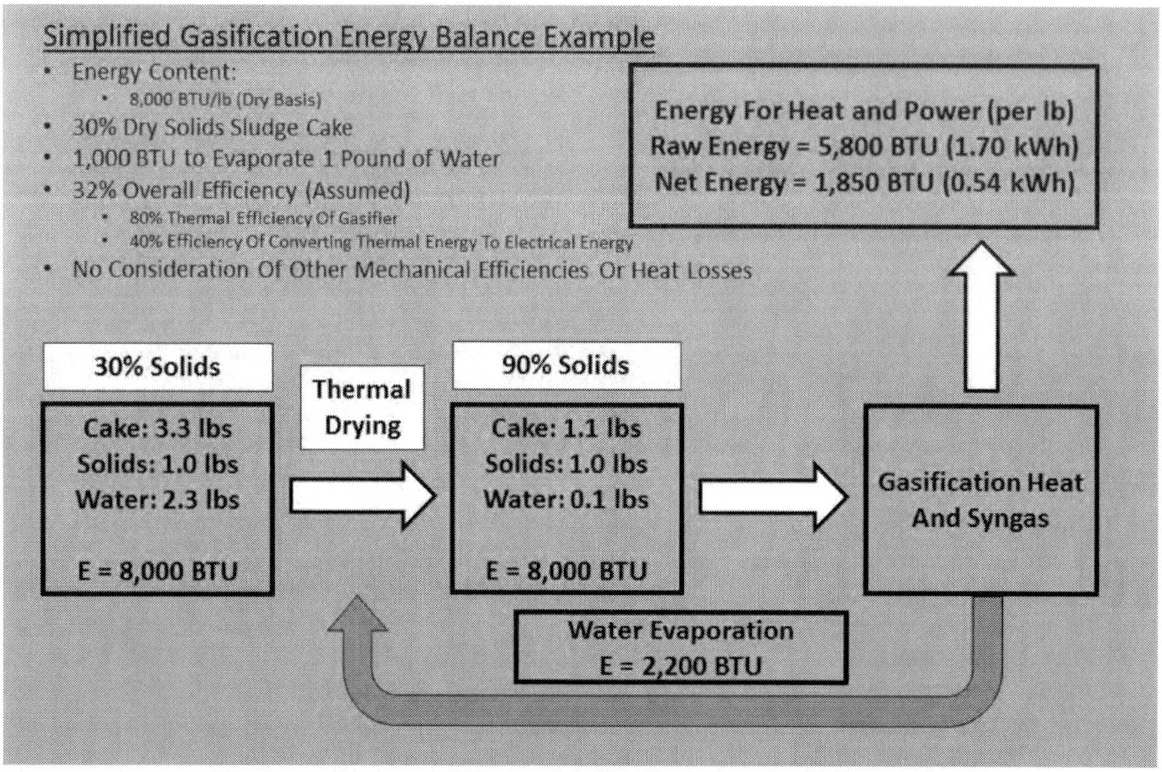

Figure 2 – Energy Balance for Sewage Sludge Gasification

This simplified example shows that for a sludge with a relatively high energy content (8,000 BTUs per pound of dry solids) that a gasifier processing 20 dry tons of material per day would be theoretically capable of producing over 20,000 kWh of electricity per day. Understanding the potential yield of heat and power requires a much more detailed analysis that what is shown here. However, this example illustrates the importance of optimizing the performance of mechanical sludge dewatering equipment. Increasing the dry solids content of the sludge by only a few percentage points can dramatically decrease the energy required for drying the sludge to a dry solids content of 85-90%.

At the time of this writing there are no gasifiers in the United States which operate solely on sewage sludge without the addition of supplemental feedstock materials (e.g. wood waste, waste plastics, solid waste). The successful implementation of a sewage sludge gasifier in Anchorage would offer significant benefits for energy savings while allowing the plant to produce cleaner air emissions. This plant might then serve as an example for other municipal wastewater plants which are seeking to eliminate their incinerators and move to a biosolids management strategy which offers a volume reduction comparable to that of incineration.

GASIFICATION BACKGROUND AND THEORY

Gasification is the name given to the series of thermodynamic processes which occur to change solid carbonaceous materials (feedstock) into combustible gases in a low-oxygen environment. These generated gases can then be combusted in order to extract useful heat and/or other forms of energy. Each of the processes described is interrelated with the other processes because the products of one reaction become the reactants for another. The specific configuration of the gasifier and the operation of the gasifier will affect the performance and the resulting gas composition.

Gasification reactions occur in the presence of high temperature. Gasification is also sometimes referred to as "choked combustion" because the amount of oxygen (or air) introduced into the process is deliberately restricted. Restricting the oxygen prevents the carbon and other materials present in the gasifier from being fully oxidized. The result is the production of partially oxidized or reduced gases which can collected and subsequently burned as a fuel for space heating or in an engine for power generation.

Sewage sludge gasification is typically preceded by multiple unit processes including dehydration, pyrolysis, and combustion. The gasification process itself can include the introduction of a water source to achieve a shift in the content of hydrogen production. These topics are described briefly below.

Dehydration

Dehydration is a process wherein the moisture content of the feedstock material is reduced. For moist feedstock materials such as sludge solids, dehydration is an essential gasification pretreatment process. Dehydration is a thermal drying process wherein the feedstock is exposed to enough heat to drive off entrained moisture. As such dehydration equipment is typically configured as a heated reactor that can maximize heat transfer to the applied sludge and withdraw moisture.

As the carbonaceous feedstock introduced into a dryer reactor rises in temperature, any water entrained in the material will begin to evaporate. The presence of some amount of moisture in the feedstock material is normal and water vapor serves as an important reactant in several of the gasification process reactions. However, water evaporation requires a significant amount of energy. In fact, dehydration is typically the largest parasitic energy demand for the entire gasification process.

If excessively wet feedstock materials are placed into the gasifier, it may interfere with the other downstream gasification reactions. Typical mechanically dewatered sludge solids feedstock is 28 to 32 percent dry solids. The dehydration process increases this to 70 to 90 percent dry solids.

Many types of thermal sludge solids dryers are commercially available. They include belt dryers, drum dryers, paddle dryers, and fluidized bed dryers. Of these, paddle dryers are operated at high temperatures (up to 750 and 1,500 °F, respectively) and both paddle dryers and fluidized bed dryers operate using an external heated fluid transfer medium such as hot oil or steam. Belt dryers and drum dryers are operated at lower temperatures (typically 300 °F or less).

Pyrolysis
The pyrolysis process – literally meaning to break with fire – occurs after the water has been evaporated from the feedstock materials. During pyrolysis, the combination of high temperatures and an oxygen limited environment cause the feedstock materials to volatilize into gaseous materials.

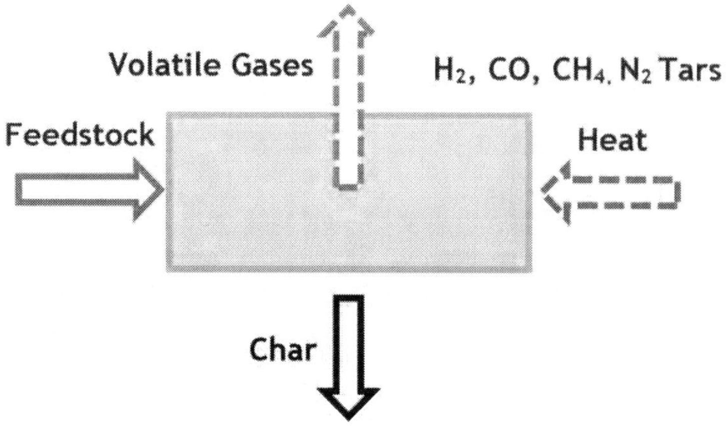

Figure 3 – Simplified Pyrolysis Process

The primary gases released in pyrolysis include hydrogen (H_2), carbon monoxide (CO), methane (CH_4), and tars. Nitrogen gas (N_2) passes through the process as largely unreacted and is of importance in that it can dilute the concentration of combustible gas components in the finished produced gas. Of these, the gases which have the most potential for recovery and reuse as an energy source are hydrogen, carbon monoxide, and methane.

The formation of tar during pyrolysis presents an operational challenge. If allowed to cool and reform, tars can build up within the system and foul process equipment. The amount of tar formed during pyrolysis depends on the configuration of the gasifier and the type of feedstock material in the gasifier. A common means to control tar formation in a gasifier is through a process referred to as 'cracking' which breaks tars into smaller compounds. (NREL, 1998)

Pyrolysis converts part of the feedstock mass into gas, but the fraction of the feedstock left behind becomes 'char'. Char consists of both inert materials and organic materials not volatilized. The type of char formed by these reactions depends largely on the type of feedstock material. For instance, pyrolysis of wood materials produces a char which typically consists of medium-sized lightweight pieces of char. However, materials like sludge tend to form much finer dust-like char.

The pyrolysis reactions generally begin to occur in the 390 to 570 °F range. The heat required for the pyrolysis reaction is obtained from the combustion of the volatilized gases.

Combustion
In the presence of oxygen and a high temperature, the feedstock material will react with oxygen to form carbon dioxide. In a gasifier this reaction produces both heat and carbon dioxide which are useful for other processes which occur in the gasifier. However, combustion reactions in gasification are limited as the availability of oxygen is controlled to provide only 25 to 30 percent

of the oxygen required for complete feedstock oxidation. What limited combustion occurs takes place in those portions of the reactor where the stoichiometric requirement of oxygen is met.

The source of oxygen for most gasifiers is air. However, steam can also be used as the source of oxygen supply. The advantages of using air as an oxygen supply are that it is readily available and therefore inexpensive. The disadvantage of air is its high nitrogen content which passes through the gasification process and dilutes the heating value of gas produced. The advantages of steam gasification are that the gas produced is higher in hydrogen (and therefore energy content) and the presence of water contributes to the water-gas shift reaction described further below. The disadvantage of steam gasification is the operation and maintenance of a steam system.

Gasification

Gasification occurs when char formed by the pyrolysis reactions reacts at high temperatures with water vapor and carbon dioxide. These reactions form the majority of the gases produced by the gasifier. The reactions are generally endothermic and consume the heat generated by the combustion reactions needed to keep the char hot and sustain gas production.

The two major reactions which occur in the gasification phase involve the reaction of water vapor and carbon dioxide with the char produced in pyrolysis. These reactions, described below, produce carbon monoxide and hydrogen gas.

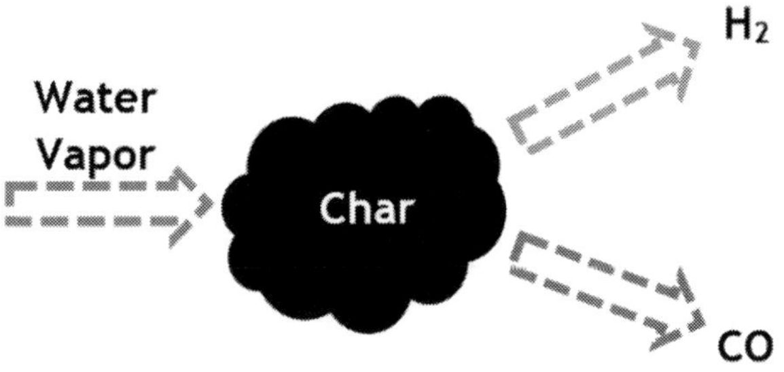

Figure 4 – Simplified Gasification Process

Steam Reforming and Water-Gas Shift Reactions

Steam reforming of hydrocarbons with a subsequent water-gas shift reaction has been a common means of commercially producing hydrogen gas. In summary, steam reforming for hydrogen production is an endothermic, high temperature and pressure process (1,500 to 1,600 °F, 20 to 30 atm) in which steam is contacted with a carbonaceous fuel feedstock in the presence of a metal catalyst (nickel) to produce carbon monoxide and hydrogen. This produced gas and steam is then cooled to temperatures of 450 to 700 °F during which the carbon monoxide becomes carbon dioxide, in what is called a water-gas shift reaction. To complete commercial hydrogen gas production, the carbon dioxide is stripped from the hydrogen using one of several processes including amine absorption or pressure swing adsorption (PSA).

Applied to sewage sludge gasification, the introduction of steam in the gasification process produces a syngas fuel of higher energy content due to increased hydrogen concentrations. Research has shown that sewage sludge gasification conducted with high temperature (1,800 °F) steam increased hydrogen production by a factor of three times over that which would be produced without the steam, resulting in an increase of gas energy yield of approximately 100 percent over that produced from an air operated gasifier.

Comparison of Gasification with Incineration

Options for disposal of sewage sludge solids is limited by law to land application, landfill disposal, or incineration/ash disposal. While the Anchorage wastewater treatment plant currently uses incineration for sludge solids reduction, the option to deploy gasification has some advantages in the quality of the emissions produced. Table 1 provides a summary of a few key differences between sewage sludge incineration and gasification:

Table 1 – Comparison of Incineration and Gasification

Incineration	Gasification
• Excess Oxygen Provided For Complete Combustion	• Limited Oxygen Provided (Choked Combustion)
• Oxidative Environment	• Reducing Environment
• Volatile Nitrogen Products: NO_2	• Volatile Nitrogen Products: NH_3, N_2, HCN
• Volatile Sulfur Products: SO_2	• Volatile Sulfur Products: H_2S
• High Particulate Matter Production (Air Scrubber Required)	• Decreased Particulate Matter Production

Fundamentally, incineration processes are designed to achieve the complete combustion of volatile materials placed into the incinerator by reacting those volatile materials with a stoichiometric excess of atmospheric oxygen. On the other hand, gasification is designed to restrict the amount of oxygen provided to the process in order to drive the high temperature conversation of carbon to useful gaseous fuels which can be used for energy generation.

Gasification is sometimes confused with biogas generated during anaerobic digestion, but the processes are not at all similar. Anaerobic digestion is a biological process which relies on anaerobic organisms to metabolize carbonaceous materials present in sewage sludge and generate methane and hydrogen gas in the absence of oxygen and at temperatures ranging from 80 to 130 °F depending upon whether the process is operated as a mesophilic or thermophilic process. By contrast, gasification is a non-biological physical chemical process which operates at temperatures on the order of 1,500 °F in order to produce a gas composed of carbon monoxide, hydrogen and methane from which useful energy can be recovered.

REGULATORY STATUS

Existing State and Federal regulations applicable to stationary combustion equipment address atmospheric gas emissions from heat and/or power generation combustion processes, discharge of liquid process wastewaters generated as a result of boiler and turbine equipment operations, and disposal of solids waste combustion residuals.

Applicable air quality regulations have been promulgated as a result of the Clean Air Act. These regulations require stationary sources including power plants, factories, and sewage sludge incinerators comply with maximum contaminant emission concentrations expressed as mass per volume of flue gas. Contaminants of concern include particulate matter, sulfur oxides, carbon monoxide, ozone, ammonia, lead and other metals, dioxins/furans, and mercury.

In addition to control of air emission contaminants, stationary combustion equipment typically is configured with closed loop heating systems which must be periodically purged or blown down to control the concentration of dissolved solids within these systems. And finally, any residual ash created by combustion/incineration of dewatered sewage sludge must be transported and disposed in permitted landfill facilities such as the Anchorage Regional Landfill.

If the incinerator continues to operate in Anchorage, new sewage SSI Rule regulations applicable to AWWU incinerator at the Anchorage wastewater plant will likely add additional operational, monitoring, and reporting requirements to existing regulatory requirements for sludge incineration. In anticipation of the forthcoming SSI Rule, gasification process equipment manufacturer/service provider MaxWest environmental (now PHG Energy) asked EPA for a ruling as to whether or not the SSI Rule would apply to sewage sludge gasification processes. The EPA responded in a December 2013 letter with a determination that the new SSI Rule would not apply to sewage sludge gasification processes.

For treatment plants like the one for the city of Anchorage, this EPA ruling could allow for the same significant volume reduction afforded by incineration while producing cleaner emissions which are not subject to the SSI Rule.

ANCHORAGE PILOT PROJECT

In order to determine the feasibility of gasification for the city of Anchorage, AWWU decided to engage in a full scale testing campaign in which Anchorage's sludge was to be used as a fuel source in an existing sewage sludge gasifier in Germany. This meant that the sludge cake produced at the Anchorage facility currently would be dried to more than 90% dry solids, bagged, and then shipped to an existing gasifier in Balingen, Germany for full scale testing.

Full scale testing was thought to be the best way to predict the performance of a sewage sludge gasifer processing Anchorage's biosolids. This testing would allow for the quantification of a gasifier's energy generation potential and a better understanding of its emissions characteristics. During the testing campaign the instrumentation on the existing gasifier will be utilized for performance monitoring. These instruments include temperature and pressure probes inside of the gasifier, load cells for feed rate and ash quantification, and online gas monitors which will track the composition of the synthesis gas produced.

Another motivation for testing Anchorage's sludge at full scale is the difference in the sludge characteristics of a secondary wastewater sludge and a primary wastewater sludge. The German sludge being processed at the gasifier in Germany is a secondary sludge with a lower energy content than the primary sludge which is produced in Anchorage. The heterogeneous nature of a primary sludge could cause the sludge to behave differently in the gasifier and might lead to operational difficulties which are as yet unknown. Therefore, it was thought that the best way to

learn how a high energy primary sludge might behave inside of a gasifier would be to test it at full scale.
To achieve this, AWWU rented a pilot scale dryer from Wyssmont and set up that equipment at the treatment plant site (Figure 5). The dryer was a multi-tray dryer in which sludge cake was fed into the top of the dryer and was then subjected to a constant stream of warm dry air as the sludge cake fell to each successive tray. By the time the sludge had reached the bottom of the dryer it was at least 90% dry solids.

This dried sludge was then loaded into plastic lined supersacks which were sealed and protected from the weather (Figure 7). Finally, the supersacks were loaded into shipping containers for shipment to Germany.

Figure 5 – Site Used for Temporary Sludge Drying

Figure 6 – Dried Biosolids Produced by Temporary Dryer

Figure 7 – Stored Dried Biosolids

SUMMARY AND CONCLUSIONS

Gasification is a non-biological thermal process which converts carbonaceous feedstock materials into a gas in a limited oxygen environment at high temperatures. Gasification offers the potential for viable biosolids volume reduction while providing a source of renewable energy. Gasification of wastewater biosolids is especially appealing for Alaska given its relatively remote location and limited applications for beneficial use of wastewater biosolids.

The intent was to have the full scale gasification testing campaign completed before the date of the conference, but unfortunately scheduling delays caused the date of the testing to be pushed back beyond the date of the conference.

At the time of this writing the sludge is enroute to Germany and a team of engineers from Anchorage will be meeting the sludge when it arrives at the plant in Germany to conduct the testing campaign.

ACKNOWLEDGEMENTS

This work would not be possible without the support of the Anchorage Water and Wastewater Utility and its continued drive to pursue the development of innovative water and wastewater solutions for the Anchorage community. The author would specifically like to acknowledge Tom Winkler, Mike Fezatte, and Steve Nuss.

The author would also like to acknowledge Dr. Jeanette Brown of Manhattan College who has been an invaluable source of support and technical expertise on sewage sludge gasification.

REFERENCES

Kumar, G. Suresh. Advances in Gasification Technology: A Basic Review. Proceedings of the 1st international conference on Advances in thermal and energy systems. 2013.

Sieger, Ron. Thermal Drying of Wastewater Solids. Water Environment Federation Residual and Biosolids Committee Biotechnology Subcommittee. January 2004.

Worley, M. Biomass Gasification Technology Assessment – Consolidated Report. National Renewable Energy Laboratory (NREL). November 2012.

Takahashi, Hajime. Study on Sewage Sludge Gasification. Tokyo Metropolitan Government Bureau of Sewarage.

Milne, T.A. Biomass Gasifier “Tars”: Their Nature, Formation, and Conversion. National Renewable Energy Laboratory (NREL). November 1998.

USEPA. Technology Assessment Report – Aqueous Sludge Gasification Technologies. June 2012.

Roos, Carolyn J. Clean Heat and Power Using Biomass Gasification for Industrial and Agricultural Projects – USDOE Clean Energy Application Center. February 2010.

Vestergaard-Hansen, Bo. Evaluating and Selecting Class A Biosolids Heat Drying Options In The Pacific Northwest. Undated.

USDOE. A Comparison of Gasification and Incineration of Hazardous Wastes – Final Report. March, 2000.

Sieger, Ron. Biogasification and Other Conversion Technologies. Water Environment Federation Residual and Biosolids Committee Biotechnology Subcommittee. April 2002.

DUFFIN CREEK EXPERIENCE TWO NEW FLUID BED MUNICIPAL SLUDGE INCINERATORS EQUIPPED WITH STATE OF THE ART EQUIPMENT

E. Ferguson – Senior Project Manager, York Region
B. Dobson – Project Superintendent, Duffin Creek Water Pollution Control Plant
K. Dangtran, Ph.D. – Vice President, SUEZ
*L. Takmaz, Ph.D. – Product Manager, SUEZ

The Duffin Creek Water Pollution Control Plant (WPCP) is located at 901 McKay Road in the City of Pickering, Ontario, Canada. The plant is jointly owned by the Regional Municipalities of York and Durham. Both regions are experiencing significant population growth. The rated plant capacity has been increased to 630 million liters per day (ML/d) through the completion of the Stage 3 Process Expansion Projects

Duffin Creek WPCP has a long experience with fluid bed Thermal Oxidizers. Two fluid bed units were built back in the late 1970's by GL&V. Two, additional reactors built by SUEZ were recently added to provide the facility with a firm solids processing capability of 270 metric dry tonnes per day (MDTPD). Identical to the original systems, the new oxidizers are also equipped with two waste-heat recovery systems producing superheated steam to drive the two steam turbines coupled with two fluidization air blowers. Unit No. 4 fluidization air blower is connected to an electric motor (447 kW) and a steam turbine through a clutch mechanism. Unit No. 3 fluidization air blower is connected to only a steam turbine. A clutch mechanism between the fluidization air blower and the steam turbine enables the switching from turbine to electric motor during normal operation.

Once both units are operating at full capacity, the fluidization air blowers are driven by steam turbines during steady state operation. This unique design should be a template for future incineration systems.

Both fluid bed units have refractory lined hot wind-box and air distributor inside the reactor. . Each fluid bed unit has the capacity to incinerate 105 MDTPD total solids. Unit No. 4 passed the performance testing in June 2013. Unit No. 3 passed the performance testing in February 2014. Both fluid bed units have satisfied the emission requirements per air permit issued by the Ministry of the Environment and Climate Change (MOECC) including mercury, dioxin, and furan removal.

Each fluid bed unit is equipped with a dedicated heat recovery system employing a primary heat exchanger and waste heat recovery boiler. Primary heat exchanger (shell and tube design) is used to preheat the fluidization air to minimize the auxiliary fuel usage during steady state operation. Each unit was designed to be autogenous, with a sludge feedstock composition of 68% volatile, 28% total solids, and 5,560 kcal/kg sludge heat value based on volatile part of dry solids (dry sludge). Based on the feedback from the plant, both units are operating autogenously even with a sludge feedstock content as low as 24%. Flue gas from the fluid bed reactor is passed through a primary heat exchanger and then discharged into a waste heat boiler to generate superheated steam. From waste heat boiler, flue gas is sent to a wet scrubber to remove particulate and acid gas (SO_2, HCl). The wet scrubber is not equipped with a caustic injection system. Due to today's more stringent air requirements, a Kombisorbon mercury-removal system is installed downstream of the multi-venturi wet scrubber to remove mercury, dioxins, and furans. The Kombisorbon system includes a conditioner having a droplet separator to remove free water droplets from the clean flue gas discharged from the wet scrubber. The conditioner also has a heat exchanger to increase the saturated flue gas temperature about 20C above the dew point temperature to prevent moisture formation inside the fixed carbon bed adsorber. Clean flue gas discharged from the wet scrubber flows through the cold side of the conditioner heat exchanger and the steam from the low pressure steam header is used as the heating medium for the hot side of the conditioner heat exchanger. Clean flue gas from the conditioner heat exchanger is sent to Kombisorbon adsorber to remove mercury, dioxins and furans. There is also an ID Fan installed before the stack to maintain vacuum conditions inside the waste heat boiler to prevent any potential flue gas leak into the building. The Kombisorbon adsorber has three layers with the first layer filled with regular carbon and the following two layers filled with activated carbon.

The text provides the detailed information on the process, the scope of equipment installed, and the information regarding the performance test results. Mercury-removal technologies available today for fluid-bed sewage-sludge incinerators are also discussed, along with their respective mercury-removal efficiency associated with each mercury removal technology.

The successful operation at Duffin Creek WPCP has shown that the improved thermal oxidizer design incorporating enhanced air pollution control and energy recovery systems to reduce the operational expenditures is an economical, environmentally friendly, and cost effective solution for sludge disposal. Duffin Creek WPCP fluid bed units have satisfied the sludge disposal needs of the plant, emission requirements of the Ministry of the Environment and Climate Change (MOECC) and the future growth needs of both Regions.

KEYWORDS

Fluidized bed incineration, sludge disposal, wet scrubber, mercury, dioxin, furan, SSI MACT

* Corresponding Author

BACKGROUND

Duffin Creek WPCP is located in the city of Pickering, Ontario (Canada) on the shoreline of Lake Ontario. Although maximum treatment capacity of the plant is 630 ML/d, the plant is operating at 520 ML/d approved by the MOE with a provision to increase the capacity to 540 ML/d subject to further approval from MOE in the near future. The plant has the capability of treating 90% of the wastewater transferred from the York Region. Duffin Creek also serves the city of Pickering and town of Ajax, both located in the Durham Region. The plant is jointly owned by the two regions, York and Durham. After the treatment, processed water is discharged to Lake Ontario. Operation of the plant is strictly monitored by the Ontario Ministry of Environment (MOE) similar to the other wastewater treatment plants operating in the province of Ontario.

SYSTEM DESCRIPTION

Each Duffin Creek unit has a reactor with a hot windbox, refractory arch dome, and teardrop-shaped freeboard (Figure 1). Reactor diameter and height is 10.44 m and 12.36 m respectively. Four sludge ports are used to feed the sludge into the sand bed. The reactor is designed to have at least 7 seconds residency time to have a complete combustion resulting in lower emissions. A primary heat exchanger is installed downstream from the reactor to heat up the fluidization air to 650 C to have autogenous combustion. Further waste heat recovery is accomplished by installing a water tube waste heat boiler to generate superheated steam at 371 C and 35 barg (3,500 kpag). Steam generation rate per train is 7,348 kg/hr. Following the waste-heat boiler, a wet scrubber and tray cooler are installed to remove particulate and acid gas from the flue gas. An ID fan (Induced Draft Fan) is installed after the scrubber-tray cooler to maintain negative pressure inside the waste heat boiler. A Kombisorbon mercury-removal system is installed on the discharge side of the ID fan, which includes a conditioner, heat exchanger, and activated-carbon-bed adsorber. The reason for having a conditioner is to ensure the flue gas is free of water droplets before entering the adsorber. Gas temperature is raised about 20 C through the heat exchanger. Steam is used as the heating medium on the hot side of the conditioner heat exchanger. After the removal of elemental / ionic mercury, dioxins, and furans, clean flue gas is sent to stack.

Figure 1: Duffin Creek fluid bed reactor.

As shown by Figure 2, Kombisorbon Adsorber is fabricated from FRP material and fitted with three layers including a dust protection layer and two adsorbent layers. The adsorbent is Sulphur-impregnated activated carbon. Currently, there are four installations in North America using the Kombisorbon technology to remove mercury, dioxins and furans from the flue gas discharged from the sewage sludge fluid bed incinerators. One of these installations (G.E. Booth WWTP) is currently the world's second largest sewage sludge incineration plant (with 4 lines of fluid bed, 100 (MDTPD) burning capacity for each unit) located outside Toronto, Ontario (CA). The second installation, which is the subject of this technical paper, Duffin Creek WPCP is also located outside Toronto, Ontario (CA). The third installation (Y.C.U.A. WWTP) is located outside Detroit, Michigan (USA). Mattabassett WPCF (Cromwell, CT, USA) is the fourth installation in North America using the Kombisorbon Technology. All four projects have been designed, executed, and built by SUEZ.

Figure 2: Kombisorbon adsorber to remove mercury and dioxins-furans from flue gas.

Figure 3 shows the fluidizing-air-blower coupled with the steam turbine, via a clutch mechanism. One of the trains (No. 3) has a fluidization air blower connected only to a steam turbine. The other train (No. 4) has a fluidization air blower connected to a steam turbine and an electric motor (447 kW). A clutch mechanism is used to switch from electric motor to steam turbine during normal operation. During switching electric motor is maintained at the design speed. In case, the turbine trips or sludge feed into incinerator is lost, fluidization air blower is switched back to the electric motor from the steam turbine. Once both units are operating at full capacity, fluidization air blowers are connected only to steam turbines.

Figure 3: Duffin Creek fluid air blower coupled with steam turbine driving the blower.

A mixture of condensate and vapor from turbines is recycled through a steam-recycling loop and sent back to waste heat boilers. The recycling loop includes condensate tanks, deaerators, chemical feed skids, heat exchangers, and pumps. Part of the steam is being used for building heating. In addition, low-pressure steam from the steam system is directed to the conditioner heat exchanger to warm up the adsorber with ambient air during cold start up. This design further improves energy efficiency through the avoidance of electrical costs, which would have been incurred to warm up the ambient air with an electric heater. When the adsorber is warm enough, sludge injection into the incinerator is permitted, airflow into the adsorber is cut off, and the flue gas is permitted to pass through the adsorber.

The ash-removal system (Figure 4) on the waste-heat boiler is a wet-ash system. Each boiler has its own ash-removal system including a quench tank and slurry pumps. Scrubber effluent is used to quench the dry boiler ash in the wet ash tank. Discharge from the wet-ash tank is sent to the dewatering system through slurry pumps.

Figure 4: Duffin Creek waste heat boiler wet ash removal system.

MERCURY CONTROL TECHNOLOGIES

Duffin Creek fluid bed incinerators are equipped with the Kombisorbon system to remove mercury, dioxins, and furans from the flue gas. Flue gas is passed through a conditioner first and then passed through the carbon layers to capture mercury, dioxins, and furans. Also, there is a heat exchanger before the carbon bed to ensure the flue gas temperature is above the dew point to prevent condensation inside the carbon bed. Operators can take carbon samples from the adsorber during normal operation and test the samples for mercury loading to determine the current condition of activated carbon. It is estimated that every two or three years, carbon bed material is replaced with fresh carbon. The actual life expectancy of the carbon is based on the mercury loading rate at each facility. The Duffin Creek units do not have sufficient operating time on them at the time of this paper to provide a definitive life expectancy. The advantage of this technology is the lower operational expenditure for the plant and higher removal efficiency meeting the most stringent mercury emission limit in the world with maximum one micron per dry standard cubic meter corrected to 7% oxygen in stack. The main disadvantage of this technology is the high-pressure drop across the entire mercury removal system requiring a larger ID fan for a push-pull system or a larger fluidization air blower for a push-only system.

There are other available technologies to remove mercury from the flue gas applicable to sewage sludge fluid-bed incinerators. One of these technologies uses activated carbon injection into the flue gas stream. A carbon bed is eliminated from the design. Dry ash is removed from the flue gas through a cyclone, bag filter, or dry electrostatic precipitator before the injection of activated carbon. Reaction temperature is about 200 C. Spent chemical is removed through a bag filter and sent to the nearest landfill. The main advantage of this technology is the low-pressure drop and being mostly applicable to dry-fluid-bed incineration systems. The dry ash from the incineration system might be marketable to reduce the cost associated with the activated carbon usage. The main disadvantage of this technology is the continuous use of activated carbon resulting in higher operational expenditure for the plant. In addition, this technology has lower removal efficiency compared to Kombisorbon technology.

A more recent mercury removal technology has been developed by W.L. Gore & Associates known as the Sorbent Polymer Composite (SPC) technology. EnviroCare International applied the SPC technology to sewage sludge fluid-bed incineration systems. With this patented new technology, sorbents and catalysts are used in capturing mercury and SO_2 from the flue gas. SPC technology does not require a heat exchanger to increase the flue gas temperature above the flue gas saturation point. SPC modules are installed either inside the scrubber after the mist eliminator section or as a standalone system after the scrubber. SPC modules remove mercury and SO_2 from the flue gas at very low gas temperatures (typically lower than 93 C). SPC modules are hydrophobic ensuring a clean operation. The main advantage of SPC technology is the low capital and operational cost, however mercury removal efficiency is low compared to Kombisorbon technology. The main disadvantages of this technology is that SPC technology can't be used to remove dioxins and furans from the flue gas and also this technology can't be used to meet the Maximum Achievable Control Technology (MACT) limits in the US for a new incineration plant. MACT limit for a new incineration plant in the US is 0.001 mg/dscm (mg=milligrams, dscm=dry standard cubic meter) corrected to 7% oxygen. This technology has been implemented in the US at three existing sewage-sludge fluid-bed incineration plants. One of these plants is Mill Creek WWTP located just outside Cincinnati, Ohio with three fluid-bed incineration lines. Mill Creek WWTP was originally designed, executed, and built by SUEZ with a start-up year of 2010. The plant was retrofitted with SPC technology to be in compliance with MACT limits for an existing fluid incineration plant. The mercury limit for an existing fluid-bed incineration plant is 0.037 mg/dscm correct to 7% oxygen. The other two plants are located in Cleveland (OH) and Cromwell (CT). All three plants have passed performance testing meeting MACT emission limits in stack.

Table 1 compares the mercury removal efficiencies for the three different technologies discussed above.

Table 1: Mercury removal efficiency comparison.

Technology	Mercury Removal Efficiency
Activated Carbon Injection	50 – 90+
Activated Carbon Bed	>99
SPC	<70

COMMISSIONING AND PERFORMANCE TESTING

Duffin Creek fluid bed incinerators have been through the start-up, commissioning, and performance testing in 2013 and 2014. Unit No. 4 passed the performance testing in June 2013. Unit No. 3 passed the performance testing in February 2014. Both fluid bed units have satisfied the emission requirements per air permit issued by MOECC including mercury, dioxins, and furans. During the performance testing, sludge, ash, and water samples have been collected and analyzed. RWDI Inc. based in Ontario, Canada has been contracted to perform the stack testing on both units. Table 2 shows the stack emission results on Unit No. 4 showing compliance with air permit. Results of stack testing on Unit No. 3 are presented in Table 3.

Table 2: Unit No. 4 stack-emission test results during the performance testing in Jun. 2013.

Parameter	Limit	Test Results (Unit No. 4)
Oxygen	Minimum 4%	8.7%
Total Hydrocarbons	100 ppm (10 min. avg.) 20 ppm (30 min. avg.)	6.3 ppm 4.9 ppm
Hydrogen Chloride	30 ppm	0.15 ppm
Dioxins and Furans	100 pg/m^3	2 pg/m3
Total Suspended Particulate	20 mg/m^3	0.6 mg/m3
Arsenic	99% Removal Efficiency	100%
Cadmium	89% Removal Efficiency	99.9%
Chromium	99% Removal Efficiency	100%
Lead	92% Removal Efficiency	100%
Nickel	99% Removal Efficiency	99.9%
Mercury	Max. 70 $\mu g/m^3$	0.5 $\mu g/m^3$

Table 3: Unit No. 3 stack-emission test results during the performance testing in Feb. 2014.

Parameter	Limit	Test Results (Unit No. 3)
Oxygen	Minimum 4%	7.7%
Total Hydrocarbons	100 ppm (10 min. avg.) 20 ppm (30 min. avg.)	4.1 ppm 3.2 ppm
Hydrogen Chloride	30 ppm	0.517 ppm
Dioxins and Furans	100 pg/m^3	3.7 pg/m^3
Total Suspended Particulate	20 mg/m^3	0.35 mg/m^3
Arsenic	99% Removal Efficiency	100%
Cadmium	89% Removal Efficiency	99.9%
Chromium	99% Removal Efficiency	100%
Lead	92% Removal Efficiency	100%
Nickel	99% Removal Efficiency	99.9%
Mercury	Max. 70 $\mu g/m^3$	0.91 $\mu g/m^3$

CONCLUSION

Overall, Duffin Creek units have met the stack emission limits at the design burning capacity. Both incinerators have a unique design with the steam turbines driving the fluidizing air blowers in lieu of electric motors. Steam driven fluidized air blowers and steam heated mercury conditioners avoid future electrical consumption as compared with non-steam recovery designs.

The project has come to a successful completion through a very strong coordination between all parties involved during the execution phase.

REFERENCES

K. Dangtran, L. Takmaz, H. Pham, J.Y. Bergel, J. Welp, P. Burrowes (2011), "Overview of municipal sludge incineration in North America – from green to greener – the Lakeview, the Duffin Creek and the Southerly experiences", Water Science and Technology, page 2213-2218, Volume, 63.10 (2011).

Duffin Creek Expansion Project: https://duffincreek.ebuild.ca/

D. Howard, K. Dangtran, L. Takmaz (2013), "Greensboro Experience - Addition of a second fluid bed sewage sludge incinerator at T.Z. Osborne Water Reclamation Facility and its compliance with the new MACT emission requirements", WE&T Magazine, Sep 2013, page 51 – 55.

William Fernandes, Levent Takmaz, Ky Dangtran (2012), "Retrofitting Lakeview GE Booth Wastewater Treatment Plant with Mercury Control System Using Kombisorbon Granulated Activated Carbon Fixed Bed Adsorber", Everything About Water Magazine, page 118-120, February 2012.

J. Kolde, P. Barilla, J. Knotts (2016), "The Gore Mercury Control System", W.L. Gore & Associates, Inc. (www.gore.com)

B. Higgins, M. Pomerleau, P. Randolph, T. Bauer, J. Kolde and P. Barilla (2014), "Demonstrated Mercury Reduction for SSI MACT Using Mercury Adsorption Modules", Residuals and Biosolids Show, Austin Convention Center, 2014, May 18-21, 2014, Austin, TX, USA.

European Commission Report (2001), "Ambient Air Pollution by Mercury (Hg) Position Paper", Chapter 7, prepared by the working group on mercury, ISBN 92-894-4260-3.

Can We Overcome Hydrolysis Limitation by Better Understanding the Impacts of Physics Within Anaerobic Digestion?

Elizabeth Manning[1,2], Adrian Romero[1,2], Baoqiang Li[1,2], Ahmed Al-Omari[2], Matthew Higgins[3], Rumana Riffat[1], Sudhir Murthy[2], and Haydee De Clippeleir[2]

[1]The George Washington University, Washington, DC 20052, USA
[2]DC Water and Sewer Authority, Washington, DC 20032, USA
[3]Bucknell University, Lewisburg, PA 17837, USA

Keywords: Hydrolysis, Diffusion, Anaerobic Digestion, Mesophilic Anaerobic Digestion, Biogas

Abstract

Investigating the mechanisms involved in the hydrolysis step in anaerobic digestion has the potential to aid in the optimization of the digestion process as a whole. Previous studies have shown that diffusion limitations affect other activated sludge processes such as nitrification (Shaw, 2013). Thus, in order to understand the physical limitations that activated sludge experiences within anaerobic digestion, the diffusion limitations are investigated. Batch-scale tests were performed with full-scale digested sludge (20-day solids retention time). Analysis was done on the gas production data collected by respirometers. The hydrolysis rates were taken as the rate during the last 7 hours of a 24-hour time period, during which time the gas production rate curve flattens to a straight line. The two tests performed were: 1) Varied total solids concentration, and 2) Varied polymer dose addition. Varying these two parameters affects the floc particle size, bulk viscosity, microbe population density, and distance between microbes and particles. The results from these tests showed that varying these parameters did not alter the hydrolysis rates of the sludge. Two conclusions were drawn from the tests: 1) the contribution of a floc's boundary layer's diffusion resistance is negligible compared to the overall diffusion resistance in the floc, and 2) mixing sludge at a high shear rate (fast mixing speed) may allow for increased solids capacity of digesters.

Introduction

Hydrolysis, the first step in the breakdown of particulate matter in of anaerobic digestion, in which particulate substrate is degraded into soluble substrate, is widely considered to be the rate-limiting step within the process and thus determines the capacity of anaerobic digesters (Pavlostathis and Giraldo-Gomez, 1991). The most common kinetic model used to describe the hydrolysis rates in anaerobic digestion is the first-order model (Vavilin et al., 2008), which is given as Equation 1. In Equation 1, dS/dt is the degradation rate of the substrate, kh is the hydrolysis rate constant, and S is the concentration of the substrate initially in the sludge.

$$\frac{dS}{dt} = -k_h S$$

Equation 1. First-order rate model describing hydrolysis (Vavilin et al., 2008)

Experiments have been performed to identify the values for the hydrolysis rate constant, kh, for various types of sludges. The kh values range between 0.005 and 2.0 day^{-1}, depending on the substrate (Vavilin et al., 2008). In the first-order model, the hydrolysis rate is only dependent on the substrate concentration and the hydrolysis rate constant thus is only applicable in substrate-limited conditions. Researchers have observed that under transient conditions where substrate is abundant, the hydrolysis rate became zero-order (i.e. independent of substrate concentration) and thus did not increase with increasing substrate addition. Additionally, by modeling the hydrolysis step using the first-order rate model, the mechanistic understanding of the process is vailed and any changes to the process to improve hydrolysis rates, other than substrate availability, will not be captured or explained. Two alternatives to using the first-order model to describe hydrolysis kinetics are available: (i) simplified mixed-order equation (Monod, 1978) or (ii) Contois Equation (Contois, 1959). The difference between these two models is that the Contois model includes the impact of surface area to the kinetic rates. Within these models, the hydrolysis rate reaches a maximum when substrate concentration becomes non-limiting. The latter approaches are limited because they use a constant apparent affinity parameter, kh, which is assumed to be affected only by the maximum hydrolysis rates and the initial substrate concentration. However, in this paper we challenge that assumption and we investigate the role of diffusion in the efficiency of the hydrolysis rates within anaerobic digesters.

In activated sludge systems, microbial reactions almost all occur under first order kinetics, as substrates are limited. The apparent half-saturation constants, K_S, have been shown to impact competition between microorganisms and determine the effluent quality and the capacity of the systems (Al-Omari et al., 2012). Shaw et al. (2013) has shown that the apparent K_S is driven by diffusion rather than by the intrinsic affinity constant of the microbial players. The former increases the apparent K_S by approximately a 10-fold compared to intrinsic kinetic parameters of single cells, demonstrating diffusion to be the process limitation. Using Fick's First Law of Diffusion (Equation 2, Shaw, 2015), three variables that impact diffusion can be identified that affect the rate of diffusion (rd): particle size (A), distance between substrate and microbe (dS/dx), and diffusivity (De).

$$r_d = -D_e \frac{\partial S}{\partial x} * A$$

Equation 2. Fick's First Law of Diffusion (Shaw, 2013)

Within activated sludge, two layers are considered to be affecting the overall, or effective, diffusivity (D_e) and thus the overall diffusion limitations. These two layers are the boundary layer and the biofilm and can be visualized in Figure 1.
Table 1 describes the various parameters within sludge that will affect diffusion and identifies which will affect diffusion only in the boundary layer, only in the biofilm, or will affect diffusion in both the boundary layer as well as the biofilm.

Table 1. Parameters That Impact Diffusion in Sludge

Boundary Layer Only	Biofilm Only	Both Boundary Layer and Biofilm
• Bulk viscosity • Shear rate • Total solids content • Bulk diffusivity • Surface area of floc	• Viscosity within floc • Floc density & porosity • Substrate diffusivity • Floc diffusivity • Surface area of substrate • EPS content	• Bulk biomass density • Active fraction of biomass • Concentration distribution of substrates, microbes, and enzymes

This research seeks to identify the mass transport limitations taking place within the boundary layer and biofilm due to the three variables identified within Fick's First Law of Diffusion (particle size, substrate-microbe distance, and diffusivity), which can be quantified in terms of surface area, varied chances of collision, and viscosity and shear rate, respectively.

Methodology
To gain an understanding of various physical factors that might affect the hydrolysis rate, batch tests based on respirometry were performed. Within each 250 mL batch bottle, conditions were varied to evaluate the impact on hydrolysis and the biogas production was monitored over a 24-hour period. Tests were performed in triplicate using sludge from full-scale digesters at DC Water's Blue Plains Advanced Wastewater Treatment and each experiment included a control condition, which contained the unchanged digestate sludge. The full-scale digesters samples for this study operate at a 20 day solids retention time (SRT) and are fed with pretreated thermally-hydrolyzed (THP) sludge.

Sludge and liquid analyses were performed at the initial and final conditions of each experiment. These analyses included total solids (TS), volatile solids (VS), total suspended solids (TSS), volatile suspended solids (VSS), total and soluble COD, and ammonium. All analytical analyses were performed according to the standard methods (APHA, 1997). The gas production data was also collected and analyzed for each experiment.

Full-scale anaerobic digestion sludge samples were incubated for 24 hours at 34 C without addition of new feed, which allowed for the assumption that biogas production rates achieved during the experiment were mainly linked to hydrolysis of particulate substrate, as most readily biodegradable organics had already been removed during the 20-day digestion time. The hydrolysis rates were determined based on the biogas production curves for the batch bottles over the 24-hour time period, more specifically as the final stable biogas production rates during the last seven hours of the 24-hour testing period, as depicted in Figure 2. The hydrolysis rates obtained from the respirometer gas production curves were then normalized based on the weight of volatile solids contained in each batch bottle.

1) ***Varying Solids Concentration***

In order to determine how varying the viscosity of the sludge by increasing its total solids (TS) content will affect the hydrolysis rate, respirometer batch tests were performed on sludge with TS contents ranging between 5% and 10%. Centrifugation and resuspension of various volumes of supernatant was used in order to vary the %TS content of the full-scale digestate. The control for these experiments was sludge that did not go through the centrifugation and resuspension process. 250 mL batch bottles were filled with the sludge of various total solids concentrations and connected to a respirometer for a 24-hour period. The bottles were continuously mixed at 300 rpm on a stir plate.

2) ***Varying Floc Size***

In order to determine how various polymer doses affect the hydrolysis rate of digestate sludge, respirometer batch tests were performed on sludge with varying polymer doses ranging between 0 to 20 lb polymer/ton TS. This test was modeled after nitrification tests showing increases in oxygen half-saturation constants from 0.2 to 1.9 mg O_2/L for nitrifiers after flocculating the sludge through addition of a flocculent (Mulder, 2016). Flocculent polymer was added to sludge originating from DC Water's full-scale digesters (20 day SRT, fed with THP). 250 mL batch bottles were filled with the sludge of various polymer doses and connected to a respirometer for a 24-hour period. The bottles were continuously mixed at 300 rpm on a stir plate.

Results and Discussion

1) ***Impact of Solids Concentration on Hydrolysis Rate***

Pretreatment of sludge by the thermal hydrolysis process (THP) decreases the average particle size of the substrate, allowing the anaerobic digesters to operate at increased loading compared to conventional mesophilic anaerobic digestion (MAD) (Wilson et al., 2011). The increased loading capacity is mainly contributed to the decrease in viscosity that the sludge experiences because of THP. The lower viscosity makes it more feasible for anaerobic digesters to pump and mix higher volumes of sludge at higher TS concentrations. However, it is not fully clear if this also results in increased hydrolysis kinetics. The goal in performing the Total Solids Test was to investigate if the increased total solids content, which increases the bulk viscosity of the sludge, would impact the hydrolysis rates.

The diffusivity is an important parameter within diffusion and is impacted by temperature as well as viscosity (Tan et al., 1992). As we already know from literature that increased hydrolysis rates can be achieved at higher temperatures (Veeken et al., 1999), the investigation of viscosity has been prioritized. The viscosity in each bottle was determined based on the shear rate that the bottles experience with a magnetic stir bar stirring at 300 rpm, which is 73 sec^{-1}. There is no statistical difference between the hydrolysis rates produced in bottles with 5% TS through 10% TS when the 250 mL bottles of sludge were mixed at 300 rpm with magnetic stir bars. As the shear rate gets to be this high, the viscosity decreases. Additionally, at such high shear rates, the differences in viscosities among various TS concentrations become negligible. Thus, it was determined from this experiment that at high shear rates, the effect of viscosity on mesophilic anaerobic digestion was negligible. The implications of this finding indicate that if the shear rate of mixing in a CSTR is high enough, the loading can be pushed higher than normally is feasible without reaching process failure. This will be an important finding in overload scenarios when there is a higher amount of solids in the influent than normal. Further testing will be needed to determine the impact of viscosity at lower shear rates.

The role that the viscosity plays in the sludge may also be analyzed with a physical mechanistic approach. The hypothesis to explain the negligible difference in hydrolysis rates in this test is that varying the bulk viscosity of the sludge only has an impact on the diffusivity of the boundary layer of flocs. Since negligible change was seen with the increased viscosity in the boundary layer, it can be hypothesized that diffusion limitations within the boundary layer are negligible compared to the diffusion limitations within the biofilm portion of the floc.

2) *Impact of Polymer Dose on Hydrolysis Rate*

In performing the Polymer Dose Test, two hypotheses were formulated regarding the impact of particle size on hydrolysis rate: (i) flocculation of sludge would increase floc size of biomass and lead to decreased biogas production due to increased diffusion limitations; (ii) decreased substrate-microbe distance within the flocs would lead to increased biogas production as colonization of substrate by microbes in enhanced.

The results of the Polymer Dose Test showed that for polymer doses ranging from 0 to 20 lb polymer/ton TS, there was no statistically significant difference in hydrolysis rates. A possible explanation for the lack of variability in hydrolysis rates seen during the Polymer Dose Test is that rather than the size of the floc being important, the density within the floc was the controlling factor. On bulk-scale, the flocs varied in particle size with the polymer addition. However, varying the floc sizes on this scale is only changing the conditions in the boundary layer, not the biofilm. The lack of variation of hydrolysis rates with increased floc sizes indicates that changing the boundary layer conditions for diffusivity has a negligible effect on the overall diffusivity within the sludge.

The Polymer Dose Test also was hypothesized to decrease the substrate-microbe distance within the sludge with increasing polymer doses. When the substrate-microbe distance was minimized, hydrolysis rates were expected to increase according to our diffusion assumption from Fick's First Law of Diffusion (Equation 2). However, the results did not reflect this expectation. Instead, the increasing polymer doses had a negligible impact on the hydrolysis rates.

The hypothesis associated with the Polymer Dose Test with respect to the substrate-microbe travel distance was that through flocculation, the particulates and the microbes would be brought closer together and would have a shorter distance to travel to reach each other. Thus, if the sludge was concentrated in a floc, the hydrolysis rates would increase due to the ease of transport to and colonization of the substrate. The lack of variation in hydrolysis seen suggested that the positive effects of bringing the substrates and microbes closer together was confounded by the increased population density within the flocs. In this perspective, the biological matrix within the floc could have become too dense for the microbes to migrate to the particulates or for the excreted hydrolytic enzymes to migrate to the particulates for degradation. The porosity and the biomass density within the flocs at varying polymer doses will need to be quantified in the future to investigate this hypothesis further.

Conclusion

This paper showed that the hydrolysis rates remained unchanged when the bulk viscosity was changed at a high shear rate and when large flocs were formed due to various polymer doses in the sludge. These findings support the hypothesis that changing the diffusion parameters within the boundary layer of the flocs (such as the bulk viscosity and the polymer dose) has a negligible effect on the overall diffusivity of the sludge. Thus, further investigations will need to vary the diffusion parameters on the biofilm layer in order to identify changes in the overall diffusivity of the sludge.

References

Al-Omari, A., Wett, B., Han, H., Hell, M., Bott, C., Murthy, S. (2012). Full-plant deammonification based on NOB-repression, AOB Seeding, Anammox Seeding and Successful Retention. Proc. IWA Conf. On Nutrient Removal, Harbin.

APHA (1997). Standard Methods for the Examination of Water and Wastewater. Intern rapport, American Public Health Association, American Water Works Association, Water Environment Federation.

Contois, D.E. (1959). Kinetics of Bacterial Growth: Relationship between Population Density and Specific Growth Rate of Continuous Cultures. *Journal of General Microbiology*. **21**, 40-50.

Monod, J. (1978). The Growth of Bacterial Cultures. *Selected Papers in Molecular Biology by Jacques Monod.* 139-62.

Mulder, C. (2016). *Impact of Intrinsic and Extrinsic Parameters on the Oxygen Kinetic Parameters of Ammonia and Nitrite Oxidizing Bacteria*. Ph. D. thesis, University of Ghent, Belgium.

Pavlostathis, S.G., Giraldo-Gomez, E. (1991). Kinetics of Anaerobic Treatment. CRC Crit. Rev. Environ. Contr. **21**, 411-490.

Shaw, A., Takacs, I., Pagilla, K., and Murthy, S. (2013). A New Approach to Assess the Dependency of Extant Half-Saturation Coefficients on Maximum Process Rates and Estimate Intrinsic Coefficients. *Water Research*. **47**, 5986-5994.

Shaw, A., Takacs, I., Pagilla, K., Riffat, R., De Clippeleir, H., Wilson, C., and Murthy, S. (2015). Toward Universal Half-Saturation Coefficients: Describing Extant Ks as a Function of Diffusion. *Water Environmental Research*. **87**, 387-391.

Tan, K.K., and Thopre, R. (1992). Gas Diffusion into Viscous and Non-Newtonian Liquids. *Chemical Engineering Science*. **47**, 3565-3572.

Vavilin, V.A., Fernandez, B., Palatsi, J., and Flotats, X. (2008). Hydrolysis Kinetics in Anaerobic Degradation of Particulate Organic Material: An Overview. *Waste Management* **28.6**, 938-951.

Veeken, A., and Hamelers, B. (1999). Effect of Temperature on Hydrolysis Rates of Selected Biowaste Components. *Bioresource Technology*. **69**, 249-254.

Wilson, C., Tanneru, C., Banjade, S., Murthy, S., and Novak, J. (2011). Anaerobic Digestion of Raw and Thermally Hydrolyzed Wastewater Solids Under Various Operational Conditions. *Water Environment Research*. **83**, 815-825.

Figures:

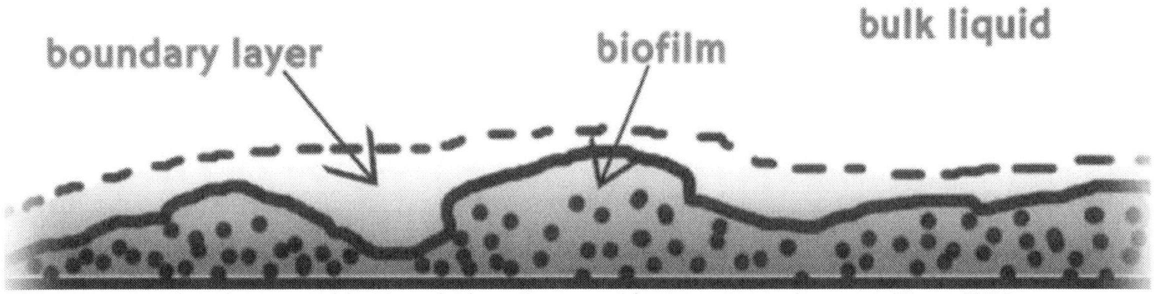

Figure 1. Boundary layer and biofilm within a floc, graphic (http://2011.igem.org/Team:ZJU-China/bio-modeling.html)

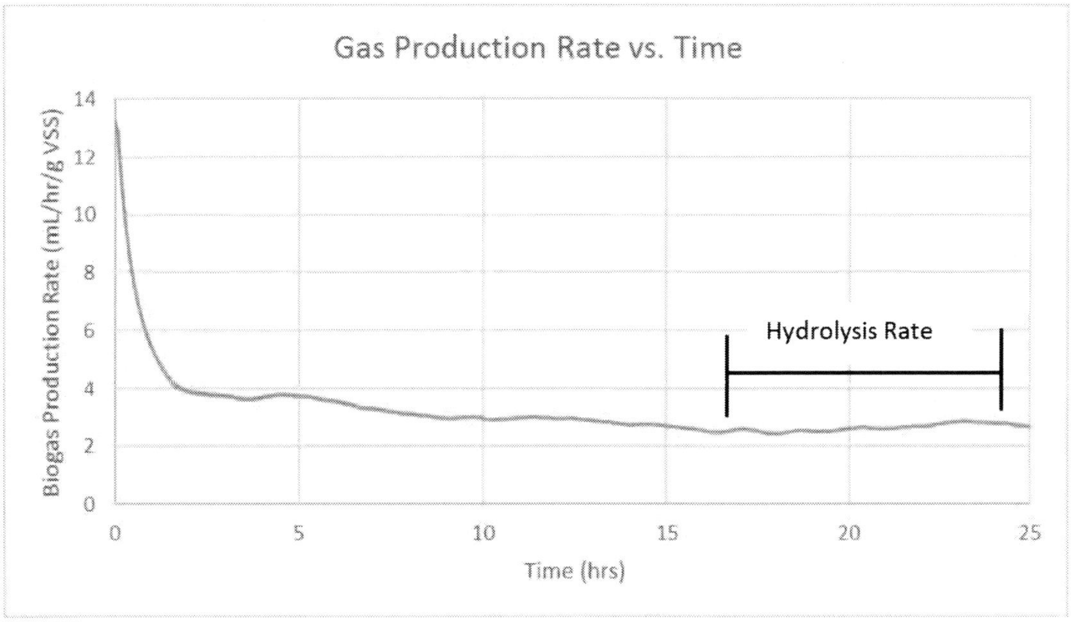

Figure 2. Determining hydrolysis rate with the biogas production rate curve

A Review of Rapid Volume Expansion Research and Development of Design Methodologies for Mitigation in Full-Scale Digesters

Steve Krugel[1*], Thomas Chapman[2], William Persich[1]

[1]Brown and Caldwell, Seattle, Washington; [2]Brown and Caldwell, Tucson, Arizona
*email: skrugel@brwncald.com

ABSTRACT

Recently, rapid volume expansion (RVE) in anaerobic digesters resulting from changes in gas holdup within the digester after sudden changes in operational situations has been postulated and demonstrated to be the cause of severe digester overflows into receiving streams, clogging of gas systems, and in some cases, severe disabling structural damage to the digesters, including cover collapse. Recent research into the mechanisms of RVE in digesters (i.e. Chapman and Krugel (2011); Chapman et al. (2014); Higgins et al. (2014); Bartek et al. (2015); and Bartek et al. (2016)), has uncovered important information to help the industry understand and mitigate the causes and extent of RVE. Although RVE has been observed in many full scale digesters, the research into this phenomenon to a large extent has been conducted on small laboratory scale reactors, supplemented with limited full scale data from only a few plants. There is a need to consider methodologies for scaling up results to full-scale digesters of different sizes and shapes and to different mixing designs. This is vitally important in designing to accommodate and mitigate RVE.

This paper discusses the theory and methodologies for estimating the potential rate and volume of volume expansion in full-scale digesters. In addition, issues of impacts of gas surge following an RVE event and impacts of gas holdup on digester solids retention time (SRT) are discussed.

KEYWORDS: Volume expansion, rapid volume expansion, rapid rise, gas holdup, anaerobic digestion, digester foaming, digester overflows, digester safety

INTRODUCTION

For decades, wastewater treatment plant operators have experienced and reported on 'foaming' events with their anaerobic digesters. In some cases, these events have caused rapid and serious sludge overflows and over-pressurization leading to filling and clogging gas system piping, large sludge spillages, and, in the most severe cases, disabling structural damage to the digesters, including cover collapse.

Recent research has defined and demonstrated the phenomenon of rapid volume expansion (RVE) that explains the causes of many of these severe events. Rapid volume expansion occurs following a sudden change in digester operation, such as loss of mixing, changing the character of the sludge and the dynamics of gas bubbles forming within the sludge and rapidly causing the volume of sludge to expand. Rates of expansion can approach the gas production rate and total expansion volume has been shown to approach and possibly exceed 20 percent of digester volume. Following the RVE event, as an example when mixing is restored, a sudden drop in volume is observed, accompanied by a large spike of gas release, several times the biological gas production rate.

As the understanding of fundamental basis of RVE has developed, the need for synthesizing the theory into models and tools that can be used for the design of full scale digesters has remained.

There is a need to consider methodologies for scaling up observations and test results to full-scale digesters of different sizes and shapes, to different mixing designs, and to sludges of different characteristics. This is vitally important in designing to accommodate and mitigate RVE. Some important digester design questions related to RVE include:

1. What is the potential rate of rapid volume expansion? - needed to design appropriate overflow flow capacity
2. What is the potential total volume of expansion? - needed to design appropriate overflow storage volume, the impact of overflow on downstream processes, or need to consider alternative design measures such as standby power for mixing, etc.
3. What is the rate of gas surge following turning mixers back on following an RVE event? - needed to appropriately size the gas system including size of piping, pressure relief, and flares.
4. What is the sludge specific gravity (s.g.) and solids retention time (SRT) under steady state mixed conditions - needed to design appropriate volume for digester to meet intended SRT.
5. How is RVE impacted by the type of sludge: rheology, total solids concentration, THP vs conventional digestion, mesophilic vs. thermophilic operating temperatures, etc.?
6. How is RVE impacted by the digester tank geometry (pancake vs. silo, etc.)?
7. How is the RVE impacted by the configuration and intensity of digester mixing?

The goal of this paper is to synthesize the theory, data, and observations generated to date with respect to RVE and propose design calculation methodologies for estimating the potential rate and volume of expansion in full-scale digesters of different sizes and mixing configurations, estimate the impacts of gas surge following an RVE event, and discuss impacts of gas holdup on digester solids retention time (SRT). These proposed design methodologies are intended for design engineers and operators to better understand the mechanisms at play and which can be controlled in RVE events. They are intended as a starting point design tool for mitigating RVE that our industry can critique, refine, and improve upon moving forward.

BACKGROUND

Recent research has defined and demonstrated a phenomenon that explains the causes of many severe digester overflow events that historically have been attributed to 'foaming'. Chapman and Krugel (2011) recognizing that in some instances digester foaming does occur, postulated that another phenomenon resulting from changes in gas holdup within the digester after sudden changes in operational situations was the cause of the most severe events. Gas holdup is a phenomenon where bubbles form within the sludge, but do rise out of the liquid until their buoyant force exceeds the yield stress within the sludge, causing an expansion of the sludge volume. They termed this phenomenon rapid volume expansion (RVE), recognizing that these events can begin to occur within minutes of the operational change. They further described the theory of gas holdup and reported on factors known to influence it, including sludge rheology, bubble dynamics, temperature and pressure, and digester configuration and design. They described known full scale cases that supported this theory and listed the operational changes that can cause these events, namely:

- Sudden change in digester mixing, the most severe being a mixer stoppage such as during a power outage
- Sudden loss of headspace gas pressure
- Sudden large increase in sludge feed
- Rapid biological activity during digester startup

The most severe triggering event has been shown to be sudden loss of digester mixing.

Chapman et al. (2014) further described this phenomenon and reported on an RVE case on a full-scale digester at the Brightwater Wastewater Treatment Plant (Brightwater) in King County, Washington that was closely monitored and supported the gas holdup theory.

Higgins et al. (2014) postulated sludge rheological relationships that influenced RVE and reported on laboratory experiments at Bucknell University that demonstrated RVE. Their research showed detailed shear rate/viscosity profiles and the effect of mixing shear rate on gas holdup. Their results suggested a critical mixing shear rate exists at the 'knee' of the shear rate/viscosity curve, below which the viscosity of the suspension increases rapidly, resulting in gas holdup and volume expansion.

Bartek et al. (2015) further developed the rheological basis for RVE and reported on additional laboratory experiments at Bucknell on 11 different sludges. This work demonstrated the relationship between sludge yield stress and apparent viscosity and the maximum amount of volume expansion and provided time-lapsed video proof that volume expansion is occurring instead of foaming.

Bartek et al. (2016) reported on additional laboratory work and developed a mechanistic model based on bubble dynamics and sludge rheology that could be used to predict RVE extent.

APPROACH

The goal of this paper is to synthesize the theory, data, and observations generated to date with respect to RVE and propose design calculation methodologies for estimating the potential rate and volume of expansion in full-scale digesters of different sizes and mixing configurations, estimate the impacts of gas surge following an RVE event, and discuss impacts of gas holdup on digester solids retention time (SRT).

While research cited above has shown a clear relationship between the rheological properties of the sludge that must be considered, other physical and chemical reactions must also be explored that contribute to the character of RVE events. In this paper, theories are postulated based on gas laws and chemical equilibria to explain gas portioning involved in RVE.

Data from previous research and full-scale observations cited above has provided the analytical footprint of RVE events. This data is analyzed to test these theories, describe the partitioning of gas during RVE, and further describe other key expansion parameters such as rate and extent. Methodologies and example calculations are presented to provide an initial starting point for the operator and designer when considering how to mitigate RVE events. Example calculations are included in Appendix A attached to this paper.

INTERPRETATION OF LABORATORY TEST RESULTS

Data provided by Higgins et al. (2014) has provided one of the clearest and best footprints of an RVE event to date. As such this paper starts with a breakdown and analysis of that data from the physical perspective of gas partitioning.

Theoretically, the maximum rate of volume expansion from gas holdup is a function of the rate of gas production. If all the gas being produced in the digester were held in suspension at measured temperature and pressure, the volume of the digester contents would be increasing at the rate of which gas is being produced. However, following a mixing power outage as an example, the actual rate of expansion will be less than the pre-event measured gas production as there are other ways the gas can partition in the digester, including:

- Some gas will continue to evolve from the surface of the digester
- Some gas will go into solution
- Some gas bubbles will form within the sludge at compressed conditions (at depth in the reactor) versus measured gas production at standard temperature and pressure or at digester headspace conditions.

Therefore, in order to determine the amount and rate at which gas is contributing to volume expansion, we need an understanding of how gas forms during a volume expansion event after the sudden loss of mixing or under other RVE-causing scenarios. Additional insights into these mechanisms can be gained by further evaluation of RVE test data from Higgins et al. (2014). The data provided by Higgins et al. (2014) can provide a better understanding of how gas partitions during a RVE event and provides a baseline that can be used to predict important RVE parameters, such as the rate of volume expansion, for full-scale digesters of various heights and even mixing configurations. Figures 1 and 2 from the 2014 Higgins work show predicted and measured digester volume change after loss of mixing and measured gas production before, during, and after the RVE event and are particularly useful to analyze.

During a RVE event in the absence of mixing, gas will continue to evolve from the liquid surface of the digester and gas bubbles held in digester contents will lead to an increase in digester volume, forcing gas out of the headspace of the digester. This fraction of the total gas production during a RVE event is what is measured by the respirometer and is denoted by the area under the gas production rate curve, as seen in Figure 2. Meanwhile, a portion of the total gas production during a RVE event will be forced into solution and some pre-event gas production volume will be lost to compression of gas deeper in the digester. This fraction of gas does not contribute to the gas production measured by the respirometer, resulting in the reduction in gas production during a volume expansion event. The region between the pre-event gas production rate and the rate measured during the RVE event represents this portion of gas, as denoted in Figure 2. Estimates of this partitioning are developed and a methodology proposed for assessing the rate of gas production from these experimental results.

Gas volumes and rates reported in Figures 1 and 2 are at 38°C and 6 inches of gas pressure above atmospheric (Higgins, 2016). When reporting gas volumes or rates at standard temperature and pressure (STP), standard temperature is taken as 15°C (288°K) from the International Organization of Standards (ISO), typically used by the gas industry and the wastewater industry when dealing with digester gas.

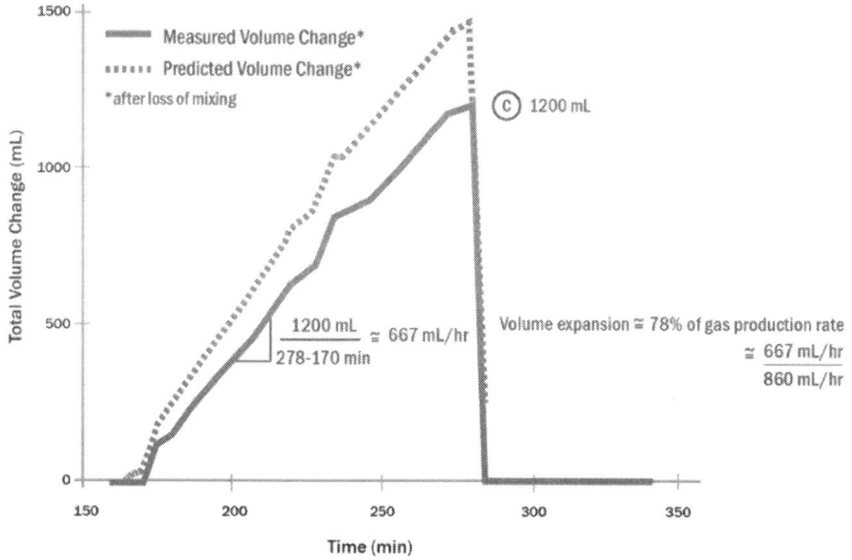

Figure 1 – Predicted and Measured Digester Volume Change After Loss of Mixing (from Higgins et al., 2014)

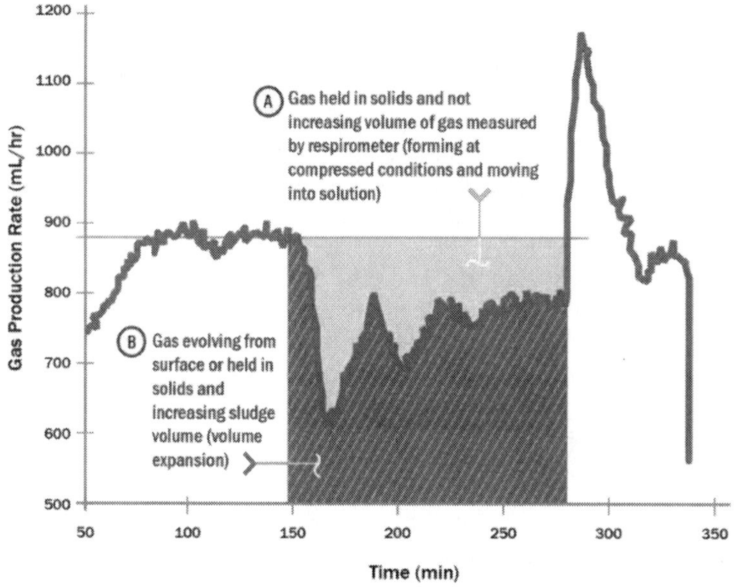

Figure 2 – Measured Gas Production Before, During, and After RVE Event (from Higgins et al., 2014)

Calculations used to interpret the data in Figures 1 and 2 are presented in Appendix A, calculations 1 through 15. The following are key results from the interpretation of these curves:

- Gas production throughout test period, **GPR_M** = 860 mL/hr (from Figure 2) at 38° C and 1 atm plus 6 inches of gas pressure = 1.015 atm; Gas production rate at standard temperature and pressure (STP) = 808 mL/hr
- Volume of Expansion, <u>C</u> = 1200 mL (compressed, held up gas bubbles produced; from Figure 1)
- Volume Expansion Rate = 667 mL/hr (Figure 1)
- **RVE_r** = Ratio of volume expansion rate to the pseudo-steady state, pre-test gas production rate at STP = 0.83 or 83%; or 78% based on gas production reported at test temperature and headspace pressure
- <u>A</u> = Produced gas at 38° C and 1.015 atm. held in solids, but not contributing to measured volume increase of the gas = 226 mL (estimated from Figure 2). (This is produced gas moving into solution, **V_{GS}**, and volume of produced gas lost due to gas compression of bubbles below the surface at higher pressure, **V_{GC}**, held up at depth and compressed because mixing has stopped and no longer moving gas bubbles formed to the surface of the reactor)
- <u>B</u> = Produced gas measured during the RVE event = gas evolving from surface of sludge + gas pushed out the headspace due to volume expansion (gas held in sludge as bubbles and increasing sludge volume) and being measured with the respirometer at 38°C and 1.015 atm. = 1,537 mL
- Total gas produced during the RVE event (mixer down time) = 1,763 mL
- Gas evolved from surface during the RVE event = measured gas - volume expansion = 337 mL
- Volume of stored gas in bubbles at STP = 1,160 mL
- Produced gas moved into solution = produced gas - evolved gas - expanded held up gas - gas volume reduction due to compression at test T and P = 191 mL

These results are further discussed and used in subsequent evaluations in the following sections.

RATE OF RAPID VOLUME EXPANSION IN FULL SCALE DIGESTERS OF DIFFERENT HEIGHTS AND MIXING CONFIGURATIONS

With different digester shapes (heights) and mixing configurations, the rate of volume expansion is expected to change. Specifically, the height of the tank and the mixing configuration will largely impact the pressure at depth and influence the portion of bubbles forming at compressed conditions or going into solution. Consequently, this will influence the volume of gas contributing to the rate of volume expansion. In the previous section, a methodology has been proposed and used to estimate the rate of RVE. Using the interpretations and relationships discussed in the previous section, an analysis is presented of how these parameters will change in full scale digesters. Table 1 provides a summary overview of the expected rate of volume expansion under different digester heights and mixing types.

Table 1. Rate of Volume Expansion under Various Mixing Configurations and Digester Depths			
Mixing Configuration	Digester Depth, ft.	Pressure at Depth [b], ft.	Ratio of RVE Rate to Gas Production (X100) [c], %
Higgins et al. (2014)	2 [a]	35.4	83
Fully Mixed	36	55.4	58 [d]
Fully Mixed	65	74.4	43 [d]
Half Mixed [e]	36	62.4	31 [d]
Half Mixed [e]	65	85.4	23 [d]

Notes: [a] Assumed
[b] Assumes absolute pressure at depth of centroid of pressure in the reactor
[c] Gas production rate expressed at standard temperature and pressure
[d] Assumes a 1.1 multiplier (safety factor) for operating s.g. below 1.0.
[e] Assumes bottom half of digester is pump mixed and top half is not

This understanding of the rate of volume expansion provides an estimate for the design of emergency overflow systems that should be installed on digesters. The following sections provide a suggested methodology and example calculations used to develop Table 1 and to provide the digester designer tools to estimate the rate of RVE in full-scale digesters.

Estimated Depth of Centroid of Pressure in Full Scale Digesters

In order to estimate the characteristics of an RVE event in a full scale digester, it is first necessary to estimate the average pressure at depth within the digester. One could consider the pressure profile and integrate through the depth. However, a simplified model is used here to approximate the depth to the centroid of the pressure. The term "centroid" is used loosely to define the depth to the point of approximate average absolute pressure in the vessel. This simplified approach calculates the total volume X pressure (VP) and then estimates the depth at which half of that value is found. Figure 3 can be helpful in visualizing this approach, representing the pressure profile in a 70-foot tall digester. Assuming atmospheric pressure of 34 ft of H_2O and headspace gauge pressure of 1 ft, approximate absolute headspace pressure is 35 ft of H_2O and the approximate pressure at full depth is 35 ft + 70 ft = 105 ft of H_2O (ft_P). This model assumes that the sludge is homogeneous with a s.g. of 1.0. Suggested corrections for lower s.g. are discussed below.

Calculating the apparent depth (ft_D) within the digester of the centroid of pressure, total volume X pressure (VP) = total volume X pressure of each pressure component illustrated in Figure 3.

From Figure 3 and where A is the surface area of the digester,

$$VP = (A)(70\ ft_D)(35\ ft_P) + (A)(\tfrac{1}{2})(70\ ft_D)(70\ ft_P) = (4{,}900)(A)\ \text{ft depth-ft pressure}\ (ft_D\text{-}ft_P),$$

If x represents the depth to the centroid where half of VP is found, then

$$(\tfrac{1}{2})(4{,}900\ A) = (A)(35x + \tfrac{1}{2}xy)$$

In this case, x depth in ft_D = y pressure in ft_P, and

$2{,}450 = 35x + \frac{1}{2}x^2$

$x = 43.3$ ft of depth; $(43.3/70)(100) = 62\%$ of depth

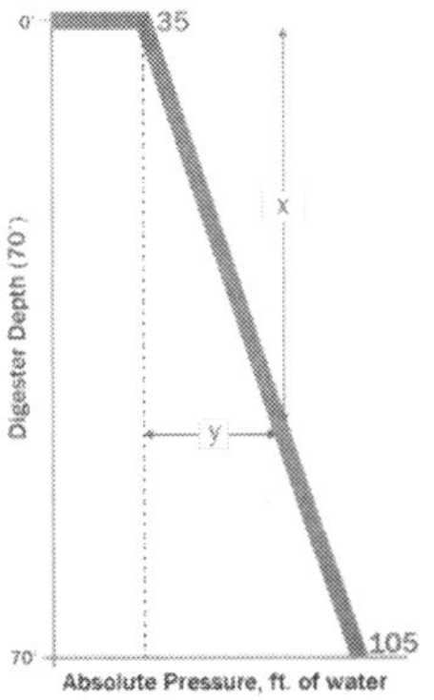

Figure 3 – Model for Estimating Depth to Centroid of Pressure in a 70-Foot Deep Full-Scale Digester (not to scale)

Using this methodology, the estimated depth to the pressure centroid for digesters of various depths are depicted in Figure 4.

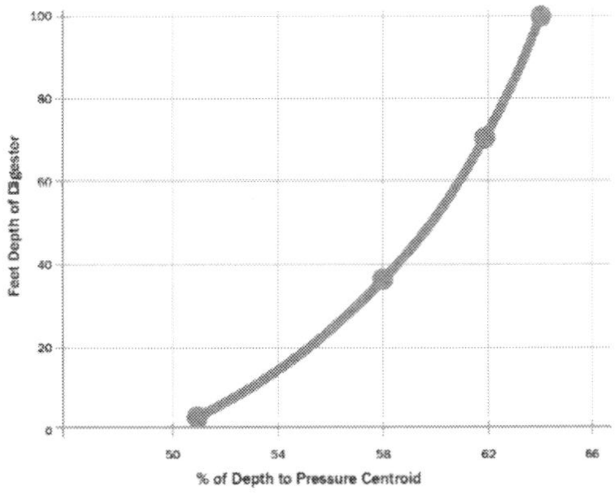

Figure 4 – Approximate Depth to Centroid of Pressure in Digesters of Various Depths

As mentioned above, this methodology assumes that the sludge is homogeneous with a s.g. of 1.0. This is close to correct for a heavily mixed digester that is "elutriating" the gas to the headspace. But as discussed below, a digester will normally operate at a s.g. somewhat less than 1.0 and, particularly in an unmixed digester, can operate with s.g. as low as 0.7 with a gradient of s.g from higher to lower as you move from the bottom to the top of the digester. This is due to the profile of held up gas bubbles. As a result, assuming a s.g. of 1.0 will slightly under predict the rate of volume expansion and slightly over predict the theoretical maximum amount of gas that can move into solution. For design purposes, a safety factor should be applied to the RVE rate; a safety factor of 1.1 is reasonably reflective of an actual starting operating s.g. of 0.9. For the gas dissolution calculation, leaving the estimate conservative is prudent and no adjustment is proposed.

Estimation of Ratio of Volume Expansion Rate to Gas Production Rate (RVE_r) Based on Gas Compression in Fully Mixed Full-Scale Digesters

To compare results of the 10 L test digester to a deep, full-scale digester, an estimation was made of the volume expansion, V_{T2}, for test conditions and duration for 36-ft tall digester mixed full height, where P_{T2} is 33.9 ft atmospheric pressure + 6 in gas pressure + 21 ft (approximate depth to centroid of pressure in 36-foot deep digester). The resulting RVE_{r2} = 0.53 or 53% (Appendix A, calculation 16). Adding a 1.1 safety factor adjusts this to **58%**.

For a 65-ft tall digester mixed full height, RVE_{r3} = 0.39 or 39% (Appendix A, calculation 17). Adding a 1.1 safety factor adjusts this to **43%**.

As a check of this methodology, data was analyzed from an RVE event in a full-scale, 59-foot diameter, 65-foot-tall digester (61.3-foot sludge depth) at the Brightwater Wastewater Treatment Plant (Brightwater) in King County, Washington, reported by Chapman et al. (2014). Figure 5 depicts monitored data during this event.

It should be noted that the level readings in Figure 5 are actually mass changes in respective tanks, as the "level" indicated is measured using a pressure-based level instrument. In reality, Digester 3 has a fixed overflow point to the storage tank and the level can be considered to be constant in the digester. Approximate specific gravity (s.g.) can be estimated assuming Digester 3 is full before and after the event and estimating the mass pushed out of the reactor during the volume expansion. It should also be noted that the "level" (mass) estimated leaving the digester and entering the storage tank are different, attributed to slight difference in calibration of the pressure-based level instruments.

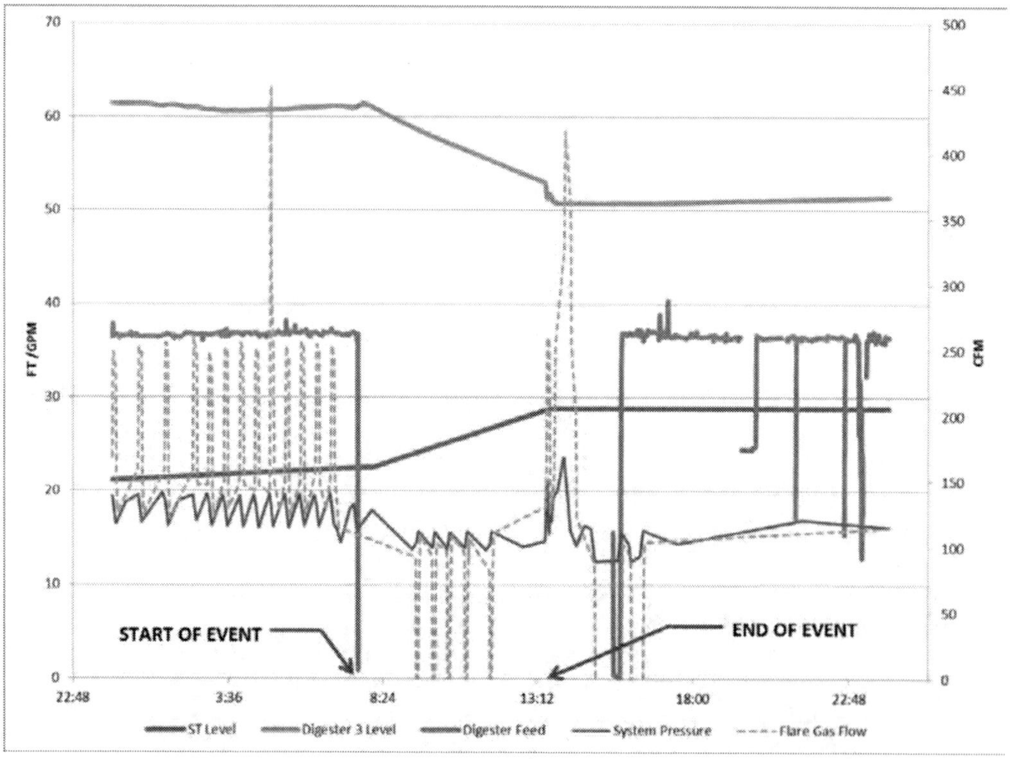

Figure 5 – Measured Parameters During 2012 RVE Event at Brightwater (from Chapman et al., 2014)

The initial gas flow rate reported before the event was approximately 167 cfm; after the event it reduced to approximately 115 cfm. At Brightwater, the gas flow meters report flows at 1 atm and 70°F (294°K). Converting these gas flows to STP at 1 atm and 15°C would result in 164 cfm and 113 cfm, respectively. The s.g. in the digester following the RVE event can be estimated assuming a start s.g. of 1.0 and using the indicated change in mass within the digester; 51-ft end level indication/61-ft start level indication = 0.84. In this case, the calculation of actual RVE rate is not straightforward as the s.g. of the overflowing sludge will be somewhat lower than average as gas is moving to and accumulating at the surface. In addition, the s.g of the overflowing sludge will vary with time, likely higher at the beginning of the event following vigorous mixing, and lower as the event proceeds. If the overflow from the digester is assumed to be at the s.g. at the end of the RVE event, the RVE rate could be estimated as = 205,000 gal OF/ (6 hrs X 0.84) ⇒90 cfm; The **RVE_r** ratio based on start gas production and digester s.g. change would then = (90 cfm/164 cfm) X100 = **55%**.

The RVE rate based on level in storage tank receiving overflow, assuming a s.g of the overflowing sludge of 0.84, would be estimated = 143,000 gal/ (6 hrs X 0.84) ⇒ 63 cfm. Based on this assumed s.g., the **RVE_r** ratio = (63 cfm/ 164 cfm) X 100 = **38%**. Pagilla et al. (1997) reported a s.g. of 0.7 at the surface of a mechanically mixed digester with an average digester s.g. of 0.94. If the RVE rate were calculated for storage tank receiving overflow assuming a s.g. of 0.7, the **RVE_r** ratio = (76 cfm/ 164 cfm) X100 = **46%.** These estimates compare to a theoretical estimate of 43% in a 65-foot tall digester from calculation 17 (Appendix A).

Estimation of Ratio of Volume Expansion Rate to Gas Production Rate (RVE_r) Based on Gas Compression in Full-Scale Digesters Mixed Half Height

Consider a pumped mixing system that draws from the bottom and discharges half way up a standard height tank. In this case the bottom half is mixed mechanically and top half of digester mixes only through natural gas rise in the digester. To simplify the RVE calculation, the top half is assumed to be at steady state; turning off mixer will not cause RVE in top half. In reality the top half has some entrained gas rising from below at the time of mixer stoppage. As this gas moves to the surface, the top half may actually exhibit a short term volume reduction, since further gas production in the bottom half is being held up below. This may result in a lower short term RVE rate as the top half reaches a new steady state after these initially-entrained bubbles release.

An estimation was made of volume expansion, V_{T4}, for test conditions and duration of a 36-ft tall digester mixed half height, where **P_{T4}** is 33.9 ft + 6 in gas pressure + 28 ft (estimated approximate depth to centroid of pressure in bottom half only of a 36-foot tall digester). Since digester is fed at the bottom, the assumption was made that 60% of gas production is in bottom half of the digester. The resulting **RVE_{r4}** = 0.28 or 28% (Appendix A, calculation 18). Adding a 1.1 safety factor adjusts this to **31%**.

For a taller 65-ft digester with pumped mixing system that draws from the bottom and discharges half way up the tank (i.e. Oceanside egg shaped digesters in San Francisco), the resulting **RVE_{r5}** = 0.20 or 20% (Appendix A, calculation 19). Adding a 1.1 safety factor adjusts this to **23%**.

As discussed above, with respect to calculations 18 and 19 (Appendix A), the bubble dynamics in the top half of the digester are considered to be at steady state. This may result in a lower short term RVE rate than calculated as the top half reaches a new steady state after initially-entrained bubbles that had risen into the top half from below are released.

EFFECTS OF GAS MOVING IN AND OUT OF SOLUTION WITH RAPID VOLUME EXPANSION

As discussed above, gas moving into solution is one route for produced gas during an RVE event, the others being gas forming bubbles at compressed conditions and expanding the sludge column and gas evolving into the digester headspace. Gas moving into solution reduces the RVE rate and gas moving out of solution is important in determining the rate of gas surge following an RVE event when the expanded column collapses after mixers are turned back on and soluble/stored gas from the event is stripped out of solution.

The effects of gas moving into solution become more significant at depth due to higher pressure; the saturation level of a given gas increases and more gas that is produced can move into solution.

Under normal steady state mixed conditions in a digester, gas that has moved into solution at depth is mechanically moved to the surface, where under less pressure, it releases as bubbles to a new solubility equilibrium at the lower pressure. At steady state under mixing, gas in solution from depth is continually "elutriated" from the solids by the mixing action moving it from the bottom to top of the digester, resulting in an unsaturated condition for the gas at depth. This unsaturated condition is gradually and partially replenished by additional gas production as it returns to depth. It follows then that when mixing stops, there is a reserve capacity for gas to go into solution at depth.

A simplified model could consider that before rapid volume expansion occurs originating from a specific depth, this reserve capacity would need to be filled. This assumes gas moves into solution as rapidly as produced; this is a reasonable assumption for gas being produced in small bubbles throughout the solution (another model considers that bacteria are producing gas in solution and that bubbles don't form until the liquid in a region becomes saturated.). It is likely that the 15-minute delay from mixer stopping to volume expanding seen by comparing Figures 1 and 2 above can be attributed in part to this (mixer stopped at time 155 minutes and volume expansion began at time 170 minutes). Part of this phenomenon can also be explained by residual gas near the surface evolving from the surface. This is a limited surface evolution however, restricted by the new lack of mixing mechanically moving more gas to the surface. The resulting increase in viscosity without mixing, leads to additional gas holdup within the sludge. Surface evolution will remain low for some time until the sludge is fully expanded and the unmixed digester reaches a new unmixed steady state with new bubbles rising to the surface.

In the experiment discussed above, with gas generated at 870 mL/hr at the start of the RVE event, there would have been 232 mL produced in first 15 minutes. During this initial 15 minutes after the mixer was turned off, there was no volume expansion and therefore the only sinks for produced gas during this period were gas evolving from the surface or moving into solution. From calculation 15 (Appendix A), the total estimated gas moving into solution over the two-hour test period is 191 mL. From Figure 2, the gas measured in the first 15 minutes is approximately 186 mL, about half of surface evolution estimated for the entire 2-hour test period from calculation 12 (Appendix A).

The estimated difference between gas produced and the gas evolved from the surface over the initial 15-minute test period is 46 mL (Appendix A, calculation 20).

This represents an estimate of the amount of gas that has moved into solution over that period. If that rate of gas moving to solution were to be maintained for the entire test period, approximately (46 mL) X (123 min/15 min) = 377 mL would move into solution. As discussed above, the estimated volume of gas that actually moved into solution, **E**, is 191 mL. This is evidence, as expected, that as the RVE event progressed, portions of the sludge reached saturation with respect to gas in solution, resulting in a slowing rate of solution and increase in gas bubble formation.

There are many examples of similar phenomena in nature that can help the reader relate to what is happening in the digester under these conditions:

- The bends - at depth (and higher saturation levels) a diver raises the soluble nitrogen concentration of his blood through respiration; when pressure reduces as he moves back to the surface, the nitrogen rapidly comes out of solution, forming bubbles in the blood - the bends.
- Opening a soda bottle - reduces pressure and bubbles come out of solution.
- Pouring a soda into a glass - agitation (mixing) and further reduced pressure and release of bubbles.
- Brewing beer - biological production of gas in a reactor, resulting in increased gas in solution

Theoretical Maximum Amount of Gas Moving into Solution in a Digester

This analysis is an approximation of the theoretical amount of gas that can move into solution using the gas laws. The amount of gas that actually moves in and out of solution will depend on the gas production rate, sludge physical and chemical conditions, the equilibrium of CO_2 in the gas phase with carbonic acid, bicarbonate, and carbonate, and be impacted by some rate-dependent conditions. In addition, some gas will remain in the gas phase in equilibrium with gas in solution even below saturation levels.

For this analysis, CO_2 and CH_4 are the principal gases of interest. CO_2 is far more soluble than CH_4. Henry's law, gas solubility data, and aqueous carbon dioxide equilibria reactions are used to estimate the amount of gas in solution (saturated) at 38°C and 55°C for CO_2 and CH_4. The impact of water vapor in the gas phase must also be considered to determine partial pressures as the digester gas will be saturated with water vapor.

For ease of comparison of gas in other states during an RVE event, this analysis will estimate the equivalent gas volume in solution. For this purpose, we define Mass Equivalent Volume (**M_{Ve}**) as the equivalent volume at STP for a given mass (or moles) of gas in solution. This will be a useful parameter for equivalent comparison of different digester regimes (temperatures and depths, etc.).

Important gas data for this analysis taken from various references are shown in Table 2.

The following gas laws were found useful for these analyses:

Henry's law proportionality constant, H = p/x where p is absolute partial pressure in atmospheres and x is mole fraction (represented by mass) in solution at a constant temperature.

$H_{T1}Mass_{T1} = H_{T2}Mass_{T2}$ and $H_{T1}/H_{T2} = Mass_{T2}/Mass_{T1}$ at constant pressure

As discussed above, Mass Equivalent Volume (**M_{Ve}**) is the equivalent volume at STP for a given mass (or moles) of gas.

Table 2. Gas Data for Determining Solubility and Partial Pressures

Parameter	Value (see Note a)
Density of CO_2	1.964 g/L at STP_{0C} (273°K) 1.8617 g/L at STP_{15C} (288°K)
Density of CH_4	0.6798 g/L at STP_{15C} 0.6682 g/L at 20°C
CO_2 gas solubility (see Note c)	179.7 cc CO_2 in 100 cc water at 0°C 90.1 cc CO_2 in 100 cc water at 20°C 0.1688 g of CO_2 in 100 g water at 20°C 0.1257 g of CO_2 in 100 g water at 30°C 0.0973 g of CO_2 in 100 g water at 40°C 0.0761 g of CO_2 in 100 g water at 50°C 0.0576 g of CO_2 in 100 g water at 60°C
Henry's law proportionality constant for CH_4 (see Notes b and c)	H = 37,600 at 20°C H = 44,900 at 30°C H = 48,600 at 35°C H = 52,000 at 40°C H = 55,100 at 45°C H = 57,700 at 50°C H = 62,600 at 60°C
Vapor pressure of H_2O at 38°C	49.7 mm Hg = 2.2 ft of H_2O
Vapor pressure of H_2O at 55°C	118 mm Hg = 5.3 ft of H_2O

Notes: [a] All values expressed at 1 atmosphere of pressure

[b] H constant = p/x where p is absolute partial pressure in atmospheres and x is mole fraction in solution at a constant temperature

[c] from Perry and Chilton, *Chemical Engineer's Handbook*, 5th Edition

CO_2 Solubility

For CO_2, one can use gas solubility tables above to interpolate mass of gas in solution (saturated) at 38°C and 55°C; and then find mass equivalent volume (**M_{ve}**) in solution in water at STP. Note these estimates are in pure water. Actual solubility will depend on alkalinity and other sludge characteristics. These calculations (calculation 21 in Appendix A) assume pure CO_2 gas in bubbles/headspace and are not yet adjusted for partial pressure of other gas constituents (e.g. CH_4 and water vapor, etc.) or aqueous carbon dioxide equilibria reactions. The estimate considers only CO_{2aq} or CO_2 molecules in solution. At 38°C the estimated CO_2 in solution considering gas solubility data is 553 mL per L (Appendix A, calculation 21). At 55°C this decreases to 359mL.

Since CO_2 is in continuous equilibrium with other carbonate species, as conditions change, the gas moving in and out of solution will be to a great degree controlled by these equilibrium reactions. The next step in the estimation of gas moving in and out of solution is to consider the equilibrium reactions between aqueous CO_2, carbonic acid, bicarbonate, and carbonate species.

Carbon Dioxide Equilibrium Chemistry and Estimating Total CO_2 That Can Move into Solution

Equations and parameters for estimating carbon dioxide in solution based on equilibrium reactions with carbonic acid, bicarbonate and carbonate are presented below.

Carbon Dioxide Equilibrium Equations

$[CO_{2(aq)}] = P_{CO2}/H$, where P_{CO2} = partial pressure of CO_2 (absolute) in atm and H = Henry's constant, atm/(mole/L); (Note: different units than proportionality constant reported in Table 2)

$K_h = [H_2CO_3] / [CO_{2(aq)}]$

$K_0 = ([CO_{2(aq)}] + [H_2CO_3])/P_{CO2}$

$K_1 = [HCO_3^-] [H^+] / [H_2CO_3]$

$K_{1(app)} = [HCO_3^-] [H^+] / ([H_2CO_3] + [CO_{2(aq)}])$, used in conjunction with K_0 and accounts for K_h and K_1

$K_2 = [CO_3^-] [H^+] / [HCO_3^-]$

$K_w = [H^+] [OH^-]$

$K = 10^{-pK}$ and $pK = -\,^{10}\log K$

$[H+] = 10^{-pH}$ and $pH = -\,^{10}\log [H^+]$

$K_h = 1.7 \times 10^{-3}$ at 25°C in pure water (Housecroft and Sharpe, 2005)

Reaction Rates

Reaction rates for CO_2 moving in and out of solution have been reported for pure water at 25°C (Wikipedia, "Carbonic Acid") as:

Rate Constant for $CO_{2aq} + H_2O \rightarrow H_2CO_3 = 0.039\ s^{-1}$

Rate Constant for $H_2CO_3 \rightarrow CO_{2aq} + H_2O = 23\ s^{-1}$

CO_2 moves into solution within seconds and effectively moves out of solution in less than a second. For the purposes of this analysis, these reactions can be considered to occur instantaneously.

Temperature Corrections

$pK_0 = -2622.38/T - 0.0178471T + 15.5873$ (Harned and Davis, 1943)

$pK_{1(app)} = (3404.71/T) + 0.032786\ T - 14.8435$ (Harned and Davis, 1943)

$pK_2 = (2902.39/T) + (0.02379\ T) - 6.498$ (Harned and Scholes, 1941)

Values for Henry's Constant and various solubility K factors for water at temperatures relevant to sludge digestion are shown in Table 3.

Table 3. Values of CO_2 System Gas Law and Equilibrium Constants for Various Temperatures

Factor	Value at 38°C (311°K)	Value at 55°C (328°K)
H, atm/mole/L	41.18	61.43
K_0	0.024845	0.018265
$K_{1(app)}$	5.005×10^{-7}	5.122×10^{-7}
K_2	5.846×10^{-11}	7.017×10^{-11}

CO_2 and Carbonic Acid Concentrations

$[CO_{2(aq)}] + [H_2CO_3] = K_0 P_{CO2}$

$[CO_{2(aq)}] + [H_2CO_3] = [HCO_3^-]\,[H^+]\,/K_{1(app)}$

$[HCO_3^-] = ([CO_{2(aq)}] + [H_2CO_3])\,K_{1(app)}/[H^+]$

$[CO_3^{-2}] = K_2[HCO_3^-]\,/[H^+]$

C_T = Total molar concentration of inorganic carbon in solution = $[CO_{2(aq)}] + [H_2CO_3] + [HCO_3^-] + [CO_3^{-2}]$

Using these relationships and equilibrium constants, the equivalent CO_2 in solution at pH 7.0 was estimated to be $M_{VeCO2@38C}$ = 3,528 mL per L (Appendix A, calculation 22). At pH 7.4 this increases to 7,980 mL per L.

Of the dissolved CO_2 and carbonate species (at pH 7.0), 587 mL of CO_2 equivalent is in the form of aqueous CO_2 and carbonic acid (H_2CO_3). Since this partitioning is predominantly $CO_{2(aq)}$ by a general factor of 1000 (see low K_h value above), this compares favorably to the 553 mL CO_2 equivalent estimated by calculation 21 (Appendix A). Under the condition in this example, about 16 percent of the C_T is as $CO_{2(aq)}$ and about 84 percent exists as bicarbonate (HCO_3^-).

Note that at these near neutral pH values, $[CO_3^{-2}]$ is insignificant and the formula for C_T can be simplified as follows:

$C_T = K_0 P_{CO2}\,(1 + K_{1(app)}/[H^+])$ in moles organic carbon dissolved per L of sludge

and M_{VeCO2} can be simplified, assuming standard gas conditions (1 atm and 15°C), to:

$M_{VeCO2} = 23{,}634\ C_T$ in mL of CO_2 gas equivalent in solution per L of sludge

A summary of M_{VeCO2} values for various pH and temperatures relevant to sludge digestion are shown in Table 4.

Table 4. Mass Equivalent Volume of CO_2 for Various pH and Temperature Values Relevant to Sludge Digestion

pH	M_{VeCO2} (mL of CO_2 gas equivalent in solution per L of sludge) [a]	
	38°C (311°K)	55°C (328°K)
6.8	2,443	1,827
7.0	3,528	2,643
7.2	5,245	3,929
7.4	7,969	5,986
7.6	12,287	9,234
7.8	19,130	14,382
8.0	29,976	22,542

Notes: [a] Expressed at standard temperature and pressure of 1 atm and 15°C; Assumes a partial pressure of CO_2 of 1 atm and sludge buffered to maintain indicated pH.

The values in Table 4 assume a CO_2 partial pressure of 1 atm, requiring adjustment for actual partial pressure when applying to a digester gas environment. In addition, these values were calculated at a fixed pH. In an unbuffered solution, one would expect the pH to decrease as CO_2 moves into solution and increase as it moves out of solution. However, sludge is typically well buffered. Therefore, the values in Table 4 assume that sludge is adequately buffered to maintain its operating pH as CO_2 moves in and out of solution.

CH_4 Solubility

For CH_4, use Henry's law and gas solubility tables above to interpolate mass (moles) of gas in solution (saturated) at 38° C and 55° C. Then find mass equivalent volume (M_{Ve}) in solution in water at STP. Note these estimates are in pure water. Solubility will be less in sludge with other dissolved salts (See Schaum et al., 2016). These example calculations (calculation 23 in Appendix A) assume pure CH_4 gas in bubbles/headspace and are not yet adjusted for partial pressure of other gas constituents (e.g. CO_2, water, etc.).

The estimated equivalent volume of CH_4 in solution at 38°C, $M_{VeCH4@38C}$ = 26.2 mL @ STP_{15C} in 1 L of water (Appendix A, calculation 23). At 55°C this reduces to 21.0 mL/L.

Comparison of CO_2 and CH_4 in Solution

Comparing the results of calculation 23 (Appendix A) with those of Table 4, it can be seen that CH_4 is far less soluble than CO_2, accounting for less than 1 percent of the soluble gas. Even after accounting for partial pressure within sludge (see below), the amount of CH_4 in solution is insignificant relative to CO_2. For further considerations related to the rate and total volume of RVE, it is assumed that produced CH_4 remains in the gaseous state. One consideration coming out of this assumption is that gas bubbles forming during an RVE event will likely be at an elevated proportion of CH_4 relative to CO_2, as CO_2 will readily move into solution; this may limit RVE rate based on a gas bubble generation rate resulting from the higher CH_4:CO_2 ratio until such time as the sludge is fully saturated with CO_2.

Corrections for Partial Pressure

The results of calculations estimating CO_2 in solution (Table 4) now need to be corrected to account for partial pressure of gas constituents. Example calculations for estimating partial pressure are presented in Appendix A, calculation 24. Results of partial pressure estimates are presented in Table 5 as follows:

Table 5. Estimated Partial Pressure of CO_2 at Various Digester Depths and Temperatures

Location within Digester	Estimated Ratio of Partial Pressure of CO_2 to Total Pressure at Stated Temperature	
	38°C (311°K)	55°C (328°K)
Headspace	0.374	0.339
Centroid of 36-Foot Tall Digester	0.384	0.363
Centroid of 65-Foot Tall Digester	0.388	0.372

Note: Assumes a 40:60 molar ratio of CO_2:CH_4 and includes water vapor.

The results of this analysis (Appendix A, calculation 24) indicate that moisture is a small component of the partial pressure at 38°C (For a 36-foottall digester: 6% at the surface and 4% at depth), but a much larger proportion at 55°C (15% at the surface and 9% at depth). To simplify the calculation of moving gas into solution with a conservative result, one could ignore the impact of moisture and assume the molar ratio of CO_2 and CH_4. However, it could be considered too conservative particularly at thermophilic temperatures. For the following example calculation, the relative partial pressures will be used to correct for moisture in the gas.

Potential for Driving Gas Into and Out of Solution with Mixing

The next step in analyzing the movement of gas into solution is to estimate the relative amounts of gas in solution at depth versus at the surface. The operating assumption is that when the digester is mixed, the gas moving into solution at depth is constantly being cycled to the surface of the digester by bulk movement of fluid, and gas is stripped into the headspace at the lower-pressure gas-liquid interface. When mixing stops, this "elutriation" of gas from depth to surface conditions suddenly stops and more gas will immediately begin moving into solution, increasing the concentration in solution at depth. This model then assumes the difference between gas in solution at depth and at the surface is the net ability or sink to move gas into solution after mixing stops, and the source of net release of gas from solution after mixing resumes.

For this analysis, use Henry's Law at constant temperature. Assume $M_{Ve1}/P_1 = M_{Ve2}/P_2$, where M_{Ve} represents gas volume equivalent of mass (moles, x) in solution; $M_{Ve1} = M_{Ve2}\ P_1/P_2$. Assume depth at the approximate centroid of the pressure and adjust for partial pressures of gas constituents.

At mesophilic temperature (38°C) and pH of 7.0, M_{VeCO2} = 3,528 mL at 1 atm in 1 L of water, assuming pure CO_2 gas in the bubbles/headspace (see Table 4).

An estimate of the potential for driving gas into solution in a 10 L laboratory-scale digester at 38°C and pH 7, adjusting for partial pressure and depth within the reactor is presented in Appendix A, calculation 25. This calculation assumes what can be driven into solution is the mass at depth minus the mass at the surface, M_{Ve} depth – M_{Ve} surface (assumes mixing drives all gas to surface solution steady state). $\mathbf{M_{VeCO2}}$ depth - $\mathbf{M_{VeCO2}}$ surface = 380 mL at STP_{15C} (Note: this is equivalent to 404 mL expressed at 38°C and at 1 atm plus 6-inch headspace pressure; see Appendix A, calculation 25).

As a check, from calculation 10 (Appendix A), 1,763 mL of digester gas were estimated to be produced during the 2-hour RVE event. In this case of the lab scale reactor, gas production rate far exceeded the estimated theoretical maximum CO_2 that could have been dissolved from the CO_2 equilibria reactions, so that the theoretical maximum was not limited by gas production.

Next, an estimate of the potential for driving gas into solution in a 36-foot tall digester at 38°C and pH 7, adjusting for partial pressure and depth within the reactor is presented in Appendix A, calculation 26. $\mathbf{M_{VeCO2}}$ depth - $\mathbf{M_{VeCO2}}$ surface for this tall digester = 8,660 mL in 10 L at STP_{15C} (Appendix A, calculation 26).

By comparing results from calculations 25 and 26 (Appendix A), it is noted that 20 times more gas per unit volume is estimated to go into solution in a 36-foot deep full-scale digester than a shallow lab-scale reactor. Even greater amounts of gas will go into solution in deeper digesters. The lab scale reactor generated an estimated 1,763 mL of gas during the RVE event (calculation 10, Appendix A), about one-fourth of the theoretical maximum. Therefore, in this deep full scale reactor at the assumed temperature and pH, gas production would limit the amount of gas that could move into solution. The theoretical maximum CO_2 that can be moved to solution using the methodology in calculation 26 (Appendix A) will become limiting in shallower digesters operating at low pH and thermophilic temperatures, during peak digester loading.

A conclusion that can be drawn is that for well mixed deep digesters, more gas can be driven into solution following a mixer stoppage, tending to reduce the rate of RVE over shallower digesters. In addition, this can lead to more gas being released from solution when the mixers are turned back on.

Calculation 15 (Appendix A) estimated 191 mL went into solution over the 2-hour period of the test RVE event based on measured gas volumes. Calculation 20 (Appendix A) estimated that 46 mL of gas went into solution and 186 mL evolved in the first 15 minutes of the test RVE event. Calculation 25 estimated the total gas that theoretically could go into solution was 404 mL (at test conditions). The difference between these estimates (404 mL and 191 mL) can be surmised from an understanding of their respective calculation methodology. The 404 mL estimate is the theoretical amount of gas that can go into solution based on gas laws, CO_2 equilibria reactions, and partial pressure of sludge gas constituents, assuming at time zero all of the gas in solution had been mixed to steady state surface concentration and after time zero all of the produced gas immediately went into solution up to the available solution capacity at depth.

The 404 mL estimate assumes that the digester contents are adequately buffered to maintain pH during the dissolution of CO_2. If sludge is inadequately buffered, as more CO_2 is driven into solution, the equilibrium will move to the right, with more CO_2 being driven to bicarbonate and so forth, releasing hydrogen ions and decreasing the pH of the sludge and restricting additional CO_2 dissolution. At lowered pH, the sludge will have lesser capacity for dissolving CO_2 than predicted by assuming a constant pH. A further influence is that the calculation of theoretical CO_2 that can be driven into solution is based on formula derived for reactions in water. Sludge is more complex and other dissolved compounds and ions will hinder the ability of carbonic ions to interact as modeled (Wisc.edu", Chemistry of Carbonic Acid in Water"). In addition, as discussed above in the section of centroid of pressure, the assumption of an operating s.g. of 1.0 slightly overstates the amount of gas that can be driven into solution. These considerations may explain all or some of the difference between the estimated 191 mL that went into solution over the entire RVE event based on measured gas and the estimated theoretical maximum of 404 mL. Other calculation methodology and measurement errors may also be partially responsible for the difference. It will be valuable to further explore and refine our model of these relationships in the future as our understanding of RVE reactions matures.

Potential for Driving Gas Out of Solution with Sudden Pressure Loss

An event at a large east coast wastewater treatment plant was recently described by a plant operator: operations had to take two of four digesters out of service for repair, doubling the load on the remaining two. Volatile solids reduction (VSR) was reduced, but acceptable with the significantly increased organic loading rate (OLR) and reduced hydraulic retention time (HRT). However, the gas production was too great for the gas piping laterals in each of the two on-line digesters. Gas pressure in the digesters increased rapidly until emergency relief hatches on the roof of the digesters began to chatter and eventually opened, resulting in an immediate pressure drop. This was followed by a large 'foaming' event, an RVE event presumed to be caused by gas being stripped from solution due to the rapid pressure drop. The event reportedly lasted for several days.

Using the calculation methodology for CO_2 equilibrium discussed above, an estimation of the volume expansion was made for a 36-foot tall digester at pH 7.0, losing 12 inches of headspace pressure (Appendix A, calculation 27). The estimated gas from solution in this case is 28 mL/L $\cong$ 3% volume expansion.

Doubling the headspace pressure loss to 24 inches doubles the estimated volume expansion to 6%. Increasing the pH assumption from 7.0 to 7.4 increases the volume expansion estimate for a 12-inch pressure drop from 3 to 6%.

If the s.g drops in a 65-foot tall digester by volume expansion from a mixer outage and the digester overflows, maintaining approximately constant level, the pressure in the digester contents will reduce as overflow occurs. A 10% drop in s.g. in an overflowing digester will be equivalent to about a 6-foot drop in pressure on the contents. Using the methodology above and correcting for the 10% lost sludge to overflow, this situation would represent a 12% volume expansion from gas coming out of solution alone, exacerbating other RVE causes.

These estimates assume sludge that is fully saturated with CO_2 at the stated starting conditions. As discussed above, a vigorously mixed digester will "elutriate" the gas from solution and be operating at less than full saturation controlled by the gas production rate, resulting in a lower volume expansion than estimated here. On the other hand, a poorly mixed digester will be operating at full saturation and experience volume expansions close to these estimates, partially ameliorated by the loss of gas to the headspace.

With respect to the rate of expansion in these instances, we have assumed above that the equilibrium reactions moving CO_2 in and out of solution occur virtually instantaneously. However, the complex ionic nature of sludge will slow these reactions. In addition, the pressure drop from say a weighted pressure relief hatch opening, for example, will occur slowly and control the rate of expansion.

COLUMN COLLAPSE AND GAS SURGE FOLLOWING RVE

Theory of Column Collapse

From Higgins work discussed above, volume collapse occurs at between 3 and 15% volume expansion during a no-mixing RVE event. The theory is that as volume expansion progresses (more bubbles being produced), more and more bubbles merge, forming larger bubbles. The bubble/liquid matrix approaches the 'edge' of bubbles rising. When bubbles in a certain region get to a critical size when their buoyant force exceeds the sludge yield stress, they suddenly rise. As they begin to rise, the sludge moves to the right on the yield stress/apparent viscosity graph below, causing a drop in viscosity and more bubbles are now in the region of being buoyant and begin to rise - less viscosity - more rising bubbles, etc. Basically the system of stable bubbles entrained in the sludge matrix "collapses" and the bubbles are released, and the sludge level drops to the previous level before RVE.

It has been observed in the laboratory (Higgins, 2016) that once collapse occurs, another RVE event can begin again, essentially cycling between RVE and collapse.

Column Collapse Under Full Scale Operating Conditions

At Brightwater, the digesters undergo column collapse during high digester feed operating conditions (Strehler, 2016). A collapse to about 1 to 2 feet below the gas dome occurs during/after a high secondary sludge wasting event. Normally the plant feeds sludge at a constant 25 gpm/digester; during the high wasting events, operators increase the feed rate to 75 gpm/digester for a day or two. This is tripling the digester solids load (mostly TWAS). Gas production goes way up; volume expands, but is overflowing to the storage tank. This occurs to certain point when the level in the digester suddenly drops (collapses). This collapse is only 2 to 3% of the total digester volume (2 feet below gas dome is 3% apparent volume, but the cover is sloped so there is not direct ratio of depth to volume). This expansion and collapse occurs while sludge feeding continues and mixers are on.

A similar event occurs during a heavy scum feed event going to one of three digesters only.

These events exhibit low volume of collapse indicating a low total volume of expansion (2 to 3%). It can be concluded that at a sudden change to high load (a load spike) will create a volume expansion, even under continuous intense mixing. However, since mixing continues in this event, shear rate will remain high and thus apparent viscosity will remain low (see Higgins et al., 2014), resulting in a lower total volume increase before collapse than under a mixer outage RVE event.

Volume of Gas Surge Following a Column Collapse After an RVE Event

In Figure 2, a large spike in measured gas occurs immediately following turning the mixers back on. Within about 40 minutes, the measured gas stabilizes to a new base gas production (about 850 mL/hr) slightly lower than before mixing was stopped. This would be expected if digester feeding was not continued during the mixer outage; gas production would be expected to begin to taper off after a couple hours without feeding. The measured gas spike happens coincident with an immediate drop in digester volume equal to the rise during the volume expansion (Figure 1). When the volume drops in the digester vessel following turning the mixer back on, room is being created for some of the released gas, equal to the prior stored (compressed) gas volume, slightly modified by the STP calculation (See calculation 13, Appendix A: 1,160 mL vs 1,200 mL). As compressed/stored bubbles are released from the liquid, bubbles expand more than the liquid level drop, which is expected as compressed gas expands. Therefore, the spike in measured gas above the new base gas production (about 850 mL/hr) is the gas being released from the sludge (from compressed bubbles and from solution) and is approximately 73 mL (estimated from Figure 2).

An estimate was made of the volume of gas surge following the laboratory test RVE event from Figures 1 and 2 (Higgins et al., 2014) that results from releasing and expanding gas bubbles. The extra gas from expanding bubbles evolving into the headspace above the volume of column collapse = 34 mL (Appendix A, calculation 28).

Therefore, of the estimated 73 mL of gas spike following the RVE event, 34 mL is estimated from bubble expansion being greater than the column collapse volume and 73 mL - 34 mL = 39 mL is from gas being released from solution. This is lower than the estimated 191 mL (calculation 15) that went into solution during the mixer outage. (Note from calculation 24: 404 mL at 1.015 atm and 38°C was estimated as possible to drive into solution.) Note that from calculation 20, an estimated 46 mL of gas went into solution in the first 15 minutes of the RVE event. The gas surge event following mixer startup lasted about 40 minutes. It could be expected that further gas will be released over extended time following the initial gas surge.

The gas being stripped from solution is estimated to be (39 mL/404 mL) X 100 = 10% of the theoretical maximum that could have been pushed into solution during the RVE event. In this case of the lab scale reactor, gas production rate far exceeded the estimated theoretical maximum CO_2 that could have been dissolved, so that the theoretical maximum was not limited by gas production.

An analysis using similar methodology was conducted for the gas surge following the 2012 Brightwater RVE event shown in Figure 5 above and described by Chapman et al. (2014). In that case, the gas production during the RVE event was far less than the estimated theoretical maximum that could have gone into solution. Therefore, the theoretical maximum that could go into solution was limited to the gas production rate, CO_2 specifically. The gas that evolved from solution during the approximately 60-minute gas surge event (following RVE) was estimated as approximately 37% of the theoretical maximum that could have been pushed into solution during the RVE event based on the CO_2 production rate.

In the case of Brightwater, expanding volume flows through an overflow pipe by gravity to a sludge storage tank. In this case the actual gas release from held bubbles is calculated from the expanded volume remaining in the digester only, as that is the only reactor where mixing resumes (the storage tank is unmixed). Based on the expanded volume in the digester only, 13.5% of the digester volume is released from compressed, held gas bubbles.

From the above discussion it can be concluded that gas moving in and out of solution is a significant component of the overall RVE mechanism and is important in determining the rate of gas surge following an RVE event when mixers are turned back on.

Methodology for Estimating Gas Surge Rate Spike Following a Column Collapse After an RVE Event

As discussed above, the time a typical gas surge following an RVE event will last, t_S, is from 40 to 60 minutes. The total volume of gas being measured during the gas surge, V_{SURGE}, will be comprised of 3 gas sources:

1. **Base gas production being generated biologically, Q_{GPROD}.** This will likely be slightly diminished from the gas production rate seen at the time of the RVE commenced because food will have been continually degraded during the potentially hours-long event. When designing a gas system to accommodate this component, estimated peak gas production for the digester should be used, considering the RVE event could occur during this condition; peak day would be a reasonable and conservative assumption for design; peak week gas production would also be a reasonable, less conservative alternative assumption.

2. **Gas being released from solution, V_{SOL}.** This is the estimated volume of gas stripped from solution on re-start of mixing (or other collapse triggering event). From calculations above, this will be some fraction (F_{SOL}) of the theoretical maximum gas in solution during the RVE event; either calculated based on CO_2 production rate or the CO_2/bicarbonate reactions, whichever is limiting. The theoretical maximum will be gas potential at depth minus gas potential at headspace pressure. From discussion above, 10% of theoretical was found for this component based on results from the tests in a 10 L laboratory digester where CO_2/bicarbonate reactions were limiting. For full scale digesters, gas production will normally be limiting and a factor of 40% of CO_2 production is prudent.

3. **Gas being released from bubbles held up during the RVE event, V_{BUB}.** These bubbles are released as the expanded volume collapses on re-start of mixing (or other collapse triggering event). The volume of the RVE is equal to held bubbles under compression. When released, these bubbles will fill the space created by the column collapse and extra gas being relieved of its compressed condition will be included in the gas surge at headspace pressure.

A simplified schematic depicting this model for calculation purposes is shown in Figure 6.

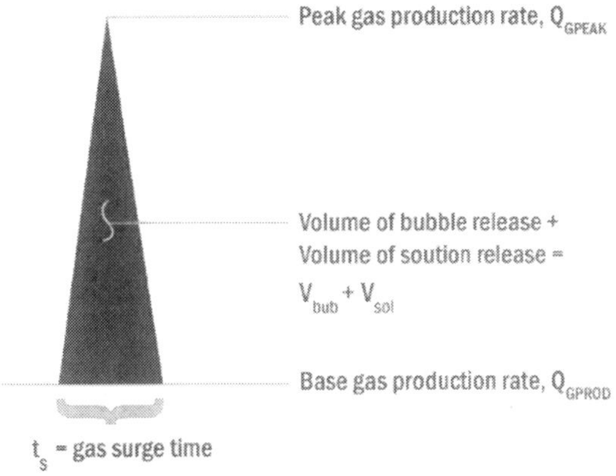

Figure 6 – Simplified Model of Gas Surge Following an RVE Event

Using this model:

$V_{SURGE} = (Q_{GPROD}/t_S) + V_{SOL} + V_{BUB}$

Q_{GPROD} = Peak gas production during the RVE event (peak day reasonable and conservative assumption for design; peak week gas production also a reasonable, less conservative alternative assumption)

t_S = assumed duration of the surge (60 minutes a reasonable assumption)

$V_{BUB} = V_{RVE}\,[(P_D/P_{EMERG}) - 1]$, where: (see calculation 28 for example)

> V_{RVE} is the total volume estimated for the rapid volume expansion within the digester prior to the surge event
>
> P_D is the absolute pressure at mid-depth of the digester prior to the surge event (includes operating headspace pressure)
>
> P_{EMERG} is the absolute pressure in the headspace under surge or emergency conditions (assumed to be at the pressure relief valve setting)

$V_{SOL} = (F_{SOL})\,(M_{VeCO2}$ depth - M_{VeCO2} surface), where: (see example calculation 26 in Appendix A)

> F_{SOL} = fraction of theoretical maximum gas that can be driven into solution that will be released during the post RVE surge; 0.4 a conservative and reasonable assumption where M_{VeCO2} is limited by CO_2 production rate
>
> M_{VeCO2} depth = mass equivalent volume of CO_2 that can be driven into solution at digester operating temperature and at centroid of absolute pressure depth
>
> M_{VeCO2} surface = mass equivalent volume of CO_2 that can be driven into solution at digester operating temperature and at digester headspace absolute pressure

Note that in this calculation, soluble CH_4 is ignored as it is far less soluble than CO_2.

Once the volume of the gas surge is estimated, a peak rate of surge, Q_{GPEAK}, needs to be estimated. From observations from the lab scale tests (Higgins et al., 2014) and at full scale at Brightwater (Chapman et al., 2014), gas release rises to a sharp peak then back down to the base gas production rate. Gas production rate will be uniform over the surge period, but the total volume from bubble release and release from solution will spike during the surge.

$Q_{GPEAK} = Q_{GPROD} + [PF\ (V_{BUB} + V_{SOL}) / ts]$, where

PF is defined as a peaking factor. If the gas rate rises steadily to a peak and then declines similarly to base rate over the surge period as depicted in the Figure 6, the peaking factor will be 2. During the Brightwater RVE event, the PF was observed to be approximately 4. However, in that event, the base gas production was well below design peak gas production rate and the gas management system was easily capable of handling the spike without a pressure increase in the headspace. For design purposes, a peak gas production rate is assumed to size the gas management system. It is reasonable under this design condition that a higher emergency headspace pressure is assumed up to the pressure relief setting. Although spikes with a PF of 4 and greater have been observed, these spikes are of short duration and represent a small volume of gas. This smaller spike volume (resulting from high PF) can reasonably be assumed to be contained in the digester headspace resulting in a pressure spike. As such, for design purposes, a PF of 2, with a pressure spike up to the gas relief setting is a reasonable assumption. The designer should verify that a small peak volume resulting from a high PF of short duration can be contained within the digester headspace and within the pressure differential between operating P and relief P.

In design, the gas management system, including the laterals, gas piping, gas flares, pressure relief valves and emergency relief vents must be designed in combination to be capable of handling Q_{GPEAK}.

DIGESTER LEVEL RISE RATE IN A RVE EVENT

For a given RVE volumetric rate and digester volume, the level rise rate will vary depending on the tank geometry. For a given tank volume, the level rise rate will be greater for a tall digester. Whereas the RVE rate and volume is important to consider in design of emergency overflows and storage volumes, the level rise rate is important to consider in designing and operational monitoring of RVE instrumentation and control systems.

Two types of level measurement are useful in digester design and operation: a radar level instrument located at the top of the digester that monitors and records the actual level of the top of the sludge, and a pressure-based level instrument at the bottom of the tank, measuring the weight of liquid over the transducer and inferring the tank level. For a digester design with reserve headspace volume that can be used to store the volume expansion, the surface radar level can detect the rising sludge and deduce the rate of rise and rate of volume expansion using the tank geometry. In some digester designs, there is little reserve headspace and instead an overflow is used to remove any rising sludge (to alternative storage or discharge, i.e. plant headworks). In this instance the surface level instrument cannot detect the rising sludge after reaching the overflow; the pressure-based level instrument will then record a drop in level. However, this is actually a drop in specific gravity as the constant volume of sludge (level locked at the overflow level) will be increasingly less dense as gas bubbles are held up in the sludge. An example of this can be seen from Figure 5 at Brightwater. The onset of this drop in s.g. while the surface radar level device detects constant level will be indicative of the RVE event.

TOTAL VOLUME OF RAPID VOLUME EXPANSION

To date, the maximum volume of RVE has been observed for a variety of sludges and full scale cases, but only limited hypothesis as to the conditions leading to the limiting maximum volume before column collapse occurs.

From Bartek, et al. (2015), another 11 different sludges with 11 different viscosities were tested until the expanded volume collapsed, when the bubbles from continuing gas production reached a size that their buoyant forces overcame the shear stress and they rose, naturally mixing the digester contents. From this data a maximum RVE volume was determined relevant to the specific sludge and this shallow 10 L test vessel (the last point on each curve). Higher viscosities clearly tended to result in higher RVE rates and maximum volumes. From this data set, the maximum RVE was 13%.

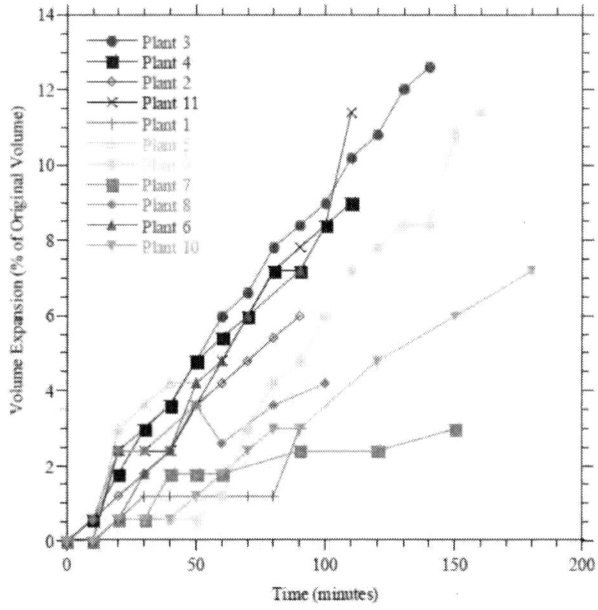

Figure 7 - Relationship Between Maximum Volume of RVE and Yield Stress (from Bartek et al., 2015)

In addition, Bartek et al. (2015) plotted the relationship between maximum RVE volume and sludge yield stress using several rheological models and demonstrated a strong correlation with each model. Their plot using viscometer data is shown in Figure 8.

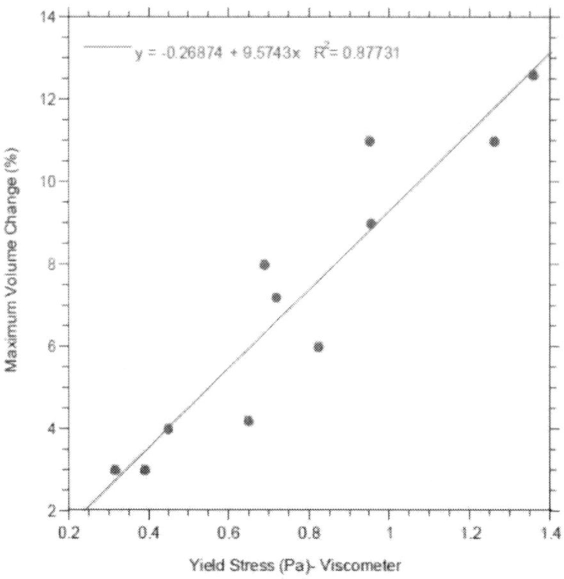

Figure 8 - Relationship Between Maximum Volume of RVE and Yield Stress (from Bartek et al., 2015)

This is the only data to data relating maximum RVE volume to rheological conditions of the sludge.

From Bartek et al. (2016) eleven additional sludge samples tested and all reached a maximum RVE before bubbles reached a sufficient size to overcome shear stress, bubbles rose and the expanded volume collapsed back to original volume. Expansions ranged from 2 to 15% before collapse. From these 11 samples sludges, the maximum RVE was 15%. As volume expansion progresses (more bubbles being produced), more and more bubbles merge, forming larger bubbles. When bubbles get to a critical size they suddenly rise. Basically the system of stable bubbles entrained in the sludge matrix "collapses" as the bubbles are released and the sludge level drops to the previous level before the RVE. Column collapse has been observed in full-scale digesters. Information on these observations will also be reported.

From Chapman et al. (2014) Brightwater RVE event lasted for 6 hours before mixers came back on; digester is submerged fixed cover with constant level overflow; Digester pressure level instrument indicated 10 feet of drop from 61 to 51 feet which is indicative of a drop in assumed s.g. of 1 to about 0.84 (or from actual 0.92 to 0.77). Under constant s.g this would have indicated a volume of about 205,000 gallons or 16 percent of tank volume; however, since the actual level was constant because of surface overflow, this only indicated change in s.g. Assuming this expanded volume had been retained in the digester (assuming sufficient head space), and the overflow amount was of comparable s.g. (0.77), and the retained expanded volume did not collapse, this would represent a volume expansion of 16% of digester contents. Since low s.g. sludge was constantly overflowing, however, it is unknown whether the OF volume had a lower s.g. or whether the contents would have collapsed before reaching this level of expansion as seen by Bartek et.al. 2016 and Higgins et al. 2014.

As the Brightwater digester overflowed, the storage tank level, as measured by a pressure transducer level device, rose by about 6.8 feet = total volume of approximately 143,000 gallons; This indicates that the digester lost approximately 11% of its solids to the storage tank (6.8 ft/61.3 ft original digester measured height). However, if the overflow volume had a s.g. of 0.6 as it left the digester (as opposed to the average digester s.g. after RVE of 0.77), this would indicate a volume expansion in the digester of about 19% (6.8 ft/ (0.6 X 61.3 ft)).

From this event the conclusion can be drawn, based on sludge overflowing and maintaining a constant digester level along with the estimated change in the digester s.g, that the maximum RVE volume is greater than 16% and may be as high as 19% or more. The total possible RVE volume cannot be determined because the mixers were restarted before the end of the event.

From a RVE laboratory experiment for a major west coast U.S. city in 2015, at a typical sludge feed rate (yielding approximately 20-day SRT), the sample sludge expanded to a maximum of about 10%. However, when the feed rate was doubled (approximate SRT of 10 days), the maximum RVE rose to about 22%. From this work, it appears sludge feed rate (and presumed gas production rate) also will impact the maximum RVE volume as well as RVE rate, and that RVE volumes as high as 22% can be seen in very high rate digesters.

Note that from this experiment, the maximum volume expansion was about 13% before volume collapse. From experiment 1 above, the volume expanded 15% before mixers were turned on; from Brightwater, the volume expanded about 16% before mixers were turned on; no volume collapse before mixers were turned on in these later two cases.

From data available to date:

1. **The maximum volume expansion varies with the sludge and tank configuration; there seems to be a direct relationship between sludge yield stress and maximum RVE volume.**

2. **The maximum volume expansion when solids are retained in the tank (adequate headspace) is greater than 15% of the original tank volume (before solids collapse) and may be greater than 22% for very high yield stress sludges in high feed rate digesters.**

3. **The maximum volume expansion when solids are not retained in the tank, but instead overflow at constant tank level is greater than 16% of the original tank volume.**

It is important to recognize the relationship suggested by Bartek et al. (2015) between the total expansion volume before collapse and the sludge rheology, specifically yield stress. It is important when considering RVE as well as other digester design features such as mixing) to have rheological testing on the sludge done to provide input to these calculations. In some instances, new digestion systems or processes are being designed where a representative sludge may not be available. In these instances, it is prudent for the designer to assume conservative values, particularly for digester safety features such as overflows and emergency relief devices.

RVE IN STAGED DIGESTION SYSTEMS

Although additional work needs to be done on staged systems, one hypothesis is that staged digestion may partially mitigate RVE compared to a single stage system. Using TPAD as an example - most of the gas is generated in the first thermophilic stage - during an RVE event, the first stage will expand at a fast rate, a function of gas production in stage 1 only, to some limit of expansion volume - say 15%. The expanding sludge overflows to the second stage mesophilic tank (or could be a storage tank or second stage thermophilic) that also has a 15% RVE volume limit (before collapse). The two stages may not be additive from the perspective of total expansion volume; potentially cutting the total RVE volume in half. The expanded overflowing sludge adds to the expansion of the second stage volume until it reaches its independent 15% maximum expansion.

The second stage expands at a rate controlled by its independent gas production rate (much slower for the second stage) plus any expansion coming from first stage, minus any off-gassing of first stage sludge in the second stage digester or storage tank. The RVE rate is reduced in second stage by the amount of off-gassing and slower gas production rate. The designer would still size a large overflow (OF) from stage 1 to stage 2. It would be prudent that the OF from stage 2 would be the same size, unless an assumption is made for a certain percent off-gassing. With this hypothesis, the maximum volume the designer will need to contend with outside the digester complex is limited to the 15% only in the second stage. As an example, if there are 3 first stage and 3 second stage digesters, the maximum RVE volume is 15% X 3 or 45% of a single digester volume. This assumes that the first stage OF flows by gravity to stage 2 and enters the body of the digester (does not enter the top where it can just overflow and not expand the body of stage 2 contents). If it enters the contents, then off-gassing cannot occur. In contrast, if there is a six-digester complex with all digesters operating in parallel, then the maximum expansion volume could be 15% X 6 or 90% of a single digester volume. This hypothesis is in need of test confirmation before relying on for design.

DIGESTER MIXING AND IMPACTS ON RVE

Mixing Intensity and Impact on Bubble Rise and RVE

Figure 9 presents estimated bubble rise rates for various yield stress/viscosities reported by Bartek et al. (2016) and compares this to rise velocities as a result of mixing at various turnover rates. The yield stresses plotted in Figure 9 represent the range of yield stresses per Bartek et al. (2016). Rise velocities for mixing of three digester heights assume mixing the digester contents from bottom-to-top (mixing contents up would be down through a draft tube system).

The conclusion that can be drawn from this comparison is that at typical high rate design turnovers (30 minutes to one hour) the impact of mixing (top-to-bottom or bottom to top) far outweighs the bubble rise rate and therefore moves bubbles to the surface far faster than they can rise naturally with the result of 'elutriating' bubbles from sludge and operation of the digester approaching a s.g. of 1. It should be noted that the bubble rise rates calculated from rheological conditions assume perfect spheres and as such likely somewhat overstate the rise velocities. Velocities will slow down as bubbles get larger and distort. In addition, pressure and temperature affect bubble drag. A more in-depth literature review and discussion of factors affecting bubble rise rate was presented by Chapman and Krugel (2011).

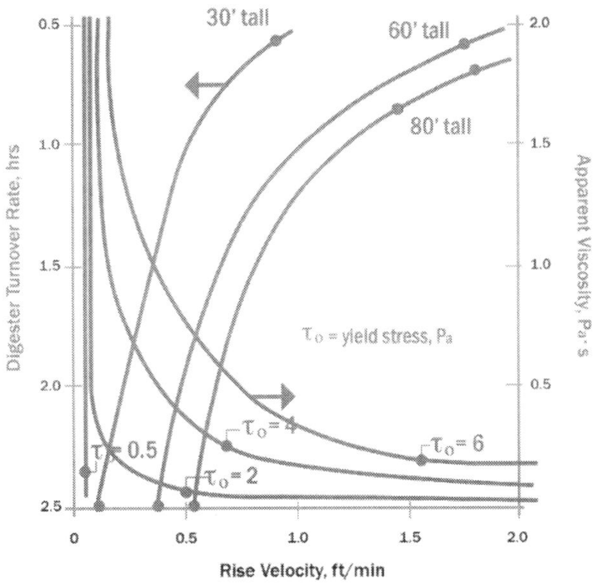

Figure 9 – Comparison of Bubble Rise Rates at Various Rheological Conditions versus Pumping Rates for Digester Mixing

Intermittent Mixing and Feeding with RVE

Bartek et al. (2015) reported on a full scale plant that had been feeding and mixing their digesters on a 4-hour cycle and were experiencing RVE events. The RVE event began immediately after the feed/mix cycle, and gas surge occurred when the feed/mix cycle resumed four hours later. As described by Bartek et al. (2015), "The addition of feed would increase gas production, and the cessation of mixing would result in the solids exhibiting a yield stress. As a result, during the periods between the mixing and feeding cycles, gas holdup would be expected." They conducted experiments where feeding/mixing interval was reduced. Reducing to one hour seemed to stop the RVE cycle.

Digester Tank Design Volume - Reduced Specific Gravity and Mixing

From experiments and operating data, heavy mixing results in moving gas out of the liquid column, "elutriating" the gas to the headspace, resulting in operation of a digester near a s.g. of 1.0. Operating at low or no mixing expands the sludge column with gas hold up, resulting in a digester operating at a much lower overall and stratified s.g. A review of the phenomenon of reduced and stratified s.g. and a discussion of RVE and foaming control measures was presented by Willis et al. (2016).

Operating at reduced s.g. influences the true SRT at which a digester is thought to be operating. As an example, at a feed and operating s.g. of 1.0, volume into a tank is equal to volume out. For a 750,000-gallon digester fed at 50,000 gpd, the SRT equals the HRT at 15 days. If this digester is fed 50,000 gpd of raw sludge at s.g. of 1.0, but is operated at a s.g. of 0.8, the withdrawal rate will be 62,500 gpd and the SRT will be 12 days. In general, a digester with an SRT of 15 days will perform better (more complete VSR) than at an SRT of 12 days.

Mixing intensity and configuration and sludge solids concentration (viscosity controlling) can be used to control operating s.g. When sizing a digester, the interplay between operating s.g., SRT, and design tank volume must be considered.

As discussed by Higgins et al. (2014), for each specific sludge, a shear arte/viscosity relationship can be determined. An example curve is shown in Figure 10. As shear rate (mixing intensity) increases beyond the knee of the curve, there is a precipitous drop in viscosity. With lower viscosity, less gas holdup is possible and less expanded volume at operational steady state. With too low of mixing, or a power outage, viscosities will be high, resulting in high gas holdup and lower s.g.

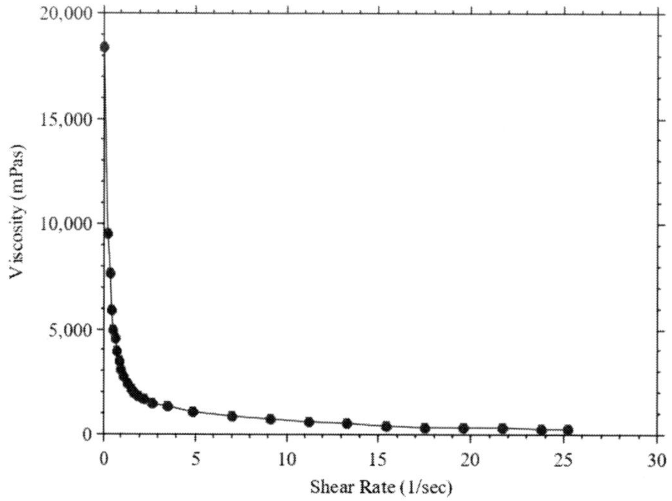

Figure 10 – Shear Rate versus Viscosity in Sludge (from Higgins et al. 2014)

There will be a clear tradeoff between tank volume and mixer intensity. If low mixing is provided, the digester will operate at higher viscosity and lower operating s.g. In this case a larger tank is needed for a desired design SRT. Also there will be less RVE, because the sludge contents would already be carrying higher volume of gas. A potential problem is evident if mixing energy (shear rate) is too low, operating on the knee of curve in viscosity vs shear rate figure where slight changes can cause digester contents to move in and out of rapid rise during normal operation. This can cause serious operational problems.

It is recommended that designs should keep mixing input (shear rate) to the right of the knee of the shear rate/viscosity curve for the specific sludge being designed for. This highlights the importance for testing the rheological characteristics of sludges if the plant sludge is available during design, or assuming very conservative rheology if the plant sludge not available (i.e. new plant or changing sludge character, etc.).

Not only is mixing energy, but also mixing configuration is important for design to mitigate RVE. Mixing top to bottom or visa versa moves bottom contents up and helps keep gas "elutriated' and a higher operating s.g. (closer to 1.0). Mixing tangentially will keep viscosity low (increasing s.g.), but will not as efficiently elutriate gas mechanically, thus lowering s.g.

Mixing only the bottom half of the tank reduces the s.g of the top half of the tank and overall tank SRT, but also reduces the RVE potential (see Appendix A, calculations 18 and 19). These tradeoffs are important to consider in digester design.

IMPACT OF SOLIDS FLOTATION ON RVE

For simplicity in calculating RVE parameters, the forgoing discussions assume a homogenous sludge matrix, with sludge character varying with depth and influenced only by gradation of gas bubble compression and concentration. In reality, the solids concentration and s.g. of the sludge will also be influenced by solids flotation within the digester.

In the design of dissolved air flotation thickeners (DAFTs), an air-to-solids (A/S) ratio of 0.01 lb air/lb solids is typical for primary and waste activated sludges; an A/S ratio of 0.1 has been used on the high end of DAFT applications. An A/S ratio of 0.1 results in about 1 cfm of air/lb solids fed. In a typical digester, gas is being produced at a rate of about 16 cu ft/lb solids leaving the digester. There is clearly sufficient gas bound in a digester sludge matrix to cause flotation and concentration of solids at the surface of the digester, subject to the more restrictive rheology in digesters compared to DAFTs.

The presence of nocardioforms and other filaments such as *Microthrix parvicella* within the reactor also effects flotation. Pagilla et al. (1997) reported data on stratification of s.g. and total solids (TS) concentrations within a digester in the presence of nocardioforms and reported greater stratification with a gas-mixed digester than a mechanically-mixed digester. Although the mixing shear rate or mixing energy inputs were not reported, Pagilla's work showed for a gas mixed digester, s.g. as low as 0.5 in the surface 1 m in a digester with average s.g of about 0.88; and for a mechanically-mixed digester, s.g. as low as approximately 0.7 in the surface 1 m of a digester with and average s.g. of about 0.94. Both digesters showed total solids concentrations of about 5 percent in the surface 1 m while the average digester TS concentration was about 2.5 percent. Parker et al. (2014) reported on the phenomenon of nocardioform filaments' hydrophobic surface exacerbating flotation and reviewed classifying selectors as a means to control foam formation. Their property of having hydrophobic surfaces gives them an accelerated attachment of gas to solids, causing nuisance foams in both activated sludge plants and digesters. Higgins (2016) observed and reported on a discernable foamy layer on top of expanding liquid volume during RVE research.

For the case of RVE, it is unlikely that the nuisance foam causing organisms are a necessary condition for the phenomenon. But given their properties, it would still be a valid research need to quantify their presence and overall impact during RVE events.

It is fair to speculate that this flotation phenomenon in an RVE episode would drop SRT faster than HRT. If the solids in the digester are swept out faster than they would be without flotation, the inventory in the digester would be reduced, and the SRT could very well be reduced. This would not only be the case for RVE events, but also in steady state continuous operation at low mixing rates and surface withdrawal, a consideration for digester performance and stability. However, with high mixing rates and appropriate mixing configurations, this would not be the case, as solids that are tending to float are reintroduced (mixed) back into the sludge matrix.

SUMMARY AND CONCLUSIONS

RVE rate is a direct function of gas production rate. A model is proposed and can be used to estimate the rate of RVE in digesters of different depths and mixing configurations.

A conclusion that can be drawn is that for well mixed deep digesters, more gas can be driven into solution following a mixer stoppage, tending to reduce the rate of RVE over shallower digesters. In addition, this can lead to more gas being released from solution when the mixers are turned back on.

In deep full-scale digesters at typical temperature and pH, CO_2 gas production will limit the amount of gas that can move into solution during an RVE event. The theoretical maximum CO_2 that can be moved into solution, based on CO_2 equilibria reactions at constant pH, may become limiting (instead of CO_2 gas production rate) in shallower digesters operating at low pH and thermophilic temperatures, during peak digester loading.

With respect to maximum RVE volume, from data available to date:

1. The maximum volume expansion varies with the sludge and tank configuration; there seems to be a direct relationship between sludge yield stress and maximum RVE volume.
2. The maximum volume expansion when solids are retained in the tank (adequate headspace) is greater than 15% of the original tank volume (before solids collapse) and may be greater than 22% for very high yield stress sludges in high feed rate digesters.
3. The maximum volume expansion when solids are not retained in the tank, but instead overflow at constant tank level is greater than 16% of the original tank volume.

It is important when considering RVE (as well as other digester design features such as mixing) to have rheological testing on the sludge done to provide input to design calculations. Digester designs should keep mixing input (shear rate) to the right of the knee of the shear rate/viscosity curve for the specific sludge being designed for.

Mixing intensity and configuration and sludge solids concentration (viscosity controlling) can be used to control operating s.g. When sizing a digester, the interplay between operating s.g., SRT, and design tank volume must be considered.

At typical high-rate mixing design turnovers (30 minutes to one hour), the impact of mixing (top-to-bottom or bottom to top) far outweighs the bubble rise rate and therefore moves bubbles to the surface far faster than they can rise naturally with the result of 'elutriating' bubbles from sludge and operation of the digester approaching a s.g. of 1.

Calculation methodologies have been proposed that can be useful in estimating the rate of RVE, the extent of gas moving into solution, and the spike of gas surge following an RVE event for different digester and mixing configurations.

Additional research and investigations are needed to further develop and refine design methodologies to mitigate RVE.

Readers should be aware that, although we have learned a lot about the very complex mechanisms at play in RVE, it is still early in our industry's understanding of these mechanisms; there will surely be additional advancements and modifications to these design approaches. The approaches used to date, even where successful, may not be appropriate in all circumstances. Each digester design is unique and, as such, the designer should not use or rely on these calculations at face value, but apply their own judgment and refinements to accommodate their unique project situation. Consultation with an experienced designer in these areas is advisable. The proposed design calculation methodologies discussed herein are intended as a starting point for further discussion of measures to help mitigate RVE that our industry can critique, refine, and improve upon moving forward.

ACKNOWLEDGEMENTS

The authors would like to thank Nick Bartek and Dr. Matt Higgins of Bucknell University for sharing their research findings and our personal discussions on RVE and Andy Strehler of King County, Washington for his willingness to provide RVE data and his review of this paper.

REFERENCES

Bartek, Nicholas and Matthew J. Higgins, Sudhir N. Murthy, Steven Beightol, Ahmed Al-Omari (2016), Development of a Mechanistic Model to Understand Volume Expansion Due to Gas Holdup in Anaerobic Digesters, *Proc. Water Environment Federation Annual Residuals and Biosolids Management Conference*, Milwaukee, WI.

Bartek, Nicholas, and Matthew J. Higgins, Sudhir N. Murthy, Steven Beightol, Travis Peaslee (2015), Causes and Cures of Rapid Volume Expansion in Anaerobic Digesters Due to Gas Holdup, *Proc. Water Environment Federation Annual Residuals and Biosolids Management Conference*.

Chapman, Tom and Andy Strehler, Phuong Truong, Steve Krugel (2014), Digester Gas Hold-Up and Rapid Rise Foam Formation, *Proc. Water Environment Federation Annual Residuals and Biosolids Management Conference*, Austin, TX.

Chapman, Tom and Steve Krugel (2011), Rapid Volume Expansion - An Investigation into Digester Overflows and Safety, *Proc. Water Environment Federation Residuals and Biosolids Conference*, Sacramento, CA.

Chemengineering (2017), https://chemengineering.wikispaces.com/Henry's+law

Harned, Herbert S. and Raymond Davis Jr. (1943), The Ionization Constant of Carbonic Acid in Water and the Solubility of Carbon Dioxide in Water and Aqueous Salt Solutions from 0 to 50°, *Journal of the American Chemical Society*, *65* (10), pp 2030–2037.

Harned, H.S. and S.R. Scholes Jr. (1941), The Ionization Constant of HCO_3^- from 0 to 50°, *Journal of the American Chemical Society*, 63, pp 1706–1709.

Higgins, Matthew J. (2016), personal communication, April 5.

Higgins, M.J., and S. Beightol, S.N. Murthy, P.J. Nielsen (2014), Insights into Rapid Volume Expansion in Anaerobic Digesters Associated with Gas Holdup, *Proc. Water Environment Federation Annual Conference*, Austin, TX.

Housecroft and Sharpe (2005), *Inorganic Chemistry*. 2nd ed., Prentice-Pearson-Hall, p.368.

Pagilla, Krishna R. and Kent C. Craney, and Wendell H. Kido (1997), Causes and Effects of Foaming in Anaerobic Sludge Digesters, *Water Science and Technology*, Vol. 36, No. 6-7, pp. 463 to 470.

Parker, Denny and John Bratby, Don Esping, ted Hull, Rick Kelly, Henryk Melcer, Rion Merlo, Rod pope, Todd Shafer, Eric Wahlberg, Robert Witzgall (2014), A Critical Review of Nuisance Foam Formation and Biological methods for Foam Management or Elimination in Nutrient Removal Facilities, *Water Environment Research*, Volume 86, No. 6, pp. 483 to 503.

Perry, Robert H. and Don W. Green, James O. Maloney (1997), *Perry's Chemical Engineers' Handbook*, 7th edition, McGraw Hill.

Schaum, C. and T. Fundneider, P. Cornel (2016), Analysis of Methane Emissions from Digested Sludge, *Water Science and Technology*, 73.7, IWA Publishing.

Strehler, Andy (2016), King County, Washington, personal communication, June 8, 2016.

Wikipedia (2017), "Carbonic Acid", *https://en.wikipedia.org/wiki/Carbonic_acid*

Willis, John and Luis Carrio, Tom Chapman, Jonathon Keaney, Gary Newman, Taft McNeal, Chris Muller, Peter Pianelli, Lance Salerno, Perry Schafer (2016), Are Your Digesters Burping, Frothing, or Otherwise Not Behaving? – Let's Talk about Solutions and Not Repeat the "Digester Foam" Paper You've Already Heard Five Times, *Proc. Water Environment Federation Annual Residuals and Biosolids Management Conference*, Milwaukee, WI.

Wisc.edu (2017), "Chemistry of Carbonic Acid in Water", *https://ay14-15.moodle.wisc.edu/prod/pluginfile.php/172623/mod_resource/content/5/Carbonate%20System%20Chemistry.pdf*

APPENDIX A – SAMPLE RVE CALCULATIONS

Interpretation of Figures 1 and 2 from Higgins et al. (2014):

(1) Gas production throughout test period: At the start of the RVE event (time mixer turned off) the gas production rate was approximately 870 mL/hr. One could assume a constant gas production rate over the test period. However, there was an actual slight reduction indicated from Figure 2 to about 850 mL/hr at time 340 minutes. The average gas production rate over test could be estimated at about 860 mL/hr, expressed at measured TP. Assume **GPR_M** = 860 mL/hr (from Figure 2); at 38° C (311°K) and 1 atm. (33.9 ft) plus 0.5 ft gas pressure = 1.015 atm. (34.4 ft).

Taking standard temperature as 15°C (288°K) from the International Organization of Standards (ISO), typically used by the gas industry and the wastewater industry when dealing with digester gas,

(2) Gas production rate at standard temperature and pressure (STP), **GPR_{stp}** = P_M**GPR_M** T_{stp}/T_MP_{stp} = (33.9 ft + 0.5 ft) X (860 mL/hr) X (288°K) / (311 °K X 33.9 ft) = 808 mL/hr

(3) Mixer downtime = 278 minutes - 155 minutes = 123 minutes = 2.05 hrs (Figure 2)

(4) Rapid Volume Expansion period = 278 min - 170 min = 108 min = 1.80 hrs (Figure 1)

(5) **C** = Volume Expansion = 1200 mL (compressed, held up gas bubbles produced at 38°C; from Figure 1)

(6) Volume Expansion Rate, **VER** = 1,200 mL/1.80 hrs = 667 mL/hr (Figure 1)

(7) **RVE_r** = Ratio of volume expansion rate to the pseudo-steady state, pre-test gas production rate at STP = **VER/GPR_{stp}** = (667 mL/hr) / (808 mL/hr) = 0.83 or 83% for 2-hour test; OR 78% based on gas production reported at test temperature and headspace pressure (667/860 = 0.78).

(8) A = Produced gas at 38° C and 1.015 atm. held in solids, but not contributing to measured volume increase of the gas = approximately (860 mL/hr - 750 mL/hr) X (2.05 hrs) = 226 mL (estimated from Figure 2). (This is produced gas moving into solution, V_{GS}, and volume of produced gas lost due to gas compression of bubbles below the surface at higher pressure, V_{GC} held up at depth and compressed because mixing has stopped and no longer moving gas bubbles formed to the surface of the reactor).

(9) B = Produced gas measured during the RVE event = gas evolving from surface of sludge + gas pushed out the headspace due to volume expansion (gas held in sludge as bubbles and increasing sludge volume) and being measured with the respirometer at 38° C and 1.015 atm.

(10) Total gas produced during the RVE event (mixer down time) = A + B = 860 mL/hr X 2.05 hr = 1,763 mL (at 38° C and 1.015 atm)

(11) B = Measured gas during the RVE event = total gas production - A = 1,763 mL - 226 mL = 1,537 mL (at 38° C and 1.015 atm)

(12) D = Gas evolved from surface during the RVE event = measured gas - volume expansion = B - C = 1,537 mL - 1,200 mL = 337 mL (at 38° C and 1.015 atm.)

(13) V_{stp} = mL volume of stored gas in bubbles at STP

$P_{stp}V_{stp}/T_{stp} = P_TV_T/T_T$; using absolute pressure and temperature

P_{stp} = 1 atm = 33.9 ft of water

P_T = Pressure (test) at depth of 1.0 ft (approximate depth to the centroid of pressure in a 2-foot deep test digester) plus 6 inches (0.5 ft) of gas pressure; test P_T = 35.4 ft

T_{stp} = 15° C + 273 = 288° K

T_T = 38° C + 273 = 311° K

$V_{stp} = P_TV_T\,T_{stp} / T_T\,P_{stp}$ = (35.4 ft X 1,200 mL X 288°) / (311° X 33.9 ft) = 1,160 mL volume of stored gas in bubbles at STP

(14) What is the equivalent volume of expanded gas in bubbles lost to compression? (Volume, V_M, at measured gas pressure, P_M, minus volume, V_C, at compressed gas pressure, P_C)

$P_MV_M = P_CV_C$

P_M = 33.9 ft + 0.5 ft = 34.4 ft;

P_C = 33.9 ft + 1.0 ft + 0.5 ft = 35.4 ft

$V_M = P_CV_C/P_M$ = (1,200 mL) (35.4 ft /34.4 ft) = 1,235 mL

Volume reduction of held bubbles due to compression = $V_M - V_C$ = 1,235 mL - 1,200 mL = 35 mL

(15) What is the estimated produced gas that has moved into solution?

E = Produced gas moved into solution = produced gas - evolved gas - expanded held up gas - gas volume reduction due to compression at test T and P = (A +B) - D - C - (V_M - V_C) = 1,763 mL - 337 mL - 1,200 mL - 35 mL = 191 mL

E = (A +B) - D - C - (V_M - V_C) = A + B - (B - C) - C - (V_M - V_C) = A- (V_M - V_C) = 226 mL - 35 mL = 191 mL

Estimate Ratio of Volume Expansion Rate to Gas Production Rate (RVE_r) Based on Gas Compression in Fully Mixed Full-Scale Digesters

(16) What would volume expansion, V_{T2}, be for test conditions and duration for **36-ft tall digester** mixed full height, where P_{T2} is 33.9 ft atmospheric pressure + 6 in gas pressure + 21 ft (approximate depth to centroid of pressure in 36-foot deep digester)?

$V_{T2} = P_{stp}V_{stp}T_{T2}/T_{stp}P_{T2}$ = (33.9 ft X 1,160 mL X 311°K) / (288 °K X 55.4 ft) = 767 mL

Gas production rate at STP, $GPR_{stp} = P_M GPR_M T_{stp}/T_M P_{stp}$ = (33.9 ft + 0.5 ft) X (860 mL/hr) X (288°K) / (311 °K X 33.9 ft) = 808 mL/hr

RVE_{r2} = Ratio of volume expansion rate for test conditions and duration to gas production rate at STP for 36-ft tall digester with 'full' mixing = (767 mL/1.80hr) / (808 mL/hr) = 0.53 or 53%. Adding a 1.1 safety factor adjusts this to **58%**.

(17) What would volume expansion, V_{T3}, be for test conditions and duration if **65-ft tall digester** mixed full height, where P_{T3} is 33.9 ft atmospheric pressure + 6 in gas pressure + 40 ft (approximate depth to centroid of pressure in 65-foot deep digester)?

$V_{T3} = P_{stp}V_{stp}T_{T3}/T_{stp}P_{T3}$ = (33.9 ft X 1,160 mL X 311 °K) / (288 °K X 74.4 ft) = 571 mL

RVE_{r3} = Ratio of volume expansion rate for test conditions and duration to gas production rate at STP for 65-ft tall digester with 'full' mixing = (571 mL/1.80 hr) / (808 mL/hr) = 0.39 or 39%. Adding a 1.1 safety factor adjusts this to **43%**.

Estimate Ratio of Volume Expansion Rate to Gas Production Rate (RVE_r) Based on Gas Compression in Full-Scale Digesters Mixed Half Height

(18) What would volume expansion, V_{T4}, be for test conditions and duration if 36-ft tall digester mixed half height, where P_{T4} is 33.9 ft + 6 in gas pressure + 28 ft (estimated approximate depth to centroid of pressure in bottom half only of a 36-foot tall digester)? Since digester is fed at the bottom, assume 60% of gas production is in bottom half of the digester.

$V_{T4} = P_{stp}V_{stp}T_{T4}/T_{stp}P_{T4}$ = (33.9 ft X 1,160 mL X 0.6 X 311°K) / (288 °K X 62.4 ft) = 405 mL

RVE_{r4} = Ratio of volume expansion rate for test conditions and duration to gas production rate at STP for 36-ft tall digester with 'half' mixing = (405 mL/1.80 hr) / (808 mL/hr) = 0.28 or 28%. Adding a 1.1 safety factor adjusts this to **31%**.

(19) Consider a taller 65-ft digester with pumped mixing system that draws from the bottom and discharges half way up the tank (i.e. Oceanside egg shaped digesters in San Francisco). In this case bottom half is mixed mechanically and top half of digester mixes only through natural gas rise in the digester. The top half is assumed at steady state, so turning off mixer will not cause RVE in top half.

What would volume expansion, V_{T5}, be for test conditions and duration if **65-ft tall digester** mixed half height, where P_{T5} is 33.9 ft + 6 in gas pressure + 51 ft (approximate depth to centroid of pressure in bottom half only of 65-foot tall digester)? Since digester is fed at the bottom, assume 60% of gas production is in bottom half of the digester.

$V_{T5} = P_{stp}V_{stp}\ T_{T5}/T_{stp}P_{T5}$ = (33.9 ft X 1,160 mL X 0.6 X 311°K) / (288 °K X 85.4 ft) = 298 mL

RVE_{r5} = Ratio of volume expansion rate for test conditions and duration to gas production rate at STP for 65-ft tall digester with 'half' mixing = (298 mL/1.80 hr) / (808 mL/hr) = 0.20 or 20%. Adding a 1.1 safety factor adjusts this to **23%**.

Estimate Difference Between Gas Produced and Gas Evolved from the Surface Over the Initial 15-minute Test Period from Higgins et.al. (2014)

(20) Estimated difference between gas produced and the gas evolved from the surface over the initial 15-minute test period is 232 mL - 186 mL = 46 mL.

Estimate CO_2 in Solution Based on Gas Solubility Data

For CO_2, one can use gas solubility tables above to interpolate mass of gas in solution (saturated) at 38°C and 55°C; and then find mass equivalent volume (M_{Ve}) in solution in water at STP. Note these estimates are in pure water. Actual solubility will depend on alkalinity and other sludge characteristics. These calculations (calculation 21) assume pure CO_2 gas in bubbles/headspace and are not yet adjusted for partial pressure of other gas constituents (e.g. CH_4 and water vapor, etc.) or aqueous carbon dioxide equilibria reactions. The estimate considers only CO_{2aq} or CO_2 molecules in solution.

(21) **Soluble CO_2 at 38° C** = 1.030 g in 1 L of water;
$M_{VeCO2@38C}$= 1.030 g X (1,000 mL/L) / 1.8617 g/L = 553 mL at STP_{15C} in 1 L of water (represents the volume of gas at STP that can be dissolved in water at 38°C)

Estimate CO_2 in Solution Based on Equilibrium with Carbonic Acid, Bicarbonate, and Carbonate

(22) **Equivalent CO_2 in solution at pH 7**

Mesophilic digester at 38°C

Assume: pure CO_2 gas at 1.0 atm; $P_{CO2} = 1$

pH = 7.0; $[H^+] = 10^{-7}$

$[CO_{2(aq)}] + [H_2CO_3] = K_0P_{CO2} = 0.024845$

$[HCO_3^-] = ([CO_{2(aq)}] + [H_2CO_3])\ K_{1(app)}/[H^+] = 0.024845 \times 5.005 \times 10^{-7} / 10^{-7} = 0.12435$

$[CO_3^{-2}] = K_2[HCO_3^-]/[H^+] = 5.846 \times 10^{-11} \times 0.12435/10^{-7} = 0.7269 \times 10^{-4}$

C_T = 0.02485 + 0.12435 + 0.00007 = 0.14927 moles/L

Density of CO_2 gas at standard conditions (15°C and 1 atm) = 1.8617 g CO_2/L gas

$M_{VeCO2@38C}$ = (0.14927 moles CO_2/L sludge) X (44 g/mole CO_2) X (1,000 mL/L)/ (1.8617 g CO_2/L gas) = 3,528 mL of CO_2 gas equivalent in solution per L of sludge

At pH 7.4

$[CO_{2(aq)}] + [H_2CO_3] = K_0P_{CO2}$ = 0.024845

$[HCO_3^-] = ([CO_{2(aq)}] + [H_2CO_3])\, K_{1(app)}/[H^+]$ = 0.024845 X 5.005 X 10^{-7} / $10^{-7.4}$ = 0.31235

$[CO_3^{-2}] = K_2[HCO_3^-]/[H^+]$ = 5.846 X 10^{-11} X 0.31235/$10^{-7.4}$ = 4.5867 X 10^{-4}

C_T = 0.02485 + 0.31235 + 0.00046 = 0.33766 moles/L

Density of CO_2 gas at standard conditions (15°C and 1 atm) = 1.8617 g CO_2/L gas

$M_{VeCO2@38C}$ = (0.33766 moles CO_2/L sludge) X (44 g/mole CO_2) X (1,000 mL/L)/ (1.8617 g CO_2/L gas) = 7,980 mL of CO_2 gas equivalent in solution per L of sludge

Estimate CH_4 Solubility

For CH_4, use Henry's Law and gas solubility tables above to interpolate mass (moles) of gas in solution (saturated) at 38° C and 55° C. Then find mass equivalent volume (**M_{Ve}**) in solution in water at STP. Note these estimates are in pure water. Solubility will be less in sludge with other dissolved salts (See Schaum et al., 2016). These calculations assume pure CH_4 gas in bubbles/headspace and are not yet adjusted for partial pressure of other gas constituents (e.g. CO_2, water, etc.).

(23) **Soluble CH_4 at 38°C**; H = 50,600 in atm/mole fraction

Equivalent $V^e_{CH4@38C}$ = moles CH_4/ (L/mole) at 38°C

Measured P = 1 atm + 0.5 ft/33.9 ft = 1.015 atm.

X = P_p / H = 1.015 atm/50,600 = 0.00002 moles CH_4/mole H_2O

1 L water = 1 kg X (1,000 g/kg) X (1 mole H_2O/18 g) = 55.6 moles H_2O

Moles CH_4 in 1 L of water = 0.00002 X 55.6 = 0.00111 moles CH_4

Conc. of CH_4 = (0.00111 moles/1 L) X (16 g/mole) X (1,000 mg/g) = 17.8 mg/L

Ideal Gas Law = 22.4L/mole at 0°C and 1 atm; 25.1 L/mole at 38°C and 1.015 atm

Equivalent $V^e_{CH4@38C}$ = 0.00111 moles CH_4 X (25.1 L/mole) X (1,000 mL/L) = 27.9 mL

$V_2 = V_1 (T_2/T_1) X (P_1/P_2)$

$M_{VeCH4@38C}$ = (27.9 mL) (288°K/311°K) X (1.015/1 atm) = 26.2 mL @ STP_{15C} in 1 L of water

Estimate Corrections for Partial Pressure

(24) What is the partial pressure of digester gas constituents and various depths and temperatures?

Assume: Gas constituents are CO_2, CH_4, and H_2O

Molar Proportion of CO_2:CH_4 = 40:60

Vapor pressure of H_2O @ 38°C = 49.7 mm Hg = 2.2 ft of H_2O

Vapor pressure of H_2O @ 55°C = 118 mm Hg = 5.3 ft of H_2O

Pressure in digester headspace = 33.9 ft atmospheric + 1 ft gas ≅ 35 ft H_2O

$P_{Total} = P_{H20} + P_{CO2} + P_{CH4}$

$P_{CH4} = 1.5 \times P_{CO2}$

$P_{Total} = P_{H20} + 2.5P_{CO2}$

Partial pressures at 38°C in headspace

35 ft H_2O = 2.2 ft + 2.5 P_{CO2}

P_{CO2} = 13.1 ft; (13.1/35) = 0.374

P_{CH4} = 19.7 ft; (19.7/35) = 0.562

P_{H20} = 2.2 ft; (2.2/35) = 0.063 (≅ 6% water vapor)

Partial pressures at 38°C at 21-foot depth (approximate centroid of pressure of 36-foot tall digester)

56 ft H_2O = 2.2 ft + 2.5 P_{CO2}

P_{CO2} = 21.5 ft; (21.5/56) = 0.384

P_{CH4} = 32.4 ft; (32.4/56) = 0.579

P_{H20} = 2.2 ft; (2.2/56) = 0.039 (≅ 4% water vapor)

Partial pressures at 55°C in headspace

35 ft H_2O = 5.3 ft + 2.5 P_{CO2}

P_{CO2} = 11.9 ft; (11.9/35) = 0.339

P_{CH4} = 17.9 ft; (17.9/35) = 0.509

P_{H20} = 5.3 ft; (5.3/35) = 0.151 (≅ 15% water vapor)

Partial pressures at 55°C at 21-foot depth (approximate centroid of pressure of 36-foot tall digester)

56 ft H_2O = 5.3 ft + 2.5 P_{CO2}

P_{CO2} = 20.3 ft; (20.3/56) = 0.363

P_{CH4} = 30.4 ft; (30.4/56) = 0.543

P_{H20} = 5.3 ft; (5.3/56) = 0.094 (≅ 9% water vapor)

Estimate Potential for Driving Gas into Solution

(25) What is the potential for driving gas into solution in a 10 L laboratory-scale digester at 38°C and pH 7?

Equivalent Volume of gas possible in solution <u>at approximate depth of the centroid of pressure</u> in 10L laboratory digester at 38°C; from calculation 24, assume partial pressure fraction of CO_2 = 0.37:

M_{VeCO2} = 3,528 mL X (35.4 ft/33.9 ft) X 0.37 = 1,363 mL in 1 L of water = 13,630 mL in 10 L @ STP_{15C}

Equivalent Volume of gas possible in solution <u>at surface</u> in 10L lab digester at 38°C at pH 7:

M_{VeCO2} = 3,528 mL X (34.4 ft/33.9 ft) X 0.37 = 1,325 mL in 100mL of water = 13,250 mL in 10 L @ STP_{15C}

Estimate Equivalent Volume of gas that is possible to drive to solution in 10 L lab digester at 38°C; assume this is mass at depth minus mass at the surface, M_{Ve} depth – M_{Ve} surface (assumes mixing drives all gas to surface solution steady state):

M_{VeCO2} depth - M_{VeCO2} surface = 13,630 mL - 13,250 mL = 380 mL at STP_{15C} (Note: this is equivalent to 380 mL X (311°/288°K) (1/1.015 atm) = 404 mL expressed at 38°C and at 1 atm plus 6-inch headspace pressure)

(26) What is the potential for driving gas into solution in a 36-foot tall digester at 38°C?

Equivalent Volume of gas possible in solution at approximate depth of centroid of pressure in 36-foot digester at 38°C and pH7; from calculation 24, assume at depth, partial pressure fraction of CO_2 = 0.38:

M_{VeCO2} = 3,528 mL X (55.4 ft/33.9 ft) X 0.38 = 2,191 mL in 1 L of water = 21,910 mL in 10L @ STP_{15C}

Equivalent Volume of gas possible in solution at surface in 36-foot digester at 38°C and pH 7; from calculation 23, assume at surface, partial pressure fraction of CO_2 = 0.37:

M_{VeCO2} = 3,528 mL X (34.4 ft/ 33.9 ft) X 0.37= 1,325 mL in 1 L of water = 13,250 mL in 10L at STP_{15C}

Equivalent Volume of gas possible to drive to solution in 36-foot digester at 38°C, M_{Ve} depth – M_{Ve} surface (assumes mixing drives all gas to surface solution steady state):

M_{VeCO2} depth - M_{VeCO2} surface = 21,910 mL – 13,250 mL = 8,660 mL in 10L at STP_{15C}

Estimate Gas from Solution with Pressure Loss of 12 Inches

(27) Estimate gas from solution with pressure loss of 12 inches

Before pressure loss, M_{VeCO2} = 3,528 mL/L X (55.9 ft/33.9 ft) X 0.39 = 2,269 mL/L @ STP_{15C}, where 3,528 mL/L is from Table 4 and 0.39 is the partial pressure of CO_2

After pressure loss, M_{VeCO2} = 2,269 mL/L X (54.9 ft/55.9 ft) X 0.39 = 2,228 mL/L @ STP_{15C}

Change in M_{VeCO2} = 2,269 mL/L – 2,228 mL/L = 41 mL/L

Convert to operating temperature and pressure = (41 mL/L) X (33.9 ft/54.0 ft) X (311°K/288°K) = 28 mL/L ≅ 3% volume expansion

Estimate Volume of Gas Surge Following an RVE Event that Results from Releasing and Expanding Gas Bubbles

(28) What is the volume of gas surge following an RVE event that results from releasing and expanding gas bubbles?

From Higgins et al. (2014) test, expanded volume from held up gas at depth, EV_D, and subsequent collapse = 1,200 mL (12% of digester volume)

The equivalent volume at expanded surface conditions, $EV_S = EV_D (P_D/P_S)$

Pressure at depth = P_D = 33.9 ft + 1 ft at depth + 0.5 ft headspace pressure = 35.4 ft

Pressure at surface = P_S = 33.9 ft + 0.5 ft headspace pressure = 34.4 ft

EV_S = 1,200mL (35.4/34.4) = 1,234 mL

Extra gas from expanding bubbles evolving into headspace above the volume of column collapse = 1,234 mL - 1,200 mL = 34 mL

Making Room for Energy Neutrality

Michael Theodoulou[1], Dorian Harrison[2], Nicholas Bonkoski[3]

[1]Senior Product Manager, Advanced Anaerobic Digestion, GE Water & Process Technologies
Author Tel: +1 905-465-3030 Email: michael.theodoulou@ge.com
[2] Technical Director, Advanced Anaerobic Digestion, GE Water & Process Technologies
Author Tel: +44 01623 429500 Email: dorian.harrison@ge.com
[3] Domain Leader, Advanced Anaerobic Digestion, GE Water & Process Technologies
Author Tel +1 484- 459-3762 nicholas.bonkoski@ge.com

ABSTRACT

There is an aspirational goal within the Municipal wastewater treatment marketplace to make energy neutral wastewater treatment plants a reality. In order to achieve this, not only does there need to be both the technology available as well as processes developed to achieve it, but also that it is financially advantageous. GE Water & Process Technologies offers the technologies to the marketplace to achieve this, and one of those key technologies is Biological Hydrolysis (BH). BH is an enabling technology that can be either retrofit into existing wastewater treatment plants, or combined with the construction of new plants to significantly increase digester efficiency. This paper will outline the application of BH at two existing wastewater treatment plants where its inclusion allowed for additional organic materials to be brought in and digested with the existing plant sludge. This practice maximized the digester assets to increase biogas production, which was converted to renewable energy, more than covering the plant parasitic load.
With the energy neutrality ideal shown in practice, the costs of implementing BH to increase digester efficiency is compared versus building new mesophilic digestion volume. As will be seen in the analysis, using BH to increase digester efficiency in a new build scenario had up to a 44% advantage versus building solely mesophilic anaerobic digester capacity, when considering the capital investment. When looking at the scenario of increasing digester capacity at an existing plant, the advantage of adding BH versus building additional mesophilic anaerobic digestion volume was up to 54%.

KEYWORDS

Advanced Anaerobic Digestion, Biogas, Biosolids, Energy Neutral, Hydrolysis, Sludge, Wastewater Treatment, Co-Digestion

INTRODUCTION

Across the United States, there are around 1,500 municipal wastewater treatment plants (WWTP) with anaerobic digestion to treat sewage sludge. Of these however, only a fraction employs combined heat and power (CHP) systems to convert the generated biogas into renewable electricity and heat. The Environmental Protection Agency estimated that only 104 wastewater treatment plants in the US had CHP in 2011 (EPA 2011). This could be partially due to two reasons: 1) The amount of potential energy in the indigenous sludge to the wastewater treatment plant is inadequate to produce enough electricity to substantially impact the WWTPs parasitic loads; or 2) The amount of biogas generated from the indigenous sludge is too low to employ CHP technology and achieve an adequate payback period. In the same EPA report, it was estimated that only plants over 30 MGD would produce enough biogas to have an attractive investment payback if CHP technology was employed.

Due to the relative dilute nature of Municipal sewage sludge, existing anaerobic digestion (AD) technology within WWTPs are inefficiently utilized. Both volume taken up by the liquid fraction of sludge, and long hydraulic retention times to achieve modest volatile solids destruction lead to this inefficiency. This AD infrastructure however, can be a tremendous asset to achieve energy neutrality in WWTPs. On the basis of maximizing the utilization of these existing assets, advanced anaerobic digestion technologies such as Biological Hydrolysis is aimed at reducing the required hydraulic retention time (HRT) of sludge within digesters, as well as enabling digesters to operate at a higher solids content. This allows for not only the indigenous sludge treatment to the WWTP, but also importation of additional organic materials such as sludge from other WWTPs, high strength liquid or food waste, and Fats oils and Grease.

Biological hydrolysis (BH) is a technology that can be retrofit upfront of existing anaerobic digestion to accelerate the hydrolysis and acidification phases of anaerobic digestion, producing high levels of Volatile Fatty Acids (VFA's) such that organic material is ready for Methanogenesis. Within the multi-phase digestion process, hydrolysis is recognized to be the rate limiting step (Appels, 2008). This limitation impacts the degree to which volatile solids (VS) can be reduced in a conventional mesophilic digester. By eliminating this limitation, organic material can be digested in a shorter HRT, as well as at higher solids content. Now more organic material can be digested in the same digester volume. Utilizing this newly discovered digester volume, energy neutral wastewater treatment is possible.

The primary motivators of plants owners to adopt a BH system are:

- Maximize digester efficiency, allowing for either plant capacity expansion or importing sludges and other organic wastes to digest in existing anaerobic digestion infrastructure;
- Minimize the capital investment for digester infrastructure on new build
- Enabling repowering of the wastewater treatment plant by increasing biogas production from existing infrastructure and producing electricity through combined heat and power systems to cover plant parasitic loads; and
- Produce an enhanced biosolids product that achieves up to a six-log reduction in indicator pathogen content.

BIOLOGICAL HYDROLYSIS DESIGN APPROACH

Biological Hydrolysis (BH) is designed to be a non-invasive solution that can be installed upfront of existing anaerobic digestion infrastructure to enable wastewater treatment plant owners to maximize the efficiency of their digester assets. The installation of BH system consists of six serial reactor vessels whereby sludge is heated to 42°C in the first reactor, represented in Figure 1. Thermal energy is typically drawn from a plant hot water loop, where the heat is provided by either a biogas boiler, or if the plant is equipped with combined heat and power (CHP), from the CHP engine.

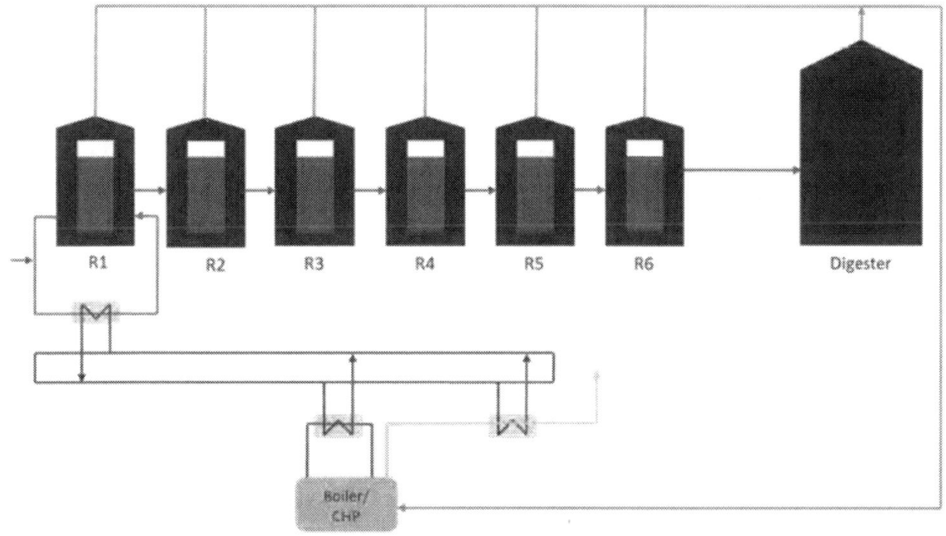

Figure 1: Biological Hydrolysis Schematic

The reactors operate in a semi-continuous reverse cascade batch system where sludge is batched once per hour from R6 to the digester, then R5 to R6, R4 to R5 and so on to R1 to R2. Following this final transfer, a fresh batch of sludge is transferred to R1. With each batch transfer, only a portion of each reactor vessel is transferred forward. This allows for concentrations of the enzymes developed to remain in the reactor vessel. The total design HRT within the BH system is approximately 3 days. As sludge is only heated in the initial reactor, over the course of progression through the remaining reactors, the temperature reduces to between 35-40°C prior to entering the methanogenic digester.

Each reactor in the BH system is mixed utilizing unconfined gas recirculation, drawing biogas from the head space of the six reactors, compressing and injecting it into the base of each reactor sequentially through mixing pipes. While each of the reactors are completely mixed, due to the method of batch flow in a reverse cascade fashion, short circuiting is minimized. This ensures that all sludge entering vessel R1 spends the majority of the design HRT within the BH vessels, being fully hydrolyzed and acidified prior to forwarding to the digester.

In addition to eliminating short circuiting, the batch transfer through the six reactor vessels provides essentially a representation of a plug flow across the entire process. As hydrolysis is a first order reaction, the kinetics rate equation is different for plug flow reactors compared to completely stirred tank reactors (CSTR). The specific hydrolysis rate equations from a first order model (Hong and Bae, 1996):

Plug Flow: N = (1-e^(-kt)*100

CSTR: N=(1-(1/(1+kt))*100

N = conversion rate, k = kinetic coefficient, and t = HRT (=V/Q)

Applying the kinetic coefficients of 0.25 for 35°C and 0.4 for 55°C (Siegrist et al., 2002), the hydrolysis conversion rates are shown in Table 1 below:

Table 1: Hydrolysis Conversion Rates

HRT (Days)	N% (35°C)		N% (55°C)	
	Plug Flow	CSTR	Plug Flow	CSTR
0.5	11.8	11.1	18.1	16.7
1.0	22.1	20.0	33.0	28.6
1.5	31.3	27.3	45.1	37.5
2.0	39.3	33.3	55.1	44.4
2.5	46.5	38.5	63.2	50.0
3.0	**52.8**	**42.9**	**69.9**	**54.5**
3.5	58.3	46.7	75.3	58.3
4.0	63.2	50.0	79.8	61.5
4.5	67.5	52.9	83.5	64.3
5.0	71.3	55.6	86.5	66.7

As observed from the above hydrolysis conversion rates, plug flow reactors have a significant advantage over CSTR reactors. In mesophilic conditions, 35°C, with a reactor HRT of 3 days, a plug flow reactor would have a 23% advantage on hydrolysis rate over a CSTR. Similarly, in thermophilic conditions, 55°C, a plug flow reactor would have a 28% advantage. As the BH process operates at 42°C for 3 days, it is expected that the advantage over a CSTR would be between 23-28%.

In addition to hydrolyzing sludge, the retention time and operational strategy the BH system is designed to promote acidification to occur. Sludges exiting a BH system are expected to have high Volatile Fatty Acid (VFA) concentrations, conditioning sludge to be converted into biogas in the downstream Anaerobic Digester. The cumulative design HRT of the BH system is important as the design intent is to transfer the sludge to the Anaerobic Digester prior to significant methanogenesis.

In addition to the described operation of a BH system at 42°C, in practice BH systems have also been designed and put into operation with an enhanced treatment step. This process called the Monsal 55 is commercially available from GE Water & Process Technologies (GE W&PT), and incorporates a pasteurization step. In the Monsal 55, reactors 4 through 6 operate at an elevated temperature of 55 to 63°C (depending on governing regulations and solids content), and operate in a true batch hold process, where vessels 5 and 6 operate in parallel, and the entire vessel contents are held to be safely pasteurized. In the Monsal 55, the minimum batch hold time is 5 hours.

Feed sludge solids concentrations to BH systems are designed up to 8% dry solids content. The natural hydrolysis process which occurs across the BH system reduces the viscosity of the sludge drastically. The sludge exiting the BH process, while still at 6-8% dry solids will show characteristics like raw sludge at 2-3% dry solids sludge.

REPRESENTATIVE PERFORMANCE

There are currently 12 BH plants that have been commissioned by Monsal Ltd., which is now part of GE W&PT. The first plant was commissioned in 2002, and in total the 12 plants process about 185,000 dry tonnes of sludge (combination of primary and waste activated) every year. Of the twelve facilities, the processing size range is from 5,000 dry tonnes per year up to 40,000 dry tonnes per year. To summarize performance of this installation base, operating data from eight operating plants are included (Theodoulou, 2016). The performance of these plants is summarized into:
a) The impact on VS reduction over the entire digestion process with the inclusion of Biological Hydrolysis;
b) The biogas yield measured in terms of m3/tonne dry solids sludge fed; and
c) Achievable digester organic loading rate.

Measuring Impact on Volatile Solids Reduction with Biological Hydrolysis Inclusion

For each respective site evaluated, the VS reduction is determined by comparing the VS mass of the sludge fed to the Biological Hydrolysis process versus the VS mass of the digested sludge exiting the mesophilic digester. The percentage VS reduction will be calculated by the following equation:

$$VSR\% = (VS_{\text{sludge fed}} - VS_{\text{digested sludge}}) / VS_{\text{sludge fed}}$$

In addition to VSR% being trended over the observed operational period, the average VSR% are plotted versus the plant average digester HRT. The cumulative plots of all of the plants evaluated generate a relationship curve of VSR% versus HRT for sewage sludge pre-treated with Biological Hydrolysis. In order to determine the benefit compared to conventional mesophilic digestion, a model developed using Simba software is utilized to compare the Biological Hydrolysis case that fits the observed VSR% from operating plants to a modelled result for a conventional mesophilic anaerobic digester with the same sludge characteristics. The characteristics used are combined primary / waste activated sludges at a 50:50 ratio, and the resultant VS content of 70%.

For the eight plants evaluated, the Volatile Solids Reduction (VSR%) have been discretely plotted versus digester HRT in Figure 2 below. In addition to the discrete points, a relationship curve has been fitted to the evaluated plants. In addition, the VSR% versus HRT modelled curve in Simba software, developed by inCTRL Solutions of Oakville Canada, for conventional mesophilic anaerobic digestion was plotted for comparison.

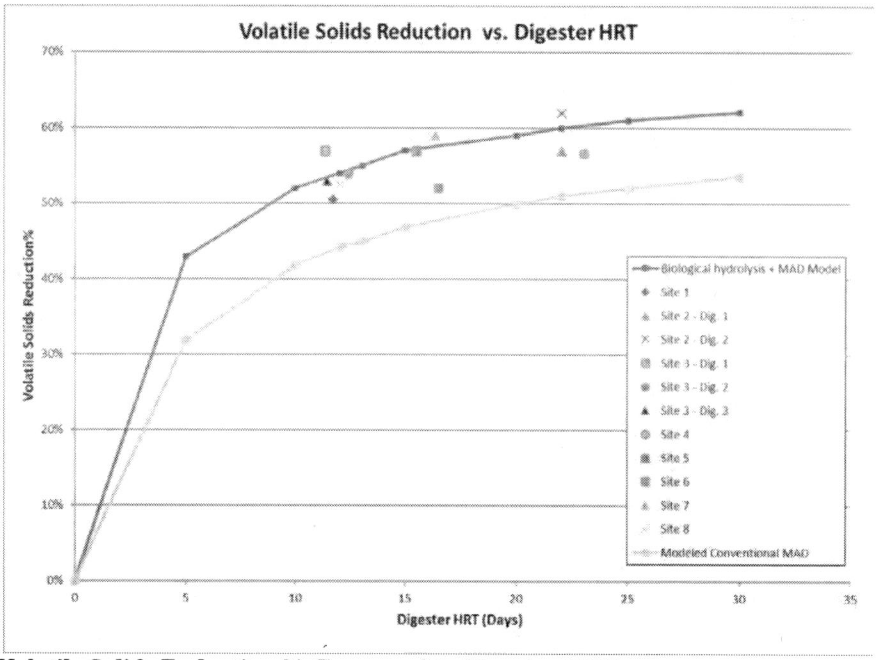

Figure 2: Volatile Solids Reduction % Compared to Digester HRT

As can be seen in the figure, multiple wastewater treatment plants with BH systems operate their anaerobic digesters with a HRT of 12-14 days. There is a variation of VSR% of these plants between 50% - 58%. This operation of these plants represents compared to the modelled conventional mesophilic anaerobic digester a higher VSR% of 14-34%. The second cluster of plants which operate with a digester HRT in the ~22 day range see the average improvement of 19%. Additionally, if the comparison was made to the conventional digester with a HRT of 24 days, incorporating BH and doubling the OLR by reduced HRT to 12 days would still result in an additional 6-10% VSR, in addition to doubling the amount of sludge treated.

Determining Biogas Yield

The efficiency of converting sludge to biogas is expressed via the biogas yield compared to the sludge mass which is fed to the overall digestion process. Within each of the evaluated plants, biogas as well as feed sludge flow is continuously measured and totalized daily. Additionally, as part of the analytical program for the selected plants, the Total Solids (TS) content and VS content of the feed sludge is analyzed. The biogas yield is calculated as:

$$BGY\ (m3/tds) = Q_{biogas(m3/d)}/(Q_{feed(m3/d)} \times \%TS)$$

(tds = tonnes dry solids)

Table 2 below summarizes the impact seen on the eight sites evaluated with Biological Hydrolysis on the digester efficiency. With respect to biogas yield, in all cases measured, the biogas yield is above 350 m3/tds, and regularly measured above 400 m3/tds. It is evident that, except for site 5, sites with longer digester HRTs have higher biogas yields.

Table 2: Biological Hydrolysis Effectiveness on Downstream Anaerobic Digesters

Site	Digester Vol. (m^3)	Digester HRT (d)	OLR (kg VS/m^3-day)	VSR%	Biogas Yield (m^3/tds)
1	16,368	11.7	3.6	51	N/A
2	3,500	22	2.5	59	458
3	8,880	11.5	3.5	55	350
4	11,358	23	3.3	57	385
5	3,960	16.5	4.0	52	406

6	20,045	15.6	3.5	57	375
7	9,000	16.4	2.4	59	396
8	7,950	12	2.5	53	360

Achievable Digester Organic Loading Rate (OLR)

It is suggested that with the application of biological hydrolysis in front of anaerobic digestion, digester organic loading rates (OLR) can be increased by 2-3 times while maintaining biogas yield. In the absence of pre-thickening, the increase of OLR which is suggested when incorporating biological hydrolysis is achieved by reducing the digester HRT. While the range is wide in the data presented in Table 2, it is indicated that higher OLR results in a reduction of VSR%. This is expected as for most sites, the increased OLR is achieved specifically by decreasing digester HRT. The BH system however, can operate up to 8% dry solids content in the feed sludge. As such, increasing the digester OLR by both decreasing digester HRT as well as increasing the feed solids concentration can result in maximum digester throughput and maintaining a respectable biogas yield. An example of this is site 6, where the plant treats both indigenous and import sludge to a total of 40,000 tonnes dry solids / year. The inclusion of the Biological Hydrolysis system allowed the plant owner to consolidate sludge treatment from other nearby wastewater treatment plants, and more than triple the sludge treated in the digesters. The resultant increase by 300% of the plant biogas production is converted in combined heat and power units and generates enough electricity to fully cover the parasitic load of the plant.

ENERGY NEUTRAL BIOLOGICAL HYDROLSYSIS CASE STUDIES

To represent the benefits of adoption of the BH technology, two cases of WWTPs that implemented BH are summarized below. In both cases, BH was implemented at the site to maximize the efficiency of their existing infrastructure. Both plants, which are owned by water utilities in the UK made the decision to convert the respective WWTPs to regional energy centres, where both indigenous sludge to the plants are treated, as well as significant volumes of import sludge and/or organic wastes.

Great Billing Sludge Energy Centre

The Biological Hydrolysis facility at Great Billing was commissioned in 2009. Prior to its start-up, the WWTP treated the indigenous combined primary and waste activated sludge. With the inclusion of the Biological Hydrolysis facility, the WWTP was enabled to accept additional undigested sludge cake from 5 neighbouring WWTPs. The summary of the digester operating conditions before and after the inclusion of biological hydrolysis are summarized in Table 3 below:

Table 3: Great Billing Impact of Biological Hydrolysis

	Organics Treated	Mass Sludge Treated	Digester OLR (kg VS/m³-day)	Renewable Electricity Produced (kW)
Prior to Biological Hydrolysis (Pre 2009)	Indigenous Sludge	12,000 Dry Tonnes/y	1.1	1,300
Post Biological Hydrolysis (Post 2009)	Indigenous + Import Sludge	40,000 Dry Tonnes/y	3.5	4,200

As shown, the amount of sludge treated at the plant more than tripled, with the utilization of existing digestion infrastructure, only now with BH. Also of note at Great Billing, the implementation of BH also was done concurrently with the construction of a cake receiving facility, where the undigested sludge cake from the 5 neighbouring WWTPs is received at 25% solids. Since this import undigested sludge cake is blended with the indigenous liquid sludge in a cake blending facility Figure 3 the resultant feed sludge concentration is 8% dry solids.

Figure 3: Great Billing Cake Blending Facility

This 8% undigested sludge is fed to the BH system, and is of high viscosity. Through the BH process, the rheology of the sludge changes with the release of moisture, CO_2 and organic acids, to a point where at the end of BH, the sludge characteristics are more like a 2-3% dry solids sludge. As such, no changes to the downstream digester infrastructure or mixing equipment was required at Great Billing. Reducing digester retention time can achieved by putting in BH, and by enabling higher solids operation up to 8% dry solids, the loading of the digester can be more than tripled. The operating digester HRT for the plant is consistently ~ 16 days.

Avonmouth WwTW

Avonmouth implemented Biological Hydrolysis in 2008. Prior to its start-up, Avonmouth was treating both the indigenous sludge from the plant as well as import sludge from neighbouring facilities. The amount that was treated through anaerobic digestion however was only 51% of the total. Through a series of upgrades at the facility, including Biological Hydrolysis, not only was the entire sludge volume (Indigenous + Import) treated through the AD infrastructure, an additional 40,000 wet tonnes/year of food waste was pre-processed on site and digested within the existing infrastructure. The summary of the digester operating conditions before and after the inclusion of these upgrades are summarized in Table 4 below:

Table 4: Avonmouth Impact of Biological Hydrolysis

	Organics Treated	Mass Sludge Treated	Digester OLR (kg VS/m^3-day)	Renewable Electricity Produced (kW)
Prior to Biological Hydrolysis (Pre 2008)	Indigenous Sludge	18,360 Dry Tonnes/y	~1.8	1,900
Post Biological Hydrolysis (Post 2008)	Indigenous + Import Sludge + Food waste	36,000 Dry Tonnes/y +40,000 Wet Tonnes/y Food Waste	3.6 – Sludge 4.5 – Food Waste	5,750

With the adoption of BH at Avonmouth, the plant increased the digester Organic Loading Rate (OLR) to 3.6 kg VS/m^3-day, roughly doubling the loading rate from their conventional practice. With the increase in sludge digestion and adding the digestion of 40,000 wet tonnes/year of food waste, the plant tripled biogas production. This is aided by the fact that the biogas yield from food waste is between 1.5 – 2 times higher than that of municipal sewage sludge. With this increased biogas production, the plant makes up to 5.75 MW of renewable electricity through site CHP engines. This allows for the coverage of the entire WWTPs electrical demand and leaves some excess that is sold to the national grid.

CAPITAL INVESTMENT BENEFITS

In cases of new WWTP construction, which would also include anaerobic digestion (AD), the benefit of incorporating BH will be the reduced digester infrastructure required, and hence a reduced investment to build digester infrastructure. Designing and operating new digesters at higher organic loading rates and solids concentrations, made possible by BH will reduce the volume of AD that needs to be built. While new build is shown to be an attractive option, as mentioned above, there are approximately 1,500 WWTPs in the US that have anaerobic digestion (AD). Add to that about 50 plants in Canada which also have existing AD.

Considering the existing AD installed base, the largest market opportunity for implementing BH technology is to retrofit existing WWTPS with AD. This enables those plants to either increase capacity without increasing existing digester infrastructure volume, and/or to maximize the utilization efficiency of those existing assets to bring in more sludge or organic wastes. Both approaches will result in increased biogas production, and the potential to convert the biogas into renewable energy.

Regardless of new build or retrofit, the capital cost of the BH technology is the same. Based on previous BH projects, Figure 4 below shows cost of the BH technology based on the mass of dry solids treated. Two technology cost curves are shown, one for the base BH technology, and one with the inclusion of pre-thickening to increase the feed solids concentration into the system to 8% solids.

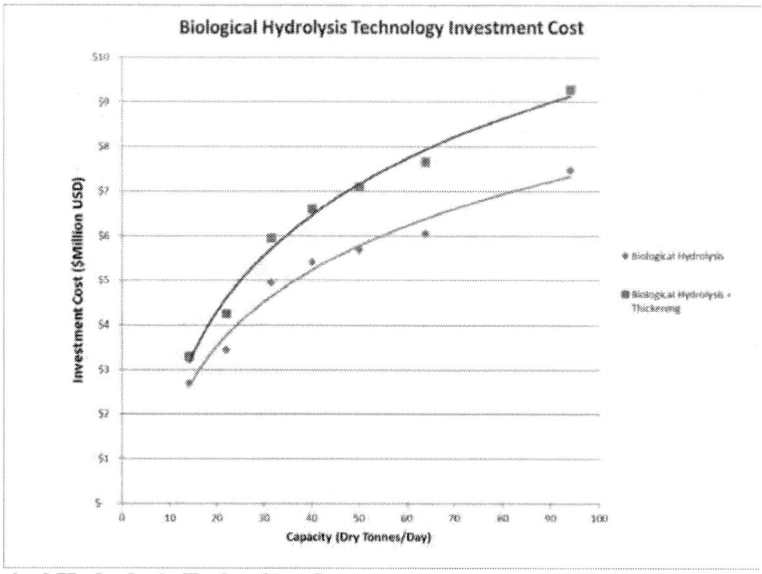

Figure 4: Biological Hydrolysis Technology Investment Cost

To estimate the total installed and constructed investment cost of the technology, cost factors are then added to the technology estimates. The technology package basis for the BH system not only includes all the equipment and reactor vessels, but also includes the installation of the reactor vessels, all the piping both internal and external to the vessels and all valves and instrumentation. As such, the total install factor for this portion of the technology package is lower than that of the remaining AD components.

For comparison purposes, the utilization of BH with downstream optimized mesophilic AD will be compared to the cost of conventional mesophilic AD sized for the treatment of the same mass of dry solids. For this comparison, a technology cost of $2 USD / gallon of digester volume is used. For all comparison purposes, the total install factor for the balance of plant (BOP) are shown in Table 5 below:

Table 5: Technology Cost and BOP Factor

Technology	Technology Cost	BOP Factor
Biological Hydrolysis	As per Figure 4	1.0
BH Pre-Thickening	As per Figure X less BH cost	1.5
Conventional Mesophilic AD	$2 USD/Gallon	1.5

To provide an example of how this is applied, consider the $2 USD / gallon for digester volume. The BOP factor is 1.5, so the additional capital cost for digestion volume is $3 USD / gallon. As such, the total installed cost for digester volume would be $5 USD / gallon.

New Build Comparison

All technology costs above are based on investment cost per tonne of dry solids for sludge capacity. To normalize this to compare based on the overall WWTP size, it is estimated that the amount of sludge handled per conventional WWTP capacity is 0.82 Tonnes dry solids/MGD plant capacity (0.217 tonnes /MLd) (Peeters, 2014). Additionally, it is assumed that the primary sludge to waste activated sludge ratio is 60%:40%.

In the new build scenario, conventional mesophilic digestion organic loading rate (OLR) is assumed to be 1.5 kg VS/m^3-day, with the volatile solids % of the total solids being 75%. For the application of BH + mesophilic AD, the OLR is assumed to be 4 kg VS/m^3-day, based on the potential shown from operating plant data. Based on these assumptions, the comparison is made for the plants defined in Table 6 below:

Table 6: Digester Capacities of Convention MAD & BH + MAD

WWTP Capacity (MGD) (MLd)	Daily Sludge (Tonnes Dry Solids / Day)	Conventional MAD – Digester Capacity (m^3)	BH + MAD – Digester Capacity (m^3)
17 (65)	14	7,000	1750
27 (102)	22	11,000	2,750
38 (145)	31	15,700	3,925
49 (185)	40	20,000	5,000
61 (231)	50	25,000	6,250
78 (295)	64	32,000	8,000
115 (434)	94	47,000	11,750

Considering the said digester capacity in Table 6, along with the technology cost defined for conventional MAD, BH and pre-thickening, and adding the BOP factor, the comparison of capital investment is shown in Figure 5 below:

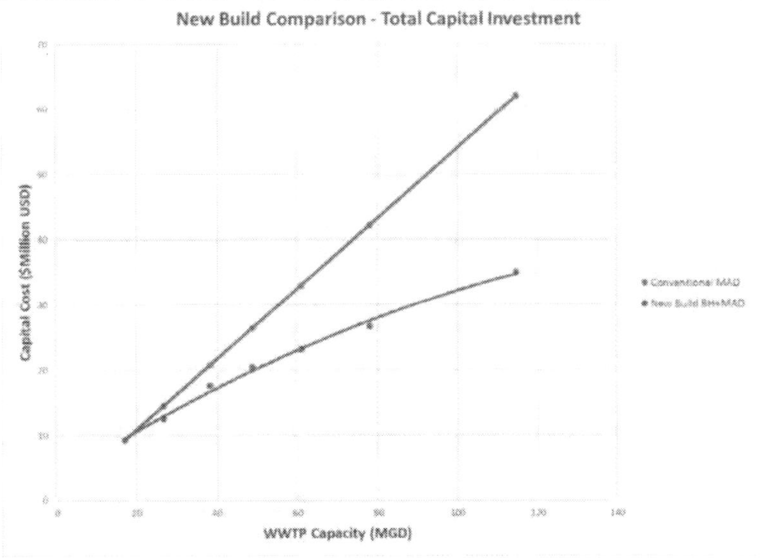

Figure 5: New Build Comparison – Total Capital Investment

As shown in Figure 5, the Total Capital Investment for BH + MAD is significantly less than conventional MAD, depending on the plant size it is applied to. Plants lower than 17 MGD (65 MLd), there is no real advantage to include BH over just conventional MAD, however as plant size increases, the advantage grows. The total capital investment cost in a 25 MGD (95 MLd) is approximately 14% lower when BH + MAD is applied, and this advantage increases to about 44% for a 100 MGD (379 MLd) WWTP. Although digester volume redundancy is not taken into account in this comparison, if a percentage of redundancy were included, the advantage of BH + MAD would be even greater.

Plant Expansion Comparison

Considering there is already such an expansive installation of digester assets across North America, the largest market impact BH will have is to enable plant owners to maximize the efficiency and utilization of those existing assets. One challenge that all plant owners are faced with is how to handle the need to expand the WWTP. From the examples of BH given earlier in this paper, it is evident that it is an enabling technology to expand the treatment capacity of existing mesophilic AD, without having to add digester volume. The question is whether this is advantageous compared to building new digestion volume. To look at this, the same technology costs and BOP factors were utilized, and the total plant expansion capital investment was considered compared to the size of the expansion. Figure 6 below compares the Capital Investment of adding conventional MAD volume versus the implementation of BH.

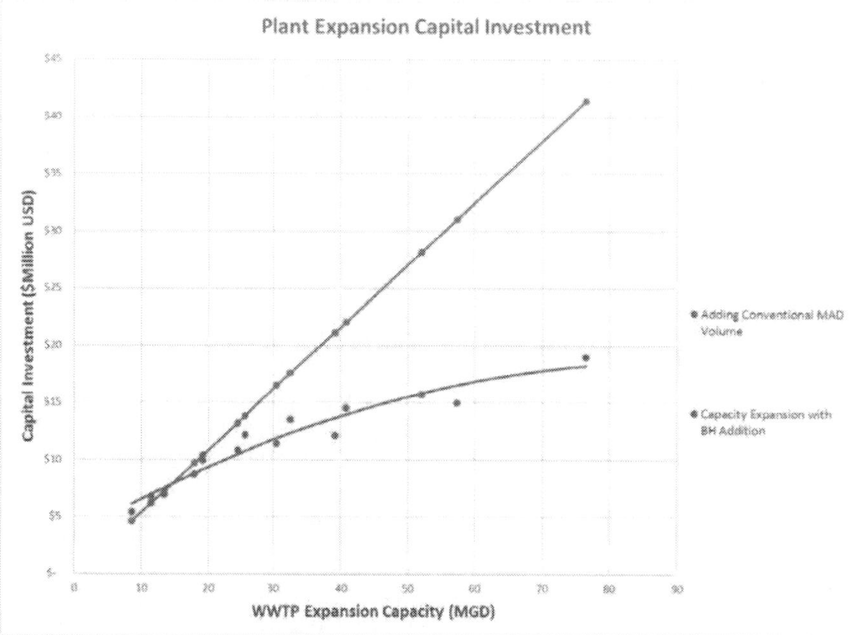

Figure 6: Plant Expansion Capital Investment Cost

As expected the benefits of implementing BH over adding conventional MAD is most pronounced the larger the expansion capacity. In fact, in a scenario of a WWTP expansion of 75 MGD, by implementing BH rather than adding more digester volume, the savings would be 54%, or over $22Million USD based on the parameters of this modeling. The advantage of implementing BH decreases as the expansion size decreases, and it is seen that if the plant expansion is less than 15 MGD (57 MLd), then BH's advantage disappears. This however does not consider the size advantage of BH and that many WWTPs are space constrained, and building more digester volume may not be an option.

One concern of WWTP owners is that the addition of BH would bring about a new process that operators need to be trained on and maintain, however, the BH system is designed to be simplistic in operation, employing standard equipment that are known to WWTP operators. Furthermore, although it is not considered in this evaluation, the energy requirements to power the BH system are much lower than the power requirements would be for a capacity expansion in digester volume. All mechanical equipment is outside of the BH reactor vessels, so maintenance is in a safe environment, and does not require the system to be taken down for service.

ENABLING OF ENERGY NEUTRALITY

While earlier sections have focused on the economic benefit of applying BH to maximize the efficiency of the utilization of mesophilic anaerobic digesters, the utilization of the technology can also be applied when the driver is to increase renewable energy generation to cover WWTP parasitic loads. As was seen in the two case examples, as well as in the plant expansion comparison, utilizing BH can enable the capacity of digesters at existing WWTPs to be tripled. This increase in capacity could be used for plant expansion, but also can be used to bring in additional sludge or organics to boost biogas production, and when converted to renewable energy eclipse the parasitic demand of the WWTP. In both the cases of Great Billing and Avonmouth, the implementation of BH allowed for the WWTP to see this benefit. Achieving energy neutrality at WWTPs not only presents a potential benefit towards plant operational costs, it also provides a reliable long term solution to ensuring the energy needs for the plant are provided.

Furthermore, the conversions of WWTPs from being energy consumers to energy producers, reduces the demand to distribution grids, but it also provides a reduction in greenhouse gas (GHG) emission reductions. This GHG emission reduction can be seen at several levels:

- Elimination of the utilization of electricity generated from non-renewable fossil fuel sources (eg. Coal or Natural Gas)
- Reduction in the overall transportation of resultant biosolids, considering the potential for higher solids reduction in a facility which implements BH
- Potential diversion of sludge and other organic wastes from landfills, where otherwise the organic material would break down and emit into the atmosphere

CONCLUSIONS

Based on the studies and analysis done in this paper, the following conclusions are drawn:

- The realization of energy neutral wastewater treatment plants was seen in the two WWTP case studies in Great Billing and Avonmouth. In both cases, BH enabled the WWTP to import either wastewater sludge and/or food waste into the WWTP and digested in existing infrastructure. The digestion of the additional organics produced enough biogas that was converted into renewable electricity, covering the parasitic load of the plant, leaving excess to export.
- When comparing the utilization of BH in a new build scenario, comparing the capital investment in the construction of a plant, a plant constructed using BH + MAD, compared to conventional MAD, the BH + MAD had a 14-44% capital investment advantage, considering plant sizes from 25 – 100 MGD.
- When comparing the utilization of BH to be retrofit in an existing WWTP to enable the existing anaerobic digestion capacity to handle 3 times the amount of organics, versus the construction new MAD, the utilization of BH had up to a 54% capital investment advantage. At plant sizes below 15 MGD however, no advantage to utilizing BH was apparent over constructing new digestion capacity.
- In addition to the economic and capacity benefits, utilizing BH to increase digestion at WWTP also provides benefits towards long term energy security for WWTPs, as well as potential GHG emission reductions which have a long term beneficial impact to the environment.

REFERENCES

1. Appels, L., Baeyens, J., Degreve, J., Dewil, R. (2008), Principles and potential of anaerobic digestion of waste-activated sludge, Progress in Energy and Combustion Science **34**, 755-781.
2. EPA (2011), Opportunities for Combined Heat and Power at Wastewater Treatment Facilities: Market Analysis and Lessons from the Field.
3. Hong, Y., Bae, J., (1996), Effects of Heat Pre-treatment and Reactor Configurations on the Anaerobic Treatment of Volatile Solids, Journal of the Korea Society of Water and Wastewater, 10(2), pp. 104-116.
4. Theodoulou, M., Harrison, D., Hong, Y., Keutgen, H. (2016), Natural Efficiency: Evaluating Digester Performance Improvements by Adding Biological Hydrolysis, Singapore International Water Week, July 10-14, 2016.

Application of Rheological Data for Non-Newtonian Sludges; Use of the "Differential Viscosity" for Mixing Simulations and System Friction Calculations

Marilyn Pine[1], Kent Keeran[1], Glenn Dorsch[1]

[1] Vaughan Company, Inc., Montesano, WA

ABSTRACT

Proper sizing of mixing and pumping systems requires a good understanding of the viscosity of the unique sludge or slurry under consideration. A Newtonian fluid is one in which the viscosity does not change with shear rate; its viscosity can be described with one value. However, because of the solid components present, many biological slurries (TWAS, digested sludges, primary sludges, etc.) are non-Newtonian in nature. By definition, a non-Newtonian fluid is one whose viscosity is not constant and which varies with shear rate. Its viscosity can no longer be described with one value and must be described as a set of values, or as an equation which calculates viscosity as a function of shear rate. This relationship between viscosity and shear rate can be used as an input into a Computational Fluid Dynamics (CFD) simulation to predict fluid velocities and mixing effectiveness or it can be used to approximate frictional losses in piping systems.

Viscosity as a function of shear rate can be derived from a fluids "flow curve". The flow curve is a set of data points which plot measurements of shear stress versus shear rate. This data can be obtained from a rotational viscometer or ideally, a rheometer. The most commonly used convention for describing the viscosity of a non-Newtonian fluid is with an expression for "apparent viscosity"; it can be calculated by taking a shear stress value at any point on the fluid's flow curve and dividing by the corresponding shear rate at that point. This is essentially the slope of a line from any given data point back to the origin. This method is typically used since it can be calculated from a single data point and can be measured in the field at a given shear rate.

Experience with simulations and actual observed results over the last 10-13 years, however, has shown that using "apparent viscosity" values or equations in CFD simulations of mixing tank applications can lead to non-realistic and overly conservative results which might predict the need for higher mixing power than necessary. This is especially true for fluids with higher yield stresses and higher viscosities (typically sludges with higher percent solids and/or with polymer thickening). This paper discusses an alternate method to calculate viscosity by defining an equation for the actual slope, or first derivative, of the fluid's flow curve to describe the fluid's "differential viscosity" at varying shear rates. Better correlation has been found between simulated results and actual observed system performance when using this method for predicting mixing system velocities. This method requires a complete set of flow curve data in order to determine the true slope, and thus viscosity, at any given shear rate.

Figure A1 demonstrates flow curve data (shear stress versus shear rate measurements) for a non-Newtonian slurry. The apparent viscosity at Point A is calculated as the slope of line A1 which extends from Point A back to the origin. The differential viscosity at point A is calculated as the slope of line A2 which is tangent to the flow curve at point A. Note that both the apparent and differential viscosity values vary at different shear rates.

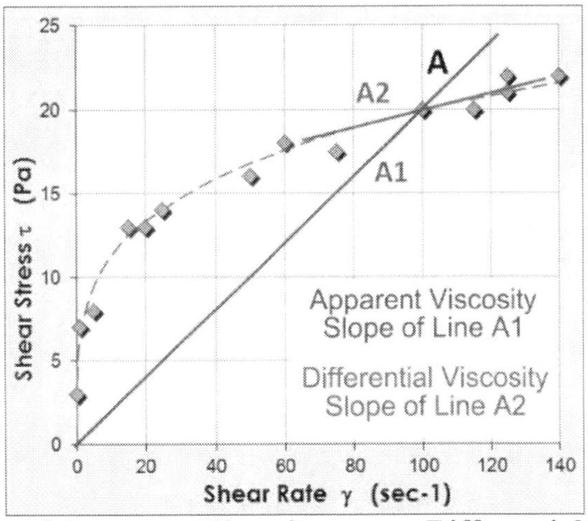

Figure A1: Comparison of Apparent Viscosity versus Differential Viscosity at Point A

Understanding the rheology of each unique application is important to predict how pumping and mixing systems will work in the field and to avoid overestimating the amount of energy usage and power required. Lessons learned over the last decade have helped to refine techniques for interpreting and applying flow curve data for sizing mixing and pumping systems.

Figure A2 below illustrates CFD results from a study where the actual viscosity of the thickened sludge being introduced into a mixing system was measured. The application of field data and differential viscosity provided the best representation of actual observed mixing velocities and system performance.

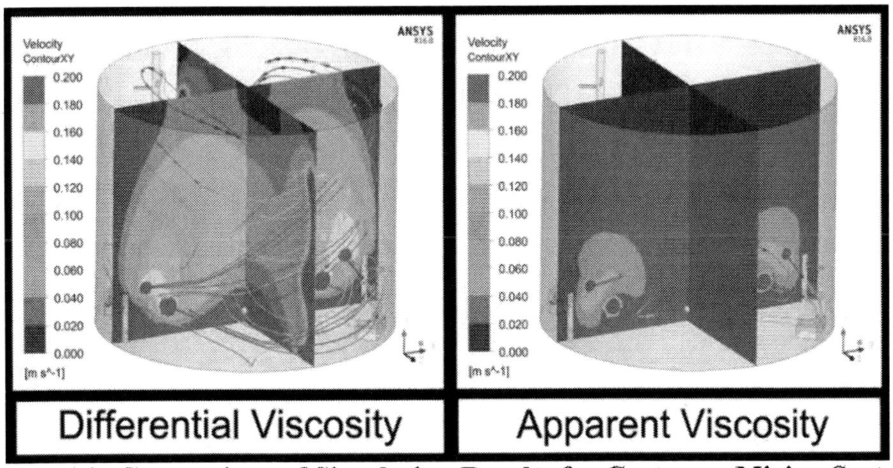

Figure A2: Comparison of Simulation Results for Customer Mixing System

This paper will discuss:

- The importance of accurate flow curve data over a wide range of shear rates.
- How apparent and differential viscosity can be calculated and an example of how the interpretation of the flow curve data can impact results, in some cases dramatically.
- Examples of real-life applications and physical validation.
- How the differential viscosity can also be applied to frictional losses in piping.

KEYWORDS Non-Newtonian, differential viscosity, apparent viscosity, mixing, CFD

1.0 Introduction

Knowing the unique rheological properties of a given sludge or slurry is important to understand when sizing pumping and mixing applications. Sludge viscosity will directly impact CFD simulations for mixing systems and system head loss calculations for pumping. Because the fluids encountered in these mixing systems tend to be non-Newtonian with shear thinning behavior, regions in the tank that are moving slowly have the highest local viscosity and therefore are the hardest areas in which to generate mixing. On the other hand, once the thick fluid starts to move and the local shear rate increases, the corresponding fluid viscosity in that region drops and it is subsequently easier to mix. Because the rheological characteristics of sludges can vary dramatically depending on percent solids, chemistry, temperature, processing, polymer addition, etc., the fluids that required to be mixed can sometimes be so viscous that they are very difficult to mix or pump. Depending upon each fluid's unique flow curve, mixing thick blended sludges can be a challenge, and one must be very careful not to undersize the mixing systems. However, oversizing can lead to increased power requirements and improper pump selection. This paper summarizes lessons learned over the last 10-13 years in applying rheological data to mixing systems and their simulation to achieve representative results.

2.0 Rheology Background

2.1 Flow Curves; Newtonian and Non-Newtonian Fluids

Viscosity is a measurement of a fluid's resistance to flow or movement. It is derived from a fluid's "**flow curve**" (or **rheogram**), which is a plot of shear stress (τ) measurements at different shear rates (γ) (see Figure 1). Viscosity is essentially the "slope" of the curve.

A constant slope throughout the entire flow curve indicates a constant viscosity at all shear rates. Fluids with this behavior are classified as **Newtonian** fluids; examples include water, light motor oils, and sludges with very low percent total solids (TS). In CFD simulations, a Newtonian fluid's dynamic viscosity can be described by one constant value for all shear rates within the fluid domain.

However, due to the presence of a higher percent solids component, many biological slurries behave as **non-Newtonian** fluids. The shape of their flow curve is no longer a straight line; the slope of the curve changes with shear rate. In this case, viscosity is not constant and is a function of how fast the fluid is moving (shear rate). For a mixing simulation of non-Newtonian fluids, the dynamic viscosity can no longer be defined by one value and it is necessary to create an expression which describes the local dynamic viscosity as a function of the local shear rate, or how fast the fluid is moving in each area of a mixing tank.

In Figure 1 below, the slope of the upper non-Newtonian flow curve rapidly increases at the left hand side of the plot, at low shear rates. This higher slope correlates with higher viscosities at low shear rates, as shown in Figure 2, and directly impacts the amount of energy required for mixing. Typically as the shear rate increases, the slope of the non-Newtonian flow curve begins to decrease, resulting in a similar decrease in viscosity. The rate at which this viscosity decreases with shear rate will impact the ability to pump the fluid and to accommodate the associated head loss and resistance to flow in the piping system.

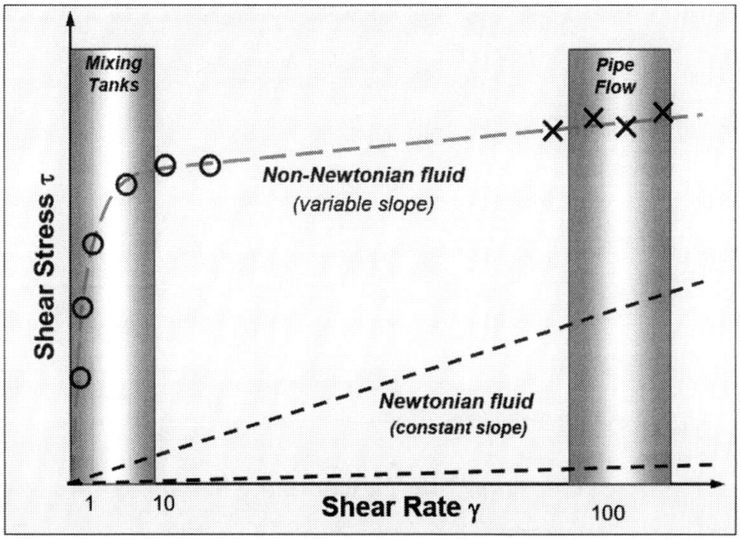

Figure 1: Schematic of Fluid's Flow Curve (Hilts *et al.,* 2014)
(Shear Stress τ vs. Shear Rate γ)

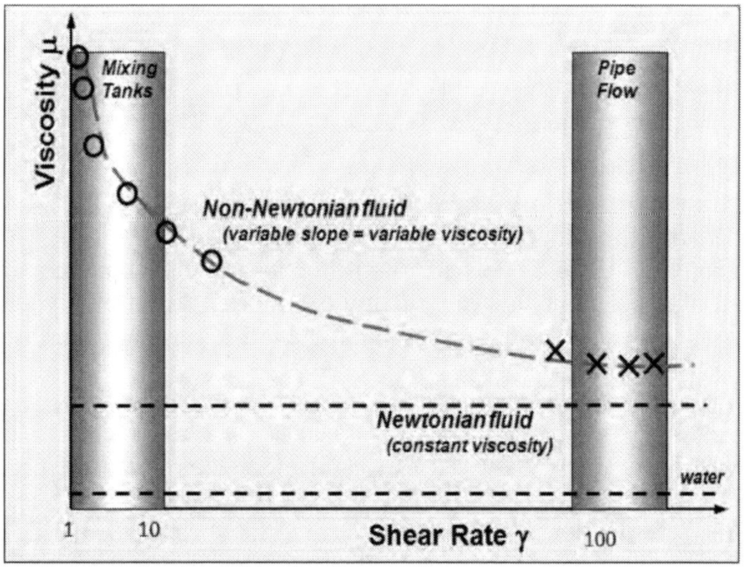

Figure 2: Viscosity μ vs. Shear Rate γ (Hilts *et al.,* 2014)

2.2 Comparison of Typical Rheological Models

Non-Newtonian fluids are often characterized depending on how well their flow curve data appears to fit a given equation or rheological model. Some of the more common models are shown in Figure 3.

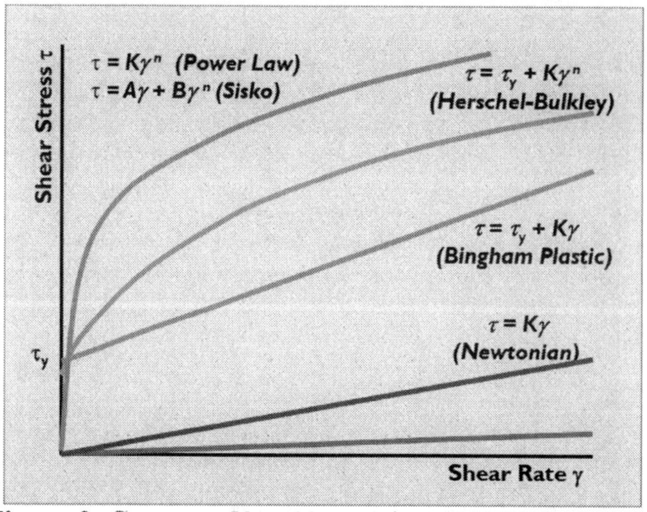

Figure 3: Common Non-Newtonian Rheological Models

Depending upon the particular type of slurry or sludge, one form of equation may best fit the data for one sludge or for a given range of shear rates, but not another. Three commonly used rheological models are the Bingham Plastic, Power Law, and Herschel-Bulkley. They have the following forms:

Bingham Plastic $\tau(\gamma) = \tau_y + K\gamma$ [1]

Power Law $\tau(\gamma) = K\gamma^n$ [2]

Herschel-Bulkley $\tau(\gamma) = \tau_y + K\gamma^n$ [3]

where: τ = shear stress (Pa)

τ_y = yield stress (Pa)

γ = shear rate (s^{-1})

K and n are fitting coefficients (with appropriate units)

It is important to note that these models are simplifications to characterize a given fluid; it can be difficult to fit a set of measurements neatly to any one model. Additionally, one form of equation may fit one portion, or shear rate range, of data and another may best describe another range of the data.

2.3 Apparent Viscosity

As discussed above, because viscosity varies for different shear rates for non-Newtonian fluids, the viscosity varies with the slope of the flow curve for different shear rates. The conventional way to describe the viscous behavior is by measuring the shear stress at a given shear rate and reporting what is defined as the **"apparent viscosity".** This is merely the measured shear stress divided by the corresponding shear rate (A) at which it was measured.

$$\mu_{A\,apparent} = \frac{\tau_A}{\gamma_A} \quad [4]$$

μ = dynamic viscosity (Pa s)

τ_A = shear stress (Pa) at a given shear rate A

γ_A = shear rate (s^{-1}) at a given shear rate A

Or, given an equation $\tau(\gamma)$ to describe shear stress as a function over a range of shear rates:

$$\mu(\gamma)_{apparent} = \frac{\tau(\gamma)}{\gamma}$$ evaluated at varying values of shear rate γ. [5]

2.4 Differential Viscosity

Another method to define viscosity is the **"differential viscosity"** which is defined as the derivative of shear stress with respect to shear rate. It is found by calculating the slope of the tangent to the flow curve at a given shear rate (the first derivative of the flow curve).

$$\mu(\gamma)_{diff} = \frac{d\tau(\gamma)}{d\gamma} \quad [6]$$

2.5 Graphic Comparison/Visualization of Apparent and Differential Viscosity

The following figure and discussion from "Rheology Modifiers Handbook: Practical Use and Applications" provides an excellent description and schematic of the apparent viscosity and the differential viscosity discussed above.

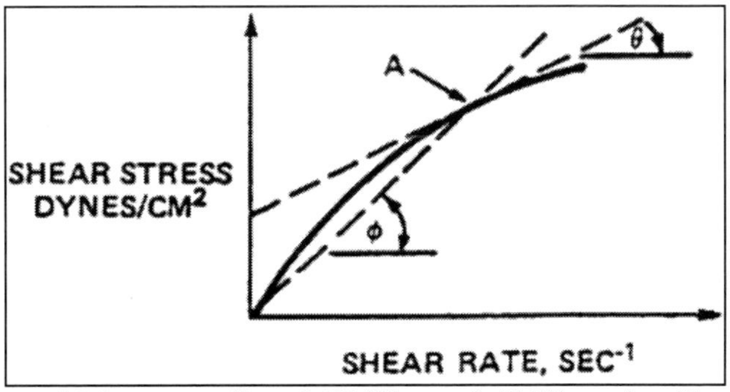

Figure 4: Definition of "Apparent" and "Differential" Viscosity (Braun & Mosen, 1999)

"The 'differential' viscosity is equal to the slope of the shear stress versus shear rate curve at some point A (or the tangent of the angle θ). The 'apparent' viscosity is equal to the slope of a line that connects the origin with a given point A on the shear stress versus shear rate curve (or the tangent of the angle ϕ). Of the two methods for expressing the coefficient of viscosity, the 'apparent viscosity' is usually chosen. This is because an 'apparent' viscosity is easily measured at one fixed shear rate while a 'differential' viscosity requires measurements at several shear rates followed by measurement of the slope at the shear rate of interest." (Braun & Mosen, 1999)

3.0 Validation of Mixing Simulation Results with Physical Observations

3.1 Early Experiences with non-Newtonian CFD Mixing Simulations

One of the first experiences with a thickened blended sludge simulation arose with a mixing system in New Zealand. The use of the apparent viscosity function (derived from non-Newtonian flow curve equations provided by the owner) resulted in simulations depicting very poor mixing velocities. As a result, the proposed pump size, mixing flow, and horsepower were increased to ensure that desired mixing velocities and flow patterns would match typical supplied mixing systems. However, subsequent startup results and testing indicated that these tanks actually were mixing at much higher velocities than originally simulated. A re-simulation, at a later time, with the differential viscosity method demonstrated a much more realistic assessment of the mixing system. In hindsight, it is generally agreed that these systems were most likely oversized based on the results obtained with the apparent viscosity method. More recent systems installed at the same location were designed with less power and flow.

3.2 Simulation of TWAS Blend Tank and Proposed Modifications

The use of a differential viscosity equation for another owner helped to provide a much more realistic assessment of their system and a realistic understanding of how proposed modifications would improve the mixing. Their installed 7 m (23 ft.) diameter x 6.1 m (20 ft.) high cylindrical blend tank mixing system was not providing visible mixing at full tank level when sludge viscosity was at its highest. Once alerted to the problem, CFD modeling was used to investigate the system and to make recommendations for improvement. This was a good opportunity to evaluate the mixing system with blended sludge rheology measurements obtained from their facility. The particular fluid that was desired to be simulated was polymer thickened waste activated sludge (TWAS); this type of sludge commonly exhibits a very high "yield" behavior at the left side of the flow curve. These blend tank mixing systems can be very challenging and may require extra mixing energy to overcome high viscosity at low shear rates.

3.2.1 Field Measurements of TWAS

The on-site consulting engineer was able to take some shear stress measurements with a rotational viscometer at varying shear rates. The results of two combined days of samples are shown in Figure 7, plotting shear stress (Pa) versus shear rate (s^{-1}). Although it would have been desired to have more flow curve data at higher shear rates, the data allowed the generation of an equation to describe the flow curve at lower shear rates. This equation was then used to input a differential viscosity formula into various CFD simulation trials to evaluate ways to improve the mixing system. These measurements from an actual process sample also helped with comparison of the rheology of this TWAS sample with other sludge data and to understand the magnitude of the resistance to flow at low shear rates.

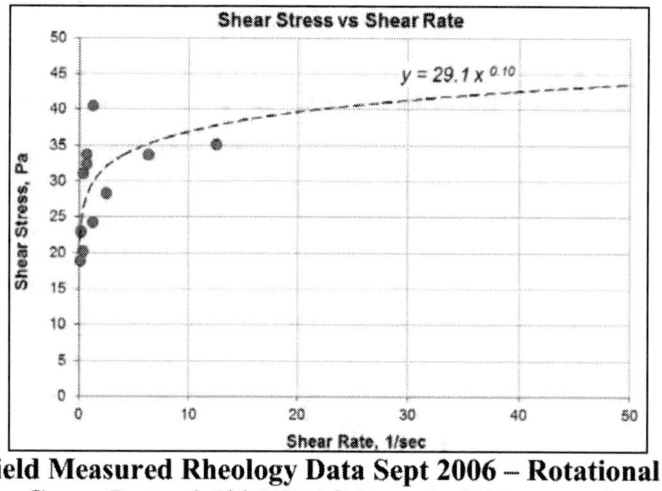

Figure 7: Field Measured Rheology Data Sept 2006 – Rotational Viscometer Flow Curve Data, 6-7% TWAS Sample (Bishop *et al.*, 2007)

In this example, the flow curve was chosen to be represented by the power law equation [2]:

$$\tau(\gamma)_{POWER} = 29.1\,\gamma^{0.1}$$

and its corresponding first derivative has the form shown in [6]:

$$\mu(\gamma)_{diff} = \frac{d\tau(\gamma)}{d\gamma} = nK\,\gamma^{\,n-1} = 2.91\gamma^{\,-0.9} \qquad \text{(Differential Viscosity, Pa s)}$$

The conventional method of deriving apparent viscosity [5] yields

$$\mu(\gamma)_{app} = \frac{\tau(\gamma)}{\gamma} = 29.1\,\gamma^{\,-0.9} \qquad \text{(Apparent Viscosity, Pa s)}$$

Depending upon the method used to derive an expression for viscosity, different equations were input into the CFD analysis for comparison.

3.2.2 CFD Analysis Results of Initial Mixing System

Initial simulations provided a baseline for comparison with the actual observed mixing system. Prior to any modifications, the tank was not able to be mixed when full of TWAS; reported surface mixing could only be seen to levels of about 2.1 m (7 ft.). Figures 8a and 8b are contour plots of fluid velocity through a vertical cross-section of an initial mixing tank configuration with a simulated fluid depth of 6.1 m (20 ft.). Even though this was a challenging thickened sludge to mix and both simulation results demonstrated poor mixing, the differential viscosity results were more representative of actual observations. Simulation results were useful in helping the owner to understand that even though they may not have seen surface movement as the tank levels increased, the fluid was still being mixed in the lower regions of the tank.

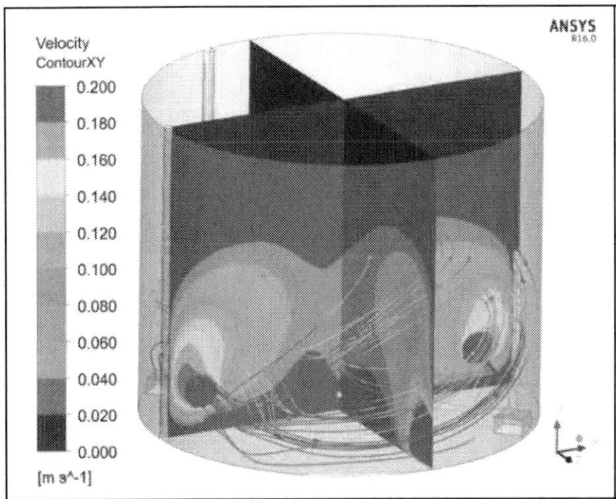

Figure 8a: Differential Viscosity

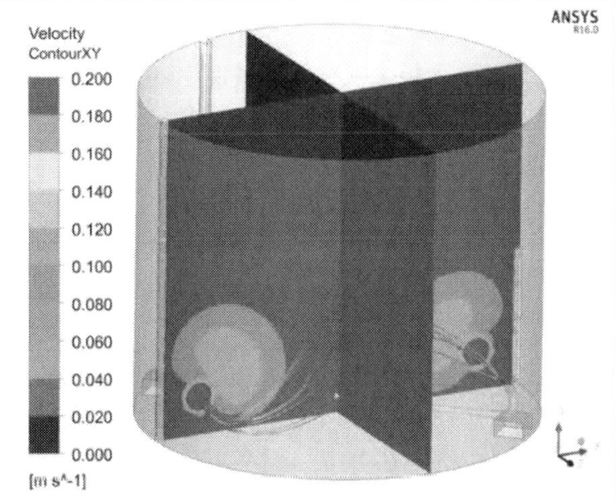

Figure 8b: Apparent Viscosity

Figure 8: Fluid Streamlines and Velocity Contour Plots through Cross-Sectional Planes.
Baseline Trials on Initial Mixing System: 193 m^3/hr, 11kW (850 GPM, 15HP).
Simulated Fluid Depth = 6.1 m (20 ft.).

3.2.3 CFD Analysis Results after Final Changes Made to Modify System

Various changes were proposed to improve mixing using the same pump and pump horsepower. Different alternatives included evaluating different numbers of nozzles, nozzle diameters, nozzle aiming, etc. Figures 9a and 9b show simulated results with the same 6.1 m fluid depth for the final modified system. Some of the proposed changes included doubling the number of nozzles in the mixing system while adjusting the nozzle exit diameter to keep the pump operating in an acceptable portion of the pump curve, as well as re-aiming nozzle angles and adding an elbow to help direct more of the flow towards the upper regions in the tank. The simulated flow and horsepower correspond to the current modified system in operation. Again, based on actual observations of the mixing system at various operating levels, the results with the differential viscosity more closely matched with actual system performance and mixing after the modifications were put into place. Surface mixing was reported at a depth of 3.7 m (12 ft.); the apparent viscosity results were not an accurate portrayal of the system in operation.

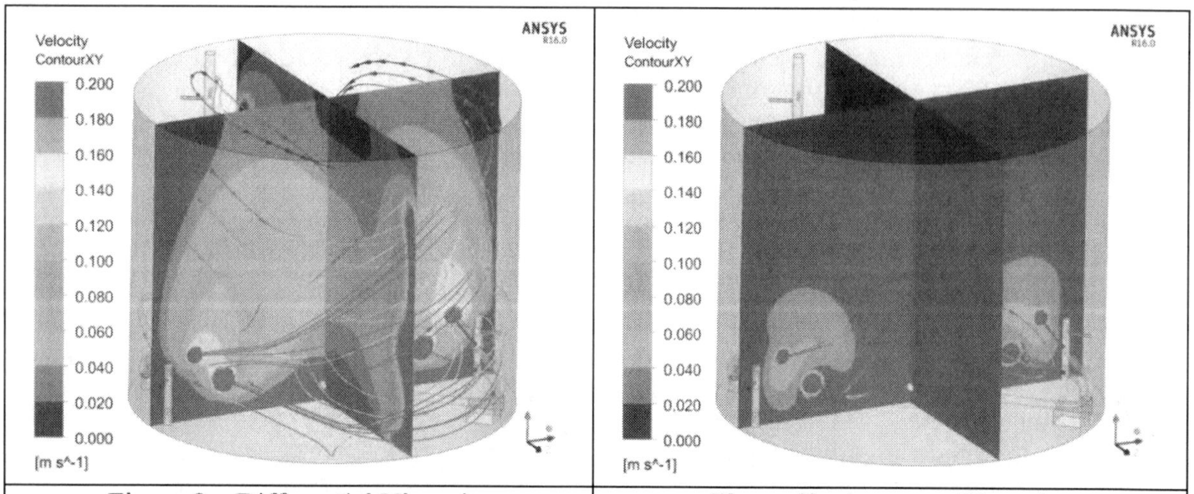

Figure 9a: Differential Viscosity	Figure 9b: Apparent Viscosity

Figure 9: Fluid Streamlines and Velocity Contour Plots through Cross-Sectional Planes.
Final Modified Mixing System: 239 m^3/hr, 11 kW (1050 GPM, 15HP).

3.2.4 CFD Analysis Results of Estimated System to Provide Full Mixing

A question also arose as to how much horsepower and flow might have been required to completely mix these tanks completely full with 100% TWAS. Figures 10a &10b below show the results of simulations with increased mixing flow and power. As expected, the result with the differential viscosity shows higher mixing velocities at all levels in the tank as the flow and power are increased. Results using the apparent viscosity show that the contents of the tank are still virtually at a standstill, except for small high velocity areas directly next to the nozzle outlets. If the apparent viscosity results were relied upon, the amount of power required to fully mix these tanks would have been significantly overestimated. While the owner did not upgrade the system as shown in Figure 10, these simulations were useful in understanding how much mixing energy and power might be required for future tanks with thickened sludges having a similar rheology.

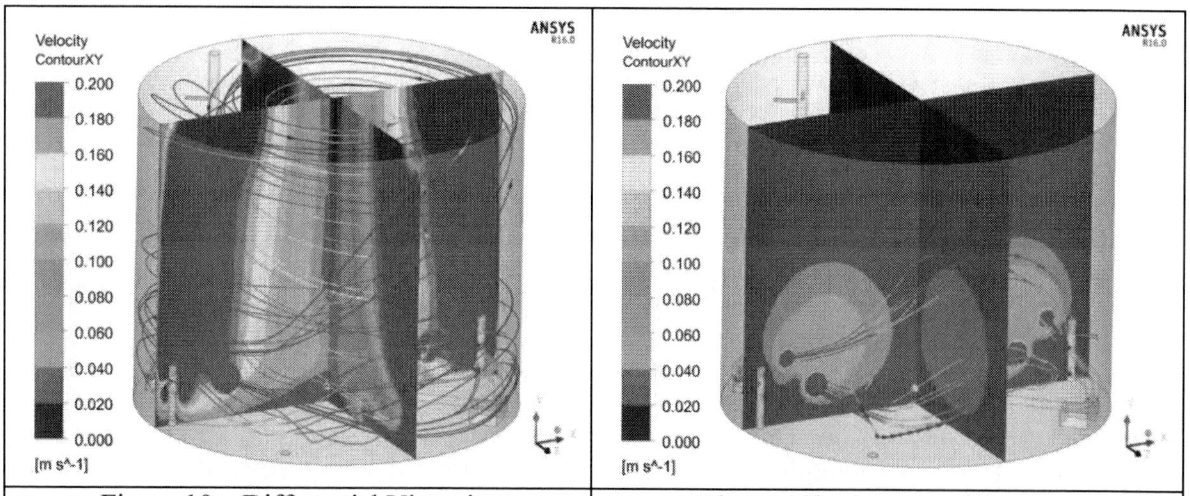

Figure 10a: Differential Viscosity	Figure 10b: Apparent Viscosity

Figure 10: Fluid Streamlines and Velocity Contour Plots through Cross-Sectional Planes
Alternate Proposed Mixing System: 427 m^3/hr, 30 kW (1880 GPM, 40HP).

3.3 Example Mixing Simulation Results for an Anaerobic Digester

From about 2006 forward, the goal has been to get the best understanding of each unique sludge to be mixed, to obtain any actual shear stress versus shear rate measurements, and to develop an expression for differential viscosity for input into mixing simulations. Continued feedback on mixing test results, videos, and field observations have helped to confirm that simulated results based on the differential viscosity method better correlate with actual performance. In a typical anaerobic digester, the range of TS and viscosity of digested sludges is typically less than those seen with thickened WAS sludges as discussed in Examples 3.1 and 3.2. Figure 11 plots a couple of representative power law flow curve equations which were chosen to qualitatively illustrate the difference in mixing simulation results for a digester application.

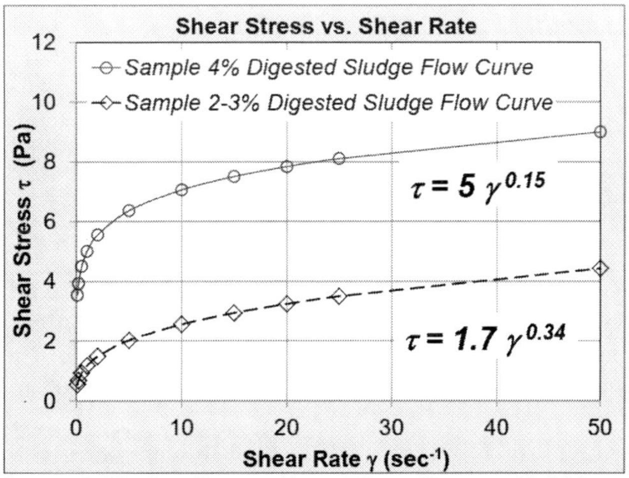

Figure 11: Sample Digested Sludge Power Law Flow Curves (2-3% and 4% DS)

The difference in simulation results when comparing the differential and apparent viscosity methods is not as pronounced for lighter sludges and fluids with lower TS as shown below for the chosen 2-3% digested sludge in Figure 11. (Power law equation where $\tau = 1.7\gamma^{0.34}$). The tank diameter chosen for this particular simulation was 33.5 m (110 ft.).

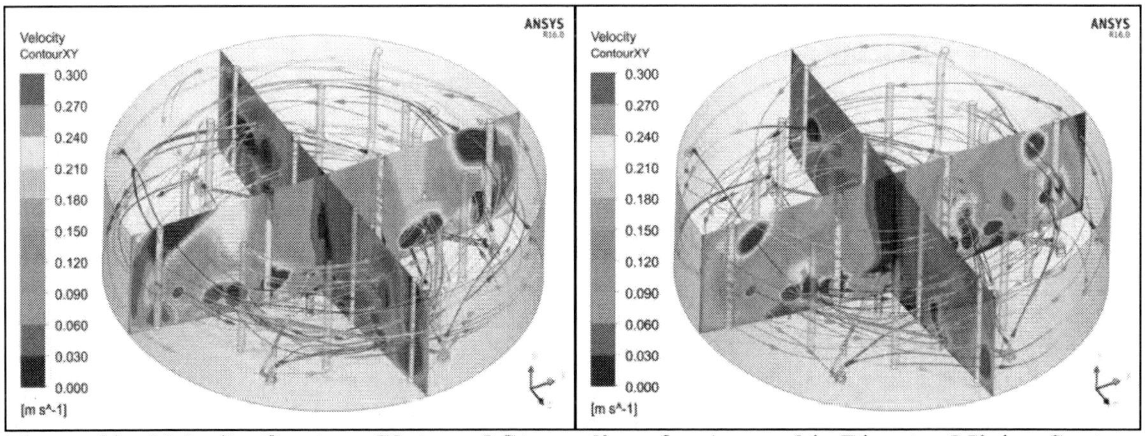

Figure 12: Velocity Contour Plots and Streamlines for Anaerobic Digester Mixing System, Estimated 2-3% Digested Sludge. (Images courtesy of the City of Calgary)
Left (12a): Differential Viscosity, Volume Average Velocity = 0.20 m/s
Right (12b): Apparent Viscosity, Volume Average Velocity = 0.15 m/s

However, because average shear rates in these mixing simulations are typically very low, on the order of 1 s^{-1} or less, a change in yield stress has a marked difference on viscosity at low shear rates and the difference between the two viscosity methods becomes more pronounced. Figure 13 illustrates how overall mixing velocities within the tank have slowed when a sludge with the sample equation for 4% digested sludge ($\tau = 5\gamma^{0.15}$) is considered, as well as how the difference in results between the two viscosity methods is now more significant.

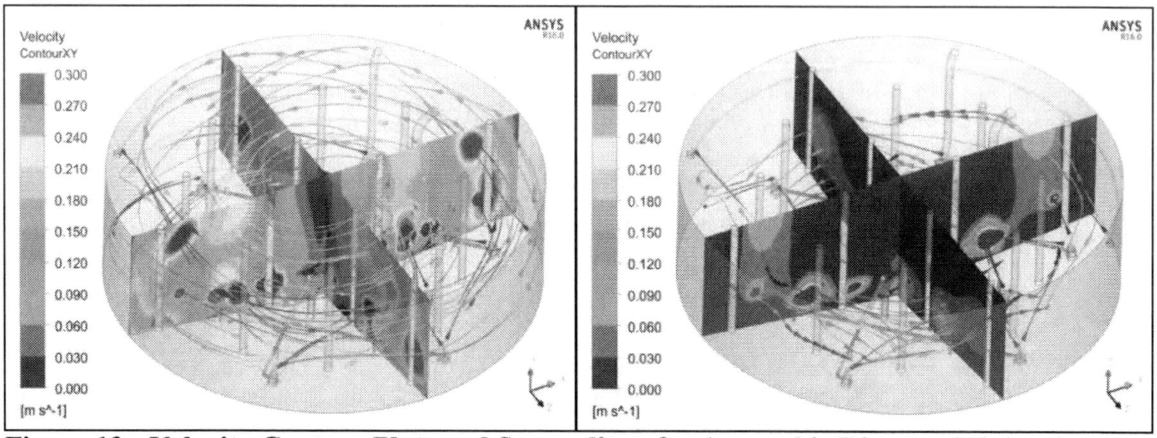

Figure 13: Velocity Contour Plots and Streamlines for Anaerobic Digester Mixing System, Estimated 4% Digested Sludge. (Images courtesy of the City of Calgary)
Left (13a): Differential Viscosity, Volume Average Velocity = 0.17 m/s
Right (13b): Apparent Viscosity, Volume Average Velocity = 0.03 m/s

Again, the different results shown in Figures 13a and 13b were derived from the **same** power law flow curve equation for 4% digested sludge. **The results only differ depending upon which method was used to derive an expression for viscosity as a function of shear rate from the original power law flow curve.** Based on experience with numerous digester mixing systems, the results simulated with an apparent viscosity derived from the 4% DS flow curve equation is not an accurate portrayal of how these systems actually perform or mix.

4.0 Obtaining Good Rheology Data for Calculation of Differential Viscosity

4.1 Non-Newtonian fluids cannot be described by one viscosity value

Most wastewater sludges with TS values equal to or greater than 2-3% typically behave as non-Newtonian fluids; their viscosity cannot be described by one value. Because their viscosity varies with shear rate, the calculation of viscosity first requires a set of shear stress versus shear rate measurements. These values must be obtained from an instrument that is able to measure shear stress for a corresponding shear rate; equipment that assumes the fluid is Newtonian and only provides one viscosity value with no reference shear rate may not adequately describe the fluid's rheology. Shear stress versus shear rate measurements may be obtained from a rotational viscometer (example data shown earlier in Figure 7) or more ideally, a rheometer which can obtain more detailed shear stress data over a wide range of shear rates. Figure 14 below demonstrates rheometer data which is a good representation of an ideal set of measurements for a TWAS sample. From these data points, one can roughly estimate the magnitude of resistance at low shear rates, evaluate the rate of curvature at higher rates, as well have enough data points to aid in deriving a flow curve expression.

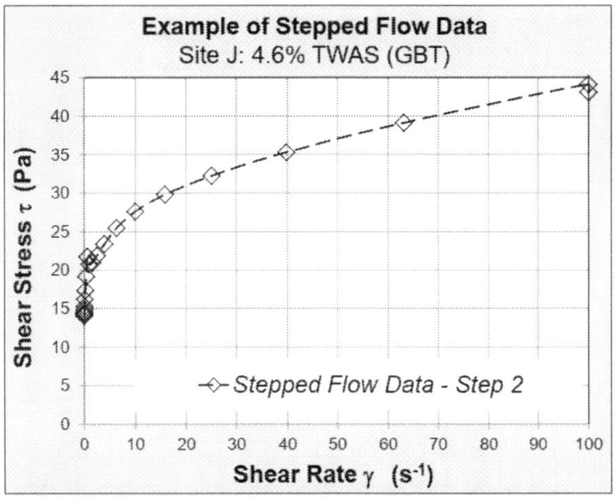

Figure 14: Sample Flow Curve Data obtained from Rotational Rheometer (Hilts *et al.*,2014)

4.2 Assume that each sludge is unique, because it probably is

In general, the yield stress and viscosity of biological slurries tend to increase with higher percent solids. All sludges are unique and their differences can become increasingly more pronounced with the type of slurry, percent solids, processing, etc. Polymer thickened sludges may tend to have the most unpredictable and variable rheologies, in terms of both yield stress behavior and resistance to flow at higher shear rates. This is especially true as the percent solids increases for > 5% thickened blended sludges (Hilts *et al.*, 2014). Whenever possible, it is best to generate a flow curve from a representative sample of the sludge under study or consideration. Because the viscosity will vary with temperature, one should also try to test at a representative process temperature as well.

Flow curves contain a wealth of information. Shear stress measurements at very low shear rates (for example, 1 s^{-1} or less) are very useful for evaluating a sludge's behavior when mixing; measurements at higher shear rates (up to 100 s^{-1}) will help one to understand how the same sludge might react in a piping system at higher velocities. Additionally, each flow curve will tend to have a different rate of increase at higher shear rates. If a power law equation has been used to describe the data, higher exponent values will result in flow curves corresponding to fluids which do not decrease in viscosity as much at higher shear rates. This will be indicative of a fluid which may be harder to pump with associated higher friction losses.

4.3 Derive representative equation to use over shear rate range under consideration

In some cases, especially as the percent solids increases or with polymer thickening, one may not be able to come up with one-size-fits-all equation that fits the whole range of data. It may be preferable to have a couple of different flow curve equations depending upon the region of shear rates that one is interested in. For example, a Bingham Plastic (BP) straight line equation is often a very good fit for data at higher shear rates, however it typically will not correctly define the viscous behavior at low shear rates; a power law or other equation might be a better choice. Figure 15 demonstrates how a BP equation fit to the sample data shown previously in Figure 14 may be a good approximation at higher shear rates but not for the lower shear rates associated with mixing. Depending upon the magnitude of a sludge's yield stress and slope of the equation, the BP equation combined with the apparent viscosity method may be overly conservative for mixing simulations. Also, one should try to keep all raw data for future reference; do not try to oversimplify the data. Once an equation has been fit to data and the original underlying data is gone, it is impossible to know how well the final flow curve equation described the original data.

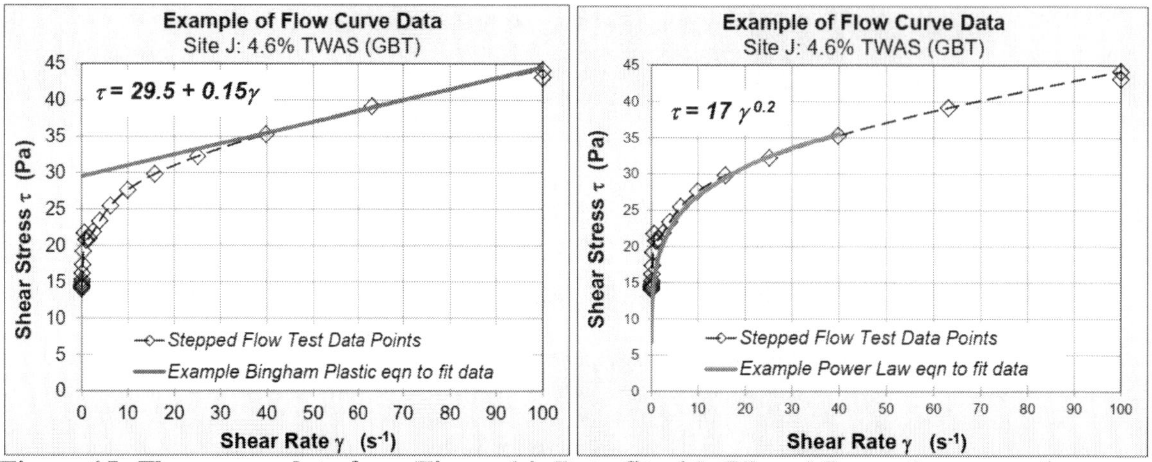

Figure 15: Flow curve data from Figure 14; Data fit with BP and Power Law equations.

4.4 Differential Viscosity is First Derivative of Flow Curve

While a value for apparent viscosity can be reported with one shear stress measurement at a particular shear rate, the calculation of differential viscosity requires a series of measurements such that the rate of change in the relevant shear rate range can be determined. The use of a "yield stress" term (such as in the Bingham Plastic [1] or Herschel-Bulkley [3] equations) may make it difficult to calculate the initial slope of the data and to accurately model the fluid's resistance to movement at very low shear rates. Ideally, the original flow curve data points can be used instead to derive an equation which passes through zero and does not contain a yield stress term, such as with a Power Law equation. Sometimes it can be challenging to find an expression which fits the data and also which can be differentiated. As an example, one form of flow curve equation which may be used and which often helps to fit non-Newtonian sludge data for a wide range of shear rates is in the form of the Sisko equation as shown on Figure 3:

$$\tau(\gamma) = A\gamma + B\gamma^{n} \quad \text{where} \quad \mu(\gamma)_{diff} = \frac{d\tau(\gamma)}{d\gamma} = A + nB\gamma^{n-1} \qquad [7, 8]$$

5.0 Application of Flow Curve Measurements and Differential Viscosity to System Friction Loss Calculations

System calculations, even for a system assumed flowing with water or a very light sludge can be very challenging. Depending on the piping material, age of piping, fittings and their placement in proximity to each other, different reported loss coefficients, etc., there are a number of sources of variation which affect friction loss calculations in a piping system. For sludges with higher TS, it is generally assumed that viscosity will start to play a larger role in head loss due to friction. If a strict multiplier (based solely upon percent solids) is applied for head loss calculations, its use may lead to overly conservative estimates because it does not consider the decrease in viscosity at higher pipe velocities with a non-Newtonian fluid. Also, just as with mixing simulations, as one starts to consider sludges with higher TS and/or the effect of polymers for thickening, the point when the viscosity will start to have a significant impact on friction loss calculations may also be when the actual sludge rheology is unpredictable and historical measurements from other sludges are less reliable. One way in which the differential viscosity approach has been used is by applying a given sludge's flow curve equation, when available, to estimate head loss due to the friction via the steps below. Using the differential viscosity rather than apparent viscosity may give a slightly lower estimate for head loss.

1. **Calculate average velocity V** in each piping segment or component based on pipe diameter D and pumping flow rate Q

2. **Estimate shear rate γ (s^{-1})** in each segment of piping
   Although the velocity profile across the interior cross-section of a pipe varies with distance from the wall, a simplified method of estimating the shear rate at the wall for a pipe of a given diameter is based on average pipe velocity:
   $$Shear\ Rate\ \gamma_N = \frac{8V}{D} \qquad [9]$$
   Note that [9] calculates average pipe shear rate γ_N based on an assumed velocity distribution for laminar flow and a Newtonian fluid. The actual wall shear rate for pipe flow is higher for non-Newtonian fluids but also very difficult to measure and quantify. The Rabinowitsch-Mooney equation adds a correction factor for non-Newtonian fluids with assumed laminar flow (Nguyen *et al.*, 2012). For a power law fluid:
   $$Shear\ Rate\ \gamma_{RM} = \frac{8V}{D}\left[\frac{3n+1}{4n}\right] \qquad [10]$$
   where n = the exponent in the power law equation $\tau = K\gamma^n$. Corrected shear rates will be higher than for those using [9] and will increase as the exponent in the power law equation becomes smaller, indicative of more shear-thinning behavior and reduced viscosities at higher flows. Estimating shear rate with [9] should result in a lower estimate for shear rate and a higher limit for predicted head loss.

3. **Calculate dynamic viscosity μ** at the shear rate γ using differential viscosity function

4. **Convert dynamic viscosity μ to kinematic viscosity ν**

5. **Calculate dimensionless Reynolds number Re** from kinematic viscosity, pipe velocity, and diameter
   $$\mathrm{Re} = \frac{VD}{\nu} \qquad [11]$$

6. **Calculate head loss in pipe segment due to friction losses**
   The Darcy-Weisbach equation is commonly used to calculate head loss h_f due to friction in piping with a length, or equivalent length, L based upon piping diameter D, velocity V, and acceleration constant g (Cameron, 2002).
   $$h_f = f_{DARCY}\left(\frac{L}{D}\right)\left(\frac{V^2}{2g}\right) \qquad [12]$$
   Equation [12] is used in conjunction with a friction factor f which varies with the type of flow (laminar or turbulent) and the relative roughness of the pipe. For assumed laminar flow (about Re <2000), the roughness of the pipe's interior surface should not affect the friction factor; the friction factor f is estimated directly from the value of Reynolds number.
   Laminar flow: $$f_{DARCY} = 64/\mathrm{Re} \qquad [13]$$

Head loss resulting from friction and fluid viscosity can be applied to sections of straight lengths of piping, fittings, and valves which have an "equivalent length" of piping. Reducers, enlargers, inlets, nozzles exits, and pipe exits which contribute to overall system head loss due to changes in velocity head are generally treated separately and not adjusted based on friction factor.

7. **Sum the total losses for all segments of piping and plot** for a range of different system flow rates to generate system curves.

8. **Compare system curves for both assumed laminar and turbulent flows for chosen pump and application.** For higher values of Reynolds number, the value of the friction factor calculated with the assumption of laminar flow gets smaller, and the friction factor for turbulent flow and the increasing effects of pipe roughness start to dominate. If the water/turbulent flow head losses begin to dominate at higher values of flow, those values are used. Where the two methods intersect the pump performance curve can serve as estimated upper and lower limits to bracket the potential head loss and flow of the final system ranging from the sludge curve estimate to the water curve estimate. This range can be compared with system requirements and capability of chosen pump.

For non-Newtonian sludges with assumed laminar flow, the shape of the predicted system curve will roughly correlate with the shape of its flow curve; it will depend upon the magnitude of yield stress behavior and the exponent controlling how the viscosity decreases at higher rates. Often an initial jump in predicted head loss will be apparent at very low GPM; this jump roughly corresponds to the initial yield stress of the non-Newtonian fluid providing more resistance at low flow and shear rates. As viscosity drops due to non-Newtonian behavior at higher shear rates, the head loss due to friction begins to drop as well. Eventually the flow transitions to turbulent flow and the system curve "merges" with the water curve at higher pumping flows. Many digested sludges tend to have system curves closer to the water estimate, even for higher 4-5% solids, due to the reduced viscosity at higher flows; this has been confirmed with actual pump pressure readings. Conversely, the combination of high yield stress behavior along with a high power law exponent n, as often seen with polymer thickened WAS sludges, can result in a fluid that is very difficult to get moving and whose friction losses do not decrease at higher shear rates, leading to very high friction losses in a piping system.

6.0 Conclusions and Summary

This paper summarizes lessons learned over the last 10-13 years; these lessons have been applied to sizing and analysis of mixing systems, pump sizing, and hydraulic calculations in viscous slurries. Mixing these fluids can be very challenging, (for example, thickened sludge blend tanks), so it is critical to properly evaluate their expected performance. The use of an inaccurate flow curve equation or viscosity model can potentially lead to oversizing or undersizing of mixing systems and decreased confidence about the expected capability of a mixing system. Both the choice of **the flow curve equation** as well as the **method for calculating viscosity** can have a significant impact on predicted results. The choice of viscosity equation may not be as significant for some applications which have fluids that are more Newtonian in behavior or are simulated at higher shear rates. However, the derivative-based differential viscosity method may give markedly different results when considering fluids or sludges with a high yield point behavior in applications that deal with low shear rates, such as mixing systems. Based on simulation results, actual field observations, and pump pressure measurements, applying the differential viscosity appears to provide much better correlation between simulated results and actual system performance.

ACKNOWLEDGEMENTS
The authors would like to thank the various contacts and owners throughout the years who have helped to collect sludge viscosity measurements and to provide feedback on installed mixing systems and pump performance.

REFERENCES
Bishop, R.; McCormick, J.; Peterson, J.; Dorsch, G.; Pine, M. (2007) "Blending Raw Sludges with Pump Mix Systems", presented at 2007 PNCWA Annual Conference in Vancouver, WA.

Braun, D.; Rosen, M. (1999) "Rheology Modifiers Handbook: Practical Use and Applications", William Andrew Publishing, 1st edition, Figure 1.2.

Cameron Hydraulic Data: A handy reference on the subject of hydraulics and steam (2002), edited by C. C. Heald, 19th edition, published by Flowserve Corporation.

Hilts, B.; Pine, M.; Dorsch, G. (2014) "Characterizing Polymer Thickened Biosolids for Pumping Applications", presented and published at WEF Residuals and Biosolids Conference 2014 in Austin, TX.

Nguyen, Q. and Nguyen, N.; (2012) Incompressible Non-Newtonian Fluid Flows, Continuum Mechanics - Progress in Fundamentals and Engineering Applications, Dr. Yong Gan (Ed.), ISBN: 978-953-51- 0447-6, InTech

Digester Cover Selection and Safety: Balancing Risk and Reward

Authors: John Maley, HDR; Benjamin Miller, Clark Construction; Matthew Williams, WesTech

ABSTRACT

Anaerobic Digesters are playing a key role in the increasing trend for Water Resource Recovery Facilities to become more energy independent. Addition of novel feedstocks for codigestion can lead to new stresses on old equipment. Resulting foaming and rapid volume expansion events can be a major nuisance in gas piping but their effects can damage or even destroy digester covers. As part of these projects, it is critical that existing digester covers are upgraded and new covers properly selected and designed with appropriate features to ensure proper and safe operation.

Unfortunately, there are many stories of digester cover problems and failures throughout the industry and some have even been fatal. While it is not possible to discuss every situation individually, some consistent vulnerabilities emerge for different cover types and materials of construction. Designers and operators should have a sound understanding of potential failure modes to ensure safety measures and procedures be incorporated to reduce the risk of failure.

In some cases, the safety equipment may provide a false sense of security. Explosions are certainly a risk in and around digesters and mitigating this risk gets a lot of attention during the design of nearby electrical and safety equipment. However, overpressurization events are more common so both risks need attention and prevention.

The Little Patuxent Water Reclamation Plant in Howard County, Maryland has recently experienced a digester cover selection process firsthand for three new digesters that will be added (in 2017) as part of a biosolids processing upgrade from lime stabilization to belt drying. In addition to a review of the cover selection process, this paper also discusses instrumentation and equipment employed to help reduce the risks that have contributed to failures at other facilities.

Through awareness of prior modes of failure and vulnerabilities (risk) associated with different cover types, designers will learn how certain additional safety features can play a key role in decreasing the instance of digester cover failures, despite the increasing number of operating digesters.

KEYWORDS

Digester, cover, safety, tank, concrete, precast, cast-in-place, biogas, foam, rapid rise, flare, gas relief, primary digester, secondary digester

INTRODUCTION

Municipal sludge digestion has been ongoing for decades and over that time many different system designs have been developed. This paper focusses on two aspects of design for common circular tank digesters: cover selection and safety. This paper first covers design considerations when selecting covers and safety equipment, including real life impacts of safety shortcomings. The second part of the paper describes how the authors implemented these design practices in the design of a digestion facility for Howard County, Maryland.

COVER DESIGN CONSIDERATIONS

Cover Types

There are three primary cover types, fixed, floating, and membrane. Fixed covers can either be steel or concrete, floating covers are steel, and membrane covers are PVC impregnated geotextile fabric. Four key considerations when choosing a cover are the gas storage requirements, cost
(both capital and O&M), process compatibility, and in certain cases, risk.

Applications where digester gas use is reliably less than digester gas production do not require provisions for gas storage, although it may be installed for process flexibility. For example, a floating or gasholder cover may be used on a secondary digester or digested solids storage tank to allow for intermittent dewatering. In systems like these the gas relief valve setting (normally at the flare) controls the digester gas pressure. Digesters that don't require gas storage can use either floating or fixed covers.

Applications where digester gas use is expected to fluctuate above or below the digester gas production rate will benefit from digester gas storage. The most common method of storing digester gas is in the head space between the digester liquid surface and the cover. There are generally two categories of gas storage covers: constant pressure or constant volume. Dual membrane and floating gasholder covers are both constant pressure whereas fixed covers are constant volume. Floating gasholder covers benefit from the simplicity; the weight of the cover itself provides the pressure to the gas system. Dual membrane gasholder covers can hold up to three times the gas volume of a floating gasholder cover for a given tank diameter. The constant pressure of the gas stored by both dual membrane and floating gasholder covers is preferred for systems where gas demand is independent of gas production and a certain feed pressure is expected such as digester heating or thermal sludge drying. In many systems the gasholding capacity is sized to stabilize fluctuations in the difference between gas demand and gas production. With a gasholder cover, pressure is kept constant, even if liquid is being removed from the tank. The Dual-membrane cover allows the most flexibility in liquid level since the gastight seal is at the top of the wall and independent of the liquid.

In general, the goal of any digester gas storage system is to retain digester gas when demand is low to offset natural gas purchase when demand is high. If the number of gasholder covers required to stabilize fluctuations is less than the total number of digesters in a system, the remaining digesters may have fixed covers. Gasholder covers, due to their easier installation are typically more economical than fixed covers, however they require energy and effort to operate. While it may be appealing to have excess gas storage, it is of little value if not utilized. In addition, there may be some degree of risk in utilizing only dual membrane covers (discussed later).

Gas storage may not be necessary and fixed covers alone could be used at plants were the gas utilization system can ramp up and down as needed to maintain the gas pressure within a set range and dewatering is relatively constant. A Combined heat and power (CHP) system would allow this to some extent. The extra provisions made for gasholding can be eliminated if the gas demand can be tailored to match the gas production.

SAFETY DESIGN CONSIDERATIONS

Overflow

One of the most prominent conditions to avoid at a wastewater treatment plant is overflow. The 10 States Recommended Standards for Wastewater Facilities (10 States) and most others require that digesters be designed with gravity emergency overflow protection. An overflow that is either recycled to the head of the plant or captured in a containment area is an operational nuisance. For this reason there are overflow protections that commonly precede a gravity emergency overflow.

The first level of protection against overflow has to do with operational strategy. Operationally, whenever a volume of flow is added to a digester, an equal volume should be taken out. This is even true for secondary digesters with fluctuating storage volumes when looked at over long span of time. When averaged over a long period of time the digester feed rate should equal the digester withdrawal rate.

Operationally it's not as simple as setting the feed and withdrawal flow to the same rate. To set the rates the same the operator would have to either rely on equaling the readings from inlet and discharge flow meters or the output of inlet and discharge positive displacement pumps. The calibration drift of meters or the wear of PD pumps can never be depended on to be equal. Eventually the variation would cause tank level to rise or fall. For this reason, level detection and control is the key variable. Normally, the feed rate will be dictated by the upstream sludge production, so the control logic will operate the withdrawal pumps as needed to maintain a digester level set point. The non-automated equivalent to this would be a gravity overflow to either a secondary digester or digested sludge holding tank. Even if automation isn't required/desired, level detection should be utilized to sound alarms at both low and high level set points.

10 States calls for a gravity emergency overflow in digester tanks. An emergency overflow should be located above both the normal operating and the alarm set point elevations. This position gives operations staff time to make corrections to prevent an impending overflow.

In the last few years light has been shed on some overflows previously thought to have been caused by foam, but now believed to be associated with a separate "rapid rise" phenomena. Rapid rise can be explained as a change in the characteristics of the digester sludge (e.g. increased viscosity as a result of mixer shutoff) that cause digester gas "holdup", which can rapidly increase the volume of the sludge (as the density drops).

A rapid rise condition is different than foam in the sense that it is the actual sludge (and gas bubbles within) reaching high levels instead of a foamy layer above the sludge. Unlike with foam (which can't be expected to flow), an adequately sized gravity overflow can be a reasonable protection against rapid rise. The fluid nature of sludge allows this as compared to the elastic semi-solid nature of foam. An overflow that is sized for rapid rise will should have a much higher capacity than an overflow sized for over feeding or regular sludge withdrawal. There are some scholarly articles that can provide some insight on how to predict the rate of a rapid rise (Chapman and Kruger, 2011). That prediction can be used to size an overflow for rapid rise. Another level of protection against overflow identified in research is to have sufficient freeboard designed into the tanks. The freeboard recommendation ranges from 4-12% of sidewater depth depending on tank shape (Chapman and Kruger, 2011).

Over Pressure

Over pressure of a digester is a serious condition that can damage the tank and covers or cause injury. A digester's first line of defense against an overpressure situation is operational; the monitoring of digester gas pressure. Under normal conditions the digester pressure is either constant (systems with either floating or dual membrane gasholder covers) or near constant (systems that withdraw gas to maintain pressure between high and low setpoints). Once an upset condition is reached though, the pressure can change rapidly. There should be both high and low level alarms of varying degrees in place.

Gasholder digester covers are most vulnerable to high or low pressure damage when they are operating at or near their extremes (i.e. full or empty). Gas storage volume monitoring provides valuable information that allows the operations staff to adjust the gas utilization to operate within safe limits and avoid extremes. On floating gasholder covers the volume of digester gas in storage can be accurately calculated using floating height of the gasholder cover. For dual membrane covers the storage volume is monitored by measuring the height of the inner membrane. This method is not very precise because the inner membrane is an unpredictably irregular surface when in mid range between full and empty. This is not much of a drawback though because much like a cars fuel gauge, a resolution that is in 1/4 tank increments is enough to make educated decisions. Designs that draw gas off the top of the cover and do not allow this fuel gauge feature have proven to be very difficult and problematic to operate and should be avoided.

In addition to monitoring to avoid overpressure, it is important to add safeguards that come into play if an overpressure event occurs. The first line of defense should be the flare and its associated pressure regulating valve. In digester systems where the biogas production exceeds the demand, the flare will be in constant use during normal conditions. Many flares have a limited turndown capability. If the expected flow rate to a flare exceeds a single unit's turndown capability, multiple smaller units will be required. Some enclosed flare designs are capable of nearly complete turndown and may offer a greater degree of flexibility. The total flaring capacity should be adequate to flare the maximum gas production rate.

If for some reason the flaring system is not able to prevent pressure from building in the digester, the next line of defense is emergency pressure/vacuum relief valves (PVRVs) at each tank that vent digester gas to the atmosphere. These devices are mounted in pairs at each tank or cover and on a "safety selector valve". The safety selector valve is a special 3-way valve constructed to reduce pressure drop, maintains a vertical orientation of the PVRVs and allows the isolation of only one PVRV at a time ensuring that the tank and cover are always protected.

Each gas relief valve (both tank mounted and at the flare) needs a flame arrestor mounted directly upstream to protect against flashback. Over pressure situations can be created by the same situations that can cause overflow situations when coupled with malfunctioning (or non-existent) overflow protection. The flame arrestors associated with the aforementioned PVRVs can cause an issue if over-filling, foam, or rapid rise conditions occur. The flame arrestor's tight grid or coil of metallic plates that are meant to cool any potential flashback will likely get blinded by any foam or sludge trying to escape, as is shown in the example below. This situation requires an additional level of protection to ensure the pressure does not exceed safe levels.

Overpressure (foam) Example 1

Chronic foaming plugged gas withdrawal lines and blinded flame arrestors of gas safety equipment (Figure 2) and caused (at different times) three very large fixed steel covers to be pulled from the top of the wall (Figure 1). The most violent event occurred during the morning commute and a "sludge geyser" was captured by a traffic helicopter and aired on the evening news.

Figure 1 – Anchor bolts pulled out from the concrete wall as cover was lifted by overpressure event.

Figure 2 – Foam blinding the Flame Arrestor on the Emergency Pressure & Vacuum Relief Valve (Partially cleaned coil shown at left).

An appropriate safety device (later installed on the example project) for relieving foam is the "emergency relief manhole cover". These are gasketed manhole covers that are either hinged or loosely tethered in place and are weighted to open (relieve) at the desired pressure. The simplicity of this device serves as the last line of defense against over-pressurization and mitigates Murphy's Law.

The emergency manhole relief pressure should be set to a higher pressure than the PVRVs to ensure that initial relief is through flame arrestors and PVRVs and the gas/foam/sludge relieves through the manhole covers only as a last resort.

Vacuum

A negative pressure can be created in the digesters from a number of means. The more obvious causes include lowering of the liquid level or gas withdrawal via compressors or blowers. Some of the less obvious causes can be temperature changes or rapid lime addition (pH increase, usually at startup) causing CO_2 in biogas to dissolve rapidly.

The loads resulting from a vacuum condition are of primary concern; however air introduced into the digester may cause explosive conditions to develop as the gas concentration in the digester drops below the upper explosive limit (UEL). Fortunately, there should be very few ignition sources in the moist interior of an operating digester, but this unsafe condition should be avoided in all cases.

Vacuum Example 1

After float testing a 100' diameter steel truss gasholder with the tank full of water, both PVRVs were isolated and the tank was left to drain at the end of the workday. The cover was resting on the corbels (in its lowest position). The sidesheets were still submerged so, as the water drained, a significant vacuum pressure formed under the cover. Figure X3 shows what the cover looked like the following morning. This was possible because the PVRVs were mounted on independent isolation valves and not safety-selector or 3-way type valves.

Figure 3 – Top of Digester Cover After Vacuum Damage. Warped ¼" Steel plate

Freezing

Freezing is a common issue that renders the various pressure relief safeguards ineffective by blocking the flow path. It is important to always keep in mind that raw digester gas is saturated with water vapor at the temperature of the gas space in the digester tank (typically about 95 degrees F.). Any drop below that temperature causes the water vapor to condense out. Sometimes water accumulating in low points is enough to interfere with flow but in cold climates, this condensation may freeze. Because the gas is warm as it leaves the digester, it is tempting to consider omitting insulating and freeze protection. This is a mistake because conditions may develop that could cause this condensation to freeze and obstruct gas withdrawal. In cases where there is little or no flow through a conductive pipe, condensation freezing can occur at ambient temperatures just below freezing. This condition may occur on the piping to the flare where condensation accumulates during intermittent flow and the no-flow conditions allow time for it to freeze and ultimately block the pipe. This is also one reason why PVRVs are often insulated and located on the digester covers.

On branches that see continuous flow of warm raw digester gas the ambient temperature will have to be below freezing to cause condensation inside the pipe to freeze. Risk increases with cooler digester gas (resulting from longer pipe runs), reduced gas flows and with lower ambient temperatures. Danger of freezing begins when the temperature of the inner surface of the pipe wall drops below 32 degrees F. There can also be a compounding effect as flow is restricted; less warm flow causes more condensation to freeze.

Freezing Example 1

During extreme cold conditions of the "Polar Vortex" in 2014, an apparently malfunctioning gas pressure sensor on a digester with a membrane gasholder caused the control panel pressure reading seemed stuck at the maximum level. The cover was designed for approximately 16" w.c. and when the display at the sensor was finally checked, it read 39" w.c. The sensor's functionality was confirmed as it correctly read lower pressures as the tank was drained slightly. It was later determined that ice blocked the gas withdrawal line running down from the top of the gasholder (shown in Figure 4) so no gas could be removed. The ice-filled pipe, shown in Figure 5 was not located and thawed until several days later.

To avoid problems with creep (stretching), membrane gasholders are designed with significant factors of safety for short-term events which is why this event was a "near-miss" and not a complete membrane failure despite the internal pressure reaching an estimated 50+" w.c. (over 4 times the design pressure).

Figure 4 – Uninsulated Gas Lines Between Digester Cover and **PVRV**.

Figure 5 – Ice formed in the horizontal pipe and isolated the **PVRVs** from the Digester.

Freezing Example 2

An uninsulated 2" gas withdrawal pipe froze. A ball valve installed in the gas line could not be turned (indicating that the pipe was full of ice). PVRVs were mounted high above the cover with no insulation or heat tracing and the flame arrestors likely became blinded with ice and/or the PVRV pallet was frozen shut. Pressure built up inside the cover and the force broke several of the holddown straps and pried off some cover supports (Figure).6) One side clearly released first and popped up and the cover was stuck in a tilted position.

Figure 6 – Fixed Cover Lifted and Damaged From Frozen Gas Piping and **PVRVs**

LITTLE PATUXENT WRP DESIGN EXAMPLE

Existing vs. New Tanks

Two digesters were built early in the millennium at the Little Patuxent Water Reclamation Plant (LPWRP) in anticipation of a new dairy related industrial contributor. The anticipated load never came to fruition while simultaneously regulations regarding the disposal of biosolids in the region were changing. The county initiated an investigation to figure out how to best adapt their biosolids processing to comply with the regulations and to maximize its nutrient value to potential users. The resulting study that was initiated looked to achieve the county's goals for the best value possible.

The study involved looking first into the most obvious option of reusing the existing pre-cast digester tanks and supplementing the tankage if needed to achieve the level of digestion required. It was found that a third tank would be required to achieve the desired volatile solids (VS) destruction. Modification of the existing tanks would also be required to achieve the necessary operability. The third tank option was compared against the construction of two or three new tanks. The evaluation found that three new tanks taking up roughly the same footprint of the existing two tanks was the most feasible and economical path forward.

Creating a new area of disturbance (that required modifications to the adjacent administration building) to build a third tank along with the limited ability to modify the existing precast tanks were the main factors that eliminated the option of reusing the existing tanks. To keep the footprint of three new tanks within that of the two existing required a "tall pancake" configuration. Each of the new tanks were designed to have a 2:1 diameter to side water depth ratio with an additional 11% of freeboard (5 feet). The 11% of sidewater depth (SWD) freeboard is at the high of the recommended 4-12% range (Chapman and Kruger, 2011).

The new digesters will digest a mix of waste activated sludge (WAS) and primary sludge. Digestion of WAS has been found to be a common factor in most documented foaming events. This is due to the ever-present nuisance of filamentous bacteria. The property of filamentous bacteria that causes foam in basins is the same as that causes foam in digesters. The property is the tendency to trap gas in bubbles. This property of filamentous bacteria slows the release of gas increasing the chances of foaming event. As earlier mentioned, it has been determined that one protection against foaming or rapid rise reaching a destructive level is a higher than typical freeboard. The lower the ratio between diameter and height causes volume increases to use more freeboard. This helps to explain why recorded instances of rapid rise tend to have a much more destructive effect on silo shaped digesters than on pancake shaped digesters. Another way of stating this is: if a common wastewater tank freeboard of 24" were applied to silo and pancake tanks equally, a rapid rise situation would have a much greater effect on the tank with a tall sidewater depth.

Tank Construction

Cast-in-place

Classic cast-in-place construction is the most flexible construction method from a design perspective. Penetrations can be made where ever the designer desires as well as can be made after the digester is built. Cast-in-place tanks tend to be thicker than the options that are post-tensioned. The extra thickness naturally retains heat better, but if the tanks are to be insulated concrete thickness plays a much less significant role in heat retention. Cast-in-place tanks were the most expensive option due to volume of concrete and labor intensive rebar work.

Precast Post-Tensioned

Precast post-tensioned tanks tend to be a more economical alternative to cast-in-place tanks. The primary factors that make them a more economical alternative are:

- The significant portion of construction benefits from the efficiency and quality control associated with casting in a factory setting.
- Post-tensioning allows a much smaller volume of concrete to be used.
- Labor onsite is minimized. The panels are erected as they arrive onsite. This reduces the tank construction time from weeks to days.

Cast-in-place, Post-Tensioned

Cast-in-place post-tensioned tanks benefit from post tensioning same as the pre-cast tanks, therefore reducing concrete volume and rebar work. Cast-in-place post-tensioned tank manufacturers also employ a mechanized casting technique that minimizes construction time and labor.

Tank Covers

Early in design, the plan was to heat the digesters using biogas and store the excess gas to offset natural gas used by a 5 days/week sludge drying operation. The drying operation would blend biogas and natural gas at a ratio that would minimize/eliminate flaring. These criteria would have required covering at least two of the digesters using membrane gasholder covers. Later, it was decided that the sludge drying would occur 24-7. This reduced the gas storage requirement to one membrane gasholder cover to accommodate an occasional dryer shutdown. The resulting design incorporated the membrane gasholder cover on the secondary digester. (Figure 7).

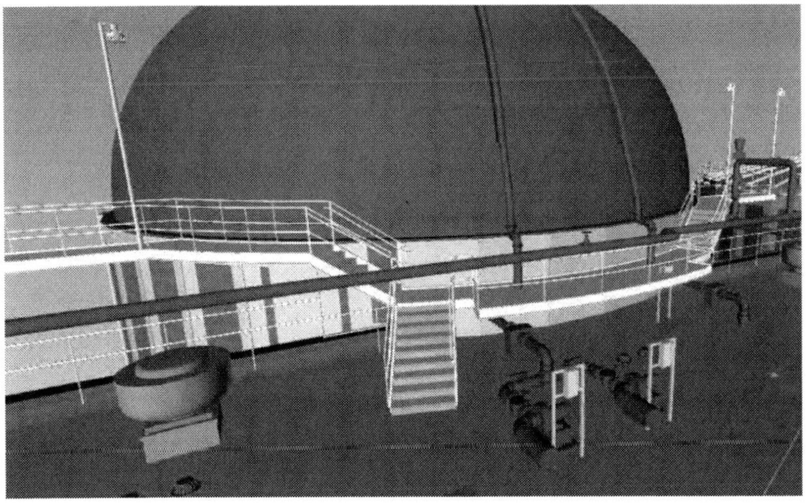

Figure 7 – Secondary digester with dual membrane gasholder cover. The fans that supply air pressure to the cover are shown in the foreground.

There were multiple options for fixed primary digester covers. The default configuration was a fixed steel cover for the primary digesters, but the two styles of post-tensioned tank each offered their own alternative. The pre-cast post-tensioned tank manufacturer offered a cover composed of precast "pie-wedge" panels supported in the center by a hollow column. The cast-in-place post-tensioned tank manufacturer offered their proprietary cover composted of built up shotcrete and reinforcing. Both of the concrete tank cover options offered cost savings over fixed steel covers.

Selection

Pre-cast post-tensioned tanks were selected because they were more economical than the other two alternatives (cast in place and cast in place post-tensioned) regardless of cover type. Evaluation of cover type continued after the tank selection. The pre-cast cover option offered enticing cost savings over steel fixed covers. There was risk though, in that this proposed pre-cast cover design had not been used in a pressurized digester application before. The pre-cast cover manufacturer took on structural design responsibility of the cover and offered an appealing warranty and guarantee. As the overall project design approached 90% completion there was still some uncertainty regarding the PVC liner use for a gas seal and corrosion prevention. This largely resulted from the difficult and extensive overhead PVC welding that was required to ensure success. These factors that drove the decision to move forward with designing around fixed steel covers since the pre-cast option's increased risk and lack of prior installations outweighed the cost savings for this project. Figure 8 shows the two primary digesters with fixed steel radial beam covers flanking the secondary digester. Table 1 shows the breakdown of costs for each alternative.

	Cast in Place Tanks	**Cast-in-Place, Post-Tensioned**	**Precast, Post-Tensioned Tanks**
Cast in Place Cover	$ 500,000	$ 470,000	N/A
Precast Cover	N/A	N/A	$ 225,000
Steel Fixed	$ 490,000	$ 490,000	$ 490,000
Gas Holder	$ 420,000	$ 420,000	$ 420,000

Table 1 – This table shows the price per cover for each tank and cover type combination

Figure 8 – This overview of the digestion facilities shows the two primary digesters flanking the secondary digester. In between the three tanks is the Digester Building which houses the digester mixing pumps, transfer pumps, and heating recirculation pump on the first floor. The second floor houses the boilers, heat exchangers, and hot water pumps. Gas handling equipment is located on the roof.

Safeguards Implemented

Overflow Protection

The first line of defense against overflows involves monitoring and feed pump interlocks. Monitoring is done via both radar liquid surface detection and pressure depth detection. The two simultaneous methods of monitoring allow the detection of a decrease in liquid density which would indicate rapid rise. The primary tanks each have (in order from lowest elevation to highest) a duty overflow to gravity feed the secondary digester, an emergency overflow, an oversized rapid rise overflow (Figure 8), and an emergency relief manhole cover (Figure 9).

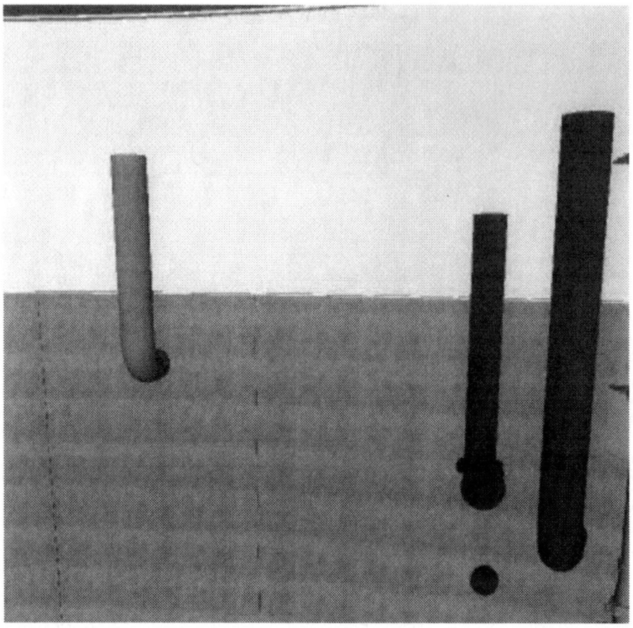

Figure 8 – From left to right this picture shows the emergency overflow, duty overflow, and rapid rise over flow in a primary digester.

Figure 9 – This view shows the emergency relief manhole cover coming off the side of the secondary digester. The emergency relief manhole covers on the primary digester are mounted on top of the fixed covers. The membrane cover on the secondary tank required a side mounted relief.

Over Pressure and Vacuum Protection

Gas storage volume is monitored by redundant laser level transmitters that detects the position of the inner membrane. There is a single enclosed flare stack with redundant regulating valves, flame trap assemblies, control panels, and pilot fuel sources. The flare is rated for the maximum gas production of the digestion system and was selected for its ability to handle lower flows.

Each tank has two locally mounted PVRVs mounted on safety selector valves (Figure 10). Each valve is sized to relieve the maximum gas production from the digester. The emergency relief manhole cover is the last line of defense against overpressure. High pressure indication, fan interlocks, and relief valve prevent the membrane gasholder system from inadvertently creating an over pressure event. There is low pressure indication that is interlocked with both the gas and digested sludge withdrawal.

Figure 10 – This view shows the PVRVs mounted on a safety selector valve on one of the primary digester covers.

Freeze Protection

All exterior piping is heat traced and insulated. The pipe going to the flare is buried below the frost line to avoid freezing. All piping in the system is sloped to various low points. All low points in the system have parallel manual and timer operated drip traps. Pressure/vacuum relief valves are mounted on top of the fixed steel covers and directly adjacent to the membrane gasholder cover (no long piping runs) and will have insulating jackets.

CONCLUSION

Throughout the history and development of digesters, various cover options and tank safeguards have been utilized with varying results. Over this time and still today, designers' investigations into the digestion process, and down to the bacterial level, have enabled the evolution of new and better technologies and approaches to combatting rapid rise and other common digestion issues while maintaining a safe treatment plant environment. The analysis in this paper illustrates some of the critical variables to control and monitor when designing a new digestion system, and the safety examples presented clearly illustrate the repercussions of not following best practices. The Little Patuxent WRP case study shows the results of considering and applying the design approach discussed here and the resulting specific precautions taken to protect the process system and the people working around it.

REFERENCES

Chapman, T. Krugel, S. (2011) "Rapid Volume Expansion – An Investigation into Digester Overflows and Safety". Proceedings of the Water Environment Federation Annual Residuals and Biosolids Management Conference, Sacramento, CA.

Parameters Influencing Anaerobic Digestion of Sewage Sludge

Giacalone, Stefano[1,2], Winter, Peter[1], Smith, Stephen R.[2]
[1]Thames Water, [2]Imperial College London, UK
Correspondence Author Tel: +44 (0) 7747 640501
Email: stefano.giacalone@thameswater.co.uk

ABSTRACT

Anaerobic digestion (AD) is one of the main technologies adopted in most large scale wastewater treatment plants (WWTPs) for sewage sludge treatment. It is also a well established technology, but significant scope remain to improve AD process performance to increase bio-methane production for energy generation, maximising solids destruction, sludge odour control and pathogen reduction for agricultural use, or conditioning for further processing of digestate. These factors have a major impact on the environmental and economic sustainability of the WWTP as a whole.

This research aims to quantify the major parameters in sludge biodegradability, as well as the interactive effects of these parameters on the sludge AD process. The objectives include characterising the major physico-chemical properties of sewage sludge and assessing the impact of up stream conditions such as metal dosing and activated sludge retention time, on anaerobic degradability. The interactive effects of different operational AD parameters were examined in continuous anaerobic reactors. These included: primary sludge (PS) to waste activated sludge (WAS) ratio (20:80 to 80:20 %DS), organic loading rate (1.5 to 4 kgVS/m^3/d) and temperature (37 to 45 °C). The outcomes will assist in benchmarking AD performance across different WWTPs, and will determine the scope for process optimisation and stable operation.

Initial results on sludge characterisation suggested that the parameters influencing biodegradability were different, for different types of sludge. For PS, the relative abundance of the different organic constituents had a major influence on the biogas production. In WAS, volatile organic matter content and sludge age were the most important contributors to bio-methane potential.

In terms of digester operational conditions, the ratio of PS:WAS was the most important parameter controlling overall biogas yield. Temperature and organic loading rate were also important in maintaining long term digester stability. Pathogen destruction was influenced by both temperature and feed regime.

KEYWORDS

Anaerobic Digestion, Operational Parameters, Process Optimisation

INTRODUCTION

Anaerobic digestion (AD) is a complex bio-chemical treatment process, in which a large number of variables interact simultaneously. Little research has focussed on the interactive effects of the operational parameters within the AD process, or on the overall system performance and stability.

This project follows two major research topics, one aims at defining the physico-chemical characteristics of sewage sludge, and how these are impacted by up-stream process; the other aims at quantifying the interactive effects of three operational parameters on the overall AD performance.

Sewage sludge is the solid fraction arising from the treatments of wastewater, and from large scale, high-rate wastewater treatment plants (WWTPs), two types of sludge are generated: primary sludge (PS) and waste activated sludge (WAS). The anaerobic digestibility of these sludge types varies greatly. Primary sludge is a highly degradable material, containing varying proportions of proteins, oil and grease, carbohydrates, fibres (mainly cellulose from toilet paper) and grit. The anaerobic biodegradability of PS varies in the range of 50-70% of the volatile solids (VS) content, and the theoretical biogas yield depends on the composition, but is estimated in range of 900-1000 $NL_{biogas}/kgVS_{destroyed}$ (Parkin and Owen, 1987; Barber, 2014). A large concentration of pathogens is found in PS (Mininni *et al.*, 2004) and their reduction or elimination is necessary for use in land restoration or agriculture (Water UK, 2006). Waste activated sludge is composed of microbial cells, extracellular polymeric substance (EPS), soluble microbial products (Shana *et al.*, 2013) and suspended solids carried over from the primary sedimentation process. The composition of WAS is more consistent compared to PS because it is produced in a biological reactor. Proteins generated from the assimilation of soluble nitrogen (N) and biological oxygen demand (BOD) in wastewater comprise the largest fraction of VS in WAS. Activated sludge is highly recalcitrant under anaerobic conditions, and only a small fraction, in the range of 20-44%VS, is biodegradable in the AD process (Winter and Pearce, 2010; Shana, 2013; Barber, 2014). Solids retention time (sludge age) in the activated sludge process has a strong impact on the biodegradability of WAS, and increasing age correspondingly reduces degradability (Bolzonella *et al.*, 2005; Batstone & Jensen, 2011). Theoretical gas yields for WAS range between 800-900 $NL_{biogas}/kgVS_{destroyed}$ (Parkin and Owen, 1987; Barber, 2014). Pathogens are also commonly found in WAS, and this sludge stream also requires pathogen removal for land application (Smith, 2014a). On the other hand, because nutrients, such as N and phosphorus (P) are concentrated in WAS (Mininni *et al.*, 2004), the fertiliser value is high, and the AD process can be exploited to mineralise a fraction of these nutrients.

Temperature is a critical control variable in all bio-chemical processes. The rate of substrate hydrolysis in AD is raised by increasing temperature and approximately doubles with an incremental rise of 20°C, following the Arrhenius relationship (Ge *et al.*, 2011). Ge *et al.* (2011) showed that the metabolic rate of other various major microbial groups (acidogens, acetogens and methanogens) are not impacted by process temperature to the same extent. This implies that, as increasing temperature can enhance the hydrolysis rate and AD acid production is much faster than acid consumption (Parkin and Owen, 1987; Gerardi, 2003; Appels *et al.*, 2008), this may lead to reactor acidification. Temperature also affects the dissociation point of ionic species and the solubility of gases (Batstone & Jensen, 2011). For instance, these have considerable impact on the carbonate and ammoniacal nitrogen (Amm-N) systems, which are of crucial importance for digester stability. Operational aspects of managing the digester temperature also include changes in the absolute humidity of gas and the thermal self-sustainability of AD process.

The organic loading rate is a mathematical expression, and is a function of the feed sludge dry solids (DS), volatile solids (VS) and hydraulic retention time (HRT). Previous studies (Hill, 1982; Gianico *et al.*, 2013; Gou *et al.*, 2013) demonstrate that the substrate loading rate to the digester impacts the absolute and specific methane (CH_4) production rate. The relationship between CH_4 productivity and OLR is non-linear (Husain, 1998), which may be explained due to biological stress and substrate inhibition. Limits on maximum OLR are also dictated by processing capabilities. Digestion operation managers typically aim to maximise the OLR to intensify the use of assets, and this can be achieved either by increasing the sludge concentration within the reactor or minimising the retention time. The impact of HRT on AD processes has been widely studied (Parkin and Owen, 1987; Miron *et al.*, 1999; Gerardi, 2003; Appels *et al.*, 2008). The lower limit of HRT depends on the rate limiting step, which can either be hydrolysis in the case of low temperature and slowly biodegradable material, or methanogenesis for soluble, rapidly degradable feedstocks (Miron *et al.*, 1999). Increasing the HRT over 20 days appears to have little impact on biogas yield, and the high capital cost of increasing digester volume to accommodate longer retention times suggests that techno-economics become unfavourable (Parkin and Owen, 1987).

In this research, the physicochemical properties of PS and WAS will be determined, including: DS and VS concentration, elemental fraction, major organic macromolecules, chemical oxygen demand (COD), metal dosage, age and bio-methane potential (BMP). The main operational conditions influencing the process that were investigated were: the ratio of PS:WAS used as feed to the AD reactor, temperature and organic loading rate (OLR). The results are used to suggest the performance boundaries of conventional AD assets and how they may be further optimised.

METHODS AND MATERIALS

Sludge source

Sludge characterisation

Sewage sludge samples were collected at approximately two month intervals from three WWTPs for the determination of physico-chemical properties; details of the WWTPs are summarised in Table 1.

Table 1: Main characteristics of the wastewater treatment plants monitored during the research on sludge characterisation

WWTP	Primary sedimentation	Chemical dosage	Secondary treatment
Reading	Lamella	No	BNR 3 stage Bardenpho with pre-anoxic and internal recirculation
Slough	Settlement tank	Ferric addition in activated sludge plant	BNR 3 stage Bardenpho
Maple Lodge	Settlement tank	Ferric addition prior to primary sedimentation	BNR Sebokeng configuration. Mainly chemical P removal.

Operational parameters trial

Sewage sludge samples were collected at weekly intervals from Reading WWTP, which receives municipal wastewater from a population equivalent of approximately 220,000 people. Primary sludge was collected after lamella sedimentation. Waste activated sludge was collected after belt thickening.

Experimental apparatus

Sludge thickening

A sludge thickening device was designed and constructed using woven drainage bags to increase control over the sludge DS concentration (Figure 1). Primary sludge, was collected from the desludging lines of the lamellae primary settlement tank at approximately 1.5% DS and was stored in an industrial bulk container. The sludge was pumped through a static mixer, dosed with polymer electrolyte, and transferred into the draining bags, to retain the solids and remove free water. The collected sludge had a DS concentration in the range 10-13%. In the case of WAS, the sludge was collected after the thickening belt at approximately 6% DS, and was pumped into draining bags to increase the DS content to approximately 9%.

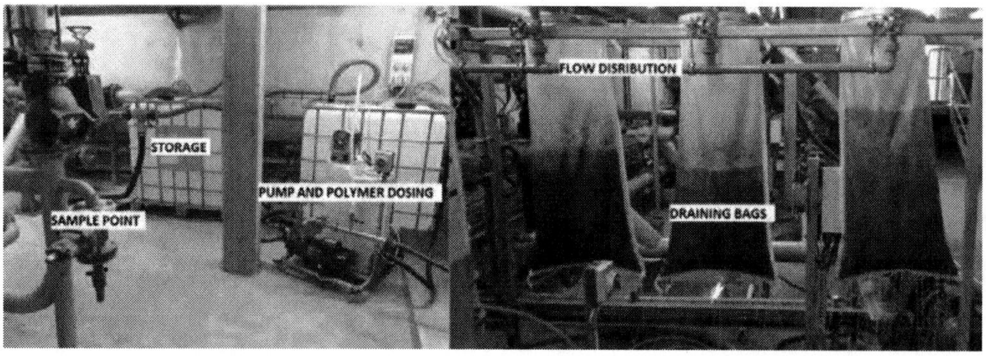

Figure 1: Sludge thickening system for raw primary sludge

Continuous anaerobic reactors

Autodigesters with 50 L working volume were used for the experimental programme (Figure 2). The reactors were fully automated, completely stirred digesters and were supplied with sludge semi-continuously. The sludge feed was supplied from a refrigerated and stirred feed tank, which was cleaned and supplied with fresh sludge every week. The recirculation of sludge in the feed tank and the feeding and withdrawal of sludge in the reactor was performed using peristaltic pumps. The bio-reactor was equipped with a heating jacket, pH and temperature probes and a level sensor to control the active volume. The waste tank was located on a digital load cell and the mass of sludge withdrawn at every cycle was measured. Sludge was fed in batches at a frequency of 3 times per day. The biogas was conditioned through a copper sulphate catchpot to strip hydrogen sulphide (H_2S) and silica gel to dry the gas. Gas was continuously monitored for CH_4 and carbon dioxide (CO_2) fractions and flowrate. The values expressed here refer to dry gas, normalised at 25°C and 1 Atm. The performance was measured as gas yield (m^3/t dry solids fed (DS_{fed})).

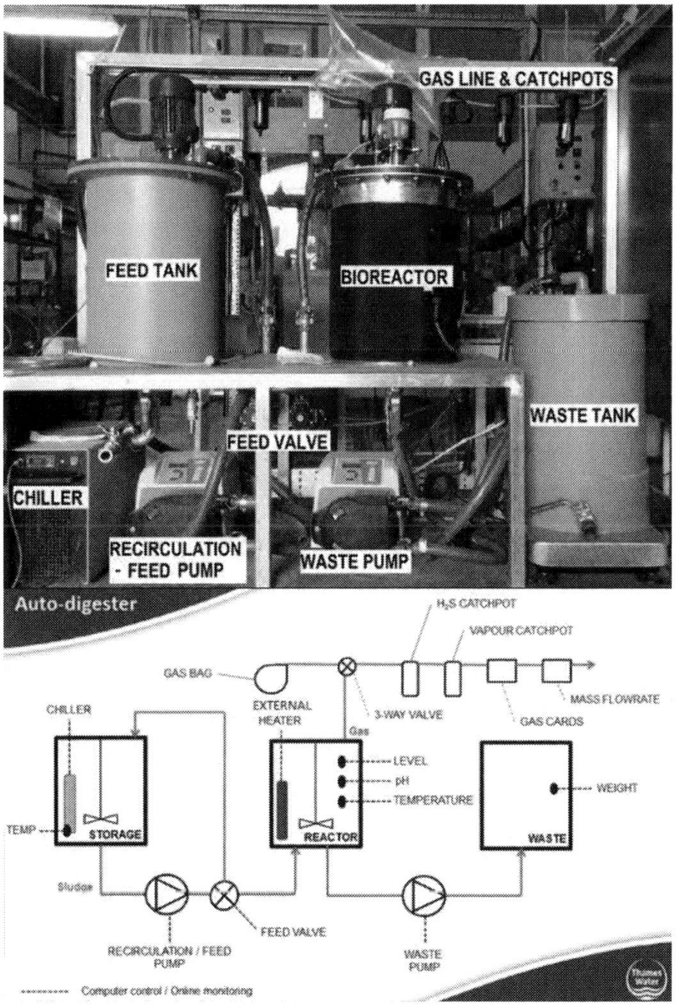

Figure 2: Image and schematic for autodigester apparatus

Batch anaerobic reactors

An automated methane potential system (AMPTS®, BioProcess, Norway) was used to determine the BMP of sewage sludge. The procedure was in accordance with Angelidaki *et al.* (2009), and the digestion tests were performance at 37°C with a 16 days retention time. The seed sludge was sourced from Reading WWTP, which operates a mesophilic AD plant.

Sampling and physico-chemical analysis

A range of different physico-chemical properties were measured in the sludge characterisation and AD operational parameters experiments and details are given in table 2.

Table 2: Routine physico-chemical analysis performed on sewage sludge

Sludge characterisation	
Analysis	**Method**
Elemental fraction	CHN determined via combustion with mass spectroscopy and nuclear magnetic resonance
Proteins	Crude protein determined via combustion by thermal conductivity cell
Fibre	Fibre determined after acid and alkali digestion by weight
Oil	Oil determined after acid hydrolysis followed by ether extraction by weight
Chemical oxygen demand	Hach Lange test kit
Metals	Aluminium and Iron measured via inductively coupled plasma by optical emission spectroscopy
Operational parameters	
Analysis	**Method**
pH	Online probe
Alkalinity	Titration and Colorimetry
Volatile Fatty Acids	Gas chromatography
Ammoniacal nitrogen	Steam distillation and titration
All experiments	
Analysis	**Method**
Dry and volatile solids	Oven at 105°C followed by furnace at 550°C

Experimental design and treatment conditions

Operational parameters experiment

The control variables selected as experimental treatments were temperature, OLR and PS:WAS. The range and rationale for the experimental conditions are indicated in Table 3.

Table 3: Range and rationale for experimental control variables

Parameter	Minimum	Median	Maximum	Reason for selected range
PS:WAS (DS basis)	20:80	50:50	80:20	Wide range of scenarios given by different plant configurations.
OLR (KgVS/m^3/d)	1.5	2.75	4	Current dry solids range achievable by conventional thickening units.
Temperature (°C)	37	41	45	Insufficient literature regarding upper mesophilic range.

Hydraulic retention time was maintained constant at 16 days in all the treatment combinations.

The experimental design to quantify the interactive effects of the treatment parameters was constructed using the Box-Behnken method (Myers and Montgomery, 2009), a surface response statistical method, which optimises the selection of treatment conditions to minimize the standard error of regression coefficients. This statistical strategy requires treatment combinations at the midpoints of the range of experimental conditions and therefore provides the greatest statistical confidence around the centre of the experimental combination cube. The Box-Behnken simulation was completed using the Minitab statistical package and indicated that 12 experiments performed in duplicate would be necessary within the specified parameter range, as well as a six replicate experiment at the centre of the experimental cube. Each treatment combination was carried out as a pseudo-replicate based on multiple hydraulic retention times.

A multiple regression analysis was performed to determine the interactive effects of control variables on different objective functions, using Minitab statistical package.

RESULTS AND DISCUSSION

Sludge characterisation

Elemental fractions

The relative abundance of carbon (C), hydrogen (H) and N for each sludge type and source is summarised in Figure 3. The results indicated that PS was chemically a more variable matrix compared to WAS, which was relatively more consistent, particularly for the C and N fractions.

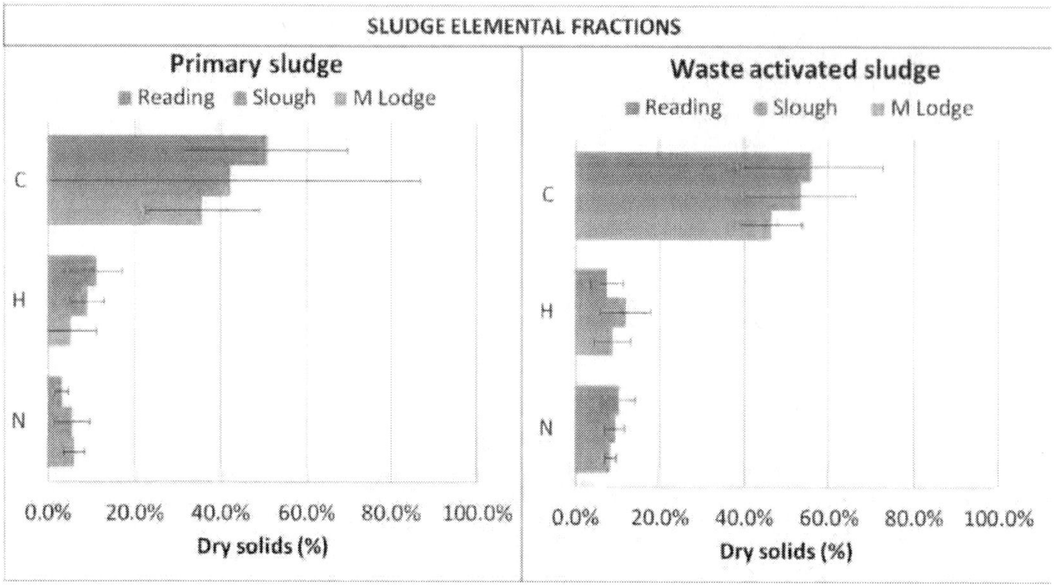

Figure 3: Mean elemental fractions of C, H, N for primary and waste activated sludge types from three WWTPs, error bars indicate standard deviation. (n=8)

The average oxygen (O) composition was calculated from the composition of organic macromolecular constituents (see later) and was estimated as 25% DS and 21% DS for PS and WAS, respectively. The average molar concentrations, normalised to N, are shown in Table 4; the C/N ratio was consistent with published values, although the H/N ratio was more variable (Mininni *et al.*, 2004, Appels *et al.*, 2008).

Table 4: Average elemental fractions expressed as normalised molar concentrations.

Site	Type	C	H	O	N
Reading	PS	18.7	49.9	6.9	1.0
Slough	PS	8.8	22.8	3.9	1.0
Maple Lodge	PS	6.8	11.7	3.6	1.0
Reading	WAS	6.4	10.1	1.9	1.0
Slough	WAS	6.8	17.9	2.1	1.0
Maple Lodge	WAS	6.6	14.8	2.3	1.0

Organics macromolecules

The sludge was fractionated for major organic constituents and the results are presented in Figure 4. A small proportion of the solids was classified as "other" and did not fall into the major constituents groups tested, potentially reflecting other minor organic cell fractions.

A further determination of the fibrous material in PS indicated the relative abundance of cellulose, hemicellulose and lignin. The results showed that for PS, 75-80% of the fibre content was cellulose and hemicellulose, which is highly degradable under AD conditions, while the remaining 20-25% was lignin, which is recalcitrant to anaerobic degradation. The results were generally consistent with other reports (Gonzalez, 2006; Barber, 2014, Smith 2014).

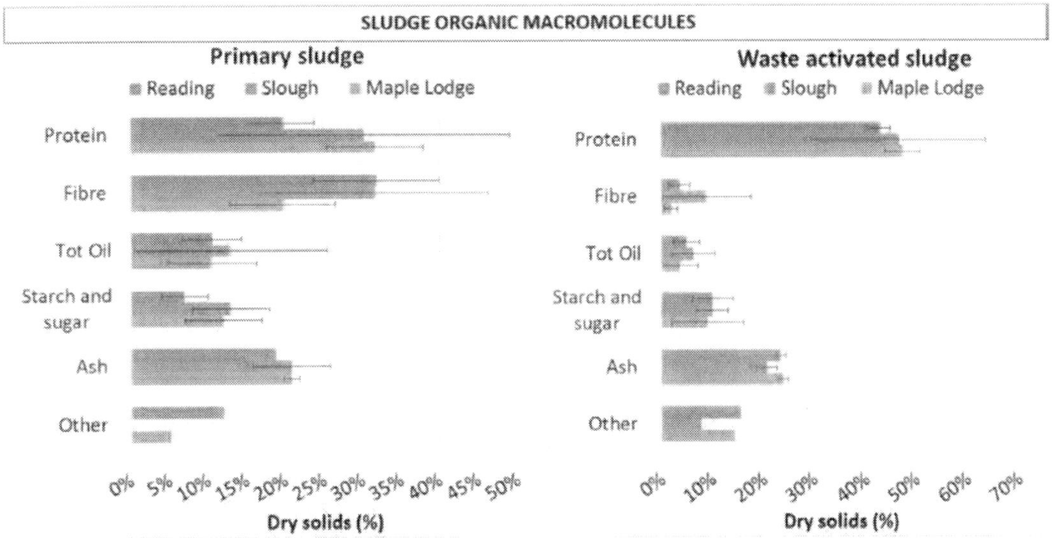

Figure 4: Mean concentration of organic fractions in primary and waste activated sludge types from three WWTPs, error bars indicate standard deviation. (n=8)

Other operational and physico-chemical parameters

Table 5 lists other properties of sludge that were determined, that could have a potential to impact sludge biodegradability under AD conditions (Kepp and Solheim, 2000).

Table 5: Average and standard deviation for physico-chemical sludge properties

Site	Type	VS (%DS)	COD (gCOD/gDS)	COD/VS (gCOD/gVS)	Iron (g/kgDS)	Alum (g/kgDS)	Age (days)
Reading	PS	81% ±	1.41 ± 0.22	1.74 ± 0.27	14.5 ± 2.1	1.7 ± 0.4	n/a
Slough	PS	79% ±	1.17 ± 0.07	1.48 ± 0.12	11.2 ± 3.1	4.9 ± 0.8	n/a
M.Lodge	PS	79% ±	1.11 ± 0.30	1.40 ± 0.38	35.0 ± 2.4	8.9 ± 0.6	n/a
Reading	WAS	77% ±	0.95 ± 0.05	1.24 ± 0.07	26.6 ± 2.9	4.0 ± 0.6	20.5 ±
Slough	WAS	79% ±	0.97 ± 0.15	1.22 ± 0.19	12.0 ±	4.2 ± 7.6	12.8 ±
M.Lodge	WAS	76% ±	0.83 ± 0.19	1.09 ± 0.25	57.7 ± 1.4	6.9 ± 0.4	15.2 ±

Bio-methane yield

The BMP was calculated and compared to the maximum theoretical yield obtained from the elemental and organic fractions measured in the sludge samples, following the assumptions indicated by Angelidaki *et al.* (2004). The biodegradable fraction was also calculated as fraction of the BMP relative to the maximum theoretical yield (Figure 5). The theoretical CH_4 yield calculated based on the elemental content or organic composition varied greatly, therefore the calculated biodegradable fractions also varied. Data for the measured BMP presented low deviation, thus the large deviation measured in the biodegradability accounted for the variability of the analytical measurement of the physico-chemical properties. On account of this, WAS appeared to present a narrower range of results, compared to PS, especially with the CHN method. Thus the theoretical values obtained with the CHN method, together with the biodegradability factor, could provide an interesting tool to forecast methane yield under AD conditions.

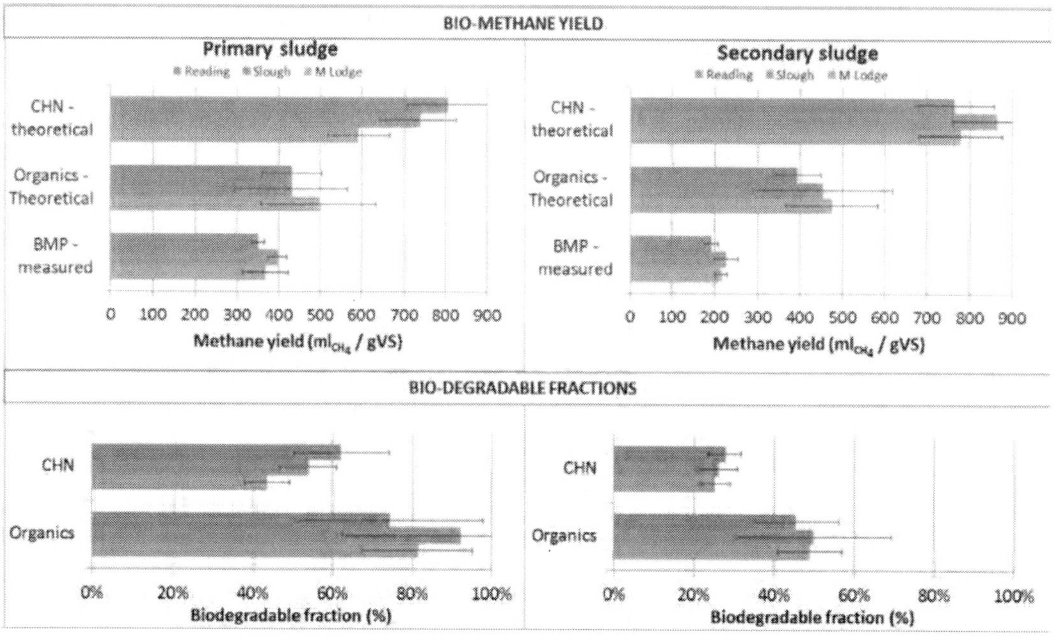

Figure 5: Measured and theoretical CH_4 yield and sludge biodegradable fractions. Gas values are normalised to STP. Error bars indicate standard deviation.

The impact of average COD/VS on BMP was tested on the available dataset and results are presented in Figure 6. It would appear that when both types of sludge were included in the analysis, a correlation could be found. On the other hand, if the two types of sludge were to be analysed separately, no correlation with BMP would be found. This suggested that COD/VS provided only a general indication of the differences in biodegradability between different sludge types, but not within sludge types.

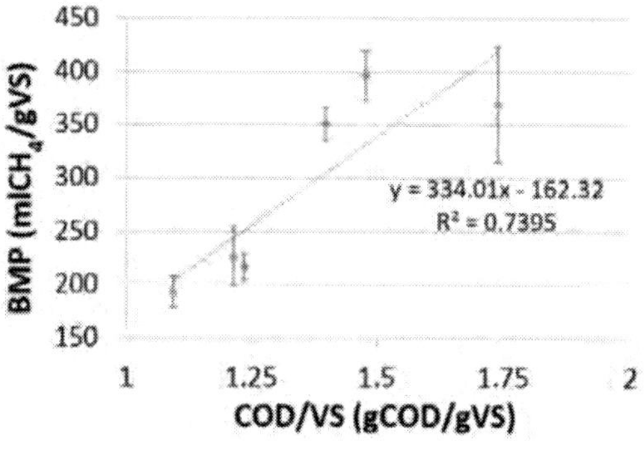

Figure 6: Correlation between average COD/VS ratio and average BMP of sewage sludge (n=6)

Another parameter that from literature (Bolzonella et al., 2005, Ge *et al.*, 2013) was found having a significant impact on WAS biodegradability was solids retention time in the activated sludge plant (age). Barber (2005) also suggested that the VS fraction of sludge could have a significant impact on the specific biogas yield of sludge. Figure 7 indicates the correlation obtained from testing age and VS, in isolation, with WAS BMP.

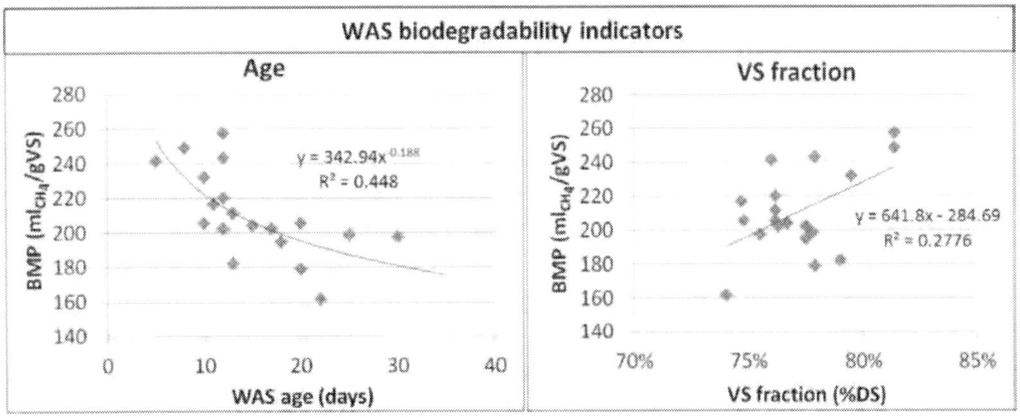

Figure 7: Impact of age and VS fraction of WAS bio-methane potential (n=19)

The two parameters were indicative of WAS biodegradability, but independently did not appear to fully explain the variability in BMP. Therefore a multiple regression analysis was conducted, utilising age and VS content as input and BMP as objective function. Data from Bolzonella et al. (2005) were included in the regression analysis to increase the range of age and VS to expand model applicability. The produced model (Figure 8) was highly statistically significant ($P<0.001$) and with a high coefficient of determination (R^2=98%).

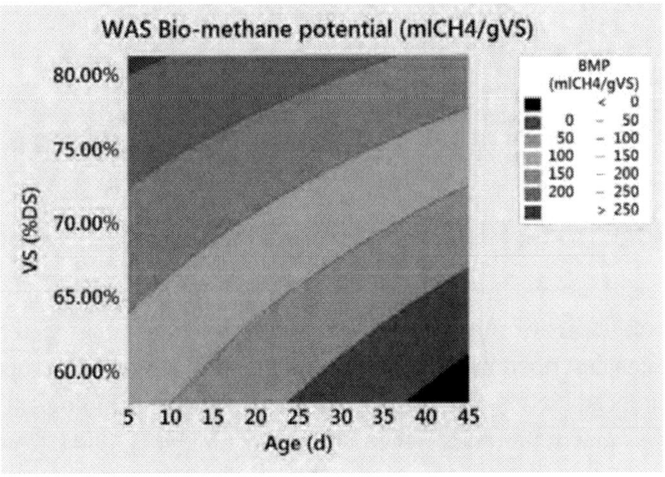

Regression equation ($P<0.001$, R^2=98.7%)

$$BMP = -8.00*Age + 404.5*VS*VS + 7.55*Age*VS$$

Figure 8: Interactive impact of Age and VS content on WAS bio-methane potential

The two parameters used were easily measureable and were routinely monitored in WWTP, thus this model provide an efficient and inexpensive way to determine the level of biodegradability to be expected from WAS, as well as a mean to optimise the performance of the plant.

Operational parameters

The results presented here include a portion of the tested matrix (Table 2), as not all the experiments were concluded at the time this paper was written. No results are presented for the maximum tested OLR of 4 kgVS/m^3/d.

Biogas and solids destruction performance

The results for biogas yield (Figure 9) indicated that the dominant control variable was the PS:WAS fraction, with higher PS corresponding to increased gas yield. However, the impact of temperature and OLR was also significant. Across all treatments increasing the temperature from 37 to 41°C produced a positive effect on gas yield. On the other hand, the further temperature rise to 45°C, gave positive effects only for high WAS treatments, when compared to 37°C. Increasing the organic loading rate from 1.5 to 2.75 gkVS/m^3/d also appeared to have a positive impact on gas yield across all treatments.

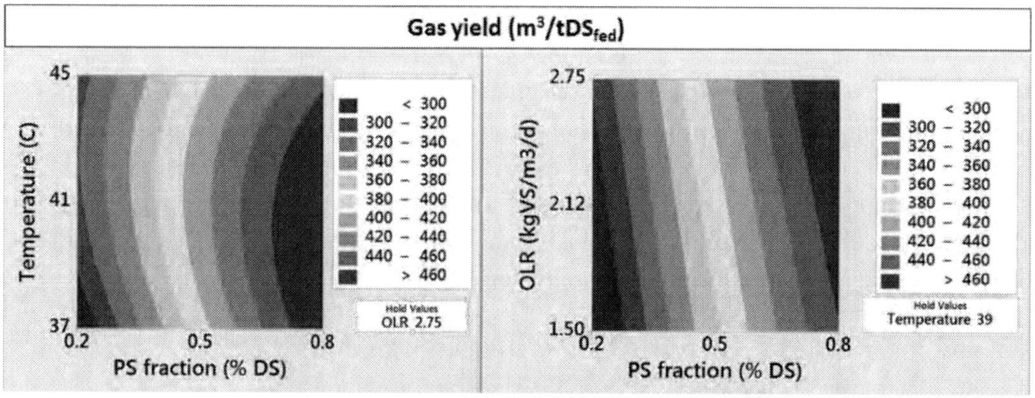

Regression equation ($P<0.001$, $R^2=73.6\%$)

Gas yield = -1892 + 914*PS% - 108.3*OLR + 102.6*Temperature - 123.0*PS%*PS% - 1.289*Temperature*Temperature - 12.26*PS%*Temperature + 3.43*OLR*Temperature

Figure 9: Contour plot for Gas yield as function of PS:WAS fraction, OLR and temperature.

The difference in gas yield between high PS fraction and high WAS fraction may be explained due to the different accessibility of substrates in the two types of sludge, as well as differences in organic constituents as described in the previous paragraphs. Rising the temperature in the reactor may have produced two separate effects, stimulating the biological activity and increasing the hydrolysis rate of the sludge. It could be suggested that 39-41°C was an optimal regime for biological activity, independently of the type of sludge fed, as the gas yield was increased compared to 37°C. Benefits of raising temperature to 45°C were only significant for high WAS regimes. This could find an explanation in the fact that WAS was more slowly degradable compared to PS, and the increased hydrolysis rate produced by the increase in temperature, led to an overall increase in performance at 45°C for high WAS regimes, compared to 37°C.

These results found partial confirmation in the solids destruction data (Figure 10).

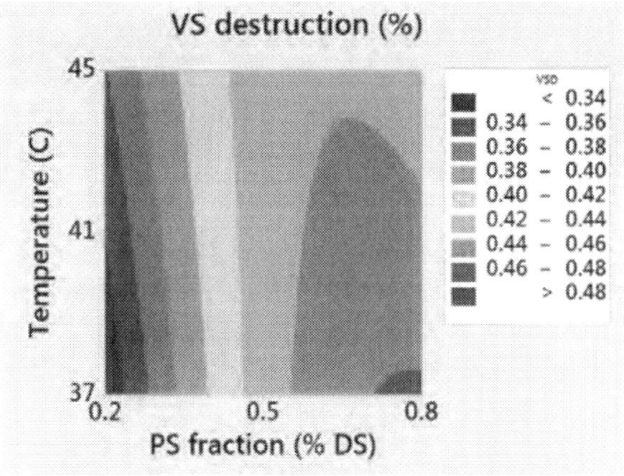

Regression equation ($P<0.001$, R^2=43.8%)

VSD = -0.0563 + 1.171*PS % + 0.00726*Temperature - 0.4019*PS %*PS % - 0.01451*PS %*Temperature

Figure 3: Contour plot for VS destruction as function of PS:WAS and temperature.

The general trend indicated in figure 10, confirmed that higher PS fraction in the feed corresponded to higher VS reduction, and thus higher gas yield. The VS reduction for the high WAS treatments confirmed the positive impact of increased temperature, with an average increase in VS reduction of approximately 10% from 37 to 45°C, for an 80% WAS treatment. As the fraction of PS increased, the impact of the temperature became less evident. Overall the R^2 of the model was low, suggesting that the model could not explain all the variability in the data. The reason for this could lie in the large variability of VS composition of PS, which resulted in larger variation of VS destruction compared to high WAS treatments.

Stability indicators

The stability of the bio-reactors was assessed by monitoring pH, volatile fatty acids (VFA), alkalinity and ammoniacal N concentration. Alkalinity is a key parameter to asses digester stability, because it gives the reactor the capacity to buffer the acids produced during the intermediate biological steps, thus reducing the changes in pH and increasing the environmental stability of the overall AD process (Parkin and Owen, 1987). Ammoniacal N has the potential to inhibit the process, if the fraction of free ammonia (NH_3) increases over the threshold tolerance of the bacterial groups (Batstone & Jensen, 2011). The trends for these two key stability indicators are summarised in Figure 11 and 12.

The concentration of alkalinity in the reactor appeared influenced primarily by the OLR of the system, which in this case was altered by increasing the feed DS and maintaining a constant HRT. For this experiment the increase in OLR from 1.5 to 2.75 kgVS/m3/d corresponded approximately to an increase in solids concentration from 3% to 5.8% DS. The PS:WAS fraction also appeared to have a significant impact, with alkalinity increasing with increasing WAS content in the feed. Temperature also contributed to alkalinity, but its contribution was minor when compared to the other two parameters.

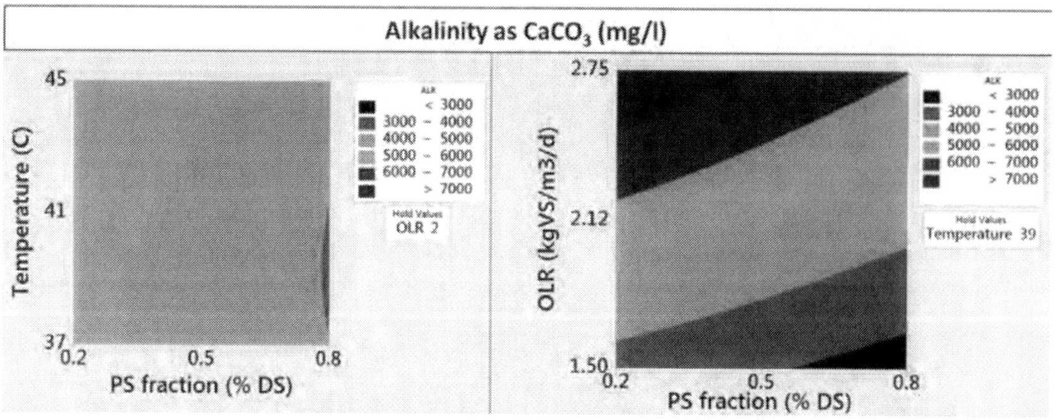

Regression equation ($P<0.001$, $R^2=97.5\%$)
Alkalinity = 15046 + 3714*PS % + 2029*OLR - 837*Temperature
+ 10.64*Temperature*Temperature - 1042*PS %*OLR - 99.3*PS %*Temperature
+ 40.0*OLR*Temperature

Figure 4: Contour plots for alkalinity as function of PS:WAS, temperature and OLR.

The explanation for increased alkalinity with increased OLR was straightforward, as higher concentration of compounds in the input, led to higher concentration of compounds in the reactor. However, its implications were important for AD management, as it suggested that effectively a higher load, up to 2.75 kgVS/m3/d, generated a more stable system and not the other way round.

The impact of PS:WAS to total alkalinity could be explained with the contribution to the production of bicarbonate alkalinity from ammoniacal N (Sotemann, 2005). As showed in Figure 12, higher WAS content generated higher ammoniacal N, thus higher alkalinity.

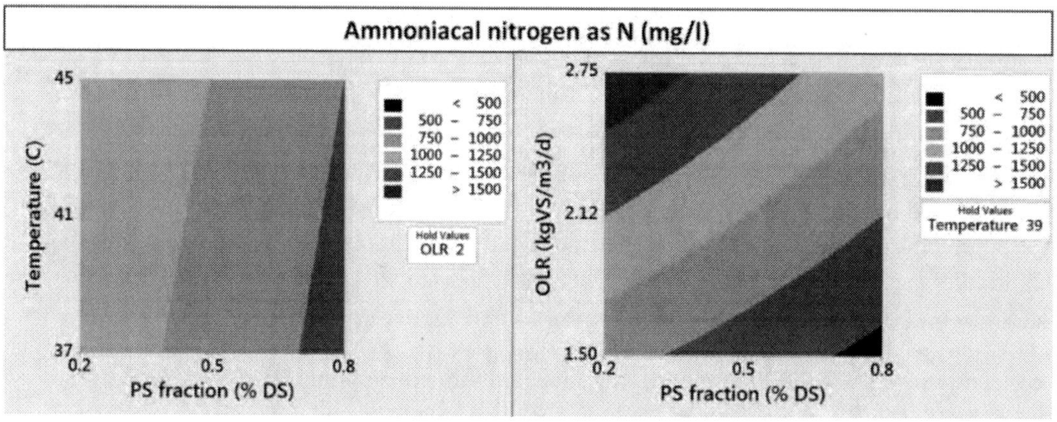

Regression equation ($P<0.001$, $R^2=97.3\%$)
Ammoniacal N = 10220 + 3711*PS % - 545*OLR - 520*Temperature
+ 6.34*Temperature*Temperature
- 550.0*PS %*OLR - 92.0*PS %*Temperature + 37.12*OLR*Temperature

Figure 5: Contour plots for ammoniacal nitrogen as function of PS:WAS, temperature and OLR.

The concentration of ammoniacal N in the reactor followed very much a similar trend to alkalinity. The production of ammoniacal N, was generated by the breakdown of proteinaceous material, thus increased with the solids concentration (OLR) and WAS content of the feed. Taking into account VFA concentration, pH and biogas CH_4 and CO_2 fractions, no sign of inhibition was found for the level of ammoniacal N indicated in Figure 12. For better reference the level of NH_3 were calculated as shown in Figure 13 (Hansen *et al.*, 1997).

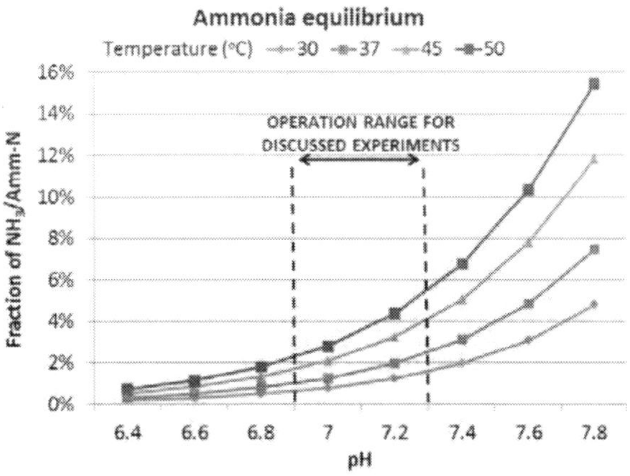

Figure 6: Fraction of free ammonia relative to temperature and pH (Hansen *et al.*, 1997)

Pathogens reduction

The level of pathogens reduction (Figure 14) was assessed by measuring the concentration of *Escherichia coli* in the feed and digested sludge. The results presented here were for an OLR equivalent to 2.75kgVS/m^3/d only, as not sufficient data were available for the other treatments. The regression model was conservative, as the effects of DS destruction was not included, and samples with no measurable concentration were set to limit of detection value (2.4 log/gDS).

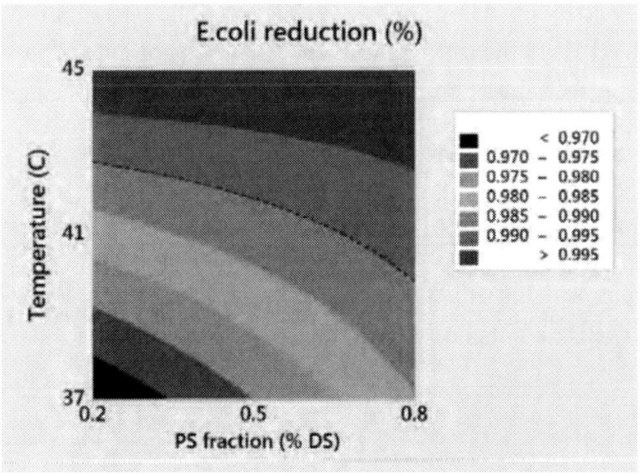

Regression equation ($P<0.001$, R^2=49.7%)

E.coli reduction = 0.7779 + 0.1559*PS % + 0.005021*Temperature + 0.0260*PS %*PS % - 0.00404*PS %*Temperature

Figure 7: Contour plot for *E.coli* reduction as function of PS:WAS and temperature.

The results indicated that the main factor affecting *E.coli* reduction was temperature. The treatments operating at 45°C achieved a level of destruction above 99%, which is the minimum requirement to obtain a CLASS B compliant biosolids product, for the current UK legislation (Water UK, 2006). In contrast, none of the treatments operated at 37°C achieved a sufficient level of pathogen removal for land application, indicating the need for secondary digestion. The regression also indicated a significant impact of the PS:WAS ratio. It appeared that treatments with high PS ratio, would have a higher level of pathogen removal, for an equivalent temperature and retention time, compared to treatments with a high WAS proportion. Figure 14 suggested that at 80% PS, a temperature of 41°C would be sufficient to reach 99% *E.Coli* removal. Although this is plausible given the other results, confirmatory studies are needed to validate the outcome of the empirical model.

The effect of temperature on pathogen removal was in line with what indicated in other literature (Smith *et al.*, 2005). The main mechanism of inactivation of enteric bacteria was the denaturation of protein caused by high temperature, which disrupted and inactivated cells. The effect of PS:WAS ratio could be explained with the presence of EPS in WAS, which acted as a protective environment for enteric bacteria, reducing the rate of inactivation and maintaining a higher level of pathogens in the digested sludge.

CONCLUSIONS

Anaerobic digestion is highly dependent on upstream processes, and an holistic view should be taken when considering process optimisation. In this chapter it will be discussed how to make best use of the results of this research, with the aim of optimising the performance of the AD process.

Upstream processes

In the previous chapter the best methods to determine sludge biodegradability were discussed.

Primary sludge is influenced by the incoming wastewater, and the best method to determine its AD potential remains direct BMP measurement. For PS there is not much scope for optimising its potential. The only relevant actions could be:

- Dosing ferric or alum compounds to aid primary sedimentation. This would increase the ratio of PS in the sludge feed, thus the overall gas yield of the feed, but would not increase the gas yield of PS.

- Minimise storage time. This can help avoiding fermentation of sludge prior to AD, and potential loss of methane outside the reactor.

For WAS, arising from BNR secondary treatments, the method suggested in this research to determine the AD potential is by characterising age and VS content of the sludge. This implies that management of the secondary plants has profound effects on the sludge properties and the following action could improve AD performance:

- Minimise sludge WAS age. This can be achieved by reducing mixed liquor suspended solids in the activated sludge plant or by increasing the fraction of wastage in order to decrease the solid retention time. The major constraints to this involve maintaining a sufficient level of biomass in the activated sludge plant, maintaining a minimum retention time in the activated sludge plant to allow for nitrification and having sufficient volume in the AD reactors to allow the extra WAS volume. Techno-economic considerations should also take into account of dewatering and sludge disposal costs of the increased WAS throughput.

Anaerobic digestion process

Following the results from the experiment on operational parameters affecting AD, a number of recommendations can be given to optimise the process performance.

- Organic loading rate should be kept between 2.5 and 3 kgVS/m3/d (5.5 – 6.5% DS at 16 days HRT). This should ensure the system has enough alkalinity to buffer the acid produced in the intermediate steps of AD.

- Temperature should be kept between 39 and 41°C, maintaining temperature changes below 2°C/HRT, this ensures the bacterial communities have time to adapt to new conditions.

- Increasing the temperature further to 45°C should be done where:

  A) The feed sludge has a high WAS ratio (WAS>50% of total DS).

  B) The WWTP does not have other barriers to pathogen reduction. If no other thermal or chemical treatment is in place, temperature can be raised to 45°C to significantly increase enteric pathogen reduction.

  C) The feed sludge is sufficiently high in solid concentration. In the event of low alkalinity (due to low OLR) and high temperature, there is the increased risk of acidifying the digester, due to the conditions of low buffering and fast hydrolysis rate.

ACKNOWLEDGEMENTS

This work was carried out as part of the Thames Water Utilities Innovation Programme and the requirements for the EngD STREAM Programme in the Department of Civil and Environmental Engineering, Imperial College London. The views expressed in the paper are those of the authors and not necessarily those of Thames Water.

This project was sponsored by Thames Water, Anglian Water, Severn Trent Water and the Engineering and Physical Sciences Research Council.

REFERENCES

Angelidaki, I. and Sanders, W. (2004) Assessment of the anaerobic biodegradability of macropollutants. *Reviews in Environmental Science and Bio-Techonology,* 3, pp.117-129

Angelidaki, I., Alves, M., Bolzonella, D., Borzacconi, L., Campos, J.L., Guwy, A.J., Kalyuzhnyi, S. Jenicek, P. and Van Lier, J.B. (2009) Defining the biomethane potential (BMP) of solid organic wastes and energy crops: a proposed protocol for batch assays. *Water Science and Technology,* 59.5, pp.927-934.

Appels, L., Baeyens, J., Degreve, J. and Dewil, R. (2008) Principles and potential of the anaerobic digestion of waste-activated sludge. *Progress in energy and Combustion Science*, 34, pp.755–781.

Barber, W.P. (2005). The effect of improving sludge digestion. *Water and Environment Journal*, 19(3), pp.214–224.

Barber, W.F. (2014) Influence of wastewater treatment on sludge production and processing. *Water and Environment Journal*, 28(1), pp.1–10.

Batstone, D.J. & Jensen, P.D. (2011) Treatise on Water Science - Anaerobic Processes Fundamentals. *Water Quality Engineering, 4, 615-639*

Bolzonella, D., Pavan, P., Battistoni, P. and Franco Cecchi (2005) Mesophilic anaerobic digestion of waste activated sludge : influence of the solid retention time in the wastewater treatment process. *Process Biochemistry*, 40, pp.1453–1460.

Ge, H., Jensen, P.D. and Batstone, D.J. (2011) Relative kinetics of anaerobic digestion under thermophilic and mesophilic conditions. *Water Science and Technology*, 64, pp.848–853.

Ge, H., Batstone, D.J and Keller, J. (2013) Operating aerobic wastewater treatment at very short sludge ages enables treatment and energy recovery through anaerobic sludge digestion. *Water Research,* 47, pp.6546-6557.

Gerardi, M.H. (2003) *The Microbiology of Anaerobic Digesters*. In Wiley-Interscience, ed. Wastewater Microbiology Series. John Wiley & Sons, Inc. Publication.

Gianico, A, Braguglia, C.M., Cesarini, R. and Mininni, G. (2013). Reduced temperature hydrolysis at 134 C before thermophilic anaerobic digestion of waste activated sludge at increasing organic load. *Bioresource Technology*, 143, 96-103.

Gonzalez, M. V., (2006), Enhancing gas production in mesophilic anaerobic digestion, *PhD Thesis, University of Surrey, UK.*

Gou, C., Yang Z., Huang J., Wang H.,Xu H. and Wang L. (2013). Effects of temperature and organic loading rate on the performance and micobial community of anaerobic co-digestion of waste activated sludge and food waste. *Chemosphere*, 105, 146-151.

Hansen, K.,H., Angelidaki, I. and Ahring, B., K. (1997) Anaerobic digestion of swine manure: inhibition by ammonia. *Water Research,* 32(1), pp.5-12

Hill, D. T. (1982). A comprehensive dynamic model for animal waste methanogenesis. Transaction of the ASAE, 25, 1374.

Husain, A. (1998) Mathematical models of the kinetics of anaerobic digestion - A selected review. *Biomass and Bioenergy*, 14, pp.561–571.

Mininni, G., Braguglia, C.M., Ramadori, R. and Tomei, M.C. (2004) An innovative sludge management system based on separation of primary and secondary sludge treatment. *Water Science and Technology*, 50(9), pp.145–154.

Miron., Y., Zeeman, G., Van Lier, J.B. and Lettinga, G. (1999) The role of sludge retention time in the hydrolysis and acidification of lipids, carbohydrates and proteins during digestion of primary sludge in CSTR systems, *Water Research*, 34 (5), pp. 1705-1713

Myers R.H., Montgomery D.C. and Anderson-Cook C.M. (2009) *Response Surface Methodology: Process and Product Optimization Using Designed Experiments*. John Wiley and Sons

Kepp, U., and Solheim O.E. (2000), Thermo-dynamical assessment of the digestion process, *CIWEM Biosolids Conference*, UK.

Parkin, B.G.F. & Owen, W.F. (1987) Fundamentals of Anaerobic Digestion of Wastewater Sludges. *Journal of Environmental Engineering* , 112(5), pp.867–920.

Shana, A., Ouki, S., Asaadi, M., Pearce, P. and Mancini, G. (2013) The impact of intermediate thermal hydrolysis on the degradation of kinetics of carbohydrates in sewage sludge, *Bioresource Technology*, 137, pp 239-244

Smith, S.R., Lang, N.L., Cheung, K.H.M and Spanoudaki, K. (2005) Factors controlling pathogen destruction during anaerobic digestion of biowastes. *Waste Management*, 25, pp.417–425.

Smith, S.R. (2014). How activated sludge has been transformed from a waste to a resource, and the implications of this for the future of the activated sludge process. *In* Horan, N.J (ed) *Activated sludge: Pasts, Present and Future.* 2-3 April, Manchester

Sötemann, S.W., Risotw, N.E., Wentzel M.C. and Ekama, G.A (2005) A steady state model for anaerobic digestion of sewage sludges. *Water SA*, 31(4), pp.511–528.

Water UK (2006) The Safe Sludge Matrix – Guidelines for the Application of Sewage Sludge to Agricultural Land [online] https://www.gov.uk/managing-sewage-sludge-slurry-and-silage [Accessed: 01/02/2015]

Winter, P. & Pearce, P (2010) Parallel digesstion of secondary and primary sludge. In *15th European Biosolids and Organic Resources Conference*, Manchester, UK.

Extracting Energy From Sludge in the UK, Recent Experience

Keith Panter* Dr Bill Barber**

* Ebcor Ltd, UK (E-mail: keith@becor.freeserve.co.uk)

** Cambi Inc (E-mail: bill.barber@cambi.com)

Abstract

In the UK, a combination of strict financial regulation, high energy costs, and various renewable energy incentives have led the Water Industry to continue investing heavily in advanced anaerobic digestion. A trend of recent years has seen the shut down and decommissioning of drying plants – deemed unnecessary with respect of best practice environmental option (BPEO) of land application of cake, expensive to operate and high carbon footprint. In addition, the number of existing incineration facilities has dwindled significantly as they have suffered increasingly expensive operation, and are being replaced by anaerobic digestion plants combined with biosolids use in agriculture. In the meantime, existing digestion facilities are being upgraded to enhance performance further and enable them to be used more intensely. As a consequence, the UK Water Industry leads the world in the use of advanced anaerobic digestion. With 21 full-scale facilities, UK has the highest proliferation of thermal hydrolysis plants that generate 60 MWs power continuously from 1500 metric dry t/day and account for nearly 30% of all sludge in UK. This is nearly half of all renewable energy produced from biogas generated from sewage sludge in the UK. In addition, there are a number of advanced biological hydrolysis systems, acid phase, Enzymic and Enhanced Enzymic Hydrolysis (EH and EEH); EH has 9 project accounting for another 250 tds/day.

This paper describes the latest findings from the UK with respect to energy recovery from sludge and looks at how this experience can be used elsewhere. What is notable from the UK is the following:

Combined Heat and Power (CHP) output has increased by 249% with a digestion capacity increase of 113%. This asset sweating has been achieved by a combination of increased VS loading, increased VSR% and additional CHP installations at sludge centres. There has been a wholesale move away from energy intensive/high carbon footprint process, such as incineration, thermal drying and lime stabilisation. 92% of all sludge is now digested with CHP and biosolids to land. There has been an increase in the quality of biosolids in terms of odour, stackability and pathogens that has assisted this switch away from high carbon footprint technologies. Nearly all dryers and incinerators in UK have been de-commissioned and lime stabilisation is now very rare except in remote communities.

Keywords: Renewable energy, anaerobic digestion, CHP

THE LAST TEN YEARS IN UK.

The rapid growth of anaerobic digestion and CHP stems from the mid-2000s when there was a rise in world energy prices due the Gulf crisis. This coincided with the introduction of Renewable Obligation Credits (ROCs), similar to RECs in USA. These ROCs still trade for about $60/MWh, though in later years the value of ROCs has been reduced by 50% for WWTPs with new projects and other forms of incentive now in place are leading to gas to grid projects (stimulated by Russian dominance of gas pipelines in Europe). The latter can be considered the German model that is leading to so many biowaste/manure projects in parts of Europe.

Therefore in the UK , a combination of strict financial regulation, high energy costs, and ROCs have led the Water Industry to continue investing heavily in advanced anaerobic digestion. A notable feature of the UK is the organisation into 10 water companies (England and Wales) and 2 Water Authorities (Scotland and N. Ireland) that have seen multiple benefits from regionalisation. In this context, the main effect has been to see the formation of sludge centres.

Sludge centres are locations where Anaerobic Digestion (AD) is centralised and raw sludge is trucked to the receiving site either as cake or liquid and digested. The advantages of sludge centres are as follows:

1. Digestion and CHP expertise is focussed on one location,
2. By doubling solids load the WWTP becomes self-sufficient in energy
3. Nutrient management is centralised also
4. Sludge pre-treatment becomes more affordable
5. Product quality can be enhanced
6. Expensive lime stabilisation systems are shut down

A trend of recent years has seen the shut down and decommissioning of drying plants – deemed unnecessary with respect of best practice environmental option (BPEO) of land application of cake, expensive to operate and high carbon footprint. In addition, the number of existing incineration facilities has dwindled significantly as they have suffered increasingly expensive operation, and are being replaced by anaerobic digestion plants combined with biosolids use in agriculture.

Surprisingly one of the major spin-offs of the advanced AD approach has been a rapid increase in cake quality with respect to both dry solids and stabilisation (particularly so following thermal hydrolysis pre-treatment) so that products are easy to store and distribute without nuisance and safety concerns. This success has been replicated recently in the US with DC Water project at Blue Plains.

SITUATION COMPARISON BETWEEN 2015 AND PRESENT TIME

The UK Water Industry leads the world in the use of advanced anaerobic digestion. With 21 full-scale facilities, UK has the highest proliferation of thermal hydrolysis plants anywhere in the world, that treat over a quarter of the UK's sludge production so that 1500 metric dry t/d, equivalent to 19 million p.e., and generating 60 MWs of electrical energy continuously. In addition, there are numerous advanced biological hydrolysis systems, of these the most popular technology is Enzymic Hydrolysis that has 9 operating facilities treating about 250 metric dry t/d, though there was a drop in uptake of this technology in this decade. These two technologies account for about one third of all UK sludge between them. See table 1 for the largest project in the UK that are in the range of 3 – 12 MWs generation.

Table 1. Top 10 renewable energy generators in the UK Water Industry

	Site	Pre-treatment
1	Davyhulme CHP at Davyhulme WWTW	Thermal Hydrolysis
2	Minworth Generating Station	Thermal Hydrolysis added 2016/7
3	Great Billing STW CHP Plant (RR)	Enzymic Hydrolysis
4	Mogden STW (RA)	Conventional
5	Avonmouth STW CHP Generation	Enzymic Hydrolysis
6	Bran Sands (RSTC Advanced Digestion) Facility	Thermal Hydrolysis
7	Cardiff East WWTW CHP	Thermal Hydrolysis
8	Howdon AAD	Thermal Hydrolysis
9	Longreach STW (RB)	Thermal Hydrolysis
10	Stoke Bardolph STW Generating Station	Separate maize digestion from attached sludge farm + conventional sludge

A notable feature of these two technologies is the ability to operate at higher organic loading. Cambi THP plants are operating at anywhere between 4 and 8 kg VS/m3/day and enzymic plants more commonly in the 3-4 kg VS/m3/day. Compared to the UK average of about 2 kg VS/m3 /day (equivalent to .155 lb VS/cuft/day) that is more typical of US digester loading rates –

PARADIGM SHIFT IN TREATMENT OF SLUDGE FOR ENGLISH WATER COMPANIES

The tables below compare the overall situation for the 10 English Water companies in 2005 and the changes made to date in the application of digestion and CHP.

see table 2 for the estimated situation in 2005 for the English water companies (telephone survey by the Author).

Table 2. Estimate of CHP from AD in English water companies 2005

Situation in 2005									
	pop M	TDA raw	digested	% digested	M3 capacity	MWs gen	Theoretical	% generated from biogas	Comments
Southern	4.4	128480	120000	93%	78000	3.2	11.2	29%	Gas used for drying
Thames	13.0	379600	250000	66%	250000	15	33.0	45%	2 raw sludge incinerators - CHP not optimised?
Anglian	5.9	172280	67000	39%	85000	3	15.0	20%	programme to digest 80000 tds more sludge under AMP4
Yorkshire	4.8	140160	77000	55%	77000	2	12.2	16%	focus on incineration
NWL	2.6	75920	10000	13%	10000	0.5	6.6	8%	Raw sludge drying
UU	7.0	204400	130000	64%	185000	10.0	17.8	56%	raw sludge incineration proposed
DCWW	3.0	87600	20000	23%	20000	1	7.6	13%	raw sludge drying and lime
ST Water	8.3	300000	300000	100%	315000	23	26.1	88%	digested sludge incineration
Wessex	2.5	73000	40000	55%	40000	4	6.3	63%	CHP not optimised, I raw sludge dryer
SWW	1.6	46720	10000	21%	10000	0.5	4.1	12%	raw sludge to lime - low cost
Total	**53.1**	**1608160**	**1024000**	**64%**	**1070000**	**62**	**139.9**	**44%**	**of total sludge**
	theoretical MWs								
		tonnes VS		**VS conversion**		**average digester loading**			
		1206120		**45%**		**1.97**	**Kg VS/m3/day**		

Now compare this with the situation that is planned for the end of the investment period in 2019 in Table 3 below.

In 10 years there has been a paradigm shift in the way the water industry treats sludge.

Table3. Estimate of CHP from AD in English water companies 2019

Planned position in 2019									
	pop M	TDA raw	digested	% digested	M3 capacity	MWs gen	Theoretical	% generated from biogas	Comments
Southern	4.6	133671	130000	97%	78000	12.9	13.4	96%	taking advantage of green credits, stopping drying
Thames	13.5	394936	390000	99%	250000	38.5	39.7	97%	diverting sludge from incineration to digestion
Anglian	6.1	179240	167000	93%	105000	17	18.0	94%	closing all raw sludge operations and maximising digestion
Yorkshire	5.0	145822	140000	96%	89000	13	14.7	89%	moving away from incineration over the next 5 years
NWL	2.7	78987	75000	95%	50000	8	7.9	101%	stopped raw sludge drying in favour of THP, digestion CHP
UU	7.3	212658	210000	99%	200000	21.0	21.4	98%	new incinerator abandoned in favour of THP, digestion CHP
DCWW	3.1	91139	80000	88%	60000	8	9.2	87%	completing regionalsiation of sludge centres withTHP/digestion CHP
ST Water	8.6	312120	312000	100%	315000	28	31.4	89%	Closed all Incinerators, adding THP to some plants and gas to grid
Wessex	2.6	75949	70000	92%	40000	7	7.6	92%	some optimisation of digestion CHP
SWW	1.7	48607	40000	82%	20000	1.5	4.9	31%	no change
Total	**55.24524**	**1673130**	**1614000**	**96%**	**1207000**	**155**	**168.1**	**92%**	**of total sludge**
		average digester loading			113%	249%	**92%**	**96%**	**of digested sludge**
VSR	**52%**	**2.75**	**Kg VS/m3/day**						

The following can be seen:

- total generation has increased from 64 MWs to 155 MWs – an increase of 249%
- but digestion capacity has only risen by 113%
- 44% of sludge was digested and biogas used in 2005 – this will rise to 92% when all current projects are complete in 2019
- Organic loading has increased from 1.97 kgVS/m3/day to 2.75 kgVS/M3/day – a 40% increase
- Comments show that there has been a wholesale move away from thermal drying, incineration and lime stabilisation that are all energy intensive and have high carbon footprints

The water industry has sweated its assets by increasing VS loading, VSR% and adding CHP to sludge centres so that MWs generated have increased by 249% while digestion stock has only increased 113%. Typical power yields for thermal hydrolysis are between 0.62 and 1.01 MWh electrical per dry metric dry ton [1,2], showing a wide variation due to feedstock differences, biogas bypass and CHP availability.

Nearly all the digested sludge is going to farmland as biosolids with a large increase in Class A and high DS% cake[2,3] with a typical content of 28 to 34% dry solids for thermally hydrolysed sludge. The result of this is that all the thermal drying plants in UK have been shut down or are about to be – though 3 still operate in Scotland. Also, nearly all the incineration plants have been shut down and the remaining few are under review. This is a triumph for beautiful biosolids!

Conclusions

There has been a paradigm shift in the UK away from energy intensive/ high carbon footprint process such as incineration, thermal drying and lime stabilisation to energy producing processes using advanced digestion. This has been achieved by sweating existing digestion assets mainly and making sludge centres. The overall cost of sludge management has gone down with the sludge as a fertiliser and the sludge centre as an energy factory.

References

1. Merry J, Oliver B, Proc of 19th European Biosolids Conference, Nov 2014, Manchester UK
2. Panter K, Fountain P, Shana A, Ringoot, D, Proc 21st European Biosolids Conference Nov 2016 Edinburgh UK
3. Panter K, Proc WEFTEC 2009, session 91, October 2009, Orlando, Fla

Innovative Biosolids Conditioning Ahead of Belt Filter Presses to Improve Cake Solids Dewatering

John H. Rickermann, PE[1*], Charles E. Volk, PE[2]

[1*] CH2M (operator, engineer), Wethersfield, CT, email: john.rickermann@ch2m.com

[2] Lehigh County Authority (owner), Allentown, PA

ABSTRACT

Solids handling costs are among the largest at a typical wastewater treatment plant, so methods to optimize these expenses are of significant interest. In particular, any approach or system that can improve cake dewatering without replacing existing dewatering equipment or dramatically increasing chemical feed rates is worth studying. A full-scale installation in Lehigh, Pennsylvania of an innovative technology to condition liquid biosolids ahead of belt filter press dewatering, using compressed air, offers one such approach. The results offer a direct comparison of dewatering performance on identical presses running side by side on the same material, with and without conditioning. Detailed changes in operating expenses were tracked to determine that the system offers an attractive payback period of the installed capital costs for the conditioning system, and operator attention and maintenance requirements are minimal.

KEYWORDS: Biosolids conditioning, dewatering, belt filter press, polymer, optimization, anaerobic digestion, operability

INTRODUCTION

Dewatering costs represent one of the largest operations and maintenance (O&M) expenses for a wastewater treatment plant, and can represent as much as 35% of the total O&M budget, depending on the processes used (WEF, 1997). For wastewater plants that need to dewater their biosolids for storage, transportation, land application or for further processing, drier cake generally means less cost. For example, hauling of dewatered cake on a $/wet ton basis leads cost conscious plant managers to reduce the amount of water content in their biosolids.

Optimizing the existing dewatering process is typically the first step. Operators, engineers and vendors can focus on polymer preparation, dosage, injection and mixing locations, along with dewatering equipment adjustments like centrifuge pond depth or belt filter press belt speeds. If the equipment cannot be further optimized, or it is beyond its useful life (and cannot be refurbished), then the usual second step is capital replacement either in kind or with a more effective dewatering technology. Capital replacements or upgrades may be the best option in many cases, depending on the objectives of the utility, but are of course expensive undertakings.

In the case of the Lehigh County Authority Wastewater Pretreatment Plant (WWPTP) in Allentown, Pennsylvania, a third option was needed. Optimization steps had been exhausted, but the dewatered cake solids content was still too low at 15.3% total solids (TS) on average. Contracted hauling and land applications costs were very competitive, but based on wet tons hauled and still added up to be a major O&M expense. Also, the low cake solids was a risk factor during winter storage (when land application was not possible) if blending with drier materials could not prevent runoff, which would force landfill disposal at three times the land application cost. The four (4) existing belt filter presses (Figure 1) were otherwise performing adequately or being refurbished, and onsite trials and studies of other dewatering methods (screw press, centrifuge, new belt filter presses) suggested the capital cost to upgrade the system, or even single units, was not feasible.

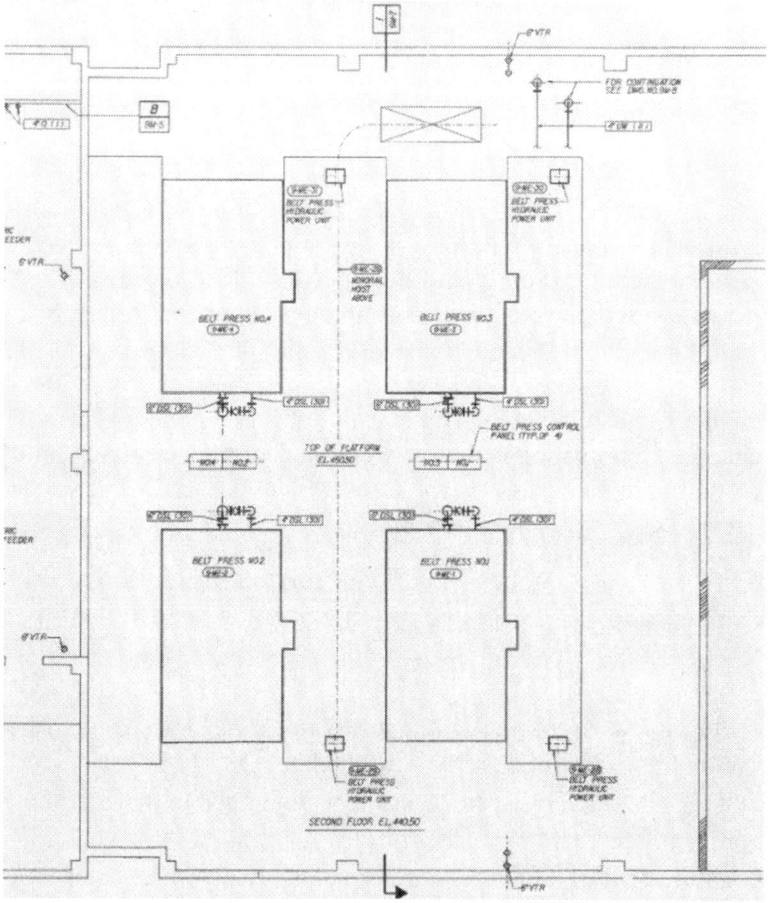

Figure 1. Existing Layout of Belt Filter Presses

METHODOLOGY

Optimization

The WWPTP receives a combination of high strength wastes from local breweries, food processors, some industries, and municipal sewage, which is then processed via preliminary treatment, primary clarifiers, a high purity oxygen activated sludge process, secondary clarifiers, and discharged back to the gravity sewer to the Allentown Kline Island WWTP. Solids are handled through gravity belt thickening of waste activated sludge, anaerobic digestion, belt filter press dewatering, and land application. The combined influent BOD loading averages 1,300 mg/L at an average daily flow of 17,000 m^3/d (4.5 mgd), but varies widely.

CH2M has been the contract operator of this plant for 23 years. The dewaterability of the anaerobically digested solids had fallen in recent years to between 14-16% TS on the 2-meter belt filter presses, perhaps as a result of changes to the influent characteristics to the WWPTP and the higher fraction of brewery waste. Digestion was stable and resulted in excellent volatile solids destruction, and loading rates were modest at 379 liters per minute/meter (100 gpm/m) or less, at 2-2.5% TS feed solids, with one of the four presses in standby. Capture rates were 97% or better, with an average filtrate concentration of 207 mg/L. Dewatering was only run during normal day shift hours to ensure adequate operator attention, and the presses were kept in good repair.

Despite these O&M practices, the polymer dosage rate was an extraordinary 30 g/kg (60 lb/DT), which was at least three times the typical dosage rate (M&E, Table 14-40) for belt filter press dewatering of anaerobically digested solids following an activated sludge process with primary clarifiers. While this high dosage had been the case for many years, the decline in cake solids content was a recent change.

The WWPTP operations staff had been using a local polymer representative for a major national supplier for years, and had gone through dozens of different polymers, both emulsion and dry (since the plant has both preparation systems available) in annual jar tests and full-scale tests. At the start of this particular optimization cycle, the vendor conducted a rigorous jar testing analysis to determine the optimum dosage, and also verified the blending and aging system was appropriate. But no improvements could be identified.

Following this work, the plant staff brought in a consulting chemical engineer with particular expertise in polymer selection to conduct an independent analysis of the polymer dosage, with a more system-wide focus that included belt speeds, feed points for the polymer ahead of the belt presses, and types of polymers. Some attention was also given to the performance of the anaerobic digesters in an additional study, including computational fluid dynamics models of the mixing regime within the digesters that had recently been upgraded. No appreciable difference in performance was achieved at the end of these efforts either.

Capital Upgrades

During the optimization activities, other dewatering technologies were being evaluated as a means of improving performance. High solids centrifuges and screw presses were considered, including onsite demonstration testing (typically 1-week trials with trailer test machines). In each case, the dewatered cake performance could be improved to within the target solids concentration (18-20% TS), but some other tradeoff was required: poor filtrate concentrations, higher operating costs, significant capital costs for modifications to the existing solids handling building and new equipment, and what appeared to be an ongoing need for very high polymer dosages. After studying these options through a business case evaluation, the capital costs could not be justified to replace the existing belt filter presses.

Digested solids samples were also sent for bench-scale study to belt filter press vendors for dewaterability analysis, and the results suggested that belt filter press dewatering could perform better than the current results indicated, by 1-2% TS. It was decided that perhaps the dewatering and polymer consumption performance could be improved with a thorough equipment refurbishment. As a result, one of the belt filter presses was taken out of service and completely rebuilt by the manufacturer and a local mechanical contractor, with new rollers, belts, drives, etc. In a subsequent side by side test using identical operating parameters (feed rates, belt speeds, polymer preparation and dosages, etc), and at the same time, the performance of the refurbished press and any of the three remaining presses was found to still be the same.

Despite significant effort and attention by plant staff, consultants and vendors, the dewatering performance was essentially unchanged and well below the 18-20% TS target, and the polymer consumption remained unreasonably high.

Biosolids Conditioning Concept

An innovative technology by Orége (pronounced *or-EDGE*) was encountered that suggested a new approach: in-line biosolids conditioning installed ahead of the dewatering equipment, to boost dewatering performance and reduce polymer consumption, without changing the dewatering equipment or feed pumps. The company was headquartered in Toussus-le-Noble, France, but had recently opened an office in the United States (400 Plasters Avenue NE #100, Atlanta, Georgia, 30324).

The process concept is as follows: liquid biosolids are pumped to the SLG™ unit, where a short detention time pressure vessel is used to inject pressurized air at up to 4 bar (60 psi). The operating principle is for the pressurized air to restructure and bind with solids and water in a homogeneous blend. Then, the air-injected solids are released from the pressure vessel through a control valve, which returns the mixture to atmospheric pressure. When the bound air is de-pressurized, it tends to separate the water and solids particles to which it had been bound, creating more free water for rapid dewatering, especially in the gravity zone of the belt filter press. Figure 2 shows a close-up view of the key components of the SLG™ unit.

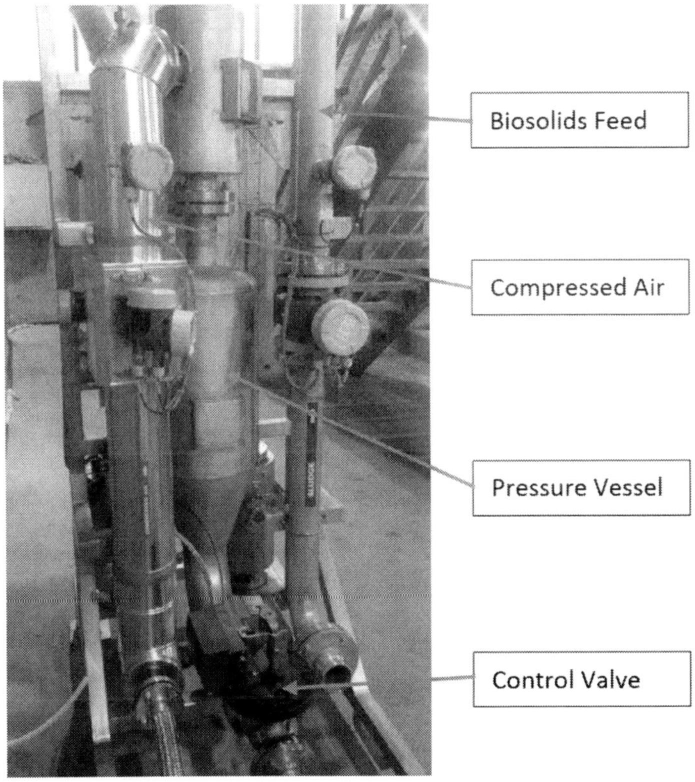

Figure 2 Biosolids Conditioning System

Some additional research was conducted by CH2M to determine if a competing technology was available using a similar concept, but none was identified. This appeared to be a unique approach.

The O&M cost impact of adding this system was considered, but it appeared to be minimal, with a 17 kW (25 hp) positive displacement blower running during press operations in additional power requirements, and almost no operator attention or maintenance requirements.

SLG™ Temporary Installation

Orége was contacted, and provided with operating data and optimization records to determine how viable their conditioning system would be for the WWPTP situation. Table 1 shows the belt filter press performance for just over a one-year period of the optimization and refurbishment activities. Several installations were in service in Europe, but none had yet been installed in North America. For this demonstration, Orége was being selective about where they would devote their resources in bringing this technology to North America, and several competing locations were under consideration for a demonstration.

Table 1 Baseline Dewatering Performance Data Prior to Biosolids Conditioning Trial

Month	BFP Cake %TS	BFP Filtrate TSS, mg/L	Digested VS %	Influent Flow, m³/d (MGD)	Influent BOD, mg/L	Influent TSS, mg/L
Aug-14	16.1%	163	72.8%	17148(4.53)	1305	784
Sep-14	16.5%	217	70.7%	15293(4.04)	1347	690
Oct-14	16.2%	259	72.6%	14801(3.91)	1257	457
Nov-14	15.3%	238	71.5%	14006(3.7)	1248	385
Dec-14	15.7%	187	72.0%	15180(4.01)	1338	356
Jan-15	15.3%	165	69.5%	14233(3.76)	1331	415
Feb-15	14.7%	209	70.7%	14915(3.94)	1478	387
Mar-15	14.1%	166	70.2%	17072(4.51)	1187	341
Apr-15	14.1%	143	71.8%	16807(4.44)	1219	415
May-15	14.6%	201	69.9%	16959(4.48)	1330	588
Jun-15	15.1%	259	73.5%	17943(4.74)	1082	447
Jul-15	15.9%	294	74.3%	17716(4.68)	1093	361
Aug-15	15.8%	195	73.7%	16618(4.39)	1258	630
Average	**15.3%**	**207**	**71.8%**	16050(4.24)	**1267**	**481**

The initial evaluation by Orége indicated a likely improvement could be achieved in dewatering performance, so a team of engineers from France came to the WWPTP and conducted a detailed evaluation of flows and loads, belt filter press operating practices, polymer usage, and historical performance. Several additional polymers were bench tested which would complement the SLG™ conditioning system.

An agreement was reached with Orége, CH2M, and Lehigh County Authority to temporarily install a full-scale conditioning system ahead of one of the four belt filter presses, and if the system could meet specified performance metrics, it would be purchased and modified into a permanent installation. Failure to meet the performance metrics would result in removal of the system at no cost to the owner or operator. The agreement stipulated a four-week test window, which was considered necessary in order to verify performance given the highly variable influent loading at the plant.

A combination of performance improvements were set as the goal: either a 5% TS cake dryness improvement alone, or a blend of improved cake dryness and polymer consumption reductions that would create enough annual savings from the baseline to pay for the unit within 4 years. The cost of additional power consumption and treating recirculated filtrate was considered in the financial analysis as well.

The system was installed using a local mechanical contractor who was already very familiar with the WWPTP, under the direction of Orége. The equipment consisted of the SLG™ unit (model SLG2-30) on a skid, a small positive displacement blower and control panel on a skid for compressed air, and a small polymer makeup unit. A PLC control interface with the existing belt filter press feed system was also required to manage pumping.

Temporary piping connections were made using flexible hoses (Figure 3), and power was routed from a spare location in the motor control center for the solids handling building (for safety reasons, this part of the installation was permanent and to code).

Figure 3. Temporary Biosolids Conditioning Installation

Orége also chose to provide a team of full time, onsite startup specialists to ensure the installation was done to their satisfaction, and to manage the biosolids conditioning system operation during the four-week performance testing period.

The system was started and optimized over a two-week period ahead of the performance testing to verify polymer selection, loading rates, and other operating parameters for the system. Data was collected and analyzed by both WWPTP staff and Orége startup experts.

RESULTS

After four weeks of performance testing, with sampling by both WWPTP and Orége teams, the results met the payback performance objective through a 3-3.5%TS cake dryness improvement (up to 18.5% TS), and a 25% reduction in polymer consumption (down to 23 g/kg or 45 lb/DT). Capture performance also improved to over 99%. This was in direct comparison to the other belt filter presses which continued to operate with the same digested biosolids, on the same operating schedule. At the unit costs for polymer, land application, and filtrate retreatment, the capital cost of the Orége system would be recovered in 3.1 years, which was well within the performance objective.

Table 2 presents the performance test results under different polymer loading scenarios. These results are the average measurements from both Orége and WWPTP staff, which were consistently in close agreement.

Table 2. Performance Testing Results

Parameter	With Inline Biosolids Conditioning			Without Conditioning
Polymer Dosage Rate, g/kg (lb/DT)	22-24 / (45-48)	25-26 / (49-52)	27 / (55)	30 (60)
Average Cake Solids, % TS	18.55 +/- 0.1	18.95 +/- 0.1	20.3 +/- 0.1	15.3
Average Filtrate Capture, %	99.3 +/- 0.2	99.3 +/- 0.2	99.3 +/- 0.2	97

As a result of this strong performance, CH2M purchased the SLG™ system as agreed, and Orége had their contractor relocate the unit into a permanent location (under the belt filter press platform to avoid safety hazards with the piping to and from the SLG™). This is shown in Figure 4.

Figure 4. Relocated Conditioning System Under Press Platform

It was observed that the gravity zone of the belt filter press plays a much greater role in dewatering performance when biosolids conditioning is added. The visible free water drains much more quickly as liquid biosolids is applied to the gravity zone, and takes on a muddy characteristic by the first or second chicane (within 1m of application). This degree of drainage is not seen on the unmodified dewatering systems until near the end of the gravity zone, which indicates that the conditioning system is creating better free water separation.

As a further improvement to the permanent installation, Orége also made changes to the small polymer blending system by adding an aging tank to improve polymer chain development before injection, since the existing polymer preparation system for the existing belt filter presses could not be divided between upgraded and current systems, and the small demonstration scale system had limited capabilities. Finally, some adjustments were made to the system control panel PLCs and programming. Orége also provided training and documentation to WWPTP staff before turning over operations.

Since August 2016, the operating history shows that the biosolids conditioning system continues to perform as intended, although changes in the influent loading to the WWPTP continues to impact performance, as shown in Figure 5.

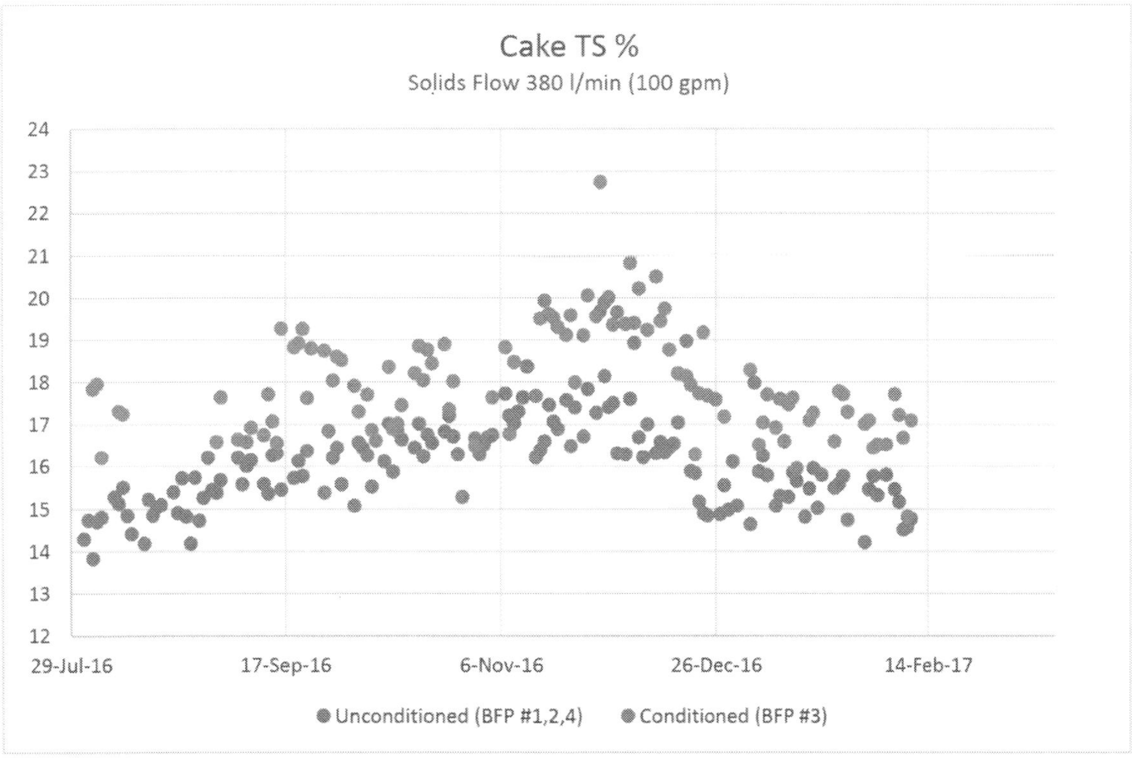

Figure 5 Conditioned Dewatered Cake Performance Since August 2016

The belt filter press with biosolids conditioning is favored for operations given the improved performance, while one or two of the unmodified presses are left in standby. At 380 l/min (100 gpm) feed rate to the belt filter press, the current polymer dosage remains 24 g/kg (47.5 lb/DT) using Clarifloc C-6296 (45% active). The SLG™ reactor pressure is set at 1.58 Bar (23 psi), with an air/solids ratio of 3.

WWPTP operations staff have been able to run the biosolids conditioning system without difficulty, and the control interface has been intuitive and simple to manage. It runs in automatic and is "care free" according to the operations supervisor. No significant maintenance issues have occurred – the 304 stainless construction tolerates the challenging atmosphere within the solids handling building (hydrogen sulfide, ammonia, and humidity), and none of the individual components have shown signs of failure or premature wear, other than a faulty pressure relief valve that was replaced without incident.

DISCUSSION

As a result of how this demonstration project occurred, Orége has discontinued the practice of making temporary installations of their biosolids conditioning system for trials. There was no cost saving by making a temporary installation, since the additional cost for a permanent install was negligible. Demonstration projects to prove performance before purchase will instead move directly to permanent installation, since Orége considers the chances of not meeting performance targets very small.

An unexpected but welcome finding for this project was the improvement in polymer consumption. Other Orége installations in Europe suggested a 10% reduction in consumption for municipal treatment systems, but the nature of the mostly food processing and brewery waste appears to be such that the compressed air conditioning makes the polymer more effective.

The inability of the existing polymer preparation system to accommodate different polymer types for each belt filter press presented an interesting challenge, which was eventually resolved with the installation of a stand-alone system just for the press with biosolids conditioning. As a test, the new polymer selected by Orége for use with their system was run on one of the un-conditioned belt filter presses, in case this particular polymer had been missed previously as a better coagulant aid. If successful, a separate polymer system would not be required. Unfortunately, the new polymer did not perform as well, and the aggregate reduction in performance by using it for all operating presses would have negated the savings from using it with the biosolids conditioning system. But this test served to prove that the performance boost was from the conditioning system, and not the change in polymer.

CONCLUSIONS

Side by side tests like this of innovative solutions are particularly helpful in showing the benefits and challenges of a new technology. After months of testing and operation, the performance of this biosolids conditioning system appears to be stable and sustainable, with clear annual cost savings at the WWPTP. Other suitable locations are currently being investigated by the CH2M contract operations division for application of this technology, especially where biosolids management costs are particularly high, and replacement of the existing dewatering equipment is not a short-term option.

REFERENCES

Water Environment Federation (1997) *Benchmarking Wastewater Operations - Collection, Treatment, and Biosolids Management,* Project 96-CTS-5; Washington, D.C.

Metcalf & Eddy (2003) Wastewater Engineering: Treatment and Reuse, 4th ed.; McGraw Hill: New York, New York.

A Laboratory Based Method for Predicting Dewaterability

Zwelani Ngwenya[1], Matthew J. Higgins[1], Steven Beightol[1], Sudhir N. Murthy[2],
[1]Bucknell University, Lewisburg, PA; [2]DC Water, Washington, DC

ABSTRACT

The goal of this research was to develop a lab-scale dewatering method that estimates dewatering characteristics such as polymer demand, cake solids and capture. The method replaces traditional centrifuge tubes with belt filter cups which are fitted with circular belt filter press fabric. Digestate is conditioned and then portions of the conditioned sludge is placed in the centrifuge cups and centrifuged at a set g-force for a defined time. The essential parameters defining the method were: (1) mixing the cake during centrifugation, (2) the time of centrifugation, and, (3) dry solids loading onto the belt filter cups. Mixing the cake during the centrifugation increased the final cake solids as did increasing the centrifugation time. However, cake solids decreases as solids loading increases. Using appropriate parameters, the method could accurately predict full scale belt filter press cake solids with only 0.15% difference relative to full scale. Overall, the method is well suited for use in the laboratory because it demands less equipment and also produces cake which can be further tested for other dewatering parameters such as odors.

Key Words: Dewaterability; centrifuge; cake solids; conditioning; sludge; CST; OPD.

INTRODUCTION

Dewatering is employed by utilities to reduce the water content of sludges and digestate. Dewatering increases dry solids which in turn helps minimize costs associated with solids management. In addition to digestate volume reduction, the dewatering process also affects the nutrient and odor levels of solids. The need for assessment of dewatering parameters such as cake solids, nutrients, polymer demand and cake odors on digestate produced from pilot scale or lab reactors has led to the development of various laboratory scale dewatering methods. The two most popular dewatering methods used in large wastewater treatment plants are the centrifuge and belt filter press. Numerous lab-scale methods have been developed to simulate performance of these full scale devices in terms of cake solids, polymer demand predictions and other dewatering parameters. Because of the small volume of digestate produced in pilot and lab scale studies, lab scale dewatering methods that produce cake using small volumes of digestate samples, are needed to help characterize digestate dewaterability at a small scale. Furthermore, researchers also make use of lab-scale methods of dewatering to compare the dewaterability of different sludges. One of the most difficult and challenging aspects of sludge dewatering at a lab scale is the absence of a universally accepted dewatering method that is easy to use, requires less equipment and can estimate cake solids while also simulating full scale dewatering methods.

While previous studies have developed different lab-scale dewatering methods that simulate full-scale mechanical dewatering devices, very few have used a relatively simple and accurate apparatus that's capable of simulating full scale methods. Previous attempts at developing lab scale dewatering methods have resulted in two distinct approaches. One group comprise methods that measure the rate of dewatering. Such surrogate methods of dewatering are the capillary suction time (CST), specific resistance to filtration (SRF) and the drying curve method. In many cases, researchers are interested in optimizing cake solids, hence methods that predict the extent of dewatering are usually used as substitute for or a complement to the surrogates of dewatering methods. Physical dewatering methods that produce cake and predict the extent of dewatering have been developed by previous researchers. These methods include, a bench scale centrifuge using centrifuge tubes, piston press and a crown press.

Dewatering surrogates

One of the earliest surrogates of dewatering previously used is the SRF, which was developed by Coakley et al., (1956). The SRF measures the permeability of a sludge layer deposited on a filter medium to which a vacuum has been applied. The method characterizes the sludge based on its filtration rate (Vesilind, 1988). A drawback of this technique is that it produces low cake solids which may not be representative of the actual extent of dewaterability of the digestate. Furthermore, while the method gives an indication of how well a sludge dewaters, it provides no information regarding solids capture efficiency.

As a response to the difficulty of conducting the SRF test, Gale and Bakersville (1968) developed the CST apparatus. The CST test measures the rate of filtration and gives information on how well a sludge dewaters. Digestate that release their water slowly have high CST values while sludges that readily release their water have low CST values (Vesilind, 1988). The CST device was deemed to be a simple, quick and inexpensive means for measuring the dewaterability potential of wastewater sludges (Vesilind, 1988). The technique has been used in wastewater and water treatment facilities to evaluate polymer demand as well as to detect changes in sludge dewaterability. Unfortunately, despite its widespread acceptance as a useful general measure of sludge "dewaterability", the CST does not predict the extent of dewaterability, namely cake solids. Instead it measures the rate of water removal. In addition, it does not result in cake for solids analysis and cannot predict capture efficiency.

Kopp & Dichtl, (2000) developed the drying rate curve method to predict digestate dewaterability. This method assumes that, of the three types of water found in sludges namely, free water, interstitial water and bound water, only the free water can be removed by mechanical dewatering means. The authors used thermogravimetric measurements to dry a sludge sample at constant temperature and airflow and produced a drying rate curve in which drying rate was plotted against moisture content, measured as mass of water per mass of suspended solids of the sludge. Using an arithmetic scale, the drying rate curve allows for the determination of the different types of water found in sludge. The curve is linear at the start of drying and then after a certain moisture content, it becomes nonlinear. The authors propose that this point of inflection marks the end of the free water content in the sludge. After determining the point of inflection and developing the equation of the tangent, the moisture content of the sample at the point of inflection is used to calculate the solids content of the sludge. The authors argue that this is the maximum cake solids achievable since the remaining bound and interstitial water cannot be separated from the sludge by mechanical dewatering. The authors showed that the method accurately predicted full cake belt filter press and centrifuge percent cake solids. While the method predicts percent cake solids accurately, it does not predict polymer requirements or capture efficiency and does not produce a cake that can be evaluated for odors or other parameters.

Physical Dewatering Methods

Using a bench scale centrifuge with centrifuge tubes has been a common method to approximate full scale dewatering process of the centrifuge. This method involves centrifuging the solids to relatively high g-force and then decanting the resultant supernatant. The cake solids are then determined by measuring the solids content of the resultant pellet in the centrifuge tube (Vesilind, 1970). While this method is quick, easy and results in cake solids for further testing, it often leads to an under prediction of cake solids as observed during preliminary phases of this work at Bucknell University. The thin water film that remains at the pellet/supernatant interface after decanting often leads to low cake solids prediction.

Baskerville et al. (1978) used a piston press as a laboratory scale approximation of the full scale belt filter press. The apparatus consists of a cylindrical filtration cell and a freely moving pistol. The application of pressure to the non- porous piston causes the filter cake in the cell to dewater. In the simulation of full scale belt press pressures Baskerville et al. (1978) found out that pressures of 700kPa are sufficient and can be achieved using a compressed-air cylinder and pressure regulator. The resultant cake solids consistencies from the piston press were comparable to full scale BFP results.

Severin and Collins (1992) used the Crown Press to simulate belt filter press (BFP) dewatering. They reported that the Crown Press produced cake solids similar to those produced on BFP.

However, Galla et. al. (1996) reported that when improper belt tensions are used in pressing regime calculations, the crown press does not accurately simulate the operation of a BFP. The process of ensuring that proper belt tensions are applied makes the method cumbersome. Emery (1994) reported that the Crown Press was able to simulate the wedge zone and high-pressure zone of the BFP but on all Crown Press simulations, severe belt blinding was observed. Belt blinding prevents further drainage of water from the sludge resulting in low cake solids predictions.

While previous studies have developed methods to simulate full scale mechanical dewatering devices, very few have used a relatively simple apparatus that's capable of producing reliable estimates of cake solids, polymer demand and capture efficiency. A reproducible laboratory dewatering method is needed to evaluate mechanisms of dewatering, and the effects of different upstream processes on dewatering at the lab scale and full-scale systems. In many cases, researchers use surrogates of dewatering such as CST or time to filter. Although these methods give an indication of dewatering rate, they do not provide information on the extent of dewatering. The development of a reliable method that is easy to use and can estimate cake solids while also simulating full scale dewatering methods would be of great value.

The goal of this research was to develop a laboratory method that:

(1) uses relatively small volumes of samples;
(2) is predictive of polymer demand for conditioning;
(3) is predictive of cake solids for a given full-scale process such as BFP;
(4) is reproducible;
(5) produces a cake that could be evaluated for parameters such as odors; and,
(6) is easy to implement.

MATERIALS AND METHODS

Experimental Approach

First, the method was developed using anaerobic digestate collected as effluent from pilot scale anaerobic digesters operated at Bucknell University. The reactors were part of a separate study and were operated under mesophilic temperatures with retention times of 10 and 15 days, respectively.

In order to test and develop the method, anaerobic digestate was obtained from three different wastewater treatment plants (WWTPs) in Pennsylvania and Washington DC. Conventional and thermally hydrolyzed sludge were used. These samples were used to evaluate different parameters in the conditioning and dewatering test as described in the overview of the method section.

Overview of the method

Figure 1 summarizes the major steps of the dewatering method. Step 1 involves conditioning using a cationic polymer. The sample is placed in baffled cylindrical container and then mixed with the polymer using a variable speed mixer. Step 2 is free draining the conditioned sludge. Sludge is drained for a minute and the mass of the wet solids measured. In step 3, wet solids are transferred to belt filter cups and dewatered in bench scale centrifuge. Finally, after centrifugation, cake is placed in a 105 °C oven and total solids are determined.

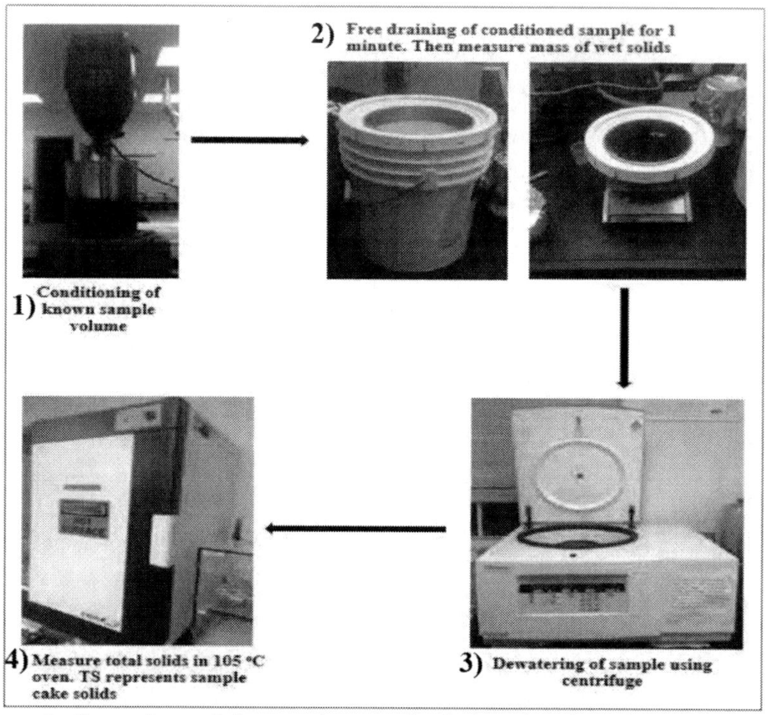

Figure 1: Overview of dewatering method showing key dewatering steps

Conditioning: A high molecular weight cationic polyacrylamide (SNF FLOPAM FO 4440) was used for conditioning. The powdered polymer was made up to a concentration of 0.5% active content. The solution was mixed at 400 rpm for 30 minutes. The prepared polymer was then allowed to age for approximately 45 minutes. Next, increasing dosages of polymer were added to 500 mL samples of digestate. The polymer was mixed at a velocity gradient, G, of 500/s for 10 s, and then a G of 100/s for 90 s. The initial high speed mixing ensures that the polymer is well mixed with the sludge for optimal coagulation. The slow speed allows for complete mixing without excessively shearing the sludge. After mixing, the CST was measured, and the optimum dose was considered as the dose that gave the minimum CST. This dosage was then used for the dewatering step.

Free draining: The conditioned wet solids are free drained through a belt filter press fabric. A spatula is used to mix the wet solids to allow for faster drainage. The wet solids are mixed for a minute for complete draining. The mass of the free drainage wet solids is measured and serves as an input to the developed spreadsheet that calculates target mass of wet solids to be added to centrifuge cups (discussed in the following section) for dewatering.

Dewatering: The dewatering method used is a modification to the bench scale centrifuge test and is primarily targeted at eliminating the thin layer of water that forms on the surface of the dewatered cake when using the traditional centrifuge tubes. Since this often leads to an under prediction of the percent cake solids, the new method was designed to keep the dewatered cake separate from the centrate. The traditional centrifuge tubes were replaced with centrifuge cups as shown in Figure 2, which had a piece of belt filter press fabric supported about half way up the cup.

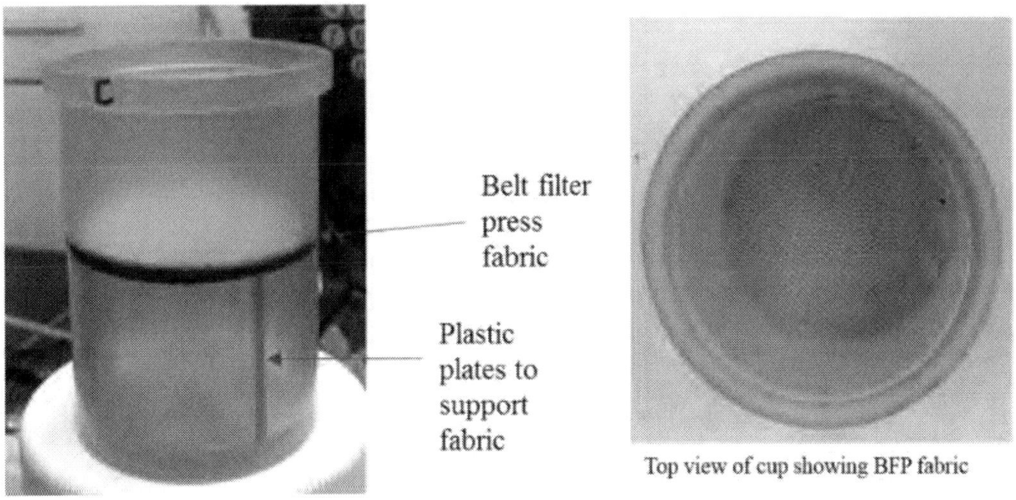

Figure 2 The centrifuge cups with the belt filter cloth fitted and supported. The dewatered cake forms on the filter cake while the centrate drains through the filter cloth to the bottom of the cup.

After free draining, a portion of the free drained solids is added to the cup and the mass of solids added to these cups depend on the total solids (TS) concentration to achieve a desired loading rate on the belt filter fabric. Table 1 shows the sample spreadsheet used to calculate mass of wet solids to be added to the centrifuge cups.

Table 1: Sample spreadsheet for the calculating of mass of wet solids to add in BFP cups

Volume of digestate (mL)	500		
TS (%) of sludge before dewatering	3.09		
Target dry solids (g)	Mass of dry solids in sample sludge[1] (g)	Mass of wet solids after Free draining (g)	Mass of wet solids to placce in BFP Cup[2] (g)
5	162	188.72	61
5	162	168.89	55
10	324	165.33	107
15	485	176.30	171
20	647	164.63	213

The mass of dry solids in sample sludge are calculated using equation 1:

$$^{1}\textit{Mass of dry solids in sample sludge} = \frac{Target\ dry\ solids\ (g) * 100}{TS\ (\%)} \tag{1}$$

Where: TS is sludge total solids prior to dewatering.

Equation 2 calculates the mass of wet solids to place in the BFP cup:

$$^{2}\textit{Mass of wet solids to place in BFP cup} = \left[\frac{Mass\ of\ dry\ solids\ in\ sample\ sludge\ (g)}{Volume\ of\ sample\ (mL)}\right] * Mass\ of\ wet\ solids\ after\ free\ draining\ (g) \tag{2}$$

The samples are then centrifuged at 140 x *g* for 2 minutes, and then at 350 x *g* for an additional 2 minutes to seat the cake on the belt fabric. Next, the cake is centrifuged at 2075 x *g* for a defined time. Centrifugation is stopped at different time intervals to mix the cake on the fabric using a spatula to redistribute the cake and allow better drainage. After centrifugation at 2075 x g for 10 minutes, the centrate drains to the bottom of the cup while the dewatered cake solids remain suspended on the filter cloth.

The different x *g* values were calculated using Equation 3 which accounts for the distance of cake from the centrifuge rotor center:

Converting rpm to times gravity (x *g*)

$$x\ g = 1.118 * 10^{-5} * R * S^2 \tag{3}$$

Where: R is the distance between the center of the centrifuge rotor and the surface of the cake in the centrifuge cup.

S is rotor speed in revolutions per minute (rpm)

Total Solids Determination: Total solids measurement was measured following Standard Methods 2540 B (APHA, 2012). The slight modification to the method was that solids were left in the oven for at least 24 hours to ensure that a constant weight was obtained. Total solids were measured twice for each dewatering cycle. TS were measured pre- and post-dewatering. First, digestate TS was calculated for use as input in the spreadsheet that calculates the mass of wet solids to be loaded on the belt filter cups, and secondly, it was measured on dewatered cake to determine the cake solids.

Cake Thickness: The thickness of the belt filter fabric was measured using a Vernier caliper. After centrifugation, the cake and the belt filter fabric was carefully removed from the cup and the thickness was measured. The actual cake thickness was calculated by subtracting belt filter fabric thickness from total thickness of belt filter fabric and the cake. Since the cake surface was not uniform, cake thickness was measured at four evenly spaced points around the cake. The average thickness was then calculated.

Parameters of Interest: The development of the method involved investigating the following impacts on cake solids and cake thickness:

1. Mixing and non-mixing during centrifugation
2. Number of times cake is mixed during centrifugation
3. Target dry solids loaded on the belt filer cup, in terms of kg dry solids per m^2 of belt fabric
4. Time of centrifugation at 2075 x g

Statistical Analysis: Microsoft excel was mainly used to generate graphs, tabulate data and perform basic statistics such as mean and standard deviation calculations. R software package was used to perform tests for: equal variances; one-way analysis of variance (ANOVA) to compare the mean values of the data sets for statistical differences at the 0.05 significance level.

RESULTS AND DISCUSSION

The development of this method was aimed at assessing the impact of four major parameters, namely;

1. Mixing and non-mixing during centrifugation
2. Number of times cake is mixed during centrifugation
3. Target dry solids loaded on the belt filer cup, in terms of kg dry solids per m^2 of belt fabric
4. Time of centrifugation at 2075 x g

Impact of cake mixing and loading rate

The first parameter studied was the impact of mixing the cake during the ten minute period of the 2075 x g force. To ensure that all other conditions remained the same during dewatering, the mixing and no mix centrifuge cups were placed in the centrifuge at the same time. The two cakes were subjected to the same temperature, time, pressure and force. Mixing was done once at the high speed centrifugation phase. The cake was mixed using a spatula at the 5th minute time interval of ten minutes. Mixing allows for the release of water that remains trapped in the cake as it compresses due to high speed centrifugation. A schematic presentation of the mixing sequence is shown in Figure 3.

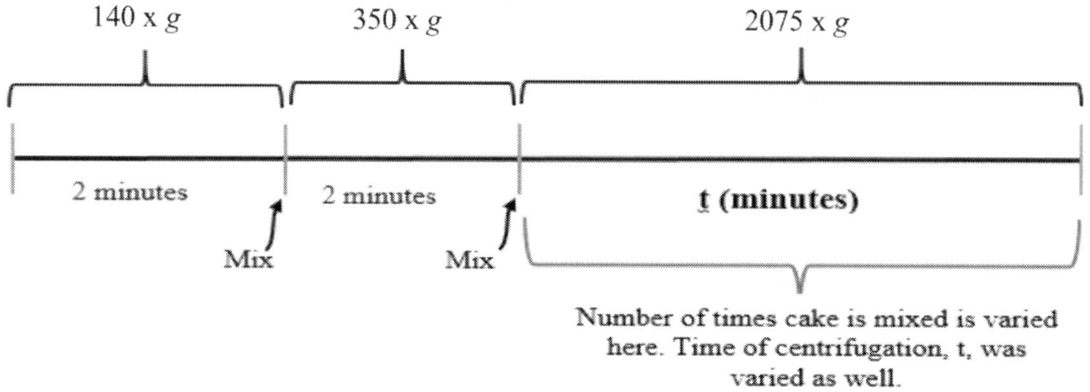

Figure 3: Overview of the mixing sequence of cake in the centrifuge. Number of times cake is mixed and the duration of centrifugation is varied only at the 2075 x g phase.

Percent cake solids were statistically different for cake mixing and no mixing, indicating that mixing does increase cake solids as shown in Figure 4.

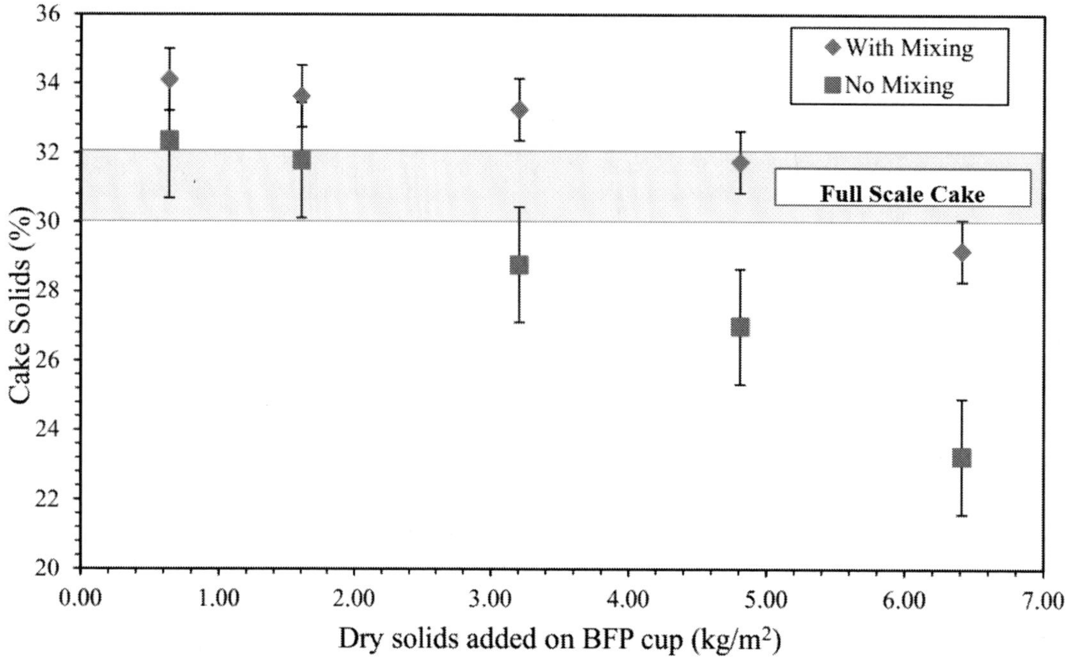

Figure 4: Effect of mixing, and increasing dry solids added onto the BFP cup on solids cake content using thermally hydrolyzed sludge from pilot digesters at Bucknell University. Error bars represent one standard deviation. Cake was mixed once at 2075 x g.

Visual inspection also showed that the cake without mixing developed a thin water layer after centrifugation at 2075 x g. The presence of this layer of water also explains the lower cake solids obtained for the no mix regime. Also higher solids loading are shown to lead to decreased percent cake solids as depicted in Figure 4. This is mostly due to the longer drainage path generated by increased thickness of cake. The cake thickness as a function of dry solids loading is shown in Figure 5.

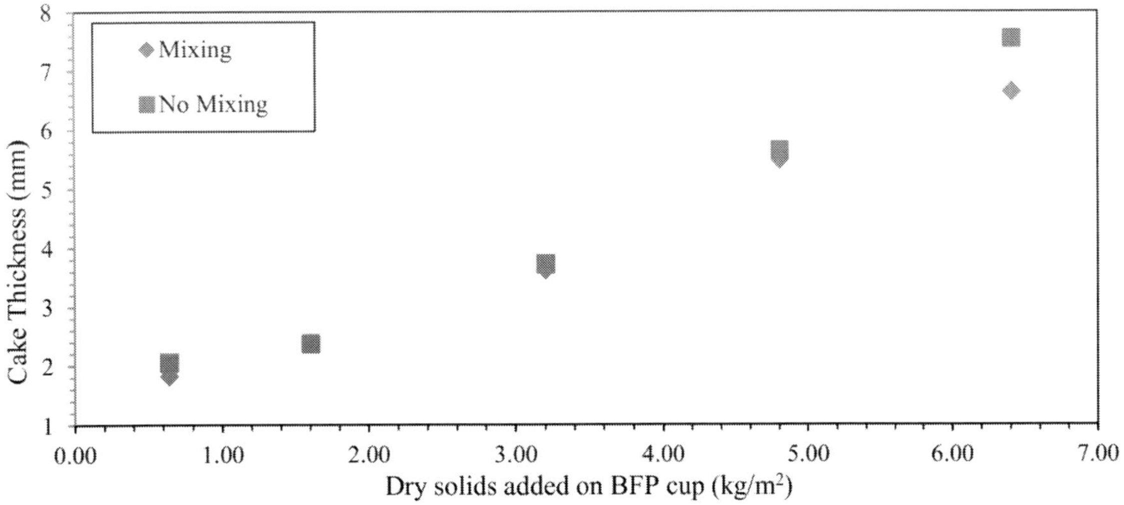

Figure 5: Effects of dry solids loading on cake thickness for thermally hydrolyzed sludge from pilot digesters at Bucknell University.

The typical range of the cake solids from the full-scale belt filter press is also shown on Figure 4, and this lab method does produce cake solids within this range depending on the loading and mixing conditions. Figure 4 also highlights the flexibility of the method. If a study is performed to simulate belt filter performance, then dewatering without mixing at low dry solids loading (between 0.6 and 2 kg/m^2) yields cake solids comparable to full scale belt filter press performance. On the other hand, if the goal of a dewatering study is to maximize cake solids, without any need to predict full scale performance, then dewatering with mixing yields higher cake solids at lower dry solids loading. To attain cake solids comparable to full scale belt filter performance using the mixing regime, high dry solids loading yield percent cake solids within the full scale range. Based on Figure 4, the appropriate solids loading to predict full scale cake solids is approximately 5 kg/m^2.

The non-mixing regime leads to lower cake solids at higher dry solids loading. The percent change between mixing and non-mixing cake solids increased linearly with increased target dry solids loading as shown in Figure 6.

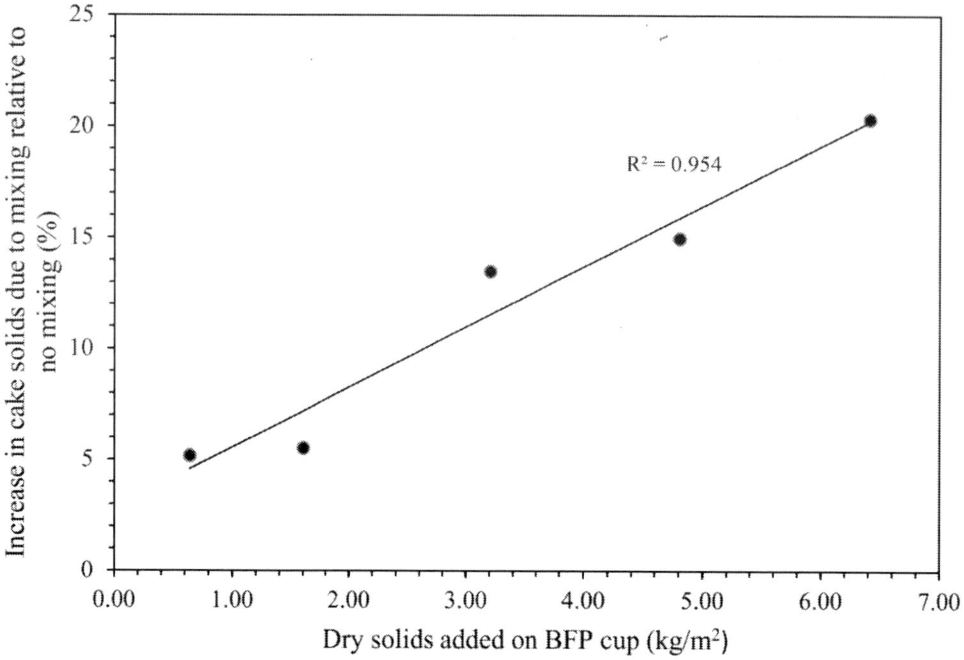

Figure 6: Percent increase in cake solids vs. dry solids loading for thermally hydrolyzed sludge obtained from pilot digesters at Bucknell University.

The effects of a longer drainage path become more pronounced at higher solids loading (between 5 and 7 kg/m^2) and therefore mixing of the cake has a greater impact.

Centrifuge Time and Cake Mixing

After demonstrating that incorporating mixing and using lower dry solids loading on the modified centrifugal method yields high percent cake solids; different times of centrifugation at high speed were tested. For this arrangement, the loading rate of 0.64 kg/m^2 and number of times the cake solids were mixed at 2075 x g was kept constant. The cake solids were mixed only once midway during the high g-force centrifugation time. The no mix regime was also employed for comparison.

The cake solids were increased with increased time of centrifugation at 2075 x g as shown in Figure 7.

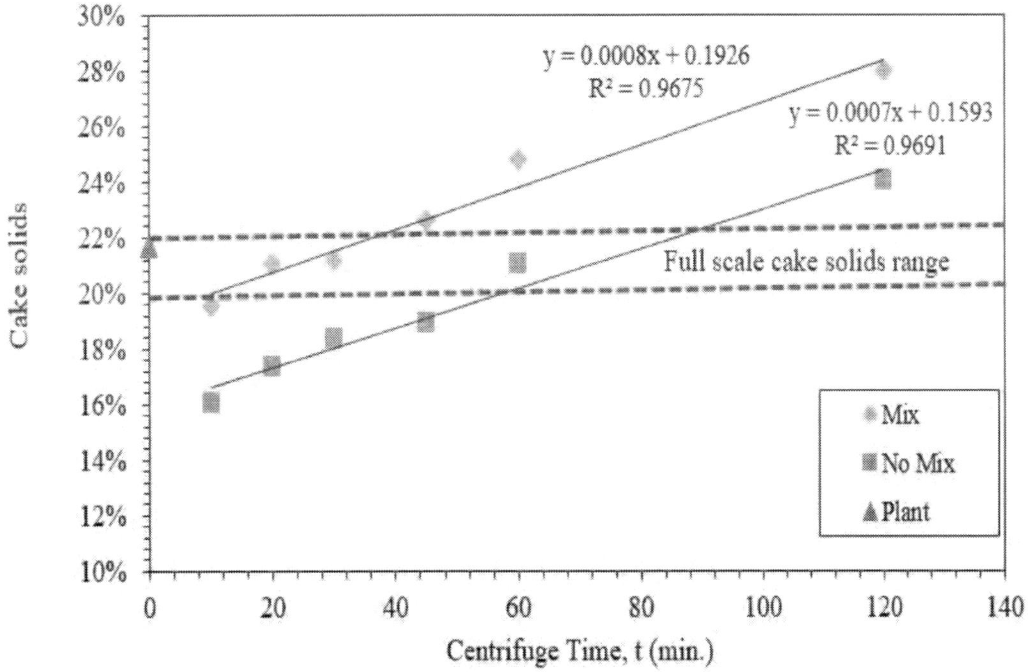

Figure 7: Effect of varying centrifuge time on cake solids at 2075 x g. Conventional sludge was used as sample.

Additionally, Figure 7 further supports the conclusion that mixing is a necessary step in order to attain higher percent cake solids. Embedded in Figure 7, is the full-scale belt filter press cake solids range for this conventional sample. With cake mixing, it takes less than 20 minutes to produce cake comparable to full scale dewatering. Without mixing, producing cake in the range of full scale cake solids would take approximately four times as long.

Frequency of Mixing

The impact of the frequency of mixing was investigated for the time interval of 20 minutes at a force of 2075 x g. The 20 minute time interval was selected because it allowed for a wider range of mixing frequency (0 to 5) with reasonable time spacing between mixing. Figure 8 shows that increasing the frequency of mixing leads to an increase in cake solids. Mixing frequently reduces the drainage path and exposes water trapped with the sludge. This facilitates maximum release of water leading to high cake solids. Additionally, mixing frequently reduces the amount of water trapped in the sludge as the cake compresses at high g force.

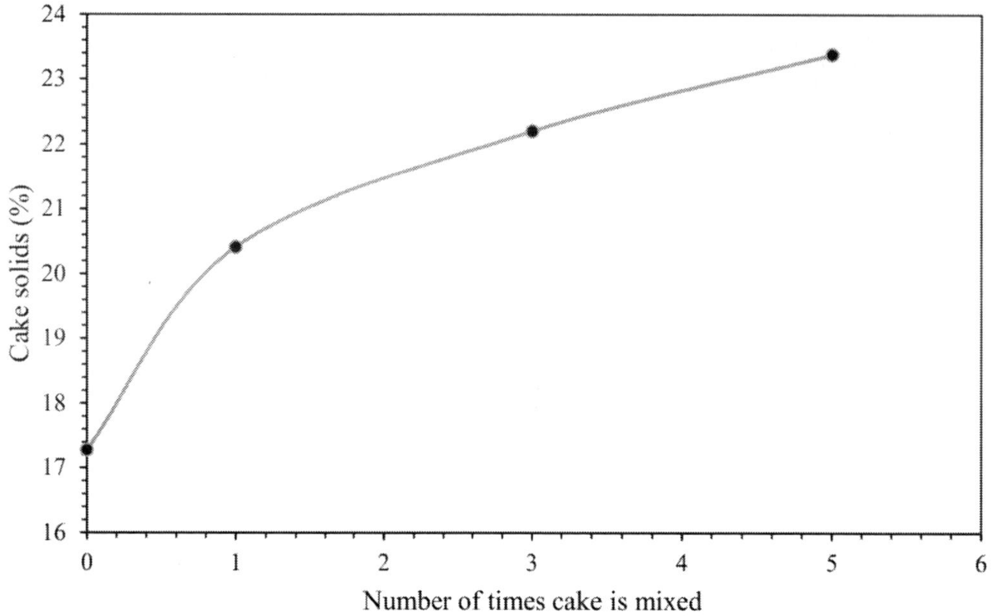

Figure 8: Impact of increasing the number of times cake is mixed at high 2075 x g for a 20 minute time interval

Mixing five times during the 20 minute time interval is rather tedious. In order to make the method shorter in terms of time, the lower end of the centrifugation time spectrum (from Figure 7) was further developed in order to produce cake comparable to full scale at minimum centrifugation time. To get high cake solids at lower centrifuge time, the frequency of mixing at 10 min and 20 min intervals was increased. Previously, the cake was mixed only once midway the high speed centrifugation time interval. The frequency of mixing was increased to 2 times within each high speed time interval. The times at which centrifugation was stopped to allow for mixing was evenly split during this time interval.

Table 2 shows that mixing twice at the 20 minute interval overestimates full scale cake solids by over 4 percent. In contrast, mixing twice at the 10 minute interval resulted in accurate predictions of full scale cake solids. On average, mixing at 10 minute interval only under predicted the cake solids by 0.36 percent and this difference is shown to be statistically insignificant at a 95% confidence level.

Table 2: Impact of mixing twice at 2075 x g for the 10-and 20-minute interval. Percent cake solids compared to full scale belt filter press results on conventional sludge

Time interval at 3000 x g centrifugation (minutes)	Number of times cake mixed	Cake solids (%)	Standard Deviation (%)	Percent difference relative to fullscale
10	2	19.47	0.38	-0.15
20	2	23.67	0.58	21.38
Full scale	-	19.5	-	-

The impact of mixing frequently at higher time interval such as 20 minutes could be used as the method's approximation of the centrifuge performance. Full scale comparison utilized in this study have drawn from plants using the belt filter press for dewatering. The 4 percent cake solids difference between the 20 minute- and 10 minute- two time mixing intervals is comparable to the full scale difference between the centrifuge and belt filter press cake solids, respectively. To assess this prediction, samples from plants with different dewatering devices such as belt filter press and centrifuge will be collected and tested in the lab as a way of applying the developed method.

Conditioning

The CST vs polymer dose profile for each sample dewatered was plotted to determine the optimum polymer dose (OPD). The OPD was considered to be the dose that resulted in the lowest CST as shown as circled data point in Figure 9.

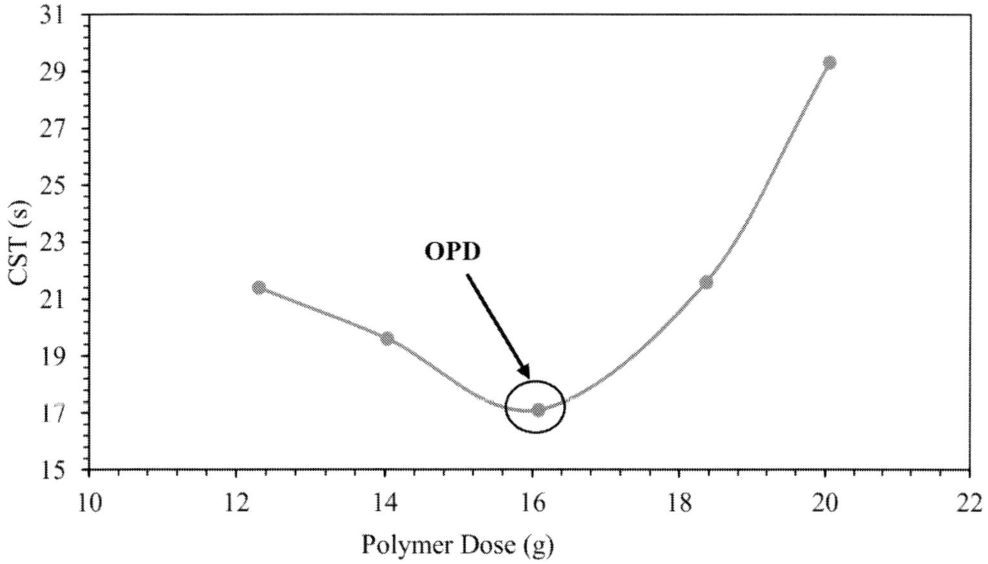

Figure 9: Capillary suction time (CST) – polymer dose profile used to determine the OPD for conventional and thermally hydrolyzed sludge during conditioning

A plot of minimum CST against cake solids was made for different samples tested during the development of the method. Figure 9 shows that the minimum CST values achieved during conditioning did not correlate with cake solids. This lack of correlation was consistent with observations made by Higgins et al., (2015). Some previous studies have used CST as an indicator of dewaterability (Ormeci & Vesilind, 1999), but CST values do not necessarily correlate with final dewaterability as measured in terms of cake solids. Figure 9 demonstrates that CST may be used to assess the rate of dewaterability but may not be reliable in predicting the extent of dewatering.

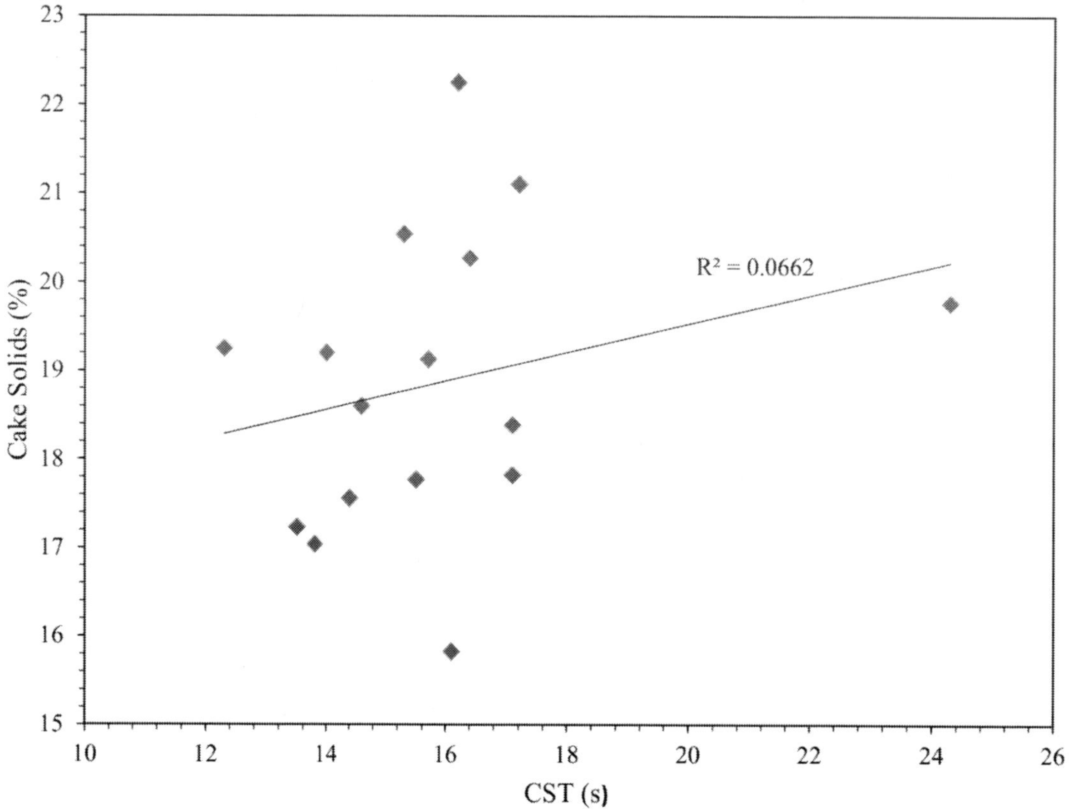

Figure 9: Minimum CST after conditioning vs cake solids for anaerobic digestate

Reproducibility of the method

The reproducibility of the method was tested by dewatering five samples of the same sludge, sampled at different dates, and analyzing the variance. The samples were dewatered for a 10 minute time interval at 2075 x g and the cake mixed twice. Table 3 shows that the confidence intervals were rather small which demonstrates that the method is reproducible.

Table 3: Reproducibility parameters of the dewatering method for conventional digestate

Conventional sludge	Cake solids (%)
Mean	20.30
Range	19.39 - 21.66
Standard deviation	0.96
Coefficient of variation	0.05
Confidence interval at 99%	1.97
Confidence interval at 95%	1.19
Confidence interval at 90%	0.91

SUMMARY AND CONCLUSIONS

The modified centrifugal method presented in this study results in the production of cake solids which can be used for further testing and analysis such as odor content. It was demonstrated that the method gives cake solids comparable to full scale when the solids loaded on the belt filter press cup are minimized and the time of centrifugation at 2075 x g set to 10 minutes. High cake solids were also produced when the cake was mixed twice during the high speed centrifugation phase. The method was proven to be reliable and reproducible, with a very small 95 percent confidence interval. The method can be used in the laboratory for accurate comparison of the dewaterability of different sludges.

ACKNOWLEDGEMENTS

This research was funded by the Water Environment and Reuse Foundation (WERF). The authors gratefully acknowledge this support.

REFERENCES

Baskerville, R. C., & Gale, R. S. (1968). A simple automatic instrument for determining the filterability of sewage sludges. *Water Pollut. Control*, *67*(3), 233-241.

Baskerville, R. C., Bruce, A. M., & Day, M. C. (1978). Laboratory techniques for predicting and evaluation the performance of a filter belt press. *Filtration & Separation*, 1-5.

Coakley, P., & Jones, B. R. S. (1956). Vacuum sludge filtration, I. Interpretation of results by the concept of specific resistance. *Sewage Ind. Wastes*, *28*, 963-976.

Emery, B.P. (1994). Predicting Belt Filter Press Performance Using Laboratory Techniques. Master's Thesis. Dept. of Civil Engineering. University of Illinois at Urbana-Champaign.

Galla, C. A., D.L. Freedman, B.F. Severin, and B.Y. Kim. (October 1996). "Laboratory Prediction of Belt Filter Press Dewatering Dynamics," Proceeding of the Water Environment Federation 69th Annual Conference, Residual and Biosolids Management Symposium, Dallas, TX.

Graham, T. M. (1999). Predicting the performance of belt filter presses using the Crown Press for laboratory simulation. CLEMSON UNIV SC COLL OF ENGINEERING.

Kopp, J., & Dichtl, N. (2000). Prediction of full-scale dewatering results by determining the water distribution of sewage sludges. *Water Science and Technology*, *42*(9), 141-149.

Örmeci, B., & Vesilind, P. A. (2000). Development of an improved synthetic sludge: a possible surrogate for studying activated sludge dewatering characteristics. *Water Research*, *34*(4), 1069-1078.

Severin, B. F., & Collins, B. H. (1992). Advances in predicting belt press performance from lab data. In *Wat Env Fed, 65th Ann Conf & Expo* (pp. 205-216).

Vesilind, P. (1988). Capillary Suction Time as a Fundamental Measure of Sludge Dewaterability. *Journal (Water Pollution Control Federation), 60*(2), 215-220.

Vesilind, P. A. (1970). Estimation of sludge centrifuge performance. *Journal of the Sanitary Engineering Division*, *96*(3), 805-818.

Paradigm Shift if Dewatering Operations Moved to the Center of the Plant Universe

William Wehner[1], Murthy Kasi[1], Mario Benisch[2], Anthony Perreira[3], Jim Wodrich[1]
[1]HDR Inc., Boise, ID; [2]HDR Inc., Portland, OR; [3]City of Boise, Boise, ID

ABSTRACT

Dewatering system influences are not largely considered when developing influent loads of phosphorus in the operation of enhanced biological phosphorus removal (EBPR) based treatment plants. A study was conducted to improve the nutrient capture efficiency in cake biosolids from a belt filter press (BFP) and reduce the nutrient recycle loading to the head of the plant. Reducing the biosolids feed rate to BFP system by 50% showed that the solids capture efficiency improved from 12.5 to 15% while concentrations and loads decreased by more than 35% for total phosphorus (TP) and total Kjeldahl nitrogen (TKN) in the filtrate and by more than 70% for TP and TKN in washwater. The resulting load reductions in TP and TKN in recycle streams constituted 7 and 18% of the total influent load to the EBPR facility, respectively. Changes made to optimize a dewatering operation demonstrated that dewatering was a central link between EBPR performance, biosolids treatment, resource recovery, and overall process stability. This new paradigm of plant operations focus will only gain more importance as resource recovery is expanded to accepting food waste and effluent nutrient limits are lowered.

KEY WORDS

Biosolids, dewatering, solids capture, total phosphorus, and nutrient recycling

INTRODUCTION

Biosolids digestion and dewatering are critical aspects of wastewater treatment plant operations. Dewatering systems are designed to maximize solids content of the biosolids in order to reduce trucking and handling costs. Wastewater biosolids handling operations typically operate dewatering systems to process material as quickly and efficiently as possible to minimize operational and disposal costs. Liquid separated from the biosolids during dewatering (filtrate) and the washwater used to clean belts and screens are typically recycled back to the head of the plant. These liquid recycle streams can carry high concentrations of nitrogen and phosphorus. In addition, treatment plants with enhanced biological phosphorus removal (EBPR) release a high concentration of monovalent cations in biosolids which can result in poor dewatering performance due to poor polymer and biosolids interactions (Higgins 2014). Despite efforts to reduce the nutrient loading via the recycle sidestream by using treatment processes, such as struvite recovery systems, the typical dewatering operation still allows for significant nutrient loads to enter the recycle stream, thus creating a compounding nutrient loop.

As plants struggle to meet increasingly difficult effluent limitations or managing operational costs, they have begun investigating sources of nutritional recycling to reduce the nutrient loading to the plant EBPR process. On the other hand, current efforts for dewatering are more focused on improving the quality of feed solids to increase the dewatering efficiency via thermal and chemical conditioning, polyelectrolyte (polymer) optimization, or enzyme treatment (Ayol, 2005). While the technologies are advancing, dewatering optimization and nutrient reduction efforts in the recycle streams are often perceived as independent unit processes in terms of plant design and operation.

Purpose of the Study

This study will shift the focus to optimizing the dewatering process in an attempt to improve the nutrient capture efficiency and measure the reduction in nutrient recycle impacts to the main-stream treatment process.

Facility Description

The City of Boise currently operates two (2) independent wastewater treatment plants, the West Boise Water Renewal Facility (WBWRF) and the Lander Street Water Renewal Facility (LSWRF). Aerial overviews of the two (2) treatment facility are shown in Figure 1.

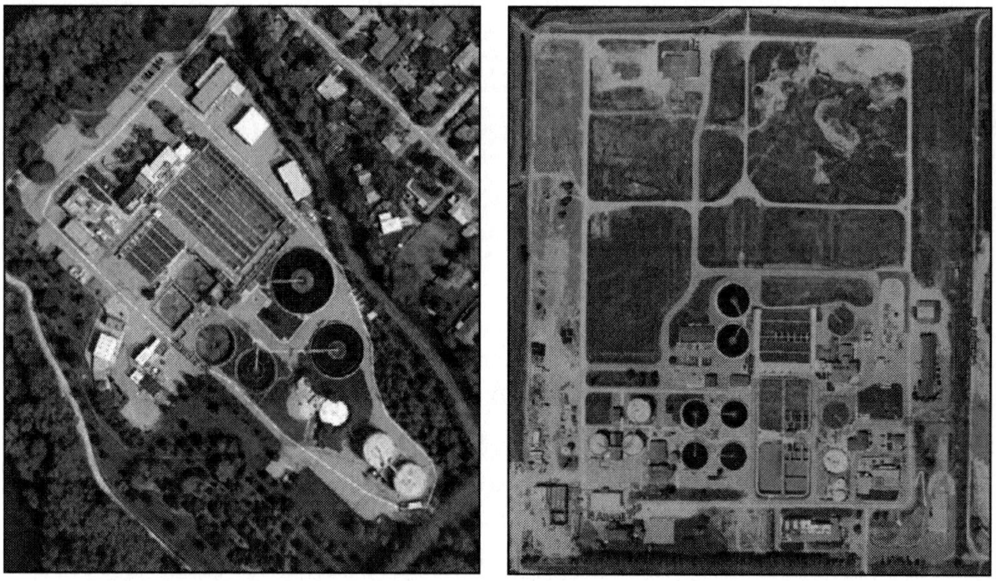

Lander Street Water Renewal Facility

West Boise Water Renewal Facilty

Figure 1: Overview of the City of Boise Water Renewal Facilities

Both treatment plants have EBPR processes to achieve effluent phosphorus limits. Biosolids from the liquid treatment stream at both facilities are processed in anaerobic digesters. The LSWRF currently does not have the capability to dewater digested biosolids. The digested biosolids are instead pumped several miles to the WBWRF. Digested biosolids from both the facilities are pumped to a 760 m^3 (200,000 gallon) steel Blend Tank where they are mixed and stored prior to dewatering. The 2015 annual average of the combined LSWRF and WBWRF digested solids loading to dewatering facility was 1.0 dry metric ton per day (11 dry tons per day). Dewatering of the biosolids is achieved using two (2), 2.8 meter wide belt filter presses (BFP). Biosolids from the two facilities vary in quality and maintaining a consistent, predictable feed biosolid stream to the BFPs is challenging.

Filtrate from the BFP flows by gravity to two filtrate storage basins (filtrate equalization) prior to being pumped to the Struvite Production Facility (SPF) where phosphorus is harvested from the filtrate. Washwater generated by the BFP operation flows separately back to the head of the plant. The dewatered BFP cake is discharged into cake storage hoppers and loaded into trucks for delivery to the Twenty Mile South Biosolids Application Site (TMSBAS) for land application as a Class B biosolid. The dewatering process flow diagram at the WBWRF is illustrated in Figure 2.

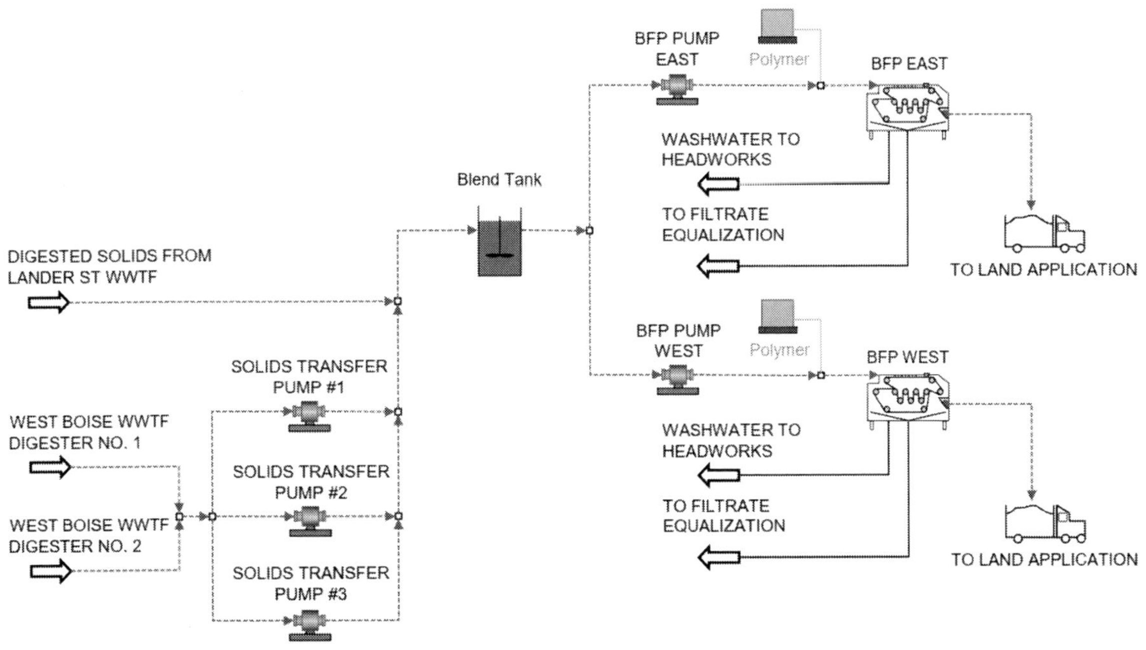

Figure 2: Dewatering Process Flow Diagram at the WBWRF

When compared to current BFP dewatering performance, the performance of the dewatering process was better prior to the installation of EBPR at WBWRF. At that time, the dewatering process achieved a higher cake solids content (15% to 16%) as well as a higher solids capture rate (95% - 97%). Over the past several years, the cake solids content has reduced to 11-12% and the capture rate reduced to below 90%.

Current Belt Filter Press Operations and Challenges

The current preferred plant operation of the dewatering facilities is to operate the two (2) BFPs in parallel for eight (8) hours a day and five (5) days a week. The operating philosophy for the plant is to dewater at night during a single shift to allow for trucking during off peak hours. The belt speed, belt tension and polymer dose are adjusted to optimize cake dryness and prevent blinding on the belt. The belt is run at a conservative belt speed to maximize cake dryness while requiring minimal supervision.

The dewatered biosolids are trucked approximately 20 miles to the Twenty Mile South Biosolids Application Site facility for land application on a City of Boise owned farm. Cake production is measured by weighing in a hopper equipped with load cells before dumping into a truck by gravity for truck transport.

METHODOLOGY

All testing of the equipment was conducted at the WBWRF in June 2016. The Blend Tank stored and mixed the influent biosolids with air at the beginning of the day before testing was conducted. The existing BFP used in the test was an Andritz 2.8 meter Model SMX S8.

Feed Biosolids Characteristics

Biosolids mixtures within the Blend Tank were varied throughout the testing to sample from a range of feed solids composition of the LSWRF and WBWRF digested biosolids. The WBWRF has the capacity to store the biosolids after digestion. Feed biosolids temperature was taken for each test. The un-insulated Blend Tank allows the biosolids to increase in temperature over the course of a summer day. Thus, the biosolids temperature increased throughout any individual testing day.

Dewaterability of Biosolids Samples

Laboratory testing was performed individually by the Andritz Company (Graz, Austria) and KB Kopp (Lengede, Germany) to determine the dewaterability of the digested biosolids from WBWRF, LSWRF, and a 50/50 mixture of the two biosolids. The actual types of tests differed between the two companies and complete test details are not disclosed. Andritz Company conducted BFP simulations to determine the maximum possible cake solids (%) and the optimum polymer dose required to achieve the maximum possible cake solids. KBKopp used the free water content concept to determine the dewaterability of biosolids by a mechanical dewatering equipment and the polymer conditioning method to determine the optimum polymer dose required. The free water content represents the freely moving water between the biosolid particles without being adsorbed by the biosolids particles or bound by capillary forces and is the only type of water content that can be removed by mechanical dewatering equipment.

Belt Filter Press Equipment Adjustment

Prior to the testing, several equipment changes were made by the BFP vendor to optimize the BFP operation. The equipment parameter changes such as belt speed and belt tension and polymer dosing were varied during all testing conditions. Additional adjustments included the alignment of head box that distributes the feed solids along the width of the BFP.

BFP Operational Goals and Optimization Testing

In addition to the equipment parameters, the impact of feed rate on cake dryness and nutrient release to recycle streams was investigated. Two (2) different feed rate conditions were tested - High Flow Mode and Low Flow Mode. The High Flow Mode was typical of the regular dewatering operations at the plant with two (2) BFPs operating for eight (8) hours a day and five (5) days a week to process enough biosolids to maintain residence time in the digesters. The Low Flow Mode represented a shift in the typical operations to develop a constantly running BFP operation throughout the week with only one BFP operating for twenty-four (24) hours a day and seven (7) days a week. Operational parameters for the two testing modes are summarized in Table 1.

Table 1: BFP Operational Parameter Changes

Parameters	High Flow Mode	Low Flow Mode
Feed Biosolids Flow Rate	700-760 L/min (185 – 200 GPM)	325-360 L/min (85-95 GPM)
Belt Speed	4.25 m/s (13.9 ft/sec)	1.50 m/s (5 ft/s)
Polymer Feed	32.6 L/min (8.6 GPM)	13.6 L/min (3.6 GPM)
Belt Tension	4.5-5.0 N/m (40-45 pounds per linear inch (PLI))	4.5-5.0 N/m (40-45 (PLI))

Sampling and Analysis

Flows for various influent and effluent streams were either measured or estimated using mass balance calculations on total suspended solids (TSS). Baseline (typical operation) samples were collected during the initial phase of the testing. Testing was conducted throughout the day with samples collected from the influent biosolids to the BFP, washwater, filtrate, cake, and the make down water for the polymer. The BFP was in operation during the week between Monday and Friday for up to eight (8) hours and was shut down during the weekend. Samples from the influent and effluent flow segregations were analyzed for a variety of physical and chemical constituents, including organic and inorganic nitrogen species, dissolved phosphorus and dissolved reactive phosphorus, and metal cations. Samples were analyzed at the WBWRF laboratory. Methods used for the analyses of samples for various constituents are summarized in Table 2.

Table 2: Testing Constituents and Measurement Methods

Constituent	Method
Hardness	EPA 200.7
pH	Hach Method 4500
Conductivity	Standard Method 2510B
Alkalinity	USGS Method I-2050-85
Sodium, Na	EPA 200.7[1]
Calcium, Ca	EPA 200.7[1]
Potassium, K	EPA 200.7[1]
Magnesium, Mg	EPA 200.7[1]
Ammonia, NH_3-N	Standard Method 4500-NH_3
Total Kjeldahl Nitrogen, TKN	Titrimetric Method PAI-DK01
Nitrate and Nitrite, NO_X	EPA 353.2(W)
Total Solids, TS	Standard Method 2540B[2]
Total Suspended Solids, TSS	Standard Method 2540D
Volatile Suspended Solids, VSS	Standard Method 2540E[2]
Dissolved Phosphorus, DP	EPA 200.7[3]
Dissolved Reactive Phosphorus, DRP	EPA 365.1
Chemical Oxygen Demand, COD	Hach Method 8000[3]
Chlorine, Cl_2 (Polymer Make Down Water)	Standard Method 4500-Cl

[1]Solid samples (feed biosolids and cake) tested with EPA 6010c
[2]Solids samples (feed biosolids and cake) tested with Standard Method 2540G
[3]Feed biosolids centrifuged for measurement

RESULTS

Feed Biosolids Characterization

Typically, the biosolids from WBWRF are warmer than the LSWRF biosolids by 5°C, since the digesters are on the site with the dewatering process for WBWRF requiring no travel time while the digested biosolids from LSWRF are pumped for several miles to WBWRF Blend Tank. The biosolids are not stored anywhere other than the digesters and the Blend Tank. On average, total Kjeldahl nitrogen (TKN) and chemical oxygen demand (COD) concentrations in digested biosolids from WBWRF are higher than in the digested biosolids from LSWRF.

One of the parameters used to evaluate dewaterability of biosolids is the monovalent to divalent cation (M/D) ratio. The ratio is indicative of a biosolids/polymer interaction known as the divalent cation bridging theory (DCBT) (Sobeck and Higgins, 2002; Higgins et al., 2014). The theory explains that divalent cations will bind to negatively charged organic material and create bridge connections to other negatively charged particles, thus resulting in destabilized flocs. A high concentration of monovalent cations will occupy available binding sites, in both the biosolids particles and polymer, which will result in net repulsive interactions within the biosolid/polymer mixture and stabilized suspended particles. M/D ratio above 2.0 would be indicative of a poor dewaterability for influent biosolids. The average M/D ratio of the mixture of biosolids from LSWRF and WBWRF was 1.44 which is indicative of moderate dewaterability potential. The LSWRF biosolids M/D ratio was around 1.6 but was still not indicative of any significant monovalent cation interference.

The WBWRF digested biosolids pH was between 7.9 and 8.1 while the typical values range between 7.2 and 7.4. Concentration of dissolved phosphorus (DP) was higher in the LSWRF biosolids than the WBWRF biosolids. The feed biosolids characteristics values are summarized in Table 3.

Table 3: Feed Biosolids Characteristics

Parameter	Unit	Feed Biosolids		
		West Boise	Lander Street	Mixed
Total Solids (TS)	%	2.3	2.3	2.3
Volatile Solids (VS)	%	73.2	73	73
Total Phosphorus (TP)	mg/kg	30,771	35,483	32,450
Dissolved Phosphorus (DP)	µg/L	202,286	296,000	242,000
Ammonia	mg/kg	59,386	58,000	62,017
Total Kjeldahl Nitrogen (TKN)	mg/kg	113,000	92,617	112,000
Chemical Oxygen Demand (COD)	mg/L	26,386	25,200	25,867
pH	-	7.9 -8.1	7.5	7.6
Temperature	°C	31	29	30
Conductivity	µS/cm	8,707	8,387	8,492
Alkalinity	mg/L	3,669	3,380	3,618
Monovalent to Divalent Cation Ratio (M/D)	-	1.4	1.6	1.4

Dewaterability Tests

Plant monitoring data indicated that the BFP produced an average cake dryness of approximately 11% TSS after the implementation of EBPR process. Independent dewaterability tests on digested biosolids conducted by the Andritz Company and KB Kopp suggested that an additional 2-3% cake solids concentration was achievable for the influent biosolids. The tests also indicated that a higher polymer dose of 10 kilograms of polymer/dry megagram of biosolids (kg/dry Mg) (20 lb/DT) should be used to optimize the solids capture. The current operation used 5.0-6.0 kg/Mg of biosolids (10 to 12 lb/DT), which was almost half of the recommended dose from the dewaterability tests.

Impact of BFP Equipment Adjustments

Adjustments made to BFP operational parameters such as belt tension, belt speed, polymer feed rate, and polymer dose resulted in cake solids of 12.5% which was 1.5% higher than the current operations. It should be noted here that the typical plant operation of BFPs as well as during the adjustment of BFP operational parameters was at a biosolids feed rate of roughly 700-760 L/min (185-200 gpm), which was considered as High Flow Mode in this study.

The polymer dosage was limited in the current feed rate (High Flow Mode) due to accumulation of flocculated solids and unreacted polymer at the head box area of the BFP causing blinding on the belt. The head box was found to be lower on one side than the other causing irregular solids distribution on the belt. Adjustment to head box allowed for better biosolids distribution across the belt width and increased polymer dose without biosolids accumulation at the head box area.

BFP Operational Goals and Optimization Testing

Polymer doses were between 5 and 6 kg/Mg and between 8 and 9 kg/Mg for High Flow and Low Flow modes, respectively. The reduced feed rate to the BFP during Low Flow Mode allowed for an increase in polymer dose. The polymer doses, operation times and the resulting weekly volumes of digested solids for each operational mode are summarized in Table 4.

Table 4: Summary of BFP Flow Operations

Flow Classification	Throughput Flow Rate	Polymer Usage	Operating Units	Operation Time	Weekly Volume of Digested Solids
High Flow Mode	665-760 L/min (175-200 gpm) (each BFP)	5.0 – 6.0 kg/Mg (10-12 lb/DT) (dry polymer)	2	8 hr/day 5 days/week	3,200,000 to 3,630,000 L/week (840,000 to 960,000 gal/week)
Low Flow Mode	325-360 L/min (85-95 gpm)	8.0 – 9.0 kg/Mg (16-18 lb/DT) (dry polymer)	1 (1 - BFP Redundant)	24 hr/day 7 days/week	3,250,000 to 3,620,000 L/week ((856,000 to 957,000 gal/week)

The three (3) effluent flow streams from the BFP are the cake, filtrate, and washwater. Maximizing the solids content of the cake ensures a dense product which can be transported at a reduced cost. Greater solids capture efficiency and a lower solids content in the filtrate demonstrate the cohesiveness of the cake and the effectiveness of the polymer at the higher dosage rates in the Low Flow Mode. The washwater will reflect the amount of blinding occurring during the BFP operation, as high blinding would result in more aggressive washing. The data results indicate that there was a significant decline in the solids passing through to the filtrate and washwater and an increase in the amount of solids captured in the cake during the Low Flow Mode. The solids content in the cake increased 2-3%. However, to appropriately compare the operational modes, the loads were extrapolated out to a weekly load. Also, due to the variability in the influent biosolids during each operational mode, the solids loadings were normalized to a single biosolids characteristic to adequately compare the values.

Overall, the data indicated that BFP Low Flow Mode operation resulted in a reduction in TS load for filtrate and washwater. A 2-3% increase in dryness of average cake solids was observed. Notably, the lowered flow rate allowed for a higher polymer dose due to increase reaction time. Previously, a high polymer dosage would be ineffective since the reaction time was shorter and the excess polymer resulted in blinding on the belt. The lowered flow rate also increased the biosolids/polymer mixture residence time on the gravity section off the BFP. The results are illustrated in Figure 3, Figure 4 and Figure 5.

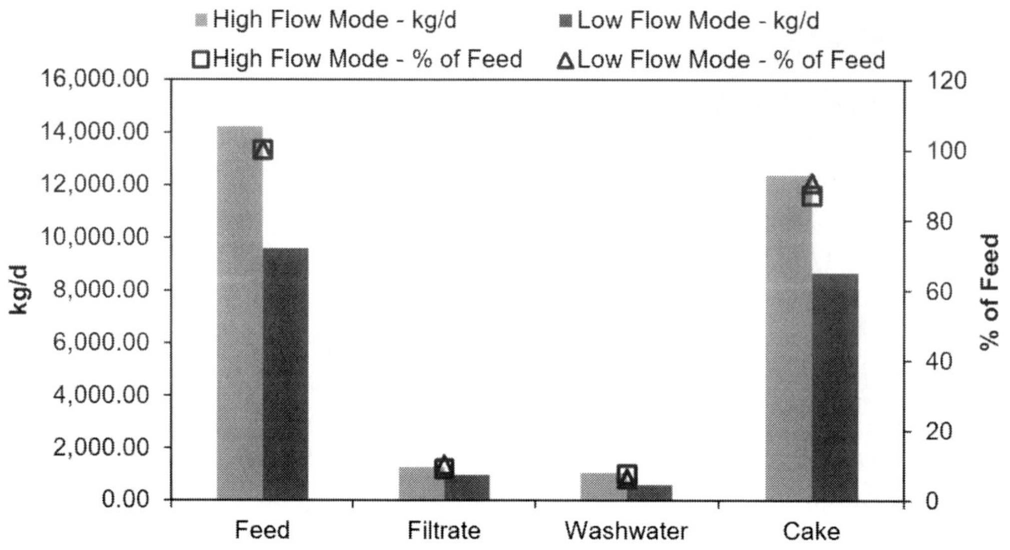

Figure 3: Total Solids Daily Load BFP Operations Comparison

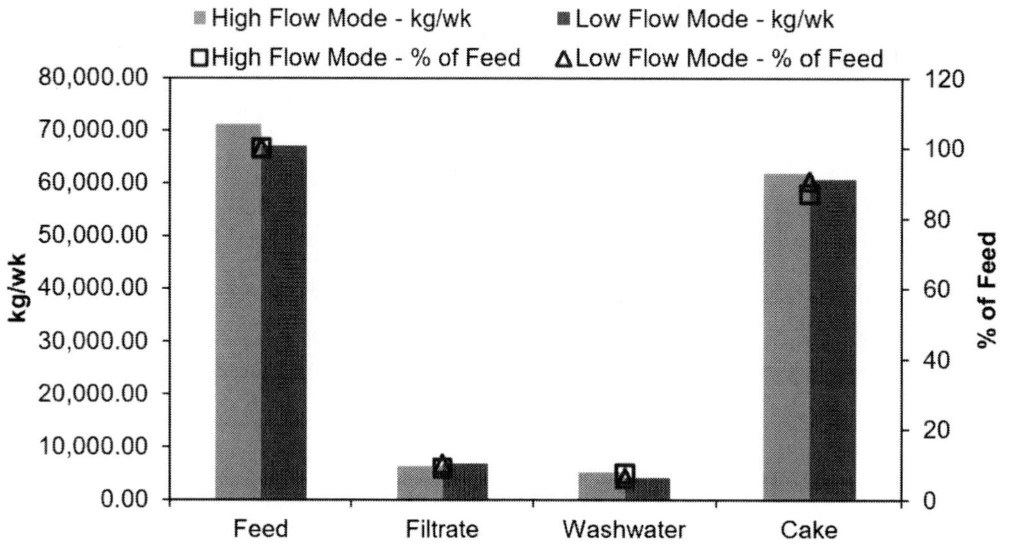

Figure 4: Total Solids Weekly Load BFP Operations Comparison

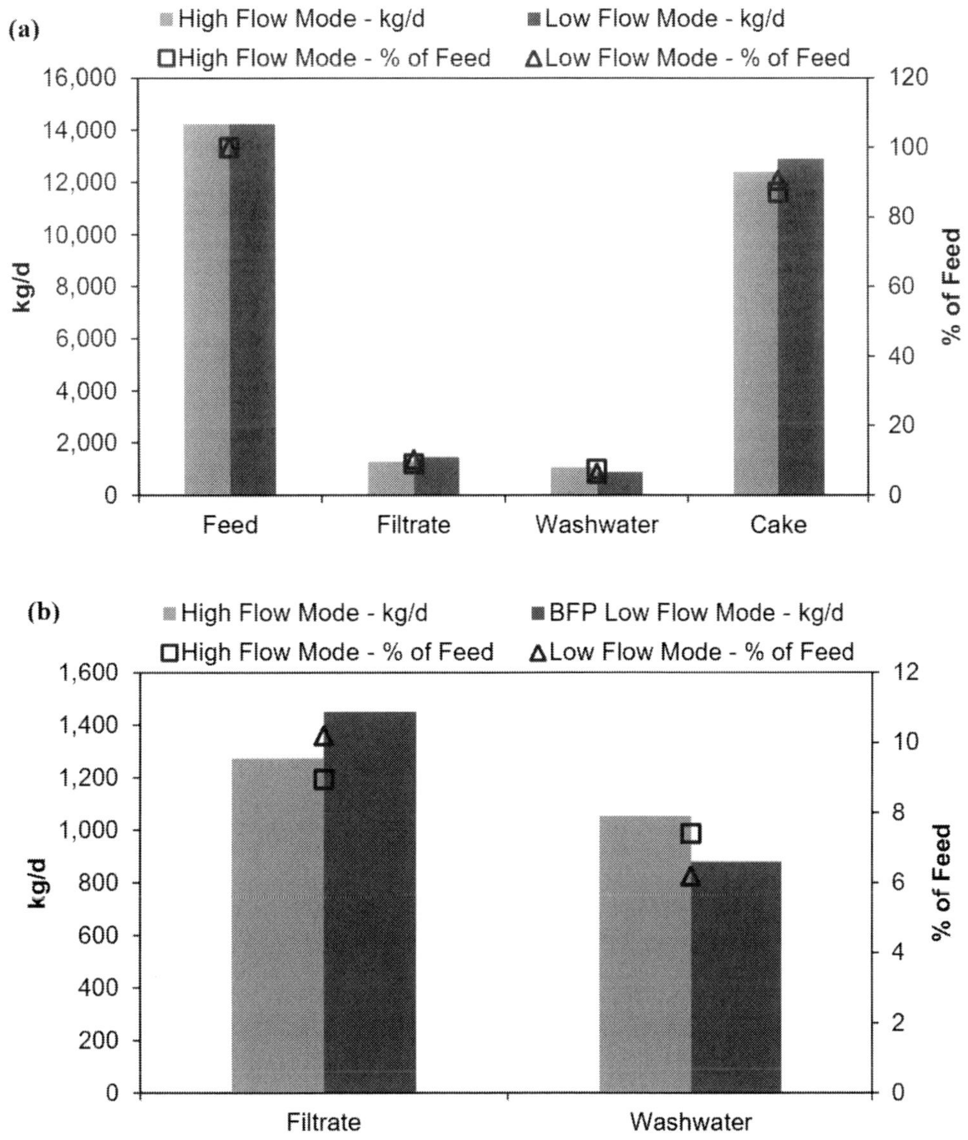

Figure 5: Normalized Total Solids BFP Operation Comparison
(a) All Components and (b) filtrate and washwater only

(Values normalized to the same influent biosolids characteristics)

Impact of Feed Rate on Quality of Recycle Streams

To determine the impact of dewatering processes throughout a plant, the nutrient makeup was analyzed for the cake, filtrate, and washwater. Optimization of the BFP process had a positive impact in terms of a reduction in recycle stream nutrient loading. TP load as a percent of feed decreased by 11% in filtrate and increased by 6% in the cake during BFP Low Flow Mode as compared to High Flow Mode. However, it should be noted here that the concentration of TP in cake remained the same for both BFP operations, which is around 25,000 mg/kg. The decrease was minimal for TKN, while a 40% decrease was observed for TP and DP loads in combined filtrate and washwater. The results are illustrated for daily loads in Figure 6, weekly loads in Figure 7, and concentrations in Figure 8. The TP recycling results were normalized to the influent biosolids TP characteristics in Figure 9.

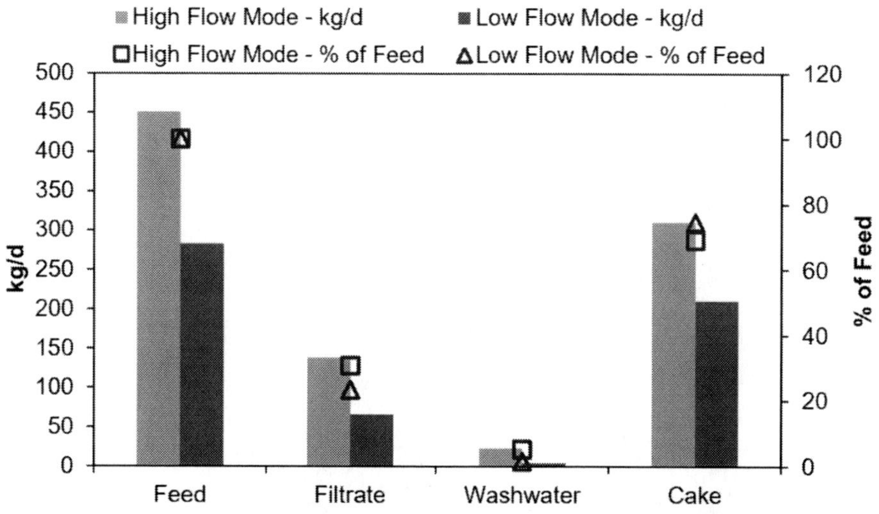

Figure 6: Total Phosphorus Daily Load BFP Operations Comparison

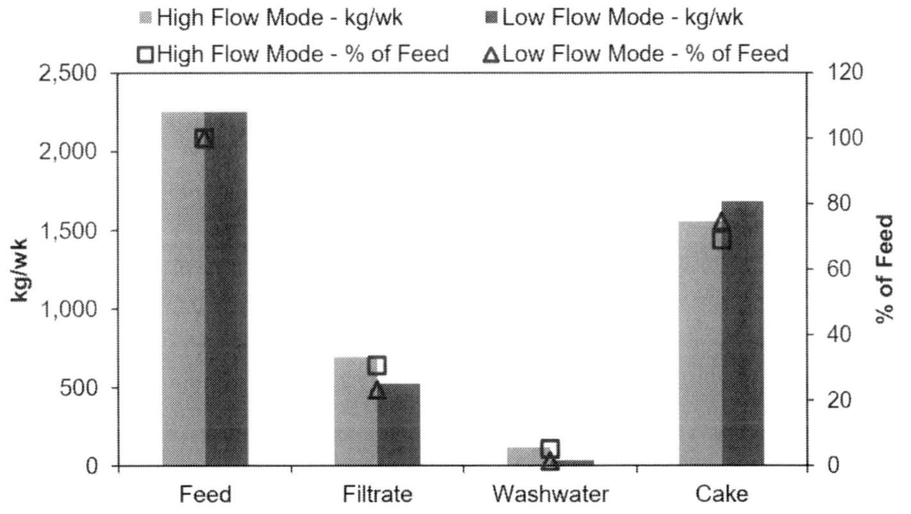

Figure 7: Total Phosphorus Weekly Loading BFP Operations Comparison

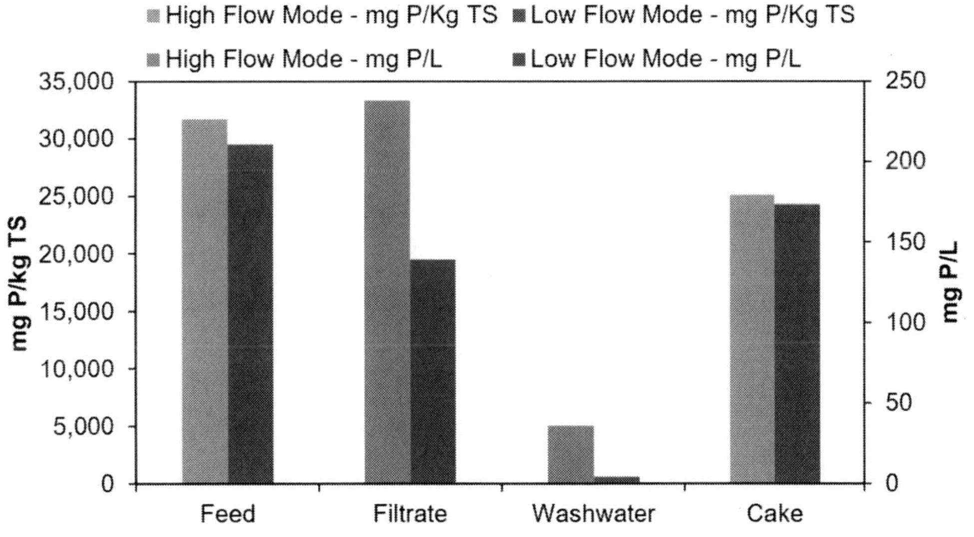

Figure 8: Total Phosphorus Concentration BFP Operations Comparison

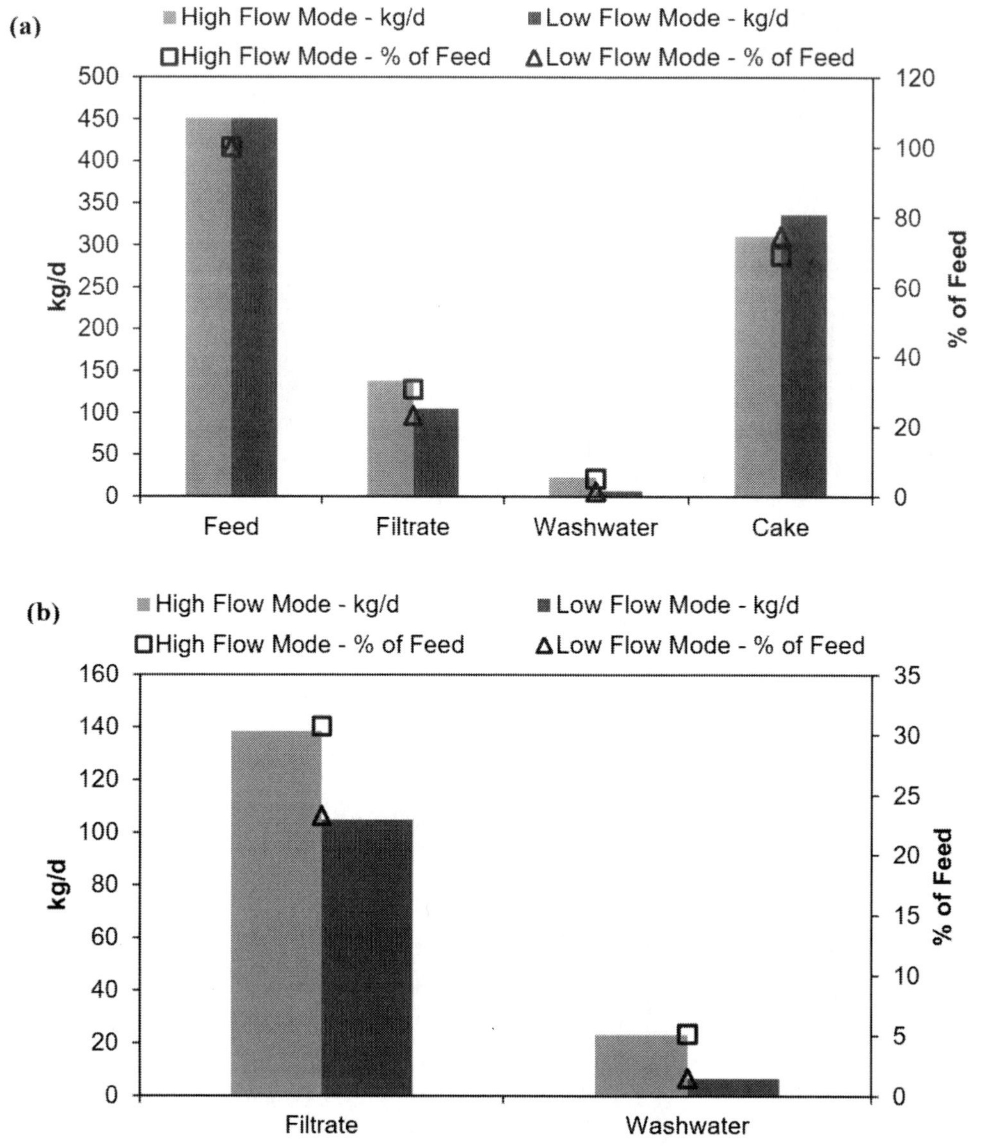

Figure 9: Normalized Total Phosphorus BFP Operation Comparison
(a) All Components and (b) filtrate and washwater only

(Values normalized to the same influent biosolids characteristics)

Impact of Feed Rate on Land Application

The impacts of BFP operations on the quality and quantity of the dewatered biosolids were analyzed to assist in evaluating the operational changes on trucking and acreage requirements at the Twenty Mile South Biosolids Application Site. Table 5 summarizes the potential TS, TP, TKN and ammonia loads and concentrations during various plant operational scenarios. The BFP

High Flow Mode (Scenario 1) produces approximately 99.0 metric tons (109 wet tons (WT)) of solids at 12.5% solids, which would equal to 3.6 trucks per day or 18.2 trucks per week. Operating the BFP in the low throughput, Low Flow Mode (Scenario 3) produces 93.5 metric tons (103 WT) at 13.8% cake solids, which would equal to 3.4 trucks per day or 17.2 trucks per week. Thus, the BFP Low Flow Mode operation could reduce the number of truck trips by 1 truck per week or 52 trucks a year. With the round trip to the Twenty Mile South Biosolids Application Site being roughly 64 kilometers (forty (40) miles) and at a cost of $200 per trip, the potential savings for the City would amount to 3,350 kilometers (2,080 miles) and $10,000 per year.

Table 5 : Summary of Cake Characteristics to Twenty Mile South Biosolids Application Site

Scenario	Conditions	TS			TP	TKN	Ammonia
		kg/d	%	wet Metric Tons/d	mg/kg	mg/kg	mg/kg
1	- Plant Flow = **102,200 m^3 (27 MGD)** - BFP Operation = **High Flow** - Plant Effluent TP = **2.5 mg P/L (current condition)**	12,400 (27,320 lb/d)	12	99.0 (109 tons)	25,100	63,735[1]	14,048[1]
2	- Plant Flow = **102,200 m^3 (27 MGD)** - BFP Operation = **High Flow** - Plant Effluent TP = **0.3 mg P/L**	12,465 (27,473 lb/d)	12	99.0 (109 tons)	36,087	65,000[2]	14,300[2]
3	- Plant Flow = **102,200 m^3 (27 MGD)** - BFP Operation = **Low Flow** - Plant Effluent TP = **2.5 mg P/L (current condition)**	12,890 (28,416 lb/d)	14	93.5 (103 tons)	24,300	63,735[1]	14,048[1]
4	- Plant Flow = **102,200 m^3 (27 MGD)** - BFP Operation = **Low Flow** - Plant Effluent TP = **0.3 mg P/L**	12,890 (28,400 lb/d)	14	93.5 (103 tons)	36,309	65,000[2]	14,300[2]

1. Average of 2006 to 2015 biosolids data
2. Assumed a 2% increase in TKN and ammonia concentrations

Discussions with plant operating staff indicated that the plant was able to achieve a WBWRF effluent TP discharge of approximately 2.5 mg-P/L to the river during the BFP pilot testing. It should be noted that recent modifications to the plant process has reduced the effluent TP to less than 1.0 mg P/L. BFP Low Flow Mode operation did not impact TP concentration in the biosolids (cake) as compared to BFP High Flow Mode (Scenario 3 in Table 5). Therefore, it was assumed that the ammonia and TKN concentrations also would not differ between the two BFP operations. The cake TP concentration will increase if the plant operational conditions are optimized for a lower WBWRF effluent TP (e.g. 0.3 mg-P/L). From mass balance calculations and BioWin 5.0® (EnviroSim Associates Ltd.) modeling, it was estimated that the TP in biosolids would increase to 36,087 and 36,309 mg/kg for BFP High Flow and BFP Low Flow operations, respectively, for the plant to meet an effluent TP of 0.3 mg P/L (Scenarios 2 and 4). Similarly, the TP in biosolids would increase to 35,900 mg/kg at plant build-out capacity of 147,000 m^3 (39 million gallons per day (MGD)) and operated to achieve 0.3 mg-P/L. Agronomic rates were evaluated for each of these scenarios and it was found that the impact of BFP Low Flow Mode compared to current BFP High Flow Mode operation on land application acreage requirements was insignificant.

CONCLUSIONS

The benefits of operating a BFP system at a reduced feed rate were demonstrated through this study with results showing an increase in BFP solids and nutrient capture efficiency and a decrease in nutrient recycling, especially through TP.

1. The BFP Low Flow Mode operation at a reduced feed rate allowed for higher polymer dose without flooding in the head box area. The polymer dose at the reduced feed rate was approximately 9 kg/dry Mg (18 lb/DT) which conformed to the recommendations from optimized laboratory testing results.
2. The BFP Low Flow Mode operation at reduced feed rates and belt speeds throughout the week would significantly reduce the nutrient cycling throughout the plant, including TP as compared to the current operation.
3. Nitrogen and phosphorus concentrations and loads were greatly reduced in filtrate and washwater during the BFP Low Flow Mode operation at a reduced feed rate. As compared to the BFP High Flow Mode, operation at Low Flow Mode with a reduced feed rate decreased concentrations and loads by more than 35% for TP and DP in filtrate and by more than 70% for TP, ammonia and TKN and by 69% for DP in the washwater..
4. The BFP Low Flow Mode operation resulted in a reduction of 27 kg/d (59 lb/d) of TP in total recycle (filtrate and washwater), which constituted 7% of the current influent load of 375 kg/d (825 lb/d) to the WBWRF.
5. The concentration of TP in cake remained consistent at 25,000 mg/kg during both modes of BFP operation.
6. Agronomic rates evaluation indicated that the impact of BFP Low Flow Mode operation on land requirements was insignificant. The most significant impact of TP solids loading to the Twenty Mile South Biosolids Application Site will occur when WBWRF effluent discharges are reduced to the 0.3 mg P/L from the interim requirement of 2.8 mg P/L between May 2016 and April 2017.

REFERENCES

Andritz (2016) Laboratory Results on Anaerobically Digested Sludge Samples from West Boise and Lander Street Wastewater Treatment Facilities, A Report Submitted to the City of Boise. January 13, 2016.

Ayol, A., Dentel, S. (2005) Enzymatic treatment effects on dewaterability of anaerobically digested biosolids-II: laboratory characterizations of drainability and filterability. Process Biochemistry 40(7): 2435-2442.

Benisch, M.; Schauer, P.; Neethling, J.B. (2014) Improving Dewaterability of Digested Sludge from EBPR Facilities, Proceedings of the Water Environment Federation, Residuals Biosolids 2014, 1-24.

Higgins, M., Bott, C.; Schauer, P.; Beightol, S. (2014) Bio-P Impact Dewatering after Anaerobic Digestion? Yes, and not in a good way! WEF Residuals and Biosolids Conference, 2014.

Higgins, M.J. and Novak, J.T. (1997) The Effect of Cations on the Settling and Dewatering of Activated Sludges: Laboratory Results. Water Environment Research, 69, 215.

KB Kopp (2016) Dewaterability of Sewage Sludge Samples from WWTP Boise West (USA) and from WWTP Lander Street (USA), A Report Submitted to the City of Boise. January 2016.

Sobeck, D. C., Higgins, M. J. (2002) Examination of three theories for mechanisms of cation-induced bioflocculation. Water Research; 36(3):527-38.

Dewatering Biosolids with a Volute Press at two Biological Nutrient Removal Wastewater Treatment Plants in Montana

Rickey Schultz[1*]

[1]HDR, Missoula, Montana
*Email: rickey.schultz@hdrinc.com

ABSTRACT

Two biological nutrient removal wastewater treatment plants in Montana recently evaluated alternative biosolids dewatering technologies. Dewatering equipment was piloted on site at both facilities prior to selection of full scale equipment. A PW Tech volute press was selected for installation at both plants. Each volute press has performed well since installation and has functioned to reduce return nutrient loading within the facility.

KEYWORDS: biosolids dewatering, volute press, biological nutrient removal, advanced wastewater treatment

INTRODUCTION

Selection of the most suitable biosolids dewatering equipment at biological nutrient removal treatment plants can be a challenge. Polymer selection and usage, dewatering performance, equipment footprint, capital costs, and operation and maintenance costs are among a few items that must be considered during the selection process. Two Montana municipalities recently evaluated alternative biosolids dewatering technologies to replace existing equipment.

Kalispell, Montana

The Kalispell, Montana Advanced Wastewater Treatment Plant (AWWTP) has an average daily flow of 9.1 ML/d (2.4 MGD) and a design capacity of 20.4 ML/d (5.4 MGD). A process diagram for the plant is shown in Figure 1. A biological nutrient removal process, called the modified Johannesburg process, is utilized for optimum biological oxygen demand, nitrogen and phosphorus removal using a combination of pre-anoxic, anaerobic, anoxic and aerobic basins. Primary effluent and RAS are mixed in the pre-anoxic zone which is followed by an anaerobic zone where the volatile fatty acids (VFAs) are introduced. Primary sludge fermentation is used to create the short chain VFAs for use by microorganisms in the anaerobic zone of the bioreactor which facilitates phosphorus removal. The anaerobic zone is followed by an anoxic zone, where mixed liquor recycle is introduced, then a swing zone, which can be aerobic or anoxic, and finally the aeration zone.

Fermenter sludge is digested in a two stage, anaerobic system and then dewatered prior to hauling. Previous plant operations involved mixing waste activated sludge (WAS) with fermenter sludge prior to digestion but this resulted in phosphorous release in the pressate conveyed to the headworks. Undigested waste activated sludge is now thickened by dissolved air floatation thickener units and then combined with digested fermenter sludge prior to dewatering. The biosolids conveyed to the dewatering equipment are proportionally 75% undigested WAS and 25% digested sludge. Two 2-meter Enviroquip belt filter presses, one duty unit and one standby unit, and dry polymer were used for dewatering.

The City has a dual approach for dewatered biosolids disposal. A local composting facility and the Flathead County Landfill are both utilized for disposal. The City hauls dewatered biosolids approximately 56 kilometers (35 miles) to Glacier Gold composting facility in Olney, Montana. The 5-year, recently-renewed agreement with the facility is for 544 dry metric tons (600 dry US tons) of dewatered biosolids annually at a cost of approximately $262.5 per dry metric ton ($238 per dry US ton). The contracted amount of biosolids that can be delivered is determined by the capacity of the equipment at the composting facility and, therefore, cannot be increased. In the past five years, the amount of biosolids hauled to Glacier Gold has averaged approximately 456 dry metric tons (503 dry US tons). The agreement with Glacier Gold also states the dewatered biosolids delivered to the facility shall be between 13-20% solids concentration; the City's historical average has been 14-15% solids concentration.

The second biosolids disposal method currently used by the City is to take one load per week of dewatered biosolids to the Flathead County landfill approximately 16.1 kilometers (10 miles) from the AWWTP. The average weekly load taken to the landfill is approximately 1.9 dry metric tons (2.1 dry US tons). Tipping fees at the landfill are $34.23 per wet metric ton ($31.05 per wet US ton).

Finally, the City is also permitted by the United States Environmental Protection Agency (US EPA) to dispose of digested sludge on a 16.2 ha (40 acre) site on the south side of the city. The disposal site is used about every 5 years for land applying digester contents when the digesters are emptied and cleaned.

Dewatering operations typically occur Monday through Thursday, Saturday, and Sunday at the Kalispell AWWTP. The dewatered biosolids are classified as Class B according to 40 Code of Federal Regulations (CFR) Part 503.

The two belt filter presses used for dewatering at the Kalispell AWWTP were a source of costly repairs for the City and staff experienced continued downtime stressing plant operations. The City elected to replace one of the belt filter presses with new dewatering equipment.

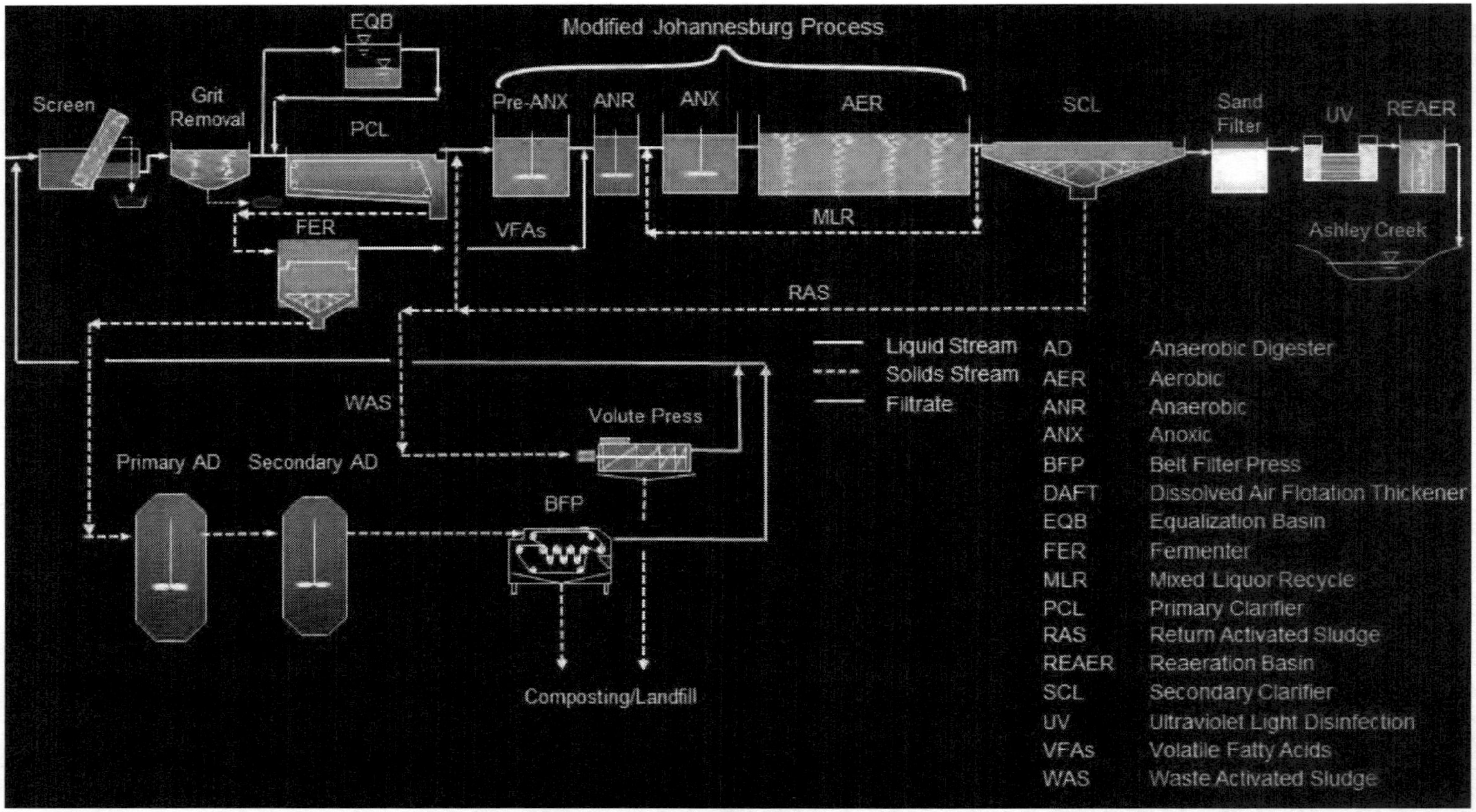

Figure 1. Kalispell AWWTP Process Flow Diagram

Missoula, Montana

The Missoula Wastewater Treatment Plant (WWTP) has an average daily flow of 25.7 ML/d (6.8 MGD) and design capacity of 45.4 ML/d (12.0 MGD). A process diagram for the plant is shown in Figure 2. The secondary treatment and biosolids handling processes at the Missoula WWTP are very similar to the Kalispell AWWTP. A modified Johannesburg process is used for biological nutrient removal at the Missoula WWTP. As is the case in Kalispell, the swing zone is normally operated as an anoxic zone, rather than an aerobic zone, as there is sufficient aeration capacity to meet ammonia discharge permit limits.

Similar to the solids handling operations at the Kalispell AWWTP, primary sludge is conveyed to a single-stage fermenter and then to anaerobic digestion. Digested sludge has an average solids concentration of 1.8% and comprises approximately 33% of the feed sludge to the dewatering equipment. A portion of return activated sludge is pumped to a dissolved air floatation thickener unit prior to dewatering. The thickened waste activated sludge has an average solids concentration of 2.6% and comprises approximately 67% of the feed sludge to the dewatering equipment. Thickened WAS is combined with the anaerobically digested sludge and then dewatered by a centrifuge prior to conveyance to the neighboring composting facility.

One, 200-hp centrifuge was operated 6 to 8 hours a day, 6 days a week. The centrate, or liquid sidestream from the centrifuge, is routed back to the headworks without equalization creating the potential for a slug nutrient load to the secondary treatment process during dewatering operations. The centrifuge is operated using 11 kg of dry polymer/dry metric ton of cake (22 lbs of dry polymer/dry US tons).

The average daily dewatered biosolids produced at the Missoula WWTP is approximately 6.6 dry metric tons per day (7.3 dry US tons per day). Dewatered biosolids are conveyed to a composting facility adjacent to the WWTP. The composting facility, which was privately owned and operated, was recently purchased by the City. Because of the neighboring composting facility there are no tipping fees or hauling costs incurred by the City for biosolids disposal and the desired cake solids is only 15-16%.

The City sought to provide dewatering equipment redundancy and reduce return nutrient loading from dewatering operations with new dewatering equipment.

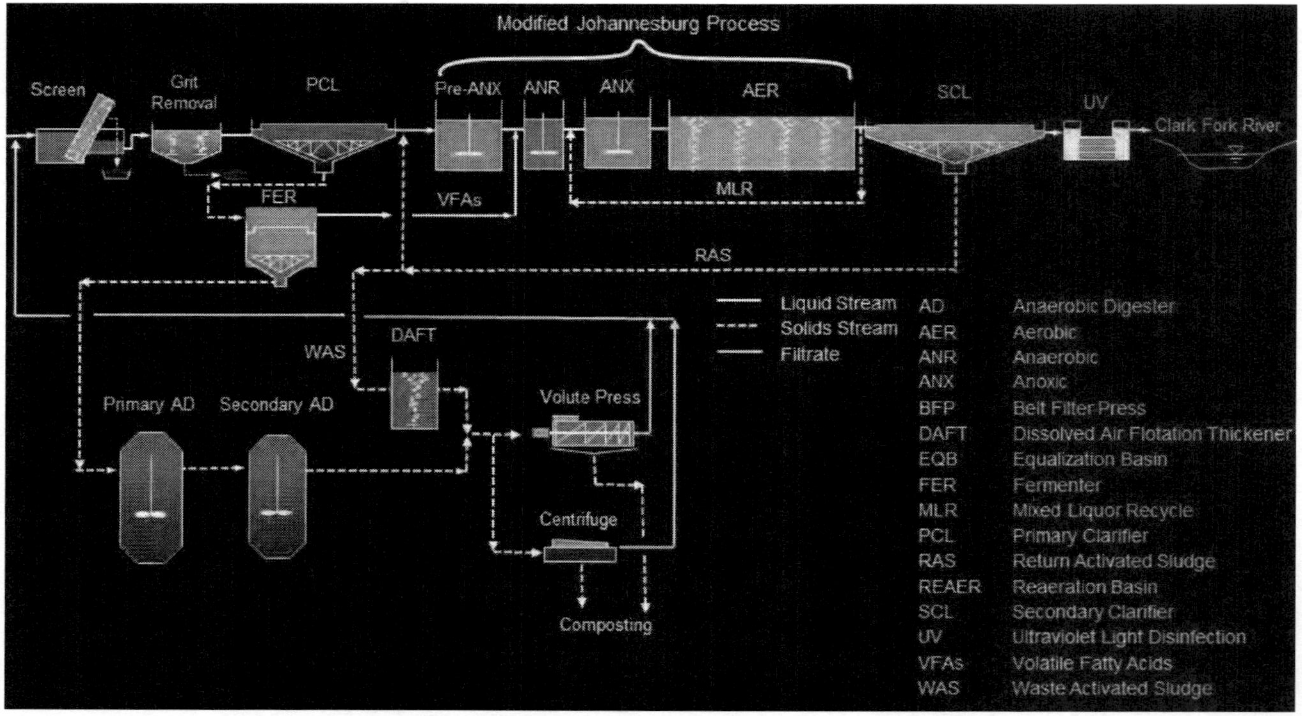

Figure 2. Missoula WWTP Process Flow Diagram

METHODOLGY

Piloting of new dewatering equipment was recommended at both facilities prior to consideration for design because site specific conditions such as sludge quality and conditioning can greatly affect dewatering performance. Water and power for pilot equipment were supplied by the City at each plant.

Kalispell, Montana

A Fukoku Kogyo Company (FKC) model SHX 200 x 1500L screw press, a Prime Solutions model RFP36D rotary fan press, and a Process Wastewater Technologies (PW Tech) model ES132 volute press were piloted in Kalispell from late August to early October 2013. The following criteria were used to evaluate dewatering alternatives and eliminate those which do not meet the City of Kalispell's needs:

- Alternative produces at least 14-15% solids concentration
- Alternative does not increase the operation and maintenance requirements of the dewatering processes
- Alternative meets future demands
- Alternative includes equipment which can be placed in the available space without requiring the construction of new facilities
- Alternative is energy efficient
- Alternative does not require auxiliary equipment to be implemented into the existing process

Screw Press

A screw press (Figure 3) uses a slow spinning auger inside a porous wire cage to compress solids and filter water. Water is released through the wire cage and the dewatered cake is pushed out through the end of the unit. Screw presses operate at a low rotating speed and have low energy and low spray water requirements. These machines are capable of unattended operation due to their use of automated control systems. Screw presses have simple maintenance, abundant municipal installations, and consistently produce a cake with high percent solids.

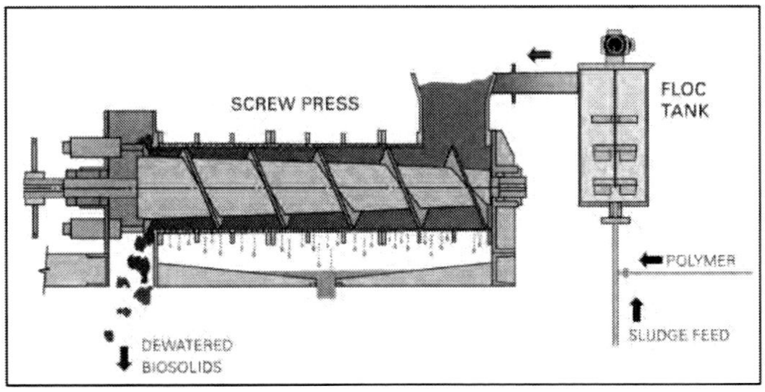

Figure 3. Screw Press

Rotary Fan Press

The rotary fan (Figure 4) press consists of two round porous metal discs that slowly rotate on a shaft. Solids are pumped between the discs and water is released through the pores in the discs as the discs rotate, compressing the sludge against a pressure bar. Rotary fan presses have a low power requirement due to the slow rotational speed. The machine is simple in design, operates with few moving parts and therefore, similar to the screw press and volute press, can be operated unattended with automatic control systems.

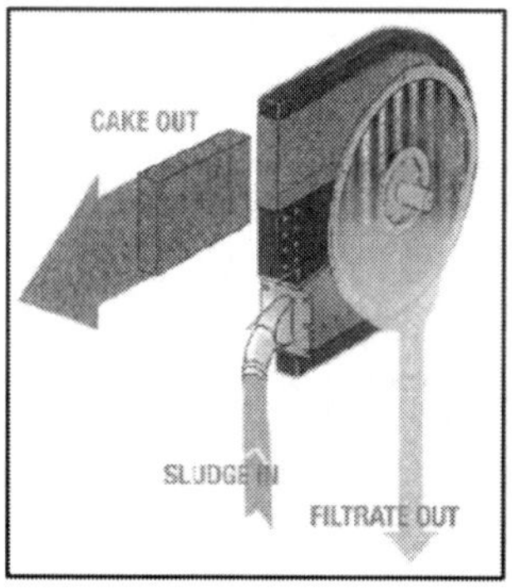

Figure 4. Rotary Fan Press

Volute Press

The volute dewatering press (Figure 5) consists of a dewatering drum, which houses spacers, fixed rings and moving rings. The initial section of the unit is the thickening zone where the threads are spaced further apart and the gaps between the rings are larger. As the sludge is moved down the drum into the dewatering zone, the distance between the screw threads decreases with the gaps between the rings. This operation squeezes water from the sludge as it travels down toward the end-plate where the solids are discharged. The volute press, similar to the screw press has a low rotational speed, low power and low spray water requirements.

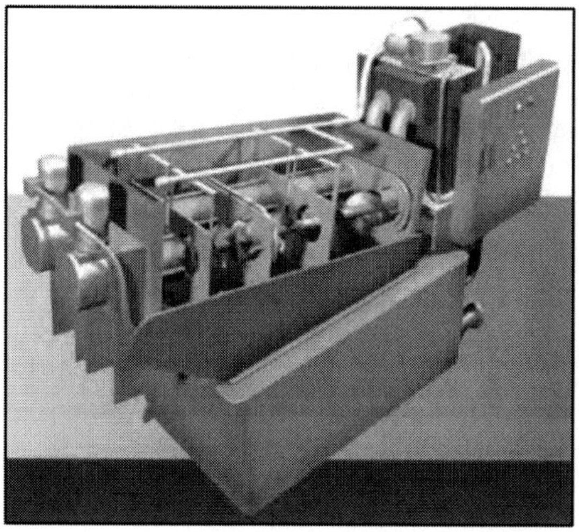

Figure 5. Volute Press

Onsite Piloting Trials

The feed sludge for the FKC screw press and PW Tech volute press was pumped from the sludge line feeding the belt filter press into a holding tank. Sludge was then fed from the holding tank into a flocculation tank. Polymer was added to the sludge just before entering the flocculation tank. Changes in level in the holding tank and polymer tanks were used to determine feed sludge volume and polymer usage for each trial run. The setup for the rotary fan press was similar but feed sludge was pumped directly to the press and polymer was mixed in-line prior to entering the dewatering channels.

Samples of dewatered biosolids were collected from the discharge of all piloting equipment, weighed, and analyzed for moisture content. Pressate gravity flowed to a nearby drain back the headworks. A piloting report was provided by each manufacturer following onsite testing.

Piloting Results

The FKC screw press was piloted over two days in late August 2013. Four trial runs were performed. The first three trials were at screw speeds varying from 0.3 to 0.75 rpm to see impact on outlet consistency and capacity of the screw press. The dry polymer currently used at the plant, T-Flock 2621, was used for the first three runs. The fourth trial run used Ashland 274 FLX polymer and demonstrated FKC's Class A solids process which involves lime and steam addition.

The solids concentration of the feed sludge during the screw press pilot ranged from 1.01% to 1.40%. The screw press manufacturer does not report solids capture rate for field trials but states a minimum capture rate of 92% will be included as a process guarantee.

Results from the first three trial runs of the FKC screw press were similar (Figure 6). Cake solids with a concentration of 15-16% were produced and polymer usage was approximately 10 kg/dry metric ton (20 lb/dry US ton). As would be expected, lower screw speeds produced slightly drier solids while the faster screw speeds produced the highest capacities.

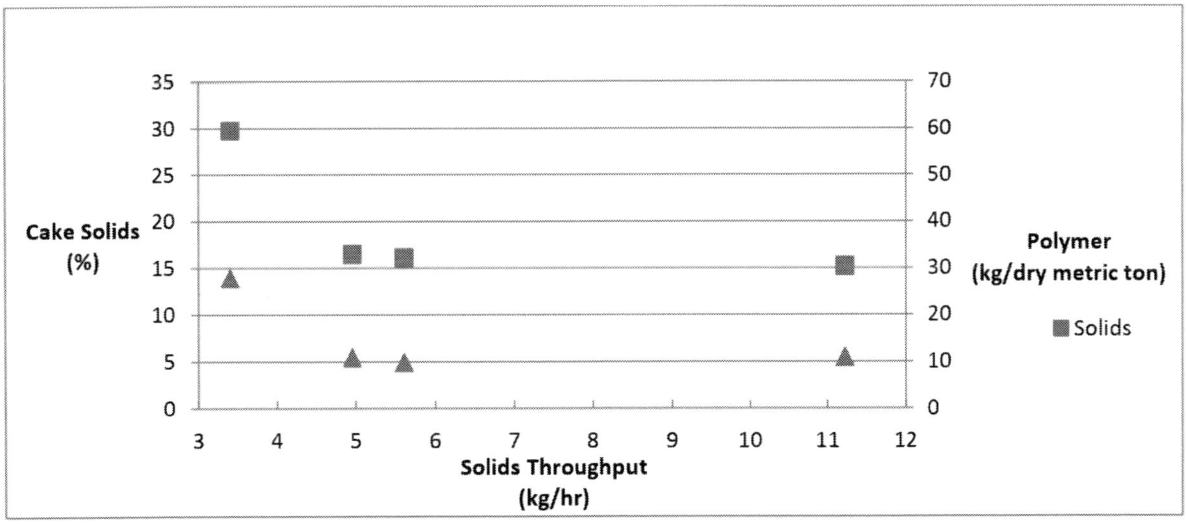

Figure 6. Screw Press Piloting Results

The fourth trial run produced solids meeting 40 CFR Part 503 requirements for Class A biosolids. Lime was added at 150 kg/dry metric ton (300 lb/dry US ton and steam was added to the screw's hollow core. Preliminary sizing by the manufacturer for full scale Class A equipment would require removal of both existing belt filter presses. The extra space would be needed for the longer press to allow for extra time to meet treatment requirements. Because the

City would only like to remove one belt filter press at this time, the equipment required to meet Class A biosolids was not further considered for final design.

The rotary fan press was only piloted for one day instead of the scheduled two days due to upsets to the solids digestion process within the plant. A total of six trial runs were performed, each at a different sludge feed rate (Figure 7). One emulsion polymer, Ashland 279 FLX, was used for all trial runs. The feed solids concentration during all fan press trials was 1.6%. Solids capture rate ranged from 76.3% at the highest solids throughput to 98.8% at lower throughputs.

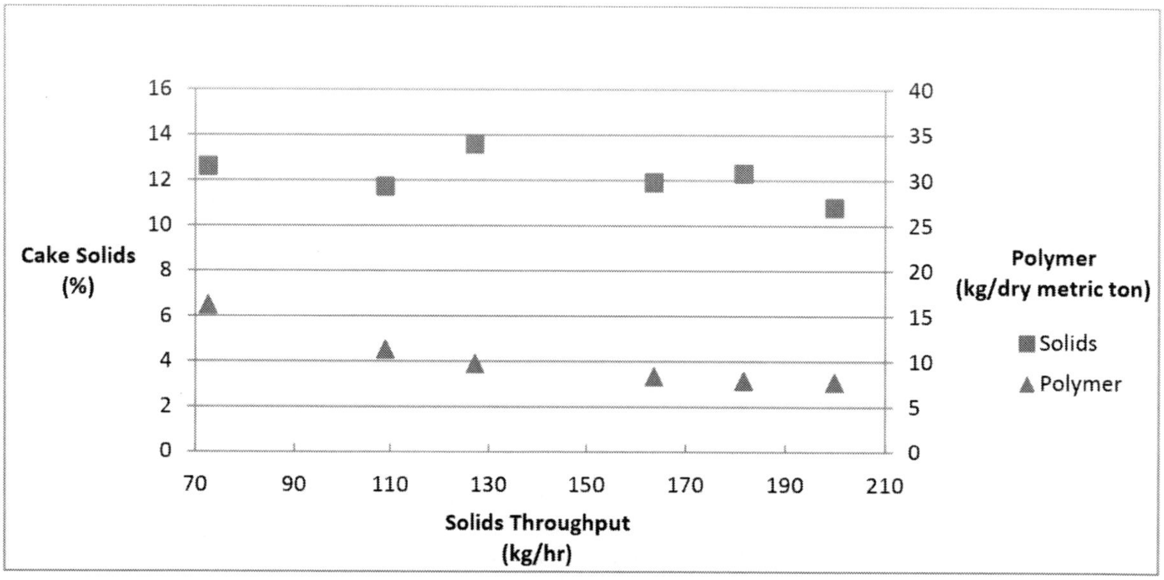

Figure 7. Rotary Fan Press Pilot Results

Overall the rotary fan press performed reasonably well. The solids throughput was high but the greatest cake solids concentration produced was 13.6%. Polymer usage ranged from 7.5 to 16 kg/dry metric ton (15 to 32 lb/dry US ton).

The final dewatering equipment technology piloted at the Kalispell AWWTP was PW Tech's volute press. Piloting occurred over two days at the beginning of October during which 15 trial runs were performed. Two emulsion polymers were used during the volute press pilot. Ashland Praestol k275FLX was used at the lower solids throughput and CIBA 8848 was used at higher solids throughput. The Ashland polymer did produce better results, however the runs at higher feed rates performed with the CIBA polymer could have negatively impacted the cake solids production. The feed solids concentration during trial runs ranged from 1.3% to 1.7%. Solids capture rate was a minimum of 99.5% for all runs.

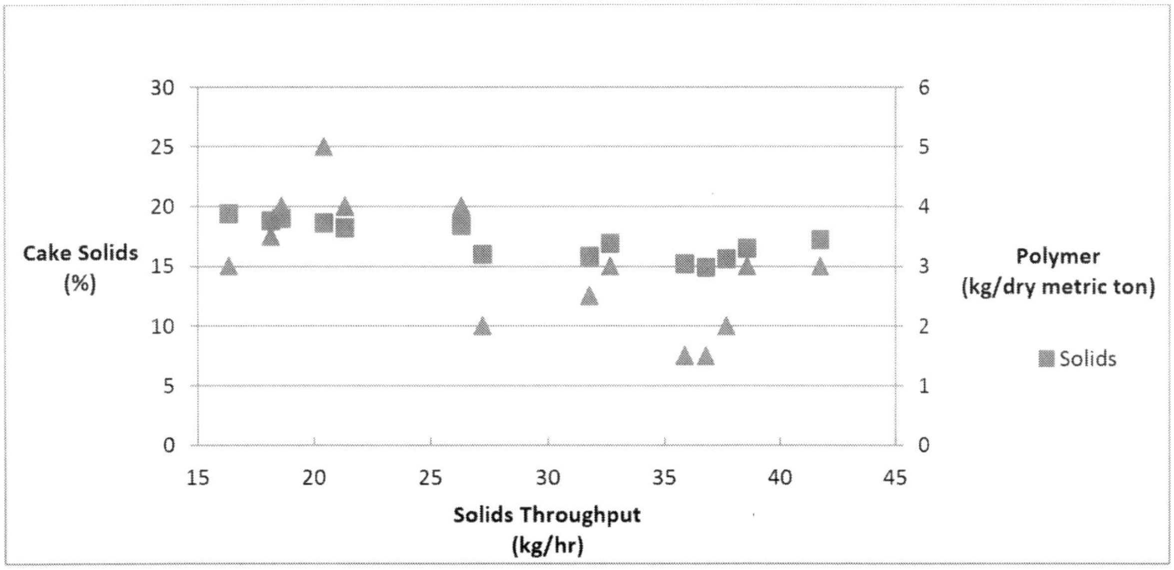

Figure 8. Volute Press Piloting Results

The PW Tech volute press performed well during piloting (Figure 8). Cake solids concentrations were a minimum of 14.9% and as high as 19.4%. Polymer usage ranged from 1.5 to 5 kg/dry metric ton (3.3 to 10 lb/dry US ton). At 1.5 kg/dry metric ton (3.3 lb/dry US ton) of polymer the cake solids concentration was just above 15%.

Dewatering Equipment Selection

Following onsite piloting efforts a proposal was received from each manufacturer for full scale dewatering equipment based on performance during piloting and future loading condition of 4,576 kg/day (10,080 lb/day). The PW Tech volute press was selected to replace one of the belt filter presses at the Kalispell AWWTP. Equipment selection was based on polymer consumption, equipment footprint, capital cost, and operator preference.

Missoula, Montana

A FKC model SHX 200 x 1500L screw press and PW Tech ES132 volute press were piloted in Missoula in August 2013 and August 2014, respectively.

Screw Press

The FKC screw press pilot included a myriad of timed trials, each designed to target a specific purpose. TWAS and digested sludge were both run separately, and a blend of 25% digested and 75% TWAS was also run. FKC utilized the dry polymer that the City currently uses for the centrifuge. FKC did note that the dry polymer did not work as well as expected for the anaerobically digested sludge. Additional bench tests were performed followed by additional runs with an emulsion polymer with marginally improved results.

The total cake solids ranged from 13.5% (TWAS only) to 29.2% (digested sludge only). This was an expected result because it is typically more challenging to dewater TWAS than digested sludge for municipal wastewater solids, particularly biological phosphorus removal TWAS. The tests on the blended sludge yielded 15-16% solids at an approximate polymer dose of 9 kg/dry metric ton (18 lb/dry US ton).

Volute Press

Piloting occurred over two days during which 15 trial runs were performed. Two emulsion polymers were used during the volute press pilot. Ashland Praestol k279FLX was used at the lower solids throughput and CIBA 8848FS was used at higher solids throughput. The Ashland polymer did produce better results, however the runs at higher feed rates performed with the CIBA polymer could have negatively impacted the cake solids production. The feed solids concentration during trial runs ranged from 2.4 to 2.5%. Solids capture rate was a minimum of 97.2% for all runs.

The PW Tech volute press performed well during piloting. Cake solids concentrations were a minimum of 14.1% and as high as 20.3%. Polymer usage ranged from 6 to 16.5 kg/dry metric ton (12 to 33 lb/dry US ton). At 6 kg/dry metric ton (12 lb/dry US ton) of polymer the cake solids concentration was 14.1%.

Dewatering Equipment Selection

Following piloting efforts HDR and City staff toured volute press installations at wastewater treatment facilities in Nevada to further confirm operational capabilities of the equipment. Proposals for full scale dewatering equipment were received from several dewatering equipment manufacturers, including technologies that were not piloted, based on performance during piloting and/or a future loading condition of 8.6 dry metric tons/day (9.5 dry US tons/day). The PW Tech volute press was selected for installation at the Missoula WWTP. Equipment selection was based on polymer consumption, equipment footprint, capital cost, and operator preference.

RESULTS

A PW Tech model ES 353 [2] volute press was installed at the Kalispell AWWTP and at the Missoula WWTP in the spring and summer of 2015, respectively. The press installed at each location is a two drum model with optional installation of a third drum for additional capacity in the future.

Kalispell, Montana

The PW Tech volute press installed at the Kalispell AWWTP has performed well since installation. Detailed performance data is not currently available but the volute press operates 12 hours/day, 7 days/week and produces cake with a solids concentration of 14-15% solids. Seasonal changes in dewatering performance have been experienced in Kalispell, but plant staff are investigating factors that may be impacting the dewaterability of biosolids.

Missoula, Montana

The PW Tech volute press has been successfully operating at the Missoula WWTP for almost 18 months. In the summer of 2016, plant staff extended the hours for dewatering operations from 6:00 am until midnight, six days a week. The extension of these operations allowed the phosphorus load to be spread out over the day with limited slug loading to the headworks of the plant. In return, the City observed steady final phosphorus levels in their plant effluent and additional ferric chloride dosing was not required for phosphorus reduction.

The volute press loading is 360 kg/hour (792 lbs/hour), equivalent to 8.6 metric tons/day (9.5 US tons/day). During start-up the press ran up to 413 kg/hour (910 lb/hour). Typical loadings in the fall of 2016 were around 286 kg/hour (630 lbs/hour) and the press ran for approximately 8 hours a day. The City does run the press up to 265 L/min (70 gpm) and at that point have reached about 368 kg/hour (811 lbs/hour) on a temporary basis. The press meets the design load of 8.6

metric tons/day (9.5 tons/day) but current flows to the plant does not require the volute press to operate at the higher design loading for any extended period of time.

During start up the press produced 17.8% cake solids and 94.5% capture rate. During the limited time for polymer testing at start up, it was determined that additional polymer testing would allow the percent solids and capture rate to be improved upon. Since comfort was gained in running the volute press, limited data is being gathered regarding the capture rate. Additionally, it is difficult to get a representative sample for the filtrate return. Regardless, the average capture rate and cake solids in October 2016 was 95.5% and 15.4%, respectively. Recent data for cake solids is shown in Figure 9 below.

It is noted that while the volute press operates above 15% solids about half the time, there is considerable time that it operates below the 15%. City operations staff have chosen to operate the volute press at less than optimal conditions because the dewatered biosolids are conveyed to the neighboring city-operated composting facility. Therefore, there are no hauling costs associated with biosolids disposal, the higher moisture content is beneficial to the composting process, and less polymer can be used to treat to the desired level, saving on chemical costs.

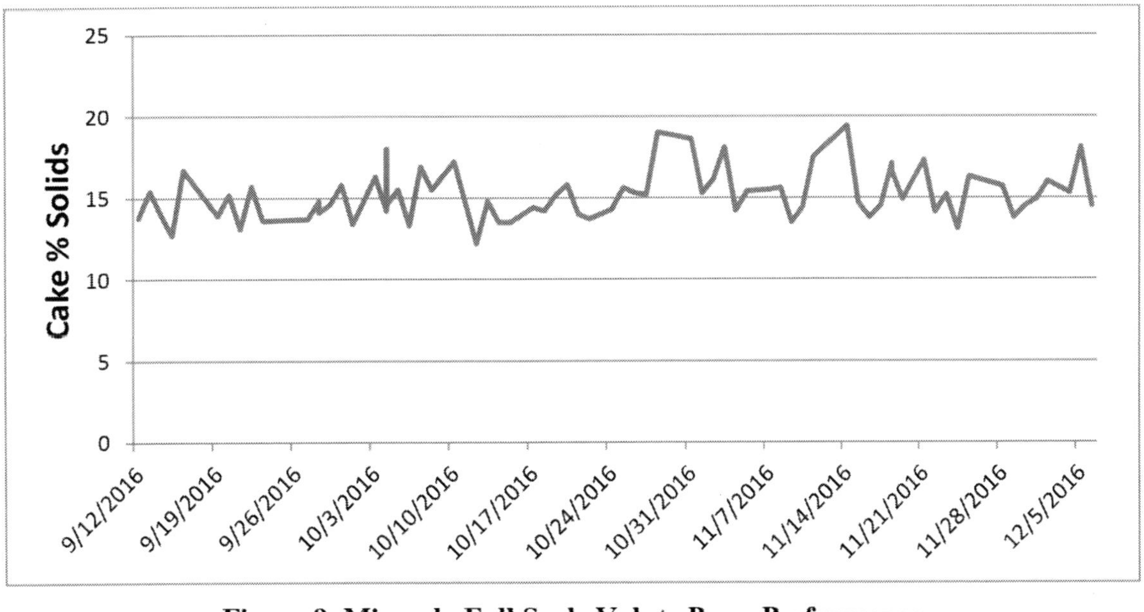

Figure 9. Missoula Full Scale Volute Press Performance

A few different polymers have been tested since the volute press has been installed in an attempt to find the least expensive polymer. The required dosing rate to reach the desired cake percent solids changes with each polymer, but each polymer tested has been within an acceptable range of 8 to 20 kg/dry metric ton (16 to 40 lbs/dry US ton), averaging 14.5 kg/dry metric ton (29 lbs/dry US ton). The City has settled on a less expensive polymer that requires a slightly higher dosage for overall cost savings.

CONCLUSIONS

The volute press model installed at the Kalispell AWWTP and at the Missoula WWTP was one of the first of its kind at a large, municipal wastewater treatment plant. The press at each facility is producing cake at the desired solids concentration. Additionally, the automation of the equipment has aided in reducing the return nutrient load within the plant by equalizing the pressate flow.

ACKNOWLEDGEMENTS

The author gratefully acknowledges the staff at both the Kalispell and Missoula wastewater treatment plants for their cooperation and assistance in collecting performance data and providing information on the operation of their respective volute presses.

Piloting to Establish Performance Criteria at Miami's South District WWTP

Terry Goss[1]; Brian, Stitt[1]; Manuel Moncholi[2]; Mohammad Abu-Orf[3]; Ismael Diaz[4]
[1]AECOM Water, [2]Miami-Dade Water and Sewer Department, [3]Formally AECOM Water, [4]Gannett Fleming

ABSTRACT

A six-month centrifuge thickening, digestion and centrifuge dewatering pilot study was conducted at Miami-Dade's South District Wastewater Treatment Plant (SDWWTP). The pilot study was set up to simulate future thickening, digestion and dewatering operating conditions to establish future thickening and dewatering performance criteria. Full scale piloting was conducted in three phases. Phase 1 operation, focusing on thickening performance began in November 2015. The testing showed that the centrifuge could reliably produce thickened solids at 5-6% total solids (TS) and achieve greater than 95% solids recovery. Testing was conducted using both dry and emulsion polymers. The dry polymer required 2.5 to 3.5 gram per dry kilogram (g/kg) or 5-7 pound per dry ton (lb/DT) active compared to 0.5-1.5 g/kg (1-3 lb/DT) active emulsion. It was also possible to thicken the sludge to 5-6% TS without the use of polymer but this reduced hydraulic throughput by about 50% to allow solids recoveries to remain above 90%. During Phase 2 operation, which began in December 2015, the sludge feed to the thickening centrifuge was switched to thickened sludge from the gravity concentrators to increase centrifuge solids throughput. The centrifuge thickened sludge was fed to Digester 9 which was isolated to receive only mechanically thickened biosolids at 5-7% TS. The solids content in Digester 9 was increased from approximately 2% to approximately 3.5% TS and the volatile solids reduction (VSR) in Digester 9 averaged 46%. Phase 3 of the pilot operation focused on the centrifuge dewatering of the digested biosolids from Digester 9. Testing was conducted using both dry and emulsion polymers with multiple polymer injection points. The testing showed that 16-18% TS cake could be achieved with 10 to 15 g/kg (20-30 lb/DT) active dry polymer. The pilot testing showed that the dewatered cake solids was lower than the preliminary design value of 20% TS with a presumed 12.5 g/kg (25 lb/DT) active dry polymer dose. Dryer cake at 20-22% TS could be produced using emulsion polymer but required higher active polymer dosages above 20 g/kg (40 lb/DT).

KEYWORDS

Biosolids, Thickening, Dewatering, Anaerobic Digestion, Centrifuges, Pilot Testing

INTRODUCTION

As part of the Consent Decree Project for Miami Dade's South District Wastewater Treatment Plant (SDWWTP), the Program Management and Construction Management (PMCM) team conducted pilot testing of centrifuge thickening prior to anaerobic digestion and centrifuge dewatering post anaerobic digestion. The pilot dewatering testing used digested biosolids from the digester receiving only mechanically thickened biosolids at high solids input concentration after steady state operation was achieved. The pilot testing established performance criteria to be used for the design build documents and associated process performance guarantee.

SDWWTP is located in the southeastern portion of Miami-Dade County and serves the southern and southwestern portion of the county. The SDWWTP is a high purity activated sludge secondary treatment facility with a permitted capacity of 112.5 million gallons per day (mgd) or 17,740 cubic meters per hour (m^3/h).

SDWWTP produces only waste activated sludge (WAS) and wasting is controlled using a modulating valve and flow meter that is tapped off the return activated sludge pumps. The WAS is mixed with polymer in the piping and sent to four 55 feet (16.8 meter) diameter gravity thickeners with a 13 feet (4.0 meter) side-water depth. The gravity thickeners thicken the solids to 2-3% total solids (TS) before being stabilized in twelve 105 feet (32 meter) diameter anaerobic digesters each with a nominal operating volume of 1.5 million gallons (5,700 cubic meters). The digesters are arranged in three clusters of four digesters per cluster, where two digesters per cluster operate as primary digesters and two operate as secondary digesters. Digester Cluster 3 normally operates with two of the digesters acting as sludge storage tanks before dewatering. Digester 9 is located in Cluster 3 and acts as a primary digester that discharges to Digester 10 which acts as a secondary digester. The digested biosolids are further dewatered using four (4) Alfa Laval PM 75000 centrifuges which achieve 18-22% TS. The sludge fed to the centrifuges is currently conditioned using a dry polymer type system.

SDWWTP is replacing the existing gravity thickeners with new centrifuge thickeners which are expected to improve the performance and increase the capacity in the existing digestion complex. The existing dewatering centrifuges are also near the end of their useful service life and the design includes replacing the system with a new centrifuge dewatering complex.

PILOT STUDY OVERVIEW AND OBJECTIVES

To better establish performance criteria for the new thickening and dewatering centrifuges, a six-month centrifuge thickening, digestion and centrifuge dewatering pilot study was conducted at Miami-Dade's SDWWTP. The pilot study was set up to simulate future thickening, digestion and dewatering operating conditions to establish thickening and dewatering performance criteria. **Figure 1** shows a plan view layout of the site and identifies the locations for the centrifuge thickening and dewatering pilot trailers. **Figure 2** provides photos of the pilot testing trailers provided by Centrisys which were selected after a competitive bidding process. The PMCM team oversaw the piloting effort with outstanding support from the SDWWTP staff.

Periodic samples collected throughout the pilot operation were all analyzed for TS. During Phase 2 and Phase 3, the PMCM team regularly monitored the volatile solids (VS) content of the thickened sludge fed to the digester and digested biosolids samples. Digested biosolids pH was also measured. The centrate samples were also analyzed for total suspended solids (TSS). College interns were trained to perform the sampling and laboratory analysis throughout the duration of pilot testing period.

The pilot operation began in November 2015, and was conducted in three distinct Phases as outlined in **Table 1**.

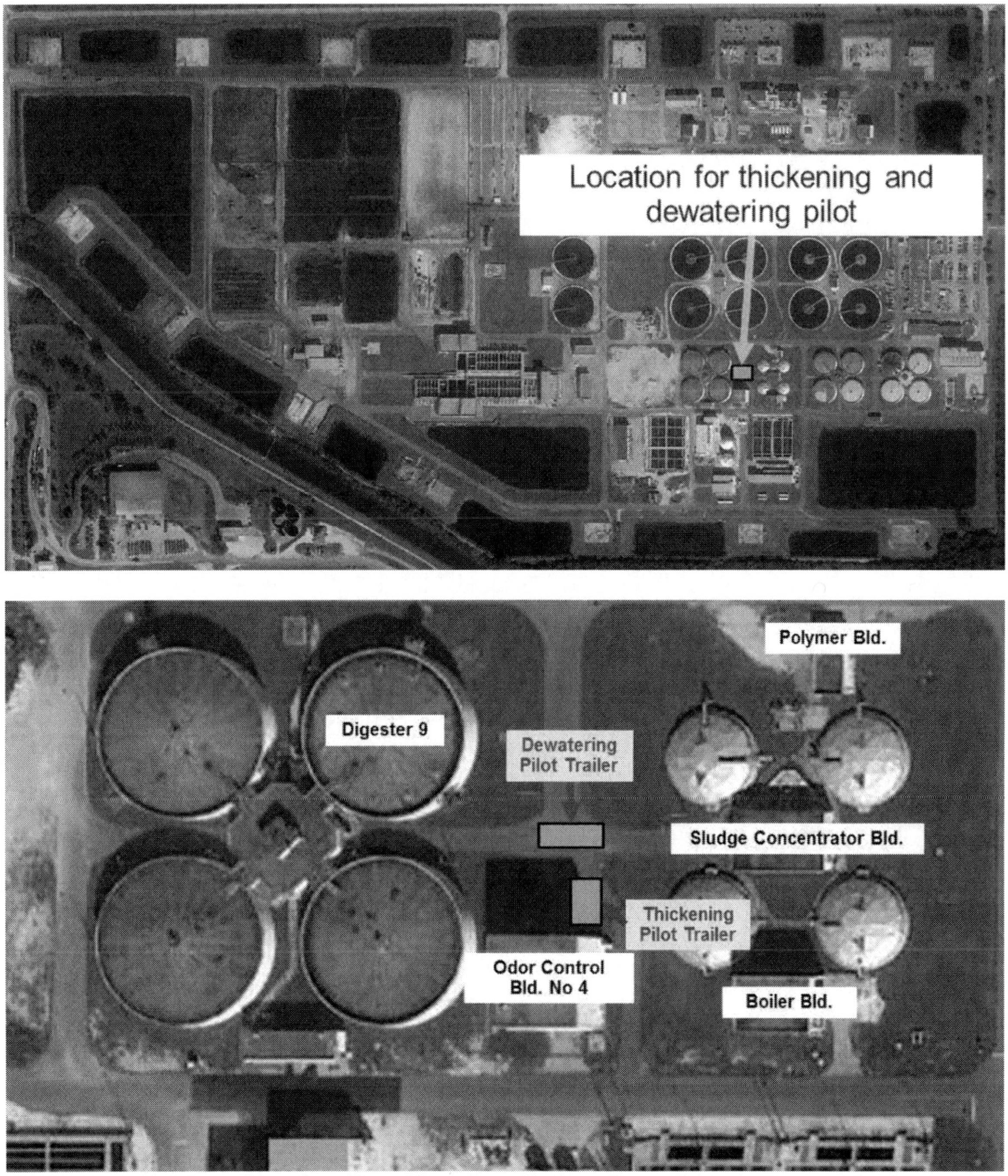

Figure 1: Site Plan Showing Centrifuge Installation, Overall (top), Zoomed In (bottom)

Figure 2: Photos of Pilot Equipment, Thickening (left), Dewatering (right)

Table 1: Summary of Pilot Testing Phases

Pilot Phase	Objective	Duration	Sample measurements
Phase 1: Thickening pilot operating with unthickened waste activated sludge (WAS)	Determine centrifuge WAS thickening performance criteria	4 weeks	TS – all TSS - centrate
Phase 2: Thickening pilot operating with gravity thickened WAS	Simulate high rate anaerobic digestion to monitor performance and prepare sludge for Phase 3	5 months	TS – all VS/TS – digester feed & digested biosolids pH – digested biosolids
Phase 3: Dewatering pilot operating with anaerobically digested biosolids following Phase 2 Thickening	Determine centrifuge dewatering performance criteria with centrifuge thickened anaerobically digested WAS	4 weeks	TS – all TSS - centrate

PHASE 1: THICKENING PILOT TESTING

Phase 1 piloting operation was based on feeding unthickened WAS to the pilot thickening centrifuge. The purpose of Phase 1 operation was to determine the optimum polymer design conditions and performance of the centrifuge thickening. The overall target for the centrifuge thickening performance, as stated in the basis of design and specifications, was to thicken the WAS to 5.5% TS while maintaining greater than 95% solids recovery. Determining the necessary polymer dose to achieve this performance is also important. Parameters that were adjusted for the centrifuge thickening included the pool depth, bowl speed and differential scroll speed.

Thickening Pilot Testing - No Polymer Operation

The system initially started up on lower sludge flows with no polymer injection. **Figure 3** summarizes the operation without polymer for a medium bowl speed equal to 2,590 revolutions per minute (RPM) and differential speed of 12 RPM with flow rates ranging from 40 to 100 gallons per minute (gpm) or 151 to 379 liters per minute (L/min).

This operation showed that it was possible to thicken the sludge up to 6% TS without the use of polymer, but at higher throughputs (above 60 gpm or 227 L/min), the solids recovery was sacrificed. The tests were conducted at a deep pool depth, a medium pool depth and a shallow pool depth. The trend shows that at the shallowest pool depth solids recovery improved but total thickened solids was sacrificed. The thickened solids concentration at the deepest and medium pool depths were nearly the same but it should be noted that the feed solid content was lower when testing the deepest pool depth (0.9% TS) while the feed solids were closer to 1.4% solids when testing the medium pool depth. If testing was done on the same feed solids concentration, it would be expected that the solids would be thicker at the deepest pool depth.

Results of the testing without polymer indicated that it is possible to maintain the desired thickened TS concentrations without the use of polymer. Although the operation without polymer may sacrifice solids recovery, future operation at this condition may be desired when sludge production is below design capacity as it could reduce operation and maintenance costs associated with polymer consumption.

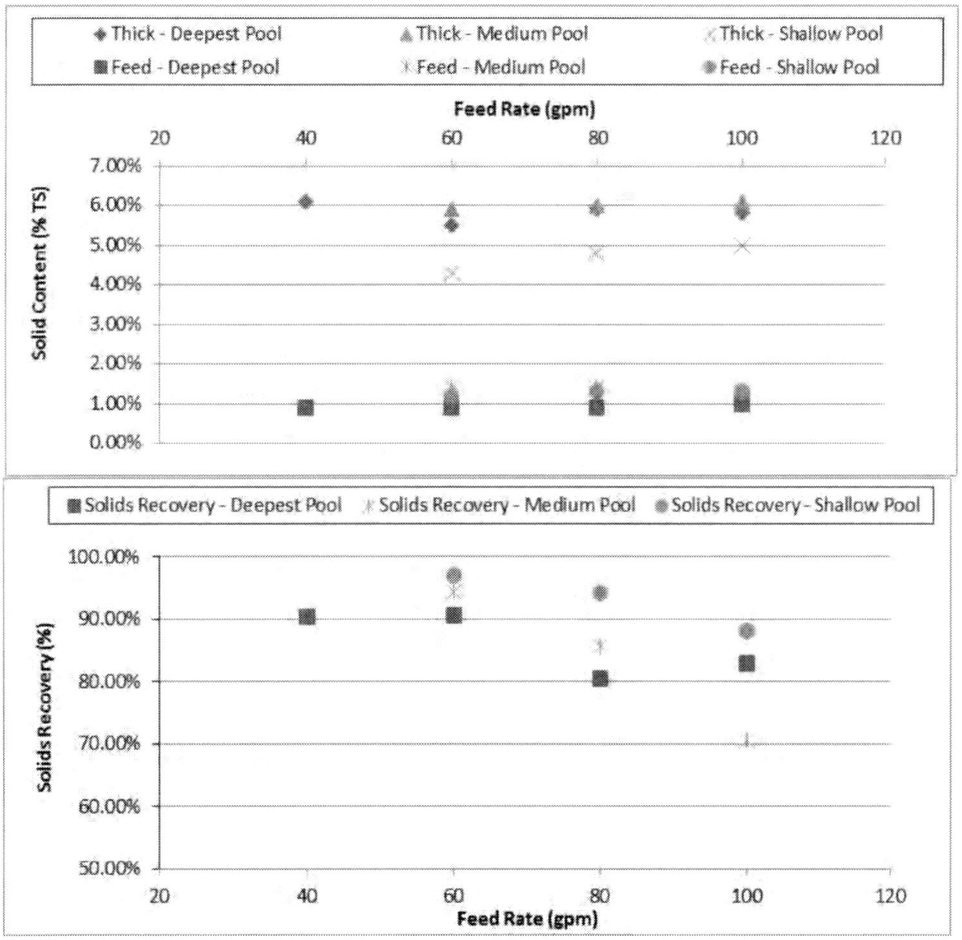

Figure 3: No Polymer Operation, Medium Bowl Speed (2,590 RPM)

Thickening Pilot Testing – Emulsion Polymer Setup and Initial Testing

The pilot unit was set up to allow injection of polymer at two locations as illustrated in **Figure 4**; either directly into the bowl of the unit (internal injection) or in the sludge feed line upstream of the centrifuge inlet (external injection). Polymer flow was measured during each sampling event using a calibration column located on the pilot trailer.

The initial polymer used was PRAESTOL® K144-L which is a cationic, high molecular weight emulsion polymer. Polymer dosage, using internal injection, was slowly increased at a constant throughput of 100 gpm (379 L/min) while visually monitoring the clarity of the centrate. The testing was conducted at the shallowest pool depth and a differential speed of 12 RPM. Once polymer was added, the bowl speeds were reduced from 2,590 RPM to between 1,900 to 2,100 RPM to keep the thickened solids from being too thick. The results showed that with the increased polymer dosages and reduction in bowl speed, the thickened solid content remained steady but the solids recovery improved.

Once relatively clear centrate was achieved, the flowrate to the machine was increased to maximize throughput while optimizing solids recovery (based on visual observations of centrate quality). In addition, other cationic, high molecular weight emulsion polymers were tested in the machine. Optimization testing showed that the PRAESTOL K144L polymer worked the best.

Figure 4: Thickening Polymer Injection Point

Thickening Pilot Testing – Dry Polymer Setup

The Centrisys thickening unit contains an emulsion polymer system and is not set up with the provisions to operate on dry polymer so a creative solution was required to facilitate testing dry polymer in the pilot Centrifuge. In order to allow the testing, one of SDWWTP polymer make up systems, which was not currently in use, was used to make up the dry polymer solution which was pumped to a tote that was connected to a dedicated portable pump to meter the dry polymer solution into the centrifuge. Photos of set up are provided in **Figure 5**. The pump on the trailer used for emulsion polymer was too small to pump the dry polymer solution.

Figure 5: Temporary Dry Polymer Set Up

A calibration curve was developed for the polymer pump by a series of bucket tests conducted at different pump speeds and the curve was compared to the theoretical zero psi pump curve showing good convergence. The dry polymer used for testing was Polydyne C-3283 which is the polymer the plant currently uses in their dewatering centrifuges.

Thickening Pilot Testing – Results

Pilot tests were run to further determine results for the centrifuge thickening operation. The results from the optimization trials showed that good performance could be achieved with medium bowl speeds of 2,400 to 2,600 RPM. At a lower flow rate of 100 gpm (379 L/min), an emulsion polymer dose of 1 pound per dry ton (lb/DT) or 0.5 grams per dry kilogram (g/kg) showed good results but as the flow was increased to 150 gpm (568 L/min), at least 2 lb/DT (1 g/kg) was needed. The higher polymer dose requirements at higher flows also correlated to the point where external polymer injection provided better results than internal polymer injection. The polymer curves and flow curves conducted during the performance testing period was to further test the limits for the operational parameters. Additional performance testing was conducted in March of 2016 using the plants dry polymer, which is Polydyne C-3283. The purpose was to repeat the November testing but with the dry polymer that more closely represents future operation.

Thickening Pilot Testing – Polymer Curve Results

Polymer curve tests were conducted by maintaining a constant volumetric throughput of sludge feed to the centrifuge but changing the polymer dose to measure the impact. With the exception of changing polymer dose, all other parameters on the centrifuge remained the same for each polymer curve test. In November 2015, three different polymer curves were generated for the 144-L emulsion polymer at 100 gpm (379 L/min), 150 gpm (568 L/min) and 170 gpm (644 L/min) WAS flow rates through the pilot centrifuge. In March 2016, an additional 144-L emulsion polymer curve was conducted at 130 gpm (492 L/min) and several dry polymer curves were conducted at 130, 150 and 170 gpm (492, 568, and 644 L/min).

Figure 6 also shows the data for emulsion polymer curves at the 130 gpm (492 L/min) and 150 gpm (568 L/min) WAS feed, conducted at medium bowl speed (2,400 – 2,600 RPM). At 130 gpm (492 L/min) WAS feed, the curve was conducted with internal polymer injection and at 150 gpm (568 L/min) WAS feed, the curve was conducted with external polymer injection. The concentration of the WAS during this testing ranged from 1.1% to 1.5% TS.

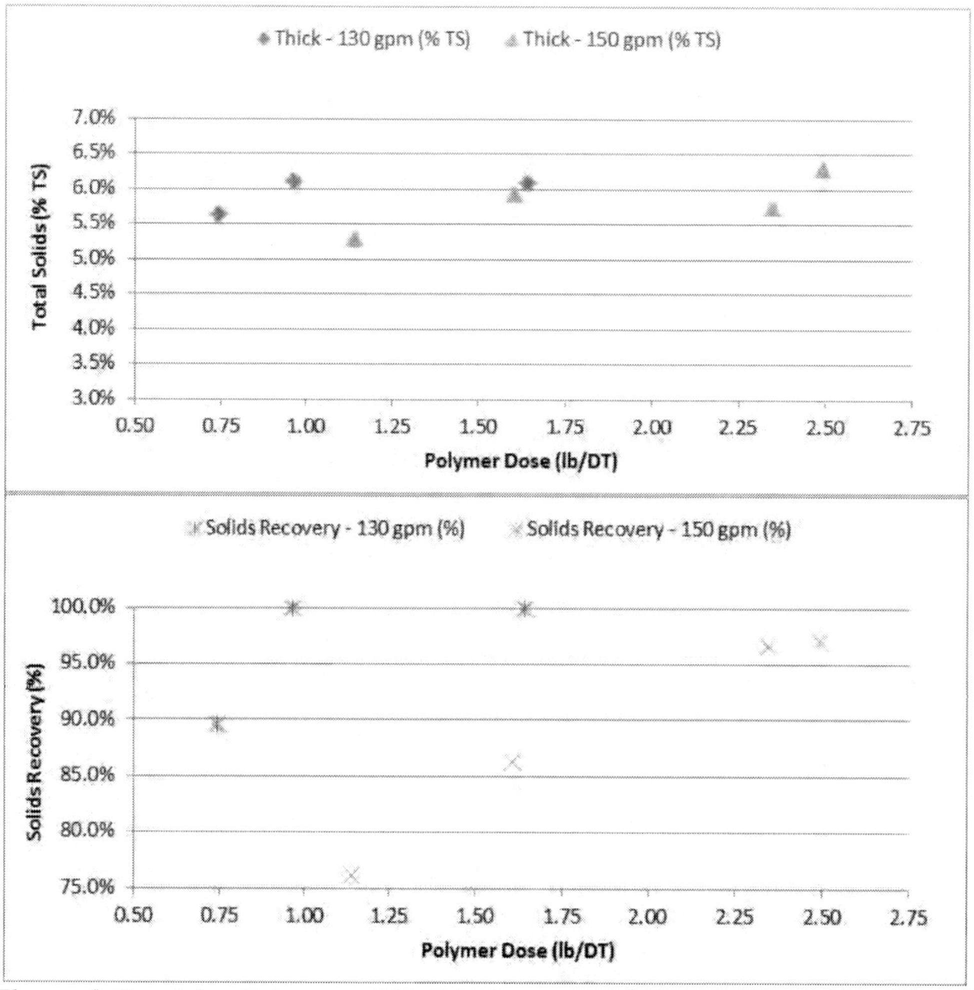

Figure 6: Emulsion Polymer - 130 & 150 gpm WAS feed, Medium Bowl Speed

As shown in **Figure 6**, at 130 gpm (492 L/min) WAS feed with internal emulsion polymer injection and a deep pool depth, good performance is obtained for polymer doses greater than 1 lb/DT (0.5 g/kg) active achieving close to 6% TS. At polymer doses greater than 1 lb/DT (0.5 g/kg) active, close to 100% solids recovery was also achieved. Exceeding 1.0 lb/DT or 0.5 g/kg (up to 1.6 lb/DT or 0.8 g/kg active) did not have much of an impact on the solids concentration or the solids recovery.

At 150 gpm (568 L/min) WAS feed with external emulsion polymer injection and a shallow pool depth, greater than 5% TS was achieved for all polymer doses tested. The solids recovery exceeded 95% for the two highest dosing points when active dosing was greater than 2.3 lb/DT (1.2 g/kg).

The dry polymer curve for the WAS flow rate of 130 gpm (492 L/min) with both internal and external dry polymer injection is shown in **Figure 7** for a medium bowl speed of 2,586 RPM and deep pool depth. The polymer concentration during these tests was approximately 0.8% TS. An additional 130 gpm (492 L/min) WAS polymer curve, also shown in **Figure 7**, used a 0.4% TS polymer solution (with internal dry polymer injection) at a bowl speed of 2,408 RPM and deep pool depth. For all tests, the feed averaged 1.2 to 1.4% TS.

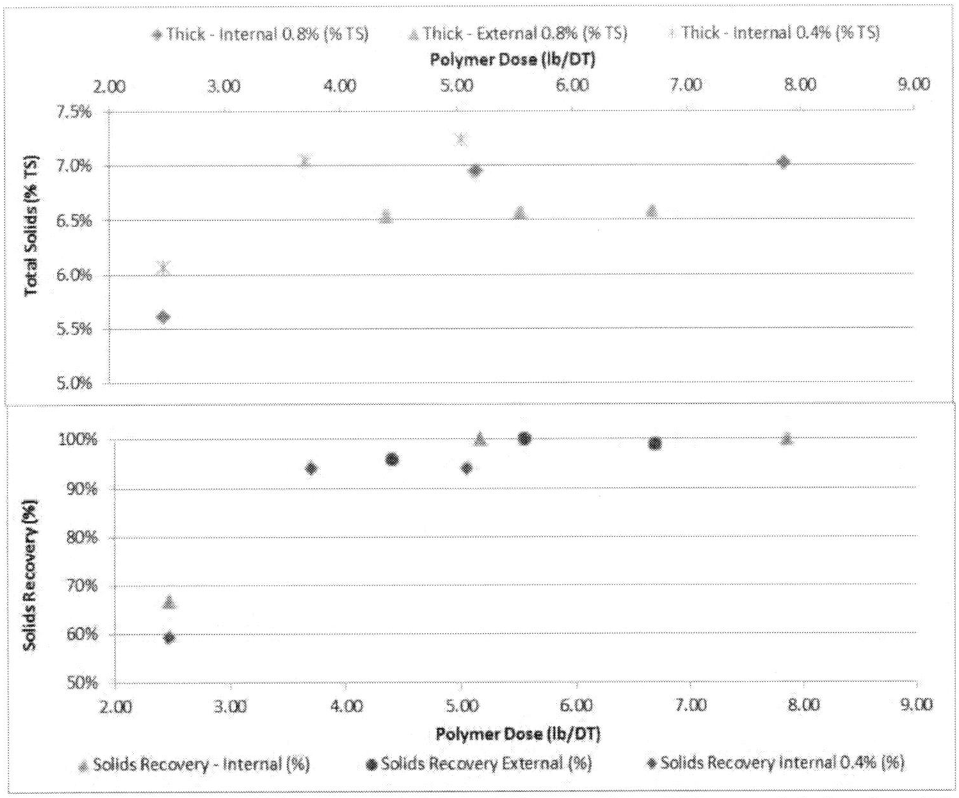

Figure 7: Dry Polymer - 130 gpm WAS Feed, Internal and External Injection, Medium Bowl Speed

For internal dry polymer injection with a 0.8% TS polymer solution, greater than 5.5 % TS was achieved for all polymer doses tested, however, only greater than 95% recovery was achieved at polymer doses greater than 5 lb/DT (2.5 g/kg). For external dry polymer injection with a 0.8% TS polymer solution, the active polymer dose ranging from 4.3 to 6.7 lb/DT (2.2 to 3.4 g/kg) did not show significant differences in centrifuge performance in terms of solids content. The thickened solids content ranged from 6.5 to 6.6 % TS and the solids recoveries were greater than 95% for all samples. The TSS sample analyzed for 4.3 lb/DT (2.4 g/kg) may also have had an error and it was noted that the centrate was visually dirtier than the samples with high polymer dosages. For internal dry polymer injection with a 0.4% polymer solution, greater than 5.5 % TS was achieved for all polymer doses tested and at polymer doses greater than 3.8 lb/DT (1.9 g/kg), the solids recovery was near the 95% target. The thickening pilot results show that at 130 gpm (492 L/min) WAS feed, approximately 5 lb/DT (2.5 g/kg) of the dry polymer is required to maintain greater than 95% recovery but at the settings tested, the thickened sludge exceeds the needed solids content and approaches 7% TS. Further optimization would be required to maintain the target of 5.5% TS such as lowering the pool depth.

The dry polymer curves for the WAS flow rate of 150 gpm (568 L/min) with both internal and external dry polymer injection is shown in **Figure 8** for a medium bowl speed of approximately 2,585 RPM and deep pool depth. The polymer concentration during these tests ranged between 0.7 to 0.8% TS. For all tests, the feed averaged 1.2 to 1.3% TS.

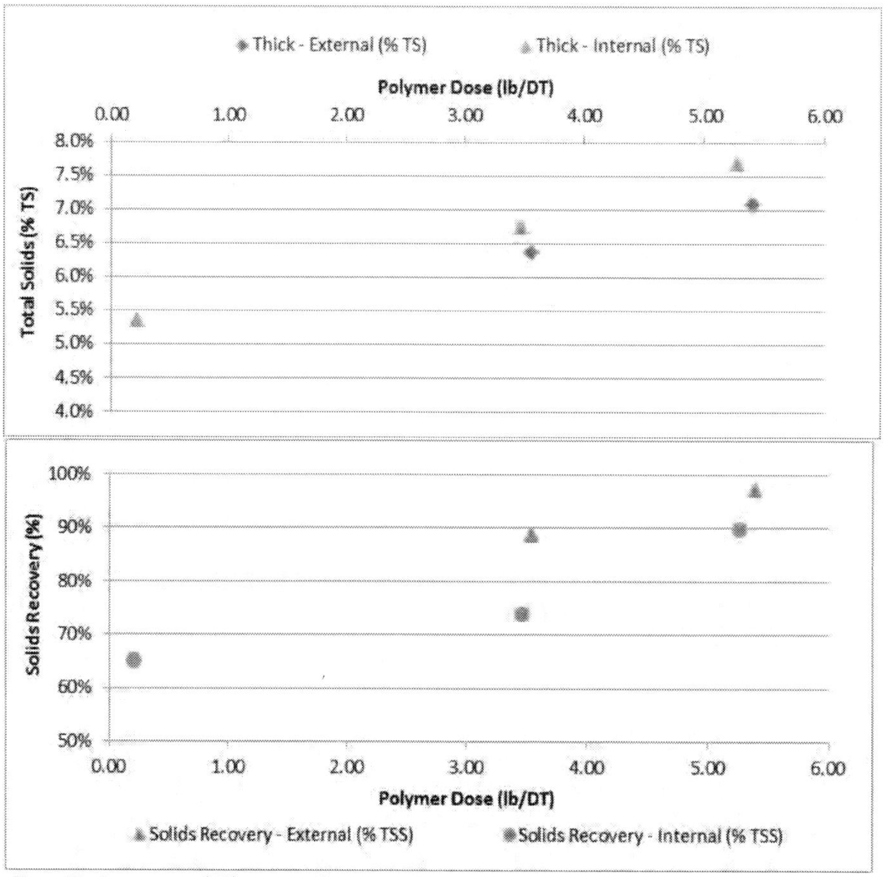

Figure 8: Dry Polymer - 150 gpm, Internal and External Injection, Medium Bowl Speed

For internal dry polymer injection, greater than 5% TS was achieved for all polymer doses tested, however, none of the points achieved greater than 95% recovery. Flocs were observed in the centrate for the external polymer injection testing so it was evident that this flow was too high for the internal polymer injection to work efficiently. For external injection, greater than 6% TS was achieved for all polymer doses tested. The solids recovery, however, was only greater than 95% for the highest active polymer dose which was 5.3 lb/DT (2.7 g/kg). It may have been possible to get more optimal performance (closer to 5.5 % TS with greater than 95% solids recovery) with a shallower pool depth.

Thickening Pilot Testing - Extended Operation Results

After generating the polymer, flow and bowl speed curves, the unit was operated several days at a constant flowrate with optimized settings to test the stability of operation throughout the course of a day. During November 2015, these tests were conducted using the 144-L emulsion polymer at a WAS flow rate of 135 gpm (511 L/min), 165 gpm (625 L/min) and 200 gpm (757). In March 2016, the extended operation testing with dry polymer was repeated twice at 135 gpm (511 L/min). Samples were collected during these trials approximately once every hour. The results with emulsion polymer were stable throughout the course of a run but the results with the dry polymer showed more fluctuation with solid recovery degradation over time (see **Figure 9)**. It was planned to again repeat the 135 gpm (511 L/min) extended operation testing using the dry polymer, however, the gearbox on the progressing cavity thickened sludge pump failed before this testing was completed. Due to the lead time to repair, it was not possible to conduct the additional testing in the schedule for the project. It is believed that limitations in the set-up and lack of mixing on the polymer tote contributed to the poor performance for the extended operation testing with dry polymer. The concentration of polymer samples collected throughout the extended run varied between 20 to 40% on both days tested.

A comparison of the extended operation at 135 gpm (511 L/min) with emulsion and dry polymer is depicted in **Figure 9**. The runs with dry polymer at 4-7 lb/DT (2-3.5 g/kg) active polymer dosages were conducted with a deeper pool depth than the run with emulsion polymer conducted at 2-3 lb/DT (1-1.5 g/kg) active polymer dosages. The deeper pool depth is likely the reason the solid content was higher with the dry polymer testing than the emulsion polymer testing. The feed WAS concentration during all three runs ranged from 1.1 to 1.5% TS.

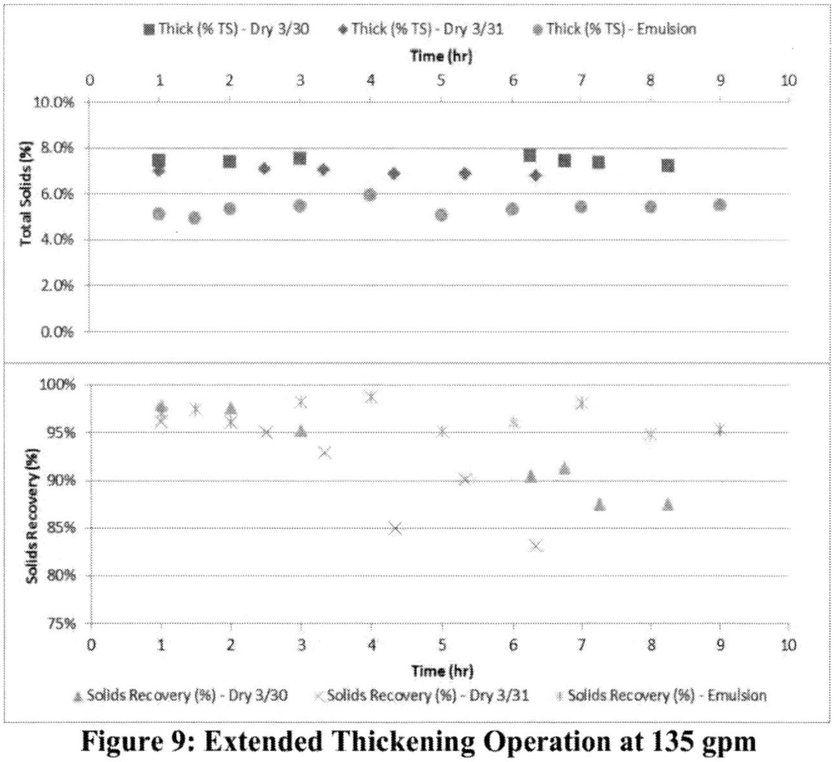

Figure 9: Extended Thickening Operation at 135 gpm

The thickening pilot testing showed that the centrifuge could reliably produce solids at 5-6% TS and achieve greater than 95% solids recovery. Testing was conducted using both dry and emulsion polymers. The dry polymer required 5-7 lb/DT (2.5-3.5 g/kg) active dosing compared to 1-3 lb/DT (0.5-1.5 g/kg) active based on the emulsion. It was also possible to thicken the sludge to 5-6% TS without the use of polymer but this reduced hydraulic throughput by about 50% to allow solids recoveries to remain above 90%.

PHASE 2 – CONTINUOUS THICKENING PILOT OPERATION

During Phase 2 operation, mechanically thickened sludge was fed to Digester 9 to simulate future high rate digestion conditions and to increase the solids content of the digested biosolids for the Phase 3 dewatering pilot operations. The feed to the thickening pilot was switched to gravity thickened sludge to increase solids loading through the thickener to increase the turnover rate in Digester 9. This mode of operation started in December 2015, continued through March 2016, and remained running in parallel with the dewatering piloting conducted in Phase 3. During Phase 2 operation, the thickening centrifuge was fed approximately 100 to 150 gpm (379 to 568 L/min) of sludge from the gravity concentrator and operated continuously. **Figure 10** summarizes the centrifuge performance during the Phase 2 operation. On March 21, 2016 the feed was switched back to the unthickened WAS when the additional thickening testing using dry polymer was conducted.

During Phase 2 operation, the feed from the gravity concentrators averaged 2.2% TS but ranged between 1.5 to 3% TS. The concentration in the gravity concentrator measured by the pilot staff compared closely with the concentration based on the plant records. The thickened solids content averaged 6.2% TS but fluctuated between 5 to 7% TS. During the Phase 1 operation, it was found that the hydraulic pressures inside the machine would increase overtime, likely due to grit building up in the machine, and would shut down on an alarm if pressures reached too high of a level. In order to mitigate this, the machine was operated at a lower speed with higher polymer doses for most of the Phase 2 operation which allowed for the unit to operate continuously. Weekly cleaning and routine maintenance was conducted throughout operation.

Throughout Phase 2 operation, the solids content in Digester 9 was increased to approximately 3.5% TS. In comparison during this period, the other operational digesters operated at 1.5 to 2% TS as shown in **Figure 11**.

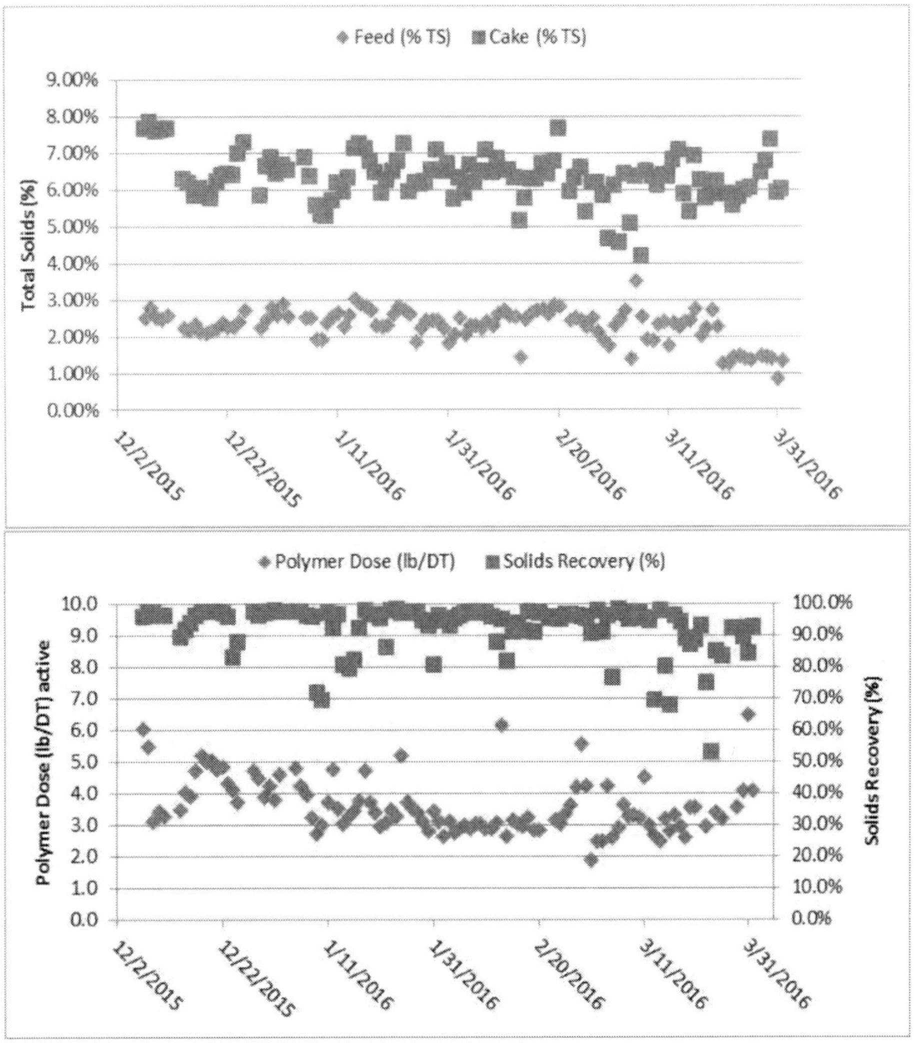

Figure 10: Phase 2 Continuous Thickening Operation

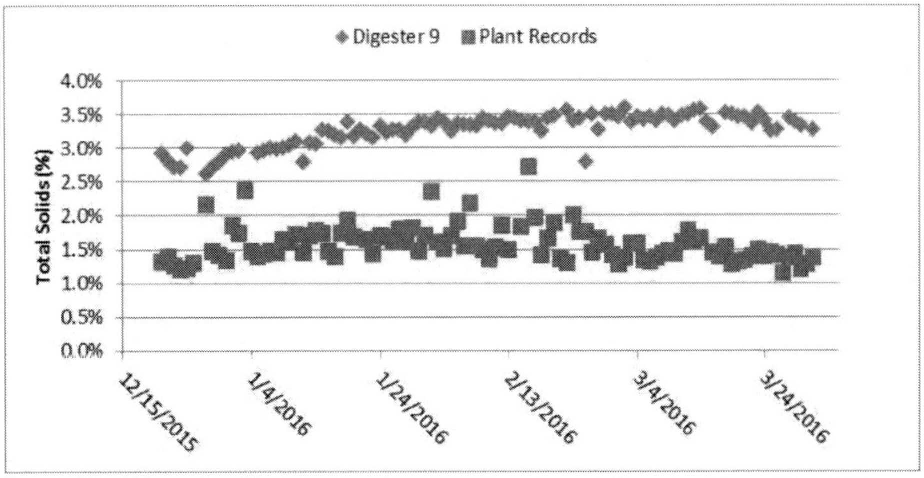

Figure 11: Digester Concentration Comparison, Digester 9 vs. Plant Records

During the Phase 2 operation period, Digester 9 averaged approximately 46% volatile solids reduction (VSR) with raw undigested sludge averaging 87% VS/TS and digested biosolids averaging 77% VS/TS. Gas production averaged 16 ft^3/lb VSR (1 m^3/kg VSR) throughout the four months of operation. The Solids Retention Time (SRT) in the digesters averaged a little more than 30 days and solids loading rate (SLR) averaged 0.11 lb VS/ft^3-d (1.8 kg VS/m^3). When comparing the pilot data to the plant's other digesters, the volatile solids feed matched the plant records but the digester SRT was shorter at approximately 20 days and digested biosolids volatile solids content was slightly lower with the plants VSR during this period averaging approximately 42%. Thus, it appears that operating digestion with longer SRT and higher thickened sludge increased the VSR by about four percentage points, reducing the biosolids for downstream dewatering and beneficial use.

It was desired to have Digester 9 at a new steady state before starting the dewatering piloting, so dewatering was targeted to start after approximately 3 digester SRT's have been achieved. **Figure 12** confirms that the dewatering performance testing was conducted after three digester turnovers were achieved in Digester 9. The data in **Figure 11** showed that the concentration in Digester 9 reached a consistent value of approximately 3.5% TS by March 2016.

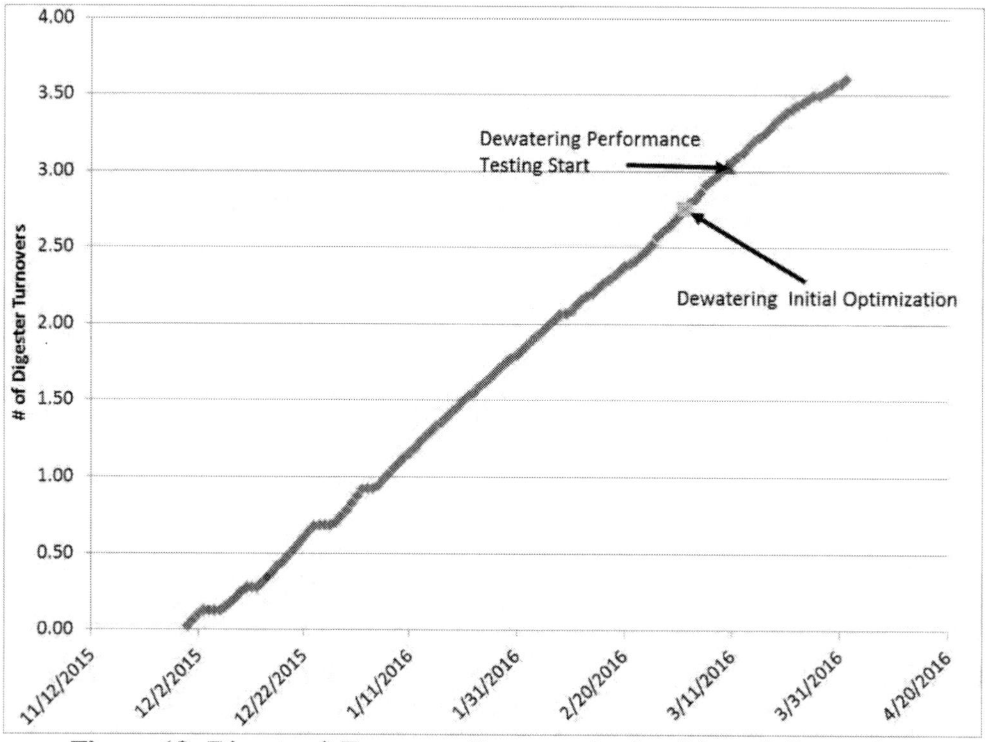

Figure 12: Digester 9 Turnover Progress during Phase 2 Operation

PHASE 3 – DEWATERING PILOT TESTING

The purpose of Phase 3 operation was to determine the optimal design conditions and performance of the dewatering centrifuge using the thickened biosolids fed from Digester 9. The overall target for the centrifuge dewatering performance as stated in the basis of design and specifications was to dewater the thickened digested biosolids to 20% TS while maintaining greater than 95% solids recovery. The necessary polymer dose to achieve this performance is also important to determine. The draft specifications indicate that active polymer dose should be less than 25 lb/DT. Textbook values for anaerobically digested WAS only biosolids are not readily available as most anaerobic digesters in industry digest WAS blended with primary sludge. Metcalf and Eddy 5th edition lists 16 to 25% TS expected for untreated WAS and lists 22 to 25% TS expected for anaerobically digested combined WAS and primary. For both untreated WAS and anaerobically digested WAS and primary, the polymer consumption is expected to be 15 to 30 lb/DT (7.5 to 15 g/kg) active and solids recoveries are expected to be 95% or greater (M&E 5th edition, 2014).

Dewatering Pilot Testing – Emulsion Polymer Initial Testing

The initial testing started with emulsion polymers on March 2, 2016. Five cationic, high molecular weight emulsion polymers were tested in order to select the most effective two polymer types for further testing. The emulsion polymers were able to achieve 22 to 25% TS. Based on visual observation of cake dryness and centrate quality, Centrisys proceeded with purchasing more of the two candidate emulsion polymers (PRAESTOL 274 FLX and 290 FLX emulsion polymers).

Dewatering Pilot Testing – Dry Polymer Setup

The centrifuge trailer has a dry polymer feeding system that was rarely used and required some effort to make it functional. The polymer blending system did not provide adequate mixing of the dry polymer and left unmixed and residual portions of polymer in the tank. The dry polymer feed pump, however, provided suitable control to deliver a dry polymer solution to the centrifuge. SDWWTP also had a dry polymer make up system that was no longer used, however, was functional. Initially, dry polymer solution was metered from the plant's make-up system directly to the dewatering centrifuge, however, the plant's polymer pumps could not adequately control the polymer feed for testing. In order to conduct the testing, a hybrid of both systems was used. Polymer solution made up using the plant's system was pumped into the dry polymer solution hopper on the Centrisys pilot dewatering centrifuge. The Centrisys pump was then used to meter the polymer to the centrifuge. The dry polymer used for the dewatering testing was Polydyne C-3283 which is currently used for the SDWWTP dewatering centrifuges.

Dewatering Pilot Testing – Polymer Injection Location Optimization

Polymer injection location in the biosolids is important in getting proper biosolids flocculation for the desired dewatering. Several injection locations were tested using the plant's dry polymer. A mixed injection system, which included dosing a portion of the polymer in a static mixer in the interconnecting hose and the rest of the polymer injected at the grinder on the pilot trailer, was initially found to be the best method for polymer injection based on visual observations of the centrate clarity. However, later testing found that injecting polymer directly into the feed tube of the centrifuge provided better centrate quality. The first few weeks of testing were based primarily on the mixed polymer injection and the later testing was conducted using primarily the feed tube injection point.

Dewatering Pilot Testing – Polymer Curve Results

Polymer curve tests were conducted by maintaining a constant volumetric throughput of digested biosolids feed to the centrifuge but changing the polymer dose to measure the impact. With the exception of changing polymer dose, most of the other parameters on the centrifuge remained the same for each polymer curve test. During some of the tests, however, the differential speed was adjusted to increase cake solids while still trying to maintain good quality centrate based on visual observations. Testing showed that reducing the differential speed would increase cake solids but could sacrifice centrate quality and solids recovery.

Polymer curve tests were conducted using both emulsion and dry polymer with both mixed and feed tube polymer injection. All of the dry polymer curves conducted on thickened biosolids from Digester 9 are presented in **Figure 13** and the data depicts the injection point and polymer concentration.

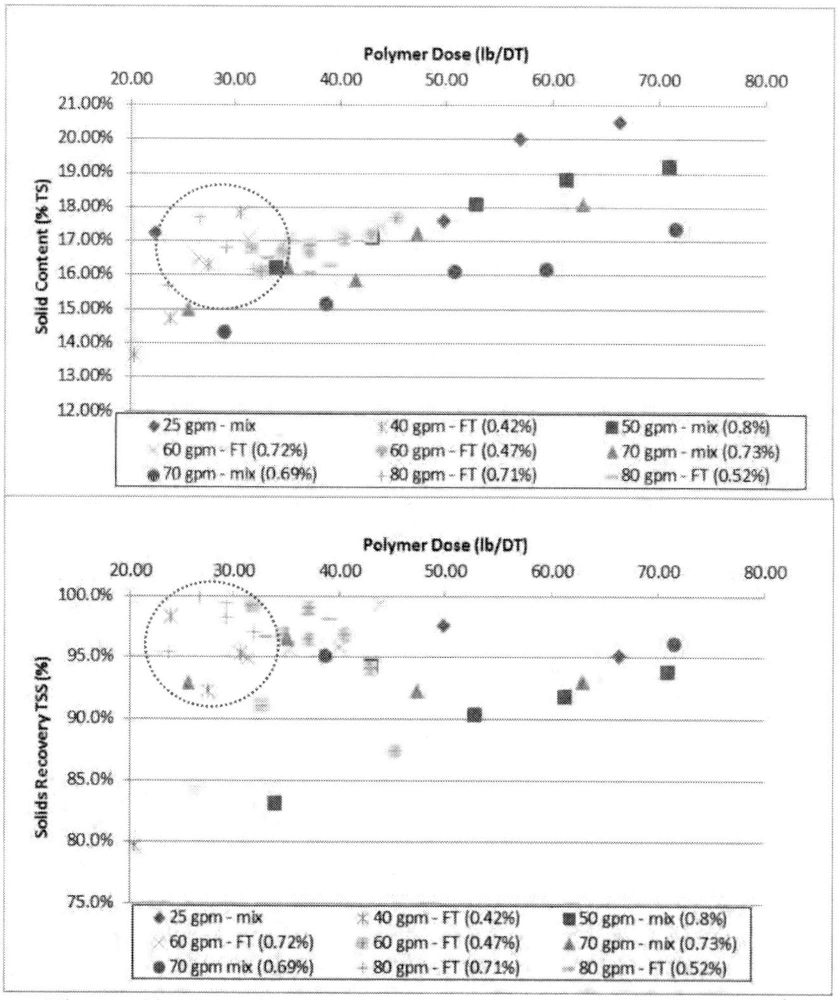

Figure 13: Compiled Dry Polymer Curves from Digester 9[1]

[1] "mix" represents "mixed polymer injection and "FT" represent "feed tube polymer injection"

For the majority of the dry polymer testing, the targeted polymer concentration was 0.8%, however, actual concentrations were measured daily throughout testing and solution concentration appeared to vary from day to day. In addition, the team conducted testing with a more dilute dry polymer concentration. The red circle on **Figure 13** denotes the area targeted for optimization with dry polymer. All of the emulsion polymer curves conducted on thickened biosolids from Digester 9 is presented in **Figure 14**.

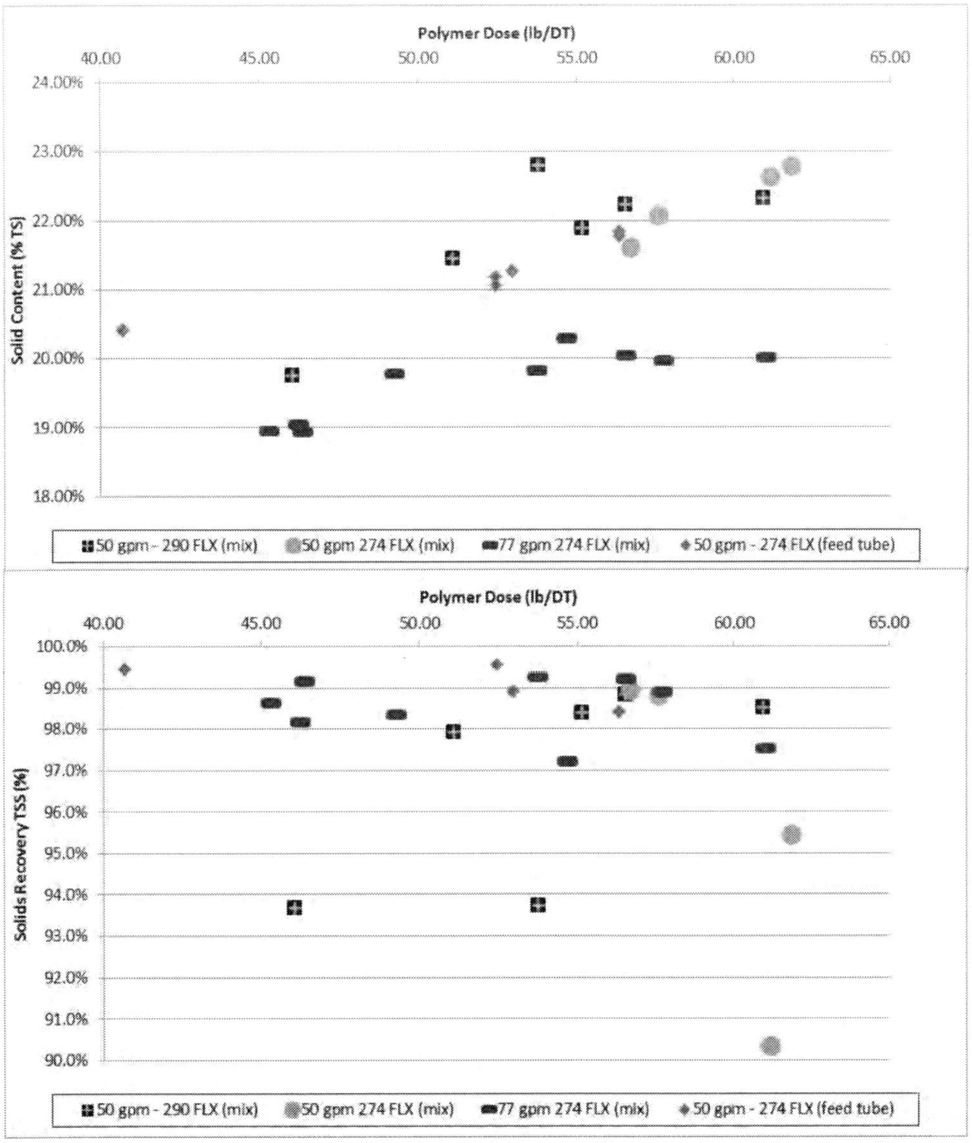

Figure 14: Compiled Emulsion Polymer Curves from Digester 9

The polymer testing curves showed that SDWWTP's digested biosolids did not dewater to the level anticipated in the preliminary design ($\geq$ 20% TS) at active polymer doses of 25 lb/DT (12.5 g/kg). Using the plant's dry polymer (Polydyne C-3283), it was difficult to dewater to greater than 18% TS unless the active polymer dose was above 50 lb/DT (25 g/kg). Dryer cake could be produced using emulsion polymer but high polymer doses were also required. For the dry polymer, using feed tube injection verses the external or mixed injection allowed for lower polymer doses while still maintaining recoveries above 95%.

Dewatering Pilot Testing - Extended Operation Results

In addition to polymer curve tests, the dewatering centrifuge was operated several days at a constant flowrate to test the stability of operation throughout the course of a day. Three tests were conducted at 50 gpm (189 L/min), two with dry polymer and one with 274-FLX emulsion polymer. Two tests were also conducted at 75 gpm (284 L/min), one with dry polymer and one with 274-FLX emulsion polymer.

The extended runs using dry polymer are shown in **Figure 15**. The data showed stable performance with high recoveries (>98%) for lower active polymer doses of 19 to 23 lb/DT (9.5 to 11.5 g/kg) as compared to results from the polymer curve testing. The dewatered cake solids during the extended operation tests averaged 17 to 17.5% TS at 50 gpm (189 L/min) feed flow with 19 to 22 lb/DT (9.5 to 11 g/kg) active polymer dose. At 75 gpm (284 L/min) with an active polymer dose of 23 lb/DT (11.5 g/kg), the dewatered cake solids averaged 16 to 16.5% TS.

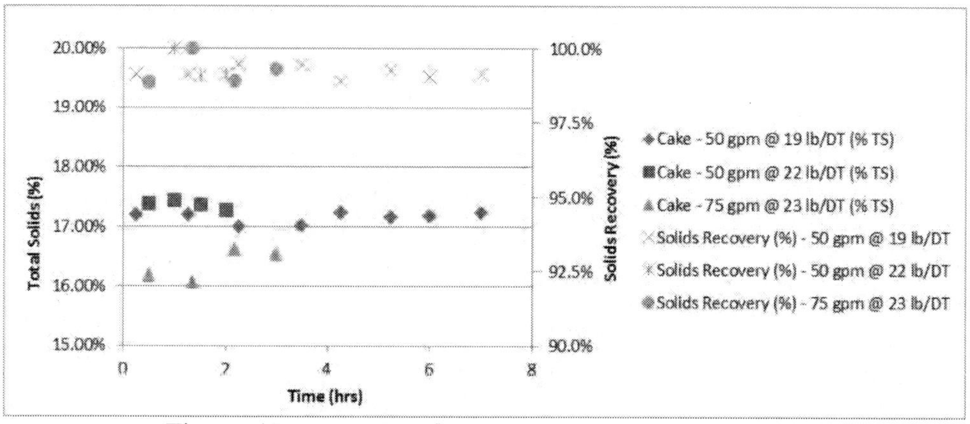

Figure 15: Extended Operation using Dry Polymer

The extended runs using emulsion 274 FLX polymer are shown in **Figure 16**. The data showed stable performance with high recoveries (>96%) for active polymer doses of 37 to 44 lb/DT (18.5 to 22 g/kg). The dewatered cake solids during this tests averages 20.5 to 21% TS at 50 gpm (189 L/min) feed flow with 44 lb/DT (22 g/kg) active polymer dose. At 75 gpm (284 L/min) with an active polymer dose of 37 lb/DT (18.5 g/kg), the dewatered cake solids averaged 19.3 to 19.6% TS.

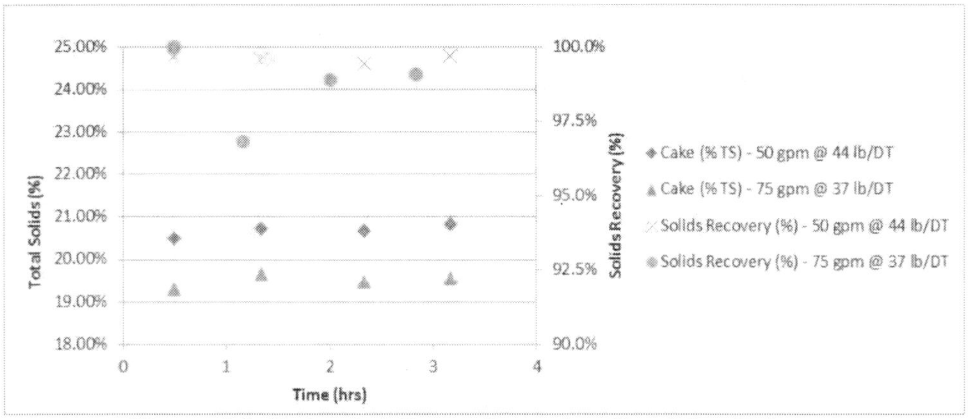

Figure 16: Extended Operation using Emulsion Polymer

The Phase 3 dewatering testing showed that 16-18% TS cake could be achieved with 20-30 lb/DT (10 to 15 g/kg) active dosing of dry polymer. The pilot testing showed that the dewatered cake solids were lower than the preliminary design value of 20% TS with a presumed 25 lb/DT (12.5 g/kg) active polymer when using dry polymer. Dryer cake at 20-22% TS could be produced using emulsion polymer but required higher dosages above 40 lb/DT active (20 g/kg).

CONCLUSIONS

The Phase 1 thickening testing showed that the centrifuge could reliably produce solids at 5-6% TS and achieve greater than 95% solids recovery. Testing was conducted using both dry and emulsion polymers. The dry polymer required 5 to 7 lb/DT (2.5 to 3.5 g/kg) active dosing compared to 1 to 3 lb/DT (0.5 to 1.5 g/kg) based on the emulsion. It was also possible to thicken the sludge to 5 to 6% TS without the use of polymer but this reduced hydraulic throughput by about 50% to allow solids recovery to remain above 90%.

During Phase 2, the solids content in Digester 9 was increased from approximately 2% to approximately 3.4 to 3.5% TS and the VSR in Digester 9 averaged 46% with a digester SRT of approximately 30 days. During the same period of time, the other digesters at the plant received gravity concentrated sludge at 1.5% to 3% TS and averaged approximately 42% VSR with a digester SRT of approximately 20 days. The increase in VSR was likely due to the longer SRT and higher solids concentration.

The Phase 3 dewatering testing showed that 16 to 18% TS cake could be achieved with 20 to 30 lb/DT (10 to 15 g/kg) active dosing of dry polymer. The pilot testing showed that the dewatered cake solids were lower than the preliminary design value of 20% TS with a presumed 25 lb/DT (12.5 g/kg) active polymer when using dry polymer. Dryer cake at 20 to 22% TS could be produced using emulsion polymer but required higher dosages above 40 lb/DT (20 g/kg) active. The pilot dewatering results highlight the importance of piloting to refine design performance criteria.

ACKNOWLEDGEMENTS

The team would like to thank all of the plant staff at SDWWTP, especially Leo Pou and Mike Garcia for the outstanding support during this pilot testing. The team would also like to thank the staff from Centrisys and CNP for their commitment to making the testing a success. We also thank the interns Andres McEwen, Yareliz Negron, and Kate Ireland who helped make this project possible by monitoring pilot operations, collecting samples and running numerous laboratory tests.

REFERENCES

S-884 - SDWWTP SLUDGE THICKENING & DEWATERING BLDG CD 1.06, 1.08, Specification Section 46 71 36 – CENTRIFUGES, MWH, January 20, 2016.

American Public Health Association, American Water Works Association, Water Environment Federation, "Standard Methods for the Examination of Water and Wastewater" 20th ed., 1999

Metcalf & Eddy. (2014). *Wastewater Engineering Treatment and Reuse.* 5th Edition.

Reducing Dewatering Costs Through an Optimization Program

Rashi Gupta, Carollo Engineers*

Steve Walker, Carollo Engineers

Mike Brehm, Bryan Stevenson, Ken Tagney, Matthew Verosik, Eastern Municipal Water District

*3150 Bristol St. Suite 500

Costa Mesa, CA 92626

ABSTRACT

The centrifuge dewatering systems operated by the Eastern Municipal Water District (District) in Perris, California, were assessed and an optimization program was developed to improve performance and reduce operating costs. The District implemented recommendations for improvements in operations, polymer, equipment, systematic optimization tasks, and recordkeeping. Sludge characteristics were tracked and monitored relative to dewatering performance so that the impacts of upstream processes could be determined. Operational observations led to changes in upstream processes that increased primary sludge content in digester and dewatering feed, which subsequently yielded better dewatering performance. Between the various operational, performance tracking, and process changes at the two plants and the successful optimization measures, the District has reduced its annual dewatering costs by approximately $200,000.

KEYWORDS

Centrifuge Optimization, Dewatering Performance, Dewatering Costs, Sludge Characteristics, Cake Dryness, Polymer Demand

INTRODUCTION

The Eastern Municipal Water District (District) in Perris, California, operates four regional water reclamation facilities (RWRF). The San Jacinto Valley RWRF (rated for 53-MLD (14-mgd) average annual flow) and Perris Valley RWRF (rated for 83-MLD (22-mgd) average annual flow) are the subjects of this paper. Both facilities include primary clarifiers that are used to thicken primary sludge, conventional activated sludge processes, waste activated sludge thickening, anaerobic digestion, and dewatering (duty centrifuges and standby belt filter presses).

During the early 2000s, population in the District's region was rapidly increasing and each facility was expanded for additional capacity and changing regulatory requirements. These expansion projects modified the plants to provide nitrification/denitrification (NdN) and Title 22 reuse effluent rather than the previously practiced carbonaceous biochemical oxygen demand (CBOD) removal only. Various projects also changed thickening equipment from gravity belt thickeners (GBTs) and dissolved air flotation thickeners (DAFTs) to rotary drum thickeners (RDTs). Dewatering centrifuges were installed as duty units with existing belt filter presses (BFPs) as standby.

Dewatering and subsequent transport/land application (hauling) of biosolids represented an annual District-wide operating cost of $3.7M from 2011-2014. The monthly dewatering costs for the San Jacinto Valley RWRF (SJVRWRF) and Perris Valley RWRF (PVRWRF) for that period, broken down between power, polymer and hauling are shown Figures 1 and 2, respectively. "Hauling" costs in these figures and throughout this paper represent the costs of hauling and land application, as provided by third party contracted firms. Because the facilities do not have the ability to separately measure power consumption by the centrifuges, power costs are conservatively estimated assuming full power draw based on nameplate horsepower on centrifuge motors. In late 2014, the District embarked on a concentrated effort to reduce dewatering costs across its facilities by 5 to 10-percent, resulting in annual savings of $185,000 to $370,000.

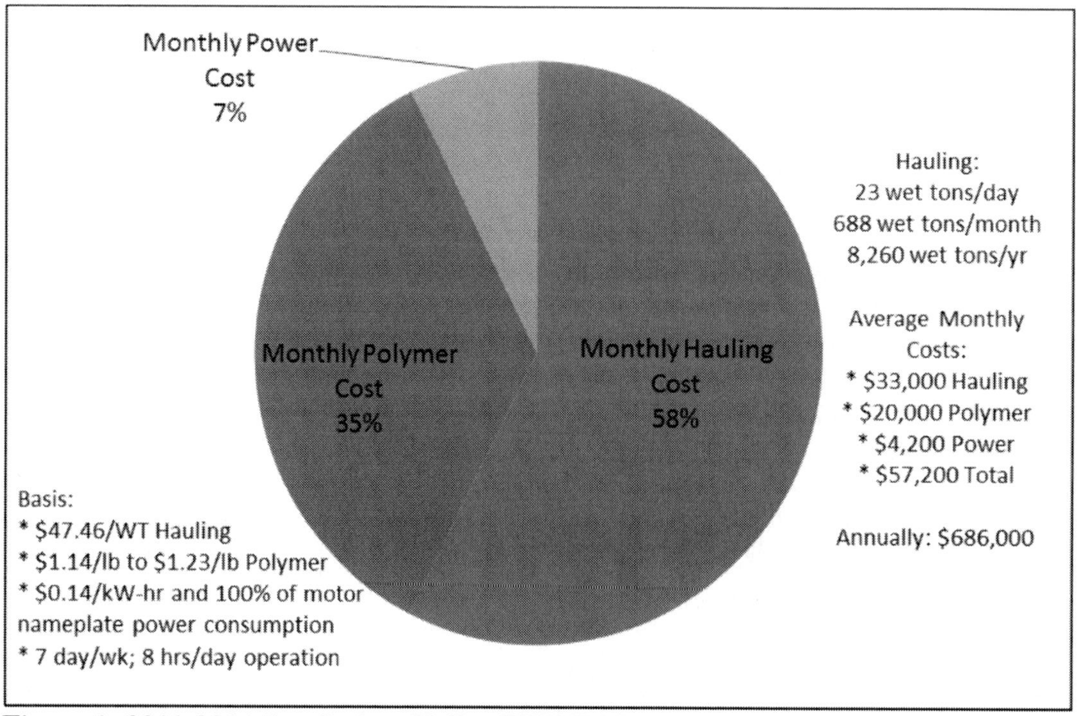

Figure 1: 2011-2014 San Jacinto Valley RWRF Monthly Dewatering Cost Breakdown

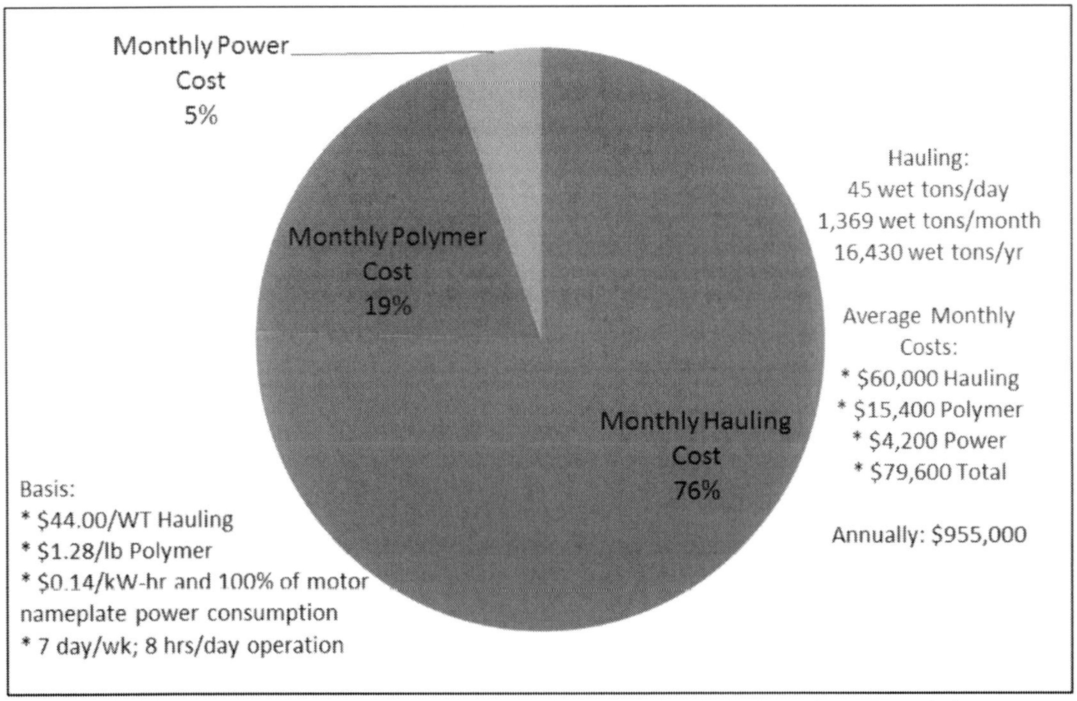

Figure 2: 2011-2014 Perris Valley RWRF Monthly Dewatering Cost Breakdown

METHODOLOGY

The initial dewatering optimization program included several steps. First, the project team studied the existing operations, system equipment, data collection and tracking, maintenance procedures and history, and performance metrics. Recommendations for changes to these items were documented and presented to the District. This was followed by onsite staff training on centrifuges, polymer systems, and field optimization. Plant staff then implemented a number of recommended changes in an ongoing effort to increase cake dryness and reduce polymer consumption and minimize dewatering costs. This paper documents the major recommendations made by the project team, those changes that were successfully implemented, and the subsequent performance metrics at the San Jacinto Valley and Perris Valley RWRFs.

RECOMMENDATIONS

The primary performance metrics for the District's dewatering program include cake dryness, polymer consumption, power, and percent solids capture by the centrifuges. The project team determined each facility's baseline performance, associated costs, and suggested changes to improve performance and tracking of metrics. As shown in Figures 1 and 2, power represents a relatively small portion of overall dewatering costs so the team's power-specific recommendation was for the District to continue operating their centrifuges during off-peak periods to the extent possible. The centrifuges consistently exhibited excellent percent capture (>95%), so the suggested modifications focused on improving cake dryness and polymer consumption.

The suggestions centered on four key parameters:

1. Feed sludge characteristics - The team emphasized the importance of primary sludge (PS) in the feed sludge. The facilities were encouraged to track the mass-based primary

sludge-to-thickened waste activated sludge (PS:TWAS) ratio and operate in a way that maximizes the PS in the digester and dewatering feed sludge.

2. Polymer - The District had a long-term contract with a polymer supplier based on jar testing from several years prior. New jar testing to identify optimal, separate polymers for thickening and dewatering was recommended to address changes in plant processes over the years. To accommodate future changes in sludge, a shorter term and the flexibility to use a primary and alternative polymer supplier were suggested for the new polymer contract. The team recommended that aging polymer blending equipment be replaced and load cells be used to track polymer consumption from totes. Additional recommendations included field optimization relative to polymer solution concentration, injection location, dose, dilution water characteristics, and pressure.

3. Systematic Optimization - A systematic approach to optimization and key parameters to test were developed with staff. In addition to the polymer-specific parameters noted above, it was suggested that regular testing be done to determine the impacts of centrifuge throughput (hydraulic and solids), bowl speed, torque/pressure, and pond depth on polymer dose and cake dryness.

4. Performance Tracking and Monitoring - Each of the District's facilities tracked dewatering performance differently, with some using written records in notebooks and others tracking specific metrics electronically. A standardized District-wide, electronic tracking system to input and monitor specific operational and performance parameters on a daily basis was recommended. This information was used to track and publish both operational targets and performance results on a monthly basis in a location that all staff regularly utilize (e.g. lunchroom bulletin boards).

The SJVRWRF and PVRWRF management implemented many of the recommended suggestions. Results and observations from their efforts are presented below.

RESULTS AND DISCUSSION

The most dramatic impact on cake dryness and polymer demand at the facilities is associated with the PS:TWAS ratio in feed sludge. Figure 3 summarizes dewatering performance metrics and the proportion of sludge that is PS for each of the District's four facilities. As shown in Figure 3, the SJVRWRF has maintained a feed that exceeds 60% PS from 2011 through 2016 and the PVRWRF increased its ratio to a similar value in the 2015-2016 period by changing its primary sludge thickening and pumping operations. The SJVRWRF maintained cake at 24% solids with polymer doses between 19 and 20.5 kg/dry solids tonne (38 and 41 pounds (lbs) active polymer/dry ton (lbs act/DT)). Once it increased its PS:TWAS ratio, the PVRWRF improved cake dryness from 20% to 23% with a polymer dose of approximately 19.5 kg/dry solids tonne (39 lbs act/DT). In comparison, the District's other two facilities with PS:TWAS ratios that favor TWAS (35% to 38% PS) require more polymer at 21.5 kg/dry solids tonne (43 lb act/DT) and achieve approximately 20% cake. The impacts of the PS proportion are further evidenced in Figure 4, which illustrates the differences in cake dryness at the PVRWRF for both centrifuge and belt filter press when the PS proportion increased from 36% (2011-2014) to 59% (2015-2016).

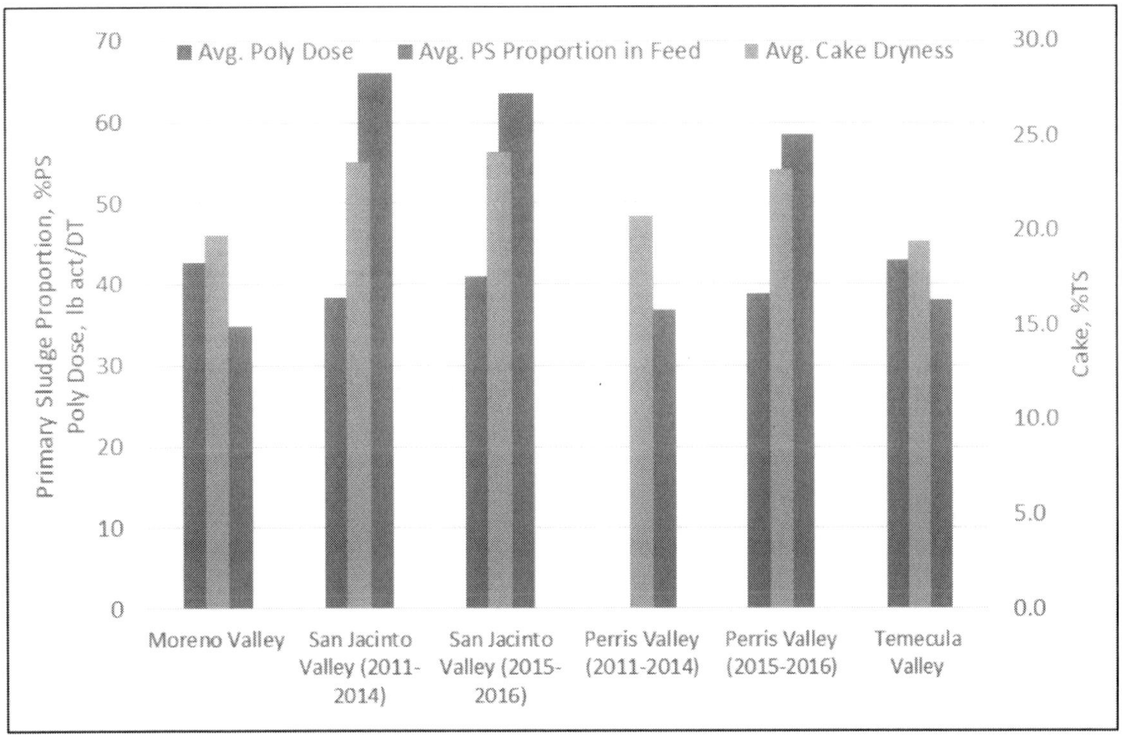

Figure 3: Facility Comparison Showing Performance Impact of PS Proportion in Sludge Feed

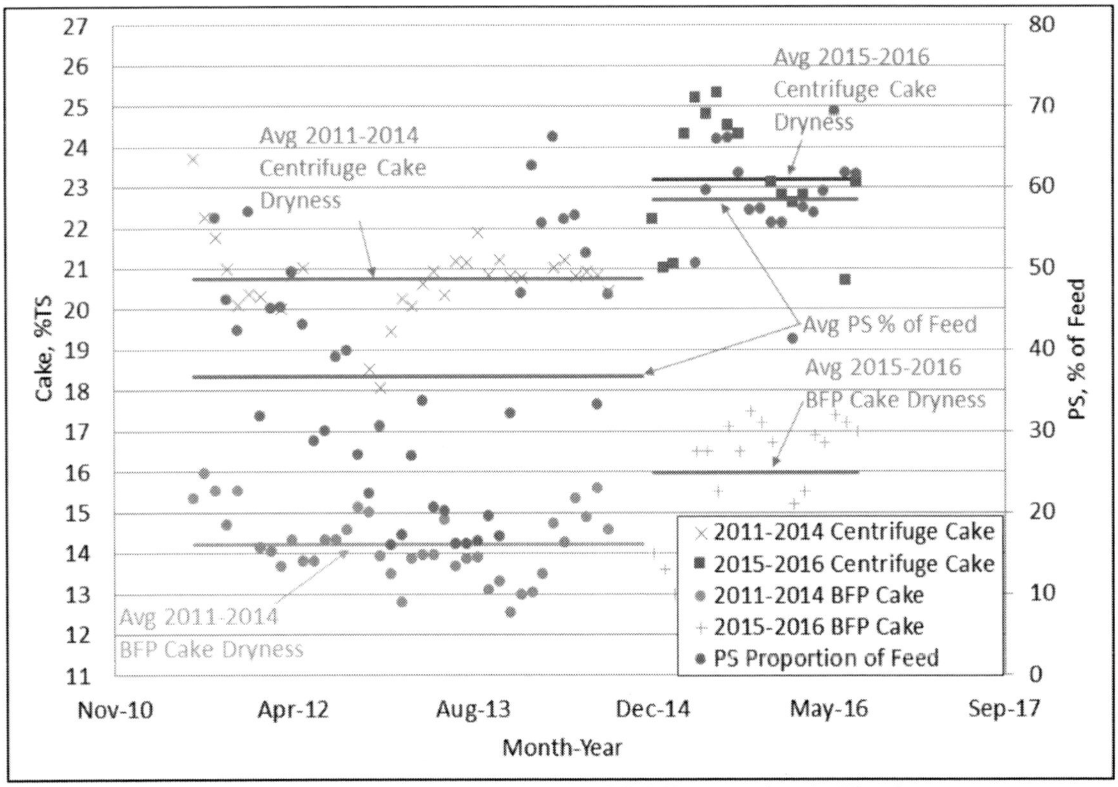

Figure 4: PVRWRF Changes in Dewatering and PS Proportion in Feed

In addition to the increase in PS:TWAS ratio, plant staff found that two other parameters impact dewatering performance: polymer solution concentration and the temperature of the sludge feed. The digested sludge at the PVRWRF is thicker than typical digested sludge - averaging 2.5% TS, which dewaters better with thinner polymer solution concentration. Dewatering sludge when it remains hot also helps its dewaterability. Plant staff at both the PVRWRF and the SJVRWRF determined that dewaterability suffers when the digesters are operated to maintain 35-degree C (95-degree F) rather than 37.8-degree C (100-degree F). Digested sludge is stored in a digested sludge storage tank prior to dewatering, but staff minimized the time the sludge is held before dewatering to minimize loss of heat. Figure 5 shows not only how cake dryness has improved between the 2011-2014 and 2015-2016 periods, but also how polymer consumption has reduced as staff has worked to optimize the dewatering system through such operational optimization.

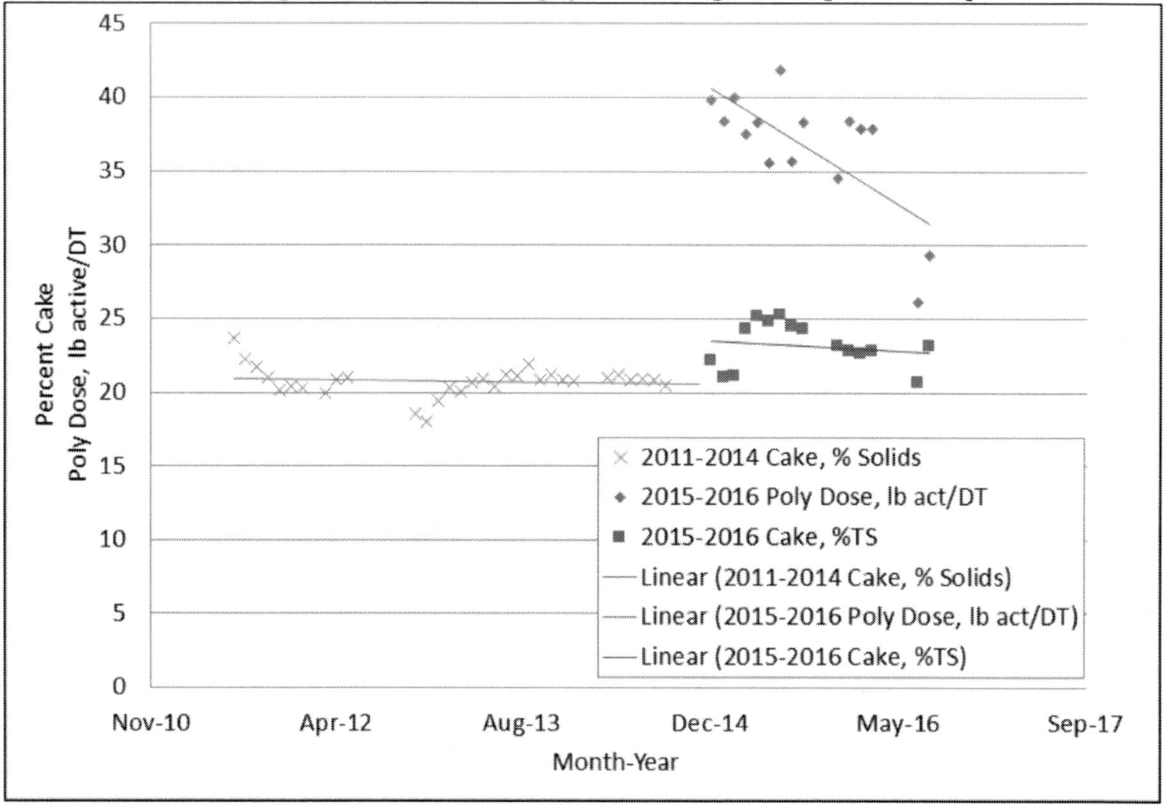

Figure 5: PVRWRF Centrifuge Dewatering Performance

As shown in Figures 6 and 7, cake dryness has remained consistent throughout the 2011 through 2016 period at the SJVRWRF. This is notable because the facility has undergone a wholescale plant expansion that changed the process from carbonaceous biochemical oxygen demand (CBOD) removal to nitrification/de-nitrification during that time. Figure 7 shows that the plant modifications have caused a drop in the PS:TWAS ratio and fluctuations in the solids load. However, plant staff has strived to maintain average cake dryness above 22% and polymer dose below 20.5 kg/dry solids tonne (41 lb act/DT). In addition to operational vigilance regarding polymer dilution and sludge characteristics, staff tried two different polymers, as indicated by the polymer active content in Figure 7. Although the 46% active polymer produced drier cake, doses were higher and the District's contract did not allow a switch between polymer suppliers at that time. The current contract has now expired and jar testing is underway at each of the District's facilities to select the most appropriate polymers for thickening and dewatering, separately for each plant. The next contract will be based on the polymers that minimize overall cost as determined through impacts on both hauling and chemical costs. It will have a 3-year term, with options to renew each year for two years. This will give the District flexibility to test and switch polymers again if needed.

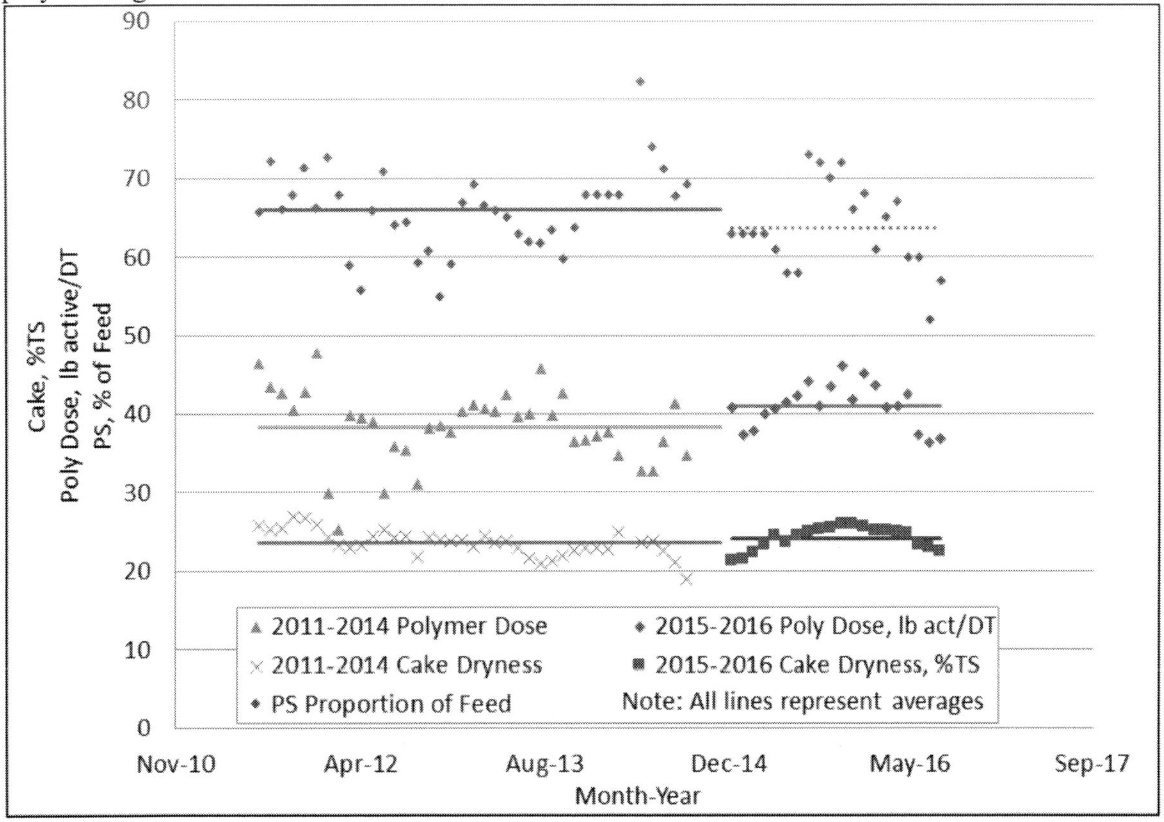

Figure 6: SJVRWRF Dewatering Performance from 2011-2016

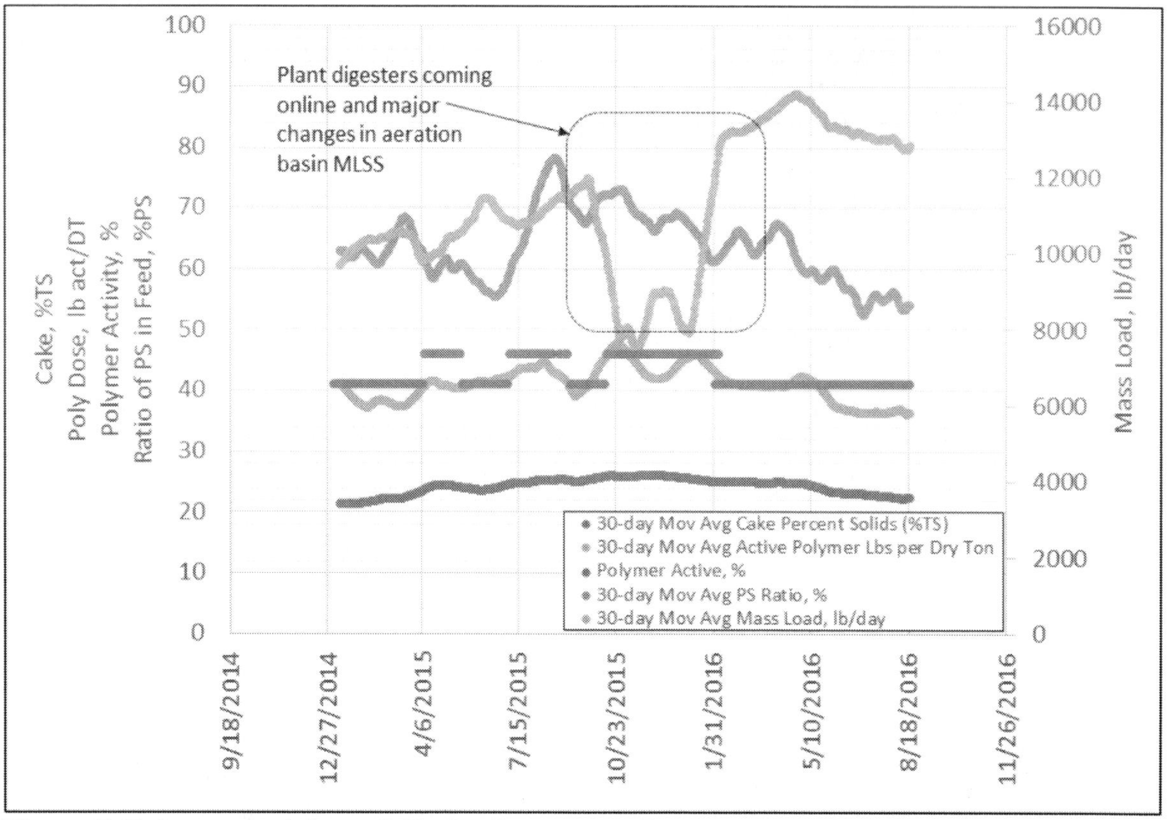

Figure 7: Polymer Activity and Sludge Impacts on Dewatering at SJVRWRF

Improvements to the facilities' dewatering systems have not been solely due to operational changes and optimization. Dewatering data are now tracked daily on standard spreadsheets that automatically calculate specific performance metrics and associated costs. These metrics are published and posted, along with the names of operators on duty during the shift. Operators that meet performance targets are recognized, which has encouraged other staff to emulate the success. Management has changed shift rotations to keep operators posted to the dewatering process for extended periods so that they gain better understanding of the entire system and optimization measures.

CONCLUSIONS

All of the District's facilities continue their efforts at reducing dewatering costs, with specific focus on primary sludge pumping strategies that maximize the PS:TWAS ratio and polymer selection. Figure 8 shows the impacts thus far at the SJVRWRF and PVRWRF between the 2011-2014 and 2015-2016 periods. Between the various process changes at the two plants and the successful optimization measures, the District has reduced its annual dewatering costs by approximately $200,000. Such success has not been based solely on equipment modifications. Rather, complete system optimization has included process and operational changes, close tracking of sludge characteristics and performance parameters, revisiting polymer selection and contract terms, and recognition of operators' success at meeting performance targets.

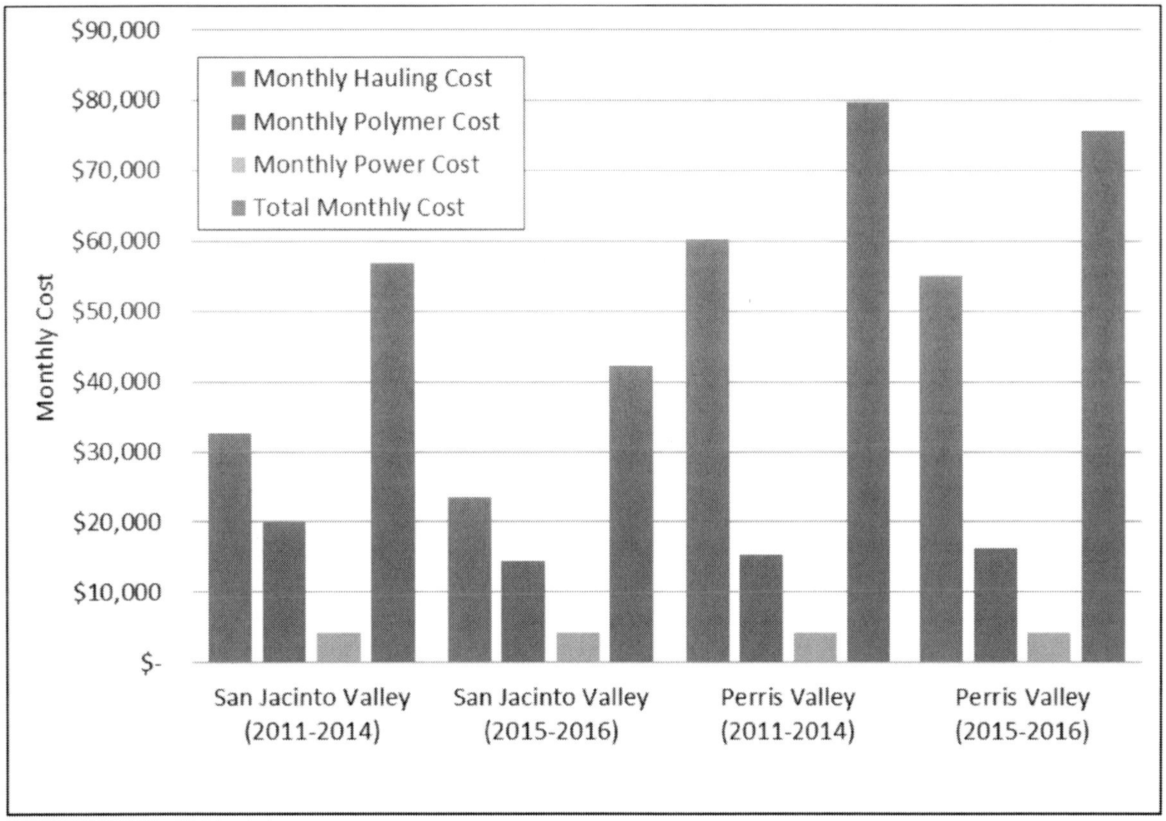

Figure 8: Changes in Monthly Dewatering Costs

AUTHOR INDEX

AUTHOR INDEX

AUTHOR INDEX

Water Environment Federation (WEF)
601 Wythe Street
Alexandria, Virginia 22314
USA

ISBN 978-1-5108-6297-5

9 781510 862975

CURRAN ASSOCIATES INC.
proceedings.com

Residuals and Biosolids 2017 Conference

Seattle, Washington, USA
8 - 11 April 2017

Volume 2 of 2

Water Environment Federation®
the water quality people™

Residuals and Biosolids 2017 Conference

Seattle, Washington, USA
8 - 11 April 2017

Volume 2 of 2

ISBN: 978-1-5108-6297-5

Printed from e-media with permission by:

Curran Associates, Inc.
57 Morehouse Lane
Red Hook, NY 12571

Some format issues inherent in the e-media version may also appear in this print version.

Printed by Curran Associates, Inc. (2018)

For permission requests, please contact Water Environment Federation (WEF) at the address below.

Water Environment Federation (WEF)
601 Wythe Street
Alexandria, Virginia 22314
USA

Phone: 1-800-666-0206
Fax: 1-703-684-1545

csc@wef.org

Additional copies of this publication are available from:

Curran Associates, Inc.
57 Morehouse Lane
Red Hook, NY 12571 USA
Phone: 845-758-0400
Fax: 845-758-2633
Email: curran@proceedings.com
Web: www.proceedings.com

TABLE OF CONTENTS

VOLUME 1

SESSION 1

SESSION 2

SESSION 3

SESSION 9

SESSION 10

VOLUME 2

SESSION 11

SESSION 12

SESSION 13

SESSION 14

SESSION 15

SESSION 17

SESSION 18

SESSION 20

ADDITIONAL PAPERS

Screw Press Maximum Dewaterability and Energy Consumption

Edward W. Fritz, P.E.[1]*, Dieter Weinert[1], Harald Neumann[2]

[1]Huber Technology, Inc., [2]Huber SE

*Ed.Fritz@hhusa.net

ABSTRACT

Huber SE has partnered with Germany's Federal Ministry of Education and Research to conduct a study of screw press technology. The study examined the screw press's ability to attain maximum dewaterability results and the energy consumption associated with operation of a dewatering screw press. A variety of sludge types were tested to examine the effect of different solids handling processes on the maximum dewaterability and energy consumption of screw presses.

KEYWORDS

Dewatering, sludge, energy, dewatering equipment performance, screw press

INTRODUCTION

The water resource recovery facility of the future will be required to meet both more stringent permit requirements and reduce their energy requirements. The dewatering process reduces sludge volume, minimizing sludge storage, transportation, and disposal costs. Energy savings can also be realized by achieving maximum dewaterability while minimizing the energy consumption of the dewatering process. Huber SE has partnered with Germany's Federal Ministry of Education and Research (BMBF) to conduct a study towards the energy optimized water resource recovery facility of the future. For this part of the project, tests were conducted to study two primary objectives: 1) evaluate screw press cake solids compared with maximum sludge dewaterability, and 2) evaluate energy consumption for each part of the press.

METHODOLOGY

Maximum Dewaterability

Sludge maximum dewaterability was determined by analysis of the sludge drying curve. This method has been shown to be accurate within ± 1.5% of maximum sludge dewaterability (Kopp, 2001). Sludge was also characterized based on properties such as sludge type, volatile suspended solids, and total solids concentration. Sludge types for this study included municipal waste activated sludge, mesophilic digested sludge, and post-thermal hydrolysis process (THP) sludge.

Samples were collected for maximum dewaterability during the screw press testing to ensure representative samples were being analyzed. Sludge was provided to the containerized system and multiple tests were performed at varying solids loading rates. Samples were collected every two hours once the screw press had achieved steady state following any changes in loading rate.

A high, middle, and low loading rate were tested at each facility and a minimum of three samples were collected for each loading rate.

Containerized Screw Press

A Huber screw press system was provided for each study. The screw press system was contained in a specialized shipping container and included the screw press, feed pump, polymer preparation system, filtrate polishing, cake conveyance, and controls. This provided all necessary equipment for the dewatering process in a single container, allowing easy measurement of all trains of the dewatering process as well as measurement of the dewatering process's energy consumption.

Figure 1 - Containerized Screw Press

Specific Energy Demand

Energy consumption of the dewatering system was recorded during a series of solids loading runs. Solids loading was adjusted between 50% and 120% of the machine's nominal loading rate to determine the specific energy demand of each loading rate. Some loading rates were repeated to confirm results at solids loading rates above the machine's nominal loading rate. Sludge types for this study included municipal waste activated sludge, mesophilic digested sludge, biological dairy sludge, thermophilic digested sludge, and post-thermal hydrolysis sludge.

RESULTS

The sludges tested were from different facilities. Tests for maximum dewaterability show that the screw press equipment is able to achieve average results within the margin of error of the maximum dewaterability test. Higher solids loadings rates showed a lower average cake solids percentage. Power consumption data at facilities showed the screw press operating at less than 70% of its full load power draw under different solids loading rates.

Maximum Dewaterability

The three types of sludges studied for maximum dewaterability had different characteristics. Table 1 shows results of the sludge testing and provides the characteristics for each facility's sludge.

Sludge Type	Sludge Concentration	Volatile Suspended Solids	Maximum Dewaterability
Waste Activated Sludge	2.3% solids	68% VSS	20.5% solids
Mesophilic Digested Sludge	4.5% solids	50% VSS	31.5% solids
Post-THP Sludge	5.2% solids	62% VSS	33.0% solids

Table 1 - Sludge Characteristics

The waste activated sludge had a tested maximum dewaterability of 20.5% solids. The average performance of the screw press was within the standard accuracy (1.5%) of this value for all loading conditions. Figure 2 shows the average results for each loading rate from the waste activated sludge facility. Dashed lines represent the ±1.5% accuracy of the maximum sludge dewaterability. Lower loading rates resulted in slightly higher dry cake solids.

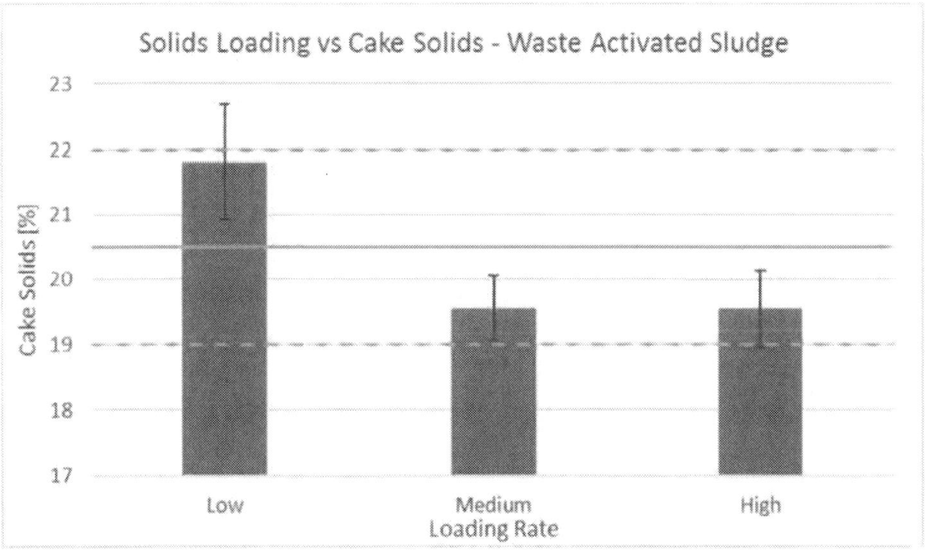

Figure 2 - Waste Activated Sludge Dewaterability

The mesophilic digested sludge had a tested maximum dewaterability of 31.5% solids. The performance of the screw press was within the standard accuracy (1.5%) of this value for all loading conditions. Figure 3 shows the average sample results for each loading rate from the mesophilic digested sludge facility. Dashed lines represent the ±1.5% accuracy of the maximum sludge dewaterability. Lower loading rates resulted in slightly higher dry cake solids.

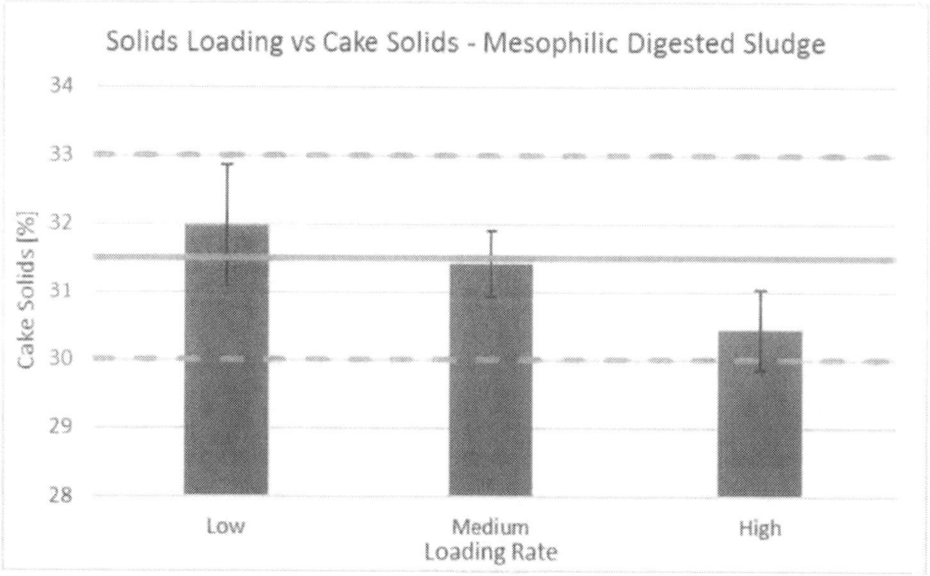

Figure 3 - Mesophilic Digested Sludge Dewaterability

The post-THP sludge had a tested maximum dewaterability of 33.0% solids. The performance of the screw press was within the standard accuracy (1.5%) of this value for all loading conditions. Figure 4 shows the average sample results for each loading rate from the post-THP sludge facility. Dashed lines represent the ±1.5% accuracy of the maximum sludge dewaterability. Lower loading rates resulted in slightly higher dry cake solids.

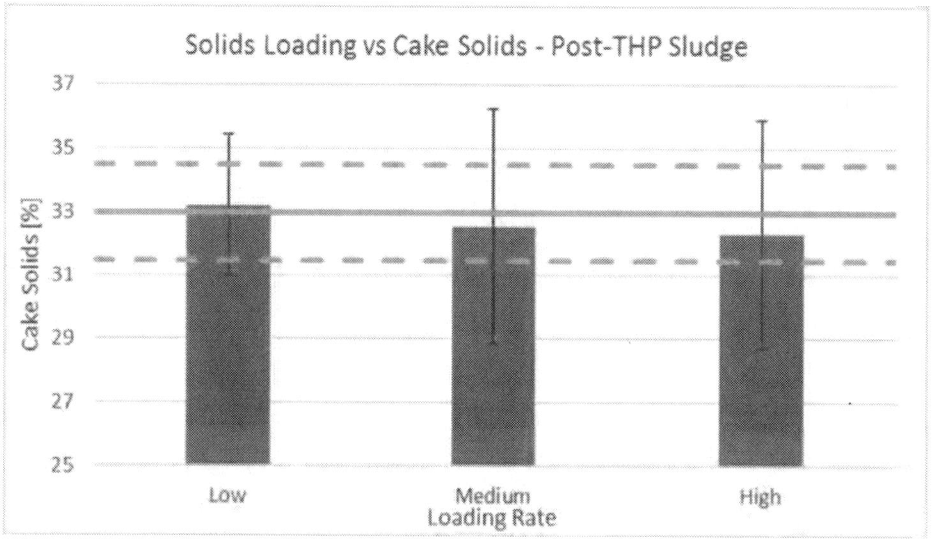

Figure 4 - Post-THP Sludge Dewaterability

Specific Energy Demand

The rated full load power demand for the Q-Press unit is 2.2 kW. In testing, the actual power demand for the Q-Press unit averaged 1.35 kW, about 60% of the full load power demand. Energy demand for the screw press increased as solids loading to the press increased, as shown in Figure 5. Also shown are the feed pump and polymer pump (other major dewatering system equipment) energy consumption. Other major dewatering system equipment did not show a significant change in power consumption with respect to nominal capacity of the screw press.

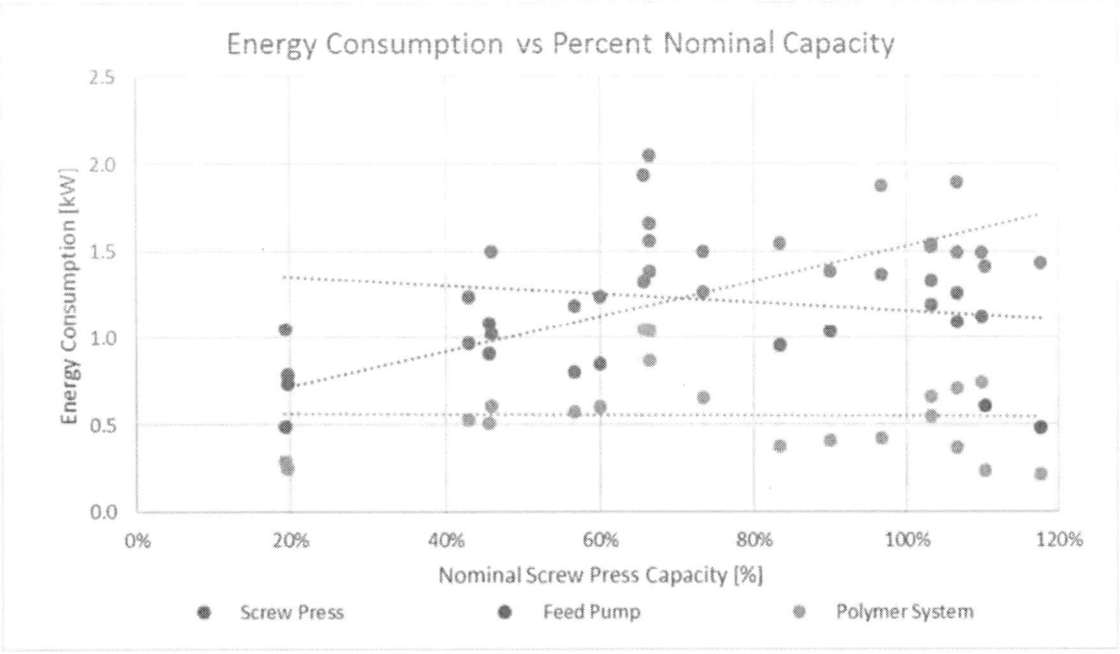

Figure 5 - Energy Consumption vs Nominal Capacity

The specific energy demand (energy required per unit mass of solids dewatered) of the screw press showed a slight decrease throughout the range of solids loadings. This is shown on Figure 6 along with the specific energy demand of other major dewatering system components, which also showed a decrease in specific energy demand as throughput increased.

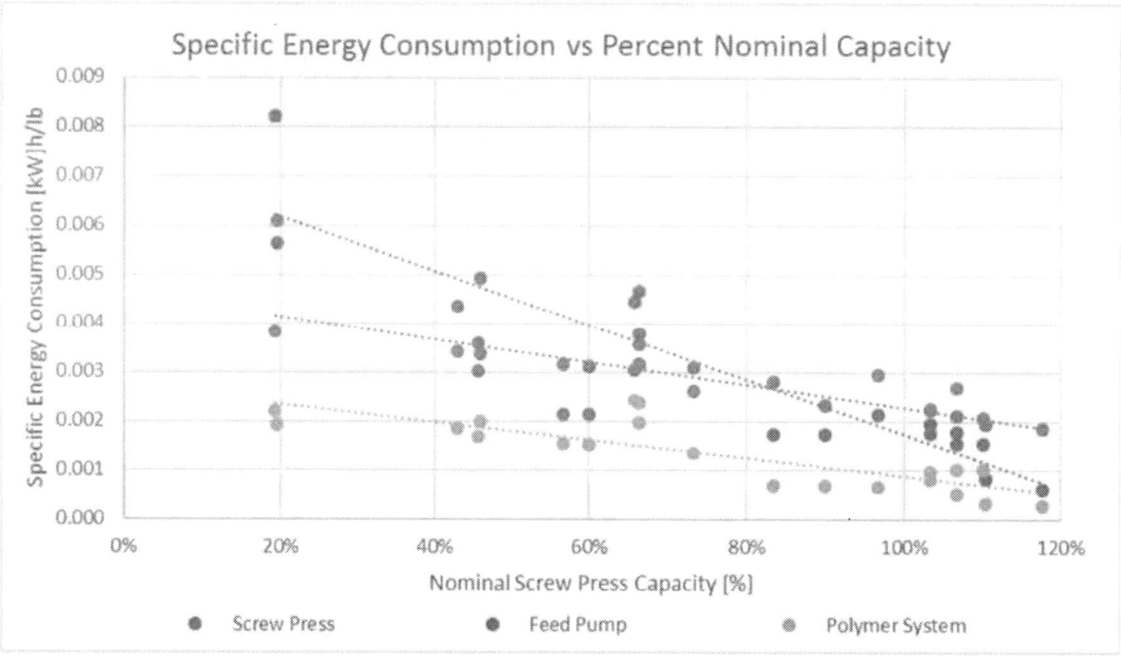

Figure 6 - Specific Energy Demand of Dewatering Equipment

Specific energy demand of the dewatering system varied significantly, depending on the type of sludge being processed. Waste activated sludge had the highest specific energy requirement, averaging 0.0081 kWh/kg (0.0037 kWh/lb) sludge processed. Mesophilic Digested Sludge and post-THP sludge were lower with 0.0053 kWh/kg (0.0024 kWh/lb) and 0.0040 kWh/kg (0.0018 kWh/lb) sludge processed, respectively. Figure 7 shows the average specific energy demand of each type of sludge. Error bars are one standard deviation.

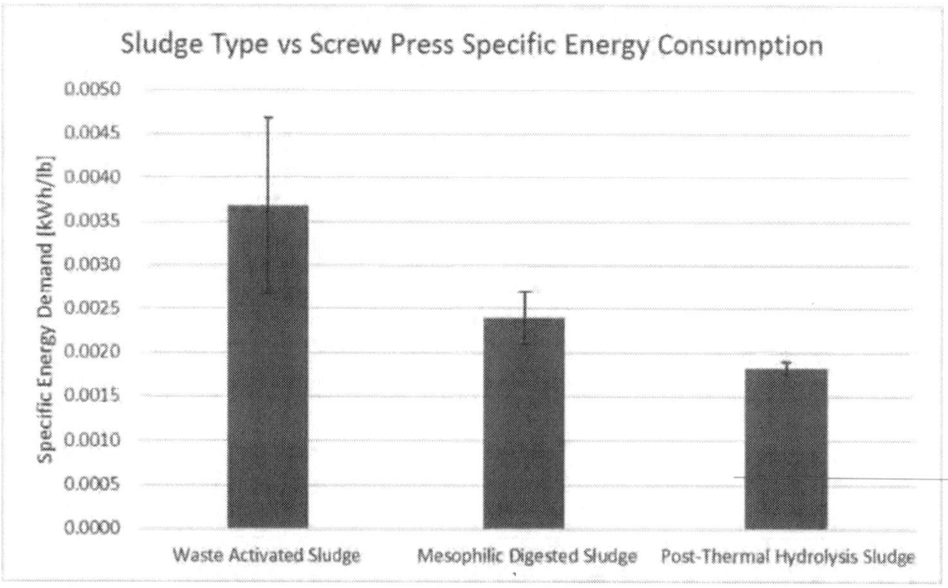

Figure 7 - Specific Energy Consumption vs Sludge Type

The screw press specific energy demand averaged between 24-46% of the total dewatering energy demand, depending on type of sludge being processed. The specific energy demand for each major component and the overall energy demand of the dewatering system is illustrated in Figure 8. Waste activated sludge showed the highest specific energy demand and was significantly higher than all other sludge types. As the overall specific energy demand decreased, the screw press percentage of the specific energy demand increased.

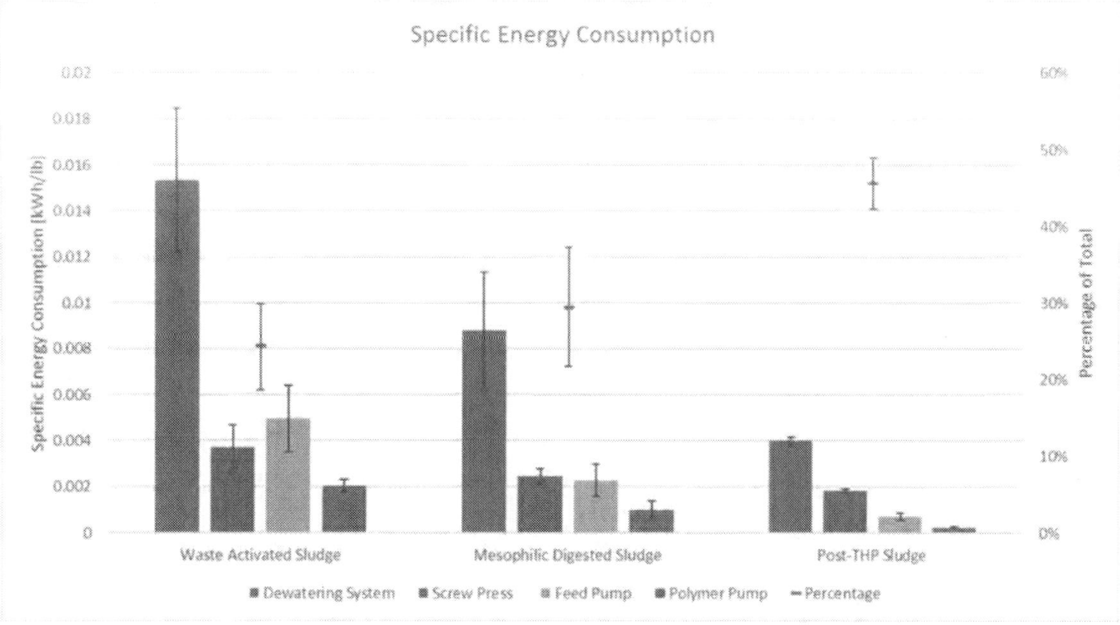

Figure 8 - Specific Energy Consumption Comparison

CONCLUSIONS

Screw press maximum dewaterability and specific energy demand were observed and recorded for different sludge types. Results show that the screw press equipment was able to achieve cake solids near the maximum dewaterability of the sludge for waste activated sludge, mesophilic digested sludge, and post-THP sludge as determined by the sludge drying curve. This indicates the Huber screw press is capable of mechanically dewatering these sludge types to near the maximum extent possible. Tests showed increasing solids loading rate resulted in a decrease in dry cake solids, as expected of mechanical dewatering equipment.

Screw press energy consumption increased with respect to solids loading rate but the specific energy demand decreased slightly with respect to loading rate to the screw press. This indicates that the energy consumption of the screw press is directly proportional to the amount of sludge processed and does not significantly gain or lose efficiency at different solids loading rates. The decreasing trend in the overall dewatering system specific energy demand was attributed to the decreasing specific energy demand of the sludge feed pump and polymer feed pump. This was due to the increased efficiency of these pumps as they approached the design point.

Specific energy consumption varied depending on the type of sludge being dewatered. Waste activated sludge required the greatest amount of energy per mass of sludge dewatered. This was because the sludge was thinner. The waste activated sludge feed solids was about 50-60% of the mesophilic digested sludge and post-THP sludge, resulting in the specific energy consumption being approximately double for waste activated sludge. Mesophilic digested sludge and post-THP sludge required less specific energy than waste activated sludge. The screw press required about one-third of the specific energy demand of the entire dewatering system, depending on the sludge characteristics.

Understanding the relationship between solids loading rate and dry cake solids of the screw press unit provides greater accuracy in predictions of screw press performance. This provides information to owners and engineers when balancing dewatering throughput with dry cake solids. The correlation between sludge types and energy consumption for screw press units allows more accurate energy consumption data when calculating life-cycle costs. Both aspects of this study, maximum dewaterability and specific energy consumption, assist engineers and owners in determining the best overall dewatering solution.

REFERENCES

D. Weinert, M. Roediger, S. Schuler. "Predicting Screw Press Performance." *Residuals and Biosolids.* Water Environment Federation, 2011. 996-1015.

J. Kopp, N. Dichtl. "Influence of the free water content on the dewaterability of sewage sludges." *Water Science and Technology* 44, no. 10 (2001): 177-183.

J. Kopp, N. Dichtl. "Prediction of full-scale dewatering results of sewage sludges by the physical water distribution." *Water Science and Technology* 43, no. 11 (2001): 135-143.

Challenges and Opportunities for Approaching Net Zero Energy and Improving Resiliency of a Unique Solids Handling Facility in California

Irina Lukicheva, CH2M, Dan Frost, CCCSD, Nitin Goel, CCCSD, Peter Burrowes, CH2M, Summer Bundy, CH2M, Jamie Pigott, Carollo Engineers

ABSTRACT

Central Contra Costa Sanitary District (Central San) is in the process of developing a Comprehensive Wastewater Master Plan for their wastewater treatment plant in Martinez, California. The District collects and treats an average of 34 million gallons per day of wastewater from nearly half a million residents within a 145 square mile service area.

Solids from primary treatment and waste activated solids are dewatered and then incinerated. Central San's incineration facility is currently one of two wastewater solids incineration facilities in California. The incineration facility became operational in the 1980s and included two multiple hearth furnaces and a centrifuge dewatering system. Waste heat recovered from the incinerator is used to drive steam turbines to generate air for secondary treatment aeration.

As part of the Master Plan, Central San is evaluating future liquid and solids treatment alternatives that will address the aging infrastructure needs, increase plant resiliency against man-made or natural disasters, increase the efficiency of energy recovery at the plant, accommodate projected capacity needs, improve operational reliability, improve regulatory resiliency, and meet increasingly stringent water quality and air pollution control regulations. In addition, one of the main goals of the Master Plan was improvement of the Central San's energy recovery portfolio. The current practice of energy recovery in the form of waste heat recovery to steam to generate air for secondary treatment is no longer the most energy efficient and due to the delicate interconnections between liquids and solids treatment, it carries risks of not being able to supply sufficient air when solids and energy recovery processes are offline, for example in the event of the earthquake.

As part of the Master Plan evaluation, Central San, CH2M and Carollo Engineers evaluated several alternatives for solids handling. These included advanced anaerobic digestion, fluidized bed incineration, drying, innovative thermal technologies, and combinations of the above listed processes. Mesophilic anaerobic digestion (MAD) of primary solids was chosen as a first processing step to generate biogas for energy recovery from the digestion of primary solids. Separate digestion of primary solids allows the plant to reduce the size of digestion required and allow for the import of high strength wastes such at fats, oils, and grease and liquid organic waste that can be co-digested in the MAD process to enhance biogas production and maximize power production and waste heat utilization. Digested primary solids combined with thickened undigested waste activated solids are to be dewatered using centrifuges. Drying of dewatered solids prior to incineration would be accomplished using a new dryer followed by incineration using a fluidized bed incinerator. The incineration process was selected to be continued as the most efficient method of managing solids that reduces energy demands, reduces greenhouse gas emissions, and does not require development of a new biosolids disposal management program, thus helping address increasing concerns over the future availability and cost of biosolids disposal and reuse for facilities in the San Francisco Bay Area.

Biogas produced from the digestion process would be used along with landfill gas in a cogeneration turbine to produce electricity. Heat recovered from the incineration in addition to the waste heat generated from the cogeneration turbine would be used to generate additional electricity using an Organic Rankine Cycle turbine. The overall energy conversion efficiency from solids into power in this case is significantly higher compared to the overall conversion efficiency for the current plant configuration.

The project will generate about 3.7 MW projected electrical power requirements using biogas, landfill gas and heat recovered from incineration. Even with the increased conversion efficiency, the plant generated electricity may not be sufficient to meet the long term future plant demand, especially if the plant demand includes energy requirements for energy intensive recycled water processes such as reverse osmosis. To meet future recycled water power demands, Central San could consider some additional imported natural gas to the cogeneration turbine or can import electricity or produce green energy using alternative renewable energy generation technologies such as solar photovoltaic panels and wind turbines.

KEYWORDS biosolids, incineration, energy recovery

INTRODUCTION

Central Contra Costa Sanitary District's (Central San) wastewater treatment plant is located in Martinez, California. Central San collects and treats wastewater for 470,000 residents and 3,000 businesses in central Contra Costa County. Central San operates and maintains over 1,500 miles of sewer, 19 pump stations, and 21 miles of force main that transport wastewater to Central San's treatment plant.

Central San's treatment plant currently treats approximately 136,400 m^3/d (30 mgd (million gallons per day)) dry weather flow. Central San's treatment plant process flow diagram is shown in Figure 1. The treatment processes include a Headworks with bar screens and grinders, influent pumping, pre-aerated grit removal, primary sedimentation, primary effluent pumping, activated sludge secondary treatment, and UV disinfection. The treated effluent is discharged into Suisun Bay or treated further for recycled water use. Tertiary filtration is used to produce Title 22 recycled water for on-site use, landscape irrigation, and other recycled water uses within Central San's service area. The current recycled water system operates at a peak day demand of approximately 6,800 m^3/d (1.5 mgd) for utility water (2W) for plant use and 9,100 m^3/d (2.0 mgd) for recycled water users, primarily for landscape irrigation.

The solids treatment process flow diagram is shown in Figure 2. Solids are byproducts of the primary and secondary treatment processes. Waste Activated Solids (WAS) from the secondary treatment process are thickened with dissolved air flotation and blended with raw primary sludge in the Sludge Blending Tank (SBT). Lime is added in the SBT prior to centrifuge dewatering to improve performance of the dewatering polymer and to reduce the potential for formation of fused agglomerations of solids (clinkers) in Central San's multiple hearth furnaces. Dewatering polymer is added directly upstream of the centrifuges to improve cake dryness and reduce water sent to the incinerators. Dewatered sludge (referred to as "cake") is pumped to one of the two multiple hearth furnaces. Incineration of the blended sludge requires supplemental fuel which is currently landfill gas. Ash remaining after incineration is re-used in soil amendment products but can also be landfilled when necessary. In the event that the cake cannot be incinerated, it is conveyed to the Emergency Sludge Loading Facility for off-loading to trucks for landfill disposal.

Central San's incineration facility is currently one of two wastewater solids incineration facilities operated in California. The incineration facility became operational in the 1980s and included two multiple hearth furnaces and a centrifuge dewatering system. The facility also included an energy recovery system to recover waste heat from incineration and generate steam to run steam-driven aeration blowers. The excess heat from the incineration process off-gases is converted to medium grade steam using waste heat recovery boilers. Energy recovery has historically been an important Central San strategy and driver.

The energy for the plant comes from four sources: heat recovered from incinerating plant solids, imported grid electricity (purchased from Pacific Gas and Electric (PG&E)), natural gas, and landfill gas. Landfill gas is primarily used as supplemental fuel for incineration, natural gas is primarily used in the cogeneration unit to generate electricity and steam, and natural gas is also used in the auxiliary boilers to generate the satisfy the remaining steam demand that is not met with the incineration and cogeneration unit waste heat recovery boilers. Approximately 95 percent of the treatment plant power demand is satisfied using onsite co-generation (~2.7 MW) with the natural gas turbine and the remaining power demand (~150 kW) is met with imported grid power (electricity).

Figure 3 shows a breakdown of the sources of energy currently used at the plant. From 2014 to 2015, the total energy consumed from those sources was approximately 2,400 MMBTU/d.

The generated electricity is mostly used for liquid treatment pumping, solids treatment and disinfection (Figure 4). In conventional treatment facilities, the secondary treatment process is the largest electricity consumer. However, because Central San relies on a steam system to power the aeration blowers, relatively little electricity is required for secondary treatment. Steam is generated from waste heat from the multiple hearth furnaces, waste heat from the cogeneration turbine, and steam made by the auxiliary boilers. Figure 5 shows the amount of steam that each steam-generating process contributes to the total steam production. Overall energy distribution through current plant is shown in the form of Sankey diagrams in Figure 6.

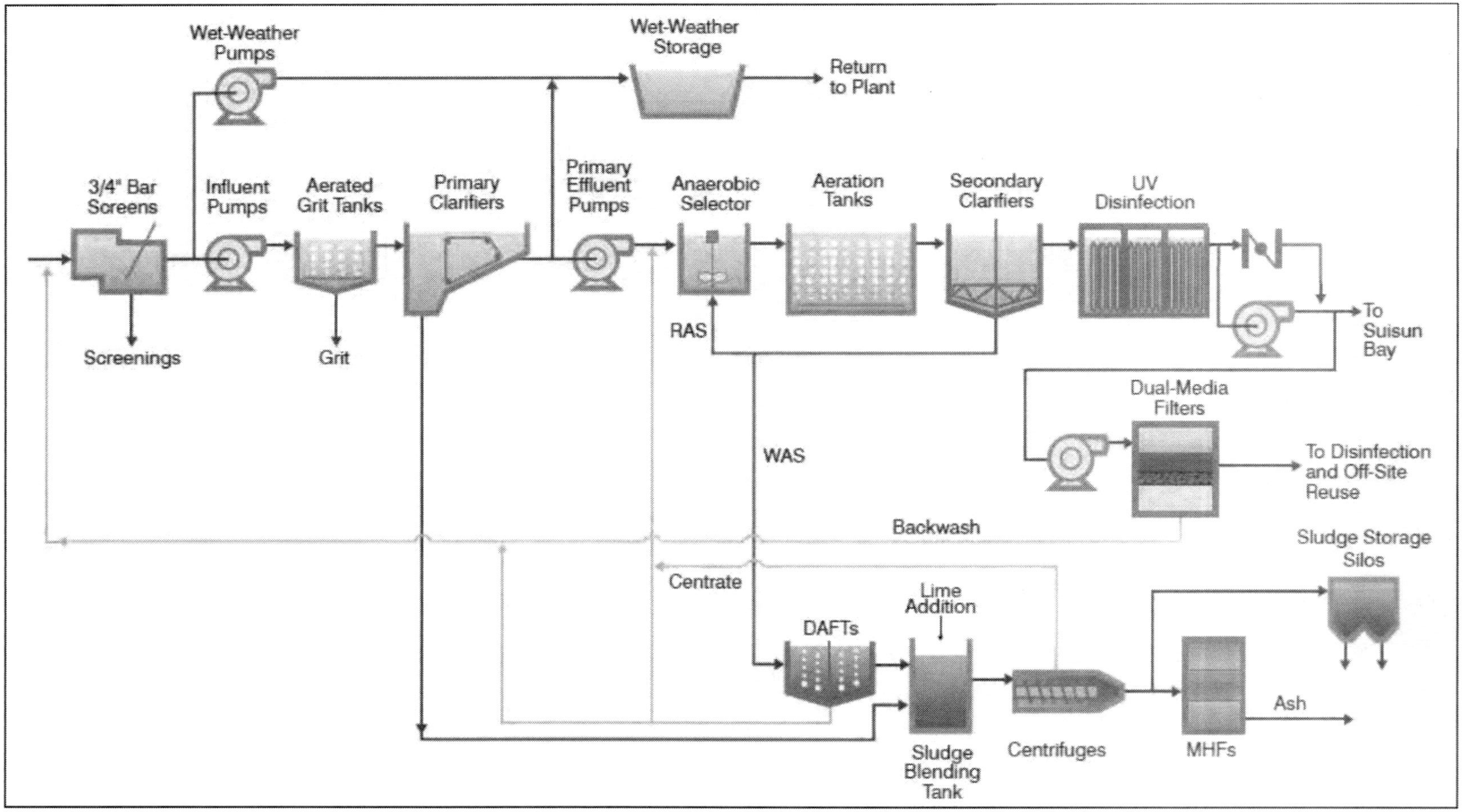

Figure 1. Plant Process Flow Diagram

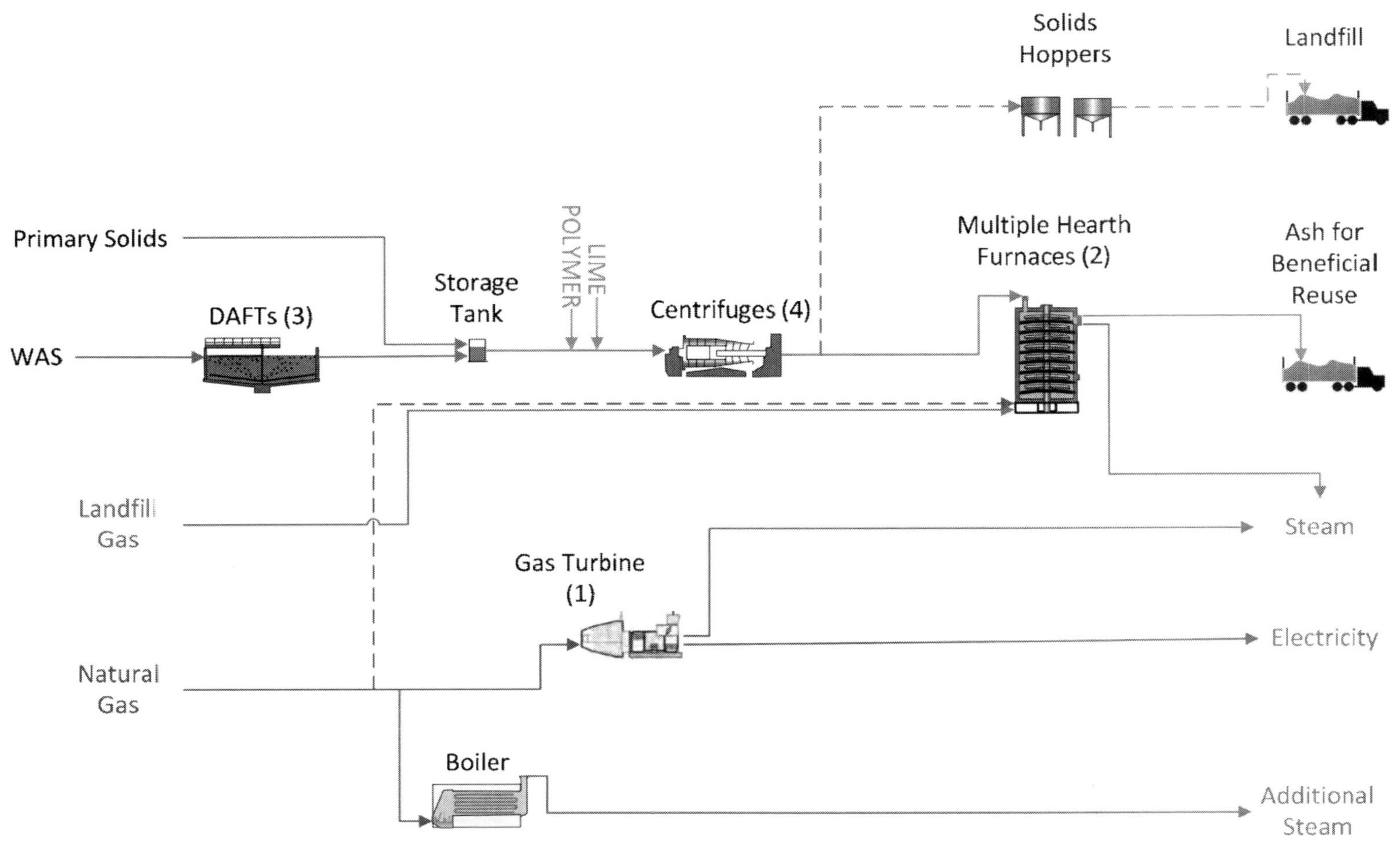

Figure 2. CCCSD solids processing process flow diagram

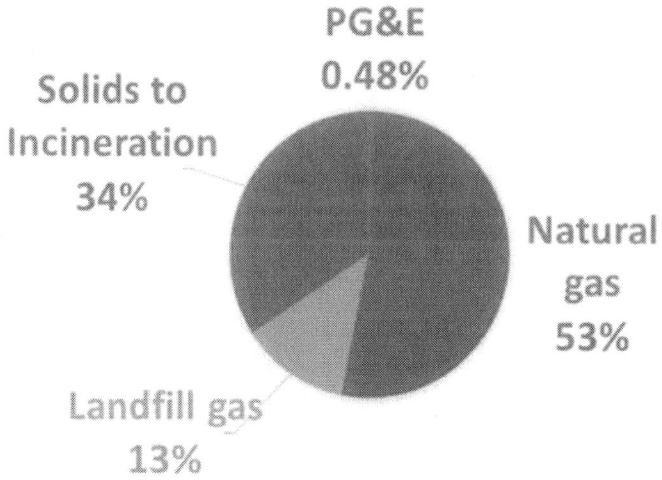

Figure 3 Central San energy source profile (heat and electricity).

(Note: Calculated based on the heating value of natural gas, landfill gas and solids in MMBTU/d, and kW of energy from PG&E also converted to MMBTU/d)

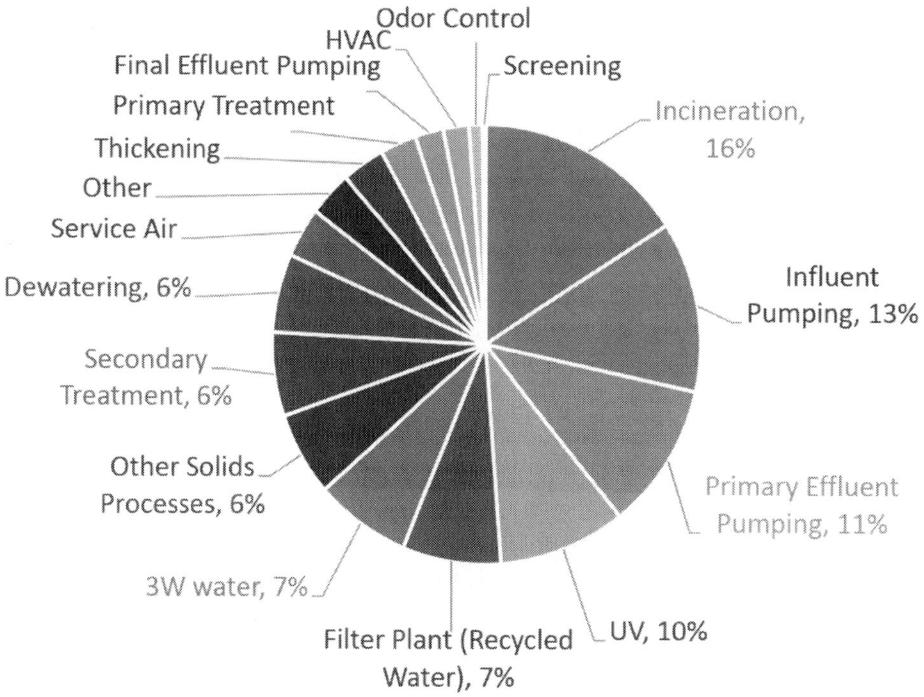

Figure 4 Electricity usage profile by process area

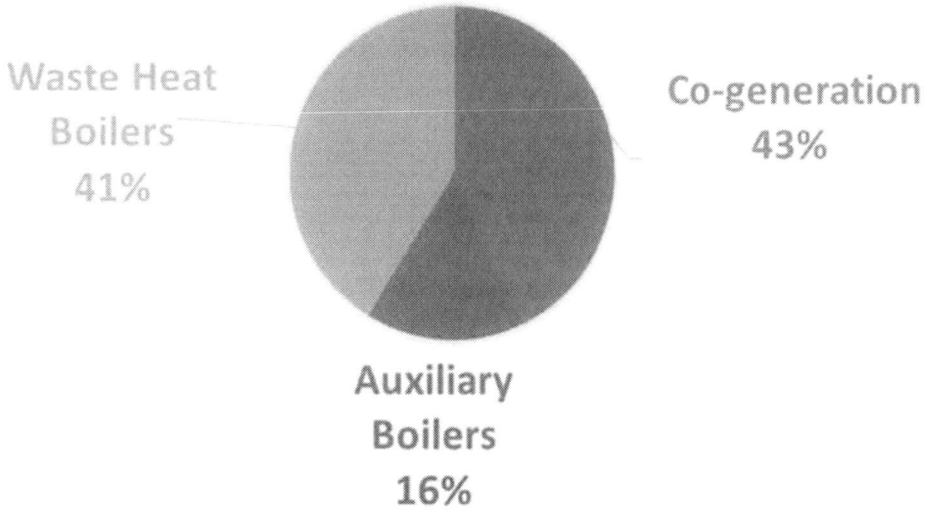

Figure 5 Steam Generation Profile

PROJECT GOALS AND DRIVERS

Recognizing the importance of protecting its assets, Central San commissioned the Comprehensive Wastewater Master Plan (CWMP) to build on its planning efforts over the past few years. The following key "drivers" were identified to help guide the plan's direction as it relates to solids handling and energy management:

Aging Infrastructure - Maintain performance and reliability of existing assets to ensure reliable wastewater collection and treatment. This includes repair or replacement (R&R) of aging equipment and some structures to extend their useful lives.

<u>Maximize Remaining Life of Incinerators:</u> Through condition assessments, it was determined that the incinerators are in a good condition due to Central San's robust maintenance program and practices and have at least 20 years remaining physical life. Central San is also able to operate one of their two furnaces per year while performing preventive maintenance on the offline furnace. Although the incinerators are in good condition, several support solids handling support facilities (e.g. dewatering equipment) are beyond their useful life and require improvements in order to realize the full life of the incinerators. The CWMP focused on developing a long-term plan for transitioning from Central San's existing incineration facility while performing near-term improvements to maximize the remaining life of the incinerators.

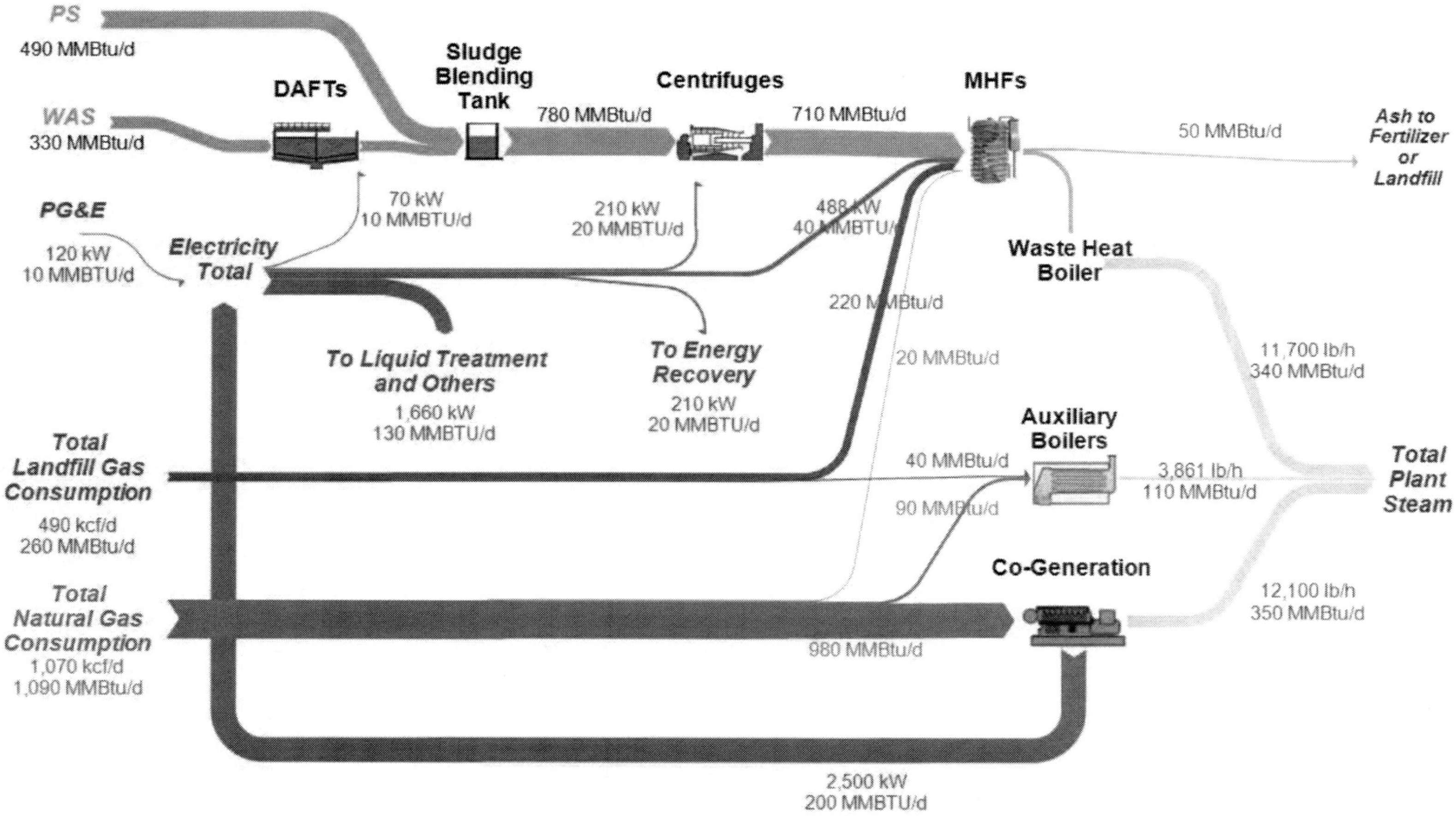

Figure 6. Central San Sankey Diagram of Current Solids Processing and Energy Recovery

Capacity - Increase capacity, if required, to accommodate planned growth for the communities served within Central San's service area. As flows and loadings increase in proportion to population growth, some treatment processes may need to be expanded.

Address Limited Incineration Capacity: The incinerators are approaching their design and permitted capacity. Currently, the solids produced during peak solids events cannot be processed through one incinerator and need to be stored for later processing. In the longer term, the average processing capacity of the incineration will also become limited resulting in the increased difficulty of solids management and the need to process some of the solids generated at the plant through an alternative parallel solids handling process or through off-site disposal. The CWMP evaluated options for reducing solids loading to the incinerators in addition to developing a phased implementation of the long-term solids handling improvements to address potential capacity triggers.

Regulatory - Comply reliably with regulatory requirements, protect human health and the environment, and plan for anticipated future regulatory requirements.

Reliably Meet Air Regulations: Over the last few years, Central San has been working on improvements to Central San's incineration air pollution control train and ash handling system to meet recent federal sewage sludge incineration rules (SSI MACT 129). To achieve reliable compliance with SSI MACT 129, the CWMP recommended considering air pollution control improvements to the wet scrubber air pollution control system.

Plan for Increasingly Stringent Air Regulations: In the coming years, federal, state, and regional regulations for air emissions may become stricter. The Bay Area Air Quality Management District (BAAQMD) is currently considering implementation of a new rule for reduction of risk from air toxic emissions at existing facilities, which may require significant plant-wide improvements including additional air pollution control equipment for Central San's incinerators.

Manage Greenhouse Gas Emissions: Currently, Central San successfully operates its existing solids handling processes and natural gas cogeneration facility such that it maintains greenhouse gas emissions to below the California Air Resources Board (CARB) Cap and Trade Threshold of 25,000 metric tons of CO_2 equivalents. However, there are state greenhouse gas emission regulations and goals that may reduce this threshold or transform the way Central San operates its facilities. Additionally, Central San has relied on local landfill gas to provide supplemental fuel to the incinerators. However, the remaining life of the landfill is questionable and a reduction in available landfill gas supply would require more import of natural gas creating a significant challenge for Central San to stay below the CARB Cap and Trade Threshold of 25,000 metric tons of CO_2 equivalents

Increasing Regulations on the Biosolids Land Application and Landfilling: The majority of the wastewater facilities in the San Francisco Bay Area generating biosolids typically land apply Class B biosolids or send biosolids to a landfill. In the future, the amount of organics that could be sent to landfills is going to be limited, which directly impacts the amount of biosolids that can be sent to the landfill for alternative daily cover. This diversion of organics and ultimately biosolids from the landfills may trigger increased competition among biosolids producers for land application and composting and lead to numerous facilities upgrading their solids processing to produce higher quality products and possibly look at thermal minimization technologies that do not produce significant amounts of biosolids that require disposal. These future limitations should be kept in mind during the planning of solids treatment facilities for Central San.

Sustainability/Optimization - Minimize life-cycle costs, maximize benefits, and achieve economic stability through optimization, resiliency, resource recovery, and energy projects. Projects include optimizing existing treatment processes, completing energy efficiency improvements, recovering resources from solids processes, generating renewable power, increasing recycled water production, and completing improvements for reliability and resiliency against man-made and natural disasters.

Improve Operational Resiliency: For many years, Central San has successfully operated a waste heat energy recovery system for its incinerators and cogeneration facility. Steam generated from the waste heat is used to drive steam-driven turbines that provide air for secondary treatment. Although this system was innovative for its time, the overall energy conversion is not as efficient as other energy recovery options that are now available. This interconnection between solids handling, energy recovery, and secondary treatment is unique but it also results in an operational vulnerability. Unexpected outages or disruptions in solids handling and/or steam production can create a ripple effect on the ability to provide oxygen for secondary biological treatment. Similarly, unexpected failures or changes in the air and steam systems can result in bypassing of the waste heat recovery system and air pollution control train resulting in a reportable compliance activity under the new SSI MACT 129 rule. Both Central San's Pump & Blower Building that house the steam-driven aeration turbines and the Solids Conditioning Building that houses the solids handling and cogeneration facility do not meet current seismic standards presenting an increased risk to Central San.

Strive to Achieve Net Zero Energy: Central San also recently adopted an energy policy that is aimed at achieving net zero energy. The CWMP evaluated several solids and energy alternatives that provide better energy conversion efficiencies and that enable Central San to import high strength wastes that will be key to achieving net zero energy. To achieve net zero energy, the CWMP determined that other energy efficiency improvements and renewable energy production would be required in addition to more energy efficient processes and imported high strength wastes.

ALTERNATIVES EVALUATION AND COMPARISON

Electrical versus Steam Blowers

The improvement of energy resiliency was the first step in solids and energy alternatives evaluation. In the current configuration, aeration air is supplied by steam-driven blowers. Steam is supplied from the waste heat recovery boilers for the multiple hearth furnaces and cogeneration system, in addition to steam from the auxiliary boilers. There were several drivers for replacing the existing steam-driven blowers:

- The steam-driven blowers cannot turn down to meet low aeration demands. During these conditions, excess air is wasted.
- The steam-driven blowers are located in the Pump & Blower Building and the boilers and steam system are located a long distance away in the Solids Conditioning Building. An extensive amount of steam and condensate piping is routed through tunnels. These piping systems are old, corroding, and will require costly replacement. Additionally, the buried air piping from the blowers to the aeration tanks has significant leaks resulting in substantial air losses.
- The linkage between secondary treatment aeration and solids handling waste heat production creates a vulnerability for Central San. Disruptions in either system can have ripple effects across both the solids and secondary treatment systems, in addition to creating air emission compliance issues by triggering boiler bypass events. Decoupling these two systems can provide a significant improvement in operational resiliency.
- The waste heat steam system cannot supply enough energy for the blowers during peak aeration demands (hot weather and weekends) while still maintaining redundancy in the system. An additional auxiliary boiler is required to meet redundancy criteria.
- The anticipated future higher aeration demand for potential nutrient removal facilities required larger blowers, thus the existing steam blowers would have to be replaced with the larger steam blowers or with electrical blowers.
- Because of hydraulics issues, the existing two steam-driven blowers and the one back-up electric blower cannot be operated together, which limits capacity and operational flexibility.
- Central San's single back-up electric blower is not sized adequately for typical daily aeration demands. It is currently used as a temporary back-up when maintenance is required on the steam system or steam-driven blowers.

An alternative considered was to use heat recovered from the incineration and co-generation facilities to generate electricity instead of the steam. Generated electricity could be used anywhere at the plant and in the case of solids treatment disruptions, electricity could be purchased from the grid or produced from on-site backup generators.

Numerous incinerator facilities in the United States and Europe have investigated and constructed energy recovery for their multiple-hearth or fluidized bed incineration systems. They all convert heat to electricity using steam turbines associated with waste heat boilers or more recently using Organic Rankine Cycle (ORC) systems. Since the ORC systems are easier to operate and maintain and are typically more efficient compared to steam turbines, the ORC was chosen for further evaluation. ORC technology is similar to the steam turbines used at Central San, but instead of using water vapor, the ORC turbine uses vaporized high-molecular-mass organic fluid. Flue gas/waste heat from the incinerators and waste heat from the cogeneration facility will be sent to a thermal oil heat exchanger to recover energy as heated thermal oil. Heated thermal oil will then be sent to the ORC turbine to generate electricity that is connected to the plant electricity grid system.

Two alternatives were developed to meet the needs of the aeration system:

<u>Continue with steam-driven aeration</u>: This alternative would entail replacing the steam-turbine driven blowers with similar steam-driven blowers, but sized efficiently to meet the necessary

turn-down requirements. Several related improvements would be required for this approach: the air piping would need to be replaced to eliminate the leaks in the current system, the steam and condensate piping would need to be replaced because it is old and corroded, and the auxiliary boiler and waste heat boiler would need to be maintained and replaced as they age. An additional auxiliary boiler and an additional electric blower would also be required to meet redundancy requirements.

Convert to electric-driven blowers: This alternative is to replace the existing steam-driven blowers and back-up electric blower with multiple small electric-driven, high-speed turbine blowers, housed in a new aeration building with new above-grade aeration air piping. Converting to electric, high-speed turbine blowers can provide several benefits: they are more efficient, have the capability to operate over the required range of air flows, and their operation would be independent of the steam system. The auxiliary boiler and waste heat boilers would be replaced with a thermal oil heat exchanger and ORC turbine designed to convert waste-heat from the furnaces into electricity for the plant.

A lifecycle cost evaluation was completed to compare the alternatives. The projected aeration demands and energy consumption were used to develop the lifecycle cost estimates. Table 1 presents the results of the comparison. As shown in the results, the capital costs for the conversion to electrical blowers is higher; however, the reduction in annual O&M costs makes the lifecycle cost comparable to the steam-driven blowers. The electric blowers option was selected because the lifecycle cost was comparable and provided several benefits to treatment plant operations and was overall a more reliable alternative.

Table 1. Net present value comparison of steam powered bowers vs. electric blowers.

Cost	New Steam Powered Blowers	Electric Blowers
Total Project Cost	$32,832,200	$37,608,400
Total Annual O&M Cost (dollars annually)	$2,691,600	$2,431,550
Net Present Value[1]	**$49,300,000**	**$47,600,000**

1. NPV is calculated as sum of the total project cost and total O&M costs throughout the 20 years

Alternatives Selection

The alternatives evaluation process was conducted through a series of workshops with Central San staff and consulting teams. The first workshop targeted evaluation of the universe of alternatives shown in Figure 7, which is the broad field of possible solids stream, and energy process alternatives currently available in the industry.

The following pass/fail criteria were used for the initial screening:

- Is the technology proven/scalable to meet liquid, solids, and air regulations?

- Does the technology fit within site constraints?
- Does the technology maximize use of existing facilities?

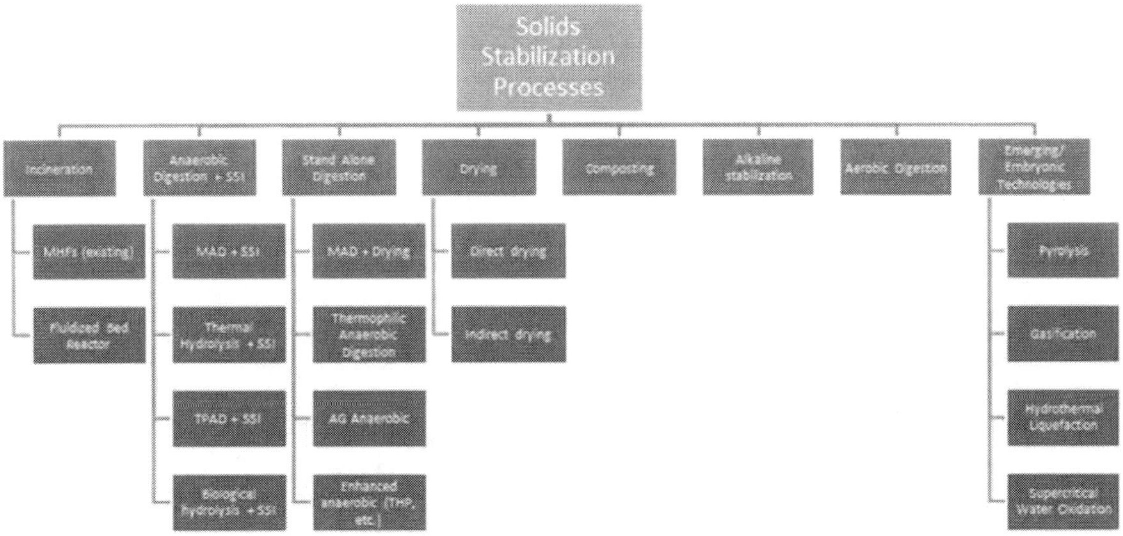

Figure 7. Universe of Solids Alternatives

The screening criteria were applied to all alternatives and the remaining viable solids stabilization technology alternatives were carried forward for final screening and alternative development and evaluation.

Assuming future solids disposal regulations will limit disposal options, only the proven technologies that result in production of Class A biosolids or ash were considered. In addition, the technologies that allow energy recovery were preferred. All emerging and embryonic high-temperature oxidation solutions were also deferred from the alternatives analysis due to the lack of proven performance. However, over the next several years, Central San will be exploring the feasibility of some of these technologies (e.g. Gasification, Hydrothermal Liquefaction, Supercritical Water Oxidation, etc.).

The following alternatives for long-term planning were carried forward for the master planning evaluation:

- Alternative 1: Replacement of the MHFs with new fluidized bed incinerators (FBIs).
- Alternative 2: Mesophilic anaerobic digestion combined with a new dryer and one FBI in place of the MHFs.
- Alternative 3: Enhanced Anaerobic Digestion (thermal hydrolysis followed by anaerobic digestion).

In addition, a baseline alternative where Central San continues to use the existing MHFs was included in the evaluation. Due to concerns regarding air pollution control requirements, investment into the existing furnaces, and the age of the furnaces, long-term use of the MHFs was not considered to be a viable alternative. Central San agreed that the existing MHFs should be used for their remaining useful life and then the selected alternative in the Master Plan should be implemented. Unless an emerging technology shows promise and is determined to be more optimal than the selected alternative, the selected alternative will serve as the basis for MHF replacement whenever MHF replacement is required – either at the end of its useful life or if the MHFs were to experience catastrophic failure.

Final Alternatives Evaluation

Following assumption were used for alternatives evaluation:

- For comparing operational costs among alternatives, it was assumed that the future electricity requirement for the plant would be 4,220 kW. Actual electricity requirement for the plant may be different due to variety of liquid treatment options that could be implemented in the future for nutrient removal and/or recycled water.
- All incineration alternatives assume that the heat recovery to generate steam currently practiced at Central San is replaced with a thermal oil heat exchanger and ORC turbine that generates electricity, as discussed previously.
- Provide an approach that minimizes greenhouse gas (GHG) production, maximizes cogeneration (to the GHG cap limit of 25,000 metric tons of CO_2 equivalents), and prioritize the use of fuel source in the following order: biogas, landfill gas, and then natural gas. The remaining electrical demand was assumed to be satisfied by imported grid electricity. A separate evaluation was included to determine the feasibility of implementing solar and/or wind renewable power production to offset imported grid and natural gas usage.
- Table 2 shows flows and loads that were used as a base for alternatives evaluation.

Table 2. Central San Projected Solids Flows and Loads

	Years and Load Conditions			
Parameters	**2035 Average Annual**	**2035 Max Month**	**2035 Max Week**	**2035 Peak Day**
Primary Solids Load, (kg/d)	33,300	30,800	36,450	62,000
Primary Solids %VS	84%	84%	84%	84%
WAS Solids Load (kg/d)	29,100	37,000	38,900	42,000
WAS Solids %VS	77%	77%	77%	77%
Total Solids (kg/d)	**62,400**	**67,800**	**75,350**	**104,000**

Alternative 1 (A1): Replacement of the MHFs with new fluidized bed incinerator (FBIs)

Alternative 1 is the replacement of the existing MHFs with two FBIs. The end product from this alternative is ash produced at an average annual rate of 17,240 wet kg/day (19 wet tons/day). The process flow diagram for this alternative is shown in Figure 8.

The following describes Alternative 1:

- The existing DAFTs would continue to thicken the WAS. Consistent with current operations, thickened WAS and primary solids would be blended and fed to centrifuge dewatering prior to incineration.
- Improvements to the solids blend tank and mixing system are required to reduce peak loadings to the incinerators and maintained a well-mixed solids feed. These improvements are necessary in the near-term for the current MHFs, which would continue to be operated while new incinerators are constructed.
- The use of new centrifuges would result in improved dewatering, which would result in a drier feed solids and reduce (or potentially eliminate) the required auxiliary fuel for incineration.
- In the future, two new FBIs would replace the existing MHFs. Two FBIs (one duty, one standby) are included to provide year-round solids stabilization, allowing one FBI to be out of service for routine and major maintenance similar to the current MHF arrangement. The evaluation assumed that both FBIs would be located inside the existing Solids Control Building. During pre-design, the construction of a new building could be further evaluated.
- Waste heat from the existing incinerators and future incinerators will be sent to the thermal oil heat exchanger and ORC.
- Cogeneration would operate with a combination of conditioned landfill gas and natural gas up to the cap and trade threshold.

Alternative 2 (A2): Mesophilic anaerobic digestion combined with a new dryer and FBI in place of the MHFs

Alternative 2 includes anaerobic digestion, thermal drying, and incineration. The end product of Alternative 2 is ash. The process flow diagram is shown in Figure 9.

The following describes Alternative 2:

- As with Alternative 1, an improved or new sludge blend tank, new centrifuges, and an ORC conversion would be included.
- The anaerobic digesters are designed to digest primary sludge and WAS, and the existing MHFs would be phased out during the next 20 years and replaced with a single FBI when the existing MHFs are demolished. The digesters would be sized to operate as a stand-alone Class B biosolids process when the FBI is offline for maintenance or in emergencies. Increasing the digester capacity to achieve Class B treatment provided similar maintenance redundancy as included in Alternative 1.

- With the conversion of volatile organics to biogas in the digesters, the FBI would likely require additional fuel as compared to processing solids with an FBI only. To reduce the need for auxiliary fuel, a dryer using a portion of the heat recovered from the incinerator would be used to partially dry the dewatered digested solids prior to incineration.
- This alternative provides the opportunity to add fats, oils, and grease (FOG) and high strength waste (HSW) receiving facilities to augment biogas production in the anaerobic digesters.
- Cogeneration would operate with a combination of conditioned biogas, conditioned landfill gas, and natural gas up to the cap and trade threshold.

Alternative 3 (A3): Enhanced anaerobic digestion (thermal hydrolysis followed by mesophilic anaerobic digestion)

This alternative is based on producing Class A biosolids for beneficial reuse (fertilizer blending and land application). The purpose for treating to Class A biosolids standards in this alternative was to address concerns regarding California organics diversion goals from landfills and the impact on the feasibility and cost of disposing of Class B biosolids. The anaerobic digesters with thermal hydrolysis would be constructed when the MHFs have reached the end of their useful life. The process flow diagram is shown in Figure 10.

The following describes Alternative 3:

- Mesophilic anaerobic digesters would be configured with thermal hydrolysis pretreatment. This process reduces the number of digesters required and has been shown to be cost competitive with conventional digestion. It is noted that other Class A treatment trains are also viable. This configuration was selected to provide a reasonable and comparable alternative. If this alternative were to be selected, the project's predesign phase would evaluate the appropriate Class A treatment train to meet Central San's goals.
- To provide the necessary pretreatment upstream of the THP units, this alternative relies on WAS co-thickening with Primary solids in three DAFTs, TWAS screening, and centrifuge dewatering.
- This alternative provides the opportunity to add fats, oils, and grease (FOG) and high strength waste (HSW) receiving facilities to augment biogas production in the digesters.
- Cogeneration would operate with a combination of conditioned biogas, conditioned landfill gas, and natural gas up to the cap and trade threshold.
- Dewatered Class A biosolids would be hauled off site and beneficially reused.

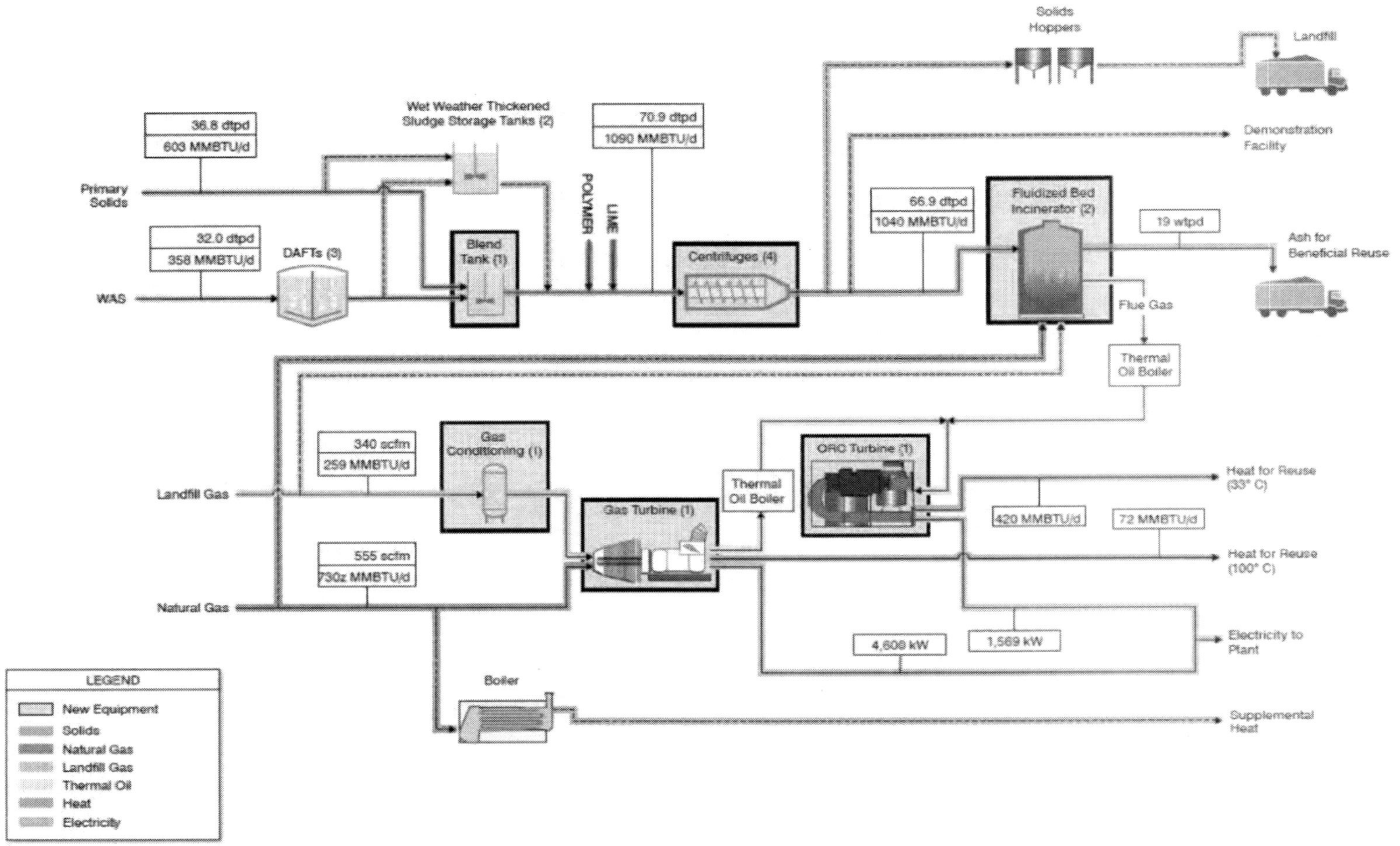

Figure 8. Alternative 1 Process Flow Diagram

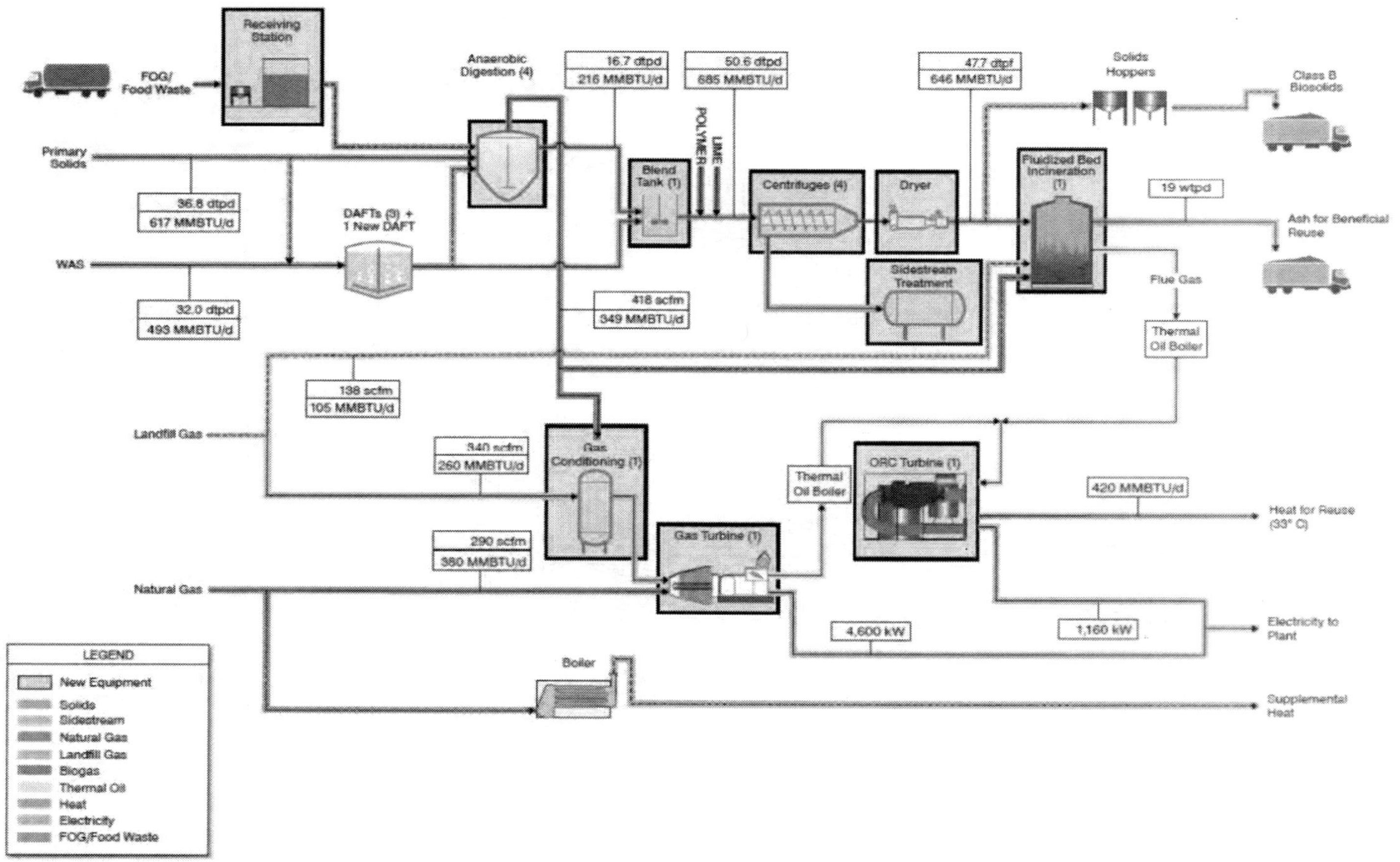

Figure 9. Alternative 2 Process Flow Diagram

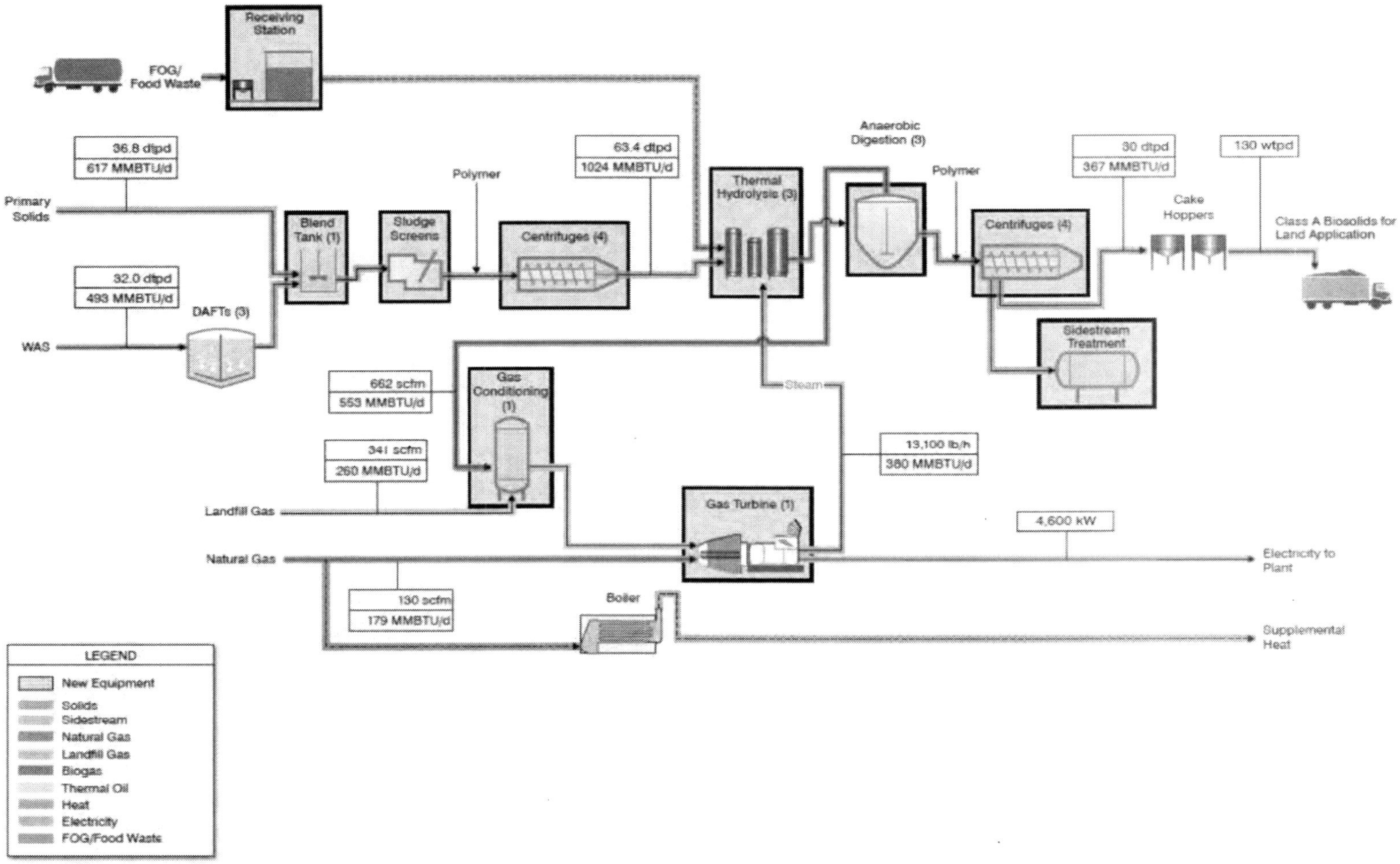

Figure 10. Alternative 3 Process Flow Diagram

Triple Bottom Line Plus Evaluation of Alternatives

To select the preferred alternative, each alternative was evaluated using the triple bottom line plus method (TBL+). The triple bottom line plus method takes technical, financial, social, and environmental factors into consideration. The following summarizes the basis of the TBL+ scoring for each of the criteria.

Technical

- Technical Objective #1 (T1) - The following reliability and performance criteria were used to evaluate each alternative:
  - Reliability/Redundancy: A2 was the highest scored, due to the ability to have two different means of off-hauling stabilized solids from the plant – either incineration with ash reuse or Class B biosolids disposal.
  - Proven technology: All alternatives equal scoring for this criterion.
  - Process stability: A1 was scored higher than A2 and A3 because it did not include anaerobic digestion, which is a biological process that does carry some additional process stability risk.
- Technical Objective #2 (T2) - The following efficiency criteria were used to evaluate each alternative:
  - Minimize expendables and number of equipment units: A1 includes the fewest number of units and received the highest score, A2 includes the second fewest, and A3 includes the most processes and received the lowest score.
  - Ease of operation: A1 was scored the highest because this process is most similar to the existing incineration process. A2 adds digestion and biogas processes. A3 is an entirely new process and includes greater operational complexity than the other alternatives.
  - Efficient use of space: A3 was scored lower than A1 and A2 due to the need for additional process space.
  - Flexibility to meet future regulations and innovation: A1 and A2 were scored higher than A3, due to less control over potential biosolids regulations that may heavily impact A3. A3 final product is digested and dewatered cake, thus it is relying on the year-round land application of biosolids.

Financial

- Financial Objective #1 (F1) - Minimize Capital Costs:

  - Capital cost: All alternatives were scored equally because the planning level capital cost estimates are relatively similar for the three alternatives.

- Financial Objective #2 (F2) - Minimize Operations & Maintenance Costs:

  - Annual O&M costs: All alternatives were scored equally because the planning level O&M costs are similar for the three alternatives.

Social

- Social Objective #1 (S1) - The following criteria for protecting public health and safety were used to evaluate each alternative:

  - Comply with regulatory requirements: All alternatives provide equivalent ability to meet this criteria.

  - Resiliency for catastrophic events: A2 scored the highest because it includes two potential end products (ash or Class B biosolids) providing more resiliency in case of emergencies.

- Social Objective #2 (S2) - The following criteria for maintaining good public relations were used to evaluate each alternative:

  - Limit odors to within fence line: A3 scored the lowest because it includes an additional solids process facility which would require additional odor control.

  - Minimize noise/visual impacts: All alternatives provide equivalent ability to meet this criteria.

Environmental

- Environmental Objective #1 (E1) - The following criteria for minimizing impact on the local environment were used to evaluate each alternative:

  - Recycled Water Production: not applicable to the solids evaluation.

  - Minimize impacts from sludge/biosolids: A3 scored lower than A1 and A2 due to the significantly increased number of truck trips required for biosolids disposal.

- Environmental Objective #2 (E2) - The following criteria for minimizing impact on the global environment were used to evaluate each alternative:

  - Minimize Greenhouse gas (GHG) emissions: Estimated GHG emissions for each solids alternative are included in Appendix B. A3 scored the highest, followed by A2, and then A1.

- Minimize energy use/maximize energy self-sufficiency: A3 scored the highest, followed by A2, and then A1.

The results of alternatives comparison are shown in Figure 11. Alternative 2 scored the highest and was recommended for implementation.

Alternatives Comparison in Terms of Energy Recovery

The energy profile was analyzed to assess the efficiency of each alternative in recovering and beneficially reusing energy at the plant. The energy profile was evaluated for the following conditions:

- **Current Operations:** This represents the current facilities at Central San, including the existing MHFs with the current steam energy recovery system. This scenario caps the existing MHFs at a feed of 54.8 dptd, consistent with the existing Title V air permit.

- **Baseline:** The baseline condition is the current operation with the implementation of the aeration system upgrade, including removal of the steam system and addition of the thermal oil heat exchanger and ORC turbine generator. This scenario caps the existing MHFs at a feed of 54.8 dptd, consistent with the existing Title V permit, and sends projected excess loading of 11.9 dtpd to a solids handling demonstration facility.

- **Alternative 1** - Replacement of the MHFs with new fluidized bed incinerator (FBIs)

- **Alternative 2** - Mesophilic anaerobic digestion combined with a new dryer and FBI in place of the MHFs

- **Alternative 3** - Enhanced anaerobic digestion (thermal hydrolysis followed by mesophilic anaerobic digestion)

The sources of energy include the solids, landfill gas, natural gas, and PG&E grid power. Currently, energy produced from the solids feed is converted to heat, steam, and electricity. Energy profiles were prepared for each solids alternative to develop the design criteria for the energy facilities. The energy profiles were estimated considering the following:

- Projected energy demand is based on operating the liquid treatment train with nutrient removal.

- The efficiency calculation for current conditions is based on an annual feed rate to the MHFs of 54.8 dtpd (limited by incineration Title V permitted capacity), while the estimate for Alternatives A1 through A3 was for a feed rate of 60 dtpd (actual projected feed rate).

- The steam-driven blowers are replaced be electric blowers. Electric power would be generated using an ORC that recovers waste heat from the incinerators.

- In all cases, the air emissions from the MHFs were held below 25,000 metric tons of CO_2 equivalents (the GHG cap). It was assumed that electricity would be preferably generated by the cogeneration system rather than purchased from PG&E.
- The existing gas turbine will be replaced with a new 4.4 MW, gas turbine, which will be fueled with natural gas. In the case of Alternative 2, the turbine needs additional landfill gas, natural gas, or biogas to run above minimum loading to the turbine.
- All alternatives were assumed to consume landfill gas at the rate of 260 MMBTU/day. If landfill gas supply diminishes, it would be replaced by biogas, natural gas, or electricity, depending on the configuration of the plant at the time of the reduced supply.

Figure 12 shows the solids conversion process and resulting efficiencies for the five conditions.

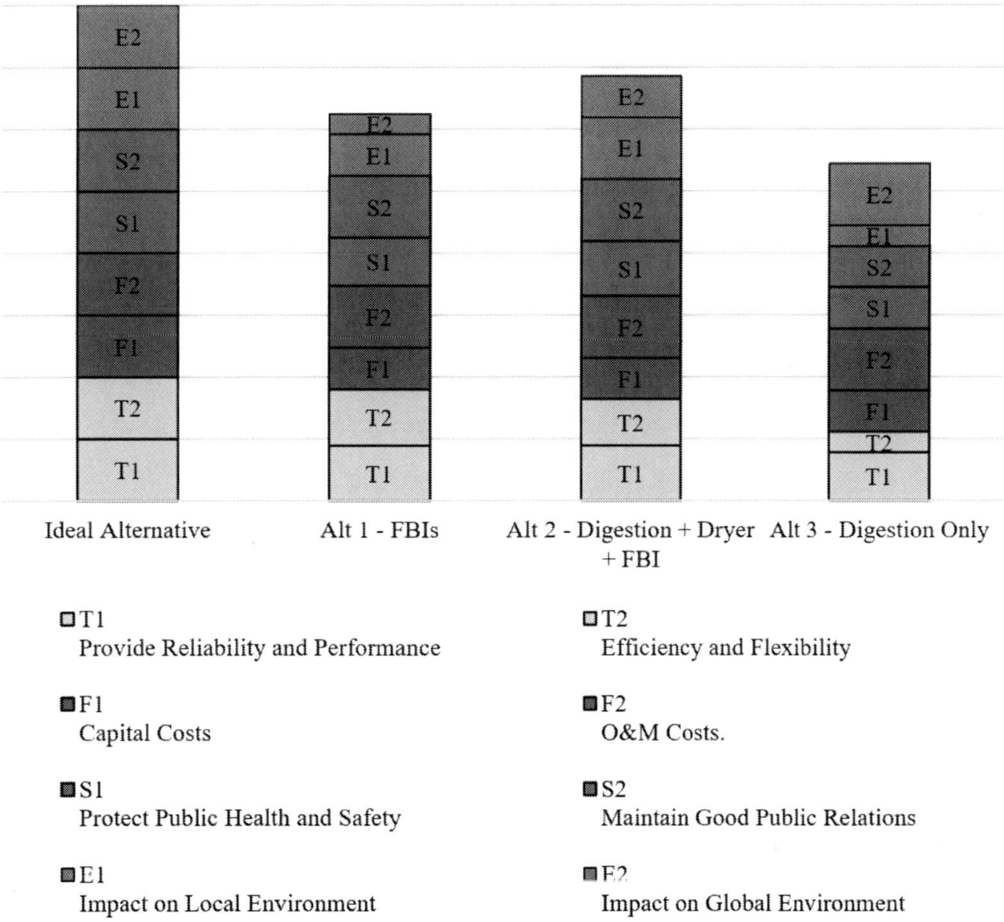

Figure 11 Cumulative criteria scoring summary solids alternatives

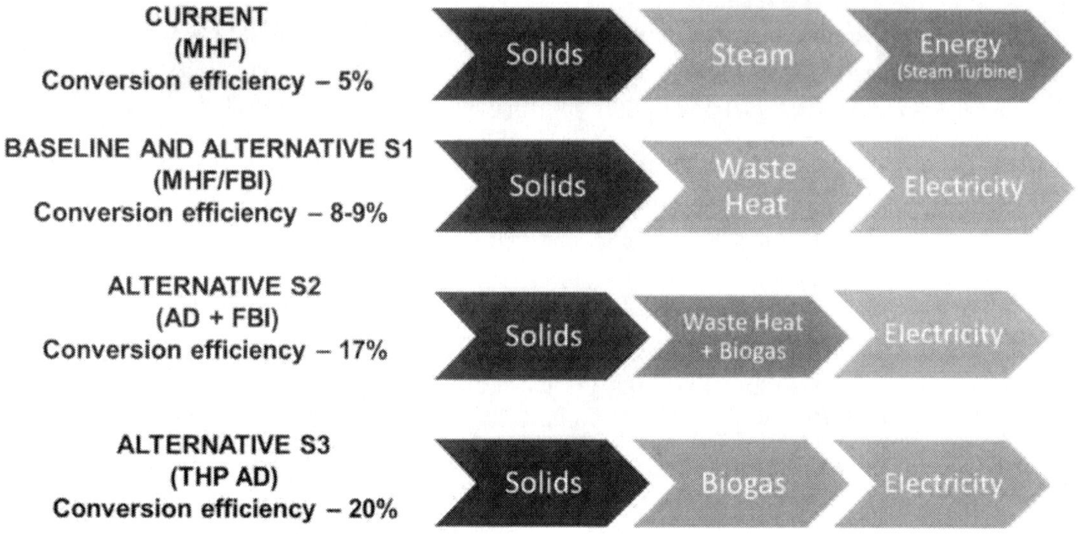

Figure 12 Solids conversion efficiency for each solids alternative.

The efficiency of the conversion depends on the solids processes used:

- Current incineration process utilizes the energy contained in the solids to produce waste heat, then steam via boilers, and then energy via the steam turbine. This process is 5 percent efficient in converting the energy in sludge to energy for plant usage.
- The Baseline and Alternative 1 also utilize incineration to generate waste heat from the solids, which is converted to electricity using the thermal oil heat exchanger and ORC turbine. These alternatives improve the solids energy conversion efficiency from 5 percent to up to 9 percent.
- Alternative 2 utilizes the anaerobic digestion of solids to generate biogas and waste heat from incineration, which are converted to electricity. This alternative raises the conversion efficiency to 17 percent due to the generation of biogas and the higher efficiency of the newer cogeneration turbine that would be required.
- Alternative 3 utilizes anaerobic digestion of solids to generate biogas, which is converted to heat and electricity. This stand-alone digestion alternative has a 20 percent conversion efficiency and is the most efficient because it maximizes biogas production and does not require supplemental fuel like the incinerators would.

Energy Recovery Comparison between Alternatives

Figure 13 presents the energy production in terms of the energy value (BTUs) produced from each energy source (solids, landfill gas, natural gas, and imported grid power).

Figure 14 presents the *electricity* produced by each energy source under each alternative, to meet an estimated 9.9 MW of total electricity demand for the future plant (all future solids and liquid upgrades included). The total demand of 9.9 MW is comprised of 4.2 MW for wastewater processing and 5.7 MW for producing 20 mgd of recycled water through membrane and reverse osmosis processes.

The following conclusions were drawn from the evaluation:

- **Baseline Conditions:** Several improvements can improve the overall efficiency of the existing MHF process such as the conversion of the waste heat recovery and steam systems with steam-driven blowers to a new electric blower system and waste heat to electricity using an ORC turbine. Additionally, more efficient cogeneration turbines can reduce the amount natural gas consumed for cogeneration. Lastly, a drier MHF feed cake would help to reduce the supplemental landfill gas required for the MHFs.

- **Alternative 1:** Alternative 1 benefits from the ability of the FBIs to process higher solids loads with less supplemental fuel, producing more energy. The reduced landfill gas amount can be conditioned and use in the cogeneration turbine and reduce natural gas combustion that produces anthropogenic greenhouse gas emissions. Compared to Alternatives 2 and 3, Alternative 1 produces the least electricity from solids and requires the most natural gas to generate electricity.

- **Alternative 2:** Alternative 2 generates more electrical energy from the same amount of solids as A1 due to the addition of anaerobic digestion upstream of incineration. Digestion plus incineration is more efficient at energy recovery than incineration alone. It requires less natural gas than Alternative 1 but requires more purchased electricity. Alternative 2 production and use of biogas results in more electricity derived from the solids and a reduce demand for imported natural gas compared to Alternative 1. It requires some additional grid electricity because the biogas and natural gas already maximize use of the cogeneration turbine capacity.

- **Alternative 3:** Alternative 3 produces the most electricity from solids due to the higher production of biogas with thermal hydrolysis pretreatment included. This alternative also requires the most imported electricity because the conditioned biogas and natural gas maximize use of the available cogeneration turbine capacity. This alternative uses the least amount of natural gas but it requires the most imported electricity. Alternative 3 does not have an ORC; therefore, the total potential for power production is lower than other alternatives.

- It is noted that since Alternatives 2 and 3 are below the greenhouse gas cap, they could be configured to include an additional turbine that could combust natural gas while remaining under the GHG cap.

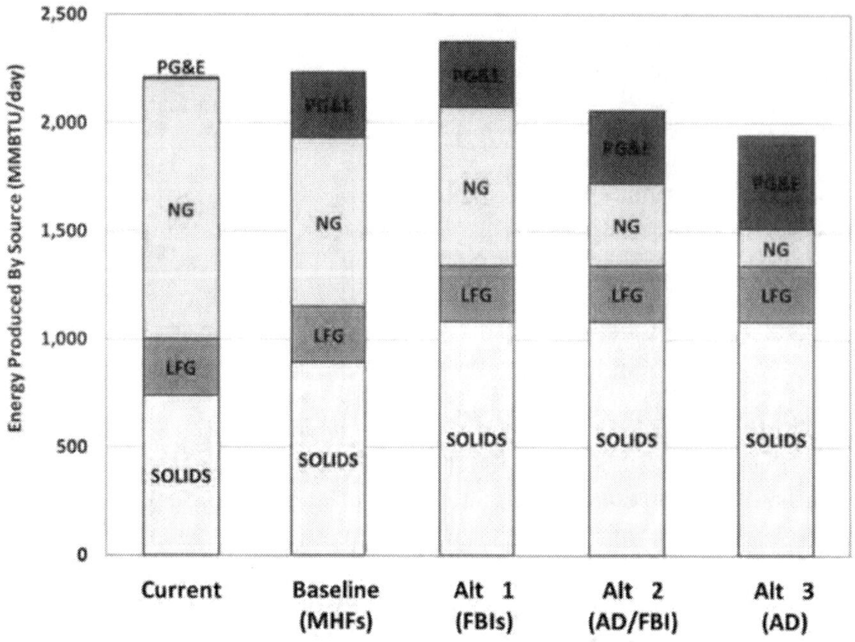

Figure 13 Comparison of solids alternatives by energy source

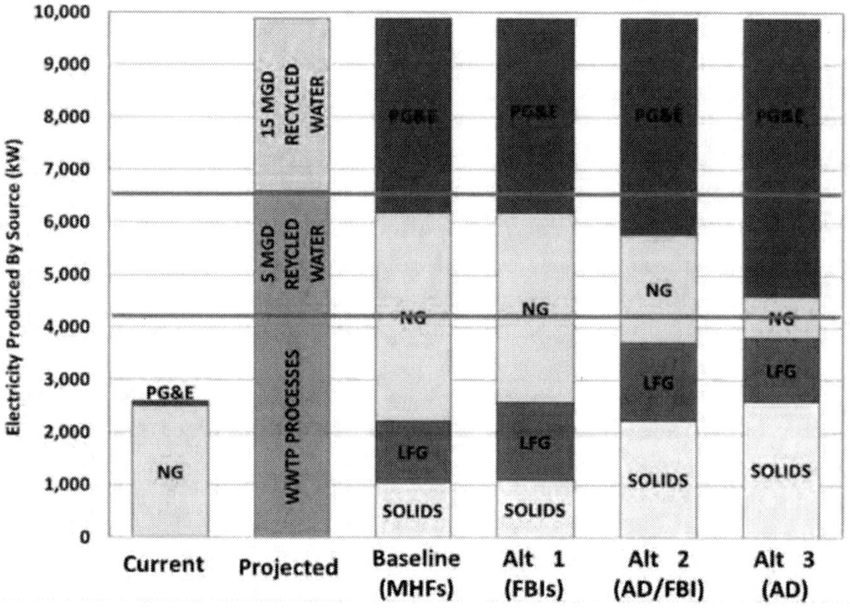

Figure 14 Electrical production by solids alternatives to meet electrical demand

Additional Energy Sources

Central San would like to strive towards net zero energy. As described above, all alternatives would require the continued import of grid electricity to meet the plant's energy needs and some use of natural gas. Renewable energy potential (wind and solar) were evaluated to reduce the Central San's reliance on imported grid electricity and natural gas. Additionally, alternatives including anaerobic digestions were evaluated for accepting imported high strength wastes for co-digestion. Figure 15 shows how each alternative could incorporate additional renewable energy sources to closer approach net zero energy.

The following are assumptions used for this evaluation:

- High-strength waste is calculated based on 20,000 gal per day of fats oil and grease (4.5% TS) imported to the plant. This amount was determined to be feasible given the service area.
- The maximum solar power potential was calculated for the installation of solar panels on all available Central San land and wet weather basins of approximately 130 acres. The peak solar production of this size of facility would be approximately 12 MW.
- The maximum wind power potential was calculated for the installation of wind turbines on all available Central San land of approximately 130 acres. The peak wind production of this size of facility would be approximately 1.9 MW.

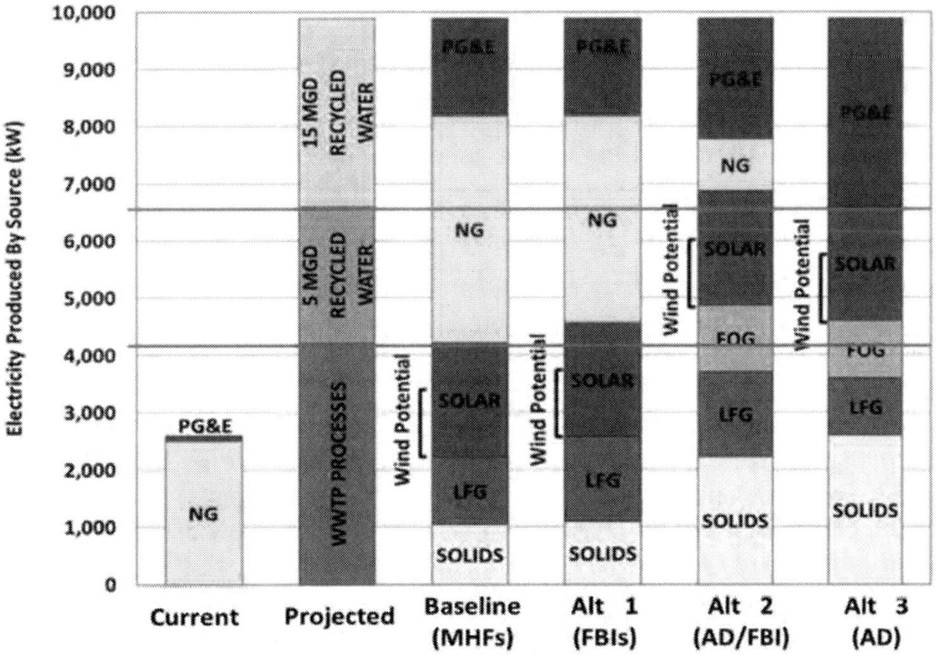

Figure 15. Electricity produced by source for additional energy alternatives

Summary of Energy Alternatives

The following summarizes the energy evaluation findings:

- Central San cannot reach energy neutrality using incineration alone alternatives (Baseline and Alternative 1).
- Stand-alone anaerobic digestion (Alternative 3) offers the highest overall energy conversion efficiency. The second most efficient configuration is anaerobic digestion followed by fluidized bed incineration (Alternative 2).
- Anaerobic digesters play a key role in reducing GHG emissions by allowing imported carbon energy source (high strength waste).
- Net zero energy can be achieved for treatment process energy demands (excluding potential future recycled water power demands) if a high strength waste facility were installed with Alternatives 2 and 3 and sufficient high strength waste is available.
- With co-digestion and solar, net zero energy can be achieved for treatment process energy demands and up to 5 mgd of potential future high quality recycled water energy demands.
- Net zero energy for all future projected energy demands does not appear to be feasible. Central San will need to weigh recycled water opportunities and goals with net zero energy goals to determine an optimal balance.
- Further investigation and tracking of net metering rules and tariffs is important to understand how much renewable power can be wheeled across the grid and to confirm the payback for renewable energy production facilities. For large renewable power installations over 1MW, Central San would need to explore other opportunities with the electrical utility and possibly consider battery storage in order to maximize the use of renewable power being generated.

CONCLUSIONS

A universe of technologies for solids processing at Central San were evaluated. The broad list of alternatives was then narrowed to an optimal alternative using a triple bottom line plus evaluation. The future recommended facilities include:

- Conversion of the waste heat recovery and steam system to thermal oil heat exchangers followed by an Organic Rankine Cycle (ORC) turbine to generate power from waste heat. Electricity generated from the ORC turbine will be used to power new, more efficient electric blowers.
- Anaerobic digesters will be added upstream of incineration to improve the resiliency of the solids processing system and provide a means to recover energy from biogas. The digesters will be designed to process solids and other high strength wastes, such as FOG

or food waste, to create more biogas. Digested biosolids will be dried and then incinerated.

- As the MHFs near the end of their useful life (within approximately 20 years), they will be replaced with a single fluidized bed incinerator (FBI). The FBI is a more advanced incineration process that produces fewer emissions than the MHFs and requires less fuel augmentation for combustion. The upstream digesters will be designed to operate independently from the FBI to allow for scheduled FBI shutdowns or to be operated for Class B biosolids for emergencies. This arrangement avoids the need for a redundant FBI train.

Energy improvements needed to strive towards net zero energy include:

- A 20,000 gal/d high strength waste facility to augment biogas production from the anaerobic digesters. A new, larger and more efficient cogeneration gas turbine will be installed to generate more electricity from the increased biogas production.

- Monitor energy incentive programs and re-evaluate net metering rules and tariffs for renewable energy. In anticipation of the solar energy facilities that would be required to offset future power demands, Central San will reserve land for potential solar and wind projects.

State of the Art Review of Digester Gas Treatment and Upgrading Technologies

Christine Polo[1], Kevin Jankowski[2], Cary Solberg[2], Jay Kemp[1], Greg Knight[1], Dave Long[1]

[1]Black & Veatch

[2]Milwaukee Metropolitan Sewerage District (MMSD)

1.0 Abstract

Beneficial use of digester gas in Combined Heat and Power (CHP) systems requires treatment for removal of contaminants including hydrogen sulfide (H_2S), moisture, siloxanes and Volatile Organic Compounds (VOC). Conversion to Renewable Natural Gas (RNG) for pipeline injection or vehicle fuel use requires removal of the same contaminants, plus digester gas upgrading to separate carbon dioxide (CO_2) from the methane fraction of the digester gas, with the methane portion representing the RNG.

Digester gas treatment and upgrading technologies can have high capital and operating costs and may make the difference between an energy recovery project that pays back or does not. Selection of cost-effective digester gas treatment technologies is thus important for successful energy recovery projects.

There are many digester gas treatment and upgrading technologies available offered by a wide range of manufacturers. Biogas treatment is a growing industry that serves different sources of biogas including landfills, municipal digesters, and agricultural digesters. The industry continues to evolve with new technologies being developed or adapted from other industries, existing technologies being improved, and technology providers starting, merging, changing names and sometimes going out of business.

This paper and presentation reviews the state of the art of digester gas treatment and upgrading technologies being used in the wastewater industry based on recent work performed for the Milwaukee Metropolitan Sewerage District (MMSD) and other clients. This review includes a comprehensive survey of the industry with lists of technology providers, and number of installations in North America and worldwide. A comparison of capital, operations and maintenance (O&M) and life-cycle costs is also presented.

Wastewater utilities considering digester gas treatment and upgrading technologies will gain a general understanding of digester gas treatment and upgrading technologies and which technologies may be the most appropriate for their applications based on treatment requirements, plant scale, contaminant concentrations and value of cleaned digester gas or RNG.

2.0 Technology Review

2.1 GENERAL CONCEPTS

2.1.1 Process Overview

Contaminants in biogas can be controlled or removed by a number of processes and technologies, which can be categorized as follows:

- Physical processes, processes that do not involve the chemical or biological transformation of the contaminant, such as:
  - **Chilling** for moisture removal or deep chilling for siloxane removal.
  - **Adsorption** of siloxane molecules onto activated carbon or other proprietary media. Adsorption is a process in which the substance adheres to the surface of the adsorbent, without chemically reacting with the adsorbent. Adsorption processes can be non-regenerable or regenerable. Regenerable systems use heat (Temperature Swing Adsorption) or pressure changes (Pressure Swing Adsorption) to regenerate the media.
  - **Membrane** separation of carbon dioxide from methane
  - **Physical absorption** of carbon dioxide into water in water scrubbers, whereby the carbon dioxide dissolves into a water solution but does not react with chemicals in the water.
- Chemical processes: processes involving chemical reactions, such as:
  - Addition of ferric salts to digesters, which results in the chemical **precipitation** of ferrous sulfide (FeS)
  - **Chemical absorption** which involves absorption into a liquid solution, followed by a chemical reaction. This is typically configured as a **chemical scrubber**, a vessel filled with packed media that maximize the contact area between the liquid adsorbent and the biogas. For example: A) iron sponge which involves the absorption or dissolution of H_2S into the liquid phase and subsequent chemical reaction with the hydrated ferric oxide. B) Chemical scrubbers for H_2S removal in which the gas dissolves in a caustic soda solution and subsequently reacts with the caustic soda. C) Amine scrubbers for carbon dioxide removal.
- Biological processes: processes involving biological activity, such as:
  - **Biological scrubber** for H_2S removal in which biogas is passed through a packed bed wet scrubber where Thiobacillus bacteria convert H_2S into elemental sulfur.

2.1.2 System and Vessel Configuration

Digester gas treatment systems must be properly configured to maximize their performance, efficiency, and reliability. For example, iron sponge systems for H_2S removal require the gas to be wet, so they are placed prior to moisture removal. Conversely, activated carbon systems for siloxane removal can also remove water, so it is critical to place these systems after moisture removal to maximize the life of the media.

Media-filled vessels (whether they act through adsorption or absorption) are often arranged in a configuration referred to as **lead/lag**. The lead/lag configuration consists of two vessels in series. The raw gas passes through the lead vessel first where it reacts with the media to remove most of the contaminant. Subsequently, the gas flows through the lag vessel, to polish or remove any remaining contaminant. Contaminant concentrations are measured after the lead vessel to detect breakthrough which indicates the media has been used up. Once breakthrough is detected, the lead and lag vessels are switched: the lag vessel is switched to lead, and the lead vessel is taken offline for media replacement. The vessel with fresh media is brought back online as the lag vessel.

2.2 HYDROGEN SULFIDE REMOVAL

H_2S is removed to avoid corrosion of metal equipment, to protect the more expensive siloxane removal media, and to reduce sulfur oxide (SOx) emissions from biogas combustion equipment.

Processes to remove H_2S from biogas include physical processes like activated carbon adsorption, chemical processes like chemical absorption onto iron sponge or iron oxide granular media and packed bed chemical scrubbers, and biological scrubbers. The H_2S concentration in the biogas can also be controlled by adding ferric salts to the sludge ahead of or in the anaerobic digesters.

The H_2S removal processes, technology providers, and number of installations are summarized in Table 2-1, and described in more detail in the sections below.

Table 2-1 H_2S Removal Processes, Technology Providers, and Number of Installations

PROCESS	PROVIDER	US & CANADA	WORLD	WRRF IN US & CANADA	WRRF IN WORLD
Iron sponge	Clean Methane Systems	70	90	60	90
	ESC	21	21	21	21
	Marcab	100	100	100	100
	MV Technologies	44	45	4	4
	Unison	21	21	19	19
	Biorem	5	5	5	5
Iron oxide granular media	Clean Methane Systems (SulfaTreat)	8	10	7	10
	ESC (SulfaTreat)	2	2	2	2
	Unison (SulfaTreat)	40	40	36	36
	Venture (SulfaTreat)	6	7	0	0
	MV Technologies (AxTrap)	2	2	1	1
Activated carbon	Hitachi Zosen Inova (HZI)	0	45	0	1
	Theia Air	8	9	4	4
	MV Technologies	2	2	0	0
Biological scrubber	Biorem	10	15	7	7
	BiogasClean	15	159	7	24
	Clean Methane Systems	3	6	3	6
	DMT	1	17	1	1
Chemical scrubber	Clean Methane Systems	1	1	0	0
	DMT (biological regeneration)	3	45	1	1
	PAQUES (biological regeneration)	27	190	15	158

2.2.1 Adding Ferric to Digesters

H_2S concentrations in the digester gas can be controlled by adding ferric salts such as ferric chloride upstream of the anaerobic digesters or directly into the anaerobic digesters. The iron reacts with the H_2S to precipitate ferrous sulfide (FeS). The dose of iron required can be estimated stoichiometrically. Ferric salts are commonly added to wastewater for other purposes such as enhanced primary treatment and phosphorus removal. Thus, many Water Resource Recovery Facilities (WRRF) already have the chemical feed system required for ferric addition, increasing the financial viability of this option.

2.2.2 Iron Sponge

In this process, biogas is passed through a permeable bed of iron sponge. Iron sponge is made up of wood chips coated in hydrated iron oxide. As H_2S passes through the iron sponge media, an exothermic (heat-producing) reaction takes place, converting H_2S to ferric sulfide (black solid) and water.

$$Fe_2O_3H_2O + 3H_2S \rightarrow Fe_2S_3 + 4H_2O$$

To increase media life and improve treated gas quality, iron sponge vessels are often arranged in a lead/lag setup, described in Section 2.1.2.

Iron sponge systems can be regenerable or non-regenerable. In non-regenerable systems, the media is used until it is spent at which point it is removed and replaced with new media. The spent media is disposed in a landfill. A disadvantage of non-regenerable iron sponge systems is that the used media can “cement up” and the process of removing the media and refilling the vessel can be messy and labor intensive. However, suppliers have developed means of breaking up the media or removing the media in large discs to facilitate media change-outs.

In regenerable systems, the iron sponge is regenerated and re-used up to 3 times before replacement. The regeneration of iron sponge produces elemental sulfur which coats the wood chips and decreases the effectiveness of the material, so the bed life is typically reduced by 50% with each regeneration cycle. The regeneration of the iron sponge is a highly exothermic reaction that uses water and air to oxidize the iron sulfide compounds forming sulfur and re-generating the hydrated ferric hydroxide. If not submerged in a flowing water bath and controlled properly, the iron chips can get very hot and combust spontaneously.

2.2.3 Iron Oxide on Granular Media (SulfaTreat and AxTrap)

Several manufacturers have developed processes whereby the hydrated ferric oxide is carried on granular media instead of wood chips. While more expensive than iron sponge, using granular media has operational advantages. The media is free-flowing and will not cement up like iron sponge media. The most common granular media process is Sulfa-Treat.

SulfaTreat is a patented process where the H_2S laden digester gas is passed through a bed of media that selectively reacts with H_2S in the gas. SulfaTreat media is based on iron-based materials similar to the iron sponge. Like iron sponge vessels, SulfaTreat vessels are often arranged in a lead/lag setup.

While the SulfaTreat processes is patented and the technology is currently owned by M-I SWACO, SulfaTreat systems are provided by several suppliers including Clean Methane Systems, ESC, Unison and Venture. Another granular media process is the AxTrap process, provided by MV Technologies.

2.2.4 Activated carbon

Activated carbon can also be used to remove H_2S from biogas. However, activated carbon is more expensive than iron sponge and granular oxide media. For this reason, activated carbon has typically only been used in specialized polishing applications where very low H_2S concentrations must be achieved, such as to remove H_2S prior to biogas upgrading for injection into the natural gas pipeline or use as vehicle fuel; or for very small or low concentration applications, where the cost of media replacements is less significant. Because activated carbon also removes moisture, moisture is typically removed prior to activated carbon H_2S removal systems to extend the media life.

2.2.5 Biological scrubber

Biogas is fed into a biological scrubber where Thiobacillus bacteria convert H_2S into elemental sulfur. The scrubber is filled with packing media which provides the surface for the bacteria to grow on. The scrubber liquid is pumped to the top of the tank where spray nozzles distribute it over the packing media. The scrubber liquid trickles down through the packed bed and provides the moisture and nutrients necessary for the bacteria. A small amount of air is injected into the reactor which is needed by the bacteria to oxidize the H_2S.

Startup can take several days to weeks for the Thiobacillus bacteria to grow on the media. The biological process requires additional monitoring to assure the bioreactor operates within specified guidelines. Because the process injects air into the gas stream in the bioreactor section, N_2 and O_2 may be introduced into the biogas stream. This is particularly a problem for pipeline injection and vehicle fuel applications, which require very low levels of N_2 and O_2.

Biological scrubbers compete with more conventional iron sponge systems in high H_2S load applications. While the capital cost is slightly higher, the operating cost is lower because it does not require media replacement.

2.2.6 Chemical Scrubber

Biogas is fed into a packed absorber column where the gas comes into contact with a caustic soda solution. The H_2S dissolves into the solution and reacts with the caustic soda. A disadvantage of chemical scrubbers is the use of chemicals, which can result in a high operating cost and may require use of corrosion resistant construction materials. To reduce the use of chemicals, some companies provide chemical scrubbers with biological regeneration. In fact, the vast majority of chemical scrubber installations include biological regeneration.

One example of a regenerable chemical scrubber system is Paques' THIOPAQ process. In this process, the spent chemical solution is fed to an aerobic bioreactor where Thiobacillus bacteria convert H_2S into elemental sulfur by the same mechanism described above for the biological scrubber. Lean or regenerated solvent is then recycled back to the reactor, while the elemental sulfur slurry produced by the Thiobacillus bacteria is pumped to a settling tank for sulfur extraction. A schematic of the process is shown in Figure 2-1.

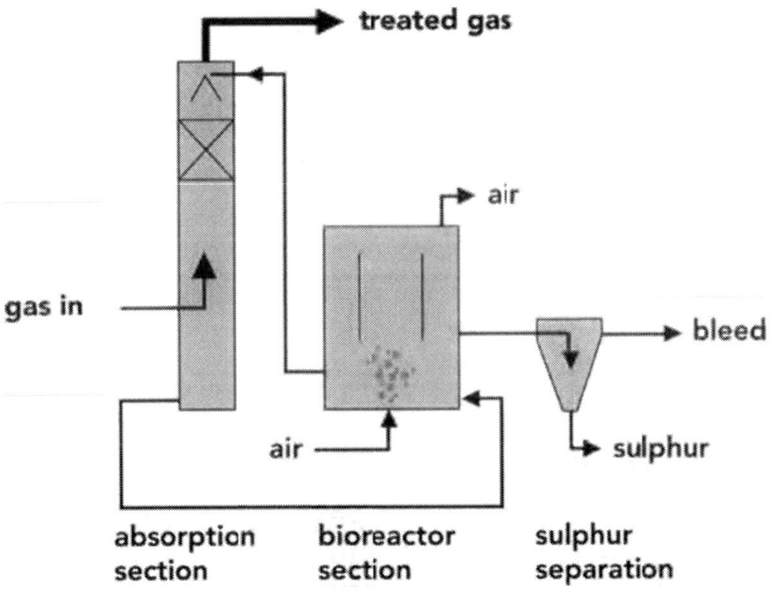

Figure 2-1 Schematic of Paques THIOPAQ H_2S Removal System (Source: Paques)

2.3 MOISTURE REMOVAL

Moisture removal is beneficial to avoid condensation of water in the gas piping and fuel trains. Condensation in the presence of even low concentrations of H_2S can create corrosive conditions. Moisture removal is generally required ahead of siloxane removal systems; the water vapor interferes with or reduces the effectiveness of adsorptive media.

The vast majority of installations remove moisture by cooling the gas in a chiller. Some specialized applications use physical absorption onto a dessicant.

The moisture removal processes, technology providers, and number of installations are summarized in Table 2-2, and described in more detail in the sections below.

Table 2-2 Moisture Removal Processes, Technology Providers and Number of Installations

PROCESS	PROVIDER	US & CANADA	WORLD	WRRF IN US & CANADA	WRRF IN WORLD
Chiller	Clean Methane Systems	75+	120+	70+	85+
	ESC	18	18	18	18
	Parker	100+	1000+		
	Perennial Energy	25	25	6	6
	Unison	151	152	125	126
	Venture	9	9	0	0
Desiccant dryer	Venture	1	1	0	0
	Nano	2	2	0	0

2.3.1 Conventional Chillers

The most common type of moisture removal process employs a chiller that uses a refrigerant to cool a liquid (glycol) which in turn is used to cool gas using a liquid-to-gas heat exchanger. Moisture is removed by cooling the gas through a heat exchanger to a dew point of 40 degrees Fahrenheit (°F) to condense the water in the gas stream. A second heat exchanger is then used to reheat the gas (usually to 70° to 80°F) using heat taken from the gas in the first heat exchanger. This reheating step provides separation from dew point that allows the gas to be piped without additional moisture dropping out as the gas cool, as long as the temperature remains above the 40°F dew pointtemperature .. The chiller is the same as used for building air conditioning systems.

2.3.2 Dessicant dryers

Dessicants are solid substances that absorb water. Dessicant dryers work by feeding the gas to a vessel full of dessicant media; the moisture in the gas absorbs to the dessicant. The saturated dessicant is typically regenerated using hot air. At least two vessels are included so that one vessel is operating while the other is regenerating.

2.4 SILOXANE REMOVAL

Siloxanes are a group of silicon containing organic compounds that turn into solid silicon dioxide upon combustion. Silicon dioxide can cause major wear of moving mechanical equipment like turbine blades, engine fuel injectors and pistons, and can coat heat recovery heat exchangers reducing their efficiency. Volatile organic compounds (VOC) are typically removed by the same processes that remove siloxanes, so their concentration must also be taken into account when sizing systems.

Siloxanes are typically removed through adsorption in vessels filled with solid media, most commonly activated carbon. Some technology providers offer adsorbent media filters with proprietary media such as activated alumina, silica gel, and polymers. Some of these proprietary medias can be regenerated by heating it up in a process called Temperature Swing Adsorption (TSA). Siloxanes can also be removed through a deep chilling process by cooling the gas to a dew point temperature -10°F or lower. Most adsorption-based siloxane removal systems require H_2S and moisture removal upstream to protect and prolong the life of the media.

The siloxane removal processes, technology providers, and number of installations are summarized in Table 2-3, and described in more detail in the sections below.

Table 2-3 Siloxane Removal Processes, Technology Providers, and Number of Installations

PROCESS	PROVIDER	US & CANADA	WORLD	WRRF: US & CANADA	WRRF: WORLD
Activated carbon; non-regenerable	Clean Methane Systems	100	150	95	100
	ESC	28	28	28	28
	Marcab	2	2	1	1
	Theia Air	5	5	2	2
	Unison	101	102	89	90

PROCESS	PROVIDER	US & CANADA	WORLD	WRRF: US & CANADA	WRRF: WORLD
Temperature Swing Adsorption	Venture	8	8	1	1
	HZI	0	1	0	1
	ESC	2	2	2	2
	DCL	5	5	1	1
	Parker	1	110		
	Venture	28	28	0	0
	Willexa	30	30	4	4
Low temperature gas drying	Pioneer Air	22	28	6	7

2.4.1 Activated Carbon Non-Regenerable Adsorption

Activated carbon non-regenerable adsorption is the most common type of siloxane removal used in the municipal wastewater industry. The biogas is passed through vessels filled with activated carbon. Siloxanes and VOCs in the gas adsorb or adhere to the surface of the activated carbon media. H_2S and moisture must be removed from the biogas prior to passing through the activated carbon vessels to protect and prolong the life of the media. A particulate filter is provided after the activated carbon vessels to capture any carbon particles that may be carried over in the gas stream.

Activated carbon vessels are typically arranged in lead-lag configuration, described in Section 2.1.2. The vessels are typically made of stainless steel. The systems are typically designed so that media replacement occurs every three to six months. Spent media is disposed in a landfill.

2.4.2 Temperature Swing Adsorption (TSA)

In the TSA process, the media is regenerated through a change in temperature. In this process, the biogas is passed through one of two vessels where the siloxane and VOC molecules adhere or adsorb to the surface of the media. Meanwhile, the other vessel is being regenerated, as shown in Figure 2-2. After some time, the vessels are switched so that the regenerated vessel is brought on line and the contaminated vessel is put on the regeneration cycle. The switching of tanks occurs automatically and continuously on an adjustable cycle to provide continuous siloxane removal with no interruption of the gas stream. The switch is usually programmed to take place every 12 or every 24 hours.

During the regeneration phase, the media can be heated by an electric heater or gas heater. The heat breaks the bond between the media and the siloxane and VOC molecules. The desorbed contaminants are carried out in the air stream. Depending on air emission regulations, this offgas may be vented to the atmosphere or disposed in a flare or thermal oxidizer.

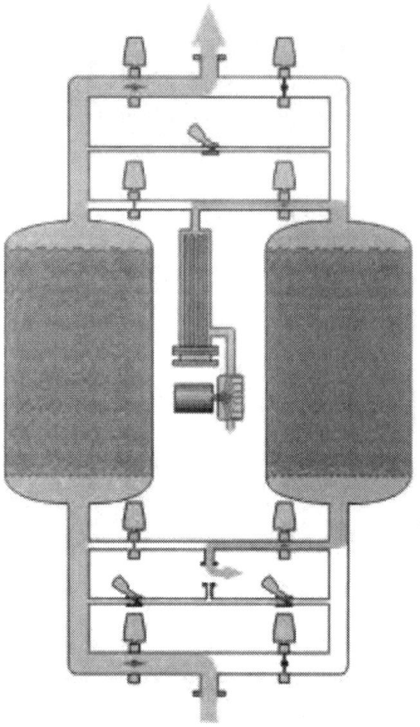

Figure 2-2 Temperature Swing Adsorption Process (Courtesy of Willexa)

A single media or a combination of medias can be used in regenerable systems. These include activated carbon and proprietary materials such as zeolites, silica gel, activated alumina, and polymers. Some of the technology providers provide a cartridge system rather than the more typical cylindrical stainless steel vessels. Some systems require moisture removal upstream while others do not. The system details for each technology provider are listed in Table 2-4.

Table 2-4 Temperature Swing Adsorption System Details and Differences

PROVIDER	MEDIA	REMOVES MOISTURE?	UPSTREAM MOISTURE REMOVAL REQUIRED?	CONTAINER	BETTER PERFORMANCE AT HIGHER PRESSURE?
ESC		Yes	Yes	Vessel	No
DCL	Polymer	No	No	Cartridge	Yes, but limit is < 10 psi
Parker	Polymeric resin	No	No, but recommended	Cartridge	Yes, but limit is < 7 psi
Venture	Mixture of activated alumina, silica gel & molecular sieve	Yes	Yes	Vessel	Yes
Willexa		Yes	Yes	Vessel	Yes

While the initial capital cost of regenerable systems is greater than non-regenerable systems, the media is replaced less frequently reducing operating costs. Thus, large biogas flow rates with relatively high siloxane concentrations typically favor regenerable systems. For this reason, most landfills use regenerable siloxane removal systems.

2.4.3 Deep Chilling

Deep chilling, or low temperature gas drying, works by cooling the gas to a very low temperature to remove all the moisture in the gas, and with it, a majority of the siloxanes. Approximately 96 to 97 percent of the siloxanes can be removed when the gas is dried to a dew point temperature of -10°F or lower. A major advantage of this system is that it removes both moisture and siloxanes in one step, potentially reducing the cost and complexity of the system.

Disadvantages of the deep chilling system are that the technology is not well proven and the system is provided by only one manufacturer, Pioneer Air. Pioneer Air has several installations using an older design which had issues with uneven cooling and freezing at the bottom of the reactor. Pioneer Air has updated its design to resolve this issue. The new design uses a glycol system similar to a conventional chiller instead of direct refrigerant chilling. There is only one installation so far using this new design located in Millbrae, CA. The system has performed well and has experienced no issues during its first year of operations, based on conversations with management and operating staff.

2.5 GAS UPGRADING: CARBON DIOXIDE REMOVAL

Conversion to Renewable Natural Gas (RNG) for pipeline injection or vehicle fuel use requires digester gas upgrading which consists mainly of separating carbon dioxide (CO_2) from the methane fraction of the digester gas, with the methane portion representing the RNG.

The gas upgrading processes, technology providers, and number of installations are summarized in Table 2-5, and described in more detail in the sections below.

Table 2-5 Gas Upgrading Processes, Technology Providers and Number of Installations

PROCESS	PROVIDER	US & CANADA	WORLD	WRRF: US & CANADA	WRRF: WORLD
PSA	Guild	24	27	6	6
	Carbotech	1	80	0	12
	Xebec	6	20	1	1
	Greenlane	2	5	0	0
Membrane	Air Liquide	25	60	2	4
	Xebec	0	0	0	0
	BioCNG (Unison)	10	10	4	4
	Hitachi Zosen Inova (HZI)	0	11	0	1
Water Scrubbing	Greenlane	20	98	2	32

PROCESS	PROVIDER	US & CANADA	WORLD	WRRF: US & CANADA	WRRF: WORLD
Amine Scrubbing	Purac	0	32	0	5
	Hitachi Zosen Inova (HZI)	0	37	0	0

2.5.1 Pressure Swing Adsorption (PSA)

Pressure Swing Adsorption is a physical adsorption process where the media is regenerated through changes in pressure. Different vendors provide different types of media including molecular sieve (zeolite), activated carbon, etc. As shown in Figure 2-3, the digester gas is passed through the bed of media at high pressure which maximizes adsorption of CO_2 onto the media. Once the media is saturated, the vessel is then depressurized and the CO_2 desorbs from the media. At least two vessels are provided so that at any given time one vessel is in adsorption mode, while the other is in desorption mode. A waste gas rich in CO_2 is produced which, depending on the local air emission regulations, may be disposed into the atmosphere, or combusted in a flare or thermal oxidizer.

Most PSA systems for carbon dioxide removal require H2S, moisture, and siloxane removal upstream to protect the PSA media. Guild's system which uses a patented molecular sieve media is unique in that it does not require pretreatment and removes H_2S, moisture, siloxanes and carbon dioxide in one step through the PSA process.

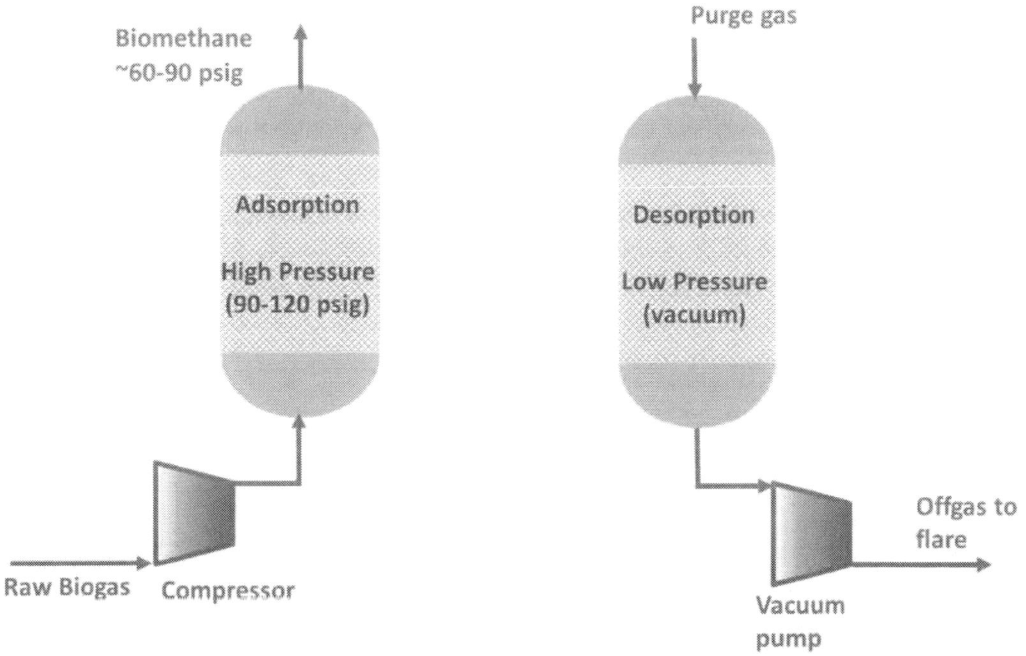

Figure 2-3 Pressure Swing Adsorption (PSA) Schematic

2.5.2 Selective Membrane

Membrane separation works by pressurizing the digester gas and passing it through selective membrane modules that separates the carbon dioxide from the methane. Pressure inside the membrane creates a concentration gradient which drives diffusion of molecules through the membrane. As shown in Figure 2-4, the membrane allows CO_2 to pass through, while the CH_4 remains inside the membrane. Membrane systems require H_2S, moisture, siloxane and VOC removal upstream. Membrane systems can have two stages or three stages of membrane modules depending on the target methane concentration.

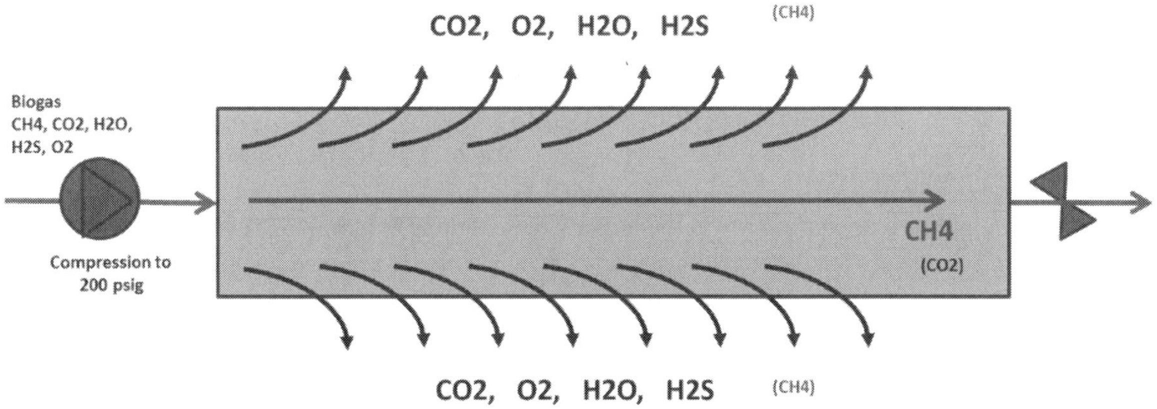

Figure 2-4 Selective Membrane Schematic

2.5.3 Water Scrubbing

Water scrubbing is a physical absorption process that works by pressurizing the digester gas and passing it through a packed bed water scrubber. The CO_2 dissolves or absorbs into the water.

Similar to PSA, the water carbon dioxide solution is then regenerated by depressurizing the vessel. As shown in Figure 2-5, the systems are typically set up with two vessels: one continuously in absorption mode and the other in regeneration mode.

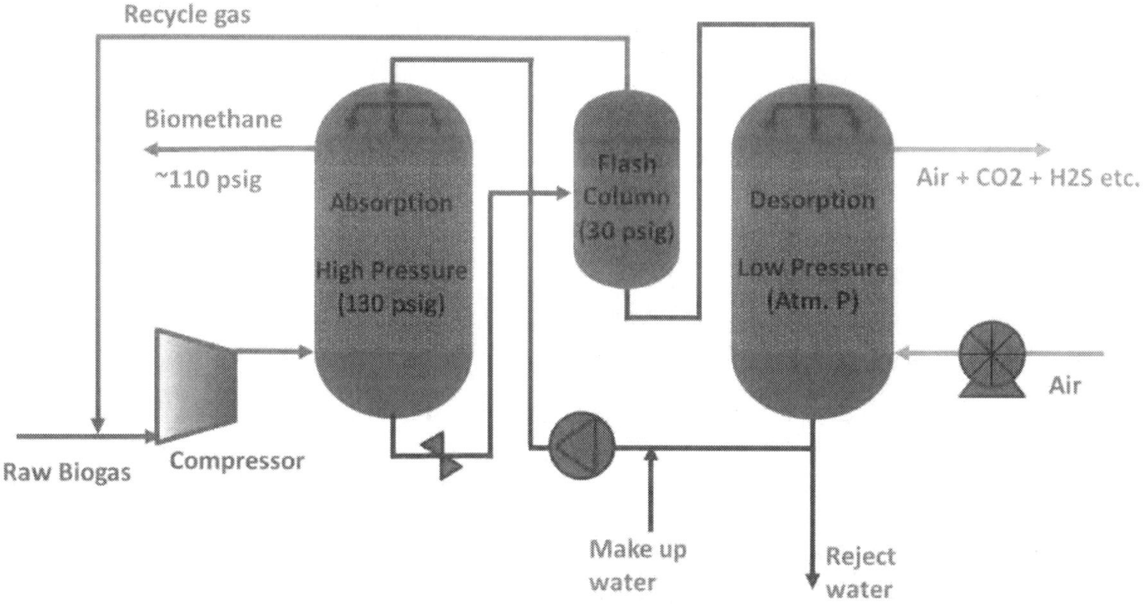

Figure 2-5 Water Scrubbing Schematic

2.5.4 Amine Scrubbing

Amine scrubbing works almost exactly like water scrubbing, with one enhancement. After CO_2 dissolves into the water solution it reacts with amines. Thus, it is a chemical absorption process. Absorption occurs at low temperature. The amine solution is regenerated at high temperature. The systems are typically set up with two vessels: one continuously in absorption mode and the other in regeneration mode.

3.0 MMSD South Shore WRF Case Study

A comprehensive analysis of digester gas treatment technologies was conducted to select the best fit gas treatment system for the South Shore Water Reclamation Facility (WRF) operated by the Milwaukee Metropolitan Sewerage District (MMSD).

3.1 BASIS OF DESIGN

Digester gas flows are presented in Table 3-1. Raw digester gas contaminant concentrations and treatment requirements for H2S, moisture, and siloxane removal are summarized in Table 3-2.

The South Shore WRF currently adds ferric for chemical phosphorus (P) removal. For this reason, H_2S concentrations over the past two years have been consistently below 100 ppmv. If current operations continue, H_2S removal is not needed. However, a wastewater process change to biological phosphorus removal could reduce or eliminate iron addition. Thus, a higher H_2S concentration of 500 ppm was used as a basis of design.

Table 3-1 Digester Gas Flows

DIGESTER GAS FLOWS	UNITS	FLOW	BASIS	USE FOR:
Flow, maximum	scfm	1,500	Engine gas utilization capacity and current peak gas production.	Equipment sizing
Flow, average	scfm	900	Current average gas production.	O&M cost estimate (power, media usage)
Flow, minimum	scfm	500	Current minimum gas production.	Turndown requirement
Note: assumed no growth in gas production over 20 year period.				

Table 3-2 Digester Gas Characteristics and Quality Requirements

DESIGN BASIS	UNITS	INLET (RAW GAS)	OUTLET (TREATED GAS)	NOTES
H_2S Removal				
Biogas temperature	F	95		
Biogas pressure	inches w.c.	7-10		
Moisture	-	Saturated		
H2S concentration, average	ppmv	500	<100	Use for equipment sizing.
H2S concentration, peak	ppmv	1,300	<100	
Moisture Removal				
Biogas temperature	F	95	70	Reheat after moisture removal.
Biogas pressure	inches w.c.	2-7		
Biogas pressure	psig		55	Higher pressure specific to this application.
H2S concentration	ppmv	<100		H_2S removal prior to moisture removal.
Moisture	-	Saturated	<40 F dewpoint	
Siloxane Removal				
Biogas temperature	F	70	<130	
Biogas pressure	psig	50		
Siloxane concentration, average	ppbv	3,000	<100 ppbv	Use for equipment sizing.
Siloxane concentration, peak	ppbv	30,000	<100 ppbv	

3.2 TECHNOLOGY SCREENING

The first step in the alternatives analysis involved a technology screening in which the wide range of biogas treatment technologies were screened down to a smaller number carried forward for a detailed life-cycle cost evaluation. The screening was done based on Black & Veatch's project experience, number of installations, and discussions with system providers. The following technologies were carried forward:

H_2S Removal

- ✓ Ferric addition
- ✓ Iron sponge
- ✓ Iron Oxide on Granular Media
- ✓ Biological scrubber
- ✓ Chemical scrubber with biological regeneration
- × Activated carbon is not recommended because the cost of the media replacement would be too high considering the digester gas flow and concentrations at the South Shore WRF.

Moisture Removal

Black & Veatch recommended that conventional chillers be selected for the South Shore WRF because of its wide, successful and reliable application to biogas treatment . Other technologies are not well proven for biogas applications.

Siloxane Removal

- ✓ Activated carbon, non-regenerable
- ✓ Temperature Swing Adsorption with vessel containers
- ✓ Low Temperature Gas Drying or Deep Chilling
- × TSA with cartridges is not recommended for two reasons. First, cartridge systems are typically viable for applications that are smaller than the scale of the South Shore WRF. And second, cartridge systems cannot be operated at higher pressures envisioned at South Shore WRF.

3.3 LIFE-CYCLE COST EVALUATION

A life-cycle cost evaluation was performed to identify the most cost effective H_2S, moisture, and siloxane removal technologies for the South Shore WRF. The capital costs and O&M costs were calculated based on information provided by system manufacturers and capital cost adders and unit costs specified by MMSD. This section presents the results and recommendations from this evaluation.

3.3.1 H_2S Removal Alternatives

The life cycle costs of each H_2S removal system are shown in Figure 3-1 and Table 3-3. The biological and chemical scrubber systems appear to have the lowest life-cycle cost. However, these systems require the treatment units to be kept warm, either using heat tracing on the treatment vessel or installing the treatment system indoors. The cost of this heating equipment was not included in the manufacturers' pricing or the cost analysis. It is expected that the heating costs would increase the life cycle costs of the biological and chemical scrubber systems to be nearly equivalent to ferric addition. Minimal capital costs are associated with ferric addition and the facility's staff has significant experience with the material. Therefore, ferric addition is the recommended alternative for H_2S removal at this facility.

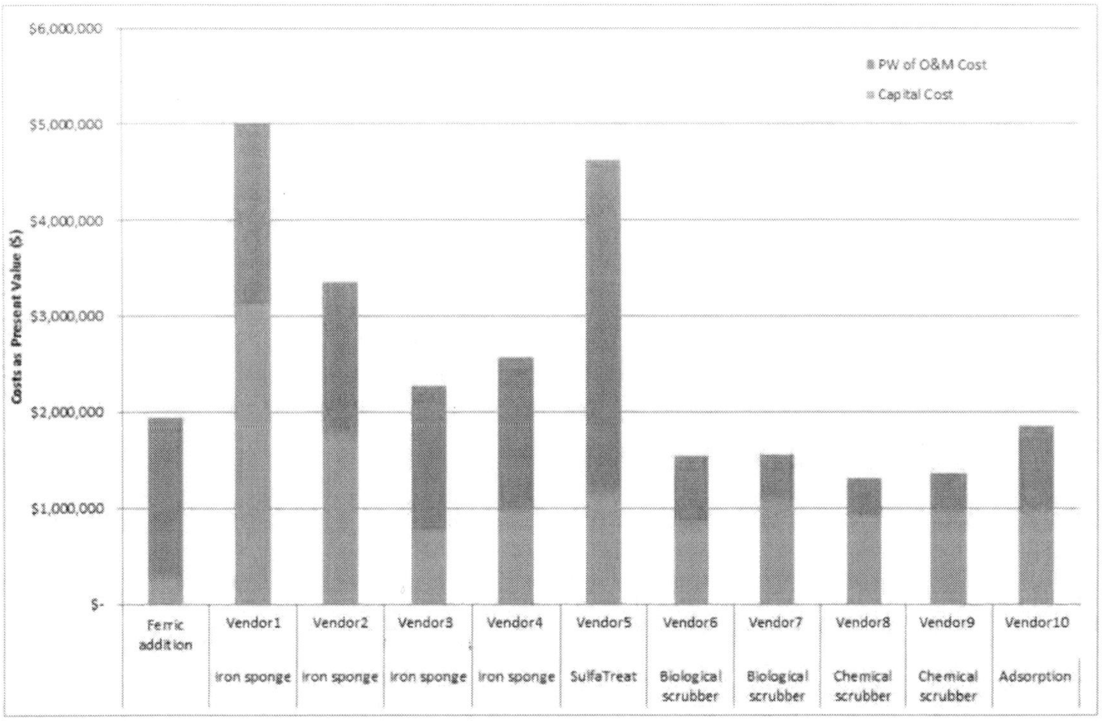

Figure 3-1 Life Cycle Costs for H_2S Removal Systems

Table 3-3 Summary of Life-Cycle Cost Results for H_2S Removal Systems

		FERRIC ADDITION	IRON SPONGE				SULFA TREAT	BIOLOGICAL SCRUBBER		CHEMICAL SCRUBBER		ADSORPTION
			VENDOR 1	VENDOR 2	VENDOR 3	VENDOR 4	VENDOR 5	VENDOR 6	VENDOR 7	VENDOR 8	VENDOR 9	VENDOR 10
Capital Cost	$	277,000	3,122,000	1,765,000	754,000	974,000	1,146,000	847,000	1,075,000	911,000	967,000	971,000
Annual O&M Cost	$/yr	106,000	121,000	102,000	97,000	102,000	223,000	45,000	31,000	26,000	26,000	56,000
PW of O&M Cost	$	1,660,000	1,890,000	1,590,000	1,520,000	1,590,000	3,480,000	700,000	480,000	400,000	400,000	880,000
Total PW Cost	$	1,940,000	5,010,000	3,360,000	2,270,000	2,560,000	4,620,000	1,550,000	1,560,000	1,310,000	1,370,000	1,850,000
Annualized Cost	$/yr	124,000	322,000	215,000	146,000	165,000	297,000	99,000	100,000	84,000	88,000	119,000
Annualized Unit Cost	$/MM Btu	0.44	1.13	0.76	0.51	0.58	1.05	0.35	0.35	0.30	0.31	0.42

3.3.2 Siloxane Removal Alternatives

The siloxane removal technologies assessed were activated carbon adsorption, Temperature Swing Adsorption (TSA), and deep chilling. Pioneer Air's deep chilling technology removes both moisture and siloxanes. An additional life-cycle cost analysis was carried out to compare deep chilling to the more conventional moisture and siloxane removal system costs from Unison.

Activated Carbon Adsorption vs Temperature Swing Adsorption

The life cycle costs of each siloxane removal system are shown in Figure 3-2 and Table 3-4. The activated carbon systems have a lower life-cycle cost than the TSA systems at this siloxane concentration.

Additional analysis was carried out to determine at what siloxane concentration the regenerable TSA system makes sense at the average design flow of 900 scfm. Figure 3-3 shows a break-even curve comparing carbon adsorption and TSA. The break-even point between carbon adsorption and TSA is 2,300 ppbv for the low cost TSA option. For the high cost TSA option, the break-even point is 3,300 ppbv.

Deep Chilling vs Conventional Moisture and Siloxane Removal

An additional life-cycle cost analysis was carried out to compare deep chilling to conventional moisture and siloxane removal. The moisture removal system using a chiller and heat exchanger and the siloxane removal system using activated carbon adsorption was used as a basis for the comparison. Table 3-4 provides a summary of the life-cycle cost results. The capital costs for the deep chilling system are lower than conventional moisture and siloxane removal. The O&M costs of deep chilling are slightly lower due to the reduced carbon usage, despite the higher power usage. Overall, the present value costs are slightly lower for the deep chilling system.

Further analysis during preliminary design revealed additional installation costs for the deep chilling system for the more complex controls, which resulted in the capital costs being slightly higher than for conventional moisture and siloxane removal. For South Shore WRF there was insufficient space in existing buildings to locate the deep chilling equipment and an outdoor installation or new building proved to be undesirable for MMSD. For these reasons, conventional moisture removal and activated carbon for siloxane removal were selected.

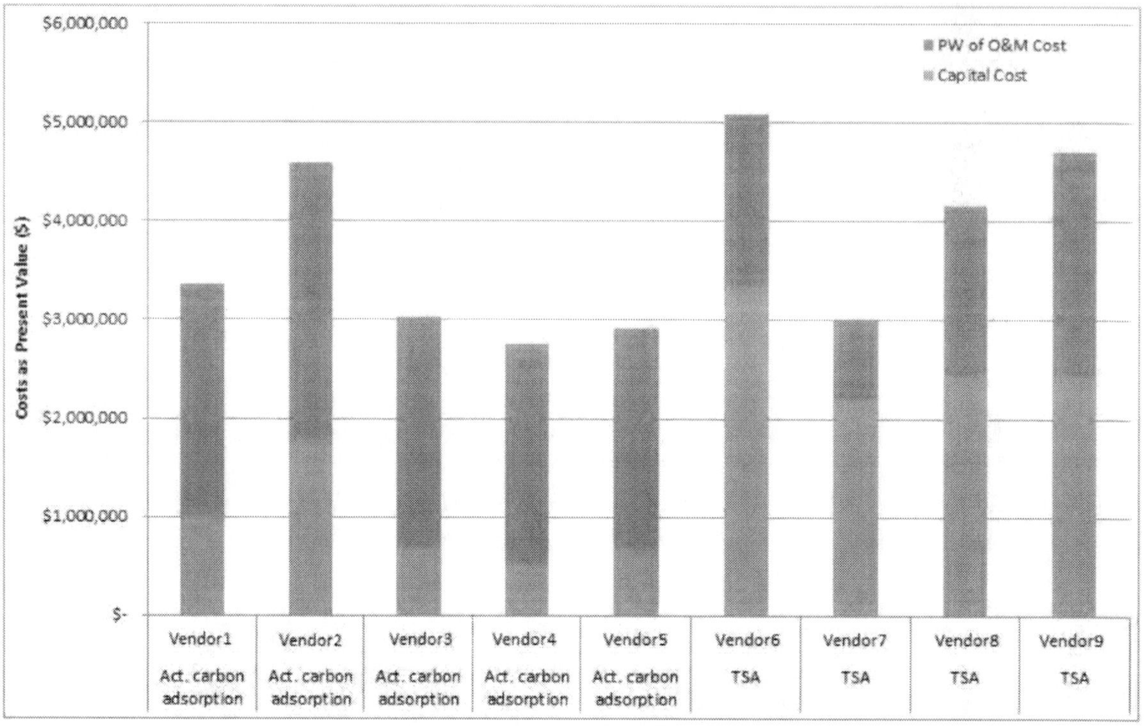

Figure 3-2 Life Cycle Cost Results for Carbon Adsorption and TSA Siloxane Removal Alternatives

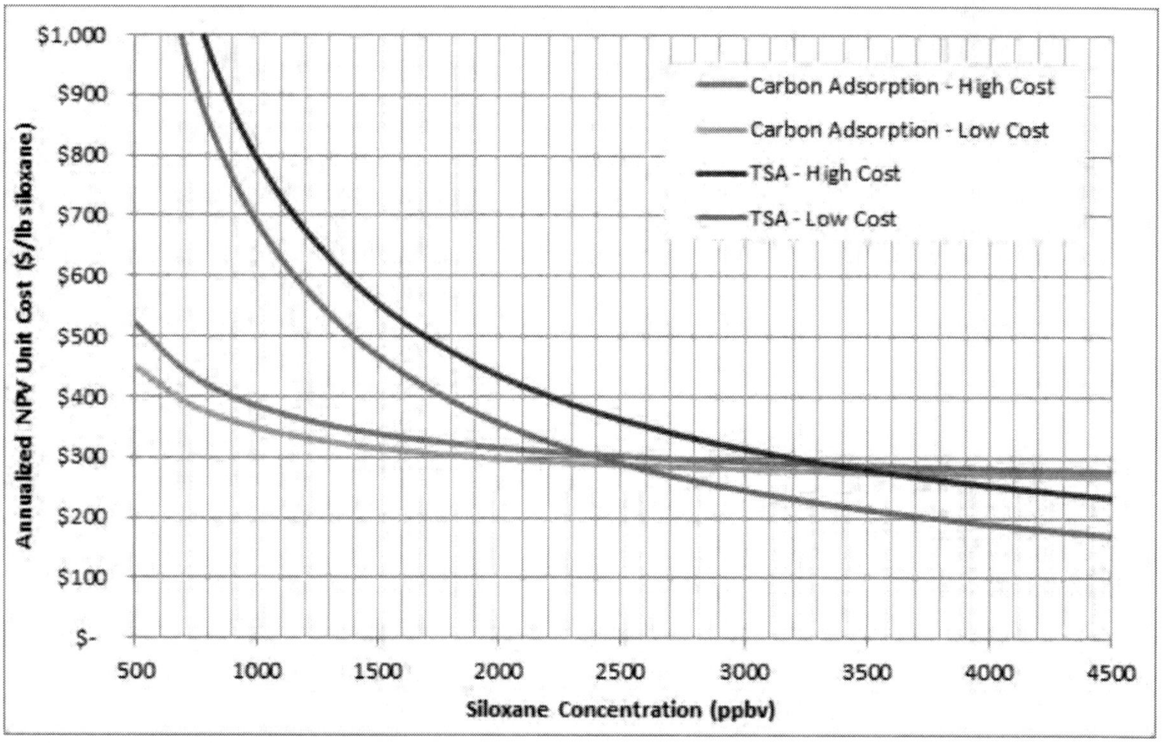

Figure 3-3 Carbon Adsorption vs TSA Break-Even Curve

Table 3-4 Summary of Life-Cycle Cost Results for Siloxane Removal Alternatives

		ACTIVATED CARBON ADSORPTION					TEMPERATURE SWING ADSORPTION (TSA)			
		VENDOR 1	VENDOR 2	VENDOR 3	VENDOR 4	VENDOR 5	VENDOR 6	VENDOR 7	VENDOR 8	VENDOR 9
Capital Cost	$	1,010,000	1,750,000	700,000	520,000	660,000	3,350,000	2,190,000	2,420,000	2,410,000
Annual O&M Cost	$/yr	152,000	182,000	150,000	143,000	145,000	111,000	52,000	111,000	148,000
PW of O&M Cost	$	2,360,000	2,830,000	2,340,000	2,230,000	2,260,000	1,730,000	820,000	1,740,000	2,300,000
Total PW Cost	$	3,370,000	4,590,000	3,040,000	2,750,000	2,920,000	5,080,000	3,010,000	4,160,000	4,710,000
Annualized Cost	$/yr	216,000	294,000	195,000	176,000	187,000	326,000	193,000	267,000	302,000
Annualized Unit Cost	$/MMBtu	0.76	1.04	0.69	0.62	0.66	1.15	0.68	0.94	1.07

Table 3-5 Deep Chilling vs Conventional Moisture and Siloxane Removal

		CONVENTIONAL MOISTURE & SILOXANE REMOVAL	DEEP CHILLING
Capital Cost	$	2,840,000	2,740,000
Annual O&M Cost	$/yr	355,000	308,000
PW of O&M Cost	$	5,530,000	4,800,000
Total PW Cost	$	8,370,000	7,540,000
Annualized Cost	$/yr	537,000	484,000
Annualized Unit Cost	$/MMBtu	1.89	1.70

3.4 DIGESTER GAS TREATMENT TECHNOLOGY RECOMMENDATIONS

Based on the capital, O&M, life-cycle cost and other factors presented in this section, the following systems are recommended for H_2S, moisture, and siloxane removal at South Shore WRF.

- No treatment for H_2S removal required at current H_2S concentrations. If concentrations increase (for example, if South Shore WRF switches to biological P removal), then ferric addition at the anaerobic digesters is recommended
- Compressor and chiller and heat exchangers for moisture removal.
- Activated carbon adsorption system for siloxane removal.

FROM BIOGAS TO BOILER FUEL: CLEANING UP DIGESTER GAS BIOLOGICALLY AT A LARGE SCALE

Melissa Woo, PE - CDM Smith
Tom Jacobs, PE - Trinity River Authority
Samir Mathur, PE, BCEE - CDM Smith

ABSTRACT

The Trinity River Authority (TRA) Central Regional Wastewater System (CRWS) serves approximately 1.2 million people in the Dallas Fort Worth area and has a rated average daily flow of 162 million gallons per day. Within the past ten years, TRA has shifted towards a more sustainable approach with regards to its biosolids planning, including the selection of thermal hydrolysis process (THP) technology. When TRA made the notable decision to install one of the largest thermal hydrolysis systems in the world at its CRWS facility, TRA set on a path to have one of the largest biological gas treatment systems as well. THP utilizes a high pressure process to sterilize sludge, while improving biodegradability, reducing volume and increasing downstream performance, such as in anaerobic digestion and dewatering. However, THP requires a great amount of heat, and to meet energy demands, the biogas produced from digesters will be used to fuel heating boilers in the THP process. The digesters will produce an average of approximately 2,000 standard cubic feet per minute (scfm), and the heating boilers will use up to 1,260 scfm (any gas not used by the boilers will be flared). To effectively use biogas for boilers, the THP project will include a digester gas conditioning system to remove moisture and impurities and to compress gas to pressures required for boiler use.

The Texas Commission on Environmental Quality (TCEQ) regulates air emissions from facilities and fuel used for combustion units, such as boilers. To meet permit-by-rule (PBR) requirements, the facility must have sulfur dioxide (SO_2) emissions below a certain threshold, and its boilers must use fuels that contain limited sulfur compounds. These PBR restrictions drove the gas handling design to include hydrogen sulfide (H_2S) removal, and a maximum H_2S concentration of 150 parts per million by volume (ppmv) was set for the biogas feed to the boilers.

During preliminary engineering, several digester gas H_2S removal technologies were investigated, including adsorption (iron sponge), biological treatment and chemical-biological treatment. A sulfur balance was developed to estimate H_2S concentrations in the gas stream from the proposed digesters and to compare the different technologies. Based on a variety of factors such as life cycle costs and operation and maintenance, the biological treatment technology was selected. The biological H_2S removal system uses cylindrical biological scrubbing reactors and media to convert H_2S in the gas stream to sulfate (SO_4^{2-}) and elemental sulfur (S^0). Controlled amounts of water and air are added to promote biological activity and to allow for the oxidation of dissolved sulfide to elemental sulfur and sulfate. Ancillary equipment includes recirculation pumps, recirculation heat exchangers, air blowers and nutrient feed equipment.

This paper provides an overview of the gas treatment and handling processes, with a focus on the biological H_2S removal system, a technology relatively new to the US. The paper covers permitting drivers, technology screenings, design considerations, design criteria, and key lessons learned.

KEYWORDS

Digester gas conditioning, biogas treatment, hydrogen sulfide, biological treatment.

INTRODUCTION

The Trinity River Authority (TRA) Central Regional Wastewater System (CRWS) serves approximately 1.2 million people in the Dallas / Fort Worth metropolitan area of Texas and has a rated annual average daily flow (AADF) of 162 million gallons per day (MGD) and a peak 2-hour flow of 405 MGD. Within the past ten years, TRA has shifted towards a more sustainable approach with regards to its facility planning, including the selection of thermal hydrolysis process (THP) technology and anaerobic digestion for its biosolids treatment. When TRA made the notable decision to install one of the largest thermal hydrolysis systems in the world at its CRWS facility, TRA set on a path to have one of the largest biological gas treatment systems as well. THP systems utilize a high pressure process to sterilize sludge, while improving biodegradability, reducing volume and increasing downstream performance, such as in anaerobic digestion and dewatering. However, THP requires a great amount of heat and energy. To sustainably meet energy demands, the biogas produced from anaerobic digesters will be used to fuel heating boilers in the THP process. To effectively use biogas for boilers, the project will include a digester gas conditioning (DGC) system to remove moisture and impurities such as hydrogen sulfide (H_2S) and to compress gas to pressures required for boiler use.

This paper covers the engineering design for the DGC area, with an emphasis on the biological treatment system. Specifically, the paper discusses permitting drivers, preliminary evaluations, final design, important design considerations and lessons learned.

PROJECT BACKGROUND

The solids improvements to the TRA CRWS facility was divided over four different phases (I – IV). The project phase described here, Phase III, consists of the design and construction of the THP, digestion, gas treatment and dewatering processes, shown in **Figure 1,** which presents the overall solids process diagram of the Phase III project improvements from the project Preliminary Design Report (CDM Smith, 2015).

As shown in the figure, gas collected from anaerobic digestion is routed to gas conditioning for H_2S removal, moisture removal and compression for feeding the THP boilers, as well as flaring any waste gas. This process is discussed in more detail later in this paper. Also shown in the figure, future gas handling facilities include siloxane removal and combined heat and power (CHP).

PERMITTING DRIVERS

A permitting assessment was performed to determine the implications of new process emissions from THP boilers, waste gas flares, a cooling tower and odor control scrubbers.

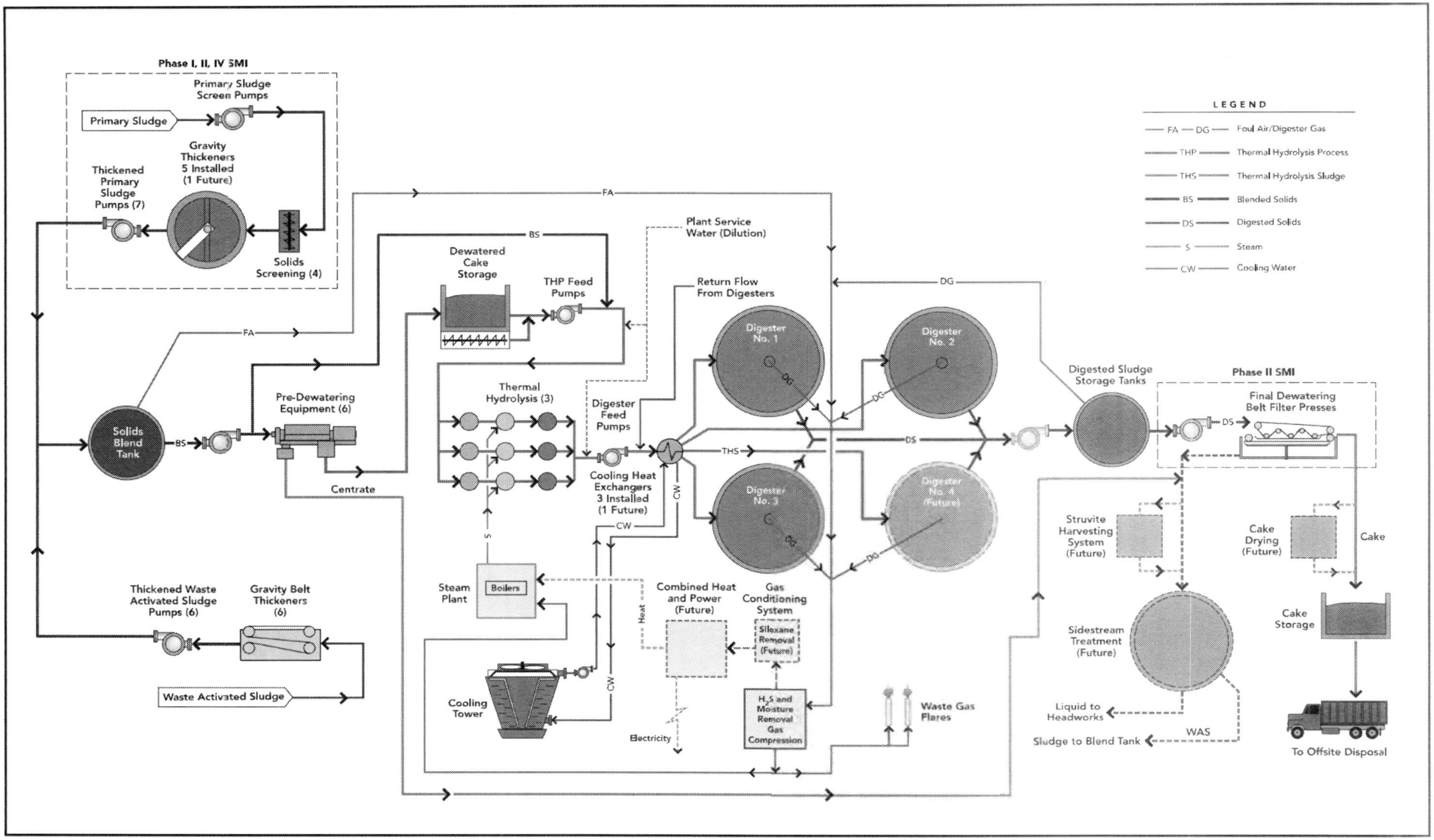

Trinity River Authority of Texas
Central Region Solids Management Improvements

Figure 1
Phase III - Solids Process Flow Diagram

The Texas Commission on Environmental Quality (TCEQ) regulates and manages air permitting in the State of Texas through the following two programs:

- Federal (Title V) Operating Permit Program
- New Source Review (NSR) Pre-Construction Permit Program

Under the federal program, a facility that will be classified as a Major Title V Source as a result of a project is required to submit a Title V permit application prior to operation of the project. The applicable major source threshold is 100 tons/year (tpy) for any air non-hazardous air pollutant (the threshold is less for hazardous air pollutants or for facilities located in certain classified nonattainment areas). If the facility is considered a "minor" or "non-major" source, then a Title V permit is generally not required.

The TCEQ NSR program requires a facility owner to obtain either a Standard Permit, Minor NSR Permit or Major NSR Permit before the construction of a project, unless the facility qualifies for a Permit-by-Rule (PBR) or is classified as a "De Minimis Source," a source with negligible air emissions that does not have to obtain an air authorization. Except for De Minimis classification, a PBR allows for a much simpler and streamlined permitting process than the other forms of permitting in the program; for example, public notice periods are not required, and the permitting processing timeframe is approximately 45 days (compared to 285 or 365 days).

For the project, the following contaminants were examined: carbon monoxide (CO), H_2S, nitrogen oxides (NO_x), particulate matter (PM_{10} and $PM_{2.5}$), sulfur dioxide (SO_2) and volatile organic compounds (VOCs). The existing and future sitewide emissions (after implementation of the project) were quantified, and it was determined that the facility will be classified as a Minor Title V Source, eliminating the need for a Title V Operating Permit. Then these projected emissions were compared with PBR thresholds under the NSR program. Upon confirming PBR status, specific PBR requirements were examined for specific technologies and equipment, such as the proposed flares, boilers, cooling water units and odor control units.

Based on permitting requirements for the PBR, H_2S and SO_2 became the key constituents of concern for the design of the DGC area, with the following limits:

- Maximum H_2S concentration for fuel (digester gas) feeding the boilers: 150 - 160 parts per million by volume (ppmv or ppm) of H_2S.
- Yearly SO_2 (uncertified) emission rate for the flares: 16 tons per year (tpy)

As a "not to exceed" value, the 150 ppm criterion is the more conservative requirement; this boiler fuel requirement stems from Rule §106.183 of Chapter 106, Subchapter G of Title 30 of the Texas Administrative Code (TAC), (which requires a maximum 0.1 grain of total sulfur compounds (calculated as sulfur) per standard cubic feet of fuel (gr/scf)). In contrast, the yearly SO_2 emission total allows for periods of higher loading or spikes. Moreover, while listed in the PBR, the SO_2 flare emission rate is "uncertified" and not enforceable. The actual enforceable limit applies to the entire facility, as 25 tpy for all source emissions per Rule §106.4 (Chapter 106, Subchapter A) of Title 30.

In addition to sulfur related compounds, the boilers and waste gas flares have other emission limits (certified and uncertified) per PBR requirements and based on yearly totals (e.g. CO, NO_X, PM and VOCs). While these requirements impact the design and specification of the flares and boilers, no additional treatment is required to meet the yearly emission limits.

MASS AND SULFUR BALANCES

Following the PBR evaluation, it became evident that the digester gas treatment system was to be designed to achieve an effluent H_2S concentration of 150 ppm or less. During preliminary engineering, a key first step involved determining the scale of treatment required based on the incoming H_2S concentration. Without any digestion processes already on site, there was no raw digester gas data available in terms of production or gas quality. Additionally, the implementation of THP technology is expected to increase the quantity of biogas, as well as change the constituents in the gas compared to conventional biogas systems. For example, the ammonia concentration is expected to increase for biogas from hydrolyzed, digested sludge.

The project team developed a preliminary overall mass balance and sulfur balance to project biogas production and estimate H_2S concentrations. Important assumptions for the mass balance included a volatile solids content of 75%, a volatile solids reduction (VSR) during digestion of 58% and a gas production rate of 18 standard cubic feet (scf) per pound of volatile solids reduced (lb/VSR). During final design a sensitivity analysis was conducted to test a more conservative VSR value of 68% to check and finalize design criteria. The mass balance accounts for year 2040, and average daily influent flows to facility are projected to be 167 MGD. **Table 1** documents key values from the overall mass balance with regards to biogas production, set for future 2040 conditions.

Table 1 - Overall Mass Balance – Gas Production

Parameter	Units	2040 Average	2040 Maximum Month	2040 Peak 2-Week	2040 Peak Week	2040 Peak Day
Total Sludge to Digesters, Dry Solids	lbs/day	375,800	450,900	469,700	488,500	601,200
Volatile Solids Content	%	75.0%	75.0%	75.0%	75.0%	75.0%
Volatile Solids, Dry Solids	lbs/day	281,800	338,200	352,300	366,400	450,900
Volatile Solids Destruction	%	58.0%	58.0%	58.0%	58.0%	58.0%
Volatile Solids Destruction	lbs/day	163,500	196,200	204,300	212,500	261,500
Gas Production Rate	scf/lb VSR	18	18	18	18	18
Digester Gas Production	scf/day	2,942,000	3,531,000	3,678,000	3,825,000	4,708,000
Digester Gas Production	scf/hr	122,600	147,100	153,200	159,400	196,100
Digester Gas Production	**scfm**	**2,043**	**2,452**	**2,554**	**2,656**	**3,269**

For the sulfur balance, TRA provided influent sulfur data over a 2-month period in 2014, which showed an average of 72 milligrams per liter (mg/l) of sulfate (SO_4^{2-}), or 24 mg/L as sulfur. It was assumed that this concentration would not change in future design conditions. Sulfur enters the facility in the influent and leaves through the plant effluent, solids handling, biogas handling or off-gassing as odorous air. Approximately 99.3% of influent flow is discharged through the plant effluent, while 0.7% is routed to solids processing. The amount of sulfur discharged in the effluent versus solids processing depends on a number of factors including:

- Assimilation of sulfur by biological processes
- Adsorption of sulfur to solids
- Solids capture through clarification

During the preliminary engineering stage, there was not sufficient data or time to analyze the above factors in detail, and therefore professional judgement was used to estimate sulfur partitioning to solids throughout plant processes. As shown in Figure 1, the headspaces of the Blend Tank (BT), digesters and Digested Sludge Storage Tank (DSST) will be connected for collecting biogas. In addition to the digesters, the BT and DSST will operate in anaerobic conditions. With a significant hydraulic retention time (16 hours) and sulfur-reducing bacteria present, it was assumed that all sulfate will be biologically converted to hydrogen sulfide species in the BT. Not every form of hydrogen sulfide is readily volatile, however; only the aqueous form of hydrogen sulfide $H_2S_{(aq)}$ is volatile, while the hydrosulfide ion (HS^-) and sulfide ion (S^{2-}) are not. The proportion of hydrogen sulfide species present in a solution depends on the solution's pH. At lower pH, the $H_2S_{(aq)}$ form dominates. The BT (pre-THP) will operate in a slightly acidic pH range (approximately a pH of 6), causing 90% of hydrogen sulfide species to exist in the $H_2S_{(aq)}$ form. The digesters and DSST (post-THP) will operate at an alkaline pH range of 7.2 to 8.4, with approximately 25% of hydrogen sulfide species in the $H_2S_{(aq)}$ form (based on a pH of 7.5). Since the BT, digesters and DSST will be enclosed with relatively long detention times, it was assumed that the liquid and gaseous phases will operate in equilibrium. Thus Henry's Law can be used to estimate the partial pressure (concentration) of H_2S in the gaseous phase based on the dissolved concentration of $H_2S_{(aq)}$ in the liquid phase. The equation below shows Henry's Law:

$$k_H = {c_a}/{p_g}$$

k_H = Henry's constant for the species
c_a = Concentration of the species in the aqueous phase
p_g = Partial pressure of the species in the gas phase

For H_2S, a value of k_H = 0.10 mole per liter per atmosphere (mole/(L-atm)) (at standard temperature of 298.15 kelvin (K)) was corrected for operating temperature and used for the sulfur balance model calculations (Sander, 1999). During preliminary design, a range of resulting H_2S concentrations was calculated. An important assumption that greatly impacted results was the partitioning factor of sulfate to solids, or the proportion of sulfate that becomes insoluble, digested or adsorbed to solids (as sulfate or other sulfur compound) throughout the treatment process. Essentially the amount of sulfate in the solids stream (feeding the BT and digesters) is comprised of both soluble sulfate in the liquid phase and partitioned sulfate (or other compound) to the solids phase. Based upon the minimum sulfur concentration expected in a biological cell, the partition factor was expected to be at least 0.2%. A preliminary selection of 1.0 to 2.5% was used for the partitioning factor, which resulted in concentrations from 1,500 to 3,000 ppm respectively. This range seemed reasonable, and the project team moved forward to technological screening and evaluation based on preliminary design criteria.

PRELIMINARY EVALUATION

During preliminary engineering, H_2S removal, gas compression, moisture removal and siloxane removal were considered for the project. While carbon dioxide removal is also an option, it is recommended specifically for refined methane applications (e.g. natural gas pipeline or vehicle fuel) and therefore was not considered for this project. This section focuses on the screening of H_2S removal system and technologies, as well as the selection.

Regarding the other processes considered, it was decided to include gas compression and moisture removal systems for the biogas feeding the boilers. The removal of siloxanes (silicon-based compounds found in personal care and other consumer products) from biogas is recommended for combined heat and power (CHP) systems with difficult-to-maintain equipment, as siloxanes can create silicon oxide compounds during combustion that can clog and erode such equipment. Siloxane removal is not critical to boiler operation, as operators can readily access and clean boilers periodically. For this reason, siloxane removal was omitted from the project to save cost. Reserved space and piping connections were included in the design to allow flexibility for siloxane vessels to be installed in the future if desired or needed.

Although permitting was the primary driver, H_2S removal has a number of other benefits. Gaseous H_2S can be corrosive and hazardous to health, and the combustion of H_2S produces sulfur oxides, a major contributor to acid rain and smog.
For removal of H_2S, the following technologies were evaluated:

- Iron sponge
- SulfaTreat®
- Biological Treatment
- Chemical-biological Treatment.

Iron sponge is a physical-chemical process, which consists of reactors filled with wood chip media impregnated with iron oxide (Fe_2O_3). As gas passes through the reactor, H_2S reacts with iron oxide to form iron sulfide (Fe_2S_3), in solid form. After a certain period, the adsorption sites become filled with iron sulfide, and the media can be regenerated by adding oxygen. Once media can no longer be regenerated, the media requires replacement, and the used media can be landfilled. This technology has been in use for 100 years, and it can consistently achieve low concentrations (25 ppm), even with varying influent gas conditions. Iron sponge units require temperature control through the addition of water, during regeneration and media change out. Adding oxygen can cause an exothermic reaction and introduce a fire risk, if temperature is not adequately controlled.

SulfaTreat® is a relatively new, proprietary physical-chemical process similar to iron sponge. Instead of wood chips, SulfaTreat uses an inert ceramic base, coated with iron oxide. Unlike iron sponge, SulfaTreat media cannot be regenerated; the media can be landfilled once it has been spent. Similar to iron sponge, since it uses adsorption processes, treatment performance is consistent. The inert nature of the ceramic media eliminates the fire risk during media change out that affects iron sponge systems. However, since the media is proprietary, it is approximately twice as expensive as iron sponge media. At the same time, SulfaTreat does not require as frequent media change-outs as iron sponge.

Biological treatment uses biological populations to convert hydrogen sulfide to sulfuric acid and elemental sulfur that can be removed in a liquid and/or solid form. These systems utilize reactors filled with synthetic media conducive to biological growth, and the media requires periodic cleaning. Water, air and nutrients are added to support biological growth. A general rule is that if gas inlet concentrations are greater than 1,000 ppm, then biological treatment systems become economically feasible. Typical outlet concentrations are 100 ppm or lower. While the biological process is not new, these manufactured biogas systems are relatively new, especially to the United States. In contrast to the adsorption type systems above, the biological systems are sensitive to varying influent loads and temperature.

Chemical-biological treatment utilizes at least two reaction vessels in series, the first being an absorption reactor and the second being a biological reactor. The absorption reactor facilitates the conversion of gaseous H_2S to dissolved sulfide, $H_2S_{(aq)}$, through the addition of water and caustic chemical, used to raise the pH. The bioreactor then uses biological populations to oxidize dissolved sulfide, similar to biological treatment units described above. Chemical-biological systems can achieve efficient treatment for high H_2S concentrations, (e.g. 3,500 ppm). However, compared to the biological system, it requires more vessels and is more operation intensive.

The above technologies were evaluated based upon multiple criteria including level of treatment required, cost, footprint, and operations and maintenance. Refer to **Table 2** for a summary of the advantages and disadvantages. The results of the economic evaluation showed the adsorption systems very expensive for the project's DGC application without other significant advantages. For example, for an iron sponge system, it was estimated that approximately nine vessels (12-ft diameter) would be required, with $1.1 million in capital costs and $1.2 million in annual costs (primarily due to media replacement). A SulfaTreat system would have similar, but slightly higher costs. In contrast, a biological system would require one to three vessels (depending on the height allowed), as well as an equipment container or skid system to house ancillary equipment. Costs were estimated to be approximately $1.5 million in capital costs and $50,000 in annual costs. A chemical-biological system was estimated to be slightly more expensive.

Table 2 - Advantages and Disadvantages of H_2S Removal Systems

Option	Advantages	Disadvantages
Iron Sponge	- Long-standing technology - Can regenerate media - Consistent performance, even with varying influent loads	- Very labor intensive media replacement - Poses a fire risk if oxygen and temperature are not adequately controlled - High annual costs for relatively high H_2S loadings
SulfaTreat	- Reduced fire risk compared to iron sponge - Less frequent media replacement compared to iron sponge - Consistent performance, even with varying influent loads	- Labor intensive media replacement - Media is proprietary, (approximately 2x that of iron sponge) - High annual costs for relatively high H_2S loadings

Option	Advantages	Disadvantages
Biological Treatment	- Low operation and maintenance costs - Relatively simple maintenance compared to other technologies - Less space requirements compared to other technologies for relatively high H_2S loadings - Better lifecycle costs for relatively high H_2S loadings compared to adsorption systems	- Relatively new in municipal applications - Varying influent loads can impact biological performance - Relatively long acclimation time to grow bacterial populations at startup - Process control is automated with minimal adjustment capability
Chemical-Biological Treatment	- Can handle very high concentrations of H_2S - Better lifecycle costs for high H2S loadings compared to adsorption systems - Better control of headloss buildup in vessel compared to biological system	- Relatively new in municipal applications - Requires addition of caustic chemical - Relatively labor intensive to maintain both biological and chemical processes. - Process control is automated with minimal adjustment capability

A biological system was recommended and accepted for implementation due to:

- The level of treatment (25 ppm) provided by adsorption systems is not needed for this project.
- Lifecycle costs and space requirements for the adsorption systems showed to be significantly greater than for the biological system.
- The ability to handle very high (>3,500 ppm) concentrations, provided by chemical-biological systems, is likely not necessary for this application.
- Chemical-biological systems are more maintenance intensive than biological systems.
- While biological systems are susceptible to varying influent loads, transient outlet spikes are expected to be infrequent or rare; additionally PBR flare requirements are based on a yearly basis, and boiler operation will have natural gas as a backup.

ENGINEERING DESIGN AND LESSONS LEARNED

At the final design stage, design criteria and process layouts were finalized. Refer to **Figure 2** for the DGC area process flow diagram and **Figure 3** for an isometric of the DGC area. As shown in Figure 2, biogas collected from the BT, anaerobic digesters and DSST will be routed through a booster fan to the biological H_2S removal system, with all gas normally treated to meet permitting obligations for H_2S and SO_2. After treatment, the booster blowers will compress and send gas through the moisture removal system (heat exchanger and coalescing filter) to the Boiler Building for use as boiler fuel. Any gas not used by the boilers will be flared.
The following sections summarize the design of each process area, with a focus on the biological H_2S removal system. Key design considerations and lessons learned are also discussed.

Biological H_2S Removal System: Design Criteria

Following preliminary engineering, the design criteria for the biological system was established. Typically gas treatment systems are designed for average daily or maximum month flows, depending on the data available and the reason for treatment such as permitting restrictions.

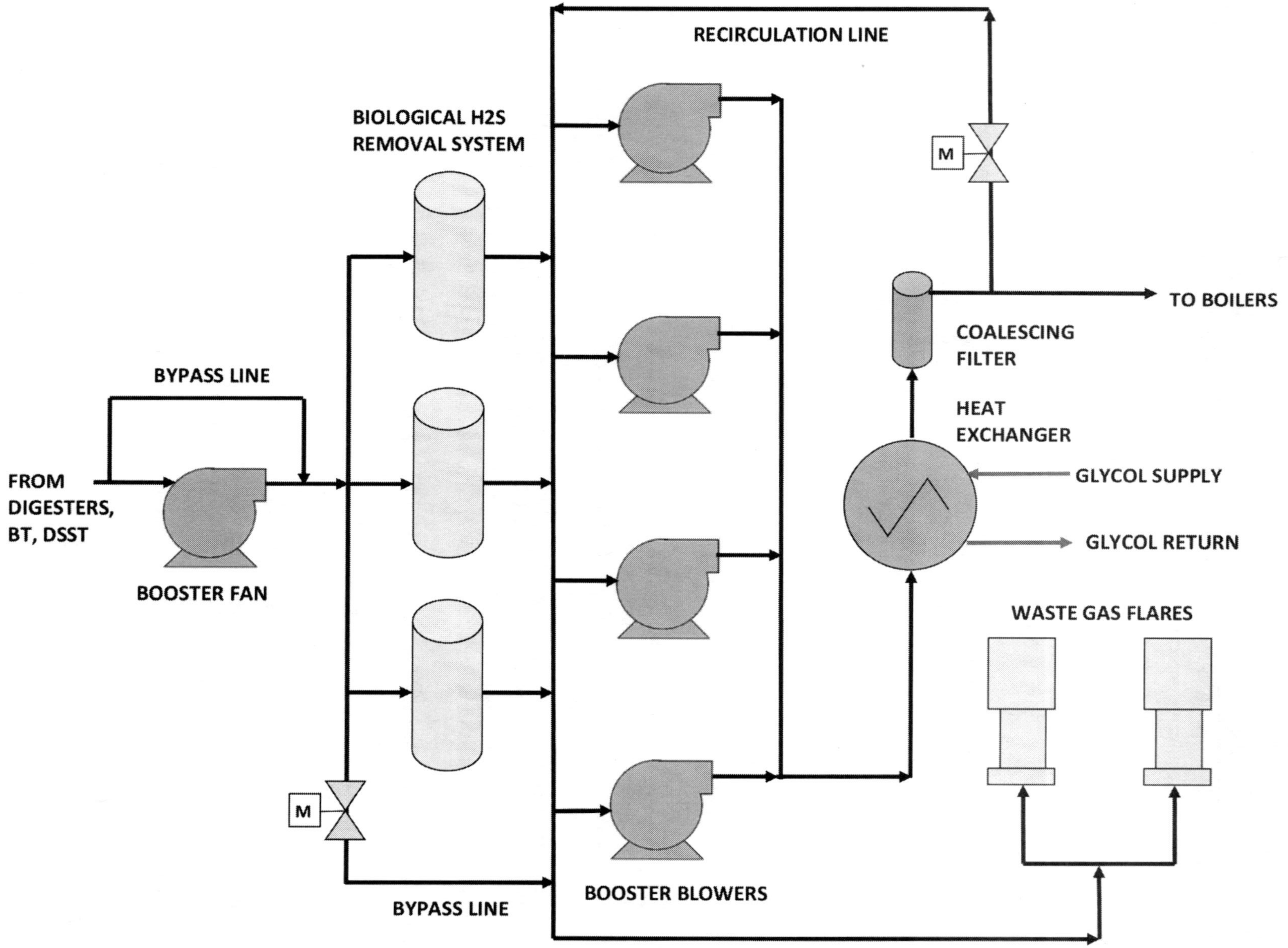

FIGURE 2
DIGESTER GAS CONDITIONING AREA PROCESS FLOW DIAGRAM

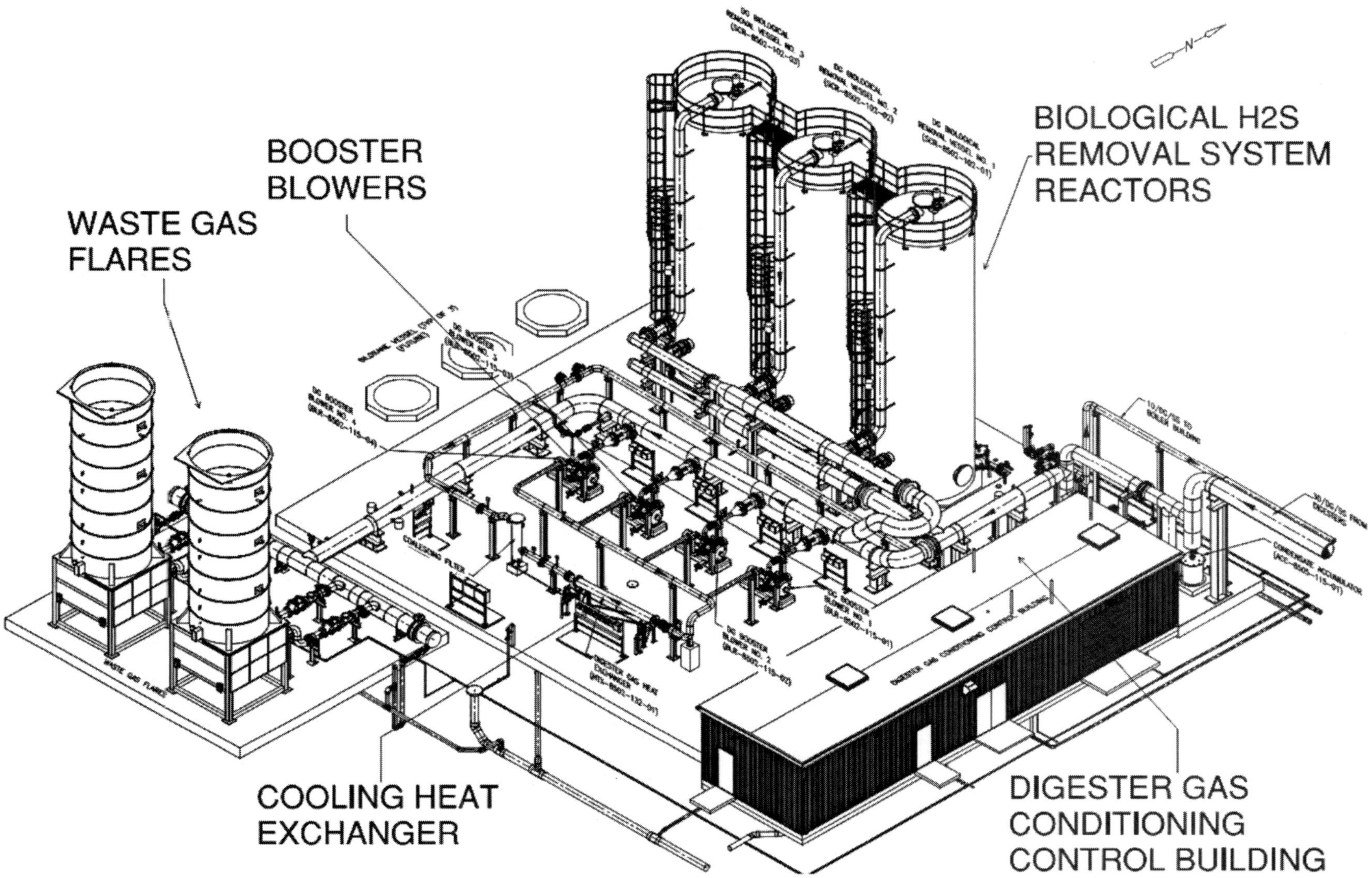

FIGURE 3
DIGESTER GAS CONDITIONING AREA ISOMETRIC

Without any data on flow and concentration, a conservative approach was utilized to set the design flow at 2,700 scfm, as based on the peak week condition (Table 1) from the overall mass balance. Peak day was deemed too conservative for treatment purposes, although the vessels were specified to pass the peak day flow.

The sulfur balance was also finalized, and using a sulfate partitioning factor of 2%, maximum H_2S concentrations were calculated at approximately 2,600 ppm. **Table 3** summarizes the design criteria.

Table 3 - Biological H2S Removal System Design Criteria

Parameter	Value
No. of Bioreactors	3
Maximum Vessel Diameter	14 ft
Maximum Vessel Height	45 ft
Peak Day	
Gas Flow Rate	3,500 scfm
Maximum Allowable Pressure Drop	6 – 7 in wc
Peak Week	
Gas Flow Rate	2,700 scfm
H_2S Influent Concentration	2,600 ppmv
Maximum Allowable H_2S Effluent Concentration	150 ppmv
Maximum Allowable Pressure Drop	4 – 5 in wc
Average Day	
Gas Flow Rate	2,000 scfm
H_2S Influent Concentration	2,600 ppmv
Maximum Allowable H_2S Effluent Concentration	100 ppmv
Maximum Allowable Pressure Drop	2 – 3 in wc

For aesthetic reasons, it was necessary to ensure the bioreactors do not exceed the height of the digesters, and thus the height was limited to 45 feet. This also set the number of bioreactors required. The system will not have full redundancy, but having multiple reactors will allow for partial redundancy during maintenance and cleaning procedures. Recommended pressure drops varied from manufacturer to manufacturer, and actual pressure drops will depend on actual loadings and the frequency of cleaning the reactors. The booster fan upstream of bioreactors (Figure 2) will provide flexibility for system cleaning by providing additional pressure to ensure biogas ultimately reaches the flares. The initial source of pressure for the system will come from the digester headspace. The fixed tank covers will build up in pressure up to 18 inch water column (in wc) under normal operating conditions. Since the two waste gas flares will be controlled by pressure, the flares will set the downstream pressure condition. A bypass line with a motorized valve (opening based on pressure) will provide an emergency bypass around the biological system in case of extreme spikes in flow.

Biological H_2S Removal System: Background

These biological systems consist of cylindrical fiberglass reinforced plastic (FRP) bioreactors, which operate in parallel, as well as ancillary equipment, including pumps and blowers. As introduced earlier in this paper, the bioreactors use synthetic media and biological processes to convert gaseous H_2S to sulfuric acid and elemental sulfur. During normal operation, to promote aerobic biological growth, water and air are added to the bioreactors. The H_2S in the gas dissolves into the water (aqueous phase), and dissolved sulfide (such as in form bisulfide, HS^-) is oxidized by *Thiobacillus* bacteria to elemental sulfur and sulfuric acid in a two-step process, as shown in the chemical equations below (Duangmanee, 2009):

(1) $2HS^- + O_2 \rightarrow 2S^0 + 2OH^-$

(2) $2S^0 + 3O_2 + 2H_2O \rightarrow 2H_2SO_4$

Refer to **Figure 4** for a schematic of a typical biological H_2S removal system. As shown in the figure, the following ancillary equipment is required:

- Air blower(s) for adding air (oxygen)
- Recirculation pump(s) for recirculating water
- Heat exchanger(s) for heating recirculated water and maintaining bioreactor temperature
- Nutrient feed and storage for adding nutrients
- Control valves for process feed control (makeup water, air, hot water)

Note that the schematic will differ depending on the manufacturer. For example, one manufacturer adds makeup water directly to the bioreactor, while another adds it to the recirculation loop. Similarly, one drains blow-down water directly from the bioreactor, while another drains such water from the recirculation loop. Whether the system operates in a counter-current or co-current design also depends on the manufacturer.

Typically, manufacturers provide the ancillary equipment listed above in skid units housed in metal (shipping container type) or FRP container buildings to protect from the elements. The additions of air, water and nutrients are automatic and regulated by at least one manufacturer-provided programmable logic controller (PLC). Instrumentation on biogas inlet and outlet lines monitor pressure, temperature and oxygen content at the bioreactors as feedback to the PLC. For the project, three different manufacturers/suppliers were identified during final design: Varec Biogas of Huntington Beach, California; DMT Environmental Technology of Joure, Netherlands; and Biorem Technologies Inc. of Puslinch, Ontario. Refer to **Figure 5, 6 and 7** for H_2S removal systems by each manufacturer.

Biological H_2S Removal System: Lessons Learned

While using biological processes for removing H_2S is not new (i.e. odor control bioscrubbers), these types of systems for treating biogas are relatively new to the U.S.; although there are a significant number of operating systems in Europe and Asia. Moreover, once this system is operational, it will be one of the largest systems in the U.S., if not worldwide. As one of the largest facilities planned in the US and world, there have been significant lessons learned throughout the planning of this project. This section summarizes some of the major lessons learned throughout this process.

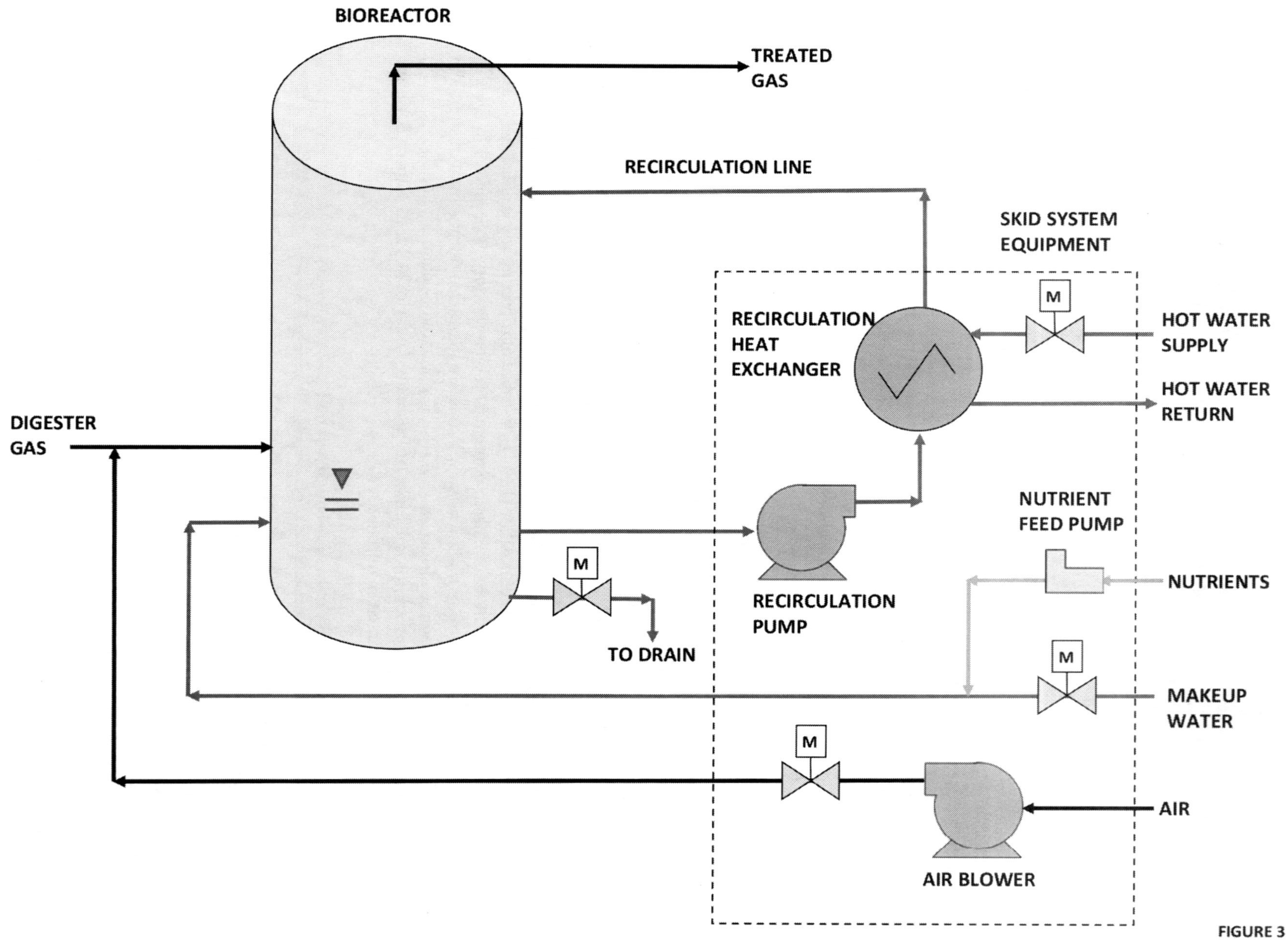

FIGURE 3
BIOLOGICAL H2S REMOVAL SYSTEM SCHEMATIC

Figure 5
Varec BiogasCleaner System (Varec Biogas, 2011)

Figure 6
DMT Sulfurex® BF System (DMT Environmental Technology, 2017)

Figure 7
Biorem Biogas Sweetening System (Biorem Technologies Inc., 2017)

Ancillary Equipment Housing

When considering this type of system for biogas treatment, the ancillary equipment requirements and the preference for housing such equipment should be considered. While manufacturers typically provide one or more pre-fabricated containers for this purpose, implemented skid systems have ranged from being enclosed under a canopy or housed in a constructed building. For the Dallas area, an outdoor canopy would require heat tracing and insulation of the equipment piping and components, and the cost savings would be minimal compared to the manufacturer standard pre-fabricated building. For this project, a separate (Contractor-provided) pre-fabricated building (herein called the Digester Gas Conditioning Control Building or DGCCB) will be provided, primarily to have a more robust structure, provide additional space and house other electrical panels for the DGC area in an isolated air conditioned room. The building will provide these benefits, but will also have an impact on cost and footprint.

In relation to the building space, electrical classification and space ventilation should be an early topic of discussion as well. Different manufacturers have different recommendations, and the National Fire Protection Association (NFPA) manual for wastewater treatment, does not define this specific scenario. In particular, a pathway of biogas to the ancillary equipment could occur if a leak occurs and gas enters through the connecting piping to the air blowers or recirculation pumps. In this case, it was decided to treat the space as if the room houses gas processing equipment, which resulted in rating the H_2S system room either as Class 1 Division 2 with continuous ventilation (12 air changes per hour) or Class 1 Division 1; the latter choice was selected. Gas monitors for detection of methane and H_2S were also included in the room design to conform with classification requirements.

Operating Temperature and Heating Capacity

Another important design factor to consider early in the design process is the operating temperature of the biological treatment system and the gas inlet temperature. Ideally, the biological populations prefer a temperature between 90 – 98 degrees F, similar to digesters. Located in Dallas, Texas, the gas pipes were designed to be above grade and uninsulated, for simplicity in construction and condensate collection. Compared to other places in the country, Dallas is relatively warm throughout the year and does not often get very cold. However due to length of pipe from the digester area to the DGC area, gas in uninsulated stainless steel piping will cool or heat depending upon the ambient temperature and sun intensity. While biological systems include recirculation heat exchangers for heating recirculation water and maintaining bioreactor temperature, the heating capacity depends on the available heating source. With no CHP included in this design, there will be no external hot water loop onsite. To address this, a separate tankless hot water heater will be installed in the DGCCB to provide a dedicated hot water loop. To save energy costs, insulation could be added to the biogas pipelines in the future to reduce temperature drop in the pipeline during winter temperatures.

Bioreactor Cleaning

As presented in the chemical reaction equations above, products of the biological processes include sulfuric acid (or sulfate) and elemental sulfur. The produced sulfate will be generally removed with the blow down (or drain) operations. Regarding elemental sulfur, a portion will be removed with the blow down, while a portion will adhere to reactor media. Over time, elemental sulfur and biomass will accumulate within the media, causing increased headloss and requiring cleaning. Manufacturers typically recommend periodic cleaning of one to two times per year. The periodic cleaning process involves filling the vessel with water, letting the vessel soak, agitating the media with air, and draining the vessel to remove the accumulated sulfur and biomass. The cleaning time for each vessel should take approximately two days per vessel. The cleaning drain flow or byproduct is expected to have a sulfur loading significantly greater than the blow down flow during normal operation. There was concern with routing this flow to the sanitary drain and to the head of the plant, due to the potential spikes in vapor phase H_2S concentrations at preliminary treatment processes, as well as at the digesters. Other facilities, with biological H_2S removal systems installed, were contacted, and one out of three that routed this flow to the head of the plant experienced a spike in H_2S concentrations at its digesters; (the other two facilities contacted send the flow to sludge or waste lagoons).

A number of location options were considered for handling the cleaning flow, including: sanitary drain and head of the plant; aeration basins for biological oxidation; final dewatering for removal with sludge; new settling tank with sludge hauling, new dewatering dumpster with sludge hauling, or plant effluent mixing. Mixing the sulfur sludge with biosolids or plant effluent was quickly ruled out due to potential biosolids classification and permitting impacts. The settling tank was determined infeasible due to the volume required to handle the cleaning flow. While a dewatering dumpster seemed promising, the materials of construction offered by manufacturers cannot handle residual sulfuric acid that may be present during cleaning. The aeration basin was feasible, but the work required would put the project significantly out of its scope of work at the plant. For these reasons, it was determined to keep the original design of routing the cleaning flow to the head of the plant. Moreover, due to the large size of the CRWS facility, the added sulfur load is expected to be absorbed without significant issues.

Adding an increased sulfur load to the head of the plant is a valid concern, especially for small plants. If necessary, other options should be considered and addressed early in the design, as such options could expand the scope of the project significantly.

Gas Compression: Design Criteria

To convey biogas to the Boiler Building, four (three duty) gas tight multistage centrifugal blowers were proposed. Due to a wide range in boiler demand of approximately 120 to 1,260 scfm, the booster blowers will be equipped with variable frequency drives (VFDs). The system will also include a recirculation line to address very low flow conditions and provide relief during surge conditions. The boilers require a pressure of 5.0 pounds per square-inch gauge (psig) to operate, and each blower will be able to produce 7.0 psig of pressure at 420 scfm. The blowers will be controlled based on a pressure set point on the gas line, and the pressure set point will dictate the turning on/off and ramping up/down of the blowers. While the gaseous H_2S concentration should not exceed 150 ppm, transient spikes may occur, and 150 ppm is still a significant concentration. Therefore, the blowers will be constructed of corrosion resistant materials and coatings (stainless steel, coated aluminum, and coated cast iron). Classified as Class 1 Division 1, the blowers will be located outdoors to save cost. The VFDs, PLC and control panel will be located within the DGCCB Electrical Room.

Moisture Removal System: Design Criteria

Downstream of the biological H_2S removal system and booster blowers, the gas will have a high water content and high temperature. The moisture removal system will provide cooling, reduce moisture and prevent condensation in the long pipeline from the DGC area to the Boiler Building. A simple shell and tube heat exchanger has been proposed to cool the biogas. The CRWS facility has a chilled water loop system, but the recently expanded loop does not have sufficient capacity for the DGC area. As a result, a chiller was included in the design, located within the DGGCB in a room separate from the H_2S equipment and the electrical rooms. The heat exchanger will receive chilled glycol to remove moisture from 1,260 scfm of biogas to a dew point of 35 degrees F. A dual electric actuate drip trap will be installed to receive the condensed water. Downstream of the heat exchanger, a coalescing filter will provide polishing by removing particulates and droplets to 1.0 micrometer at an efficiency of 95%, well below the requirements needed by the boilers.

CONSTRUCTION CONSIDERATIONS

The project specifications for digester gas conditioning equipment included requirements for functional, start-up, and performance testing. Careful consideration must be taken with regards to the order and conditions of start-up and performance testing. For example, as soon as the digesters receive sludge, the waste gas flares must be ready to receive biogas. Prior to starting up the booster fan, the digesters must be producing a sufficient amount of biogas for the fan to operate in a stable condition on its curve. Note that it is not expected that the booster fan will be needed in low flow conditions.

To meet permit requirements, the H_2S removal system must be operating satisfactory (with microbial populations acclimated), with effluent concentrations less than 150 ppm, prior to starting up the booster blowers, utilizing the heat exchanger and coalescing filter, and sending digester gas to the boilers on a continuous basis.

For performance testing of the booster blowers and heat exchanger, biogas production should be equal to or greater than the design maximum capacity of the blowers and heat exchanger (1,260 scfm). Also, for performance testing of the waste gas flares and H_2S removal system, the plant's solids processing must have achieved stable operations for at least 45 days, and biogas production must have reached a minimum rate (approximately 1,500 scfm). Since it is not expected that biogas production will reach the Peak Day and Peak Week flow rates specified (Table 3) within a reasonable period for performance testing, bioreactors can be tested individually or in pairs so that the flow to a given reactor reflects these higher flow rates.

CONCLUSION

In summary, this paper provides an overview of the gas treatment and handling processes for the TRA CRWS Phase III Solids Improvements Project, with a focus on the biological hydrogen sulfide removal system, a technology relatively new to the U.S. The paper presents permitting drivers, initial technology screenings, design criteria, design considerations and key lessons learned.

Based on findings throughout the design process, biological treatment of biogas for H_2S removal has great potential for applications in the U.S., especially when annual costs for adsorption systems are high.

As the project moves into the construction phase, CDM Smith and TRA expect to continue to learn about the biological system technology, especially during start-up and performance testing.

ACKNOWLEDGEMENTS

The authors of this paper gratefully acknowledge the staff of TRA for working with CDM Smith during various phases of permitting, planning, preliminary and final design on this project. The authors also commend TRA's leading role in the industry for using new, innovative, and sustainable technologies. Additionally, the authors acknowledge Chris Campbell and Cynthia Hibbard of CDM Smith for leading the permitting process for the project, as well as Engin Guven and Cameron Clark, both formerly at CDM Smith, for their roles in preliminary engineering and design of the digester gas conditioning system.

REFERENCES

Biorem Technologies Inc. The Benefits of Biogas Conditioning. http://www.biorem.biz/?portfolio=the-benefits-of-biogas-conditioning (accessed February 2017).

CDM Smith. *Trinity River Authority of Texas Central Regional Wastewater System Preliminary Design Report for Phase III Solids Improvements*. CDM Smith: Dallas, Texas, 2015.

DMT Environmental Technology. Gas Desulphurization. http://www.dmt-et.com/products/biogas-and-syngas-deshulphurization (accessed February 2017).

Duangmanee, T. (2009). Micro-aeration for hydrogen sulfide removal from biogas. Graduate Thesis and Dissertations, Paper 10748. 2009. http://lib.dr.iastate.edu/cgi/viewcontent.cgi?article=1765&context=etd (accessed February 2017).

Sander, R. Compilation of Henry's Law Constants for Inorganic and Organic Species of Potential Importance in Environmental Chemistry. 1999. http://www.henrys-law.org/henry-3.0.pdf (accessed February 2017).

Varec Biogas. BiogasCleaner. 2011. http://www.varec-biogas.com/en/products/show/45 (accessed February 2017).

To Treat, Or Not To Treat (Post Combustion Exhaust), That Is The Question

Jason Wiser[1], James Schettler[1], Don Trueblood[1], Philip Heck[3]

[1]Brown and Caldwell, 201 E Washington Street - Suite 500, Phoenix, AZ 85004
[2]Central Valley Water Reclamation Facility

ABSTRACT

Digester gas, produced from the anaerobic stabilization of sewage sludge, is largely comprised of methane. As such, digester gas has long been used as a fuel for combined heat and power (CHP), also known as cogeneration systems that provide process heat and produce renewable electricity. Internal combustion engines are the most mature and common prime mover used for digester gas fueled CHP applications. One of the challenges with utilizing digester gas as fuel is the need to treat the gas to remove impurities in order to meet the engine manufacturer's fuel quality requirements. Air emissions present another challenge for digester gas fueled CHP systems, especially for internal combustion engines.

Air emission limits vary by geographic jurisdiction. They are constantly evolving and generally become more stringent as emission control technologies continue to improve. While exhaust quality is an important design and air permitting consideration for any digester gas fueled combustion system, this paper will focus solely on exhaust treatment for lean-burn internal combustion engines. In just the last few years it has become more common for digester gas fueled internal combustion engines to require some form of post combustion exhaust treatment in order to comply with air emission limits.

Selective catalytic reduction (SCR) systems and oxidation catalysts are commonly used to treat engine exhaust. These exhaust treatment systems operate according to the principal of catalysis to reduce the concentration of harmful and regulated air pollutants found in combustion exhaust gases.

Despite the best efforts to treat digester gas for use as engine fuel, operating post combustion exhaust treatment equipment will eventually result in the catalysts becoming deactivated and ineffective. Due to the sensitive nature of the catalyst equipment, more thorough digester gas treatment is required than what is normally needed to meet typical fuel quality requirements. A robust and properly operating digester gas treatment system should prolong SCR system and oxidation catalyst life.

In addition to providing a detailed description of SCR systems and oxidation catalyst technologies, this paper will also present case studies of two water resource recovery facilities (WRRFs) that are in varying stages of modifying existing CHP systems. Air permitting requirements for both projects will be presented as well as the system modifications required to meet the more stringent permit limits. Digester gas treatment to meet engine fuel requirements and to protect the new catalysts will be presented.

KEYWORDS
Internal combustion engine, exhaust emissions, SCR, oxidation catalyst, digester gas.

INTRODUCTION

In general, air permitting agencies utilize two methods for imposing emission limits. First, an agency can adopt and implement performance standards as is the case with the South Coast Air Quality Management District's (AQMD) oxides of nitrogen (NOx) limits for digester gas fueled engines. South Coast AQMD, is widely regarded as having the strictest air emissions limits in the United States. Performance standards can be applied to new and existing emissions emitting processes, equipment, or facilities. Second, an air permitting agency can impose what is known as Best Available Control Technology (BACT). BACT levels vary and are categorized based on the combustion source and type of fuel. Under BACT the most stringent technologies available for reducing and or limiting air emissions must be implemented so long as the technology is deemed to be both cost effective and technologically feasible. Of course, cost effectiveness is open to interpretation across the various air districts and is typically evaluated based on the cost of air pollutant removal in terms of dollars per pound or ton of pollutant removed. The Bay Area AQMD is another strict air district and it imposes BACT on emission sources within its jurisdiction. The Bay Area AQMD's BACT requirements have been adopted from the South Coast AQMD's performance standards for new gas engines. However, BACT is required at the time of permitting and is not applied to existing processes, equipment, or facilities. Other jurisdictions around the country typically require a top-down BACT analysis to be performed as part of the permitting process. Such an analysis generally includes the review of the most stringent air emissions limits imposed around the country for similar processes, equipment, or facilities. For a top-down BACT analysis, most jurisdictions exclude air emissions requirements imposed by California air districts, which are generally perceived to be too stringent. Other jurisdictions consider these stricter air emissions limits to be the Lowest Achievable Emission Rate (LAER). While BACT requirements vary from jurisdiction to jurisdiction, the general nationwide trend has been for BACT requirements to become ever stricter. Air emissions limits must be considered on a case by case basis per the local jurisdiction and based on the process, equipment, or facility in question.

NOx is one of the six criteria air pollutants as defined by the Clean Air Act (CAA) and regulated by the United States Environmental Protection Agency (EPA). NOx is a regulated air pollutant because it contributes to the formation of ozone. NOx is emitted by virtually all combustion sources due to the high temperature conversion of nitrogen (N_2) and oxygen (O_2) present in combustion air. Most NOx emissions originate as NO, which accounts for up to 95 percent of the total, with the remainder primarily originating as NO_2. Upon entering the atmosphere, NO is quickly oxidized to NO_2. Collectively, NO, NO_2, and several other minor oxides of nitrogen are referred to as NOx.

Carbon monoxide (CO) is also defined as a criteria air pollutant per the CAA. In general, CO forms as a result of incomplete combustion, which is commonly caused by insufficient combustion air (lack of O_2) and low combustion temperature. In internal combustion engines, CO also forms due to the short residence time in the cylinders. Hydrocarbons (HC), including volatile organic compounds (VOC) and non-methane hydrocarbons (NMHC), also form as a result of incomplete combustion.

Internal combustion engines, by virtue of their inherent combustion technology, pose unique air permitting challenges. Intermittent combustion, as occurs repeatedly within each cylinder, generally produces more NOx and CO than continuous combustion. Further, NOx forms at high temperatures. The dilemma facing the control of air emissions from internal combustion engines is that reducing CO and HCs can be achieved by increasing combustion temperature, which would then result in higher NOx emissions. Thus, combustion zone temperatures must be lowered while not causing an increase in CO and HC emissions. The most common method to accomplish this delicate balance is to adjust the engine's air/fuel ratio to be fuel lean. The influence of air/fuel ratio on NOx and CO emissions is clearly depicted in Figure 1.

Figure 1 – Gas Engine Emissions in Relation to Air/Fuel Ratio (Credit: EPA)

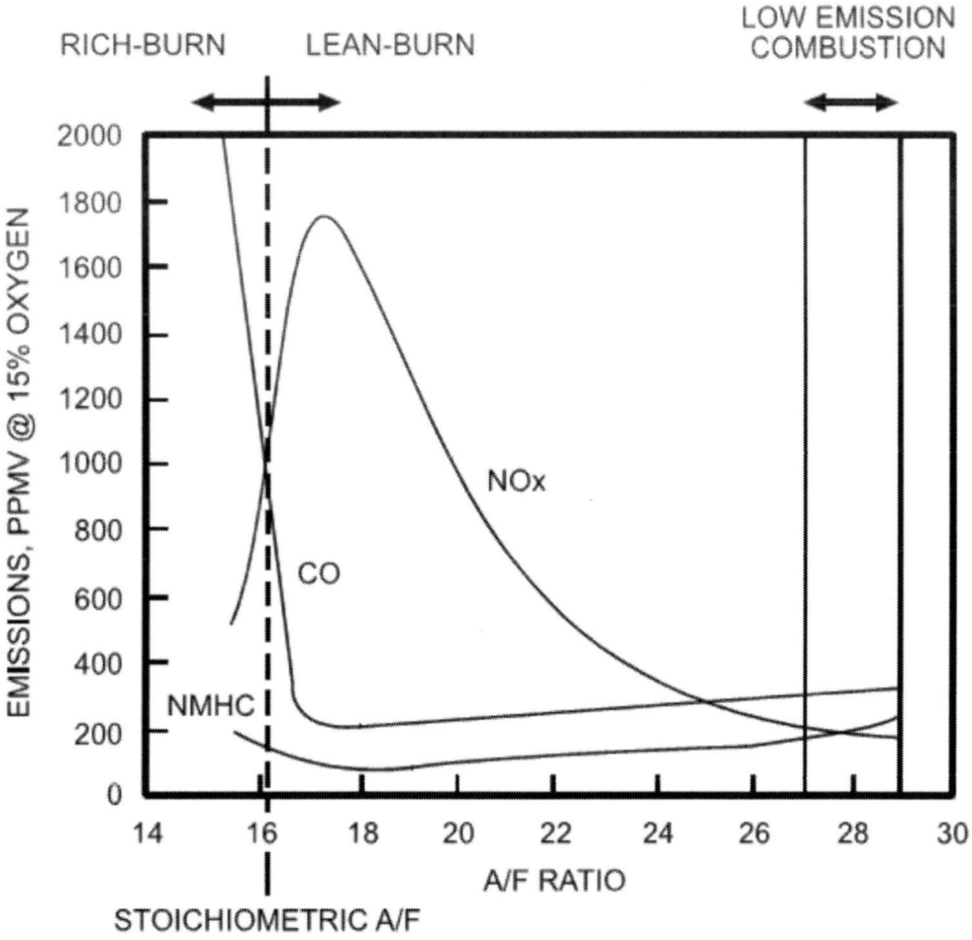

Historically, internal combustion engines were operated at near stoichiometric air/fuel ratios to maximize horsepower per the amount of combustion air consumed. The stoichiometric air/fuel ratio is defined as that ideal air/fuel ratio at which all fuel is completely combusted and all O_2 is consumed per the chemical reaction. However, for actual combustion application and to ensure optimal combustion, excess air is provided. When excess combustion air is provided, the air/fuel ratio is referred to as Lambda (λ), which is defined as the operating air/fuel ratio divided by the ideal stoichiometric air/fuel ratio. It follows then that λ is 1.0 at stoichiometric air/fuel ratio.

As noted on Figure 1, air/fuel ratios less than stoichiometric are referred to as rich-burn, whereas air/fuel ratios greater than stoichiometric are referred to as lean-burn. To the left side of stoichiometric, or rich-burn, NOx emissions decrease rapidly due to a lack of O_2 in the combustion chamber and lower combustion temperatures. Conversely, to the right side of stoichiometric, or lean-burn, NOx emissions increase rapidly due to the high combustion temperature and the presence of excess oxygen. However, as the air/fuel ratio, or λ, continues to increase NOx emissions begin to rapidly decline. This is due to lower combustion temperatures despite the presence of excess O_2.

Again referring to Figure 1, CO is high when operating with a rich air/fuel ratio. This is due to insufficient O_2 needed to achieve complete combustion. In general, CO emissions are low throughout the lean-burn region due to the presence of sufficient O_2. CO emissions are lowest when λ is slightly greater than the stoichiometric air/fuel ratio because combustion temperatures are high and sufficient O_2 is available for complete combustion to occur. However, as λ continues to increase, the CO concentration increases slightly due to less efficient combustion caused by lower temperatures.

Until about 1985 to 1990 rich-burn engines were very common for use with digester gas due to their stable fuel combustion at or near the stoichiometric air/fuel ratio. Rich-burn engines eventually gave way to more modern lean-burn engines, which were developed to produce much lower emissions and are also more fuel efficient. Lean-burn engines can be characterized as having a λ in excess of about 1.1 or more with more than about 8 percent O_2 in the exhaust. However, to achieve the necessary low levels of emissions required in certain jurisdictions, lean-burn engines must be equipped with SCR systems and oxidation catalysts.

POST COMBUSTION EXHAUST TREATMENT

SCR systems and oxidation catalysts are required to comply with South Coast AQMD's performance standards for all gas engines and Bay Area AQMD's BACT emission limits for new gas engines. SCR systems and oxidation catalysts both operate according to the principal of catalysis to reduce harmful and regulated air pollutants found in combustion exhaust gases. A catalyst is a substance that significantly increases the rate of a chemical reaction. SCR units and oxidation catalysts both consist of a substrate featuring a honeycomb like lattice, which provides a large surface area to which a catalyst coating is applied. The catalyst impregnated wash coat significantly increases the surface area, as can be seen at the microscopic level in Figure 2.

Figure 2 – Microscopic View of High Surface Area Catalyst Wash Coat (Courtesy of DCL)

In many cases the catalyst coating consists of a combination of platinum group metals (PGM), typically, typically platinum and palladium. The chemical reactions that drive the reduction of exhaust emissions occur at the catalyst surface, as depicted in Figure 3. The large surface area serves to bring the chemical molecules into close contact while simultaneously reducing the chemical reaction's activation energy thus driving the reaction at a greatly accelerated rate.

Figure 3 – High Surface Area Coating within Substrate Passageway

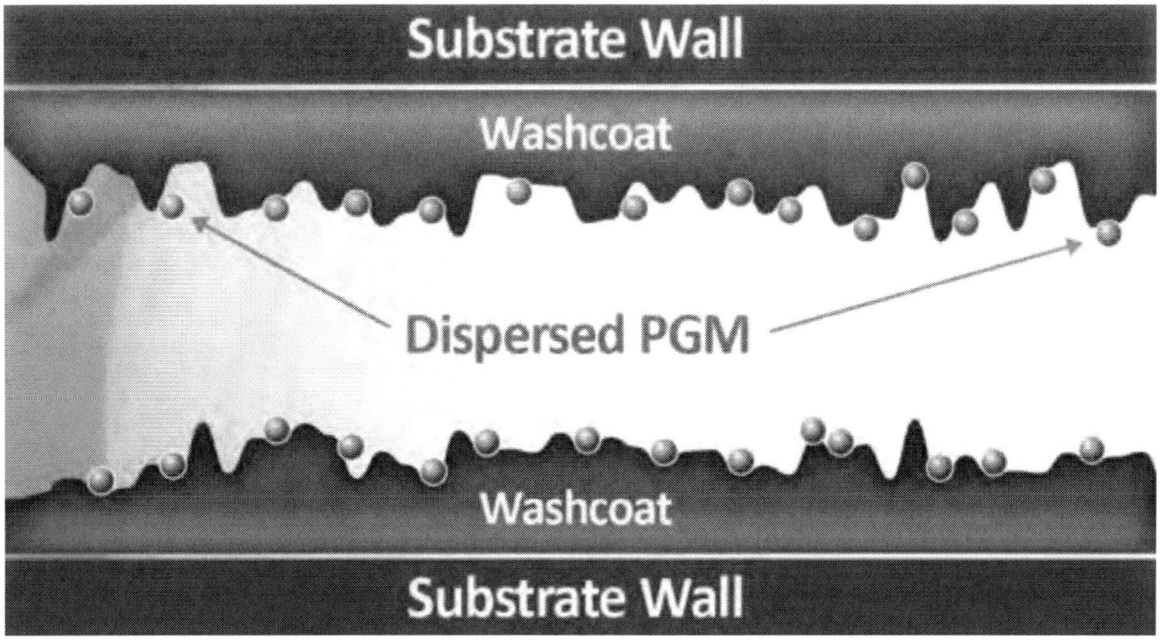

Figure 4 depicts a typical layout of post combustion exhaust treatment equipment. The particulate filter depicted is standard for diesel engines and is generally not included with a gas engine application. SCR and oxidation catalysts are often provided in separate housing units, but can also be integrated into a single housing unit.

Figure 4 – Post Combustion Exhaust Treatment (Courtesy of DCL)

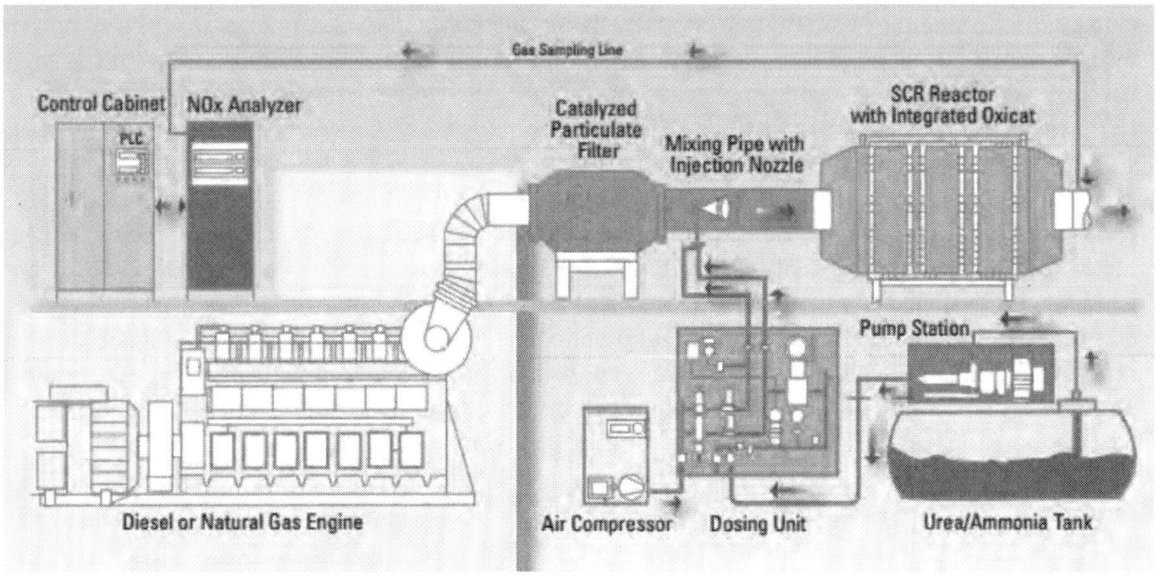

The primary elements depicted in the Figure 4 are described below.

SCR Catalyst. In the case of SCR systems, NO and NO_2, collectively known as NOx, are converted to N_2 gas and water vapor when reacted with ammonia (NH_3) at the surface of the catalyst coated substrate. NOx removal efficiency is determined by the exhaust temperature, design and configuration of the substrate, catalyst activity, and the amount of NH_3 injected into the exhaust stream. NOx reduction by SCR systems occurs according to the following chemical reactions:

$$4NO + 4NH_3 + O_2 \Rightarrow 4N_2 + 6H_2O$$
Equation 1

$$6NO_2 + 8NH_3 \Rightarrow 7N_2 + 12H_2O$$
Equation 2

There are at least two types of SCR catalysts available for NOx reduction and use with lean-burn internal combustion engines. The first utilizes a vanadium pentoxide (V_2O_5), also known as vanadia, catalyst embedded in a ceramic substrate. The second type of SCR catalyst utilizes a zeolite-based catalyst coating on a ceramic substrate.

Ammonia Reactant. Ammonia required for the reduction of NOx can be supplied as a gas (anhydrous ammonia) or as an aqueous solution. Water-based urea solutions are most common because it is easier to handle and store and is less toxic than ammonia gas. Urea is available at a concentration of either 32.5 or 40 percent. These urea concentrations have freezing points of negative 11 and 0 degrees C (12 and 32 degrees F), respectively. The 40 percent urea concentration has a slightly greater NOx reduction capacity per unit volume.

With the engine operating and once the SCR elements have reached their normal operating temperature, typically 315 – 427 degrees C (600 – 800 degrees F), aqueous urea is injected into the exhaust gas piping upstream of the SCR housing where NH_3 is dispersed and mixed with the engine exhaust. Compressed air is supplied with the urea to atomize the solution to promote the efficient mixing and contact of the NH_3 molecules with NO and NO_2 molecules. This also helps prevent the injection nozzle from plugging. Vanes are located throughout the mixing section to create sufficient turbulence in the exhaust gas to achieve adequate mixing with the urea prior to reaching the SCR elements. Heat from the exhaust gas vaporizes water in the urea solution causing the formation of small urea crystals, which then breakdown into ammonia. NOx present in the exhaust gas then reacts with the ammonia at the SCR catalyst per Equations 1 and 2, above.

Control systems are available to automatically optimize urea dosing so that post SCR NOx emissions comply with permit limits while simultaneously limiting ammonia slip (caused by excessive urea dosing). These control systems include a post SCR NOx analyzer. Based on the measured NOx, a signal is sent to the urea dosing system to increase or decrease the urea dose as needed to simultaneously comply with NOx emissions requirements while limiting urea dosing and subsequent ammonia slip to the extent practical. This method helps reduce urea costs and routinely limits ammonia slip to less than 10 parts per million by volume (ppm_v). If additional ammonia slip control is desired, a dedicated ammonia slip catalyst can be added to the SCR catalyst housing. An ammonia slip catalyst is designed to specifically target and reduce ammonia emissions. In this case the ammonia slip catalyst would be located downstream of multiple SCR catalyst elements.

Oxidation Catalyst. The purpose of an oxidation catalyst is to reduce CO, formaldehyde (CH_2O), other un-combusted HCs, particulate matter (PM), and hazardous air pollutants (HAP). An oxidation catalyst consists of a substrate of rolled and corrugated steel foil coated with a high surface area primer. A PGM based catalyst coating is then applied over the primer.

Figure 5 – Rolled and Corrugated Steel Foil Catalyst Substrate

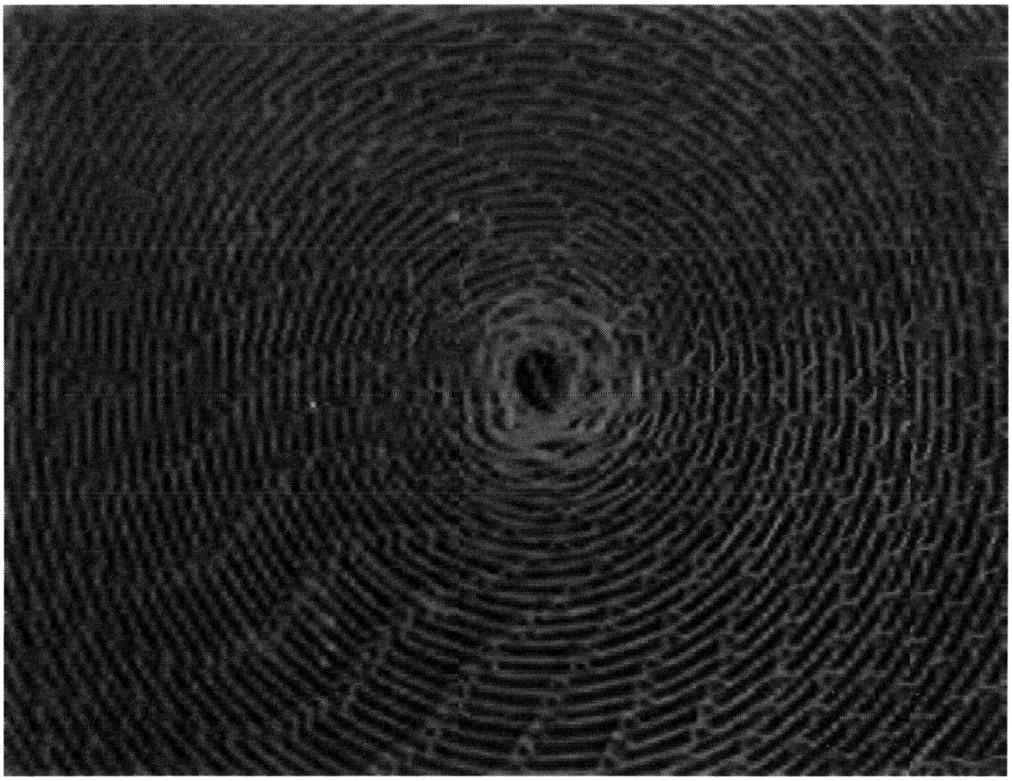

These catalysts utilize excess oxygen from lean-burn combustion to convert CO and HCs to water vapor and carbon dioxide (CO_2) according to the following reactions:

$2CO + O_2 \Rightarrow 2CO_2$
Equation 3

$CH_2O + O_2 \Rightarrow H_2O + CO_2$
Equation 4

$HC + O_2 \Rightarrow H_2O + CO_2$
Equation 5

Continuous Emissions Monitoring System. To ensure compliance with established performance standards or BACT, many air quality jurisdictions require the installation of a continuous emissions monitoring system (CEMS) to continuously measure and record emissions thus demonstrating air permit compliance. In addition to NOx, CO, and O_2, CEMS units can also monitor other emissions including sulfur oxides (SOx), CO_2, VOCs, and PM. Additional emissions monitoring beyond NOx, CO, and O_2 would be determined by the local air permitting authority during the design and permitting process.

Despite the best efforts to treat engine fuel, operating post combustion exhaust treatment equipment will eventually result in the catalysts becoming deactivated and ineffective. Catalyst deactivation can result from the combustion of engine lubricating oil or compressor oil that carries over into the fuel stream. Because lean-burn engines require so much combustion air some facilities have experienced catalyst deactivation as a result of contaminated combustion air. Some engine manufacturers now have detailed combustion air quality specifications for use with exhaust catalysts. There are several mechanisms by which a catalyst can become deactivated and are described as follows:

Masking. This type of deactivation occurs when elements are deposited on the surface of the catalyst thus preventing exhaust gas contaminants from contacting the catalyst at that location.

Fouling. Fouling occurs when solid deposits buildup within the passageways of the catalyst substrate thus restricting exhaust gas flow and limiting catalyst effectiveness. In severe cases fouling can cause significant back pressure on the engine which can result in reduced fuel efficiency and/or cause the engine to stall.

Masking and fouling of catalysts can be reversed with proper maintenance, typically performed by the catalyst manufacturer.

Poisoning. Catalyst deactivation caused by poisoning cannot be reversed and is caused when certain elements combine with the catalyst. Because this process cannot be reversed, the only solution is to replace the catalyst elements. Catalyst poisons can be present in the engine's lube oil, coolant, and fuel as well as the combustion air.

Deposits known to poison catalysts include combustion products, such as silicon dioxide (SiO2, from siloxanes) and sulfur (from hydrogen sulfide [H_2S]) and sulfated ash from engine lube oil. Coordination between the engine and catalyst manufacturers is imperative to select an appropriate low ash engine lube oil that meets both the engine's needs but will not lead to premature deactivation of the catalysts.

Thermal Deactivation. In the case of oxidation catalysts, excessively high exhaust temperatures (over 1,350 degrees F) can cause significant problems. At high temperatures, metal foil substrates have a tendency to weaken and PGM metals can agglomerate reducing catalyst performance. Thermal deactivation is not reversible and is most often caused by engine misfires.

Due to the concerns and costs associated with catalyst deactivation, manufacturers will guarantee catalyst performance only when engine exhaust temperature and fuel quality requirements are met. Catalyst life, under proper operating conditions, is typically guaranteed for 8,000 to 16,000 operating hours depending on the manufacturer.

Operations and maintenance (O&M) activities for post combustion exhaust treatment equipment are primarily accomplished via engine O&M activities, including fuel treatment. Periodic inspection of the catalyst elements is suggested to provide visual confirmation of catalyst condition and the degree of buildup of deposits. Aside from periodic visual inspections, automated controls monitor exhaust temperature and pressure and post SCR NOx emissions. These parameters provide an indication of overall system performance. When these monitored parameters or visual inspections indicate deactivation of the catalysts, cleaning and/or replacement is recommended to be performed by the catalyst supplier. These systems have remote monitoring capability which allows the system supplier to remotely observe and track system operation and performance. Remote monitoring is valuable in that it enables the system supplier to provide operations staff with immediate feedback on system performance and to recommend corrective actions. Remote monitoring also allows the system supplier to more quickly mobilize its field staff when catalyst cleaning or replacement is needed.

ENHANCED DIGESTER GAS TREATMENT

Over time, operating exhaust treatment catalysts become deactivated. Deactivation can occur rapidly when catalysts are exposed to exhaust gases produced from digester gas fueled combustion equipment. Therefore, it is absolutely critical that digester gas fuel be thoroughly pre-treated to remove harmful contaminants such as sulfur compounds and siloxanes.

Catalyst elements have exceptionally large surface areas and thus are extremely prone to fouling. Thus catalyst manufacturers have stringent requirements regarding the quality of the exhaust gases. A survey of several leading manufacturers of exhaust treatment equipment indicates that in order to meet exhaust emission requirements, digester gas fuel must be treated prior to combustion to limit sulfur species to no more than 20 ppm_v and siloxanes to no more than 20 parts per billion by volume (ppb_v).

Typical digester gas treatment required to meet both the engine's fuel standards and the post combustion exhaust treatment equipment requirements includes H_2S removal, typically iron sponge or similar; gas cooling for moisture removal; pressurization; and siloxane removal. Most analytical laboratories are unable to detect siloxanes at the 20 ppb_v level prescribed by the catalyst manufacturers. Due to the challenges inherent with monitoring siloxane concentrations in real time, it is recommended that activated carbon vessels be arranged in series so that breakthrough can be detected downstream of the first vessel. While this technique provides for sufficient time to replace the media in the first vessel before breakthrough can occur at the second vessel, it also requires frequent digester gas sampling and laboratory analysis, which can be quite expensive if performed at necessary frequency. However, the consequence of siloxane breakthrough is that a catalyst can become deactivated within just a few hours of siloxane exposure. Catalyst replacement is far more expensive than preventative maintenance associated with regular and frequent digester gas analysis. Alternatively, modern siloxane removal systems feature media that can be regenerated in-vessel every few days thus ensuring that media remains fresh and able to remove siloxanes to a very low level.

RESULTS

This section presents the ongoing, air permitting, engineering and construction related activities at two WRRFs that are currently in the process of implementation of modifying existing CHP systems.

Case Study #1: Laguna Treatment Plant; Santa Rosa, CA

The Laguna Treatment Plant (TP) is located in Santa Rosa, CA and is within the jurisdiction of the Bay Area AQMD. The Laguna TP's CHP system was installed in 2009 and includes four 1,100 kW Cummins engines. The internal combustion engines were originally permitted to operate with a blend of digester gas and up to 10 percent natural gas. At the time the engines were permitted the Laguna TP did not produce enough digester gas to operate more than one engine at a time with digester gas. However, a new co-digestion facility at the Laguna TP has been operational since 2016 and has resulted in a significant increase in digester gas production but not enough to operate two engines solely on digester gas.

In an effort to move closer to energy neutrality, the City now intends to increase the amount of natural gas blended with digester gas for fueling the engine-generators. Because the engines are currently permitted to operate with digester gas and up to 10 percent natural gas, an increase in the amount of natural gas will require a new air permit (Authority to Construct and Permit to Operate) as well as a modification to the existing Title V permit. Consequently, re-permitting the engines will require the affected engines to be equipped with post combustion exhaust treatment equipment, including SCR systems and oxidation catalysts, and associated support systems.

The project will include the design and installation of two SCR systems, two oxidation catalysts, a CEMS unit, and associated support systems for two CHP engines. The project includes the design and implementation of improvements to the existing digester gas conditioning and natural gas blending systems. With these project enhancements in place, the CHP capability and operation will be as follows:

Table 1 – Laguna TP Engine Configuration and Operation

Engine	Configuration	Normal Operation	Alternate Operation
Engine 1	SCR	Optional operation with 100% natural gas.	Engines 1 and 2 can alternate duty/standby.
Engine 2	SCR	Continuous operation with blended digester gas and natural gas.	
Engine 3	No modification	Continuous operation with 90 to 100 percent digester gas.	Engines 3 and 4 can alternate duty/standby.
Engine 4	No modification	Continuous operation with 90 to 100 percent digester gas.	

During the design of the new CHP facility in 2007, BACT emissions limits for natural gas were much lower in the Bay Area AQMD than for digester gas. Post combustion exhaust treatment for digester gas was not required because its contaminants are known to foul and poison catalysts which would result in cost-prohibitive replacements. However, since that time the Bay Area AQMD has revised the BACT limits for digester gas fueled internal combustion engines. The new and stricter air emissions can be achieved by thoroughly treating the digester gas fuel to remove contaminants and implementing post combustion exhaust treatment. These changes to the BACT limits for digester gas fueled internal combustion engines do not affect the existing CHP facility at the Laguna TP because the facility obtained its operating permit under the previous BACT rules.

However, increasing the natural gas content of the CHP facility's blended fuel to more than 10 percent, as is planned for this project, will require the engines to be re-permitted by Bay Area AQMD for natural gas operation and require the existing Title V permit to be modified. The following table compares the emissions limits per the CHP facility's current Permit to Operate with the current BACT limits for internal combustion engines fueled by either digester gas or natural gas.

Table 2 – Internal Combustion Engine Exhaust Emission Limits per Bay Area AQMD

Pollutant	Current Permit g/bhp-hr	Natural Gas – BACT g/bhp-hr	Digester Gas – BACT g/bhp-hr
NOx	0.75	0.071	0.151
CO	2.55	0.102	0.892
Notes: SCR and oxidation catalysts are required to comply with current BACT per the Bay Area AQMD.			

Based on information provided by the engine and catalyst manufacturers and anticipated Bay Area AQMD emissions limits, the following performance criteria have been established for the Laguna TP project.

Table 3 – Catalyst Performance Criteria

H_2S limit in the engine fuel, maximum, ppm_v	20
Silicon limit in the engine fuel, maximum, ppm_v	0.02
Engine exhaust temperature, maximum, degrees F	858
Engine exhaust temperature, minimum, degrees F	779
Exhaust pressure drop, maximum, inches WC	12
NOx emissions limit, g/bhp-hr	0.07
CO emissions limit, g/bhp-hr	0.10
Ammonia slip, maximum, ppm_v	10

Figure 6 – Engine-Generators at the Laguna TP

The existing digester gas treatment system at the Laguna TP includes an iron sponge system that reduces H_2S from approximately 100 ppm_v to less than 10 ppm_v. Thus the existing H_2S removal system is believed to provide the level of treatment required for the future exhaust treatment equipment. The existing digester gas treatment system also includes activated carbon-filled vessels, arranged in parallel, to remove siloxanes from the digester gas fuel. The siloxane removal system will be modified to operate in series and a third vessel will likely be provided for redundancy. Operating the siloxane removal vessels in series is absolutely critical given the facility's acceptance of high strength waste via its new co-digestion program. Depending on the source and type of material, alternate digester feedstocks have been known to contribute unusually high levels of siloxanes to the digester gas.

Case Study #2: Central Valley Water Reclamation Facility; Salt Lake City, UT

The Central Valley Water Reclamation Facility (WRF) has successfully operated a CHP facility since it was initially constructed in the early 1980's. Digester gas is currently used as fuel for five engine-generators. The existing engine-generators are relatively old and inefficient compared to new more advanced models. The existing Waukesha engine-generators include two VHP model units, each rated for approximately 1,150 kW and installed in 1983 and three model AT27 units, each rate for approximately 1,340 kW and installed in 2002. While the newer Waukesha AT27 engines have proven to be suitable for continued operation they lack sufficient redundancy for their role. Meanwhile, the 33-year-old Waukesha VHP engines have reached the end of their useful life and are ready for immediate replacement.

The total CHP nameplate power generation capacity is approximately 6,300 kW. However, due to certain limitations with the CHP system in general and the physical condition of the Waukesha VHP units, the practical or reliable power generation capacity is limited to approximately 3,300 kW.

Central Valley WRF is now in the process of replacing the existing engine-generators with new more reliable advanced reciprocating engines with high electrical efficiency and low exhaust emissions. Four GE Jenbacher 1,812 kW engine-generators have been pre-purchased to replace the existing Waukesha engine-generators.

The primary objectives of the engine-generator replacement project are to provide new, modern engine-generators that maximize electrical output by more efficiently using the available digester gas and to maintain or reduce emissions in order to comply with current and known future emissions requirements.

The four new GE Jenbacher engine-generators were selected after a careful and detailed evaluation of several engine-generator options in the size and capacity range needed to most efficiently utilize the available digester gas to be produced at the Central Valley WRF. The new lean-burn, spark-ignition engines selected for installation at the Central Valley WRF have special state-of-the-art features, such as dual aftercoolers with distinctive higher pressure turbochargers and sophisticated combustion controls to achieve both extremely high fuel-to-electrical efficiency and very low exhaust emission rates.

The Central Valley WRF has submitted an application for a Title V permit, which is currently under review. Future plant-wide emissions, after the new engines are installed, are not expected to drop below the Title V threshold of 100 tons per year. Therefore, the current Title V permit application will continue to be processed and is expected to be approved, albeit with necessary modifications to reflect the removal of existing engine-generators and installation of new engine-generators. The Central Valley WRF is currently close to triggering the need for a Prevention of Significant Deterioration (PSD) permit. This would be triggered by the emission of 250 tons per year of any pollutant for which the facility is in attainment. The Central Valley WRF is located within a CO attainment area and current facility-wide CO emissions are approximately 237 tons per year.

BACT might be required regardless of the permitting level and would essentially require that new equipment meet the lowest emission levels of similar equipment that has been permitted. Within the last few years, Brown and Caldwell conducted a BACT analysis for similar internal combustion engines in the same geographic area (within 50 miles of the Central Valley WRF), under the same regulatory jurisdiction. At that time BACT was established as 0.5 grams per brake horsepower-hour (g/bhp-hr) for NOx.

By comparing BACT emissions levels for the new engine-generators to emission rates for the existing engine-generators contained in the air permit, it is clear that replacement of the existing engine-generators will reduce facility-wide emissions for NOx and CO (emission rates for other pollutants emitted from the existing engines were not available). Therefore, PSD is not considered to be a practical issue with respect to this project.

Due to wintertime temperature inversions, the Salt Lake Valley has more stringent PM2.5 emissions standards than most areas in the United States. PM2.5 are typically caused by the following:

- Combustion of H_2S and its subsequent conversion to SOx
- Carryover of lubricating oil from gas compressors to the fuel stream
- Contaminated combustion air

The Utah Division of Air Quality (DAQ) could require PM2.5 emissions as low as 0.0003 g/bhp-hr. While gas fueled internal combustion engines do not normally produce any significant PM2.5 emissions, this possible 0.0003 g/bhp-hr limit is below what most engine manufacturers are willing to guarantee. However, H_2S removal from the digester gas and the engine exhaust oxidation catalysts would both help to minimize PM2.5. Further, the upgraded CHP facility will not include oil-flooded gas compressors and combustion air will be thoroughly filtered prior to entering the engines.

At project startup in early 2016 it was believed that oxidation catalysts would be required in order to comply with PM2.5 limits. The new digester gas treatment system was originally designed to remove impurities to the necessary levels required to protect the oxidation catalysts. However, more recent communications with the Utah DAQ revealed that oxidation catalysts will not be required for this project. Because the project will reduce facility-wide emissions for all criteria pollutants, including PM, a Notice of Project Exemption has been submitted to the Utah DAQ indicating that the installation and operation of the new high efficiency and cleaner burning engines will be exempt from ordinary permitting processes.

Despite the recent design change to eliminate oxidation catalysts from the project, design of the previously planned and very robust digester gas treatment system design will not be modified. The existing gas treatment system, in particular the siloxane removal system, does not have sufficient capacity to treat the projected future design digester gas flow. The following modifications to the gas treatment system will be provided with the expansion of the CHP facility:

- Add a granular iron oxide-based system to remove H_2S from the digester gas. This new treatment system will be required to limit SOx emissions from the engine exhaust and will likely be required by the new air permit. Granular iron-oxide type media is the preferred choice of H_2S removal media, in lieu of iron sponge, due to its non-binding characteristics, which is expected to simplify media removal efforts and associated labor.

- Replace the existing activated carbon-based siloxane removal system with a new activated carbon-based siloxane removal system. A total of six new vessels will be required to treat the full 1,500 scfm of digester gas flow. Vessels will be arranged in pairs with each pair of vessels operating in series to achieve exceptionally low levels of siloxane in the engine fuel. By arranging the vessels in series siloxane breakthrough can be detected between the vessels via frequent gas sampling and laboratory analysis. The downstream vessel will continue to provide siloxane removal during the time that the upstream vessel is undergoing media replacement.
- Add higher capacity gas cooling/moisture removal equipment to meet the future design digester gas flow.
- Increase gas blower capacity. For the new engines, new larger capacity gas blowers will be needed to pressurize the gas to about 41.4 kilopascal (6 pounds per square inch [gage pressure]). This pressure is necessary to overcome pressure losses through the gas treatment system while providing sufficient residual gas pressure to the engine fuel train.

Figure 7 – Existing CHP Units to be Replaced at the Central Valley WRF

CONCLUSIONS

Large stationary internal combustion engines remain the most mature and proven CHP technology for use with digester gas. However, exhaust emissions represent the primary limitation of internal combustion engines. Engine manufacturers are continually seeking to improve combustion characteristics and engine performance. Lean-burn engine technology has been available for over 25 years. As air emission regulations have evolved and become ever stricter in recent years, it is now commonly required in many jurisdictions to include post combustion exhaust treatment equipment on digester gas fueled engines. In turn, post combustion exhaust treatment equipment systems require the engines to receive exceptionally clean fuel so that the combustion exhaust does not prematurely deactivate or harm the catalysts. However, many jurisdictions do not require post combustion exhaust treatment for digester gas fueled internal combustion engines. Utilities and their engineers should communicate clearly, early and often with their local air permitting agency to ensure a clear understanding of emission limits is achieved as this will dictate the level of post combustion exhaust treatment that may or may not be required.

Enhanced digester gas treatment can be costly. In some cases, the addition of post combustion exhaust treatment equipment and the requisite enhanced digester gas treatment equipment may make internal combustion engine based CHP systems prohibitively expensive, not to mention the additional air permitting challenges associated with engines. However, despite the technical, economic, and air permitting challenges, digester gas fueled CHP systems with post combustion exhaust treatment have been successfully demonstrated at several WRRFs and continue to be implemented at other WRRFs such at the Laguna TP. Utilities should proceed with caution when considering post combustion exhaust treatment for internal combustion engines. Case in point: The Central Valley WRF will likely avoid including post combustion exhaust treatment in its new CHP project by clearly communicating with the local air district and demonstrating the CHP upgrades will result in lower plant-wide emissions.

ACKNOWLEDGMENTS

The authors appreciate the support and facility access provided by the case study subjects; Laguna Treatment Plant and Central Valley Water Reclamation Facility.

REFERENCES

Doyle, B.; (2003) *Combustion Source Evaluation*, Third ed.; United States Environmental Protection Agency

Emission Standards Division; (1993) *Alternative Control Techniques Document – NOx Emissions from Stationary Reciprocating Internal Combustion Engines*; United States Environmental Protection Agency

Schettler, J.; Harris, C.; Stutz-McDonald, S.; Lindstrom, K.; (1987) *Associates Committee Report on Internal Combustion Engines Using Digester Gas and Related Air Emission Requirements*; California Association of Sanitation Agencies

Evaluation of Biogas-to-Energy Options for Edmonton's Gold Bar Wastewater Treatment Plant

Nicholas Szoke[1*], Alfredo Suarez[2]

[1] Stantec Consulting Ltd.
[2] EPCOR Water Services Inc.
*Email: Nick.Szoke@Stantec.com

ABSTRACT

EPCOR is developing near- and long-term buildout options for Edmonton's Gold Bar Wastewater Treatment Plant (GBWWTP). Identified upgrades consider integrated resource recovery opportunities where possible to increase operational efficiency. Digester gas (biogas) generated at GBWWTP is an on-site renewable energy resource. Currently, biogas not used in GBWWTP's boilers is flared, and represents an opportunity to improve operating efficiency through a more cost-effective Biogas-to-Energy (BTE) technology. A review of BTE technologies, their benefits/drawbacks, biogas conditioning requirements, advanced digestion considerations, operational requirements, and major capital and operating costs was conducted for informed decision making purposes. The review found that a reciprocating engine generator combined heat and power (CHP) solution to be the best BTE technology for possible implementation at the GBWWTP.

KEYWORDS: Biogas; biogas-to-energy; heat; power; advanced digestion; gas conditioning

APPROACH

A co-generation (co-gen) feasibility study was recently completed by Stantec for EPCOR. The study assessed the economics and other merits of a reciprocating engine or turbine combined heat and power (CHP) system fed primarily on biogas. The study found a positive business cases exists to implement a CHP system. In addition, the CHP system could provide resiliency to electricity supply and costs, generate "green" carbon credits, has flexibility for future expansion, has a high capacity factor, and is capable of generating steam for future advanced digestion upgrades. Notwithstanding the positive findings of the feasibility study, EPCOR identified the need to investigate alternative biogas-to-energy (BTE) technologies to confirm that a reciprocating engine CHP system is the best BTE solution for GBWWTP.

The following list of BTE technologies represents possible technical options practicable to GBWWTP.

1. Boilers to generate hot water for process heat and occupied spaces
    a. Conditioned biogas (H_2S reduction only).
2. Combined heat and electrical power
    a. Reciprocating engines
    b. Fuel cells
3. Biomethane "pipeline quality gas" for use as renewable natural gas (RNG)
4. Compressed-RNG (C-RNG) for use in fleets vehicles
5. Liquefied-RNG (L-RNG) for efficient storage, transportation, and export

SUPPORTING INFORMATION

Currently, Edmonton's GBWWTP treats approximately 280 ML/d (74 MG/d) of wastewater generated from a service area population of 900,000 people. This wastewater stream generates approximately 100 dry Tonnes/day of sludge and stabilized by conventional mesophilic anaerobic digestion (MAD).

The primary purpose of conventional MAD of sludge at GBWWTP is to biochemically degrade the organic matter in the sludge to stabilize the volatile solids (VS) content before disposing of the treated digestate (Biosolids) to the Clover Bar storage lagoons. The biochemical degradation of VS results in the generation of off gases, mainly methane (CH_4) and carbon dioxide (CO_2) with lessor concentrations of hydrogen sulfide (H_2S), moisture (H_2O), and siloxanes. The sludge composition (i.e., feed stock) along with the degree of VS degradation strongly influences the resulting biogas constituents and their concentrations. Increased VS degradation will result in increased biogas production. Since the mesophilic operating temperature of the digesters is held at a near constant temperature of 37 °C, solids retention time (SRT) is the main process control parameter currently influencing VS stabilization and the biogas produced. A schematic of the existing solid removal and treatment processes at GBWWTP is shown in **Figure 1.**

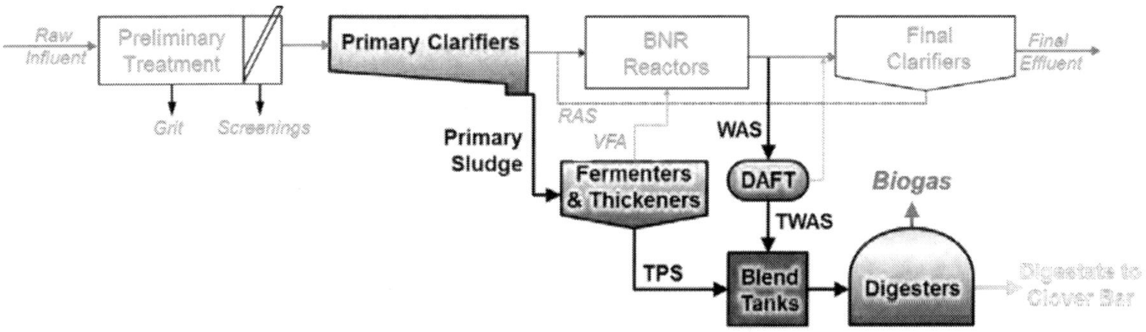

Figure 1: GBWWTP Solids Handling and Treatment Processes

Untreated biogas is used in the plant's boiler systems to provide process heat year-round and seasonal heating of occupied spaces during colder months. The boiler system was upgraded to use natural gas (NG) to meet peak heating load demands during the winter months when the supply of biogas was insufficient. Biogas not used in the boilers is flared to regulate digester headspace pressure, and represents an opportunity to improve operating efficiency through more cost-effective and practicable use of this renewable resource.

Based on 2013 data, the average daily discharge of blended sludge into the digesters was 2.1 ML/d at a solids concentration of 4.7%. The total working volume of all digester combined is 58 ML. With the largest digesters offline, the firm digestion capacity is approximately 49 ML and has a solids retention time (SRT) during dry weather conditions between 24 to 28 days. Based on the Van Kleek equation, the volatile solids reduction (VSR) resulting from current digester operation is estimated at 50.4 % on average. However, using the same data but applying a mass balance approach, the estimated VSR would be 55.3 %. VSR directly influences the amount of biogas generated. The difference between VSR estimates relates to the mass of inert solids in the feed sludge and biosolids.

VSR can be correlated to the volume of biogas generated and its CH_4 content. The expected biogas volume can be estimated from the mass of VS destroyed using a 0.75 to 1.12 Nm^3/kg relationship, typically 0.95 for convention wastewater MAD. The amount of CH_4 generated can be estimated based on 0.68 kg/m³ CH_4 density relationship. The estimated biogas volume is expected to range between 28,300 to 42,200 Nm³/day, with a typical volume of 35,900 Nm³/day. The estimated biogas volume agrees very well with monitored conditions at GBWWTP: average production is measured to be 37,900 Nm³/day. Based on biweekly samples, the MAD process at the GBWWTP produces near stable methane (CH_4) content in the biogas measured at an average of 61.4% and correlates very well based on the average biogas density of 1.132 kg/m³. CH_4 content of biogas from GBWWTP digesters based on biweekly sampling was relatively constant at between 61 to 62 %, with an average content of 61.4 %, as shown in **Figure 2**. A calibrated and verified BioWin model of the GBWWTP was applied to 2013 conditions and predicted a CH_4 content of 63.7 %, VSR of 54.0 %, and a biogas flow rate of 1279 m³/hour (field rate), which all agree very well with monitored 2013 conditions.

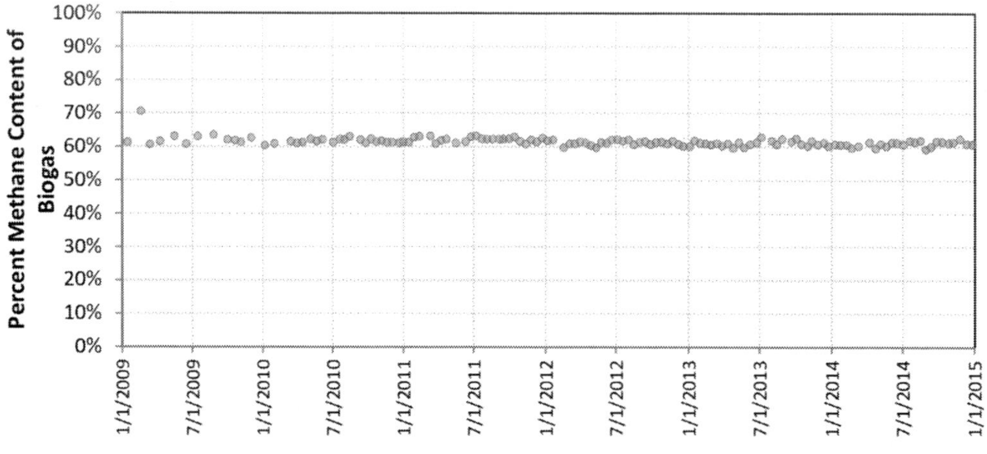

Figure 2: Monitored Percent Biogas Methane Content (2009 to 2014)

The average field measured volume produced was approximately 1280 m³/hour (2009 to 2014). Statistical analysis of hourly data found that the production can vary significantly over the course of a day, week, and month. **Figure 3** plots biogas production and illustrates its variability. The variation in biogas production underscores the need for temporary storage to attenuate the variation and maximize the amount available for subsequent use in a BTE technology.

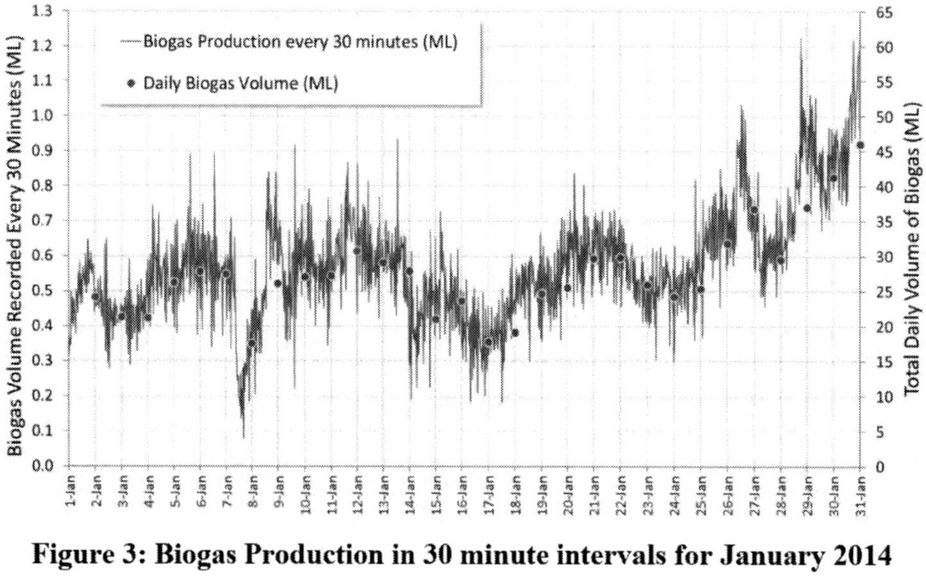

Figure 3: Biogas Production in 30 minute intervals for January 2014

Biogas volume is a field recorded value at a temperature of approximately 37 °C and a pressure of 134.7 ± 10.9 kPa. Adjusting the field biogas production of 11,200 ML to SATP (i.e., 101.325 kPa and 15 °C) yields a normalized biogas volume equal to13,830 NML/year or 1580 Nm³/hour on average. At 61.4% CH_4 the biogas has a gross and net heating value of 23.2 and 20.9 MJ/m³ (on a dry basis).

Contaminants in biogas are of concern when used as a fuel source for certain applications. Biogas produced between 2009 to 2014 was used to determine representative quality characteristics.

- Hydrogen Sulfide (H_2S) ranged from a high of 3400 to a low of 400 mg/L. Historical information indicated that H_2S was as high as 5000 mg/L and represent the upper limit for gas conditioning purposes.
- Carbon Dioxide (CO_2) content was quite stable between 29 to 40 % with a peak content of 45%. A peak CO_2 content of 45% was used as a conservative upper limit for calculating the Methane Number (MN) and for gas conditioning purposes.
- Biogas at 37 °C will be saturated with water and have a moisture content of approximately 44 mg/L @ 1 atm. Dehydration to the required moisture content is BTE specific.
- Siloxane, measured as Si, ranged from 0.5 to a high of 11.3 mg/dscm (milligrams per dry standard cubic meter). Limited data suggests that Si levels could exceed 12 mg/ dscm. To be conservative, a value of 15 mg/dscm was used as the upper range for gas conditioning purposes.
- Nitrogen (N) content, including ammonia, ranged from a high of 20,900 mg/L-N to a low of zero. To be conservative, it was assumed that 90% of the measured N was in the form of NH_3 (i.e., 22,840 mg/L as NH_3) and represent the upper limit for gas conditioning purposes.

GAS CONDITIONING

While raw biogas can be used in boilers, the H_2S, moisture and siloxanes negatively impact their operation and performance. Reducing the H_2S, moisture and siloxanes concentrations will significantly reduce boiler maintenance costs and reduce the rate of heat transfer efficiency degradation. Conditioning of biogas to reduce particulates, H_2S, moisture and siloxanes is typically required to meet "engine quality" requirements to minimize the degradation of engine components and reliable stable operation.

Further upgrading of biogas from engine quality to Biomethane, referred to a "precision cleaning", is required to achieve pipeline natural gas (NG) quality. Biomethane and renewable natural gas (RNG) are interchangeable terms in this paper. Precision cleaning or upgrading to RNG involves removal of carbon dioxide (CO_2), oxygen (O_2), nitrogen (N), H_2S, moisture, particulates, and siloxanes to very low levels. Some BTE technologies require a higher level of gas conditioning than other BTE technologies, along with specialized systems for stable and reliable operation.

Figure 4 schematically illustrates the interaction between digestion and biogas uses for the range of BTE technologies evaluated for GBWWTP.

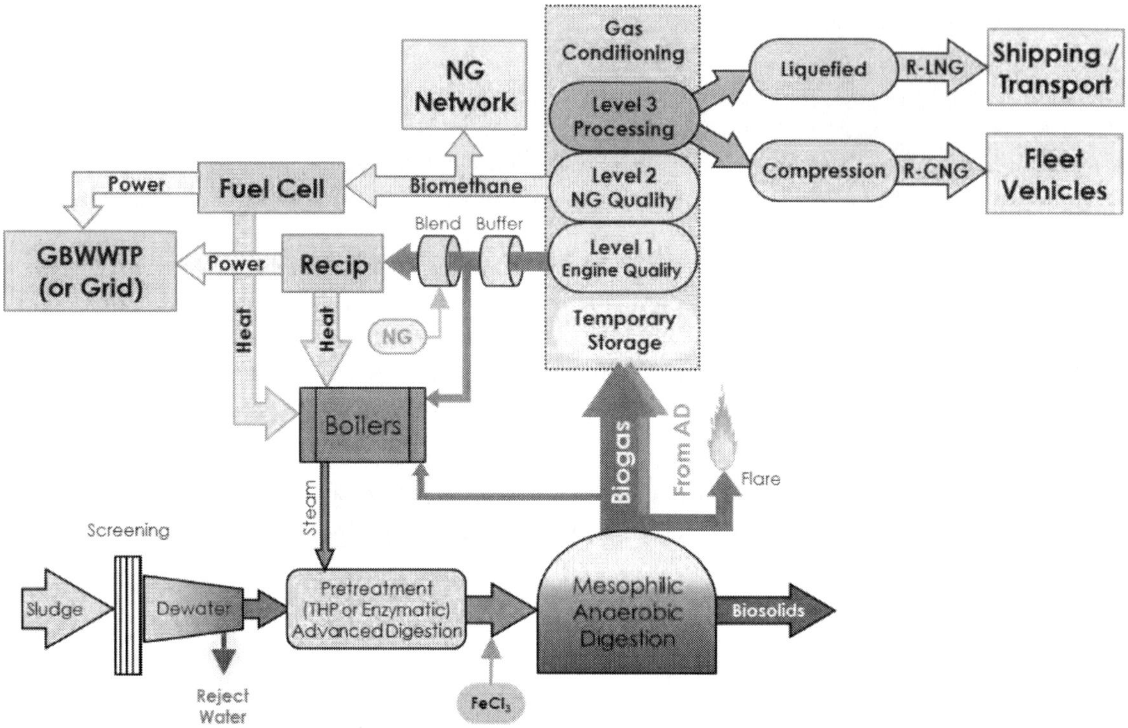

Figure 4: BTE Technology Options and Gas Conditioning Requirements

A wide range of technologies are available for gas conditioning. To meet engine quality requirements, it is essential to quantify the concentrations for the contaminants of concern, along with regulatory limits on the exhaust emissions. Several gas conditioning suppliers were contacted to obtain representative costs for the level of conditioning required to meet the tolerance limits for each BTE technology. A plethora of technologies exist to achieve the moisture, H_2S, and siloxane limits for engine quality requirements. Precision cleaning technologies are less common and typically employ pressure swing adsorption (PSA), water scrubbing, organic physical scrubbing, and chemical scrubbing. A limited number of vendors were contacted to obtain planning level capital and operating costs to process 1580 Nm^3/hour of biogas quality for specific BTE technologies, as presented below. Note, all costs presented in this paper are in 2016 Canadian dollars (CAD),

- **Eco-Tec BgPur™ Plus**
  - H_2S removal only
  - Equipment cost, $1.34 million (plus applicable taxes)
  - Operating cost, utility + chemicals, $75 K per year

- **Parker, PpTec**
  - Removal of moisture, particulates, and siloxanes
  - Equipment cost $1.06 million (plus applicable taxes)
  - Operating costs, utility + chemicals, $15 K per year

- **Puregas Solutions LLC**
  - Precision cleaning after H_2S, siloxane and moisture removal to achieve pipeline quality, Renewable Natural Gas (RNG)
  - Equipment cost, $4.0 million (plus applicable taxes)
  - Operating cost, utility and chemicals, $150 K per year

- **Molecular Gate**
  - Single process, full treatment to achieve pipeline quality, Renewable Natural Gas (RNG)
  - Removes: H_2S, Siloxane, CO_2, NH_3, VOCs, H_2O
  - Equipment cost, $4.4 million (plus applicable taxes)
  - Operating cost, utility and chemicals, $400 K per year

- **Mequipco Ltd.**
  - Gas Compression/Moisture Removal System (Provided by Unison)
    - Equipment cost, $ 860 K (plus applicable taxes)
    - Operating cost, utility and chemicals, $8 K per year
  - Regenerative Siloxane Removal System (Provided by Willexa/Built By Unison)
    - Equipment cost $ 980 K (plus applicable taxes)
    - Operating cost, utility and chemicals, $70 K per year

REVIEW OF BTE TECHNOLOGIES

1.0 Boilers

Untreated biogas generated from GBWWTP MAD digesters are collected and sent to fire-tube boilers to provide heat to appliances or equipment. To regulate digester pressure, biogas not used in the boilers is flared. The existing boiler systems were recently upgraded to operate on natural gas (NG) to provide an alternative fuel source when biogas supply was not sufficient to meet peak heating load demands. A dedicated piping system is used to convey hot water from boiler house #2 to heat exchangers to preheat the sludge and to keep the digesters operating at a near constant mesophilic temperature of 37 °C. Boiler house #1 provides the hot water needed to heat occupied spaced (e.g., buildings and tunnels).

The existing supply of biogas is saturated with moisture and contains high concentrations of H_2S and other trace contaminants (e.g., siloxanes) that can negatively impact the operation and performance of the boilers. Water and H_2S form an acid that can aggressively corrode metallic fire-tubes. Combustion of the raw biogas results in a silica-based particulate that coats the fire-tubes and over time reduces their heat transferred efficiency. The combination of corrosion and silicate build-up in the boilers necessitates annual servicing and maintenance repairs to keep the boilers operating reliably and efficiently. GBWWTP spends approximately $500 K per year to service the boilers. There are two (2) boiler facilities on site, each housing five (5) boilers, each with tank volume of 75 m^3 (265 ft^3). Each boiler has a capacity of 2.94 MW and can operate on biogas or NG.

Many wastewater treatment plants add a metal salt to the sludge fed to digesters to reduce H_2S concentration in the biogas. The amount of metal salt required to reduce H_2S can be approximated from stoichiometric relationships. The following example is based on Ferric Chloride ($FeCl_3$) since it is commonly used to reduce H_2S in biogas.

- $2\ FeCl_3 + 3\ H_2S \rightarrow Fe_2S_3 + 6\ HCl$
- 3.17 mg/L of $FeCl_3$ required per 1.0 mg/L of H_2S, or 1.16 mg/L Fe^{3+} per 1.0 mg/L of S

Iron (Fe) dosages estimated from stoichiometric relationships are theoretical values. Field trials are required to determine site-specific dosages since there are many other substances in the raw sludge that could react with the $FeCl_3$. Based on experience elsewhere, approximately 670 Tonne/year would be required to reduce H_2S from 2000 mg/L to 200 mg/L, on average. At a cost of $400/Tonne of chemical addition of $FeCl_3$ translates to approximately $270,000 per year.

Since GBWWTP does not have a dedicated dosing system for digesters H_2S reduction, a system would need to be added. The major components of the system would consist of:

- A receiving and transfer station;
- Appropriate temperature controlled storage system;
- Chemical dosing system;
- H_2S biogas monitoring equipment;
- HVAC for indoor equipment, appropriate health and safety equipment;
- Electrical, instrumentation and controls; and
- Staffing and administration requirements

The financial assessment of a metal salt dosing system is summarized below:

- Capital Cost: $1008 K
- Annual Chemical costs: $ 270 K
- Annual staffing and administration: $ 50 K
- Avoided cost of boiler maintenance: $ 400 K per year
- The payback in approximately 12.6 years
- Using a discount rate of 6.57 %, the net present value (NPV) over a 20-year period would be approximately $ 110 K indicating a weak but positive business case

The key merits and challenges for a metal salt dosing option are summarized below:

- Merits
  - Baseline for comparison
  - Optimization of boiler efficiency can reduce NG supplement
  - Reduced time and maintenance costs
  - Fe adds value to biosolids as a land applied product
  - Modest improvement in operational efficiency
  - Potential increase in biogas production
- Challenges
  - Addition of a $FeCl_3$ dosing system
  - Chemical acceptance, storage, handling, safety, staffing, and administration
  - Potential to overdose and negatively impact Ostara nutrient recovery process
  - Additional sludge mass and pumping to Clover Bar Sludge Lagoons

2.0 Combined Heat and Power (CHP)

A feasibility study was previously completed for GBWWTP and a reciprocating engine over turbines was found to have the best business case. CHP is the simultaneous production of electricity and heat from a single fuel source. Heat is produced as a by-product of electricity generation from the combustion of the selected fuel, conditioned biogas. Often the heat produced is equal to or greater than the electrical power being generated. It is the recovery of this heat that makes cogeneration a highly energy-efficient process. **Figure 5** illustrates the energy efficiency of a CHP system over conventional power generation (Source: IEA analysis, USEPA, 2008).

Reciprocating engines operating on biogas typically require pretreatment to reduce contaminants of concerns to meet engine manufacturers' fuel specifications. While the CH_4 content is sufficient for use as fuel, moisture, particulates, H_2S, and siloxanes are often present in elevated concentrations and can cause engine reliability problems. **Table 1** summarizes typical manufacturers' requirements for key contaminants and compared to recent biogas sample analysis (2009 to 2014 inclusive).

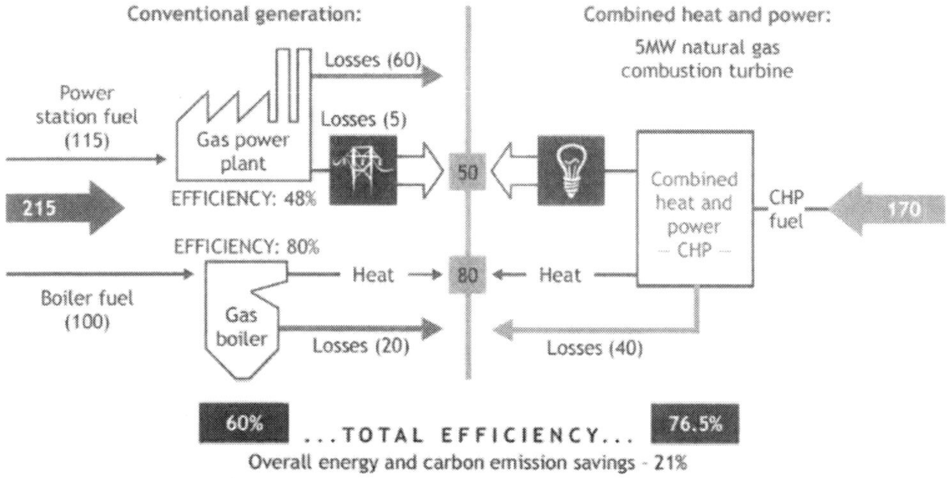

Figure 5: Example of CHP Efficiency Gains over Conventional Power Generation

Table 1: Maximum Allowable Biogas Fuel Contaminant Levels

Engine or Source	H_2S limit	Siloxane limit (as Si)	Water limit
Reciprocating Engines[a]	≤ 57 mg/MJ	≤ 0.56 mg/MJ	≤ 80% Relative humidity
GBWWTP Biogas	400 to 3400 ppm Ave. = 2882 mg/m³ Ave. = 138 mg/MJ[b]	0.4 to 11.3 mg/m³ Ave. = 9 mg/m³ Ave. = 0.43 mg/MJ[b]	Saturated

Notes: a - Requirements based on a Caterpillar G3500 Reciprocating Engine
b - Based on 61.4% methane in the biogas, LHV of 20.9 MJ/m³

Based on measured concentrations for contaminants of concern, gas conditioning for moisture, H_2S, and Siloxane will be required. For the feasibility study, representative costs were assembled for the required equipment as given in **Table 2,** and consist with planning level quotes provided by different vendors.

Table 2: Gas Conditioning Equipment and Costs

Item	Units	Equipment Cost	Install Cost	Total
LP gas compression, aftercooler, separator	2	$ 665,000	$ 240,000	$ 905,000
Chiller	1	200,000	90,000	290,000
Dehydration skid	1	378,750	72,000	450,750
Piping (304 SS)	1	425,000	220,000	645,000
Natural Gas and biogas mixing station	1	78,750	7,500	86,250
Analyzer (mass spec)	1	123,750	20,000	143,750
Instrumentation & Controls	1	62,500	25,000	87,500
H_2S bio-scrubber	1	1,487,500	120,000	1,607,500
Acid drain neutralization (skid)	1	75,000	2,000	77,000
Siloxane removal (regenerative)	1	781,250	120,000	901,250
Totals		**$ 4,277,500**	**$ 906,500**	**$ 5,194,000**

Several reciprocating engine options were investigated in the feasibility study. Containerized package system quotes were provided by vendors for MWM and Jenbacher engines. The MWM engine operating at 100% capacity can produce 1.56 MWe and has an electrical and thermal efficiency of 40.9% and 43.0%, respectively. At 50 % capacity this system can produce 0.78 MWe with an electrical and thermal efficiency of 37.7% and 45.8%, respectively. The broad efficiency range for this engine is well suited to ramp up and down with biogas fuel supply fluctuations. Similar performance was stated for the Jenbacher engine. The cost of three containerized package systems delivered to GBWWTP was estimated at $4.9 million. The cost for new switchgear/motor control centers and cabling was estimated at $2.5 million. The estimated annual operating costs for a CHP system are summarized in **Table 3.**

Table 3: Estimated Annual Operating Cost for a CHP system

Item	**Cost**
Labor Cost	$ 75,000
Service Contract by Generator Manufacturer	500,000
Owner's Maintenance Cost	150,000
Insurance/Property Tax	175,000
Management/Administration	25,000
Other Operating Expenses	15,000
Total	**$ 940,000**

An analysis of NG demand before and after the implementation of a CHP was conducted to provide an estimate of the change in annual NG consumption. The analysis indicated that a reduction of ~ 7,700 GJ of NG energy could be realized with the implementation a CHP system. The implementation of temporary storage (i.e., methane storage spheres) along with better controls to minimize flaring are responsible for the reduced NG consumption. Based a retail supply cost of $3.95 per GJ of NG, the NG savings would be approximately $30 K annually. Based on a total power cost of $100 per MWh, the estimated power savings from generating approximately 35.5 GWe of power per year would be ~ $ 3.55 M. Anticipated upgrades to the transmission and distribution lines is expected to increase the price of grid power. A 2 % annual increase was allowed for in future power costs. Conversely, the abundance of NG on the market along with its sustained production is expected to stabilize NG pool prices and not increase supply costs as quickly as power costs. That is, the "spark gap" will grow favoring onsite power generation. **Table 4** summarizes the items used to develop the business case for reciprocating engine.

Table 4: Summary of Construction Cost for a Reciprocating Engine CHP System

Item	**Cost**
Site work	$1,332,000
Building	2,738,000
Gas Conditioning	5,194,000
Mechanical and gas storage	3,365,000
Instrumentation & Controls	486,000
Electrical Install & Wiring	2,901,000
Subtotal	**$ 16,016,000**
General Conditions	2,499,000
Allowances for profit, contingency, escalation, bonds, admin., etc.	4,782,000
Subtotal	**$ 23,297,000**

Reciprocating Engines and Switch Gear (Owner furnished)	7,473,000
Total Construction Costs	**$ 30,770,000**
Approvals and site investigations	785,000
Engineering Design	3,400,000
Total	**$ 34,955,000**

The financial assessment of a reciprocating CHP system is summarized below:

- Capital Cost: $34.995 M
- Annual operating cost: $ 940 K
- Power savings: $ 3.55 M per year, escalating at 2% per year
- NG savings: $ 30.4 K per year
- Greenhouse gas credits, $ 227 K for the first eight years
- The payback is approximately 11.0 years
- Using a discount rate of 6.57 %, the NPV over a 20-year period would be approximately $ 2.9 M indicating a positive business case.

The key merits and challenges for a CHP system are summarized below:

- Merits
  - Effective utilization of a renewable resource to improve operational efficiency
  - Can reduce NG consumption and biogas flaring
  - Offset power costs
  - Positive and sound business case
  - Connect to boiler house #2 to provide sustained heat for digester operations
  - Increased resilience to NG and power costs escalations
  - Expandability of system
  - Can outsource equipment maintenance
  - Option to sell excess power to the power grid
  - Proven technology widely practiced elsewhere
  - Containerized systems are designed with service in mind for minor and major overhauls.
- Challenges
  - Limited space near digesters to implement a CHP system
  - Reliable biogas conditioning system to consistently meet engine manufacturer contaminant tolerance limits
  - Temporary biogas storage
  - Supplemental supply of NG to meet power demands
  - Power generation verse heat loads trade-off
  - Transformation and switch gear to seamlessly integrate power
  - Buildings to house certain equipment
  - Availability of containerized package systems for desired CHP size
  - Maintenance, administration, noise, emissions, approvals, regulatory, public

3.0 Biomethane, Renewable Natural Gas (RNG)

An alternative to on-site CHP generation via reciprocating engines fueled by conditioned biogas is to precision clean the biogas to as higher standard and inject into the NG pipeline. Biogas upgraded to NG quality is referred to as Biomethane, a renewable natural gas (RNG). **Table 5** provides a quality summary of the NG supplied by ATCO Gas to the GBWWTP based on their material safety data sheet and Gas Quality Specifications for the TransCanada pipeline and other pipelines.

The biogas treatment process used to condition the biogas for use in a reciprocating engine as configured for the GBWWTP would require additional purification to meet pipeline NG quality standards. Specifically, after the H_2S, moisture and siloxanes have been initially removed from digester biogas, there are several processes that that can be added to remove CO_2 and moisture. In addition, depending on the biogas moisture, H_2S, N and O_2 concentrations, these substances might also need to be reduced and could influence the choice of precision cleaning technology to meet pipeline NG quality standards, as presented in **Table 5**.

Table 5: ATCO Gas NG Quality Standards

Components	% by Volume
Methane (CH_4)	95
Ethane & Heavier Hydrocarbons	2
Heating Value	Min. 36 MJ/m^3
Nitrogen	Max. 2% by volume
Oxygen	Max. 0.4 % by volume
Carbon Dioxide	Max. 2% by volume
Water	Max. 65 mg/m^3 (1.8 mg/MJ)
Hydrogen Sulfide	Max. 23 mg/m^3 (0.64 mg/MJ)
Total Sulphur	Max. 115 mg/m^3 (3.2 mg/MJ)
Siloxane	Not reported

One of the purification methods with the longest proven history and performance is compressing the biogas and then absorbing the carbon dioxide in water in a high pressure tank. This method requires high pressure equipment and has a CH_4 recovery rate of about 90%. Pressure swing absorption (PSA) is another possible technology offered. The Molecular Gate PSA system is a single stage absorption system and has recovery rate of 92%. An amine absorption process from Purac claims to have a methane recovery rate > 99 % and a CO_2 removal rate > 99 %. The costs provided by the vendors do not reflect the same level of detail provided in **Table 2**. Elements from **Table 2** have been added to the Molecular Gate PSA technology to provide a more complete cost, along with allowances as present in **Table 6**.

Table 6: Biogas Conditioning to NG Quality.

Item	Units	Equipment Cost	Install Cost	Total
Piping (304 SS)	1	$ 425,000	$ 220,000	$ 645,000
Natural Gas and biogas mixing station	1	78,750	7,500	86,250
Analyzer (mass spec)	1	123,750	20,000	143,750
Instrumentation & Controls	1	62,500	15,000	77,500
Molecular Gate PSA system	1	2,900,000	1,450,000	4,350,000
Shipping	1		100,000	100,000
Tail Gas Flaring	1	200,000	100,000	300,000
Subtotals		**$ 3,790,000**	**$ 1,912,500**	**$ 5,702,500**
Estimating allowance (30%)		1,137,000	573,750	1,710,750
Contingency (10%)		379000	191,250	570,250
Totals		**$ 5,306,000**	**$ 2,677,500**	**$ 7,983,500**

GBWWTP is close to a large network of refineries. Precision cleaning biogas to pipeline quality and selling it to either the local utility (ATCO Gas) or a local customer willing to purchase RNG as an alternative to NG is a possibility, but would require negotiations, long-term contracts, and reliable precision processing of the biogas to meet pipeline quality specifications. If ATCO Gas were willing to accept RNG produced at GBWWTP, an injection system along with quality and flow monitoring system would be required. A potential obstacle to accepting RNG is the willingness of the utility or customer to accept new trace contaminants in the supply, (e.g., siloxane).

Moreover, imprecise conditioning of the biogas could result in RNG being off-spec and unacceptable by the consumer. This situation would require flaring of off-spec RNG, and represent a revenue loss reflected as an additional operating cost. The cost of negotiations has not been included since it is not possible to gauge the level of effort required to realize a long-term agreement.

For this option it has been assumed that onsite storage is not required but rather a small buffering tank employed to help control the injection of RNG into the pipeline at the proper pressure. An injection station will need to be constructed to monitor the quality, compress the RNG, and accurately measure the volume and quality injected into the pipeline. Off-spec RNG will be directed to a flare system for destruction. The estimated cost to construct an injection station is ~ $1.4 million.

A fundamental consideration in the production of RNG for sale is the consideration of onsite heating needs. At GBWWTP, heating loads increase in the colder months leaving only a portion of the year when biogas production exceeds onsite demands. Year 2014 was a lower than average biogas production year based on the last five years and represents a conservative case for economic assessment. After biogas systems optimization, 53 % of the annual biogas produced could potentially be available for upgrading to RNG based on 80% utilization efficiency in the boilers. For current conditions, an average hourly volume of 1580 Nm^3 of biogas is produced with a net heating value of 20.9 MJ/m^3. Processing 53% of the annual production would correspond to 7.33 million Nm^3, and 150 TJ of energy. Assuming 95 % precision performance and a 95% CH_4 capture efficiency, 138 TJ could be sold for profit. Assuming it is possible to get at least 90% of the current retail cost NG priced of $ 3.95/GJ, a net annual revenue of approximately $ 490 K could be realized.

Table 7 summarizes estimated annual operating costs to inject RNG into the NG pipeline.

Table 7: Estimated Annual Operating Cost for a RNG system

Item	**Cost**
Labor Cost	$ 75,000
O&M + Power[1]	240,000
Insurance/Property Tax	175,000
Management/Administration	25,000
Residuals Handling and Disposal	25,000
Total	**$ 540,000**

*Notes: 1. Cost prorated based on amount of biogas processed (344*0.53+58 ~240)*

Since the annual operating cost exceeds the potential revenue generation, a negative return on investment would result from the implementation of this BTE option. The main operating cost impacting the feasibility of a RNG system is the electrical cost to process the biogas to NG quality. With the anticipated escalation in power cost over NG cost, an RNG system could become even less attractive from an economic perspective. Due to the poor business case for this option, the injection of RNG into the NG pipeline is not be recommended for consideration.

4.0 Compressed RNG

Review of experience elsewhere finds that compressed RNG (C-RNG) has been effectively used in vehicles as an alternative fuel where the cost of petroleum-based fuels is significantly more costly than C-RNG. Government incentives (or penalties) along with environmental stewardship initiatives towards the use of renewal fuels has commonly been considered in "triple bottom line" analyses. The analysis requires a scoring system to be established prior to the ranking of alternatives to prevent a bias and skewing the scoring process. The scope of the current assignment did not allow for such an analysis, only economics and the major benefits and drawbacks were considered in BTE options comparisons.

Biogas must be conditioned to NG standards for its use in spark ignition engines used in fleet vehicles. Additional factors to consider in the use of the RNG in fleet vehicles.

- Compression
- Onsite storage
- Hauling to a central station for dispensing
- Engines designed to use R-CNG
- Maintenance garage to service the fleet vehicles using R-CNG
- Fuel consumption, and distance per tank of R-CNG
- Cost of diesel avoided.

Many challenges exist with the construction a dispensing station at GBWWTP. Assuming all challenges were successfully resolved, biogas not used in the boiler system could be made available for fleet vehicle use. The mostly likely and plausible scenario would be to haul the C-RNG to a central location where the C-RNG could be dispensed efficiently to fleet vehicles. Edmonton's "sister" city Calgary, recently invited prequalified parties to bid on the Stoney CNG Bus Storage and Transit Facility. The facility will be capable of housing up to 400 Compressed Natural Gas buses, including indoor vehicle storage, 40 maintenance bays, two steam bays, on-site natural gas compression, Compressed Natural Gas storage distribution infrastructure, and administrative facilities. The estimated capital cost for the project is $200 million. A new smaller CNG facility in Edmonton to house 100 C-RNG buses could conceivable range between $ 80 to 100 million. In such a scenario, all labour and maintenance costs would be borne by a third party (e.g., City of Edmonton) to own and operate new C-RNG facility.

For this option, a dispensing system at GBWWTP consisting of a high pressure compressor, storage tanks, and a loading system onto tanker trucks would be employed to process and delivery C-RNG to a central C-RNG facility. The installed cost for a dispensing system at GBWWTP would be in the order of $2.2 million with an operating and maintenance costs of approximately $350,000 per year. For infrastructure development, an allowance of $500 K was included. Depending on the size of onsite storage, facility might need to be operated seven days per week. It is envisioned that at least two full time workers would be required to operate and maintain the new storage and dispensing facility

A large number of C-RNG fleet vehicles would be required for this option. Since a large number of busses and refuse collection trucks exist in Edmonton, an opportunity exists for them operate on R-CNG. Diesel engines can be converted to operate on C-RNG and reported to cost in the $ 90 K range per engine. Conversions can be problematic and not a preferred solution. New vehicles can be purchased with engines that are designed to operate on C-RNG but incur a cost premium of approximately $ 60 K per vehicle. The premium for buses or refuse truck operating on C-RNG would be borne by the owner and costs recovered by the fee charged in their operation.

The energy value of C-RNG is lower than both diesel and regular gasoline which requires more C-RNG to be used to travel the same distance. As such, the driving range of the vehicle would be reduced if the same tank size was used and require more frequent refueling, or replaced with larger tanks to increase the driving range. To assess the cost-competitiveness of C-RNG, it needs be converted to fuel equivalents, as follows.

- Diesel equivalent (DE), 1 L of C-RNG L = 0.729 L of diesel
    - The cost of 1 L of diesel peaked at ~ $1.40 and currently sells at $ 0.92
        - Diesel has an energy value of 35.5 MJ/L
        - The current cost of NG is $ 3.95/GJ,
        - Assume the compression of NG adds a 50% premium, CNG = $ 5.90/GJ
    - After processing, 138 or 261 TJ is available as C-RNG
    - 138 TJ = 3.88 ML (DE) (53 % of biogas produced)
        - CNG cost for 138 TJ = $ 0.81 M
        - Diesel cost for 3.88 ML = $ 3.57 M
        - Savings, $ 2.76 M per year
    - 261 TJ = 7.34 ML (DE) (100 % of Biogas produced)
        - CNG cost for 261 TJ = $ 1.54 M
        - Diesel cost for 7.34 ML = $ 6.76 M
        - Savings, $ 5.22 M per year

The current CNG pump price in Edmonton is $ 0.75/L with pricing closely related to the energy values of the fuels. The compression to CNG in the foregoing assessment assumed cost of $ 5.90/GJ, which is the equivalent of $ 0.20/L of C-RNG. Using $5.90/GJ for C-RNG produced at the GBWWTP, the annual revenue would be $ 816 K or $ 1.54 M, as noted above. Assuming it is possible to sell the product to the vendors at $0.40/L, the annual revenue generated would be $ 2.13 M or $ 4.03 M, respectively. It is important to note that the economics is very sensitive to the sale price of C-RNG. At or below $0.30/L, the revenue would not be sufficient to provide a positive business case.

The estimated annual operating costs to produce C-RNG are summarized in **Table 8.**

Table 8: Estimated Annual Operating Cost for a C-RNG system

Item	Cost
Labor Cost (2 FTE)	$ 200,000
O&M + Power[1]	752,000
Insurance/Property Tax	175,000
Management/Administration	25,000
Residuals Handling and Disposal	25,000
Total	**$ 1,052,000**

Notes: 1. Cost based on processing all the biogas (344+58+350)

The financial assessment of C-RNG systems are summarized below:

Case 1: Only process and sell biogas not used in boilers.

- Capital Cost: $10.7 M
- Annual operating cost: $ 890 K
- Power escalating at 2% per year
- Revenue from C-RNG sale: $ 2.13 M per year
- The payback period is approximately 8.7 years
- Using a discount rate of 6.57 %, the NPV over a 20-year period would be approximately $ 2.87 M indicating a positive business case, on par with the CHP option.

The estimated annual cost to use only NG in the boilers is $514 K if all biogas was process to C-RNG and included in Case 2.

Case 2: Process all the biogas and use NG in the boilers as required.

- Capital Cost: $10.7 M
- Annual operating cost: $ 1.57 M
- Power escalating at 2% per year
- Revenue from C-RNG sale: $ 4.03 M per year
- The payback period is approximately 4.4 years
- Using a discount rate of 6.57 %, the NPV over a 20-year period would be approximately $ 16.2 M indicating a strong and positive business case, better than a CHP option.

While Case 2 presents a strong business case, the successful resolution of several challenges would be required to implement a C-RNG system at the GBWWTP. The business case is premised on the assumption that a third party would implement a CNG facility and bear the full cost of implementation and operation. A long-term agreement would need to be established with a stable third party for the supply of C-RNG. The processing of C-RNG on site will require specially trained staff to operate and maintain C-RNG facility. The C-RNG would be hauled to the CNG facility. Moreover, the loss of resilience and increased exposure to NG price escalation. Many of these factors are not within the control of EPCOR and represent major risks to the business case.

The key merits and challenges for a C-RNG system are summarized below:

- Two cases assessed for this option:
  1) Process biogas not used in boilers,
  2) Process all biogas and buy NG for all onsite heat loads.
- Assumptions:
  - A third party would buy the product and dispense it to fleet vehicles
  - Sell C-RNG at $ 0.40/L (approximately half of current pump price)
    - below $ 0.30/L business case becomes negative.
  - C-RNG hauled from GBWWTP to CNG facility owned and operated by others
- Merits
  - Case 2) economics suggests a strong business case but risky
  - Case 1) economics positive but not as strong as reciprocating engine CHP.
  - Renewable fuel displacing regular gasoline or diesel
  - Cleaner burning with lower emissions
  - Longer engine life, less frequent oil changes
  - Lower fuel operating cost for same distance travelled

  - Better engine performance than petroleum-based fuels, higher octane rating, less knocking and superior starting under severe weather conditions, hot or cold.
- Challenges
  - Conditioning to NG standards, off-specification product flared
  - Onsite storage and transfer to tanker trucks to haul C-RNG to dispensing station
  - Land availability at GBWWTP to implement a conditioning and compression facility, tank farm for temporary product storage, transfer system to load product on tanker trucks.
  - Operate 7 days per week, 24 hours per day to process biogas as it is produced.
  - Fill time for tankers or individual fleet vehicles increases as the supply is depleted
  - Safety considerations for onsite storage of a compressed (20 to 25 MPa) explosive gas, tank sizing, weatherizing, inspections, etc.
  - Operating staff and security
  - Increased truck traffic to and from GBWWTP
  - Road restrictions, hours of operations
  - Resistance by the local community.
  - Permitting, zoning, bylaws, regulation, certification, etc.
  - Stable third party to purchase and dispense C-RNG, long-term agreement
  - Exposure to NG price escalation if all biogas is processed and sold as C-RNG

5.0 Liquefied RNG (L-RNG)

An alternative to C-RNG is to liquefy RNG to a denser state. There are several factors that influence choosing liquefied RNG (L-RNG) over C-RNG system.

- Vehicle type and application
- Operating requirements (e.g., range, speed, hours of service, idle/down time)
- Payload and storage tank weight
- Fueling infrastructure, and route planning.
- Shipping to remote locations or overseas

It is important to distinguish between L-RNG and C-RNG storage requirements. L-RNG is in a cryogenic liquid form stored at –160 °C and stored at 1 atmosphere of pressure. C-RNG is stored in pressurized tanks (20 to 25 MPa) that can operate within a broad range of ambient temperatures. Since more energy is contained in L-RNG, it is better suited to heavy-duty vehicles travelling > 400 km/day and operating 5 or more days per week. L-RNG can be rapidly dispensed to fill large tankers, or refuel trucks. Moreover, L-RNG can be transported great distances, including overseas shipments to deliver the product where pipeline transport is not possible.

Low oil prices have negatively impacted the demand for L-RNG, which has caused the selling price of L-RNG to drop sharply. As of May 2016, the landed price of L-RNG on a global scale has ranged between $2.0/GJ ($1.51 USD per Million BTU) and $5.8/GJ ($4.32 USD per Million BTU). The average landed price was $ 4.9/GJ (3.63 USD per Million BTU). Costs will be incurred to get the product to market. To allow for this expense 75% of the average landed price, $3.7/GJ was used as the selling price of L-RNG. Since the selling price is less than the current retail cost of NG at $3.95/GJ only the biogas flared is considered for this option. The annual revenue from L-RNG would be approximately $ 511 K.

The cost to construct a liquefaction facility based on recent projects is \$400 to \$1200 per Tonne of CH_4 per year. Based on a measured CH_4 density of 0.68 kg/m³ and an average daily production of 37.9 ML/d, the estimated capital cost for a liquefaction system would be approximately \$3.8 to \$ 11.5 million, use \$7 million was used. The annual operating and maintenance cost of a liquefaction facility was assumed to be on the order of 2 % of the capital cost, or \$140,000. Based on a recent benchmarking report (LNG Emissions Benchmarking Prepared for: BC Climate Action Secretariat March 2013) approximately 450 to 500 GWh of energy is required per million Tonnes of LNG produced. Approximately 9500 Tonnes of LNG could be produced from all the biogas currently generated from GBWWTP digesters. The corresponding annual energy requirement would be approximately 4.8 GWh. The prorated energy cost to process the biogas that would otherwise be flared is 2.6 GWh, or approximately \$ 216 K/year. The estimated annual operating costs to produce LNG at GBWWTP are summarized in **Table 9**.

Table 9: Estimated Annual Operating Cost for a L-RNG system

Item	Cost
Labor Cost (2 FTE)	\$ 200,000
O&M + Power[1]	414,000
Insurance/Property Tax	175,000
Management/Administration	25,000
Residuals Handling and Disposal	25,000
Total	**\$ 839,000**

*Notes: 1. Cost prorated based on amount of biogas processed (4800*0.53*85+58 +140)*

Since the annual operating cost exceeds the potential revenue generation, a negative return on investment would result for this BTE option. The implementation of an LNG system at the GBWWTP is not recommended since it has a negative business case and other non-financial drawbacks

6.0 Fuel Cell CHP

Fuel cell technology is typically not cost competitive with other CHP technologies due its high capital cost, gas conditioning requirements, and life cycle (i.e., replacement timeframe). Due to limited installation history and operation on RNG, use of fuel cells is not considered a fully proven technology. Nonetheless, for completeness it is included in the evaluation of possible BTE technologies.

A 2015 study (Catalog of CHP Technologies, U.S. Environmental Protection Agency Combined Heat and Power Partnership, March 2015), reported that the capital cost to implement fuel cells was \$5000 to 6500 USD/kWh, and had a maintenance cost of \$ 0.032 to 0.038 USD/kWh. While the reciprocating engine had a lower electrical efficiency their overall efficiency was between 75 to 80 % indicating a consistently higher degree of overall performance than fuel cells. Fuel cell systems have inherently low emissions because the process does not involve combustion. Since fuels cells produce power via electrochemical reaction, the power generation process is quiet, unlike reciprocating engines which generate considerable noise and vibration. The capital and operating costs for the reciprocating engine option was adjusted as noted below and used in the economic analyses to assess its business case.

- Capital cost:
  - Increase of \$ 16.1 M for fuel cells over reciprocating engine cost
  - Increase of \$ 2.8 M for Gas conditioning to achieve NG quality
- Operating Costs
  - Annual Maintenance cost
    - Increase of \$ 860 K for fuel cells

- Increase of $ 58 K for Gas conditioning equipment
- Increase of $ 214 K for power associated Gas conditioning

The financial assessment of a Fuel Cell CHP system is summarized below:

- Capital Cost: $53.9 M
- Annual operating cost: $ 1850 K
- Power savings: $ 3.34 M per year, escalating at 2% per year
- NG savings: $ 30.4 K per year
- GHC Credits, $ 227 K for the first eight years.
- The payback period is > 20 years
- Using a discount rate of 6.57 %, the NPV over a 20-year period would be ~ $ -29.1 M indicating a very negative business case.

While fuel cell technology has positive attributes of very low emissions and quiet operation, it has a very negative business case. Its high energy capacity, clean and quiet power are its key attributes. Reduction of capital costs in the future could make fuel cell CHP an attractive option but at present it is not recommended due to its poor business case.

SUMARRY

An evaluation of alternative BTE technologies finds that a reciprocating engine CHP solution to be best option to improve operational efficiency at GBWWTP. The evaluation confirmed that a reciprocating engine option is the most practicable solution and has a positive business case. However, processing all the biogas to NG standards and compressing it for possible use in fleet vehicles has the potential to have a stronger business than a reciprocating engine option, but does not appear as practicable, would face major implementation challenges (e.g., safety, staffing, approvals, and public acceptance), and relies on a third party to purchase and dispense the compressed natural gas. All other options had poor or negative business cases and accordingly were not recommended. **Table 10** summarizes the economic evaluation of the BTE alternatives reviewed.

Table 10: Economic Summary of Alternatives Reviewed.

Biogas-to-Energy (BTE) Technology	Capital Cost	O&M Cost	Revenue	Payback (years)	NPV[5]
Boilers[1]	$ 1.0 M	$ 270 K	$ 500 K	12.6	**$ 110 K**
Reciprocating Engine CHP[2]	$ 35.0 M	$ 940 K	$ 3580 K	11.0	**$ 2,900 K**
Renewable natural gas (RNG)[3]	$ 9.4 M	$ 540 K	$ 490 K	never	$ -10,300K
(a) Compressed RNG[4] (C-RNG)[3]	$ 10.7 M	$ 890 K	$ 2100 K	8.7	**$ 2,900 K**
(b) Compressed RNG[4] (C-RNG)[3]	$ 10.7 M	$ 1050 K	$ 4030 K	4.4	**$ 16,200 K**
Liquefied RNG (L-RNG)[3]	$ 15.0 M	$ 840 K	$ 510 K	never	$ -19,000 K
Fuel cell CHP[3]	$ 53.9 M	$ 1850 K	$ 3370 K	> 20	$ -29,000 K

Notes:

1. *H_2S treatment only*
2. *H_2S, H_2O, and Siloxane treatment*
3. *Full treatment to achieve NG quality standards (e.g., CO_2 reduction)*
4. *(a) Use only biogas that would otherwise be flared, (b) Use all biogas*
5. *NPV based on a discount rate of 6.57% and 20 years, electricity cost escalated at 2% per year*

Based on the merits and challenges associated with all options reviewed, a reciprocating engine CHP solution is the most-cost effective option that can be practically implemented at GBWWTP for current conditions. In addition, a reciprocating engine CHP system has the flexibility to be expanded readily in the future to capitalize on increased biogas production resulting from increased sludge generation due to population growth or implementation of advanced digestion options.

Stirling Engine for Biogas Cogen – A Case Study of Helena MT

Brian Mitchell, Keith Albretson, Michael Moe
WesTech Engineering, Inc.
3665 S. West Temple
Salt Lake City UT 84107

ABSTRACT

This presentation will highlight Helena MT WWTP, which is the only WWTP in North America with a stirling engine generator. Helena is running the engine on unscrubbed biogas. The presentation will provide an overview of what a stirling engine is and further discuss how it was implemented in Helena. Small facilities would normally be deterred by the cost of gas scrubbers and the significant O&M costs of traditional CHP solutions. As a stirling engine does not require scrubbing it allows facilities a more productive option than simply flaring the gas. The electricity can be used to power part of the plant, or be connected back to the grid. The hot water output can be used to supplement the digester boiler or as radiant building heating. This technology is ideal for plants with a renewable energy focus and can take advantage of existing incentives and grants.

There are many small wastewater treatment plants that would like to convert their biogas into heat and power, but due to economies of scale have been unable to do so. Gas scrubbing was cost prohibitive for the small plant of Helena MT but the Cleanergy Stirling engine is less sensitive to H_2S and siloxanes than traditional CHP systems which allowed Helena MT to move forward with the technology. The Cleanergy Stirling engine also has the potential to burn low content methane gas. Waste streams from biogas scrubbing or aerobic lagoons have methane concentrations at 20-25%, which requires natural gas supplementation to flare. The Cleanergy Stirling engine can burn the waste stream without the need of a natural gas supplement.

KEYWORDS

biogas, waste to resources, renewable energy, CHP, sustainability, Cogen

STIRLING ENGINE BACKGROUND AND INTRODUCTION

The stirling engine was invented in 1816 by Robert Stirling. It was originally designed as a solar thermal electric machine, but has since been used with many heat sources including biogas. [1.]

Figure 1. M. Frey (2014). "Pump Design Stirling Engine" [2.]

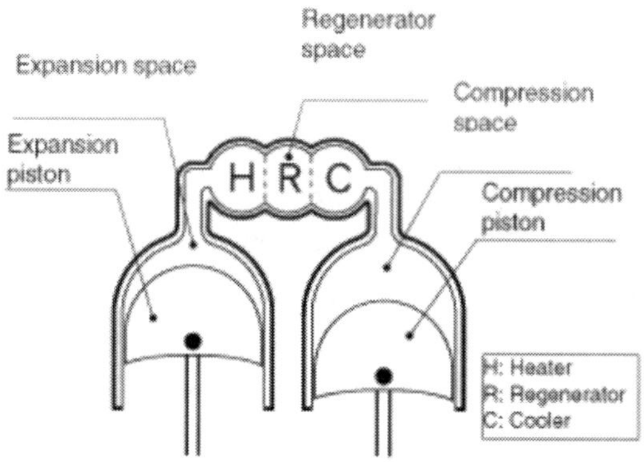

Fig 2. K. Hirata (1997). "Schmidt Theory For Stirling Engines" [3.]

The external combustion heats up the expansion space "H" and depresses the expansion piston. The compression piston is pushed back which introduces the working fluid through the regenerator space "R" and then to a cooling source "C" which will in turn compress the gas. The compression piston pushes the expansion piston back to repeat the cycle again. [3.]

The working gas between the expansion space, regeneration space, and compression space is usually Hydrogen or Helium. The system is a closed system, meaning the combustion and air fuel mixer will never touch the pistons or moving parts of the engine. [4.]

As an external combustion engine, it has the advantage of using a wide variety of fuel sources of varying quality and heat content. Since the internal combustion engine was more practical for automobiles, the stirling engine hasn't been as popular in recent history. However, its unique properties including reliability, efficiency, and low-emissions make it an ideal solution for small to mid-size CHP facilities using biogas.

Wastewater Treatment Application

Digestion facilities produce biogas as part of their solids reduction process. The biogas is made up of mainly methane (CH4) at 65-70% and carbon dioxide (CO2) at 25-30%, with trace amounts of nitrogen, hydrogen, hydrogen sulfide (H2S), siloxanes (Si-O-Si compounds), and oxygen. [5] When biogas is run in an internal combustion engine or microturbine the hydrogen sulfide and siloxanes are particularly damaging. They cause corrosion and buildup in the intricate moving parts of the engine that require engine overhauls and head replacements. For internal combustion engines and microturbines to use biogas as an energy source the biogas must be scrubbed and dried to protect their inner workings. This is an involved and expensive process and is cost inhibitive for small plants. As a stirling engine is less sensitive to contaminants like hydrogen sulfide and siloxanes, eliminating expensive gas scrubbing is a possibility. In Wastewater Treatment plants the cooling source for the stirling engine is water, which has a byproduct of hot water. This hot water can be used to heat the digester sludge, a warehouse, or another process that requires hot water.

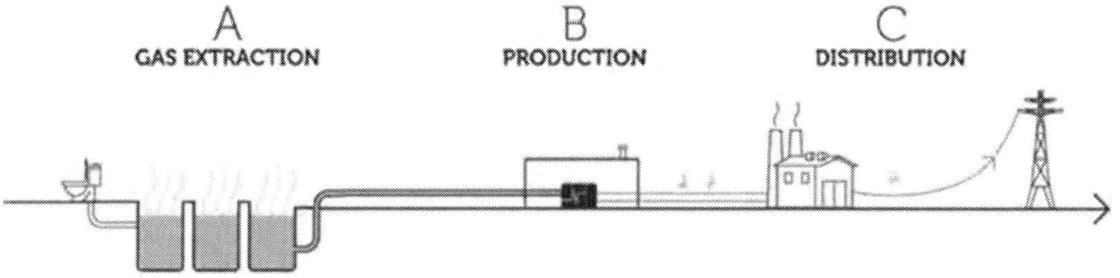

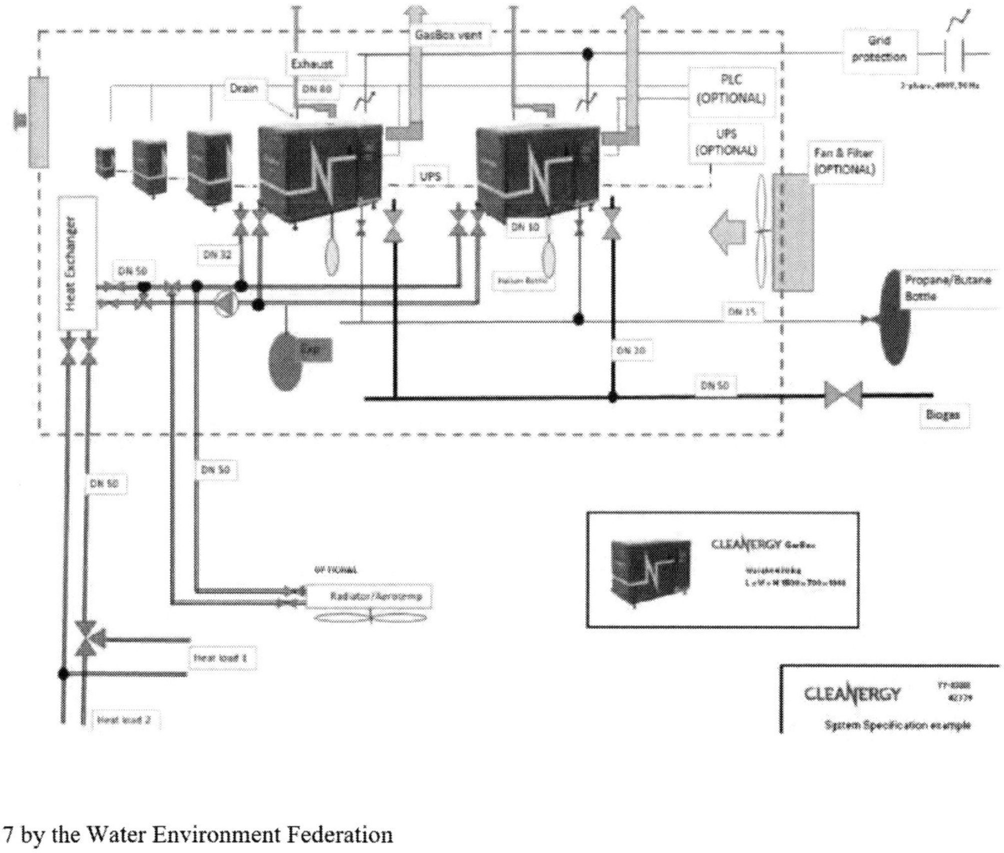

Dimensions and Performance

The Cleanergy units are relatively small, less than 5 feet long and 3 feet wide. They are self-contained units with built in control panels that can run alone or paired with others depending on the amount of gas flow. Each unit produces 7.2 kW of electricity on 2.6 SCFM of biogas. The burner principles allow for methane concentrations as low as 18% which allow the unit to be used in unique applications like covered lagoons and waste gas streams from gas scrubbers.

GasBox General Dimensions	
GasBox Dimensions (Length x Width x Height)	59"x27"x40"
Clearances (Front, Side, Top, Back)	40"x32"x36"x16"
GasBox Weight	992 lbs

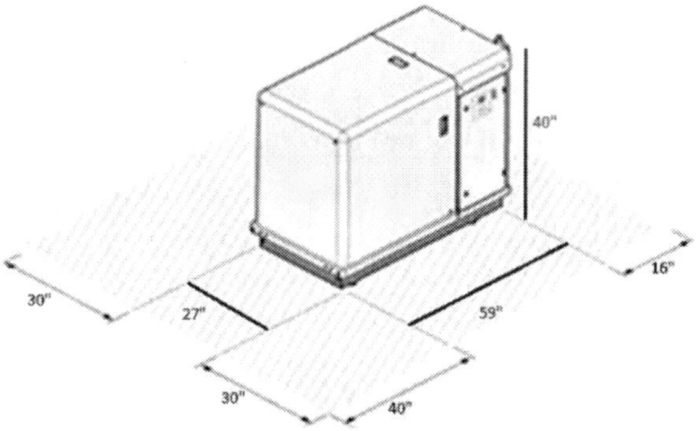

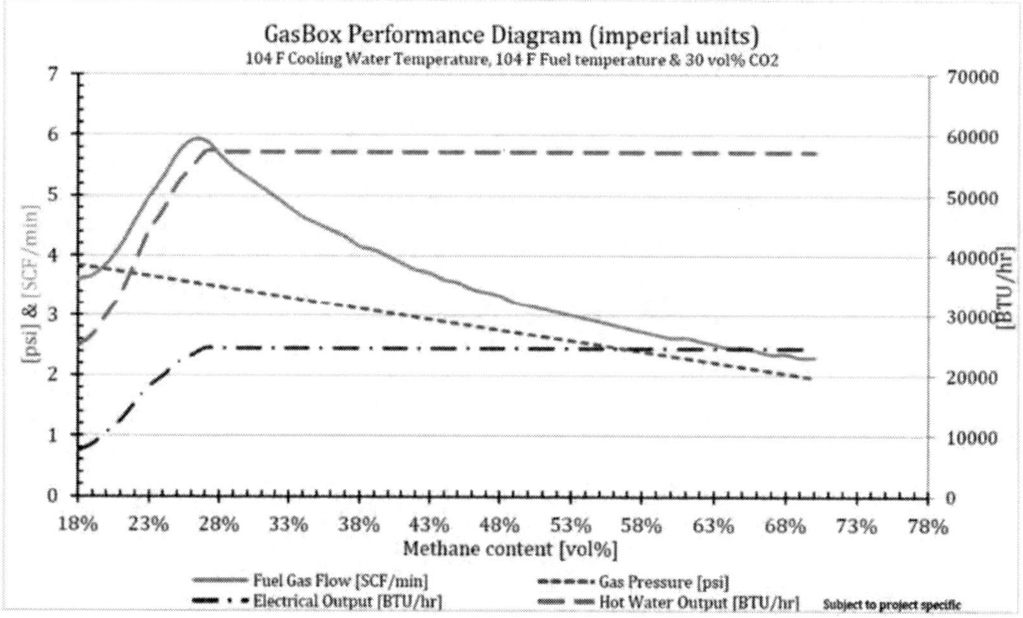

Combustion Principals

The burner is optimized for combustion of different gaseous fuels, even with low energy content. The basic principle is MILD combustion (Moderate or Intense Low oxygen Dilution). The burner has low oxygen and CH4 levels because of a higher inlet temperature which is achieved through Combustion Gas/heat Recirculation (CGR). The inlet temperature is between 900-1200k and the CGR is between 1100-1400K. The MILD combustion and CGR result in a high efficiency engine (70% overall efficiency compared to 35-40% efficiency for conventional combustion), low NOx and Co emission, and low noise (65 dB).

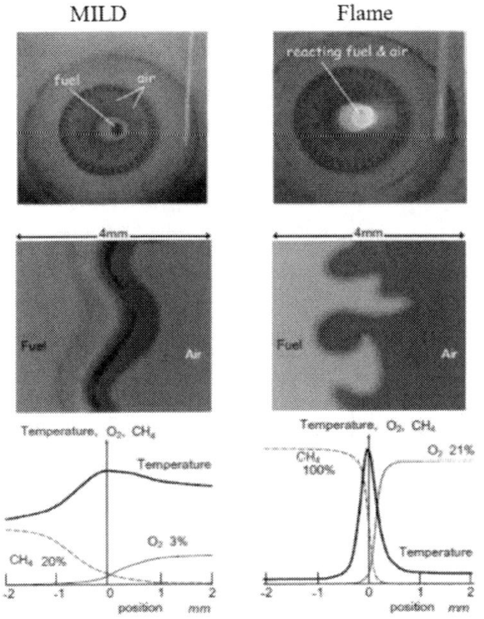

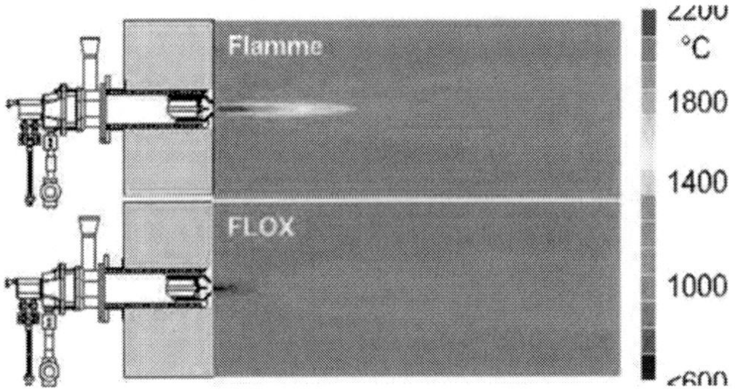

CASE STUDY AT HELENA WWTP

The Helena WWTP processes approximately 2 million gallons per day of wastewater. It services a community of approximately 30,000 people. The digester at the plant produces approximately 10,000,000 cubic feet of biogas a year.

The WWTP in Helena was the first US installation for Cleanergy's GasBox system. While there are 40 Cleanergy Stirling engines in Europe, they are new to the US market. They generate electricity and thermal power from the dirty biogas being emitted from the digesters. The gas was previously flared, but the plant was motivated to harness that alternative energy source. The plant partnered with WesTech and installed a single Gasbox with an electrical capacity of 7.2 kW and hot water output of 50,000 BTU/hr. The plant has plans to expand to 7 total units which would give 50.4 kW of electricity and 350,000 BTU/hr of hot water.

The engine requires cooling water to be between 86F and 120F. As this water cools the helium in the engine its temperature is increased by approximately 50 degrees F. To utilize the waste hot water Helena uses an Alpha Laval plate heat exchanger to supplement the boiler in the sludge heating process. A temperature control valve and shell in tube heat exchanger are used to ensure the water returning back to the stirling engine are within the manufacturers recommended 120^{0} F limit.

The Helena WWTP runs their digesters around 10 inches of water column. As a stirling requires 4 psi for optimal efficiency a small gas booster pump is needed. The pump was installed on the outside wall of the digester building to conform to the Class 1 Div 2 explosion proof requirements. For safety purposes the Helena plant uses the inert gas Helium as the working gas in the engine.

FUTURE OF THE HELENA WWTP CHP

The Helena WWTP is partnering with WesTech/Cleanergy to upgrade from 1 unit to 7 units. The current engine is running on raw biogas which has siloxanes levels of 15,205 ppbv. This is high compared to a typical WWTP which would have siloxane levels around of 5,000 ppbv. At the current siloxane levels, the engine requires cleaning and maintenance every 3 months. The plant would like to increase this interval to 9 months. This will be accomplished with a small activated carbon filter to reduce the hydrogen sulfide and siloxanes. This filter will not act as a full gas scrubber, but rather a roughing type filter to reduce the siloxanes to a more manageable level.

CONCLUSION

As small anaerobic digestion plants can't afford to scrub their biogas in order to use it, they are forced to flare it. Flaring is unsightly and frustrates sustainably minded constituents. A stirling engine is a great way for these small plants to capitalize on the renewable energy source they produce. This technology has been proven in Europe and is showing itself as a viable option to US plants.

REFERENCES

1. P. Gevorkian (2010). Alternative Energy Systems in Building Design. HISTORICAL TIME LINE OF SOLAR ENERGY, (McGraw-Hill Professional)
2. M. Frey (2014). "Pump Design Stirling Engine". Retrieved 2017-1-13
3. K. Hirata (1997). "Schmidt Theory For Stirling Engines". Retrieved 2017-1-13
4. W.R. Martini (1983). "Stirling Engine Design Manual (2nd ed)" NASA.
5. 1987 Anaerobic Sludge Digestion, Manual of Practice No. 16 Second Edition, Water Pollution Control Federation

Update on Biogas Upgrading Project for City of Portland

"From Flare to Fuel"

Vu Han[1], Paul Suto[1], Jeremy Holland[2], Danny Grady[3], Muriel Gueissaz-Teufel[1]

[1]City of Portland, Bureau of Environmental Services (BES)

[2]HDR Portland, Oregon

[3]City of Portland, Bureau of Planning and Sustainability (BPS)

ABSTRACT

The City of Portland, Bureau of Environmental Services (BES), has been working on a project to upgrade biogas at the Columbia Boulevard Wastewater Treatment Plant. The project was originally started in 2013 as a cogeneration expansion project, with the intention of evaluating other options of biogas use. The project ultimately became an RNG (Renewable Natural Gas) to vehicle fuel project due to the economic and sustainable benefits of offsetting fossil based vehicle fuel. As a result, the City has been working on a variety of project development issues over the last three years in order to move toward completion of a project to treat biogas to pipeline quality RNG and sell the RNG as a transportation fuel. The following paper will describe the steps the City has taken once the decision was made to produce RNG, the obstacles that have been overcome to move the project forward, and the future plans for completing the project and selling RNG as a transportation fuel.

INTRODUCTION

Project Background:

The Columbia Boulevard Wastewater Treatment Plant (CBWTP) is located in north Portland, Oregon. The plant started up in the early 1950's and serves most of the city of Portland, with population of more 600,000 people. See Figure 1 for aerial photo of the CBWTP.

Figure 1: Columbia Boulevard Wastewater Treatment Plant

The CBWTP treats an average of 65 mgd (million gallons per day) of dry weather flow, with peak weather flow capacity of 450 mgd. CBWTP currently produces about 1.6 million cubic feet of biogas a day, or about 1,100 scfm, with methane content ranging about 55% to 65%. Of the biogas production, more than three quarters is beneficially used and about a quarter of it is flared. The current beneficial uses include: boilers for digester heating and various heating uses onsite, cogeneration to offset power usage, and sale of untreated biogas to a roofing manufacturing facility nearby. Figure 2 and 3 below illustrate the current biogas utilization at the CBWTP.

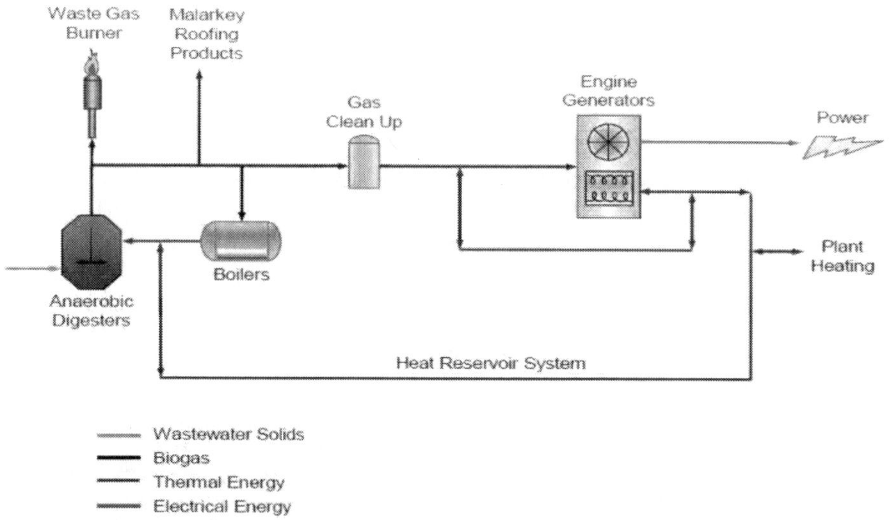

Figure 2: Current Biogas Utilization Schematic

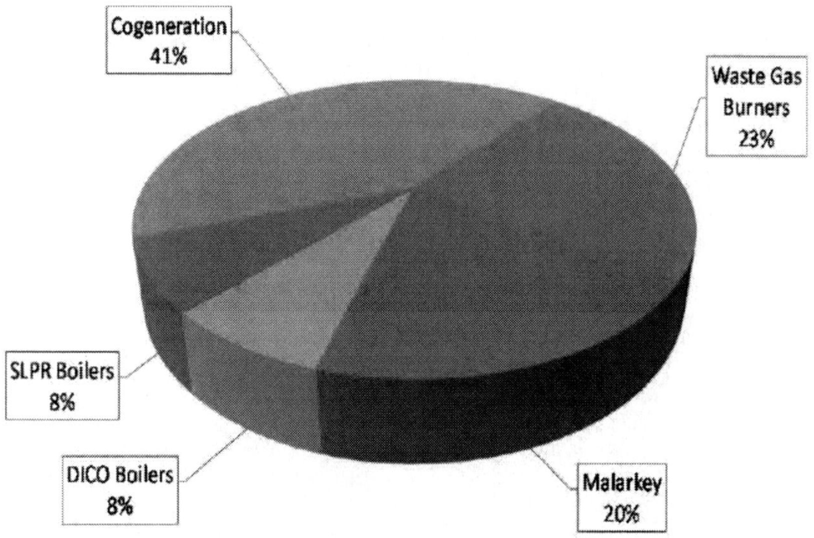

Figure 3: Current Percent Biogas Utilization

Biogas used for the boilers and sales to the nearby Malarkey Roofing Company are raw digester gas; hence, no gas treatment is required. Biogas used for the Cogeneration has to be treated for removal of hydrogen sulfide, moisture and siloxane, to meet engine specifications.

The original scope of the project started as cogeneration expansion. However, through alternative analysis during the project preliminary design, and further development and market opportunities of the renewable energy for transportation fuel, the project scope was changed to producing renewable natural gas (RNG) for vehicle fuel.

Overall Project Goals:

The overall project goals include:

1. Best beneficial use of the remaining biogas.
2. Maximize ratepayers' benefits.
3. Optimize return on investment with best balance of risk and reward
4. Align with City's goals, policies, and Climate Action Plan

With those stated goals, HDR (consulting engineering firm) performed a triple-bottom line analysis to include monetized values of the environmental and social values, for various alternative uses of the biogas. The results clearly indicated that the best beneficial use of the remaining biogas is to be used for vehicle fuel, which has highest benefits from the financial, environmental and social aspects.

Steps, Obstacles and Future plans:

After figuring the benefits for vehicle fuel and determining the direction of the project, many challenges and obstacles still have to be worked out, such as who will be the off-taker of RNG, where to deliver RNG and how to deliver it. It is especially challenging for the Pacific Northwest, as there is currently no infrastructure and market for RNG. Early concepts for transporting RNG include building commercial fuel station(s) and trucking RNG to retailers and wholesalers. Those concepts were not materialized due to logistical and access challenges.

The City approached NW Natural (NWN), a local distribution company (LDC), for a partnership on pipeline injection and delivery of the RNG. There have been numerous discussions with NWN in the last couple years, but BES now has an approved tariff with NWN to inject RNG into its pipeline for the purpose of transporting the gas. Once injected into the NWN pipeline, RNG can be transported through the vast gas pipeline network, even to an out of state market. An illustration of pipeline injection concept is shown in Figure 4.

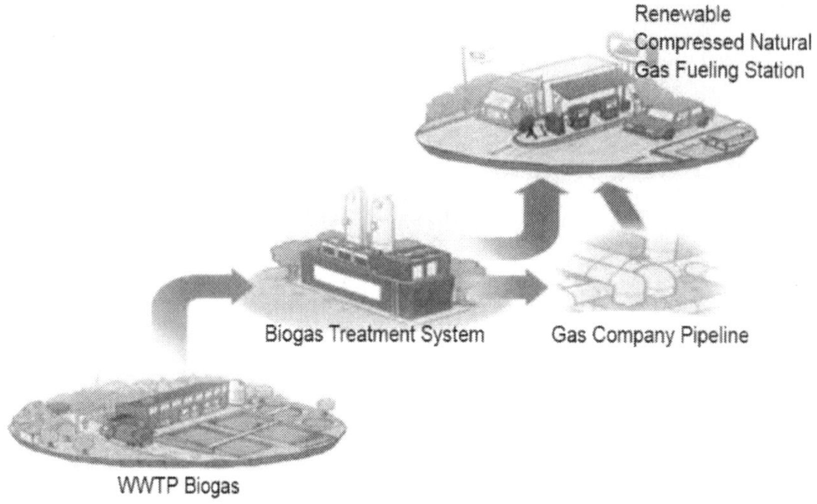

Figure 4: RNG Pipeline Injection Concept
Source: Greenlane Biogas

The RNG produced will comply with the specifications by NWN for pipeline injection. At the writing of this paper, an RFP for sale and purchase of RNG with off-taker(s) is advertised. It is anticipated that construction of the biogas treatment facility (producing RNG) to start summer 2017, and RNG available in the fall 2018.

With current lack of an RNG for vehicle fuel market and infrastructure in Oregon, it is anticipated that RNG will be sold to out of state markets, such as California, in the early years. However, it is the City's interest, as well as our pipeline partner's (NWN), to develop a local market in the long term to keep the consumption of a renewable and sustainable energy, as well as the environmental and social benefits locally. To help jump start the local development plan, the City will install a Renewable Compressed Natural Gas (RCNG) fueling station onsite for some vehicles in its fleet, and possibly contractor vehicles with operations onsite.

The City is also considering organic wastes, primarily food waste, as additional feedstocks to CBWTP to augment biogas production, which will improve the project return on investment and payback. The incorporation of food waste will also help provide a solution to another waste, and keep the waste to fuel ecosystem completely local.

Approaches to the Project:

The following approaches are used in this project and are discussed in this paper.

1. Sustainable Business Approach
2. Technical Requirements
3. Biogas Management Plan

SUSTAINABLE BUSINESS APPROACH

The City's goals with the biogas upgrade/utilization project were driven primarily by developing a sustainable revenue stream, and secondarily by sustainability-producing a renewable fuel. That is, the project needed to have a business pay back for the capital expense, and not just provide environmental benefits. Using the recently completed cogeneration project (2008) as a model and 'baseline project', the project team moved forward with confirming the preferred end use of the biogas based on financial and social and environmental factors, traditionally called a triple bottom line analysis- financial, environmental, and social.

Although the project was first planned as expanding cogeneration, City staff were interested in other potential uses. The main reasons for expanding cogeneration had a limited benefit financially, compared to the initial cogeneration project. That is, while the first project had beneficial uses for both the electricity and heat produced, the new project would primarily benefit from the electricity. This is because the plant could generate sufficient heat with the existing cogen and boilers. Another concern with electricity production was that the plant had generation size limit of 2 MW. Above this limit, the plant would be classified as a major power producer and would be subject to different rate schedules, which would have significant impact to the project financials. Expanding cogen would easily exceed 2 MW as the plant was currently at 1.7 MW. A legislative process to change the 2 MW limit was not successful.

Feasibility Analysis of Biogas Use Alternatives

In late 2011, the City participated in an Oregon Department of Energy (ODOE) biogas feasibility study. The report evaluated the feasibility for biogas use at four Oregon wastewater treatment plants, including CBWTP. The evaluation included biogas upgrading (or cleaning/conditioning) technologies and requirements for various potential end uses. The end uses included electricity production, natural gas pipeline injection, and renewable CNG for a vehicle fuel. The analysis was simplified to screen alternatives by considering sending most of the gas (less what is needed for the boilers) to any one of the alternatives, as this would provide the best economy of scale for each alternative. Alternative 2 -Expanding Cogen included two sub alternatives- 2A continuing to sell biogas to a local customer, and using all of the remaining gas in cogeneration and expanding cogen based on that size; -2B sending all available gas to cogen and expanding cogen. The recommendation for the City of Portland's CBWTP from the final report, Bioenergy Optimization Assessment at Wastewater Treatment Plants, March 2012, was to consider adding cogeneration capacity and also consider producing renewable CNG. See Table 1 that presents the financial benefits for the alternatives, with only alternatives 2 and 5 being financially viable. The study did show that GHG emissions benefits of each options as well, and the two recommended alternatives had the lowest net GHG emissions. The study did note the logistical challenges with the renewable CNG vehicles: finding a fleet, converting vehicles, storing rCNG, getting access to the local gas utilities pipeline. These analysis was helpful for bureau staff to further refine and consider alternatives for the biogas utilization/upgrade project.

Table 1: CBWTP Biogas Utilization Alternatives Financial Results Comparison from ODOE study

Performance Metric	**Alternative 2 Cogeneration**	**Alternative 3 Pipeline Injection**	**Alternative 4- CNG (all gas)**	**Alternative 5- CNG (partial gas)**
Total Construction Cost	$3,979,725	$3,849,768	$5,166,720	$4,113,720
Annual Net Profit (or Loss)	$413,724	($947,594)	($493,988)	$729,865
Net Present Value (20 year)	$4,606,659	($8,762,142)	($4,320,906)	$6,823,475
Simple Payback	Year 7	N/A	N/A	Year 7
Notes: From Bioenergy Optimization Assessment at Wastewater Treatment Plants, March 2012, Oregon Department of Energy				

Biogas Upgrade Project Alternatives Analysis

The project was formally revised to be called the biogas utilization project as expanding cogen was now just one option. Based on the ODOE study, the alternatives analysis did not consider phasing out the existing cogen unit. Basically, the alternatives would consider uses for the gas currently being flared and also part of the gas currently sold to a local user. The primary alternatives included: Expanding Cogen (with and without storage), Renewable CNG, Natural Gas Pipeline injection, and Biosolids Drying. The project consultant team used a Sustainable Return on Investment (SROI) Analysis to provide a monetized triple bottom line analysis. That is, the societal and environmental impacts were monetized using economic models to show the benefits in terms of dollars. This analysis would provide a two phased level of screening to support both the financial business case and overall benefit to the public and environment. For the vehicle fuel alternative, the assumption used was that the biogas would be upgraded to fuel a vehicle fleet directly (via pipeline and/or trucking to storage).

Biogas Utilization Alternatives

Figure 5 below shows both the existing biogas uses and proposed alternatives considered in the SROI analysis.

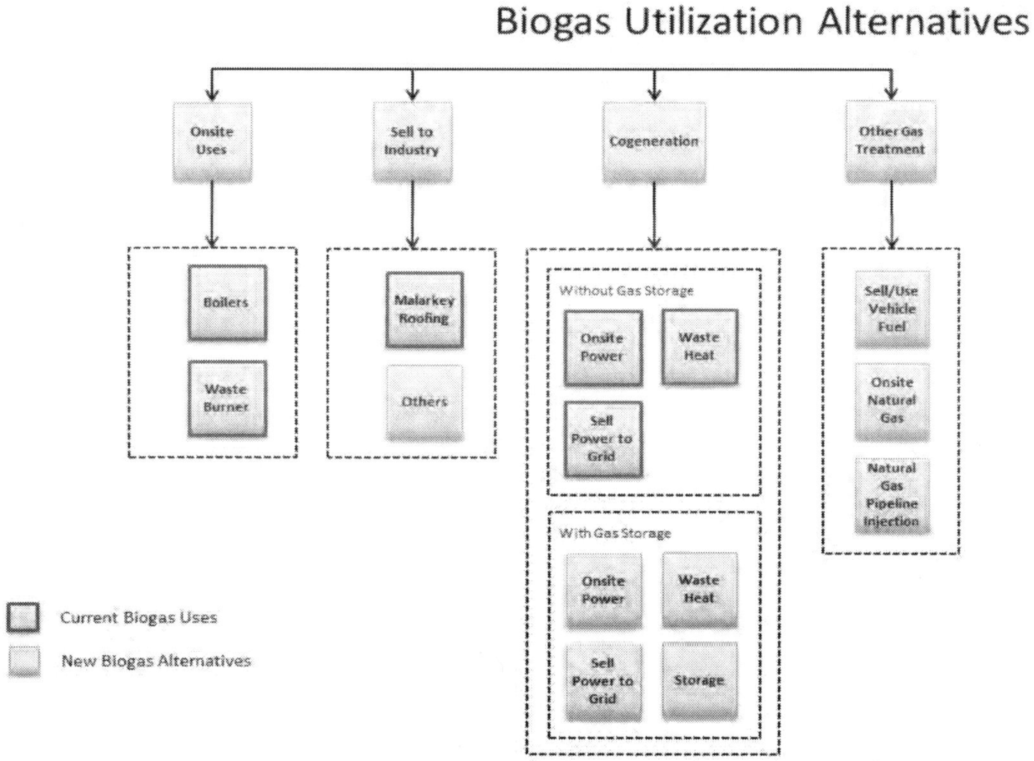

Figure 5: City of Portland CBWTP summary of existing uses and alternatives considered for SROI Analysis.

Below are examples of the Structure and Logic diagrams to develop the models.

Examples for the selected alternative are shown Figure 6 and 7 (biogas to vehicle fuel).

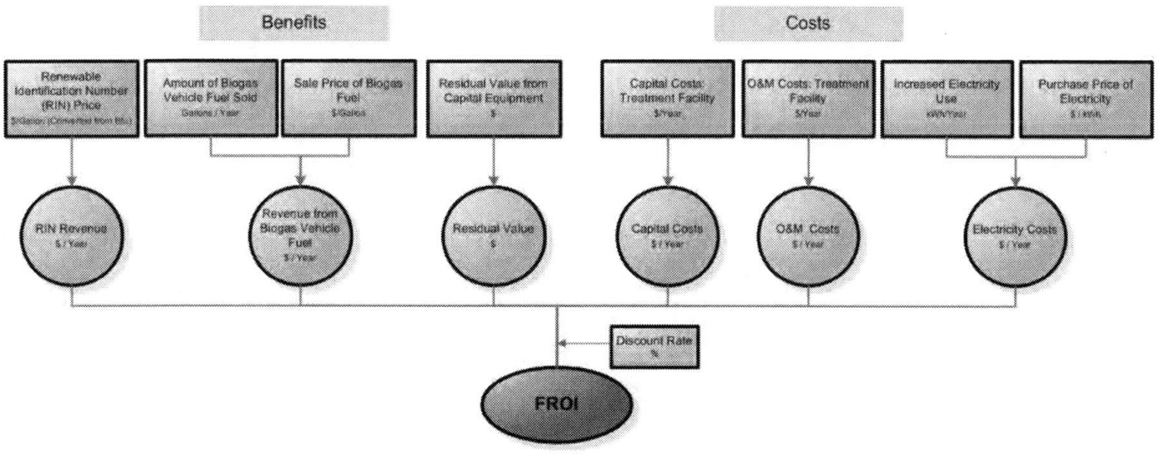

Figure 6: City of Portland CBWTP summary of specific financial impacts considered in development of the biogas treatment for vehicle use alternative.

The financial impacts include' capital costs, replacement costs, annual operating and maintenance costs, power costs. Financial benefits include sale as a vehicle fuel (commodity price), the RIN (Renewable Identification Number) revenue, and residual value. Note that RINs are only included in the financial analysis, and not the SROI analysis.

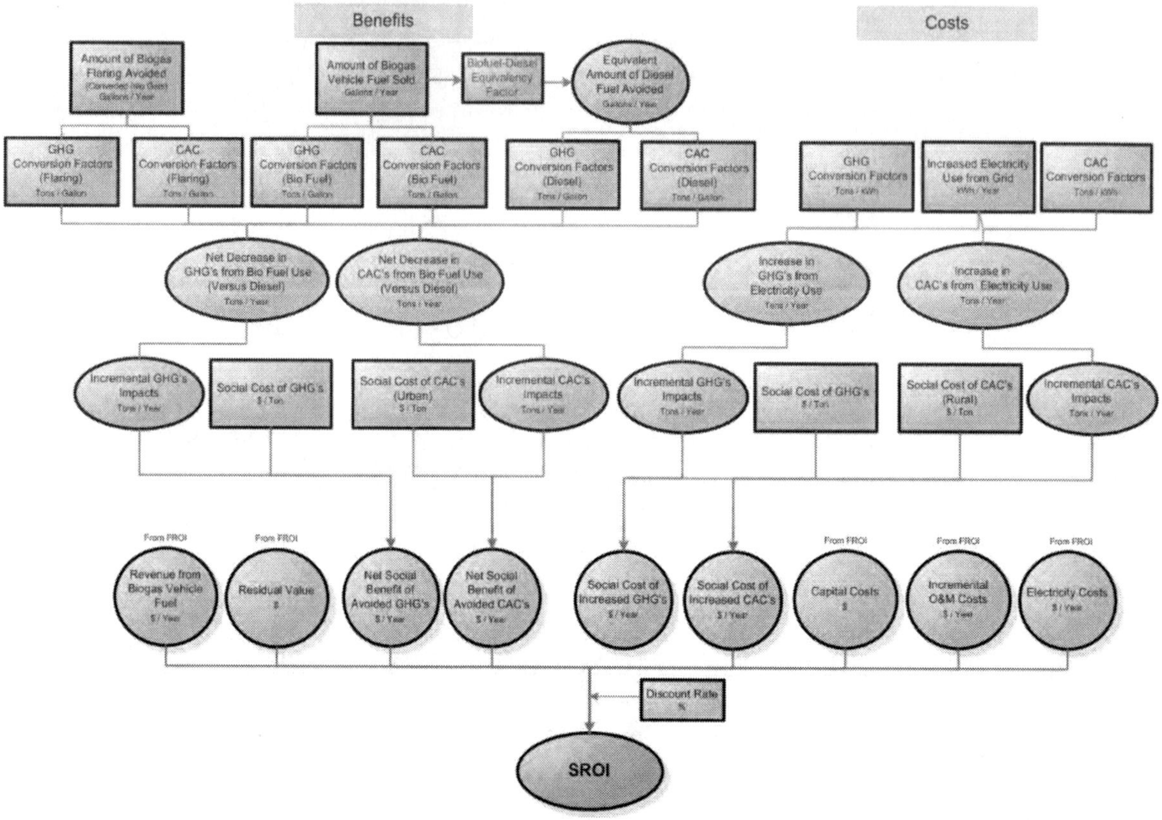

Figure 7: City of Portland CBWTP summary of specific impacts considered for the SROI Analysis for the biogas treatment for vehicle use alternative.

The societal benefits and costs relate to produced and avoided Criteria Air Contaminants (CACs) from the incremental difference in biogas fuel versus diesel fuel (from fossil fuels), and avoided flaring of biogas, purchase of additional power needed.

Table 2: City of Portland Biogas Utilization Financial and SROI Analysis Results

	Alternative 1: Cogeneration Expansion without Biogas Storage	**Alternative 2: Cogeneration Expansion with Biogas Storage**	**Alternative 3: Biogas Treatment for Vehicle Use**	**Alternative 4: Biogas Treatment for Sale to Industry**	**Alternative 5: Biogas Used in Biosolids Drying**
Metrics	Mean Values	Mean Values	Mean Values	Mean Values	Mean Values
Financial Return on Investment					
Net Present Value Benefits	**($1,827,153)**	**($13,824,271)**	**$8,244,500**	**($3,845)**	**($37,177,765)**
Discounted Payback Period	**N/A**	**N/A**	**10 y 7 m**	**N/A**	**N/A**
Sustainable Return on Investment					
Net Present Value Benefits	**($999,891)**	**($12,610,227)**	**$82,642,711**	**$5,717,842**	**($35,248,952)**
Discounted Payback Period	**N/A**	**N/A**	**2 y 11 m**	**11 y 8 m**	**N/A**
Net Monetized Societal Impacts	**$1,239,586**	**$1,862,860**	**$83,955,695**	**$5,721,687**	**$1,928,813**

Based on the financial analysis (cash only impacts), only one alternative has a positive financial scenario, in terms of both financial payback and NPV-Biogas for sale as a vehicle fuel. The financial payback is estimated to be just under 11 years. The financial analysis showed the other alternatives carried a high financial risk to move forward. It was interesting to see cogen expansion to not have a positive return on financial investment. A key factor was that no financial incentives were available (such as tax credits) to reduce the capital cost burden for the City.

Looking at the SROI results show positive societal benefits for all alternatives, but one alternative has a significant increase-over $83 M. Alternative 4 resulted in a net positive NPV after factoring in societal benefits. Based on both the financial and societal benefits, the selected alternative was biogas treatment for vehicle use. A regional factor that affected the SROI analysis included the fact that Oregon's electrical grid emission intensity is low compared to other regions because of the availability of hydropower and renewables (wind and solar). This resulted in renewable electricity not being as compelling compared to renewable fuel.

The SROI results are reassuring, as they provide added environmental and societal benefits, and align with other City's goals, such as the Climate Action Plan. However, at the end of day, green benefits alone are not sufficient for the City to proceed with the project; it has to be economically viable.

The SROI analysis used the risk analysis and Monte Carlo simulation techniques to account for uncertainty in both the input values and model parameters. With all elements converted to monetary values and discounted to present value terms, the probable ROI and SROI are shown Figure 8 below.

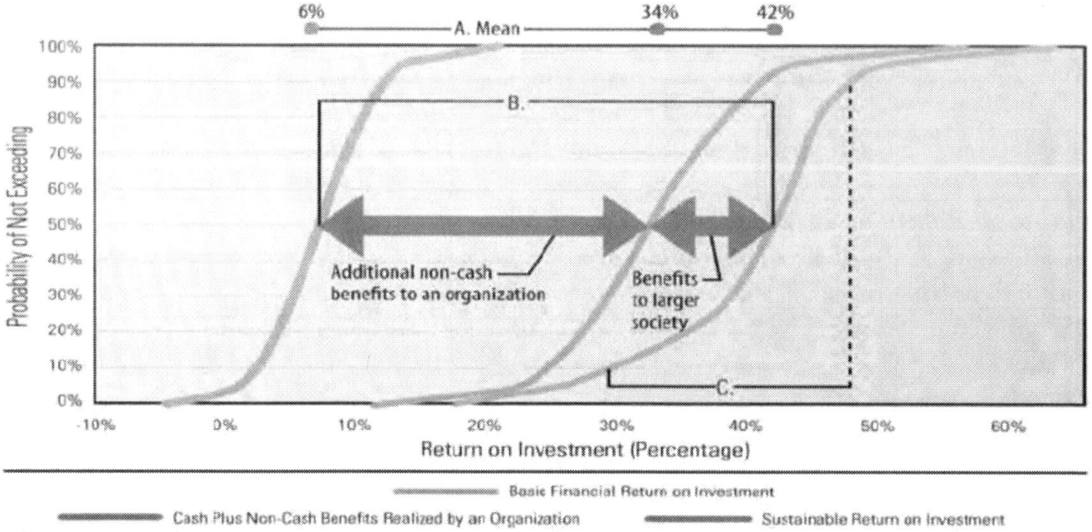

Figure 8: The Sustainability "S" Curve to Optimize the Project Total Value

Reducing Project Capital Costs -Will Our Wishes Be Granted?

City staff were also interested in reducing the financial burden of the costs and pursued looking for grant and tax credit opportunities. Early during project design (near 30% complete), City staff submitted grant and tax credit applications based on the project model of producing vehicle fuel for a local fleet. In 2015, the City was successful in securing an ODOE Alternative Vehicle Fuel Infrastructure tax credit for approximately $1.6 million. The tax credit set a timeline for project completion by March 2018. The key elements of the project included gas upgrading equipment and a fueling station.

Who Will Purchase the Fuel? RFP/Buyers

Based on the financial analysis, it became clear that a fleet or fuel purchaser is needed. The City was concerned with building a facility and not having anyone to sell the fuel too. The project had the classic chicken and egg scenario- local fleets were not interested in converting without a readily available fuel source. Fuel suppliers were reluctant to invest in infrastructure without an available fleet. The notion of "build it and they will come" was not a comforting situation, nor would it gain support from City leadership to move forward with the project.

After numerous discussions with the local gas utility over a period of over a year, the gas utility was interesting in partnering with the City to allow RNG onto the gas grid. This would be the first project in the state of Oregon to achieve this goal. The City was interested in eliminating the logistical challenges of storing and hauling the RNG, and connecting the local gas line would achieve that goal. Furthermore, by connecting to the gas grid, the customer options would increase significantly because of the ability to sell in markets where CNG fleets exist and are interested in RNG. City staff contacted many potential buyers while the discussions with the local gas utility were occurring, to test the viability of that approach. One of the key lessons was the benefits of economies of scale-the more gas, the better. (add in volumes- 100,000 dekatherms per year minimum)

Since a local RCNG market does not exist in the Portland Metro area, the expectation based on outreach to potential off-takers (or buyers) is that that the RNG will be sold out of state into existing markets, such as California. This would be under a model of "offset in displacement", which is acceptable under the EPA's renewable fuel standards program. The key requirements are that the gas as a commodity (no RINs included) are sold to the local gas utility, and there has to be physical connectivity between the RNG connection point and the NG transmission system. Under this model, transmission fees are not collected by the gas utilities. The City will encourage the buyers to develop local users over the length of the contract. Figure 9 shows a key part of the increase in RIN values-the increase in mandates volume levels for the RFS program. The figure also shows the various classes of RINs. Of note is that fuel produced from municipal wastewater solids meets D-3 category requirements, and that category has the highest increase. This reinforced the importance to the City to find a buyer to take advantage of the current and coming market.

The City's long term goal is to have the societal benefits be realized locally; however, given that there isn't an immediate opportunity with a local fleet, including the City's fleet, the City is preparing to plan the project so that over time, local fleets are developed. By selling the fuel to the most active and financially lucrative market, financial risks are reduced, and the new facilities are planned to realize a good payback. If a local fleet develops over time, the volume of RNG needed for that fleet can be redirected to meet that demand.

Before contracting with buyers, the City needed certainty that the RNG could be connected to the local gas grid. The City entered into a Memorandum of Understanding (MOU) with the local gas utility in 2016, which set the stage for accessing customers. With the MOU in place, the City next turned to connecting with fuel purchasers by issuing an RFP in December 2016. The RFP results will confirm the financial analysis complete to date, to help the project move forward into construction.

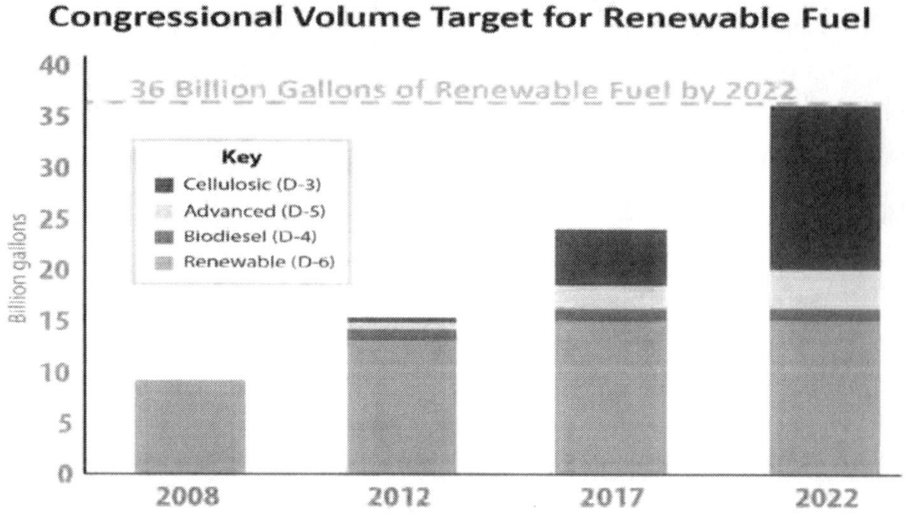

Figure 9: EPA Renewable Fuel Standards Trend of Volume Targets for Fuel, and categories of RINs. Source: EPA.

The More Gas the Better

Based on the sizing of the gas upgrading equipment at 1,100 SCFM peak, the City (project team) evaluated opportunities for increasing gas production to utilize the available capacity. After considerable analysis, two plans were developed:

Plan A: Partnering with a local waste hauler and bringing a source separated and processed commercial municipal food waste slurry to CBWTP to increase gas production. This plan is not certain, as the food waste feedstock is not certain, and will need further partnerships to develop.

Plan B: Managing the existing gas produced so that the gas is sent to the most beneficial end use, which was determined to be RNG as a vehicle fuel. Figure 10 below shows a comparison of the net financial value of the gas uses. Since RNG as a vehicle fuel is considerable higher in value, one of the cogen engines could be shutdown and the gas be sent to the RNG production facility.

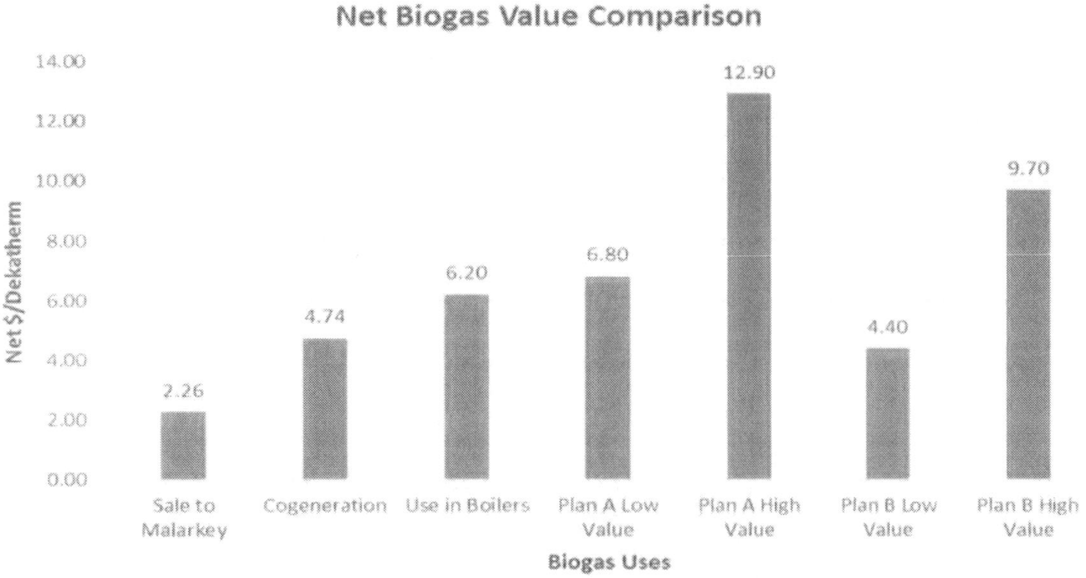

Figure 10: Comparison of the relative values of all the biogas uses at CBWTP, Portland, Oregon.

TECHNICAL REQUIREMENTS

1. **Confirming gas quality requirements for pipeline injection**
In order to inject RNG in to NWN pipeline, it has to meet specifications by NWN. Pipeline injection specifications are set by local gas utilities, and they can vary from one gas utility to another. The City started discussions on pipeline injection concept with NWN in 2014. The City now has an MOU with NWN, and approved tariff to inject RNG into NWN gas pipeline. The City has received draft pipeline gas specifications from NWN. Draft specifications include three categories of gas constituent requirements, similar to requirements by SoCal Rule 30. The three categories of requirements include:
    a. Gas quality specifications, e.g. methane content, invert gas and moisture limits.
    b. Health protective constituent specifications, e.g. P-dichlorobenzene, ethylbenzene, vinyl chloride, hydrogen sulfide.
    c. Pipeline integrity constituent specifications, e.g. ammonia, hydrogen, mercury, siloxane.

Part of the specifications also include testing and monitoring requirements. The methodology and frequency of testing and monitoring depend on the parameters, ranging from online monitoring by instrumentation to periodic grab samples. The specifications also outline trigger levels for alarm and shutoff, as well as shutdown and restart procedures.

NWN will install a gas monitoring station upstream of the interconnection to the gas utility pipeline. This station will also serve as custody transfer of RNG from the producer (City) to the gas utility and the off-taker. A site map of the RNG interconnection is shown in Figure 11.

Figure 11: Site Map of RNG Interconnection.

2. **Selecting a gas treatment technology**

The biogas technology was evaluated by the project team, consisting of HDR (engineering firm), and City engineering and O&M staff. The following three technologies were considered by the project team:

a. Membrane Treatment
b. Pressure Swing Adsorption
c. Water Scrubbing

The process of evaluating the technologies included collection of installation records, previous engineering consultant experiences with technologies, engineering requests from the technology providers, and site visits of actual installations. Site visits were particularly helpful so that City staff could discuss performance and operations with equipment operators and see physical installations to better understand the technologies. See Figures 12, 13, and 14 for examples of site visit locations. The various technologies were compared against one another considering capital cost, operating cost, and biogas yield and associated revenue. The selected technology through this process was the water scrubbing technology due to the size of system the City is constructing, the capital costs versus biogas yield, and the operations considerations of the water wash versus other technology choices.

Figure 12: Membrane Treatment Site (Air Liquide Membranes)

Figure 13: PSA System Site Visit (PSA)

Figure 14: Water Scrubbing System (Greenlane Biogas)

3. **Sizing of gas treatment system and storage**

The current average biogas production is 1,100 scfm. However, biogas production fluctuates daily and seasonally. The peaking factor is established at 1.5; hence, the biogas production range can vary from 550 scfm to 1,650 scfm. See Figure 15 for current typical biogas production range.

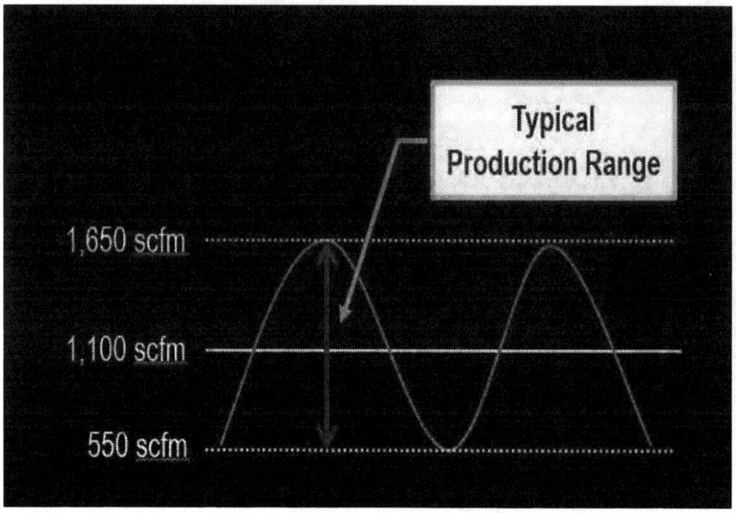

Figure 15: Current Typical Biogas Production Range.

There is small amount gas storage capacity in the current biogas system. However, if the RNG gas treatment system were sized specifically for the biogas volumes that are flared and sold to the roofing company, additional biogas storage would be needed to address the issues with the biogas production fluctuation. Biogas storage is expensive. Also, the project design includes provisions for future organic waste intake, which is estimated to be another 300 to 500 scfm. To address the concern of biogas production fluctuation and include future provisions, the project team concluded on larger RNG gas treatment system, with 1,100 scfm capacity (up to 340,000 dekatherms a year output). The system will have turn down up to 70%, which can match the minimum biogas volume in the typical biogas production range.

The larger capacity system is a more cost effective in terms of unit cost. With the water scrubbing equipment, the incremental of cost of production is significantly less. The larger system will also provide flexibility to manage and direct gas to higher uses, and hence, maximize revenues, such as the Plan A and Plan B mentioned above.

4. **Considering system redundancy versus cost**
   This biogas project is driven primarily by political and economic factors, and is not a regulatory required project. Hence, redundancy is not a requirement and such decision is based solely on economics. The manufacturer guarantees 95% uptime of their gas treatment system. The City analyzed the cost and benefits for uptime greater than 95%, with the criterion of less than 10 years of simple payback on the investment. The conclusion was that it is not worth the cost to improve the system redundancy and reliability greater than 95% uptime. The decision was to design and install a single unit system. Figure 16 shows an example of redundancy costs for various options.

Approx. Capacity (scfm)	Models	Number of units	Reliability Factor	Annual Lost Production (MMBTU)	Annual Lost Revenue ($)	Capital Cost Difference ($)	Simple Payback (years)
400	Rimu	1	95.0%	5,782	$57,816	-	-
400	2 x Kanuka	2	97.5%	2,891	$28,908	$720,000	25
600	Matai	1	95.0%	8,672	$86,724	-	-
600	Rimu + Kanuka	2	97.5%	4,336	$43,362	$945,000	22
600	3 x Kanuka	3	98.3%	2,891	$28,908	$1,665,000	29
800	Totara	1	95.0%	11,563	$115,632	-	-
800	2 x Rimu	2	97.5%	5,782	$57,816	$1,112,500	19
800	2 x Kanuka + Rimu	3	98.3%	3,854	$38,544	$1,832,500	24

Figure 16: Redundancy Costs of Various Options

BIOGAS MANAGEMENT PLAN

Need for Biogas Management Plan

The existing biogas system constantly flares gas, and significant amount at times. Hence, there isn't a critical need to prioritize the gas uses. With the new RNG system, all biogas will be 100% beneficially used. Therefore, it will be crucial for the City to prioritize and optimally manage the gas uses to maximize revenues. Furthermore, the gas will need to be carefully managed to maintain stable system operations, balance internal supply and demand, as well as external contract obligations, such as volumes nominated to the off-taker(s).

Onsite RCNG Fueling Station

As part of the received tax credit, the City will install an onsite RCNG fueling station. The station will fuel City Fleet converted vehicles that operate on CNG. It will also potentially fuel contractor vehicles that have operations at CBWTP, such as contactor truck use for biosolids land application program and/or solid waste haulers to landfills. The RCNG fueling station will help jumpstart the development for a local market and infrastructure for RNG. When the local market for RNG has taken off, more RNG would be retained locally to keep the environmental and societal benefits locally.

POLITICAL DRIVERS AND IMPORTANCE OF PARTNERSHIP

The City of Portland is striving to become one of the leading cities in green energy and sustainability. It has a world award winning Climate Action Plan and recently passed a fossil fuel export ban to outlaw coal, oil and propane export terminals in Portland. The RNG for vehicle fuel project aligns with these renewable and sustainability goals, in addition to the financial benefits to the City budget and ratepayers.

This project crosses boundaries of a number of industries, from waste to gas utility to then end user of fuel. Therefore, it is important to form and maintain good partnerships for the entire concept to work. It also provides the City opportunities to form public/private work relations, and allows it to focus on its core business, which is wastewater treatment and resource recovery and leave the marketing and sales of the fuel to someone else.

CONCLUSION

The RNG to vehicle fuel project has tremendous benefits the City from the financial, environmental and society aspects. It has potential to keep the entire ecosystem regionally, from waste collection, fuel generation, consumption, and retention of the benefits locally. With that, there are challenges, risks and uncertainties. As presented, there are many external factors to consider and partnerships to form in order for to work. It has been a long in the making and many steps the City has gone through to this point, in finishing design and prepare for construction. The projected revenues for RNG rely on RIN value and it fluctuates with market demand and supply. The City has performed numerous financial analyses, including sensitivity analyses, to ensure the project is economically viable. At time of the writing of this paper, the City has an RFP advertise for the RNG offtake. The proposals received from the RFP will verify the City's projections and confirm the City's decision to proceed with construction.

A GREENER FUTURE FOR THE CITY OF RALEIGH TURNING BIOSOLIDS INTO VEHICLE FUEL

Greg Knight[1,4], Michael Bullard[2], Kent Lackey[1], Christine Polo[1], Jay Kemp[1], Aaron Brower[3], T.J. Lynch[3]

[1] Black & Veatch
[2] Hazen & Sawyer
[3] City of Raleigh
[4] 1120 Sanctuary Pkwy #200, Alpharetta

ABSTRACT

The City of Raleigh is progressing with a Bioenergy Recovery Program to convert biosolids primarily generated at the Neuse River Resource Recovery Facility (RRF) into a Class A product for beneficial use and resource recovery via land application and to make beneficial use of the biogas generated in anaerobic digestion for use as vehicle fuel.

Historically, biogas produced from anaerobic digesters at wastewater treatment facilities has predominantly been used in boilers, with any surplus flared, or used in reciprocating engines (or turbines) to produce electricity. In some cases it has been used to drive plant equipment such as pumps and aeration blowers.

Recent advances in gas upgrading technology mean that conversion of biogas to renewable natural gas (RNG) for pipeline injection or, following compression, to compressed natural gas (CNG) for vehicle fuel are now viable alternatives to more traditional combined heat and power (CHP) applications.

For the Nesue River RRF, various options were evaluated for biogas use including engine driven blowers, CHP with engine generators, conversion to CNG vehicle fuel and conversion to RNG for pipeline injection as well as hybrid options involving a combination of solutions. Following the evaluation, the option selected was to convert the biogas to RNG for pipeline injection, combined with conversion of the City's bus fleet to CNG to realize the green credentials of the RNG via the Renewable Fuels Standard.

Given selection of RNG to pipeline with a CNG vehicle fuel end use as the preferred option, various technologies needed to be considered for biogas upgrading at the Neuse River RRF. A technology evaluation was conducted and life cycle cost results are presented for each option.

SELECTION OF BIOGAS UTILIZATION APPROACH

Potential biogas utilization options needed to be considered for the Neuse River RRF which could maximize the value of the biogas while also providing sufficient heat (in the form of steam) for the thermal hydrolysis process.

Initially, a screening level evaluation was conducted to identify gas utilization options which could be expected to be financially viable over a 20 year project life cycle. The screening evaluation identified the options shown in Table 1 for further evaluation to determine their financial viability for implementation at the Neuse River RRF. As well as producing an end product from the biogas which would have financial value to the City, provision of steam for the Thermal Hydrolysis Process (THP) was also a key requirement of the gas utilization system. Also shown in Table 1 is the method by which each option would be able to produce steam for the THP.

Initially an option of cleaning up biogas to produce a renewable natural gas product for injection into the natural gas network was ruled out during the screening evaluation due to the significant distance between the plant and the nearest natural gas main with a large enough size to receive the gas.

Table 1 – Gas Utilization Options Evaluated for implementation at the Neuse River RRF

Option	Prime Use of Biogas	Steam Production for Thermal Hydrolysis
1	Fired steam boilers	Fired steam boilers
2	Engine driven blower	Combination boilers using waste heat from engine and biogas firing
3	Combined heat & power (CHP) engine Generators	Combination boilers using waste heat from engine and biogas firing
4	Conversion to Compressed Natural Gas (CNG) Vehicle Fuel for City Buses	Fired steam boilers
5	Conversion to CNG Vehicle Fuel for City Garbage Trucks	Fired steam boilers
6	Conversion to CNG Vehicle Fuel for City Buses and Garbage Trucks	Fired steam boilers
7	2 MW CHP engine generator followed by Conversion to CNG Vehicle Fuel for City Buses	Combination boilers using waste heat from engine and biogas firing
8	1 MW CHP engine generator followed by Conversion to CNG Vehicle Fuel for City Buses	Combination boilers using waste heat from engine and biogas firing

Further detail regarding the options considered is provided in the discussion below.

Option 1 (Base Option) – Steam Generation Only

Biogas would be used to fire a conventional shell steam boiler to supply steam to the THP. Excess biogas would be flared. This option was used as a baseline against which to compare all other options.

Gas supplied to the boiler would need to be cleaned to remove moisture, hydrogen sulfide (H_2S) and possibly siloxanes. Without siloxane removal the efficiency of the boiler would decrease over time as silicon deposits build up in the boiler tubes. A small amount (0.5 hours) of low pressure gas storage was also included with this option. A conceptual schematic is provided in Figure 1.

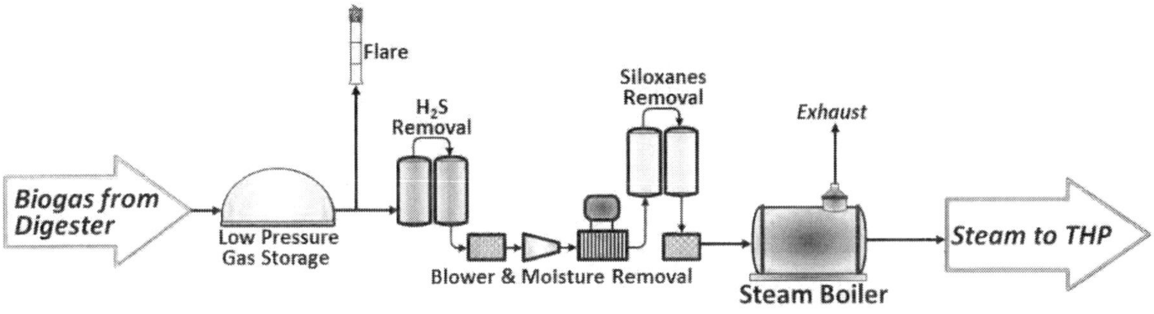

Figure 1 - Gas utilization schematic, steam boiler only

Option 2 – Engine Driven Aeration Blower

Biogas would undergo basic cleaning and be combusted in a single engine which would drive an aeration blower. Air produced would be transferred via new pipework to the existing air pipework main and would offset costs associated with electricity consumption of the existing aeration blowers. Heat recovered from the engine associated with the blower would be recovered for steam production. This would require a combination boiler to produce steam from either biogas firing or engine exhaust heat. Digester gas would require low pressure storage, H_2S removal, moisture removal and siloxane removal. A conceptual schematic showing the engine driven blower option is provided in Figure 2.

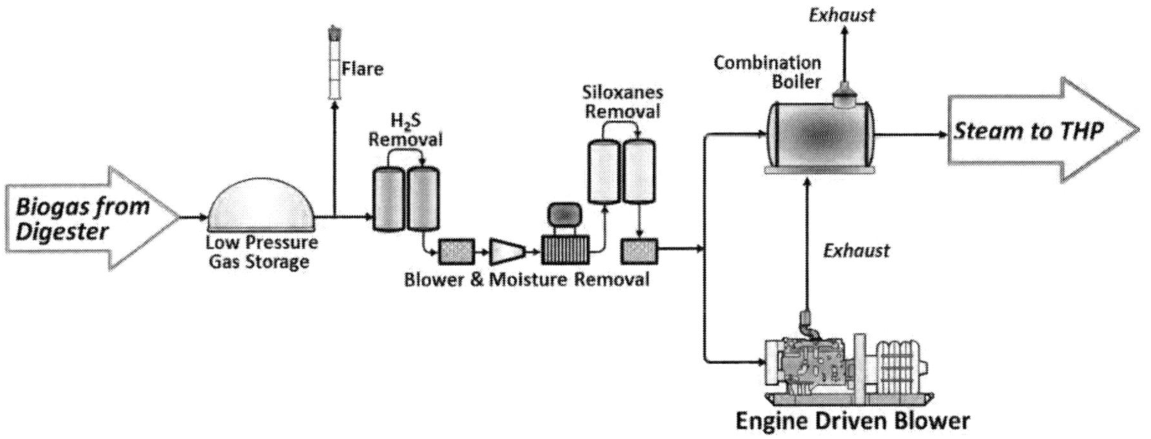

Figure 2 - Gas utilization schematic for engine driven blower option

Option 3 – Combined Heat & Power with Engine Generators

Biogas would undergo basic gas treatment including moisture, H_2S and siloxane removal and would be used to fire three engines which would drive generators to produce electricity. Energy in the engine exhaust gas would be used in a combination boiler to raise steam for the THP. Jacket heat from the engines would be partially recovered for heating the boiler feedwater. The engine generators would also serve as backup power and were assumed to offset the cost associated with new diesel generator capacity required on site. During an electrical outage, the engine generators could be fueled with natural gas if needed. Figure 3 provides a schematic for the engine generator option.

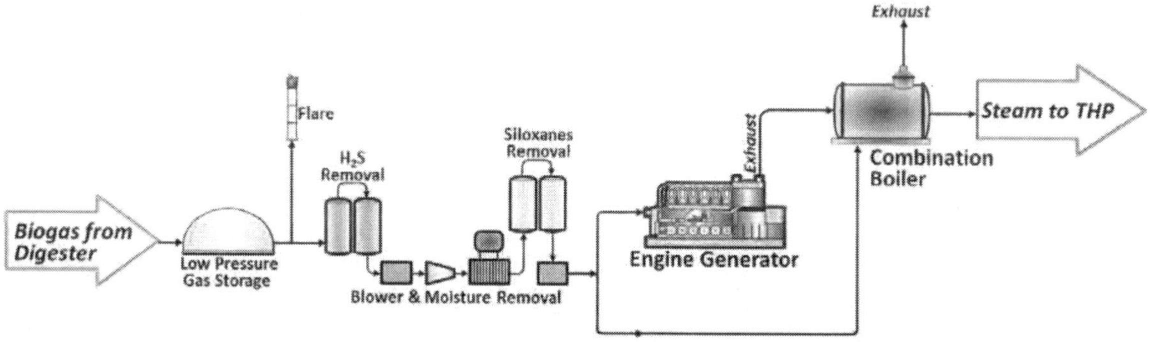

Figure 3 – Gas utilization schematic for combined heat and power with engine generators

Option 4– Compressed Natural Gas for Bus Use (Phased Purchase)
Biogas would undergo advanced treatment or upgrading to produce renewable natural gas and would be compressed to 3,500 pounds per square inch gage (psig). Compressed natural gas (CNG) would be stored on site and delivered under pressure to CNG trailers for transfer to the City's bus depot. A new fueling station would be constructed at the bus depot. The fueling station would include an offload compressor system to allow CNG to be transferred from the CNG trailers to on-site CNG storage vessels. CNG supply to the fueling station would be supported by on-site compression of piped natural gas. This option was based on the phased purchase of buses using replacement frequency / age of fleet information provided by the City. Figure 4 provides a conceptual schematic of a gas upgrading solution for vehicle fuel which is relevant to all the vehicle fuel options.

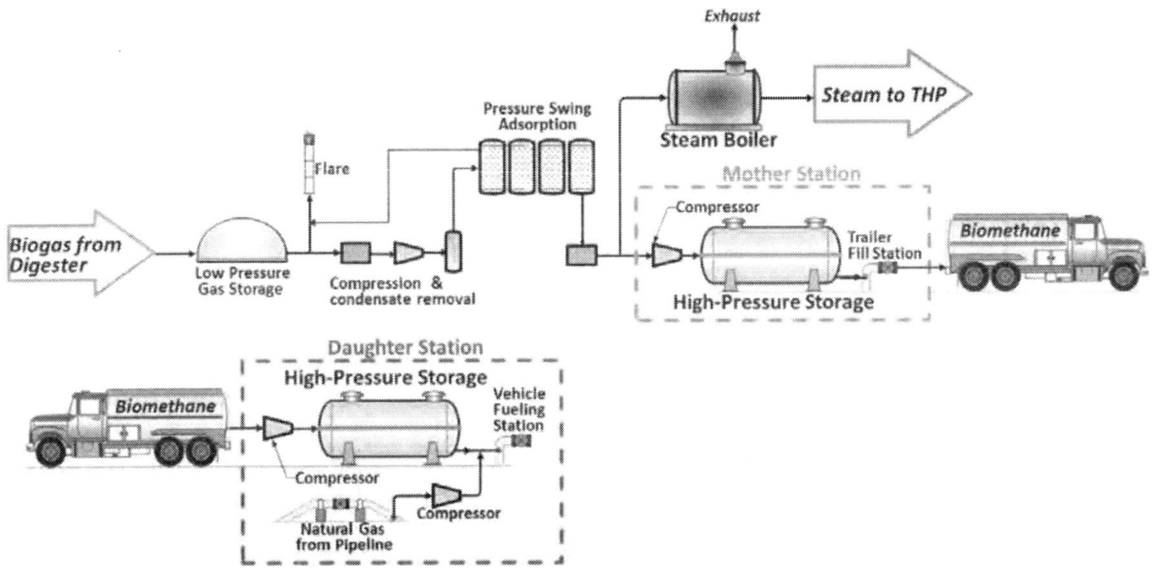

Figure 4 - Gas utilization schematic for vehicle fuel options (note that for options using both buses and garbage trucks it is likely that two daughter stations would be required – one at the bus depot and one at the garbage truck depot)

Option 5 – Compressed Natural Gas for Garbage Truck Use (Phased Purchase)
This option is essentially identical to option 4 except for the fact that the City's garbage trucks would be used instead of buses and the fueling station would be located at the garbage truck depot.

Option 6 – Compressed Natural Gas for Bus and Garbage Truck Use (Phased Purchase)
This option was a combination of options 4 and 5 and included fueling stations at both the bus and garbage truck depots.

Option 7&8 – Partial use of CHP or engine driven blowers followed by future conversion to vehicle fuel
During the screening stage evaluation, the City expressed an interest in a phased solution using CHP initially followed by longer term implementation of conversion to vehicle fuel. The logic behind this solution was that initially, the CNG vehicle fleet capacity would be too small to make full use of the available biogas. The option would therefore allow the City to avoid (or reduce) flaring of gas initially by implementing a short term solution, pending a longer term solution whereby the City's vehicles would be converted to CNG. For option 7, a 2MW CHP facilitywas considered. Under the requirements of the local electricity Utility, this would result in the City selling the power produced to the grid as 'purchased power'. For option 8, a 1 MW CHP facility was considered. This would allow the power produced to be 'net metered', allowing the City to use the power produced to offset site power consumption and achieve a higher effective rebate

for the energy produced. Options 7 and 8 are summarized in the conceptual schematic in Figure 5.

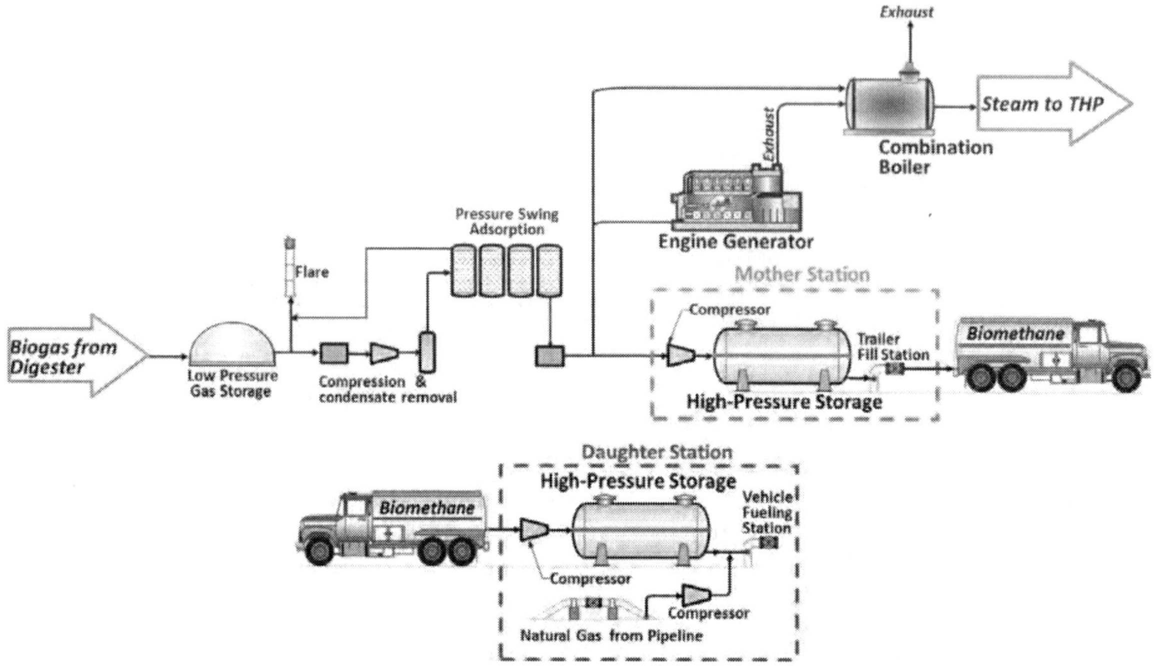

Figure 5 Gas utilization schematic for hybrid option using engine generators initially followed by gas clean-up for vehicle fuel.

METHODOLOGY

The basis of the evaluation was a year by year net present worth assessment intended to identify significant differences between the options being considered in terms of expected life cycle cost outcome. The evaluation was based on the biogas production rates shown in Table 2.

Table 2 – Projected Biogas Production

Parameter	Units	Current	Near Term	Ultimate
Year	-	2015	2032	2040
Biogas flow	SCFM	634	1,158	1,512
Biogas energy (lower heating value)	million BTU/h	21	39	50

Key assumptions relating to each option considered are provided in Table 3.

Table 3 – Key assumptions associated with each gas utilization option evaluation

Parameter	Units	Value
Engine Driven Blower		
Engine driven blower availability	%	95%
No. of units	No.	1
Capacity	SCFM	44,361
Blower mechanical power input per unit air flow	kW/SCFM	0.035
CHP Engine Generators		
CHP engine availability	%	95%
CHP engine parasitic load	%	5%
Capacity	kW	2,000
No. of units	number	3
Total capacity	kW	6,000
Gas Cleaning to CNG Vehicle Fuel		
Methane loss for advanced gas cleanup	%	5%
Bus fleet	no. of vehicles	98
Garbage truck fleet	no. of vehicles	109
RNG Usage - Buses	DGE per day	96
RNG Usage - Garbage Trucks	DGE per day	23
Rate of Replacement of Bus Vehicles	Vehicles per year	10.0
Rate of replacement of garbage truck vehicles	Vehicles per year	12.5
Useful lifespan of CNG bus	years	12
Useful lifespan of CNG garbage truck	years	7
Number of CNG tube trailers required	number	3
Incremental cost of CNG bus	$ per vehicle	$45.1 k
Incremental cost of CNG garbage truck	$ per vehicle	$38.1 k

Unit rates used to calculate operating costs are summarized in Table 4.

Table 4 – Operating Cost Assumptions

Parameter	Units	Value
Electricity Purchase Rate	$ per kWh	0.076
Electricity Sales Rate	$ per kWh	0.053
Diesel fuel cost	$ per gallon	2.60
Gas upgrading power use	kWh/scfm	0.46
Electrical consumption for CNG fuelling station	kWh/DGE	0.87
Operating man hours per day for CNG fuelling station	h/d	24
Cost of labor	$/h	34.09
Value of RINs	$/DGE	0 to 2.00
Alternative fuel excise tax credit	$/GGE	0 to 0.50

Equipment costs were estimated based on a combination of vendor quotes, previous project costs (adjusted for inflation and plant capacity) and literature information. Project costs were calculated for each option. Maintenance costs were estimated based on experience using a fixed percentage of equipment costs for each piece of equipment and using vendor information where available.

For each option, a life cycle evaluation was carried out. Financial rates and project term used for the life cycle evaluation are listed in Table 5.

Table 5 – Financial Rates

Item	Units	Value
Discount rate (cost of capital)	%	4.0%
Inflation	%	2.0%
Term of evaluation	Years	20

Regarding the technology evaluation which was conducted to evaluate gas cleaning technologies, key assumptions are defined in Table 6.

Table 6 – Assumptions for Gas Upgrading Technology Evaluation

Cost Item	UNITS	PSA 1	PSA 2	PSA 3	Amine Scrubbing 1	Amine Scrubbing 2	Membrane 1	Membrane 2	Membrane 3
Capacity per train	scfm	625	589	625	727	625	625	625	625
Methane capture	%	95.1	97.1	96.1	99.9	99.9	99.5	99.6	98.0
Methane in product gas	%	98.0	97.1	97.0	99.0	98.9	98.5	99.4	97.5
System uptime	%	95.0	95.0	95.0	96.0	96.0	96.0	96.0	96.0
O&M labor	hrs/day	4	4	4	2	2	2	2	2
Power use	kW/scfm	0.52	0.42	0.38	0.170	0.142	0.410	0.941	0.352
Product gas pressure	psig	90	90	110	1	1	235	320	246
Thermal oxidizer power use	kW/scfm	0.009	0.009	0.009	~	~	~	~	~
Amine chemical usage	lb/yr/scfm	~	~	~	0.7	16.2	~	~	~
Heating energy	kW/scfm	~	~	~	0.187	1.09	~	~	~

RESULTS

The options identified in Table 1 were subjected to a detailed technical evaluation. For each option, a full energy balance was calculated for each year of operation of the facility being evaluated over a 20 year life cycle. An example of the energy balance for Option 8 is provided in Figure 6 which shows the balance of biogas utilized by each component of the process during each year of the plant operation. As shown, for this option, initially a portion of the biogas would be flared due to the 1MW CHP not having sufficient capacity to make full use of the biogas. In year 5, conversion to CNG vehicle fuel would commence and would ramp up as the City converted buses to operate on CNG.

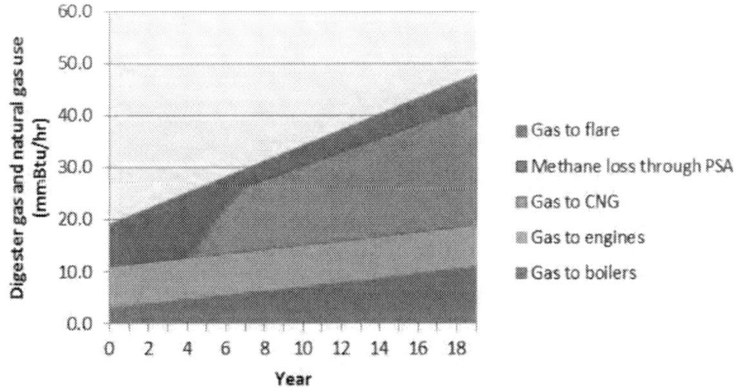

Figure 6 – Energy balance for Option 8 – 1MW CHP followed by conversion to CNG vehicle fuel

Figure 7 provides an overview of the results, showing how much biogas was converted to various uses on an average basis over 20 years.

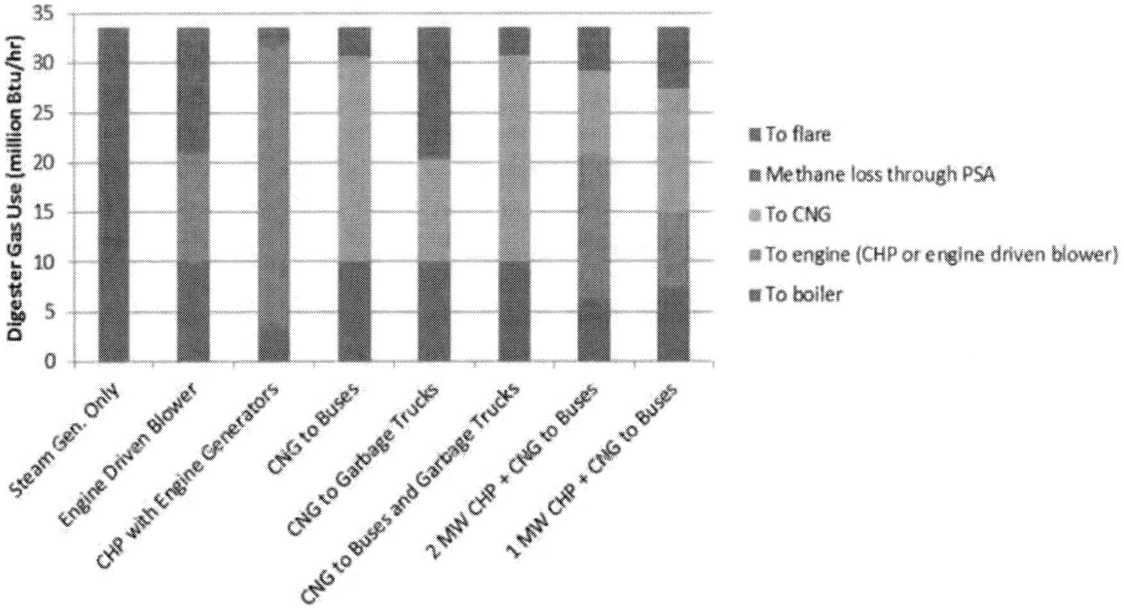

Figure 7 - Summary of biogas utilization from the energy balances associated with each option (values shown are averages over the life of the plant)

The economic viability of each of these options was evaluated on a life cycle cost basis to determine the most financially advantageous option for the City. Results are presented in Figure 8.

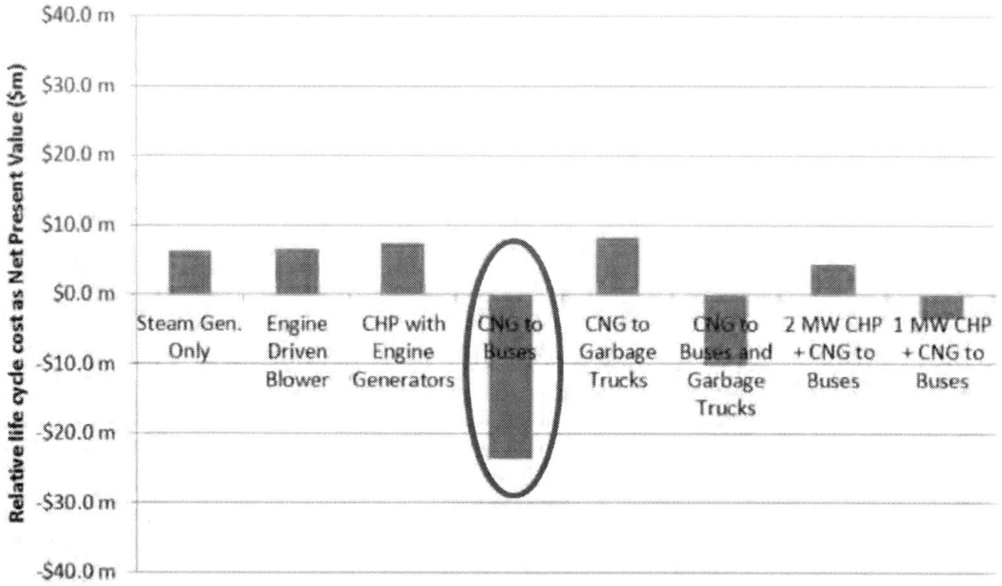

Figure 8 – Net life cycle costs for each gas utilization option shown as present values (negative is a saving)

The results show clearly that converting the biogas to CNG for use in the City's buses is by far the most cost advantageous option for the City. On top of the raw financial benefits which are shown in Figure 8 above, there are also currently various funding and green incentives in place which have the potential to further improve the life cycle cost outcome. Figure 9 shows the impacts of the following incentives and tax credits on the life cycle cost outcome.

- Federal funding which currently covers 80% of bus purchase costs (applied to the incremental cost difference between a diesel and a CNG vehicle)
- Renewable Identification Numbers (RINs) which are a tradable commodity associated with the production of renewable vehicle fuel under the Renewable Fuels Standard.
- The alternative fuel tax credit, which currently incentivizes the use of CNG vehicle fuel.

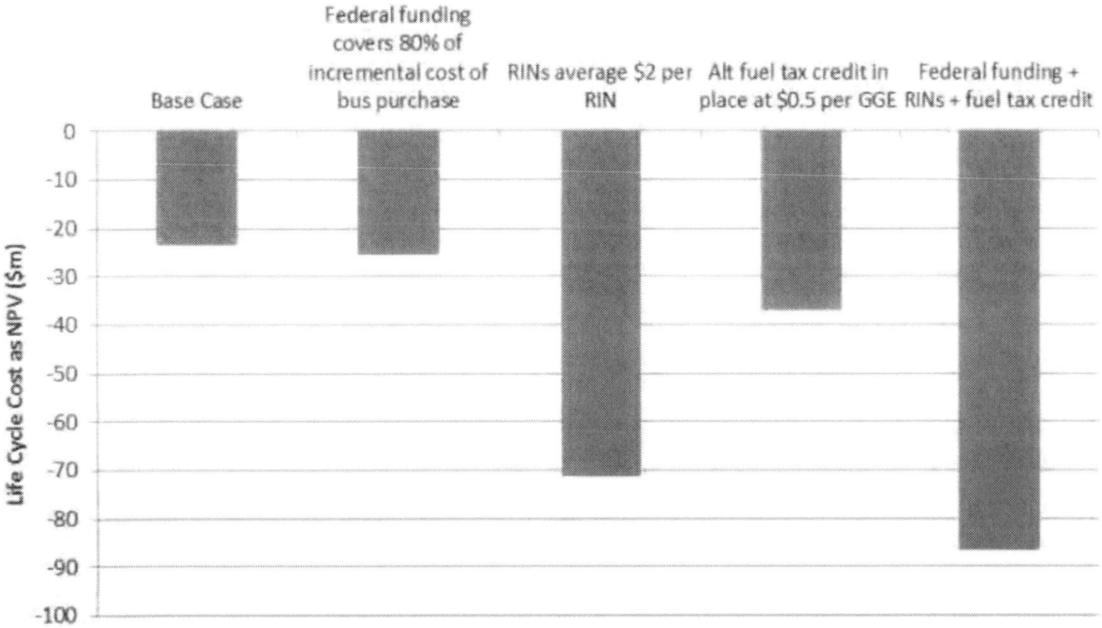

Figure 9 – Impact of federal funding of bus purchase, RINS and alternative fuel tax credits on the life cycle cost outcome of the option to convert biogas to CNG for use in the City's buses

The results in Figure 9 show that federal funding of bus purchase, RINS and alternative fuel tax credits could lead to a significant improvement in the life cycle cost outcome compared to the raw cost results, if these incentives continue to remain in place over the 20 year life cycle of the project. Obviously it is important to consider that there can be no certainty in this, as all of these impacts could potentially be affected by government policy.

Based on the evaluation, conversion of biogas to vehicle fuel for use in the City's buses was identified as the best long term option for the City. However, during the course of the study, it transpired that the local gas utility had a requirement to build a new large natural gas main through the land owned by the City adjacent to the Neuse River RRF. This presented an opportunity to the City to achieve more value for the gas during the early phases of the project (pending conversion of the City's bus fleet to CNG) and to provide a more straightforward means of transferring gas in future from the Neuse River RRF to the Vehicle Depot.

The previous evaluation work was updated, confirming that utilizing the proposed gas main for transfer of the RNG product from the Neuse River RRF was an economically attractive option and this option was selected by the City. Results showing the addition of the gas pipeline option are provided in 10.

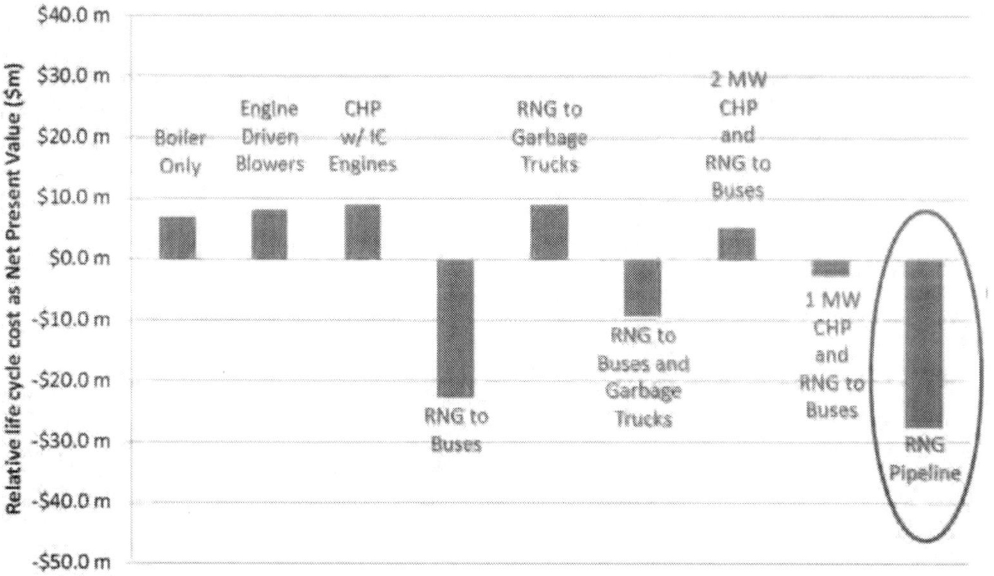

Figure 10 – Life cycle costs updated with RNG to pipeline option

SELECTION OF BIOGAS UPGRADING TECHNOLOGY

Following the selection of a biogas utilization option which required cleaning of biogas from the Neuse River RRF to pipeline quality, it was necessary to determine the most appropriate technology for achieving the quality required by the local gas utility.

Various options were evaluated including selective membranes, pressure swing adsorption, water scrubbing and amine scrubbing. Some of the options considered are shown in Figure 11. While water scrubbing was initially considered, it was later ruled out of the evaluation because the leading vendors of water scrubbing technology in the USA recommended a PSA system instead.

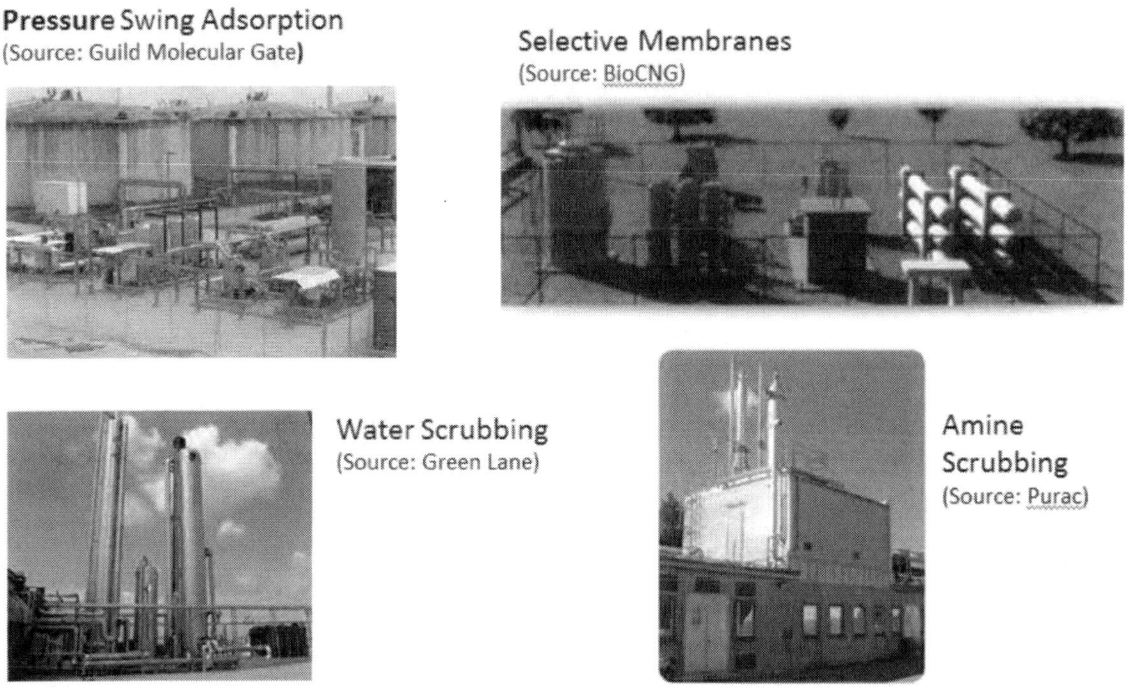

Figure 11 – Some of the biogas upgrading technologies considered

Capital costs associated with the various options are summarized in Figure 12. Due to the phased installation of additional gas trains as the load on the Neuse River RRF increases, a residual value was assigned to gas trains installed part way through the period covered by the life cycle cost evaluation. It was found that the amine scrubbing options generally had a higher capital cost. The membrane and PSA options generally had a similar capital cost.

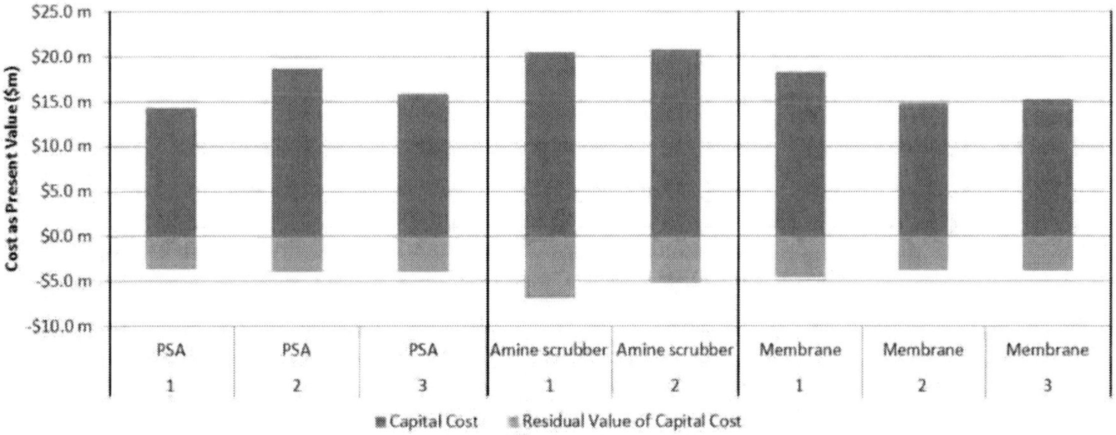

Figure 12 – Capital costs associated with each gas upgrading solution

Results of the evaluation showing a breakdown of operating costs are provided in Figure 13. As shown, one of the PSA options had significantly lower operating / maintenance costs than the other options. This was primarily due to the type of media used in the system, which does not require prior H_2S removal to prevent media poisoning.

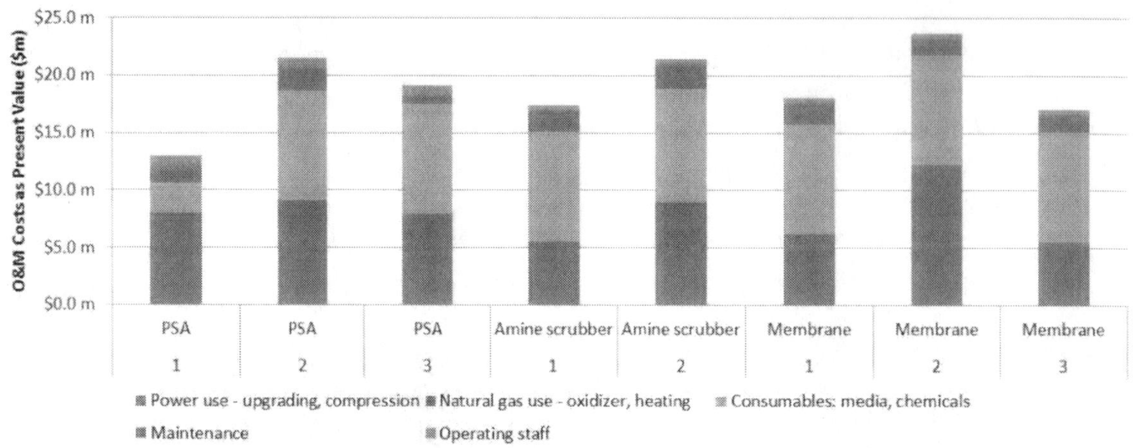

Figure 13 – Operating cost breakdown for each of the gas cleaning technologies

Life cycle costs for each technology option are presented in Figure 14. Due to the lower operating costs associated with PSA option 1, this option came out more favorably than the other options on a life cycle cost basis.

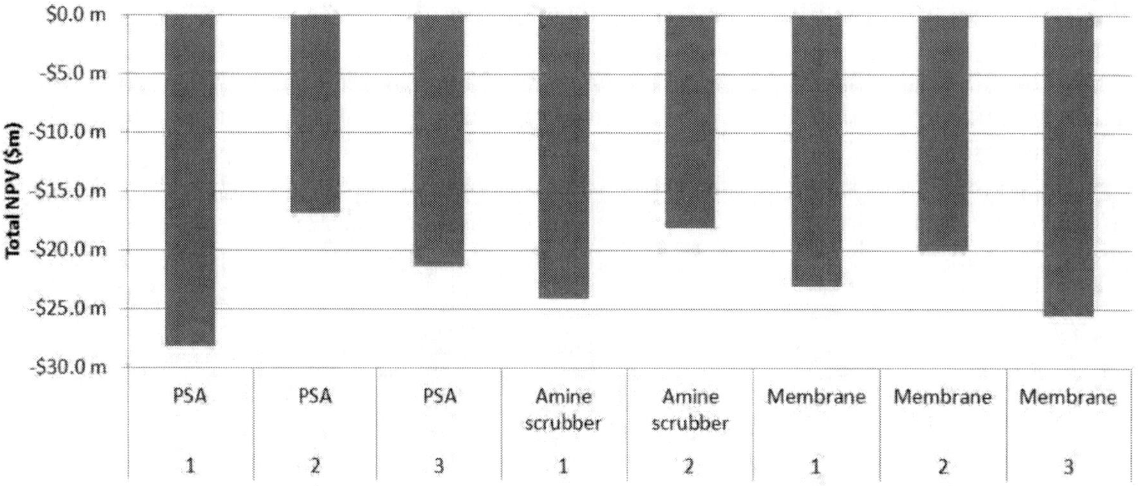

Figure 14 – Life cycle cost shown as net present value over 20 years (negative is favorable)

THE PATH FORWARD

Following the evaluation work conducted during this study, the City has continued to explore the use of CNG in the City's bus fleet. The City commissioned a separate feasibility study which looked in more detail at the cost benefit of using CNG in the City's buses. This supported the findings in this study and confirmed that use of CNG offers significant cost advantages over diesel fuel for the City bus fleet, even if the City were to purchase natural gas directly from the grid.

The local gas utility has confirmed that routing of the natural gas transfer main will be local to the Neuse River RRF and discussions are ongoing between the City and the utility.

The Bioenergy Recovery project is now in detailed design, a Principal Contractor for the scheme has been selected. The Project will be administered as a Construction Management at Risk (CMAR) Contract. Construction is expected to start in the first quarter of 2018.

Selection of the gas cleaning vendor will be achieved via the CMAR contract with the vendor selected based on life cycle cost, with hard costs and guaranteed performance factored into the determination.

Key Words

Anaerobic digestion, combined heat and power, compressed natural gas, gas utilization, vehicle fuel, pipeline injection, biogas, digester gas, gas upgrading.

What's the Best Use of Your Digester Gas? Well, it Depends

Ian McKelvey[1]*, Eron Jacobson[1], Peter Zemke[1], John Smyth[2]
[1]Brown and Caldwell, 701 Pike St, Suite 1200, Seattle, WA 98101
[2]King County Wastewater Treatment Division, 201 South Jackson Street, Seattle, WA 98104
*To whom correspondence should be addressed. Email: imckelvey@brwncald.com

ABSTRACT

A series of biogas utilization studies were performed for King County's Wastewater Treatment Division to determine the performance of the County's existing systems and identify preferred alternatives for utilizing biogas in the future. Although the County's treatment plants operate within close proximity to each other (within 25 miles) and are subject to the same financial and policy pressures, the results of the studies demonstrate how sensitive the preferred use of biogas at wastewater treatment facilities can be to site-specific conditions at a given facility. By completing the reviews using the same standardized approach for each treatment plant, the results for each plant can be compared to each other and an understanding of why a utility may select one means of biogas utilization over another can be understood.

KEYWORDS

Digester gas, biogas, biomethane, combined heat and power (CHP), biogas upgrading, renewable identification numbers (RINs), pipeline quality gas, effluent source heat pumps

INTRODUCTION

Increasingly, wastewater treatment agencies are seeking opportunities to reduce energy consumption at their wastewater treatment plants (WWTPs), both by reducing the consumption of energy and by more efficiently recovering energy from the wastewater treatment process. For those facilities that include anaerobic digestion of their residual solids, the methane rich digester gas (or biogas) produced during the solids stabilization process is a prime source of renewable energy available for recovery.

One such agency seeking to maximize resource recovery is King County's Wastewater Treatment Division (WTD). King County owns and operates three regional wastewater treatment plants in the greater metropolitan Seattle area that each utilize anaerobic digestion for solids stabilization. The County has a strong tradition of beneficially using the recovered biogas at these plants but recently hired Brown and Caldwell to determine the performance of the existing biogas utilization systems and identify a preferred alternative for optimizing biogas usage in the future at each facility. These studies were performed sequentially with the first taking place in 2012, the next in 2016, and the last still in progress.

The process used to generate and evaluate alternatives is not unlike the process used by many other communities. The baseline condition and performance of existing systems was established, alternatives were brainstormed and screened to a short-list, and the short-listed alternatives were evaluated in detail based on established financial, environmental, and operational objectives. What is unique about these studies is differences in the results can be attributed to factors outside of the evaluation process or the objectives used for the evaluation. In other words, because the same process was used to evaluate alternatives, the same agency was involved at all three plants, and the same objectives for a future biogas utilization program were used, the differing results from each study indicate that the preferred use of biogas at a facility is due to site-specific conditions.

This paper documents the existing biogas utilization systems at each of the WTD's three regional wastewater treatment plants and describes the approach and methods used to evaluate biogas utilization alternatives. The results are presented and conclusions are drawn indicating that there is no biogas utilization alternative that is more universally beneficial over all others; the preferred use of biogas at wastewater treatment facilities is ultimately dependent on site-specific conditions.

DESCRIPTION OF EXISTING SYSTEMS

King County's wastewater treatment system serves 1.7 million people in the metropolitan Seattle region. The County's service area covers 424 square miles, providing trunk sewer conveyance and regional wastewater treatment to 34 wholesale customers. King County has been providing regional wastewater treatment services since 1965 and in 2015 treated an average of 178 million gallons of per day (mgd) of sewage. Regional wastewater treatment occurs at three regional wastewater treatment plants: South Treatment Plant (South Plant) in Renton, Washington, West Point Treatment Plant (West Point) in Seattle, Washington, and Brightwater Treatment Plant (Brightwater) near Woodinville, Washington. Figure 1 is a map showing the WTD's service area and treatment plants.

These facilities are different sizes, utilize different treatment approaches, and are fed from different types of sewer systems. One similarity between all three facilities is that they all use mesophilic anaerobic digestion to produce Class B biosolids and biogas. A summary of the characteristics of the three regional treatment plants is presented in Table 1.

To initiate a study of biogas utilization alternatives at each plant, a baseline of the existing biogas utilization and heating systems at each facility was completed. This included a site visit, interviews with plant operators, and evaluation of data provided from King County. This baselining effort established the performance and condition of the existing systems and provided biogas production and heat demand parameters needed to develop alternatives. The historical and current use of biogas at each facility is summarized in the following sections.

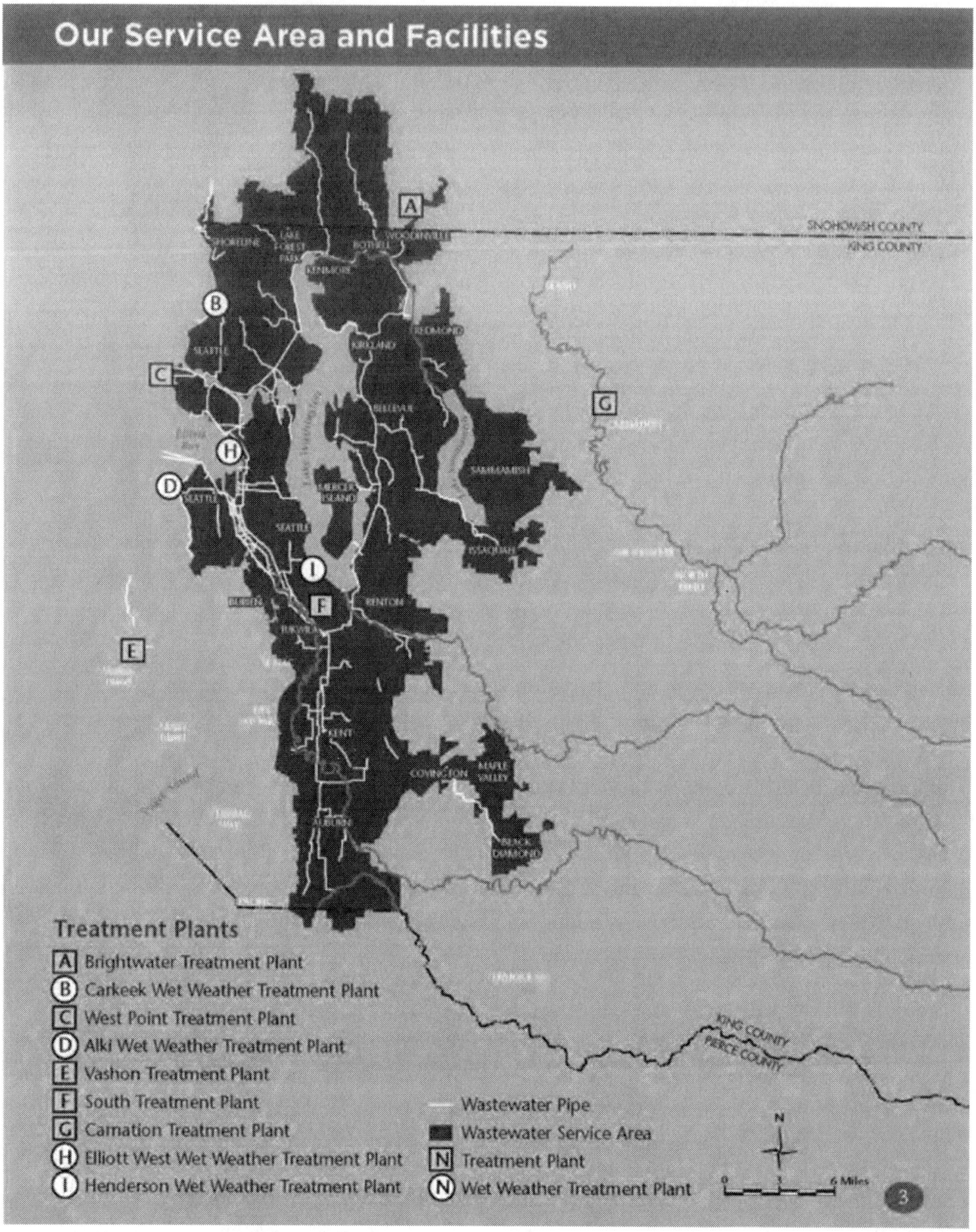

Figure 1. King County WTD's Service Area and Treatment Plants.

Table 1. King County WTD's Regional Treatment Plant Characteristics

Characteristic	South Plant	West Point	Brightwater
Location	Renton	Seattle	near Woodinville
Sewer type	separated	combined	separated
Design capacity			
- **Maximum month**	144 mgd	215 mgd	41 mgd
- **Peak capacity**	325 mgd	440 mgd	100 mgd
Average annual flows (2015)	69 mgd [a]	92 mgd	17 mgd
Secondary treatment method	conventional activated sludge	high-purity oxygen activated sludge	membrane bioreactor
Solids thickening process	dissolved air flotation	gravity belt	gravity belt
Solids stabilization process	mesophilic anaerobic digestion	mesophilic anaerobic digestion	mesophilic anaerobic digestion
Number of active digesters (active digestion volume)	4 (10.4 million gallons)	5 (9.5 million gallons)	3 (3.8 million gallons)
Biosolids dewatering process	centrifuge	centrifuge	centrifuge
Annual biosolids produced (2015)	13,400 dry tons [b]	11,600 dry tons	3,300 dry tons

a – South Plant also treated 24 million gallons of hauled septage in 2015.
b – South Plant also treated raw sludge from the Vashon and Carnation WWTPs in 2015, totaling 140 dry tons.

South Plant

South Plant has provided secondary treatment since it began operation in 1965 but has only included solids treatment since 1987 (prior to that, residual solids were transferred to West Point). As part of the enlargement that implemented solids treatment, a biogas scrubbing system was implemented which upgrades the plant's biogas to a pipeline-quality biomethane product for sale to the local natural gas utility. To maximize the amount of biogas available for use in this system, South Plant originally used effluent-source heat pumps to meet its process and space heating demands. These systems were expanded when the plant underwent a third enlargement in 1995.

Due to a number of factors, the County implemented a hot water boiler and a cogeneration system in 2002 and 2004, respectively. The two primary drivers were the dramatic increase in electricity prices in the western states in the early 2000's and the loss of efficacy of the plant's heat pumps when the original R12 refrigerant was banned and replaced with R134a refrigerant.

As a result, the plant now uses the gas-fired hot water boiler as its primary source of heat. Table 2 lists the biogas and heating equipment at South Plant.

Table 2. South Plant Biogas Utilization and Heating Equipment

System	Type	Installed	Capacity
Gas scrubbing	High-pressure water solvent	1987 and 1995	1,667 scfm [a]
Heat pumps	Effluent source, centrifugal compressor	1987 and 1995	5 at 6.0 MMBtuh each [b]
Hot water boiler	Fire tube	2003	11.7 MMbtuh
Cogeneration	Combined cycle turbines	2006	2 at 3.5 MW [c]

a – Installed capacity. Operators indicate actual capacity may be up to 30% lower.
b – Five units are installed but only one remains operational.
c – The system also has a 1.0 MW steam turbine that has not been used to date.

In terms of the size of the biogas utilization system, hourly biogas production at South Plant averaged 920 standard cubic feet per minute (scfm) in 2015 with 95% of biogas production ranging between 700 and 1150 scfm (see Figure 2). Figure 3 shows the plant's heat demand between 2010 and 2015. Average daily heat demand was 5 million British thermal units per hour (MMBtuh) during this period with a peak average week of 8 MMBtuh and a minimum average week of 1 MMBtuh. Specific end uses of the plant's biogas and sources for the plant's heat demand were documented during the study in 2012. Figure 4 indicates that the majority of the gas produced at South Plant was either scrubbed and sold to the natural gas utility (Puget Sound Energy, PSE) or burned in the boiler, while Figure 5 shows that most of the plant's heat needs are met by the hot water boiler.

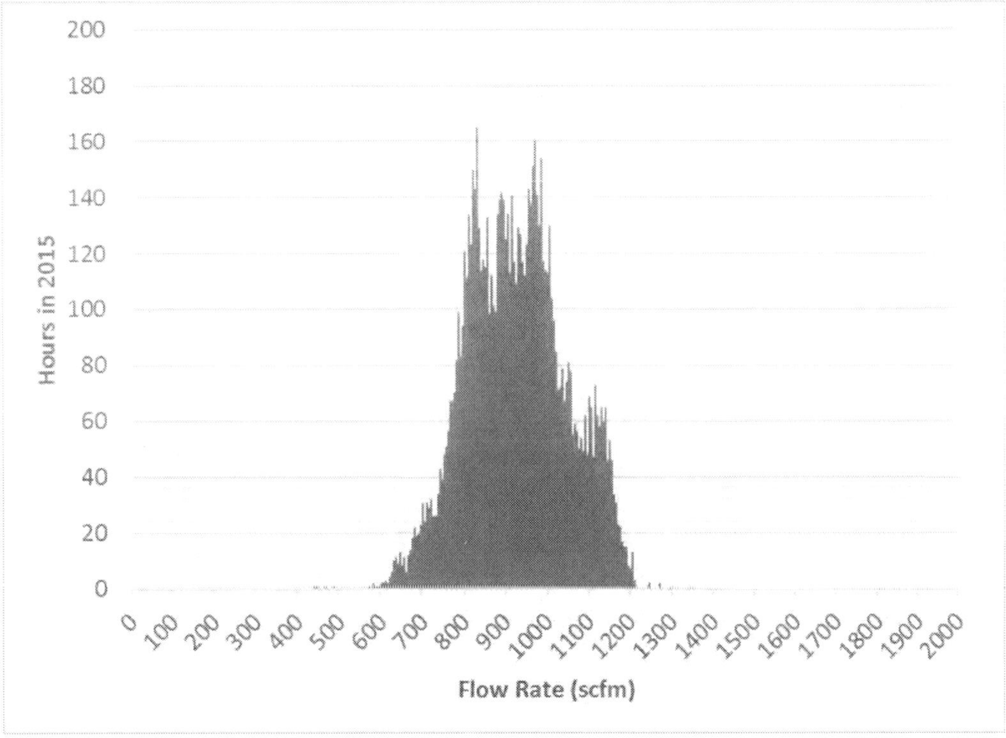

Figure 2. Histogram of hourly biogas production at South Plant in 2015 (Bin size: 5 scfm).

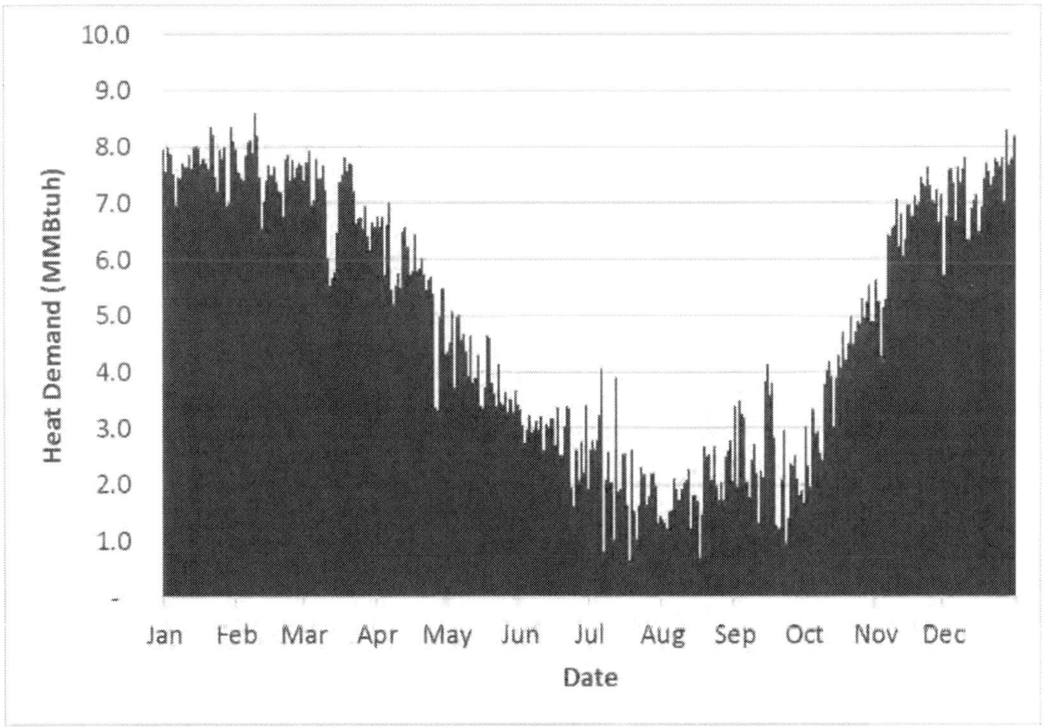

Figure 3. Estimated average daily heat demand at South Plant between 2010 and 2015

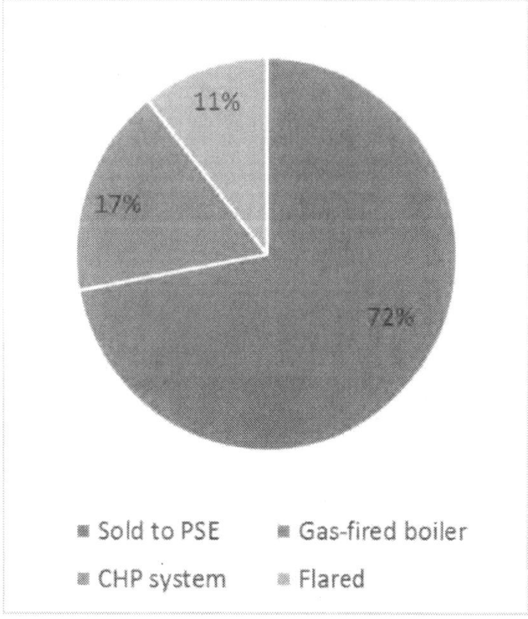

Figure 4. Biogas use at South Plant in 2012.

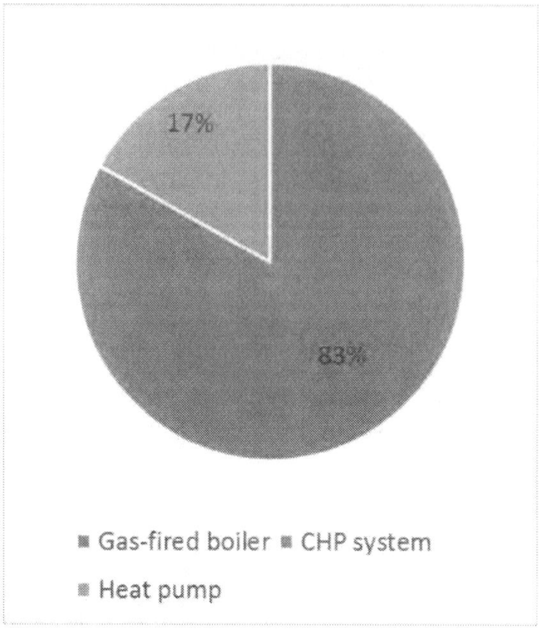

Figure 5. Heat supply sources at South Plant in 2012.

Due to the age of the existing gas scrubbing system and the lack of redundancy in the existing heating system, the biogas utilization study for South Plant focused on the best system for utilizing all of the plant's biogas while reliably satisfying the plant's heat demand.

West Point

West Point was constructed as a primary-only treatment facility with anaerobic digestion in 1966. The unreliable power supply to the plant led to the plant historically using biogas-fired engines to drive the plant's raw sewage pumps (RSP) and effluent pumps; heat from the engines was recovered to meet the plant's heat demands, supplemented by boilers. In the 1980's, with the additions of Digesters 4 and 5, the County also added a combined heat and power (CHP) system consisting of three 1.2 megawatt (MW) reciprocating internal combustion (IC) engines.

With the addition of the secondary treatment process in 1995, the plant's energy profile was changed and the biogas utilization systems were modified. The effluent pumping station engines were replaced with electric motors (but the RSP engines were replaced in-kind) and gas-fired hot water boilers were installed as the plant's primary source of heat. In 2011, a new CHP facility was installed to replace the aging existing system. Table 3 lists the current biogas and heating equipment at West Point.

Table 3. West Point Biogas Utilization and Heating Equipment

	Type	Installed	Capacity
Raw sewage pump engines	Reciprocating internal combustion	1996	4 at 440 bhp
Combined heat and power	Reciprocating internal combustion	2011	2 at 2.3 MW [a]
Hot water boiler	Fire tube	1992	2 at 6.9 MMBtuh, 1 at 15.5 MMBtuh
a – Upstream gas conditioning is only sized for operation of one engine at a time			

Hourly biogas production at West Point averaged 1050 scfm in 2014 with 95% of biogas production ranging between 820 and 1250 scfm (see Figure 6). Figure 7 shows the plant's estimated heat demand in 2014 with an average heat demand of 9 MMBtuh, a peak average week of 15 MMBtuh, and a minimum average week of 4 MMBtuh. Based on this gas production and heat demand, the plant's biogas consumption in 2014 is shown in Figure 8 while the sources of heat are indicated in Figure 9. These figures indicate that the plant primarily utilizes the CHP facility as a source of heat and beneficial end use for the plant's biogas.

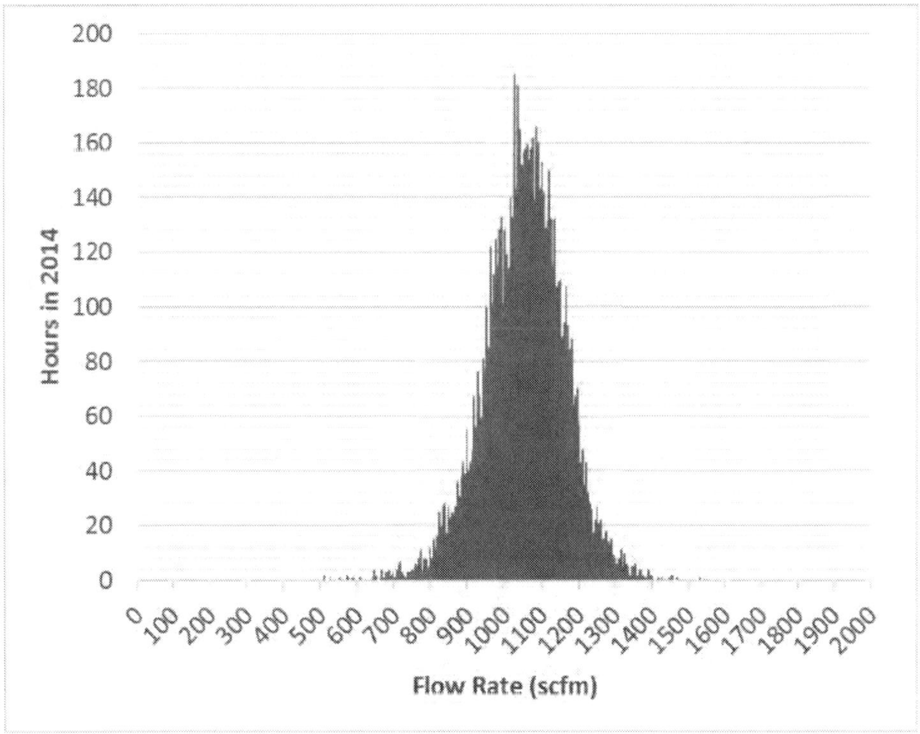

Figure 6. Histogram of hourly biogas production at West Point in 2014 (Bin size: 5 scfm).

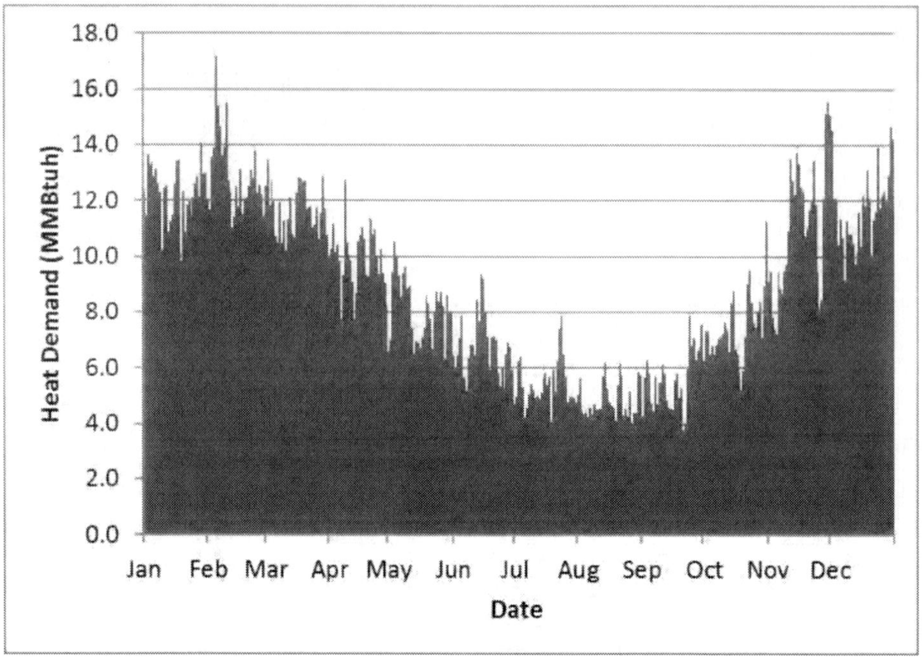

Figure 7. Estimated daily heat demand at West Point in 2014

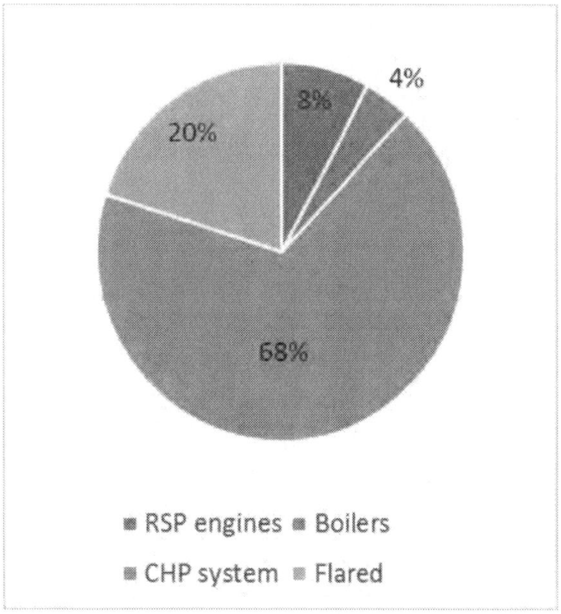

Figure 8. Biogas use at West Point in 2014.

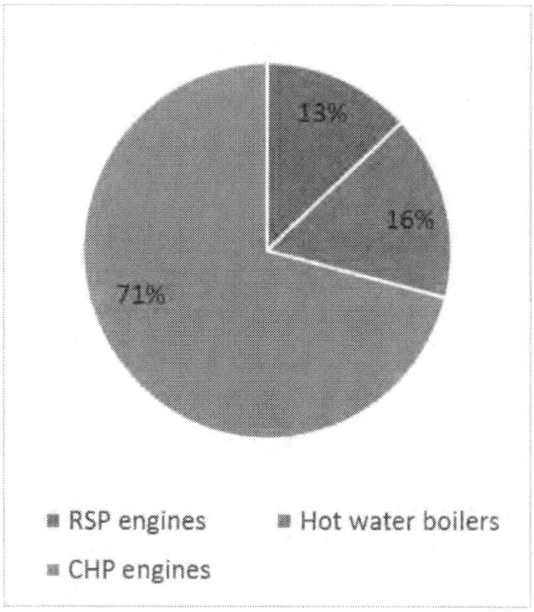

Figure 9. Heat supply sources at West Point in 2014.

Due to the need to operate the CHP engines to meet contractual requirements and the decision to continue to operate the raw sewage pump engines to eliminate concerns regarding backup electrical power, the biogas utilization study at West Point focused on best utilizing the surplus gas being flared in enclosed waste gas burners.

Brightwater

The Brightwater plant is the newest treatment plant in WTD's system, beginning full operation in 2012. The plant meets its process and space heating demands utilizing gas-fired hot water boilers (two at 17.2 MMBtuh). A CHP system was removed from the design of the plant to save capital costs and as a result, any biogas produced in excess of the boiler fuel demand is flared.

Between November 2015 and November 2016, average hourly biogas production at Brightwater was 290 scfm with 95% of biogas production falling between 210 and 390 scfm (see Figure 10). During this same period, the average heat demand was 5 MMBtuh with peak average week of 9 MMBtuh and a minimum average week of 2 MMBtuh (see Figure 11). Figure 12 shows the breakdown of biogas consumption at Brightwater indicating that just over half of the biogas produced is used to fuel the boilers.

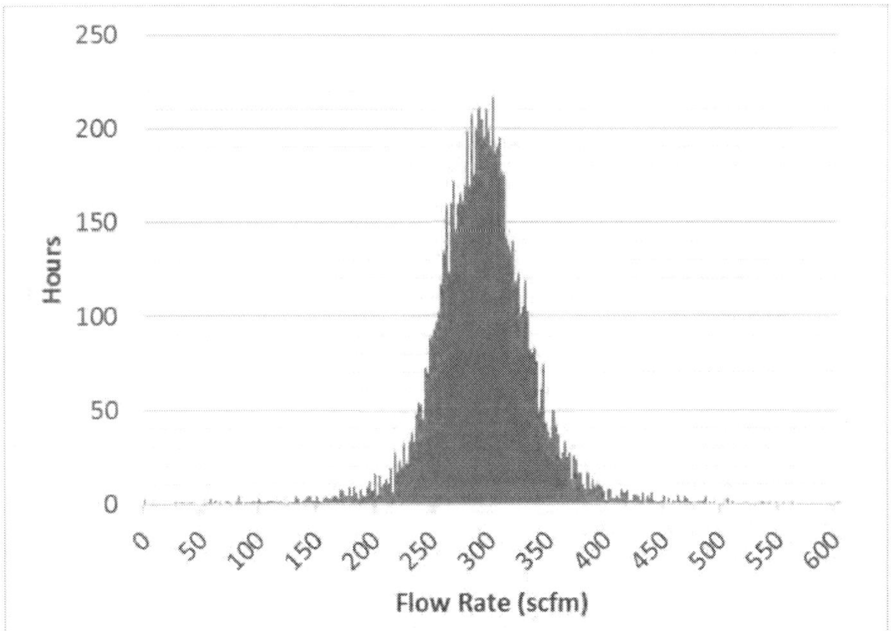

Figure 10. Histogram of hourly biogas production at Brightwater from November 2015 to November 2016 (Bin size: 2 scfm).

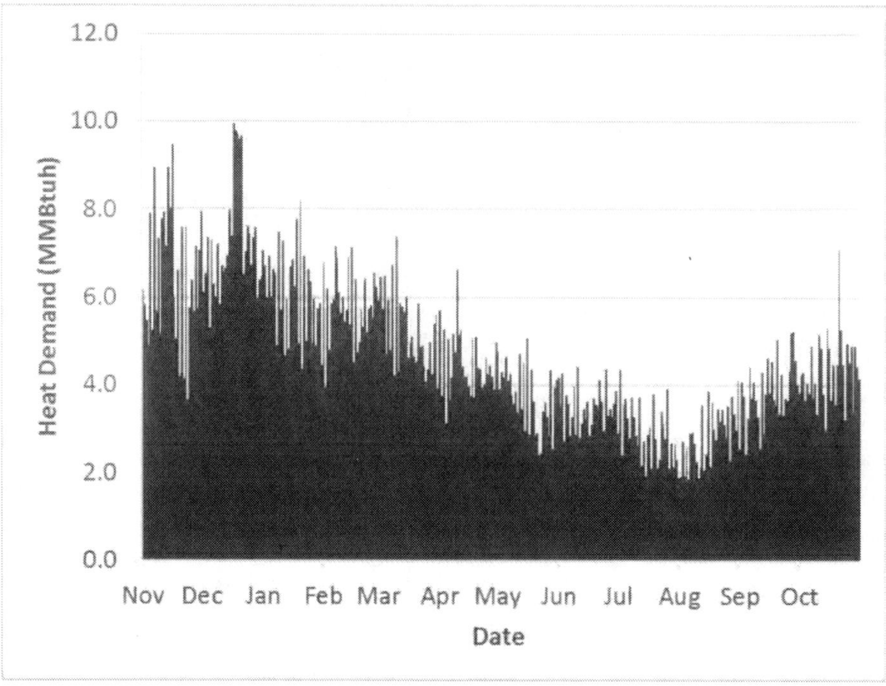

Figure 11. Estimated daily heat demand at Brightwater between November 2015 and November 2016.

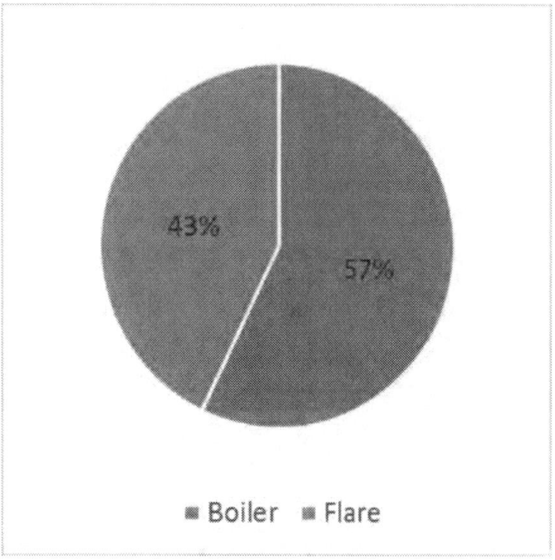

Figure 12. Biogas use at Brightwater Between November 2015 and November 2016.

With the lack of a beneficial use outside of the gas fired boilers, the study at Brightwater is focused on the best use of all of the plant's biogas while meeting the plant's heat demand.

Table 4 summarizes the biogas production and heat demand parameters for all three treatment plants. With this data and the baselining of the existing systems complete, the studies next turned to the development and evaluation of alternatives.

Table 4. Biogas Production and Heat Demand at WTD's Three Regional WWTPs

Parameter	South Plant	West Point	Brightwater
Average gas production (scfm)	920	1050	290
Gas production range [a] (scfm)	700-1150	820-1250	210-390
Average heat demand (MMBtuh)	5	9	5
Minimum average week heat demand (MMBtuh)	1	4	2
Peak average week heat demand (MMBtuh)	8	15	9
a – 95% of total production observed			

EVALUATION OF BIOGAS UTILIZATION ALTERNATIVES

To compare the biogas utilization alternatives for each facility, it was first necessary to identify the County's biogas utilization objectives. Objectives for each plant were developed in a workshop setting with plant operations, process, and reliability-centered maintenance staff; WTD resource recovery, project planning and delivery, and project management staff; and Department of Natural Resources and Parks (DNRP) policy staff. The objectives were divided into three categories: financial, environmental, and operational. To reflect the relative importance of the objectives compared to each other, a weighting was applied such that objectives considered more important than others received a higher weight.

The objectives identified were largely the same for each plant. Financial objectives were focused on minimizing the life-cycle cost of any end use. Environmental objectives were based on a compilation of goals identified in the County's energy plans and strategic climate action plans. Operational objectives focused on minimizing risk while maximizing system reliability and flexibility. A summary of the objectives used in all three studies is provided in Table 5.

Table 5. Objectives Used for Biogas Utilization Alternatives Analysis

Objective	Relative Weighting
FINANCIAL	
Maximize the program's net present value [a]	8
Minimize sensitivity to commodity price changes [b]	0
ENVIRONMENTAL	
Reduce use of and expenditures for energy	1.6
Reduce greenhouse gas emissions	1.6
Increase production of renewable energy	1.6
Increase consumption of renewable energy	0.6 (not in the WP or BW list)
Reduce quantity of digester gas sent to the flares	0.6 (not in the SP or BW list)
Invest in alternative fuel transit and fleet vehicles	0.6

OPERATIONAL	
Maximize system redundancy and reliability	1.3
Maximize system operational flexibility	0.6
Minimize WTD labor requirements	1.6
Minimize reliance on outside contracts	0.3
Minimize technical risk	1.6
Minimize air quality treatment requirements	0.6
a – South Plant study broke this objective out into its individual components (i.e., capital costs, operation and maintenance costs, revenues, and grants) b – Provided for information only. A detailed sensitivity analysis for each alternative was performed. WP = West Point, BW = Brightwater	

When used with a 1 through 5 scoring scale, the weighting of objectives allowed for a maximum score of 100 points. The relative weighting reflected the reality that financial objectives held more sway over the environmental and operational objectives but was not the only important factor. Within the environmental factors, the weighting shows the County's commitment to reducing energy consumption and greenhouse gas emissions. Lastly, the operational objectives were weighted such that technical risk, the reality of limited staffing, and system reliability were more critical than operational flexibility, the need for third party maintenance, or the risk of air quality regulatory changes.

As Table 5 shows, the objectives and their relative weighting were largely the same for all three studies. This reflected that the County's goals for a biogas utilization system were largely independent of the specific treatment plant generating the biogas.

Once the objectives for a biogas utilization system were defined, alternatives were brainstormed with plant staff and evaluated based on these objectives. The alternatives considered for each plant are listed in Table 5 below. Ultimately, the alternatives for each plant were screened down to three alternatives for a more detailed evaluation using the objectives identified in Table 5. The alternatives carried forward for a more detailed consideration are identified in bold text in Table 6.

Table 6. Biogas Utilization Alternatives Considered at WTD's Three Regional WWTPs

South Plant	**West Point**	**Brightwater**
Status quo (boiler + refurbished scrubbing system)	**Status quo (flare surplus gas)**	**Status quo (boilers)**
Status quo to produce renewable compressed natural gas (rCNG)	Status quo + improved boiler controls	**Existing boilers + gas scrubbing for pipeline injection**

Status quo with sale to a third party	Status quo + upsize existing CHP conditioning	Existing boilers + gas scrubbing for vehicle fuel
New gas scrubbing system	**Additional CHP with one IC engine**	Heat pumps + gas scrubbing for pipeline injection
New gas scrubbing system to produce rCNG	**Additional CHP with two IC engines**	**Heat pumps + gas scrubbing for vehicle fuel**
New gas scrubbing system with sale to a third party	Additional CHP with microturbines	Heat pumps + gas scrubbing, trucked off site
New boilers + refurbished scrubbing system	Biogas upgrading for pipeline injection	**IC engines**
New boilers + refurbished scrubbing system to produce rCNG	**Biogas upgrading, trucked off site**	IC engines + heat pumps
New boilers and gas scrubbing system	Biogas upgrading, trucked off site with storage	Microturbines
New boilers and gas scrubbing system to produce rCNG	Convert RSP engines to electric motors	
New heat pumps + refurbished scrubbing system	New heat pump	
New heat pumps + refurbished scrubbing system to produce rCNG	Convert electric motors to gas-fired engines	
New heat pumps and gas scrubbing system	Add biogas storage	
New heat pumps + gas scrubbing system to produce rCNG	Organic Rankine cycle	
Operate existing CHP facility full-time (refurbished gas scrubbing system)	Convert standby generator to CHP	
Operate existing CHP facility full-time (new gas scrubbing system)		
Convert CHP system to biogas operation and run full-time		
New biogas IC engines		
New biomethane IC engines + refurbished gas scrubbing system		
New biomethane IC engines and gas scrubbing system		

The alternatives considered generally focused on a few groups of technologies:

- Combined heat and power: Systems that generated both electrical power and heat were considered at all three plants. In particular, reciprocating internal combustion (IC) engines were considered at all three plants but microturbines and modifying South Plants existing combined-cycle turbines were also considered.
- Biogas upgrading: Systems that separated the methane in the biogas from other constituents to produce a biomethane product were also considered at all three plants. Specific technologies included high-pressure water solvent systems, pressure swing adsorption systems, and membrane separation systems. End uses for the biomethane that were considered included pipeline injection and renewable compressed natural gas (rCNG) vehicle fuel use
- Heating: Because meeting plant heating demands is inextricably linked to biogas utilization options, various means of meeting heating needs were considered, both in terms of using the biogas directly for heating (e.g., hot water boilers) or using another energy source in order to maximize the availability of biogas for another end use (e.g., electric heat pumps).

Detailed evaluation of the three selected alternatives at each plant (plus the status quo alternative kept for baselining purposes) was completed next. This detailed evaluation included preliminary siting, a Class 5 construction cost estimate, a greenhouse gas emissions inventory, life-cycle cost analysis, and a sensitivities analysis for each alternative. The results from these efforts were used to reevaluate the scoring for each objective and select a preferred alternative.

A brief summary of the analysis for the South Plant and West Point plants follows. The Brightwater study is currently ongoing and this detailed analysis is in progress.

<u>South Plant</u>

Because of the age of the existing systems, the status quo for South Plant required the significant rehabilitation of the existing gas scrubbing system and the addition of new boilers to meet the plant's redundancy needs (to be located in a new building). This resulted in a relatively high capital cost for a status quo alternative. In addition, this alternative assumed that the new boilers would continue the plant's existing practice of using high-Btu fuel (i.e., natural gas or biomethane) to improve system reliability.

In one alternative, a new gas scrubbing system would be paired with boilers. For the new gas scrubbing system, a pressure-swing adsorption system was assumed as the base technology based on the design team's familiarity with the system and the relatively low cost compared to other biogas upgrading technologies considered (e.g., water solvent, membranes). To conserve construction cost, the study assumed a new PSA system could be installed in the same space occupied by the plant's existing gas scrubbing system (see Figure 13). To minimize greenhouse gas emissions and avoid the high energy consumption required to upgrade biogas, the boilers selected would operate on low-Btu (i.e. minimally conditioned) biogas.

For the alternative in which heat pumps would be used rather than boilers to meet the plant's heat demands, a new building would not be needed. Instead, the new heat pumps would be installed in the same location as the plant's existing heat pumps. High-temperature dual-stage heat pumps with screw compressors were assumed to minimize the operational challenges the plant had experienced previously with their single-stage centrifugal compressor heat pumps.

The CHP alternative at South Plant assumed three 1.2 MW IC engines would be installed in a new building. To fuel the new CHP system, a new gas conditioning system would be included to remove moisture, hydrogen sulfide, and siloxanes. The existing CHP system at South Plant runs on only high-Btu fuels and therefore does not already include this equipment.

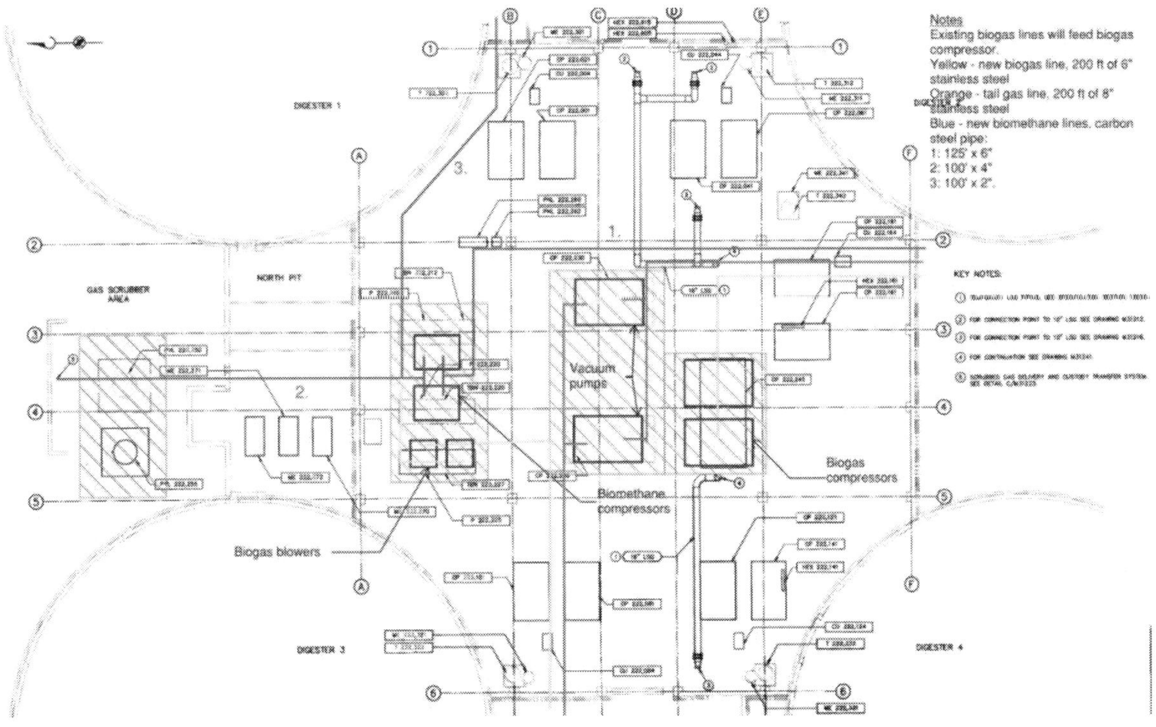

Figure 13. Preliminary equipment layout of a new biogas upgrading system at South Plant

Table 7 summarizes the major results from this analysis. The sensitivities analysis performed found that the results were most sensitive to the price of the generated biomethane or power (including associated Renewable Identification Numbers [RINs] or Renewable Energy Certificates [RECs]). In particular, the sales price the County could negotiate with the purchasing utility could ultimately determine whether any selected alternative would have a positive net present value.

Table 7. Results from detailed analysis of South Plant Biogas Utilization Alternatives

Parameter	Status Quo	Scrubbing + Boilers	Scrubbing + Heat Pumps	CHP System
Construction cost (2013 dollars)	$10 million	$11 million	$12 million	$19 million
Net present value (2013 dollars)	-$5 million	-$2 million	-$3 million	-$2 million
Greenhouse gas emissions (metric tons of eCO_2 per year)	-3,000	-8,000	-10,000	-2,000

Figure 14 graphically demonstrates the results when comparing the alternatives using the objectives outlined in Table 5. Both gas scrubbing alternatives scored highly within 1 point of each other.

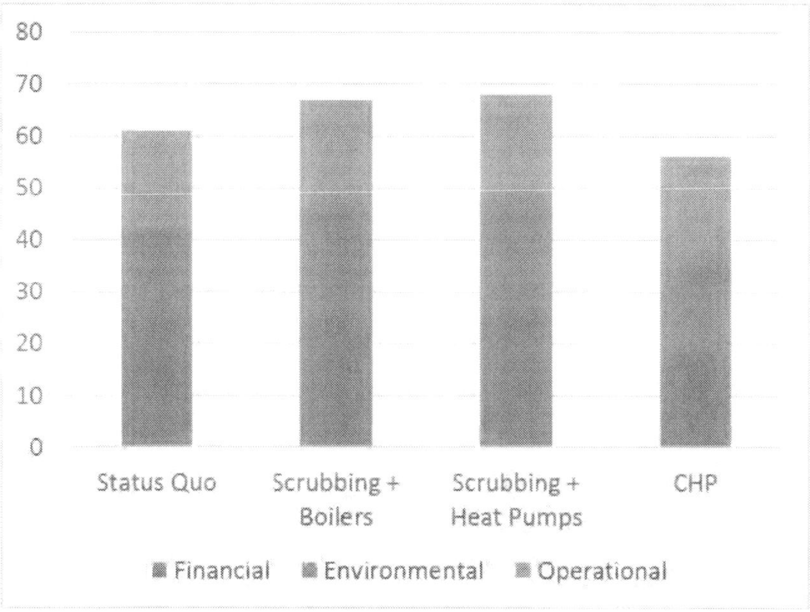

Figure 14. Results from the South Plant Biogas Utilization Study

West Point

At West Point, the selected alternatives focused on how to best utilize the surplus gas leftover after the needs and capacity of the existing systems had been met. As a result, the status quo did not carry any capital cost with it; the plant would continue to use biogas in the RSP engines, boilers, and CHP system, and would flare any surplus biogas.

The first alternative considered adding one additional engine to the plant's existing CHP facility. The new engine would be smaller than the existing engines (1.5 MW versus 2.3 MW) allowing for improved flexibility in performing with variable fuel quantities. Figure 15 provides an example of how a third engine could be added to the existing CHP facility. The lack of any major structural work made this a relatively cost effective alternative.

The second alternative consisted of two new engines to further improve the plant's ability to consume biogas as the gas production and heat demand fluctuated. Two 700 kW engines would not be able to fit within the space for the existing CHP system however, so an additional space for the system would need to be constructed within the plant.

The third alternative was for a new biogas upgrading system (i.e. gas scrubbing system) that would upgrade the surplus biogas to pipeline quality biomethane for sale to a third party. Because of the plant's location, injection into the local natural gas utility grid is not viable. As a result, this alternative assumed the upgraded gas would be compressed and trucked via CNG tube trailer to the third-party purchaser directly. To facilitate trucking off-site, the system would be located in the old Chlorine Handling Building, near the existing truck parking area. This allows for the County to also consider converting a portion of their biosolids hauling trucks to CNG-fueled engines and running them on the rCNG.

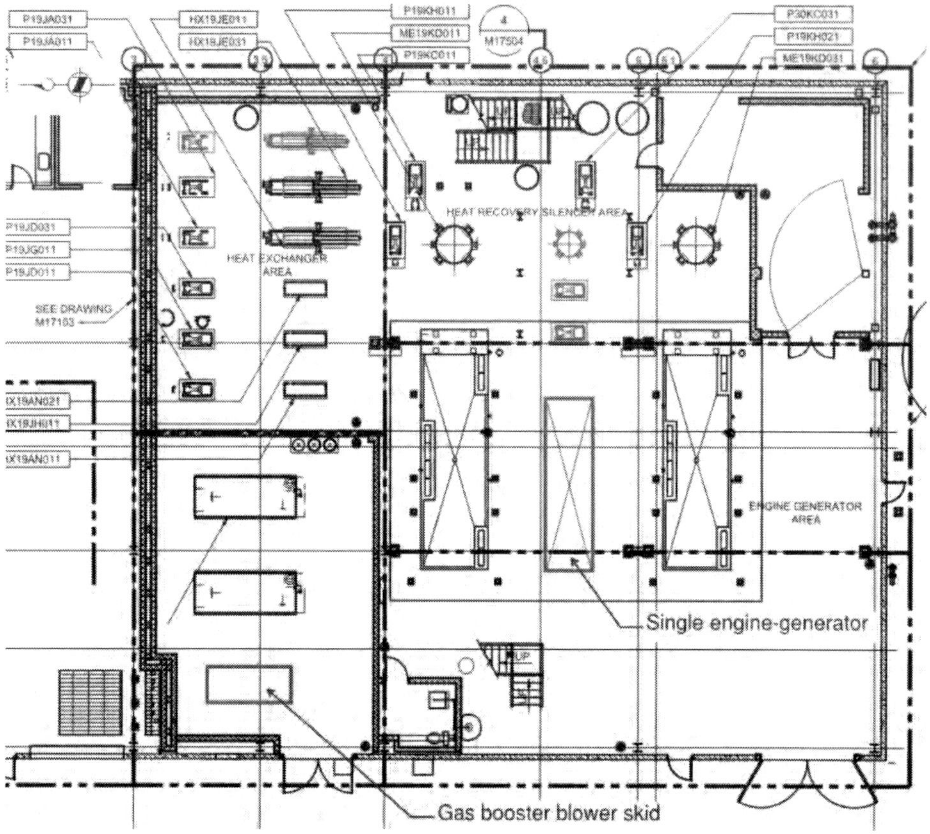

Figure 15. Preliminary layout for one additional IC engine at West Point

For the candidate technology, a 400 scfm BioCNG membrane system was assumed based on the appropriate size for the plant's needs, the low height profile (as compared to a water solvent system), and the convenience of having the entire biogas upgrading system provided by a single vendor.

Table 8 summarizes the major results from this analysis. One may note the extremely small values for greenhouse gas emissions reductions for the CHP alternatives. This is because the electrical utility that serves West Point (Seattle City Light) has a very small carbon profile (less than 2% of power supply is from traditional fossil fuels).

The sensitivities analysis performed found that the NPV results were extremely sensitive to the value of the electricity produced by the CHP system and the amount of surplus biogas available. In particular, the County's current agreement with Seattle City Light has very favorable rates and if these rates can be extended to the new system, the NPV of one additional engine would be positive. But if lower rates are assumed, the value of this alternative quickly becomes negative. Similarly, if the amount of surplus biogas varies significantly, the alternatives would either become cost prohibitive (at lower than expected biogas flow rates) or would all become net positive in terms of cost (if more biogas is available than expected).

Table 8. Results from detailed analysis of West Point Biogas Utilization Alternatives

Parameter	Status Quo	One IC Engine	Two IC Engines	Gas Upgrading
Construction cost (2016 dollars)	-	$5 million	$8 million	$7 million
Net present value (2016 dollars)	-$2 million	$1 million	-$3 million	-$8 million
Greenhouse gas emissions (metric tons of eCO_2 per year)	-	-200	-200	-3,200

Figure 16 graphically demonstrates the results when comparing the alternatives using the objectives outlined in Table 5. The single IC engine alternative scored the highest.

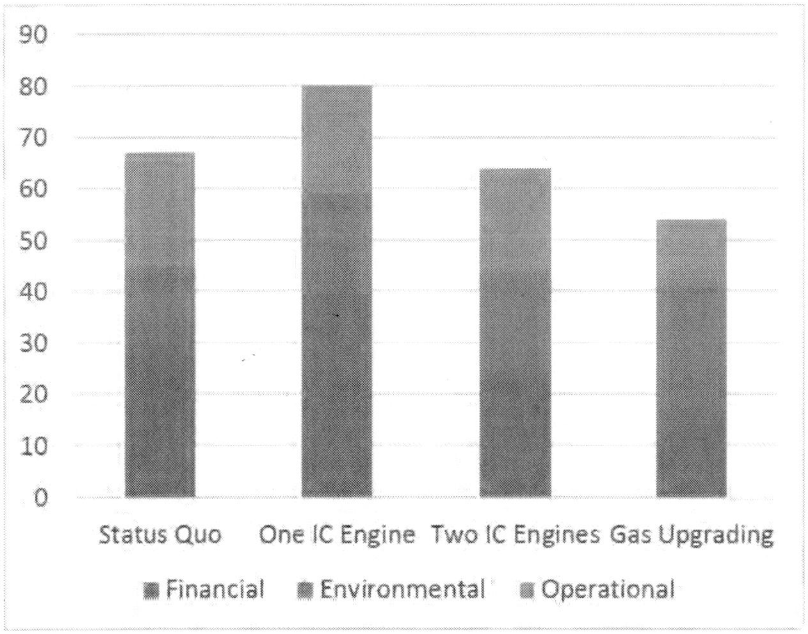

Figure 16. Results from the West Point Biogas Utilization Study

Brightwater

Detailed evaluation of the selected alternatives at Brightwater is still in progress but a description of the alternatives can be provided. Because the boilers at Brightwater are still in good working condition and meet the plant's needs, the status quo for this study would be to continue utilizing biogas to meet the plant's heating needs and flaring the remainder. As a result, there would be no capital cost associated with the status quo but also no benefit in terms of the County's energy goals.

One alternative under consideration would add a biogas upgrading system to produce a biomethane product from the gas remaining after the plant's heat demand had been satisfied. The biomethane would be injected into the local natural gas utility's transmission grid but sold to a third-party buyer for the value of the associated RINs.

Another alternative would install a similar biogas upgrading system but would pipe the biomethane directly to a nearby refuse hauler CNG fueling station instead of injecting biomethane into the natural gas transmission system. By selling the biomethane directly to the end user the County would avoid the contractual and quality requirements imposed by the natural gas utility but would pay for this in terms of additional market risk (e.g., only having one possible user for the product) and capital cost for rCNG storage. This alternative will also consider the use of effluent-source heat pumps to reduce the heat required from the boilers, increasing the amount of biogas available for upgrading. The layout of the treatment plant lends itself to low-temperature (i.e., 130-140 degrees Fahrenheit) heat pumps because the higher temperature process heating needs (e.g., sludge heating) can easily be separated from the lower temperature space heating needs.

The last alternative under consideration at Brightwater is to install a new CHP facility as was originally planned during detailed design of the plant. Because the original plan for the plant included this facility, the expectation is that integration of this type of facility into the plant should be more straight-forward than that for the other alternatives under evaluation.

There are no results to report to date on the detailed evaluation; the evaluation is expected to be completed during the summer of 2017.

RESULTS AND CONCLUSIONS

Despite the fact that the same evaluation process was used in both studies and that the County's objectives for biogas utilization is largely the same at both facilities, the results of the studies have led the County to move forward with differing approaches for South Plant and West Point. At South Plant, the study found that biogas upgrading with direct utility pipeline injection was the preferred approach for the County and has resulted in a project to replace the existing scrubbing system with a new biogas upgrading system and to install both hot water boilers and effluent source heat pumps. The high value for the RINs associated with the biomethane and the improved greenhouse gas emissions profile led the County in this direction but the plant's familiarity with the biogas upgrading technology and the availability of existing infrastructure for pipeline injection made this alternative viable. For West Point, the value of the RINs and the environmental benefits would still be present for a biogas upgrading alternative, but ultimately the lack of natural gas infrastructure at the plant led to this alternative not being very viable. Instead, the plant's long history of operating biogas-fired engines and the current rate the plant receives for RECs made an expanded CHP system the preferred alternative for the County.

Thus, site specific and historical factors end up playing a significant role in selecting a preferred use of biogas at any given wastewater treatment plant. The study for Brightwater has not been completed yet but site-specific factors can already be seen playing an effect on the direction of this study too. Brightwater's location is such that injecting biomethane into the natural gas utility is a viable alternative and the proximity of a large CNG refuse hauler refueling station could make a dedicated vehicle fueling alternative possible. In addition, the site-specific layout of process and space heating needs at the plant makes the use of a "hybrid" heating supply system that utilizes boilers for higher temperature needs and heat pumps for lower temperature needs a possibility.

The results of these studies shed light on how a consistent approach to evaluating biogas utilization alternatives with similar objectives can still result in widely different outcomes, even for facilities in relatively close physical proximity operated by the same utility. For utilities considering options for implementing a biogas utilization system, this conclusion emphasizes the importance of site-specific evaluations rather than applying a one-size-fits-all solution to every facility. There are many good uses for digester gas but these studies show that ultimately the best use for your digester gas will depend on your specific situation.

ACKNOWLEDGEMENTS

The authors gratefully acknowledge the cooperation of King County's WTD staff that provided invaluable input, review, and direction for these studies:

- South Treatment Plant: Mike Wohlfert, Curtis Steinke, Pete Carter, Rick Butler, Dave Jurgens, Lou Broadhead, Lester Van Gelder
- West Point Treatment Plant: Eugene Sugita, Showell Osborn, Dave Truman, Al Williamson, Steve Huang, Jim Belcher, Steve Zamperin, Charles Wenig, Chris Boyle, Kate Osborn
- Brightwater Treatment Plant: Carol Nelson, Andy Strehler, Bruce Kessler
- King Street: Bob Bucher, Pardi Sukapanpotharam, Carl Grodnik, Jessie Israel, Dave Broustis, Felix Brandli, Sue Hildreth, Amanda Connell

Biogas testing: Critical techniques to minimize capital and operating costs

Jeremy Holland, HDR Portland, OR

1 INTRODUCTION

Biogas is a renewable resource generating a lot of interest. While it has traditionally been used at wastewater treatment plants in boilers to generate heat for digestion or in engines to generate electricity and heat, it is increasingly being looked at to do more, such as to fuel vehicles, be injected into natural gas pipelines, and used as feedstock for bioplastics. These technologies place a heightened importance in the quality of the final biogas product being delivered and as such, biogas treatment systems are more important than ever before.

While there are many efforts focused around the treatment systems that can be used for the various end product requirements, less attention has been given to understanding the true condition of the untreated "raw" biogas. Sampling methods vary widely at different treatment plants and laboratories for testing biogas use different analysis methods and equipment. The end result is that data from testing is unreliable and this lack of good data costs significantly in capital and operating costs.

2 GAS SAMPLING METHODS

There are two different sampling methods that are used for biogas facilities- grab samples and real time sampling. Grab samples are taken with discrete volumes of biogas in various vessels and delivered to analytical laboratories for analysis. Real time sampling collects gas samples at regular intervals from the system and delivers the samples to analytical tools on site for analysis. Grab samples are the typical method for understanding biogas quality and are useful in order to evaluate a large number of analytes under a controlled laboratory setting. Real time sampling is used primarily as process control to target specific constituents in the gas that are critical to meeting a fuel specification for the end use or to protect emissions control equipment.

2.1 Grab Sampling Techniques

Grab sampling is done through four different means- methanol impinger method, tedlar bags, summa canister, and dual valve cylinder. Technically the methanol impinger is not explicitly a grab sample, however, it is included in this grouping because it does collect a discrete sample that is shipped to a lab for analysis. The methanol impinger method involves using tubes containing a methanol solution to bubble biogas through in order to adsorb gas contaminants. Impingers are used for siloxane and VOC sampling of the biogas. Because gas is bubbled through the solution, it is necessary to measure flow through the impingers and monitor the length of time the biogas is measured through the impinger.

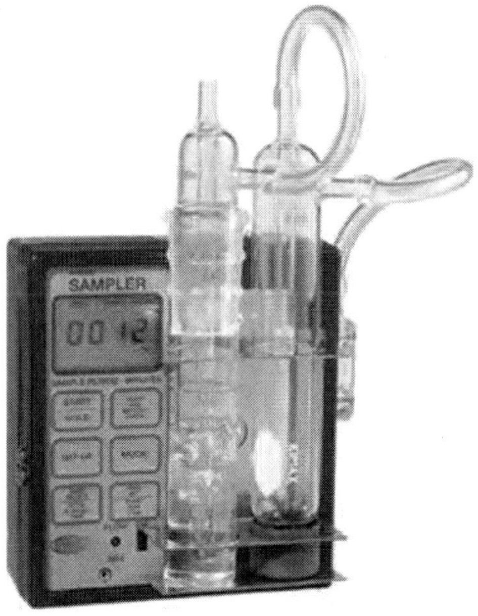

Figure 1. Example of Impinger Set, Picture Courtesy SKC Inc.

True grab sample vessels include the tedlar bag, summa canister and sample canisters. Tedlar bags are sample bags made from Polyvinyl Flouride and patented by DuPont. Tedlar bags are non-reactive so there is no concern about the constituents of biogas reacting with the bag and

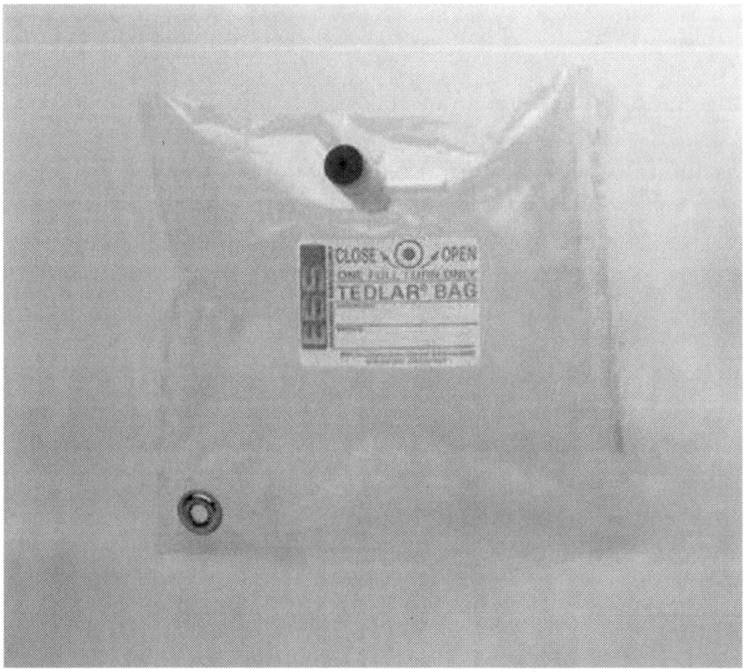

Figure 2. Tedlar Bag, Photo Courtesy Cole Parmer

impacting the results of the sample. The bags do have a small amount of permeability which should be considered. The following is gas permeability of some major gases over a 24 hour period. While the amounts of gas loss would be low, it is still worth recognizing that there can be an impact to some consituents over time. Another challenge with Tedlar bags

is the difficulty of accurately sampling without air contamination. Because Tedlar bags have a

Gas Permeability	
Carbon Dioxide	11.1 cc/(100in^2)(24 hr)(atm)(mil)
Helium	150 cc/(100in^2)(24 hr)(atm)(mil)
Hydrogen	58.1 cc/(100in^2)(24 hr)(atm)(mil)
Nitrogen	0.25 cc/(100in^2)(24 hr)(atm)(mil)
Oxygen	3.2 cc/(100in^2)(24 hr)(atm)(mil)

Figure 3. Gas Permeability of Tedlar Material, Courtesy DuPont

single filling valve with no way to purge gas through the bag, very careful connection to biogas piping is necessary to avoid air contamination. Review of Tedlar bag samples from several biogas facilities has shown that sampling variability of oxygen and nitrogen are significant and indicate air intrusion into the sample.

Another type of grab sample vessel is a summa canister. Summa canisters have been used for many years in air sampling. The canisters are made of stainless steel so are not reactive to most constituents, although H2S can be reactive to the vessels. If H2S concentrations are expected to be high, it is recommended to use glass lined canisters to avoid any possible reactivity of the H2S with the metal. Summa canisters are usually provided in a vacuum and have more robust valving for sampling connection. This allows for more reliable sampling results, although careful connection to the biogas system is still necessary.

Figure 4. Summa Canister

The third sampling vessel for grab samples is a dual valve sample cylinder. Dual valve cylinders offer the highest degree of sample confidence due to the fact that biogas can be purged through the tank with the second valve, allowing for removal of any air intrusion that may have occurred initially. These cylinders are also stainless steel so as with Summa canisters, H2S can react to the cylinder walls and therefore they should be glass lined or otherwise protected against H2S.

Figure 5. Dual Valve Sampling Cylinders

2.2 Real Time Gas Monitoring

Real time gas monitoring is necessary to measure gas quality for process control of upgrading equipment to produce renewable natural gas and for monitoring gas quality at pipeline interconnections. The most common monitoring methods are gas chromatographs for major gas analysis, such as methane, carbon dioxide, oxygen, and nitrogen. Other methods for analysis include tunable diode laser absorption spectroscopy (TDLAS), non-dispersive infrared (NDIR), and fourier transform infrared (FTIR).

In Gas Chromatography, the sample is injected into the analyzer and the gas and associated constituents pass through a column and can be analyzed. The constituents move through the column at different rates, there fore the different elements of the biogas can be detected and analyzed to determine what chemical is present and the amount. TDLAS, FTIR, NDIR all use similar principles of passing light through a gas and the phase change of the light can be used to determine the types and amounts of various chemicals in biogas. Companies such as Qualvista and Protea offer

Figure 6. Gas Chromatograph, Courtesy ABB

Figure 8. Gas Chromatograph, Courtesy ABB

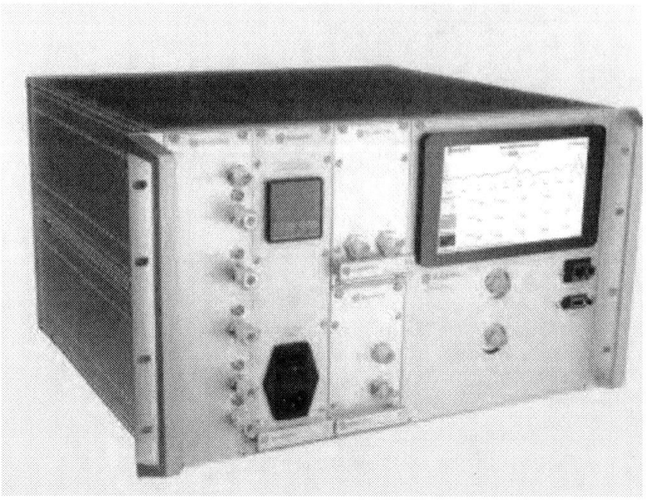

Figure 7. Qualivista NDIR monitor, Courtesy Qualvista

Companies such as Spectrsensors produce TDLAS instruments w hich can be used to measure in the field such parameters as hydrogen sulfide and moisture. NDIR sensors are included in monitoring packages from Qualvista and FTIR from Protea.

While upfront costs for these real time monitoring systems can be quite high, over time the benefits of real time sampling and reducing the need for grab samples and outside labs will become more important, particularly as biogas upgrading becomes more common.

ANALYSIS

The results of three rounds of sampling were completed at the Columbia Boulevard Wastewater Treatment Plant. The facility has traditionally used a specific laboratory for analysis and used tedlar bags for taking gas samples. Changes to sampling procedure included a first step to use a summa canister instead of a tedlar bag. A second step was added to use a dual valve cylinder and also to vary the analysis methodology. Results from the sampling and testing show significant differences in the Oxygen (O2) and Nitrogen (N2) concentration across the samples. Table 1 presents the results of O2 and N2 concentrations across the three sampling technique changes.

Table 1 – CBWTP Sampling Results

Summa Canister	**N2 %**	**02 %**	**CO2 %**	**CH4 %**
Raw Gas	0.66	0.1	36.4	62.8
After Media Tanks	0.49	0.07	36.2	63.2
After Siloxane Tank #1	0.54	0.08	36.7	62.6
After Siloxane Tank #2	1.57	0.36	37.8	60.2
Dual Valve	**N2 %**	**02 %**	**CO2 %**	**CH4 %**
Raw Gas Sample 1	0.502	0.106	30.6	41.9
Raw Gas Sample 2	21.44	6.12	40.1	59.3
Tedlar Bag	**N2 %**	**02 %**	**CO2 %**	**CH4 %**
Raw Gas Sample 1	3.06	0.86	34.8	60.8
Raw Gas Sample 2	1.9	0.518	32.9	67.4

Tedlar bag sampling results in O2 and N2 concentrations of .5% O2 v/v and 2% N2 v/v. Summa canister sampling results in concentrations of .1% O2 v/v and .5% N2 v/v. The dual valve cylidner sampling method produced similar results to the summa canister. One critical finding in the summa and dual valve canister sampling results are the deviation samples that are present in both cases. Given that multiple samples were taken and sampled, the deviation samples indicate clearly that air contamination occurred during sampling or testing. O2 and N2 are of particular concern because pipeline injection projects must meet natural gas utility fuel specifications which have extremely low concentrations. If the initial data from tedlar bag sampling or the two deviation samples were used as the basis of design for a biogas treatment system, an additional process would be required to remove oxygen with significant capital and operating costs.

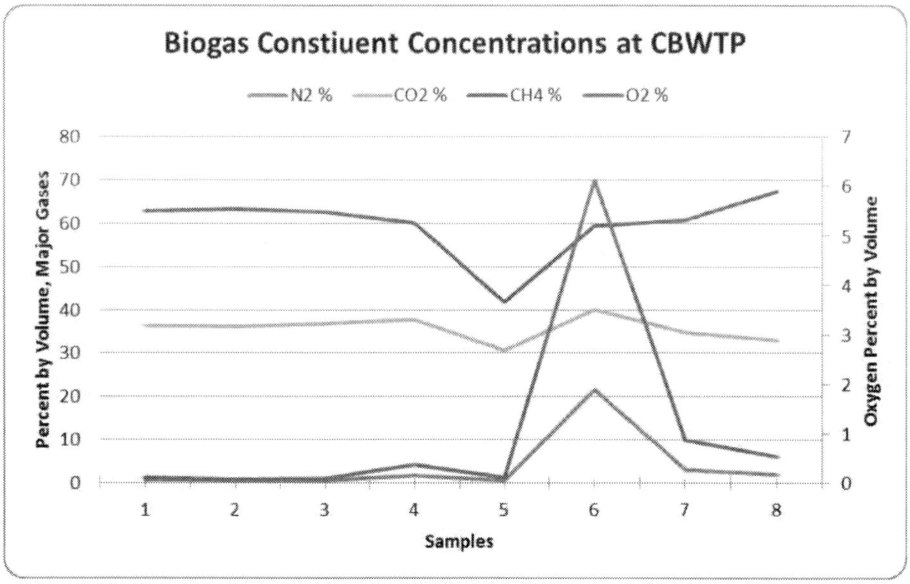

Figure 9. Results of different gas sampling methods

If it can be shown consistently that the biogas does not have off-specification concentrations of O2 and N2, this will reduce capital costs and operating costs. Oxygen removal equipment can be as much as 20% of the overall equipment costs for a biogas treatment system and has ongoing O&M costs associated with the system.

CONCLUSION

There is significant variability in the sampling techniques and testing approaches used at wastewater treatment plants to analyze untreated biogas. The most common techniques for sampling are using tedlar bags for grab samples. Tedlar bags are inherently problematic for sampling due to the propensity of the sampling technique to allow air intrusion into the sample. Summa canisters provide an improved sampling method and dual valve canisters and further improve the sampling procedure to avoid contamination of the sample. Laboratory equipment and testing methods must be understood in order to select the right method and equipment with enough sensitivity to detect low levels of contamination. Data from the Columbia Boulevard case study show significant improvement in the data when sampling methods are changed.

Biogas treatment systems are a significant cost and the ability of a treatment plant to limit the cost of treatment increases the return on investment of a project and increases the likelihood of more projects being completed. When the choice between flaring a renewable resource and beneficially reusing that resource is based on bad data, then the odds of a bad choice being made go up. If we can educate wastewater treatment plants and testing facilities on the importance of the sampling procedure and the testing techniques we can increase the confidence in the data itself and greatly improve on the economics of biogas treatment systems.

TITLE:

Next Generation Resource Recovery: Co-Digestion to Renewable Natural Gas (RNG) Pipeline Injection at the Des Moines WRF

ABSTRACT:

The Des Moines Metropolitan Wastewater Regional Authority (WRA) is a national leader in transforming wastewater treatment into a revenue generating resource recovery operation. Recent upgrades to the anaerobic digestion facilities of municipal sludge and FOG and other organic wastes processed in the co-digestion process has more than doubled WRA's biogas production, and this upward trend is expected to continue. WRA is moving ahead with plans to process up to 2,250 scfm of anaerobically digested biogas into pipeline quality renewable natural gas (RNG). This approach offers the opportunity to generate revenue from the sale of RNG to the local utility, as well as revenue from sale of Renewable Information Numbers (RINs) via compliance with EPA's Renewable Fuel Standards for a D5 advanced biofuel credit. Final design is in process now with construction completion expected in 2017.

CDM Smith will present this case study describing the biogas treatment technology alternatives considered, evaluation process, and recommended design criteria for new WRA facilities. Background information regarding RNG quality requirements, utility constraints and perspectives, financial expectations, and planned approach to sell RINs will be offered.

Commercially available treatment technologies used to separate methane from carbon dioxide and other contaminants include pressure swing adsorption (PSA), liquid scrubbers, and membrane scrubbing systems. Technology Evaluation highlighted many differences between these technologies, including: ability to handle high hydrogen sulfide concentrations, methane capture efficiency, cost, ease of operation and maintenance (O&M), and ability to consistently meet product purity requirements.

The presentation will include a discussion of the Bioenergy model developed by CDM Smith. This tool models various process conditions, financial scenarios, and technology options impacts on the revenue and anticipated payback of the project. A screenshot of one page of the Bioenergy model is included in Figure 1.

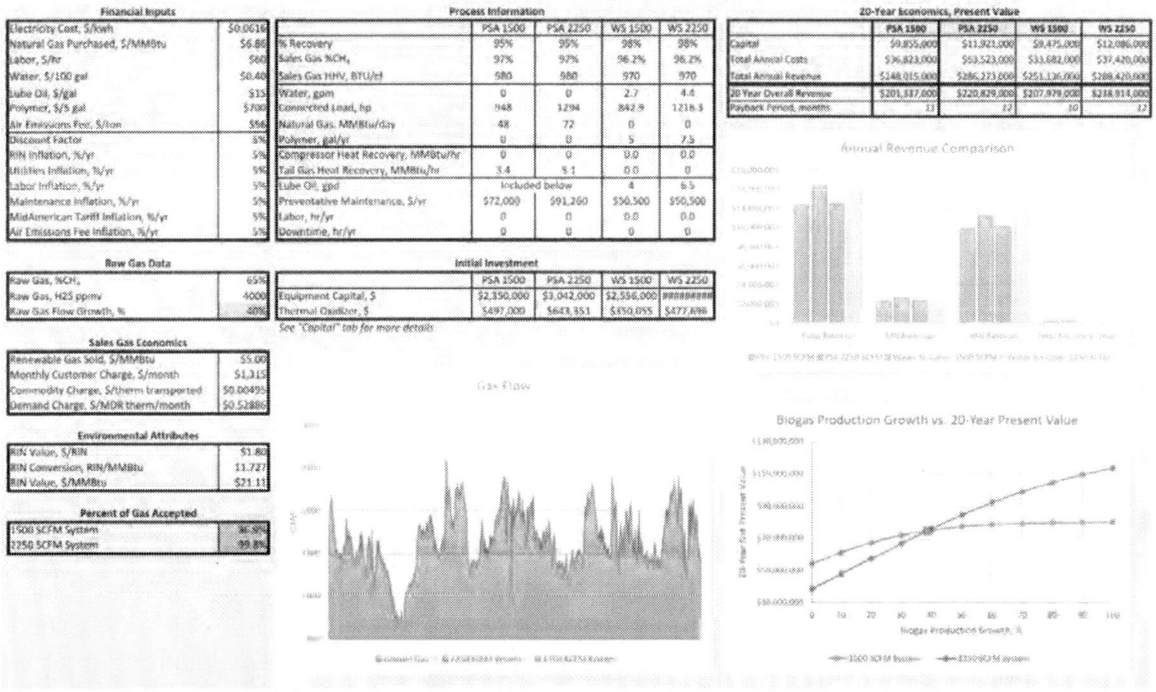

Figure 1: Bioenergy Model

The design documents are anticipated in September 2016 with construction starting in Q4 of 2016. CDM Smith will present the design documents, including the facility code modifications required to install the biogas treatment technology within an existing unclassified building. CDM Smith will also share lessons learned and negotiating tips for working with municipalities, gas utilities, and RIN partners for other wastewater agencies considering this approach to resource recovery.

Converting Biogas into Energy and Vehicle Fuel

Adam Klaas - Unison Solutions, Inc.
adam.klaas@unisonsolutions.com
5451 Chavenelle Road
Dubuque, IA 52002
Phone: 563-585-0967

INTRODUCTION

Utilizing biogas as a source of fuel for producing electricity, heat and vehicle fuel is not a new concept, but it can be confusing. This presentation will discuss all facets of a biogas to energy project, from the suitability factors needed to start a project to the equipment and technologies available. We will also present case studies of different sites that have been operational for many years.

Site suitability, local utility rates and the utilities needed for renewable energy are all important components of any project. Every municipal plant uses significant amounts of electricity and natural gas in the day to day operations. Defining these requirements and the actual costs will help determine payback with a cogeneration system. For sites that have CNG vehicles as part of their municipal fleet, the biogas to RNG may be a good option. Many municipal sites are using a combination of these technologies to produce electricity, recover heat and produce RNG for vehicles.

STEPS TO DESIGN A SYSTEM

The first requirement for a biogas to energy system is to capture your biogas. This can be done with a variety of technologies, but all include anaerobic digestion of the solids to produce biogas. This gas typically is 100% saturated and contains methane, carbon dioxide, nitrogen, oxygen. Depending on the waste source and the treatment protocols at the wastewater it may also contain hydrogen sulfide, siloxanes and volatile organic compounds. All of these components effect the treatment method needed for the biogas so the next step in the process is testing the biogas for all of these components.

Determining BTU content can be determined by utilizing EPA Method 3C and ASTM 3588 for the BTU calculation. Methane (CH_4), Carbon Dioxide (CO_2), Oxygen (O_2) and Nitrogen (N_2) are reported in terms of their concentrations in the sample. The BTU content is an important component in determining output from cogeneration equipment.

Sulfur content in biogas includes hydrogen sulfide levels. These compounds are quantified using either EPA Method 3C or ASTM Method D6968. Hydrogen sulfide is a corrosive compound that can cause excessive wear on internal components of cogeneration equipment.

Siloxane speciation can be determined with the EPA Method TO-15 or a GA-AED method using standards for the most common siloxanes found in biogas. Siloxane compounds are manmade chemicals used in all types of personal care and cleaning products. Their chemical structure is similar to volatile organic compounds with the addition of silica molecules. During combustion, the silica forms a hard white deposit on internal surfaces of the cogeneration equipment and vehicles. Depending on the type of cogeneration equipment being used, siloxane tolerances can vary greatly. However, they increase the maintenance cost for all equipment, regardless of the manufacturer's specification so identifying their concentrations is important.

Volatile organic compounds are also quantified using EPA Method TO-15. Methods TO-14 and 17 are sometimes used in addition to capture more compound information. These compounds are important at sites when siloxane removal is necessary as they compete for the same porosity as the siloxane compounds. For systems producing RNG for vehicle fuel, they are especially important because they can harm the carbon dioxide removal membranes.

SYSTEM EVALUATION AND SIZING

Once the gas testing is completed and the available gas flow rates calculated, choosing the best cogeneration system for the site is the next step. We'll review the variations between biogas conditioning systems for engines vs. turbines or just utilizing the gas for boiler systems.

Initial equipment costs, along with operating and installation costs per kW should be reviewed before making a buying decision. Additional factors to consider are space availability, integration into existing thermal and electric systems, along with grid interconnection requirements and state regulations. Many sites have multiple engines or turbines so that one unit can be shut down for maintenance without losing the entire capacity of the installation.

The appropriate sizing for a cogeneration system can be either electric or thermal limited, and is dependent on the size of the facility and the equipment on site. Defining the current electric and thermal energy usage is the first step in determining which type of cogeneration equipment would be best suited for a facility. Then the site's potential future requirements for both electricity and thermal energy can be reviewed. Electricity cost (demand and energy) and the amount the local utility will pay per kWh for renewable energy are factors to consider whether producing excess electricity is a viable alternative.

The volume of biogas produced is one of the variables that will help in determining which cogeneration product best meets the needs of a site. For many sites, the biogas production varies during a typical week. For these sites, the turn down capacity of the cogeneration equipment becomes a factor. Most cogeneration equipment has a 50% turn down. This requires biogas conditioning systems to be designed for the higher biogas volume, while still having the ability to effectively operate at lower biogas production levels. For example, if a site is designed for 180 scfm and has two engines that use 90 scfm each, one engine can be serviced while the second one produces electricity at its 50% turndown. If this same site (designed for 180 scfm), began operation at 90 scfm and one engine, an additional engine of the same size could be added in the future when the biogas production increased.

Wastewater treatment plants with anaerobic digesters are sized by the "million gallons/day" (MGD) capacity for waste, and the respective amount of biogas that it produces. Many facilities use co-digestion of high strength waste (HSW) and the base sludge, to increase biogas production in digesters with excess capacity. In this process, the energy-rich organic waste materials such as waste from food processing will maximize the production of biogas per unit of mass. The anaerobic digestion of these biodegradable substrates provides approximately three to four times more biogas production per volume of digester capacity than municipal waste alone.

HSW from other types of manufacturing may also be used. The composition of the waste and how it reacts in the digester will determine whether it is of sufficient quality to boost the biogas production. There may also be more impurities in manufacturing waste that could negatively affect the operating and maintenance cost of the entire cogeneration system.

Following is a generic example of the payback for a small digester using a 65 kW turbine with and without high strength food processing waste.
<u>Option 1</u> – Municipal sludge only: The site takes in 1.0 MGD of waste and produces 38,000 ft^3/day biogas with the cost of electricity at $0.075/kWh and the value for heat at $0.80/therm. The total estimated cost for the biogas conditioning system, turbine, site engineering, and installation would be approximately $500,000. Based on the value of both power and heat produced by the system, the payback would be 10 years.
<u>Option 2</u> – Municipal sludge plus high strength food processing waste: The biogas produced would increase to 76,000 ft^3/day. The new, total estimated cost, including an additional 65 kW turbine and other small incremental costs would increase to approximately $655,000. But, the payback would decrease to 6 years due to the increased power and heat produced.

This will also include the various types of hydrogen sulfide and siloxane removal technologies available. The final technology discussed will be the additional equipment needed for the biogas to vehicle fuel systems.

CASE STUDIES

Case studies for both biogas to energy and biogas to vehicle fuel systems will be discussed in detail. The sites reviewed will include both municipal and industrial digesters from around the country. We will review the size of each site and the equipment being utilized. A summary of their project with timelines, project cost and payback information will also be discussed.

<u>Case Study Overviews</u>

Hornsby Bend Biosolids Facility, Austin, TX

- 450 scfm of biogas produced
- (1) GE Jenbacher IC engine produces electricity, and recovers heat
- System includes H_2S removal, compression/ moisture removal, and siloxane removal systems and has been installed since December 2012

Dubuque WRRC, Dubuque, IA

- 40 MGD municipal plant (Avg flow 14 MGD)
- 350 scfm of biogas produced
- (1) CR600 Capstone turbine produces electricity, and recovers heat for the boiler system
- System includes H_2S removal, compression/ moisture removal, and siloxane removal systems
- Part of a $38 M plant renovation
- October 2012, Initial system conditioned biogas for boiler fuel only
- August 2014, a compression skid for 600 kW Capstone turbines. Electricity produced is utilized at the plant only
- 2016 – plant is producing excess gas and is evaluating putting conditioned biogas into the natural gas pipeline

Potawatomi Casino, Milwaukee, WI

- Food waste from casino and local grocery stores
- High strength waste is also used to supplement the food waste
- 700 scfm of biogas produced
- (2) 1,000 kW IC engines produce electricity and recover heat
- System includes compression/moisture removal and has been operational since October 2013

Persigo WWTF, Grand Junction, CO

- 12.5 MGD design flow (Avg. 8 MGD)
- 100 scfm of biogas is collected from site
- Time fill and fast fill stations for CNG fueled collection trucks and city buses
- 5.8 mile dedicated pipeline from the wastewater facility to the fueling facility
- 142,000 gallons of gasoline diverted=CO_2 emissions reduction of 3 million pounds/year
- Challenges at the plant have included high H_2S and low moisture conditions
- Utilizing a new H_2S removal media in two large vessels to help reduce operating costs
- BioCNG™ - 100 scfm, has been installed since spring 2015

Dogfish Head Brewery, Milton, DE

- 80,000 gallons/day waste stream treated by a Tribrid Bioreactor
- 200 scfm of biogas is produced
- Brewery was founded in 1995 and produces 175,000 barrels/year
- Effluent water is treated by GE, reverse osmosis membranes
- GE Jenbacher IC engine for electricity and heat recovery. All power produced is used at the plant.

Danville Sanitary District, Danville, IL

- 16 MGD municipal plant (Avg. flow 5-8 MGD)
- 50 scfm of biogas produced
- (1) Tech 3 IC Engine, produces 150 kW of electricity and 778,000 BTU/hour of thermal energy

- Engineering was done in-house.
- Total project cost was $750,000 and with their $200,000 grant from the Illinois Department of Commerce and Economic Opportunity, the projected payback is expected to be approximately 5 years.
- Long term goals include net zero
- Challenges at the plant have included high H_2S
- Site utilizes a H_2S/siloxane removal media on the dry gas portion of the system
- Project planning started in 2008 with system start up in October 2012

SSF- Blue Line Transfer, South San Francisco, CA

- 100 scfm of biogas is collected from site
- 8 dry anaerobic digesters
- 11,200 tons of food and green waste diverted per year
- Time fill and fast fill station for CNG-fueled collection trucks
- Challenges at the site have included extremely high volatile organic levels and widely fluctuating H_2S
- Utilizing a new VOC removal media to help with operating costs
- BioCNG™ - 100 scfm, has been operational since spring 2015

Plymouth Utilities, Plymouth, WI

- 1.01 MGD municipal plant
- 50 scfm of biogas produced
- (2) Capstone CR65 turbines produces electricity
- System includes compression/ moisture removal, and siloxane removal systems and has been operational since December 2014
- Challenges at the site included tight space requirements. Skid was divided into 2 sections so that installation could be achieved within the desired space constraints.

Loyd Ray Farms, Yadkinville, NC

- Agricultural – covered lagoon/digester
- 50 scfm of biogas produced
- (1) 65 kW Capstone turbine produces electricity
- System includes compression/moisture removal and has been operational since December 2013

Sierra Nevada Brewery, Mills River, NC

- 85,000 gallons/day average waste stream processed by an anaerobic contact sequencing batch reactor
- 250 scfm biogas produced
- Fuel supplies 2 – 200 kW Capstone Turbines
- Solar panels are also used on site to produce 600 kW
- All power produced is used on site

Power or Fuel: Renewable Natural Gas a Feasible Alternative

Trung Le[1], Joe Marino[1], Trent Montemayor[1], Jacob Klosinski[2]

[1]Greeley and Hansen, USA
[2]City of South bend, Indiana

Introduction

The recovery and utilization of biogenic gas is no longer an economic-only driven deliberation. Social, political, and regulatory factors often drive these projects as much as economics. As municipalities continually move towards a vision based on sustainable principles, resource recovery has become a large part of this vision to reduce cost. Gas utilization has become a high priority and major component of energy recovery in large-scale and small-scale wastewater treatment facilities. The general direction of wastewater digester gas energy recovery has often gone towards combined heat and power also known as co-generation. However, an alternative to electricity generation is renewable natural gas (RNG) production through gas cleaning technologies. This method was implemented in the City of South Bend, Indiana's Wastewater Treatment Plant due to the existing natural gas infrastructure and RNG's greater flexibility of end uses.

The most basic form of gas utilization is to use an unprocessed or low grade biogas in boilers. As processing increases, the quality of digester gas improves as well as its value. This biogas can be used in a multitude of different turbines to generate useable heat and/or electricity. The highest level of cleaning and processing can turn digester gas into pipeline quality RNG which can directly be used as a fuel source or sold to natural gas vendors. Gas can be utilized internally or externally depending on the level of processing and the economics that drive that demand. One incentivizing program that has made generating and selling renewable energy appealing stems from state Renewable Portfolio Standards (RPS).

Twenty-nine states including the District of Columbia have enacted energy mandates as part of the RPS which have influenced how municipalities approach the use and generation of energy. Municipalities can produce renewable energy in the forms of electricity or fuel. Each can qualify for a renewable energy credit program such as Renewable Energy Certificates (REC) or Renewable Identification Number (RIN). Therefore, an energy price and a green premium can be derived from these renewable energies.

The City of South Bend, Indiana embarked on a mission to maximize the use of its biogas through a phased approach. This case study evaluates the City's decisions to move towards RNG production rather than the industry norm of cogeneration and the methods to get there.

Case Study: The City of South Bend, Indiana: Gas Cleaning and Utilization

The City of South Bend (City), Indiana Wastewater Treatment Plant (WWTP) is a Class IV activated sludge treatment facility with an average design flow of 48 MGD and two (2) mesophilic anaerobic digesters. Previous operation of the WWTP, utilized digester gas to heat digester sludge in a boiler system. However, the amount of digester gas energy produced far exceeded the amount of energy required to heat the sludge and additional excess gas was flared. Due to this waste of gas, South Bend sought to maximize the use of its digester gas through different pathways. The implementation of resource recovery in small to mid-sized WWTPs can often be cost prohibitive. To reduce risk, the City decided on a multi phased-approach to allow for a lower year-to-year capital investment and finer tuning of the design as components of the project are finalized and come into operation.

The multi-phase project focused on optimizing and intensifying gas production, capturing and cleaning of the gas and then utilizing the gas. An evaluation was made on the existing infrastructure and possible end uses for the gas. This initial evaluation found that upgrades to the existing system could improve gas production to as much as 20 to 30%. The issue in recovery of additional digester gas can fall under the principles of diminishing returns as capital, infrastructure requirements, and implementation cost for small-scale facilities can exceed the recovery benefits. With the goal of the City to maximize the potential of its gas generation, low-capital alternatives were considered. The City decided that producing a natural gas fuel rather than electricity through cogeneration was the best option as it would allow them to directly supply part of their CNG facilities fuel needs.

Design and Considerations

To implement any gas utilization plan, upgrades to the current digester system were needed and included upgrades of one (1) mesophilic anaerobic digester and one (1) sludge storage tank. Additionally, improvements were needed to the process control equipment, installation of high rate mixers (external draft tubes), a buffer storage tank, a new digester cover, and heat exchangers.

Gas Cleaning System

A gas cleaning system with a raw gas input of 82 CFM and clean gas output of 43 CFM was installed to handle producing higher grade RNG (assuming 40% of carbon dioxide, water vapor, siloxanes, other contaminants and losses through the cleaning process). The technology used to clean the digester gas to the higher grade renewable natural gas is a pressure swing absorption (PSA) system.

This design is a Guild system rated to accept gas flow rates between 50,000 to 310,000 ft^3/day of digester gas.

The gas cleaning process is as follows:

- Digester gas is compressed to 100 psig
- The compressed gas is cooled to the ambient air temperature to condense the moisture in the gas
- The gas is then heated to 140 °F
- CO_2, N_2, Siloxanes, and VOCs are removed using a proprietary adsorbent media in a process called Guild Pressure Swing Adsorption (PSA)

Media acts like a molecular sieve and is sized to adsorb compounds larger than CH_4

- Consists of three sieve beds that are rotated in service then cleaned under a vacuum

- Cleaning process produces a waste tail gas
- H_2S is removed in the waste tail gas that is flared

The gas cleaning system is composed of the following equipment:

- Compressor
- Main Pressure Swing Adsorption and Filter
- Tail Tank

Adsorbent media is regenerative and is regenerated approximately every minute. The expected media lifespan is approximately 10 years and media replacement can be completed using large buckets. The only other components of the gas cleaning system requiring maintenance are the compressor and vacuum (oil and filter changes).

Digester and Infrastructure Upgrades

The city operated an anaerobic digestion system consisting of two primary digesters and two secondary digesters. Each digester is 110-foot in diameter with a 25-foot side water depth and a 9-foot cone. Each digester has a volume of 1.99 million gallons. All of the digesters were constructed in 1954. The primary digesters were upgraded in 1988 but new upgrades are required to the mixing, covers and internal piping of the digesters. Digester No. 1 and Digester No. 3 were cleaned and are currently inactive, pending the need for improvements and upgrades due to deterioration of internal piping and cover structures. With the increased mixing upgrade, the City only needed to operate one digester based on current sludge production rates and stay within suggested volatile solids loading rates for mesophilic digesters (100-300 lb VS/1000 cfd)$^{-1}$ and SRTs (15-20 days) based on current flow and loadings. Future flow rates would require operation of a second digester to achieve the recommended SRT. It was decided that Digester No. 2 would undergo upgrades to limit capital cost.

Digester No. underwent the following upgrades

- New gas holding cover
- External draft tube mixes equipped with water jackets
- New centrifugal Flygt N-technology pump to act as a transfer pump
- New flex type broiler
- New sludge and digester gas piping, valves and gas appurtenances

Gas Utilization Strategy

End use alternatives were evaluated based on a lifecycle cost assessment to connect to the current system. Each option required the recovery of a high grade natural gas (pipe line quality) to supply different components of the plant or to be sold to a natural gas company.

Three alternatives were evaluated in addition to the use of digester gas for heating sludge:

1. Supply RNG to two natural gas engines used in the operation at the plant. This included a 250 HP Waukesha engine that powers a raw main sewage pump (MSP) and a 1,000 HP Caterpillar engine used to power one of the blowers.
2. Produce RNG for used in the plant fleet vehicles. Cleaned gas was routed to the vehicle fueling station where natural gas is then compressed to fill up to fourteen City Vehicles. Each truck contains a 60 gallon diesel equivalent compressed natural gas tank.
3. Sell natural gas to gas utility. Currently, this alternative is not feasible due to the minimum production rate requirement that exceeded South Bend's capacity to produce gas. However, South Bend's production of digester gas is not fully optimized and is still a plausible future option to take advantage of the appeals of RINs.

Alternatives 1 & 2 were selected and incorporated into the final phase of the project. In this configuration, digester gas will be cleaned and then used first to fuel the boiler system. The boiler has a capacity of 3,780 MBH; however, the normal demand ranges between 1,000 MBH to 2,400 MBH. On the most extremely cold (but infrequent) winter days the demand is estimated to be closer to the capacity of the boiler. Surplus RNG will be then used to supply the two natural gas engines and the remainder of the RNG will be used to fuel the plant fleet vehicles (Figure 1).

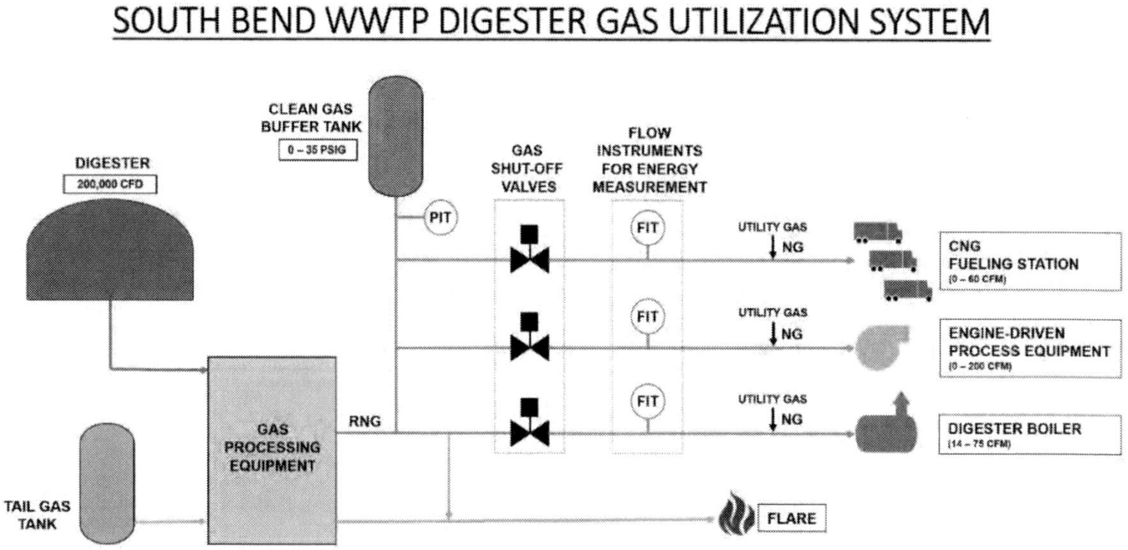

Figure 1: Schematic of Gas Utilization System

Vehicle Fueling Station

The vehicle fueling station located on the wastewater treatment plant site compresses natural gas for filling seven City vehicles: six trash trucks and one roll-off truck that hauls solids to the Organic Resource Facility. Filling is typically performed weekdays during the night. The station is rated for 48 CFM at a pressure of 3,600 PSI. Each truck contains a 60 gallon diesel equivalent compressed natural gas tank.

With the boiler running and no storage, digester gas will provide 90 diesel equivalent gallons of gasoline per night on average, which is equivalent to seven trucks filling up approximately 20% of their tank nightly.

Future Construction

The ongoing construction of the final planned phase is scheduled to be constructed by September 2017 and includes the following elements:

- Independent control of gas flow to allow the City to select the end use of the gas
- Energy monitoring / recovery for each end user integrated into the plant wide SCADA
- Plant wide gas distribution of manufactured RNG
- Distributed pressure regulating valves to automatically transition between RNG and purchased NG

Conclusions

The decision for South Bend's current operation is to fuel its CNG fleet and natural gas engines. Implementation of this strategy would utilize 88-100% of the current renewable natural gas production with a payback period of 20-28 years. Upgrades to the digesters provided to be beneficial with an increase of 33% in gas production. On average, RNG generated from the digesters satisfies the MSP process engines, the digester boilers, and admin domestic heating needs during the daytime. During the night, the system can satisfy the digester boilers, admin heating, and 50% of the vehicle fueling station.

Whereas electricity generation from digester gas is limited in its use, generation of renewable natural gas has flexibility as a commodity to be used, sold or traded. Renewable natural gas has several economical viable end product routes. If future thresholds are set such that the City of South Bend may benefit from the RIN program, it would make economic sense to maximize the amount of cleaned digester gas being used to collect RINs (i.e. vehicle fuels) and supplement natural gas demands on-site (i.e. for boilers and engines) using utility grade natural gas. The feasibility of clean digester gas sale externally to capture RINs should revisited on a periodic basis as it has the potential to improve on the already favorable economics of cleaned digester gas use to offset natural gas use for boiler, engine, and fleet fueling.

Using a Chemically Hydrolyzed Biosolids for Co-digestion

Joshua Registe*, Joshua Perez*, Jeanette Brown*, Robert Sharp*, George Nakhla**, and Ajay Singh***
*Manhattan College, Department of Civil and Environmental Engineering, **Western University ***Lystek

ABSTRACT

Manhattan College performed an independent study on the use of a chemically hydrolyzed biosolids product for co-digestion studies. The biosolids material is produced by Lystek, Cambridge, Ontario Canada. The objective of this study was to determine if the hydrolyzed Lystek product (Lystek) could be used in anaerobic digestion (AD) to potentially increase gas yields and methane content. The AD study was divided into two phases; Phase 1, an initial study using a bioassay procedure to determine the amount of Lystek to add in the anaerobic digestion process for optimal gas yield and methane concentration and Phase 2, a more in-depth study of Lystek using bench scale anaerobic digesters. The results of the bioassay indicated that gas yields increase within a Lystek feed volume range of 15-25% with 20% to 25% yielding optimal results. This information was then use in the bench scale anaerobic digesters. The digesters were operated at an SRT of 15 days. Primary sludge was used in both the control and experimental reactor. The results indicated that the addition of Lystek increased biogas production as well as increased methane content of biogas. In addition, there appears to be little or no hydrogen sulfide in the biogas when using Lystek. There was an increase in ammonia and orthophosphate concentration possibly due to the higher volatile solids reduction in the Lystek reactor. Further research will include evaluating higher percentages of Lystek, dewaterability of the digestate, confirmation of the reduced hydrogen sulfide concentration, and determination of the cause of increased ammonia and orthophosphate concentration.

KEYWORDS

Co-digestion, Lystek, Hydrolyzed Product, Improved Digestibility, Biogas Enhancement

INTRODUCTION/BACKGROUND

Project Description

The goal of water resource recovery facilities (WRRFs) is to recover resources embedded in wastewater. Anaerobic digestion is a process that allows for recovering of energy in the form of methane. The purpose of this independent study was to determine if the addition of biosolids produced by the Lystek process (Lystek) when added to an anaerobic digester would enhance biogas production. The study was performed in two phases, with the first phase being a bioassay to determine the optimal percentage of Lystek to add to digesters, and the second phase being a bench scale study. The goal was to determine if the addition of Lystek would increase methane concentration and biogas production.

Anaerobic Digestion

Anaerobic digestion is a well-established and understood process used for the stabilization of raw primary and waste activated sludge (Parkin 1986). Essentially, it is a series of processes where microbes break down and digest biodegradable material present in a system void of oxygen. This process ultimately reduces the volatile solids in to produce biogas that can be used as an energy source for heat, or electricity. There are 4 steps in anaerobic digestion (Figure 2): hydrolysis, acidogenesis, acetogenesis, and methanogenesis. There are two beneficial by-products - biogas and digestate.

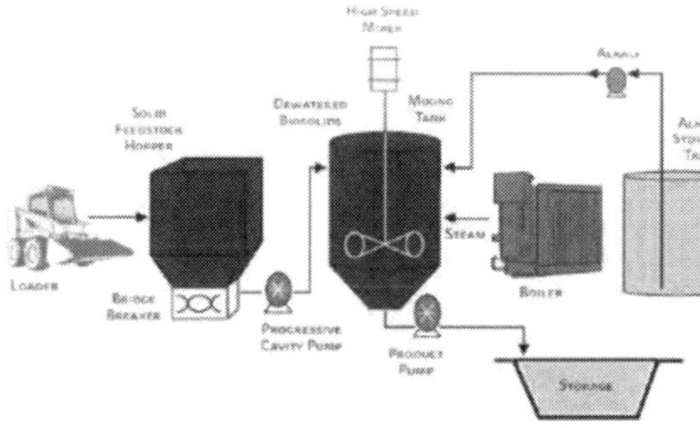

Figure 1. Schematic of Lystek Process

Biogas is typically composed of methane (~60-65%) and carbon dioxide (~35-40%) (Abdeshahian et al. 2016). The production rate of biogas varies between 0.7-1.12 m^3/kg (12 to 18 ft^3/lb) of volatile solids (VS) destroyed and has an energy content of 22 kJ/m^3 (~600 BTU/ft^3) HHV. The biogas yield is dependent on the digestibility of organic matter in the system, as well as, temperature, pH, kinetics, and solids retention time. Chemical Oxygen Demand (COD) is generally used to quantify substrate energy content and therefore predict biogas production. Biogas production can also be estimated from the percent reduction of volatile solids. The amount of volatile solids in a system and therefore the amount destroyed can cause the gas production to fluctuate. The ultimate effectiveness and success of anaerobic digestion is measured by the amount of biogas produced and percent VS or COD reduction.

The intent of WRRFs is to optimize and enhance anaerobic digestion to increase biogas production to recover energy embedded in the waste, thus reducing their energy footprint. In many cases this enhancement is accomplished through co-digestion. Co-digestion is anaerobic digestion of two or more substrates and can be considered a modification to the conventional anaerobic digestion of a single substrate (Mata-Alvarez 2014). Essentially, the theory behind co-digestion is to add a secondary substrate to wastewater residuals in an anaerobic digester. Currently, the main substrates that are used, besides wastewater sludge, are fats, oils, and grease (FOG), and food waste as well as other products with high COD. The hypothesis for this study is that the addition of Lystek to the digester will enhance biogas production and methane concentration.

Lystek

Lystek biosolids treatment process (Figure 1) is an innovative, energy efficient and economical low temperature thermal-chemical hydrolysis technology which uses a proprietary combination of heat to 70-75° C (158 – 167° F), pH 9.5-10.0 using alkali / KOH, and high shear mixing treatment to convert biosolids into a high solids (15% -17%), homogeneous, pathogen-free, and nutrient rich Class A EQ liquid product. This process is applicable to undigested and digested biosolids. The process conditions make the residual recalcitrant organics in the digested biosolids more amenable to further biodegradation after re-feeding to anaerobic digesters. Compared to conventional thermal hydrolysis technologies that operate at high temperatures and pressures,

Lystek technology uses a combination of milder processing conditions to create a lysed product with multiple beneficial applications. The liquefaction process causes disintegration of cells and hydrolysis of complex particulate complex organic molecules into simpler and rapidly degradable compounds.

PROJECT DESCRIPTION

Introduction

The goal of this project was to determine if the addition of the Lystek product to an anaerobic digester would enhance biogas production, to what extent, and at what feed rate. The study was divided into two phases. Phase 1 was a bioassay to quickly determine optimal feed rates and the Phase 2 used bench scale reactors over long SRTs.

The sludge used for both the bioassay and the anaerobic digestion study was obtained from Middletown, New York. The following are the characteristics of the sludge:

- Type: Primary
- Total Solids: 2-4%
- Volatile Solids: 76%,
- Chemical Oxygen Demand: 60,000 - 80,000 mg/L
- Soluble Chemical Oxygen Demand: 2000 – 4000 mg/L
- Ammonia: 100-350 mg/L
- Phosphorus: 20-50 mg/L.

Phase 1-Anaerobic Bioassay Study

The anaerobic Bioassay (ABA) was used to determine the optimal feed rate for Lystek addition to anaerobic digesters when mixed with primary sludge on a volume percent basis. The procedure for the anaerobic bioassay was taken from a toxicity assay method developed from "Bioassay for Monitoring Biochemical Methane Potential and Anaerobic Toxicity" by Owen et al. (Owen, 1978). By using this test, the optimal dosage can be quickly determined and then used in the anaerobic reactors. Since the anaerobic reactors are operated at a 15 day SRT, ABA reduces the time required to determine optimal feed rates to one day.

ABA's were performed in 150 ml glass serum bottles sealed with butyl rubber caps under mesophilic anaerobic conditions. The serum bottles were dosed with digested sludge from Middletown Wastewater Treatment plant and either raw primary sludge or an acetate-propionate feed solution was added in varying dosages. The bottles were prepared at ambient air temperature and then sealed and placed in an incubator at 35°C. Gas flow readings were measured after 1 day with an Agilent Technologies ADM1000 Universal Gas Flow meter with a range of 0 to 1000 mL/min.

Table 1. Acetate-Propionate Feed Protocol

Acetate-Propionate Feed		
Sample Volume (mL)	Sample Type	Percent Lystek
50	Acetate-Propionate	0% (Control 2)
50	Digestate	
0	Lystek	
42.5	Acetate-Propionate	15%
50	Digestate	
7.5	Lystek	
40	Acetate-Propionate	20%
50	Digestate	
10	Lystek	
37.5	Acetate-Propionate	25%
50	Digestate	
12.5	Lystek	

Table 2. Primary Sludge Feed

Primary Sludge Feed		
Sample Volume (mL)	Sample Type	Percent Lystek
30	Primary	0% (Control 1)
50	Digestate	
0	Lystek	
25.5	Primary	15%
50	Digestate	
4.5	Lystek	
24	Primary	20%
50	Digestate	
6	Lystek	
22.5	Primary	25%
50	Digestate	
7.5	Lystek	

The acetate-propionate solution was prepared with 37.5 g/L sodium-acetate and 13.25 g/L sodium-propionate to act as substrate for the anaerobic microorganism populations present in digesting sludge. Primary sludge served as a secondary substrate to mimic real-life feed conditions. The feed protocol for each substrate is shown in Table 1 and Table 2. Two Lystek products were used; one recently obtained using a Canadian Lystek facility (< 6 months) and one prepared from sludge from Hunt's Point WWTP (an older product, > 6 months). All tests were performed in triplicate on the ABA for each Lystek product.

Phase 1-Results and Discussion

Hunt's Point Lystek

The results using Lystek from Hunt's Point are shown in Figures 2 and 3. The ABA using acetate-propionate plus Lystek showed poor results from Lystek addition, almost as if the acetate-propionate were inhibiting the methanogens as concentration increased (Figure 3). The serum bottle with 0% Lystek feed showed the highest gas production with a decline of about 9% at 15% Lystek feed. There was an increase of about 1% at 20% Lystek as compared to 15% and then a sharp decrease of about 6 % at 25% Lystek. However, the ABA using primary sludge showed encouraging results (Figure 3) with about a 5% increase in gas production from no Lystek to about a 20% Lystek feed rate.

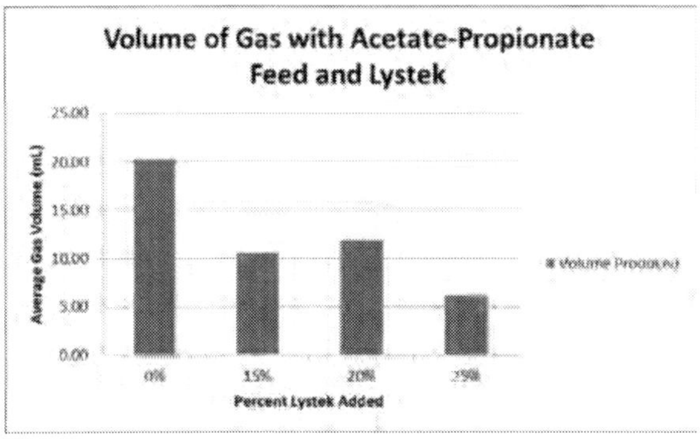

Figure 2. Gas volume produced from various concentrations of Hunt's Point Lystek with Acetate-Propionate

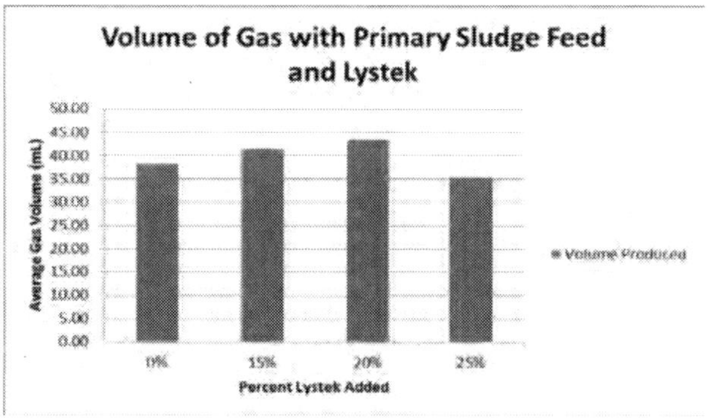

Figure 3. Gas volume produced from various concentrations of Hunt's Point Lystek with primary sludge.

Lystek product from a Canadian facility

The results using Lystek from Canada are shown in Figures 4 and 5. The optimal dosage once again appears to be 20% Lystek feed and shows a similar pattern to Experiment 1. The Acetate-Propionate results with the Canadian Lystek were different than that from Hunt's Point. In this case, 20% Lystek feed had the same gas production as the 0% Lystek serum bottle, but the volume was lower for the 0% Canadian Lystek as compared to the Hunt's Point Lystek.

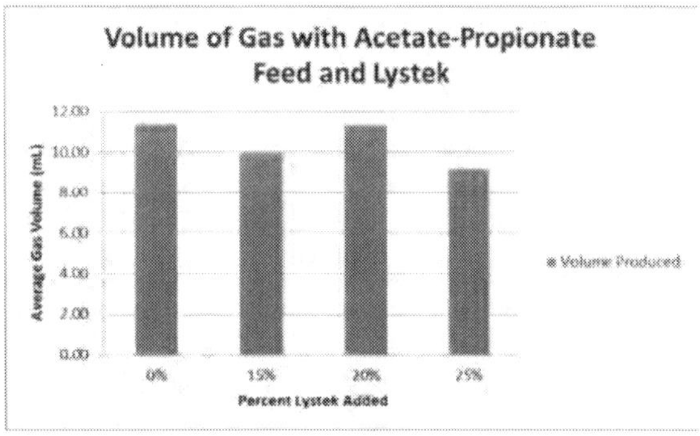

Figure 4. Gas volume produced from various concentrations of Hunt's Point Lystek with Acetate-Propionate

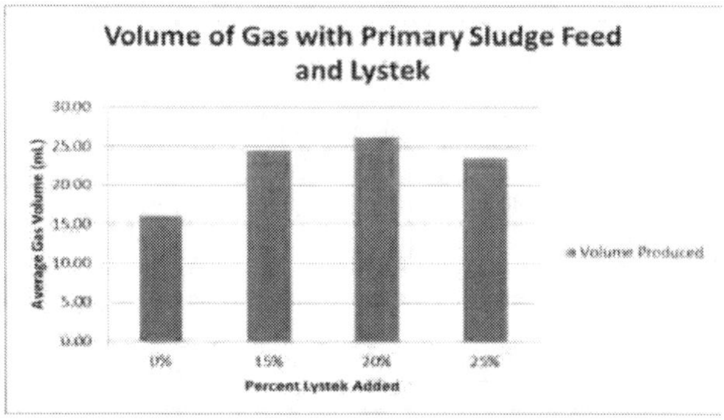

Figure 5. Gas volume produced from various concentrations of Canadian Lystek with Primary Sludge

However, for the primary sludge ABA, there was a much larger increase from no Lystek to 20% Lystek feed and much less of a decline at 25%. Overall in the tests using the Canadian Lystek, there was a significant increase in gas production for all concentrations as compared to Hunt's Point Lystek. This may be due to the Hunt's Point Lystek being more than 6 months old and may have undergone some degradation while in storage. The results from the Acetate-Propionate bottles are relatively consistent for both Lystek products but are hard to understand and need to be analyzed further. Figure 9 shows a side by side comparison of the performance of the two Lystek products.

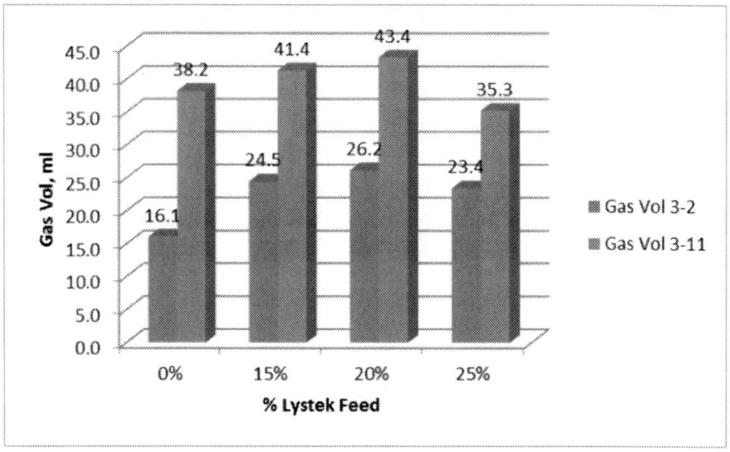

Figure 9. Gas volume produced from various concentrations of Hunt's Point Lystek with Acetate-Propionate

Phase 2-Anaerobic Digester Study

Bench Scale Reactors

Bench scale reactors were used to simulate performance of full-scale anaerobic digestion using a 15-day SRT. The goal was to determine if Lystek would enhance biogas production under the typical conditions used at New York City WRRFs. Two batch reactor systems were constructed for this phase of the experiment. A schematic of the reactor is shown in Figure 6. The 5-liter custom reactors were wrapped with digitally controlled heating tape and bubble wrap to maintain a temperature of 35°C (Figure 7). To maintain anaerobic conditions, the reactors were sealed using rubber stoppers. Biogas exited from the reactors through a one-way flow valve. Gas volume was continuously measured using HOBOware data loggers attached to wet-tip gas meters (Wet Tip Gas Meter Co., Nashville, TN), which measured gas production by volumetric displacement of water. Biogas composition was determined using a Landtec 5000 biogas meter. The meter was calibrated before each reading using a standard gas mixture of methane, CO_2, and H_2S. The biogas unit measures the percentage of methane (CH_4), carbon dioxide (CO_2), and oxygen (O_2) in the biogas and then subtracts those values from 100%. If the sum does not equal 100%, the difference is typically assumed to be nitrogen. The meter was also used to measure hydrogen sulfide (H_2S) concentration.

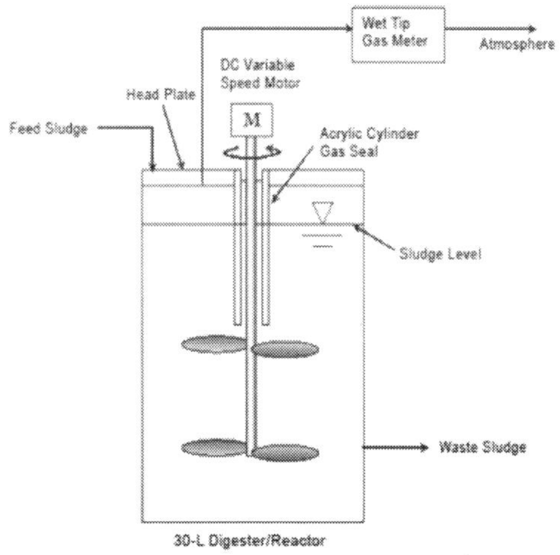

Figure 6. Schematic of bench scale

Figure 7. Experimental Setup using 5-L Reactors

The control reactor (Reactor A) was fed primary sludge and (Reactor B) was fed the same primary sludge plus varying amounts of Lystek on a volume-volume ratio with a 15 day SRT. Primary sludge collected from the Middletown, NY, Wastewater Treatment Plant was characterized after collection and before being fed to the reactors. TS and VS analyses were performed using the standard pan method. Hach kits were used to determine the COD (TNT 822), NH4-N (TNT 832), Volatile Acids (TNT 872), Total Phosphorous (TNT), Ortho Phosphorous (TNT 845), Alkalinity (TNT 870). This same characterization process was used to determine the characteristics of the reactors as well on every feed day.

Figure 8. Experimental Setup using 3.5-L

From the beginning of the test, there was difficulty in creating anaerobic conditions in the reactors originally thought to be due to air leakage. Although a major effort was made to completely seal the reactors and appurtenant equipment, it was impossible to establish steady state anaerobic digestion determined by low biogas production, low methane concentration, and poor volatile solids destruction. Because of these problems, this system was shut down and a literature search performed which resulted in the development of a new system. (Note: We later determined the problem with the reactors was due to too large a head space entrapping too much air.) This new system was a smaller scale version of the previous reactors and uses 3.5-liter vacuum filter flasks with a working volume of 3 liters and headspace of 0.5 L. Other than the change in physical reactors, the same materials and methods explained earlier were used with this arrangement as well.

The flasks were placed in a water bath set at 35°C and were attached to tipping buckets that recorded gas production. The same gas reading method as previously explained was used. Reactor feeding was performed every Monday and Thursday with feed volumes of 600 mL and 800mL respectively. The working SRT was 15 days. Primary sludge was characterized in the same manner as for the first reactor setup. Initially both reactors were fed primary sludge to establish identical, steady-state conditions before adding Lystek to the experimental reactor. Once steady-state was achieved, primary sludge alone continued to be fed to the control reactor (Reactor A) and primary sludge plus varying amounts of Lystek was fed to the experimental reactor (Reactor B) on a volume-volume ratio. The amount of Lystek was increased over time starting at 15% and ending at 25%. Primary sludge feed solids concentration was maintained between 2-4% solids.

Phase 2-Results and Discussion

The following results and discussion are based on the reactor performance from March 31 to June 16. The Control Digester was fed 100% primary sludge throughout the entire test period. The feed cycle to the Lystek reactor was:

- 03/31/16 - 04/21/16: Baseline Phase, 0% Lystek, 100% primary Sludge (by Volume)
- 04/21/16 - 05/05/16: 15% Lystek, 85% Primary Sludge
- 05/05/16 - 06/06/16: 20% Lystek, 80% Primary Sludge
- 06/06/16 - 06/23/16: 25% Lystek, 75% Primary Sludge

Volatile Solids Reduction

For the first SRT, the volatile solids reduction in both reactors was lower than expected, but started increasing through the second SRT. The Lystek reactor had a greater volatile solids reduction than the control reactor except for the first SRT of the 20% feed period (Figure 9). The Lystek reactor did reach the typical volatile solids reduction expected from anaerobic digestion towards the end of the 20% feed period and continuing throughout the 25% feed period. At a 15-d SRT in a full-scale mesophilic digester, the typical volatile solids reduction would be about 45%. The higher

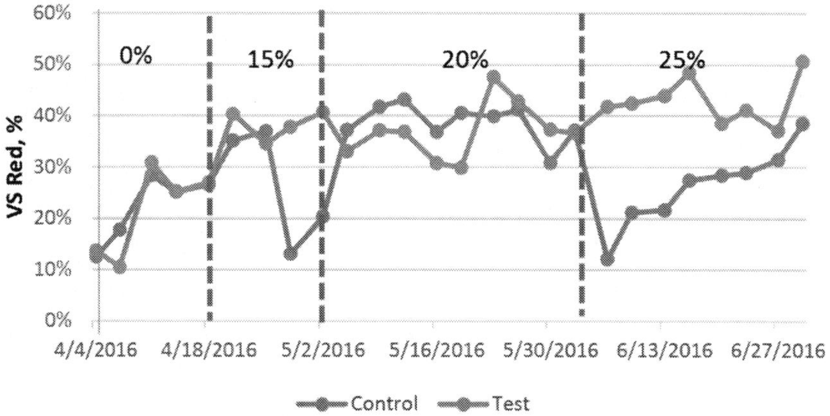

Figure 10. Volatile Solids Reduction

volatile solids reduction in the Lystek reactor correlates fairly well with the biogas yield at the same feed rate.

Biogas Yield
Throughout the entire testing period biogas yield was higher in the Lystek reactor than in the control reactor except for a brief period at the beginning of the 15% Lystek feed (Figure 11), likely due to acclimatizing of the microorganisms to Lystek.

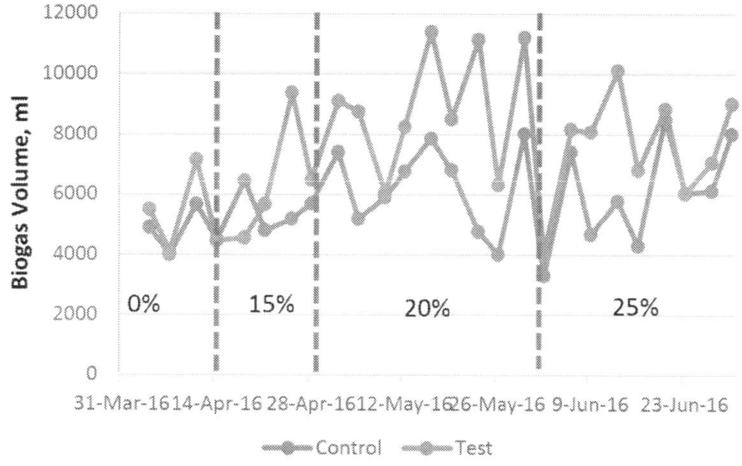

Figure 11. Biogas Yield

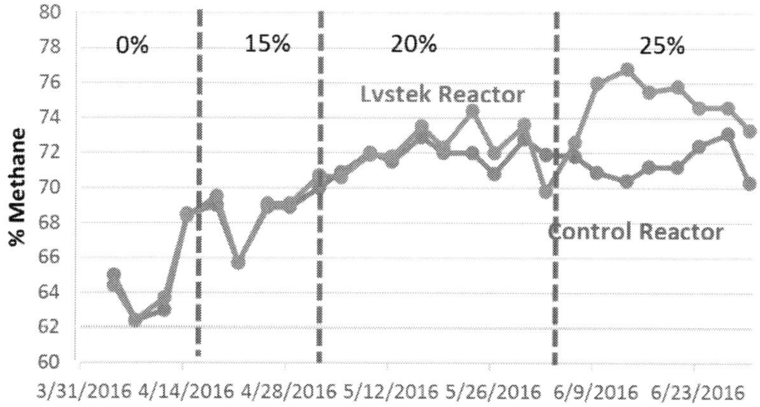

Figure 12. Methane concentration in biogas

Biogas Composition
Methane, carbon dioxide, oxygen, and hydrogen sulfide were measured in the gas twice per week. At 0% to 15% Lystek feed, the methane concentration in both reactors was very similar and ranged from 62% to 72% (Figure 12).

For the first few days of the 20% Lystek feed, both reactors produced similar concentrations of methane; however, after about 10 days, the biogas from the Lystek reactor has a much higher methane concentration that did the control reactor. Towards the end of the 20% feed period, there was a significant drop in methane concentration in the Lystek reactor which cannot be explained. When 25% Lystek was fed, the methane concentration significantly increased while the control reactor remained relatively constant. With additional time, the methane concentration in the Lystek reactor decreased but was still higher than the control reactor. The decline at the end was due to no longer feeding the digester in preparation for shutdown. It should be noted that the methane concentration of biogas can range from 55 to 70%, but is typically considered to be between 65 and 70% for wastewater sludge (Metcalf and Eddy, 5th ed). Both reactors had concentrations significantly higher than typical at certain times during the project. To ensure accuracy, the meter was calibrated each day measurements were made, so there is confidence in the results. Moreover, there is certainly a difference in the methane concentration from control to Lystek on a relative basis. In addition to methane, carbon dioxide and hydrogen sulfide concentrations were measured. Carbon dioxide concentrations average about 29% and O_2 remained at less than one percent (~0.3%). Hydrogen sulfide concentrations (Figure 13) increased in the control reactor but decreased to almost zero in the

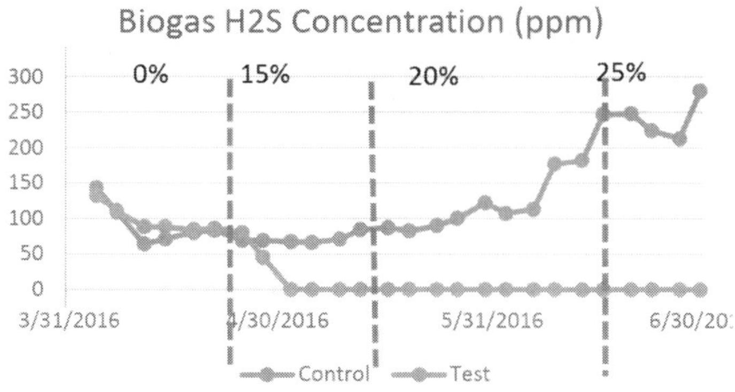

Figure 13. Hydrogen Sulfide concentration in biogas

Lystek reactor. The reasons for this are not known at this time. However, there is certainty that these results were not due to an instrument malfunction or calibration error since the measurements for both reactors were done within minutes of one another. One possibility may be a reaction with metal ions which form a sulfide salt instead of hydrogen sulfide. It is important to understand this and requires further study since if there is little to no H_2S in the biogas, there would be a significant reduction in the cost for gas cleaning equipment.

Since digester supernatant or filtrate/centrate from dewatering anaerobically digested sludge is returned to the head of the plant, it is important to quantify the concentration of nitrogen and phosphorous entering the plant through these sidestreams.

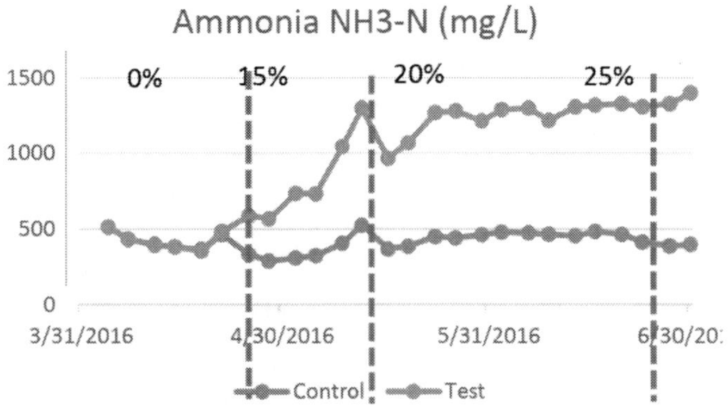

Figure 14. Supernatant ammonia concentration

Ammonia

Ammonia is a concern since it adds to the influent nitrogen concentration and impact facilities that have a nitrogen permit limit. Ammonia is produced in the anaerobic digestion process and the concentration is a function of SRT; the lower the SRT, the lower the concentration. The typical concentration at a 15-d SRT is between 1200 and 1400 mg/L. The ammonia concentration from the Lystek reactor fell within those averages (Figure 14); however the control reactor had a considerably lower concentration. The reason for this is most likely due to the concentration of ammonia in Lystek product coupled with the increased volatile solids reduction within the Lystek reactor. Further studies are needed to confirm and better understand these results

Phosphorous

Phosphorous is released under anaerobic conditions which exist in the digester which can then be returned in plant sidestreams adding to the influent concentration. This is a concern for facilities that have a permit limit for phosphorus. For this project, reactive orthophosphate was measured. Figure 15 shows the concentration of orthophosphate at various concentrations of Lystek feed relative to the control. There appears to be a greater release of orthophosphate in the Lystek reactor which increases at higher concentrations of Lystek feed. This may be due to the increase in volatile solids reduction or the initial concentration of phosphorus in Lystek, but needs further study to understand the reasons for this higher concentration.

Conclusion

The results of this project indicate that adding Lystek to an anaerobic digestion process appears to increase both methane production as well as gas yield as compared to a digestion process with only primary sludge. Hydrogen sulfide production in the digester gas appears to be very low with Lystek as compared to primary sludge. If this can be replicated, this is a significant advantage since it reduces the need for gas cleanup prior to use as an energy source. Ammonia production is in the normal range so there is no negative impact by adding Lystek. Volatile solids reduction is also higher than the control. However, there is an increase in ammonia and orthophosphate in the Lystek reactor. The next steps for this research are to use a mixture of primary and waste activated sludge (WAS) to determine if the presence of WAS changes biogas production, H_2S production, etc.; increase the concentration of Lystek to determine optimal concentration, confirm and understand the reasons for little to no H_2S in the biogas, evaluate the reasons for the increase in ammonia and orthophosphate concentrations, and to determine the effects of Lystek addition on dewaterability.

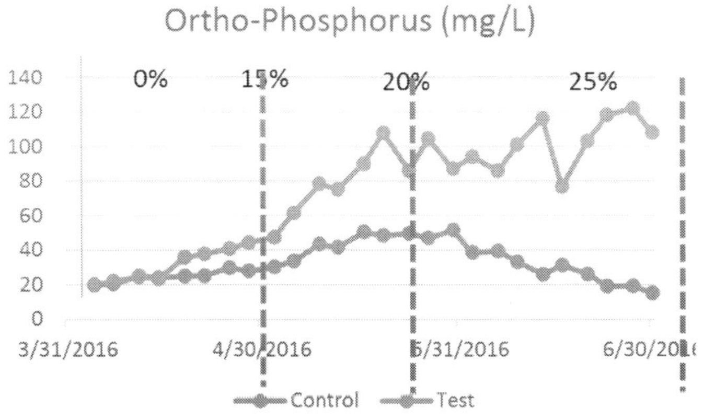

Figure 15. Supernatant ortho-phosphate concentration

References

Abdeshahian, Peyman et al. 2016. "Potential of Biogas Production from farm animal waste in Malaysia". Renewable and Sustainable Energy Reviews, 60: 714-723.

Elbeshbishy, Elsayed et al., 2014. Impact of alkaline-hydrolyzed biosolids (Lystek) addition on the anaerobic digestibility of TWAS in lab and full-scale anaerobic digesters. Waste Management, 34: 2090-2097.

Fernandez-Rodriguez, J. et al. 2015. Temperature-phased anaerobic digestion of Industrial Organic Fraction of Municipal Solid Waste: A batch study. Chemical Engineering Journal, 270: 597-604.

Mata-Alvarez, J. et al. 2014. A critical review on anaerobic co-digestion achievements between 2010 and 2013. Renewable and Sustainable Energy Reviews, 36: 412-427.

McCarty, P. & McKinney, R. 1961. Volatile Acid Toxicity in Anaerobic Digestion. Journal Water Pollution Control Federation, 33: 223-232.

Owen, W., Stuckey, D., Healy, J. J., Young, L., & McCarty, P. (1979). Bioassay for Monitoring Biochemical Methane Potential and Anaerobic Toxicity. Water Research 13: 485-492.

Parkin, G. and Owen, W. (1986). "Fundamentals of Anaerobic Digestion of Wastewater Sludges." Journal of Environmental Engineering 112: 867-920.

Park, Ki Young, et al. 2016. Combination of different substrates to improve anaerobic digestion of sewage sludge in a wastewater treatment plant. International Biodeterioration & Biodegradation 109: 73–77

Wu, Li-Jie, et al. 2015. Upgrading of anaerobic digestion of waste activated sludge by temperature-phased process with recycle. Energy 87: 381-389.

Kenosha's Energy-Optimized Resource Recovery System

Authors
Joseph Hughes, Centrisys Corporation
Dr. -Ing Andreas Duennebeil, PONDUS Verfahrenstechnik GmbH

Abstract
The intent of this paper is to convey the story behind Kenosha Water Utility's (KWU's) quest to become less energy reliant and more environmentally sustainable. Faced with an aging treatment plant in need of anaerobic digester repairs and a valuable biogas resource that was underused, KWU issued a Request for Proposals (RFP) for the design, installation, start-up and commissioning of an energy-optimized resource recovery system.

The goals of the project were to: increase the generation of methane gas from the biosolids; generate greater than 500 kW of continuous power from the methane gas to run the new process and send usable power back to the existing plant power network; use the electric and thermal energies generated to dry the biosolids to 90% dry solids and produce Class A biosolids, reducing the volume and cost of biosolids disposal; maintain existing effluent quality of the plant; maintain or reduce the ambient noise, odor, and particulate levels beyond the waste water treatment plant; obtain a performance warranty for the system; and, provide a payback period of 8 years or less.

This paper reviews the design/build construction phase of the project and the following one-year optimization phase of the process. Preliminary process results are presented and lessons learned during the project are described.

Keywords: Thermo-chemical hydrolysis, combined heat and power (CHP) units, biosolids, biogas

Introduction

The Kenosha wastewater treatment plant has an average daily capacity of 28.6 MGD with nine (9) primary clarifiers followed by six (6) activated sludge basins, four (4) secondary clarifiers, disinfection, and dechlorination prior to discharge to Lake Michigan (See Figure 1 below). The plant went online in 1940 and has been expanded and upgraded in 1967 and 1985.

Figure 1: Kenosha Wisconsin Wastewater Treatment Plant

The six (6) anaerobic digesters operated fairly well for their age, but the frequency of needed repairs was increasing and available capacity was decreasing. In addition, there was an opportunity to leverage the biogas potential of the digesters to reduce the plant's overall operating costs. As a result, KWU issued a Request for Proposals (RFP) for the design, installation, and the start-up and commissioning of an energy-optimized resource recovery system.

The goals of the project outlined in the RFP included:

- Increasing the generation of methane gas from the anaerobic digesters.
- Generating greater than 500 kW of continuous power from the methane gas to run the new process and send usable power back to the existing plant power network.
- Using the electric and thermal energies generated to dry the biosolids to 90% dry solids concentration and produce a Class "A" biosolids, reducing the volume and cost of biosolids disposal.
- Maintaining existing effluent quality of the plant.
- Maintaining or reducing the ambient noise, odor, and particulate levels beyond the WWTP.
- Obtaining a performance warranty for the system.
- Providing a payback period of 8 years or less.

On October 1, 2013 KWU entered into a design/build contract with the manufacturer of a thermal hydrolysis process for the Kenosha Energy-Optimized Resource Recovery System (KE-ORRS) project. The project was initiated and the Engineer of Record, Donohue and Associates, Inc. (Donohue), prepared 113 sheets of design/build level plans for the installation of the system. In February 2014, the design/build manufacturer entered insolvency proceedings in Germany and became unable to provide the thermal hydrolysis process or otherwise support the project.

Centrisys Corporation (Centrisys) is a centrifuge manufacturer and service company located in Kenosha, Wisconsin. Centrisys has built a relationship with KWU through two (2) mutually beneficial equipment installations that paid for themselves in less than 2 years. With the KE-ORRS project, Centrisys saw the opportunity to engage in another mutually beneficial project with KWU, and signed a contract to take over as the Design/Builder in early October 2014. On October 27, 2014, KWU kicked off the revamped KE-ORRS design/build biosolids project.

There was an immediate push to get the GC under contract and the project underway because of a grant application that KWU had applied for with We Energies, the local electrical utility. The $500,000 Focus on Energy Grant was available to Wisconsin utilities that implemented sustainable energy-saving projects. However, in order to be eligible, the projects had to attain Substantial Completion before the end of 2015.

Methodology

In an effort to get the proposed General Contractor (GC) J. F. Ahern Company and their subcontractors under contract, the project team focused on the parts of the project that were not changing from the original design. These parts of the project included the cleaning of and modifications to the existing anaerobic digesters, demolition work, structural concrete repair work to the solids handling building, and the construction of a temporary building to house the dewatering centrifuge. After multiple on-site meetings between KWU, Centrisys, Donohue, and the team of contractors, a 32 point proposal was signed on November 20, 2014 as the initial contract with the GC.

As soon as the contract was signed, the GC mobilized equipment and site trailers, and began working on the digester cleaning, concrete repair work described above. Three (3) plate and frame presses, which were replaced by the decanting centrifuge years earlier, used lime to dewater the biosolids. Over time, the lime had eaten away at the concrete structure of the solids handling building. As a result, a temporary structure was built to house the decanting centrifuge so that the structural repairs to the solids handling building would not interfere with the dewatering operations at the plant (See Figure 2 below).

Figure 2: Deteriorated & repaired structural concrete beams and temporary dewatering structure

As previously mentioned, the lysing technology needed to be replaced when Centrisys took over as the Design/Builder for the KE-ORRS project because the supplier of the previously selected technology had gone insolvent. Centrisys selected the PONDUS thermo-chemical hydrolysis process (See Figure 3 below), which was substantially different than the previously selected technology. As a result, the 113 sheets of design/build level plans for the installation of the insolvent system required numerous equipment layout and design changes.

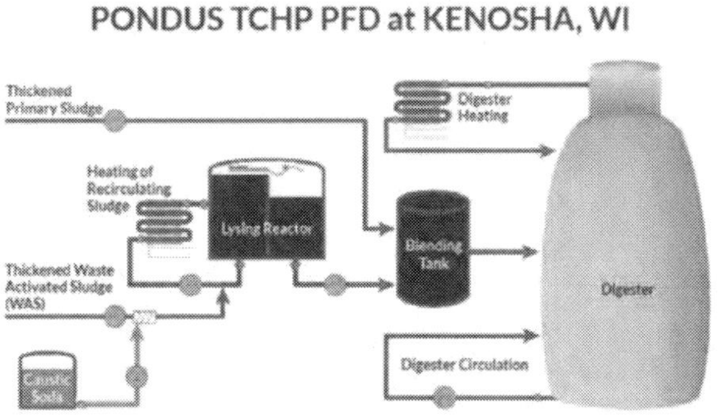

Figure 3: PONDUS Process Flow Diagram

In an effort to keep the GC and subcontractors productive, the procedure of Donohue preparing a limited set of drawings and associated specifications with specific construction items for the GC to bid on was repeated 12 more times between November 20, 2014 and July 15, 2015. Of the 113 original contract drawings based on the original hydrolysis system and associated unit processes, 12 were deleted, 55 new drawings were added, and 322 drawing revisions were made during the construction of the project. Because of the fluid nature of the project, the GC consolidated work items from multiple drawing sets to prepare proposals, resulting in eight project phases (the initial contract and 7 CO's). The individual technologies included in the KE-ORRS project are summarized as follows (See Figure 4 below):

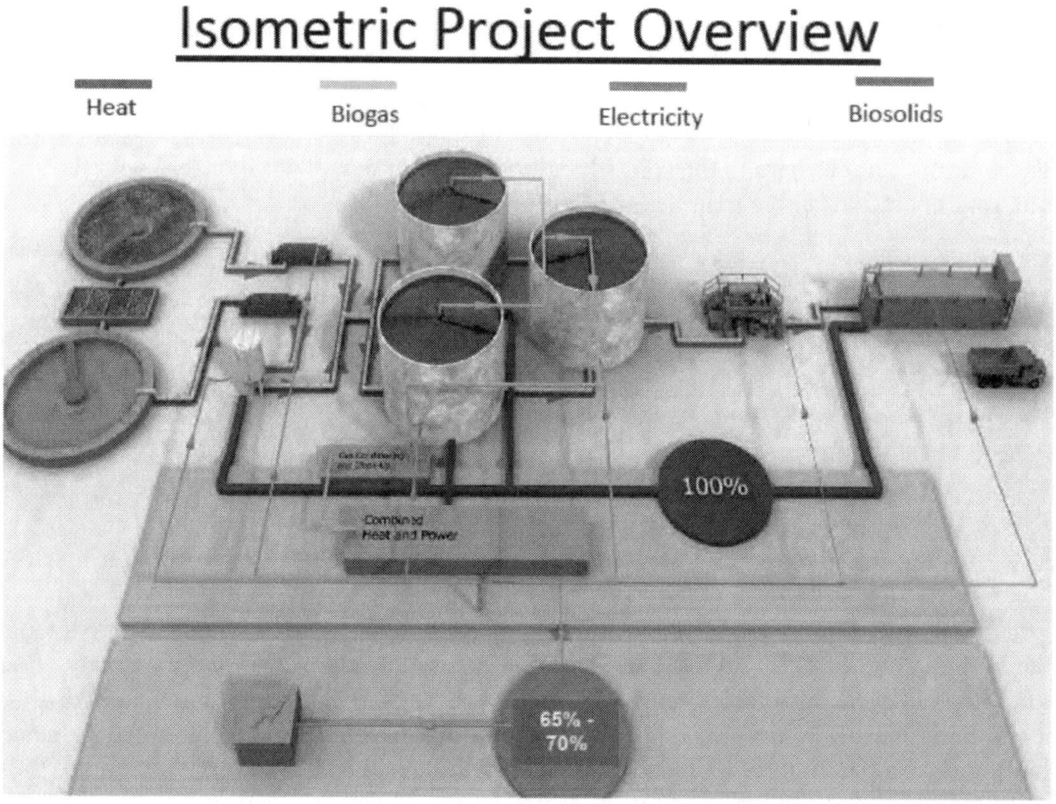

Figure 4: KE-ORRS Schematic Layout and Energy Distribution

Primary Sludge Thickening: The sludge discharged from the primary clarifiers typically ranges between 3% and 4% solids concentration. In the new high-solids concentration anaerobic digestion process, the primary sludge is sent to a Centrisys thickening centrifuge to increase the solids concentration up to 7%. The Thickened Primary Sludge (TPS) is then discharged to a blend tank, where it is mixed with hot Hydrolized Thickened Waste Activated Sludge (HTWAS) resulting in a warm mixture of TPS and HTWAS to feed to the digesters.

Waste Activated Sludge (WAS) Thickening: The WAS typically ranges between 0.9% and 1.2% solids concentration. In the new high-solids concentration anaerobic digestion process, the WAS is sent to a Centrisys thickening centrifuge to increase the solids concentration up to 7%. The Thickened WAS (TWAS) is then discharged to the PONDUS TCHP.

PONDUS Thermo-Chemical Hydrolysis Process (TCHP): PONDUS TCHP consists of a sodium hydroxide storage tank and feed pump, a heat exchanger, an annular reactor with approximately 2 hours of detention time, recycle and discharge pumps, and instrumentation. The TWAS is dosed with 50% sodium hydroxide solution as it is pumped to the PONDUS system heat exchanger. A recycle stream withdrawn from the reactor is combined with the chemically treated TWAS to preheat the feed and reduce the viscosity of the combined flow into the heat exchanger. The combined influent /recycle flow is pumped through a heat exchanger designed to maintain the temperature of the reactor vessel between 60° and 70°C (140° to 160°F), with a target temperature of 65°C (150°F). The hot, Hydrolyzed TWAS (HTWAS) is pumped from the PONDUS reactor to the blend tank, where it is mixed with the cooler thickened primary sludge. The resulting warm combination of TPS and HTWAS is then pumped to the two (2) high-solids concentration anaerobic digesters.

Digester Mixing: KWU has six (6) anaerobic digesters. The new high-solids concentration anaerobic digestion process uses three (3) of the 70 ft diameter digesters. Each of the digesters has a 26 ft side water depth comprising approximately 700,000 gallons of capacity. The two (2) heated primary high-solids concentration anaerobic digesters and the single unheated secondary digester were equipped with new mixing equipment. Each digester was equipped with a 50 hp horizontal chopper pumps capable of recirculating 3,200 gpm through three (3) separate double-nozzle discharge assemblies to homogenize the digester contents. The pumps are typically operated at 75% capacity for 22 hours per day and at 100% for 2 hours per day.

Biogas Conditioning System: The biogas conditioning system consists of a compressor and heat exchanger with a glycol chiller for moisture removal, particle filtration, and activated carbon filters for siloxane removal.

Combined Heat & Power (CHP) units: Electric and thermal energy is generated by two (2) CHP units that each include biogas fueled 1,050 cubic inch V-8 engines and brushless synchronous generators. Each CHP is capable of generating 330 kW of electrical energy and 1,467,200 BTU/hr of thermal energy with approximately 38% and 50% efficiencies, respectively.

Backup Boiler: The hot water system includes a 1,704,000 BTU/hr natural gas boiler that can provide supplemental hot water if a CHP unit is taken out-of-service.

Biosolids Dewatering: The digested biosolids discharged from the secondary digester typically ranges between 2% and 3.5% solids concentration. The digested biosolids are sent to a Centrisys decanting centrifuge to increase the solids concentration up to approximately 30%. The dewatered biosolids are then discharged to a conveyor that feeds a progressive cavity pump which directs the biosolids to the dryer.

Biosolids Drying: The digested and dewatered biosolids are discharged to a low temperature hot air dryer capable of processing 1,050 kg/hr of solids and evaporating 700 kg/hr of water. The biosolids are extruded through a strainer onto the upper belt which transports the biosolids through the dryer. The semi-dried biosolids are discharged into a crusher and onto the lower belt for further drying. The temperature and wetness of the dried biosolids are continuously monitored, and they are discharged to a conveyor system that loads a dump truck for hauling.

Two-Stage Odor Scrubber: Approximately 90% of the air used in the dryer is recycled while 10% is discharged to a two (2) stage odor scrubber. The odor scrubber has a heat exchange plenum to reduce the temperature of the exhaust air and preheat the raw intake air, followed by a cooling spray prior to entering the acid dosed media zone. The acid-dosed (H_2SO_4) first stage is followed by an alkaline-dosed (NaOH) second stage.

The new design not only incorporated a new lysing technology (the first PONDUS TCHP in the United States), it also consolidated the system's technology in the existing solids handling building. In order to enable that consolidation, the existing plate and frame presses, and a 12 ft diameter, 35 ft tall lime silo had to be demolished and removed from the building.

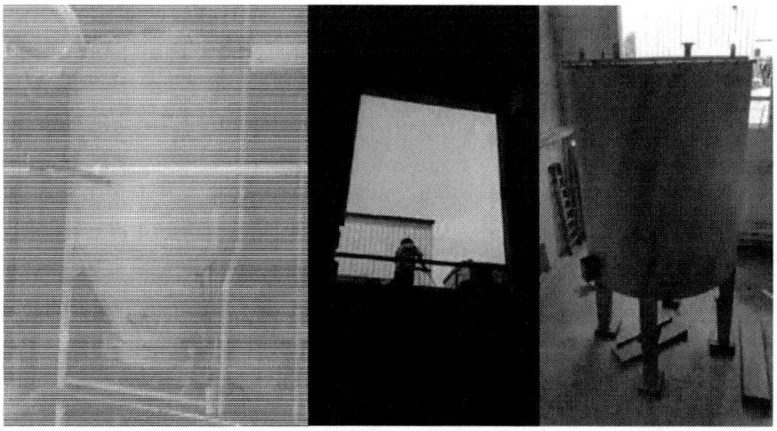

Figure 5: Existing lime silo that was demolished to make room for the PONDUS reactor

The lime silo was cut into pieces to remove it from the basement of the solids handling building; however, the PONDUS reactor could not be cut into pieces to install it in the building. The only way to get the approximately 8 ft. diameter by 12 ft. high reactor into the basement was to remove a section of the wall. That presented the opportunity to replace the brick with a large window that would bring more natural light into the basement, and enable the removal of the reactor in the future, if necessary (See Figure 5 above). The removal of the plate and frame presses from the solids handling building provided room for the new primary sludge thickening centrifuge to be located in the same room with the existing WAS thickening centrifuge and the existing dewatering centrifuge (See Figure 6 below).

Figure 6: The new thickening and dewatering centrifuge room at the Kenosha WWTF

As mentioned earlier, one of the goals of the project was to qualify for the Focus on Energy Grant sponsored by We Energies by reaching Substantial Completion before the end of 2015. For the KE-ORRS project, Substantial Completion was defined as both the PONDUS TCHP and the biosolids dryer being operational. The project team was determined to meet this goal. So we prepared an aggressive start-up schedule that would be simultaneously implemented along with the remaining construction schedule. Figure 7 below shows the start-up schedule for the numerous control panels on the KE-ORRS project

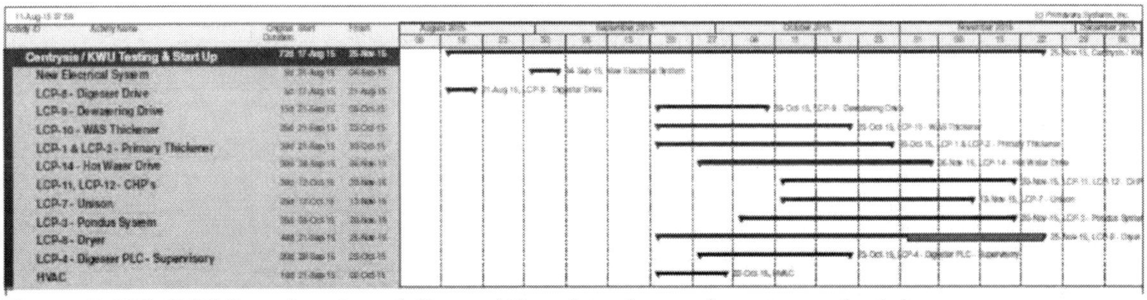
Figure 7: KE-ORRS project Local Control Panel testing and start-up schedule

The start-up was two-fold. There was the electrical/mechanical start-up, and then there was also the process start-up. The electrical/mechanical start-up involved 12 control panels, 27 pumps, 48 instruments including gas and liquid flow meters, temperature, pressure, and level monitors, 42 VFD's, mixers, two (2) CHP's, a Dryer, an odor scrubber, and a truck loading bay. The process start-up involved a conversion from a batch feed anaerobic digestion process using four (4) primary digesters and two (2) secondary digesters, to a continuous feed high-solids concentration anaerobic digestion process using (2) primary digesters and one (1) secondary digester.

The aggressive electrical/mechanical start-up schedule was adopted in mid-September that laid the pathway for the concurrent start-up of many individual pieces of equipment and unit processes. The dewatering centrifuge was started-up the second week of October, 2015, followed by the boiler, the hot water system, and the primary sludge thickening centrifuge the following week. The WAS thickener, the CHP's, and PONDUS Process were started-up next. Once the CHP's were up and producing heat and electricity, the Dryer could be started-up.

The process start-up required a transition from a batch feed mode that used four (4) heated primary digesters followed by two (2) unheated secondary digesters to a continuous feed mode with two (2) heated primary digesters and one (1) unheated secondary digester.

In the batch mode, KWU would feed one (1) primary digester for 8 hours per day. When one of the primary digesters became full, the secondary digesters were settled, supernatant was decanted, and then biosolids were pumped to the decanting centrifuge for dewatering. This provided room for the transfer of sludge from the primary digesters to the secondary digesters, enabling the primary digesters to be fed again.

In late August 2015, the process of converting the digester operating strategy from batch-feed mode to continuous-feed mode was initiated. Rather than flowing directly from the primary clarifiers to the digesters, the primary sludge had to be redirected to a buffer tank feeding a progressive cavity pump feeding the primary sludge thickening centrifuge. The TPS was directed to a blend tank where it was mixed with hot HTWAS, and the mixture was pumped to the digesters (See Figure 8 below). In addition, the contents of two (2) of the primary digesters that were being taken out of service had to be emptied. Ultimately, the continuous feed mode of the high-solids concentration anaerobic digestion process would operate as shown in Figure 8.

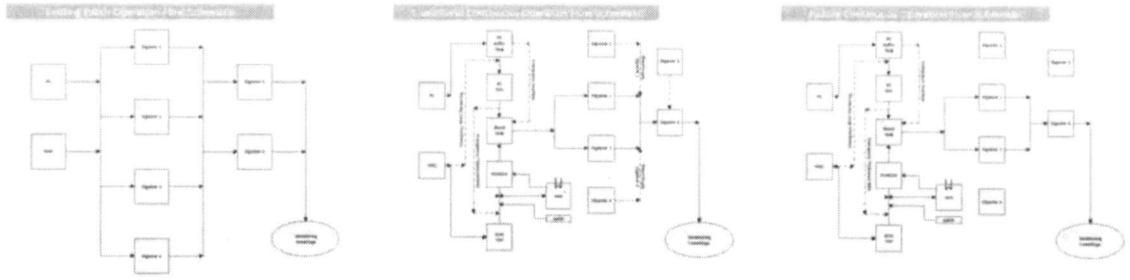

Figure 8: Batch feed mode, transitional feed mode, and continuous feed high solids concentration mode of operation for anaerobic digesters

After the system had been started-up, KWU reviewed the overall project and issued a 184 item Punch List of items that needed to be addressed by the Contractor and/or the Design/Builder on November 19, 2015. On November 30, 2016, Representatives from We Energies visited the facility to inspect the operations of the PONDUS TCHP and the biosolids dryer. Based on the increase in biogas production along with the generation of thermal energy and electricity to operate the new equipment and provide additional electrical energy to other equipment at the plant, We Energies awarded KWU a $500,000 Focus on Energy grant for the renewable energy project.

After Substantial Completion was achieved and the Focus on Energy grant was awarded, some of the unit processes were taken offline to enable the completion of the punch list items. The Contract included a one-year optimization period prior to the beginning of the contractual ROI period, so temporarily taking unit processes offline did not result in any financial penalties. By January 15th, 2016, approximately 80% of the 184 items identified in the Punch List had been addressed. The Contractor submitted updated versions of the Punch List at the February and March monthly project meetings, and on April 4, 2016, the Contractor submitted the completed Punch List.

From a process perspective, the entire System was put back into continuous operation on February 22, 2016. The System operated for approximately a month when primary sludge thickening had to be stopped due to an accumulation of rags on the shaft and propellers of the mixer in the primary sludge buffer tank (See Figure 9 below). Because of the extent of the rag issue, it was decided that the primary sludge would be sent directly to the primary digesters without additional thickening until a grinder could be installed prior to the buffer tank. Unfortunately, the rag problem was too severe to be resolved by a second grinder, so the primary sludge continued to be sent directly into the digesters without further thickening.

Figure 9: Primary sludge buffer tank with and without rags wrapped around mixer

KWU had been aware of the screening problem at the plant and was already moving forward with plans to install new influent perforated plate screens. The new influent perforated plate screens were started-up in late February 2017

Results

Initial results during the Optimization Phase of the project also show how robust the digesters are when operated with the PONDUS Process. In April of 2016, the PONDUS TCHP was turned off for almost five days, during which the biogas production levels dropped. When the PONDUS TCHP was brought back online, the biogas production almost immediately recovered to the previous level and then surpassed the previous level of gas production (See Figure 10 below).

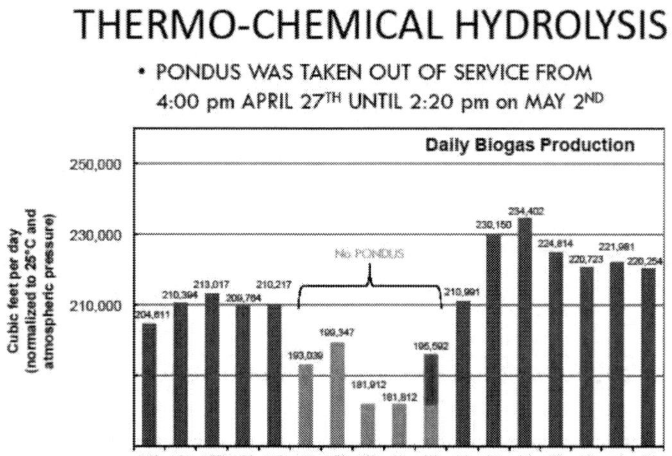

Figure 10: Biogas production drop off without PONDUS feed and increase with PONDUS feed

Initial results also indicate a 15% to 25% increase in biogas production. The electricity generated by the biogas is used to power the equipment for the new system, after which approximately $200,000 worth of electricity was sent to the plant in the first 12 months of operation. Part of the reason for the increase in biogas production may be attributed to the increase in soluble COD produced by the PONDUS Thermo-Chemical Hydrolysis Process. Marquette University evaluated the soluble COD concentrations in lysed and non-lysed sludge and found a 172% increase in the soluble COD concentration.

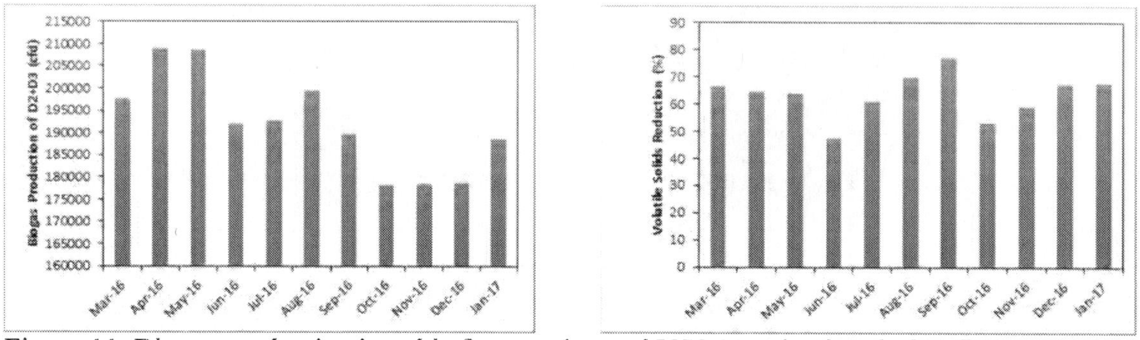

Figure 11: Biogas production in cubic feet per day and VSS % reduction during first year of operation

KWU operates a drinking water treatment plant in addition to the wastewater treatment plant. Two or three time per year the water plant sends alum sludge to the wastewater plant for disposal. In 2016, the wastewater treatment plant received alum sludge during the last week of May and the first two weeks of October. In Figure 11 above, a drop off in biogas production can be seen in June after the alum sludge was sent to the plant in the last week of May. Similarly, the biogas production drops off in October after receiving alum during the first two weeks of the month. VSS reduction also drops off after receiving alum sludge from the water plant (See Figure 11 above).

Initial results also indicate that the PONDUS TCHP system increases the dewaterability of the digested biosolids. It may be due to the higher VSS reduction rate achieved by the PONDUS TCHP system. More VSS reduction results in less organics in the biosolids, resulting a lower charge repelling biosolids particles. As a result, an increase in dewatered solids concentration of 2 to 4 percentage points has been observed (See Figure 12 below).

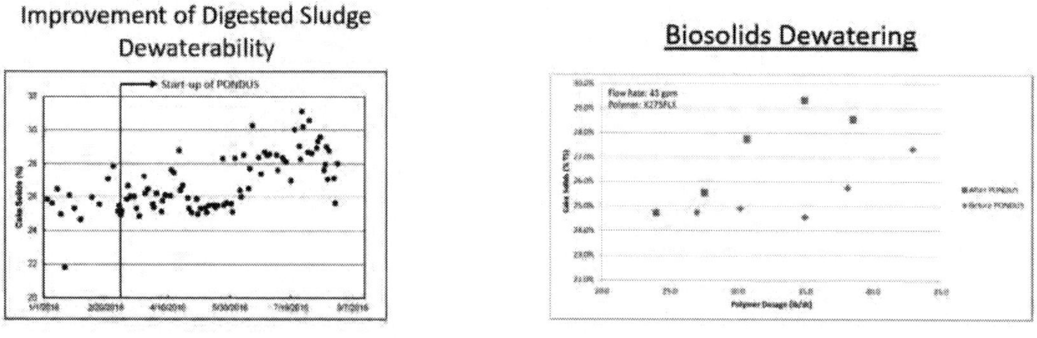

Figure 12: Increase in dewaterability of PONDUS fed digested sludge

Lessons Learned

The most important lesson resulting from the KE-ORRS project may not have been learned, but it was confirmed. That lesson is that this type of project has many inter-related aspects and is complicated. The KE-ORRS project included a one year optimization period following the construction phase of the project and prior to starting the return on investment phase. The optimization period provides time to fine tune the operation of the process, but it also allows the operators to stress the process to find limitations without the concern of meeting certain performance targets. The optimization period also provides time for the plant operating staff to become familiar with the process without the concern of meeting certain performance targets. Be sure to include an optimization phase after construction of complex inter-related systems.

CHP's provide the opportunity to generate thermal power and electricity from a natural resource. However, before initiating in a project where electricity generation is involved, one should evaluate their current electricity payment structure. Electrical utilities don't like losing customers, or having customers become competitors. Water and wastewater utilities sometimes have reduced electricity rates that are grandfathered from years ago. Implementing an energy generating process may result in the loss of a grandfathered rate, thereby reducing the realized benefit resulting from a project.

Pipes clog and spills have to be cleaned up. Another lesson learned is that you cannot have too many cleanout connections for sludge piping systems, and you can always use a few more drains.

The value of little things should not be under-estimated. Spare ring seals are inexpensive, but if one fails on a flow meter or a chemical feed pump, entire systems may have to be shut down. Similarly, rags can go from being an eyesore to a daily nuisance, accumulating around rotating shafts of pumps and mixers. Known existing problems like rags need to be addressed when engaging in a complex project. They rarely become less of a problem.

Testing is important to establish performance levels that can indicate when maintenance is required on a piece of equipment. By establishing performance levels when unit processes and pieces of equipment have recently started, one has a basis for identifying when wear levels or accumulated deposits need to be addressed.

Conclusions

The KE-ORRS project has shown that the PONDUS TCHP system offers a number of benefits to wastewater treatment plants with anaerobic digestion. Some of those benefits include the following:

- 15% to 25% increase in biogas production
- 2% to 4% increase in dewatered sludge solids concentration
- A potential reduction in polymer usage
- Less potential for foaming sludge in the digester

Now that the new influent bar screens have been installed and the entire process can be operated as intended, KWU will continue to collect data on the PONDUS process and the KE-ORRS project in the coming years.

Reducing Sludge Viscosity and Improving Dewaterability through a Thermochemical Hydrolysis Process

Peiyu Tan[1] and Zhongtian Li [1, 2*]

[1] Centrisys Corporation, 9586 58th Place, Kenosha, WI 53144
[2] Centrisys Corporation – CNP Division, 9535 58th Place, Kenosha, WI 53144
*Email: zhongtian.li@centrisys.us, zhongtian.li@cnp-tec.com

Abstract
Thermal hydrolysis pre-treatment of sludge for anaerobic digestion increases biogas production, enhances volatile solids reduction, and improves dewaterability of digested sludge. This study reports the performance of a thermochemical hydrolysis process (TCHP) that utilizes heat (60°C-70 °C) and sodium hydroxide for cell lysis. The objective of this study is to evaluate the impacts of implementing the TCHP on viscosity and floc morphology of thickened waste activated sludge (TWAS), dewaterability of anaerobically digested sludge, and biogas production. Cell lysis of TWAS was confirmed by microscope examination. Dewaterability of digested sludge was improved on full-scale dewatering centrifuges. The current data suggest at least 5-fold reduction of dynamic viscosity of WAS by thermal chemical hydrolysis treatment. Overall, the TCHP effectively hydrolyses TWAS and increases dewatered cake solids.

1. Introduction

Thermal hydrolysis pretreatment of sludge for anaerobic digestion increases biogas production, enhances volatile solids reduction, and improves dewaterability of digested sludge (Abelleira-Pereira et al. 2015, Schieder et al. 2000, Ruiz-Hernando et al. 2013). Conventional thermal hydrolysis treatment applies high temperature and high pressure for achieving adequate hydrolyzing effect of treated sludge (Schieder et al. 2000, Feng et al. 2014). Thermochemical hydrolysis combines both thermal treatment and chemical treatment for sludge hydrolysis. Both sodium hydroxide (Zhang et al. 2015, Neyens et al. 2003) and hydrogen peroxide (Abelleira-Pereira et al. 2015) have been applied for thermal chemical treatment.

One of the well documented benefits of thermal hydrolysis is improvement of dewaterability. Dewaterability of sludge has a direct impact on the economic viability of wastewater treatment plants (WWTPs). Maximizing sludge dewaterability is especially important when WWTPs are facing high sludge disposal cost (Aichinger et al. 2015, Takaoka et al. 2014). Thermal hydrolysis pretreatment of sludge for anaerobic digestion is reported to improve sludge dewaterability in respective to dry cake solids and polymer dosage (Neyens and Baeyens 2003).

Another agreed benefit of thermal hydrolysis is reduction of sludge viscosity. Viscosity is an important design parameter for anaerobic digestion at elevated total solids concentration (Dai et al. 2014). Sludge is usually thickened before subjecting to thermal treatment. Excessively high viscosity requires higher energy input for mixing (Russ et al. 2015). Thermal hydrolysis pretreatment significantly decreases sludge viscosity (Feng et al. 2015, Bougrier et al. 2008) and enables better mixing in anaerobic digester for high solids digestion (Baudez et al. 2011).

Despite of all the benefits, most of commercially available thermal hydrolysis process requires steam injection and limiting its application to large utilities. Thermochemical hydrolysis combines both thermal treatment and chemical treatment for sludge hydrolysis. Thermochemical hydrolysis can be achieved below the boiling temperature and could be applied to mid-size utilities. PONDUS is one of the commercially available technologies for thermochemical hydrolysis process. PONDUS thermochemical hydrolysis process utilizes heat (60°C- 70 °C) and sodium hydroxide. Though there are 6 full scale PONDUS installations, thermochemical hydrolysis is still at the early stage of technology adaptation and little technical information are available beyond improvement in biogas production.

The objective of this study is to evaluate the impact of a thermochemical hydrolysis process on viscosity and floc morphology of thickened waste activated sludge (TWAS), and dewaterability of anaerobically digested sludge. Dynamic viscosities of WAS and lysed thickened wasted activated sludge (LTWAS) were monitored under different NaOH dosages. Cell lysis of TWAS was confirmed by microscope examination. The results of this study will provide the necessary information to include the effect of thermochemical hydrolysis on sludge viscosity and dewaterability for future facility planning.

2. Materials and Methods

2.1. Thermochemical hydrolysis process

A PONDUS thermochemical hydrolysis process (TCHP) was implemented in Kenosha, WI in 2016 (**Figure 1**). The system includes a 14.3 m^3 reactor, a heat exchanger, and a mixing pump (**Error! Reference source not found.**). The average hydraulic retention time is 2.0 hr – 2.5hr. The reactor is heated to 65 °C - 70 °C. Sodium hydroxide solution (50% w/w) is dosed into TWAS before heat exchanger. Series dosages of sodium hydroxide solution (50% w/w) have been tested on the TCHP reactor in the range of 0.5 L/m^3 TWAS to 2.5 L/m^3 TWAS.

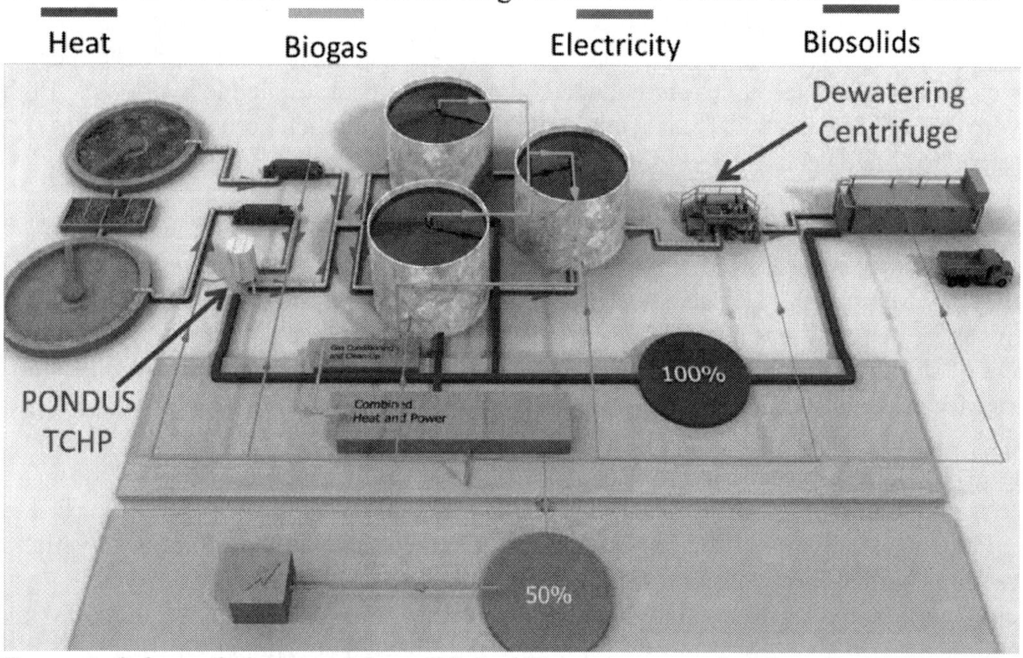

Figure 1. Schematic layout of the solids handling process at the Kenosha WWTP. The placements of TCHP and dewatering centrifuge were labeled by red arrows.

Figure 2. Installation of the PONDUS TCHP at the Kenosha WWTP.

2.2. Sludge dewaterability test

Dewatered sludge samples were collected from a dewatering centrifuge installed at the site plant (**Figure 3**). Two kinds of data were collected by Kenosha WWTP. The plant personnel take samples every day to monitor daily dewatering performance hence a time based operational curve can be generated to assess long term impact of TCHP on dewaterability of the treated and then digested sludge. Besides that, the several centrifuge optimization trials were conducted before and after the installation of TCHP to evaluate the relationship between polymer dose and cake dryness. Solids cake was measured according to U.S. EPA Method 1684.

2.3. Sludge viscosity test

Rotational viscometer (Thermo Fisher Scientific, CA) is used for dynamic viscosity test of WAS, TWAS, LTWAS, and digested sludge at 6, 12, 30, and 60 rpm. The changes of dynamic viscosities of LTWAS during cooling process are evaluated at 47.5°C, 38.2°C, and 25.0°C. Floc morphology of TWAS and LTWAS are evaluated under microscope (AmScope, IL) at 40X, 400X, and 1000X magnifications. Floc images are captured by a CCD camera installed on the microscope.

Figure 3. Centrisys dewatering centrifuge was installed at the Kenosha WWTP.

3. Results

3.1. Improvement of digested sludge dewaterability

Before thermochemical hydrolysis treatment was adopted, the average cake dryness was around 26% TS, ranging from 22-28% TS. During the optimization phase, the cake dryness improved gradually and after three months, 2% - 3% increase in cake solids was observed after the TCHP was in full operation and the highest cake dryness was approaching 31% TS (**Figure 4**).

It should be noted that at Kenosha WWTP, TCHP was not the only unit that was incorporated into the new operational design. A belt dryer was also implemented to dry dewatered sludge to 90% solid content. Due to issues with sludge discharge pump carrying sludge from centrifuge to the newly installed drier the plant intentionally kept dewatered sludge at around 27% instead of aiming for driest cake possible (**Figure 5Figure 4**).

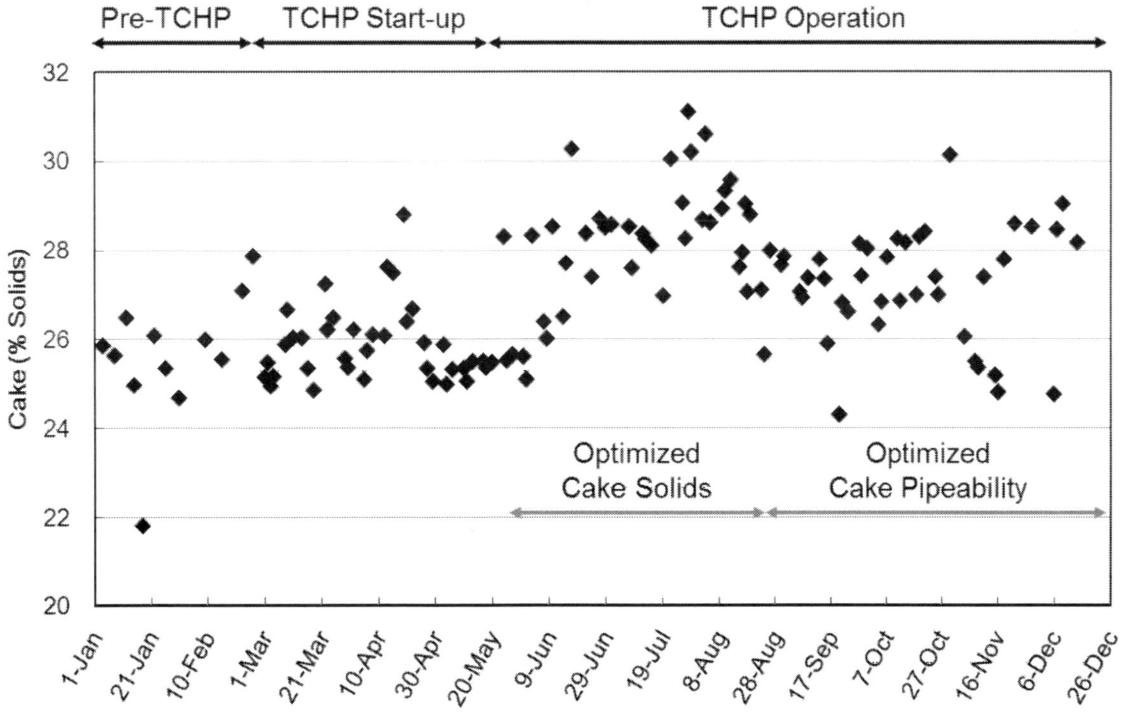

Figure 4. Historic trend of dewatering performance at Kenosha WWTP.

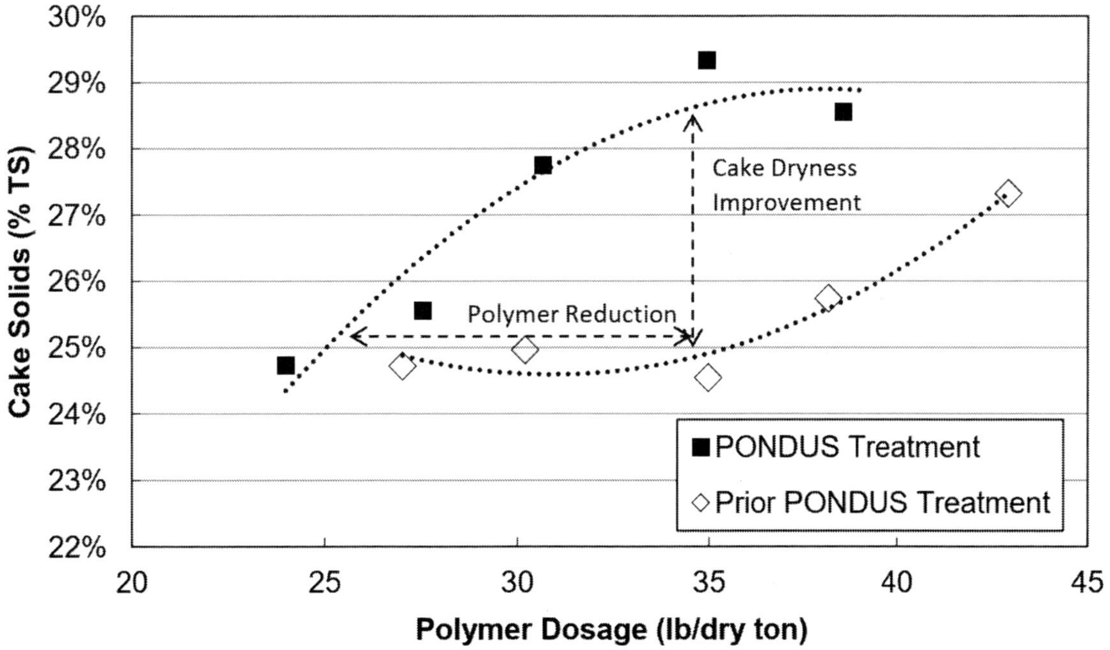

Figure 5. Polymer curves with and without thermochemical hydrolysis treatment.

3.2. Reduction of TWAS viscosity
In order to increase the capacity of anaerobic digester, waste activated sludge is thickened to 6-7% TS at the Kenosha WWTP. These increased the apparent viscosity by 1000X, as shown *in* **Figure 6**. Higher than 80% reduction of dynamic viscosities was observed after TWAS went through TCHP treatment. Both TWAS and LTWAS were non-Newtonian fluid. However, the change of dynamic viscosity of LTWAS was less significant compared with that of TWAS.

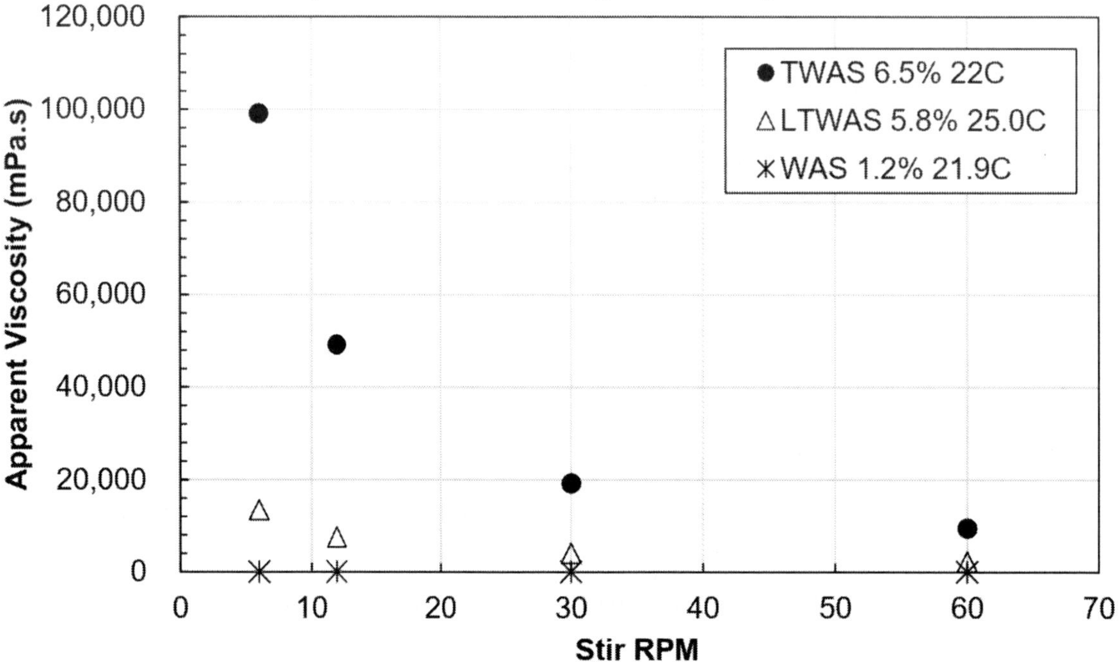

Figure 6. Comparing dynamic viscosities of WAS, TWAS, and LTWAS.

Figure 7 evaluated the dynamic viscosity of LTWAS under different temperatures. During the cooling process from 47.5°C to 38.2°C, no apparent changes of dynamic viscosity were observed at different stir rate. It was noted that the hydrolysis process continued as the sludge was cooled down, resulting a lower total solids value and lower dynamic viscosities after 4 hours cooling process from 47.5°C to 25.0°C.

3.3. Microscopic analysis of cell lysis
Floc morphology of TWAS and LTWAS were compared in **Figure 8**. Rotifer was selected as a target for comparison. Floc image at 1000X magnification clearly indicate the damage of cell structure on Rotifer. The combining effects of thermal treatment and chemical treatment on microorganisms provided pathogen reduction. Additional dynamic viscosities measurements and floc image analyses will be conducted under different dosages of sodium hydroxide solution in the range of 1.0 L/m^3 TWAS to 2.5 L/m^3 TWAS. The completion of this study will provide guidelines for the optimization of TCHP.

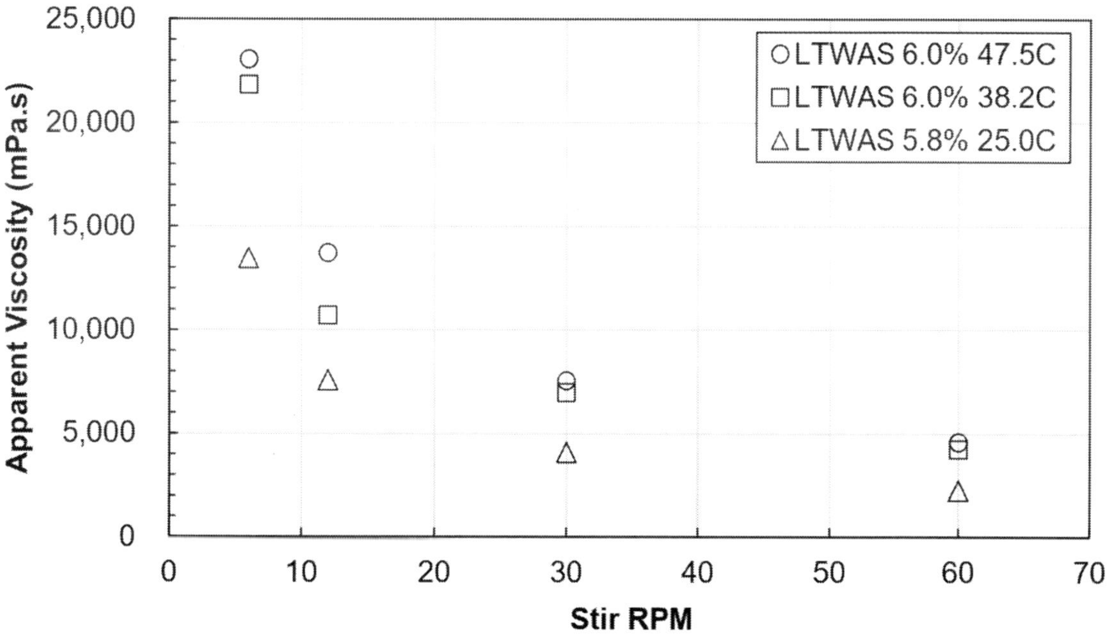

Figure 7. Comparing dynamic viscosities of LTWAS during cooling process.

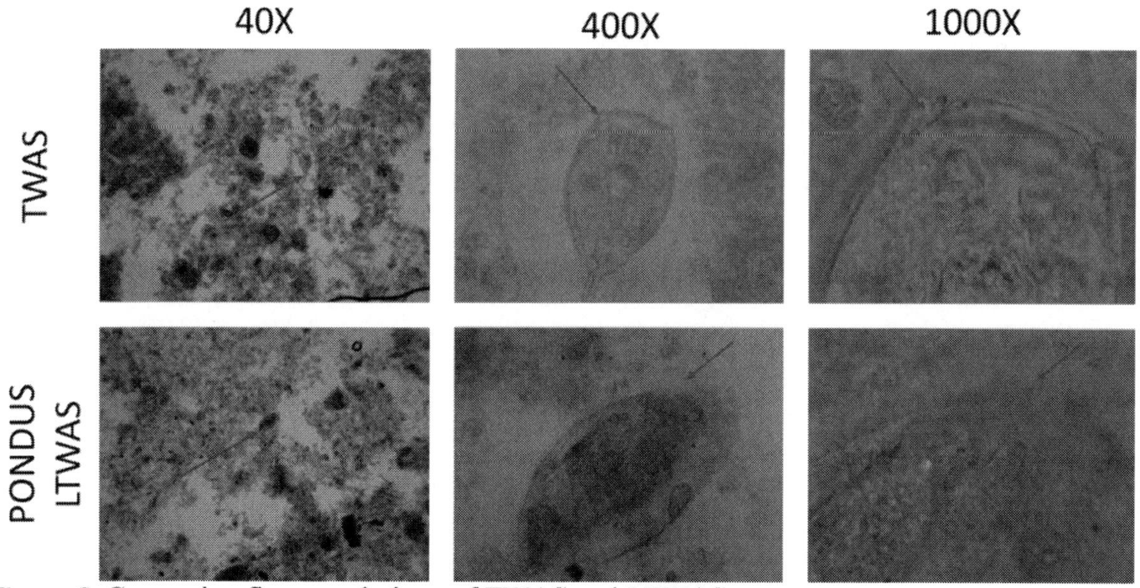

Figure 8. Comparing floc morphology of TWAS and LTWAS.

4. Discussion

4.1. Impact of TCHP on digested sludge dewaterability

Studies have shown that thermal hydrolysis impact a variety of parameters that will influent sludge dewaterability, including extracellular polymeric substances (EPS) and promotion of flocculation by reducing the amount of fine flocs (Neyens et al. 2004, Barber 2016). This this study, the centrifuge optimization trial was conducted by Centrisys process engineers in 2015 and 2016. During the trials, the polymer curve was created to illustrate the obtainable cake dryness at given polymer dosing rate. The polymer curves showing the impact of thermo-chemical hydrolysis treatment are presented in **Figure 5**. The polymer curve is of most interest since polymer consumption is directly linked to capital cost and dewatering performance. Both curves were created at 45 gpm of sludge using the same polymer at the same concentration.

Figure 5 shows two distinctively shaped curves, which means that the critical dosage for polymer activation was lowered due to TCHP treatment. The optimum dosing range for treated sludge was in between 33 to 38 lb/dry ton while the optimum dosage was above 43 lb/dry ton for untreated sludge. Within the optimum dosing range for treated sludge, a constant 3.5% to 4% increase in cake dryness was observed. At 25% cake solids in compliance with dryer operation, up to 34% polymer reduction was achieved. Please note that those polymer curves were optimized for daily operation.

During the testing for driest cake, sample collected was about 31% TS, shown in **Figure 4**. Barber reviewed the effect of thermal hydrolysis on sludge dewaterability and reported that full scale installation of WAS only thermal hydrolysis produce 28-30% cake dryness (Barber 2016). Yet the limitation came from the sludge handling after sludge dewatering. Other constraints such as vibration and hydraulic pressure of decanter centrifuge could limit for long term operation to generate highest cake solids. The improvement of dewaterability should be included in the overall facility planning when installation of TCHP was considered.

4.2. Impact of TCHP on TWAS viscosity

Reduction of viscosity is suggested to be the most important consequences of thermal hydrolysis as it allows higher solids loading rate for anaerobic digesters (Stickland 2015). A number of studies have shown that increasing the hydrolysis temperature decreases sludge viscosity between temperatures approx. 130°C - 180°C (Urrea et al. 2015, Higgins et al. 2014). Sludge become more Newtonian fluid as hydrolysis temperature increased (Bougrier et al. 2006). The decrease of viscosity was attributed to irreversible thermal denaturing of protein which might release bound water. Increase of sludge viscosity was observed during sludge cooling from 80 °C to 33 °C, which was termed as "thermal history".

In this study, we measured the dynamic viscosity of LTWAS during cooling process from 47.5°C to 25.0°C. The sludge viscosity was slightly reduced after cooled to 25.0°C. We speculate the hydrolysis continues during the cooling process and the dosage of caustic soda benefited the relatively stable sludge viscosity at decreased temperature. This observation is beneficial since the data suggested that LTWAS could be held for longer time or transported to long distance at lower temperature with even lower viscosity.

5. Conclusion

Impact of thermochemical hydrolysis on sludge viscosity, floc morphology, and dewaterability was evaluated at the PONDUS TCHP full scale installation. The major conclusions are:

- Implementation of thermochemical hydrolysis improves the dewaterability by 2.5% points with maximum cake dryness of 31%.

- At least 5-fold reduction of dynamic viscosity of TWAS by thermochemical hydrolysis and LTWAS exhibits more close to Newtonian fluids compared to TWAS. This brings benefits of more accurate modeling of mixing condition. The thermal hydrolysis reactor effectively hydrolyzed TWAS with present total solids of 6-8.5%. The results of this analysis suggest that the thermochemical hydrolysis can produce the similar effect on sludge viscosity and dewaterability as thermal hydrolysis.
- No gel formation or regaining of viscosity was observed of the hydrolyzed TWAS treated by the TCHP. Sludge holding tested verified that the hydrolyzed TWAS by the TCHP was suitable for long-distance piping.
- Sludge morphology tests verified the cell lysis effect of the TCHP. The hydrolyzed TWAS was stable in tested sealed containers.

6. References

Abelleira-Pereira, J.M., Perez-Elvira, S.I., Sanchez-Oneto, J., de la Cruz, R., Portela, J.R. and Nebot, E. (2015) Enhancement of methane production in mesophilic anaerobic digestion of secondary sewage sludge by advanced thermal hydrolysis pretreatment. Water Research 71, 330-340.

Schieder, D., Schneider, R. and Bischof, F. (2000) Thermal hydrolysis (TDH) as a pretreatment method for the digestion of organic waste. Water Science and Technology 41(3), 181-187.

Ruiz-Hernando, M., Martinez-Elorza, G., Labanda, J. and Llorens, J. (2013) Dewaterability of sewage sludge by ultrasonic, thermal and chemical treatments. Chemical Engineering Journal 230, 102-110.

Feng, G.H., Tan, W., Zhong, N. and Liu, L.Y. (2014) Effects of thermal treatment on physical and expression dewatering characteristics of municipal sludge. Chemical Engineering Journal 247, 223-230.

Zhang, S.T., Guo, H.G., Du, L.Z., Liang, J.F., Lu, X.B., Li, N. and Zhang, K.Q. (2015) Influence of NaOH and thermal pretreatment on dewatered activated sludge solubilisation and subsequent anaerobic digestion: Focused on high-solid state. Bioresource Technology 185, 171-177.

Neyens, E., Baeyens, J. and Creemers, C. (2003) Alkaline thermal sludge hydrolysis. Journal of Hazardous Materials 97(1-3), 295-314.

Aichinger, P., Wadhawan, T., Kuprian, M., Higgins, M., Ebner, C., Fimml, C., Murthy, S. and Wett, B. (2015) Synergistic co-digestion of solid-organic-waste and municipal-sewage-sludge: 1 plus 1 equals more than 2 in terms of biogas production and solids reduction. Water Research 87, 416-423.

Takaoka, M., Oshita, K., Iwamoto, T. and Mizuno, T. (2014) Effect of co-managing organic waste using municipal wastewater and solid waste treatment systems in megacities. Water Science and Technology 69(6), 1159-1166.

Neyens, E. and Baeyens, J. (2003) A review of thermal sludge pre-treatment processes to improve dewaterability. Journal of Hazardous Materials 98(1-3), 51-67.

Dai, X.H., Gai, X. and Dong, B. (2014) Rheology evolution of sludge through high-solid anaerobic digestion. Bioresource Technology 174, 6-10.

Russ, D.C., Thomas, J.M.D., Miller, Q.S. and Berson, R.E. (2015) Predicting Power for a Scaled-up Non-Newtonian Biomass Slurry. Chemical Engineering & Technology 38(1), 53-60.

Feng, G.H., Guo, Y.B. and Tan, W. (2015) Effects of thermal hydrolysis temperature on physical characteristics of municipal sludge. Water Science and Technology 72(11), 2018-2026.

Bougrier, C., Delgenes, J.P. and Carrere, H. (2008) Effects of thermal treatments on five different waste activated sludge samples solubilisation, physical properties and anaerobic digestion. Chemical Engineering Journal 139(2), 236-244.

Baudez, J.C., Markis, F., Eshtiaghi, N. and Slatter, P. (2011) The rheological behaviour of anaerobic digested sludge. Water Research 45(17), 5675-5680.

Neyens, E., Baeyens, J., Dewil, R. and De heyder, B. (2004) Advanced sludge treatment affects extracellular polymeric substances to improve activated sludge dewatering. Journal of Hazardous Materials 106(2–3), 83-92.

Barber, W.P.F. (2016) Thermal hydrolysis for sewage treatment: A critical review. Water Research 104, 53-71.

Stickland, A.D. (2015) Compressional rheology: A tool for understanding compressibility effects in sludge dewatering. Water Research 82, 37-46.

Urrea, J.L., Collado, S., Laca, A. and Díaz, M. (2015) Rheological behaviour of activated sludge treated by thermal hydrolysis. Journal of Water Process Engineering 5, 153-159.

Higgins, M.J., Beightol, S., Mandahar, U., Xiao, S., Lu, H.-W., Le, T., Mah, J., Patha k, B., Novak, J., Al-Omari, A. and N. Murthy, S. (2014) Effect of Thermal Hydrolysis Temperature on Anaerobic Digestion, Dewatering and Filtrate Characteristics. Proceedings of the Water Environment Federation 2014(15), 2027-2037.

Bougrier, C., Albasi, C., Delgenès, J.P. and Carrère, H. (2006) Effect of ultrasonic, thermal and ozone pre-treatments on waste activated sludge solubilisation and anaerobic biodegradability. Chemical Engineering and Processing: Process Intensification 45(8), 711-718.

The sludge ozonation for different types of mixed liquor under high and low pH conditions by a plug-flow reactor

Xiaoyu Zheng and Eric R. Hall

Department of Civil Engineering, The University of British Columbia, 6250 Applied Science Lane, Vancouver, BC Canada V6T 1Z4,craeszheng@gmail.com, ehall@civil.ubc.ca

ABSTRACT

Sludge disintegration by ozone treatment with a high degree of solid solubilization at a low dosage is desirable. However, sludge ozonation is a complicated process and many factors affect its efficiency. In this study, a plug-flow ozone reactor was used to test the performance of sludge ozonation by different TSS concentrations, three types of sludge, the carbonate or bicarbonate addition, under high and low pH conditions. It was found that higher degree of sludge disintegration was achieved with lower TSS concentration. There was an optimum TSS concentration at which the maximum soluble COD concentration could be achieved. The types of sludge had a significant impact on the efficiency of sludge ozonation. Both the size and the structure of particles in sludge had significant impact on the efficiency of sludge ozonation. As one of the main scavengers, the carbonate and the bicarbonate inhibited sludge disintegration while causing high cellular inactivation due to pH increase. The highest degree of sludge solubilization was achieved under high pH condition, while more than 90% cellular inactivation was observed under low pH condition during sludge ozonation.

KEY WORDS

Sludge, Ozonation, Disintegration

INTRODUCTION

As a strong chemical oxidant, ozone can be used to disintegrate excess sludge and high degree of solid solubilization can be achieved [1]. The main drawback for sludge ozonation is its high energy consumption and associated high operating cost. The ozone dosage, generally expressed by g O_3/g $TSS_{treated}$, is one of the most important parameters to evaluate the efficiency of sludge ozonation. A low dosage with a high degree of disintegration is desirable. However, the dosage reported in literature varied greatly. For example, it was reported that an ozone dosage lower than 0.015g O_3/g VSS was not enough to cause cell rupture [2]; no alternation in bacteria DNA was detected at the dosage of 0.02 g O_3/g TSS and destruction of bacteria occurred mainly at ozone dosage above 0.08 g O_3/g TSS [3]. By measuring Oxygen Uptake Rate (OUR), at an ozone dosage of around 0.02 g O_3/g TSS, 80% of microbial respiration activity was lost [4]. Another study [5] reported that around 70% of sludge bacteria activity was lost at ozone dosage of 0.03-0.04 g O_3/g TSS.

The discrepancy of the ozone dosages for sludge disintegration may be due to many other factors which may be overlooked. Sludge is a three-phase complex matrix of flocs, living cells and organo-mineral materials, and the interactions between ozone and activated sludge are complicated and poorly understood[6]. Except for the ozone dosage, the efficiency of sludge ozonation may depend on many factors including (1) sludge characteristics, (2) ozone mass transfer efficiency in the reactor and the configuration of the ozone reactor, (3) flow rate and concentration of ozone in the gas-phase, (4) flow rate and solid concentration in the treated sludge, (5) contact time between ozone and sludge, (6) scavengers in sludge, and (7) pH.

In this study, three different types of sludge were tested to investigate the sludge disintegration efficiency. Three different solid concentrations of mixed liquors were also tested with the same amount of ozone applied.

The effects of the carbonate and bicarbonate (the scavengers) addition and pH adjustment on the performance of sludge ozonation were also tested. Hydroxyl radicals (OH°) generation during ozonation was assumed to be one of the critic factors for sludge solubilization and cellular inactivation [3, 6]. The carbonate and bicarbonate ions are the most common scavengers in sludge, which react with OH°, stop producing superoxide radicals ($O_2^{\circ -}/HO_2^{\circ}$), and terminate the chain reaction. At high pH condition, the main reaction in sludge ozonation is the hydroxyl radical reaction; and at low pH condition, the ozone reaction is dominated by direct ozone oxidizing reaction. In this study, batch tests were conducted to investigate sludge disintegration with addition of the carbonate/bicarbonate, base and acid.

MATERIALS AND METHODS

A plug-flow ozone reactor was built, which included a pump (Moyno® 331), a Venturi injector (Mazzei® Model 283) and a coil mixer. The coil mixer was made of 1 foot PFA tube with ½ inch internal diameter, and the coil diameter was 4 inch. The flow rate for sludge was about 3.0 L/min, and the ozone injection rate was about 80 mg/min. The retention time of sludge in the reactor was less than 15 seconds.

Three types of sludge were tested: (1) sludge from the aerobic tank in UBC MBR (membrane bioreactor) pilot plant at 40 days SRT, (2) the mixed liquor from aerobic phase of the control SBR (20 L lab-scale) at SRT = 50d, and (3) the mixed liquor from aerobic phase of the SBR (20 L lab-scale) combined with a sludge ozone treatment unit at nominal SRT = 50 d.

Sludge from UBC MBR pilot plant had high solids concentration (TSS=11580 mg/L). It was diluted by 75% and 50% with its effluent to composite two different solid concentrations for sludge ozonation test.

The sludge samples from aerobic phase of the control SBR were prepared and completely mixed as: (1) 2 L mixed liquor + 500 mL distilled water, (2) 2 L mixed liquor + 500 mL 0.4 N Na_2CO_3, (3) 2 L mixed liquor + 100 mL 0.2 N NaOH + 400 mL distilled water, (4)2 L mixed liquor +100mL 0.8N HCl + 400mL distilled water. The pH values for sludge samples were 7.4, 10.4, 11.1, and 2.0, respectively.

The sludge samples from aerobic phase of the ozonated SBR were prepared and completely mixed as: (1) 2L mixed liquor + 500mL distilled water, (2) 2 L mixed liquor + 500 mL 0.4 N Na_2CO_3, (3) 2 L mixed liquor + 100mL 0.2 N NaOH + 400 mL distilled water, (4) 2 L mixed liquor +100 mL 0.8 N HCl + 400 mL distilled water. The pH values for sludge samples were 7.2, 10.2, 11.0, and 2.0, respectively. Each sludge sample was then treated by the plug-flow reactor with and without ozone injection. Table 1 lists types, composition and characteristics of sludge for ozone treatment.

Table 1 Types, composition and characteristics of sludge for the batch test

Types of Sludge	Sludge prepared for treatment by the plug-flow reactor	Solids concentration TSS(VSS) (mg/L)	VSS to TSS ratio (%)	Particulate Size by Volume
Mixed liquor from aerobic tank in UBC MBR Pilot Plant	2 L no dilution	11580 (8280)	71.5	D(0.1)=10.6 um, D(0.5)=23.0 um, D(0.9)=116.6 um
	2 L 75% diluted by effluent	8740 (6300)	72.1	
	2 L 50% diluted by effluent	5700 (4300)	75.4	
Mixed liquor from aerobic phase of the control SBR (SRT=50 d)	2 L mixed liquor + 500 mL distilled water	4920 (4100)	83.3	D(0.1)=33.7um D(0.5)=128.1um D(0.9)=339.8um
	2 L mixed liquor + 500 mL 0.4 N Na_2CO_3			
	2 L mixed liquor + 100 mL 0.2 N NaOH + 400 mL distilled water			
	2 L mixed liquor +100 mL 0.8 N HCl + 400 mL distilled water			
Mixed liquor from aerobic phase of the SBR combined with sludge ozonation (SRT=50 d)	2 L mixed + 500 mL distilled water	4613 (3680)	79.8	D(0.1)=31.9um D(0.5)=87.0um D(0.9)=181.3um
	2 L mixed liquor + 500 mL 0.4 N $NaHCO_3$			
	2 L mixed liquor + 100 mL 0.2 N NaOH + 400 mL distilled water			
	2 L mixed liquor +100 mL 0.8 N HCl + 400 mL distilled water			

Each time, about 2 L sludge was treated in a once-through way by the plug-flow reactor. Sludge was sampled before and immediately after treatment. Triple samples were taken from each sludge for mixed liquor suspended solids (MLSS) and mixed liquor volatile suspended solids (MLVSS) measurement.

The soluble constituents were sampled after filtration through 0.45 μm filters of the supernatant from sludge centrifuge treatment. The sludge samples were analysed on the same day of the batch tests. The soluble constituent samples were stored at 4 C until the analyses were performed at the Environmental Engineering Laboratory at UBC.

MLSS, MLVSS, the chemical oxygen demand (COD) were measured according to the Standard Methods [7]. Particle size distribution measurement was measured by a Malvern Instrument (Westborough, MA, USA) Mastersizer 2000 analyzer, with a Hydro S (Malvern) automated sample dispenser unit. The pH value was measured by a portable DO meter (Hach® HQ30d). The sludge ATP concentration was measured by Luminultra® QG21W test kits.

RESULTS

Solids Concentration

The efficiency of sludge ozonation was related with solids concentration. Figure 1 shows the efficiencies of sludge disintegration from MBR process by ozonation at three different solid concentrations. It seems that the higher degree of sludge solubilisation and cellular inactivation ratio were achieved with lower solids concentration.

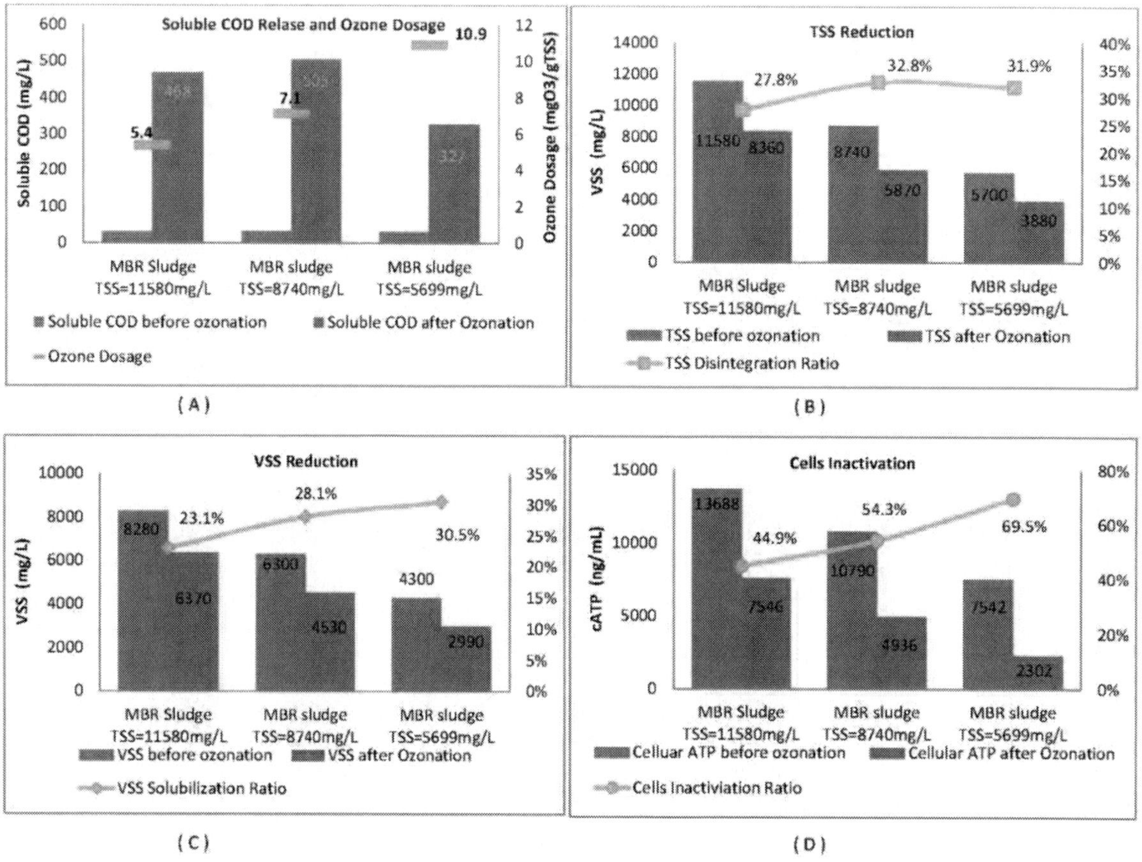

Figure 1 Sludge disintegration by ozone treatment at three different solids concentration

Since the same amount of ozone was applied to each sludge sample, the ozone dosage, if defined as the mass of ozone applied per mass of treated TSS, increased from 5.4 to 10.9 mg O_3/g TSS when TSS concentration decreased from 11580 mg/L to 5699 in Figure 1(A). However, the highest soluble COD was not achieved at the lowest TSS concentration. The highest concentration of the soluble COD after ozone treatment was measured at TSS = 8740 mg/L. It seemed that there existed an optimum TSS concentration at which the maximum soluble COD could be achieved.

The degree of TSS and VSS reduction/solubilization were generally increased when the solids concentration of the treated sludge decreased, which illustrated in Figure (B) and (C). However, the percentage of solid solubilization was not significantly increased even though the concentration of solids dropped (the ozone dosage increased).

Figure (D) shows that sludge from MBR process was highly susceptible to ozone treatment and about 44.9% of living cells in sludge were inactivated even at the low dosage of 5.4 mg O_3/g $TSS_{treated}$. The cellular inactivation ratio increased when the TSS concentration decreased.

Types of Sludge

The types of sludge had a significant impact on sludge solubilisation and cellular inactivation in sludge ozonation. Figure 2 shows the performance of sludge ozonation for three different types of sludge at different ozone dosages. At the same ozone dosage, more cells were inactivated than the particles were solubilized. In addition, sludge from the SBR process had far less degree of sludge solubilisation and cellular inactivation ratio than sludge from the MBR process even at lower TSS concentration. Sludge from the SBR combined with sludge ozonation achieved the lowest degree of sludge solubilisation and cellular inactivation ratio even its TSS was lowest and its ozone dosage was highest.

The particle size distribution of sludge may also be a factor affecting the performance of sludge ozonation. Figure 3 shows the particle size distribution of three types of sludge by volume, respectively. The median diameters of the particles in sludge from MBR, the control SBR and the ozonated SBR were 23.0 μm, 128.1 μm and 87.0 μm, respectively. The particle size of MBR sludge was significantly smaller than that of sludge from SBRs.

The ozone mass transfer mechanism in gas/liquid systems is generally explained by the "fast kinetic regime" theory proposed by Danckwerts [8]. The rate limitation occurs in the liquid film, where the ozone can react fast with the various dissolved organic substances and be consumed. This causes the apparent rate of ozone mass transfer to exceed the maximum rate of physical gas-liquid mass transfer. Normally, the thickness of the liquid film is between 15 to 20 μm. During sludge ozonation, with the enhancement by the fast reaction, the effective thickness of the film would be reduced to less than 10 μm [9]. The thin liquid film allows only smaller particles and/or periphery of large flocs to be oxidized by ozone, while the cells inside the flocs can be viable due to aggregates physical protection [10]. The theory may explain why the efficiency of ozonation of MBR sludge was higher than that of sludge from SBRs. However, it cannot explain why lower sludge disintegration ratio was achieved for sludge from the ozonated SBR than from the control SBR even though it had lower particle size than that of the control (87.0 μm versus 128.1 μm).

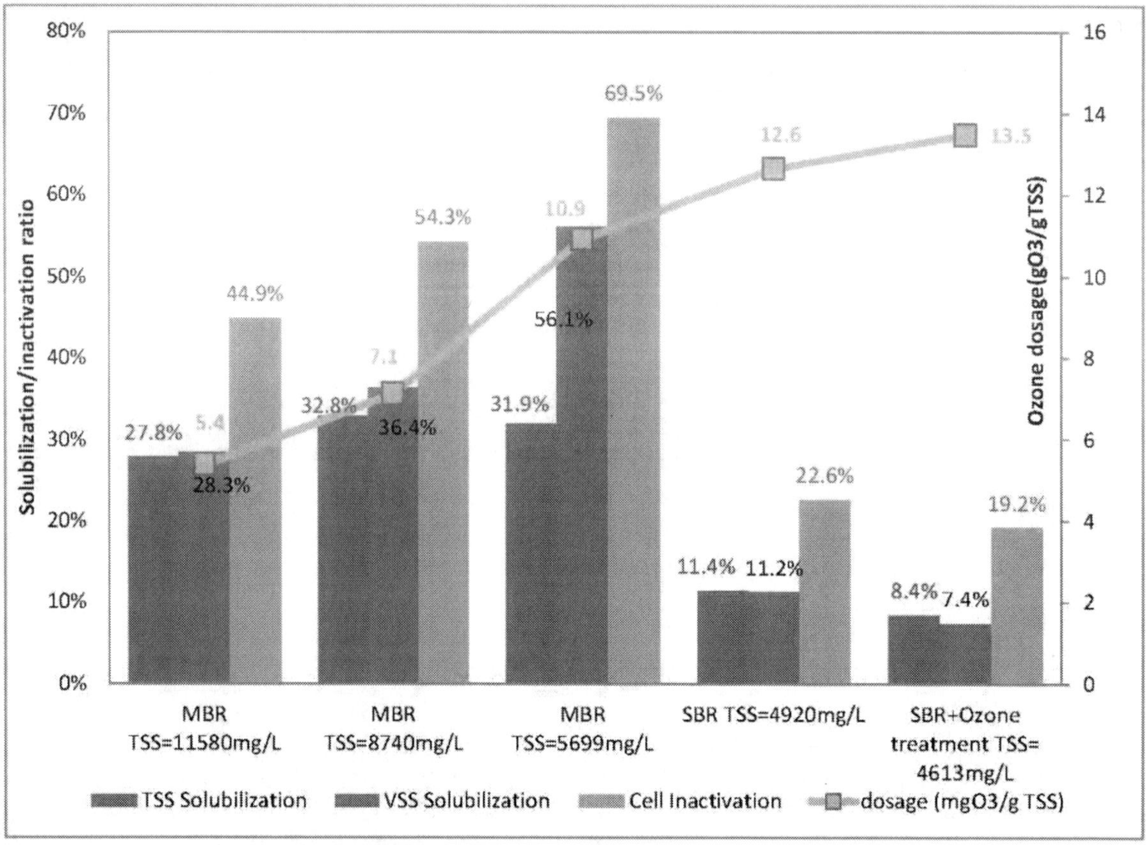

Figure 2 Sludge solubilization and cellular inactivation vs. ozone dosage for different TSS concentrations and different types of sludge

The photos of three types of sludge from microscopic observation (Figure 4) may explain the discrepancy. Figure 4 illustrates the structure differences among three types of sludge. The MBR sludge was composed of many small loose flocs, fine particles and filamentous bacteria (Figure 4(A)). The flocs with large sizes were observed in sludge from both SBRs (Figure 4 (B) and (C)). The filamentous bacteria were obvious in sludge from the MBR and the control SBR. No filamentous bacteria were found and the flocs were compact in sludge from the ozonated SBR, which was similar with many articles on sludge ozonation[11]. The loose structure of flocs and larger surface areas of filamentous may make it more chances for microorganisms to contact and to be exposed to ozone and/or hydroxyl radicals; and thus high degree of sludge disintegration and cellular inactivation ratio could be achieved. The microorganisms from the ozonated SBR may be adapted to long-term ozone exposure and formed solid and compact flocs structure which may resist the attack from ozone. Therefore, the lowest degree of TSS and VSS solubilization and the lowest cellular inactivation ratio were measured for sludge from the ozonated SBR.

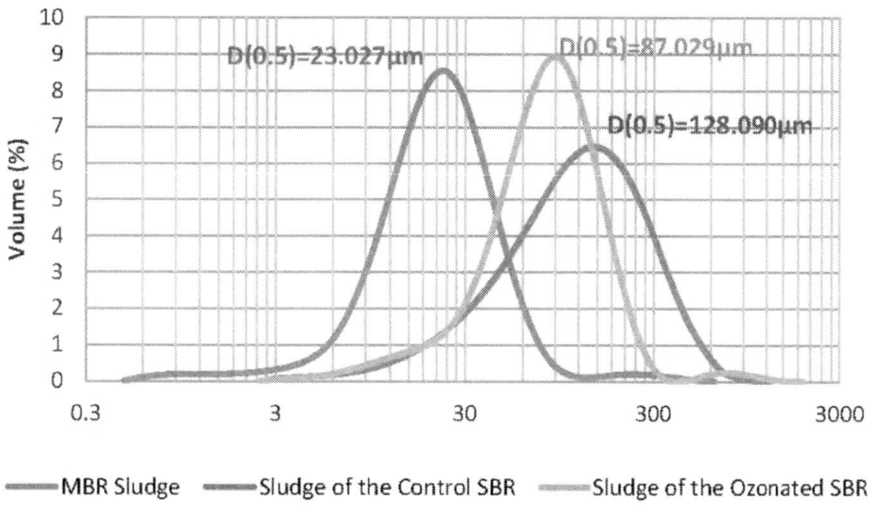

Figure 3 The particle size distribution by volume for three types of sludge

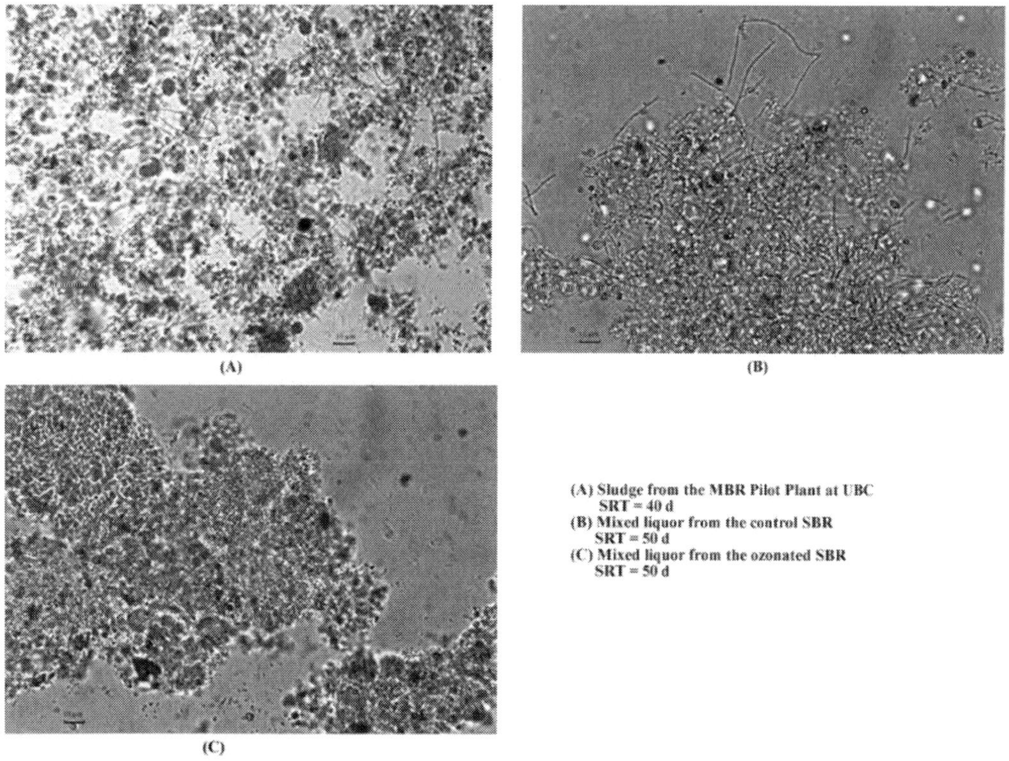

Figure 4 The microscopic photos of three types of sludge

Sludge Ozonation with Carbonates/bicarbonates addition, high pH and low pH adjustment

The results of sludge disintegration for the control SBR by adding carbonates, sodium hydroxide and hydrochloric acid with and without ozone were shown in Figure 5. Figure 6 shows the similar results of sludge disintegration for the SBR combined with the ozone treatment by adding bicarbonates, sodium hydroxide and hydrochloric acid with and without ozone.

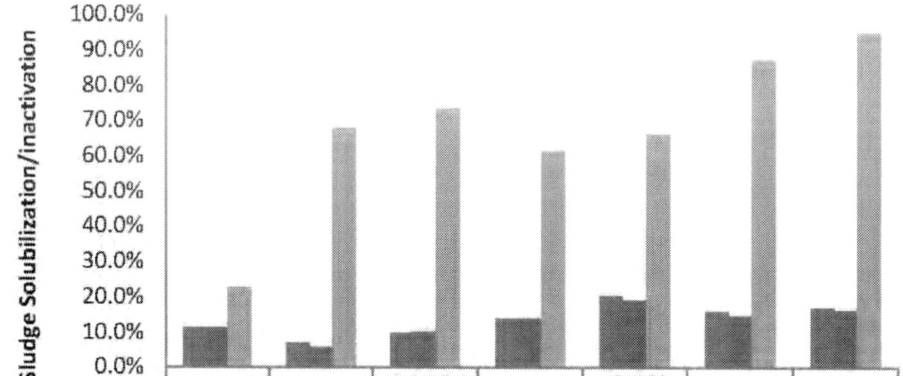

	O3	Na2CO3	Na2CO3 + O3	NaOH	NaOH +O3	HCl	HCl +O3
TSS Solubilization	11.4%	6.9%	9.8%	13.8%	20.5%	16.1%	17.1%
VSS Solubilization	11.2%	5.5%	9.9%	13.8%	19.0%	14.6%	16.3%
Cell Inactivation	22.6%	67.8%	73.4%	61.5%	65.9%	87.1%	94.9%

Figure 5 Sludge solubilization and cellular inactivation of the sludge from the control SBR by carbonate, base, and acid addition with and without ozone treatment

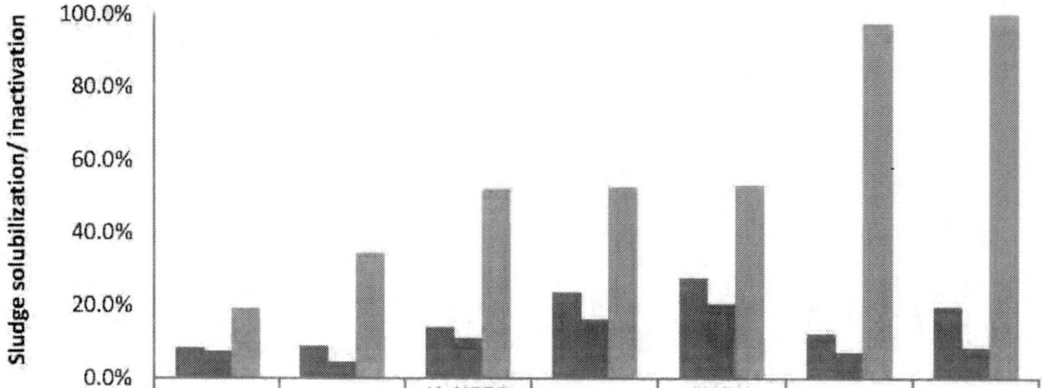

	O3	NaHCO3	NaHCO3 + O3	NaOH	NaOH +O3	HCl	HCl +O3
TSS Solubilization	8.4%	9.0%	14.0%	23.7%	27.6%	12.3%	19.6%
VSS Solubilization	7.4%	4.5%	11.1%	16.1%	20.5%	7.1%	8.4%
Cell Inactivation	19.2%	34.5%	52.2%	52.7%	53.0%	97.6%	100%

Figure 6 Sludge solubilization and cellular inactivation of the sludge from the ozonated SBR by adding bicarbonate, base, and acid with and without ozone treatment

The carbonates/bicarbonates addition had impact on sludge ozonation. In Figure 5, sludge reacted with high concentration sodium carbonates due to pH increase. About 6.9% TSS and 5.5% VSS were solubilized. The cellular inactivation ratio was 67.8% which was significantly higher than sludge ozone treatment only. When carbonates were added to sludge for ozonation, TSS solubilization ratio reduced from 11.4% to 9.8% and VSS solubilization ratio decreased from 11.2% to 9.9% compared with ozone treatment only. The carbonates/bicarbonates normally act as one of the hydroxyl radical scavengers and reduce ozone radical reaction, which reduced the efficiency of sludge ozonation. However, the cellular inactivation ratio increased from 22.6% to 73.4% which may be due to the increased pH (10.4) by carbonates addition.

Either base or acid addition alone can be used to disrupt sludge. When they were applied to sludge alone, high degree of TSS and VSS solubilization can be achieved (Figure 5). Compared with sludge ozone treatment, high cellular inactivation ratio was observed either with base or acid treatment. From Figure 5, 61.5% and 87.1% of microorganisms were inactivated by sodium hydroxide and by hydrochloric acid, respectively; while only 22.6% by sludge ozonation. If combined with ozone treatment, the efficiency of sludge disintegration was generally improved.

High or low pH treatment may inactivate higher percentage of cells than ozone treatment. In Figure 6, ozone treatment alone inactivate only 19.2% cells, while addition of bicarbonates, sodium hydroxide and hydrochloric acid separately inactivated 34.5%, 52.7%, and 97.6% of cells, respectively. It seems that sludge may be more tolerant of low dosage ozone treatment than under high pH (≥10) or low pH (≤2) condition.

When sludge ozonation combined with sodium hydroxide addition, the degrees of TSS and VSS solubilization was significantly improved. The high pH favours the hydroxyl radicals generation and sludge ozonation may be dominated with ozone chain reactions, which promotes efficiency of sludge disintegration. From Figure 5 and Figure 6, sodium hydroxide addition combined ozone treatment achieved the highest degrees of TSS and VSS solubilization. It is possible that ozone treatment at high pH condition, which favours ozone indirect hydroxyl radical reaction, improves the efficiency of sludge disintegration.

When sludge ozonation coupled with acid addition, more than 90% cells were inactivated. The degrees of TSS and VSS solubilization also increased but less than those under high pH condition. In Figure 5 and Figure 6, hydrochloric acid addition coupled with ozone treatment obtained the highest cellular inactivation ratio. It is possible that low pH may be more lethal to microorganisms than under high pH condition during sludge ozonation.

CONCLUSIONS

The efficiency of sludge ozonation depends on many factors including TSS concentration, sludge type, carbonates and bicarbonates, and pH.

The lower TSS concentration, the higher degree of sludge solubilisation and cellular inactivation ratio were expected to be achieved. An optimum TSS concentration may exist at which the maximum soluble COD could be achieved.

The types of sludge had a significant impact on sludge solubilisation and cellular inactivation in sludge ozonation. Both particle size and the structure of flocs had significant impact on sludge disintegration.

The carbonate or bicarbonate addition during ozone treatment may reduce the efficiency of sludge disintegration.

The degree of solids solubilization was significantly improved under high pH condition; while 90% cellular inactivation ratio was achieved under low pH condition when sludge ozonation combined with acid addition.

REFERENCE

1. Muller, J. A., Pretreatment processes for the recycling and reuse of sewage sludge. *Water Science and Technology* **2000,** *42* (9), 167-174.

2. Albuquerque, J. S.; Domingos, J. C.; Sant'Anna, G. L.; Dezotti, M., Application of ozonation to reduce biological sludge production in an industrial wastewater treatment plant. *Water Science and Technology* **2008,** *58* (10), 1971-1976.

3. Yan, S. T.; Chu, L. B.; Xing, X. H.; Yu, A. F.; Sun, X. L.; Jurcik, B., Analysis of the mechanism of sludge ozonation by a combination of biological and chemical approaches. *Water Research* **2009,** *43* (1), 195-203.

4. Chu, L. B.; Yan, S. T.; Xing, X. H.; Yu, A. F.; Sun, X. L.; Jurcik, B., Enhanced sludge solubilization by microbubble ozonation. *Chemosphere* **2008,** *72* (2), 205-212.

5. Saktaywin, W.; Tsuno, H.; Nagare, H.; Soyama, T.; Weerapakkaroon, J., Advanced sewage treatment process with excess sludge reduction and phosphorus recovery. *Water Research* **2005,** *39* (5), 902-910.

6. Dziurla, M. A.; Salhi, M.; Leroy, P.; Paul, E.; Ginestet, P.; Block, J. C., Variations of respiratory activity and glutathione in activated sludges exposed to low ozone doses. *Water Research* **2005,** *39* (12), 2591-2598.

7. APHA; AWWA; WEF, *Standard Methods for the Examination of Water and Wastewater*. 22nd ed. ed.; American Public Health Association: Washington D.C., 2012.

8. Danckwerts, P. V., *gas-liquid reaction*. McGraw-Hill: New York, 1970.

9. Paul, E.; Debellefontaine, H., Reduction of excess sludge produced by biological treatment processes: Effect of ozonation on biomass and on sludge. *Ozone-Science & Engineering* **2007,** *29* (6), 415-427.

10. Chu, L. B.; Wang, J. L.; Wang, B.; Xing, X. H.; Yan, S. T.; Sun, X. L.; Jurcik, B., Changes in biomass activity and characteristics of activated sludge exposed to low ozone dose. *Chemosphere* **2009,** *77* (2), 269-272.

11. Nilsson， , F.; Marinette Hagman; Artur Tomasz Mielczarek; Per Halkjær Nielsen; Jönsson, K., Application of Ozone in Full-Scale to Reduce Filamentous Bulking Sludge at Öresundsverket WWTP. *Ozone: Science & Engineering* **2014,** *36*, 7.

A Tale of Two Cities: How Climate, Drivers, and Budget Affect Thermal Hydrolysis Design

Thomas Nangle[1*], Kristina Warren[2]

[1]CDM Smith, Raleigh, North Carolina
[2]CDM Smith, Bellevue, Washington
*Email: nanglet@cdmsmith.com

ABSTRACT

Thermal hydrolysis is a proven anaerobic digestion pretreatment process that is well established in Europe and quickly gaining popularity across the United States. In addition to the installation at Washington DC (DC Water – 450 dtpd), thermal hydrolysis is in design or construction at other large water reclamation facilities (WRFs) in San Francisco (SFPUC – 200 dtpd), Dallas (Trinity River Authority 245 DTPD), and Virginia Beach (Hampton Roads Sanitation District – 92 dtpd). Franklin, Tennessee (29 dtpd) and Pontiac, Michigan (26 dtpd) are examples of smaller WRFs that are planning to utilize thermal hydrolysis to meet their long-term biosolids goals.

This paper explores the different design approaches two specific cities took to integrate thermal hydrolysis into their facilities. Many of these differences in design are the result of different drivers, climates, and budgets for the two cities. Some of the key design differences that will be covered in this paper are outlined below:

- Drivers Leading to the Selection of Thermal Hydrolysis
- Strategies to Address System Redundancies
- Steam Production and Biogas Utilization
- Approaches to Accepting Hauled in Wastes
- Biosolids End Use

Franklin, Tennessee is a rapidly growing, affluent city that is overhauling and expanding its WRF. Franklin selected thermal hydrolysis as a way to achieve Class A Biosolids while maximizing energy recovery from its wastewater treatment process. The Franklin design incorporates a significant level of redundancy in most systems and allows for future expansion to more than double the existing solids loading capacity. Franklin will accept hauled in fats, oils and greases (FOG) for co-digestion in its digesters. The plant will utilize the biogas in CHP engines with heat recovery to generate steam for the THP process, electricity to be consumed on site, and hot water for space and process heating needs.

Pontiac, Michigan is city with stagnant growth that is seeking to increase its biosolids treatment capacity to meet existing demands. Pontiac has a Class B land application program, but only has anaerobic digester capacity for half of its biosolids; the remaining half is disposed of in a landfill. Pontiac chose THP to achieve treatment of all the plant's biosolids without expansion of digester capacity. The Pontiac design incorporates less redundancy than Franklin, with lower capital cost. The upgrades include installation of a new septage receiving facility. Pontiac will use a significant part of its generated biogas to power steam boilers that will provide steam for THP and provide process and space heating for the new biosolids handling building.

KEYWORDS: Thermal Hydrolysis Pretreatment, Sustainability, Climate, Redundancy, CHP

INTRODUCTION

Franklin is a wealthy city located 20 miles south of Nashville, Tennessee. The city has seen tremendous population growth over the past few decades, with the population increasing 6-fold since the 1980 census (see **Figure 1**). The City of Franklin owns and operates a 12 mgd WRF, and currently trucks all its dewatered biosolids to a landfill over 100 miles away. Franklin's WRF is being upgraded to meet future permit limits and increase treatment capacity to 16 MGD. These upgrades include improvements to the secondary treatment facilities, a new headworks facility, a 10 MG equalization basin, an upgraded UV disinfection facility, and a new biosolids facility. The solids portion of the new biosolids facility will include thermal hydrolysis pretreatment (THP) and anaerobic digestion, a fats, oils, and grease (FOG) receiving facility for co-digestion, pre- and post-dewatering, sludge screening, and a solar dryer pilot. The gas portion of the new biosolids facility includes biogas treatment, CHP engines with heat recovery steam generators (HRSGs), backup steam boilers, and a waste gas burner. The existing biosolids facility, consisting of dissolved air flotation (DAF) thickening and belt filter press (BFP) dewatering, is past its useful life and is becoming an operational risk for Franklin. The design of the WRF upgrades, which includes the new biosolids facility, is complete and construction is expected to begin in 2017.

Pontiac is located 10 miles north of Detroit, Michigan. The city sits in the heart of the former General Motors manufacturing center, and as the auto manufacturing industry declined in Michigan, so did Pontiac's population (see **Figure 1**). No growth is projected in the service area in the next 20 years. The Oakland County Water Resources Commission (WRC) owns the City of Pontiac's WWTP. The Pontiac WWTP consists of two geographically separate treatment facilities, East Boulevard WWTP and Auburn WWTP, but the Auburn WWTP treats the biosolids for both facilities. The Auburn WWTP is rated for 22 mgd and the East Boulevard WWTP is rated for 8.6 mgd. Auburn WWTP will upgrade its biosolids facility to add thermal hydrolysis pretreatment, digester improvements including a membrane gas holder cover on the secondary digester, cake storage upgrades, a new septage receiving facility, new pre- and post-dewatering, sludge screening and an additional waste gas burner. The existing sludge facility includes waste activated sludge thickeners, primary and secondary anaerobic digesters, belt filter press dewatering, and onsite biosolids storage, which support a land application program for Class B biosolids. The digesters only have the capacity to process approximately half of the plant design capacity, with the excess sludge being dewatered and landfilled. Construction of the upgrades is expected to begin in 2017.

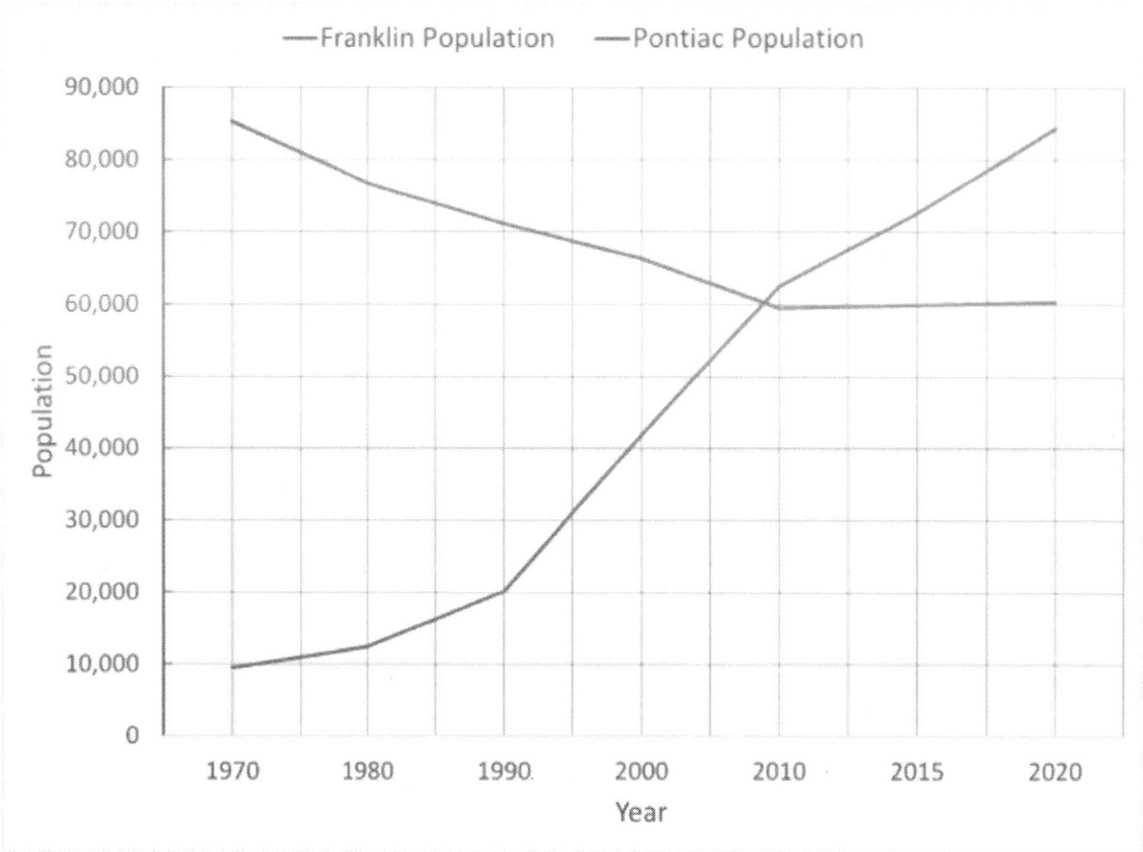

Figure 1: Franklin and Pontiac Census Data and Estimated Population Growth

TWO PATHS TO THERMAL HYDROLYSIS

Before discussing the design features of these two facility upgrades, let's explore how these two different utilities – with different drivers, budget, and climate – both selected thermal hydrolysis pretreatment as the basis of their biosolids upgrades.

Franklin needs a complete overhaul of their biosolids facility, which is located adjacent to residential communities in a warm climate. Franklin chose a facility that reduces risk by reliably generating Class A biosolids with minimal odor, maximizing the recovery of renewable energy from the biosolids, and meeting their capacity needs for the fast growth Franklin is experiencing. The main drivers for choosing thermal hydrolysis over other Class A technologies for Franklin were how it maximizes biogas production while minimizing potential for offsite odors.

Franklin's new biosolids facility will be sized to meet the projected growth the area is experiencing. **Table 1** compares the solids projections for the design year (2040) and the estimated startup year (2020) for the new biosolids facility.

Table 1: Franklin WRF Range of Projected Solids Production

	Average Day			Maximum Month		
	Solids Loading (dtpd)	%TS	Flow (gpd)	Solids Loading (dtpd)	%TS	Flow (gpd)
Startup WAS Loading (Year 2020)	12.25	0.84%	349,600	15.90	0.84%	454,500
Design WAS Loading (Year 2040)	22.65	0.84%	646,600	29.45	0.84%	840,600

The average day solids loading during startup (12.25 dtpd) is approximately 40% of the ultimate design loading (29.45 dtpd). The design will allow the new biosolids facility to be constructed in phases, allowing its capacity to grow with the population. This will improve the operating efficiency of installed process equipment, and will allow Franklin to push some of the capital cost into the future when there are more rate payers to split the debt service. **Figure 2** highlights the capacity of the new biosolids facility associated with each stage of construction, and the year when the phases' capacity is expected to be reached.

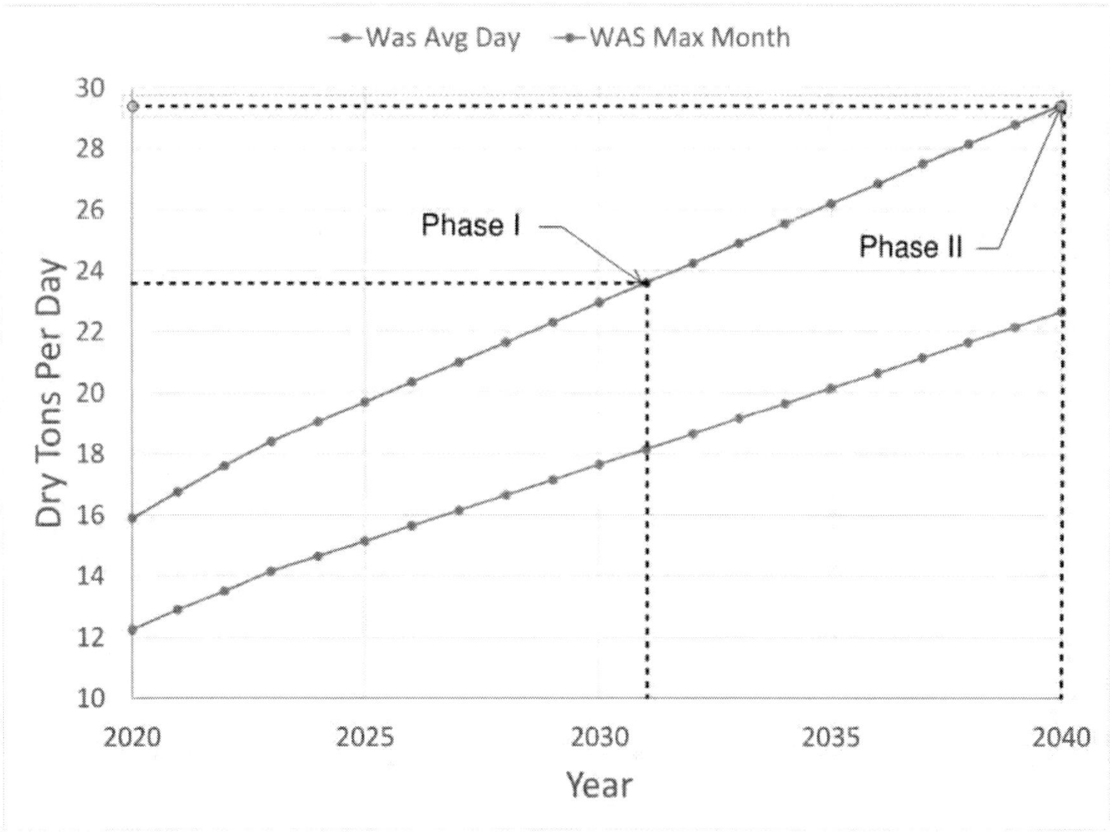

Figure 2: WAS Loading Projections and Design Capacity for Phase I and II

Conversely, Pontiac anticipates stagnant growth over the next 20 years, and instead needs to increase the biosolids handling capacity of the plant to address existing capacity limitations. Pontiac currently treats half its biosolids in its two anaerobic digesters and beneficially reuses these biosolids through its Class B land application program. Auburn's digesters do not have sufficient capacity to treat the biosolids generated by the Auburn and East Boulevard plants, so approximately half of the generated raw sludge is dewatered by the belt filter presses and landfilled. Constructing additional digester capacity on the limited available space on the plant would be costly. The primary goal of Pontiac was to process all biosolids to achieve at least Class B biosolids through digestion or other means. Secondary considerations were to reduce hauling volume through improved dewatering and the ability to generate additional biogas. While Class A biosolids was not an initial goal of the facility, it is a significant advantage for Pontiac as this will allow them to reduce the size of their cake storage pad by 30 percent. Class A biosolids will also have reduced odor potential.

The new biosolids handling facility at Pontiac is designed for a solids loading rate of 26.5 dtpd. This design loading rate will treat existing maximum month loading of the biosolids from the Auburn and East Boulevard WWTPs. The plant is adding a new septage receiving facility, and will be able to accept septage during average flow conditions but not during maximum month solids loading conditions. See **Table 2** for Pontiac biosolids loading projections.

Table 2: Pontiac Solids Loading Table comparing Auburn, E. Blvd and Septage sources

	Average Day				Maximum Month			
Source	Solids Loading (dtpd)	%TS	% of Total	Flow (gpd)	Solids Loading (dtpd)	%TS	% of Total	Flow (gpd)
Auburn Primary	8.0	4.0	39%	47,506	12.5	4.0	47%	74,228
Auburn WAS	5.1	5.0	25%	24,171	7.9	5.0	30%	37,441
East Boulevard Primary + WAS	3.9	3.6	19%	25,763	6.1	3.6	23%	40,296
Septage	3.6	1.6	17%	53,957	0	1.6	0%	0
Total	20.6	3.75	100%	151,396	26.5	4.21	100%	151,965

Comparing Basis of Designs

Cambi's B-2 system was pre-selected as the thermal hydrolysis basis of design for both Franklin and Pontiac. The B-2 system consists of four 2 cubic meter reactors sandwiched between a pulper and a flash tank. The B-2 system has similar configuration, operation, and automation as the larger B-6 and B-12 models. The B-2 model is equipped with the new Mark II control system and has additional pneumatic control valves, including one on each of the steam lances, to better pace the injection of steam into the system. Whenever thermal hydrolysis is mentioned in this case study, it is referring to Cambi's B-2 system.

Even though both Franklin and Pontiac are designed for approximately the same capacity, the identical treatment processes begin and end with the B-2 skid. Pontiac does not anticipate increased loading to the plant in the next 20 years, and the treatment process sizes can be optimized to a relatively stagnant flow range. This is because Pontiac will accept septage during average flow but not during maximum flow; therefore, the average and maximum flow rates are similar. Conversely, Franklin is constructing its facility in phases so it can grow with the demand, which requires smaller and more numerous process units that, in combination, can efficiently process the varied flow ranges. **Table 3** compares the number and capacity of the main unit processes for the Franklin and Pontiac design.

Table 3: Number and Capacity of Unit Processes at Franklin WRF and Pontiac WWTP

Process	Franklin WRF	Pontiac WWTP
Sludge Screen	440 gpm @ 0.8 - 1.2% TS Two for Phase I – 1 duty, 1 standby Three for Phase II – 2 duty, 1 standby	105 gpm @ 3.7 - 4.3% TS 1 duty, 1 standby
Pre-dewatering	820 lb/hr Centrifuges Three for Phase I – 2 duty, 1 standby Four for Phase II – 3 duty, 1 standby	2,208 lb/hr Centrifuges 1 duty, 1 standby
Thermal Hydrolysis	One B-2 system (Cambi)	One B-2 system (Cambi)
Digesters	375,000 gal. concrete tank Two for Phase I, Three for Phase II	Primary Digester 719,000 gal. Secondary Digester 662,000 gal.
Post-Dewatering	1,200 lb/hr Screw Presses Two for Phase I – 1 duty, 1 standby Three for Phase II – 2 duty, 1 standby	2,368 lb/hr Centrifuges 1 duty, 1 standby

ADDRESSING SYSTEM REDUNDANCIES – UPSTREAM OF THP

Pontiac's Auburn WWTP and the Franklin WRF biosolids upgrades are designed for 26.5 dry tons per day (dtpd) and 29 dtpd, respectively. All the sludge at this design throughput can be processed through a single B-2 thermal hydrolysis train. Installing a single B-2 system keeps capital costs low, but introduces the risk of losing the ability to process sludge if the thermal hydrolysis system is offline. Raw (unhydrolysed) sludge cannot bypass the thermal hydrolysis system to the digesters because:

1. The digesters are sized for a higher solid loading rate (fed at 10% solids after thermal hydrolysis), and a more readily digestible sludge. Feeding raw sludge to the digester will overload it and likely cause an upset.
2. Introducing raw sludge to processes downstream of thermal hydrolysis risks introducing pathogens to that process equipment, which may contaminate the hydrolysed Class A biosolids after thermal hydrolysis comes back online.

The tanks of the B-2 system require annual pressure vessel inspections, which can last between a few days to a week. During this time, Franklin and Pontiac must find a way to store and/or divert incoming sludge away from the thermal hydrolysis system.

Franklin's approach to managing sludge during THP downtime is a combination of sludge storage, diversion, and the potential for alternate treatment and cake storage. **Figure 3** shows Franklin's WAS process flow to the THP system. Waste activated sludge (WAS) is pumped from the final clarifiers through sludge strain presses and into a WAS storage tank. The screened WAS is pumped from the WAS storage tank through pre-dewatering centrifuges, which discharge the dewatered cake down a chute into the pre-dewatered cake storage bin. The cake also can be diverted out of the chute to a separate screw conveyor that can transport it to truck loading. In addition to bypassing thermal hydrolysis, this diverter also allows for continued sludge processing if the cake bin is out of service. Diverted cake will normally be transported to a landfill, but Franklin also has the option of treating the raw sludge in a dedicated solar dryer bay. This option will likely not be used due to the potential for odor and contamination of the solar dryer bays, as well as the bays not having the adequate capacity to process the 18% solid sludge before it has been reduced in the digesters and not dewatered to over 30% solids.

Franklin's 280,000 gallon pre-stressed concrete WAS storage tank provides 8 to 19 hours of detention time at max month design flow rates and average day startup flow rates, respectively. The City of Franklin will waste sludge from the clarifiers and process sludge at a faster rate prior to the annual pressure vessel inspection to lower the sludge blankets in the clarifiers and the water level in the WAS storage tank. This will allow Franklin to reduce the amount of pre-dewatered cake that needs to be diverted for landfill disposal.

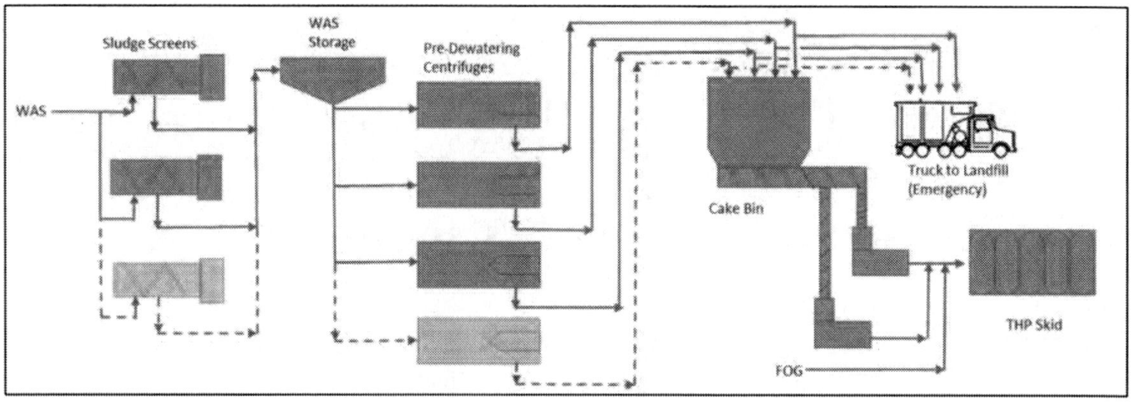

Figure 3: Franklin Biosolids Handling Equipment Pre-THP

Pontiac's approach to managing sludge during THP downtime is also a combination of sludge storage and diversion, but its budget constraints resulted in more cost-conscious solutions. **Figure 4** shows Pontiac's raw sludge process flow to the THP system. Primary sludge and thickened WAS from the Auburn WWTP are combined with hauled in septage and co-thickened PS/WAS from the East Boulevard WWTP in Sludge Blend Tank No. 1. The blended sludge is then pumped through sludge strain presses and discharged into Sludge Blend Tank No. 2. The screened sludge is then pumped to the pre-dewatering centrifuges, which discharge the dewatered cake down a chute into the pre-dewatered cake storage bin. The pre-dewatered cake can then be pumped to either the thermal hydrolysis system or to truck loading for landfill disposal. Diverting pre-dewatered cake from the process after the cake bin by pump slightly increases risk, because if the cake bin is down, so is the plant's ability to process sludge; but it results in significant cost savings in extra screw conveyors, as well as overall building height.

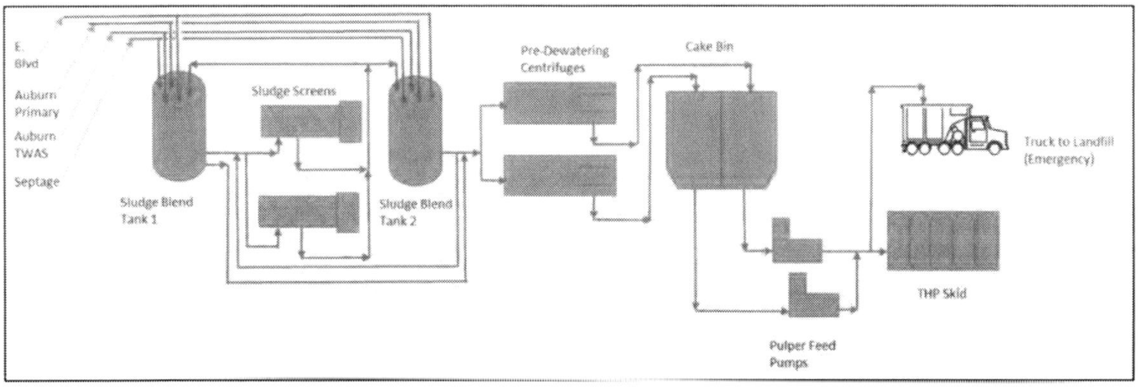

Figure 4: Pontiac Biosolids Handling Equipment Pre-THP

At Pontiac, Sludge Blend Tanks 1 and 2 provide storage like the WAS storage tank at Franklin, however each tank only provides 2 hours of detention time. The blend tanks have three design functions: raw sludge storage, blending and homogenizing sludge feed to the centrifuges and THP, and more reliable control strategy for the eleven separate pumps that transfer raw sludge and septage to the biosolids handling building. The four separate sludge sources have different concentrations of solids and are all pumped separately and intermittently throughout the day. The septage transfer pumps will be new, and the Auburn Primary sludge pumps will be replaced with new pumps. Three existing Auburn TWAS pumps and two existing East Boulevard sludge pumps will be retained to pump the sludge to the Blend Tanks. Feeding raw sludge and septage directly to a blend tank instead of directly to sludge screens simplifies the control of the transfer pumps and allows for the reuse of some of the existing pumps. Splitting the storage into two blend tanks provides redundancy for when the one tank needs to be out of service for cleaning.

An innovative approach taken at Pontiac is diverting the cake after the pre-dewatered cake bin. The cake bin provides an additional 5 hours of storage. By utilizing the cake storage bin, and substituting two smaller FRP tanks for the large pre-stressed storage tank, Pontiac achieved the same amount of storage at $250,000 less cost than Franklin. In addition to cost savings, this approach maximized the storage in the cake bin, and minimized the footprint required for the raw sludge storage and blending, as summarized in **Table 4**. Pontiac's cake pumping diversion offers operational advantages over Franklin's cake conveyance diversion in that the centrifuges would be able to continuously operate even when the plant is waiting for the next truck to discharge into. Centrifuges have operationally intensive shutdowns and startups, and are most efficient when they can run uninterrupted for extended periods of time. Since over half of the storage is provided in the cake bin, it allows the centrifuges to operate continuously between truck loading cycles.

Table 4: Comparison of Raw Sludge Storage Capacity, Footprint, and Costs

Facility	Franklin WRF	Pontiac Auburn WWTP
Raw Sludge Storage	One 280,000 Gallon Concrete Tank	Two 12,700 Gallon FRP Tanks
Footprint Required	1,600 square feet	450 square feet
Raw Sludge Storage Capacity	8 Hours (19 at startup)	4 hours
Raw Sludge Storage Capital Cost	\$370,000	\$120,000
Equivalent Raw Cake Diverted from Landfill	49 WTs @ 18% TS or 65 cubic yards	24.5 WTs @ 18% TS or 33 cubic yards
Raw Cake Storage	One 50 cubic yard bin	One 40 cubic yard bin
Raw Cake Capacity	6 hours (14 at startup)	5 hours
Total Storage Capacity Upstream of Truck Diversion	8 Hours (19 at startup)	9 hours

Sludge Storage Redundancy

Pontiac can achieve the same detention time in less storage due in large part to its ability to leverage existing sludge thickening infrastructure. Pontiac's raw sludge is approximately 5 times thicker than Franklin's; 4.0% to 0.8% total solids, respectively. Pontiac's raw sludge storage also has a less robust mixing system than Franklin's; pumped recirculation verse a Rotamix / Jetmix type mixing system, respectively.

The higher percent solids, less robust mixing system, and introduction of varied waste streams (including septage) into Pontiac's sludge blend tanks increases the potential need of tank cleaning. Pontiac built in partial redundancy to its raw sludge blending/storage system. The system is operable with one sludge blend tank out of service, but with reduced storage capacity. If one sludge blend tank is out of service, raw sludge will be fed into the other tank. Raw blended sludge will be fed to the sludge screens from the in-service tank and screened sludge returned to the same tank. When one blend tank is out of service, the pre-dewatering centrifuges will receive a blend of screened and unscreened blended sludge. Centrifuges and the Cambi system do not require screened sludge, so this partial redundancy is acceptable. Eliminating a third blend tank saved approximately \$100,000.

Cake Storage Redundancy

Another area of partial redundancy for Pontiac is that each pre-dewatering centrifuge discharges to a dedicated side of the cake bin, and each side of the cake bin discharges to a dedicated pulper feed pump (**Figure 3**). The system can operate at full design capacity if one pre-dewatering system and its associated side of the cake bin or associated pulper feed pump is out of operation. The loss of one side of the cake bin reduces system storage capacity by 2.5 hours. However, if a pre-dewatering centrifuge is down and its opposite cake bin side or its opposite pulper feed pump is down, the biosolids handling system will not be operational. In this case, raw, pre-dewatered sludge would not be able to be processed or trucked off-site, leaving the WWTP to rely entirely on sludge storage.

Conversely, Franklin's pre-dewatered cake bin can feed cake from anyone of the four centrifuges' discharge to either of the pulper feed pumps. This redundancy comes at an additional capital cost, with the cake bin including a long live bottom hopper with duty/standby augers, each feeding a duty/standby pulper feed pump. Each auger and the pulper feed pump it feeds to is powered and controlled by a separate control panel and MCC. This dual electrical connection provides an additional level of redundancy to the critical process.

ADDRESSING SYSTEM REDUNDANCIES – DOWNSTREAM OF THP

After THP at Franklin, hydrolysed sludge is injected into a one of three digester sludge recirculation loops, one for each digester. During Phase I, there will only be two digesters and two sludge recirculation loops (see **Figure 5**). These digester recirculation loops each have a vertical chopper pump to convey sludge from the digesters, to the discharge of the THP skid, through its dedicated heat exchanger, and back into its dedicated digester. There are a few advantages to injecting hydrolysed sludge into these recirculation loops:

- Reduces the risk of sludge "baking" onto the tubes inside the heat exchanger by lowering the sludge inlet temperature and diluting the FOG content of Franklin's feed sludge.
- Increases the homogeneity of the sludge due to increased mixing time with digested sludge upstream of digestion.
- Reduces hydraulic issues associated with conveying and dosing low flow rates (20-50 gpm) of high solids hydrolysed sludge into the digesters in small diameter pipes.

Each digester at Franklin has a dedicated heat exchanger to cool the hydrolysed sludge to mesophilic temperatures. Common suction and discharge headers for the digester recirculation pumps provide the ability to use any pump and any heat exchanger to control the feed and temperature of any digester. When the THP skid is offline for pressure vessel inspection, hot water can be circulated through the heat exchangers to maintain mesophilic temperatures in the digesters.

The digested sludge is pumped to a digested biosolids storage tank, which provides 10 to 25 hours of mixed storage at design maximum month and average day startup conditions, respectively. Biosolids are pumped from this storage tank to the post-dewatering screw presses. Dewatered cake is then transferred to a post-dewatered cake hopper by a duty and standby inclined screw conveyor. The hopper is elevated above the truck bay to add an additional 15 minutes of storage; enough time for trucks to be switched out.

The class A cake will either be collected from the truck loading bay by a third-party distributor, or will be transferred to the solar dryer pilot installation in the back of the biosolids facility. Two bays of solar dryers are being added to dry a portion of the 30% Class A cake to 75% total solids to evaluate the local market demand between Class A cake and Class A pellets.

Franklin is expecting to partner with a third party to distribute the Class A biosolids.

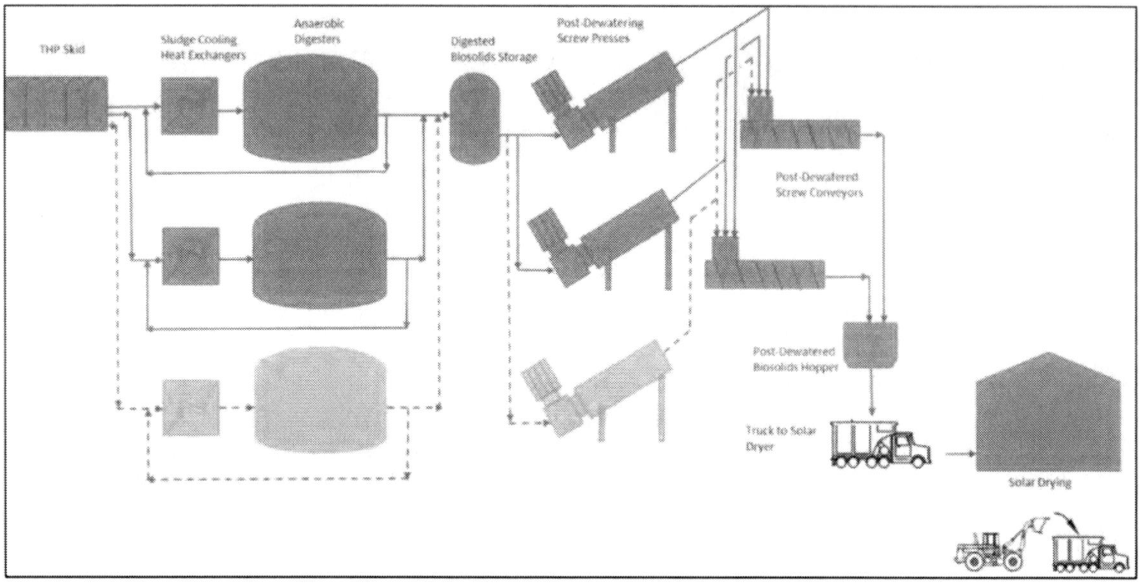

Figure 5: Franklin Biosolids Handling Equipment Post-THP

After THP at Pontiac, hydrolysed sludge is injected into a sludge recirculation loop, which passes through a single sludge cooling heat exchanger prior to discharging in the primary digester. Pontiac has no redundancy in its sludge cooling heat exchanger. This heat exchanger uses single-pass plant water as the cooling liquid. Heat exchangers can foul and lose heat exchanging capacity over time. Pontiac's took fouling into account in the heat transfer design. Although fouling and reduced heat transfer will require increased maintenance, heat exchanger failure is uncommon, as there are no moving parts aside from actuated valves, which will have shelf spares. Eliminating the redundant sludge cooling heat exchanger saved approximately $100,000.

Sludge in the primary digester, which is designed for a 15 day SRT (average conditions) is mixed and heated. Primary digester sludge overflows to the secondary digester, which will have a new flexible membrane cover. The secondary digester acts primarily as storage for sludge and biogas. Hydrolysed, digested sludge is typically fed to the post-dewatering centrifuges from the secondary digester. These centrifuges discharge to the cake storage pad using a series of conveyors.

The Pontiac cake conveyance system to the cake storage pad incorporates only partial redundancy. There are six total conveyors taking cake from the biosolids handling building to the cake storage pad (see **Figure 6**). Each post-dewatering centrifuge has two dedicated conveyors taking the cake outside the building and across the road to the pad. Post dewatering centrifuge No. 1 has dedicated Conveyors 1 and 3; Post dewatering centrifuge No. 2 has dedicated Conveyors 2 and 4. The belt conveyors on the storage pad, Conveyors 5 and 6, are not redundant. If Conveyors 5 and/or 6 are out of operation, Conveyors 3 or 4 can drop cake directly onto the cake pad. With these provisions, maintenance of the conveyors can be planned without interruption of biosolids treatment.

If one post-dewatering centrifuge is not operational, then the system can still run at full design capacity. However, if a post-dewatering centrifuge and one of its opposite conveyors is not operational, then post-dewatered cake cannot be transported to the pad. In this case, the secondary digester can provide storage of hydrolysed sludge without transferring sludge out of the digesters. If the secondary digester started from empty, it would provide up to two weeks of storage of hydrolysed sludge overflow from the Primary Digester. The secondary digester is expected to operate half full to maintain buffering capacity for sludge and biogas. The system is also designed so that both post-dewatering centrifuges can operate simultaneously to draw down the secondary digester. This will allow the drawing down of the secondary digester to increase storage capacity in preparation for the THP system to be taken offline for pressure vessel inspections.

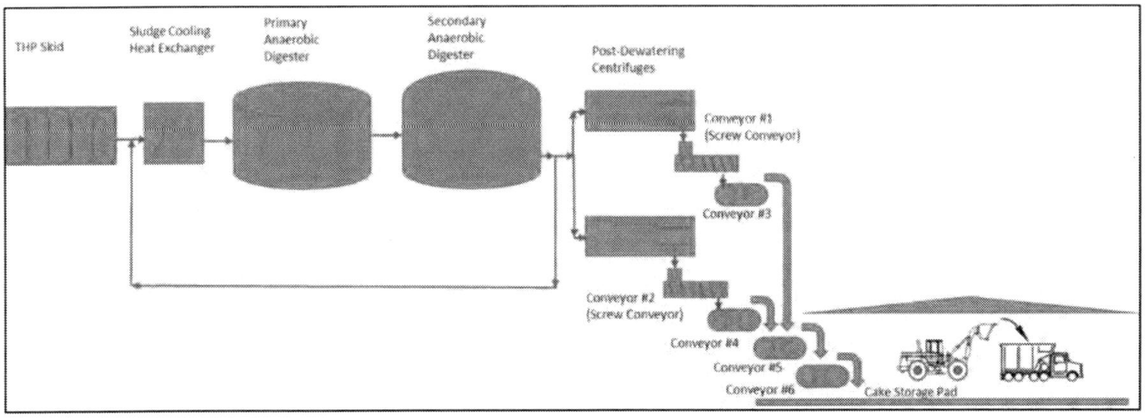

Figure 6: Pontiac Biosolids Handling Equipment Post-THP

Redundancy Effects on Building Cost

Franklin's new Solids Processing Building (SPB) and Pontiac's new Biosolids Handling Building contain most unit processes associated with each facility's upgrades. Both two story buildings include:

- Sludge Screens and Screenings Bins
- Pre-Dewatering Centrifuges
- Pre-Dewatered Cake Bin and Pulper Feed Pumps
- Post-Dewatering Centrifuges and Screw Presses (For Pontiac and Franklin, respectively.)
- Pre- and Post-Dewatering Polymer Systems
- Electrical and Control Rooms
- Numerous Process Pumps

The main drivers that informed the decisions in designing the SPB at Franklin include odor and aesthetic considerations for the neighboring residential neighborhood, a desire for the SPB to be the center of the new biosolids facility, and to provide ease of maintenance and automation of treatment processes. The most important drivers for the Biosolids Handling Building at Pontiac were space and budgetary constraints. **Table 5** compares the square footage and opinion of probable construction cost (OPCC) for the two buildings.

Table 5: Franklin and Pontiac Main Building Comparison

Facility and Building	Franklin WRF Solids Processing Building	Pontiac WWTP Biosolids Handling Building
Construction Description	2-story concrete and masonry	2-story steel frame
Above Grade Square Footage	39,400	16,700
Below Grade Square Footage	4,650	0
Total Square Feet	44,050	16,370
Building Height	33.5 – 55.5 feet	38 – 45 feet
OPCC of Building Structure	$7.4M[1]	$2.2M
Cost per Square Foot	$169	$134

[1] Cost adjusted to account for inflation that occurred between the two OPCCs and difference in cost of construction between the Nashville and Detroit metro areas.

One of the differences between the buildings was the overall height. To divert the pre-dewatered cake before it enters the cake bin, Franklin needs to provide greater distance between the pre-dewatering centrifuges and cake bin to fit a diverting gate. Pontiac was able to reduce the overall height of the Biosolids Handling Building by removing the diverter gate from the pre-dewatering cake chute, as well as the cross connection for post-dewatering cake conveyance. This allowed for a four-foot reduction in the building height on the first floor, which alone is estimated to save approximately $400,000 on the cost of the building.

Pontiac was able to reduce their building square footage in part because the facility utilized less units per treatment process than Franklin, but also because the building didn't include many of the spaces that Franklin's SPB did. For instance, Franklin's SPB included a full laboratory, storage room, enclosed truck loading bay, and a room for odor control equipment. The single largest cost difference between the two buildings was the difference in square footage.

However, another half million dollars can be attributed to the reduction in height and square footage, combined with a more economical steel frame construction (Franklin utilized concrete and masonry). It should be stressed that the savings from reduced footprint and height of the building were achieved at the cost of full redundancy, which will limit Pontiac's operational flexibility.

STEAM PRODUCTION AND BIOGAS UTILIZATION

A Cambi B-2 system requires up to 2,300 pounds of 160 psi steam per hour at maximum throughput. The actual real-time steam consumed is controlled by the B-2 skid control valves and is a function of the mass flow-rate through the system. The most common ways to provide this steam is through a steam boiler system or a combined heat and power (CHP) system coupled with a heat recovery steam generator (HRSG).

Steam Boiler Options
A steam boiler is more thermally efficient than a CHP system, and can be fueled by natural gas or biogas. Fueling with natural gas, as shown in **Figure 7**, is advantageous because it provides a steady fuel supply to the boiler, which simplifies the boiler's steam output control, and it makes biogas available to be consumed in a CHP engine to generate electricity and hot water. CHP systems operate best with a consistent steady fuel supply; when using biogas as the sole source of fuel, biogas storage is typically required to shield the CHP system from the peak and troughs of biogas production rates.

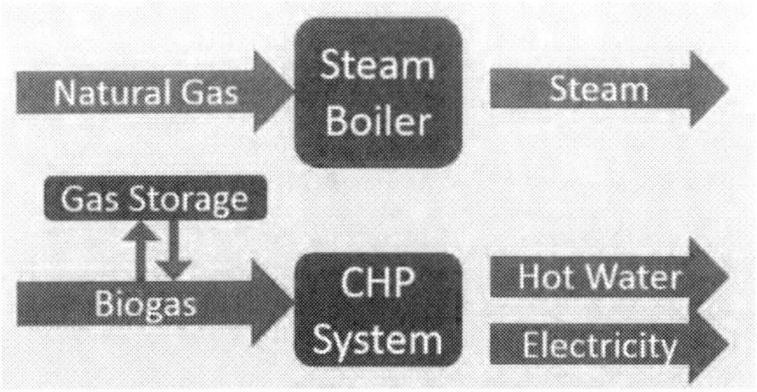

Figure 7: Steam Boiler on Natural Gas

Figure 8: Steam Boiler on Biogas

If a CHP system isn't desired, fueling a steam boiler with biogas can significantly reduce capital costs (**Figure 8**). In addition to eliminating the cost of the CHP system, steam boilers don't require the complex biogas treatment systems that CHP systems do. Running a steam boiler on biogas also cuts down on the operational costs of using natural gas, as there is typically enough heating value in the biogas to fuel a steam boiler (to provide the required steam for a thermal hydrolysis system) without having to supplement with natural gas. However, a gas storage membrane would need to be provided to ensure a steady supply of fuel is available for the steam boiler. Another advantage of this option is that a CHP system operating on biogas can be added later when more funding is available.

CHP System Option
Another common approach to producing the required steam for the thermal hydrolysis process is to use a CHP system with a HRSG. A CHP system, though not as thermally efficient as a steam boiler, typically has a similar overall fuel to energy (electrical + thermal) efficiency as steam boilers, and can be fueled by natural gas, biogas, or a blend of both. Since CHP systems are not as thermally efficient as steam boilers, they require more fuel input to produce a given amount of steam; typically, there isn't enough biogas produced to meet this fuel requirement. Two ways to address this are:

1. Size the CHP system with a HRSG to produce the required steam, and supplement the biogas fuel with natural gas to achieve the required fuel input **(Figure9)**.
2. Size the CHP and HRSG to burn the available biogas to provide a base load of steam, and fire a steam boiler operated on natural gas to supplement as needed **(Figure 10)**.

Option 1's real time blending of gases requires more complex gas feed controls for the CHP system, which comes at a premium. This option does however eliminate the need for a biogas storage membrane, as the natural gas supplement is providing the real-time buffer to changes in biogas production. Option 1 also generates more electricity and low grade heat in the form of hot water than option 2.

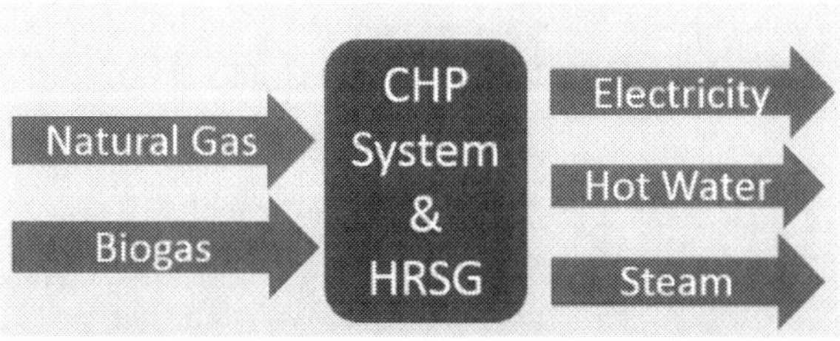

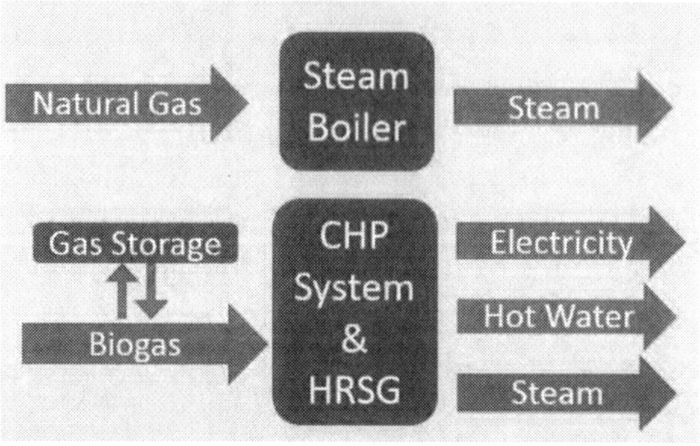

Figure 9: CHP Sized for Steam Demand **Figure 10: CHP Sized for Biogas Supply**

Franklin, with an eye on maximizing resource recovery, opted to utilize a CHP system to meet the steam demands of the thermal hydrolysis system. Two types of CHP systems were considered; gas turbines and reciprocating engines. Turbines can produce more steam from its exhaust gas, but engines produce a secondary source of low grade heat from its jacket water; this low-grade heat can be used for other process needs. Engines have a greater turn down ratio than turbines, and have better efficiencies through the turn down range; this is particularly important to Franklin, whose steam demands will be growing along with the city's population. Franklin chose to use reciprocating engines as the basis of design for the CHP system.

Franklin's steam production and gas utilization system consists two (2) GE Jenbacher JMC-312 engines, each with a HRSG would provide the required steam. Two auxiliary 50 HP steam boilers were also provided with this option. The auxiliary boilers allow for steam production to continue when the CHP engines are out of service, and the combined capacity of the boilers could produce all the steam required under maximum design conditions.

The CHP engines will be operated to produce the steam required by the THP system. The amount of fuel the engines will need to produce the steam is greater than the estimated biogas production. Natural gas will be blended with the biogas to provide the required fuel energy for the CHP system to provide the required steam. **Figure 11** shows the estimated proportions of biogas and natural gas to be consumed in the engine.

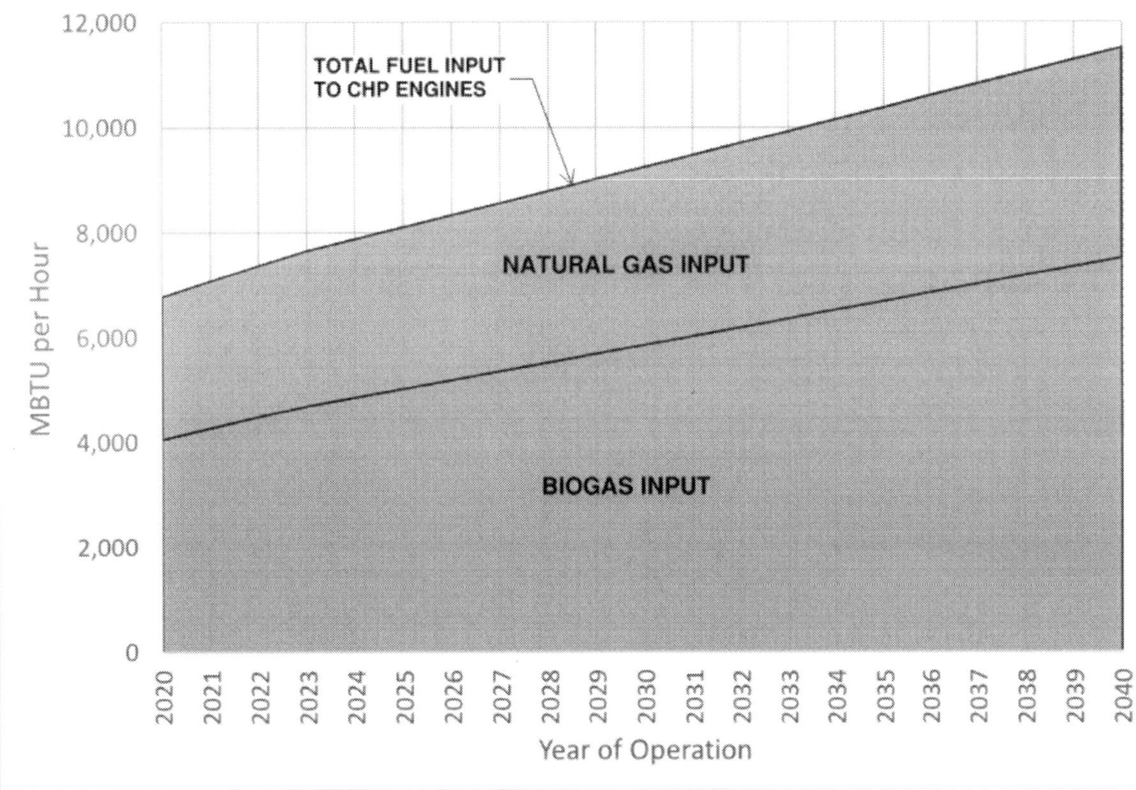

Figure 11: Estimated Fuel Input Proportions to CHP Engines

Adding natural gas incurs an additional operational cost for steam generation. However, that additional natural gas is producing additional power; the savings from using the additional power generated on-site more than covers the cost of the purchased natural gas.

In addition to operational cost savings due to increased electrical generation, sizing the CHP engines to require more fuel than is available in the biogas results in all the biogas being consumed on-site, without the need for biogas storage. Biogas storage is typically provided to engines running on biogas to dampen the peaks and valleys of biogas production, thus provide the engines with a steady, continuous fuel input. In this application, natural gas is actively blended with the biogas to make up for the shortage of fuel. Therefore the natural gas flow is adjusting to ensure a steady, continuous fuel input to the engines. Eliminating the biogas storage from Franklin's design opened up space on site and saved over $600,000 in capital costs.

The power generated will be consumed on-site to reduce the amount of power Franklin buys from the grid. The steam generated from the heat captured from the engine exhaust, via the HRSG, is consumed in the THP process. The hot water recovered from the engines jacket water system is used for heating of the FOG receiving process, and for space heating of the Digester Building. **Table 6** outlines how Franklin utilizes the energy captured from the biogas.

Table 6: Summary of Biogas Utilization for the Franklin CHP System

	Average Day Startup	Max Month Design
Biogas Production (SCFM)	90 SCFM	209 SCFM
Biogas Energy Content (MBTU/hr)	3,234 MBTU/hr	7,515 MBTU/hr
Supplemental Natural Gas (MBTU/hr)	1,658 MBTU/hr	3,997 MBTU/hr
Blended Gas to CHP Engine (MBTU/hr)	4,892 MBTU/hr	11,512 MBTU/hr
Electrical Production (kW)	535 kW	1,288 kW
Electrical Efficiency (%)	37.3%	38.1%
Steam Production (MBTU/hr)	1,053 MBTU/hr	2,531 MBTU/hr
Steam Efficiency	21.5%	22.0%
Hot Water Production (MBTU/hr)	1,193 MBTU/hr	2,869 MBTU/hr
Hot Water Efficiency	24.4%	24.9%
Combined Thermal Efficiency	45.9%	46.9%
Total System Efficiency	83.2%	85.0%

This configuration maximizes the beneficial use of biogas by converting it to beneficial electrical and thermal energy. Captured thermal energy is only beneficial when it can be used. Franklin has limited thermal loads outside of the THP process. This CHP system captures 85% of the energy from the biogas, and 38% of it is converted to electricity and 22% is converted to steam. The electricity and steam will be in demand year-round, resulting in a baseline 60% efficiency in beneficially using the biogas. The hot water will heat incoming FOG and provide space heating of the Digester Building. In the winter, all of the thermal energy in the hot water will be consumed in these processes; during the summer the space heating needs will be zero and the FOG thermal demand will be lower. Therefore, the overall estimated efficiency in utilizing the biogas is in the range of 60-85%.

Pontiac, with a tighter budget and greater heating needs, opted for steam boilers. They are much more efficient at converting the available biogas to steam than a CHP system, have a significantly lower capital cost, can utilize untreated biogas, and provide a very efficient steam system for building and process heating. Pontiac is expected to utilize most of the digester gas produced after the installation of the new thermal hydrolysis system. Digester gas production rates and storage capacity are shown in **Table 7**.

Table 7: Pontiac Biogas Production and Storage Table

Biogas Production	Equivalent Power (MBH)	Biogas Heating Value (BTU/sf)	Biogas Production (cfm)	Biogas Storage (cubic feet)	Biogas storage detention time (hours)
Ave annual	6832	590	193	81,000	7.0
Max month	8779	590	248	81,000	5.4

Digester gas is used by the three 80 hp steam boilers (2 duty, 1 standby) to provide steam to the THP system, for space heating of the new 16,700 square foot biosolids handling building, and to a steam-to-water heat exchanger to heat water for neat polymer dilution and activation. Heating of neat polymer dilution water increases the polymer efficiency and decreases polymer consumption; this heat is not critical to operation of the process, and therefore can be suspended if there is insufficient biogas to meet critical demands. **Table 8** lists the biogas demands in order of priority, with a range of biogas demands for each use.

Table 8: Summary of Biogas Utilization for the Franklin CHP System

	Summer Average Day Demand Scenario	Winter Maximum Month Demand Scenario
Thermal Demands		
System Makeup and Blowdown (MBTU/hr)	418	418
THP Steam (MBTU/hr)	2,119	2,589
Building Space Heating (MBTU/hr)	0	1,800
Polymer Makeup Water (MBTU/hr)	1,982	1,982
Total Thermal Demands (MBTU/hr)	4,519	6,789
Steam Production		
Steam Production (MBTU/hr)	4,519	6,789
Boiler Efficiency	80%	80%
Biogas to Steam Boilers (MBTU/hr)	5,649	8,486
Biogas Consumed (SCFM)	160	240

If there is insufficient pressure in the biogas feed line the boilers will use natural gas as a backup fuel. Under unusual operating circumstances, it is possible that there will not be enough steam produced to heat all the polymer dilution water while meeting THP demands and winter space heating demands. This will occur if both post-dewatering centrifuges are in operation. During normal operation in winter months, Pontiac will have sufficient biogas to meet all demands, with little to no excess biogas. During summer months when there is no projected space heating demand, Pontiac is projected to have between 65 cfm and 82 cfm of excess biogas. This project will install a new biogas storage membrane on top of the secondary digester, which will provide buffering capacity and storage to maximize biogas use and limit the amount of excess biogas that needs to be flared.

Steam Production Approach Comparison

One of Franklin's main drivers for the project was to maximize resource recovery through generation of electricity. Franklin is also located in a warmer climate, and has limited heating demands outside of the thermal hydrolysis system. These two factors pushed Franklin to utilize CHP engines with HRSGs to maximize the capture of the energy potential of the biogas. The CHP engines provide electrical generation to offset part of the plant's consumption, high pressure steam for the thermal hydrolysis process, and low grade hot water that is used for space heating, and process heating.

The primary driver for Pontiac is upgrading the biosolids facility so that all the sludge can be processed into biosolids for land application; generating on-site electricity was not a stated goal. The additional capital costs associated with providing a CHP system would have made it difficult to achieve the upgrades required for the given budget. Additionally, Pontiac experiences a colder climate than Franklin, and has greater space heating demands. These factors resulted in steam boilers being selected because of their comparatively lower operating costs, and unrivaled efficiency at creating useful thermal energy.

An interesting comparison between the two facilities was in their approach to designing steam boilers. Franklin's design includes two 50 HP steam boilers for backup steam generation for when the CHP engines are out of service. These two 50 HP boilers combined can provide all the steam generation needed for the thermal hydrolysis process. Pontiac's design includes three 80 HP steam boilers (2 duty, 1 standby), which is designed to consume the available biogas during winter months. During winter months, approximately a third of the steam created from the steam boilers is needed for the thermal hydrolysis system. Another third will be used for space heating of the new Biosolids Handling Building, and the final third for heating polymer make up water to improve the efficiency of the pre- and post-dewatering process. During summer months, space heating demands for the new biosolids handling building will be zero. When space heating demands are zero, about a third of the biogas generated will be unused and go to flare.

APPROACHES TO ACCEPTING HAULED IN WASTES

Franklin's and Pontiac's upgrades include new facilities to accept FOG and septage, respectively. Franklin is attempting to co-digest with grease from local restaurants with the goal of increasing biogas production. Pontiac is setting up a septage receiving facility to provide a service to the surrounding areas. Franklin will meter the grease into the thermal hydrolysis process to optimize digester performance. Pontiac will blend the septage with the other sludge sources further upstream, prior to sludge screening.

The Septage Receiving Facility at Pontiac includes a Septage receiving system (JWC Honey Monster) enclosed in a building. Septage trucks discharge to the septage receiving system, which processes the septage and discharges liquid to a 55,000-gallon underground storage tank. The storage tank provides approximately one day of storage. Screenings and solids are discharged into a screenings bin inside the building. The building is sized to house a second redundant septage receiving system. The redundant system is a bid alternate even though the building will be sized for the redundant system to reduce capital costs. One set of pumps is used for both mixing and transfer from the septage tank to the sludge blend tanks. The system is sized for consistent transfer of septage to maximize homogeneity of the mixed sludge and not overwhelm the mixed sludge with a slug of septage. Under maximum solids loading from the other sludge sources, the facility will not accept septage.

The FOG Receiving Facility at Franklin has duplicate receiving systems that will be enclosed and odor controlled. FOG will be discharged into one of two receiving boxes which include a bar rack and hot water spray. The screened FOG will drop into one of two 10,000-gallon FOG storage tanks. These tanks will be mixed with a pump and nozzle system (Rotamix/Jetmix). FOG will be pulled off the FOG mixing pump suction line and pumped through a FOG recirculation loop. This loop is heated in tube in tube heat exchangers with hot water from the CHP system, and the solids concentration of the FOG will be continually monitored by a microwave solids analyzer on the loop. FOG is pulled off this recirculation loop and metered into the thermal hydrolysis feed via progressing cavity pumps. Space is provided for a third FOG storage tank and receiving station for future buildout.

BIOSOLIDS END USE

Franklin currently thickens and dewaters WAS using DAF thickeners and BFP dewatering. The dewatered cake is trucked to a landfill 113 miles away. In the proposed facility, dewatered, Class A cake will be conveyed to an above head loading hopper inside a truck loading bay. The truck loading bay is included within the proposed Solids Processing Building to allow the truck loading operations to be odor controlled. The cake can then be hauled off-site for land application or further processing into other products by a third-party contractor. Alternatively, Franklin staff can also truck a portion of the class A cake to the solar dryers.

Preliminary research into Franklin's regional area indicates that a dried biosolids product may be more valuable than a dewatered product. Franklin has enough land to build 12 bays of solar dryers, which would be enough space to dry all the biosolids produced on site. Only two bays are being installed in phase 1 to evaluate the difference in demand between a dried and a dewatered Class A product. On one extreme, if there is a high demand for the dried product, Franklin would weigh the costs and benefits of investing in more solar dryer bays. On the other extreme, if there is no demand for the dried product, the two bays could be used as storage or treatment for raw dewatered sludge that is generated when the THP system is down.

Pontiac currently stores cake on an approximately 28,000 square foot uncovered cake storage pad. They currently land apply approximately half their biosolids, since half of the generated sludge is treated by the digesters to achieve Class B status. Half of the sludge is just dewatered by the belt filter presses and goes directly to landfill. There is no conveyance system currently in place from the belt filter presses to the cake pad, and all biosolids must be transported by truck to the cake pad for storage. Pontiac is required to store cake for 180 days during certain months of the year due to land application regulations. Though this requirement won't apply to the Class A biosolids, the storage will be beneficial since land application can only occur part of the year in the surrounding region. Having Class A biosolids will also be beneficial to Pontiac because of reduced odors, reduced storage footprint, and a much less labor-intensive process to handle and store biosolids.

Cake pad improvements include a cake pad cover, a series of conveyors to carry the biosolids directly to the cake pad, push walls around the cake pad, and four separate storage bays with nine-foot-high push walls. Because of the increased biosolids treatment with thermal hydrolysis and digestion of all the sludge, and the cake pad improvements, the stored cake will have higher solids content and can be stored in taller piles resulting in a 30% reduction in cake pad footprint to 20,000 square feet. Additionally, operators will no longer be required to truck the biosolids to the pad for storage, and the conveyors will be set up so that cake can be discharged onto different areas of the pad. The Class A biosolids should have less odor potential than the current Class B biosolids, and the new cake pad cover is also expected to reduce odor potential.

A key driver for Franklin is to create a Class A product that can be beneficially reused. Pontiac did not choose thermal hydrolysis to achieve a Class A biosolids product, but to increase the capacity in their digesters so they can continue their Class B land application and stop landfilling sludge. Franklin added solar dryers to test its regional market demand for dried versus dewatered Class A biosolids, and is in negotiations with a third-party distributor to market the Class A product. Pontiac plans to continue land application, and is content that thermal hydrolysis results in less and drier biosolids to dispose of.

SUMMARY AND CONCLUSIONS

Franklin WRF and Pontiac WWTP are two facilities installing thermal hydrolysis pretreatment systems to meet their biosolids processing needs. Thermal hydrolysis pretreatment of biosolids can meet a range of end goals, including increased biosolids treatment capacity, increased biogas production, reduction in odor potential of biosolids, and production of Class A biosolids for beneficial reuse. The two case studies presented herein illustrate a range of ways thermal hydrolysis technology can be applied to help a facility meet its unique biosolids challenges and goals.

An identical Cambi B2 THP skid will be used at both the Franklin and Pontiac plants, but the THP support systems were uniquely designed to address the cities' climate, budget and drivers for the project. Franklin, with its rapid growth and focus on sustainability, is motivated to generate Class A biosolids and maximize biogas production and use. The Franklin case study showed that THP subsystems like sludge screens and dewatering equipment can be incrementally implemented, illustrating one way a smaller facility that will experience future expansion can incorporate THP into its long-term plans. Franklin will benefit from the increased biogas production, and has selected CHP engines that can utilize all the biogas produced. Franklin's CHP engines will generate steam for the THP process, hot water for process and space heating, and will also generate electricity for use at the plant.

Pontiac has stagnant growth, but needs to double its biosolids treatment capabilities at the Auburn plant to meet current needs, and discontinue landfilling of raw dewatered sludge. THP allows for this expansion without increasing digester capacity, which is ideal for a plant with limited space to add another digester. Pontiac will utilize most of the generated biogas with new steam boilers to provide steam to the THP system and heat the new biosolids building. Space heating needs are significant during Pontiac's winter months, and will use up to a third of the steam generated from biogas. During summer months this biogas will go to flare. The anticipated excess biogas did not warrant a CHP system for the current design, but future addition of a CHP engine is a possibility. Pontiac had a limited budget compared to Franklin, but was able to use existing equipment, and have partial redundancy in some THP support equipment to reduce capital costs.

The Franklin and Pontiac case studies provide insights into how thermal hydrolysis pretreatment could be incorporated at smaller facilities with varying drivers, climates, and budgets.

REFERENCES

Pontiac Wastewater Treatment Plant 2014 SRF Project Plan for the Water Resources Commissioner, HNTB, July 2014

Pontiac WWTP Solids Handling and Septage Receiving Project Basis of Design Report, CDM Smith, August 2016

Pontiac Wastewater Treatment Plant 2015 State Revolving Fund Project Plan, CDM Smith, June 2015

Planning Your Class A Biosolids Program: Design Considerations and Lessons Learned in Pierce County, Washington

Melissa Newell, Pierce County Public Works

ABSTRACT

Biosolids managers are increasingly considering Class A alternatives to traditional Class B land application as their primary management strategy. Because of situational complexities, Class A program design may require thinking beyond simple "copycat" implementation of admired programs that work well for other wastewater utilities. Pierce County has successfully managed the SoundGRO® biosolids fertilizer program since 2007, and along the way, has gained experience that has refined the program from its original design. These lessons learned highlight six types of will highlight six types of planning considerations essential for developing a new Class A biosolids management program: 1) political and operational fit; 2) product quality; 3) product distribution; 4) marketing and communication; 5) regulatory compliance; and 6) ongoing business practices.

KEYWORDS: Biosolids, planning, program design, Class A "Exceptional Quality," fertilizer

More and more, biosolids managers are considering Class A alternatives to traditional Class B land application as their primary management strategy. Such alternatives promise to meet sustainability, public outreach and other goals in new ways, while also effectively providing for beneficial use in an increasingly sensitive political climate. However, Class A biosolids programs can take many forms – with products ranging from compost to dried pellets to custom soil amendments and operations varying widely in terms of distribution, marketing and other business practices. Because of this complexity, what works well for one utility may not be a good fit for another.

Pierce County has successfully managed a Class A "Exceptional Quality" biosolids program since 2007. Although now well established, the program is in many ways substantially different from what we envisioned during initial planning more than a decade ago. Changes in market conditions, first-hand experience operating treatment equipment and evolving regulations have driven changes to our program from its original design. While we continue to meet our overarching goals and regulatory obligations, we have also experienced growing pains and challenges associated with responding to unanticipated program influences.

Pierce County's lessons learned point to six types of planning considerations essential to developing any new Class A biosolids management program. These include: 1) political and operational fit; 2) regulatory compliance; 3) product quality; 4) product distribution; 5) marketing and communication; and 6) ongoing business practices. Although not a comprehensive list for every context, these key questions may help operators and planners think more broadly about potential design elements and implications for your new program.

HISTORY OF CLASS B LAND APPLICATION

Pierce County Public Works has provided high-quality, cost-effective sanitary sewer services within Pierce County, Washington, for more than 40 years. We currently serve more than 62,000 residential, commercial and light industrial accounts within our 117-square mile service area, which includes all or part of eight cities and towns, as well as unincorporated areas within the county. Our 700-mile long collection system conveys an average of 20 mgd to one regional treatment facility.[1]

The Chambers Creek Regional Wastewater Treatment Plant is located in University Place and was originally constructed in 1984. Initially designed as a conventional activated sludge process with selectors, the plant is currently rated at 28.7 MGD capacity. However, construction is underway to improve treatment processes, add biological nutrient removal and expand capacity to approximately 43 MGD, with substantial completion expected in the first half of 2017. Capital improvements include construction of new aeration basins, clarifiers and digesters; revisions to solids handling; addition of side-stream biological nutrient removal capabilities; installation of reclaimed water systems; and other improvements. Total project cost is currently estimated at $353 million.

Prior to initial plant startup, County staff explored a variety of sludge management strategies, including ocean dumping, lagooning, land application, land reclamation and landfill disposal. Analysis completed during the mid-1970s as part of pre-construction sewerage planning and environmental impact requirements recommended landfill disposal as the most economic sludge management strategy. However, rapidly evolving state laws, local policies and environmental concerns caused the County to reexamine our program criteria and shift to an approach emphasizing beneficial use, rather than disposal. Instead of focusing solely on cost and regulatory compliance, the County added environmental soundness (what we might today call "sustainability") and program flexibility and reliability to our design criteria.

As such, we initially managed Chambers Creek Regional Wastewater Treatment Plant solids by way of land application, as documented in the 1987 Pierce County Utilities Department Sludge Management Program. Pierce County biologically stabilized and treated all solids to the applicable state and local standards[2] through anaerobic digestion, then dewatered to approximately 15% solids using centrifuges before transferring them to a third party hauling contractor. The contractor transported our solids approximately 60 miles by truck to a southwest Washington strip mine, where they were beneficially used for land reclamation purposes by a permitted commercial sludge disposal company.

Pierce County continued to land apply our biosolids for the next 32 years, but the sites, purposes and contracted partners evolved. Although we continued to utilize small quantities for mine reclamation, we moved largely to agricultural use over time.

TRANSITION TO CLASS A

Our shift toward a Class A biosolids management program started in 1992, with the County's purchase of an 886-acre former gravel mine immediately adjacent to the Chambers Creek Regional Wastewater Treatment Plant.[3] As a former mine, land reclamation and site restoration were required, to which biosolids land application would be well suited. The size and nature of the property also allowed brief consideration of such land intensive biosolids strategies as lagooning or composting. However, these options did not gain significant traction due to the high interest in the property's use for public and recreational purposes.

Pierce County's local leaders increasingly valued sustainability principles and called for use of Class A biosolids for reclamation during development of the Chambers Creek Properties.[4] At the same time, our operational experiences were raising questions about long-term risk management and concerns about the threats of poor public perception associated with Class B biosolids practices. Most significantly, a large-scale digester upset incident in 2001 and other policy shifts influenced creation of a Sewer Utility management objective to move away from Class B biosolids.

Our Class A "Exceptional Quality" biosolids program launched in 2006, with completion of our biosolids drying facility. Constructed at a cost of $19 million, the new facility includes a triple bypass Andritz drum dryer[5] and all of the associated infrastructure needed to produce a pelletized biosolids product – other elements of the dryer train, air quality scrubber systems, SCADA controls, storage silos and associated assets (~ 300 total assets). The facility also houses equipment used to support Class B biosolids management practices, including centrifuges, cake loadout chute and certified truck scale. Since December 2006, we have produced SoundGRO® biosolids fertilizer continuously, running 24-hours per day, approximately five days per week and yielding a 90% solid pelletized product.

Pierce County produces approximately 2500 tons of SoundGRO® fertilizer annually. Like many wastewater utilities with Class A "EQ" programs, we market and sell our biosolids product to a wide range of customers for home and commercial use. But unlike most others, we opted to register as a fertilizer in five states.[6] At the time, fertilizer registration was seen as a way to convey value and establish demand during the early years of our program. Our registration status continues to be a benefit, but has also brought some unforeseen challenges.

SoundGRO® fertilizer sells out reliably and consistently each year, generating approximately $200,000 annually in sales revenue, and the product is well received by customers and policy makers alike. However, the program has also experienced some significant growing pains as we've refined and revised nearly all program elements, including marketing, distribution, operations and maintenance. These lessons learned will be substantially integrated into future strategic biosolids management planning efforts.

KEY PLANNING QUESTIONS

While in the early stages of facility design and program planning, Pierce County staff conducted extensive research about the technological and program design options currently in use at the time. Staff attended numerous professional conferences focused on biosolids management, consulted with successful and established utilities, and conducted site visits to see various biosolids drying technologies in action. Ultimately, we modeled our program design on that of The Ocean Counties Utility Authority (New Jersey), a similarly-sized municipality who has been successfully producing, marketing and distributing OceanGRO® biosolids fertilizer since 1997. However, despite extensive planning, we have sometimes been surprised and challenged by the evolving and emergent nature of our Class A biosolids program.

In the decade since we began Class A operations, we've benefited from the real-world practical learning that can only come from doing. In some cases, our experiences helped us to refine our already good plans. For example, our decision to register as a fertilizer has generated positive perception amongst local leaders and strong demand from customers. Our experiences have also enabled us to develop a strong operational record, effectively anticipating and addressing maintenance needs so as to minimize unplanned shutdowns. In other respects, our experiences have caused substantial changes to our program design as was envisioned during initial planning stages. For example, early customer demands influenced final product characteristics, quality control and basic operations, such that our process now optimizes for pellet quality rather than dewatering. Most significantly, our customer base is substantially different from initially expected, with much less focus on residential home users and the majority of sales to large volume commercial customers. Marketing activities that were initially thought essential are now not needed to ensure continued sales.

Pierce County's lessons learned fall into six general categories: 1) political and operational fit; 2) product quality; 3) product distribution; 4) marketing and communication; 5) regulatory compliance; and 6) ongoing business practices. By focusing on these and other key questions early in the planning process, you ensure greater success of your own biosolids management program over the long term.

Political and Operational Fit

1) Organizational values – What are your political and operational values? How are they reflected in your new program?
2) External perceptions – What trends in public and policy-maker perspective might positively or negatively affect your program?
3) Program scalability – Can your program design expand and contract in response to normal variability? Can it flexibly meet your long-term planning needs to accommodate future growth?
4) Fit within your biosolids portfolio – How does your new program complement your other biosolids management strategies? What contingency plans will you need?
5) Resource availability – Do you have space and staffing available to operate your new program under tough conditions?

Product Quality

1) Initial guarantees – What initial quality guarantees do you wish to make? What guarantees might you be willing to make, if pushed?

2) Customer requirements – What initial guarantees do your customers require? Will different customers require different products? How will you adjust to changing customer demands?
3) Quality control – How will you monitor your process for quality? When a quality issue arises, how will you respond?
4) Organizational constraints – How do your existing policies support or challenge your ability to maintain a desirable product quality over time? How might changes within your organization affect your biosolids product?

Product Distribution

1) Established demand – Is there established demand for Class A biosolids product in your area? How flexible or resilient is this demand?
2) Potential new demand – If needed, could you create new demand? What would it look like?
3) Existing distribution channels – Do you have a distribution system already in place? How can you utilize it?
4) Distribution policies – What are your distribution-related values, goals and performance measures? How do these affect your product design? What outside influences might cause a change in your policies?
5) Ownership of distribution stream – What's your desired level of ownership over the distribution of your product?

Marketing and Communications

1) Communications values – What are your critical marketing and communication goals? What goals are optional?
2) Product advocates – Who will champion your product from within your organization? Who are your external champions? How will you engage them?
3) Established communications channels – What established marketing outlets can you utilize? What new resources will you need?
4) Marketing scalability – How will you scale your marketing to meet changing demand for your product or changes in the product itself?
5) Level of engagement – To what level do you want to educate or satisfy all of your customers? How will you address negative public inquiries? How will you track changes in perception and monitor emerging regional trends?

Regulatory Compliance

1) Current requirements – What are your current biosolids regulatory requirements? How comfortable is your organization with compliance?
2) Additional or "voluntary" requirements – What additional regulations and requirements will apply to your new program? Which agencies will be involved? How will you monitor rule changes?
3) Regulatory resource needs – What existing resources will you engage? What new partnerships might be required?

4) Response to rule changes – How will you respond to emerging requirements? What changes will you proactively influence and what changes might drive a shift in your program design?

Ongoing Business Practices

1) Customer service values – What are your customer service values and goals? How will your new program integrate with existing customer service programs?
2) Policy integration – How will your existing policies support or hinder your new program? What new policies might be required?
3) Process integration – How might existing business practices and processes benefit or otherwise impact your new biosolids program? How will you integrate existing business functions with other aspects of your biosolids management program?
4) Data management – What established data management tools will you engage? What additional tools are needed?
5) Daily operations – What day-to-day business practices are essential? How will you flex these in response to changes?

Careful consideration of these and other key questions will enable you to craft a more complete understanding of how a new Class A biosolids program might fit into and complement your existing operations. Every regulatory and operational context is different, requiring unique program design elements to completely address situational challenges. But by calling upon the experiences of established Class A programs like Pierce County's SoundGRO® fertilizer program, you can set your new program up for good success and positively advance biosolids beneficial use within the local and broader community.

[1] As of March 2017, we are also planning for the addition of a second wastewater treatment facility, which will serve a master-planned community in the mid-County area. Pierce County plans to begin ownership and operations of the Cascadia Wastewater Treatment Plant at Tehaleh in 2018.
[2] At the time, municipal sludge was regulated by the Washington Department of Ecology under Revised Code of Washington (RCW) chapters 90.48 ("Water Pollution Control") and 70.95 ("Solid Waste Management – Recovery and Recycling"), as well as by Tacoma-Pierce County Health Department's 1985 "Sludge Utilization Policy." In combination, these rules set treatment standards and guided land application practices. U.S. EPA rules specific to municipal biosolids were still under development.
[3] Today known as the Chambers Creek Properties, the site is now home to a major regional park and the Chambers Bay Golf Course, with plans for additional development to include hotel and restaurant facilities.
[4] Chambers Creek Properties Master Site Plan, 1997
[5] The Model DDS-20 Andritz drum dryer is designed to evaporate up to 4400 lbs. of water per hour.
[6] Initial fertilizer registrations were filed in Washington, Oregon, Idaho, Montana and California. However, since 2007, these registrations have contracted and expanded in response to changing market conditions. Fertilizer registrations are typically filed with the state departments of agriculture.

Beginning a Complete Renewal of San José-Santa Clara Regional Wastewater Facility's Biosolids Processing Facilities

Adam Ross[1*], Mariana Chavez-Vazquez[2], Lloyd Slezak[1], Alicia Alba[2]

[1] Brown and Caldwell, Walnut Creek, California.
[2] City of San José, California.
* Email: ARoss@brwncald.com

ABSTRACT

Many wastewater treatment plants face the task of rehabilitating aging digestion infrastructure while also planning for new challenges – whether for Class A digestion, changes in biosolids disposition pathways, advancing resource recovery, or improving operability and maintainability. However, very few wastewater treatment plants deal with all of these challenges at once. This is the reality that the City of San José-Santa Clara Regional Wastewater Facility (Facility) is facing as it transitions from its existing process to an entirely new biosolids treatment strategy.

This paper details the decision drivers in the Biosolids Transition Study and the scope of the Digester and Thickener Facilities Upgrade Project, which, in addition to temperature-phased anaerobic digestion (TPAD), is implementing co-thickening, sludge screening, upgraded digester feed/transfer/withdrawal systems, and a new gas management system. Innovative, but proven, processes were combined to deliver a complete renewal of San José-Santa Clara's biosolids processing facilities. Whereas many facilities are using patented or proprietary biosolids equipment and processes, San José-Santa Clara is an example of an "open source" solution to high-solids, high-performance, Class A biosolids management.

KEYWORDS: Anaerobic digestion, temperature-phased anaerobic digestion, TPAD, biosolids.

INTRODUCTION

Unlike other facilities that have multiple biosolids disposition options, San José-Santa Clara, for several decades, has transported its air-dried biosolids to the adjacent landfill where they are used as alternative daily cover. At the time of planning, the landfill was expected to close by 2025, so although the Facility's sole biosolids pathway is convenient and cost-effective, the City wanted other disposal options. In addition to diversifying its disposal options, the Facility's sludge lagoons and drying bed operation must be discontinued to meet new odor containment goals. Between the potential landfill closure, changes prompted by odor containment, and major improvements needed at older plant solids infrastructure, San José-Santa Clara's entire solids processing system had to be redefined.

The San José-Santa Clara Regional Wastewater Facility treats an average of 110 million gallons of wastewater per day (mgd). The largest tertiary treatment plant in the western United States, the Facility serves 1.4 million residents and over 17,000 businesses in an eight-city area.

Figure 1 shows the Facility's service area in California's southern bay area. Figure 2 shows an aerial view of the Facility. Figure 3 shows the Facility's current operational footprint and process areas.

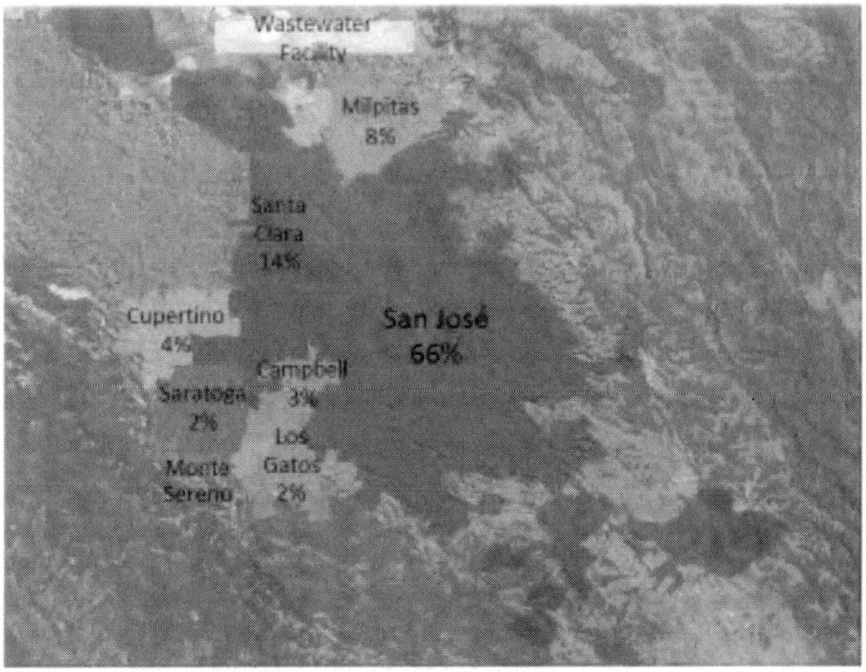

Figure 1. The San José-Santa Clara Regional Wastewater Facility's Service Area showing contributing cities. San José is located at the southern end of the San Francisco Bay.

Figure 2. Aerial view of the Facility looking from west to east. Facility operations are in the foreground. Biosolids lagoons are beyond.

Figure 3. Facility's footprint shown by process area. Biosolids processing – consisting of lagoons and drying beds – takes up 770 acres and is adjacent to development.

BACKGROUND AND PLANT MASTER PLAN

In 2008, San José-Santa Clara embarked on a master planning process to provide overall direction for rehabilitating and upgrading its facilities including potential process changes. The Plant Master Plan (PMP) used an extensive community engagement process to develop overarching environmental, economic, social, and operational goals.

One area of focus for the master planning process was biosolids management since treating wastewater at the RWF produces about 85 dry tons of solids each day. The Facility's current biosolids management practices produce a Class A biosolids product and include:

- Mesophilic Digestion
- Lagoon Stabilization – where digested solids are stored for about three years in open-air lagoons allowing further biological treatment and concentration of the solids.
- Drying – where stabilized biosolids are allowed to air dry in a series of drying beds.
- Disposition at Newby Island Landfill – where the dried and stabilized biosolids are used as daily cover in landfill operations.

Figure 4 shows the existing biosolids practices at the Facility.

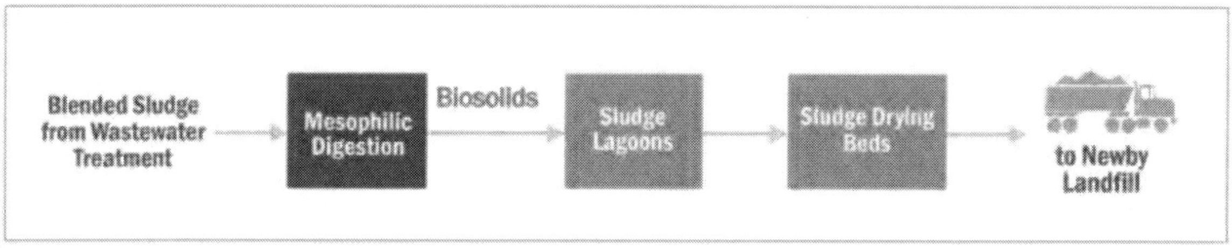

Figure 4. Existing biosolids treatment process.

This current system is land-intensive and has historically been one source of odors in the area. Because of these issues and because of the planned closure of Newby Island Landfill in 2025, the PMP recommended a new Biosolids Management Program involving a variety of enclosed, odor controlled treatment processes with the resulting treated biosolids used in a variety of off-site processing and beneficial reuse applications. The PMP envisioned a program that produced a mix of Class A and Class B biosolids products. Specific PMP recommendations related to the future Biosolids Management Program included:

- Rehabilitation of the existing thickening facilities and mesophilic digesters and an evaluation of whether or not a different type of digestion process should be implemented.
- Mechanical dewatering for all biosolids in an enclosed, odor-controlled facility to concentrate digested biosolids, which reduces the volume and weight of material requiring transport to off-site processing and beneficial reuse locations.
- Drying a portion of the dewatered biosolids using both thermal drying (20% of the biosolids) in an enclosed facility utilizing waste heat from a planned cogeneration facility and solar drying (10% of the biosolids) in enclosed greenhouses.
- Decommissioning the existing open sludge lagoons and drying beds.

- Additional processing and beneficial reuse at off-site composting facilities, land application sites and landfills.

Figure 5 shows the recommended biosolids process from the PMP.

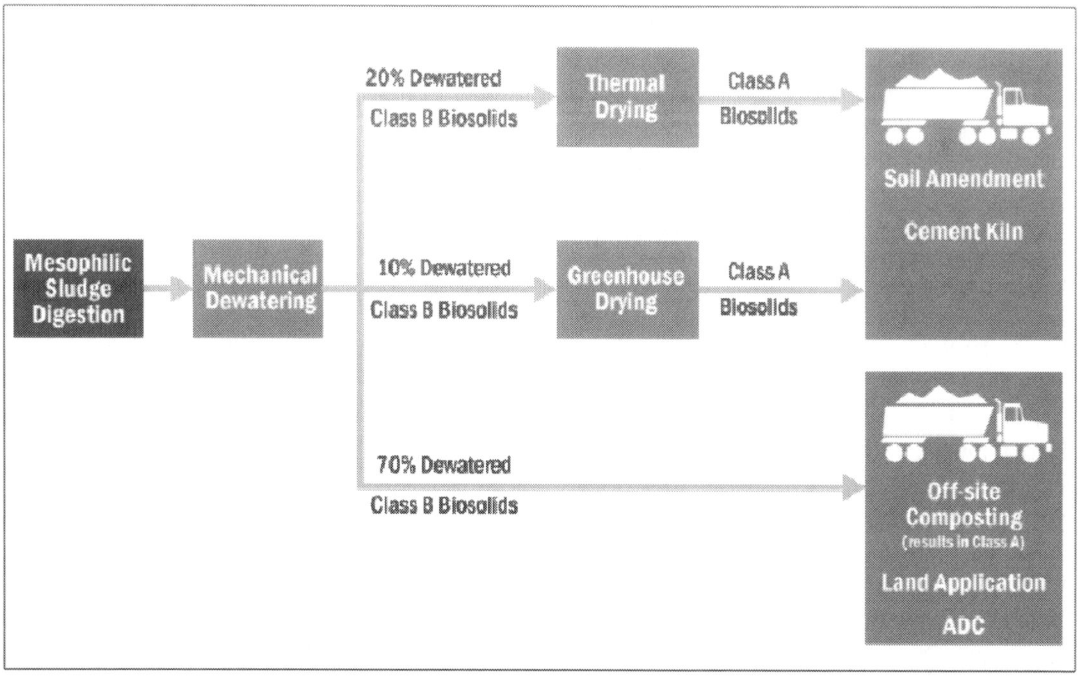

Figure 5. Recommended Biosolids Management System: Plant Master Plan.

Although the Plant Master Plan was officially adopted in 2013, the technical component of the PMP was completed in 2010. During the three-year environmental review process that occurred between 2010 and 2013, there were a number of changes in conditions potentially affecting the assumptions, recommendations, and implementation strategy recommended in the PMP. These changes included:

- In May 2011, in response to community concerns about odors emanating from the lagoons and drying beds, the San José City Council directed acceleration of the transition to the new biosolids management system and specifically called for the Facility to cease discharging biosolids to the existing lagoons by 2018, followed by emptying the lagoons and drying beds by 2024. The PMP had envisioned a three phase approach that would have decommissioned the lagoons and drying beds by 2030.
- During the environmental impact review process for the PMP, it was determined that the planned location for recommended future biosolids facilities contained potential wetlands and habitat. Siting facilities in the recommended location would likely trigger extensive environmental mitigation and a lengthy permitting process. The resulting schedule delays would push completion of those new facilities required to cease discharge to the existing lagoons well beyond the 2018 target date, therefore, alternative sites needed to be evaluated.

- In April 2014, the Facility's Treatment Plant Advisory Council provided feedback to staff to evaluate the possibility of producing Class A instead of Class B biosolids. The existing lagoon and air drying process results in a Class A-equivalent product and the Council wanted to keep the same overall level of treatment.
- In early 2014, the Facility's Capital Improvement Plan team conducted a detailed project validation review process of all projects recommended in the PMP. This validation effort led to a change in assumption from a large, open biosolids storage area near the lagoons (sized for 180 days of storage) to a managed, enclosed four-day storage facility located at the Biosolids Dewatering Facility, which is more in line with best practices in the wastewater industry.

In response to these changes, San José-Santa Clara initiated a Biosolids Transition Strategy Evaluation to evaluate end use options and to develop a processing system that provides readiness and flexibility for the future.

BIOSOLIDS TRANSITION STRATEGY EVALUATION

The Biosolids Transition Strategy Evaluation addressed certain specific issues regarding implementation of the transition from the current biosolids management system to the PMP's recommended system considering changes that occurred since the technical aspects of the PMP were developed. It included both near-term and long-term recommendations for the Biosolids Transition Strategy, taking into consideration the goals identified in PMP. The Biosolids Transition Study focused on answering several key questions related to the transition including:

- Should San José change from its current practice of mesophilic digestion to a temperature phased anaerobic digestion (TPAD) process in order to optimize solids stabilization and increase biogas production?
- Should San José accelerate the on-line date for planned thermal drying and greenhouse drying facilities and add a blending facility to take maximum advantage of low disposition costs at Newby Island Landfill until it closes?
- Should San José focus on installing treatment processes to achieve Class B biosolids at this time while preserving the ability in the future to achieve Class A biosolids? Is it economically feasible to meet the Council's direction for Class A biosolids?
- Should San José preserve the potential for other on-site biosolids processing should it be warranted by future industry, market, and regulatory conditions?
- What areas should be reserved for locating biosolids processing facilities?
- Can the 2018 target date for ceasing discharge to the lagoons be met? And if not, what can be done about that?

Answers to these questions and evaluation of alternatives entailed a two-step process. First, Brown and Caldwell's Solids-Water-Energy Tool (SWET) Model was used to help screen out less favorable alternatives. For example, based on the SWET analysis, alternatives involving 100% thermal drying and 100% composting were eliminated. Team workshops were then conducted to select and refine three alternatives for comparison against the recommendations in the PMP. The alternatives were developed considering the PMP's objective of providing a cost-effective program with diverse outlets for biosolids and included:

- Alternative 1 - Modified Base Case with TPAD (20% thermal drying, 10% solar drying) to improve solids stabilization and increase gas production.
- Alternative 2 – Base Case with a Blending Option (Accelerated on-line date for drying technologies, smaller thermal dryer, and added blending facility to allow dried biosolids to be blended with dewatered biosolids in order to maximize benefits of low disposition costs at Newby Island Landfill until it closes in 2025, while still meeting a minimum dryness (percent total solids).)
- Alternative 3 – TPAD with Future Batch Tanks. (Limit facilities to TPAD and mechanical dewatering while providing future flexibility to achieve Class A biosolids through the addition of batch tanks.)

Figures 6, 7 and 7 show the process configurations for Alternatives 1, 2 and 3 respectively.

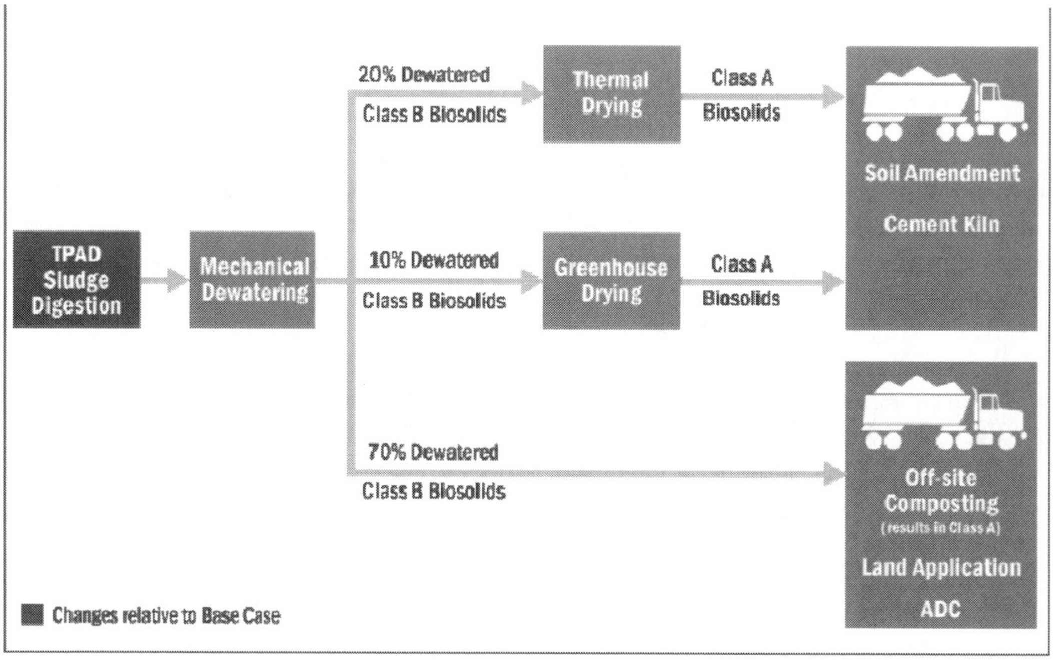

Figure 6. Biosolids Transition Study, Alternative 1: Modified Base Case with TPAD.

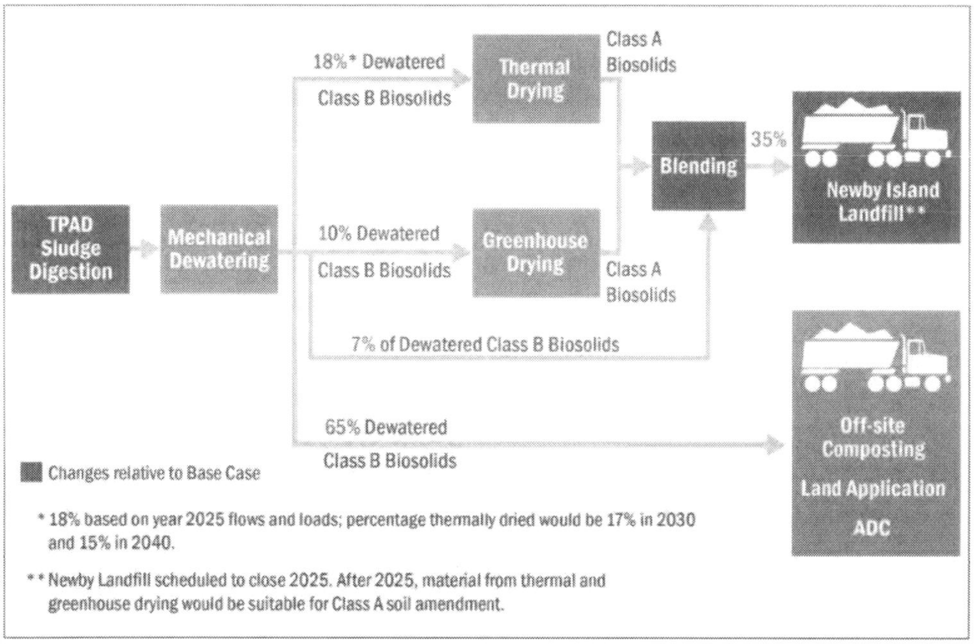

Figure 7. Biosolids Transition Study, Alternative 2: Base Case with a Blending Option.

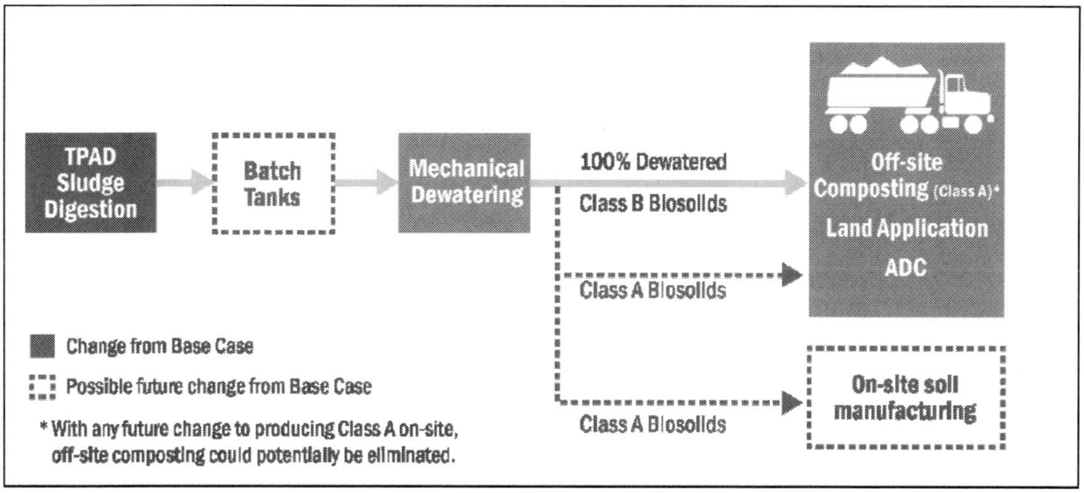

Figure 8. Biosolids Transition Study, Alternative 3: TPAD with Future Batch Tanks.

The alternatives were then reviewed and refined in a series of comprehensive and interactive workshops which involved program management, engineering and O&M staff. Alternatives were evaluated in terms of economic considerations and a Triple Bottom Line Plus (TBL+) evaluation that also considered non-economic factors. Results for each alternative include a TBL+ Performance Score" and a "Value Score." The final business case evaluation results for each alternative are summarized in the tables below.

Alternative 1 (Modified Base Case with TPAD) had TBL+ Performance and Value Scores that were comparable to the Base Case. Present Value Life Cycle Costs (PV LCC) were also equivalent. These results suggested that TPAD is comparable to mesophilic digestion. TPAD is recommended because it also provides a pathway to Class A biosolids via future addition of batch tanks, and because it results in enhanced solids stabilization and biogas production. (TPAD is discussed in the following section.)

Parameter	Base Case: PMP with Mesophilic Digestion	Alternative 1: Modified Base Case with TPAD
PV Life Cycle Cost, $ Millions	520	520
Capital Cost, $ Millions	298	306
O&M Cost, $ Millions	14.5	14.1
TBL+ Performance Score	5.3	5.4
Value Score	0.12	0.12

Alternative 2 (Base Case with a Blending Option) would have higher TBL+ Performance and Value Scores than the Base Case and would result in PV LCC savings. However, any potential savings would be highly schedule-dependent and there was substantial risk that this alternative could not be implemented soon enough to capture all savings.

Parameter	Base Case: PMP with Mesophilic Digestion	Alternative 2: Base Case with a Blending Option
PV Life Cycle Costs, $ Millions	520	490
Capital Costs, $ Millions	298	270
O&M Costs, $ Millions	14.5	14.1
TBL+ Performance Score	5.3	6.3
Value Score	0.12	0.14

Alternative 3 (TPAD with Future Batch Tanks) showed significantly higher TBL+ Performance and Value Scores compared with the Base Case, as well as substantial PV LCC savings ($140 M). Product diversity goals with this alternative will be met through multiple biosolids disposition contracts including off-site composting (Class A product), land application, and landfill disposal or ADC. Choosing Alternative 3 keeps options open for future process additions including Class A batch tanks, soil blending, partial thermal drying, and partial solar drying.

Parameter	Base Case: PMP with Mesophilic Digestion	Alternative 3: TPAD with Flexibility for Future Batch Tanks
PV Life Cycle Cost, $ Millions	520	380
Capital Cost, $ Millions	298	166
O&M Cost, $ Millions	14.5	12.3
TBL + Performance Score	5.3	8.5
Value Score	0.12	0.20

If batch tanks were installed in the future, PV LCC for Alternative 3 would increase by $10 M. If a soil manufacturing facility was also installed, PV LCC would be the same as if only batch tanks were installed. This is because the additional capital costs of soil manufacturing would be off-set by savings in disposition costs and by the revenue generated from the sale of manufactured soil. If manufactured soil was "given away" rather than sold, PV LCC for Alternative 3 would increase by $30 million.

Parameter	Alternative 3 with future Batch Tanks	Alternative 3 with future Soil Manufacturing	Alternative 3 with future Soil Manufacturing but no Revenue from Sale of Soil
PV Life Cycle Cost, $ Millions	390	390	410
Capital Cost, $ Millions	181	209	209
O&M Cost, $ Millions	11.9	10.5	11.7

Based on the comparisons included in the evaluation of alternatives, proceeding with TPAD and dewatering was recommended with use of a variety of disposition contracts to achieve the PMP's diversification goals. Installation of additional on-site processing facilities to achieve Class A biosolids were deferred pending market or regulatory need for Class A biosolids. Multiple disposition contracts need to be developed, negotiated, procured and potentially renewed to meet the PMP multiple end product and contract diversification goals.

TPAD EVALUATION

One of the main recommendations of the Biosolids Transition Study was to transform the Facility's conventional mesophilic digestion system to a temperature-phased anaerobic digestion (TPAD) process with future Class A batch tanks. This digestion process, coupled with mechanical dewatering, gives the Facility the ability to produce Class A biosolids without an energy-intensive or space-intensive post-digestion process such as drying or composting. The change from mesophilic digestion to TPAD was evaluated in-depth in a separate study, summarized below, with the goal of providing an evaluation of the best approach for the digestion process, including rehabilitation of existing digesters. Temperature (thermophilic) phased anaerobic digestion (TPAD) is a process that uses thermophilic anaerobic digestion and mesophilic anaerobic digestion in a series configuration to maximize biosolids stabilization and the production of biogas and that allows for a relatively easy upgrade to a recognized Class A digested biosolids production process. TPAD produces a Class B biosolids product, while TPAD Class A includes batch tanks between the thermophilic and mesophilic phases to produce a Class A biosolids product.

The Facility Plant Master Plan called for rehabilitation of the 16 existing digesters. This work is to occur in three phases, with the initial phase including work in four digesters (later determined to be digesters 5, 6, 7 and 8). The baseline assumption was that the rehabilitated digesters would continue to perform as mesophilic digesters and stabilization would maintain status quo. At the time the PMP was completed, it was anticipated that the existing lagoons and drying beds would continue to be in service for the next two decades. (As noted above, this assumption was subsequently changed drastically.) Instead of upgrading four digesters for parallel mesophilic operation, the TPAD evaluation considered a process in which the four upgraded tanks would serve at the first-stage thermophilic reactors, while the remaining digesters would serve as second-stage mesophilic reactors. This approach offers the following benefits:

- Allows the initial project to deliver a complete/new first-stage digestion process instead of just rehabilitation of one third of the parallel mesophilic process.
- Eliminates concerns about pumping thicker sludge through long lengths of piping to non-rehabilitated digesters.
- Reduces future rehabilitation cost of mesophilic digesters operating in less critical second-stage duty
- Produces biosolids lower in pathogen content with a low-cost, future pathway to 100 percent Class A biosolids as well as a path for multiple beneficial uses of the biosolids material in California.

Digesters 5 through 8 were evaluated for operation as thermophilic digesters in a TPAD configuration. The TPAD process thermophilic phase has a significant potential to increase process capacity. One of the primary benefits of thermophilic digestion is additional loading capacity relative to mesophilic digestion. Based on the planned digester volume, digesters 5 through 8 would be capable of providing sufficient capacity for TPAD operation and allow the ability to take one thermophilic digester out of service under all loading conditions. In addition, TPAD digestion processes offer several other advantages for the Facility:

- Approximately double the capacity of mesophilic digestion, reducing the risk of process upsets
- 5 percentage point increase in volatile solids reduction (VSR)
- 9 percent increase in useful gas production
- 7 percent reduction in biosolids production

- Relatively easy upgrade to TPAD Class A with the addition of Class A batch tanks to produce Class A biosolids
- Reduced complexity of digester feed system as raw thickened solids digester feed is required only for the thermophilic digesters 5 through 8 rather than all digesters 5 through 16
- The ability to handle significant quantities of fats, oils, and grease (FOG) to maximize gas production from the digestion system

The TPAD process requires additional equipment for heat transfer relative to mesophilic digestion to allow the digesters to operate at the higher thermophilic temperatures. Other additional features for TPAD are required to accommodate the higher feed rates and gas production rates. Additional equipment and features required to operate in a TPAD configuration are listed below:

- Digester heating system: One additional heat exchanger and associated circulation pump per digester to allow higher operating temperatures. Modifications to the hot water supply piping system are also required as greater demands for hot water are needed by the thermophilic digester heat exchangers.
- Gas piping: Larger digester gas piping to handle increased gas production rates.
- Emergency overflow system: Larger emergency overflow system required to handle increased rapid volume expansion potential of the thermophilic digesters 5 through 8.
- Digester structure: Insulation to prevent thermal stress cracking of the walls during cold weather periods. Insulation will also reduce heat losses from the digesters, reducing heat demand.
- Electrical requirements: Additional motor control center (MCC) capacity for additional equipment.
- Sludge cooling system: The process requires sludge cooling between the first and second stages, with associated heat exchangers and cooling water pumps. Facility effluent would be used for process cooling and would be routed to these heat exchangers for this purpose.

Figure 9 shows the process flow diagram for TPAD digestion at the Facility.

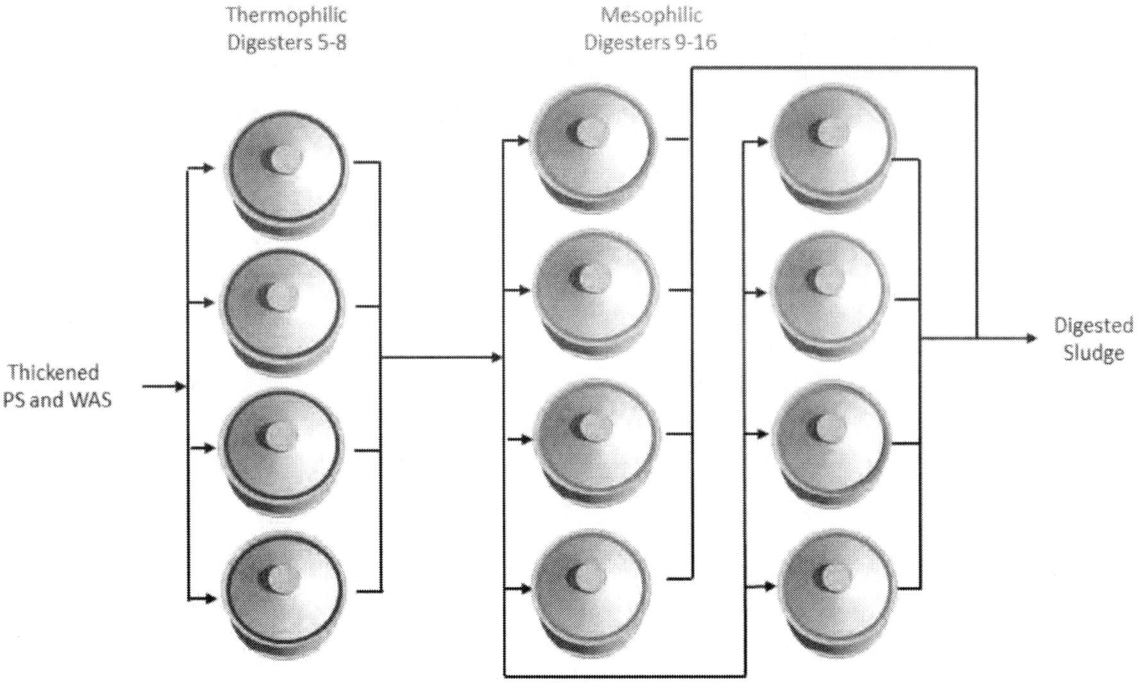

Figure 9. TPAD process flow. Digesters 5-8 serve as first-stage thermophilic reactors. Digesters 9-16 serve as second-stage mesophilic reactors. Not shown: Digesters 1-4, which are no longer required with TPAD operation and are abandoned.

Class A TPAD

TPAD can be operated as a Class B process or a Class A process. A conventional Class B TPAD process, as shown in Figure 9, can be upgraded to a Class A process by including batch hold tanks between the thermophilic and mesophilic phases of the digesters. At 131°F, the required detention time to meet Code of Federal Regulations (CFR) Title 40 Part 503 regulations requirements for Class A is 24 hours. Following the batch tanks, material treated and discharged from the mesophilic second-stage digesters is considered a Class A biosolids with odor products from the thermophilic process mitigated. In the absence of the batch tanks the resulting biosolids from the TPAD system are considered Class B. The Biosolids Transition Study recommended that batch tanks be constructed in the future in response to Class A market drivers. Figure 10 shows the TPAD process flow with Class A batch tanks.

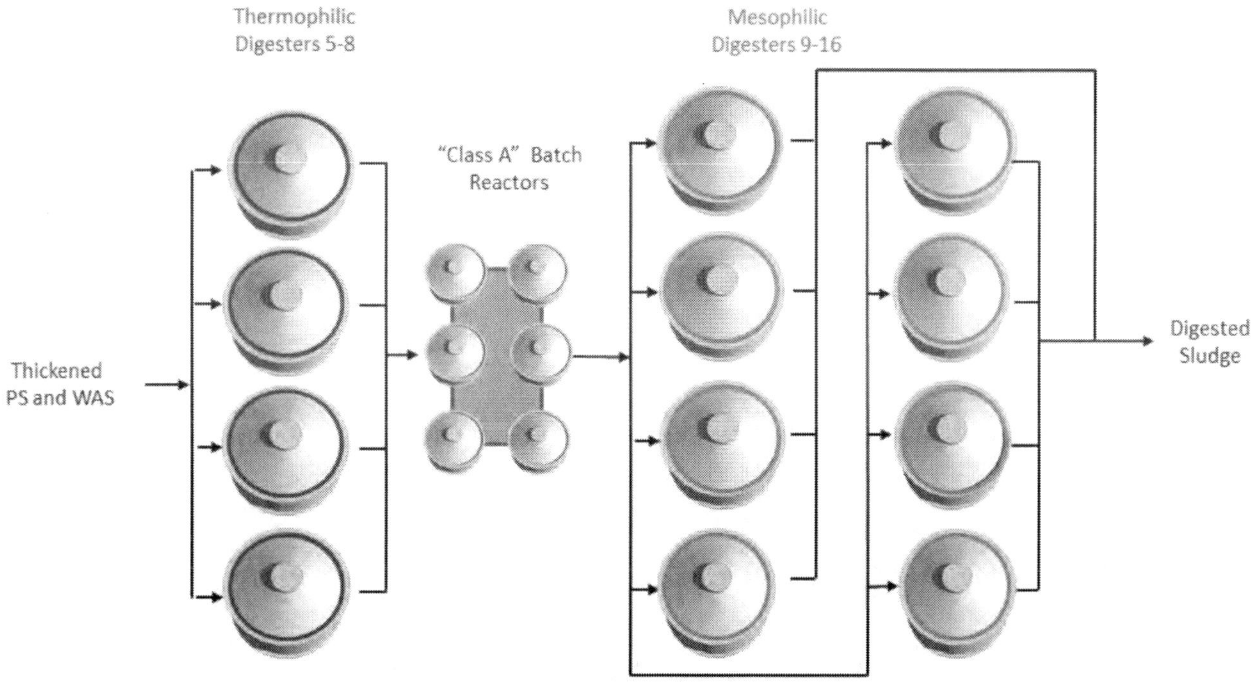

Figure 10. TPAD process flow configured for Class A biosolids – includes batch tanks to meet time-temperature hold requirements.

As part of the TPAD evaluation, project costs were considered to determine capital, annual and net present value costs for each digestion alternative: status quo / PMP baseline recommendation (mesophilic), TPAD, and TPAD Class A.

Capital Costs

Additional equipment, piping, and appurtenances are required to upgrade to TPAD. The incremental capital cost to implement TPAD to the four upgraded digesters 5 through 8 includes larger gas piping, additional pumps, addition of heat exchangers, tank insulation, and electrical allowances for the new equipment and instrumentation. TPAD also requires the added cost of a new heat loop to the thermophilic digesters. The mesophilic digesters 9 through 16 will require provisions for digester cooling from thermophilic to mesophilic temperatures. It is anticipated that the existing heat exchanger system can be repurposed as a cooling heat exchanger system, but piping and an effluent-cooled water heat exchanger loop is required to route cooling water to the mesophilic heat exchangers.

Incremental costs to implement batch tank operation (to achieve Class A operation) would include six batch tanks with insulation, circulation and transfer pumps, grinders, heat exchangers, piping, pressure vacuum relief valve (PVRV) assemblies, a gas sediment trap, and electrical allowances for the new equipment and instrumentation. A capital cost summary of these options is presented in the following table.

Digestion Alternative Capital Cost Comparison	
Alternative	Capital Cost, $M
Mesophilic	58.2
TPAD	65.1
TPAD with batch tanks	76.8

DAFT and digester upgrades and gas pipeline included in all capital costs.

Annual Costs
Annual costs are impacted by the process change to thermophilic digestion in the four upgraded digesters and are described below. Annual costs are estimated assuming a midpoint load between Day 1 and 2030.

- Net power: TPAD and batch tanks will have additional equipment and will require more electrical power than the mesophilic option; however, these alternatives result in greater gas production and thus additional revenue through cogeneration. The unit cost for power assumed is the 2013 rate of $0.108 per kilowatt-hour (kWh). Higher VSR with TPAD will create additional digester gas, which can be used at the Facility cogeneration facility to generate a credit. A cogeneration electrical efficiency of 42 percent is assumed. TPAD will likely increase VSR to a level 5 percentage points higher than mesophilic digestion; gas production will increase with thermophilic operation; and solids production will decrease with thermophilic operation. Gas production rates for TPAD and mesophilic digestion are 16.5 and 15 scf/lb VS destroyed, respectively. To account for the TPAD water vapor volume, only the increase in gas energy production is included. A supplementary 9 percent additional gas production will be attributed to the thermophilic digestion as this rate will be equivalent to the increase in gas energy available from thermophilic digester gas. The digester gas energy is assumed to be 550 Btu/ft3 for mesophilic digestion and 500 Btu/ft3 for TPAD. The additional energy from TPAD reduces the quantity of natural gas that will be needed in the future to meet the Facility's requirement of 12 megawatts (MW).
- Dewatering polymer: Solids from thermophilic digestion have been reported to require greater quantities of dewatering polymer, typical of processes that achieve greater VSR. For this analysis, it is assumed that dewatering mesophilic digested sludge will require 20 lb polymer/dry ton solids, and dewatering of thermophilic digested sludge will require 22 lb polymer/dry ton solids. A 10 percent increase is assumed for TPAD, and the polymer unit cost assumed is $1.70/lb polymer. Overall pounds of solids are lower for TPAD, however.
- Water: Double mechanical seals are assumed to be used for sludge pumping, and each sludge pump was assumed to require a continuous 1 gpm of seal water. There will be additional sludge pumps for TPAD and batch tank operations, which will increase seal water usage. Recycled water may be used and a unit cost of $1.33/100 ft3 will be applied. This cost is taken from the City's Web site (http://www.sanjoseca.gov/index.aspx?nid=2279).
- Natural gas: A cogeneration facility capital improvement project is in design and expected to be in service by 2019. This facility is designed to meet the entire Facility's power demand of

12 MW. For this study, it was assumed that digester gas and natural gas, if necessary, would be supplied to the cogeneration facility to produce 12 MW of power. Therefore, all of the produced digester gas will be used, and natural gas will also need to be purchased to create 12 MW of power. Based on heat balance estimates, mesophilic digestion will require more natural gas than TPAD because the energy production rate is lower. For estimates, a $0.50 per therm unit cost for natural gas is assumed. Approximately 47.8 million Btu per hour (MMBtuh) of natural gas will be required for mesophilic digestion and 43.3 MMBtuh natural gas for TPAD to produce 12 MW at the cogeneration facility.

- Solids hauling and disposal: Fewer solids are produced from increased VSR associated with TPAD operation. Annual hauling and disposition costs are based on an average unit cost of $35/WT. It is also assumed that dewatered cake is 25 percent solids for both mesophilic and TPAD. There is a possibility that TPAD cake will be slightly drier than mesophilic cake and is a result of the increased polymer use and improved dewaterability associated with the higher temperature processing, however, extensive laboratory testing would be required to evaluate improvements in dewatering performance. To be conservative, the same cake percent solids is assumed in this evaluation.
- Heating requirements: Sufficient digester gas will be produced from TPAD operation to meet the heating requirements of the thermophilic digesters. With the cogeneration facility operation at 12 MW, the heat recovered will be in surplus of heat requirements for TPAD, and an external heat source is not required. There will be no additional cost for the heating parameter.
- Operations: Operations cost is assumed to be equal for mesophilic and TPAD. While TPAD will require more labor for the heat exchangers and additional pumps, the feed system is simpler than mesophilic. Adding batch tanks to TPAD Class A will require an additional operator, estimated at $100,000 per year.
- Maintenance: Maintenance costs are associated with the operating equipment and gas utilization. Digester gas will be split between cogeneration and a fuel cell. The Facility has a contracted fuel cell that produces 1.4 MW of energy using digester gas. Different proportions of digester gas will be used by the fuel cell for mesophilic digestion and TPAD due to the difference in VSR and gas production. For mesophilic digestion, 79 percent of digester gas will go to the cogeneration facility and 21 percent will go to the fuel cell. For TPAD, 81 percent of the gas will go to the cogeneration facility and 19 percent will go to the fuel cell. The annual maintenance cost is assumed to be the sum of 1 percent of the capital cost, $0.031/kWh produced in the cogeneration facility, and $0.01/kWh for the fuel cell operations and maintenance. The fuel cell will run 85 percent of the time and utilize digester gas to produce 1.4 MW while the cogeneration facility will run continuously on a combination of digester gas and natural gas. Natural gas will supplement digester gas to produce 12 MW at the cogeneration facility.

A summary of the incremental annual costs required to change the process from mesophilic to TPAD and from mesophilic to TPAD Class A with batch tanks is presented in Table 21.

Annual Costs Comparison			
	Mesophilic	TPAD	TPAD with Batch Tanks
Net power production from cogen (less power consumption from equipment)	($6.75M)	($6.72M)	($6.59M)
Chemicals	$3.73M	$3.76M	$3.76M
Natural gas	$2.32M	$2.10M	$2.10M
Solids hauling and disposal	$4.50M	$4.18M	$4.18M
Operation	$1.0M	$1.0M	$1.1M
Maintenance	$3.58M	$3.65M	$3.77M
Net annual operating costs (unrounded)	$8.38M	$7.97M	$8.32M

The TPAD Class B digestion process is estimated to require the lowest annual operating and maintenance costs. While TPAD Class B requires more equipment, this process generates more power with cogeneration from a 5 percent increase in VSR and digester gas energy production. A greater polymer demand is required for the TPAD processes. With more equipment, greater seal water demand is also associated with TPAD and TPAD Class A. Natural gas demand is required for both alternatives to supplement digester gas in producing 12 MW at the future cogeneration facility; however, less natural gas is needed for the TPAD processes. With increased VSR, there is also a reduction in solids production and associated hauling costs. TPAD Class A will require greater operating labor for the Class A batch tank facility. Similarly, TPAD and TPAD Class A will require more maintenance due to the greater amount of equipment.

Net Present Value

NPV costs were developed on a conceptual level using the capital costs and annual costs. The present-worth evaluation is based on a 30-year period, a 5 percent discount rate, and a 3 percent escalation rate. This evaluation represents the total capital costs and annual operating costs for mesophilic digestion, TPAD, and TPAD with batch tanks.

Digestion Alternative Net Present Value Comparison	
Alternative	NPV Cost
Mesophilic	$246M
TPAD Class B	$244M
TPAD Class A	$263M

While greater capital costs are associated with TPAD relative to mesophilic digestion, TPAD is estimated to have reduced annual costs and thus a slightly lower NPV. TPAD Class A has the highest capital cost, and this option represents the greatest NPV cost.

TPAD Recommendation

TPAD provides the lowest NPV, and while this process had a slightly greater capital cost than mesophilic digestion, the reduced annual costs are significant. The TPAD configuration also provides several other process benefits, including a simplified raw sludge feeding system and greater digestion capacity. Given these benefits and the estimated NPV, the TPAD process is recommended for Class B digestion.

If a driver for Class A biosolids is realized, the addition of Class A batch tanks can be implemented relatively easily to provide Class A biosolids. Given the greater capital and annual costs, Class A digestion should not be implemented until there is a biosolids utilization requirement.

DIGESTER AND THICKENER FACILITIES UPGRADE PROJECT

The Digester and Thickener Facilities Upgrade Project (Project) is the initial phase of a comprehensive upgrade to the Facility's biosolids processing facilities. It follows the recommendations of the Plant Master Plan (upgrading four digesters), but leverages the work of the Biosolids Transition Study and the TPAD Evaluation to transform the digestion process to TPAD and deliver a lower lifecycle cost to the City.

The objective of the Project is to initiate a complete renewal of the Facility's sludge and biosolids processing facilities. In addition to rehabilitating the first four of the existing 16 digesters, the Project will transform the fundamental sludge thickening and anaerobic digestion processes to a more efficient and advanced regime. To achieve the objective of processing more solids and less water in the biosolids process train, six of the existing DAFT units will be rehabilitated and reconfigured to thicken a combined stream of primary sludge (PS) and waste activated sludge (WAS) to a target solids concentration of 5.5 percent. To meet the objective of reducing debris in the thickening, digestion, and dewatering processes, sludge screens will be installed to remove residual debris from the PS, upstream of thickening.

To meet the objective of efficient feed to digesters with smooth, even digester gas production, a thickened sludge distribution system will be provided that equally distributes continuous flow to every first-stage digester. To meet the objective of a digestion process that can accept higher loadings per vessel, provide greater stabilization of volatile matter, produce more digester gas energy, and result in a product with very low pathogenic organism counts, the Project will transform the digestion process to a TPAD process with first-stage thermophilic digesters to be followed by second-stage mesophilic digesters.

To achieve the objective of digester vessels with optimized volume, efficient mixing, and managed removal of scum and floatable matter, the first four rehabilitated digesters will be converted from floating-cover reactors to submerged fixed-cover reactors. To meet the objective of removing hazardous digester gas piping systems from the Facility tunnels, a new outdoor elevated pipe network will be installed through the entire digester campus that will carry uncompressed and compressed digester gas, natural gas, and fuel oil to points of utilization in the

Facility. Figure 11 shows a preliminary rendering of the digester, DAFT and new screening facility upgrade areas.

Figure 11. Digester and Thickener Upgrade Project preliminary rendering.

The vision for the Project is completion of the backbone of a major anaerobic digester process conversion. With completion of the Project, new sludge screening and thickening facilities will evenly distribute thicker sludge to four newly rehabilitated digesters that accept high organic loadings and process sludge at thermophilic temperatures. The completely new first stage of the process will perform the majority of volatile solids stabilization and produce most of the digester gas. New sludge cooling facilities will reduce the temperature of the sludge to mesophilic temperatures and the sludge will be distributed to the existing south campus digesters that will complete the second-stage digestion process. Digester gas will be safely conveyed from digesters to compressors and from compressors to uses in a new outdoor pipe network that will result in a safer campus tunnel system. The remaining digesters can be more simply renovated in subsequent rehabilitation projects that will improve the physical characteristics of the second-stage digesters. The renewed process will provide a better-stabilized product for the planned new digested sludge dewatering and load-out facilities. Additionally, the TPAD process will position the Facility for an easy conversion to full Class A biosolids production with insertion of some simple batch tanks at a future date.

The Digester and Thickener Facility Upgrade project design was completed in 2015 and the project is currently in construction. When this $108 million project is complete (expected in 2019), San José-Santa Clara will have the world's largest TPAD facility.

REFERENCES

Slezak, L.; Chow, G. (2014) *Biosolids Transition Strategy Report, prepared for the City of San José.*

Chapman, T.; Ross, A; Slezak, L. (2014) *Temperature (Thermophilic) Phased Anaerobic Digestion Process Conceptual Development, prepared for the City of San José.*

Sustainable and Cost Effective Biosolids Management Planning at OCSD

Dan Buhrmaster[1], Jim Clark[2], Tom Chapman[3], Natalie Sierra[4], Sharon Yin[5], and Tom Meregillano[5]

[1]Black & Veatch (B&V), Phoenix, Arizona
[2]Black & Veatch (B&V), Los Angeles, California
[3]Brown and Caldwell (BC), Tucson, Arizona
[4]Brown and Caldwell (BC), Andover, Massachusetts
[5]Orange County Sanitation District (OCSD), Fountain Valley, California

ABSTRACT

The Orange County Sanitation District (OCSD) in Southern California recently identified the need for a new Biosolids Master Plan (BMP). The BMP technical evaluations are complete, and a preferred biosolids management alternative has been identified. This paper presents the findings and recommendations associated with the offsite and onsite biosolids management evaluations, alternatives screening, life cycle cost analysis, and triple bottom line selection process that led to the recommended alternative. The lessons learned for this BMP will be useful for utilities that are considering updating their residuals treatment, reuse, and disposal program.

The planning effort began with the biosolids end use in mind to ensure that any product generated, whether onsite or offsite, has an associated reliable, cost-effective, and sustainable end use market. In order to better define available markets for the potential biosolids products and onsite processing options, as well as determine the product's value and marketability, comprehensive market research on the regional horticultural (lawn/garden), agricultural, and fertilizer industries was completed . This included overall industry data evaluations coupled with telephone surveys and a series of in-person meetings with companies in the local targeted outlets. Two sets of evaluation criteria were developed – one market-based set for the evaluation of product-market pairs and a second set for evaluation for products only (and associated processing technologies). A total of 20 final product alternatives and 10 final market options were evaluated using predetermined evaluation criteria and weighting values.

The project team then identified potential onsite processing alternatives for solids digestion and post dewatering that met the preferred product/markets that had been identified. An initial screening of these technologies was performed using the pre-established evaluation criteria noted above. The technical evaluation team then performed concept designs for the preferred onsite alternatives identified. The offsite (end-use) and onsite evaluations were merged to form end-to-end alternatives and 26 alternatives were identified. The final stage in the decision making process consisted of a Solids Water Energy Evaluation Tool (SWEET) life cycle cost evaluation followed by a comprehensive triple bottom line evaluation using Criterium Decision Plus™ (CDP). This paper focuses on the evaluations performed and the recommendations made for OCSD Plant No. 2, one of two wastewater treatment plants (WWTPs) owned and operated by OCSD.

KEY WORDS: Biosolids, anaerobic digestion, thermal drying, composting, land application

INTRODUCTION AND PLANNING BASIS

OCSD has been confronted with many challenges associated with limited biosolids management options imposed by strict local land application regulations in California and with structural integrity issues of its solids handling facilities. Faced with these challenges, OCSD made a decision to develop a BMP that would address these issues. OCSD issued a Request for Proposal in July 2015, the study started in January 2016, and the Draft BMP was completed in March 2017.

OCSD is the third largest wastewater treatment agency west of Mississippi River. Currently OCSD treat nearly 200 million gallons per day (mgd) of wastewater at two treatment plants (Plant No. 1 and Plant No. 2) for approximately 2.5 million people in north and central Orange County. Figure 1 shows the OCSD service area and key data. Figure 2 shows the main treatment processes for both wastewater treatment plants (WWTPs). The solids generated and recovered from primary and secondary treatment processes are stabilized in mesophilic anaerobic digestion systems. The digester gas generated is treated before being fed to Central Power Generation Systems (CenGen) to produce electrical power and heat for plant operations. The biosolids are dewatered and hauled offsite for beneficial use. The current contractor beneficial use sites and biosolids volume allocations are given in Figure 3. Currently, approximately 750 wet tons per day of Class B biosolids are produced from both plants.

The goal of the BMP is to implement sustainable, flexible, and cost-effective biosolids management options, which align with the appropriate biosolids products and markets, level of treatment, and proper solids handling facility improvements. The major work elements of the BMP consist of establishing a planning base, reviewing the current outlook of biosolids management, evaluating the markets for potential biosolids products, evaluating digestion and post-dewatering technology alternatives, and evaluating co-digestion of high-strength organic waste opportunities to develop a comprehensive capital improvement projects portfolio ready for construction.

The OCSD BMP evaluated both Plant No. 1 and Plant No.2. However, due to structural integrity issues with existing anaerobic digesters at Plant No. 2, there is a need to replace these structures over time. This paper reviews the evaluations performed to identify the new preferred digestion and/or post-dewatering facilities that are recommended for Plant No. 2. Since Plant No. 1 anaerobic digesters do not have structural integrity issues that require replacement, no new digestion alternatives were evaluated at that plant. Figure 4 illustrates OCSD's overall BMP planning approach. The focus of this paper is Task 3 – Market Evaluation and Task 4 – Technology Evaluation that led to a recommended new anaerobic digestion alternative at Plant No. 2.

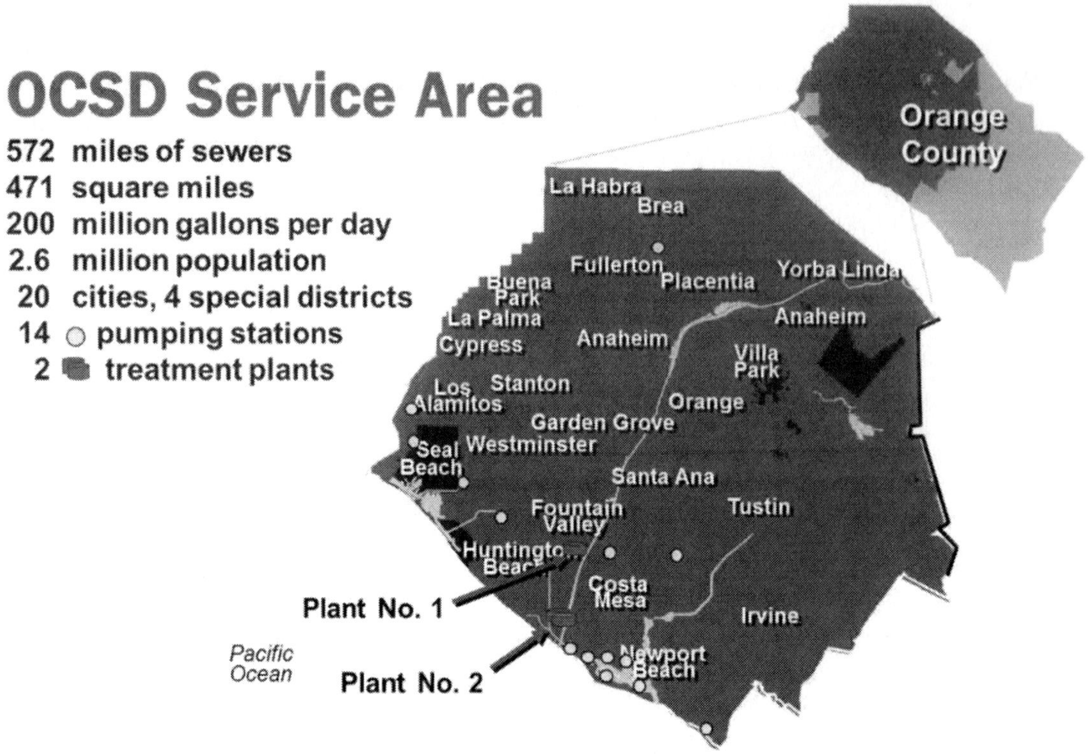

Figure 1. OCSD Service Area.

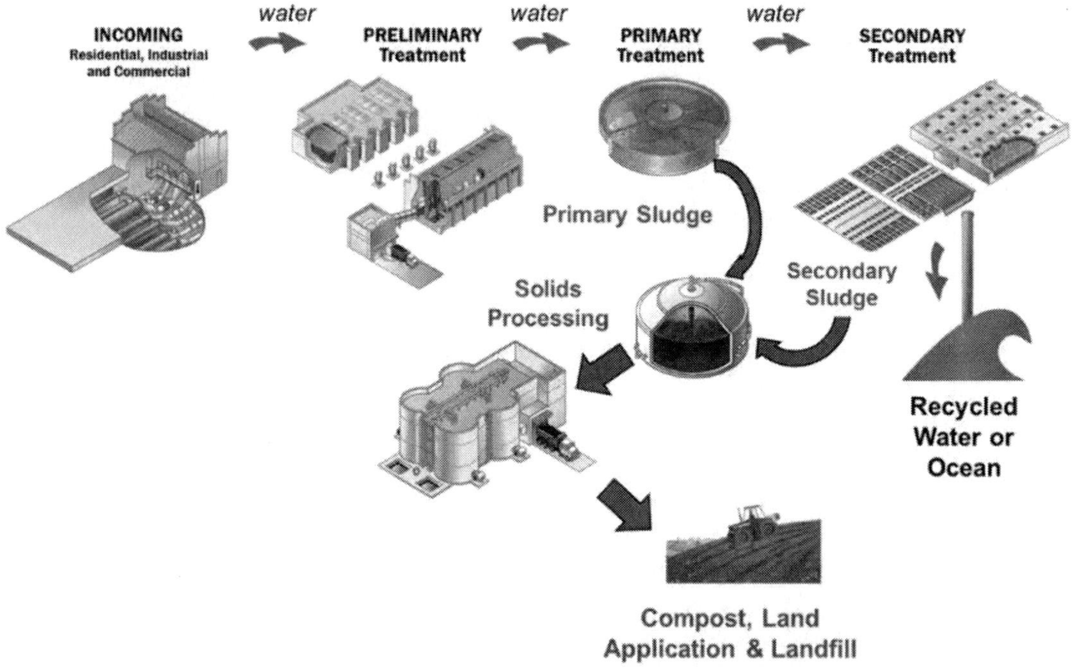

Figure 2. OCSD Wastewater Treatment Process.

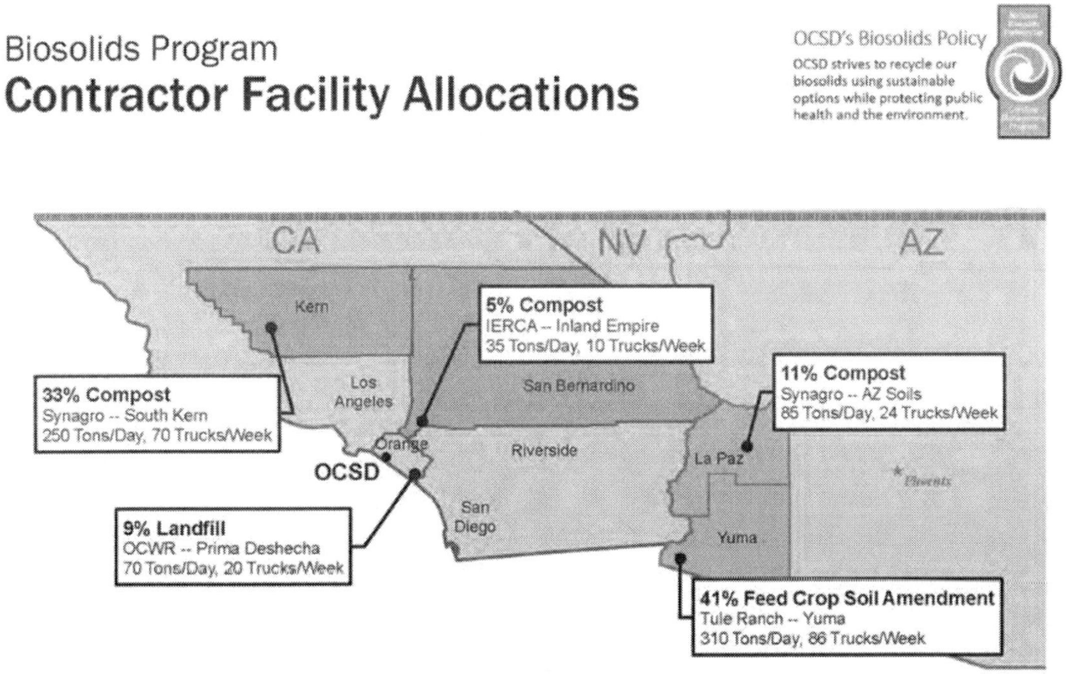

Figure 3. Biosolids Program Contractor Facility Allocations.

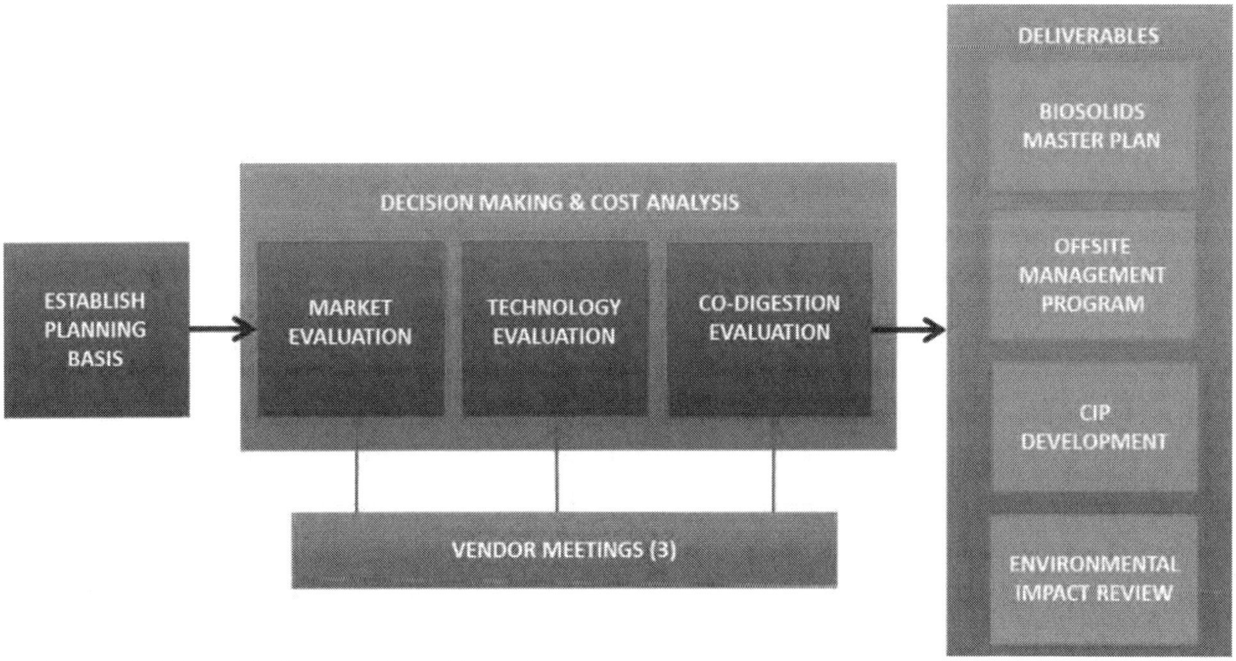

Figure 4. Biosolids Master Plan Overview.

Early in the project, a structured decision making process was established to guide the project team (OCSD, B&V, and BC) through the evaluation of complex and interrelated onsite and offsite alternatives, as illustrated in Figure 5. The offsite (end-use) biosolids management evaluation and onsite technology evaluation affect one another, and so an interactive approach was selected. After parallel evaluations were conducted and alternatives were screened, these separate options were then merged into specific end-to-end alternatives that encompass the entire biosolids management alternative. Non-economic criteria for initial screening and final evaluations were identified up front that reflected OCSD goals, values, and objectives.

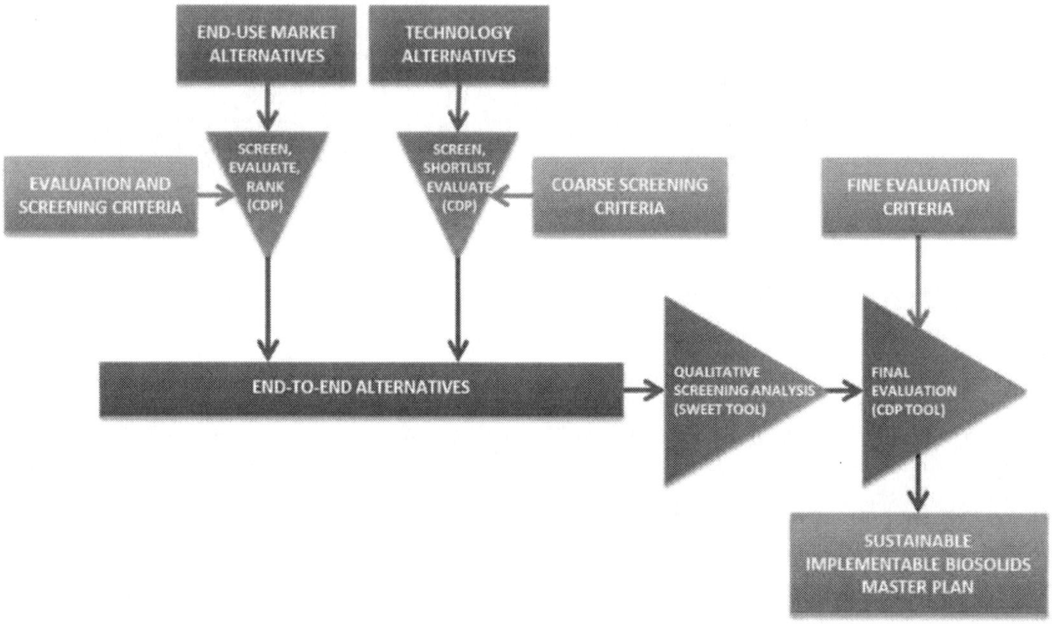

Figure 5. Structured Decision Making Process.

OFFSITE BIOSOLIDS MANAGEMENT ALTERNATIVES EVALUATION (TASK 3)

The work described herein was performed to help select biosolids products with market potential to support OCSD's Biosolids Policy (OCSD 2013). Of particular relevance to this work is OCSD's commitment to a diverse portfolio of offsite management options that support the recycling of biosolids. Thus, the planning efforts under the BMP begin with the biosolids end use in mind to research potential markets for any product generated, whether on or offsite.
Early in the execution of this task, the B&V/BC (consultant) team recommended extending the Task 3 evaluation to biosolids products in addition to end use markets. In doing so, onsite management alternatives, such as thermal drying, were included in the analysis. Products and markets were screened using evaluation criteria tailored to OCSD's goals. The screened products and markets were paired with onsite processing technologies under Task 4.

The analysis considered base products generated at OCSD's facilities (biosolids cake products), as well as further processing steps used to generate new products. The range of products was then paired with feasible markets, which were evaluated individually. Products and markets considered are presented in Table 1.

Table 1. End Use Product and Market Alternatives.

Base Product	Additional Processing	Final Product	Final Market(s)									
			Landfill Disposal	Bulk Agriculture	Bulk Horticulture/ Landscaping	In-District Projects	Soil Blending	Golf Courses and Other Specialty Customers	Distribution as Bagged Product	Land Reclamation	Fertilizer Blending	Direct Use in Energy[1]
Raw, Dewatered Cake	*None*	Raw, Dewatered Cake	Realistic									
	Thermal Drying followed by SCWO	Steam and Recoverable/ Marketable CO_2										Possible[2]
	FBI	Inorganic Ash	Realistic									
Sub-Class B Cake[7]	*None*	Sub-Class B Cake	Realistic									
	Offsite Composting	Class A Compost		Realistic	Realistic	Realistic	Realistic	Realistic	Realistic	Possible[2]		
	Lime Stabilization	Lime Stabilized Class B Cake		Realistic								
Class B Cake	*None*	Class B Dewatered Cake		Realistic								
	Offsite Composting	Class A Compost		Realistic	Realistic	Realistic	Realistic	Realistic	Realistic	Possible[2]		
	Partial Drying[4]	Class B Partially Dried Product		Realistic								Possible[2]
	On or Offsite Thermal Drying	Class A Dried Granule (low quality)		Realistic	Possible[3]	Possible[3]	Possible[2]			Possible[2]		Realistic
	On or Offsite Thermal Drying	Class A Dried Granule (high quality)		Realistic	Realistic	Realistic	Possible[2]	Realistic	Realistic	Possible[2]	Realistic	Realistic
	Thermal Drying followed by Pyrolysis	Biochar[5], Synthetic Fuel or Gas		Emerging (Biochar)	Emerging (Biochar)	Emerging (Biochar)	Emerging (Biochar)	Emerging (Biochar)	Emerging (Biochar)	Emerging (Biochar)	Emerging (Biochar)	Emerging (Syngas or Synfuel)
Class B Cake	Thermal Drying followed by Gasification	Inorganic Ash and Syngas										Emerging
	Lystek Process	Class A Lystegro		Realistic	Emerging		Emerging			Possible[2]		

Table 1. End Use Product and Market Alternatives. (cont.)

Base Product	Additional Processing	Final Product	Final Market(s)									
			Landfill Disposal	Bulk Agriculture	Bulk Horticulture/ Landscaping	In-District Projects	Soil Blending	Golf Courses and Other Specialty Customers	Distribution as Bagged Product	Land Reclamation	Fertilizer Blending	Direct Use in Energy[1]
Class A Cake	*None*	Class A Dewatered Cake		Realistic			Possible[3]			Possible[2]		
Class A Cake	Partial Drying[4]	Class A Partially Dried Product		Realistic	Possible[3]	Possible[3]	Possible[3]			Possible[2]		Possible[2]
Class A Cake	Offsite Soil Blending	Class A Soil Blend			Possible[3]	Possible[3]	Possible[3]	Possible[3]	Possible[3]			
Class A THP Cake[6]	*None*	Class A THP Dewatered Cake		Realistic			Realistic			Possible[2]		
Class A THP Cake	Partial Drying	Class A THP Partially Dried Product		Realistic	Possible[5]	Possible[5]	Possible[5]			Possible[2]		Possible[2]
Class A THP Cake	Offsite Soil Blending	Class A THP Soil Blend			Realistic	Realistic	Realistic	Possible[3]	Possible[3]			

Notes:

1. Direct use may include cement kilns, biomass plants, etc. as identified through market research.
2. Economics of option need to be evaluated and may create challenges toward implementation.
3. Further market research needed to determine whether product quality is compatible with application.
4. Partial drying to reduce volume for hauling; estimated TS content of 40–50 percent.
5. Further market research required to assess status of biosolids-based product.
6. THP cake has been shown to have less odor, better dewaterability, and less potential for regrowth when compared to traditional Class A cake.
7. Sub-Class B cake results when the requirements for Class B quality biosolids are not met.

Current Biosolids Management Alternatives and Regulatory Environment
Analyses regarding nutrient content were conducted by OCSD to support the market research conducted under this task. OCSD's biosolids quality supports beneficial use. As a result, approximately 50 percent of OCSD's biosolids are land-applied to farm land in Arizona, with much of the remaining tonnage sent to three regional compost facilities to produce Class A compost. This beneficial use is reflective of the region at large, where significant capacity exists in both biosolids processing (primarily composting, but also thermal drying) and land application (largely in western Arizona). Although it is difficult to assess the total excess capacity for land application, the consultant team estimated excess biosolids management capacity at about 1.77 million wet tons per year. This is approximately 368,100 dry tons per year based on 20 percent total solids cake product for contract composting within about 100 miles of OCSD. This total does not include some larger facilities, such as the Los Angeles County Sanitation District's Tulare Lake Compost facility, for which the agency - for market competition or other reasons - was not willing to state available capacity. Additional capacity is available for other products and markets.

California's regulatory structure can create challenges for biosolids beneficial use. In particular, a number of county ordinances ban or restrict the use of Class B and, in some cases, Class A biosolids. The California Association of Sanitation Agencies (CASA) is optimistic, however, that Class A land application will be permitted in most counties in the near future. In addition, western Arizona has ample acreage to manage California biosolids. A number of large agencies send at least a portion of their biosolids to farms in western Arizona. Pending bans on organics in landfills are another significant regulatory trend in California, potentially eliminating the use of biosolids and other organics as alternative daily cover (ADC) in the near term and resulting in a ban on biosolids and other organics disposal in landfills in the long term. Of significance to the marketplace, diversion of organics from landfills into alternative markets such as compost, creates more competition for composting capacity and biosolids-based products. Because few agencies in Southern California employ landfill ADC as an option, the possibility of competition in the compost market is a more salient market driver with respect to organics.

Comprehensive Market Evaluation
OCSD wished to better define available markets for the potential biosolids products and processing options, as well as determine the products' value and marketability. Therefore, market research on the regional horticultural (i.e., lawn/garden), agricultural, and fertilizer industries was completed. Demographic data indicate that significant regional capacity exists within these markets. A review of emerging markets for environmentally beneficial applications (e.g., dewatered cake for land reclamation, compost for erosion control, etc.) was also conducted, although these markets were considered to be less developed with respect to biosolids products. Market surveying was initially completed via telephone by staff experienced in data collection and/or biosolids product sales. Then, to improve data collection and long-term market development efforts, several companies were surveyed in person (face to face). During the process, 80 entities and individuals were surveyed including 42 landscape supply/resellers/topsoil companies (with 21 face-to-face meetings), 21 fertilizer companies (with 4 face-to-face meetings), 16 composters (with 11 face-to-face meetings), and the local solid waste entity (Orange County Waste and Recycling).

Generally speaking, experience with biosolids products in the agricultural and horticultural markets is strong in the region, which encompasses Southern California and the westernmost part of Arizona. Agriculture can absorb large volumes of fertilizer-type products, whether thermally dried biosolids or compost. Use of biosolids in agriculture is expected to continue. Interest in use of biosolids for land reclamation can further expand available acreage. Marketing and outreach by either OCSD or its contractors will help support regional use of biosolids products.

Ranking of Biosolids Management Alternatives
A comprehensive evaluation of both products and markets was performed under this task. Two sets of evaluation criteria were developed—one market-based set for evaluation of the product/market pairs and a second set for evaluation of the products and associated processing technologies. Each criterion was assigned a weighted value by the project team. The consultant team identified the product/market pairs based on the market research conducted and the project team then evaluated the products and their associated processes. The resulting scores were presented and refined in a workshop format. End-use markets were scored in accordance with the following criteria developed by the project team:

- Realistic, proven market
- Market size
- Proven value
- Future market capacity
- Resiliency to regulatory change
- Year-round dependability

The project team jointly developed triple-bottom-line criteria to help assess potential biosolids products. The criteria were:

- Minimize life-cycle management costs
- Provide broad marketability
- Provide proven, safe, and reliable technology
- Support OCSD's biosolids management goals, policy, and operations
- Provide regulatory resilience
- Minimize net carbon footprint
- Minimize impacts from negative side streams and emissions
- Enhance community relationships

Products and markets were assigned a score for each criterion from 1.0 to 5.0. Products receiving a score below 3.0 were eliminated from further consideration. Products receiving a score of 3.0 or higher were paired with their most promising markets (i.e., those receiving a score of 3.5 or higher). It should be noted that other markets could be developed with appropriate marketing, education, and outreach; however, for the purposes of further analysis, only the most likely markets were selected.

Preferred Management Alternatives
This market-based evaluation resulted in recommendations for products and markets to be carried forward into Task 4 – Technology Evaluation. In addition, these products relate back to

on and offsite technologies that were identified and screened in Task 4. For most products, several different post-dewatering technologies can achieve the appropriate product quality. Table 2 summarizes the general process flow necessary to generate the products screened in this analysis, including whether a post-dewatering process is assumed to occur on or offsite. Because of the space limitations at OCSD WWTPs, only drying processes can be located on site. Partial drying was assumed to occur on site, because the primary reason for implementing such an option is to save on truck hauling costs. Composting and soil blending were assumed to take place offsite.

Table 2. Task 3 Processes, Products, and Markets.

Stabilization Process	Post-Dewatering Process	Product	Highest-Scoring Markets
Class B Digestion	None	Class B Cake	Bulk Agriculture, Arizona
Class B Digestion	Partial Drying (On Site)	Class B Partially Dried Product	Bulk Agriculture, Arizona
Class B Digestion	Class A Composting (Offsite)	Class A Compost	Soil Blending
			Bulk Agriculture, California
			Bulk Agriculture, Arizona
			Bulk Horticulture/Landscaping
			Distribution as Bagged Product
			Golf Course and Other Specialty
Class B Digestion	Thermal Drying (On or Offsite)	Class A High-Quality Granule	Bulk Agriculture, California
			Bulk Agriculture, Arizona
			Bulk Horticulture/Landscaping
			Distribution as a Bagged Product
			Fertilizer Blending
Class A Digestion	None	Class A Cake	Bulk Agriculture, California
			Bulk Agriculture, Arizona
			Land Reclamation
Class A Digestion	Partial Drying (On Site)	Class A Partially Dried Product	Bulk Agriculture, California
			Bulk Agriculture, Arizona
Class A Digestion	Soil Blending (Offsite)	Class A Soil Blend	Bulk Horticulture/Landscaping
Class A Thermal Hydrolysis Process (THP)	None	Class A THP Cake	Bulk Agriculture, California
			Soil Blending
			Land Reclamation

Table 2. Task 3 Processes, Products, and Markets. (cont.)

Stabilization Process	Post-Dewatering Process	Product	Highest-Scoring Markets
Class A THP	Partial Drying (On Site)	Class A THP Partially Dried Product	Bulk Agriculture, California
			Bulk Agriculture, Arizona
			Soil Blending
			Bulk Horticulture/Landscaping
			Land Reclamation
Class A THP	Soil Blending (Offsite)	Class A THP Soil Blend	Bulk Horticulture/Landscaping

The alternatives recommended were then carried forward into Task 4, to be paired with specific thickening and stabilization technologies for the creation of end-to-end alternatives.

PLANT NO. 2 DIGESTION AND POST-DEWATERING EVALUATIONS (TASK 4)

Identification of Biosolids Process Technologies

Table 3 shows these processing technologies which were presented in an initial Task 4 workshop. This list was used as the starting point for identifying specific technologies that could be implemented at Plant No. 2.

Table 3. Processes to Produce End-Use Product Alternatives.

Task 3 Product Alternatives	Task 4 Process Technology Options
Class A Compost (*current product*)	Class A or B digestion + composting
Class B Cake (*current product*)	Class B digestion
Class A Cake	Class A digestion
Class A THP Soil Blend	THP/digestion + soil blending
Class A THP Cake	THP/digestion
Class A Soil Blend	Class A digestion + soil blending
Class A Dried Granule	Class A or B digestion + drying
Partially Dried Class B Cake	Class B digestion + partial drying
Partially Dried Class A Cake	Class A digestion + partial drying
Partially Dried Class A THP Cake	THP/digestion + partial drying

Specific thickening, digestion, and post-dewatering technologies identified are summarized in Tables 4 through 6. Biosolids to energy technologies identified as emerging in Task 3 included pyrolysis, gasification, and supercritical water oxidation (SCWO). Pyrolysis and gasification were discussed with OCSD, but were considered too emergent to justify current capital improvements for this BMP. Recent interest in SCWO by both OCSD (as a demonstration project) and the Bay Area Biosolids to Energy (BAB2E) coalition warrants further study of the process, but this will be conducted in parallel with the capital planning process. Lime stabilization and incineration were also briefly considered, despite the fact that they do not generate any of the products selected for inclusion under Task 3.

Table 4. Thickening Technology Alternatives.

Thickening Technologies
Existing Dissolved Air Flotation Thickening (DAFT) of Waste Activated Sludge (WAS)
Existing Primary Clarifier Thickening of Primary Sludge (PS)
GBT (WAS)
Rotating Drum Thickener - RDT (WAS and/or PS)
Gravity Thickening (PS)
Centrifuge Thickening (WAS and/or PS)
Membrane Thickening (WAS)

Table 5. Digestion Technology Alternatives.

Digestion Technologies	
Class B	Class A
• Mesophilic	• Acid/Gas
• Staged Digestion	• Thermophilic
• Acid/Gas	• TPAD
• Thermophilic	• THP
• Temperature Phased Anaerobic Digestion (TPAD)	• Enzymatic Hydrolysis
• THP	• Three Phase
• Enzymatic Hydrolysis	• Aerobic
• Mesophilic High Solids	• Intermediate THP
• Three Phase	
• Aerobic	

Table 6. Post-Dewatering Technology Alternatives.

Post-Dewatering Technologies
Thermal drying
Solar drying
Partial drying
Deep well injection
Incineration
Gasification
Pyrolysis
Supercritical Water Oxidation

Initial Screening of Onsite Treatment Technology Alternatives
Given the list of potential onsite solids treatment technologies, the project team conducted a screening process to short list the preferred technologies for further evaluation. A list of nine preliminary screening criteria was developed for this purpose, as follows:

- End use market compatibility
- Proven technology performance
- Energy/resource recovery
- Operation and Maintenance (O&M) impacts
- Safety impacts
- Environmental impacts
- Community impacts
- Flexibility for project phasing or future process changes
- Project site compatibility

The applicable evaluation criteria were applied to each technology in the following categories: (1) thickening alternatives, (2) digestion alternatives, and (3) post-dewatering alternatives, in three separate evaluations. Each technology was given a score of 1-5 (higher number is more favorable) for each evaluation criteria. End use compatibility represented a pass-fail criteria for the alternatives. If the technology was not compatible with the results of the Task 3 – End Use Evaluation, it was eliminated from further consideration and scoring. For example, it was determined in Task 3 that lime stabilization did not have a sufficient end use market. Also, incineration, gasification, pyrolysis, and deep well injection were eliminated due to the lack of a marketplace or due being too emergent for further consideration. Each criterion was given a weighting factor based on its significance to achieving OCSD's overall biosolids management goals. Scores for each technologies were multiplied by the associated weighting factor, and the products added together to provide an overall score for each technology. High scores represented technologies that were perceived to more closely achieve OCSD's overall biosolids management goals. The scores for these evaluations are given in Tables 7, 8, and 9. This led to the selection of technologies for further detailed evaluation as shown in Table 10.

Table 7. Thickening Technology Scoring and Weighting Results.

Criteria No.	Evaluation Category	Criterion	Weight %	Score						
				Gravity Thickener (PS)	GBT (PS and WAS)	RDT (PS and WAS)	Centrifuge (PS and WAS)	DAFT	Membrane Thickener	PC Thickening
1	Digestion Process Compatibility			YES	YES	YES	YES	YES	YES	YES
2	Proven Technology Performance	Ability to demonstrate proven performance and reliability	30	5	5	4	5	5	3	5
3	Energy/Resource Recovery	Recovery of renewable energy and resources	10	5	4	4	2	2	2	5
4	O&M Impacts	Complexity impacts to plant O&M	15	2	4	5	5	5	2	5
5	Safety Impacts	Safety concerns of technology	15	3	3	3	3	3	3	4
6	Odor Control	Requirements for odor control	10	1	1	3	3	3	2	3
7	Project Site Compatibility	Compatibility of technology with existing Plant No. 2 site	20	1	3	3	5	5	2	5
			100	**3.1**	**3.7**	**3.7**	**4.2**	**4.2**	**2.5**	**4.7**

Table 8. Digestion Technology Scoring and Weighting Results.

Criteria No.	Evaluation Category	Criterion	Weighting %	Score												
				Single Stage Mesophilic Digestion	Staged Digestion	Acid/Gas Digestion	Thermophilic Digestion	TPAD	THP	Enzymatic Hydrolysis	Recuperative Thickening/ High Solids Mixing	Three Phase	Lime Stabilization	Aerobic Digestion	Intermediate THP	Chemical Hydrolysis
1	End Use Market Compatibility			YES	YES	YES	YES	YES	YES	YES	YES	YES	NO	YES	YES	YES
2	Proven Technology Performance	Ability to demonstrate proven performance and reliability	25	5	5	4	4	4	4	2	2	2		2	1	1
3	Energy/ Resource Recovery	Recovery of renewable energy and resources	10	3	4	4	4	4	4	4	3	4		1	4	4
4	O&M Impacts	Complexity impacts to plant O&M	5	5	3	3	3	3	2	2	3	3		5	1	3
5	Safety Impacts	Safety concerns of technology	15	4	4	4	4	4	2	4	4	4		5	2	3
6	Environmental Impacts	Impact of technology to environment	5	3	4	4	4	4	4	4	3	4		1	4	4
7	Community Impacts	Enhance community relationships	15	3	3	2	3	3	4	4	4	2		1	4	3
8	Flexibility	Flexibility of technology for phasing or process changes	10	4	4	3	5	4	2	2	4	3		2	2	2
9	Project Site Compatibility	Compatibility of technology with existing Plant No. 2 site	15	3	2	2	4	4	4	3	3	2		1	3	3
			100	3.9	3.8	3.3	3.9	3.8	3.4	3.1	3.2	2.8	0.0	2.2	2.5	2.6

Table 9. Post-Dewatering Technology Scoring and Weighting Results.

Criteria No.	Evaluation Category	Criterion	Weighting %	Score						
				Inciner-ation	Thermal-Drying	Gasification	Pyrolysis	Solar Drying	Partial Drying (40% cake)	Deep Well Injection
1	End Use Market Compatibility			NO	YES	NO	NO	YES	YES	NO
2	Proven Technology Performance	Ability to demonstrate proven performance and reliability	25		5			3	4	
3	Energy/Resource Recovery	Recovery of renewable energy and resources	10		3			5	3	
4	O&M Impacts	Complexity impacts to plant O&M	5		4			3	2	
5	Safety Impacts	Safety concerns of technology	15		3			3	3	
6	Environmental Impacts	Impact of technology to environment	5		2			4	3	
7	Community Impacts	Enhance community relationships	15		3			5	4	
8	Flexibility	Flexibility of technology for phasing or process changes	10		3			1	3	
9	Project Site Compatibility	Compatibility of technology with existing Plant No. 2 site	15		3			1	4	
			100	**0.0**	**3.5**	**0.0**	**0.0**	**3.1**	**3.5**	**0.0**

Table 10. Preferred Plant No. 2 Onsite Technology Alternatives.

Thickening	Stabilization	Post Dewatering
Existing DAFT	Conventional Mesophilic Digestion (CMAD)	Rotary Drum Dryer – High Quality Granule
Existing Primary Clarifiers	Staged Mesophilic Digestion (SMAD)	Partial Drying with Paddle Dryer – Lower Quality Granule
New Centrifuge Thickening	Thermophilic Anaerobic Digestion (TAD)	
	TPAD	
	THP	

Creation of End-to-End Alternatives for Evaluation

Onsite technology alternatives were then paired with their appropriate products and markets, as determined under Task 3. These end-to-end alternatives are presented in Table 11. In general, end-to-end alternatives were developed with OCSD's guiding principles in mind, with an emphasis on diversification in particular. Class B biosolids from Table 11 only have two available outlets – Arizona land application and regional composting, although several individual contractors are available for each of these. For some alternatives, an onsite soil blending pilot was envisioned in order to support larger marketing efforts in the region. In general, the size and interest of the market for a given product, as assessed in Task 3, informed the makeup of the end use program included in the alternative.

Table 11. End-to-End Alternatives for Evaluation.

Thickening[1]	Digestion	Onsite Post-Dewatering Process	Resulting Biosolids Product(s)	End Use						
				Land Application, Arizona	Land Application, California	Bulk Horticulture/ Landscaping	Soil Blending	Fertilizer Blending	Class A Regional Compost[2]	Off-Site Thermal Drying
PC/DAFT	Mesophilic	None	Class B Cake	50%					50%	
		100% Drum Drying	Class A Dried Granule[3]	5%	5%	30%		60%		
		75% Drum Drying	Class B Cake and Class A Dried Granule[3]	4%[4]	4%[4]	22%[4]		45%[4]	25%[5]	
		50% Drum Drying	Class B Cake and Class A Dried Granule[3]	28%	2%				25%[5]	
		25% Drum Drying	Class B Cake and Class A Dried Granule[3]	50%[4]	2%[4]	8%[4]		15%[4]	25%[5]	
PC/DAFT	Mesophilic	Partial Drying[6]	Partially Dried Class B Cake	100%						
PC/DAFT	Mesophilic	None	Class B Cake	50%					25%	25%
Centrifuge	Mesophilic	None	Class B Cake	50%					25%	25%
Centrifuge	Mesophilic	Partial Drying[6]	Class B Cake	50%					50%	
Centrifuge	Mesophilic	100% Drum Drying	Class A Dried Granule[3]	5%	5%	30%		60%		
PC/DAFT	Series Mesophilic	None	Class B Cake	50%					50%	
Centrifuge	Series Mesophilic	None	Class B Cake	50%					50%	

Table 11. End-to-End Alternatives for Evaluation. (cont.)

Thickening[1]	Digestion	Onsite Post-Dewatering Process	Resulting Biosolids Product(s)	End Use						
				Land Application, Arizona	Land Application, California	Bulk Horticulture/ Landscaping	Soil Blending	Fertilizer Blending	Class A Regional Compost[2]	Off-Site Thermal Drying
PC/DAFT	Class B Thermophilic	None	Class B Cake	50%					50%	
Centrifuge	Class B Thermophilic	None	Class B Cake	50%					50%	
PC/DAFT	Class A Thermophilic	None	Class A Cake	50%	50%					
		Pilot Onsite Soil Blending[7]	Class A Soil Blend, Class A Cake							

Notes:
1. Thickening abbreviations as follows: PC – primary clarifier thickening, DAFT – dissolved air flotation thickening.
2. Class A regional compost encompasses any number of regional composting facilities and their associated final end-use markets.
3. High quality dried granule, as assessed in Task 3.
4. High quality dried granules in this end use only.
5. Class B cake in this end use only.
6. Partial drying indicates 25 percent of digestate is dried on a paddle dryer and blended with 75 percent dewatered cake for purposes of volume reduction.
7. 5 percent of cake production to small onsite pilot. Additional cake production to off-site soil blending.

SWEET Evaluation
The SWEET model, a proprietary software tool by BC, was used in this investigation to determine the net present value (NPV) of each end-to-end alternative. This economic screening step allows for a high level assessment of total alternative costs, with the expectation that higher cost alternatives would not represent the most preferred alternative for OCSD. Following this initial economic screening step, the CDP model assessed the remaining end-to-end alternatives on selected non-economic criteria.

Capital Costs
Capital costs for each end-to-end alternative were developed based on the summation of cost estimates for each associated facility. Table 12 provides a total construction capital cost for each end-to-end alternative.

Table 12. End-to-End Alternative Construction Capital Cost.

Alternative	Description[1]	Capital Cost	Alternative	Description	Capital Cost
ALTS 1 & 7	ET-CMAD-ND	\$426 M	ALT 14	CT-TAD-ND (CLASS B)	\$473 M
ALT 2	ET-CMAD-TD (100%)	\$426 M	ALT 15	ET-TAD-PD (CLASS A)	\$492 M
ALT 3	ET-CMAD-TD (75%)	\$487 M	ALT 16 & 17	ET-TAD-ND (CLASS A)	\$448 M
ALT 4	ET-CMAD-TD (50%)	\$479 M	ALT 18	CT-TAD-ND (CLASS A)	\$494 M
ALT 5	ET-CMAD-TD (25%)	\$468 M	ALT 19	ET-THP-PD	\$492 M
ALT 6	CT-CMAD-TD	\$480 M	ALTS 20 & 21	ET-THP-ND	\$448 M
ALTS 8 & 10	CT-CMAD-ND	\$480 M	ALT 22	ET-TPAD-ND (CLASS B)	\$465 M
ALT 9	ET-CMAD-PD	\$470 M	ALT 23	CT-TPAD-ND (CLASS B)	\$513 M
ALT 11	ET-SMAD-ND	\$456 M	ALTS 24 & 26	ET-TPAD-ND (CLASS A)	\$486 M
ALT 12	CT-SMAD-ND	\$476 M	ALT 25	CT-TPAD-ND (CLASS A)	\$535 M
ALT 13	ET-TAD-ND (CLASS B)	\$427 M			

Note:
1. ET: existing technology (DAFT), CT: centrifuge thickening, CMAD: conventional mesophilic digestion, SMAD: series mesophilic digestion, TAD: thermophilic anaerobic digestion, TPAD: temperature phased anaerobic digestion, THP: thermal hydrolysis process, ND: no drying, PD – partial drying, TD: thermal drying.

Capital costs, operating costs, and annual benefits were imported into the SWEET model to assess the 20-year net present value (NPV) of each end-to-end alternative. The net present value shown in Figure 6 is based on the following assumptions:

- Escalation rate = 3.5 percent
- Discount rate = 4 percent
- Midpoint of construction = 2023
- Life cycle duration = 2025-2045 (20 years)

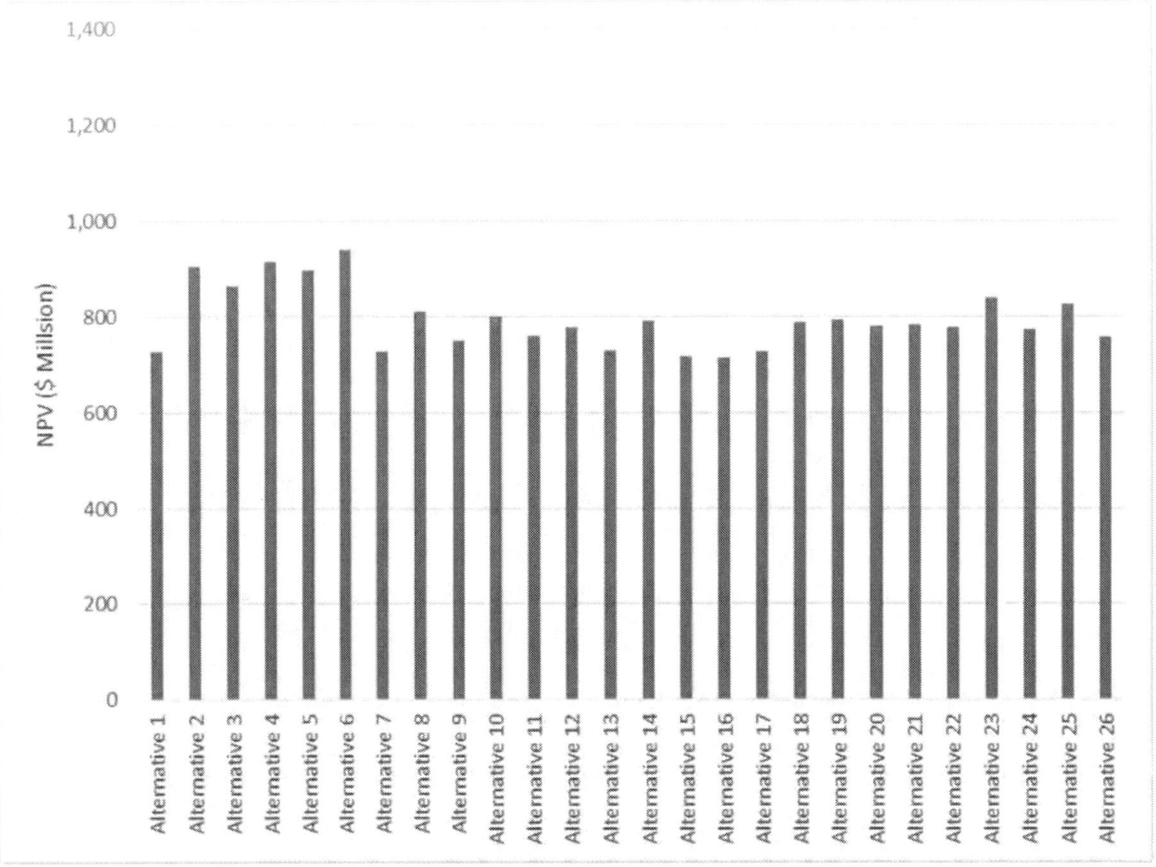

Figure 6. End-to-End Alternative Net Present Value from SWEET.

Based on the SWEET NPV results (Figure 6), the TAD end-to-end alternative using the existing thickening and Class A end use alternatives represented the lowest NPV. However, the majority of the results from the alternatives considered in this study were reasonably similar for this conceptual level of analysis. For example, Class B mesophilic digestion with the existing thickening and Class A thermophilic digestion with partial drying had almost identical NPV results. The NPV results for most alternatives presented in Figure 6 (based on a Class 5 engineering estimate) were within a 15 percent of the average value. It was noted that thermal drying alternatives generally had a higher NPV than no drying or partial drying alternatives. Therefore, thermal drying alternatives would need to score relatively high on non-economic criteria in order to be recommended for implementation at Plant No. 2.

Through an isolation of specific alternatives for comparative analysis, the following conclusions were developed.

- Centrifuge thickening did not reduce operating costs sufficiently to pay for the capital investment of the thickening facility. Therefore, centrifuge thickening alternatives were not considered further in the investigation.
- Thermal drying alternatives were considerably higher in cost than other process alternatives, but thermally dried products provide unique end use market opportunities not available with other technologies. For this reason, the thermal drying alternative was considered further in the CDP evaluation.
- Partial drying produced a product that had the potential for biosolids management cost savings that resulted in a NPV that was comparable with other alternatives. Therefore, partial drying was considered in more detail in the CDP evaluation.
- Class A and B alternatives have comparable NPV results indicating that the cost savings associated with managing a Class A biosolids product was able to offset the additional capital and operating costs associated with Class A technologies. Therefore, for alternatives that warrant further consideration, a Class A configuration was preferred over a Class B configuration.

The alternative results were presented to OCSD in a workshop format and the following six onsite strategies were selected for final CDP evaluation:

- CMAD Class B (ET-CMAD-ND-Class B)
- TAD Class A (ET-TAD-ND-Class A)
- TPAD Class A (ET-TPAD-ND-Class A)
- THP Class A (ET-THP-ND-Class A)
- CMAD with 100 percent thermal drying (ET-CMAD-TD-Class A)
- Partial drying with TAD, Class A (ET-TAD-PD-Class A)

These alternatives represented similar NPV results in the SWEET evaluation (except for thermal drying), as shown in Figure 7 below. CMAD Class B had a low NPV and represents the current process employed at Plant No. 2. CMAD with 100 percent thermal drying allows OCSD to diversify the current biosolids management portfolio since Plant No. 1 would continue to produce Class B biosolids. Combining partial drying with TAD Class A allows Plant No. 2 to produce a Class A partially dried biosolids product. This product has the potential to be marketed locally and offers operating cost savings. Thermophilic digestion alternatives (TAD and TPAD) represent a conventional and cost effective approach to produce a Class A biosolids product. In particular, the TPAD Class A options provide several unique opportunities for project phasing. Finally, the THP alternative produces a unique and desirable Class A product. This alternative was somewhat higher in NPV compared to conventional alternatives, but the process provides advantages associated with reduced footprint and high processing capacity.

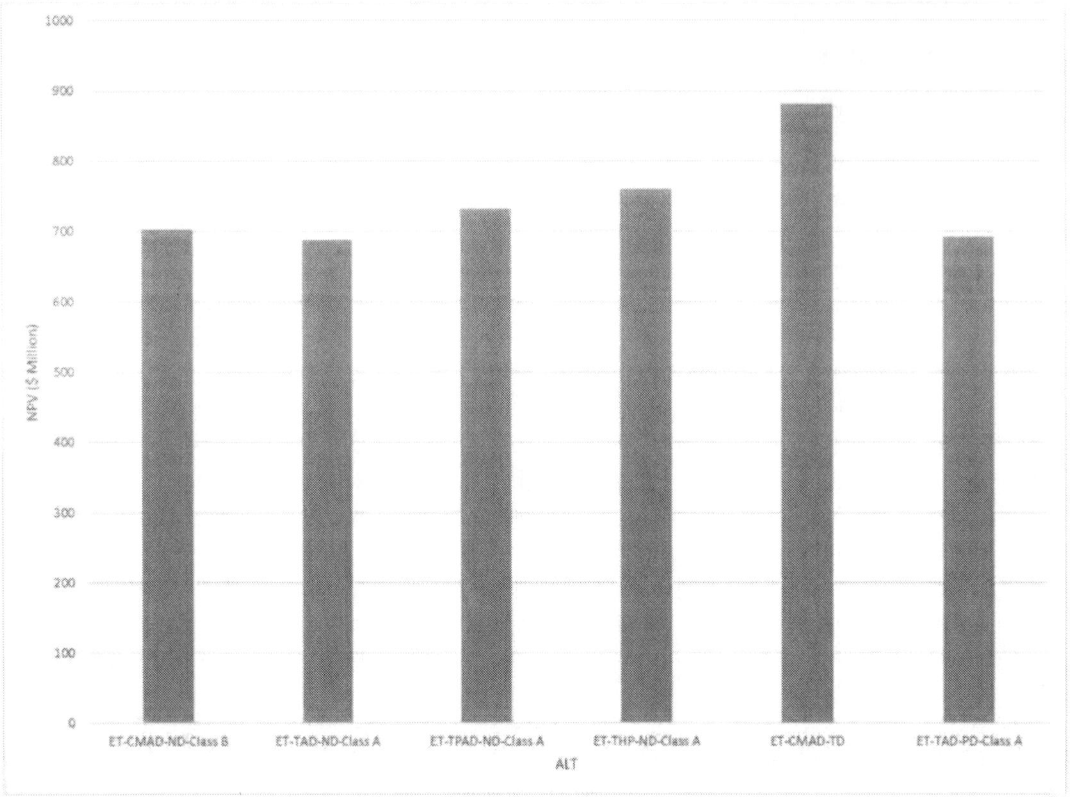

Figure 7. SWEET Results for Short Listed Alternatives.

CDP Evaluation and Selection of On-Site Processing Technologies

The CDP tool for this project used a suite of evaluation criteria which were identified and developed by the project team in a workshop format, as shown below. Figure 8 presents a breakdown of the CDP scores based on each evaluation criteria.

- Operability, maintainability, and reliability.
- Safety.
- Supports OCSD energy policy
- End use market resiliency and reliability.
- Facility and project site compatibility.
- Neighborhood benefits.
- Greenhouse gas emissions.

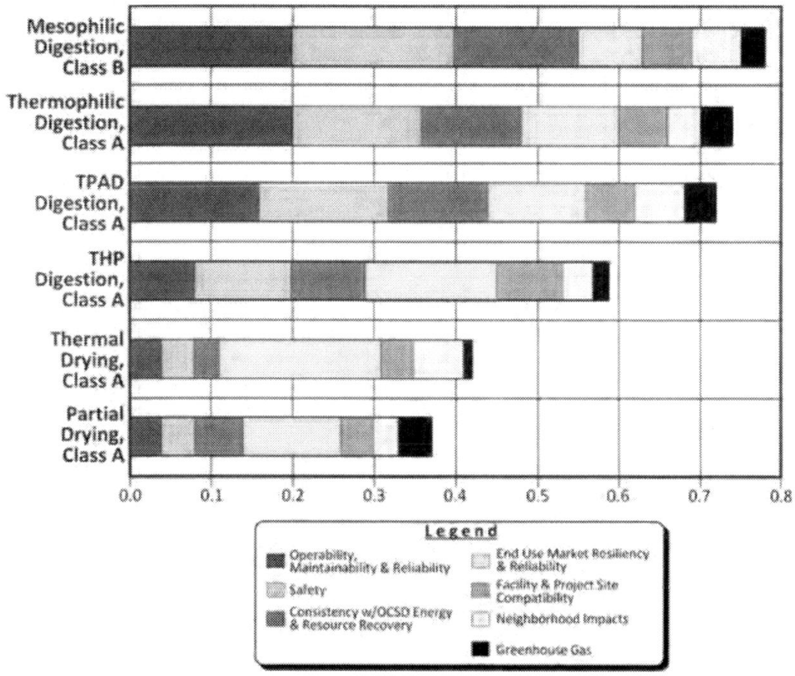

Figure 8. Task 4.5 CDP Model Scoring Breakdown.

Mesophilic digestion, TAD, and TPAD included high scores related to operability, maintainability, and reliability and safety. Class A technologies, especially thermal drying, provided high scores related to end use market diversity and resiliency. When considering non-cost scoring, mesophilic digestion, TAD, and TPAD are very similar, and these results were considered very similar at this level of analysis. However, the scoring for thermal drying alternatives was considered significantly lower than the other alternatives. Based on the CDP evaluation, it was decided that the thermal drying alternatives, both partial drying and thermal drying, would be removed from further consideration. The CDP results were not sufficiently different for the remaining alternatives to make a preferred process selection, and there was insufficient group consensus to make a decision on a preferred process technology between mesophilic digestion Class B, TAD Class A, TPAD Class A, and THP Class A. Thus, a third category of criteria was developed to compare the technology to facilitate a decision on a preferred technology, and included:

- Minimize initial project construction cost.
- Seismic risk mitigation: The existing digester structures are at risk of failure from a seismic event. The selected technology should be able to quickly and cost effectively mitigate the seismic risk.
- Diversity for the biosolids program: The selected technology should provide long term biosolids management options to provide flexibility and reliability.
- Maximize use of existing infrastructure: A significant investment has been made into the existing process infrastructure. Where possible, the implemented technology should provide the option to reuse existing infrastructure.

- Minimize odor impacts to Plant 2 neighbors and end use sites: Neighbor relations are an important concern, and the implemented technology should minimize the impact of new odors to the nearby neighbors. Likely, end use alternatives should minimize the impact of odors at the application sites to ensure long term viability of these application sites.

The results of this final evaluation are shown in Table 13. A plus (+) sign indicates a positive score, a zero (0) indicates a neutral score, a minus (-) sign indicates a negative score.

Table 13. Final Digester Technology Comparison.

Criteria	Mesophilic Class B	Thermophilic Class A	TPAD Class A	THP Class A
Initial Construction Cost	$426 M	$448 M	$286 M	$413 M
Seismic Risk Mitigation	-	0	+	+
End Use Diversity	-	0	+	+
Maximum Use of Existing Infrastructure	-	-	+	-
Minimize Odor Potential	0	-	0	-

Based on this comparison, TPAD provided a clear advantage relative to the other alternatives. The TPAD alternative provides the lowest initial construction cost and shortest schedule to implement an alternative that mitigates the seismic risk at Plant 2. In addition, the TPAD alternative provides the best opportunity to reuse existing infrastructure. Through the implementation of Class A batch tanks, the TPAD process can easily be configured to provide either Class B or Class A biosolids. Finally, the TPAD process is not expected to cause any increase in odor concerns related to the nearby neighbors or at the end use sites relative to the existing mesophilic process and product.

SUMMARY AND CONCLUSIONS

This paper presents the results of a systematic planning evaluation conducted by the project team in order to prepare a BMP that would meet the key objectives of a sustainable, flexible, and cost-effective biosolids management plan which aligns with the appropriate biosolids products and markets, level of treatment, and proper solids handling facility improvements. A structured decision making process was established early in the project and non-economic criteria were identified and developed collaboratively by the team at each alternatives evaluation stage.

The planning effort began with the biosolids end use in mind to ensure that any product generated, whether onsite or offsite, has an associated reliable, cost-effective, and sustainable end use market (Task 3 – offsite biosolids management evaluation). This market-based evaluation resulted in recommendations for products and markets to be carried forward into the onsite solids treatment technology evaluation (Task 4) and helped determine what onsite alternatives would be considered.

Onsite alternatives were identified, screened, and a short list of alternatives was carried forward for concept design in order to estimate planning level capital and operating costs. These costs were then input into the SWEET model to determine NPV results for the end-to-end alternatives that had been identified allowing further screening of alternatives. Finally CDP was used for a non-economic evaluation of the end-to-end alternatives. It was recommended that no post-dewatering facilities be implemented based on these results. However, the evaluation of digestion technologies continued with the identification of final, additional criteria. After this final evaluation step, TPAD was recommended as the preferred digestion technology. OCSD is now moving forward to complete the BMP and implement TPAD for Plant No. 2.

Comprehensive Biosolids and Bioenergy Planning

Authors: Cameron Clark*[1], Irina Lukicheva[1], Anna James[1], Kathy Rosinski[1], Dave Parry[1]

[1]CH2M

*Email: Cameron.Clark@ch2m.com

KEYWORDS: biosolids, bioenergy, market assessment

BACKGROUND

A comprehensive biosolids and energy plan was prepared for a municipal wastewater treatment plant in California. The wastewater treatment plant (WWTP) provides tertiary treatment for a service area is approximately 143,100 people based on the 2010 census. The WWTP has a permitted capacity of 59.5 million liters per day (ML/d) (15.7 million gallons per day (mgd)) for average dry weather flow (ADWF). The current ADWF is approximately 41.5 ML/d (11 mgd). Anaerobically digested, dewatered biosolids are hauled offsite. The WWTP is currently undergoing upgrades to the liquid treatment train that will improve the effluent water quality and increase the capacity of the plant.

Capacity limitation issues, aging infrastructure, and lack of redundancy led the municipality to perform a Biosolids and Energy Management Plan (BEP) to evaluate the solids treatment and bioenergy processes at the WWTP. The BEP evaluated biosolids issues to improve plant operations, such as solids reduction technologies, resource recovery, beneficial use markets, and odors.

The municipality defined BEP drivers and their vision of transforming the WWTP into a Water Resource Recovery Facility that will be a model for resource recovery from wastewater and provide cost-effective, diversified, and power-positive energy recovery. The goals of the BEP generally focused on reliability, energy recovery/independence, biosolids end use diversification, and cost effectiveness.

The BEP first focused on treatment capacity, redundancy, performance, and condition of the existing biosolids treatment facilities. A market assessment was conducted to review regulatory trends and to identify opportunities for biosolids use, organic waste, and energy markets for biosolids and energy recovery.

Market-driven alternatives for biosolids and energy management at the WWTP were then developed and evaluated using a "triple bottom line plus" multi-attribute model that includes economic, environmental, social, and operations criteria that were developed and prioritized with input from municipality staff. The preferred alternative was elected based on the cost-to-benefit ratio, which captured both non-monetary criteria and life-cycle costs.

Finally an implementation plan was developed for the preferred alternative. The plan was based on preliminary facilities and equipment sizing, estimated project costs, and site layouts. A phasing plan containing a list of projects and a preliminary schedule was developed.

EXISTING AND FUTURE CONDITIONS

Solids treatment at the plant is shown in Figure 1. It includes primary sludge screening, grit removal, gravity thickening (GT), and waste activated sludge (WAS) thickening in a dissolved air flotation tank (DAFT). Thickened sludge is combined and anaerobically digested in two egg-shaped digesters. Digested solids are dewatered by centrifuges to 20 to 23 percent total solids (TS). Class B cake is land applied six months per year during the dry season, and landfilled as alternative daily cover for the remainder of the year.

Digester gas is currently used in the boilers to heat water for digester heating and flared. A new compressed natural gas (CNG) facility will use a portion of the digester gas to fuel the City's vehicle fleet.

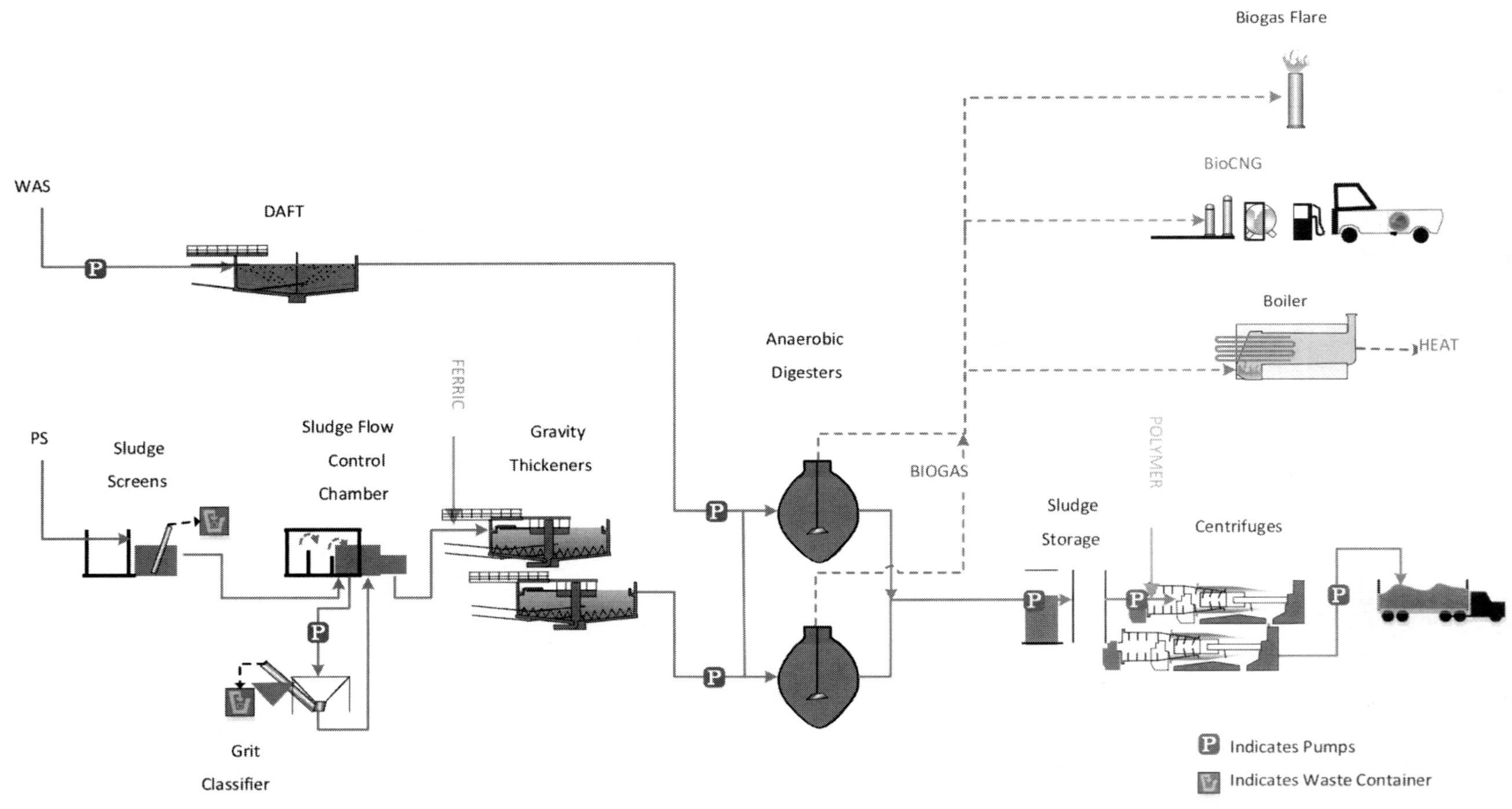

Figure 1. WWTP solids treatment process flow diagram

New preliminary, primary, and secondary treatment facilities are being constructed at the WWTP. These facilities will begin operation in 2023 and are anticipated to result in less grit passing through and accumulating in the biosolids treatment as well as improved sludge thickening.

The evaluation identified improvements needed to address capacity limitations, aging equipment, or redundancy issues with the existing biosolids treatment facilities (Table 1).

Table 1 Solids treatment processes condition and capacity assessment

	Capacity Limitation?	Aged Equipment?	Lack of Redundancy?
Gravity Thickening	NO	YES	NO
Dissolved Air Flotation Thickener	NO	NO	YES
Mesophilic Digestion	YES	NO	YES
Dewatering Centrifuges	YES	NO	YES
Cake Transfer and Storage	NO	NO	YES
Boiler	NO	NO	YES
Waste Gas Burner	NO	NO	YES

The following recommendations were made to improve reliability:

- Plan for rehabilitation/replacement of existing GT
- Plan for redundant WAS thickening capacity or implement a redundant workaround
- Modify digester operation to parallel and add digester capacity
- Install a third centrifuge (in existing building)
- Rehabilitate existing cake conveyor or add redundant conveyance capacity
- Install a second boiler or implement a redundant workaround

MARKET ASSESSMENT

A market driven approach to biosolids planning was used to identify the solids processing technologies. The ultimate goal of this approach is to identify biosolids products that can be easily distributed to multiple end uses.

Biosolids regulatory update reviewed current and future regulatory trends for management and distribution of biosolids. The biosolids market assessment focused on trends and opportunities for beneficial use. The most common biosolids end use practices are land application of predominantly Class B biosolids during dry season and landfilling or landfill daily cover during the winter months or all year long. Due to the regulatory restrictions currently being placed on disposing of organics in the landfill, the latter option will most likely become limited in the future. This is expected to put pressure on wastewater facilities to find land application sites for year-round land application sites or identify alternative ways to manage biosolids. While land application is likely to continue in the future, more stringent restrictions on Class B biosolids for land application are expected, which will likely increase distribution costs of Class B biosolids. In addition, counties where land application is taking place are increasing the demand for higher quality biosolids, and Class A biosolids are expected to have the most versatile distribution options.

The organic waste survey identified opportunities to increase energy recovery from biosolids by augmenting digester feed with local or imported organic wastes. Restrictions on landfilling organics that will limit biosolids landfilling are also expected to increase the amount of food scraps that could be used for co-digestion at wastewater treatment plants. Due to uncertainty of how much organic waste would be available to import to the plant, a moderate co-digestion program was assumed (2-3 trucks per day).

The energy assessment identified and described different beneficial uses of renewable energy and energy efficiency. The energy market assessment showed that excess digester gas is expected to be available for beneficial uses, even after the new CNG vehicle fuel production facility achieves full design capacity. Furthermore, there is an opportunity to improve energy recovery with the implementation of a co-digestion program (which would increase digester gas production) and with a combined heat and power (CHP) process (Figure 2).

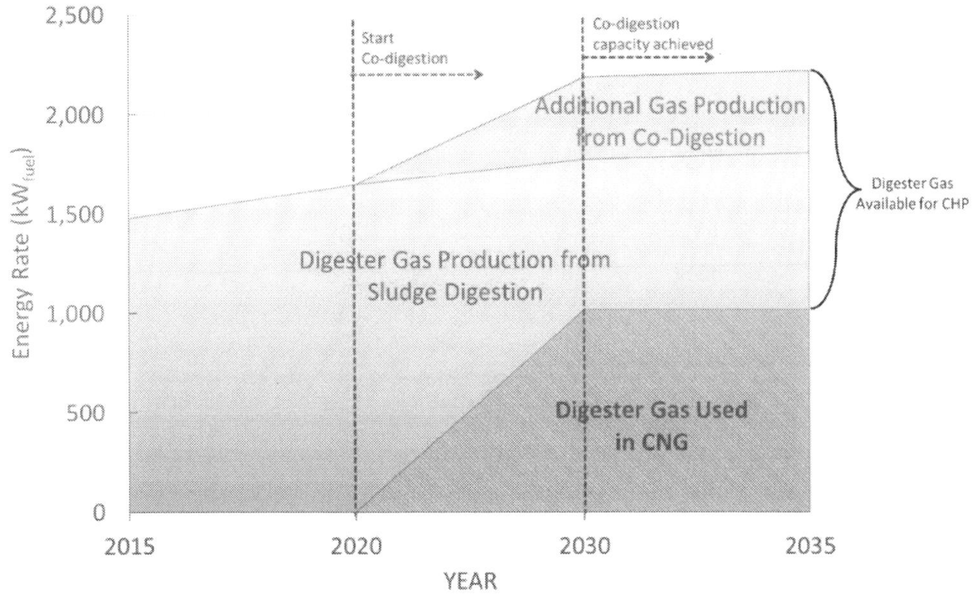

Figure 2. Projected digester gas production and usage (shown as total heating value of biogas)

The projected power demand after secondary treatment upgrades is significantly higher than current levels. To achieve an energy neutral facility using co-digestion, an aggressive (>5 trucks per day) program would be required, requiring increase digestion and CHP equipment. Therefore, an aggressive co-digestion program was not considered in this project.

Based on the market assessment, the following recommendations were made:

- Identify a path for the WWTP to produce Class A biosolids
- Explore implementation of a moderate co-digestion program
- Explore implementation of CHP for improved energy recovery

ALTERNATIVES DEVELOPMENT AND EVALUATION

A number of alternatives were developed for initial screening. Initial screening reduced the number of biosolids and energy management alternatives to the following:

Alternative 0 - Existing – intended to capture the "do nothing" scenario, this alternative does not meet any of the City's goals, but was used to compare alternatives from a financial and logistical perspective.

Alternative 1a – Class B Cake (Base Case + CHP) – includes critical reliability and redundancy improvements needed to continue mesophilic digestion and Class B cake production through the planning period. The reliability and redundancy upgrades would include the following:

- New primary and secondary solids mechanical thickening using three belt filter presses in a new building
- One new mesophilic digester (same size as existing)
- Addition of one new centrifuge installed in the existing building (same size as existing)
- Addition of a cake transfer pump in the existing building (same size as existing)
- Addition of one new cake storage hopper (same size as existing)
- One new digester gas flare (same size as existing)
- A moderate co-digestion program
- One new CHP engine installed in a new enclosure

This alternative was developed to demonstrate that achieving the City's core redundancy, capacity, and energy recovery goals for reliable operation will require investments in several unit processes.

Alternative 1b – Class B Cake (Base Case) – includes reliability and redundancy improvements, similar to Alternative 1a, but without co-digestion or CHP. This alternative reflects a minimum investment required to maintain reliable operation. The municipality requested that this option be included in the evaluation, but it showed poor financial performance compared to Alternative 1a, so it was eventually removed from consideration.

Alternative 2a – Class A Cake via Thermophilic Digestion – This alternative includes the same reliability and redundancy improvements as discussed for Alternative 1a. It also assumes that the two existing digesters will be converted to operate at thermophilic temperatures (135 degrees Fahrenheit [°F]). The conversion will require replacement of existing heat exchangers and other equipment upgrades. Additional digesters would be needed and would operate in a batch mode to meet Class A by time and temperature.

This alternative meets the City's reliability and redundancy goals, as well as energy and resource recovery goals. This alternative can be phased in to maximize the use of existing infrastructure, is adaptable to future regulations, and offers diversity of end uses. The operational considerations for this option that have to be kept in mind are operation of the digester process at higher temperatures, which could be mitigated by implementing standard operating procedures (SOPs) and staff training; potential for increased nutrient loadings from sidestream, which could be mitigated by sidestream treatment; and in some instances, reported reactivation of pathogens and odorous cake, which could be addressed by odor control.

Alternative 2b – Class A Cake via Thermal Hydrolysis Process (THP) – The advantage of this option is reliable, high-quality Class A Cake product with broad end-use opportunities in local markets. Furthermore, THP would increase the allowable solids loading to the existing mesophilic digesters. Therefore, no new mesophilic digester capacity would need to be constructed. However, THP would add significant complexity to the biosolids treatment process because it would require addition of pre-dewatering, an additional cake hopper/transfer system, and steam boiler. It would also require an increased level of staffing, and likely a significant learning curve for plant staff. WWTP staff had safety concerns related to steam system operation, and the need for a full-time certified steam operator.

The large heat demand for THP would reduce the amount of digester gas available for CHP. Therefore, microturbine units were assumed for this alternative because they are available at lower capacities, have higher heat efficiencies, and would be better suited compared to engines. Alternative CHP technologies could also be evaluated.

This alternative meets the City's reliability and redundancy goals, as well as energy and resource recovery goals.

Alternative 3b - Class A Dried Product – This alternative assumes one drum dryer will be installed to process mesophilically digested and dewatered solids. If the dryer needed to be out of service, the plant would still be able to produce Class B cake through the existing digestion process. For this alternative, a new digester would be needed to provide reliability for the digestion process. This alternative could be phased-in and would integrate well in the existing mesophilic treatment processes, taking advantage of all of the sludge treatment processes that are currently in place.

A key advantage of this option is a reliable, high-quality Class A pelletized product with broad end-use opportunities in local markets. This alternative also presents an opportunity for the WWTP to undertake marketing and distribution of its own biosolids product.

Addition of a dryer would add some complexity to the biosolids treatment process, and would require an increased level of staffing. The risks associated with operation of a drum dryer, specifically operation at high temperatures and the potential for thermal events, were a concern to WWTP staff during workshop discussion. It was noted that the risk of thermal events can be managed and mitigated through design of safety systems, training, and operating protocol. The WWTP was interested in other ways to achieve a uniform product without the high temperature associated with a drum dryer. It was noted that a uniform granular product can be produced with a belt dryer that operates at lower temperatures followed by a pelletizer. However, the final product would likely be less desirable compared to uniform pellets.

From an energy recovery standpoint, a dryer would use the large portion of digester gas, so less gas would be available for CHP. In addition, a dryer would also use natural gas, contributing to an increased carbon footprint for this alternative.

Alternative 4 – Third-Party Processing - captures opportunities for onsite third-party processing of the Class B cake into a Class A product, biochar, or other marketable material in exchange for a tipping fee. Based on a survey of local solids processing outfits, the tipping fee would likely be higher than the current fee that the WWTP is paying for solids hauling and disposal.

This alternative includes the upgrades to the solids handling system included in the Base Case, but a third-party contractor would process all or a portion of the dewatered cake and market/distribute the final product. It was assumed that the third-party processing facility would be co-located at the WWTP.

This alternative offers numerous benefits, including adaptability to future regulatory changes and diversification of biosolids end use. It could be easily phased-in and integrated into the existing treatment processes, and would have minimal impact on existing operations staffing.

However, depending on the technology selected, third-party processing may have large land requirements and is also likely to increase truck traffic to the WWTP if the third-party facility processes feed from other facilities. Risks associated with using a new technology also need to be considered, although technologies and their track record are improving. In addition, third-party contracting may have limited operational flexibility. Excess digester gas produced by the facility would be used by the third party and there would be limited opportunity for CHP. Therefore, CHP was not included in this alternative.

The evaluation of this alternative indicated that there is currently no economic benefit to the engage other third-party biosolids management entities when compared to the current practice.

Third-party processing opportunities should continue to be tracked and evaluated for economic and non-economic benefits because these opportunities are likely to become more accepted and economically viable in the future, and could provide an important path to diversification of biosolids end uses at the WWTP.

Alternative 5 – Combination of Class A Digestion and Thermal Drying - This alternative combines a Class A digestion process with drying capacity for one-half of the solids produced at the facility. Thermophilic digestion was assumed as the Class A digestion process, although THP could also be used. One dryer unit was used for evaluation purposes, with the assumption that if the dryer is out of service, the plant can still produce Class A biosolids.

This is the most flexible and diverse option that combines the benefits and concerns that have already been discussed for the Class A cake and drying options. This alternative would provide the WWTP with the flexibility to be able to respond to future regulations and also to diversify the end product. It would also allow the WWTP to develop its biosolids marketing program as well as a variety of products to support it. However, it would also be most complex from the standpoint of implementation, staffing, operation and cost.

Non-Monetary Evaluation

Alternatives evaluation based on non-monetary criteria using the "triple bottom-line" model is shown in Figure 3.

Alternatives that produce Class A product achieved the highest benefit scores as these provide the desired diversity of biosolids end use.

Continuing with mesophilic digestion to produce Class B cake with added reliability and redundancy improvements (Alternative 1) had a lower benefit score. While this alternative would achieve reliable operation, the main concern is its inability to adapt to anticipated regulatory changes and alignment with the City's biosolids recovery goals.

Alternative 0 – do nothing, achieved the lowest score. This alternative represents current conditions with significant reliability and redundancy deficiencies that would result in risky and unsustainable operation through the planning period.

Of the alternatives that meet the City's reliability and redundancy goals, as well as the biosolids and energy recovery goals (2 through 5), Alternative 2a – conversion to thermophilic digestion to produce Class A cake had the lowest life-cycle costs (Figures 4 and 5). These costs were comparable to alternatives that only addressed redundancy and reliability issues (1a and 1b), which made this alternative particularly attractive.

In comparison, other alternatives that required addition of new treatment processes, in addition to improvements for reliability and redundancy (Alternative 2b, 3b, and 5), had higher life-cycle costs. Third-party processing, while requiring low capital investment, had the highest O&M cost due to higher tipping fees.

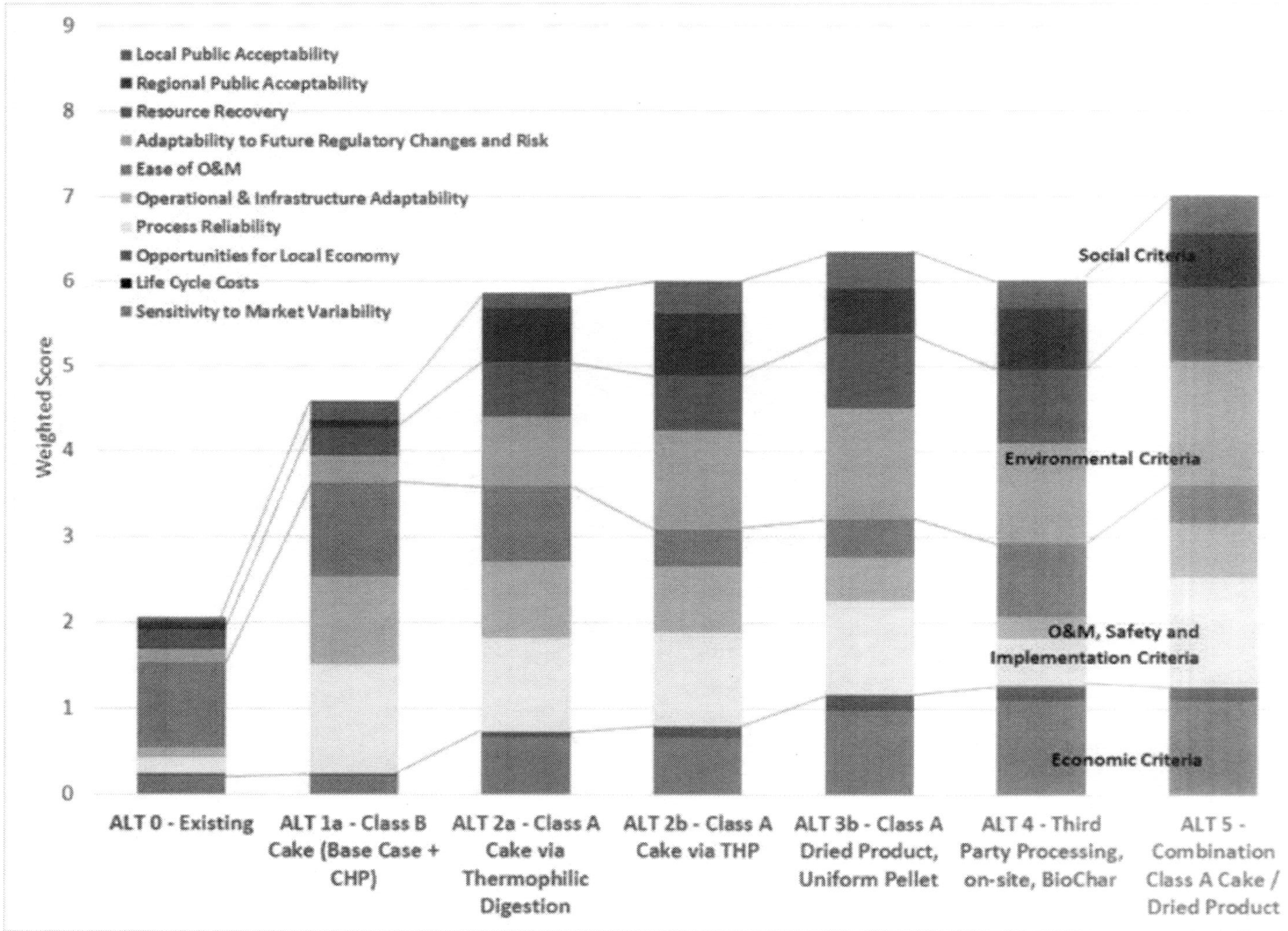

Figure 3. Comparison of alternatives based on non-monetary evaluation

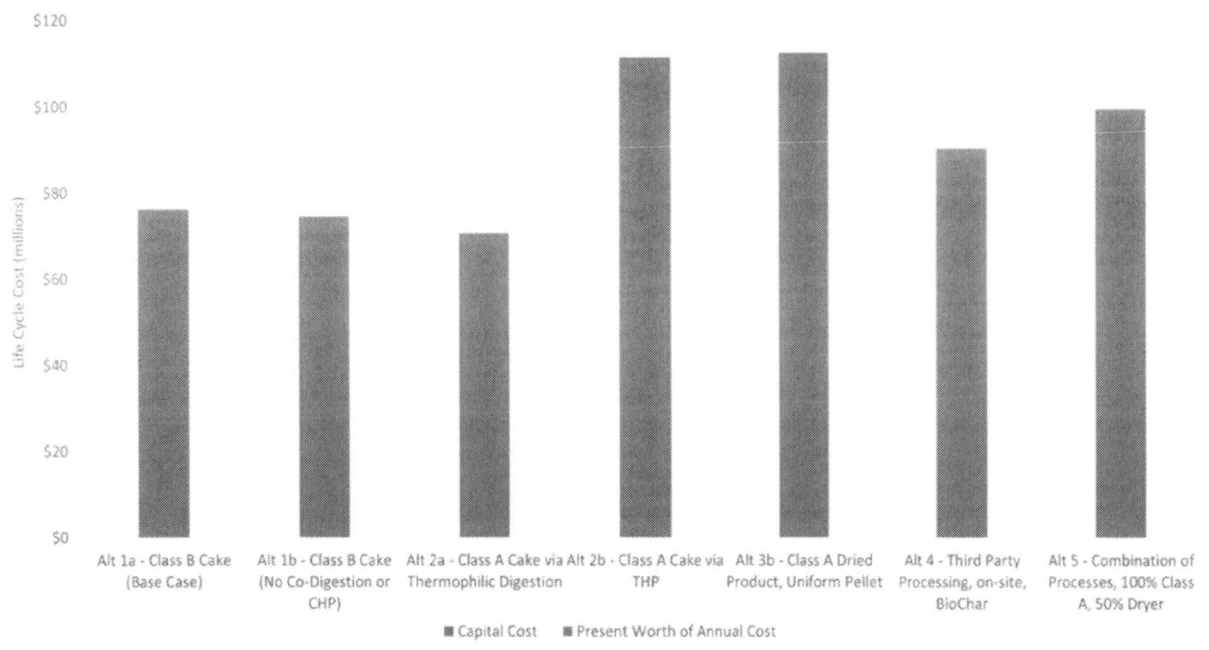

Figure 4. Life cycle cost for evaluated alternatives

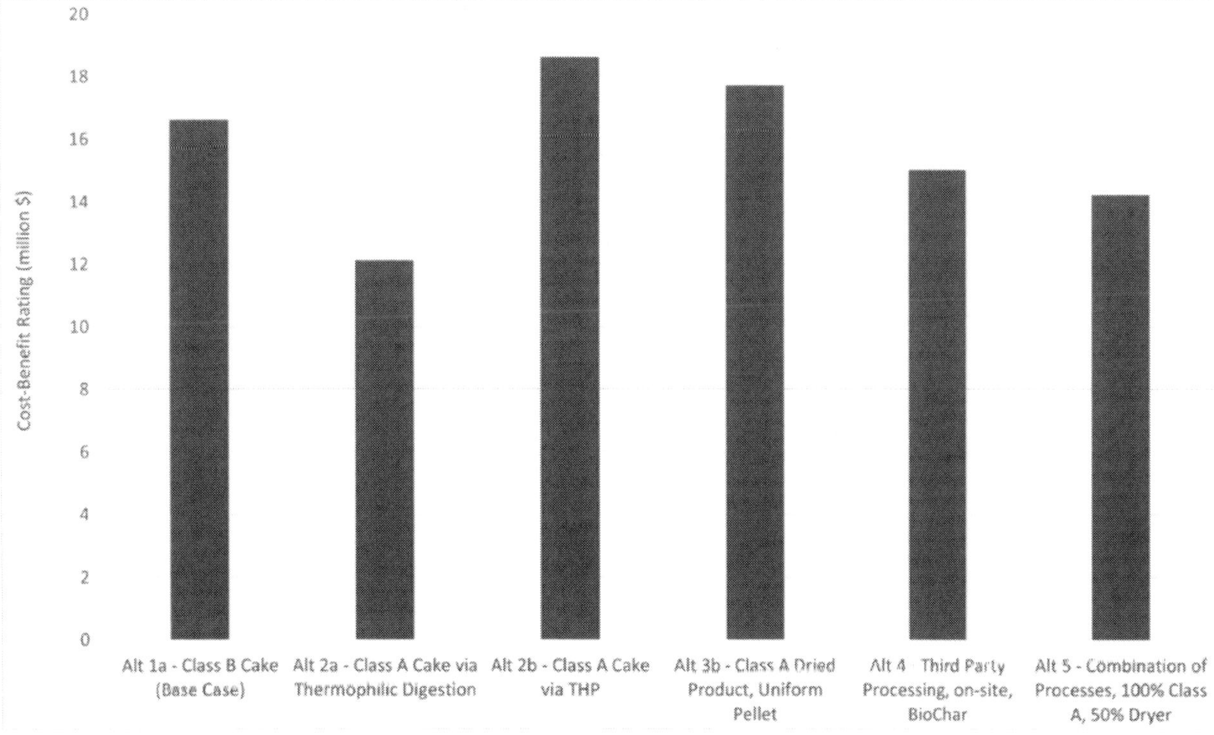

Figure 5. Cost-to-benefit ratio for evaluated alternatives

Based on the above evaluation, Alternative 2a was selected as the preferred alternative. This alternative will provide a Class A biosolids product, which facilitates resource recovery and minimizes future biosolids distribution challenges. Because this alternative also includes a co-digestion and CHP, it will also enhance energy recovery by offsetting power purchased from the utility. Investments in process equipment for conversion to thermophilic digestion were also lower than the other alternatives, resulting in a lower overall life-cycle cost. Considering the economic results and non-economic criteria scores, this alternative is expected to provide the biosolids and energy management approach that most effectively achieves the City's goals.

IMPLEMENTATION PLAN

Proposed upgrades to the WWTP biosolids processing and energy recovery systems required to implement conversion to thermophilic digestion are shown in Figure 6. Upgrades that would be implemented within the BEP planning period are indicated in orange, while future or optional processes that could be considered based on triggers are shown in blue.

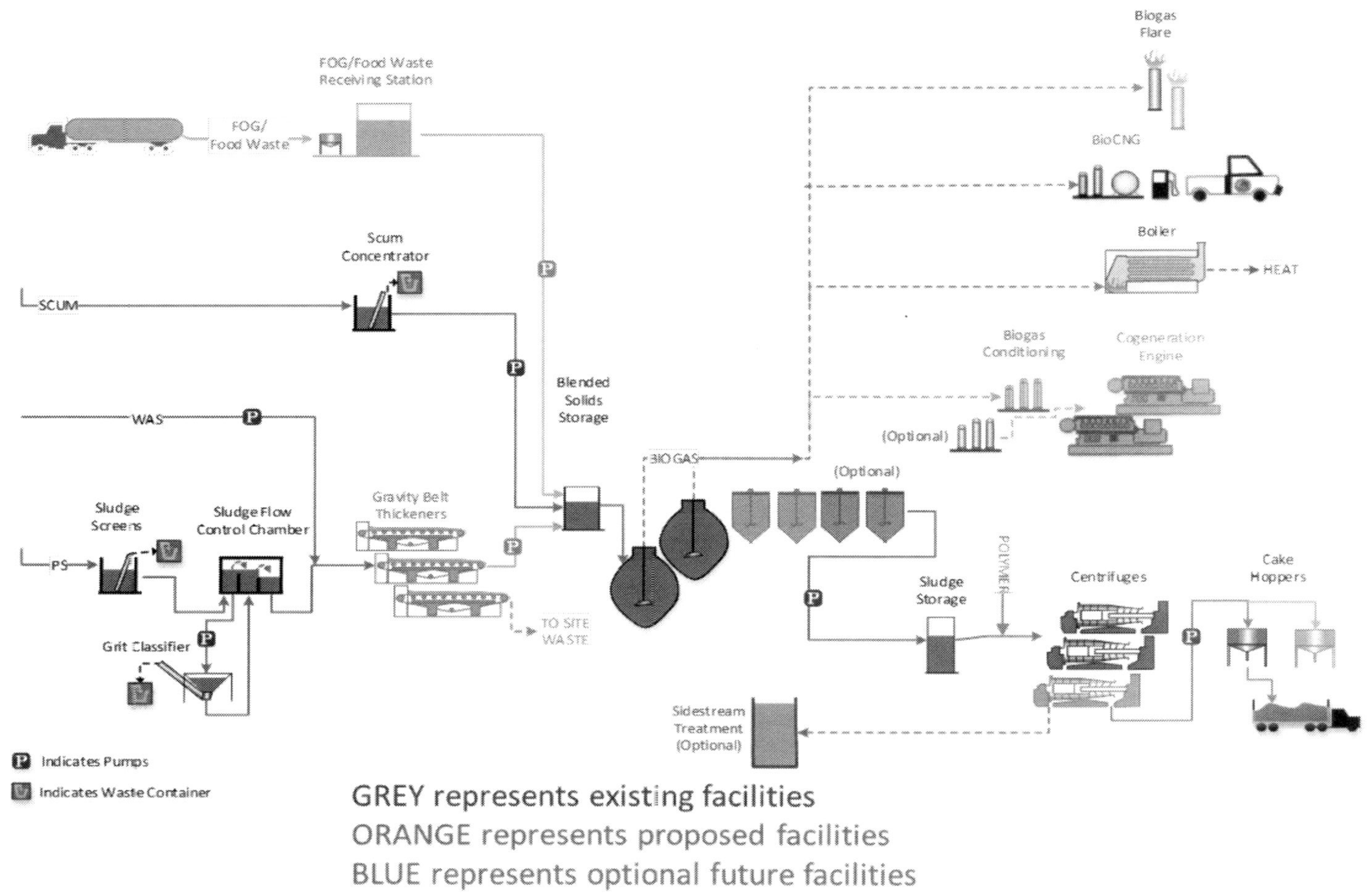

Figure 6. Process flow diagram for the preferred biosolids management alternative – conversion to thermophilic digestion

Four phases were identified to facilitate the transition and to integrate the BEP with the existing CIP:

- **Immediate Needs** include a new centrifuge, upgraded cake conveyance system, and new cake hopper. These projects address urgent reliability and redundancy needs. It should be noted that other workaround solutions to address redundancy may be considered.

- **Phase 1** includes a new odor control facility and thickening system. These projects address reliability and redundancy concerns with the thickening system by providing a fully redundant and upgraded thickening process. Phase 1 also facilitates conversion to thermophilic digestion by reducing the hydraulic loading to the digester.

- **Phase 2** includes a FOG receiving station, a new gas treatment and CHP equipment, and a new waste gas burner. These projects enhance energy recovery by increasing digester gas production and providing additional options for digester gas beneficial use. This phase also addresses redundancy concerns with the plant hydronic heating and waste gas burner system by providing standby units for both processes.

- **Phase 3** involves modifications to the existing digesters to be able to operate at higher temperatures and construction of two new small digesters to convert the digesters from mesophilic to thermophilic operation. This phase increases resource recovery options by producing a Class A biosolids product, which have more distribution options than Class B biosolids. Phase 3 also increases the loading tolerance of the digesters to allow for higher hydraulic and solids loading rates compared to mesophilic digestion.

Biosolids Trends in Tennessee: A Roadmap for the Future

Robert G. O'Dette

State of Tennessee, Nashville, Tennessee
Email: Robert.Odette@tn.gov

ABSTRACT

Biosolids activities and projects in Tennessee over the last decade are reviewed and evaluated showing that the trends for treating, handling and beneficial utilization of biosolids in Tennessee are a roadmap for the future, not only in Tennessee, but for the United States. The Tennessee case histories indicate that the trend by municipal utilities is towards beneficial recycling and producing an exceptional quality (EQ) biosolids product. All of the case histories have several common factors associated with the selection of a municipal utilities biosolids management option. These factors include sustainability, regulatory compliance, energy recovery, costs, and the potential to generate revenue. The types of biosolids treatment include, thermal drying, alkaline stabilization, autothermal thermophilic aerobic digestion and thermal hydrolysis.

KEYWORDS: Biosolids, beneficial use, recycling, exceptional quality, thermal hydrolysis.

INTRODUCTION

There are very few certainties in life. Death and taxes are the two most commonly recited, but the proper handling, treatment, use and disposal of waste is another that should be near the top of that list. Wastewater and wastewater solids are a reality of life on this planet and have to be dealt with properly. If not handled properly, the result has been and will be serious environmental and public health conditions. For the purposes of this paper, the term "biosolids" is being used to represent treated municipal wastewater treatment plant solids (sewage sludge) that can be beneficially recycled.

The land application of biosolids in the State of Tennessee is regulated under the authority of state rules, Chapter 0400-40-15, which became effective on June 30, 2013. Exceptional Quality (EQ) Biosolids (biosolids meeting the most stringent federal and state standards), are exempt from the requirements contained in Chapter 0400-40-15 pending certification by the State Biosolids Coordinator. Permitting options under the state biosolids rules include an individual permit or coverage under the general permit (GP).

The GP for the land application of non-EQ biosolids was issued on March 31, 2014, with the effective date of May 1, 2014. An important fact about the GP is that the application process utilizes the notice of intent (NOI), followed by notice of coverage (NOC), there is a provision for self-certification for those agencies that are certified via the National Biosolids Partnership (NBP). All of the regulatory provisions in Tennessee rules and permits are geared to encouraging the beneficial land application of biosolids and rewarding those municipal utilities that produce an EQ biosolids or are a certified agency by the NBP. Additionally, compliance with all applicable state and federal laws and regulations was the goal to provide liability protection to Tennessee municipal utilities that are beneficially recycling their biosolids.

Whether around the world, in the United States, or in particularly the State of Tennessee the problems associated with municipal wastewater treatment plant solids handling are basically the same. However, there are many, many different approaches to solving the problems and finding the best alternative for biosolids recycling.

The focus of this paper will be to explore case histories from Tennessee that show there are several common factors associated with the selection of a municipal utility's biosolids management option. These factors include:

- Sustainability,
- Regulatory Compliance,
- Energy Recovery,
- Costs, and
- Potential to Generate Revenue.

These factors are also reflected in biosolids management decisions nationwide and worldwide. However, one trend in Tennessee over the past few years has been that some decisions on biosolids management are not totally driven by cost (initial capital or life-cycle). In the recent past, costs were usually the driving force behind the decision and still are in about half of all municipal utilities in Tennessee. However, in the last four to five years, a number of municipal utilities have been evaluating their options with heavy emphasis on what type of biosolids product they need? Ultimately, the answer comes down to producing a biosolids product that is marketable and/or its use is sustainable for the future.

Several case histories will show that municipal utilities opted for an EQ biosolids process, but they still had problems because there were no markets or places to go with the product. Conversely, it will be shown by numerous recent case histories that municipal utilities that evaluated markets and assessed local surroundings made wise decisions with their biosolids management plan.

The chicken or the egg analogy can be applied to the selection of the biosolids management option. The question is: "What comes first the market or the product?" While there are case histories where an EQ biosolids product came first and a market was found for the product, the best approach is to determine the market and then make the biosolids product that is the best fit for that particular market. The final decision must also evaluate the reasonable potential for that market to be sustainable.

It is also important to point out that there are many municipal utilities located in rural areas where Class B biosolids is a very viable option and is likely to be for the foreseeable future. Furthermore, in areas where land application of Class B is viable, the landfill tipping fees also tend to be low. Therefore, about one-half of the wastewater treatment plant solids generated in the State of Tennessee go to a landfill because it is the cheapest and easiest option.

CASE HISTORIES

Jackson Energy Authority (JEA), Jackson, Tennessee
The JEA over the past five to ten years has utilized several different solids handling, use and disposal options for their two wastewater treatment plants, ranging from landfilling to land application and composting. A comprehensive evaluation was completed for a centralized biosolids stabilization and disposal facility for JEA by Griffey (2015).

The objectives of this evaluation for JEA were:

- Fiscally responsible.
- Minimize exposure to regulatory compliance action.
- Minimize regulatory monitoring, reporting, and record-keeping.
- Maximize existing capacity.
- Compatible with existing processes.
- Maximize biogas (free energy) production.
- Environmentally sustainable.

The primary drivers were capacity, sustainability and regulatory compliance. The secondary drivers were energy recovery, optimize processes and generate revenue. Based upon these drivers the following list of available stabilization and disposal methods were among those selected for evaluation:

- Autothermal Thermophilic Digestion (ATAD)
- Windrow Composting
- Thermal Drying
- Thermal Hydrolysis Process (THP)
- Lime Stabilization
- Gasification
- Solar Drying
- Landfill Disposal
- Land Application(1)
- Beneficial Reuse(2)

  (1) Land application means the land application of non-EQ biosolids on the generator's property. Under this scenario land application is strictly an option that does not provide an additional revenue stream to the generator.

  (2) Beneficial Reuse means distributing EQ biosolids to outside entities who pay for the material they receive. The EQ biosolids product sold would be "beneficially used" in agricultural applications as a fertilizer and/or soil amendment.

This list was boiled down to the following five alternatives for detailed evaluation:

- Thermal Drying
- Gasification
- THP

- Landfill Disposal
- Beneficial Reuse

The unit cost factors for these selected alternatives are shown in Table 1.

Table 1 Unit Cost Assumptions

Parameter	Value
Electrical, kWh	\$0.097
Natural Gas, Therm	\$0.60
Polymer Cost, lb. active	\$4.22
Plant Utility Water Cost, 1,000 gallons	\$0.19
Tipping Fee, wet ton	\$31.98
Truck Trip to Landfill, trip	\$367.92 per 30 cu yd. hopper
Class A, EQ Biosolids Product sale, ton	\$26.00

The bottom line for JEA was that going to a landfill was the cheapest option and that was the chosen approach. However, Griffey's (2015) evaluation provided a key take home point that showed of all of the options evaluated THP was the most costly option and thermal drying was the next cheapest option compared to landfilling. Additionally, the fact that one of the primary drivers was "Sustainability" it seemed difficult to see how landfilling could be considered for anything other than a short-term or interim solution. However as the Griffey pointed out, *"Landfill disposal, as well as each of the other alternatives, was* ***more*** *sustainable than what they (JEA) were doing at the time (i.e. monofilling)."* A critical point to remember is that the cheapest and easiest option is going to be the one selected unless there are local constraints due to location where factors such as odor, noise, and truck traffic make the option unsustainable.

City of Maryville, Tennessee

The City of Maryville is a very progressive utility and has experimented over the past 5 to 10 years with several biosolids management options, including:

- Composting,
- Alkaline Stabilization, and
- ATAD

Several factors were considered by the City of Maryville when they were evaluating potential biosolids management options:

- Size
- Regulations
- Power costs
- Manpower
- Current infrastructure
- Final biosolids plan
- Budget

Autothermal Thermophilic Aerobic Digestion (ATAD) is used at Maryville, Tennessee to produce an EQ biosolids product. In the ATAD process, "Autothermal" means that it is self-heating (no external heat source needed) with thickened material feed to the reactor. Thermophilic means that the reactor temperatures are maintained in the range of 50 to 65 degrees centigrade. Aerobic means that conditions in the reactor maintain the presence of oxygen. ATAD provides greater than 60% volume reduction and enhances the dewaterability of the biosolids product. However, odor considerations play a key role with this process. In aerobic decomposition dimethyl disulfide is the prevailing cause of foul odors. Dimethyl disulfide is very pungent and has a detection limit similar to hydrogen sulfide.

The ATAD process has significant money saving features as discussed by Westphal (2016):

- High Volume Reduction. 55-70% volatile solids destruction and greater than 50% total solids destruction which cuts hauling and end use and/or disposal costs by 50%.
- Improves Dewatering (even on Waste Activated Sludge). The higher percent solids (25 to 33%) translates to less material to manage and reduced polymer demand.
- Reduced Manpower. The high degrees of automation reduces operator time.
- Energy Efficient. Oxygen Reduction Potential (ORP) control strategy allows the system to match oxygen delivery to uptake demand.

The ATAD process offers a strategically sound, forward-thinking biosolids management process for municipal wastewater treatment utilities to consider. The important point to be made with discussing ATAD that is applicable across the board with biosolids products is that if you can produce an odor free and pathogen free product many more biosolids management options become available. For the City of Maryville their EQ biosolids via ATAD (with odor control) has placed them in a strong position for long-term sustainablity with relatively low costs. In this case, there is no revenue from sales as the product is given away; however, they have no problem finding outlets with little marketing effort. This is true primarily because the biosolids product is EQ and odor free.

Metropolitan Government of Nashville and Davidson County (City of Nashville), Tennessee

The City of Nashville started experiencing problems with wastewater treatment plant solids handling and disposal options in the early 1980's. For a period of time Nashville was land applying liquid biosolids in the adjacent counties of Wilson and Williamson, but public opposition in these counties eventually forced Nashville to explore other options. Over the last quarter of a century, Nashville has used the following options:

- Land Application
- Landfilling
- Composting
- Alkaline Stabilization
- Thermal Drying

Problems with public acceptance and product acceptance ultimately led to Nashville biting the bullet and building a thermal dryer that produces a very marketable product. While perhaps not the cheapest option, employing state of the art thermal drying with anaerobic digestion, has proven to be a very sound and sustainable option for Nashville with product outlets in Tennessee and Arkansas. Product sales revenue is about $10 per ton with virtually zero cost for market development and/or sustainability. It should be noted that EQ biosolids products from thermal drying operations have been placed in lucrative citrus markets in Florida at about $50 to $60 per ton (dating back 15-20 years). Additionally, valuable soil products markets in California using biosolids compost for many blends, both bagged and bulk, have been ongoing for almost 80 years.

Alkaline Stabilization

There are ten municipal utilities in the State of Tennessee that achieved EQ biosolids status and are thus exempt from coverage under the state biosolids rules. These municipal utilities are listed as follows:

1. Athens – Thermal Drying
2. Cookeville – Alkaline Stabilization (RDP)
3. Crossville – Alkaline Stabilization
4. Dickson – Alkaline Stabilization
5. Gallatin – Alkaline Stabilization
6. Lewisburg – Alkaline Stabilization
7. Maryville – ATAD
8. Nashville – Thermal Drying
9. Newport – Thermal Drying
10. Pulaski – Alkaline Stabilization

Six of these ten EQ biosolids producers utilize alkaline stabilization. The questions are: Why use alkaline stabilization? Why is it so widely used in the United States? And why has it been the most selected process for producing EQ biosolids in the State of Tennessee? The short answer is because of low cost and location with regard to acidic soils (pH less than 6.0).

Historically, lime has been used extensively at wastewater treatment plants for decades because it is simple, easy to use and low cost. Furthermore, a large portion of the United States, particularly east of the Mississippi River (especially true in Tennessee) have soils where the pH is less than 6.0 (see Figure 1).

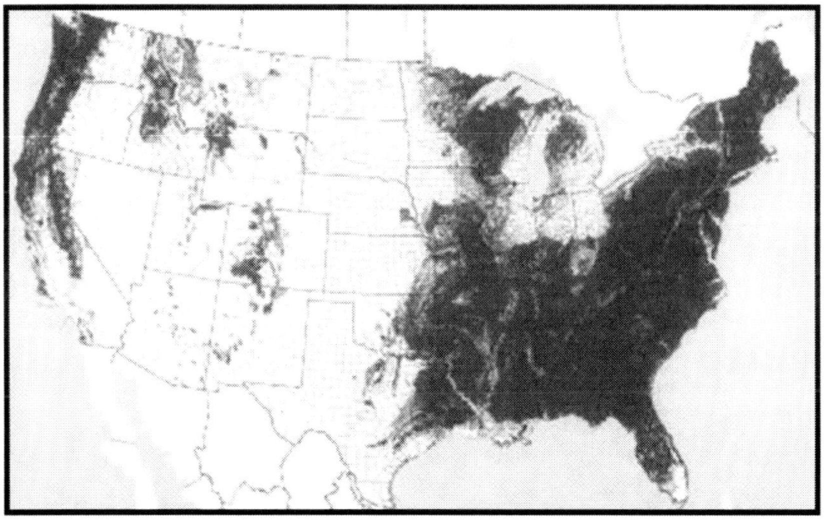

Figure 1 – Soils with pH less than 6.0 (shown in red) by Christy (2016)

It is a consensus among Tennessee farmers that the single most important thing one must do is get your soil pH correct, because that optimizes the uptake of nutrients into the plants. Having the correct pH increases crop yield and makes more efficient use of fertilizer.

Thus, the alkaline stabilized biosolids product is suitable for application in many situations, such as agriculture, and mine reclamation. The product serves as a lime substitute, source of organic matter, and a specialty fertilizer. The addition of alkaline stabilized biosolids results in more favorable conditions for vegetative growth by improving soil properties such as pH, texture, and water holding capacity.

Additionally, alkaline stabilization systems are generally quite reliable and flexible. The same equipment can often be used to produce either Class A or Class B biosolids with minor process modifications, such as a larger dosage of alkaline material and/or the addition of supplemental heating (pasteurization).

Those municipal utilities that ultimately selected alkaline stabilization to produce EQ biosolids did their homework, and because of the widespread acidic soil conditions across the State of Tennessee chose this option. For this same reason, more than fifty percent of all non-EQ biosolids land applied in Tennessee are alkaline stabilized.

Thermal Hydrolysis Process (THP)

The thermal hydrolysis process (THP) is a process in which sewage sludge is treated via elevated temperature and pressure to produce a product that is pathogen free, more biodegradable, less viscous and easier to dewater. Thus, some of the benefits of the THP are significantly smaller digester sizes (reduced by as much as 50 percent or more) and improved ability to dewater (greater than 30 percent via a belt filter press).

The **City of Franklin, Tennessee** has spent a considerable amount of time over the past 2-3 years trying to make a final determination as to which sewage sludge treatment option to utilize. The State of Tennessee, Dr. Jim Smith (EPA-NERL, Cincinnati, Ohio) and Mike Hom (EPA-Region 4) have provided comments to the city and their consultants relative to an evaluation of the CAMBI THP.

The regulatory compliance problem for a CAMBI THP system can be understood via Figure 2.

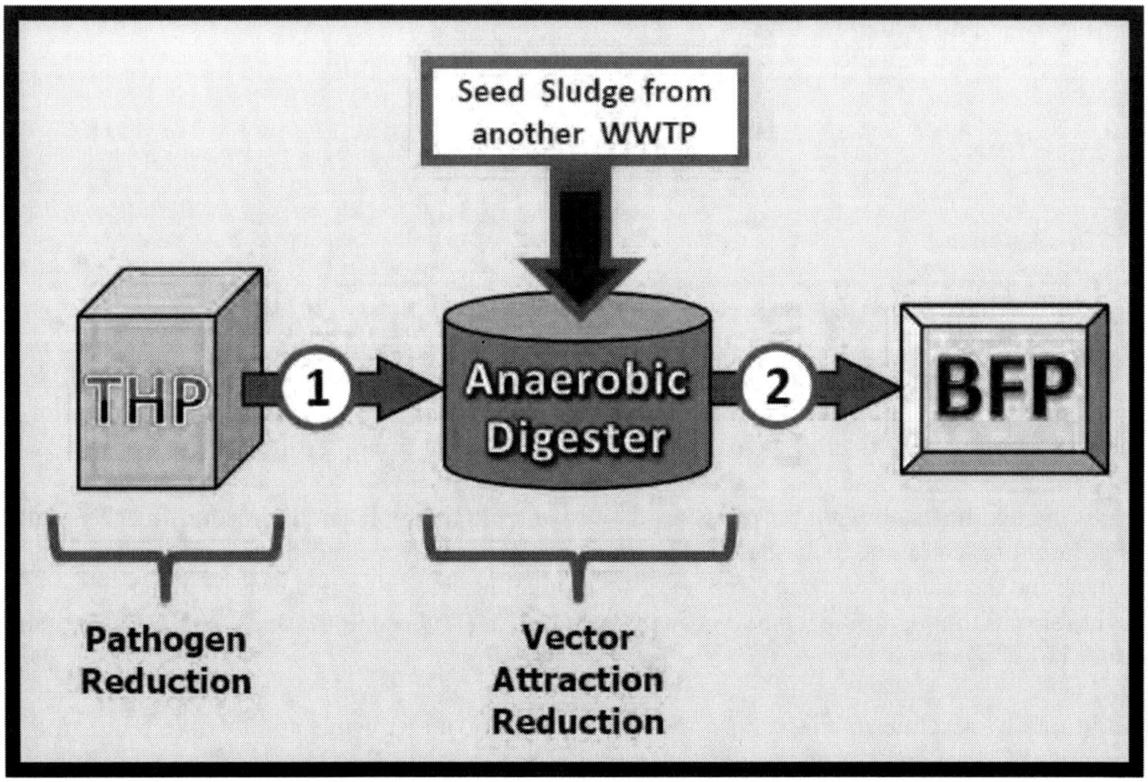

Figure 2 – Flow Schematic for THP

When sewage sludge is treated in the THP it kills everything in terms of microorganisms and is Class A at Point 1. However, at Point 1, there are no microorganisms available to seed an anaerobic digester, so seed sludge must be brought in from another WWTP. Once another sludge is added to the digester, the entire product, by regulation, becomes a "product derived from sewage sludge" and the achievement of pathogen reduction (PR) via Class A by the THP is nullified because the "product derived from" must now start from scratch to achieve PR via Class A, vector attraction reduction (VAR) to become Exceptional Quality (EQ) biosolids (assumes that the City of Franklin's biosolids meets the pollutant concentrations levels of Tables 1 and 3 of 40 CFR Part 503.13) after Point 1.

Subsequently, the City of Franklin developed a plan of operation that addressed the regulatory compliance issues of concern with CAMBI THP. A plan of operations was eventually developed that was acceptable to the state, EPA-Region 4 and Dr. Smith. This plan of operation proposed isolating the anaerobic digester after the seed sludge is added to achieve PR via time and temperature per Class A - Alternative 1. One of the major concerns for the State of Tennessee with THP was that compliance with all applicable laws and regulations could be shown if there was ever a challenge to this system by concerned citizens/environmental activist groups. To date, no THP biosolids system has been certified as EQ by the "*permitting authority*" under 40 CFR Part 503.9(p). The only operational THP system in the United States is at the Blue Plains WWTP in Washington, D.C., but neither the state of Virginia or Maryland is a delegated state for biosolids and thus, not the permitting authority per Part 503. This is a potentially huge liability issue if there is problem with a product claiming to be EQ, and it has not been permitted as such by the permitting authority per Part 503.

40 CFR Part 503 is a self-implementing rule which means that virtually everyone's opinion or interpretation is just as valid as anyone else's, including the permitting authority. When there are questions or differing interpretations relative to Part 503, it is best to have the permitting authority (EPA or one of eight delegated states) issue a permit. The permitting process would provide a public notice and possibly a public hearing and at the end of the process the utility would hold a valid permit from the permitting authority which would provide some level of liability protection or shield.

There are several other major problems associated with the THP, such as having to seed the digester from another source and thus turning the entire digester into a "product derived from sewage sludge." This then requires the biosolids processing, for regulatory compliance, to have to start over from that point forward to meet PR and VAR. Furthermore, the process is very complex requiring a heat exchanger following THP to reduce the hot sludge (160 degrees C) temperature prior to entering the anaerobic digester.

The argument has been put forward that with THP the size of the digester is cut in half and dewaterability is significantly increased. The evidence supports these two conclusions, but with a thermal dryer it is not necessary to have a digester to produce an EQ biosolids product. One example of this is the former dryer in the Bronx used by New York City with the product marketed and sold in bulk to citrus markets in Florida. Additionally, ATAD as noted by Westphal (2016), provide the same or enhanced benefits compared to THP.

Under Section 405(e) of the Clean Water Act the option for biosolids use or disposal is a "local determination" and the choice can be made regardless of cost. For THP it is not only a high cost, complex option compared to other EQ processes, but the marketability of the end product has yet to be determined.

The discussion relative to THP drives home two points: (1) it is imperative for municipal utilities to develop markets <u>first</u> and then make the product that will match those markets, and (2) to ensure that the final product is compliant with all applicable state and federal laws and regulations. Whenever either of these principles is violated, costs go up, there are liability issues and the overall sustainability of the selected option is placed in jeopardy.

SUMMARY AND CONCLUSIONS

The biosolids trends in the State of Tennessee are a road map or predictor of what is happening or soon will happen across the United States and the world. What should be clear is that there are more than one right answer to question, what is the best option for managing wastewater treatment plant solids? This is true, because whatever option is selected depends on the local situation, primarily their location and size.

In terms of relative costs comparison (dollars per dry ton of biosolids basis) for biosolids management options researched in this paper the following ranking is shown with least cost to most costly:

1. Landfill or Land Application[(a)]
2. Composting or Alkaline Stabilization[(b)]
3. ATAD
4. Thermal Drying
5. THP
   (a) Depends on location, tipping fees and haul distances.
   (b) Depends upon size and type of composting. Large windrows will be cheaper, small envessel or aerated static pile systems will be more costly.

In the final assessment, the realities of this world have to be faced and they play the ultimate role in the selected biosolids management option. In economically strapped communities, or communities with little growth in population or revenue, dollars and cents are going to be the primary driver for the disposal option selected. "Cheap" and "easy" are attractive particularly in locations such as these. Likewise it is challenging for treatment providers to elect to increase the rates of their customers to pay for "expensive" and "difficult" without other local, compelling pressures to do so. In hard economic times or locations, it is hard for officials to be "green" for the sake of being "green."

For larger municipal utilities in large urban areas, it is imperative for them to look at producing marketable biosolids products that provide the potential for a sustainable present and future use option. The biosolids products that have the highest potential for marketability are pellets from thermal drying and compost. Additionally, as shown in Tennessee, alkaline stabilized products, with high calcium carbonate (lime) equivalency, have viable, sustainable outlets where there are acidic soil conditions. The bottom line in these latter situations reinforces the fact that having a viable sustainable option trumps cheap and easy.

ACKNOWLEDGEMENTS

The author offers special thanks to Joe Griffey, P.E. (Smith Seckman Reid, Inc.), Dick Christy (RDP Technologies) and Dr. Jim Smith, P.E. (EPA-NERL) for their invaluable contributions to the content of this paper.

REFERENCES

Christy, Richard (2016) EnVessel Pasteurization (PowerPoint). RDP Technologies, Inc. Conshohocken, Pennsylvania.

Eckhardt, Gregg (2015) SAWS (San Antonio Water System) Case Study. World Environmental & Water Resources Congress. Austin, Texas.

Federal Water Pollution Control Act (33 U.S.C. 1251 et seq.), as amended through P.L. 107-303, November 27, 2002.

Gautam, T. (2015) Houston's Biosolids Production To Market. World Environmental & Water Resources Congress. Austin, Texas.

Griffey, Joseph (2015) Planning for the Future-Evaluating a Centralized Biosolids Stabilization and Disposal Facility for the Jackson Energy Authority. Kentucky-Tennessee Water Professionals Conference, Covington, Kentucky.

Hintz, Angela M. (2015) Biosolids Management Planning. World Environmental & Water Resources Congress. Austin, Texas.

Moss, Lynne H. (2015) The Road To Recovery: Regulation and Policy Considerations. World Environmental & Water Resources Congress. Austin, Texas.

Milorganite® Annual Report for 2016, Milorganite-MMSD, 260 W. Seeboth Street, Milwaukee, WI 53204 (January 30, 2017).

Nangle, Thomas; Hatcher, Michelle; Daniel, Zack (2016) Thermal Hydrolysis at a Small WRF, Challenges & Lessons Learned. Kentucky-Tennessee Water Professional Conference, Knoxville, Tennessee.

Sierant, Brian (2015) State of Texas Biosolids Program Perspectives. World Environmental & Water Resources Congress. Austin, Texas.

Slagle, Jody (2015) Blazing a resource Recovery Trail-Hornsby Bend Biosolids Management Plan. World Environmental & Water Resources Congress. Austin, Texas.

Smith, James (2015) The 503 Regulations and Biosolids Use and Disposal 20 Years Later. World Environmental & Water Resources Congress. Austin, Texas.

State of Tennessee (2013), *Chapter 0400-40-15-Biosolids Management*, Rules of the Tennessee Department of Environment and Conservation, Nashville, Tennessee (June 30, 2013).

State of Tennessee (2014), *General State Operation Permit for the Land Application of Non-Exceptional Quality Biosolids*, Permit Number TNB000000, Tennessee Department of

Environment and Conservation, Division of Water Resources, Nashville, Tennessee (Effective Date: May 1, 2014; Expiration Date: April 30, 2019).

Tsang, Richard K. (2016) Emerging Biosolids Processing Technologies-Potential Game Changers. Kentucky-Tennessee Water Professionals Conference, Knoxville, Tennessee.

U.S. EPA (1975). *Lime Stabilized Sludge: Its Stability and Effect on Agricultural Land.* National Environmental Research Center. Washington, D.C. 20460. EPA-670/2-75-012. April.

U.S. EPA (1979). *Process Design Manual for Sludge Treatment and Disposal*, EPA/625/1-79/011, Cincinnati, Ohio, September.

U.S. EPA (1989) *Environmental Regulations and Technology: Control of Pathogens in Municipal Wastewater Sludge*, EPA/625/10-89/006, Center for Environmental Research Information, Cincinnati, OH.

U.S. EPA (1992) *Technical Support Document for Reduction of Pathogens and Vector Attraction in Sewage Sludge.* EPA/822/R-93-004. Washington, DC.

U.S. EPA (1993) 40 CFR Parts 257, 403, and 503 (FRO-4203-3): Standards for the use and disposal of sewage, final rule. *Fed. Register*, 58, 9248. US Government Printing Office, Washington, D.C., February 19, 1993.

U.S. EPA (2003). *Environmental Regulations and Technology: Control of Pathogens and Vector Attraction in Sewage Sludge*, EPA/625/R-92/013, revised July, 2003, Cincinnati, OH.

Westphal, Sherri (2016) Proven Technologies Offer An Exceptional Biosolids Management Process. Kentucky-Tennessee Water Professionals Conference, Knoxville, Tennessee.

A Big Picture Approach to Addressing Multiple Improvement Triggers for both Short- and Long-term Multi-Facility Master Planning

Vera Gouchev[1,*], Paul Knowles[1], Robert Sharp[1], Emanuel Psaltakis[1], Thomas Lauro[2]

[1] Hazen and Sawyer, New York, NY
[2] Westchester County Department of Environmental Facilities
* Email: vgouchev@hazenandsawyer.com

ABSTRACT

As regional utilities are faced with tighter regulations and simultaneous goals for economic and environmental sustainability, it is advantageous if they can identify synergies among multiple facilities to improve efficiency and performance. This approach was utilized to develop a County-wide short-term (5-Year) and long-term (25-Year) Residuals Management Master Plan for Westchester County (NY) Department of Environmental Facilities (DEF). Westchester DEF owns and operates seven Water Resource Recovery Facilities (WRRFs) with a combined capacity exceeding 190 MGD. These facilities currently produce 120,000 cubic feet of screenings and grit, and approximately 19,000 dry tons of biosolids annually for beneficial use and disposal.

The 5-Year plan proposed strategies for optimizing existing facilities with a focus on reducing operational costs through reduction in residuals volume and more efficient energy usage. Private liquid sludge and dewatered cake hauling and disposal contracts are a large component of annual operating costs for DEF amounting to $7.4 million annually, which makes volume reduction a priority for the County. In addition to providing facility specific recommendations to improve operations, it was recommended that the County consolidate several facilities (Blind Brook, Port Chester, Ossining, and Peekskill WRRFs) by utilizing the additional capacity at the Peekskill WRRF to thicken, digest, and dewater. The recommendations presented in the 5-Year plan are expected to save the County $1.1 million annually on operations costs.

The 25-Year plan employed the strategy of identifying specific triggers that would necessitate improvement. For Westchester County, four triggers were identified: (1) Class A biosolids requirements; (2) limited disposal options; (3) sustainability mandates; and (4) nutrient reduction requirements. A 7-step process was used to evaluate the recommended strategies for each trigger, with a focus on strategies that could support multiple triggers. A multi-facility Residuals Handling Evaluation and Optimization Tool was utilized to perform a triple bottom line analysis and identify the most suitable approach for meeting a trigger requirement in a streamlined manner. Out of 20 different scenarios evaluated, 7 scenarios made it through the screening process. Based on this analysis, it was determined that retrofitting the existing digesters at the Yonkers Joint WRRF to operate Temperature Phased Anaerobic Digestion (TPAD) for processing cake from the New Rochelle, Peekskill, and Yonkers Joint WWRFs would be the most suitable option. In this scenario, three triggers are met: Class A biosolids requirement, limited disposal options, and sustainability mandates. Additional annual operations costs savings are anticipated to be $3-4 million per year after debt service for fully financed capital improvements.

This paper describes our approach to multiple-facility planning in order to develop a County-wide short-term (5-Year) and long-term (25-Year) Residuals Management Master Plan for Westchester County DEF.

The key elements of this manuscript include:

- Approach to evaluating current operations and establishing baseline conditions for short-term and long-term planning.
- Optimization of existing infrastructure to provide low-to-no cost short-term improvements and achieve substantial savings in annual operations and maintenance (O&M) costs.
- Multi-facility decision making process for long-term planning.

KEYWORDS: Biosolids, Residuals, Multi-Facility Planning, Decision Making Tool

INTRODUCTION

BACKGROUND

Westchester County DEF owns and operates seven Water Resource Recovery Facilities (WRRFs), with a combined capacity exceeding 190 MGD. A brief description of each facility is provided below:

- **Blind Brook (BB) WRRF** - 5 MGD Biological Nitrogen Removal (BNR) facility. The current treatment process includes headworks with screening and grit removal, influent pumping, primary sedimentation, activated sludge treatment (modified MLE process), final clarification, and sodium hypochlorite disinfection. Settled secondary sludge is pumped via Force Main to the Port Chester WRRF.
- **Port Chester (PC) WRRF** – 6 MGD rotating biological contactor (RBC) facility. The current treatment process includes influent pumping, preliminary screening, grit removal, primary clarification, secondary treatment, gravity thickening, disinfection, and effluent pumping. Thickened sludge and scum are disposed via a private liquid hauling contract for beneficial use.
- **Mamaroneck (MM) WRRF** – 23.2 MGD BNR facility. The current treatment process includes partial influent pumping, screening, grit removal, chemically enhanced primary treatment (CEPT), primary clarification, integrated fixed film activated sludge (IFAS)

aeration for BNR, secondary clarification, disinfection, and effluent pumping (for high flows and high tide only). Scum and sludge is sent via force main to New Rochelle WWTP for processing.
- **New Rochelle (NR) WRRF** – 20.6 MGD BNR facility. The current treatment process includes screening, influent pumping, grit removal, primary settling, pure oxygen intermediate clarification, Biological Aerated Filter (BAF) for BNR, sludge thickening, sludge dewatering, and ultraviolet (UV) disinfection. Dewatered sludge is disposed of via a cake hauling contract for beneficial use.
- **Ossining (OS) WRRF** – 7 MGD facility. The current treatment process includes preliminary screening, grit removal, primary clarification, activated sludge secondary treatment, final clarification, gravity thickening, and disinfection. Thickened sludge is disposed via a liquid sludge hauling contract for beneficial use.
- **Peekskill (PK) WRRF** – 10 MGD facility. The current treatment process includes: preliminary screening, grit removal, primary clarification, activated sludge secondary treatment, gravity thickening, anaerobic digestion, disposal of digested sludge by private liquid hauling contract.
- **Yonkers Joint (YJ) WRRF** – 120 MGD facility. The current treatment process includes preliminary screening, grit removal, primary clarification, activated sludge secondary treatment, final clarification, gravity thickening, anaerobic digestion, and dewatering. Dewatered sludge is disposed of via a cake hauling contract for beneficial use.

A map of all seven DEF operated WRRFs can be found in **Figure 1**. **Figure 2** provides a summary of the residuals related treatment processes available at each facility.

Figure 1: Westchester County DEF WRRF Map

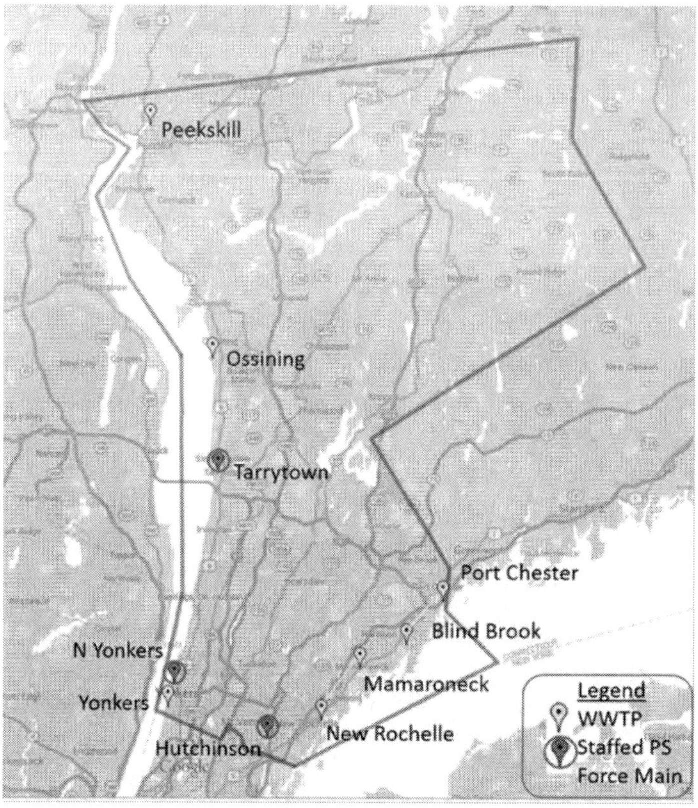

Figure 2: Westchester County DEF WRRF Residuals Process Matrix

Facility	Screen	Grit	Prim Settle	Prim Scum	Sec Settle	Sec Scum	Sludge Thick	Dewat 1	Anaer Digest	Dewat 2	Sludge FM
Blind Brook	✓	✓	✓	✓	✓	✓					✓
Port Chester	✓	✓	✓	✓	✓	✓	✓				
Mamaroneck	✓	✓	✓	✓	✓	✓					✓
New Rochelle	✓	✓	✓	✓	✓	✓	✓	✓			
Ossining	✓	✓	✓	✓	✓	✓	✓				
Peekskill	✓	✓	✓	✓	✓	✓	✓		✓		
Yonkers	✓	✓	✓	✓	✓	✓	✓		✓	✓	

- Dewat 1 – Post-thickening dewatering
- Dewat 2 – Post-digestion dewatering
- Sludge FM – Sludge force main

PROJECT APPROACH
As regional utilities are faced with tighter regulations, and simultaneous goals for economic and environmental sustainability, it is advantageous that they identify synergies among multiple facilities in order to improve efficiency and performance. This approach was utilized to develop a County-wide short-term (5-Year) and long-term (25-Year) Residuals Management Master Plan for Westchester County (NY) Department of Environmental Facilities (DEF).

This study was approached in three distinct phases:

- PHASE 1 – Evaluation of Current Operations
- PHASE 2 – 5-Year Improvement Planning
- PHASE 3 – 25-Year Modification Planning.

PHASE 1 - EVALUATION OF CURRENT OPERATIONS

APPROACH
A detailed evaluation of the current operations and equipment conditions at each facility was conducted to establish baseline conditions, account for areas for performance improvement, and identify the potential for consolidation of residuals handling processes between multiple County facilities.

The existing processes and specific performance parameters evaluated are listed below:

- Screening – volume and/or mass produced
- Grit Removal – grit production per MG treated, total grit volume, hauling trips
- Primary Treatment – surface overflow rate, primary sludge production, solids capture
- Secondary Treatment – surface overflow rate, solids loading rate, WAS production
- Thickening – solids capture, thickened sludge production, thickener recycle quality, chemical usage (e.g., polymer)
- Anaerobic Digestion – total solids reduction, volatile solids reduction, solids retention time, gas production
- Dewatering – solids capture, dewatering recycle quality, cake production

RESULTS

Each facility was provided with a dedicated technical memorandum that provided a summary of current residuals related equipment, current operational practices (e.g., units in service, pump run time, chemical doses, etc.), performance of each process, and general conclusions.

The evaluation of current operations provided a strong basis for recommendations in the 5-year and 25-year planning horizons.

PHASE 2 - 5-YEAR IMPROVEMENT PLANNING

APPROACH

The main focus of the 5-Year plan was to develop strategies for optimizing existing facilities. Process recommendations were made for the individual WRRFs and pumping stations to reduce operating costs through reduction in residual volume and more efficient energy usage. Private liquid sludge and dewatered cake hauling contracts are a large component of annual operating costs for DEF, which makes volume reduction a priority for the County.

RESULTS

Examples of facility specific recommendations to achieve these goals include the following:

- Automation of scum removal systems (MM, OS, YJ)
- Continuous polymer addition for sludge thickening processes (OS, PK, PC)
- Separate sludge thickening of PS and WAS via gravity belt thickeners (GBTs) (OS, PK)
- Installation of a combined-heat and power (CHP) system for beneficial use of digester biogas (PK)
- Co-digestion of HSOW (food-waste) (PK, YJ)
- Disposal of dewatered cake from YJ and NR WWTPs at RESCO facility in Peekskill, NY for waste-to-energy incineration process

In addition to providing facility specific recommendations to improve operations, it was recommended that the County consolidate several facilities (Blind Brook, Port Chester, Ossining, and Peekskill WRRFs) by utilizing the additional capacity at the Peekskill WRRF to thicken, digest, and dewater. Additionally, it was determined that this option becomes more financially feasible when including CHP, co-digestion of high-strength organic waste (HSOW), and dewatering operations at the Peekskill WRRF.

As shown in **Figure 3**, the recommendations presented in the 5-Year plan are expected to save the County $1.1 million annually on operating costs.

Figure 3: Source of savings for 5-Year scenario when compared to Current Operation

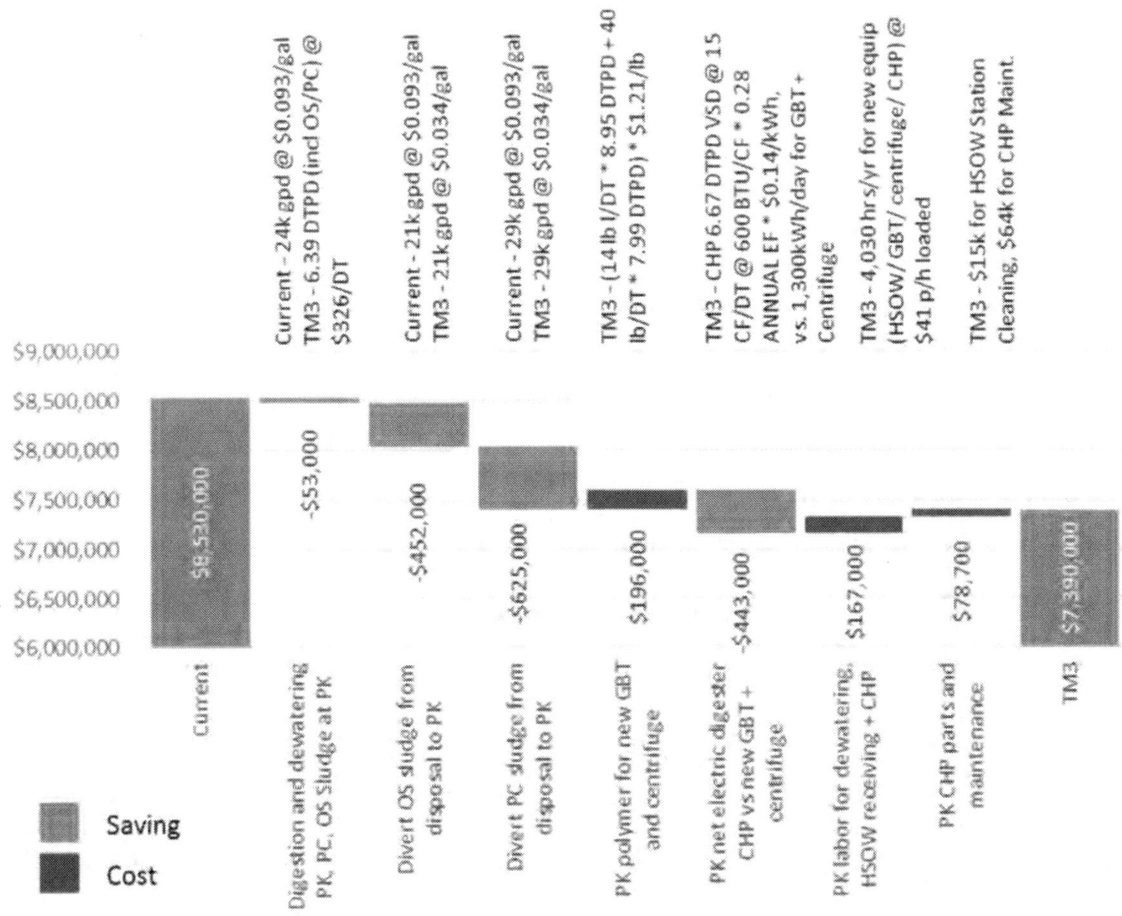

25-YEAR MODIFICATION PLANNING

APPROACH

The 25-year Modification Plan was designed to provide the County with the flexibility to adapt to realistic changes to market, societal, and regulatory trends. The following four 'triggers' were identified that would necessitate improvement in the 25-year planning horizon:

1. **Class A Biosolids Requirements:** Either regulatory (US EPA or DEC) or County programmatic mandates require that biosolids must be treated to Class A standards, per 40 CFR Part 503.
2. **Limited Disposal Options:** Changes to current disposal options (either access or cost) for residuals (including the potential use of RESCO) requires that alternative strategies are considered to reduce the mass and volume of disposed biosolids and thereby mitigate cost risk.
3. **Sustainability Mandates:** Biosolids management must be done sustainably (either based on County targets or applicable mandates).

4. **Nutrient Reduction Requirements:** Permitted limits for nutrient discharge in WRRF effluents (ammonia and phosphorous) become more stringent, requiring plant nutrient loads associated with biosolids operations to be minimized. This issue is particularly applicable to BNR plants but may also effect non-BNR plants in the future.

A 7-step process was used to evaluate the recommended strategy for each trigger, with a focus on strategies that could support multiple triggers. Briefly, the steps included:

1. Identifying the County's planning priorities with respect to technology, economics, environment, and society
2. Using a fatal-flaw analysis to short-list six technologies for further consideration
3. Proposing scenarios that described which technologies would be applied at which plant to support each trigger
4. Scoring the scenarios according to 13 qualitative parameters
5. Scoring the scenarios according to 8 quantitative parameters. A decision making tool was created to assist with calculation of quantitative parameters. An operating cost sensitivity analysis was also part of this step.
6. Comparing how scenarios scored against County priorities
7. Making final recommendations for each trigger

Twenty different scenarios were evaluated that build on the recommendations proposed in the 5-Year plan. In the 5-Year plan, sludge from OS, PC, and BB is first digested and dewatered at PK, and sludge from MA is first dewatered at NR. All scenarios are based on wastewater flows corresponding to population projections for the year 2040. Scenarios were compared to the performance of Current Operations and the 5-Year plan, which both assumed hauling and disposal of dewatered cake to either landfill or incineration for $75/DT. Factors used for comparison include operating cost (which included debt service for fully financed capital improvements) carbon footprint, truck traffic and jobs required. The ability for the existing YJ and PK facilities to accommodate increased biogas production and the ammonia load from digestate return was evaluated for each scenario. An example of the 'dashboard' built into the multi-facility Residuals Handling Evaluation and Optimization Tool is displayed in **Figure 4**.

Figure 4: Dashboard for the 5-Year plan compared to Current Operations

RESIDUALS MANAGEMENT MASTER PLAN PROCESS FLOW AND MASS BALANCE Scenario: TM3

The TM3 Scenario provides modifications from the baseline scenario to partially centralize County sludge management at Peekskill:
- Separate thickening of primary sludge in the existing gravity thickeners at Peekskill
- Thickening of Peekskill WAS and hauled combined sludge from Ossining and Port Chester in new Gravity Belt Thickeners at Peekskill
- Recommissioning of the Peekskill dewatering building, using Centrifuges
- Acceptance of HSOW in the digesters at Peekskill
- Installation of a CHP system at Peekskill to beneficially use digester biogas

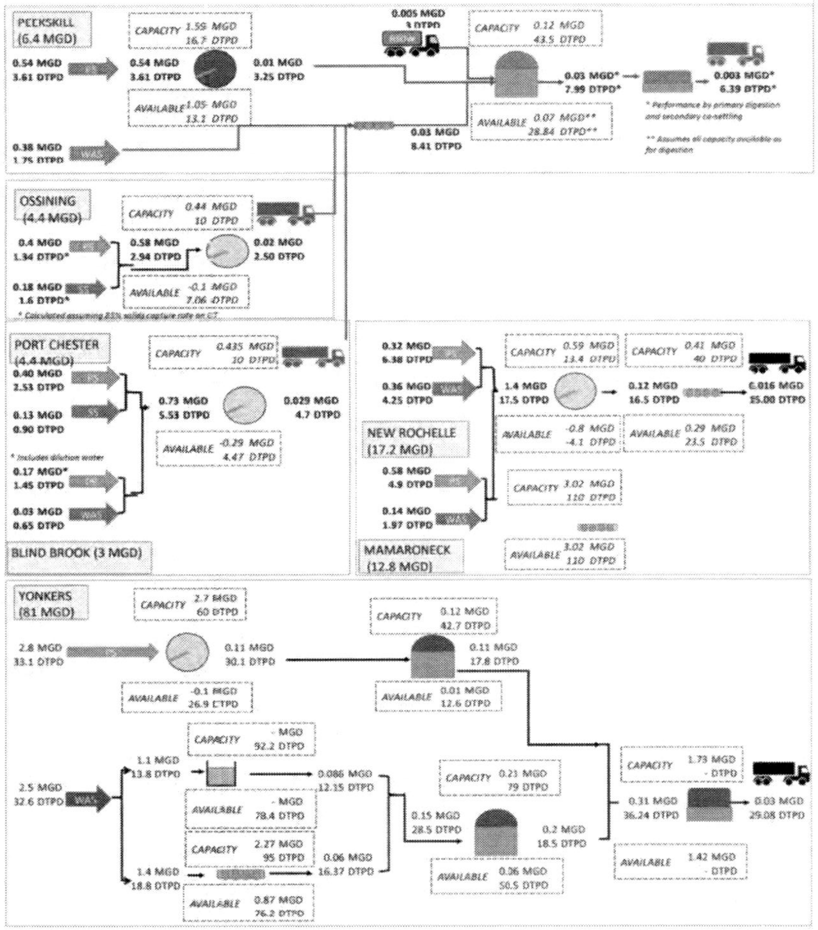

RESIDUALS MANAGEMENT MASTER PLAN MODELING RESULTS Scenario: TM3

The county wide results for residuals operations are summarized below for both this scenario and the baseline (current) scenario. The simple payback time on the capital improvements proposed under the scenario is calculated based on the estimated annual operational cost savings. The six bar charts compare plant performance under the baseline and proposed scenario on an annual and per dry ton basis for a) operating cost, b) carbon footprint c) additional labor hours required.

SUMMARY TABLE	Baseline	Scenario	Change
Operating Costs ($/MM/yr)	10.13	8.92	-1.21
Carbon Footprint (MT CO2.e/yr)	5,137	4,051	-1,087
Labor Hours (Hrs/yr)	6,125	9,588	3,463
Dry Tons of Material Disposed (DT/yr)	19,740	18,424	-1,316
Wet Tons of Material Disposed (WT/yr)	183,977	76,190	-107,787
Vehicle Miles Traveled (per year)	614,534	381,890	-232,644
Biogas Production (MMCF/yr)		73	73

Capital Cost	$MM	14.7
Simple Payback Time	Years	12.2

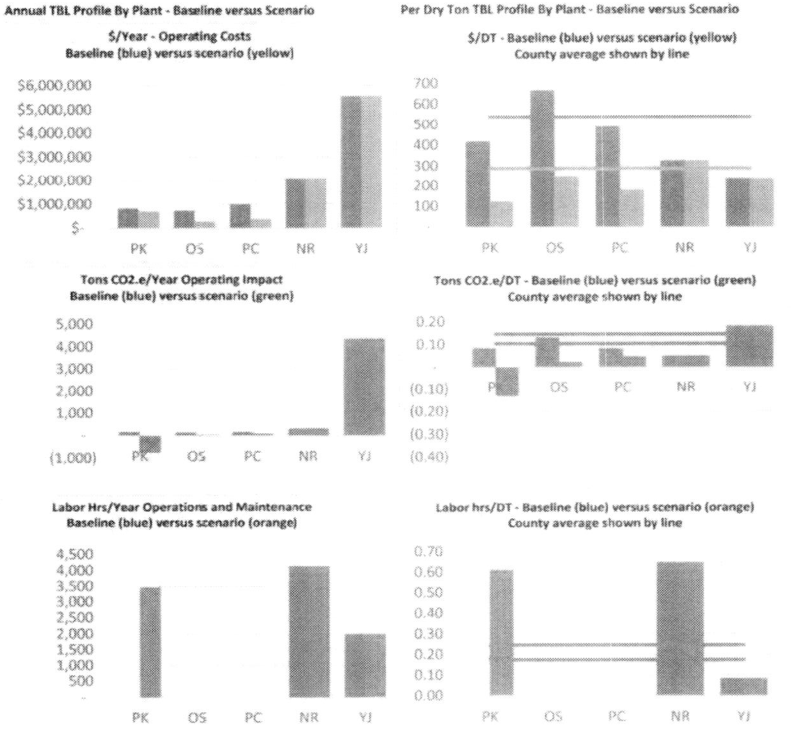

From these twenty scenarios, the seven below were shortlisted based on projected cost savings:

- **YJ Co-Digestion:** Co-Digestion of High Strength Organic Waste (HSOW) at the Yonkers Joint (YJ) WRRF.
- **YJ + PK Hydrolysis:** Thermal hydrolysis at the YJ WRRF and Thermo-Chemical Hydrolysis at the Peekskill (PK) WRRF. Class A biosolids go to beneficial reuse.
- **YJ Central Hydrolysis:** Centralized hydrolysis at YJ WRRF for processing of dewatered cake from the New Rochelle (NR), PK, and YJ WRRFs. Class A biosolids go to beneficial reuse.
- **PK Central Drying:** Centralized drying of dewatered cake from the NR, PK and YJ WRRFs, either at the PK WRRF or at a site nearby the RESCO incinerator to take advantage of waste heat. Class A biosolids go to beneficial reuse.
- **YJ TPAD**: **Conversion of the existing digesters at YJ to operate as Temperature Phased Anaerobic Digesters for processing of cake from NR, PK and YJ.** Class A biosolids go to beneficial reuse.
- **YJ Struvite Recovery:** Recovering phosphorous as struvite from the dewatering side-streams at YJ WRRF.
- **PK Central Drying + YJ Co-Digestion:** Co-Digestion of HSOW at the YJ WRRF and Centralized drying of dewatered cake from the NR, PK and YJ WRRFs, either at the PK WRRF or at a site nearby the RESCO incinerator to take advantage of waste heat. Class A biosolids go to beneficial reuse.

The summary results of the priority matching analysis are presented in **Figure 5**, which illustrates that YJ TPAD is the best matched scenario. The lower scores for YJ Struvite and YJ Co-Digestion reflect the fact these scenarios each only benefit one trigger: Nutrient Reduction Requirements and Sustainaility Mandates, respectively. Results from the decision making tool for the scenarios with the three highest priority matchings are compared in **Table 1**, along with results for the Current Operation and the 5-Year plan. Results for the current operation indicate $8.5 million per year operating cost associated with residuals management, approximately $4.5 million of which is associated with disposal of residuals. The scenarios modeled for the 5-Year plan and 25-Year plan mainly achieve savings by reducing this annual disposal cost while simultaneously increasing biogas production for use in a Combined Heat and Power (CHP) system.

Figure 5: Normalized and priority weighted aggregate quantitative and qualitative scoring for each scenario, providing a measure of how well a scenario matches the County's priorities

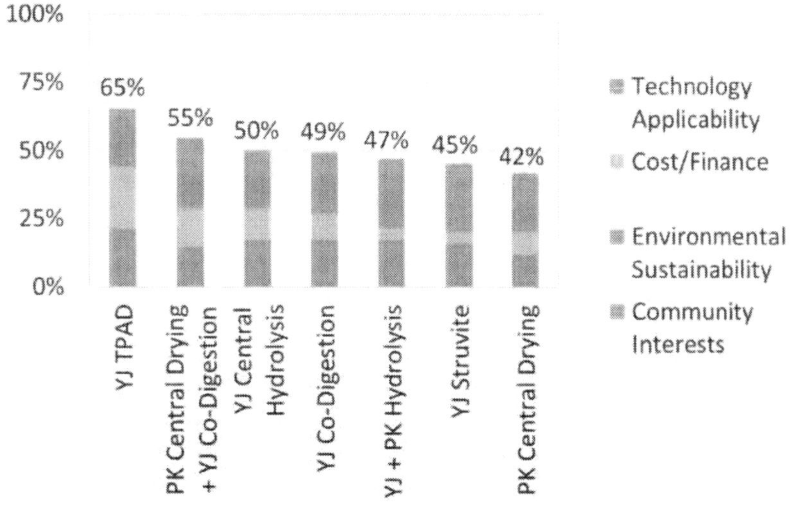

Table 1: Summary of Quantitative Modeling Results

Trigger			Nutrient Removal		
Scenario Name	Current Operation	5-Year Plan	YJ Central Hydrolysis	PK Central Drying + YJ Co-Digestion	YJ TPAD
Dry Tons per Year Processed	35,400	39,100	39,100	42,000	39,100
Total Annual O&M ($MM)*	$8.5	$7.4	$2.3	$2.9	$1.7
Range Estimated Capital Cost ($MM)	$-	$14.7 - $22	$42.2 - $63.3	$35.7 - $53.5	$30 - $45
Annual Saving ($MM)	$-	$1.13	$5.10	$4.49	$5.70
Range Estimated Payback Time (yrs)	0.0	13 - 19.4	8.3 - 12.4	7.9 - 11.9	5.3 - 7.9
Processing Cost ($/DT)	$241	$189	$59	$69	$43
Total Wet Tons per Year Disposed	184,000	80,500	32,800	18,900	46,800
Total Dry Tons per Year Disposed	19,700	18,400	12,200	17,400	14,600
Estimated YJ ammonia load from dewatering (mg/L NH_3)	3.23	3.23	6.31	3.27	4.31
Estimated PK ammonia load from dewatering (mg/L NH_3)	3.38	6.70	6.70	6.70	6.70
Carbon Footprint (MT CO_2 e/yr)**	650	(350)	(1,600)	(1,500)	(2,300)
Equivalent Number of Cars***	124	(67)	(305)	(286)	(438)
Processing Intensity (MT CO_2 e/DT)	0.02	-0.01	-0.04	-0.04	-0.06
FTEs Required	5	7	13	15	9
Processing Intensity (hrs/DT)	0.29	0.36	0.66	0.73	0.45
Trucks per Day at YJ	5	5	7	5	9
Effective Processing Cost incl Loan Repayment ($/DT)**	$241 - $298	$217 - $279	$138 - $177	$132 - $166	$100 - $127

*** Annual operating cost + annual loan payment amortized over 20 years at a 4% interest rate. Operating cost includes all fully loaded labor, polymer, electrical, transport, maintenance, and disposal costs.**
**** Includes all truck transport between plants**
***** Based on a standard passenger vehicle driving 10,000 miles per year**

Achieving County-wide sludge dewatering in the 5-Year plan will save up to $840,000 per year after debt service and reduce operational carbon footprint by 1,000 MT CO_2/yr. The YJ TPAD scenario provides additional projected annual savings of approximately $3-4 million after debt service for fully financed capital improvements with a Class 5 estimate of $30-$45 million. The majority of savings are achieved by avoiding landfill disposal costs because the Class A bisolids produced by the TPAD scenario can be beneficially reused. Estimated carbon footprint reduction of approximatly 2,000 MT CO_2 per year compared to the 5-Year plan is equivalent to removing over 350 cars per year from the road based on passenger cars driving 10,000 miles per year. The anticipated increase in ammonia loading at YJ from the dewatering return is estimated to be equivalent to only 1 mg/L NH_3. Anticipated drawbacks to YJ TPAD include traffic increase of four trucks per day at YJ and the need to replace the centrifuge dewatering system with belt filter presses to prevent sudden bacterial reactivation in Class A bisolids.

The recommendation of YJ TPAD is based on the balanced weighted priorities of the County and the goals of this study to plan for the four triggers. However, should the identified drawbacks make YJ TPAD unacceptable to the County, this study has provided two technically viable alternative scenarios that also project significant operational savings: YJ Co-digestion + PK Central Drying; and YJ Central Hydrolysis. Each scenario has it's own pros and cons and varying ability to provide a positive or negative impact on the future triggers considered by this study, as summarized in **Table 2.**

Table 2: Strategies Considered for Recommendation and their Impact on Future Triggers

Scenario and Rank	Major Pros	Major Cons	Trigger*			
			C	D	S	N
YJ TPAD **1st**	• Relatively reliable savings projection of $100/DT to $140/DT compared to the current scenario. • Only a small increase in centrate ammonia load at YJ • Relatively simple technology to operate compared to thermal dryers and thermal hydrolysis	• Requires change in dewatering technology at YJ from centrifuge to belt filter press • Increases YJ truck traffic by up to 4 trucks per day	+	+	+	0
YJ Co-Digestion + PK Central Drying **2nd**	• Avoids additional ammonia load at YJ • Only a small increase in truck traffic at YJ • Co-digestion improves economic and environmental performance of Central Drying alone	• Additional 4,000 MT/yr. carbon footprint above current operation if a suitable offsite location for free heat cannot be identified. • Risks associated with dryer operation	+	+	-	+
YJ Central Hydrolysis **3rd**	• Ability to use additional projected biogas for either heat or electric beyond that used by two engines would make this option economically competitive with and environmentally superior to TPAD	• Additional 3-4 mg/L of ammonia load to plant anticipated from centrate return • Relatively new technology	+	+	+	-

*** Key to Triggers: C = Class A Bisolids Requirements, D = Limited Disposal Options, S = Sustainability Mandates and N = Nutrient Reduction Requirements**

The risk analysis indicated that the economic performance projected for the 5-Year plan is highly sensitive to future landfill disposal costs. Conversely, the recommendations proposed under 25-Year plan provide more robust projections for economic savings that mitigate the risks present under the Current Operation and the 5-Year plan. The key to providing economic savings is to begin producing and successfully distributing Class A biosolids for beneficial reuse. Hazen recommends that the next step to support this 25-Year-Plan would be a Class A Biosolids Marketing Study, to identify potential local outlets for Class A biosolids, understand regulatory and permit requirements, and to confirm the minimum economic assumptions made by this study. Assuming the Class A Biosolids Marketing Study supports the strategies proposed by this study, the next step would be to develop a concept design for the desired scenario.

Conclusions

This study has proposed several technology scenarios that would allow the County to respond to multiple future triggers while reducing operating costs from current conditions. The three best ranked scenarios according to the Priority Matching step were YJ TPAD, YJ Co-Digestion and PK Central Drying and YJ Central Hydrolysis. The final selection by the County will depend on weighted consideration of the various pros and cons of each scenario.

Mechanistically Understanding the Dewatering Fundamentals: Impact of Biological Systems and Thermal Hydrolysis on Cake Total Solids and Polymer Demand

Mahmudul Hasan[1, 2], Qi Zhang[2,3], Rumana Riffat[1], Ahmed Al-Omari[2], Sudhir Murthy[2],
Matthew J. Higgins[4], Haydee De Clippeleir[2]
[1]*Department of Civil & Environmental Engineering, The George Washington University, 800 22nd Street, NW, Washington, DC 20052, USA*
[2]*DC Water, 5000 Overlook Ave. SW, Washington, DC 20032, USA*
[3]*Laboratory of Microbial Ecology and Technology (LabMET), Faculty of Bioscience Engineering, Ghent University, Coupure Links 653, 9000 Ghent, Belgium*
[4]*Dept. Of Civil and Env. Eng., Bucknell University, Lewisburg, PA, 17837, USA*

Abstract

A study was performed for a variety of different sewage sludges in order to determine the factors that impact polymer demand and cake total solids (TS), along with to investigate the optimum polymer dose of each sludge type. This study hypothesized that polymer demand is directly correlated with soluble protein and polysaccharide concentration of the sludge, whereas the cake total solids TS are related to the extracellular polymeric substances (EPS) and structure water. Results indicated that sludges from the thermal hydrolysis process (THP) required more polymer dosage to dewater, compared to primary sludge, waste activated sludge (WAS) and the mixture of primary and WAS due to high content of protein and polysaccharide release in the bulk liquid of the sludge. A strong linear correlations were found between optimum polymer dose (OPD) and soluble protein (R^2=0.99), and between OPD and soluble polysaccharides (R^2=0.95) of the sludge. However, THP treatment increased the cake TS by 88% for the mixed sludge and 19% for the anaerobic digested sludge. The study also found that the cake TS was negatively correlated with bound water (R^2= 0.97) and EPS (R^2=0.76) in the sludge. This suggested that the improvement of cake TS were mostly achieved by the THP causing release of bound water from the sludge.

Key words

Biosolids, Dewaterability, Sludge Characterization, Optimum Polymer Dose, Capillary Suction Time, Extracellular Polymeric Substance,

Introduction

Wastewater treatment plants generate a significant amount of sludge in the form of primary and secondary solids which require further treatment. Many resources have been devoted to reduce the cost of treatment and disposal by decreasing the sludge production, and improving the sludge dewatering process. In most cases, the cost of excess sludge management and handling, including dewatering as well as transportation, drying and incineration, comprises a significant fraction of operational costs for the wastewater treatment plants (Mikkelsen and Keilding, 2002). Among all these activities, sludge dewatering alone contributes to 30 to 50% of the annual operating cost (Sørensen, 1996).

In the treatment plant, sludge dewatering is optimized in such a way that the total solids of dewatered cake are high, whereas the polymer needed for dewatering is low. Usually, due to the biological process like activated sludge process and anaerobic digestion, the cake total solids are reduced significantly. But the factors affecting the cake total solids of dewatered sludge cakes are not well known. The biological process actually changes the composition and properties of the sludge, such as the presence of the extracellular polymeric substrates (EPS), the strength of floc structure, particle size distribution, bound water, and surface charge which eventually influences the cake total solids and the overall dewaterability (Fitria, 2014). Moreover, the presence of organic contents in sewage sludge leads to the formation of colloidal structure in sludge floc particles. These particles are very hydrophilic and make it difficult to remove water from the sludge structure (Li, 2007). However, that structure water has a good relation with EPS in sludge. EPS refers to an insoluble organic substances adhered to the surface of the sludge bacterial cell and accounts for 50% to 90% of total organic matter in activated sludge (Urbain, 1993). Besides, EPS mainly consists of protein and polysaccharides, which contribute 70% to 80% of total EPS (Dignac, 1998). Due to the strong water binding capacity, EPS accounts for water binding to the sludge surfaces and capture water inside the sludge flocs (Ding, 2014 and Keiding, 2001). Thus it was hypothesized that the structure water and EPS could play important roles on cake total solids.

For the improvement of sludge conditioning and dewaterability of primary and activated sludge, thermal hydrolysis process (THP) was introduced, and hence significant solid-liquid separation was achieved (Haug, 1978). Apart from sludge dewatering, THP also improves the disintegration of organic substances leading to the release of cell contents. Later, it was found that anaerobic digestion of sludge combined with thermal hydrolysis can dramatically improve better dewatering (Kim, 2003). Although THP has been shown to greatly impact the dewaterability of sludges, the extent of the impact and the mechanisms for the improvement have not been well studied.

Poor dewatering of sludge results excessive polymer conditioning and hence increases the operating cost. Several researchers tried to find out the reasons for the poor dewatering. Zhou et al. (2002) found a direct correlation between the dewatering rate and the amount of EPS. Later, Higgins et al. (2004) showed a very good linear correlation between protein concentration and the optimum polymer dose (OPD) measured for a number of different digested sludge types. It is generally accepted that charge neutralization plays an important role for sludge conditioning by the polymers, but it was not well proven. However, no study was conducted to determine the OPD for different types of sludges from the different treatment processes.

In this context, the objective of this study was to evaluate the dewaterability of sludges through the treatment processes that included biological treatment and thermal hydrolysis along with other conventional treatments. Comprehensive analysis of sludge characteristics in combination with determination of optimum polymer dose, cake solids and capture efficiency was performed. Specifically the impact of biological processes (activated sludge systems and anaerobic digestion) as well as the impact of THP were studied in detail to understand the mechanisms behind its dewatering behavior.

Materials and Methods

Sludge Sample: Six sludge samples were studied throughout the wastewater treatment process at the Blue Plains Advanced Wastewater Treatment Plant (Blue Plains AWTP) in Washington, DC, USA. The samples included the primary sludge, biological activated (secondary) sludge, the blend of primary and secondary sludge (1 to 1 ratio on mass basis) which is referred to as the mixed sludge, the thermal-hydrolyzed mixed sludge (referred to as the THP sludge), the anaerobic digested sludge after thermal hydrolysis (referred to as the AD sludge), and the thermally-hydrolyzed anaerobic digestion sludge. The thermally-hydrolyzed anaerobic digestion sludge was obtained by thermally-hydrolyzing the AD sludge at 160° C for 30 minutes in a THP pilot. The initial TS of all types of sludge was adjusted to around 4-5%.

Before the dewatering, all the initial characterization of all types of sludge was done. The protein concentration was measured using the Hartree (1972) modification of the Lowry et al. (1951) method, whereas the polysaccharide concentration was measured using the Dubois et al. (1956) method. The cake total solids (TS) was determined by centrifuging at 3000 x g for 10 minutes to simulate the belt filter press. The surface charge was estimated by colloid titrations according to the procedure reported by Mikkelsen and Keiding (2002). The EPS was extracted by the heating method described by Li (2007) and the structure water was determined by Lee (1995).

Dewatering tests: Lab-scale dewatering tests were performed following the methodology described by Higgins et al. (2014) and capillary suction time (CST) was used to assess the conditioning polymer demand by APHA et al. (1995). All the tests were performed in triplicate. Filtrate was analyzed for total suspended solid (TSS), volatile suspended solids (VSS), total chemical oxygen demand (COD), particulate COD, colloidal COD, soluble COD, NH3-N, soluble protein, soluble polysaccharide and surface charge. The OPD range was determined by the change in slope of CST curve which has range of -2 to 0.

Results and Discussion

The primary sludge was easily dewatered, achieving around 99% capture efficiency and producing a 42% cake TS at the OPD of 6.1 g polymer/kg sludge-TS. The secondary sludge only achieved around 15% cake solids and 97% capture efficiency, at an OPD of 5.2 g polymer/kg sludge-TS. The mixed sludge, which was a mixture of 50% of primary sludge and 50% of secondary sludge, had cake solids of 24%, and higher polymer dose could not improve the dewatering.

After the thermal hydrolysis process, the cake solids of the mixed sludge increased by 88%, from around 24% to 45%, while the polymer demand increased by 4.4 g polymer/kg sludge-TS. Anaerobic digestion decreased the cake solids by 29% (from 45% to 32% TS). After the AD sludge went through THP, the cake solids increased by 19%, from around 32% to 38% with an increase in polymer dose by 5 g polymer/kg sludge-TS.

OPD had a strong linear correlation with soluble protein ($R^2 = 0.99$), with soluble polysaccharides ($R^2 = 0.95$), and with the sum of soluble protein and polysaccharides together ($R^2 = 0.98$), whereas there is no clear correlation between OPD and soluble COD ($R^2 = 0.19$). In all cases, the THP sludge sample didn't fit the correlation curve and regarded as the outlier. Higgins et al., (2004) observed the similar correlation between the polymer demand and the soluble protein and polysaccharide concentrations, but in our study, the ranges of soluble proteins and polysaccharides were wider, from 0 to 8000 mg/L, compared to from 0 to 1400 mg/L in Higgins et al. (2014) study. OPD also showed a good correlation ($R^2 = 0.6$) with the sludge surface charge.

It was hypothesized that cake TS was limited by the structured water content of the sludge, and negatively related to sludge EPS content. Initial measurement on structure water indeed showed a good linear correlation with cake TS ($R^2 = 0.97$), and cake TS also negatively related to the EPS content in sludge ($R^2 = 0.76$). The latter might be a good indicator of structure water content (Mikkelsen and Keiding, 2002) and is more straightforward. Biological sludge had higher EPS content and lower cake TS, as the case of secondary sludge and AD sludge. Moreover, the role of THP in improving cake TS might be related to the release of structure water from cells due to the impact of high pressure and temperature, resulting in lower bound proteins and thus EPS. After THP treatment, bound EPS decreased from 167.5 to 94.3 mg COD/g VSS for the mixed sludge and 70.5 to 60.4 mg COD/g VSS for AD sludge, and thus decreasing the water retaining capacity of the sludge (Neyens, 2004). Similarly, the structured water content decreased by 31% from 7.4 to 5.1 gH_2O/g TS resulted from freeing of the cell structure water due to higher energy input (Farno, 2014).

Conclusion

This study focused on understanding the dewatering characteristics of different sludge types including biological sludge and the THP sludge in terms of polymer demand and cake TS. The polymer demand related to the soluble protein and polysaccharide content of the sludge. Cake TS related to the structure water and EPS content. Biological systems such as secondary systems decreased the cake TS due to the increase of EPS whereas improvements in cake TS were achieved by the thermal hydrolyzed process due to releasing of structure water.

Acknowledgement

Financial support for this research project was provided by the District of Columbia Water and Sewer Authority, Washington, DC, USA. The assistance of all the personnel at the Blue Plains AWTP research laboratory is gratefully acknowledged.

** Due to licensing and copyright regulation by Water Environment Federation (WEF), more detailed and additional information will be given in the upcoming journal publication. For any query the readers are encouraged to contact mahmudul@gwmail.gwu.edu.

References

1. Dignac, M. F., Urbain, V., Rybacki, D., Bruchet, A., Snidaro, D., & Scribe, P. (1998). Chemical description of extracellular polymers: implication on activated sludge floc structure. *Water Science and Technology*, *38*(8-9), 45-53.
2. Ding, A., Qu, F., Liang, H., Guo, S., Ren, Y., Xu, G., & Li, G. (2014). Effect of adding wood chips on sewage sludge dewatering in a pilot-scale plate-and-frame filter press process. *RSC Advances*, *4*(47), 24762-24768.
3. Farno, E., Baudez, J. C., Parthasarathy, R., & Eshtiaghi, N. (2014). Rheological characterisation of thermally-treated anaerobic digested sludge: Impact of temperature and thermal history. *Water research*, *56*, 156-161.
4. Fitria, D., Scholz, M., Swift, G. M., & Hutchinson, S. M. (2014). Impact of sludge floc size and water composition on dewaterability. *Chemical Engineering & Technology*, *37*(3), 471-477.
5. Hasan, M., Zhang, Q., Riffat, R., Al-Omari, A., Murthy, S., Higgins, M., De Clippeleir, H. "Optimization of Dewatering Process by Mechanistically Understanding the Dewater Fundamentals". Manuscript in preparation.
6. Haug, R. T., Stuckey, D. C., Gossett, J. M., & McCarty, P. L. (1978). Effect of thermal pretreatment on digestibility and dewaterability of organic sludges. *Journal (Water Pollution Control Federation)*, 73-85.
7. Higgins, M. J., Chen, Y. C., Maas, N., Troxell, J., & Murthy, S. N. (2004). WERF Phase 1: Factors affecting conditioning during dewatering: Sludge and Polymer Characteristics. *Proceedings of the Water Environment Federation*, *2004*(14), 356-366.
8. Higgins, M.J., Y.C. Chen, S.N. Murthy, D. Hendrickson, J. Farrell, P. Schafer. (2007) Reactivation and growth of non-culturable E. coli in anaerobically digested biosolids after dewatering. Water Research, 44, 665-673.
9. Higgins, M., Bott, C., Schauer, P., & Beightol, S. (2014). Does Bio–P Impact Dewatering after Anaerobic Digestion? Yes, and not in a good way!. *Proceedings of the Water Environment Federation*, *2014*(2), 1-11.
10. Keiding, K., Wybrandt, L., & Nielsen, P. H. (2001). Remember the water-a comment on EPS colligative properties. *Water Science and Technology*, *43*(6), 17-23.
11. Kim, J., Park, C., Kim, T. H., Lee, M., Kim, S., Kim, S. W., & Lee, J. (2003). Effects of various pretreatments for enhanced anaerobic digestion with waste activated sludge. *Journal of bioscience and bioengineering*, *95*(3), 271-275.
12. Lee, D. J., & Hsu, Y. H. (1995). Measurement of bound water in sludges: a comparative study. *Water Environment Research*, *67*(3), 310-317.

13. Li, X. Y., & Yang, S. F. (2007). Influence of loosely bound extracellular polymeric substances (EPS) on the flocculation, sedimentation and dewaterability of activated sludge. *Water Research*, *41*(5), 1022-1030.
14. Mikkelsen, L. H., & Keiding, K. (2002). Physico-chemical characteristics of full scale sewage sludges with implications to dewatering. *Water research*,*36*(10), 2451-2462.
15. Murthy, S. N., Novak, J. T., & Holbrook, R. D. (2000). Optimizing dewatering of biosolids from autothermal thermophilic aerobic digesters (ATAD) using inorganic conditioners. *Water Environment Research*, *72*(6), 714-721.
16. Neyens, E., Baeyens, J., & Dewil, R. (2004). Advanced sludge treatment affects extracellular polymeric substances to improve activated sludge dewatering. *Journal of hazardous materials*, *106*(2), 83-92.
17. Sørensen, B. L., & Wakeman, R. J. (1996). Filtration characterisation and specific surface area measurement of activated sludge by rhodamine B adsorption. *Water Research*, *30*(1), 115-121.
18. Urbain, V., Block, J. C., & Manem, J. (1993). Bioflocculation in activated sludge: an analytic approach. *Water Research*, *27*(5), 829-838.

Phosphorus Release and Thermal Hydrolysis Pretreatment Effects on the Dewatering Properties of a Bio-P Biosolid in Anaerobic Digestion

J. D. Nicholson[1,3*], S. Beightol[2], M. J. Higgins[2], C. B. Bott[3];

[1]Virginia Tech & HRSD, VA, [2]Civil & Environmental Engineering, Bucknell University, PA, [3]Hampton Roads Sanitation District, VA., [*]jeffnich@vt.edu

ABSTRACT

The use of biological phosphorus removal by activated sludge has been shown to have adverse effects downstream in the solids handling processes of anaerobic digestion. From struvite formation problems within the digesters to increased polymer dosage and dewatered biosolids solids content. The fundamental cause is unknown however phosphorus removal prior to anaerobic digestion has been shown to improve cake solids. Thermal hydrolysis pretreatment has also been shown to increase cake dewaterability. This study employed both strategies and has shown that while phosphorus release did not improve cake dewaterability thermal hydrolysis pretreatment has.

KEYWORDS: Dewatering, biosolids, biological phosphorus removal, thermal hydrolysis pretreatment

INTRODUCTION

From the discovery and integration of the natural phenomenon of biological phosphorus release and uptake into municipal activated sludge processes in the late 1960s and early 1970s until the mid-2000s the impact on anaerobic digester solids characteristics has not been well documented. Early studies focused on digester amendments of aluminum and iron salts to prevent phosphorus release in the digested sludge from returning to the mainstream plant after dewatering (Knocke 1992). These studies also found that the addition of these metal salts improved dewaterability of the bio-p biosolids; however, no comparison or significant change was noted between dewaterability of the activated sludge before and after the bio-P treatment was initiated.
More recent experiences with WRRFs converting to bio-p have shown more detrimental effects on the anaerobic digester biosolids polymer dosage and sludge dewaterability of the associated waste activated biosolids (Alm 2015, Benisch 2014). Plants in both Europe and North America have seen considerable decreases, 3-4 % TS, in dewatered cake solids with double or triple the amount of polymer sometimes needed to obtain an optimal dosage. This decreased concentration of solids leads to larger volumes of cake and higher treatment costs either for more hauling of the solids for landfill or composting or higher fuel costs for incineration.

Additionally, the increased loading of phosphorus as polyphosphate granules is ultimately solubilized in the digester which in the presence of an overabundant concentration of ammonia and a limiting amount of magnesium crystallizes to form an insoluble mineral known as struvite. Struvite may deposit and accumulate in the digester solids as well as any process downstream of the digesters solids stream; including pumps, pipes, centrifuges and other dewatering equipment causing blockages or damage to equipment. Recent studies have correlated monovalent to divalent cation ratios, higher orthophosphate concentrations, and higher interstitial water concentrations to decreased dewaterability in bio-P biosolids. Releasing the phosphorus from the activated sludge has been shown to provide a slight increase in cake solids (Higgins 2014). However, a solution has still yet to be found for increasing cake solids equal to or above pre bio-P treatment levels.

The Cambi thermal hydrolysis process (THP) has been shown to improve the rate and extent of digestion as well as improve the dewaterability of solids after anaerobic digestion (Wilson, 2009). THP pretreats the primary and secondary solids with high temperature using pressured steam. Typically operating temperatures are around 165 °C which is high enough to essentially sterilize the solids as well, producing a Class A product. THP is thought to hydrolyze a portion of the organic fraction in the solids, such as proteins, lipids and carbohydrates (Wilson, 2009). This decreases viscosity and reduces the solids that typically bind water which is difficult to remove during dewatering.

This research is intended to identify pathways for improving bio-P biosolids dewaterability, decreasing polymer demand and mitigating problematic struvite precipitation equal to or above pre bio-P levels.

METHODOLOGY

Four 10 L digesters operated at 38°C and a 15 day SRT were semi-continuously fed evaluating 4 treatment conditions. A baseline case evaluated the typical operation of a mesophillic digester fed a bio-P thickened waste activated sludge (TWAS) along with thickened primary sludge (TPS) while a second digester evaluated the effects of phosphorus release (PR) on the TWAS fed with the same TPS. PR was induced through the anaerobic addition of acetate to the WAS before thickening and dewatering in order to induce the release of phosphorus along with magnesium and potassium from the phosphorus accumulating organisms (PAO). The TWAS/TPS and PR-TWAS/TPS solids were also combined prior to being thickened to a cake and being fed to a pilot scale THP reactor and pressurized with steam for 1 hour at 6 bar and then depressurized into a flash tank at atmospheric pressure. These solids were then fed to the control THP reactor and the THP+PR reactor.

Soluble phosphorus as well as soluble cations was measured in order to correlate any effects of phosphorus concentration or monovalent to divalent cation ratios to dewaterability, polymer demand and struvite formation potential. Gas production, solids and COD loading, volatile fatty acids as well as pH and alkalinity were all monitored to insure proper reactor operation.

Once reaching steady state the anaerobically digested biosolids were dewatered according to a protocol described in Higgins 2012 in order to ascertain polymer demand and dewatered cake solids.

RESULTS

The characteristics of the feed solids fed to each reactor are summarized in table 1.

Table 1: Average values and standard deviations in parentheses of the feed solids to each reactor

Reactor Name	Control	PR	THP	THP+PR
TS, % (w/w)	4.34 (0.08)	4.33 (0.07)	8.77 (0.14)	8.77 (0.16)
VS, % (w/w)	3.40 (0.12)	3.61 (0.11)	6.98 (0.23)	7.43 (0.28)
Total COD, mg/L	64572 (2051)	69011 (2041)	133776 (4531)	142356 (5202)
Soluble COD, mg/L	3127 (288)	3497 (222)	23480 (961)	28216 (1225)
TP, mg/L	1707 (133)	1011 (99)	3178 (412)	1785 (177)

The reactors steady state operation parameters are shown in table 2.

Table 2: Average and standard deviation values of the digesters operating parameters.

Reactor Name	Control	PR	THP	THP+PR
pH	6.91 (0.04)	7.18 (0.04)	7.27 (0.05)	7.48 (0.03)
Alkalinity, mg/L as $CaCO_3$	3455 (321)	4571 (427)	8817 (1000)	11053 (369)
TS, % (w/w)	2.73 (0.04)	2.61 (0.04)	5.51 (0.15)	5.18 (0.11)
VS, % (w/w)	1.81 (0.03)	1.90 (0.04)	3.69 (0.13)	3.80 (0.12)
VS Reduction, %	46.9 (1.8)	47.5 (1.6)	47.0 (2.6)	48.4 (0.9)
Total COD, mg/L	30120 (839)	32822 (2063)	65564 (1556)	69710 (1400)
Soluble COD, mg/L	4690 (1075)	4162 (957)	17640 (2561)	17432 (2540)
Total COD Reduction, %	53.06 (2.71)	51.96 (3.83)	50.70 (2.47)	50.71 (2.15)
Digester TP, mg/L	1629 (119)	968 (102)	3283 (178)	1917 (185)
Digester OPO_4-P, mg/L	787 (97)	319 (58)	1361 (135)	579 (90)
Digester NH_4-N, mg/L	1606 (130)	1730 (289)	4090 (252)	4614 (688)
Free Ammonia (mg/L as N)	17.5 (1.1)	32.9 (4.5)	100.2 (10.9)	178.9 (39.2)
Na^+, mg/L	77.8 (5.0)	195.0 (31.0)	64.5 (5.3)	156.5 (12.6)
K^+, mg/L	760 (64.6)	434.6 (71.3)	1240.5 (361.7)	865.3 (44.6)
Mg^{2+}, mg/L	Below Detection Level			
Ca^{2+}, mg/L	9.2 (4.3)	17.4 (1.3)	15 (6.9)	27.5 (9.9)

The optimal polymer dose and cake solid results of the dewatering method are shown in Figure 1.

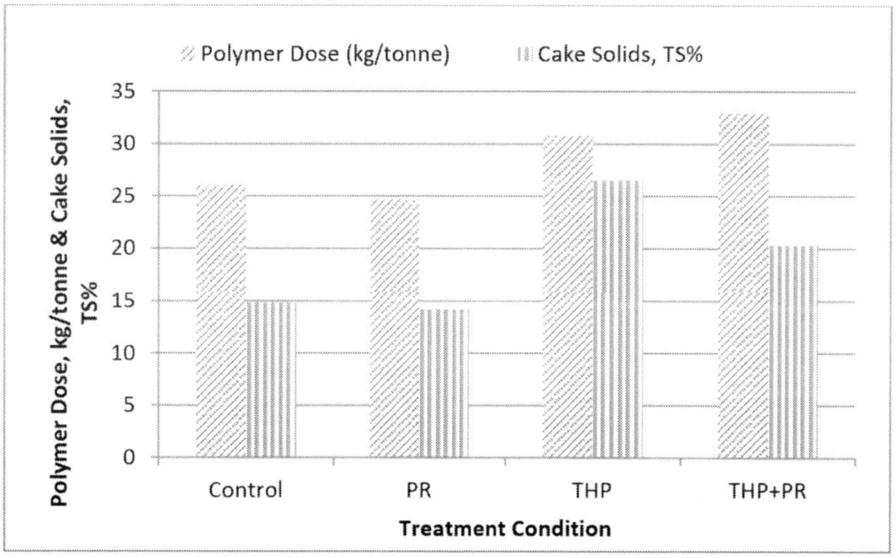

Figure 1: Polymer demand and dewatered cake solids for each reactor condition.

DISCUSSION

The PR reactors show similar differences in both the conventional and THP reactors. They both have higher pH values as compared to their respective controls that may be attributable to the reduced concentrations of phosphorus, about 40% lower, and the buffering effect it has in the system as well as the reduced production of struvite that would decrease the pH. This also confers higher alkalinity to these reactors and may make them more stable, but also increases the amount of free ammonia present. In the PR process inorganic polyphosphate is removed and replaced by organic polyhydroxy alkanoates (PHA) formed from acetic acid. For a similar total solids loading the PR reactors have a higher volatile solids fraction and COD loading. They have similar COD and volatile solids reduction so the PR reactors are also producing more methane and CO_2. During PR the PAOs also release the cations magnesium and potassium to balance the charge of the phosphorus being released. The PR reactors have lower potassium levels and higher sodium due to the sodium acetate added. However the magnesium is unknown as it was below the detection level of 4 mg/L, but the calcium levels are higher due reduced levels of complexation by orthophosphate and this pattern would also be expected for magnesium. This reduction of monovalent cations and increase in divalent cations provides a lower monovalent to divalent ratio in the PR reactors which would be predicted to improve dewaterability.

The PR reactor showed a minor decrease in polymer demand as well as dewatered cake solids of 14.22% TS compared to 14.8% TS of the conventional control reactor. The PR+THP reactor showed an increased polymer demand and a much decreased cake solids of 20% compared to the 26% achieved with the control THP reactor. The results of this study do not agree with the previous results found in Higgins 2014 showing phosphorus release to provide an increase in dewatered cake solids.

CONCLUSION

THP has shown to be a much more advantageous treatment process allowing for much improved dewatering of bio-P solids when compared to phosphorus release alone.

REFERENCES

Alm, R., Sealock, A. W., Sprouse, G. (2015) Investigations into Improving Dewaterability at a Bio-P/Anaerobic Digestion Plant. *Proceedings of WEF Residuals and Biosolids* , Washington D.C.

Benisch, Mario; Schauer, Peter; Neethling, JB. (2014) Improving Dewaterability of Digested Sludge from EBPR Facilities. *WEF Residuals and Biosolids Conference. Austin, TX.*

Higgins, M.J., P. Fountain, S.N. Murthy, K. Kelleher, P. Schafer, A. Cooper, E. Kasirga, K. Lee, J. Machisko (2012) Belt Filter Press Dewatering Trials with Cambi Thermally Hydrolyzed Biosolids – Evaluation of Critical Performance Criteria. *Proceedings of WEF Residuals and Biosolids* , Raleigh, NC.

Higgins, M.J., Bott, C.B., Schauer, P., Beightol, S. (2014) Does Bio – P Impact Dewatering After Anaerobic Digestion? Yes, and N ot in a Good Way! *WEF Residuals and Biosolids Conference. Austin, TX.*

Knocke, W., Nash, J., and Randall, C. (1992). "Conditioning and Dewatering of Anaerobically Digested BPR Sludge." J. Environ. Eng., 10.1061/(ASCE)0733-9372(1992)118:5(642), 642-656.

Wilson, C.A. (2009) Mechanisms of Methanogenic Inhibition in Advanced Anaerobic Digestion. Ph.D. Dissertation, Virginia Tech, Blacksburg, VA.

The role of soluble P & EPS on dewatering performance

Ester Rus*, Paul Fountain*, Nick Mills*, Achame Shana*, Obinna Molokwu**, and Manocher Asaadi**.

* Thames Water, ** AD Technologies

Corresponding author: *Ester.rusperez@thameswater.co.uk*

Abstract

Crossness and Beckton wastewater treatment plants have experienced issues with dewatering of their final digested sludge, requiring double the polymer demand for floc formation than that expected and achieving up to 25% less dry solids in the final cake. This paper looks into two possible root causes for the poor dewatering performance observed onsite: (1) the impact that high concentrations of soluble phosphorous on the monovalent to divalent ions on the sludge and its effects on dewaterability, and (2) the effects of erratic feeding of the digesters and the bacterial enhanced extra-polymeric substances excretion due to shock loads.

Analysis and lab scale dewatering done on digested sludge samples from other 9 THP sites across the UK showed a clear relationship between soluble P and monovalent to divalent ratio, which also showed a relationship with polymer demand and final cake dry solids. Nevertheless, both Crossness and Beckton have seen an improvement in dewaterability with no change or increasing monovalent to divalent ratios, which goes against the observed trend. Furthermore, a set of lab scale digesters seeded with Crossness and Beckton digested sludge and fed with hydrolysed sludge from the two sites under controlled conditions (constant feeding and similar iron dosing) show an improvement in dewaterability with time compared to site. This suggests other factors may be driving this improvement.

Extra-polymeric substances were considered a possible second factor having a negative impact on dewaterability. Both bound and soluble protein related extra-polymeric substances showed a weak relationship with polymer demand. Beckton site showed higher concentrations with higher polymer demand. This was not the case with Crossness.

The results seem to show that both factors explored here play a role in digested sludge dewaterability. Nevertheless, other factors are also most likely having an impact on polymer demand and final cake dry solids and further work to explore other causes is recommended.

Introduction

Thames Water has built a number of Thermal Hydrolysis Plants (THP) to meet new capacity demands, improve efficiency and reduce operating costs. Two of these plants, Beckton & Crossness, located in East London utilise a two stream Cambi B12 THP system, 6 x 4,000m^3 digesters, 3 x 2MW CHP units and 5 x Bucher press dewatering units. The guarantee for the final stage dewatering cake quality was 38% dry solids (DS). Initially the plant dewatering achieved 38%DS with a polymer consumption of 10-11 kg/TDS. Nevertheless, after 2 hydraulic retention times (HRT) the cake DS dropped below 20% and the polymer consumption increased dramatically, averaging 27kg/TDS. This poor dewatering timed with the loss or washout of the seed sludge which had come from a different STW. During this period volatile solids destruction (VSD) and biogas production remained stable and as expected (Rus *et al.*, 2015). The dewatering issue seemed to affect the rate of release of water as well as polymer demand and cake DS. The nature of the dewatering problems is very typical of dewatering issues seen on other Biological Nutrient Removal (BNR) sites. Beckton and Crossness were not designed to operate in BNR mode; nevertheless it was subsequently found that both sites were operating effectively as BNR. However, initial lab investigations suggested that the poor dewatering performance cannot be explained by the BNR \ sludge chemistry issue alone, and suggested the biology could be having a direct impact on dewaterability. Both approaches are reviewed in this paper.

Work at lab scale:

Some research has been done on the effect of soluble phosphate on dewatering of digested sludge, finding that high concentrations of soluble P in the digester could indeed lead to poor dewatering. Higgins and Novak (1997b) and Higgins *et al.*, (2014) concluded that the monovalent to divalent (M/D) ion ratio was a good indicator of dewaterability, being positively correlated with the specific resistance of sludge to filtration, and therefore, cakes with higher DS could be achieved with sludges showing lower M/D ratios. This was also found by Alm *et al.*, (2015) and Park (2002). Higgins *et al.*, (2014) suggested that biological phosphorous is key to understanding the dewaterability issues in digested sludge. According to this research, high concentrations of soluble P in the digester can result in a decrease in available magnesium and calcium ions, which play an important role in bio-floc formation prior to digested sludge dewatering. Furthermore, it can cause the release of monovalent ions such as potassium which are known to be detrimental to dewatering. Since both Crossness and Beckon sites showed signs of opportunistic BNR, this theory was initially explored and an experiment with different concentrations of ferric chloride and sulphide to lab scale digesters was performed. The results showed a drop in sol P and a better dewaterability, both in terms of polymer demand and final cake solids for those lab digesters or chemostats with higher ferric chloride dose (Rus *et al.*, 2015).

Furthermore, it is common knowledge in the wastewater industry that anaerobic digesters perform better under stable conditions. That includes constant temperatures, feed flows, working volumes, loading rates, hydraulic retention times, etc. Data shows that both Crossness and Beckton were seeing a variability in their digester feed of more than 30% during a one given day. Identifying a method to quantify the effects of erratic/ inconsistent feeding or shock loads of a digester has been the real challenge and would be the first attempt for THP+AD sludges. What is proposed in this paper is that under sustained inconsistent feeding of the digesters, bacteria would have to addapt to new, sub-optimal conditions. The biogas generation, methane production, and volatile solids reduction (VSR) may be close to what would be expected from a THP site, nonetheless, bacteria would be using a defence methanism against the unpredictable environment and excreted higher quantities of extracelular polymeric substances (EPS).

EPS is a mix of different classes of organic macromolecules excreted by bacteria as part of their metabolism and to protect themselves from the environment (Aquino and Stuckey, 2004; Aquino and Stuckey, 2007; and Shana, 2015). The main macromolecules present in EPS are polysaccharides and proteins, but there are other polymeric compounds that could be found in the intracellular space or outside the cell surface. In literature, EPS is commonly classified as:

- Soluble EPS - soluble macromolecules, slimes (Sheng *et al.*, 2010).
- Bound EPS – sheaths, capsular polymers, condensed gels, etc. (Sheng *et al.*, 2020).
    - Loosely bound EPS (LB-EPS): Outer layer with no well-defined edge.
    - Tightly bound EPS (TB – EPS): Inner layer that has a structure and it is stable around the cell surface.

The bound EPS has been identified as a gel like material that is able to hold water inside (4-5 kg water/kg bound EPS), which keeps the environment around the cell hydrated, crucial for survival. This also means that releasing water from this gel is difficult and therefore higher content of bound EPS may result in higher polymer demand for dewatering and lower cake dry solids (DS).

Soluble EPS on the other hand can also have a negative impact on dewatering since it is negatively charged and can act as a sink for polymer during conditioning (Higgins *et al.*, 2016). Higgins *et al.*, (2016) found a strong correlation between Sol EPS and optimum polymer dose ($R^2 = 0.88$) for different AD sites.

Both bound and soluble EPS are explored here.

Methodology

1. Lab experiment set up:

Six laboratory or bench scale semi-continuous anaerobic digestion (AD) rigs were set up. Each unit consists of a 10L spherical glass chemostat with 8L working volume (WV), with overhead stirrers for continuous mixing, aspirators with graduated biogas collection bottles, and a common water bath (see figure 1).

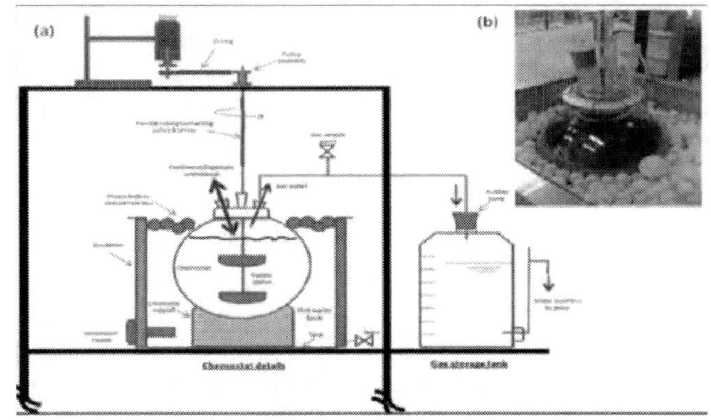

Figure 1: (a) schematic of semi-continuous AD rig and (b) picture of chemostat in a water bath (Shana, 2015).

Chemostats 1, 2, and 3 were seeded with digested sludge from Beckton and chemostats 4, 5, and 6 with digested sludge from Crossness. Ferric chloride was also dosed at different % by weight in the chemostats(see table 1):

Table 1: Sludge procedence, OLR and Fe dose for each chemostat:

Chemostat	1	2	3	4	5	6
Site	**Beckton**	**Beckton**	**Beckton**	**Crossness**	**Crossness**	**Crossness**
OLR (kgVS/m3/d)	3	2	3	3	1.7	3
Fe dose (%w/w) before Nov 2015	-	0.2	2	-	0.2	2
Fe dose (%w/w) after Nov 2015	-	0.4	0.4	-	0.4	0.4

The dosing was changed in November 2015 after the initial trial (Rus *et al.*, 2015). After November 2015, ferric chloride was added at 0.4% w/w for the dosed chemostats as that was considered the maximum dosing possible onsite.

Chemostats 1, 3 ,4, and 6 where set at the organic loading rate (OLR) that would be expected onsite, and chemostats 2 and 5 where set up at the actual OLR onsite.

Every day, hydrolysed sludge from both sites was imported to TW Innovation facilities to feed the chemostats. The feeding was done once a day.

Volatile fatty acids (VFA), alkalinity, pH, ammonia, and gas production were checked 3 times a week.

2. **THP sites sampling:**

Digested sludge was imported in 1 L bottles from different THP sites in the UK: Crossness, Beckton, Oxford, Crawley, Chertsey, Riverside, Cardiff, Brand Sands, Cotton Valley, Afan and Seafield. Crossness and Beckton were sampled regularly to monitor both the dewaterability of the digested sludge at lab scale and in some occasions, digester stability. Chemical analysis was also done.

3. **Chemical analysis:**

Chemical analysis was done by Thames Water accredited labs every 2 to 3 weeks on all chemostats. The analysis included DS, VS, soluble phosphorous, soluble magnesium, soluble calcium, soluble potassium, sulphate, and total iron. The same chemical analysis done for the chemostats was done for the site samples.

4. **Dewatering of samples:**

Dewatering trials with a bench scale piston press and a lab scale Bucher sock were done every 2 to 3 weeks. In some occasions, the frequency of dewatering tests was higher (5 times a week) to observe the effect of reseeding on dewatering. The bench scale piston press is able to mimic a full scale belt press and the Bucher socks are able to give an estimate of full scale Bucher Press performance, remaining a couple of perceptual points below the actual full scale results.

Polymer preparation: The chosen powder polymer (Flopan 4440 CT from SNF) was prepared with water and mixed with a magnetic stirrer for 45 minutes. Using a 500ml beaker, a 200ml sample of digested sludge was collected. The application of polymer to the digested sludge was done with a syringe, adding 5-10ml of polymer at a time and stirring the mix gently until stable flocs formed. Depending on the sludge quality more or less polymer was be added. It has been observed that over dosing polymer does not improve floc quality. Figure 2 shows a sample with a stable floc ready to be dewatered in the Bucher sock.

Figure 2: Flocculated sludge ready to be dewatered.

The flocculated sludge was then dewatered with two different methodologies to mimic the Bucher press and belt press technologies.

(a) **Bucher Press:** the 200ml sample is poured into the Bucher socks with no previous decantation to replicate the Bucher press as closely as possible. The methodology used to dewater in the Bucher sock was provided by Bucher. Appropriate training was provided. Figure 2 shows an example of a Bucher sock:

 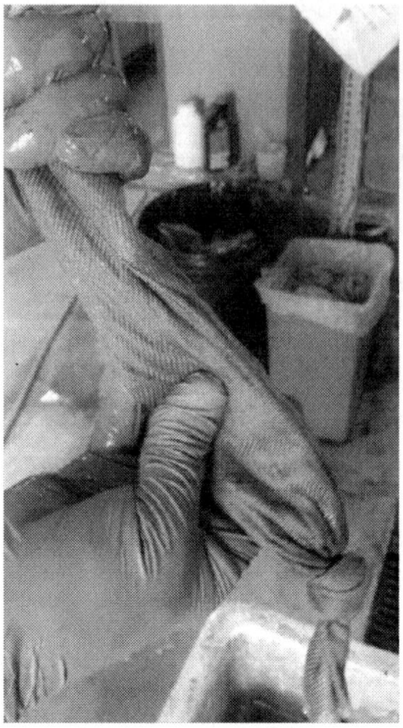

Figure 2: Lab scale Bucher Sock: (left) flocculated sludge added inside the sock, (right) redistribution of sludge in between twisting cycles to simulate full scale Bucher torsion.

(b) **Piston press:** the 200ml sample flocculated sample is added into the piston chamber. The device (seen in figure 3) is then pressurised with an air compressor in 4 different steps: (1) 5 min under 2 barg, (2) 5 min under 5 barg, (3) 10 min under 7.5barg, (4) 10 min under 10barg, and (5) min under 11 barg. During that period water release is recorded every minute.

Figure 3: Piston press.

5. Cation analysis and M/D Ratio:

The calculation for the monovalent to divalent ratio was done using the methodology followed in Higgins and Novak (1997) and Higgins *et al.,* (2014). Na^+, K^+, Mg^{2+}, and Ca^{2+} where analysed using ion chromatography and the M/D ratio was calculated on an eq/eq basis:

$$\textit{M/D Ratio (eq/eq)} = \frac{Na^+ + K^+}{Mg^{2+} + Ca^{2+}}$$

The ammonium ion was not included at this point, since most of the digesters have similar concentrations of 2.8-3gNH_4-N/L.

6. **EPS and SMP:**

The soluble fraction of EPS is obtained by centrifuging the sample for 15 minutes at 3000 rpm at 5°C. The supernatant is then double filtered through 1.5μm filters.

To measure the bound part of EPS, distilled water is added to the pellet after centrifugation and the suspension is transferred into a flask. This is heated for an hour at 80°C and left to cool. The suspension is then centrifuged at 3000rpm for 30 minutes and the supernatant is subjected to double filtration using a 1.5 μm filters.

Results and discussion

1. **Stability of chemostats, specific gas production and volatile solid reduction:**

The average and standard deviation (SD) for alkalinity, VFA, pH, NH_4-N, methane, specific gas production (SGP), and VSR across all six chemostats are shown in table 2.

Table 2: Average and SD for alkalinity, VFA, pH, NH_4-N, methane, SGP, and VSR.

Chemostat	Alkalinity (mg/L)	VFA (mg/L)	pH (-)	NH_4-N (mg/L)	CH_4 (%)	SGP (m^3/TDS)	VSR - VK (%)
1 - Beckton	7943±1068	1205±73	8±0.1	2493±476	64±9	418±40	56±0.1
2 - Beckton	7943±1068	1181±96	8±0.1	2710±420	63±5	439±54	57±0.1*
3 - Beckton	7800±1041	1147±127	8±0.1	2647±367	65±5	408±39	58±0.1*
4- Crossness	8188±839	1302±197	7.9±0.1	2509±317	65±4	414±47	55±0.1
5- Crossness	7600±397	1150±181	7.9±0.1	2457±342	62±5	403±43	56±0.1*
6- Crossness	7615±1015	1071±140	7.9±0.1	2952±309	63±5	408±56	55±0.1*

*Corrected for iron addition.

Some variability on the alkalinity and VFA values was observed due to instrumentation issues. Nevertheless, both parameters remained stable for most of the recorded period. Ammonium levels stayed below 3,000mg/L and the methane content in the biogas remained between 62-65% for all chemostats. The SGP and VSR with the Van Kleeck method were very similar to those observed onsite. Therefore the overall performance of AD seen at lab scale was comparable to that seen at full scale. In terms of gas production and VSR, what is observed at both sites (both at lab and full scale) is what would be expected from a THP +AD site.

2. **Dewaterability of lab scale and full scale:**

Figure 4 shows that eventhought the overall performance of the digesters onsite in terms of gas yield and VSR is similar to that observed at lab scale, the dewaterability is very different, particulartly in the polymer demand.

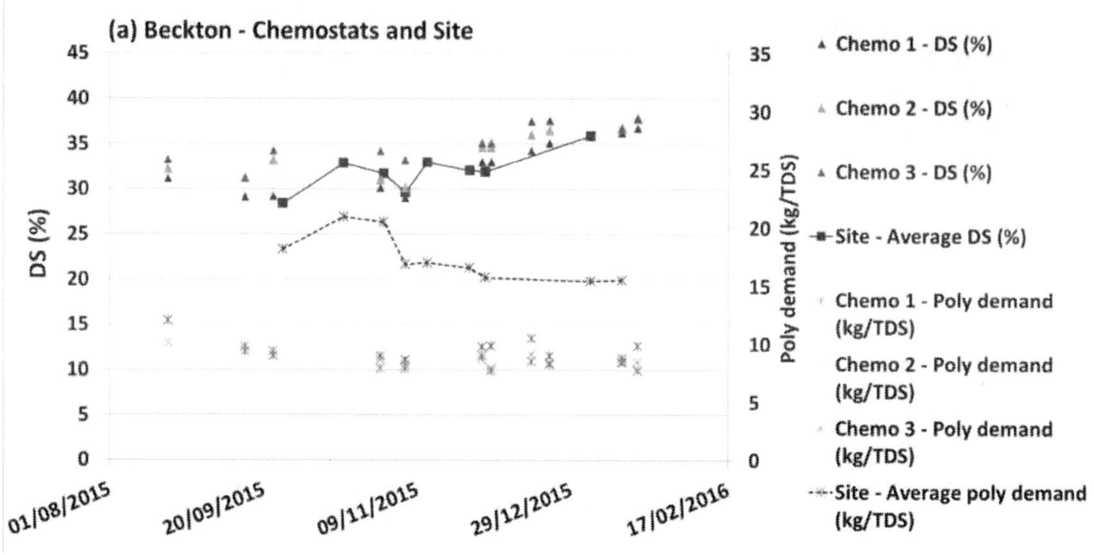

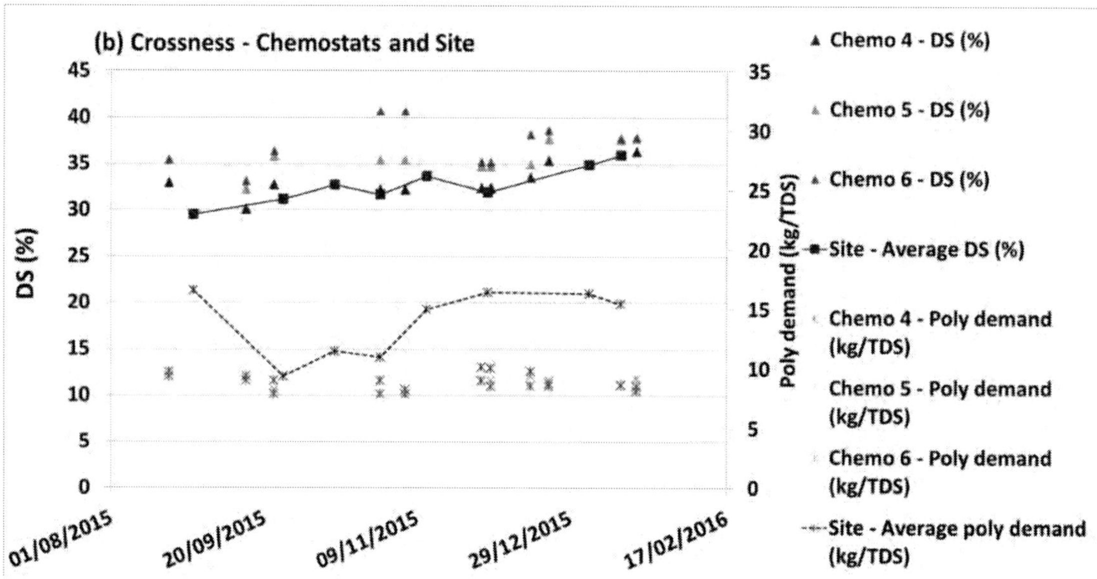

Figure 4: Evolution of full scale dewaterability compared to lab scale. All samples dewatered with a lab Bucher sock.

3. Soluble Phosphate and the monovalent to divalent ratio:

Figure 5 shows the relationship between the M/D ratio done with four main ions: Mg^{2+}, Ca^{2+}, Na^{+}, and K^{+}.

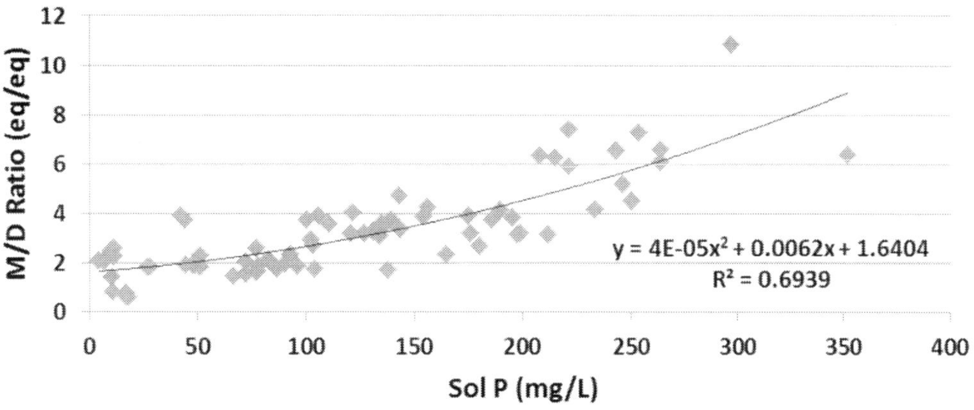

Figure 5: Relationship between sol-P and M/D Ratio (no ammonium) for all THP sites.

A strong correlation was found between both, with an R^2 of 0.7. Looking at the relationship between soluble P and the cations involved in the M/D ratio (figure 6 (a) and (b)), a downward trend was observed for the two divalent ions (Mg^{2+} and Ca^{2+}). Magnesium seems to show a stronger relationship than calcium.

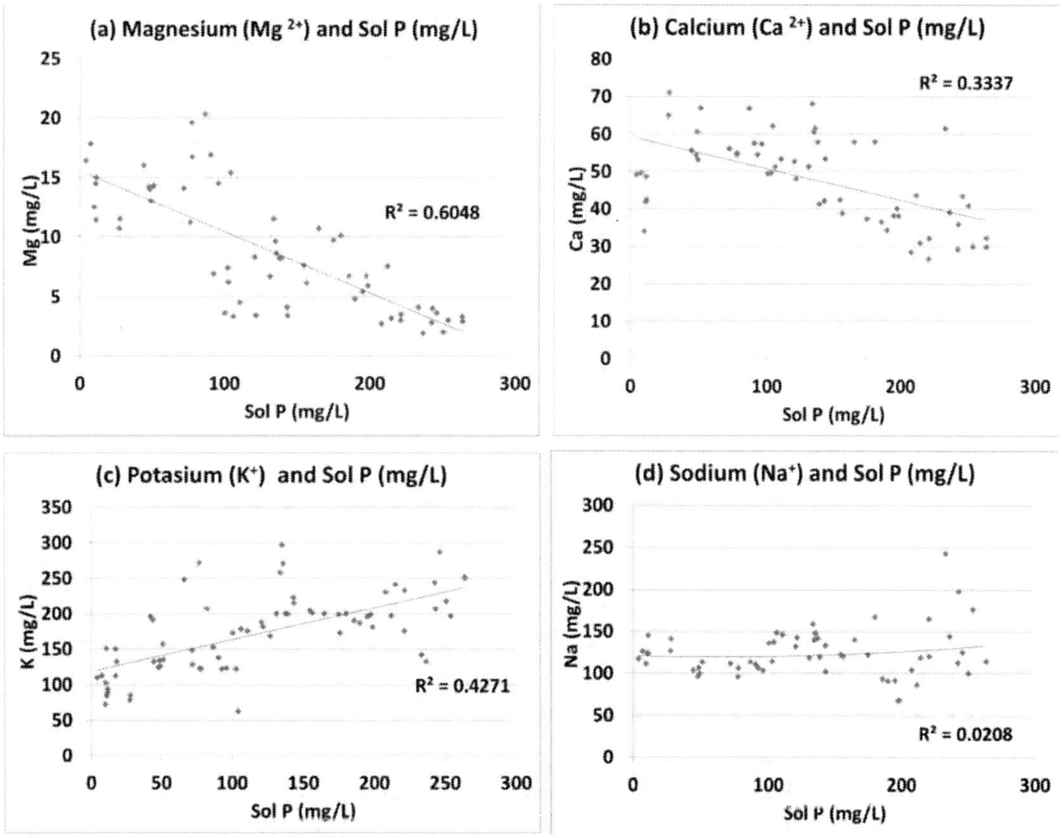

Figure 6: Relationship between (a) magnesium ion, (b) calcium ion, (c) potassium ion, and (d) sodium ion and soluble P.

According to the divalent ion bridging theory, high concentrations of soluble P are accompanied by a decrease in both magnesium and calcium ions, which are associated with bio-floc formation (Higgins and Novak, 1997). Phosphorous seems to be able to form complexes and precipitates with these two cations as observed by Higgins *et al.*, (2014). These results seem to indicate that there is a better linear fit with magnesium. In terms of soluble calcium, only Oxford sludge seemed to not follow the linear trend, with very low levels of soluble P (between 4-10mg/L) and corresponding soluble calcium concentrations between 30-50mg/L.

Figure 6 (c) and (d) show the relationship of the two monovalent ions potassium and sodium. Potassium follows an upward trend with increasing soluble P which was also found by Higgins *et al.*, 2014. The relationship of soluble sodium with soluble P is not clear at the observed concentrations. Data seems to show slightly higher concentrations of soluble Na after soluble P reached 200mg/L.

Figure 7 shows the results on the M/D ratio of the listed sites and their polymer demand during the dewatering trials in the lab. All sites depicted are THP-AD sites. Seafield is clearly driving the trend line with most of the sites remaining at an M/D ratio of 1-2.5. This may mean that even for these sites, other factors aside from the soluble P concentrations may be influencing the polymer demand for floc formation, including shock loads or sudden changes in temperature, in the digester or right before dewatering. It may also mean that our methodology (manual polymer injection and mixing + lab scale Bucher sock test) is not sensitive enough to give us significant differences across sites with similar dewatering performance.

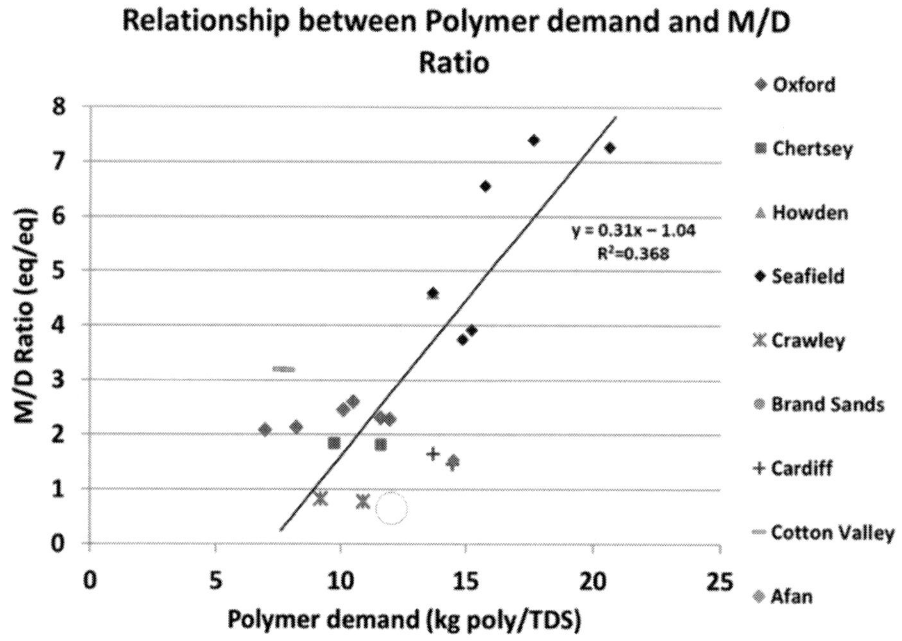

Figure 7: Relationship between polymer demand (kg/TDS) and M/D ratio (All sites apart from Crossness and Beckton).

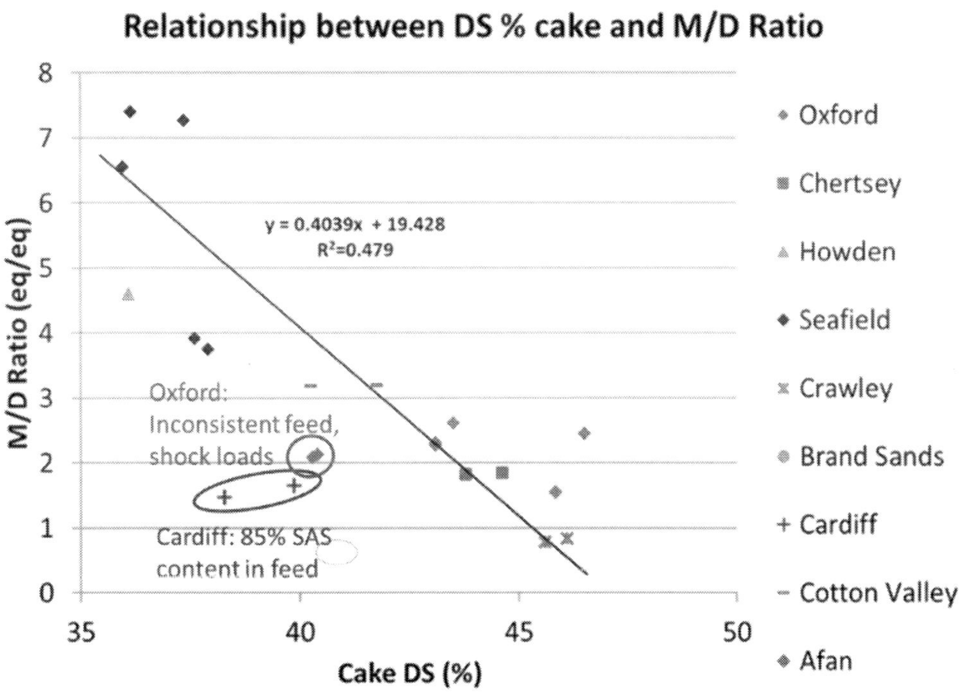

Figure 8: Relationship between cake DS (%) and M/D ratio (All sites apart from Crossness and Beckton)

Figure 8 shows the final DS of the digested sludge from the different sites and their M/D Ratio. The scatter is significant with an R^2 of 0.48 indicating a high percentage of variation in the data. This could be due to other variables affecting the dewaterability of the sample. For instance, Cardiff did not achieve more than 40% DS even though a high M/D ratio is observed. This may be related to the surplus activated sludge (SAS) content being very high (85%) compared to the other sites plotted with a similar M/D ratio. In the case of Oxford, the samples corresponding to a final cake DS of circa 40% where taken during a period where the site was feeding very inconsistently, with constant start and stops which would shock load the digesters.

(a) **Beckton M/D Ratio and final cake DS (%)**

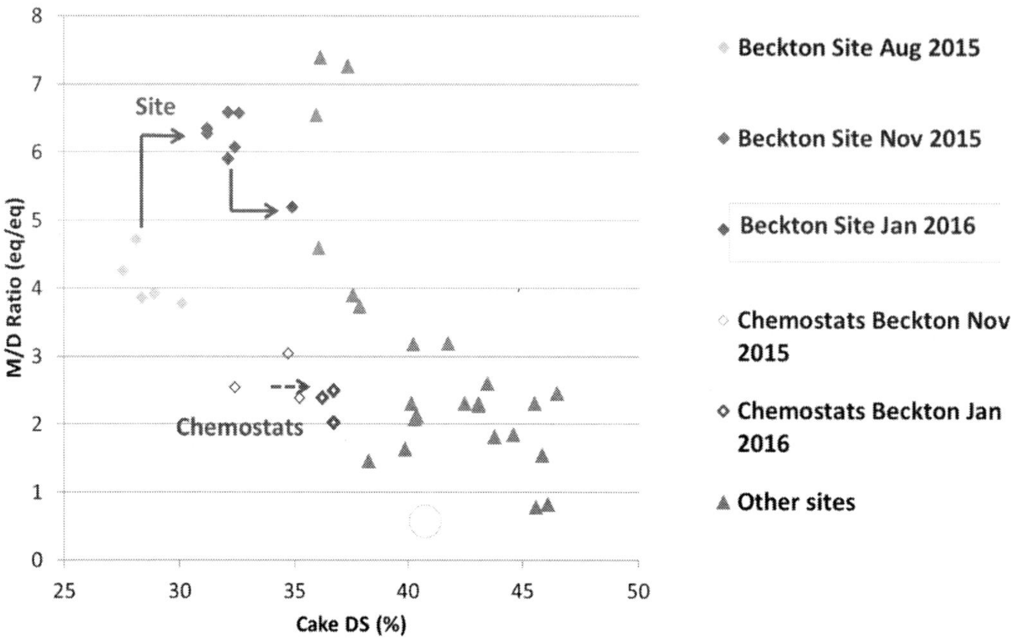

(b) **Crossness M/D ratio and Final cake DS (%)**

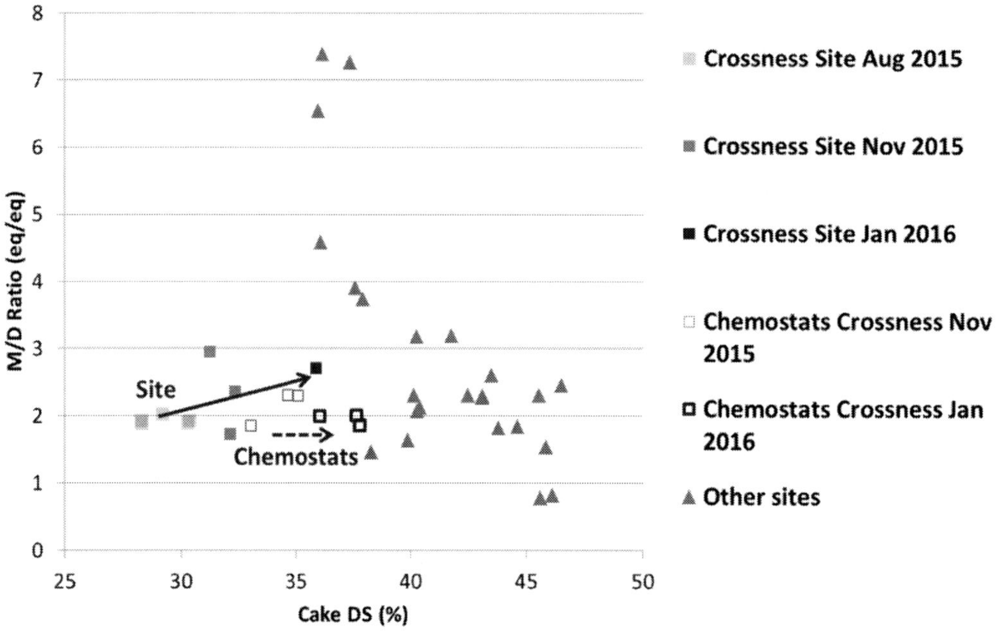

Figure 9: M/D ratio and final cake DS (%) for (a) Beckton and (b) Crossness compared to other THP sites (shown separately on figure x).

Figure 9 shows the evolution of the M/D ratio and final cake DS compared to the rest of the THP sites. Where most sites achieved DS between 35 to 46%, both Crossness and Beckton struggled to achieve 30%DS back in

August 2015. At that time both sites were seeing extreme erratic feeding, with more than 60% variability in the feed observed in the 15 min interval data collected. In terms of soluble P, only Beckton had a significant concentration of 149mg/L whereas Crossness only had 56mg/L. The higher M/D ratio for Beckton is therefore expected according to the divalent ion bridging theory (Higgins et al., 2014). This difference in sol P values across sites could be due to the higher amount of SAS treated at Beckton (80% SAS compared to 40% in Crossness).

In late November 2015, soluble P levels in Beckton went up to 247mg/l which corresponds to the increase in the M/D ratio. Nonetheless, we see higher final cake DS. Looking at the feeding regimes for that month, the feed flows into the digesters seem to be more stable, with a variability of 40%. This improvement may partly explain the better dewaterability obtained even though the sol P increased. Other improvements onsite may have contributed to it. Crossness saw a similar improvement in DS with not much change in sol P in the month of November.

In late January 2016, both sites continued to follow an upward trend in final cake DS observed at lab scale. At that point, Beckton had much more work done around the feeding system and therefore the variability of digester feed was lower than at Crossness. This may explain why even though Crossness only had 102mg/l of Sol P compared to the 246mg/l seen at Beckton, the dewatering was very similar at both sites.

The Beckton chemostats showed a significantly lower M/D ratio than that observed onsite. This also correlates to the lower soluble P seen in the chemostats (149 mg/l compared to 250mg/l seen onsite). The possibility of some mechanism changing the chemistry in the full scale digesters compared to the chemostats is being studied. An initial theory is the possibility of higher amounts of struvite formation at full scale. Magnesium ion seemed to be slightly lower for sites samples (3.4mg/l compared to 9mg/l in the chemostats). Nevertheless, whether or not this difference is significant is to be determined. The concentration of calcium ions was found to be much lower onsite as well, with values close to 40mg/l compared to 70mg/l in the chemostats. In terms of dewatering, from November to January there was a slight improvement that could be associated with a long stable period with optimum AD conditions.

The Crossness chemostats showed a similar M/D ratio to that seen onsite. The DS obtained where in all cases slightly higher than those seen for site samples in the lab scale dewatering trials also showing an improvement with time.

(a) Beckton M/D Ratio and polymer demand (kg/TDS)

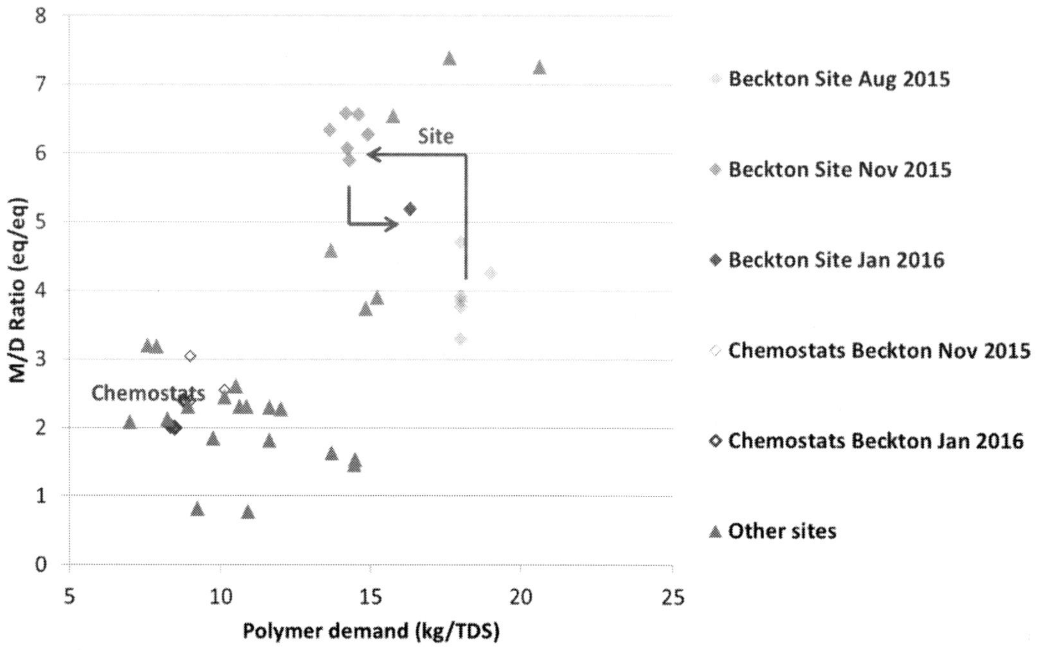

(a) Crossness M/D ratio and polymer demand (kg/TDS)

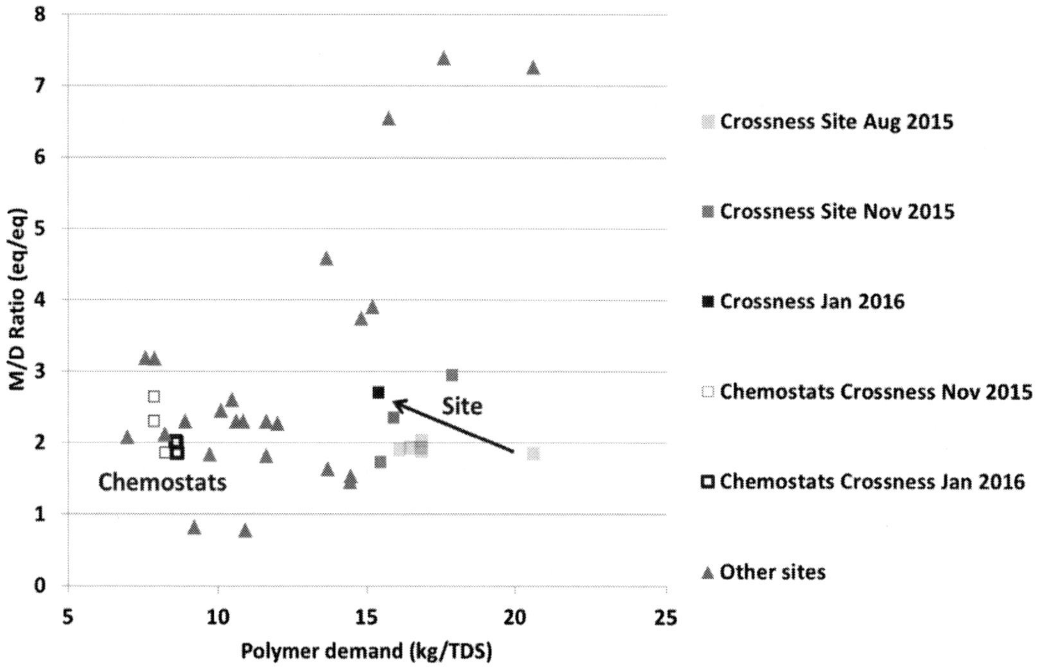

Figure 10: M/D ratio and polymer demand (kg/TDS) for (a) Beckton and (b) Crossness compared to other sites (shown separately on figure x).

Figure 10 shows the relationship of the M/D ratio with the polymer demand of Crossness and Beckton site and chemostats compared to other THP sites.

Beckton site digested samples had a significantly higher polymer demand compared to the chemostats. The M/D ratio observed onsite was higher than in the chemostats.

Crossness sees a similar difference in polymer demand between the chemostats and site, but in this case the M/D ratios are very similar, staying circa 2.5.

4. **Effects of reseeding chemostats 3 (Beckton) and 6 (Crossness):**

Two chemostats (Chemostat 3 from Beckton and Chemostat 6 from Crossness) were reseeded and dewatering tests with the piston press where done twice a week for a period of 4 weeks. Both DS and polymer dose was recorded. The results can be seen in figure 11:

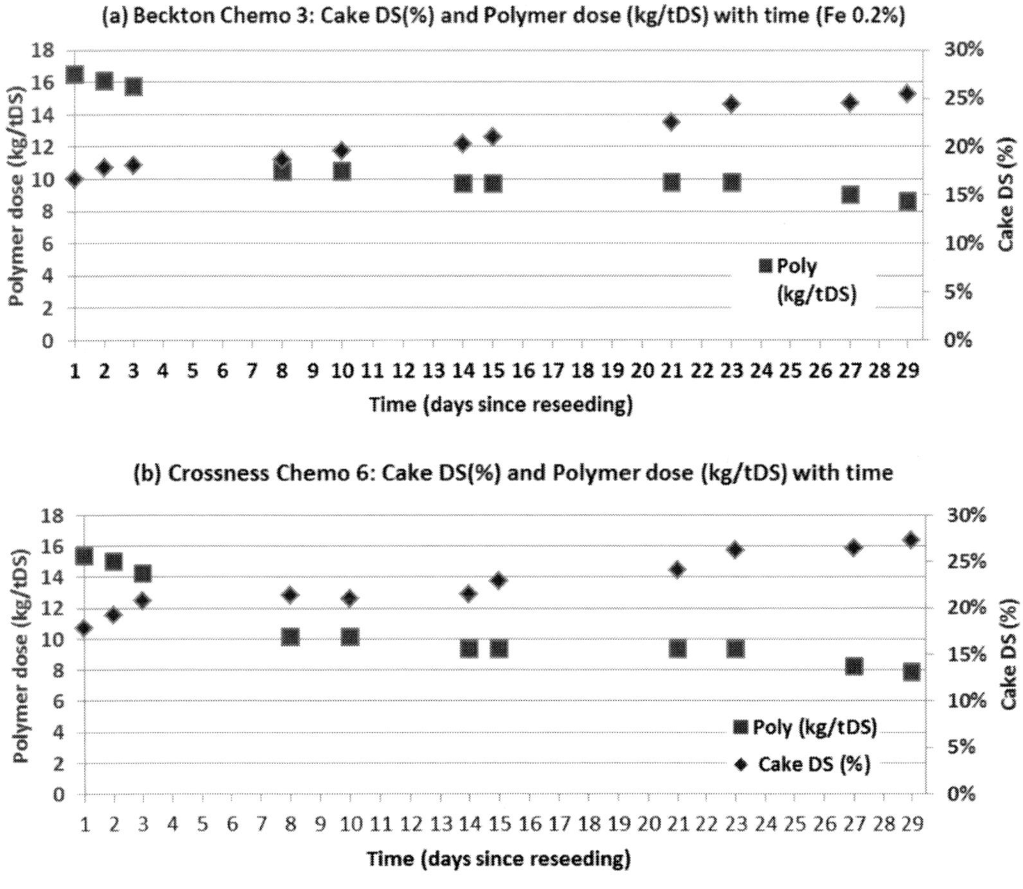

Figure 11: Reseeding of chemostats (a) 3 with digested sludge from Beckton STW and (b) 6 with digested sludge from Crossness STW and monitoring of dewaterability and polymer demand.

The iron dose remained constant at 0.2% (w/w) to match what was then dosed onsite. Therefore, under same iron conditions, a clear improvement in dewatering as well as a gradual decrease in polymer demand was observed in both cases. This would indicate that something else apart from the iron addition was affecting the dewatering performance.

5. **Extra-Polymeric Substances:**

Figure 12 shows the relationship between polymer demand (kg/TDS) and soluble protein related EPS (mg/l) for all the THP sites sampled including Crossness and Beckton. Chemostats 2 (Beckton) and 5 (Crossness) are also plotted for being the most representative of both sites.
There seems to be a relationship between these two parameters (R^2 = 0.5107). Some scatter may be expected since the presence of soluble P seems to be another variable affecting dewatering.

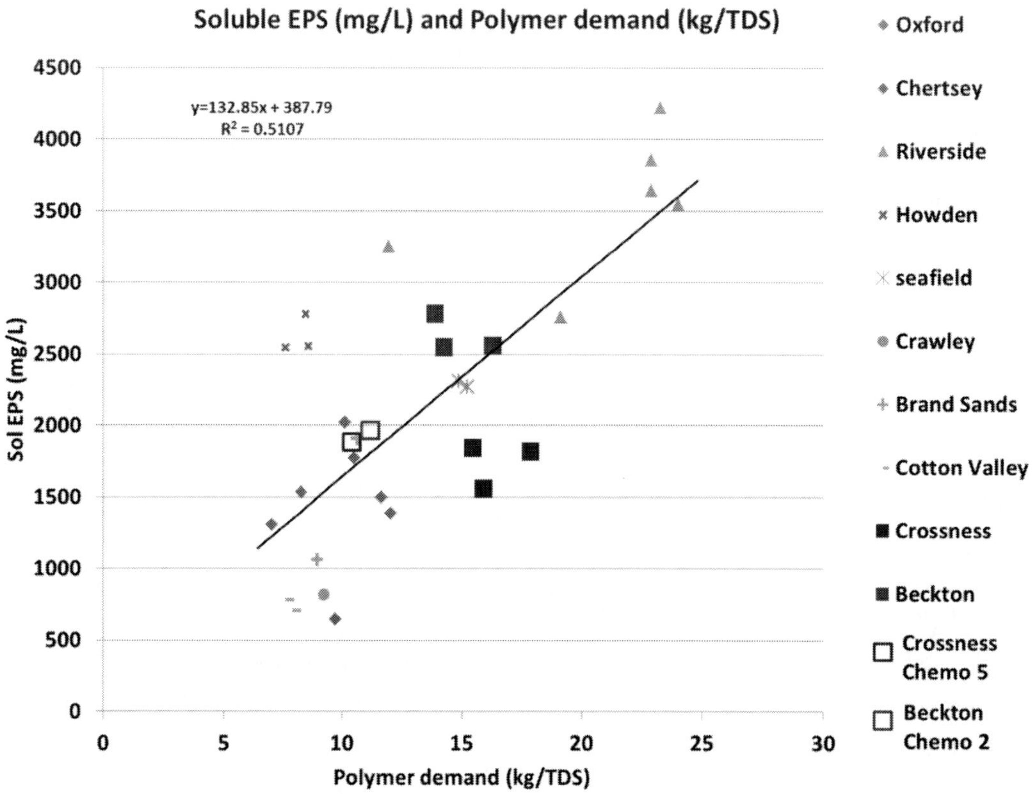

Figure 12: Relationship between soluble protein related EPS (mg/L) and polymer demand (kg/TDS).

In terms of the relationship between chemostats and sites, only Beckton followed the general trend of higher polymer demand with higher soluble EPS. The higher EPS onsite may be related to the more erratic or irregular feeding regime observed there compared to the consistent feeding regime of the chemostats. This is not observed for Crossness, where the EPS is similar in both the chemostat and full scale. The variance in sludge feeding flows into the digesters for a given day is higher at Crossness and therefore higher EPS was expected at full scale.

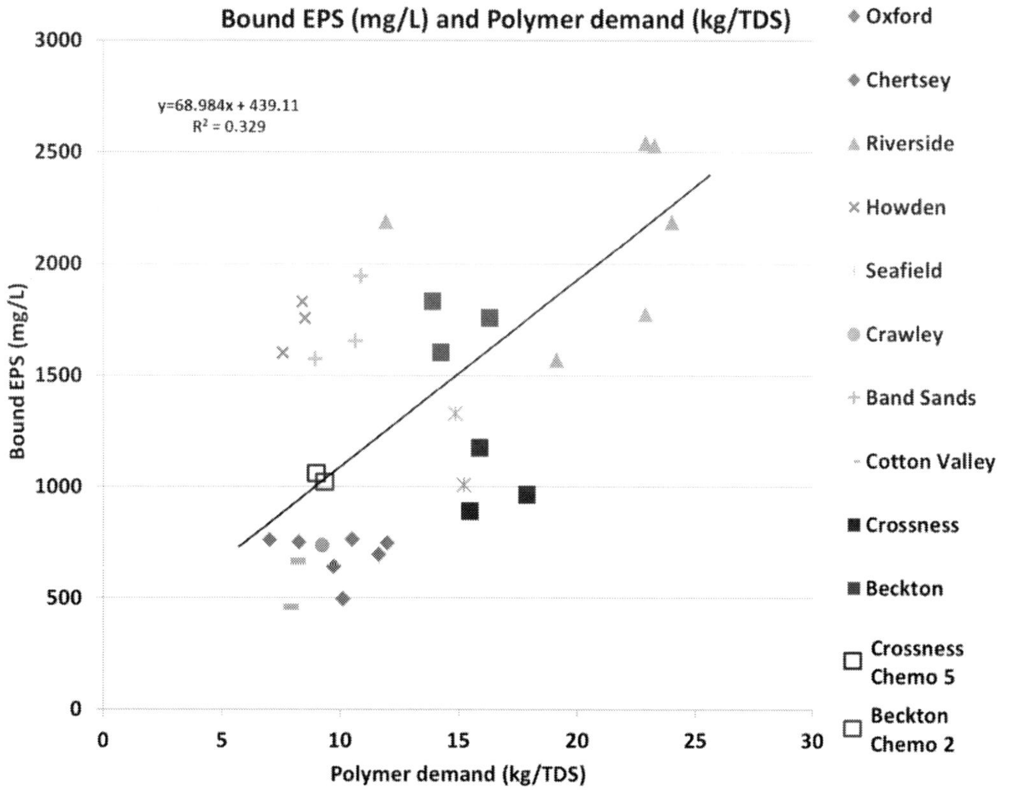

Figure 13: Relationship between bound protein related EPS (mg/L) and polymer demand (kg/TDS).

Figure 13 shows the relationship between the bound EPS and polymer demand of THP sites and chemostats 2 (Beckton) and 5 (Crossness). The relationship is weaker for this parameter, with an R^2 = 0.329. Beckton chemostat 2 shows a lower concentration in bound EPS and polymer demand before dewatering compared to site as observed in figure 12 for soluble EPS and polymer demand. Nevertheless, Crossness does not follow the linear trend, with higher polymer demand onsite compared to chemostat 5 even though the bound EPS is similar in both cases. This is also the case for soluble EPS and polymer demand depicted in figure 12.

It is worth noting that there is currently no standard methodology to follow to be able to compare our own findings with other papers.

Other variables that have not been assessed in this paper may also be having an impact on polymer demand and final cake DS which would increase the scatter for both the EPS and M/D ratio relationships with polymer demand.

Conclusions

Poor digested sludge dewaterability was observed at Crossness and Beckton sites with no other signs of digester underperformance. Two possible causes for poor dewaterability were identified for Crossness and Beckton sewage treatment plants.

One theory proposed by Higgins *et al.*, (2014) suggest that soluble P is an indirect driver for dewaterability issues, both in terms of polymer demand for flocculation and final cake DS. Results from 9 THP sites showed a strong correlation between soluble P and the M/D ratio. This ratio also correlated to dewaterability, particularly for polymer demand. Plants with higher M/D ratios showed poorer sludge dewaterability, confirming previous work (Higgins and Novak, 1997b; and Higgins *et al.*, 2014). Nevertheless, both Beckton and Crossness saw an improvement in dewaterability with increasing final cake DS and decreasing polymer demand with time even though the M/D ratio did not change for Crossness and saw an increase for Beckton. This may indicate some other factor is influencing dewaterability.

Additionally, chemostats 3 and 6 were re-seeded with digested sludge from Crossness and Beckton respectively to observe the evolution from full scale to lab scale of dewaterability with time. The results for the first 30 days after seeding showed a steady improvement of dewatering with time with no change in M/D ratio. This also suggested that soluble P may not be only factor having an impact on dewaterability.

The second factor considered was EPS. According to the data from both Crossness and Beckton, the digesters from both sites were being fed erratically. The hypothesis proposed in this paper suggests that the shock loads observed at both sites could trigger a defence mechanism in the bacterial population in the form of higher EPS excretion. Since EPS is known to be difficult to dewater, higher EPS concentrations would likely translate into higher polymer demand for bio-floc formation. Initial results showed a weak relationship between the protein related EPS (both soluble and bound parts) and polymer demand which indicates EPS is likely partly driving digested sludge dewaterability. When comparing chemostats to site, Beckton showed higher soluble and bound protein related EPS onsite compared to the chemostat. This was expected given the better feeding conditions provided at lab scale compared to that at full scale. Nevertheless, this result was not found for Crossness, with both bound and soluble EPS figures remaining very similar at lab and full scale. This may suggest some other factor driving the poor dewatering at this particular site.

Challenges around the methodology for EPS extraction were found since there is no one particular standard method to extract the polymeric substance across the business. This means no comparison can be done with other research. Therefore the next steps to take would be expand our understanding on EPS extraction methodologies by comparing results with different methods available in the literature. Furthermore, work around the observation of EPS under the microscope is already in place with encouraging initial results. Other factors that may be having an impact on dewaterability will also be looked into. Some preliminary work on particle size distribution suggests that site and chemostats may show differences in particle sizes. Sludge viscosity and bacterial population are two other parameters to investigate.

References:

Alm, R., Sealock, A.W., Koo, A., and Sprouse, G. (2015) Investigations into improving dewaterability at a bio-P/anaerobic digestion plant. *Residuals and Biosolids Conference 2015*, Washington DC, US.

Aquino, S.F. and Stuckey, D.C. (2004) The effect of organic and hydraulic shock loads on the production of soluble microbial products (SMP) in anaerobic digesters. *Water Environ. Res.* 76 (6), 2628–2636.

Aquino, S.F., and Stuckey, D.C. (2007) Integrated model of the production of soluble microbial products (SMP) and extracellular polymeric substances (EPS) in anaerobic chemostats during transient conditions. *Biochemical Engineering Journal*. 38, 138-146.

Higgins, M. and Novak, J. (1997) Characterization of Exocellular Protein and Its Role in Bioflocculation. *J. Environ. Eng.* 123:5(479), 479-485.

Higgins, M., Bott, C., Schauer, P., and Beightol, S. (2014) Does bio-P impact dewatering after anaerobic digestion? Yes, and not in a good way! *WEF Residuals and Biosolids Conference 2014*, Austin TX. US.

Higgins, M., Murthy, S., Bott, C., and Rajagopalan, G. (2016) Bioflocculation Mechanisms and Implications for Understanding the Effects of Co-Digestion and BioP on Dewatering of Anaerobically Digested Biosolids. *WEF Residuals and Biosolids Conference*, Milwaukee. US.

Park, C. (2002) Cations and activated floc structure. Master Thesis. Virginia Polytechnic Institute and State University.

Rus, E., Perrault, A., Fountain, P., and Shana, A. (2015) East London THP commissioning and laboratory investigations. *European Biosolids Conference 2015*. Manchester, UK.

Shana, A. (2015) Application of an innovative process for improving mesophilic anaerobic digestion of sewage sludge. PhD Thesis. University of Surrey.

Sheng, G-P., Yu, H-Q., and Li, X-Y. (2010) Extracellular polymeric substances (EPS) of microbial aggregates in biological wastewater treatment systems: A review. *Biotechnology advances*. 28 (6), 882-894.

Development of Kodiak, Alaska's Successful Composting Program

Todd Williams[1*], Bud Alto[1], Floyd Damron[1], Mark Kozak[2], Lori Aldrich[3]

[1]CH2M, Richmond, VA and Anchorage, AK
[2]City of Kodiak, AK
[3]Alaska DEC, Anchorage, AK
*Email: todd.williams@ch2m.com

ABSTRACT

The City of Kodiak (City) on Kodiak Island is located 400 kilometers (250 miles) southwest of Anchorage at the westernmost edge of the Gulf of Alaska. As such, it is an isolated community, only reachable by boat or by plane. The Island's main local government consists of two bodies, the City of Kodiak and the Kodiak Island Borough (KIB). In addition, the island houses the largest Coast Guard Base in the United States. The combined population of the City, KIB and the Coast guard base which generate wastewater solids that must be managed is approximately 14,000 persons. The City's existing 17.8 mega-liter/day (4.7 MGD) capacity secondary wastewater treatment facility (WWTF) currently treats an average flow of 6.8 mega-liters per day (1.8 MGD). Faced with dwindling landfill space and skyrocketing fees at the island's largest landfill owned and operated by the KIB, the City of Kodiak embarked on a Biosolids Management Study in 2007 to evaluate solids management options to landfilling dewatered wastewater residuals. Incineration, thermal drying, the Cannibal™ process and composting were the four alternatives evaluated. Composting was selected by the City based on costs and the consistency of the technology with the City and KIB's philosophies, goals and objectives. This paper provides information on the background used for selecting composting to manage the City's wastewater solids. In addition, planning, permitting, public outreach as well as design features and operating experience are provided.

BIOSOLIDS MANAGEMENT OPTIONS EVALUATIONS

In 1997-2000, the City of Kodiak built $8 million of treatment plant upgrades required due to capacity issues and sludge disposal issues at KIB landfill. In 2007 the City contracted with CH2M HILL (CH2M) to conduct a biosolids management feasibility study to consider alternatives to continuing landfilling of lime treated biosolids at the Kodiak landfill. The study identified three treatment alternatives:

- Drying & incineration of all solids with hauling of residual ash to landfill
- Cannibal treatment process to reduce the quantity of solids to be landfilled
- Composting all solids and beneficial use of compost

The result of this study in 2008 was that composting was projected to have the lowest capital and operating cost of the three options considered. At the same time, KIB completed delivery of a solids management report for the landfill known as the 2008 KIB Bell Report. This report projected disposal costs in the required new lined landfill to be $212 to $258 per ton which was much more than the 2008 cost of $75/ton. In addition, the KIB Bell Report strongly recommended encouraging diversion and recycling to extend the useful life of the landfill. Therefore, composting of the wastewater treatment plant solids potentially with other waste stream materials such as wood wastes, cardboard and brush was determined to be aligned with City's efforts to identify an affordable, environmentally acceptable disposal method. The study indicated that the feasibility of acquiring a suitable site and the availability of suitable amendments needed to be investigated further to determine the overall feasibility of composting. With the availability of land in a reasonably convenient and well buffered location, and the availability of suitable amendments and bulking agents on Kodiak Island, biosolids composting could be a sustainable, long-term alternative for the City of Kodiak. It was decided that a compost pilot study should be performed to demonstrate the efficacy of the process to City staff, the public and to the Alaska Department of Environmental Conservation (ADEC). Further, compost produced from such a pilot would allow for testing compliance with US EPA and ADEC requirements, allow testing the use of various bulking agents, and to perform a limited compost market assessment. In 2009 such a compost pilot study was initiated.

COMPOST PILOT STUDY METHODOLOGY

The City and CH2M initiated a Biosolids Composting Pilot Program at the Kodiak WWTF during the summer of 2009 to demonstrate the aerated static pile method of composting to the wastewater operations team, City staff, City Council members and the public. The pilot program purpose was to prove biosolids composting efficiency with locally provided wood chips, forced aeration, and reusable aboveground piping in Kodiak's coastal climate. The pilot program also intended to demonstrate that suitable odor controls and processing capacity could be achieved with this composting method. Secondary pilot program goals were to collect baseline data that would be required for the City to develop a permanent composting facility, including data about:

- Wood chip type and volume requirements
- Sufficient temperatures to achieve pathogen kill
- Aeration rates to manage moisture and pile temperatures
- Weather protection
- Mixing equipment capabilities
- Operating costs

Figure 1 - Completed Pilot Compost Pile at City WWTF

The pilot program was conducted using a portable composting system designed by CH2M. The portable composting system used is a variation of the aerated static pile composting method. The system facilitates composting by using either positive aeration or negative aeration with an odor control system. During the Kodiak pilot program, the composting system was operated in negative aeration mode, meaning that air was pulled through the compost pile and discharged through a biofilter for odor control. A centrifugal blower and engineered pipe manifold network was used to move air through the compost pile. Aeration piping was designed to drain to a low point for condensate collection. Figure 1 shows the final assembled compost pile which contained 30 cubic yards of wastewater solids. Temperature monitoring showed requirements were achieved to meet Class A process to further reduce pathogen requirements and vector attraction requirements. Testing of the finished compost confirmed pathogen elimination and metal concentrations easily met exceptional quality (EQ) standards. The pilot test demonstrated that high-quality compost, defined as exceeding US EPA and ADEC regulatory requirements for a Class A EQ biosolids product suitable for unrestricted use, can be produced using the aerated static pile method in Kodiak's coastal environment. Wood chip supply and processing, site preparation, and initial setup required a significant effort, but once the compost pile had been constructed, the day-to-day monitoring requirements were minimal. Pilot program conclusions were:

- A reliable wood source is needed to provide bulking agent
- Odor control is important
- Marketing of the product poses challenges in a small community
- Capital costs were estimated at $4.5 million for full scale facility design
- O&M costs were projected to be comparable to landfill disposal costs at that time

Based on the success of the pilot composting program, the City began taking steps to develop a full scale composting operation to manage wastewater solids.

INTERIM COMPOSTING PROGRAM

After this successful City-run pilot program, a temporary low-technology contractor-run composting operation was permitted by the ADEC and began at the KIB landfill in late 2012. This action was necessary because in June 2012 KIB provided written notice to the City that the KIB landfill would no longer accept wastewater solids for disposal as of Dec 15, 2012 due to operational needs and closure process of the old portion of the landfill. A partnership was begun between a private contractor and the City to compost biosolids on space atop the existing landfill. The contractor had ready access to bulking agent materials (wood chips) in large enough volumes needed to produce a Class A EQ finished compost product. The contractor also expressed interest in marketing and use of the finished compost product and owned equipment and resources to operate the composting facility. An agreement was made between this contractor and the City to process wastewater solids using the static pile composting process to produce a class B product that would be designated for use as intermediate cover at the KIB landfill until a permanent compost facility could be designed and built. Challenges arose with the temporary outdoor operation due to poor site conditions, exposure to rainfall, and inability to adequately control the composting process due to lack of aeration and excessive moisture.

Public Outreach and Regulatory Engagement

Meetings were held between CH2M and the ADEC regulators early on in the development of the full scale composting facility concept to review the technology being proposed and to obtain feedback from the regulators as to their concerns needing to be addressed in the design and operation of a composting facility. These meetings proved crucial as public feedback in the form of phone call complaints regarding the temporary compost facility operation and its contractor practices were being lodged with ADEC prior to the full scale facility design. CH2M was able to provide technology information to ADEC and answer questions about the process that had arisen. CH2M developed preliminary permit documents for filing with ADEC regarding storm water management, odor control, and planned product disposition. Public concern continued to grow when a small portion of finished product was applied by the contractor on a private site to demonstrate the product's utility without City knowledge. The ADEC was notified, and the offsite land application of Class B compost was halted. Additionally, public concern was raised over where the permanent compost facility would be sited, how the product would be used and how the process would be run by a private contractor basically threatening the confidence of the public and from the public's vocal opposition, created doubts for City Council members about the technology selection. CH2M worked closely with the City Manager, Public Works Department, City Council members and the ADEC in 2013 and 2014 to educate the public on the technology of composting and the uses of the product as well as the history of successful composting of biosolids in the US and Canada.

A tour of several operating biosolids composting facilities in the Pacific Northwest was conducted to demonstrate the technology to the City Manager, Council members and the City Engineer to allow this group to see what other communities had done to develop successful composting programs. Key findings from that trip were:

- These biosolids compost facilities have been operating for up to 25 years
- Public concern during planning faded away after facilities began successful operations
- Facilities effectively controlled odors
- Facilities produced high quality biosolids compost products (Class A exceptional quality)
- Condensate and runoff were well controlled
- Covered facilities had best control of the process and minimal impact from precipitation

- Compost product markets were strong (compost was not readily available at these sites)
- Knowledgeable operators were committed to making their operations a success
- Facilities are considered to be valuable assets to the communities they serve

The outcome of this trip along with supporting engineering documentation of the alternatives and their costs ultimately convinced City leaders of the viability of the biosolids composting technology and confirmed the City had made the right choice in selecting composting to manage their wastewater solids. A public meeting was held in August 2014 with City Council members and with the ADEC permitting representative to answer questions about the planned facility, the permitting of the operation, and to address product usage issues. This cooperation between all parties ultimately garnered the support needed to move the facility concept from design into construction. Ultimately, CH2M was retained by the City to design the facility in late 2013, assist in construction management and start-up. The design was completed in 2014, a contractor selected, and the facility was built in 2015 with operations commencing in December 2015.

Basis of Design

Figure 2 - City f Kodiak Compost Facilities showing Mixing Building on the Left and 6 Compost Bays on the Right.

The City of Kodiak Composting facility is designed to process 360 dry tonnes per year (400 dry tons) of wastewater solids with an average total solids content of 18%. 1,900 cubic meters (2,500 cubic yards) of wood chips are purchased annually to be used as the bulking agent. The facility is located adjacent to the KIB landfill with the plan to divert waste wood material and other segregated carbon-rich materials from the landfill to the compost operation in the future. Because of the extremely wet conditions typical of this Northwest environment, the receiving and mixing operations are housed indoors and six covered compost bins are used to process the aerated composting and curing processes. The mixing operation consists of a batch mixer with weigh scales so that a recipe of wastewater solids, new bulking agent and recycled bulking agent can be accurately blended in batches. A front-end loader moves the material from the covered mix bunker outside the mixing building to one of six compost bays. Each bay handles one week's worth of mix (Figure 3). Each bay includes two 15 centimeter (6 inch) diameter perforated high density polyethylene pipes on the floor which are connected to a dedicated blower. Continuous negative air flow is used to provide aeration, control pile temperatures and to capture odorous emissions for subsequent treatment in a biofilter system. Blower speed is manually adjusted using a variable frequency drive (VFD) based on automated pile temperature recordings to meet USEPA and ADEC requirements for process

Figure 3 - Compost Bay with Aeration Laterals

to further reduce pathogens (PFRP) and vector attraction reduction (VAR) to achieve Class A Biosolids standards. The six compost blowers discharge into a common header that is connected to a booster fan to treat the exhaust air through an above ground wood chip based biofilter. The booster fan also has a VFD and is sized to add cooling ambient air for optimal inlet temperature to the biofilter. The biofilter system provides over 90% reduction in compost and curing process odors. Atmospheric odor dispersion modeling was performed for the design of the entire operation to ensure the compost facility odors are properly controlled and would not be noticeable at the site boundary. The EPA approved AERMOD model was used with local terrain and meteorological data taken from Kodiak's State Airport. Five recent years of meteorological data were screened and the most conservative year (with respect to producing the highest predicted offsite concentration) was selected. Odor source data was based on measurements taken by CH2M at other aerated static pile composting facilities using a similar process to the one planned for the Kodiak. Dispersion modeling results predicted that perceptible odors would be contained within the property boundary with greater than a 99.9% confidence factor, and any slight predicted incursions would be into the adjacent Monashka Bay Road far from any residences or other potential receptors. These results indicate a minimal risk of the compost operation causing offsite odor impacts. Storm water is managed on-site in an infiltration control structure. Contaminated condensate and wash water from the operation is collected in an underground storage tank and pumped and hauled to the City WWTF. A bulking agent and compost storage building was also added to the facility in 2016 to keep materials dry from the over 180 centimeters (70 inches) of precipitation received annually.

Figure 4 - Engineered Wood Media Odor Control Biofilter

Facility Start-up and Operation

The City of Kodiak Compost facility began operation in December 2015. After facing public scrutiny about the capability of the private contractor, the City decided to take ownership of running the facility. CH2M worked closely with City staff to aid in start-up and detailed operator training regarding process control, monitoring, and record keeping. All WWTF solids produced have been processed through the composting operation since start-up. All compost batches have successfully achieved PFRP and VAR temperature requirements. After three weeks of composting, the material in each compost bay is tested in four locations for the pathogen indicator salmonella. Once successful test results are received back from the laboratory, the material is moved to an adjacent bay for continued aerated curing for 3 more weeks. Recently, because of the success of the operation in demonstrating reduction of pathogen indicators, the number of samples and tests required has been reduced to more suitable levels for this size of operation (quarterly). The ADEC

Figure 5 - Screening of Compost Product in Front Left

has reviewed all compost testing data and has declared the finished compost material has met all the requirements to be considered Class A EQ and may be distributed. Finished compost is currently managed by the original compost contractor who sells the material for unrestricted use. The City is in the process of developing its compost distribution program to sell the product and further offset operating costs. Test results for the compost are provided in Table 1.

Table 1 - Kodiak Compost Test Results

Parameter	EPA Ceiling Conc.	12-Month Average	% of Ceiling
METALS (Dry Weight)	Mg/Kg	**Mg/Kg**	
Arsenic	41	**11**	**26%**
Cadmium	39	**1**	**3%**
Chromium	-	**13**	-
Copper	1500	**148**	**10%**
Lead	300	**9**	**3%**
Mercury	17	**1**	**6%**
Molybdenum	-	**2**	-
Nickel	420	**14**	**3%**
Selenium	100	**2**	**2%**
Zinc	2,800	**260**	**9%**

Parameter	12-Month Average
NUTRIENTS (Wet Weight)	
Total Nitrogen (N) (%)	**1.1**
Nitrate N (ppm)	**160.5**
Ammonia – N (ppm)	**905.0**
Total Phosphorus (P) (%)	**0.4**
Total Potassium (K) (%)	**0.10**
pH	**7.1**
CO2 Evolution mg/g OM/day	**1.5**
Seedling Emergence (%)	**95.0**
Total solids (%)	**50.3**
Volatile Solids/Organic Matter (%)	**42.0**
C/N Ratio	**21.8**

Costs

The capital cost of the compost facility was $3.45 million in 2015 plus a new covered storage area was added in 2016 at a cost of $1.1 million. Operating costs were tracked in 2016 and continue to decrease as more operational experience is gained. Operating costs include labor, bulking agent, electricity, fuel, laboratory testing and maintenance. 1755 tons of dewatered solids were processed in 2016 at an overall cost of $386,425 amounting to a unit cost of approximately $220 per wet ton of solids processed. This may seem high, but actually is low compared to the cost of landfilling at $258/wet ton which will escalate in 2017 to $265/ton. Figure 6 shows the percentage of cost attributable to the various components. Labor is the highest component since two operators are required at this remote site. These two personnel could easily process significantly more material but for safety and operational performance, two full time staff run the facility. Bulking agent is the next largest cost and is currently purchased. Plans are in place to begin recovering wood wastes from the landfill operation to use as bulking agent once grinding equipment can be secured to do so.

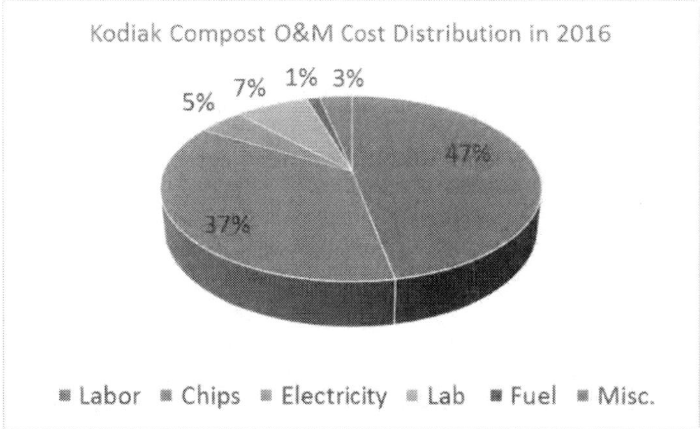

Figure 6 - O&M Cost Breakdown by Category

To date compost has been utilized by the previous contract operator to finish contractual obligations, but in 2017, compost will begin to be sold by the City further decreasing overall O&M costs.

CONCLUSIONS

The development of the City of Kodiak's composting facility demonstrates that biosolids composting is an effective solution for rural Alaska and other remote Northwest North American communities in a very wet environment. The process produces a high quality Class A EQ compost product that is a valuable product for many uses. The importance of coordination and ongoing cooperation between the City, the Borough, the state regulator and the engineer to address public concern was essential to the successful implementation of this program.

Using Lime to Beneficially Manage Wastewater Treatment Plant Residuals: A Review and Assessment of the Practice for Producing an Exceptional Quality Product

James E. Smith, Jr.[1*], A. Robert Rubin[2], Harry L. Francis[3], and Robert S. Reimers[4]

[1]USEPA-ORD-NRMRL-Ci (Retired), 5821 Marlborough Drive, Cincinnati, Ohio 45230
[2]Biological and Agricultural Engineering (Retired), Box 7625, Weaver Laboratories, North Carolina State University, Raleigh, North Carolina 27695
[3]Former Technical Manager, National Lime Association, 2851 Sarver Rd., Elliston, VA 24087
[4]Tulane School of Public Health and Tropical Medicine, New Orleans, LA 70112
[*]Email: JesmithjrDSc@aol.com

ABSTRACT

Lime is used to prepare residuals for safe beneficial use by preventing the biodegradation of organic material and either reducing the numbers (of) or eliminating any pathogenic organisms present. Lime acts in several ways including: elevating the pH and holding it at a high level for a period of time; and in some instances elevating the temperature of the residuals to a high level and maintaining it there for a specified period of time. The performance of any lime treatment system is, however, only as good as the quality of alkaline material selected, the dosage level applied, the mixing method selected and applied, the time allowed for all reactions to reach completion, and the monitoring methods employed for pH and temperature.

KEYWORDS: Lime treatment, alkaline treatment, exceptional quality biosolids, beneficial use.

INTRODUCTION

Lime stabilization of wastewater treatment plant residuals or sludge came into common use in the United States during the 1970s. Then as a way of providing a high quality product it gained popularity in the 1980s. Its popularity, however, began to decline in the early 2000s because of issues with odor, product quality, and marketability. Alkaline treatment is viewed by many in the wastewater treatment industry as a low cost, easy way to meet the federal, state, and local requirements for beneficial use of sludge in agriculture, horticulture, and reclamation. While alkaline treatment can effectively disinfect and stabilize sludge when applied correctly, it is not a simple process, but rather one that must be properly designed and operated. This paper provides a better understanding of the practice's critical issues and an assessment of the current practices for producing an exceptional quality biosolids by lime/alkaline treatment. Then drawing upon that information it makes recommendations on how to move forward with the practice. The paper is organized into three sections: 1) A review and discussion of the key factors to consider in any lime/alkaline treatment system, 2) A detailed review of practices at numerous facilities utilizing different approaches to lime treatment including the N-Viro – like, RDP, and Bioset Processes, and 3) An assessment of the study's findings with recommendations for improved design and operation of lime/alkaline treatment systems.

KEY FACTORS / ISSUES

Prime reference materials reviewed include the USEPA's 2004 Workshop and its later proceedings (USEPA, 2007), WEF's 2009 MOP (WEF, 2009), WEF's 2012 Solids Process Design and Management publication, and a relatively recent and comprehensive lime treatment of residuals presentation (WEF, 2012; Nicholson, 2013). The factors in the box below apply to all approaches to alkaline treatment and will be discussed in detail.

Need for EQ Product	Mixing Process
Regulatory Requirements	Monitoring Needs
Residuals to be Treated	Product Requirements
Lime/Alkaline Additive	Storage
Introduction of Lime into Residuals	Odor

Need for EQ Product

Exceptional Quality (EQ) Biosolids is sewage sludge or material derived from sewage sludge that meets the following criteria: 1) high quality pollutant concentration limits; 2) has undergone one of the Class A pathogen reduction alternatives; and 3) has undergone one of the vector attraction reduction options. If the biosolids are classified as EQ, they may be distributed for land application without site restrictions. Land application of EQ biosolids is not regulated by the US 40CFR Part 503 Regulation once the biosolids leave the control of the biosolids preparer. Therefore, soil blenders or other (non-preparer) users who take EQ biosolids may store the biosolids or mix the EQ biosolids with other (non-sewage sludge) materials without resampling the product. Conversely, if EQ biosolids remain within the control of the preparer, they are still considered biosolids and are still covered by the Part 503 Regulation. Like all Class A products, they must undergo microbiological testing at the last possible point before being distributed. In addition, if the preparer mixes the EQ biosolids with non EQ materials or otherwise changes the quality of the biosolids, the new material must again comply with pathogen reduction, vector attraction reduction, and pollutant concentration limits. EQ Biosolid derived product may be transported to soil blending operations to enhance the quality of the product.

Regulatory Requirements

In 1993, EPA promulgated the US 40CFR Part 503 Regulation, "Standards for the Use or Disposal of Sewage Sludge" as minimum requirements (USEPA, 1993). Two classes of disinfection were described: A and B. Class A disinfection reduces the number of pathogenic microorganisms to below the detection level, while Class B disinfection accomplishes a two log reduction of bacterial pathogens and a one log reduction of enteric viruses and helminth ova. One lime based option was called out for Vector Attraction Reduction (VAR).

One of three alternatives are usually chosen for achieving Class A disinfection with lime treatment. Alternative One establishes time-temperature relationships as shown below.

Sludge	Temp, °C	50	55	60	65	70	75	80
Moisture > 7 %	Time, hours	316	63	13	2.5	0.5	0.10	0.020
Moisture < 7 %	Time, hours	120	24	4.8	0.95	0.19	0.04	0.008

Alternative Two is a pH-time, temperature-time, drying procedure. Basically, the sludge is held at a pH above 12 for 72 hours with a 12-hour period in which the temperature exceeds 52°C, followed by air drying to a solids content exceeding 50 %. Then under Alternative 6, sludge is treated by one of the proprietary processes listed below that were recommended as equivalent to a Class A, Alternative 5, PFRP Process by the USEPA (USEPA, 2003).

- ATW, Inc., Santa Barbara, CA - Manchak process uses quicklime to simultaneously stabilize and pasteurize sludge. Quicklime, or a combination of quicklime and fly ash, is mixed with dewatered sludge at a predetermined rate in a confined space. An instant exothermic reaction is created in the product wherein the pH is raised in excess of 12 after two hours of contact, in addition, the temperature is raised in excess of 70^0C for > 30 minutes.
- N-Viro Energy Systems Ltd., Toledo, Ohio - Fine alkaline materials (cement kiln dust, lime kiln dust, quicklime fines, pulverized lime, or hydrated lime) are uniformly mixed by mechanical or aeration mixing into liquid or dewatered sludge to raise the pH to > 12 for 7 days. If the resulting sludge is liquid, it is dewatered. The stabilized sludge cake is then air dried (while pH remains > 12 for ≥ 7 days) for > 30 days and until the cake is ≥ 65 % solids. A solids concentration of ≥ 60% is achieved before the pH drops below 12. The mean temperature of the air surrounding the pile is > 5°C for the first 7 days.
- N-Viro Energy Systems Ltd., Toledo, Ohio - Fine alkaline materials (cement kiln dust, lime kiln dust, quicklime fines, pulverized lime, or hydrated lime) are uniformly mixed by mechanical or aeration mixing into liquid or dewatered sludge to raise the pH to > 12 for ≥ 72 hours. If the resulting sludge is liquid, it is dewatered. The sludge cake is then heated, while the pH > 12, using exothermic reactions or other thermal processes to achieve temperatures of ≥ 52°C throughout the sludge for ≥ 12 hours. The stabilized sludge is then air dried (while pH > 12 for ≥ 3 days) to ≥ 50% solids.
- Schwing Bioset, Inc., Danbury, CT - The "Bioset" process (See below.) is to be used to

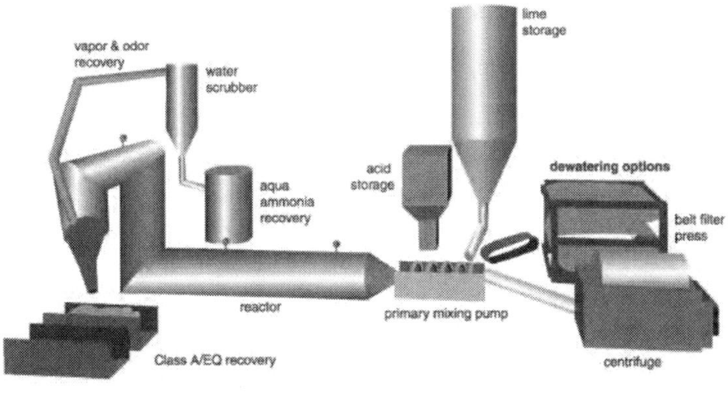

Figure 1. The Bioset Process

treat municipal wastewater sludge with a total solids concentration between 6 and 35% by weight and with a minimum ammonium concentration in the reactor discharge of 0.5 mg $NH^+{}_4$/g dry weight. Dewatered sludge solids must be mechanically mixed with calcium oxide (quicklime) to achieve a pH of equal to or greater than 12 standard units.

Sulfamic acid must be mixed with the sludge/quicklime mixture to maintain the temperature of the mix at equal to or greater than 55°C. The process must be operated in a plug flow regime with a minimum operating pressure of 27 kPa (4 psi) and a minimum solids retention time of 40 minutes at a minimum temperature of 55°C.

Comments. An example of a process meeting Class A, Alternative 1 is RDP's EnVessel Pasteurization Process which is shown below in Figure 2. It is unique because it partially uses electrical heat to achieve the required 70°C for 30 minutes. Lime is added to sludge cake in an electrically heated mixer called a ThermoBlender. The 70°C and pH of 12.0 material is discharged from the mixer into an insulated and heated plug flow pasteurization vessel. Temperature is recorded at the inlet and discharge ends. This insures that every particle of sludge is at or above 70°C for 30 minutes. Class A, Alternative 2 is essentially the process offered by N-Viro Energy Systems Ltd., Toledo, Ohio and deemed equivalent by USEPA to a PFRP Process under Class A, Alternative 6.

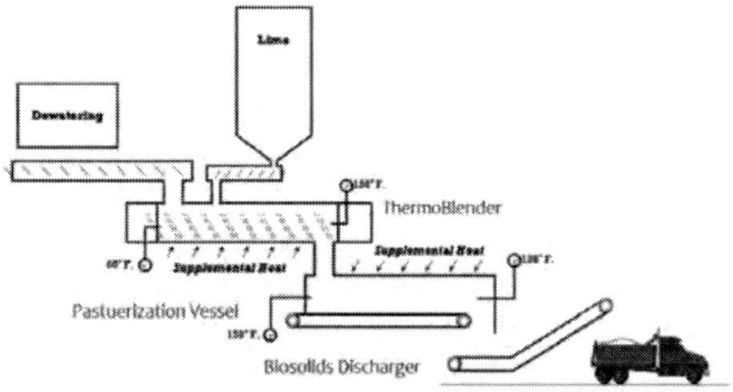

Figure 2. RDP EnVessel Pasteurization

The Bioset Process shown previously in Figure 1, in addition to providing elevated pH and temperature, takes advantage of another disinfectant, ammonia. The high pH conditions of lime treatment convert NH_4^+ to NH_3, a non-charged chemical species known to inactivate many organisms including *Ascaris* eggs. Further research showed that ammonia at temperatures near 50°C and a high pH near 11 could inactivate many pathogens (Pecson and Nelson, 2005). Studies at Tulane observed removal of viruses, pathogenic bacteria and *Ascaris* ova under closed systems at 55°C instead of 70°C due to the containment of ammonia in a closed system (Fitzmorris-Brisolara and Reimers, 2013). In the Bioset process, quick lime helps with both temperature and pH elevation while the small addition of sulfamic acid renders the lime sludge mixture thixotrophic insuring complete mixing within the closed reactor. Data demonstrated the process' effectiveness at inactivating helminth ova within 30 – 40 minutes resulting in a greater than 2 log reduction.

All alkaline treatment systems meet VAR requirements with Option 6: Sewage sludge is considered to have undergone adequate vector attraction reduction if sufficient alkali is added to: a) raise the pH to at least 12, b) maintain a pH of at least 12 without addition of more alkali for 2 hours, and c) maintain a pH of at least 11.5 without addition of more alkali for an additional 22 hours. pH should be measured in a slurry

The American Society of Testing Materials (ASTM) issued a guide for the Alkaline Stabilization of Wastewater Treatment Plant Residuals in compliance with 40CFR503 (ASTM, 1998).

Residuals to be treated

These can be primary, secondary or a combination of primary and secondary sludges. They can also either be raw or digested sludge, recognizing that anaerobically digested sludge is likely to contain a large amount of ammonia. Typically the sludges are dewatered to around 15-25 % solids when they enter the process.

Lime/Alkaline Additive

Both CaO and $Ca(OH)_2$ are frequently referred to as lime. CaO is manufactured from limestone ($CaCO_3$) by heating it to a temperature of about 1093°C to separate and drive off carbon dioxide (CO_2). CaO reacts with water to form $Ca(OH)_2$ and emit significant heat. The exothermic heat produced is 27,500 BTU/lb. mol (National Lime Association, 1990). A medium to highly reactive lime is needed to treat sludge and insure that CaO is fully converted to $Ca(OH)_2$. Because of the cost of lime, other alkaline materials that contain small percentages of available or free CaO are frequently employed. For example cement kiln dust which contains about 3 to 5 percent available CaO and generally has low reactivity with water can be used. Similarly lime kiln dust with 25 to 40 percent available CaO and a low reactivity with water can be used as even can fly ash from the combustion of coal which contains anywhere from 1 to 40 percent CaO.

Ideally, sludge should be treated with a moderately or highly reactive lime to ensure that the CaO fully converts to $Ca(OH)_2$. For the reaction to generate a high pH that migrates throughout the solids, there must be a continuous film of water throughout the material. Otherwise, the lime may not be fully hydrated or the hydroxide ions may not migrate throughout the solids. This can and does result in improper pH measurements, improper lime doses, and therefore, unstabilized solids. If calcium oxide must be pulverized, it is best to do the pulverizing at the point of application to prevent air slaking and ensure the desired reactivity.

Lime requirements vary greatly on type of sludge, anaerobically digested vs waste activated sludge for example, the temperature needed, and the solids concentration. Class "A" processes using the pasteurization criteria of 70°C for 30 minutes require a dosage of CaO of 13 to 20% on a wet weight basis. The RDP EVP process uses about 7 % on a wet weight basis with supplemental electric resistance heat, while the Bioset process uses 9-18 % CaO on a wet weight basis with a very small amount of sulfamic acid. The N-Viro Advanced Alkaline Stabilization with Subsequent Accelerated Drying (AASSAD) process uses 12-20 % CaO (or equivalent amount of CKD or LKD) on a wet weight basis to achieve 52°C for 12 hours, and additional by-product alkaline materials (CKD or LKD or fly ash) are added to achieve a final solids concentration of 55 to 65%, the amount added depending on whether air drying or heated-air drying is used. Usually where lime dose is high, the end product is best used as a liming agent and the calcium carbonate equivalency or CCE should be used to optimize load to the field.

Introduction of Lime into Residuals

Dry CaO is added to dewatered cake. The lime typically is mixed with the cake via a pug mill, plow blender, paddle mixer, ribbon blender, screw conveyor, or similar device. Mixing is discussed in the next session. Figure 3 shows a typical lime feeding scenario. Quicklime,

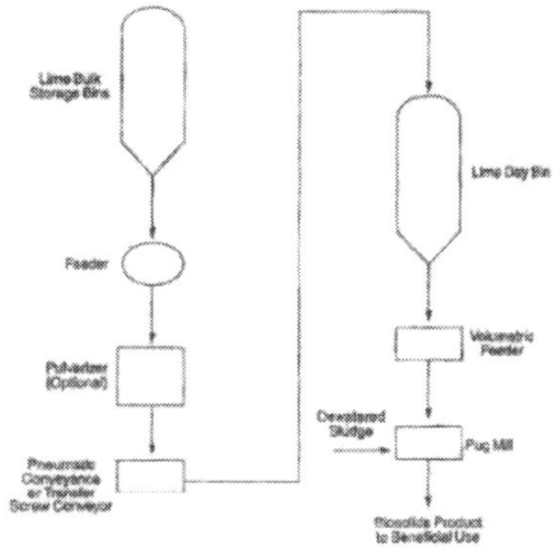

Figure 3. Typical dry lime feeding scenario (Oerke et al. 1990)

hydrated lime, or other dry alkaline materials can be used in this process. Quicklime is less expensive than hydrated lime, and the heat of hydrolysis released when quicklime is added to dewatered cake can enhance pathogen destruction.

Mixing Process

The heart of a sludge/lime treatment system is the mixing procedure. The mixing is generally done with a Pug Mill. The first known use of a pug mill dates back to 1824 where they were used to mix clay and water. Pug Mills are twin horizontal shafts that are rotated in concert to impart the required mixing action. The shafts are arranged to pull the product up through the middle and down on the outside, as shown in Figure 4. The counter rotating action is only effective when the proper bed depth is maintained in the live zone. Sludge/lime mixing pug mills need at least 60 seconds of mixing time in the live zone. They should be provided with a variable frequency drive to allow for changes in speed and mixing energy.

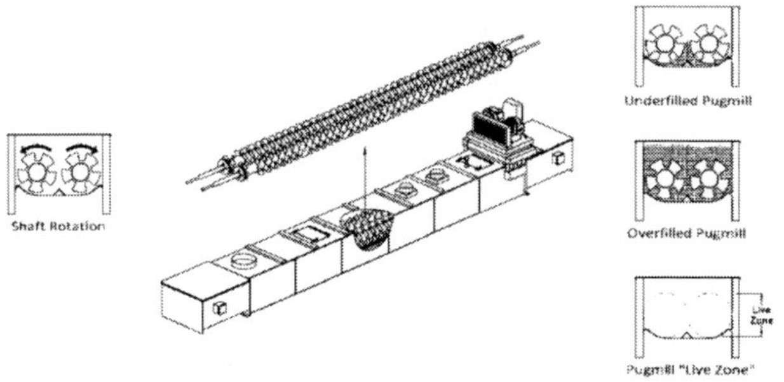

Figure 4. Dual Screw Pug Mill Mixer (Christy, 2017)

Dewatered cake and chemicals are added together at the "head" of the mixer. The proportioning of the alkaline material dosage rate to the rate at which cake is introduced is important. Good thorough mixing of the lime and sludge is critical to insure uniform distribution of calcium and pH throughout the sludge mass. Inadequate mixing or low dose rate can result in loss of pH control in storage and formation of putrescent odors. Over mixing can destroy the granular structure of the product. Uniform mixing of CaO or $Ca(OH)_2$ is much more difficult with dewatered sludge. Non-uniform mixing will change the Calcium Carbonate Equivalence (CCE) of the product and impair its acceptance for beneficial use.

Factors to be considered in striving for good mixing are sludge solids content, type and grade of lime additive, mixer design, mixing energy and mixing time. Achieving the desired pH is not alone an adequate criterion for mixing efficiency. The procedure for measuring pH involves adding water to the treated sludge and then inserting the pH probe. Any unreacted CaO or $Ca(OH)_2$ will react with the added water and pH will increase. An indirect measure of mixing efficiency is NH_3 release. If NH_3 is not detected by the nose or by Dragger tube, then the bulk of the sludge is not experiencing high pH. DCWSA has developed a Ca test that has been shown to be effective in monitoring the completeness of mixing, and has allowed them to reduce their CaO dose significantly.

Lime type and grade can greatly affect mixing uniformity. The basic reactions between CaO and $Ca(OH)_2$ and residuals are hydration and hydrolysis, both requiring interaction with free water in the sludge. The CaO is hydrolyzed to $Ca(OH)_2$ by combining with water, and the $Ca(OH)_2$ dissolves in water to raise pH. Both CaO and $Ca(OH)_2$ are relatively insoluble in water and since dewatered residuals/sludge do not contain much free water, intimate and immediate contact between the CaO or $Ca(OH)_2$ and them is critical. It is also important to keep in mind, that for Class A processes, which use CaO to generate heat, solubility of the formed $Ca(OH)_2$ decreases with temperature.

Temperature and pH Monitoring

To measure the pH of an alkaline sludge cake is not easy and the procedure shown in the box below should be followed in addition to the following considerations. To get a proper pH reading at the high end of the pH scale, the reading must be taken at the standard temperature, 25^0C, or corrected to 25^0C, using the formula shown below.

Correction factor = [0.03 pH units x ($T_{measured}$ – 25 ^{0}C)]/1.0^{0}C where T is in °C.
Actual pH = Measured pH + correction factor (factor can be + or -)

While pH meters often have temperature correction controls. These controls correct for probe temperature and conductivity only. Changes in the concentration of the test sample due to solubility that changes with temperature are not compensated by these controls. Also when testing a liquid/solid mixture with a system designed for liquids (pH probe), care must be taken to keep flow of sample moving across the pH probe surface for accuracy. Clean the electrode frequently, paying particular attention to the reference electrode junction. Bits of adherent biosolids or alkaline agent can strongly alter the measured pH, resulting in erroneous values not representative of the entire mass of material.

The sample and the fresh standard solution used for calibration of the meter must both be at 25^0C in order to obtain a correct reading. The pH electrode should be one that is recommended for alkaline condition (USEPA, 1995). If one reads the pH of a solution at other than standard temperature, then the pH meter reading will either be too high, or too low, depending on the temperature of the test sample, leading to an incorrect assumption that the levels needed for disinfection and stabilization have been achieved. In sludge containing significant amounts of

1. Weigh out 20 to 30 grams (wet weight) of limed sludge into a beaker. There should be no smell of NH_3 in the sample before water is added. Ammonia will give a false high pH if present. So if ammonia can be smelled, then allow additional time for the reaction between the alkaline material and the solids to complete.
2. Add water to the sample in a 1:1 ratio (grams of sample to mL of water). Distilled or deionized water should be used.
3. Stir the sludge/water sample thoroughly and allow to settle. If all of the water is absorbed by the sludge, repeat the procedure with more water until there is a supernatant solution above the settled solids. Allowing the sample to stand for a period of time will also allow any NH_3 to volatilize.
4. Place the electrode in the supernatant solution and record the pH after the reading has stabilized. If necessary correct the pH reading to a temperature of 25 C.
5. Rinse the electrode with distilled or deionized water between readings.

nitrogen, ammonia gas will be released when the pH rises above about 9, with almost complete conversion and release by pH 10. The pH reading of ammonia as it is "gassing off" may exceed 12, leading an uninformed operator to believe that sufficient lime has been added to meet regulatory requirements. But, as soon as the ammonia conversion is completed, the pH of the sludge mass will be at about 10 and not meeting regulatory requirements. At pH 10, bacteria will regrow and consume the unstabilized organic matter, possibly resulting in offensive odors. Measuring the pH of the fully mixed sludge, after ammonia has gassed off, assures that sufficient lime material has been added to achieve the regulatory pH of 12 or above, assuring stabilization and inhibiting regrowth.

Temperature monitoring is important. It is critical to know, for example, if you are achieving Class A status with a temperature of 70°C in all sludge particles for $\geq$ 30 minutes, when the temperature reaches 70°C and how long it stays at $\geq$ 70°C. So significant monitoring is needed for Alternatives 1 and 2 as well as Alternative 6.

Beneficial Use Plan

Beneficial use implies a material is used as a supplement to a mined or manufactured product and that the application rate is consistent with crop need based on realistic yield expectations for a given crop on a specified soil in a defined geographic setting. EQ products are considered a product and site specific permitting for the end use application is not required, but land appliers should maintain records of distribution. If the preparer is the applier, farm maps are necessary to assure the product is applied onto appropriate sites. Information on beneficial use and appropriate loadings is available from local or county extension staff, from the natural resources conservation service, from the local Soil and Water Conservation Service office or from crop consultants.

Beneficial use projects have been implemented throughout the country. Results from one project in Coastal North Carolina are provided to demonstrate the benefits of optimizing the quality of products applied with local needs. One custom soil blender compost producer has requested the lime stable product for several years and the crop yields and quality have improved.

Morehead City, North Carolina is located in eastern North Carolina and is surrounded by sensitive areas. Biosolids management has historically been through a Class B program. In 2011, the City developed an Enhanced Pasteurization Process (EPP) to blend lime and supplemental heat with dewatered biosolid to produce a Class "A" product. That was achieved through pH adjustment for VAR and heat for PFRP. The material achieved the PFRP designation, but did not receive as widespread a use as initially planned. In 2014, the lime dose to the product was reduced yet still complied with VAR requirements and the dried product was blended off site with Class "A" compost. The commingled material was tested and found to contain moderate levels of essential plant nutrients. This commingled product was applied to the established test plots of Bermuda grass, Sorghum, and turf fields. The plots have been used since 2011 and this allowed time series comparison between high lime and low lime material.

Soil materials and crops on each of the designated test plots or sites were tested prior to application of biosolid derived products to assure compliance with nutrient and lime requirements and at the end of annual growing seasons to assure crop quality and yield was assessed.

Based on the soil test results the biosolid derived product loads have been modified through time. The initial product contained lime only as the VAR compliance tool. Subsequent products contained a lower volume of lime when transported off-site for blending into a Class A compost. The material transported off site maintained the VAR/PFRP compliance at the time of use. The commingled biosolid derived products (Lime only or Enhanced Pasteurization Process and Class A compost) contained higher relative levels of N, P, K and secondary nutrients than the lime only material. Prior to the commingling operation, the biosolid derived product was applied at a rate of 3 to 4 tons per acre to satisfy the lime requirement of the soil, but this resulted in failure to meet other nutrient requirements. Following the commingling with compost, were applied at rates of 6 and 9 tons/acre on the Bermuda grass and Sorghum plots and at a rate of 6 tons/acre on the athletic fields to satisfy both lime and a more balanced nutrient requirement. The load to the athletic field was most sensitive because of the public contact on the sites. One of the fields is a "summer league" baseball field linked to the local semi-professional baseball team. On each of the athletic fields turf quality was critical because of the public scrutiny on these sites.

Plant tissue quality and yields have been measured through several growing seasons and correlated with soil test results and the biosolid derived product loadings. The soil test data show dramatic improvements in soil fertility indicators, particularly soil organic matter and other nutrients. Significant improvements in crop yield and quality associated with the beneficial use of these societal by-products are evident through assessment of the plant tissue quality and yield data.

Results from the 2011 through 2014 growing seasons are summarized in the tables below. These represent the changes in the quality of the biosolid applied when optimum beneficial use is considered as quality of the biosolid derived product applied, impacts of product application on crop yield and crop quality and finally improvements in soil fertility associated with optimization of the biosolid materials applied. The optimum quality resulted from addition of composted agribusiness residues containing high levels of essential plant nutrients with the lime-stabilized products (LSP or Enhanced Pasteurization Process). The LSP generated initially utilized lime only for PFRP and VAR compliance, modification to the process initiated in 2014 utilized supplemental heat to achieve PFRP and required lower levels of lime to achieve VAR. This is the Enhanced Pasteurization Process material. The LSP and Enhanced Pasteurization Process when blended with compost yielded a more balanced source of essential plant nutrients than the LSP only. The commingled Enhanced Pasteurization Process material with the composted agribusiness residues resulted in the most beneficial impacts to crop yield and quality as evidenced through the doubling of yield depicted in Table 3.

The changes in processing and the commingling resulted in significant improvements in crop quality and yield. Utilization of the lime only amendment did result in elevation of soil pH to levels that may adversely impact crop management.

Table 1. Biosolid derived product quality generated through commingling societal by-products (as percent except pH).

Product	pH	CCE	N	P	K	Ca	Mg
LSP only	11	43	1.7	.3	.3	18	0.4
LSP+comp	10.2	35	2.1	.4	.9	12	0.4
EPP only	11	24	3.2	.8	.5	12	0.6
EPP+comp	9.8	21	3.0	.8	1.5	8	0.6

All waste, soil and plant tissue testing accomplished at NCDA Agronomic Service Laboratory.

The LSP materials met the Class A requirements at the time they were blended with the compost. The addition of the compost generated from agribusiness residues resulted in a plant nutrient source much higher in potassium than the lime only material. The potassium levels in these commingled products are well balanced with nitrogen and phosphorus.

Table 2. Soil Test Results Following Application of Biosolid Derived Product (as NCDA index value) onto Biofuel Crops in Craven County, NC.

Site	CEC	pH	PI	KI	Ca%	MG%	ZnI	CuI
LSP	16.8	7.4	79	67	92	7	189	101
LSP+Comp	24.2	7.1	113	115	90	6	172	90
EPP only	20.0	7.0	91	86	91	6	198	115
EPP+Comp	26.8	6.9	126	131	91	8	180	105
Control	13.6	5.5	57	41	80	9	129	72
Background	9.2	5.7	57	41	81	9	118	67

Addition of the lime-stabilized materials only did result in significant elevation of soil pH. The biosolid derived product generated through the Enhanced Pasteurization Process required lower levels of lime to achieve the Class A designation and the addition of this material to test plots had less adverse impact to soil pH than the lime only LSP. The addition of the blended lime products with agribusiness compost resulted in dramatic improvement in soil quality.

Measurable Benefits were documented in selected soil fertility levels as well as crop yield and quality following addition of the various biosolid derived products. Sample data collected during the initial phase of project and subsequently as the commingling of compost was added to the biosolid management scenario suggest the commingling results in production of a high quality soil conditioner and nutrient source which improved quality of plant materials and soils in the community.

Table 3. Biofuel Vegetation (Sorghum) Quality (as % or PPM) and Yield (as Tons/ac) Associated with Application of Biosolid Derived Product and from Control Area.

Site	Yield	N	P	K
Lime only	7	2.3	.25	1.8
Lime+Comp	12	2.8	.3	2.3
EPP only	12	2.9	.3	1.7
EPP+Comp	15	3.1	.35	2.6
Control	5	1.9	.22	1.2

The beneficial use demonstrated has resulted in commingling product with soil and compost. Although the volume of material handled by the blender is greater than the volume of the lime only or the compost only, the mixing and blending has resulted in a custom blend most beneficial to plants and soil and least detrimental to underlying groundwater and adjacent surface water.

Storage

Federal and state regulations address storage issues in terms of the length of time that biosolids are stored. The Guide to Field Storage of Biosolids (USEPA, 2000) provides detailed technical information on storage methods and procedures. The Guide deals separately with the storage of Class A and Class B biosolids and addresses pathogen regrowth and transmission from stored biosolids, odor monitoring and mitigation, runoff control, and restriction of storage site access by humans, livestock and wildlife.

Lime and other alkaline materials can be stored in steel or concrete silos. Although standard steel can be used, silos must be both air tight and water tight. The steel silo with cone bottom is the most popular and comes in a variety of sizes. Silo capacity can be calculated from the dose rate of the process used and the biosolids produced. The decision to use several smaller silos rather than one larger one is a function of the process used. Processes using commercial CaO only need one silo, while the N-Viro process uses at least two silos to accommodate a variety of alkaline materials. A significant reserve capacity should be designed for alkaline sludge treatment to avoid process disruptions in the event of a supply breakdown. A recommended reserve is at least seven (7) processing days.

Fine lime materials have inherent flow problems. Lime may absorb water and form a non-flowing cake in the silo, or may arch or bridge. Silos are equipped with external vibrators or air pads, as well as internal structures, to prevent bridging. Hopper bottoms must have a minimum slope from the horizontal of 60^0 to overcome the high angle of repose of CaO ($50\text{-}55^0$) and $Ca(OH)_2$ (70^0) (National Lime Association, 1990).

Odor

Lime treated sludge can be particularly odorous, and this has caused problems for use of this method with respect to processing and beneficial use. Ammonia is the major process odor associated with lime treatment, but others include dimethyl amine (DMA) and trimethylamine (TMA), as well as odor compounds associated with sludge itself (hydrogen sulfide (H_2S) and mercaptans, among others). Ammonia is odorous in the gaseous form, and the gaseous form is prevalent at pHs greater than 9.5. At lower pHs, ammonia reacts with water, as shown below, to form ammonium ion (NH_4^+), which is highly water-soluble. The equilibrium pH for this reaction is 9.5, so at pH 9.5, 50 % of the ammonia is NH_4^+ and 50 % is NH_3 gas. Since pH is a

$$NH_3 + H_2O = NH_4^+ + OH^-$$

logarithmic scale, it means that, for every unit increase in pH the concentration of NH_3 gas increases 10 fold. Because of the high reactivity of NH_3 with water, NH_3 is not very stable in the environment. Ammonia is more of a worker issue during alkaline stabilization, when most of the NH_3 in the sludge is released into the working environment. Ammonia is rarely detected beyond distances of 10-20 feet at processing sites, so odor control measures at the processing site are rarely used. The most obnoxious odors associated with alkaline biosolids are the methyl amines, which have a strong fish-like odor, are highly persistent, and will absorb to clothing. The methyl amines can be easily detected hundreds of yards from the source and are not readily dissipated. Factors affecting the generation of odors with alkaline stabilization include:

- pH – as discussed above, primarily affects the generation of gaseous NH_3.
- Temperature – escape of gaseous compounds from sludge increases with temperature, and very high temperatures can result in formation of burnt odors.
- Residuals/sludge type – the type of sludge (raw or digested) can affect the odor compounds formed. Anaerobically digested sludge has orders of magnitude higher NH_4^+ concentrations than raw, waste activated or aerobically digested sludge, and NH_3 processing odors are most severe with anaerobically digested sludge. Anaerobically digested sludge also contains higher concentrations of reduced compounds like H_2S and mercaptans, while raw sludge contains higher concentrations of odor compounds associated with putrescence.
- Use of polymer – certain polymers that contain amino ($-NH_2$) groups are chemically unstable under alkaline conditions and hydrolyze to form methyl amines. There are no hard and fast rules as to which polymers will form methyl amines under alkaline conditions, so all polymers proposed for use in lime treatment should be tested for their stability under alkaline conditions. Also sludge/biosolids should be treated as soon after dewatering as possible. Undue storage of dewatered sludge may lead to anaerobic conditions, with NH_3 and other odor generation, and to breakdown of polymer.
- Incorporation of biosolids by plowing or by injection - is recognized in 503 as a non-treatment option to meet VAR. With alkaline biosolids, however, VAR is usually met by high pH and incorporation is not required. Incorporation of alkaline biosolids is recommended, however, to more uniformly mix the alkaline material into the soil and avoid any odors that may develop as the pH of the residuals falls and anaerobic decomposition of the organics commences.

Specific odor compounds are usually not routinely monitored. However, measurement of NH_3, methyl amines, and reduced S compounds (H_2S, methyl sulfides, mercaptans) can be used to optimize the alkaline process and anticipate field odor problems. Disposable probes (Dragger tubes) are available for many of these compounds. They are not analytically quantitative, but can be used to give relative values. Air sampling, together with gas chromatography measurements, is the most quantitative means of assessing odor.

Some regulatory authorities have had issues involving Class B and Class A alkaline biosolids. These issues are primarily associated with odors with material stockpiled in the field. In some cases, NH_3 was voiced by the public as being objectionable when alkaline biosolids were stockpiled too close to residences and public areas like schools. In other cases, methyl amines (fish odor) were identified as the cause of complaints. Causes cited for the issues included the failure to dose adequately with CaO or other alkaline products, and to uniformly mix these materials with the residuals. Frequently the putrescible odors occurred when pH dropped. The minimum dose rate is determined by the process parameters. Class A processes that use the pasteurization criteria of 70 ^{0}C for 30 minutes require a range of CaO or its equivalent of 15 to 75 % on a wet weight basis.

REVIEW OF FACILITIES PRACTICES

The authors visited and spoke with personnel at approximately 50 facilities across the country, taking care to look at both *do it yourself type processes* and *proprietary ones* like N-Viro, RDP, and Schwing Bioset. An effort was made to collect the following facility and lime treatment information.

- Size of facility, types of wastewater received and method of processing
- Method of sludge dewatering and solids concentration achieved
- Method of feed conveyance
- Type lime added (CaO, Lime Kiln Dust, etc.) and method of addition (Storage silo, Rotary Vein, Transfer Augers, VFD
- Mixing of lime and sludge: Plowshare Blender, Screw Blender, Pugmill, Heated Screw
- Curing practiced? Heat Pulse Bunker, EnVessel, Open Floor, Storage Bins
- Method(s) of Storage and Odor Control
- How beneficially used: Agriculture, Land Reclamation, Top Soil Blending Class A, or Landfill Cover

Examples of information provided by facilities utilizing the N-Viro, RDP, and Bioset Processes are shown in Figures 5, 6, and 7. Of the sludge processing facilities investigated and regulated as EQ biosolids producing ones, many were achieving pathogen reduction under Class A, Alternative 2, that is by a) Elevating the pH to greater than 12 and maintaining the pH for more than 72 hours; b) Maintaining the temperature above 52°C (126°F) throughout the sewage sludge for at least 12 hours during the period that the pH is greater than 12; and c) Air drying to over 50% solids after the 72-hour period of elevated pH. This is basically the N-Viro Process. Facilities using the RDP Process achieved pathogen reduction under Class A, Alternative 5, maintaining 70°C for ≥ 30 minutes. Bioset Processes achieved pathogen reduction under Class

A, Alternative 6, which meant maintaining 55°C for ≥ 40 minutes in addition to other conditions which were mentioned previously. All facilities inspected using an EQ alkaline treatment process met vector attraction reduction by using Option 6, adding sufficient alkali to: a) Raise the pH to at least 12, b) Maintain a pH of at least 12 without addition of more alkali for 2 hours, and c) Maintain a pH of at least 11.5 without addition of more alkali for an additional 22 hours. Further of the study's findings follows.

Treats 2.2 of a possible 6 MGD of domestic and industrial wastewaters. Has primary and trickling filter processes. Sludges are 90 % primary, 8-9 % biological, 1 % septage, plus some activated sludge from another facility. All go to anaerobic digestion. Lime kiln dust is added to achieve EQ biosolids via Class A, Alternative 2. LKD is added via auger at about 50 %. Good mixing is said to be assured via pH and temperature measurements as well as visually. Untreated sludge is 3% and final is 60-80%; Temperatures achieved are 55-90°C with pH of 12.2. Product is stored prior to marketing. Marketing was a problem resulting in some storage of up to 3 years. That is now being resolved with a PR specialist on board. Biosolids are free to try and then $15/scoop. See photos above.

Figure 5. Information Obtained from Facility using N-Viro Process

Plant is attractive, very clean and has highly automated operation. Treats 2.1 to 3.57 mgd wastewater with primary and activated sludge processes. Effluent goes to a Creek and Reservoir and is P limited. Sludge goes into an aeration mixed holding tank. Sludge at 2-2.5 % solids is belt filter press dewatered to 19-21 %. 2300 # of lime kiln dust are then mixed with the 523 # of dry solids to produce via the RDP Process a 55 % solids product. Plant pays $2/ton to farmers to take the product for corn, soy beans and landscaping. Spreading is largely done in Fall and Spring. RDP's Process employs a thermo blender, twin screw mixer and insulated pipe heated to 650°F. This movement of the sludge mixture through 20 feet of heated pipe guarantees the mixture will be at ≥100°C for 30 minutes. See photos. There was little dust and no lime-sludge balls visible in the soil like product. ……..

Figure 6: Information Obtained from Facility using RDP Process

Lime Alkaline Addition

Several plants using Alternative 2 used lime kiln dust to raise the pH and temperature and in addition some also used cement kiln dust and/or fly ash to either assist with temperature elevation or more often to help with product quality, making the product more granular and thus better in appearance and easier to use. Facilities using the RDP and Bioset Processes largely only added quicklime.

At some facilities visited introduction of the alkaline additive did not closely match, time wise, with introduction of the dewatered residuals. As such it raised questions about the level of treatment occurring. Mixing is always a major concern and is obvious upon both witnessing the process and looking at the product. You could see balls of material where the outside of the ball is coated with lime while the inside still contains raw or unreacted residuals. Similarly you could see unreacted lime particles.

Treatment Plant is a 48 mgd activated sludge one that utilizes pure oxygen. Waste sludge from the clarifiers is pumped to an equalization tank and then to belt presses where most of the remaining liquid is removed. The "cake" that comes off the belt presses at ~ 20 % solids is transferred by conveyor and high-pressure pumps to the Bioset Process. This process mixes dewatered biosolids with calcium oxide (typically 0.6 kg quicklime/1kg dry sludge) to achieve a pH of greater than or equal to twelve standard units. Sulfamic acid in solid granular form added to, and mixed with the sludge/quicklime to promote an exothermic reaction which increases the temperature of the mixture to equal to or greater than 55°C (131°F). The sludge/quicklime/sulfamic acid mixture is then continuously fed to a pressurized plug flow compact reactor (35' long X 4' dia.) using a twin-cylinder piston pump for a minimum solids retention time of forty minutes at a minimum temperature of 55°C (131°F). A pressure of 20 psig is maintained.

Figure 7: Information Obtained from Facility using Bioset Process

Monitoring Needs

The requirements for insuring pathogen and vector attraction reduction are clear and apply to **ALL PARTS OF THE RESIDUALS**. For most of the facilities visited it was unclear how if at all assurance could be given that all parts of the residuals were meeting the temperature and pH requirements. See relevant photos in Figure 5. A conveyor belt was bringing the sludge cake to a lime or alkaline material feeder where the lime was deposited with the cake and both moved into a mixer with the combination then moving along with a conveyor to a windrow pile. The introduction of the lime material to the sludge, the mixing, and the movement out of the mixer all happens quite fast. How quickly the temperature and pH rose to required levels is questionable. In some instances steam was noticeable as well as the odor of ammonia, However, more often than not all that was noticeable was a fog of dust. pH and temperature measurements, if taken, were taken by sticking a probe into the center of a windrow and/or perhaps taking a sample of material from a single location in the windrow. So how uniform the pH or temperature was throughout the windrows is not known.

Dust / Cleanliness
It was not uncommon to see at the facilities visited considerable dust in the air, sometimes accompanied by the odor of ammonia and wet residuals on the floor. From a public relations standpoint, these situations can be problematic.

DISCUSSION OF STUDY'S FINDINGS WITH SUGGESTIONS FOR IMPROVING DESIGN AND OPERATION

Three major issues were identified in the course of the study: mixing, assuring that pH and temperature requirements are achieved, tackling problems of odor and dust, and planning for beneficial use of the biosolids. Each issue will now be addressed.

Mixing
The introduction of the alkaline material to the sludge cake starts the process. It is critical that that introduction is tied to the movement of the sludge cake. Usually this means the dosing is tied to the wet solids, typically ~ 18%, passing the dosing unit. The lime must be added as the wet cake moves under the dosing unit, not before or after.

Every opportunity must be presented for reactions to occur between the alkaline (Ca) product and the residuals/sludge. For this to occur available lime should be in small particles (promoting maximum surface area) but not so small as to create dust and the moisture content of the residuals should be between 80 and 85 %. The preference with most facilities is to buy a pulverized product. Seldom does pulverization occur at the facility. Reactivity of the alkaline material (available Ca) is also important and should be between medium and high (ASTM, 1998).

Manner of mixing including vigor as well as length of time mixing occurs are critical. Some research work has recommended a mixing period of 40 to 60 seconds with an operating speed of about 30 rpm. Ultimately, it is incumbent on the operator to conduct the necessary tests to assure that adequate mixing is occurring. Some guidance developed from mixing studies at DCWASA, USEPA-Region 3 and other locations (Murthy et al, 2000; Mangus et al. 2006) follows:

- Total Ca measurements throughout the lime-treated sludge is a good measure of mixing efficiency.
- Increased CaO dose over and above the minimum required to raise pH above 12 will guarantee that pH does not decline over a period of weeks.
- CaO or $Ca(OH)_2$ grades should be selected to give the highest possible lime content and fineness.
- Increased mixing time increases mixing efficiency. Increased mixing time is required for sludge with higher solids contents and for centrifuged sludge.
- Other measurable indicators of mixing efficiency include NH_3 and amine generation, fecal coliform reduction, and reduction of specific odor compounds such as sulfides and mercaptans.

- Screw conveyors can produce sludge masses that require longer mixing times to achieve uniform mixing.

An approach is recommended for evaluating the distribution pattern of lime in the final mix and thus being assured of a proper mix. Take at least ten (10) 0.5 gm samples from the final sludge/lime mixed product and evaluate them for Ca levels. The variance can be used to evaluate if proper lime distribution/mixing and optimization of the equipment has occurred.

Assuring that pH and Temperature Requirements

The previous section addressed the need to insure a thorough distribution of Ca so that contact was made with all sludge particles. The difficulty occurs in knowing how fast all reactions occur. In most cases, though, with Class A alkaline treatment, enough lime will be added to insure that VAR is achieved.

This leaves as the critical concern that all particles of sludge meet the Class A time requirements for a 70ºC, 55ºC or 52°C temperature. In many of the utilities visited, following mixing, the combined lime-sludge cake product was on an open conveyor, where you could see steam coming off (cooling) as it moved to an open windrow. It is difficult to know what the temperature of the material is at any point in time. At least two equipment providers resolve this issue by a) having the conveyor move through an insulated channel where they can monitor when the temperature rises to the required level and note how long it stays at that level, or b) having the lime-sludge cake mixture move into a closed reactor where they can monitor the temperature and time.

An example of how temperature in a windrow can vary is shown in Figure 10. This temperature profile for this windrow was created with lime treated residuals and with daytime temperatures

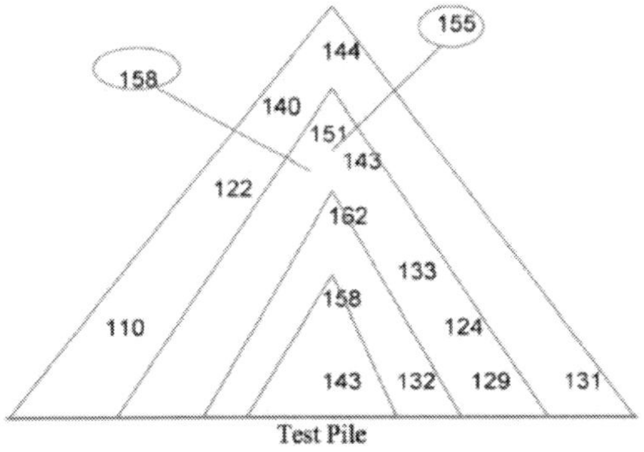

Figure 10. Open Pile Temperature Profile

of 50 to 70°F (10 to 21°C). Facility wanted to achieve 158°F (70°C) throughout the windrow. You can readily see where this did and didn't occur. In an open environment it is very difficult to maintain temperature. Similarly it is difficult to maintain pH, when CO_2 from the air can enter the material and lower the pH.

This temperature profile is typical of many open pile systems including compost windrows. The profile demonstrates the need for effective contact with the high temperatures documented above to achieve pathogen reduction necessary to have a product designated as an EQ material suited for beneficial use.

Tackling Problems of Odor and Dust as well as Temperature in a Windrow

The authors suggest consideration of an insulating membrane cover for the windrow(s). This would minimize heat loss to the environment and insure a more uniform temperature distribution throughout the windrow. Membrane technology has developed to the point where membranes are available that provide facilities a greater than 90% reduction in (escaping) process odors and VOC emissions including that of the beneficial disinfecting contaminant ammonia when compared to uncovered systems. Other attributes can include a greater than 99% containment of bio-aerosols. Membranes are now successfully used in composting systems.

Figure 11. a) Membrane Covered Windrow, and b) Benefits of using Membrane (After GORE® Cover System-North America)

Beneficial Use of the EQ Biosolids

As was discussed previously in the paper, with any sludge management plan it is important to know where the final product will go. Stock piling is seldom an answer. A critical aspect of the program is securing the involvement of farmers who are interested in utilizing the EQ biosolids. The critical player for this job is the biosolids manager. Understandably his/her involvement will likely depend on the individual communities involved and their past experience with sludge application systems; and the overall public acceptance of the concept. The manager or his/her assistant has to be a good communicator and make the public aware of how the sludge is treated and the product of that treatment can be beneficially used by farmers/ranchers. The project could even be described in the local newspaper and the public invited to tour the operation on a particular day. Through the help of the local Soil Conservation Service or agricultural extension service agent the manager needs to get the word out to the agricultural community about the benefits of biosolids and at the same time enlist customers and see if they might be willing to pay anything for the product and or pick it up and spread it. Interested farmers could be invited to visit the sludge treatment operation and talk about what they are seeking most, for example lime or nutrients or perhaps both.

It was obvious in visiting facilities and talking with managers that where there was no plan, the well treated material just accumulated and accumulated, resulting in an unsightly mess.

ACKNOWLEDGEMENTS

Authors wish to recognize the significant assistance of Richard W. Christy, RDP Technologies; Robert O'Dette, State of Tennessee; Timothy J. Nicholson, Organic and Mineral Systems; Chuck Wanstrom, Schwing Bioset; and the many State Biosolids Coordinators and Utilities who provided information and data for this paper.

REFERENCES

ASTM (1998) Standard Guide for Alkaline Stabilization of Wastewater Treatment Plant Residuals, Designation: D 6249 – 98, September 1998, ASTM International, 100 Barr Harbor Drive, PO Box C700, West Conshohocken, PA 19428-2959

Christy, R.W. (2017), Mixing of Alkaline Materials and Sludge, Personal Communication, RDP Technologies, Inc. 960 Brook Rd. Conshohocken, PA 19428, February 14, 2017.

Fitzmorris-Brisolara, K..B. and Reimers, R.S.(2013) "Developments in Disinfection Using Advanced Alkaline Stabilization," WEF Specialty Conference Disinfection and Public Health, Water and Environmental Federation, Alexandria, Va CD-ROM

Mangus, J.J.; Li B.; Burns B. R.; Butler H. G; Cole C. A. (2006), Impact of Lime Dose and Mixing Quality on Odor Generation by Lime-Stabilized Biosolids, Presented at the Water Environment Federation Annual Conference in Dallas, TX, October 2006.

Murthy, Sudhir; Strawn, Mary; Peot, Chris; Tolbert, Dorian; Bailey, Walter; McGrath, Michael (2000). Mitigation of Odors from Lime Stabilized Biosolids. Proceedings of the Water Environment Federation, WEFTEC 2000: Session 51 through Session 60, pp. 574-574(1)

National Lime Association (1990) Lime: Handling, Application, and Storage in Treatment Processes. Bulletin 213, Arlington, Virginia.

Nicholson, M. (2013) Lime Stabilization, History, Trends and Lessons Learned, presented at Mid-Atlantic Biosolids Association's Biosolids Symposium, July 16-17, 2013 in Lancaster, Pennsylvania.

Oerke, D.W. and Rogowski, S. M. (1990), Economic Comparison of Chemical and Biological Sludge Stabilization Processes, AWWA-WPCF Joint Residuals Management Specialty Conference, New Orleans, Louisiana, December 2-5, 1990.

Pecson, B.M., Nelson, K.L., 2005. Inactivation of Ascaris suum eggs by ammonia. Environ. Sci. Technol. 39 (20), 7909–7914.

USEPA (1993) 40 CFR Parts 257, 403, and 503 (FRO-4203-3): Standards for the Use and Disposal of Sewage Sludge, Final Rule. Fed. Register, 58, 9248. US Government Printing Office, Washington, DC. February 19, 1993.

USEPA (1995) Laboratory Procedures and Test Methods, under Methods 9040B, and 9045C, SW-846, revisions published in the Federal Register (pg. 17003, FR 60, Vol. 64, Tues., April 4, 1995). Also as of 2015 SW-846 is available online at http://www.epa.gov/epawaste/hazard/testmethods/sw846/online/index.htm

USEPA (2000) Guide to Field Storage of Sludge. Office of Wastewater Management. EPA/832-B-00-007.

U.S. EPA (2003). Environmental Regulations and Technology: Control of Pathogens and Vector Attraction in Sewage Sludge, EPA/625/R-92/013, revised July, 2003, Cincinnati, OH.

USEPA (2007) Alkaline Treatment of Municipal Wastewater Treatment Plant Sludge: Technical Guide, Unpublished Report, National Risk Management Research Laboratory, Cincinnati, Ohio

WEF (2009) Design of Municipal Wastewater Treatment Plants: MOP 8, Fifth Edition, Water Environment Federation, Alexandria, Virginia

WEF (2012) Solids Process Design and Management, Water Environment Federation, Alexandria, Virginia

PRODUCING HIGH VALUE CARBON PRODUCTS FROM MUNICIPAL SOLIDS GENERATED FROM CHEMICALLY ENHANCED PRIMARY TREATMENT

Wendell Khunjar[1*], Eirene Pavlakis[2], Kartik Chandran[2], Christopher Wilson[3]

[1]Hazen and Sawyer, 4035 Ridge Top Road, Suite 400, Fairfax VA 22033 USA.
[2]Columbia University
[3]Hampton Roads Sanitation District

*Email: wkhunjar@hazenandsawyer.com

ABSTRACT
In this work, we document the recovery of high value carbon products such as volatile fatty acids (VFA) and polyhydroxyalkanoates (PHA) from municipal sludge augmented with high strength organic waste - food waste (FW).

KEYWORDS: volatile fatty acids, bioplastics, carbon recovery, PHA

INTRODUCTION
A key element of low energy nutrient removal technologies like mainstream deammonification is the reliance on carbon diversion strategies, like chemically enhanced primary clarification (CEPT), to help control heterotrophic and autotrophic metabolism in mixed liquor. Solids generated from CEPT are typically anaerobic stabilized, resulting in the generation of biomethane, which can be captured and utilized as an energy source.

While energy recovery from biomethane can provide immediate benefits to many utilities, there are other valuable carbon products (e.g., volatile fatty acids (VFA), polyhydroxyalkanoate (PHA)) that have higher revenue potential than that associated with biomethane energy recovery. Therefore there is the opportunity to demonstrate that coupling carbon diversion with low energy nutrient removal can facilitate high value resource recovery.

To this end, we document the recovery of high value carbon products such as volatile fatty acids (VFA) and polyhydroxyalkanoates (PHA) from CEPT generated municipal sludge augmented with high strength organic waste - food waste (FW). Specific objectives were to: (a) maximize the VFA production from co-fermentation of municipal solids with FW, (b) maximize the PHA production from the VFA rich fermentation effluent (c) Perform an engineering assessment of implementing carbon recovery platform at full-scale WRRFs.

METHODOLOGY

Anaerobic Fermentation
Anaerobic fermentation was performed at 37°C (mesophilic condition [2]) and pH 6.5 ± 0.1. The hydraulic retention time (HRT) and solids retention time (SRT) was set to 2 days to limit methanogenesis [3]. The feed of the fermenter was 1:1 COD based mixture of primary sludge (PS) and high strength food waste (HSW) and the organic loading rate (OLR) of the fermenter was set to 10 or 25 kg COD/m^3-day depending on the experimental conditions [1].

Total COD (tCOD), soluble COD (sCOD), and total carboxylic acids (TCA) of the effluent and influent of the fermenter were determined using colorimetry (Hach kits 2125915, and 2244700, Loveland, CO). Total suspended solid (TSS) and volatile suspended solid (VSS) content in the influent and effluent was determined gravimetrically using standard methods (APHA 2005). VFA composition was determined using Ion Chromatography (Dionex ICS-2100) using a Dionex IonPac™ column (AS11-HC-4μm, RFIC™ & HPIC™ 2 X 250 mm). In addition, biomass was collected once a week for 16S-RNA based metagenomics analysis.

PHA production

Fermentation effluent was used as the influent for downstream PHA enrichment reactor after solids separation. The enrichment reactor was operated in SBR mode, at room temperature (average temperature 23°C), at a 12-hour cycle (Decant 1L, Settling, Decant 1L, Feeding 2L, Anaerobic zone, Reaction) applying feast and famine conditions (HRT=1 day, SRT=1.6 ± 0.2 days). Quantification of PHA was obtained using Gas Chromatography [5,6] coupled with flame ionization detector (GC-FID, SRI 8610C) and MXT-Wax column (30 m 0.53 mmID 0.5 μmdf) (Restek, Bellfonte, PA). Pure Sigma poly[(R)-3-hydroxybutyric acid] (PHB) and poly(3-hydroxybutyric-co-3-hydroxyvaleric acid) (PHBV, 12% mol HV) were used as reference standards. Also, PHA observation under the microscope (Nikon Eclipse 80i, Chroma JC-1 non ratiometric) was performed after staining biomass samples from the enrichment reactor with Nile Blue A.

RESULTS

Co-Fermentation of Municipal Solids and High Strength Food Waste (CofermR)

The co-fermentation system was operated to achieve OLRs of 10 and 25 kg-COD/m3-day using a 1:1 COD mass basis combination of PS and HSW. Results from this phase of research indicated that particulate COD hydrolysis was similar for both conditions (Figure 1). However, the total carboxylic acid yield decreased at the higher OLR (Figure 2). This indicates that while particulate hydrolysis efficiency may be similar, a larger fraction of the soluble COD produced at the higher OLR condition was not present as carboxylic acid. The specific nature of this soluble COD fraction was not determined in this work. Short chain volatile acid distribution in the effluent is provided in Table 1. Results indicated that operating at the higher OLR resulted in higher fraction of acetic, butyric, valeric and succinic acids than at the lower OLR.

Since prior research has demonstrated that PHA production is dependent on several factors including the amount and type of carbon organic carbon (Gilmore, Zuidervliet et al. 2015), overall productivity of a PHA production system is dependent on the mass of readily biodegradable carbon provided during the feast phase. From a practical perspective, this means that maximizing VFA production during fermentation would benefit PHA production. Consequently, fermentate from the co-fermentation system operating at an OLR of 25 kg/m^3-day was used for PHA accumulation experiments.

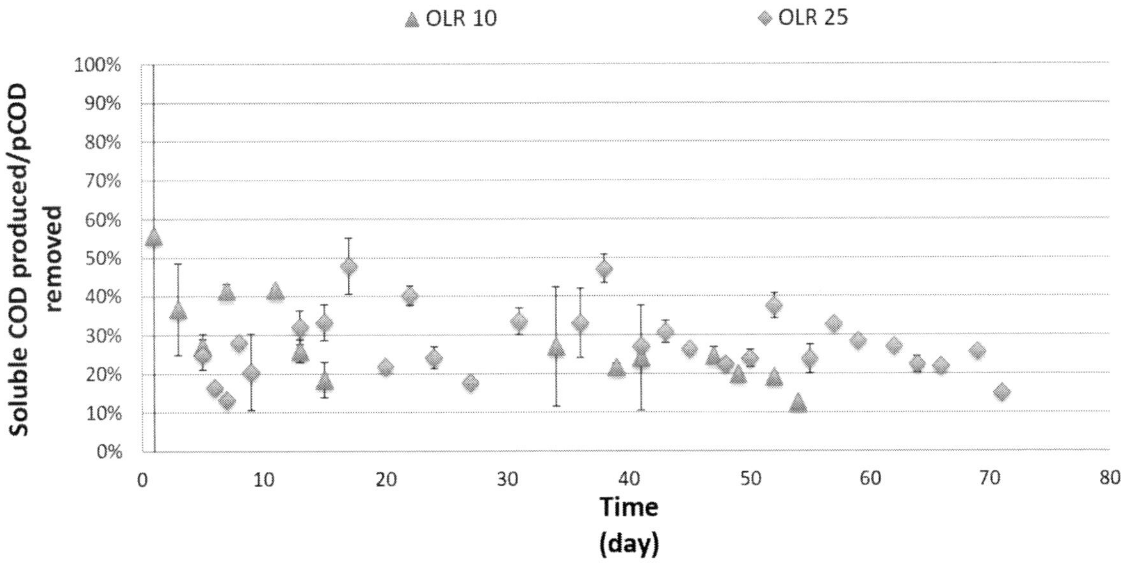

Figure 1: Particulate COD Hydrolysis Efficiency

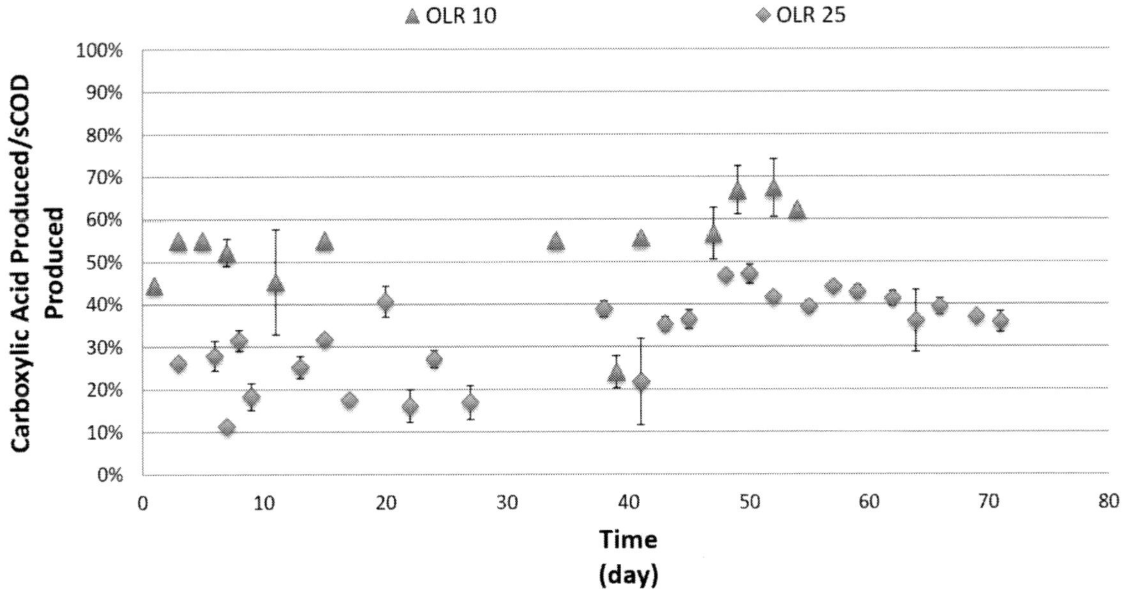

Figure 2: Carboxylic Acid Production as a Function of sCOD produced

Table 1: Distribution of VFAs

		OLR 10 kg/m3=day		OLR 25 kg/m3=day	
		Feed	Effluent	Feed	Effluent
Acetic Acid	mg COD/L	170± 130	910± 200	1,100± 740	2,600±1,900
Propionic Acid	mg COD/L	97± 85	990± 280	310± 35	820± 200
Formic Acid	mg COD/L	4.1± 5.3	260± 280	0.02± 0.0	36±44
Butyric Acid	mg COD/L	44± 52	190± 100	65±20	290±190
Valeric Acid	mg COD/L	29± 23	150± 90	43±12	250±12
Succinic Acid	mg COD/L	nr	nr	21±0.0	250±0.0

nr – not reported

PHA production

A two reactor was employed to stimulate PHA production. The first reactor served as a continuous enrichment system. This reactor was fed fermentate generated from the co-fermentation system. The goal of this system was to generate a seed culture with a high propensity for carbon storage. Biomass from the enrichment system was harvested and used within a dedicated production reactor. The production reactor was also fed fermentate generated from the co-fermentation system. Employing a two reactor strategy allowed for decoupling of enrichment and production, while also ensuring that sufficient enriched seed culture is available.

Enrichment of PHA accumulating organisms has been observed. Preliminary results indicate that some PHA accumulation is occurring during the feast phase. During the feast phase, it is estimated that PHB and PHV accounted for 3.2% (g PHB/g VSS) and 1.3% (g PHV/g VSS) of the biomass. Visualization of the PHA granules was also performed using Nile Blue A staining (data not shown). Further enrichment is underway. Enrichment strategies being utilized include increasing the organic load to the system and shortening the feast and famine phases in an attempt to prevent mineralization of organic carbon.

REFERENCES

APHA (2005). Standard Methods for the Examination of Water and Wastewater. Washington DC, USA, American Public Healt Association.

Gilmore, K., B. Zuidervliet, et al. (2015). Beyond Nutrients: Recovering Carbon And Other Commodity Products From Wastewater. Report 1, WERF: 65.

Class B Land Application: A Long-term, Highly Successful Public-Private Partnership

Jake Finlinson[1*], Dave Ruud[2], Andy Bary[3], Maile Lono-Batura[4]

[1]King County Wastewater Treatment Division
[2]Boulder Park Incorporated
[3]Washington State University
[4]Northwest Biosolids
*Email: jake.finlinson@kingcounty.gov

ABSTRACT

The King County Wastewater Treatment Division (King County) and Boulder Park, Inc. (BPI) have engaged in a public-private partnership to provide high quality class B biosolids applications for farmers in eastern Washington since 1992. The project, known as the Boulder Park Project Beneficial Use Facility (Boulder Park Project), applies biosolids to primarily dryland agriculture lands in Douglas County, Washington. The Boulder Park Project is jointly permitted, operated, and managed by King County and BPI: King County providing application equipment and project management, and BPI providing project operations and application services. Over thirty agencies from across Washington now participate in this unique and highly successful program for beneficial use of their biosolids. The Boulder Park Project is one of the largest farmer-owned and operated, multi-farmer biosolids recycling projects in the United States. Since 1992, project participants have increased from 3 to more than 100 farmers and landowners due to the measureable benefits exhibited in their soils and in crop yields they have witnessed either first-hand on their own farms or on their neighbors' farms. The project currently has over 90,000 acres permitted for biosolids land application, mostly for dryland winter wheat crops.

INTRODUCTION

The Boulder Park Project provides biosolids and application services to farmers to enhance and enrich agricultural soils by adding organic material and nutrients. Biosolids are a nutrient-rich, organic product that has proven to be an effective alternative to synthetic fertilizers. Unlike synthetic fertilizers, biosolids not only provide all essential micro- and macro-nutrients needed by the crops, but also improve the physical properties of the soil by adding organic matter. Increasing soil organic matter provides a food source for soil microbes, improves soil structure, increases soil stability and moisture holding capacity, and promotes carbon sequestration in soil. Farmers pay the participating agencies for the soil amendment value of biosolids delivered and BPI is compensated for application services.

The majority of biosolids applied at the Boulder Park Project are used to fertilize dryland wheat, with lesser applications to triticale, canola, sunflowers, and grasses. Anecdotally farmers participating in the Boulder Park Project have reported better crop yields, improved soil tilth, increased biological activity, and expedited biodegradation of standing wheat straw, as well as reduced runoff and wind erosion effects. Scientific research corroborates many of these claims. The Project's long-term research plots comparing three different rates of biosolids applications to synthetic fertilizer and no fertilizer treatments were established over 20 years ago by a participating farmer and Washington State University (WSU). Results from those research plots indicate as good or better yields when compared with synthetic fertilizers, increases in carbon sequestration, and improvements in grain quality from biosolids applications [1].

Continuous research is only one of the keys to the success of the Boulder Park Project to demonstrate the unequivocal efficacy of biosolids as a fertilizer substitute and soil amendment; just as important is your project partners, customers, community relations, and project location.

HISTORY

In 1987, KC joined with other biosolids management agencies to form the Northwest Biosolids Management Association (NBMA), currently known as Northwest Biosolids. Membership grew to also include universities, consultants, vendors, and regulators and eventually became the forum for King County to network with farmers for biosolids agriculture.

Over the next several years, King County, with the help of third-party allies and spokespersons, developed a strong agriculture program. King County biosolids staff worked long hours with farmers, county extension agents, private contractors, and farming communities to provide information about biosolids. Key to their success was the pro-biosolids message that was being delivered to the public by their allies. The message to the public was clear, transparent, logical, and easily understood [2].

WSU, Washington's land grant university and a member of NBMA, provided scientists and researchers to conduct on-farm biosolids research. WSU scientists installed demonstration and research plots on farms to measure the benefits of biosolids. Results were published and they provided presentations at public meetings to farmers, community members, health district staff, and regulatory officials.

King County began working with a local extension agent who became a strong ally and proponent of agricultural biosolids land application. He made presentations to more than 3,500 people over a two-year period and promoted biosolids use on his weekly radio program.

King County also began to work with a prominent, third-generation farmer, and an official of the state wheat growers association. This farmer used biosolids on his own farm, which became the ideal stage to provide King County's agricultural outreach – a blend of presentations and field tours by farmers, researchers, biosolids managers, and regulators to any interested parties. The farmer brought an outside perspective and worked actively with potential biosolids users, enthusiastically sharing his experiences using biosolids.

> *[A frequent question from the public was "if biosolids are so good why aren't you using them in King County?" – The fact that biosolids was indeed used in King County – in forestry and composting programs – helped minimize concern that Seattle was "dumping" their waste on rural areas.]*

BOULDER PARK PROJECT

The Boulder Park Project was first initiated in 1991 when another prominent, well-known, well-respected farmer in Mansfield, Washington was contacted by a person attending one of these outreach events and was made aware of biosolids, the benefits of biosolids, and recommended he give them a try. One main reason the recommendation was made was location. The town of Mansfield is located on the Waterville Plateau in Douglas County, Washington. The plateau is largely a rural, sparsely populated agricultural community. The plateau is split in half by the Winthrow Moraine, a ridge of glacial deposits from the Cordilleran Ice Sheet about 17,000 years ago. North of the moraine is covered by a discontinuous blanket of gravelly, sandy loam glacial till, which makes this area an ideal candidate for biosolids soil amendments.

In 1992, three Mansfield area farmers began applying biosolids to their dryland crops. The three farmers immediately saw the value of biosolids and created a company known as Boulder Park Incorporated (BPI). In 1994, King County and BPI officially became a public-private partnership known as the Boulder Park Project. Over 25 years, the Project has grown from one participating agency to over 30.

PUBLIC OUTREACH

The Boulder Park Project is a model for a successful land application program. BPI and King County are dedicated collaborators and spokespersons for the biosolids industry. BPI, as the local sponsors and landowners, takes the time and energy to talk to the public, their neighbors, farmers, local government agencies, contractors, regulators, researchers, and treatment plant managers about their project and the safety and benefits of using biosolids in agricultural applications. The fact that the company is owned and operated by local farmers who embrace and support their community, live and work in their community, and are available to their community members – along with the geographic location of the project - is really what makes this project successful.

BPI truly believes that biosolids are a valuable resource. They can see and feel the improvement in their soils in addition to their increased crop yields. Past years of intensive farming practices and repeated applications of synthetic fertilizers have taken its toll on the areas soils. After just one application of biosolids a person can walk across a field and feel the added body (tilth) from the organic matter in contrast to the thin, powdery and easily wind-eroded soils which are only fertilized with synthetic fertilizer. Adding organic matter back to the soils is slowly building soil tilth, making the soils more productive, and promoting sustainable farming for generations to come. BPI distributes biosolids to as many farmers as possible, rather than concentrating on a few ownerships. This practice promotes a strong client base, which in turn makes the project much stronger and more sustainable in the future.

Since 1992, considerable effort has been made by BPI to involve their communities, local government agencies and elected officials in their project. Their efforts have been successful, as demonstrated by a county that is well informed about biosolids with a growing demand among neighboring farmers. Information is made available to the general public and local agencies about biosolids recycling and the Project. Public involvement is encouraged by providing opportunities for public interaction with project participants via informal open houses and project tours. Tours give interested persons the opportunity to learn about biosolids, biosolids land application, and crops grown in biosolids amended soils; to see the project area; observe field applications; and get answers to their questions. BPI has an open door policy at their office for visitors wishing to learn about the project.

Tours of the Boulder Park Project's research plots and various biosolids application sites have been a part of BPI's operations since the beginning of their project. In the early days of the project, tours were popular and well attended by neighbors, farmers, researchers, and conservation agencies. Tours remain the most effective way to acquaint people with the benefits of biosolids recycling. Tours have included groups such as Douglas County (DC) Transportation & Land Services, DC Master Composters, Department of Ecology, Chelan-Douglas Health Department, eastern WA chapter of the PNCWA, National Biosolids Partnership representatives, Evergreen Rural Water of Washington, Northwest Biosolids, as well as agency representatives.

BPI has been proactive in coordinating with the media and other groups in Washington and other states explaining the benefits of biosolids use, cultivating and gaining allies, sharing relevant research, and simply talking with neighbors and friends. They are dedicated spokesmen and advocates for biosolids recycling.

Above: Large, well-attended tours allow neighbors and interested parties the chance to see how biosolids are managed by BPI.

ECONOMIC BENEFITS

The Boulder Park Project was one of the first biosolids projects in the United States to bill farmers for the fertilizer value of the biosolids and return that revenue to the participating agencies supplying the biosolids as well as to the tax payers. Charging farmers for biosolids instills the notion that biosolids are a valuable commodity and not a waste – an idea that supports stewardship and sustainability of biosolids land application in Washington State.

Farmers and the community realize economic benefits from the Project. The Boulder Park Project brings jobs to the community. BPI staffs about 10 employees. As mentioned above, farmers pay for the soil amendment value of biosolids, however, biosolids are the economical choice since biosolids provide all micro and macro-nutrients needed for their crops at about half the cost of synthetic nitrogen fertilizers. In addition, biosolids nitrogen is in a slow release form, which means farmers typically get 1-1/2 to 2 crop cycles from one biosolids application. This saves farmers money in synthetic fertilizer expenses and is especially important when wheat prices are depressed. Other economic benefits can include, but are not limited to, include local sales fuel revenues and their associated tax revenues, repair and service provider fees for biosolids equipment, and purchases of biosolids application equipment.

RESEARCH

BPI has been an enthusiastic supporter of research by supplying equipment, labor, and suitable fields to assist WSU Cooperative Extension and University of Washington (UW) scientists in dryland wheat and biosolids research over the last 20 years. University scientists and researchers have installed numerous research plots on BPI-owned project sites that are designed to evaluate the effect of biosolids on nutrient and soil conditioning properties, crop quality and yield, pollutant uptake, lateral phosphorous movement in soil and carbon sequestration. Results have shown that biosolids increase soil organic matter, moisture retention, and carbon storage; improve soil structure; enhance soil fertility; increase crop residues and yields; and, maintain grain quality.

Above: Research plots with and without biosolids at various application rates.
Below: Research plot wheat comparison (left to right): Biosolids treatment, Synthetic Fertilizer Treatment, and Control-No Fertilizer Treatment

The majority of biosolids applied at the Boulder Park Project are used to fertilize dryland (without irrigation) wheat, with lesser applications to canola, sunflowers, and grasses. Anecdotally farmers have reported better crop yields, improved soil tilth, increased biological activity, as well as reduced runoff and wind erosion effects. Scientific research corroborates many of these claims. The Project's long-term research plots comparing three different rates of biosolids application to commercial fertilizer and no fertilizer were established over twenty years ago by a participating farmer and Washington State University. Results from those research plots indicate as good or better yields when compared with commercial fertilizers, increased soil carbon sequestration, and improved quality grain quality from biosolids applications [2].

Preliminary results from the August 2016 research plot harvest (chart 1) compliment the historic results (chart 2) supporting beneficial use of biosolids in land application. Triticale yields show first crop application rates of 2 dry ton (dt) and 3 dt biosolids per acre significantly yielded more grain than conventionally fertilized grain treatments. All biosolids plots had significantly equivalent grain yield. Conventionally fertilized treatments yielded significantly more than unfertilized treatments. *(Preliminary results and chart provided by Andy Bary, Washington State University, August 2016.)*

Chart 1. 2016 grain yield results from Boulder Park Project Research Plot, GP17.

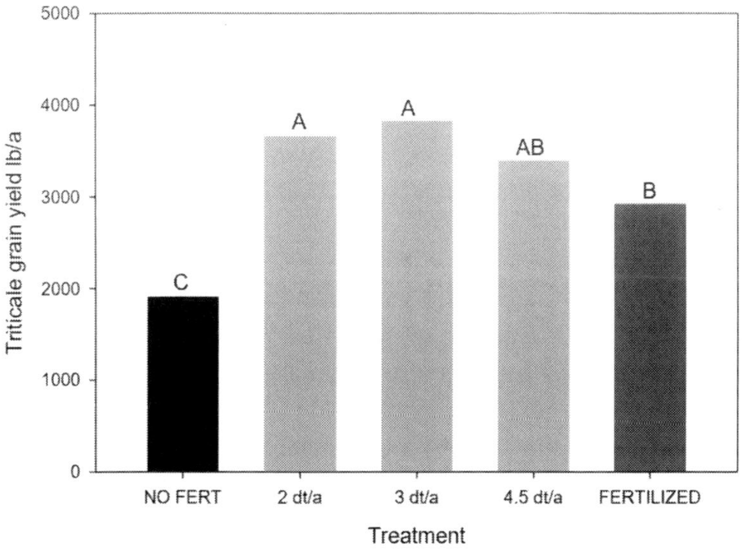

Chart 2. Historic (1994-2010) grain yield results from Boulder Park Project Research Plot, GP17.

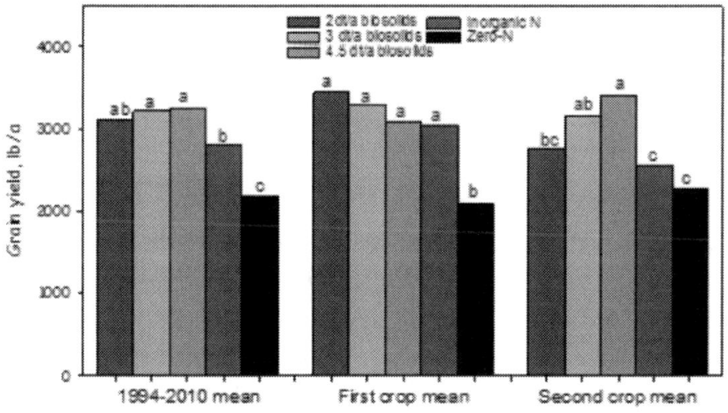

Biosolids had equal or greater grain yields than inorganic N treatment

Continuous research is key to the success of the Boulder Park Project to demonstrate the unequivocal efficacy of biosolids as a fertilizer substitute and as a soil amendment, however, the project owes its greater successes to farmers paying attention to their neighbors using biosolids and observing the tangible difference in crop yields and soil quality.

THREE-PARTY CONTRACT

Ecology has issued a mandate for the beneficial use of biosolids, which is consistent with state and local government's zero waste goals, beyond waste initiatives, and growing emphasis on landfill diversion stressing reduce, reuse, and recycle practices. Agricultural land application of biosolids has become one of the primary beneficial uses of biosolids in Washington State. As a result, the Boulder Park Project has grown to meet not only the needs of biosolids management agencies that wish to have their biosolids land applied at the Boulder Park Project, but also the growing farmer demand. Agencies producing biosolids could see that the Boulder Park Project had an excellent reputation and operated a cost-effective, reliable project. Since farmer demand had increased, the Project needed more biosolids to meet that challenge of keeping customers satisfied.

The Boulder Park Project developed the Three-Party Contract to make the Project available to other agencies. This was a win-win as the Project received more biosolids for its customers and the agencies found a cost-effective, reliable land application project for their biosolids recycling.

The Three-Party Contract has worked well for defining the parties' roles and responsibilities, standards and requirements for biosolids quality, rates for land application services and use of equipment, requirements for insurance and liability coverage, contract terms, and establishes language that allows for contract revocation if the contract terms are not adhered to as defined.

The Boulder Park Project encourages participants to visit the BPI office and talk to staff so the agencies know who is managing their biosolids; to inspect project sites and know where their biosolids are going; and to go to the research plots and see the results of 20 years of scientific research.

REGULATORY ENVIRONMENT

One of the first steps in developing a safe, socially, and environmentally sound biosolids program is to identify and comply with all federal, state and local regulations. One of the first steps to building a successful program is to determine the classification of your biosolids so that you ensure proper end use. BPI and King County developed best management practices so that all regulatory requirements, public needs and customer desires could be met. One component to our ongoing success with BPI has been the relationship built with the regional biosolids regulator.

It is vital to build mutual trust and respect with your biosolids regulator; communication is key. We communicate with our regulator at the beginning of each new project to ensures we meet all regulatory requirements as our project progresses, providing time to modify objectives and targets when necessary. We have clearly communicated our goals, guiding principles and practices, enabling our regulator to support our program bringing long term credibility to our project.

CONCLUSION

The Boulder Park Project is one of the largest farmer-owned and operated, multi-farmer biosolids recycling projects in the United States. Since 1992, project participants have increased from 3 to more than 100 farmers and landowners due to the measureable benefits exhibited in their soils and in crop yields they have witnessed either first-hand on their own farms or on their neighbors' farms with over 90,000 acres permitted for biosolids land application. BPI is not only on the forefront of beneficial land management but are proactive, becoming dedicated spokesmen and advocates for biosolids recycling.

REFERENCES:

1. Cogger, Craig G. 2013. Long-Term Crop and Soil Responses to Biosolids Applications in Dryland Wheat. Journal of Environmental Quality: American Society of Agronomy, Crop Science Society of America, and Soil Science Society of America.

2. Johnson, M. and Vogel, L. 2013. Building a Successful Biosolids Management Program in the Northwest. Northwest Biosolids.

Biosolids Products for Urban Agriculture

Sally Brown[1], Kristen McIvor[2] and Dan Thompson[3]

[1]University of Washington
[2] Harvest Pierce County
[3]Tagro, Tacoma, Washington
* Email slb@uw.edu

Key words: Urban agriculture, biosolids, community gardens, Tacoma

The traditional end use for biosolids has been agronomic crops. We have seen increased crop yields, improved soil physical properties and enhanced soil carbon sequestration as a result of applying biosolids to meet the nitrogen needs of crops (Brown et al., 2011; Cogger et al. 2013; Wuest and Reardon, 2016). As municipalities transition to Class A biosolids, there is a potential to expand use of these materials to urban areas. Within cities potential uses for biosolids products include urban tree plantings, home gardens, highway right of way and urban agriculture. In urban areas, soils are often highly compacted and nutrient poor. In many cases, legacy contaminants such as Pb can also be a concern. Biosolids based soil products have been shown to be effective for improving soil properties in urban areas and for reducing the risk associated with contaminants in urban soils (Attanayake et al., 2014, 2015, McIvor et al., 2012). A recent publication recommended use of biosolids and composts as the most effective tool to limit concerns about contaminants in urban soils for growing food (Brown et al., 2016). Use of biosolids in urban agriculture, in particular in community gardens can offer the potential for rapidly increasing knowledge of and demand for biosolids based products within urban areas.

Community gardens

Community gardens typically consist of small plots that are cultivated by individuals or families within a shared space (McIvor, 2016). These plots have traditionally been found in high- density urban areas where individuals do not have their own gardens to grow food or flowers. However, they are also now commonly found in areas where single family homes are common. The American Community Gardening Association provides a broad definition for community gardens (https://communitygarden.org/). The primary criteria are that it is a community managed open space with areas set aside for growing plants. Because these gardens are comprised of many individuals or groups, they provide a space for sharing information as well as for growing food. Many of the gardeners who are interested in community gardens are new to growing plants and join a garden in the hope of gaining knowledge. Information exchange in community gardens was seen as one of the primary places that people learn about soils and methods of cultivation (McIvor, dissertation). In many cases, one or more gardeners will come to be recognized as a source of information including information on soils. In a survey of community gardeners in Tacoma and Seattle, WA and Boston, MA, a majority of the gardeners reported getting their information on soils from other gardeners (McIvor, 2011)(Table 1).

Table 1. Sources of information on soils from community gardeners (n=15) in Seattle, Tacoma, and Boston (McIvor, 2011)

Type of information	**Number of total gardeners reporting**	**Percentage**
Fellow Gardener	12	80%
Books	9	60%
Workshops (total)	8	53%
Non-profit agency workshop	6	
Cooperative Extension workshop (Master Gardener program)	2	
Internet	5	36%
Other family, friends	4	29%
Garden shop/Vendors	3	21%
Volunteer experts	2	14%
Local media	2	14%
Phone resource (hotline)	1	7%

Biosolids products in community gardens

Developing and using biosolids products for community gardens has the potential to expand usage outside of the gardens. If the products are shown to be effective in the garden, the members of the garden are likely to use the products in their home gardens and also tell their neighbors. This has the potential to be a means to increase an urban customer based for biosolids based products. In deciding to develop biosolids based products for use in community gardens it is important to recognize that the products need to have a consistency that allows for surface spreading and incorporation into the soil. They must also look like recognizable soil materials and have a smell that is associated with soil and cultivation rather than human waste. Ease of use and availability for use are also important factors. Community gardens can be considered as a means to disseminate information on biosolids products and develop a customer base. Active participation in the gardens should be viewed as a critical step to achieving large- scale acceptance and demand for biosolids based products.

Tacoma, WA case study

The history of biosolids use in community gardens in Tacoma, WA can serve as a guide for both developing products for use in community gardens and for how to integrate these products into the day to day function of community gardens. The Tacoma WWTP began producing Class A biosolids in the early 1990s through a duel digestion system that includes a combination of aerobic thermophilic and anaerobic mesophilic digestion (Mendrey, 2014). The resulting biosolids were unacceptable for home use. A blend was developed by mixing sawdust and sand with the material and was marketed as Tagro. Tagro was readily available to residents and was broadly used. A second product- designed to be less nutrient rich and stable was developed in 2003. The Tagro potting soil consists of biosolids, aged bark and sawdust (Mendry, 2014). The program generates between 3500- 4000 dry tons per year of biosolids and use this to produce 25,000 yards of the classic Tagro mix and 20,000 yards of potting soil each year.

Use of Tagro in community gardens was initially part of a PhD research project by Kristen McIvor, then a student at the University of Washington. McIvor tested plant response to Tagro in three community gardens, then run by the City Parks Department (McIvor, 2011). As part of her work she became increasingly involved in the burgeoning community garden movement in the city. This drew the attention of the mayor, Marilyn Strickland (McIvor, K. and S. Brown, 2016). Expanding community gardens became a goal of the Mayor and use of materials that were readily available within the city was a critical tool to make this happen. Available city owned land was offered to potential gardeners. At each site, water for irrigation was made available. The City provides soil testing, construction materials to build raised beds, Tagro potting soil and/or yard waste compost to fill the beds, and cardboard and mulched wood to cover paths between the beds. These materials are provided free of charge to the gardens. There is also a program to manage the gardens. Harvest Pierce County (http://piercecd.org/190/Urban-Agriculture) is housed within the Pierce County Conservation District and runs a gleaning program in addition to the community garden program. The number of gardens in the area has grown from about 5 in 2007 to over 70 in 2016. A map of the gardens is shown in Figure 1.

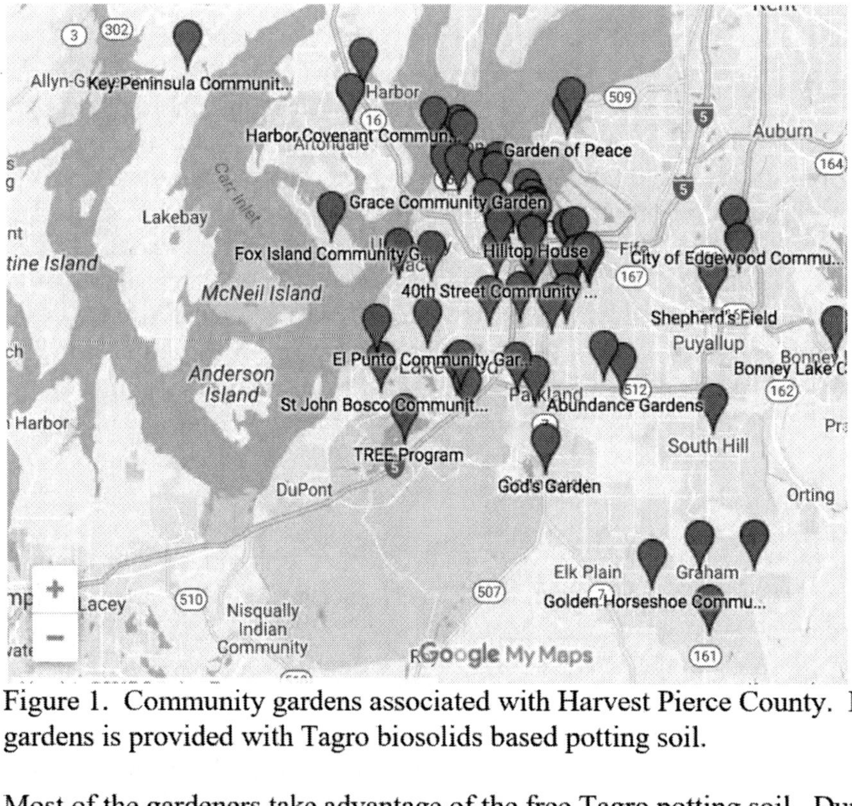

Figure 1. Community gardens associated with Harvest Pierce County. Each of the gardens is provided with Tagro biosolids based potting soil.

Most of the gardeners take advantage of the free Tagro potting soil. During initial meetings to set up a new garden some gardeners will have questions about the Tagro potting soil. These questions are typically answered by other gardeners who are familiar with the material. The Tagro soil is a highly fertile growing substrate. This makes it easier for first time growers to succeed and so makes it more likely that they will continue to garden. A row of plots in a community garden that used the Tagro potting soil is shown in Figure 2.

Figure 2. Raised bed plots in a community garden in Tacoma, WA. Each of the beds was filled with Tagro potting soil, provided free to each garden every year.

The potting soil provided to the gardens makes up less than 5% of the total biosolids produced. However, in tandem with supplying biosolids to community gardens, the Tacoma wastewater program has seen increased demand for biosolids products (Mendry, 2014). The biosolids program is now operated as an independent profit center within the utility. Sales of biosolids related products have increased every year as associated costs with managing biosolids have decreased. The utility is not yet at a break- even point but they have seen per wet ton costs fall from $60 below $30. Sales revenue was over $650,000 in 2013 and over $850,000 in 2015. The utility sold out of biosolids in the early summer of 2015. The Tagro biosolids soil products are used by over 4500 customers each year.

<u>Conclusion</u>

The use of biosolids based soil products in urban agriculture offers a potential for increased local use of these materials. Biosolids based products can enhance gardens by improving soil properties and decreasing concentrations and hazards associated with urban soil contaminants. They also make it easier for first time gardeners to be successful. Supplying biosolids to community gardens has become a central component of the Tacoma, WA program. As this program has grown, sales of the Tagro soil products have also increased. While community gardens are an excellent place for biosolids based soil products- introducing them to gardens and providing them annually requires a strong partnership with the gardens and a willingness to maintain the relationship and the service over time.

References:

Attanayake, C.P., G.M. Hettiarachchi, A. Harms, D. Presley, S. Martin, and G.M. Pierzynski. 2014. Field evaluations on soil plant transfer of lead from an urban garden soil. J. Environ. Qual. 43:475–487. doi:10.2134/jeq2013.07.0273

Attanayake, C.P., G.M. Hettiarachchi, S. Martin, and G.M. Pierzynski. 2015. Potential bioavailability of lead, arsenic, and polycyclic aromatic hydro- carbons in compost-amended urban soils. J. Environ. Qual. 44:930–944. doi:10.2134/jeq2014.09.0400

Brown, S., Kurtz, K., Bary, A., Cogger, C.. 2011 Long-term effects of organic amendments on soil carbon storage and physical properties. Environ. Sci. & Tech. dx.doi.org/10.1021/es2010418.

Brown, S.L., R.L. Chaney, and G.M. Hettiarachchi. 2016. Lead in urban soils: a real or perceived concern for urban agriculture? J. Environ. Qual. doi:10.2134/jeq2015.07.0376

Cogger, C.G., Bary, A.I., Myhre, E.A., Fortuna, A.M. 2013a. Biosolids applications to tall fescue have long-term influence on soil nitrogen, carbon and phosphorus. J. Environ. Qual. 42, 516-522.

Cogger, C.G. ., Bary, A.I., Kennedy, A.C., Fortuna, A.M. Long-term crop and soil response to biosolids applications in dryland wheat. 2013b. J. Environ. Qual. 42 (6) 1872-1880.

McIvor, K. 2011. Soil Quality and the Dynamics of Information Sharing in Community Gardens. PhD dissertation, University of Washington, Seattle, WA.

McIvor, K., C. Cogger, and S. Brown. 2012. Effects of biosolids based soil prod- ucts on soil physical and chemical properties in urban gardens. Compost Sci. Util. 20:199–206. doi:10.1080/1065657X.2012.10737049

McIvor, K and S. Brown 2016. A Case Study: Integrating Urban Agriculture into the Municipal Infrastructure in Tacoma, WA pp. 385-400 In Sowing Seeds in the City: Human Dimensions. E. Snyder, K. McIvor, and S. Brown (eds). Springer Publishers

Mendrey, K. 2014. Building profitable markets for biosolids markets. Biocycle June 55:5:41

Wuest, S.B., and C.L. Reardon. 2016. Surface and root inputs produce different carbon/phosphorus ratios in soil. Soil Sci. Soc. Am. J. 80:463-471

The Effect of External Substrate (Pure Glycerol), SRT, pH and Mixing Intensity on VFA Production in Bench-scale Prefermenters

Marzieh Ghasemi[1], Andrew A. Randall*[2]

1: Ph.D. candidate at University of Central Florida; 12800 Pegasus Drive, Suite 211, zip code: 32816-2450;

email: m.ghasemi@knights.ucf.edu; phone: 407-925-9931

*2: Professor, University of Central Florida; 12800 Pegasus Drive, Suite 211, zip code: 32816-2450, Orlando, Florida;

email: Andrew.Randall@ucf.edu; phone: 407-823-6429

ABSTRACT

Controlling the operational conditions of fermentation processes enable us to optimize the performance of prefermenters with respect to volatile fatty acid (VFA) quantity and composition. In this research the effect of solids retention time (SRT), pH and mixing intensity on co-fermentation of primary solids and an external substrate (pure glycerol) was studied in 3 phases. In Phase 1, 3 identical bench-scale reactors with 1500 ml active volume were run under controlled conditions in which every two reactors isolated an experimental variable. It was observed that adding pure glycerol to the prefermenter increased the average total VFA production from 5090 ± 4236 mg COD/L to 9639 ± 5731 mg COD/L. Between the two pure glycerol dosed-reactors with SRT as the sole experimental variable, the reactor with a SRT of 4 days produced more VFAs (12511 ± 5829 mg COD/L) than that of the reactor with an SRT of 2 days (9639 ± 5731 mg COD/L). In Phase 2, the effect of initial pH on co-fermentation of primary solids and pure glycerol was evaluated. For this reason, 3 identical bench-scale reactors were run under the same environmental conditions except for different initial pH values (5.5, 7, 8.5). The most consistent result in this phase was that the initial acidic pH inhibited the VFA production significantly. Adjusting the initial pH to both neutral (pH = 7) and basic (pH = 8.5) pH values improved the total VFA production. However, in the reactor with the initial pH of 7 propionic acid was the dominant VFA whereas in the reactor with the initial pH of 8.5 acetic acid was the primary fermentation product. The effect of mixing intensity on co-fermentation of glycerol and primary solids were studied in Phase 3 through running 2 identical reactors which were dosed with pure glycerol at the same concentration as in Phase 1 and Phase 2. Mixing intensity was the sole experimental variable between the reactors in Phase 3 (0 and 100 rpm). Experimental results revealed that the VFA production in the un-mixed reactor (9821 ± 3597 mg COD/L) was much higher than that of mixed reactor at 100 rpm (4196 ± 1156 mg COD/L). Consistent with the majority of the reactors in Phase 1 and Phase 2, propionic acid was the main fermentation end-product followed by acetic acid.

Key words: glycerol fermentation, volatile fatty acids, mixing intensity, solids retention time, primary solids, enhanced biological phosphorus removal, wastewater.

1 Introduction

Enhanced Biological Phosphorus Removal (EBPR) is known as a powerful mechanism to remove phosphorus from wastewater. EBPR removal efficiency depends both on the wastewater VFA concentration and composition. It is reported that for 1 mg/L P removal about 7-9 mg/L VFA as COD is needed (Barnard, 1993). Although acetic acid was once thought of as the most effective substrate for P removal, in long-term cultivation, the EBPR removal efficiency in reactors with higher propionic acid concentrations resulted in a higher EBPR removal efficiency. The optimum propionic acid to acetic acid ratio was reported as a 1:1 (molar ratio) by Chen et al. (2004). However, wastewater often has a limited amount of VFAs and it is mostly in the form acetic acid. Therefore, it is important to increase the VFA concentration with the desired composition (e. g. propionic:acetic ratio) to optimize EBPR. Due to high propionic acid and acetic acid production-costs it is expensive to purchase supplemental acetate and propionate in full-scale wastewater treatment plants.

On the other hand, it is reported that glycerol (the main component of biodiesel waste) fermentation results in considerable VFA production, mostly in the form of propionic acid. Barbirato et al. (1997) reported that propionic acid was the dominant glycerol fermentation end-product (by pure culture) followed by succinic acid, acetic acid, and propanol. Himmi et al. (2000) showed that propionic acid formation by propionic acid bacteria (pure culture) was about 2 times greater on glycerol than glucose. This leads to the idea of using glycerol as the external substrate in prefermenters in order to produce propionic acid which can be subsequently used as a carbon source to drive phosphorus removal in EBPR systems.

Solid retention time (SRT) is a term referring to the time that solids or microorganisms stay in a reactor. Increasing the SRT up to a certain point increases the quantity of microorganisms in the prefermenter which results in greater VFA production. However, if SRT is increased too far methanogenesis will take hold and VFA will be consumed with methane being the major end product (Rössle and Pretorius, 2001). Therefore, it is important to keep the SRT value in a proper range. Danesh and Oleszkiewicz (1997) reported that increasing the SRT in the range of 4 to 13 days at room temperature (20 ± 2°C) increased the VFA volumetric production rate in bench scale prefermenters. Elefsiniotis and Oldham (1991) found that at ambient liquid temperature (18 °C to 22 °C) increasing the SRT value from 5 to 15 days (5 d, 10 d and 15 d) enhanced the VFA specific production rate.

Gonzalez-Barcelo and Gonzalez-Martinez (2007) reported that decreasing the pH from 7.7 to 5.5 increased the acidification of primary solids in a sequencing batch reactor. In contrast, Zeng et al. (2006) investigated the effect of pH on acidification of primary solids in a batch system. It was observed that the acidification of primary solids decreased significantly when the pH decreased from 7 to 5.5. The optimum pH was reported between 6.5 and 7. Danesh and Oleszkiewicz (1997) observed that regardless of the SRT, decreasing the pH from 7-7.6 to 6.1-6.4 resulted in a lower VFA volumetric production rate. Bengtsson et al. (2009) examined the effect of pH on fermenting 4 different types of wastewater in a batch system designed based on activated primary tank prefermenters (i. e. a primary clarifier with a recycle). They observed that the optimum pH with respect to total VFA production depended on the type of wastewater but generally ranged between 5.25 and 6. A dramatic VFA production drop was seen in a lower pH environment. In addition, VFA composition was also affected by the pH and regardless of the type of wastewater increasing the pH led to more propionic acid and less acetic acid formation.

Mixing intensity is another factor that can affect the fermentation process. Danesh and Oleszkiewicz (1997) reported that decreasing the mixing period from 6 hr/cycle to 0.25 hr/cycle enhanced both the volumetric VFA production rate and the specific VFA production rate in bench-scale prefermenters.

Given the importance of operational conditions on the fermentation process, the objective of this research was to investigate the effectiveness of adding glycerol to prefermenters as an external substrate and to identify the optimum SRT, pH and mixing intensity to maximize and control the VFA production and composition.

2 Materials and methods

2.1 Phase 1: Type of substrate and SRT

In the first phase of the study, 3 identical side-stream bench-scale prefermentation reactors were operated at room temperature 22 °C with no pH change or adjustment. The liquid volume inside the reactors was 1500 mL with 500 ml headspace. Mixing energy was applied to all reactors with mechanical mixers at 50 rpm. Depending on the SRT value the proper amount of solids inside the reactor was wasted, and the reactor was reloaded with the same volume of fresh primary solids on a daily basis. Fresh primary solids were obtained from the Glendale Wastewater Treatment Facility (Lakeland, Florida) weekly and kept in a cooler at 4°C. Reactor # 1 and Reactor # 2 were dosed with 1500 mg pure glycerol/cycle. Therefore, the initial increase in glycerol concentration in Reactor # 1 and Reactor # 2 was 1000 mg/L. Reactor # 3 was run as a control without any substrate addition. Table 1 shows the experimental conditions of the 3 reactors. Sampling was conducted over 7 weeks. Since after 4 weeks the VFA production suddenly increased significantly the presented values are the average of 6 sampling events conducted in the last 3 weeks when the reactors seemed to achieve steady-state conditions.

Table 1: Operational conditions of 3 reactors in Phase 1

Reactors #	External Substrate	Dosage[2] (mg COD/L)	Temp. (°C)	pH[1]	SRT (days)	Mixing (rpm)	MLSS[1] (mg/L)	MLVSS[1] (mg/L)
R1	Pure glycerol	1217	22	4.68	4	50	16153	14620
R2	Pure glycerol	1217	22	4.79	2	50	13947	12613
R3	-	0	22	5.05	2	50	17227	14880

1: pH, MLSS and MLVSS data show the average values inside the reactors at the end of each sampling cycle.
2: Theoretical oxygen demand of glycerol was calculated as 1.217 mg COD/mg glycerol. Therefore, the substrate dosage in Reactor # 1 and Reactor # 2 was 1000 mg/L × 1.217 mg COD/mg glycerol = 1217 mg COD/L.

2.2 Phase 2: pH effect

In the second phase the reactors were run under the same environmental conditions except for the different initial pH values (Table 2). The initial pH of the reactors was adjusted to 5.5, 7, and 8.5 using the 1M HCl or 1 M NaOH solutions. The solid retention time (SRT) in all reactors was set at 4 days and they operated as semi-continuous fed batch reactors. Therefore, 375 ml of mixed liquor was removed and the reactors were loaded with the same amount of fresh primary solids approximately every 24 hours. The reactors were mixed continuously with mechanical mixers at 50 rpm. Reactor #1, 2, and 3 were dosed with pure glycerol at 1000 mg/L (1217 mgCOD/L). The experiment was conducted at room temperature (22°C). Table 2 shows the experimental conditions of the 4 reactors.

Table 2: Operational conditions of 3 reactors in Phase 2 (MLSS and MLVSS are the average of 5 to 8 data points)

Reactor #	External Substrate	Dosage (mg COD /L)	Temperature °C	Initial pH	SRT	Mixing (rpm)	MLSS (mg/L)	MLVSS (mg/L)
R1	Pure glycerol	1217	22	5.5	4	50	12467	10693
R2	Pure glycerol	1217	22	7	4	50	13217	10358
R3	Pure glycerol	1217	22	8.5	4	50	12725	10333

2.3 Phase 3: Mixing intensity

In Phase 3, two 1500 ml identical batch-reactors with 500 ml headspace were run under the controlled operational conditions. The reactors were isolated with respect to the mixing intensity. One of the reactors was mixed at 100 rpm and the other one was left un-mixed during the entire experimental period except for the sampling time when a slight mixing was provided to obtain a homogenous sample. Both reactors were dosed with 1000 mg/L glycerol (1217 mg COD/L) According to the experimental results from the first two phases, the SRT value was kept constant at 4 days and no pH adjustment was conducted during the entire experiment. The reactors were run for 3 weeks to reach the steady-state condition. Sampling started in week 4 and was conducted twice a week for 5 weeks. The experimental condition of Phase 3 is shown in Table 3.

Table 3: Operational conditions of 2 reactors in Phase 3 (pH, MLSS and MLVSS are the average of 5 to 8 data points)

Reactor #	External Substrate	Dosage (mg COD /L)	Temperature °C	Final pH	SRT	Mixing (rpm)	MLSS (mg/L)	MLVSS (mg/L)
R1	Pure glycerol	1217	22	4.29	4	0	19774	15700
R2	Pure glycerol	1217	22	4.42	4	100	19281	15283

2.4 Analytical methods

2.4.1 VFA measurement

VFAs were measured by gas chromatography (GC) using a Shimadzu GC14-A which was equipped with a Supelco Nukul Column (30m × 0.25mm I.D. × 0.25μm; Supelco, St. Louis). The column initial temperature was increased from 110°C to 190°C at the rate of 5°C/min and stayed at the highest temperature for additional 10 min to remove all the residuals from the column. The injection port and detector port both were kept at 220°C. The carrier gas helium was provided at 20 cm/min linear velocity. Samples were centrifuged and then filtered with 0.45μm membrane filters. Samples were acidified before injection with 5% formic acid in 1.5 ml GC vials (the pH must be 3 or less). 1 μl acidified sample was injected into the GC by an auto-injector AOC-20i (Shimadzu, Columbia, Maryland). To calculate the VFA production in the reactors, it was assumed that the reactors achieved steady state conditions. Therefore, the VFA production was calculated by subtracting the VFA concentration in fresh primary solids going inside the reactor from the VFA concentration in the reactors at the end of each cycle.

2.4.2 Solids

Total suspended solids and volatile suspended solids were measured according to Standard Methods section 2450 D and E (1995).

2.4.3 Chemical oxygen demand

Chemical oxygen demand was determined by using the high range (0-1500 mg COD/L) colorimeter COD vials (Lovibond, Sarasota, Florida). The absorbance of the samples were then measured using a Hach DR5000 spectrophotometer at 620μm (Hach, Loveland, Colorado).

2.4.4 Glycerol

High purity Glycerol (assay > 99.5%) purchased from Fisher Scientific was used in this study.

3 Results

3.1 Glycerol addition

Comparison of Reactor # 2 versus Reactor # 3 isolates pure glycerol as the sole experimental variable in Phase 1 (Table 1). From Table 4 it can be seen that most of the glycerol was fermented to propionic acid. The difference in total VFA production (in terms of mg COD/L) between Reactor # 2 and Reactor # 3 was higher than the COD added to Reactor # 2 in the form of pure glycerol which implies that some of the carbon from the primary solids was fermented to propionic acid in Reactor # 2 that was not fermented to propionic acid in Reactor # 3. One theory for explanation of this would be that glycerol favors the bacteria that produce propionic acid as their fermentation end product. Therefore, their population will be larger in Reactor # 2 than in Reactor # 3. As a result, some of the primary solids are fermented by this larger population and end up as propionic acid in Reactor # 2. Table 4 summarizes the VFA production and composition in Reactor # 2 and Reactor # 3. As it is shown in Table 5 Reactor # 2 has a higher VFA production rate in comparison with Reactor # 3 which is related to the higher VFA production in the glycerol dosed-reactor. According to Table 5 and Table 6, the VFA volumetric and specific production rates in this study were much higher than that of former studies. This might be related to the higher solid concentration of the reactors in the current study. This assumption is consistent with the experiment of Zeng et al. (2006) in which higher VFA production, primarily in the form of propionic acid, was observed in the reactor with a higher solids concentration. Another possible reason could be the higher COD value of the external substrate added to the reactors. The COD value of glycerol was 1217 mg COD/mg glycerol. Therefore, adding 1000 mg glycerol per liter resulted in 1217 mg COD enhancement per liter in glycerol fed reactors.

Table 4: Fermentation composition and VFA production for Phase 1, Reactor # 2 vs. Reactor # 3 (from December 2nd-2015 to December 18th-2015).

Reactors	Glycerol dosage (mg COD/L)	HAc[1] (mg COD/L)[2]	HPr (mg COD/L)	HBu (mg COD/L)	HV (mg COD/L)	VFA (mg COD/L)
R2	1217	3707 ± 2838	4596 ± 2015	492 ± 1105	845 ± 1003	9639 ± 5731
R3 (Control)	-	2866 ± 2318	1636 ± 920	241 ± 848	346 ± 657	5090 ± 4236

1. HAc: Acetic acid; HPr: Propionic acid; HBu: Butyric acid; HV: Valeric acid.
2. The COD of acetic acid, propionic acid, butyric acid and valeric acid was calculated by multiplying the mg/L values by 1.06, 1.51, 1.82 and 2.039 (their theoretical CODs per unit mass) respectively.

Table 5: VFA production rates in Phase 1, Reactor #2 vs Reactor #3 (control)

Reactors	Glycerol dosage	VFA (mg COD/L)	Volumetric rate (mg COD/L/hr)	Specific rate (mg COD/gVSS/L)
R2	1217	9639 ± 5731	200.8 ± 119.4	24.5 ± 21.5
R3 (Control)	-	5090 ± 4236	106.0 ± 88.3	9.5 ± 8.1

Table 6: VFA specific and volumetric production rate in the literature

Author	External Substrate	Dosage (mg COD/L)	VSS (g /L)	Specific rate (mg COD/g VSS/hr.)	Volumetric rate (mg COD/L/hr.)
Danesh et al. (1997)	-	-	2.7-3	3	38
Zeng et al. 2006	-	-	8.43	2.95	24.84
Zeng et al. 2006	Molasses	750	2.99	11.98	NR[1]
Bengtsson et al. (2008)	-	-	NR[1]	NR[1]	55 ± 6.3[2]

3.2 Solids retention time

Experimental results in Phase 1 showed that changing the SRT from 2 to 4 days increased the total VFA production from 9639 ± 5731 mg COD to 12511 ± 5829 mg COD (Table 7). Further analysis of the experimental data revealed that this increment resulted from higher propionic acid and valeric acid production which occurred in Reactor # 1 with the higher SRT (4 days). The average propionic acid and valeric acid production (in order) were 6187 ± 2229 and 2328 ± 1378 in Reactor #1 and 4596 ± 2015 and 845 ± 1003 in Reactor #2. Both reactors had almost the same acetic acid and butyric acid production. It should be noted that the VFA production of each cycle was calculated by subtracting the VFA concentration of the inflow at time zero from the VFA concentration of the corresponding reactor at the end of the cycle (24 hr. fermentation time). The average specific VFA production rate in Reactor #1 was 27 mg COD/g VSS/hr which was higher than that of Reactor #2 with 24.5 mg COD/g VSS/hr. It is possible that SRT has a significant impact on VFA and propionic acid production outside the SRT range analyzed (i.e. 2 to 4 days), however, due to time constraints the effect of SRTs outside this range were not studied. Table 7 summarizes the fermentation composition in Reactor # 1 and Reactor # 2. The numbers presented in Table 7 are the average of the 6 data points (from December 2nd-2015 to December 18th-2015).

Table 7: Fermentation composition and the average VFA production per cycle (from December 2nd-2015 to December 18th-2015) in Reactor # 1 vs. Reactor # 2 (SRT as the isolated experimental variable).

Reactors	SRT (days)	HAc (mg COD/L)	HPr (mg COD/L)	HBu (mg COD/L)	HV (mg COD/L)	VFA (mg COD/L)
R1	4	3526 ± 2412	6187 ± 2229	470 ± 1099	2328 ± 1378	12511 ± 5829
R2	2	3707 ± 2838	4596 ± 2015	492 ± 1105	845 ± 1003	9639 ± 5731

3.3 pH effect

Experimental results in Phase 2 revealed that the optimum pH in terms of VFA production was between 7 and 8.5. In some observations the VFA production was dominant at neutral pH and in some observations a basic environment (pH=8.5) was the most effective. Zeng et al. (2006) reported the same trend between VFA production vs. the pH value. They identified the neutral pH as the optimum pH for producing VFAs. Regarding the VFA composition for the current study, propionic acid was produced effectively in neutral and basic environments. In some observations propionic acid production was at its maximum when the initial pH was adjusted to 7 and in some observations, pH 8.5 resulted in a higher propionic acid production. On the other hand, the maximum acetic acid production occurred in Reactor # 3 where the pH was adjusted to 8.5. In Reactor # 1 and Reactor # 2 acetic acid was consumed in some cycles which suggests that there might have been minor acetocalstic methanogenesis active in them, but the amounts consumed were very small and may indicate that there was no significant change in the amount of acetic acid.

3.4 Mixing effect

Experimental results revealed that the VFA production in the un-mixed reactor was much higher than that of mixed reactor. Table 8 shows the VFA production with respect to the mixing intensity. The average VFA production in Reactor # 1 and Reactor # 2 (Phase 3) were 9821 ± 3597 mg COD/L and 4196 ± 1156 mg COD/L in order. Since only 1217 mg COD/L pure glycerol was added most of the VFAs had to originate from the primary solids, and the results with respect to mixing can be generalized to primary prefermenters with no glycerol addition. This is supported by the literature (Danesh et al., 1997, Zeng et al., 2006). Acetic acid production was inconsistent, and in some cases acetic acid consumption was observed. However, the amount was small, implying that acetoclastic methanogenesis was not significant. TCOD data showed little decrease/cycle which was consistent with low methanogenesis. Consistent with the first two phases, propionic acid was the dominant fermentation product.

Table 8: VFA production and composition in Phase 3

Reactor	Mixing (rpm)	HAc (mg COD/L)	HPr (mg COD/L)	HBu (mg COD/L)	HV (mg COD/L)	VFA (mg COD/L)
R1	0	487 ± 607	5397 ± 2276	1275 ± 356	2662 ± 615	9821 ± 3597
R5	100	271 ± 969	1773 ± 305	728 ± 207	1424 ± 168	4196 ± 1156

4 Conclusions

- It was observed that adding external substrate increased VFA production. Adding the pure glycerol at the initial dosage of 1000 mg/l increased the VFA production about 89% on average compared to a control reactor.
- Regardless of the SRT, propionic acid was the dominant fermentation end-product in the glycerol fed reactors of Phase 1 whereas in the control reactor acetic acid was the main fermentation product.
- It was revealed that increasing the SRT value from 2 to 4 days increased the VFA production by 29%.
- The most consistent results in Phase 2 was the inhibitory influence of initial acid pH of 5.5 on VFA production. In the reactor experiments the degree of inhibition was greatest for acetic acid.
- Both initial neutral and basic environments favored VFA production, however, the average propionic acid to total VFA production was higher at pH 7 than at pH 8.5.
- No mixing favored production of VFAs (particularly in the form of propionic acid).

Acknowledgements

The authors wish to thank Glendale Wastewater Treatment Plant (Lakeland, Florida) and Iron Bridge Water Reclamation Facility (Oviedo, Florida). This work was funded by SAKAL LLC-GROUPO LAKAS (Panama City, Panama).

References

Abu-Ghararah, Z. H., Randall C. W. (1991) The effect of organic compounds on biological phosphorus removal. *Water Science & Technology,* 23.4-6, 585-594

Barbirato, F., Chedaille, D., Bories, A. (1997) Propionic acid fermentation from glycerol: comparison with conventional substrates. *Applied Microbiology and Biotechnology,* 47.4, 441-446.

Barnard, J. L. (1993) Prefermentation in biological nutrient removal plants. *Proceedings of the Joint CSCE-ASCE National Conference on Environmental Engineering*; Montreal, Quebec, Canada, July, 12 – 14; 1767 – 1774

Banister, S. S., Pretorius W. A. (1998) Optimization of primary sludge acidogenic fermentation for biological nutrient removal. *Water S. A.*, 24(1), 35-42

Bengtsson, S., Hallquist, J., Werker, A., Welander, T (2008) Acidogenic fermentation of industrial wastewaters: effects of chemostat retention time and pH on volatile fatty acids production. *Biochemical Engineering Journal,* 40(3), 492-499.

Bondioli, P., Bella L. D. (2005) An alternative spectrophotometric method for the determination of free glycerol in biodiesel. *European Journal of Lipid Science and Technology,* 107(3), 153-157.

Bouzas, A., Ribes, J., Ferrer, J., Seco, A. (2007) Fermentation and elutriation of primary sludge: Effect of SRT on process performance. *Water Research,* 41(4), 747-756

Chen, Y., Randall A. A., McCue, T. (2004) The efficiency of enhanced biological phosphorus removal from real wastewater affected by different ratios of acetic to propionic acid. *Water Research,* 38(1), 27-36.

Dancsh, S., Oleszkiewicz, J. A. (1997) Volatile fatty acid production and uptake in biological nutrient removal systems with process separation. *Water Environment Research,* 69(6), 1106-1111.

Elefsiniotis, P., Oldham, W.K. (1991) The effect of operational parameters on the acid-phase anaerobic fermentation in the biological phosphorus removal process.

Proceedings of the ASCE National Conference on Environmental Engineering, New York, New York, 325-330

González-Barceló, O. G., González-Martínez, S. (2007) Anaerobic prefermentation and primary sedimentation of wastewater in a sequencing batch reactor. *Water S. A.,* 32(4), 577-583.

Himmi, E. H., Bories, A., Boussaid, A., Hassani, L. (2000) Propionic acid fermentation of glycerol and glucose by Propionibacterium acidipropionici and Propionibacterium freudenreichii ssp. shermanii. *Applied Microbiology and Biotechnology*, *53*(4), 435-440.

Hood, C. R., Randall, A. A. (2001) A biochemical hypothesis explaining the response of enhanced biological phosphorus removal biomass to organic substrates. *Water Research,* 35(11), 2758-2766.

Lopez-Vazquez, C. M., Oehmen, A., Hooijmans, C. M., Brdjanovic, D., Gijzen, H. J., Yuan, Z., van Loosdrecht, M. C. (2009) Modeling the PAO–GAO competition: effects of carbon source, pH and temperature. *Water Research*, *43*(2), 450-462.

Rössle, W. H., Pretorius, W. A. (2001). A review of characterisation requirements for in-line prefermenters: Paper 1: Wastewater characterisation. *Water S. A.*, *27*(3), 405-412

Wu, H., Yang, D., Zhou, Q., Song, Z. (2009) The effect of pH on anaerobic fermentation of primary sludge at room temperature. *Journal of Hazardous Materials*, *172*(1), 196-201.

Zeng, R. J., Yuan, Z., Keller, J. (2006) Effects of solids concentration, pH and carbon addition on the production rate and composition of volatile fatty acids in prefermenters using primary sewage sludge. *Water Science and Technology*, *53*(8), 263-269

Vermicomposting of Biosolids and Beneficial Reuse

-

New Zealand Commercial Case Studies from 4 Communities over 8 Years

Michael Quintern[1] and Max Morley[2]

[1] Quintern Innovation Ltd., New Zealand / Germany
[2] Noke Limited, Tauranga, New Zealand
Keywords: vermicomposting, biosolids, pulpmill solids, land utilisation, case studies
Presenting author email: max@mynoke.co.nz

ABSTRACT

Earthworms are the emerging work force behind the up-cycling of dewatered sludge from municipal and industrial wastewater treatment plants and of wet organic residues from food processing industries in New Zealand. Vermicomposting is increasingly becoming the strategy of choice, transforming these sludge's and wastes into a safe and highly valued fertiliser and soil conditioner known as vermicast for use in agriculture, horticulture, nurseries, and recreation areas. Acceptance of the technology has seen vermicomposting grow from processing some 2,000 tonnes of sludge in 2008 to close to 200,000 tonnes in 2015. Globally the traditional means of disposing processing wastes to landfill is under close scrutiny due to high capex and opex costs and negative environmental impacts. Increasingly stringent legislation means communities and industries are seeking the security of alternative solutions when planning their long-term waste biosolids disposal strategies. Important to these are excellence in environmental performance, ecological sustainability and the need to be cost effective. Vermicomposting technologies achieve this through the conversion of biosolids to a safe fertiliser and soil conditioner suitable for sustainable land utilisation. Vermicomposting is natural process, is highly cost effective for industrial and municipal organic wastes streams and overcomes cultural resistance of human waste utilisation to land. A number of case studies from municipal wastes and industrial are discussed. Municipal wastes from Maketu, a small community of 250 households producing 0.5 tonnes biosolids per day. Te Puke, a semi-rural community of 8,000 people producing 900 tonnes p.a., and Rotorua and Hamilton with combined populations of 218,000 and diverse industrial footprints producing 10,000 and 13,500 tonnes p.a. of biosolids respectively. Industrial sludge and wet organic residues vermicomposting operations are presented for Miraka (dairy) (2,500 t/a), Fonterra (dairy) (30,000 t/a), Seeka (horticulture) (5,000 m3/a) and Oji Fibre Solutions (100,000 t/a), (formally CHH Pulp and Paper), who have diverted their organic waste sludge's away from traditional landfill disposal to vermicomposting up-cycling.

BACKGROUND

Earthworms - or more specifically, compost worms - are the emerging work force driving the up-cycling of dewatered sludge from municipal and industrial wastewater treatment plants and of wet organic residues from food processing industries in New Zealand (Quintern, 2015) and globally (Edwards et al., 2011). Vermicomposting is increasingly becoming the strategy of choice, transforming these sludge's and wastes into a safe and highly valued fertiliser and soil conditioner known as vermicast for uses in agriculture, horticulture, nurseries, and recreation areas (Lazcano and Domínguez, 2011; Zandonadi and Busato, 2012; Hernandez et al., 2015; Joshi et al., 2015; Sinha et al., 2013).

The traditional way of disposing organic wastes from municipalities and industries to landfill is globally close scrutiny due to high capex and opex costs and negative environmental impacts. Increasingly stringent legislation means communities and industries are seeking the security of alternative solutions when planning their long-term organic waste disposal strategies. Important to these are excellence in environmental performance, ecological sustainability and to be cost effective. Vermicomposting technologies achieve this through the conversion of biosolids to a safe fertiliser and soil conditioner suitable for sustainable land utilisation. Vermicomposting is natural process, is highly cost effective for industrial (Quintern et al., 2013; Quintern et al., 2015) and municipal organic wastes streams (Quintern, 2014a; Quintern, 2014b; Quintern, 2015) and overcomes cultural resistance of human waste utilisation to land.

ORGANIC RESOURCES

Biosolids

The nitrogen and phosphate rich biosolids from municipal wastewater treatment plants are low in carbon as presented in Table 1 for the two biosolids samples form Hamilton and Te Puke in comparison to the carbon rich and low in nutrient recycled paper solids. The received biosolids are mixed with the pulpmill solids at the MyNoke vermicomposting site to achieve the preferred C/N ratio for vermicomposting (Figure 1)

Table 1. Volumes and analysis results of organic wastes such as sewage sludge, waste activated sludge from milk plants, and pulpmill solids received at various vermicomposting sites. For comparison Aa-grade limits for land applications of biosolids (sewage sludge) in New Zealand

	Sewage sludge Hamilton	Sewage sludge Te Puke	Limits for AA-grade (NZWWA, 2003)	WAS	Recycled paper solids, example
Received (t/a)	13,500	1,000		30,000	100,000
Dry Matter (%)	19.0	13.5		16	26.9
Carbon (%)	30	41		43	37
Nitrogen (%)	4.7	6.5		6.6	0.17
C/N ratio	6.4	6.3		6.5	217
pH	7.4	-		5.8	6.9
Phosphorus (mg/kg)	25,000	12,100		18,700	153
Sulphur (mg/kg)	8,800	7,200		9,000	< 5,000
Potassium (mg/kg)	1,400	3,400		9,800	240
Calcium (mg/kg)	17,900	7,500		13,500	58,900
Magnesium (mg/kg)	2,700	2,200		2,700	1,900
Sodium (mg/kg)	700	1,350		4,900	820
Boron (mg/kg)	24	57		54	< 50
Arsenic (mg/kg)	129	< 4	20	< 5	< 5
Chromium (mg/kg)	83	11	600	7	7
Cadmium (mg/kg)	0.7	0.6	1.0	< 0.3	0.2
Copper (mg/kg)	260	162	100	7	46
Lead (mg/kg)	17.2	18.1	300	0.8	15.2
Mercury (mg/kg)	0.49	1.8	1.0	< 0.3	< 0.3
Nickel (mg/kg)	28	7	60	< 5	5
Zinc (mg/kg)	330	350	300	67	77

Figure 1. Loading biosolids from Hamilton City WWTP at MyNoke vermicomposting mixing site prior to mixing with pulp and paper mill solid

Pulp and paper mill organic wastes

Oji Fibre Solutions Ltd. (Oji) operates four Pulp & Paper mills located in New Zealand 's North Island. All four sites produce organic wastes generally described as pulpmill solids or sludge's. The fibrous pulpmill solids are blended with the Fonterra milk plant sludges. The characteristics of pulpmill solids can be highly variable depending upon the source of fibre e.g from wood chips or recycled cardboard or paper (Table 1), the particular pulp mill wastewater treatment processes used and the sludge separation technology.

Pulp and paper sludge from Kinleith Pulp & Paper mill

Losses to the Pulp & Paper mill's wastewater systems from routine cleaning of the pulp & paper machine, average around 15 t/d (dry basis) of rejected fibre material. This product is referred to as 'primary solids' and is removed in the primary clarifier. The pulp & paper mill's wastewater treatment system produces approximately 12 t/d (dry basis) of secondary biological solids from the oxidation ponds which are combined with the primary solids prior to being pumped into two gravity-dewatering ponds (Figure 2). Table 2 gives an overview of the characteristics of the primary and secondary solids from Kinleith Pulp & Paper mill. By blending the different solids the limits for organic certification are met and an organic certified product is produced. The process of producing the pulp and paper sludge from the Kinleith Pulp & Paper mill is described in more detail by (Quintern et al., 2013).

Figure 2. Primary pulp and paper mill sludge in gravity dewatering pond at Oji FS Kinleith Pulp & Paper Mill before transporting to vermicomposting site

Table 2. Characteristics of pulp and paper mill sludge and recycled paper solids from various Oji Fibre Solutions' Pulp & Paper Mills in New Zealand

Parameter	Primary & secondary solids	Recycled paper solids	Primary solids	Limits organic certification*
Pulp & Paper Mill	Kinleith		Tasman	
Dry Matter (%)	17.8	26.9	34.1	
Carbon (%)	37.6	37	29.7	
Nitrogen (%)	0.5	0.17	0.4	
C/N ratio	75	217	75	
pH	7.4	6.9	7.9	
Phosphorus (mg/kg)	509	153	939	
Sulphur (mg/kg)	3,200	< 5,000	1,321	
Potassium (mg/kg)	1,060	240	2,180	
Calcium (mg/kg)	24,200	58,900	114,900	
Magnesium (mg/kg)	2,440	1,900	2,120	
Sodium (mg/kg)	1,130	820	3,640	
Boron (mg/kg)	0.28	< 50	20	
Arsenic (mg/kg)	1.0	< 5	2.6	20
Chromium (mg/kg)	2.9	7	62	150
Cadmium (mg/kg)	0.1	0.2	0.26	1.0
Copper (mg/kg)	9	46	55	60
Lead (mg/kg)	1.72	15.2	5.0	250
Mercury (mg/kg)	0.04	< 0.3	< 0.11	1.0
Nickel (mg/kg)	1.3	5	6.1	60
Zinc (mg/kg)	43	77	63	300

* (NZWWA, 2003)

Recycled paper solids from Kinleith Pulp & Paper mill

The recycled paper solids show similar carbon and nitrogen concentrations to the primary solids. The slightly higher metal concentrations of the recycled paper solids (Table 2) compared to the primary pulp and paper mill sludge are still well below the limits for safe land application (NZWWA, 2003).

Pulp and paper solids from Tasman Pulp & Paper mill

At Tasman Mill, the primary pulp and paper solids are separated in the wastewater treatment plant clarifier by adding a polymer to coagulate the solids which are then de-watered mechanically resulting in a much dryer product than the pulp and paper solids from the sedimentation ponds at Kinleith Mill (Table 2). The pulp and paper solids are generated in a continuous flow process, which enables a consistent supply to the vermicomposting operation.

CASE STUDIES

A number of case studies from municipal sewage sludge and industrial organic wastes are described and discussed. The case studies range from small communities with on-site and decentralised vermicomposting to centralised full scale vermicomposting operations where organic wastes are received from municipalities, a number of food processing industries and the pulp and paper sector.

Maketu

Maketu is a small costal community of 250 households producing 0.5 t/d of sewage sludge at a centralised wastewater treatment plant (WWTP). Sludge production doubles during the peak summer holiday months. The dewatered activated sludge is collected in a feedstock-mixing wagon. Approximately once a week dewatered pulpmill solids are mixed with the sewage sludge and applied to the on-site vermicompost area of 0.5 ha. Annually a total of 250 t sewage sludge, 300 t pulpmill solids are converted to approximately 150 t of AA-grade vermicast. Vermicast is tested according to (NZWWA, 2003) prior to utilisation on local farmland at application rates of 10 t/ha.

Te Puke

Te Puke, a semi-rural community of 8,000 population is producing 900 t/a of sewage sludge at their centralised WWTP. See Table 1 for analysis results of the sewage sludge. Dewatered sewage sludge is collected on a truck and transported to a centralised regional vermicomposting operation near a pulp and paper mill twice a week. At the centralised vermicomposting site at Kawerau the sewage sludge is combined with sewage sludge from Rotorua WWTP.

Rotorua

Rotorua is an urban community with a population of 70,000; dominantly tourism and timber processing industries and producing close to 10,000 t/a sewage sludge at their centralised WWTP. The dewatered primary and activated sewage sludge is transported daily to a centralised regional vermicomposting operation near a pulp and paper mill at Kawerau. At the Kawerau vermicomposting site both sewage sludge from Te Puke and Rotorua municipalities are mixed with approximately 14,000 t pulpmill solids. The analysis results of the pulpmill solids are given in Table 1. During a 12 months vermicomposting process, annually 8,000 t vermicast are generated. Vermicast is tested according to (NZWWA, 2003) and (NZS, 2005) prior to land application to maize crops at rates of 20 t/ha. The Kawerau vermicomposting operation has a footprint of 12 ha and is operated 3 to 5 days per week.

Hamilton

Hamilton City has a population of 153,000 with a diverse industrial footprint producing 13,500 tonnes p.a. of biosolids respectively. The anaerobic digested, dewatered sewage sludge (see Table 1 for analysis results) is transported daily to Tokoroa vermicomposting site where it is blended with other industrial organic wastes plus pulpmill solids from a cardboard recycling plant owned by Oji Fibre Solutions. The Tokoroa site operates in forestry blocks and is vermicomposting some 90,000 t/a of combined pulpmill solids, sewage sludge, sludge from local milk processing plants, and fruit wastes. The site is producing 30,000 t of various vermicast products per year on approximately 45 ha. General vermicast is tested according to (NZWWA, 2003) and (NZS, 2005) before applied to cropping and nurseries, organic certified vermicast is tested according (AsureQuality, 2013) prior to application to pasture and orchards.

Taupo

In 2017 the new Taupo site has been commissioned with a capacity for vermicomposting 3,000 t/a biosolids from Taupo WWTP, 1,000 t decanted waste activated sludge (WAS) (Figure 1) and dissolved air flotation sludge (DAF) from local milk plants. Biosolids, WAS, and DAF is combined with some 5,000 t/a pulpmill solids from various pulp and paper mills operated by Oji Fibre Solutions (Figure 2). The vermicomposting site is operating daily and has a footprint of 5 ha and is integrated in 70 ha agricultural cut and carry farm. The vermicomposting windrows are rotated on the farm to maximise benefits from the remaining vermicast after harvesting.

INDUSTRIAL VERMICOMPOSTING IN NEW ZEALAND

The first laboratory scale vermicomposting trials using paper wastes as a carbon source and blending agent for nutrient rich organic wastes have been published by Edwards, 1988). From the 1990's scientists started using pulp and paper mill sludge also referred as pulpmill solids for vermicomposting as bulking agents (Butt, 1993; Elvira et al., 1996; Elvira et al., 1997; Elvira et al., 1998; Lazcano et al. 2008a; Lazcano et al. 2008b). Vermicomposting of purely pulpmill solids has not been conducted at any significant scale because the wide C/N ratio of pulpmill solids makes this product unsuitable for vermicomposting. Laboratory trials have shown that best reproduction of earthworms is achieved when the C/N ratio is adjusted to 25 with nitrogen rich waste streams (Elvira et al., 1998; Ndegwa and Thompson, 2000; Aira et al. 2006). Various industrial organic wastes have been studied for the past 40 years using vermicomposting technology to produce a high quality soil conditioner or fertiliser (Edwards and Neuhauser, 1988). Paper wastes with a C/N ratio of up to 200 and higher were used as a carbon rich blending agent for nutrient rich wastes such as biosolids, food wastes (Edwards, 1988, Arancon et al., 2005), manure (Arancon et al., 2005), and other industrial wastes (Tucker, 2005). In recent years, Quintern et al. have demonstrated that a sub-optimal C/N ratio in the earthworm feedstock can be successfully applied in commercial vermicomposting operations and that nitrogen sources captured from the wastewater treatment plant of a pulpmill can be used as nitrogen source for blending with primary pulpmill solids (Quintern et al., 2009; Quintern, 2011; Glasner and Quintern, 2011). Vermicomposting of DAF sludge from milk processing industries in combination with pulpmill solids has first been studied and published by (Quintern, 2009).

Industrial scale vermicomposting of sludge from Oji's Pulp & Paper mills and other industrial and municipal organic waste streams commenced in 2007 and currently totals 150,000 t/a (Quintern et al., 2009; Quintern, 2011; Quintern et al., 2013; Quintern et al., 2014a; Quintern et al., 2014b). In 2010 Noke Ltd. undertook feasibility studies on commercial vermicomposting of the organic waste stream from pulp and paper industries in combination with dewatered WAS and DAF sludge from Fonterra's milk plants. The aim was to produce a valued added product, vermicast, for beneficial land utilisation. In 2014 Noke Ltd established the efficacy of operating as a full-scale vermicomposting operation and a new site is being developed to receive only these products. This site commenced operating at full industrial scale in September 2015 with a capacity of 70,000 t/a.

Vermicomposting process - general

Thermophilic composting of wet organic sludge with little or no structure as well as wet pulpmill solids with a wide C/N ratio proves both difficult and economically challenging. Both organic wastes require either mixing with suitable agents to provide oxygen flow or require adding nutrients, mainly nitrogen and phosphorus for rapid decomposition. Thermophilic composting requires mechanical turning thus adding complexity and costs to the process. The high temperatures achieved through the thermophilic process can create significant odour issues, which can be problematic if processed close to populated areas. Thermophilic composting achieves a volume reduction of approximately 1/3.

Vermicomposting is a mesophilic process, which uses epigeic compost earthworms to ferment and decompose organic matter producing a specific earthworm casting called vermicast or vermicompost (Edwards and Neuhauser, 1988). Compost worms can 'operate' in moist conditions between 60 to 90% depending on the structure of the feedstock, so no pre drying is required. Vermicomposting requires no mechanical turning since the feedstock waste is applied in smaller quantities as earthworms are mixed with the feedstock through their activity. This action is described as 'bio-turbation' in soil science or could be described as 'vermi-turbation' in the process of vermicomposting. During the digestion of the organic waste a volume reduction of up to 80% can be achieved, this reduces transportation costs of the vermicast to its end users.

Vermicomposting technology used in New Zealand

The vermicomposting technology developed by Quintern Innovation Ltd. is based on the Windrow Vermicomposting System described as low-technology vermicomposting systems (Edwards, 2011). Windrow vermicomposting technology in general is characterised as low capital expenditure (CAPEX) and easy to manage. The drawbacks of conventional Windrow Vermicomposting Systems are a large footprint, labour-intensive, slow processing time, considerable nutrient losses, and need for separation of earthworms from vermicompost. Most of these drawbacks have been resolved by Quintern (Table 3). The key improvements have been achieved by avoiding continuous feeding practices in order to reduce labour costs and processing time. The footprint / net production ratio for vermicomposting has been significantly reduced by minimising the non-productive areas of the vermicomposting site. The Quintern Windrow Vermicomposting Technology has been developed and proven over the last decade and is successfully operating at an economic industrial scale.

Integrated vermicomposting in farm management systems

To manage the nutrient losses, such as leaching and gaseous losses from vermicomposting sites, the C/N ratio of the feedstock is carefully adjusted, which reduces the potential of enriching the underlying topsoil with nutrients and humic acids. Operating at the same location over multiple years could create a hot spot of nutrient loss into the soil ecosystem. To address this potential risk, the author adopted the strategy of integrating vermicomposting into farm and forest management practices. Similar to integrating outdoor pig ranging, with extreme high nutrient accumulation in hot spots, the vermicomposting sites will rotate either on the farm (Quintern, 2005; Quintern and Sundrum 2005) or within the forest. After a certain time of vermicomposting on one specific site the vermicomposting windrows will be placed on a new paddock or forest block. On the former vermicomposting site, the vermicast will be harvested and a so-called 'catch crop' with high nutrient demand, such as maize, will be planted. In forest management systems the new plantation benefits from the vermicast residues as the demand for nutrient of trees is highest in the first years of plant growth.

Table 3. Characteristics of the Quintern Windrow Vermicomposting Systems in comparison to conventional Windrow Vermicomposting Systems described by (Edwards, 2011)

	Quintern Vermicomposting	Conventional Vermicomposting
Benefits	Low CAPEX*	Low CAPEX*
	Easily managed	Easily managed
	Not labour intensive (mechanised)	
	Reduced footprint	
	Retention of nutrients through wider C/N ratio	
	Harvesting vermicompost without earthworm separation	
Drawbacks		Labour-intensive
		Large footprint
	6 to 12 months processing time	6 to 16 months processing time
		Loss of nutrient through leaching and volatilisation
		Impossible to harvest vermicompost without earthworm separation

*Capital expenditure.

VERMICAST

Vermicast and vermicompost are used in the literature often synonymously. In New Zealand vermicast is defined as mature pure earthworm casting whereas vermicompost may contain some material that has not been processed by compost worms (NZS, 2005). There are hundreds of references characterising vermicompost in relation to the degree of earthworm activity, vermicomposting technology used, and the parent wastes or 'feedstock' used. Arancon and Edwards (2011) providing a review of the recent findings of beneficial use of vermicast.

In comparison to commonly advised vermicomposting technology, the vermicast produced in this commercial operation has received a much larger portion of fibrous and carbon rich feedstock source as a blending agent. As a result, the vermicast has a more advanced peaty structure with lower nitrogen content and therefore a slightly wider C/N ratio. This provides the MyNOKE vermicast with various advantages when applied in nutrient sensitive ecosystems, which are lacking in soil humus required for maintenance of a high nutrient and water holding capacity. Specific characteristic of the vermicast are described in Table 4.

Table 4. Volumes and analysis results of various MyNOKE® vermicast estimated for 2016. For comparison Aa-grade limits for land applications of biosolids (sewage sludge) in New Zealand and limits for organic certified vermicast

	MyNOKE® organic certified vermicast	MyNOKE® Black Gold PLUS vermicast	MyNOKE® Black Gold vermicast	Limits organic certification (AsureQuality, 2013)	Limits for AA-grade (NZWWA, 2003)
Produced (t/a)	12,500	20,000	17,500		
Dry Matter (%)	49.0	49.0	44.0		
Carbon (%)	21	21	16.7		
Nitrogen (%)	0.85	1.6	1.28		
C/N ratio	24.7	13.5	13.0		
pH	7.5	7.0	6.8		
Phosphorus (mg/kg)	1,458	4,720	7,670		
Sulphur (mg/kg)	2,810	5,100	4,880		
Potassium (mg/kg)	709	1,055	1,748		
Calcium (mg/kg)	56,700	136,100	86,600		
Magnesium (mg/kg)	1,775	2,060	1,990		
Sodium (mg/kg)	766	2,180	1,222		
Boron (mg/kg)	< 6	23	27		
Arsenic (mg/kg)	8.1	6.2	9.5	20	20
Chromium (mg/kg)	24	23	28	150	600
Cadmium (mg/kg)	0.44	0.4	0.38	1.0	1.0
Copper (mg/kg)	44	39	46	60	100
Lead (mg/kg)	24	38	35	250	300
Mercury (mg/kg)	0.09	0.08	0.11	1.0	1.0
Nickel (mg/kg)	11.7	12	8.6	60	60
Zinc (mg/kg)	131	174	197	300	300

Beneficial land utilisation of vermicast in New Zealand

Currently 50,000 tonnes of vermicast are produced in New Zealand per year (Table 4) and applied back to land to improve soil fertility and to restore soil organic matter. Vermicast is applied to pasture and cropping in agriculture, horticulture, orchards, nurseries, and to forestland with the aim to improve natural soil functions such as water and nutrient holding capacity and erosion protection. Vermicomposting operations are integrated into agricultural, horticultural and forestry management systems to involve end users into the production of vermicompost and to maximise the environmental and economic benefits of the vermicomposting technology. This means that worm beds are considered as a 'crop' and are integrated into on-farm crop rotation planning. After the vermicomposting process is complete, the resultant vermicast is harvested and exported from the site but with some vermicast remaining on top of the soil. Either a catch-crop that has a high nutrient demand or a new plantation follows the vermicomposting process and benefits from the activity through improved plant growth. This 'vermicomposting-crop-rotation' concept, developed by Quintern Innovation, also mitigates potential nutrient accumulation during vermicomposting and minimises operational and environmental monitoring costs. In fruit production systems such as olives and citrus, vermicomposting could be operated within orchards and plantations.

Vermicomposting is the ideal technology for regions with moderate climates characterised by milder winters and modest summers such as New Zealand, the southern regions of Australia and in many states of the United States. Over the last decade the technology has evolved to accept continuous waste streams as from municipal and industrial waste water treatment plants as well as for discontinuous organic residue production e.g. from fruit, wine and olive oil production. The highly "mobile", low capex, and opex technology offers great potential for small communities and processors, but can also be up-scaled to full industrial scale. Most important it is a regional solution whereby the up-cycling of organic resources to a high quality vermicast gains strong support from communities and end users. Table 4 gives an overview of the estimated production and quality criteria of different MyNOKE vermicast produced in 2016.

Application of vermicast

Vermicast is applied in bulk to various crops at rates of 2.5 t/ha to pasture, 10 t/ha to kiwi-fruit orchards, and up to 20 t/ha to maize and to other crops according to the specific soil quality and nutrient demand of the crop. Screened vermicast can be applied with standard compost or manure spreaders.

Benefits to plant growth

Vermicast produced by compost earthworms (Eisenia foetida) contains humic substances described as humic acids, humates, gibberelins, auxins, 3-indole acetic acid, and various other substances. These humic substances promote plant growth in multiple ways starting with faster germination (Lazcano et al., 2008b; Talarposhti, 2008; Bachman, 2008; Hidalgo and Agricultural, 1999), increased root development of more and longer lateral and vertical roots, increased area of root hairs and even higher root activity (Canellas et al., 2002; Canellas et al., 2010; Canellas et al., 2013; Chaoui et al., 2003; Lazcano et al., 2010; Lazcano et al., 2011; Rima et al., 2011; Roy et al., 2010; Zandonadi, et al., 2007; Zandonadi, et al., 2010). Increased root area, root growth depth, and activity leads to increased nutrient uptake and access to more available soil water during drier seasons.

Positive effects were measured in higher numbers of blossoms, flowers, and fruits, increased photosynthesis which overall leads to higher yields (Joshi et al., 2013; Lazcano et al., 2011; Roy et al., 2010), and fruit harvests (Arancon et al., 2004). Recent studies have shown that vermicast has the potential to suppress plant diseases and to control pests such as insects and nematodes (Chaoui et al., 2002; Arancon et al., 2002; Arancon et al., 2005; Serfoji et al., 2010; Edwards et al., 2004; Edwards et al., 2007; Edwards et al., 2010).

Benefits to soils

The beneficial uses of vermicompost to various crops and forests on different soils have been widely documented. Little is known about the continuous application of vermicast on intensive dairying land, especially on pasture and on intensive monoculture maize cropping systems. Intensive dairy farming is currently criticised for increased nitrate leaching and phosphate runoff. Improved topsoil functions in regards to water and nutrient retention would most likely mitigate nutrient losses, such as nitrate in the groundwater. Higher humus content and more intense root systems would improve soil function, as well as increasing nutrient uptake including nitrogen. Intensive maize cropping may lead to a reduction in soil organic matter and therefore carbon loss from these soils (Sparling & Schipper, 2004). In a recent trial, vermicast applied at a rate of 20t/ha and at a constant nitrogen fertiliser application rate of 400 kg N/ha has increased the production of maize silage by 3 to 5 t/ha dry matter. The shoot to root (S/R) ratio for maize is 5.6 (Bolinder et al., 2007) and would increase the root production by 0.54 to 0.89 t/ha dry matter. In addition to the extra humus accumulation through extra root growth, carbon is applied at a rate of 2.06 t/ha.

OUTLOOK

Vermicast application to soil has the potential to mitigate nitrate losses, increase soil organic matter directly as a carbon source and indirectly by increasing root production.

Vermicomposting of combined industrial and municipal organic wastes offers multiple benefits on the carbon footprint for industry, the community and for the agribusiness sectors. Of significance is the reduction in greenhouse gas emissions originating from the land filling of organic wastes and the avoidance of the high economic penalties associated with the design, construction and ongoing monitoring of landfills. For the farming and horticulture sectors, soil carbon would be increased and higher carbon sequestration by increasing root mass and crop yields. A reduction in mineral fertiliser application would reduce the carbon footprint and a better soil structure would reduce nitrous oxide (N_2O) emissions from pastoral soils.

REFERENCES

1) Aira, M., Monroy, F. and Domínguez, J. (2006). C to N ratio strongly affects population structure of Eisenia fetida in vermicomposting systems. *European Journal of Soil Biology (42): 127-131*
2) Arancon, N. Q., Edwards, C. A., Lee, S. S. and Yardim, E. (2002). Management of plant parasitic nematode populations by use of vermicomposts. *Brighton crop protection conference pests and diseases 2, 705-710*
3) Arancon, N. Q., Edwards, C. A., Bierman, P., Welch, C. and Metzger, J. D. (2004). Influences of vermicomposts on field strawberries: 1. Effects on growth and yields. *Bioresource Technology 93: 145-153*
4) Arancon, N. Q., Galvis, P. A. and Edwards, C. A. (2005). Suppression of insect pest populations and damage to plants by vermicomposts. *Bioresource Technology 96, 1137-1142*
5) Arancon, N. Q., Edwards, C. A., Bierman, P., Metzger, J. D. and Lucht, C. (2005). Effects of vermicomposts produced from cattle manure, food waste and paper waste on the growth and yield of peppers in the field. *Pedobiologia (49), 297-306*
6) Arancon, N. Q. and Edwards, C. A. (2011). The use of vermicomposts as soil amendments for production of field crops - Chapter 10 in *Vermiculture Technology: Earthworms, Organic Wastes, and Environmental Management* (eds Edwards, C. A., Arancon, N. Q. & Sherman, R.) CRC Press, Taylor & Francis Group, Boca Raton, London, New York, 129-151
7) AsureQuality (2013). *Organic Standard For Primary Producers.* AsureQuality
8) Bachman, G. R. and Metzger, J. D. (2008). Growth of bedding plants in commercial potting substrate amended with vermicompost. *Bioresource Technology (99): 3155-3161*

9) Bolinder, M. A., Janzen, H. H., Gregorich, E. G., Angers, D. A. & VandenBygaart, A. J. (2007). An approach for estimating net primary productivity and annual carbon inputs to soil for common agricultural crops in Canada. *Agriculture, Ecosystems & Environment 118, 29-42*

10) Butt, K. R. (1993). Utilisation of solid paper-mill sludge and spent brewery yeast as a feed for soil-dwelling earthworms. *Bioresource Technology 44: 105-107*

11) Canellas, L. P., Olivares, F. L., Okorokova-Façanha, A. L. and Façanha, A. R. (2002). Humic acids isolated from earthworm compost enhance root elongation, lateral root emergence, and plasma membrane H+-ATPase activity in maize roots. *Plant Physiology 130, 1951-1957*

12) Canellas, L. P., Piccolo, A., Dobbss, L. B., Spaccini, R., *et al.* (2010). Chemical composition and bioactivity properties of size-fractions separated from a vermicompost humic acid. *Chemosphere 78, 457-466*

13) Canellas, L. P., Balmori, D. M., Médici, L. O., Aguiar, N. O., *et al.* (2013). A combination of humic substances and Herbaspirillum seropedicae inoculation enhances the growth of maize (Zea mays L.). *Plant and Soil 366. 1-14*

14) Chaoui, H., Edwards, C. A., Brickner, A., Lee, S. S. and Arancon, N. Q. (2002). Suppression of the plant diseases, Pythium (damping-off), Rhizoctonia (root rot) and Verticillium (wilt) by vermicomposts. *Brighton crop protection conference pests and diseases 2: 711-716*

15) Chaoui, H., Zibilske, L. and Ohno, T. (2003). Effects of earthworm casts and compost on soil microbial activity and plant nutrient availability. *Soil Biology and Biochemistry 35, 295-302*

16) Edwards, C. A. (1988). Breakdown of animal, vegetable and industrial organic wastes by earthworms. in *Earthworms in Waste and Environmental Management* (eds Edwards, C. A. and Neuhauser, E. F.), SPB Academic Publishing BV, The Hague, The Netherlands: 21-31

17) Edwards, C. A. (2011). Low-technology vermicomposting systems - Chapter 7 in *Vermiculture Technology: Earthworms, Organic Wastes, and Environmental Management* (eds Edwards, C. A., Arancon, N. Q. & Sherman, R.) CRC Press, 79-90

18) Edwards, C. A. and Neuhauser, E. F. (1988). *Earthworms in waste and environmental management* (SPB Academic Publishing, The Hague, The Netherlands

19) Edwards, C. A., Dominguez, J. and Arancon, N. Q. (2004). The influence of vermicomposts on plant growth and pest incidence in *Soil zoology for sustainable development in the 21st century* (eds Shakir Hanna, S. H. & Mikhail, W. Z. A.) Cairo, 397-420

20) Edwards, C. A., Arancon, N. Q., Emerson, E. and Pulliam, R. (2007). Suppressing plant parasitic nematodes and arthropod pests with vermicompost teas. *BioCycle 48, 38-39*

21) Edwards, C. A., Arancon, N. Q., Vasko-Bennett, M., Askar, A., *et al.* (2010). Suppression of green peach aphid (Myzus persicae)(Sulz.), citrus mealybug (Planococcus citri)(Risso), and two spotted spider mite (Tetranychus urticae)(Koch.) attacks on tomatoes and cucumbers by aqueous extracts from vermicomposts. *Crop Protection 29, 80-93*

22) Edwards C., Arancon, N. and Sherman, R. (2011). Vermiculture Technology: Earthworms, Organic Wastes, and Environmental Management. Edwards, C.A., Arancon, N.Q., & Sherman, R. eds. CRC Press, Taylor & Francis Group, Boca Raton, London, New York

23) Elvira, C., Goicoechea, M., Sampedro, L., Mato, S. and Nogales, R. (1996). Bioconversion of solid paper-pulp mill sludge by earthworms. *Bioresource Technology, (57): 173-177*

24) Elvira, C., Sampedro, L., Dominguez, J. and Mato, S. (1997). Vermicomposting of wastewater sludge from paper-pulp industry with nitrogen rich materials. *Soil Biology and Biochemistry (29): 759-762*

25) Elvira, C., Sampedro, L., Benitez, E. and Nogales, R. (1998). Vermicomposting of sludges from paper mill and dairy industries with Eisenia andrei: a pilot-scale study. *Bioresource Technology (63): 205-211*

26) Glasner, U. and Quintern, M. (2011). Win Win Win: Biosolids + Pulpmill solids + Compost worms = Fertile soils (The Western Bay Way). *New Zealand Land Treatment Collective: Proceedings for the 2011 Annual Conference (32): 96-103*

27) Hernandez, O.L., Calderín, A., Huelva, R., Martínez-Balmori, D., Guridi, F., Aguiar, N.O., Olivares, F.L. & Canellas, L.P. (2015). Humic substances from vermicompost enhance urban lettuce production. *Agronomy for sustainable development, 35(1): 225-232*

28) Hidalgo, P. and Agricultural, M. (1999). *Earthworm castings increase germination rate and seedling development of cucumber* (Office of Agricultural Communications, Mississippi State University

29) Joshi, R., Vig, A. P. and Singh, J. (2013). Vermicompost as soil supplement to enhance growth, yield and quality of Triticum aestivum L.: a field study. *International Journal Of Recycling of Organic Waste in Agriculture* **2,** 1-7

30) Joshi, R., Singh, J. & Vig, A.P. (2015). Vermicompost as an effective organic fertilizer and biocontrol agent: effect on growth, yield and quality of plants. *Reviews in Environmental Science and Bio/Technology, 14(1): 137 159*

31) Lazcano, C., Sampedro, L., Zas, R. and Domínguez, J. (2008a). Enhancement of pine (Pinus pinaster) seed germination by vermicompost and the role of plant genotype. *Compost and digestate: sustainability, benefits, impacts for the environment and for plant production, 253-254*

32) Lazcano, C., Sampedro, L., Nogales, R. and Domínguez, J. (2008b). Paper sludge vermicomposts as amendments into the potting media of peppers (Capsicum annium L. var longum). *Compost and digestate: sustainability, benefits, impacts for the environment and for plant production. Proceedings of the international congress CODIS 2008. February 27-29, 2008, Solothurn, Switzerland: 211-213*

33) Lazcano, C., Sampedro, L., Zas, R. and Domínguez, J. (2010). Assessment of Plant Growth Promotion by Vermicompost in Different Progenies of Maritime Pine(Pinus pinaster Ait.). *Compost Science & Utilization 18, 111-118*

34) Lazcano, C. and Domínguez, J. (2011). *The use of vermicompost in sustainable agriculture: impact on plant growth and soil fertility.* Soil Nutrients. Nova Science Publishers, New York,: 230-254

35) Lazcano, C., Revilla, P., Malvar, R. A. and Domínguez, J. (2011). Yield and fruit quality of four sweet corn hybrids (Zea mays) under conventional and integrated fertilization with vermicompost. *Journal of the Science of Food and Agriculture 91, 1244-1253*

36) Ndegwa, P. M. and Thompson, S. A. (2000). Effects of C-to-N ratio on vermicomposting of biosolids. *Bioresource Technology (75): 7-12*

37) NZS (2005). Composts, Soil Conditioners and Mulches. New Zealand Standard, NZS 4454:2005

38) NZWWA (2003). Guidelines for the safe application of biosolids to land in New Zealand. New Zealand Water & Wastes Association, Wellington, New Zealand

39) Quintern, M. (2005). Integration of organic pig production within crop rotation: Implications on nutrient losses. *Landbauforschung Völkenrode. 31-40*

40) Quintern, M. (2009). Vermicomposting of lake weeds and pulp & paper solids for carbon resource recovery for primary sectors. *Report LC08/017. Sustainable Farming Fund*

41) Quintern, M. (2011). Organic waste free pulpmill through vermicomposting - The Kinleith way. *New Zealand Land Treatment Collective: Proceedings for the 2011 Annual Conference (32): 84-88*

42) Quintern, M. (2014a). Full scale vermicomposting and land utilisation of pulpmill solids in combination with municipal biosolids (sewage sludge). *7th International Conference on Waste Management and the Environment 7: 65-76*

43) Quintern, M. (2014b). Industrial scale vermicomposting of municipal biosolids by blending with fibrous industrial wastes. Eurasia 2014 Waste Management Symposium: 1-9

44) Quintern, M. (2015). Sardinia 2015 15th International Waste Management and Landfill Symposium. Case studies of industrial scale vermicomposting of organic wastes from primary industries in combination with municipal sewage sludge. CISA Publisher, Italy, Sardinia, Italy, 1-10

45) Quintern, M. and Sundrum, A. (2006). Ecological risks of outdoor pig fattening in organic farming and strategies for their reduction - Results of a field experiment in the centre of Germany. *Agriculture, ecosystems & environment, 117: 238-250*

46) Quintern, M., Wang, H., Magesan, G. and Slade, A. (2009). *Vermicomposting primary and secondary solids from the pulp and paper industry*. Scion, Rotorua, New Zealand, Report

47) Quintern, M., Seaton, B., Mercer, E. and Millichamp, P. (2013). Industrial scale vermicomposting of pulp and paper mill solids with municipal biosolids and DAF sludge from dairy industries. *Appita,. 66: 290-295*

48) Rima, J. A. H., Martim, S. A., Dobbss, L. B., Evaristo, J. A. M., *et al.* (2011). Citric acid addition improve humic acids action and change proteins profile from plasma membrane of maize roots. *Ciência Rural 41, 614-620*

49) Roy, S., Arunachalam, K., Dutta, B. K. and Arunachalam, A. (2010). Effect of organic amendments of soil on growth and productivity of three common crops viz. Zea mays, Phaseolus vulgaris and Abelmoschus esculentus. *Applied Soil Ecology 45, 78-84*

50) Serfoji, P., Rajeshkumar, S. and Selvaraj, (2010). T. Management of root-knot nematode, *Meloidogyne incognita* on tomato cv Pusa Ruby. by using vermicompost, AM fungus, *Glomus agcoagulans* and mycorrhiza helper bacterium, *Bacillus coagulans. Journal of Agricultural Technology 6, 37-45*

51) Sinha, R.K., Soni, B.K., Agarwal, S., Shankar, B. and Hahn, G. (2013). Vermiculture for Organic Horticulture: Producing Chemical-Free, Nutritive & Health Protective Foods by Earthworms. *1(1): 17-44*

52) Tucker, P. (2005). Co-composting paper mill sludge with fruit and vegetable wastes. *Thesis*

53) Talarposhti, R. M., Liaghati, H. and Borojeni, M. R. (2008). Effect of organic fertilizers on germination and seedling growth of tomato and suppression of soil borne diseases. *Compost and digestate: sustainability, benefits, impacts for the environment and for plant production 27, 259-260*

54) Zandonadi, D. B., Canellas, L. P. and Façanha, A. R. (2007). Indolacetic and humic acids induce lateral root development through a concerted plasmalemma and tonoplast H+ pumps activation. *Planta 225, 1583-1595*

55) Zandonadi, D. B., Santos, M. P., Dobbss, L. B., Olivares, F. L., *et al.* (2010). Nitric oxide mediates humic acids-induced root development and plasma membrane H^+-ATPase activation. *Planta 231, 1025-1036*

56) Zandonadi, D.B. and Busato, J.G. (2012). Vermicompost humic substances: technology for converting pollution into plant growth regulators. *International Journal of Environmental Science and Engineering Research, 3(2): 73-84*

Know what you don't know – using social media and market research to inform Loop® communication strategies

Ashley Mihle[1]

[1]King County Wastewater Treatment Division, Seattle, Washington
*Email: Ashley.Mihle@kingcounty.gov

ABSTRACT
As scientists and communicators, we often make assumptions about the public's opinions, beliefs, and interests. We use these assumptions to guide most of our decision making and communication strategies, but are these assumptions correct? In 2015 and 2016, King County used both social media research and market research to test these assumptions. The social media research did not have geographic boundaries, but the market research surveyed only King County residents. The social media and news assessment found that interest in biosolids has been decreasing over time and that outside of a few areas of organized resistance, anti-biosolids activists do not have a large online presence or reach. The first market research survey found low levels of biosolids awareness, but high levels of support and interest. A market segmentation was used to profile respondents into key audience groups and analyze their demographics, lifestyles, and preferences to better understand how to successfully communicate with and message to these groups. The market segmentation identified three key audience groups: the Poop Champions, the Poopsuadables, and the Pooh Poohs. The Poopsudables are the target audiences, the Poop Champions are the target messengers, and the Pooh Poohs are the group to avoid intentional engagement with. The demographics, values, and preferences of these groups is specific to King County, but any agency could perform a similar data analysis to profile target audiences. The second survey was a media test of the Loop® biosolids videos, which found that visual communication, personal stories, and the positive messaging of the Loop brand was an effective communication strategy. Sixty-four percent of respondents indicating there were very supportive of biosolids after seeing a Loop video, while only 35% of respondents indicated they were very supportive after reading an unbranded text definition and seeing a wastewater treatment diagram. If agencies can identify which audiences to target, which to use as messengers, and which to avoid, and how to message to each in alignment with a specific communications goal, then their communications efforts will be more efficient and strategic.

KEYWORDS: biosolids, Loop®, communication, market research, social media, marketing, branding, public perception

INTRODUCTION
As scientists and communicators, we often make assumptions about the public's opinions, beliefs, and interests. We use these assumptions to guide most of our decision making and communication strategies, but are these assumptions correct? In 2015 and 2016, King County used both social media research and market research to test these assumptions.

In the face of legislative challenges and in an effort to ensure the Loop® biosolids brand is relevant and accessible to target audiences, King County conducted both social media research and market research. This research resulted in a snapshot of general biosolids perceptions, support, and salience nation-wide and also within King County, as well as a better understanding of how target audiences relate to Loop brand messaging. The King County Biosolids team wanted to know what they don't know so they could better focus their communication tactics, align Loop brand messaging with specific audiences, and increase support for Loop.

METHODS

Social media and news assessment

The social media and news assessment was part of an in-depth review of materials related to potential challenges to the King County biosolids program that was conducted to help provide strategic direction in the face of potential legislative challenges. The process for this research involved the use of a number of paid subscription social and news intelligence platforms. These platforms analyze billions of daily conversations on the Internet. This research also examined news stories on biosolids and the commentary those stories generated. In addition, social mentions of biosolids on Twitter, public Facebook profiles, Instagram, forums, blogs, Google+, and YouTube from the past year were reviewed and assessed. As part of the social media and news review, a number of keywords and search queries were used, such as: biosolids, biosolids, #biosolid, and sewage sludge. These search queries included common misspellings.

Market Research Surveys

The social media and news review provided a general snapshot of national views and salience of biosolids. However, a more detailed, local assessment was needed, especially as King County reviews strategic program options and considers the possibility of increased local uses for Loop. In order to measure awareness and understanding within King County, two market research surveys were conducted to supplement existing data from the King County Wastewater Treatment Division biennial Water Quality Index Survey conducted by EMC Research.

The first survey was an un-branded survey about biosolids awareness, interest and support. In addition to demographics, this survey included questions on values, lifestyles, and behaviors that aided the data analysis, including performing a market segmentation of the survey respondents. The survey instrument was designed by PRR. Frequency tables, cross tabulations, and a market segmentation was also performed by PRR. The survey was conducted online by Research Now, with demographic quotas representative of King County demographics, for N=379 respondents. King County conducted the data analysis and reporting.

The second survey was a branded media test of the existing Loop videos. In addition to some of the same biosolids awareness and perception questions as the first survey, this survey was designed to test the Loop videos and inform the Loop communication team about future video topics, messages, and style. The survey instrument was designed by PRR, and frequency tables and cross tabulations were also run by PRR. The survey was conducted online by Research Now with demographic quotas representative of King County demographics, for N= 277 respondents. King County conducted the data analysis and reporting.

Market research survey instruments are displayed in Attachment A.

RESULTS

Social media and news assessment

King County's social media and news assessment, conducted by Mammoth, found that knowledge and awareness of biosolids is extremely low, and that interest has been declining over time. Relative popularity of the search term "biosolids" is shown in Figure 1. In addition to analyzing the volume and context of how the term biosolids is used in news and social media, the assessment measured the strength and impact of organized resistance and known detractors.

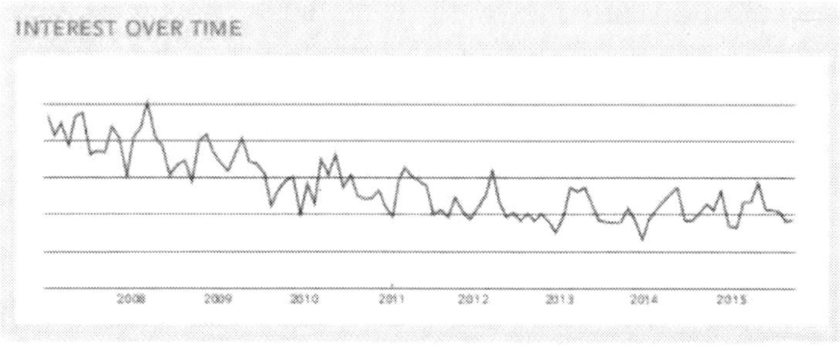

The chart (from Google Trends) shows the popularity of the search term "biosolids" on Google in the United States from January 2007 through October 2015.

A line trending downward means that a search term's relative popularity is decreasing.

The chart shows the relative popularity of a search term to itself over a specific period of time. Google Trends does not reference specific numerical values. The top search queries on Google from the past few years are "EPA biosolids" and "biosolids management."

Figure 1. Google search term "biosolids" popularity over time

The news assessment founds that stories mentioning biosolids also tended to mention sewage, wastewater, sludge, compost, and the environment. A review of Twitter found 6,500 tweets that included the word "biosolids" in 2015. The major themes included mention of areas with organized biosolids resistance, such as Kamloops in British Columbia, and Raleigh, North Carolina. Common themes included emotional perceptions and words such as cancer, health, and sewage. Overall, findings show that outside of a handful of pro GMO labeling bloggers, spam accounts, and vocal individuals, there is little evidence online of any large, organized resistance (in the United States). In summary, resistance was not well organized and had a limited audience and reach.

Market Research and Stakeholder Interviews

Despite a lack of organized online resistance, individual detractors often create issues for biosolids managers, from time consuming public information requests to disruptive public meetings to negative local media. King County needed to know more about its local community and its own Wastewater Treatment Division ratepayers. King County knew biosolids and Loop awareness was low but that once provided with basic information, most of the King County public supports Loop. However, King County did not know how this support and awareness breaks down among demographic groups and King County geographic areas. King County only had anecdotal evidence of what groups to target for outreach and education, and which messages resonated with each group.

The overwhelming finding of the market research was the extent to which King County citizens did not know about biosolids. King County knew that only 9% of citizens are aware of Loop, as demonstrated in the 2015 Water Quality Index Survey by EMC Research, but testing of deeper biosolids awareness and knowledge had not been conducted. The first survey found that 65% of respondents had never heard of biosolids. Of the 33% who indicated they were somewhat familiar with biosolids, few could accurately define the term. While not a surprising finding, this indicates the significant work that still needs to be done to inform King County citizens about wastewater treatment and the renewable resources created.

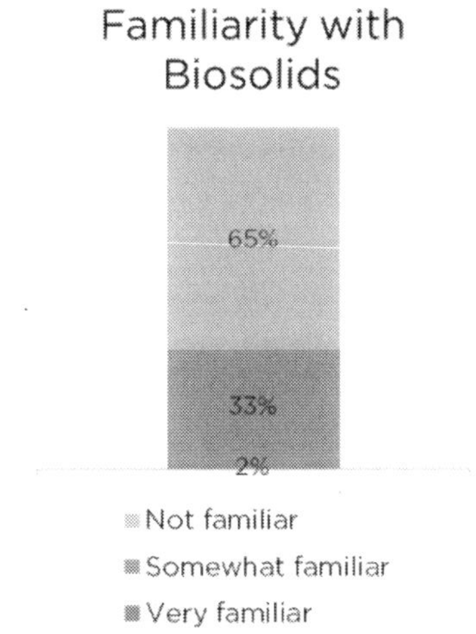

Figure 1. Familiarity with Biosolids

In addition to questions on awareness, support, and interest, in order to test anecdotal evidence and provide more data to inform future outreach strategies, King County segmented the market research data into three clusters or groups:

1. **Pooh Poohs** (20%) – Environmental issues are a very low priority, they are environmental detractors/science skeptics, and are very opposed to the use of biosolids.
2. **Poopsuadables** (44%) - Environmental issues are a lower priority, but they are <u>not</u> environmental detractors/science skeptics, and they are *possibly* open to the use of biosolids.
3. **Poop Champions** (36%) - Environmental issues are a high priority, they are environmental/science believers, and they are very supportive of biosolids use.

Familiarity, interest, and support for biosolids by each group is shown in Figure 2.

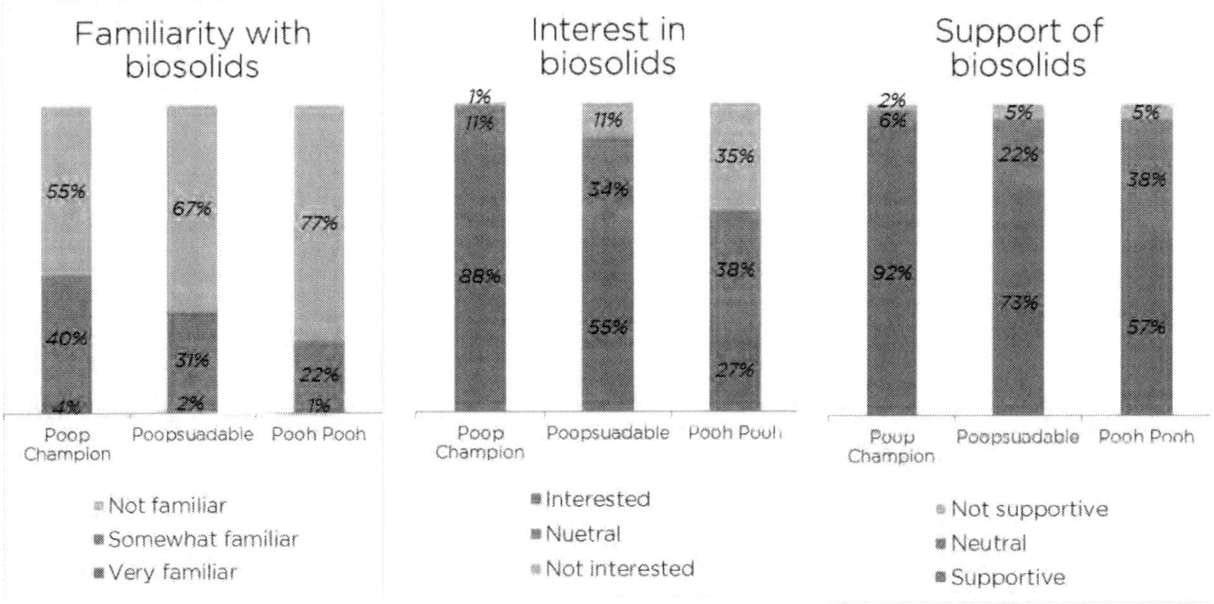

Figure 2. Familiarity, interest, and support of biosolids in King County, WA.
These three groups were extensively profiled and messaging was tested for each. While these are specific to King County and would vary considerably in other areas, these groups demonstrate how the survey instruments were used to help inform target audience and messaging.

Who are the Poopsuadables?

Demographics

Poopsuadables are present in all demographic groups but:

- More likely to be female;
- Ages 35-54;
- More likely to have children under 18 living at home; and
- No other demographics varied strongly.

Lifestyle and preferences

- Hikers, environmentalists, often engage in environmental behaviors;
- Sometimes purchase organic, local, natural, sustainable, recycled - these labels are somewhat important to them; and
- Local is the most important of these labels.

What do they want to know

- General information;
- Safety – do we do any testing; and
- Research – has there been any?

Who are the Poop Champions?

The Poop Champions include many groups identified as target audiences, including our partners and our customers.

Demographics

Poop champions are present in all demographic groups but:

- More likely to be female; and
- No other demographics varied strongly.

Lifestyle and preferences

- More likely to be gardeners;
- Hikers, environmentalists, always engage in environmental behaviors;
- Sometimes, often or always purchase organic, local, natural, sustainable, recycled - these labels are very important to them; and
- Local and sustainable are the most important of these labels.

What do they want to know

- Nothing more;
- How to get involved;
- Safety; and
- More general information about biosolids or Loop.

Who are the Pooh Poohs?

Demographics

It is important to note that the Pooh Poohs are not just anti-biosolids activists, or strongly anti-biosolids citizens. Only a very small percentage (<3%) of respondents strongly opposed biosolids. The Pooh Pooh category does include those who do not support biosolids, but this group also includes respondents who are not interested in biosolids, for whom environmental concerns are not a priority, and/or do not share the same values as WTD and Loop.

Pooh Poohs are present in all demographic groups but are:

- More likely to be male;
- More likely to be ages 55+;
- Less likely to have children under 18 living at home;
- More likely to have <$50,000 annual income; and
- No other demographics varied strongly.

Lifestyle and preferences

- NOT Hikers or environmentalists or gardeners, not worried about climate change, and less likely to engage in environmental behaviors (generally apathetic about these issues and values); and
- Rarely or only sometimes purchase organic, local, natural, sustainable, recycled - these labels are not very import (or neutral) to them.

Why do they support/not support biosolids?

- Most do not care or are not interested; and
- Need more information.

The second survey was a media test. The survey found that respondents were very interested in basic information about Loop – what it is, how it is made, and what it is used for. The media test found that for topics related to safety, public health officials were the most trusted source of information in King County, while trusted sources varied widely for more general biosolids information. In addition to informing specific strategies for future videos and identifying messaging gaps, this survey found that biosolids support was significantly stronger when respondents watched a video about biosolids and then indicated their support versus when respondents read an unbranded text definition and viewed a wastewater treatment diagram, as shown in Figure 3. Verbatim answers indicated that the positive tone, personal stories, and beautiful visuals displayed in the Loop videos created this positive response.

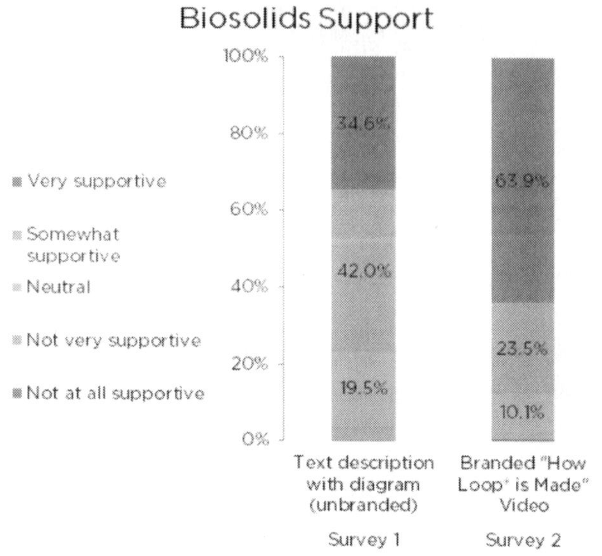

Figure 3. Biosolids support with text vs. visual definitions

CONCLUSIONS

As biosolids managers, anti-biosolids activists take up much of our time. The social media and news assessment found that outside of a few areas of organized resistance, many of these activists did not have significant online reach. Both the social media and news assessment as well as the market research demonstrated widespread lack of knowledge and awareness on the topic of biosolids. However, while the social media and new assessment showed little interest in biosolids as a whole, King County citizens indicated they are very interested in biosolids. This distinction is important, because it demonstrates the strong environmental ethic that is somewhat unique to the Pacific Northwest, and especially to the Seattle metro area. Big picture messages around sustainability, recycling, turning wastes into resources, and creating renewable resources (including recycled water and biogas) resonated strongly with King County citizens. Segmenting the survey data into three groups: Poopsuadables, Poop Champions, and Pooh Poohs will allow King County to focus messaging and outreach strategies on the Poopsuadables, focus community partnerships on the Poop Champions, and avoid engaging with the Pooh Poohs. Tailoring messaging and outreach to the Poopsuadables – so they can become Poop Champions – will build stronger public support for King County's biosolids program and help insulate it against possible legislative challenges in the future.

While the social media and news assessment was a broad, general assessment with results that are applicable to any biosolids manager, the market research survey results are specific to King County and not directly applicable to other agencies or managers. However, the analysis conducted by King County can be replicated and tailored to a specific agency's needs. Every biosolids producer has Poop Champions, Poopsuadables, and Pooh Poohs. They may have different demographics, values, and behaviors – but this overarching framework is still applicable. If agencies can identify which audiences to target, which to use as messengers, and which to avoid, and how to message to each in alignment with a specific communications goal, then their communications efforts will be more efficient and strategic.

REFERENCES

Mammoth (2015) *Biosolids Strategic Communications Plan.* Seattle, Washington.
King County (2016) *Loop Brand Evaluation.* Seattle, Washington.

Survey 1: Survey Instrument

Project ID:	**348404**
Wave Number:	**1**
SURVEY NAME:	**King County Study - Project 1**
CLIENT NAME:	**King County Wastewater Treatment**
ESTIMATED INTERVIEW LENGTH (MINS):	**10 min.**
ESTIMATED PROGRAMMING LENGTH (MINS):	
VERSION/DATE:	**V1 / 03-01-2016**

TYPE OF PROJECT: (select those that apply)

- **STS using STS Session Tracking and Universal Redirects – <u>specify with resume</u>**

SAMPLE SOURCE/COUNTRY/DETAILS:

COUNTRY	LANGAUGE(S)	SOURCE(S)
USA	English	ER 50
USA	English	VOP 22
USA	English	PL 68

REQUIREMENTS	QUESTION NUMBERS/ NOTE
ADVANCED IMAGE/VIDEO SECURITY	**NO: only default security**
RICH MEDIA TOOLS TO USE	**Confirmit / Research Now**
OTHER	

PROGRAMMING NOTES ARE IN BLUE.

DP NOTES ARE IN GREEN

ORIGINAL CLIENT INSTRUCTIONS ARE IN GREY – PLEASE IGNORE

QUOTAS:

Quota Description	Question Number	Pre-code Name OR Number(s)	Percentage	Quota	Quota set in ConfirmIT (Over quota)
Total - Quota stop			100%	400	0
Gender	**D1**	1. Male	50%	200	0
		2. Female	50%	200	0

			100%	**400**	**0**
Age	**S4**	2. 18-24	999%	999 (min n=46 –soft quota)	0
		3. 25-34	20%	80 (n=126 total from 18-34y.o)	0
		4. 35-44	19%	76	
		5. 45-54	19%	75	
		6. 55-64	15%	61	
		7. 65+	16%	62	
			100%	**400**	
Hispanic	**D2**	1. No	91%	364	
		2. Yes	9%	36	
			100%	**400**	
Race	**D3**	1. White or Caucasian	70.1%	279	
		2. Black or African American	6.7%	27	
		3. Native American or Alaska Native	1.1%	4	
		4. Asian	16.4%	66	
		5. Native Hawaiian or Pacific Islander	0.8%	4	
		6. Two or more races	999%	999 (MAX N=20 BETWEEN PRE-CODE 6 AND 98)	
		98. Other	999%	999	
			100%	**400**	
Income	**dINCOME**	1. Under 50k	35%	141	
		2. 50k to $99,999	30%	119	
		3. 100k+	35%	140	
			100%	**400**	
Gardeners	**dGARDENERS**	1. Gardeners & Landscapers	999	999	
		2 Gardening (home or community)	999	999	
		3. Farming	999	999	

		4. Yard work/Lawn care	999	999	
			100%		
D6	**D6**	5. Farming	999	999	
		6. Yard work/lawn care	999	999	
			100%	**0**	**0**

Reporting link with age (all age brackets, not just quota brackets) and D1-D6.

Screening questions

[ASK ALL]
[OE NUM – ZIP CODE - VALIDATE USING ZIP LIST IN PROJECT FOLDER]
27. What is your home zip code? *(Please enter your zip code)*
____________ *(00000)*

[ASK ALL]
[PUNCH OFF OF Q27 – SEE PUNCHING IN FOLDER]
[TERMINATE IF NOT PUNCHED BELOW]
dCOUNTY

1	**Seattle / North King County**
2	**East King County**
3	**South King County**

[ASK ALL]
[SC]
S4. Which of the following categories includes your age? *(Select one)*

1. Less than 18 **[TERMINATE]**
2. 18-24
3. 25-34
4. 35-44
5. 45-54

6. 55-64
7. 65+

Quotas: Demographics

[ASK ALL]
[SC]
D1. What is your gender? *(Select one)*

1. Male
2. Female

[ASK ALL]
[SC]
D2. Are you of Hispanic or Latino origin? *(Select one)*

1. No
2. Yes

[ASK ALL]
[SC]
D3. What is your race? *(Select one)*

1. White or Caucasian
2. Black or African American
3. Native American or Alaska Native
4. Asian
5. Native Hawaiian or Pacific Islander
6. Two or more races
98. Other, *please specify* **[OE CHA]**

[ASK ALL]
[SC]
D4. What was your total annual household income (before taxes) for 2015? *(Select one)*

1. Less than $25,000
2. $25,000 to $49,999
3. $50,000 to $74,999
4. $75,000 to 99,999
5. $100,000 to $149,000
6. $150,000 to $199,000
7. $200,000+
96. Prefer not to answer

[ASK ALL]

[PUNCH OFF OF D4]
[SC-HIDDEN]
dINCOME

1	**Under 50k**	**IF D4=1 OR 2**
2	**50k to $99,999**	**IF D4=3 OR 4**
3	**100k+**	**IF D4=5 OR 6 OR 7**

[ASK ALL]
[MC]
D5. Do you work in any of the following industries? *(Select all that apply)*

1. Advertising
2. Agriculture
3. Banking
4. Government
5. Construction
6. Education
7. Insurance
8. Landscaping
9. Market research/marketing
10. Manufacturing
11. Retail
12. Transportation
13. Utilities
99. None of the above **[EXCLUSIVE] [ANCHOR]**

[ASK ALL]
[MC]
D6. Do you do any of the following? *(Select all that apply)*

1. Biking
2. Gardening (home garden)
3. Gardening (community garden)
4. Hiking
5. Farming
6. Yard work/lawn care
7. Raising livestock (e.g. chickens, goats, or cows etc.)
8. Trail running
9. Volunteering
99. None of the above **[EXCLUSIVE] [ANCHOR]**

[ASK ALL]
[HIDDEN]
[PUNCH OFF OF D5 AND D6]
dGARDENERS

1	**Landscaping**	**(D5=8)**
2	**Gardening (home or community)**	**(D6= 2 OR 3)**

3	**Farming**	**(D6=5)**
4	**Yard work/Lawn care**	**(D6=6)**

Behaviors, Values

[ASK ALL]
[USE CONFIRMIT RATEMENT]
[MAKE PRE-CODES TEXT BLUE AND LARGER FONT]
[ADD PALE BLUE SHADING TO EVERY OTHER ROW]

1. Please indicate whether you agree or disagree with each of the following statements... (*Select one per row)*

	[RANDOMIZE ROWS]	**Completely disagree**	**Disagree**	**Neither agree nor disagree**	**Agree**	**Completely agree**
1	It is important to me to think up new ideas and be creative; to do things my own way	1	2	3	4	5
2	It is important to me to use safe products; to avoid anything that might put myself, my family, or my pets at risk.	1	2	3	4	5
3	It is important to me to do something for the good of society.	1	2	3	4	5
4	It is important to me to look after the environment; to care for nature and save resources	1	2	3	4	5
5	It is important to me to find out the facts before making decisions that affect my life, rather than just taking someone's word for it; to thoroughly	1	2	3	4	5

	research my decisions					

[ASK ALL]
[USE CONFIRMIT RATEMENT]
[ADD PALE BLUE SHADING TO EVERY OTHER ROW]
2. Please indicate how much you agree or disagree with the following statements…. (*Select one per row)*

	[RANDOMIZE ROWS]	**Completely disagree**	**Disag ree**	**Neither agree nor disagree**	**Agr ee**	**Completely agree**
1	Science and technology are making our lives healthier, easier and more comfortable.	1	2	3	4	5
2	Because of science and technology, resources can be protected or recycled for future generations	1	2	3	4	5
3	It is not important for me to know about science in my daily life	1	2	3	4	5

[ASK ALL]
[SC PER ROW GRID]
[USE LEGACY GRID]
[ADD PALE BLUE SHADING TO EVERY OTHER ROW]
3. How often do you, or anyone in your household, participate in the following activities? (*Select one per row)*

		Neve r	**Rarel y**	**Sometime s**	**Ofte n**	**Alway s**
1	Recycle	1	2	3	4	5
2	Purchase recycled products	1	2	3	4	5
3	Compost food scraps	1	2	3	4	5

10	Separate food scraps for curbside pick – up (to be composted)	1	2	3	4	5
4	Compost yard waste	1	2	3	4	5
11	Separate yard waste for curbside pick-up (to be composted)	1	2	3	4	5
5	Use natural or toxic-free cleaning/personal care products in your home	1	2	3	4	5
6	Purchase USDA certified Organic food	1	2	3	4	5
12	Purchase "organic" food (not certified)	1	2	3	4	5
13	Purchase natural food	1	2	3	4	5
7	Purchase locally grown food	1	2	3	4	5
8	Purchase energy efficient appliances	1	2	3	4	5
9	Use less fossil fuel intensive modes of transportation (e.g. walking, biking, public transport, carpooling, etc.)	1	2	3	4	5

[ASK ALL]
[SC PER ROW GRID]
[USE LEGACY GRID]
[ADD PALE BLUE SHADING TO EVERY OTHER ROW]

4. How important are the following labels when you purchase food or other products for your home? *(Select one per row)*

		Not at all importan t	**Not very importa nt**	**Ne utr al**	**Somewha t importan t**	**Very import ant**
1	USDA certified Organic	1	2	3	4	5
6	"organic"	1	2	3	4	5
2	Natural	1	2	3	4	5
3	Local	1	2	3	4	5
4	Recycled	1	2	3	4	5
5	Sustainable	1	2	3	4	5
98	Other, *please specify* **[OE CHA] [ANCHOR] [RESPONSE NOT REQUIRED]**	1	2	3	4	5

[ASK ALL]
[MC]
[1 RESPONSES MIN AND 2 RESPONSES MAX]
5. Which are the most important? *(Select up to two)*

1. USDA certified organic

6. "organic"
2. Natural
3. Local
4. Recycled
5. Sustainable
98. Other, *please specify* **[OE CHA] [RESPONSE NOT REQUIRED]**
99. None of the above **[EXCLUSIVE]**

[ASK ALL]
[SC]
6. How much of a problem do you believe climate change (also known as global warming) is? (*Select one*)

1. Not at all a problem
2. Not a very big problem
3. Neutral
4. Somewhat of a problem
5. A very big problem
97. Don't know
6. I do not believe in climate change/global warming

[ASK IF Q6 = 1-5 OR 97]
[RANDOMIZE 1 AND 2]
[SC]
7. What do you think causes climate change? (*Select one*)

1. Natural causes
2. Human activity
3. Natural causes and human activity **[ANCHOR]**
97. Don't know**[ANCHOR]**
98. Other, *please specify* **[OE CHA] [RESPONSE NOT REQUIRED] [ANCHOR]**

[ASK ALL]
[SC]
8. How familiar are you with the term BIOSOLIDS? (*Select one*)

1. Not familiar
2. Somewhat familiar
3. Very familiar

[ASK IF Q8 = 2 OR 3]
[OE CHA – BIG BOX]
9. To the best of your knowledge, describe what you think BIOSOLIDS are? *(Please enter your response in the box below)*

[SHOW ALL]

[INFO NODE 1]
[ATTCHED IMAGE FROM PROJECT FOLDER]
[SHOW NEXT BUTTON AFTER 20 SECONDS]
10. Please take a minute to read the definition and view the diagram below:

BIOSOLIDS are the nutrient-rich organic product of treating wastewater (from toilets and drains) from homes, businesses and industries.

At the treatment plant, the solid, organic portion of wastewater is separated from the liquids and sent to big heated tanks (called digesters) where beneficial microorganisms break down the material. When this biological process is complete, excess water is removed (like in the spin cycle of a washing machine), leaving a soil-like product that contains essential nutrients and organic matter.

BIOSOLIDS and compost made with BIOSOLIDS are used as a fertilizer and soil conditioner on farms, forests, gardens and landscapes.

BIOSOLIDS and compost made with BIOSOLIDS that are available to the general public have been treated to kill 100% of pathogens (called Class A). All composts made with BIOSOLIDS are in this category.

BIOSOLIDS that have been treated to kill 95-99% of the pathogens (the rest die quickly from sunlight, heat, and competing microorganisms) are called Class B and are available only to permitted users on permitted sites - NOT to the general public.

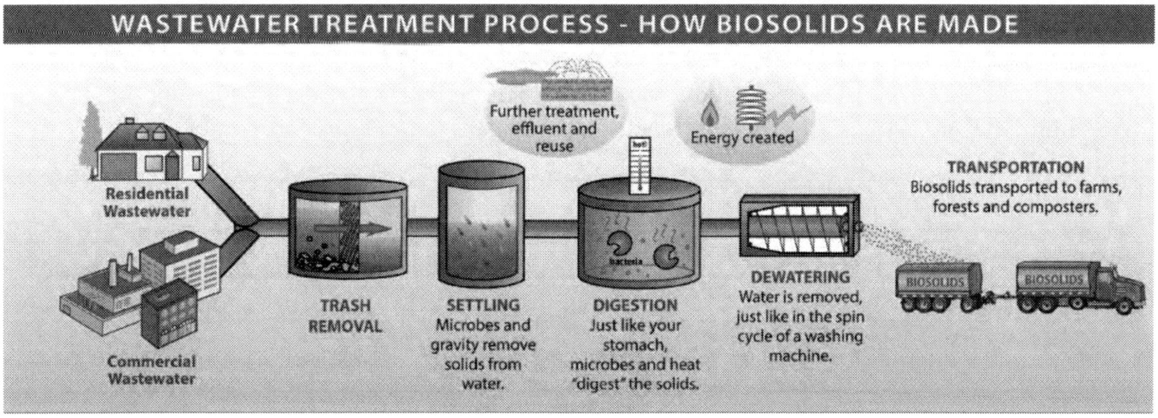

[ASK ALL]
[SC]
[DISPLAY THUMBNAIL OF IMAGE – MAKE SURE TO INCLUDE IMAGE AND DEFINITION FROM INFO NODE 1]
11. Based on what you now know about BIOSOLIDS, what is your overall level of interest in the topic of BIOSOLIDS. (*Select one*)

(Click image above for definition)

1. Not at all interested
2. Not very interested
3. Neutral
4. Somewhat interested
5. Very interested

[ASK ALL]
[SC]
[DISPLAY THUMBNAIL OF IMAGE – MAKE SURE TO INCLUDE IMAGE AND DEFINITION FROM INFO NODE 1]
12. Based on what you now know about BIOSOLIDS, what is your overall reaction to the process of recycling BIOSOLIDS as a fertilizer and soil conditioner? *(Select one)*

(Click image above for definition)

1. Not at all supportive
2. Not very supportive
3. Neutral
4. Somewhat supportive
5. Very supportive

[ASK ALL]
[USE CONFIRMIT RATEMENT]
[ONLY SHOW PRE-CODE 1-8 50% OF THE TIME AND PRE-CODES 9-16 THE OTHER 50% OF THE TIME – RANDOMLY USING LEAST FILL]
[ADD PALE BLUE SHADING TO EVERY OTHER ROW]
13. How much do each of the following facts about BIOSOLIDS increase your support for recycling BIOSOLIDS as a fertilizer and soil conditioner? (*Select one per row*)

	[RANDOMIZE ROWS]	Does not at all increase my support	Does not increase my support much	Somewhat increases my support	Very much increases my support
1	Biosolids are a sustainable way to recycle a waste product.	1	2	3	4
2	Biosolids create benefits for plants and soils such as increased plant growth and crop yields, improved soil quality, and the recycling of carbon and nutrients.	1	2	3	4
3	Biosolids recycling mitigates climate change and decreases carbon footprints by storing	1	2	3	4

	carbon in the soil and reducing use of fossil fuel intensive synthetic fertilizer.				
4	Biosolids are a safe and low risk product for farmers, foresters, gardeners, and landscapers.	1	2	3	4
5	Biosolids recycling creates economic benefits for their users.	1	2	3	4
6	Biosolids recycling is the most cost effective way to manage biosolids.	1	2	3	4
7	Biosolids are continually monitored and tested for safety and quality.	1	2	3	4
8	The Department of Ecology strongly supports recycling of biosolids as a fertilizer and soil conditioner.	1	2	3	4
9	Biosolids contain a full suite of micro and macro-nutrients – they are like a multivitamin for plants.	1	2	3	4
10	The treatment process that creates biosolids also creates recycled water, which can be used for things like irrigation or street cleaning.	1	2	3	4
11	The treatment process that creates biosolids also creates renewable energy.	1	2	3	4
12	Biosolids contain tiny amounts of some widely used chemicals, usually in lower levels than in household dust.	1	2	3	4
13	Biosolids recycling creates economic benefits for wastewater treatment plants.	1	2	3	4
14	Biosolids recycling keeps rates low for ratepayers.	1	2	3	4
15	Biosolids are regulated by the Washington Department of Ecology.	1	2	3	4
16	Biosolids products are often in high demand by customers - there's a waiting list for farmers who want to use biosolids and	1	2	3	4

	many biosolids products sell out each year.				

[ASK ALL]
[SC]
14. Based on what you now know about BIOSOLIDS, what is your overall reaction to the process of recycling BIOSOLIDS as a fertilizer and soil conditioner? (Select one)

1. Not at all supportive
2. Not very supportive
3. Neutral
4. Somewhat supportive
5. Very supportive

[ASK ALL]
[OE – LARGE BOX]
15. Why would you say you are **<PIPE IN PRE-CODE SELECTED AT Q14>** of recycling BIOSOLIDS as a fertilizer and soil conditioner? (Please enter your response in the box below)

Gardening/product preference/compost

[ASK IF (D5= 8) OR (D6 = 2 OR 3 OR 5 OR 6]
[MC]
16. Which of the following do you use? *(Select all that apply)*

1. Soil conditioner or soil amendment (to build or improve soil)
2. Fertilizer (to provide nutrients)
3. Herbicide (to kill weeds)
4. Pesticide (to kill pests)
5. Mulch (to cover the soil)
98. Other, *please specify* **[OE CHA]**
99. None/Don't use **[EXCLUSIVE]**
97. Don't know **[EXCLUSIVE]**

[SHOW Q17 AND Q17B IN SAME PAGE AS Q16]
[MAKE Q17/Q17B APPEARED AFTER Q16 IS ANSWERED]

[ASK IF Q16 = 1]
[OE]
17. What soil conditioner or soil amendment do you use? *(Please enter your response in the box below)*

[ASK IF Q16 = 2]

[OE]
17b.What fertilizer do you use? *(Please enter your response in the box below)*

[ASK IF Q16= 1 OR 2]
[RANKING – DRAG AND DROP - RANK 3 ITEMS MIN AND MAX]
[RANDOMIZE – KEEP 1 AND 2 AND 3 TOGETHER, 4 AND 5 TOGETHER, 6 AND 7 AND 8 TOGETHER, 9 AND 10 AND 11 TOGETHER]
18. What are the top three features that you look for in….*(Rank up to three features where 1 is the most important feature, two is the second most important feature, and three is the third most important feature. To rank the features, drag a feature from the boxes on the left into the numbered boxes on the right.)*

[SHOW IF Q16=1- CENTER, BOLD AND 2 POINTS FONT BIGGER] Soil conditioner or soil amendment
[SHOW IF Q16=2- CENTER, BOLD AND 2 POINTS FONT BIGGER] Fertilizer

1. Convenient to use
2. Available for purchase at a convenient location
3. Time savings/ doesn't take much time to apply
4. Organic
5. Natural
6. Safe for family
7. Safe for my pets
8. Safe for environment
9. Does what it is designed to do
10. Works consistently
11. Affordable/low cost
98. Other, *please specify* **[OE CHA] [ANCHOR]**
100. Other, please specify [OE CHA] [ANCHOR]
97. Don't know **[OE CHA] [ANCHOR] [EXCLUSIVE]**

[ASK IF Q16=1 OR 2]
[SC]
19. Where would you prefer to purchase soil conditioners/soil amendments or fertilizers? *(Select one)*

1. At a large hardware store (such as Home Depot, Lowe's)
2. At a small local hardware store
3. At the grocery store
4. At a large retail store (such as Fred Meyer, Target, Wal-mart)
5. At a gardening store or plant nursery

6. At a topsoil and landscape supply depot (such as Sawdust Supply, Pacific Topsoil, Ballard Dirt Exchange)
7. At a wastewater treatment plant or recycling and disposal station (transfer station)
8. Online, *please specify* **[OE CHA]**
98. Other, *please specify* **[OE CHA]**

[ASK IF Q16=1 OR 2]
[SC]
20. What packaging do you prefer? *(Select one per row)*

1	Bagged – small bags (approximately 5 to 10 pounds)		
2	Bagged – medium bags (approximately 10 to 20 pounds)		
3	Bagged – large bags (approximately 20 or more pounds.)		
4	Bulk/Loose		

[ASK IF Q16=1 OR 2]
[SC]
[SHOW ON SAME PAGE AS Q20 – show after Q20 is filled out]
32. How you prefer to obtain the material? *(Select one)*
1. Pick up in store
2. Delivered to home
3. No preference

[ASK IF (D5=8) OR (D6=2 OR 3 OR 5 OR 6)]
[SC PER ROW GRID]
[USE LEGACY GRID]
[ADD PALE BLUE SHADING TO EVERY OTHER ROW]

21. How likely would you be to use COMPOST made with BIOSOLIDS on your... *(Select one per row)*

		Extremel y unlikely	**Unli kely**	**Neu tral**	**Li kel y**	**Extreme ly likely**	**Not applica ble**
1	Yard	1	2	3	4	5	6
2	Flower garden	1	2	3	4	5	6
3	Vegetable garden	1	2	3	4	5	6
4	Farm	1	2	3	4	5	6
5	Landscaping	1	2	3	4	5	6
98	Other, *please specify* **[OE CHA] [RESPONSE NOT REQUIRED]**	1	2	3	4	5	6

[ASK IF (D5=8) OR (D6=2 OR 3 OR 5 OR 6)]

[SHOW IN SAME PAGE AS Q21]
[MAKE Q22 APPEAR AFTER Q21=1-5 IS ANSWERED]
22. Why would you be likely or unlikely to use COMPOST made with BIOSOLIDS on your …
(Please enter your response in the box/boxes below)

1. Yard **[OE CHA] [SHOW IF Q21_1=1-5]**
2. Garden **[OE CHA] [SHOW IF Q21_2 OR 3=1-5]**
3. Farm **[OE CHA] [SHOW IF Q21_4=1-5]**
4. Landscaping projects **[OE CHA] [SHOW IF Q21_5=1-5]**
5. **[PIPE IN ANSWER FROM Q21_98=1-5] [OE CHA] [SHOW IF Q21_98=1-5]**

[ASK ALL]
[SC PER ROW GRID]
[USE LEGACY GRID]
[ADD PALE BLUE SHADING TO EVERY OTHER ROW]

23. How do you feel about a neighbor using the following on their garden or yard? *(Select one per row)*

	[	**Strongly oppose**	**Somewhat oppose**	**Neutral/don't care**	**Somewhat favor**	**Strongly favor**
1	Manure (chicken, cow, horse, etc.)	1	2	3	4	5
2	Biosolids	1	2	3	4	5
3	Compost made with biosolids	1	2	3	4	5
4	Chemical/synthetic fertilizer	1	2	3	4	5

[SHOW Q24a, Q24b, Q24c IN SAME PAGE AS Q23]

[ASK IF Q23_1= 1 OR 2 OR 4 OR 5]
[OE CHA-BIG BOX]
24a. Why do you **[PIPE IN COLUMN TEXT SELECTED AT Q23_1]** a neighbor using manure? *(Please enter your response in the box below)*

[ASK IF Q23_2= 1 OR 2 OR 4 OR 5] [OE CHA-BIG BOX]
24d. Why do you **[PIPE IN COLUMN TEXT SELECTED AT Q23_2]** a neighbor using biosolids? *(Please enter your response in the box below)*

[ASK IF Q23_3= 1 OR 2 OR 4 OR 5]
[OE CHA-BIG BOX]
24b. Why do you **[PIPE IN COLUMN TEXT SELECTED AT Q23_3]** a neighbor using compost made with biosolids? *(Please enter your response in the box below)*

[ASK IF Q23_4= 1 OR 2 OR 4 OR 5]

[OE CHA-BIG BOX]
24c. Why do you **[PIPE IN COLUMN TEXT SELECTED AT Q23_4]** a neighbor using chemical/synthetic fertilizer? *(Please enter your response in the box below)*

[ASK ALL]
[OE CHA – BIG BOX]
25. What questions or comments do you have about biosolids? *(Please enter your response in the box below)*

Additional demographics

[ASK IF D5=8]
[MC]
26. Are you a… (*Select all that apply)*

1. Landscaper
2. Landscape designer
3. Landscape architect
98 .Other, *please specify* **[OE CHA]**

[ASK ALL]
[SC]
27a. What is the primary language spoken in your home? (Select one)

1. English
2. Spanish
3. Vietnamese
4. Russian
5. Somali
6. Chinese
7. Korean
8. Ukrainian
9. Amharic
10. Punjabi
98. Other, please specify **[OE CHA]**

[ASK ALL]
[SC]
28. What is the highest level of education you have completed? (*Select one)*

1. Less than high school
2. High school graduate or the equivalent (example GED)
3. Some college, no degree
4. Associate degree or trade/technical/vocational training
5. Bachelor's degree
6. Some graduate school, no degree

7. Graduate degree

[ASK ALL]
[SC]
29. Do you own or rent your home? (*Select one)*

1. Own
2. Rent
98. Other

[ASK ALL]
[SC]
30. Do you have children under 18 years of age living at home? (*Select one)*

1. No
2. Yes

[ASK ALL]
[SC]
31. Would you be willing to participate in a discussion group or other further research efforts to help the King County Wastewater Treatment Division learn more about opinions of citizens like you regarding biosolids? (*Select one)*

1. No
2. Yes

Survey 2: Survey Instrument

Project ID:	**353159**
Wave Number:	**2**
SURVEY NAME:	**King County Study - Project 2**
CLIENT NAME:	**King County Wastewater Treatment**
ESTIMATED INTERVIEW LENGTH (MINS):	**10 min.**
ESTIMATED PROGRAMMING LENGTH (MINS):	
VERSION/DATE:	

TYPE OF PROJECT: (select those that apply)

- **STS using STS Session Tracking and Universal Redirects – specify with resume**

SAMPLE SOURCE/COUNTRY/DETAILS:

COUNTRY	LANGAUGE(S)	SOURCE(S)
USA	English	ER 50
USA	English	VOP 22
USA	English	PL 68

REQUIREMENTS	QUESTION NUMBERS/ NOTE
ADVANCED IMAGE/VIDEO SECURITY	NO: only default security
RICH MEDIA TOOLS TO USE	Confirmit / Research Now
OTHER	Duplicate p1197307917 and apply changes

PROGRAMMING NOTES ARE IN BLUE.

DP NOTES ARE IN GREEN

ORIGINAL CLIENT INSTRUCTIONS ARE IN GREY – PLEASE IGNORE

QUOTAS:

Quota Description	Question Number	Pre-code Name OR Number(s)	Percentage	Quota
Total - Quota stop			100%	300
Gender	**D1**	1. Male	50%	150
		2. Female	50%	150
Age	**S4**	2. 18-24	999%	999 (min n=33 –soft quota)
		3. 25-34	20%	60 (n=93 total from 18-34y.o)
		4. 35-44	19%	57
		5. 45-54	19%	57
		6. 55-64	15%	46
		7. 65+	16%	47
Hispanic	**D2**	1. No	91%	273
		2. Yes	9%	27
Race	**D3**	1. White or Caucasian	70.1%	212
		2. Black or African American	6.7%	20
		3. Native American or Alaska Native	1.1%	3
		4. Asian	16.4%	48
		5. Native Hawaiian or Pacific Islander	0.8%	2
		6. Two or more races	999%	999 (MAX N=14 BETWEEN PRE-CODE 6 AND 98)
		98. Other	999%	999
Income	**dINCOME**	1. Under 50k	35%	106
		2. 50k to $99,999	30%	89
		3. 100k+	35%	105
Gardeners	**dGARDENERS**	1. Landscaping	999	999
		2 Gardening (home or community)	999	999
		3. Farming	999	999
		4. Yard work/Lawn care	999	999
			100%	

D6	**D6**	5. Farming	999	999
		6. Yard work/lawn care	999	999

Screening questions

INSERT VIDEO SCREENER
INSERT AUDIO SCREENER

[ASK ALL]
[OE NUM – ZIP CODE - VALIDATE USING ZIP LIST IN PROJECT FOLDER]
S1 [Q27].
What is your home zip code? *(Please enter your zip code)*

[ASK ALL]
[PUNCH OFF OF S1 [Q27] – SEE PUNCHING IN FOLDER]
[TERMINATE IF NOT PUNCHED BELOW]
dCOUNTY

1	**Seattle / North King County**
2	**East King County**
3	**South King County**

[ASK ALL]
[SC]
S4 [S2].
Which of the following categories includes your age? *(Select one)*

1. Less than 18 **[TERMINATE]**
2. 18-24
3. 25-34
4. 35-44
5. 45-54
6. 55-64
7. 65+

Quotas: Demographics

[ASK ALL]
[SC]
D1. What is your gender? *(Select one)*

1. Male
2. Female

[ASK ALL]
[SC]
D2. Are you of Hispanic or Latino origin? *(Select one)*

1. No
2. Yes

[ASK ALL]
[SC]
D3. What is your race? *(Select one)*

1. White or Caucasian
2. Black or African American
3. Native American or Alaska Native
4. Asian
5. Native Hawaiian or Pacific Islander
6. Two or more races

98. Other, *please specify* **[OE CHA]**

[ASK ALL]
[SC]
D4. What was your total annual household income (before taxes) for 2015? *(Select one)*

1. Less than $25,000
2. $25,000 to $49,999
3. $50,000 to $74,999
4. $75,000 to $99,999
5. $100,000 to $149,000
6. $150,000 to $199,000
7. $200,000 and above

96. Prefer not to answer

[ASK ALL]
[PUNCH OFF OF D4]
[SC-HIDDEN]
dINCOME

1	**Under 50k**	**IF D4=1 OR 2**
2	**50k to $99,999**	**IF D4=3 OR 4**
3	**100k+**	**IF D4=5 OR 6 OR 7**

[ASK ALL]
[MC]
D5. Do you work in any of the following industries? (*Select all that apply)*

1. Advertising
2. Agriculture
3. Banking
4. Government
5. Construction
6. Education
7. Insurance
8. Landscaping
9. Market research/marketing
10. Manufacturing
11. Retail
12. Transportation
13. Utilities
99. None of the above **[EXCLUSIVE] [ANCHOR]**

[ASK ALL]
[MC]
D6. Do you do any of the following? *(Select all that apply)*

1. Biking
2. Gardening (home garden)
3. Gardening (community garden)
4. Hiking
5. Farming
6. Yard work/lawn care
7. Raising livestock (e.g. chickens, goats, or cows etc.)
8. Trail running
9. Volunteering
99. None of the above **[EXCLUSIVE] [ANCHOR]**

[ASK ALL]
[MC]
[HIDDEN]
[PUNCH OFF OF D5 AND D6]
dGARDENERS

1	**Landscaping**	**(D5=8)**
2	**Gardening (home or community)**	**(D6= 2 OR 3)**
3	**Farming**	**(D6=5)**
4	**Yard work/Lawn care**	**(D6=6)**

[ASK ALL]
[SC]
Q8. How familiar are you with the term BIOSOLIDS? (*Select one*)

1. Not familiar
2. Somewhat familiar
3. Very familiar

[SHOW ALL]
[INFO NODE]
[SHOW NEXT BUTTON AFTER VIDEO IS COMPLETE]
[DISPLAY "How Loop is made" VIDEO 1]
Now, we're going to show you a short video and ask you a few questions about what you think. The play button will appear once the file has buffered.

[ASK ALL]
[OE – LARGE BOX]
[DISPLAY THUMBNAIL OF VIDEO – allow respondent to view again if they click on it]
Q32. What did you take away from this video?

[ASK ALL]
[SC]
[CONFIRMIT RATEMENT]
Q33. How much did you like this video? *(Please select your response)*
1. Dislike a lot
2. Dislike a little
3. Neutral
4. Like a little
5. Like a lot

[ASK IF Q33=4 OR 5]
[OE – LARGE BOX]
[SHOW ON SAME PAGE AS Q33, show after Q33 is filled out]
Q34. What did you like about this video? *(Please enter your response in the box below)*

[ASK IF Q33=1 OR 2]
[OE – LARGE BOX]
[SHOW ON SAME PAGE AS Q33, show after Q33 is filled out]
Q35. What did you dislike about this video? *(Please enter your response in the box below)*

[ASK ALL]
[OE – LARGE BOX]
[SHOW ON SAME PAGE AS Q33, show after Q33 is filled out]

Q36. What would make this video more appealing to you? *(Please enter your response in the box below)*

[ASK ALL]
[SC]
Q11. Based on what you now know about Loop biosolids, is Loop a topic you would want to know more about? Please select your level of interest in Loop: (*Select one*)

1. Not at all interested in knowing more about Loop
2. Not very interested in knowing more about Loop
3. Neutral
4. Somewhat interested in knowing more about Loop
5. Very interested in knowing more about Loop

[MOVED POSITION]
[ASK ALL]
[SC]
Q12. Based on what you now know about Loop biosolids, what is your overall reaction to the process of recycling Loop as a fertilizer and soil conditioner? *(Select one)*

1. Not at all supportive
2. Not very supportive
3. Neutral
4. Somewhat supportive
5. Very supportive

[SHOW ALL]
[SELECT AND SHOW ONE VIDEOS BETWEEN #2-#8 USING LEAST FILL]
[SHOW NEXT BUTTON AFTER VIDEO IS COMPLETE]
[DISPLAY VIDEO]
Q37. Now, please watch one more short video about Loop.
The play button will appear once the file has buffered.

[ASK ALL]
[OE – LARGE BOX]
[DISPLAY THUMBNAIL OF VIDEO – allow respondent to view again if they click on it]
Q38. What did you take away from this video?

[ASK ALL]
[CONFIRMIT RATEMENT]
Q39. How much did you like this video? *(Please select your response)*

1. Dislike a lot
2. Dislike a little
3. Neutral
4. Like a little
5. Like a lot

[ASK IF Q39=4 OR 5]
[OE – LARGE BOX]
[SHOW ON SAME PAGE AS Q39, show after Q39 is answered]
Q40. What did you like about this video? *(Please enter your response in the box below)*

[ASK IF Q39=1 OR 2]
[OE – LARGE BOX]
[SHOW ON SAME PAGE AS Q39, show after Q39 is answered]
Q41. What did you dislike about this video? *(Please enter your response in the box below)*

[ASK ALL]
[OE – LARGE BOX]
[SHOW ON SAME PAGE AS Q39, show after Q39 is answered]
Q42. What would make this video more appealing to you? *(Please enter your response in the box below)*

[ASK ALL]
[SC]
Q11a. After watching this video, is Loop a topic you would want to know more about? Please select your level of interest in Loop: (*Select one*)

1. Not at all interested in knowing more about Loop
2. Not very interested in knowing more about Loop
3. Neutral
4. Somewhat interested in knowing more about Loop
5. Very interested in knowing more about Loop

[ASK ALL]
[SC]
Q14. After watching this video, what is your overall reaction to the process of recycling Loop as a fertilizer and soil conditioner? (Select one)

1. Not at all supportive
2. Not very supportive
3. Neutral
4. Somewhat supportive
5. Very supportive

[ASK IF Q14 = 1,2,3,4]
[OE – LARGE BOX]
Q42a. What specific information would you need to know to increase your support of Loop biosolids? *(Please enter your response in the box below)*

[ASK ALL]
[GRID – SC PER COLUMN]
CHANGE TO LEGACY GRID FORMAT
[RANDOMIZE blocks 1-3, 4-6, 7-9 but keep answer options together in this order within the block]
Q43. Who would you trust the <u>most</u> to give you accurate information about Loop biosolids and about the safety of Loop for the health of you and your family? *(Select one information source for each of the columns)*

		General information about Loop biosolids	Safety for human health	Safety for the environment
1	University scientists or academics	1	2	3
2	Educators or teachers	1	2	3
3	Biosolids managers or industry professionals	1	2	3
4	Local wastewater treatment agency	1	2	3
5	Public health agencies (e.g. Local health department, Center for Disease Control and Prevention)	1	2	3
6	Environmental regulatory agencies (e.g. Environmental Protection Agency, Department of Ecology)	1	2	3
7	Environmental or community organizations	1	2	3
8	Testimonials from everyday people that use biosolids (e.g. farmers, Master Gardeners, landscapers)	1	2	3
9	Family/friends	1	2	3

98 Other, *please specify* **[ANCHOR]** 1 2 3

[ASK IF Q14 = 1,2,3,4]
[OE – LARGE BOX]
Q47. What specific information do you need to know to be assured that Loop is safe?

[ASK ALL]
[CLICKABLE RANKING: MIN AND MAX 3]
[RANDOMISE LIST]
Q44. Which of these Loop video topics would you find most interesting? *(Please rank your top three)*

1 How the King County Wastewater Treatment Division cleans water, makes Loop, and protects public health and the environment.
2 History of Loop
3 Safety of Loop
4 Loop and climate change/reducing carbon footprints
5 Loop and sustainability
6 Using Loop as a fertilizer (and the benefits) compared to other options (such as landfilling or burning)
7 How your community uses Loop
8 Loop and recycled water
9 Loop and bioenergy
10 Loop and its potential uses (agriculture, forestry, reclamation, gardening, landscaping, etc.)
11 Stories about Loop's customers (gardeners, foresters, farmers)
98 Other, *please specify* OE TEXT, FIXED
99 None of the above – Not interested in watching videos about Loop **[EXCLUSIVE]** FIXED

[ASK ALL]
[SC]
Q45. For a video about Loop biosolids, what video **style** of video would you find most compelling? *(Select one)*

1. Stories illustrated via marker drawing (whiteboard or chalkboard videos)
2. Animated videos
3. Interview based story telling with relevant shots of the subject matter
4. Voice-over based story telling with relevant shots of the subject matter
98. Other, *please specify* OE TEXT
99. None of the above – Not interested in watching videos about Loop **EXCLUSIVE**
97. Don't know **EXCLUSIVE**

[ASK ALL]
[SC]
Q46. For a video about Loop biosolids, what video **approach** of video would you find <u>most</u> compelling? *(Select one)*
1. Education/informational videos
2. Humor-based video
3. Dramatic
98. Other, *please specify* OE TEXT
99. None of the above – Not interested in watching videos about Loop **EXCLUSIVE**
97. Don't know **EXCLUSIVE**

[ASK ALL]
[SC]
Q28. What is the highest level of education you have completed? (*Select one)*

1. Less than high school
2. High school graduate or the equivalent (example GED)
3. Some college, no degree
4. Associate degree or trade/technical/vocational training
5. Bachelor's degree
6. Some graduate school, no degree
7. Graduate degree

[ASK ALL]
[SC]
Q29. Do you own or rent your home? (*Select one)*
1. Own
2. Rent
98. Other

[ASK ALL]
[SC]
Q30. Do you have children under 18 years of age living at home? (*Select one)*
1. No
2. Yes

[ASK ALL]
[SC]
Q31. Would you be willing to participate in a discussion group or other further research efforts to help the King County Wastewater Treatment Division learn more about opinions of citizens like you regarding biosolids? (*Select one)*
1. No
2. Yes

Class "A" Part Two. After Achieving Class A Designation, Our Path To Building A Sustainable Class A Program

Shannon Ostendorff[1], Scott Thompson[1], Tim Truax[1]

[1]City of Bend, Oregon,

ABSTRACT

The City of Bend utilized a unique partnership with the United States Forest Service, to designate their biosolids as Class A Exceptional Quality (EQ). This designation of Class A EQ biosolids paved the way for a new biosolids program for the City of Bend. This new designation opened many options for biosolids application and left City staff with the challenge of building a sustainable Class A program. This presentation will discuss some unique partners and the path the City utilized to garnish community support, develop new customers, and update their Biosolids Management Plan.

The City of Bend, Oregon's Water Reclamation Facility is an 8.5 MGD Class IV wastewater reclamation facility. Biosolids are anaerobically digested, and dewatered using a belt filter press prior to drying on 12 acres of asphalt drying beds.

The City typically dries a year's accumulation of biosolids on asphalt drying beds. In this manner, moisture is reduced to less than 10% total moisture by August each year. The dried solids are then stored under roof to prevent further moisture accumulation. The facility has historically applied Class B biosolids to local area farms using City labor and equipment. In 2014, the City worked with the United States Forest Service (USFS) to apply biosolids as a test fertilizer on forest service land reclamation projects. In order to facilitate this biosolids application, the USFS paid for testing of the City's dried biosolids to determine that they met Class A status. The material met Class A biosolids requirements under EPA Part 503 Subpart D Class A Alternative 4 and the USFS applied the Class A EQ biosolids on test plots of land on a mining reclamation site in Eastern Oregon. This application was to study plant growth rates for biosolids application compared to the current forest management practice of using bio char as a fertilizer. This study is ongoing and will be monitored for success until 2018. With newly designated Class A EQ biosolids, the City became determined to expand their existing Class B program. This next step required community education, development of a new customer base, and a comprehensive update of the Biosolids Management Plan. The Class A EQ program was developed to be cost-effective and sustainable to future use of Class A EQ biosolids.

With the new Class A EQ designation the City worked with Kennedy/Jenks to comprehensively update the Biosolids Management Plan. This plan was updated to reflect the recent expansion of the secondary treatment process, and detail the process utilized to achieve Class A status.

Critical to the success of a Class A EQ program is the development of Community support. The City started with a partnership with the local Environmental Center to help with this effort. This program started initially with Environmental Center educators providing in-classroom education on the water cycle and wastewater treatment process. The Environmental Center also helped facilitate tours at the Water Reclamation Facility. This partnership led the development of a Victory Garden for public biosolids education at the Environmental Center location, and grew a local advocate and partner for biosolids education.

The City also worked with Sally Brown of the University of Washington and the Oregon State Master Gardeners to educate Master Gardeners on the quality of biosolids as a soil amendment and promote use of Class A EQ biosolids in gardens. This partnership helped build not only local, but state support for biosolids. The Victory Garden will be a critical demonstration site for master gardeners of biosolids effectiveness.

With the Class A EQ biosolids designation, the City next conducted a survey of large scale uses of material and marketing. Although an ongoing effort, the development of this new market for Class A EQ biosolids has fully transitioned the City from a Waste to Commodity Utility. This paper details the steps taken to develop this program and foster the partnerships to fully develop a sustainable, cost effective Class A EQ biosolids program.

KEYWORDS

Biosolids, Class A, Drying Beds, land reclamation, biochar, fertilizer, nutrients, mining reclamation, partnerships, Master Gardeners, marketing

INTRODUCTION

The City of Bend, Oregon's Water Reclamation Facility (WRF) is located in Central Oregon approximately three hours southeast of the City of Portland, Oregon. The WRF is located on 1600 acres of mixed juniper and sagebrush land as shown in Figure 1.The WRF is an 8.5 MGD Class IV wastewater reclamation facility. Biosolids are anaerobically digested, and dewatered using a belt filter press prior to drying on 12 acres of asphalt drying beds.

The Bend WRF biosolids program consists of three mesophilic anaerobic digesters with a capacity of one at 800,000 gallons and two at 375,000 gallons for a total of 1,550,000 gallons. The anaerobic digestion process consistently meets the vector attraction reduction (VAR) requirements for temperature and volatile solids reduction.

The City typically dries a year's accumulation of biosolids on asphalt drying beds, pictured in Figure 2. In this manner, moisture is reduced to less than 10% total moisture by August each year. The dried solids are then stored under roof to prevent further moisture accumulation. The facility has historically applied Class B biosolids to local area farms using City labor and equipment.

The City produces on average 3 dry tons of biosolids per day. The digested solids are then dewatered using a belt filter press to approximately 13 percent solids concentration. These dewatered solids are then stored in a silo before being discharged by gravity into a truck. The truck hauls the digested, dewatered solids across the site to 12 acres of drying beds as shown in Figures 2 and 3. The solids are applied to the drying beds year round, where they are dried and turned over using a brown bear by the operations staff. The 12 acres of drying beds take advantage of Central Oregon's dry, sunny environment to dry the solids throughout the year. Figure 4 shows the brown bear operation where the operations staff works the biosolids to get the most evaporation of moisture from the solids throughout the year. The solids continue to dry throughout the year until they are 88-90 percent solids. The final turn over is very dry and is a dusty operation as shown in Figure 5.

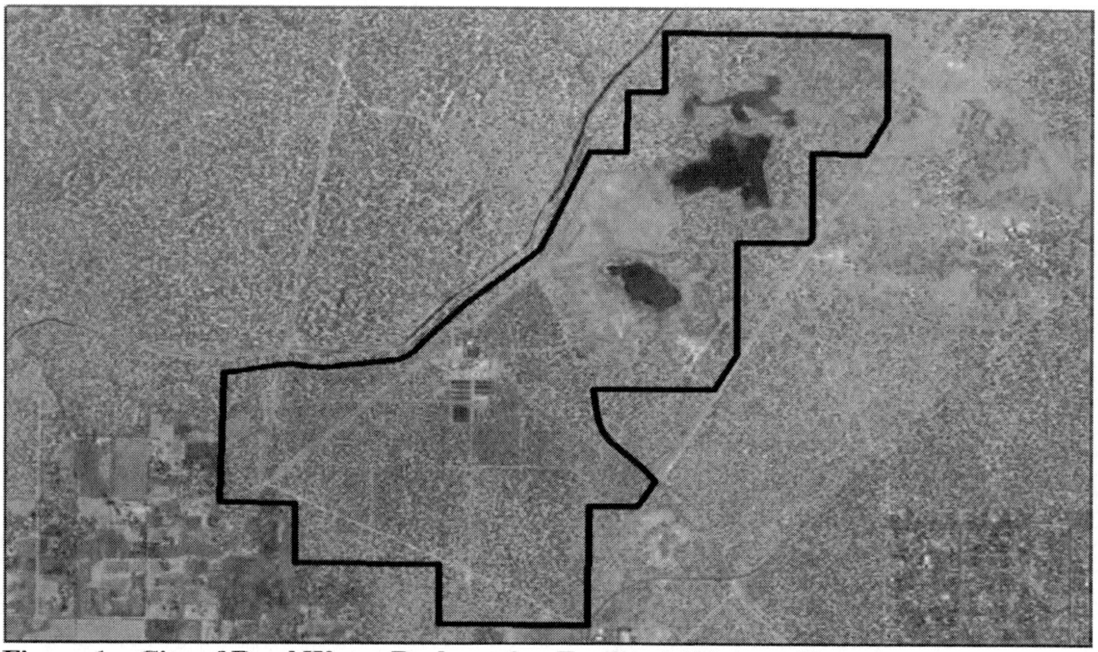

Figure 1 – City of Bend Water Reclamation Facility 1600 Acre Site

Figure 2 - City of Bend Water Reclamation Facility Drying Beds

Figure 3 – Drying Bed Operation (The liquid in the concrete tank is filtrate storage)

Figure 4 – Brown Bear Biosolids Operation

Figure 5– Dry Material Handling

In 2014, the City worked with the United States Forest Service (USFS) to apply biosolids as a test fertilizer on forest service land reclamation projects. In order to facilitate this biosolids application, the USFS paid for testing of the City's dried biosolids to determine that they met Class A status. The material met Class A biosolids requirements under EPA Part 503 Subpart D Class A Alternative 4 and the USFS applied the Class A EQ biosolids on test plots of land on a mining reclamation site in Eastern Oregon. This application was to study plant growth rates for biosolids application compared to the current forest management practice of using bio char as a fertilizer. This study is ongoing and will be monitored for success until 2018. A summary of results for pollutant concentrations, pathogen reduction, and Vector Attraction Reduction is provided in Tables 1, 2 and 3 respectively.

Table 1 shows the City meets the pollutant concentration limits set forth in 40 CFR Part 503.13 (Tables 1 and 3) and the City's biosolids data for 2014 and 2015. Based on this information the City's biosolids can be classified as Exceptional Quality (EQ).

Table 1: Historical Biosolids Pollutant Concentrations

Pollutant	Pollutant Concentrations (mg/kg) by year	
	2014	2015
Arsenic (As)	3.8	5.0
Cadmium (Cd)	1.1	1.1
Copper (Cu)	337.5	347.5
Lead (Pb)	29.5	29.3
Mercury (Hg)	<1.0	<1.0
Molybdenum (Mo)	8.1	9.9
Nickel (Ni)	19.0	16.3
Selenium (Se)	3.4	4.2
Zinc (Zn)	478	598

Table 2 summarizes the *Salmonella sp.*, enteric virus and helminth ova densities in samples collected from the drying beds for 2014 and 2015. These results illustrate the City achieves the densities specified in 40 CFR Part 503.32(a)(6) Alternative 4 for Class A biosolids.

Table 2: Pathogen Reduction: *Salmonella sp.*, Enteric Virus and Helminth Ova Data

	Salmonella sp.	Enteric Virus	Helminth Ova
Sample Date	**MPN per gram of total solids**	**MPN per gram of total solids**	**MPN per gram of total solids**
2015 (8 samples)	<3.0	<1	<1
2014 (7 samples)	<3.0	<1	<1

The City complies with the vector attraction requirements for Class A biosolids by meeting 40 CFR Part 503.33 (b)(1) Option 1. This requires at least 38% reduction in volatile solids. Table 3 summarizes the volatile solids reduction in biosolids from 2014 and 2015.

Table 3: Vector Attraction Reduction: Volatile Solids Reduction

	Volatile Reduction (%)	
	2014	**2015**
January	48	54
February	52	49
March	58	47
April	49	50
May	41	46
June	51	45
July	51	48
August	43	40
September	42	35
October	51	36
November	48	37
December	50	43

With newly designated Class A EQ biosolids, the City became determined to expand their existing Class B program. This next step required community education, development of a new customer base, and a comprehensive update of the Biosolids Management Plan. The Class A EQ program was developed to be cost-effective and sustainable to future use of Class A EQ biosolids.

With the new Class A EQ designation the City worked with Kennedy/Jenks to update the Biosolids Management Plan (BMP). This plan was updated to reflect the recent expansion of the secondary treatment process, and to detail the process utilized to achieve Class A EQ status. The City met with Oregon Department of Environmental Quality (DEQ) staff and determined the best approach for the BMP update was to prepare a plan amendment. This approach was based on the status of the WPCF permit, a construction project underway, and plans for additional solids handling improvements and facility plan modifications in the near future. The plan amendment was also drafted to include the ability for the City of Bend to act as a regional facility for other Central Oregon municipalities.

Critical to the success of a Class A EQ program is the development of Community support. The City started with a partnership with the local Environmental Center to help with this effort. This program started initially with Environmental Center educators providing in-classroom education on the water cycle and wastewater treatment process. The Environmental Center also helped facilitate tours at the Water Reclamation Facility. This partnership led the development of a Victory Garden for public biosolids education at the Environmental Center Kansas Avenue Learning Garden location. Biosolids were used to plant a pollinator crop, and information provided on the kiosk for information to community members about biosolids. This partnership helped grow a local advocate and partner for biosolids education.

Figure 6– The Central Oregon Environmental Center Mission

Figure 7– The Central Oregon Environmental Center Kansas Avenue Learning Garden location of demonstration garden for biosolids. Biosolids were used to plant a pollinator garden.

Figure 8– Kiosk at the Learning Garden which provides information about biosolids used as part of the pollinator garden

The City also worked with Sally Brown of the University of Washington and the Oregon State Master Gardeners to educate Master Gardeners on the quality of biosolids as a soil amendment and promote use of Class A EQ biosolids in gardens. The Master Gardeners helped devise the mix design for a biosolids soil amendment and helped foster a partnership with the local landscaping community. A one page flyer on the biosolids soil mixture was developed, specifying the Carbon to Nitrogen ratio and the benefits of this biosolid product. The City then worked with Deschutes Recycling to develop and package the soil mixture. The mixture is in the process of being branded and a marketing strategy developed. The City is also evaluating onsite composting to develop the composted blended soil/biosolids mixture. This partnership with the Master Gardeners helped build not only local, but state support for biosolids. The Victory Garden (pollinator garden) and educational kiosk will be a critical demonstration site for master gardeners of biosolids effectiveness.

With the Class A EQ biosolids designation, the City next conducted a survey of large scale uses of material and marketing. Survey through industry connections found large scale users. A large agricultural site in Hampton Oregon was found with 600 acres. The facility self-hauled 1,000 dry tons of Class A material in 2016. It is anticipated that the City will save up to $50,000 per year in application of material with the ability of large customers to self-haul and apply the product.

SUMMARY AND CONCLUSIONS

Although an ongoing effort, the development of this new market for Class A EQ biosolids has fully transitioned the City from a Waste to Commodity Utility. The Class A EQ program was developed to be cost-effective and sustainable to future use of Class A EQ biosolids. The focus of partnerships with the local environmental center and master gardeners has helped the City increase the value of the product both locally and to larger scale industrial users.

Figure 9 – City of Bend Class A Biosolids Product

ACKNOWLEDGEMENTS

- The authors would like to express their thanks to the following individuals: Paul Kennedy Kennedy.Paul@deq.state.or.us, Mark Cullington MarkCullington@KennedyJenks.com, and Sally L. Brown slb@u.washington.edu

A Strong Network: The Success of Regional Biosolids Groups in North America

Maile Lono-Batura[1*] and Ned Beecher[2]

[1] Northwest Biosolids
[2] North East Biosolids & Residuals Association (NEBRA)
[*] Email: maile.lono@nwbiosolids.org

ABSTRACT

Regional biosolids groups are stakeholders banded together to serve as a voice for biosolids management in their regions. While regulations, concerns, and interests may differ from state to state, regional associations have served as hubs for leveraging valuable research initiatives, sharing biosolids news, and answering member and community questions on biosolids use. By focusing attention on their respective regions, their efforts are rooted at a local level where they have the most impact. There are five active groups across North America, including the California Association of Sanitation Agencies, Mid-Atlantic Biosolids Association, North East Biosolids & Residuals Association, Northwest Biosolids, and the Virginia Biosolids Council. These groups have witnessed their own unique challenges and successes specific to their regions, yet have come to rely on the power of an extended network for continent-wide initiatives and staying informed on biosolids news across the continent. Where regional networks don't currently exist, WEF Member Associations have helped to address local challenges and offered educational opportunities for their members. Today, after three decades of experience, the resiliency of the regional groups continues to be tested by challenges such as adequate funding, the level of volunteer engagement, a transitioning workforce, and the changing dynamics of biosolids and residuals management.

KEYWORDS: biosolids, sewage sludge, residuals, management, land application, regional groups, communications, outreach

INTRODUCTION

Twenty years ago, the National Biosolids Partnership (NBP) was formed by the Water Environment Federation (WEF), the National Association of Clean Water Agencies (NACWA), and U. S. EPA. That same year saw the formation of three new *regional* biosolids associations, built upon the 10-year success of Northwest Biosolids (then known as Northwest Biosolids Management Association or NBMA). At meetings convened by the NBP, biosolids stakeholders envisioned the power of a distributed network of information sources and communications nodes across North America, linked together (and linked to WEF and NACWA) – but each one focusing locally.

Today, in 2017, that network is strong and efficient. But it is not complete. Thirty years since it started, NW Biosolids remains the leader among the biosolids groups, with dynamic effective vision, active involvement of members, and strong funding. Twenty years since they started, the North East Biosolids and Residuals Association (NEBRA) and the Mid-Atlantic Biosolids Association (MABA) continue, but a Great Lakes Byproducts Management Association has faded away. While there has been talk of forming a Southeast or other regional associations, it has not happened. Instead, state WEF Member Association biosolids committees have helped address local challenges - especially in the Carolinas, Texas, and Colorado. In California, a strong voice for biosolids was created by the California Association of Sanitation Agencies (CASA), which, through a fully-funded position, has led significant proactive advancement of biosolids resource recovery in that state. In another large state with extensive biosolids recycling, stakeholders formed the Virginia Biosolids Council (VBC). Northwest Biosolids and NEBRA extended their reach into western and eastern Canada, respectively, and now stakeholders across that country are considering creating a nationwide organization to further their network.

Altogether, these organizations and committees are a formidable structure in support of biosolids utilization and best practices. They share information through strong websites; produce numerous technical workshops, documents, and conferences; support research; provide training; and advocate for best practices, science-based regulations, and proactive communications. And they currently do it with a total of less than *four* full-time equivalents (FTEs) - and less than six when WEF & NACWA staff are included.

In 2013, the WEF *Enabling the Future* report summarized the status of the various public and non-profit biosolids support groups (Moss et al, 2013):

> These organizations form a distributed support network for biosolids professionals that has considerable strength and resiliency, which supports and facilitates the exchange of accurate information. Today, as the focus on resource recovery from biosolids intensifies, the importance of the distributed network of support for biosolids professionals becomes even greater. The increased complexity of biosolids management and the need for increased communications with more diverse audiences requires that these support mechanisms continue to grow and evolve to meet future needs. Enabling the future will require enhancing the capacity, skills, and knowledge in the public utility and private sectors involved in biosolids management.

Below are summaries of the 30-year history of Northwest Biosolids, the up-to-20 years of other regional biosolids groups, and the strong distributed network they create together. The discussion includes several major successes wrought by these groups, successes that have significantly impacted the history of North American biosolids management, helping bring the profession to where it is today.

For more about the importance of regional biosolids groups, see https://www.nebiosolids.org/why-biosolids-organizations-are-needed .

THE REGIONAL BIOSOLIDS GROUPS & THEIR SUCCESSES

Northwest Biosolids

Northwest Biosolids was incorporated as a non-profit organization in 1993 with the purpose of

- advancing environmental sustainability through the beneficial use of biosolids in the northwest region of the United States and Canada;
- fostering cooperation among governmental agencies, private corporations, and the public in the management of biosolids; and
- sharing technical information, research results, and general knowledge about biosolids management with members; local, state, and federal agencies; and the public.

Prior to incorporation, Northwest Biosolids Management Association (NBMA) was organized as the Regional Sludge Management Cooperative in 1988 and led some of the early collaborative research initiatives that helped to inform the national Part 503 biosolids regulations – as well as statewide regulations – and to serve as some of the most significant long-term biosolids research projects in existence. For nearly 30 years, Northwest Biosolids has prided itself in its ingenuity to leverage regional funding for critical research that serves as technically based knowledge to inform regulations and answer questions around safe and sustainable biosolids use. The Northwest Biosolids region currently spans Alaska, Idaho, Oregon, Washington, British Columbia, and Alberta.

Maile Lono-Batura has served as Executive Director of Northwest Biosolids for the past 18 years, leading the organization's programs, overseeing membership management, developing informational products, and coordinating all works necessary to managing the non-profit.

In 2015, Northwest Biosolids surveyed its membership on biosolids management and found that biosolids recycling remains alive and well in the Pacific Northwest, with 90% of the nearly 226,000 dry tons of biosolids generated being recycled to agriculture, forestry, land reclamation, and Class A production.

Following is an example of how this nexus of research, regulation, and outreach came together when Northwest Biosolids challenged a local biosolids ban.

Being a Voice for Biosolids Agencies in Legal Advocacy

Biosolids agencies have learned that what happens in another town, another county, state, region, or country can directly impact your program at home. The ability for news to travel has become infinitely more expedient with social media and news alerts. For Washington State, this news came in the Spring of 2011 when Wahkiakum County in southwest Washington issued a ban on Class B biosolids.

What prompted the County Commissioners to enact an ordinance that directly challenged Washington State's statute supporting beneficial use of biosolids? The Commissioners aimed to protect their County against the unknown – a proposed septage application. While animal manure application is seen as a common agricultural practice, septage and biosolids can be met with concern and questions. Washington's Department of Ecology (Ecology) and researchers from local universities met with the Commissioners to help answer questions. In response to Commissioners' concerns, Ecology included additional permit conditions to the proposed application. Finally, Ecology highlighted the State's directive (RCW Chapter 70.95J) that, 'properly managed municipal sewage sludge is a valuable commodity and can be beneficially used in agriculture, silviculture, and in landscapes as a soil conditioner' and that the biosolids program, 'shall, to the maximum extent possible, ensure that municipal sewage sludge is reused as a beneficial commodity.' An ordinance would directly oppose this law, and Ecology would need to file suit. The Commissioners decided to enact an ordinance anyway, but only on Class B, stating that Class A application was still welcome in their County. At the time, 88% of the state's biosolids were Class B, with the remaining 12% being Class A product that was sold or given away.

And so ensued the beginning of the battle for beneficial use, with a Trial Court hearing held the Fall of 2011. Following review of an extensive cost analysis prepared by Ecology on the difference between Class A vs. B, the Trial Court upheld the biosolids ban one year after the initial trial. While the analysis detailed that it costs utilities 2.5 times more to operate a Class A program, the Judge stated that, since the County was willing to accept Class A, it was not an outright ban. Until a conclusion of law and findings of fact were finalized, the ordinance was in place, and Ecology was not able to file appeal. Ecology and the County finally agreed on concluding findings. Shortly after, Ecology's appeal stated that it was their "understanding that state legislation takes precedence over local ordinances, and Ecology is tasked with ensuring that state law is followed."

When Ecology decided to appeal the trial court ruling to ban biosolids, Northwest Biosolids was faced with a critical decision. Do we trust that our voice will be represented by Ecology? Or do we ask for a seat at the table to represent the voice of biosolids customers and generators in WA and the Pacific NW? Although legal advocacy was a road Northwest Biosolids had never traveled, we chose the later. If Northwest Biosolids didn't stand for this, what did we stand for? The return on this investment was an unknown; however, the implications of upholding the ban and not having our voice heard could be detrimental and far-reaching. It was our intention to make this known.

Northwest Biosolids reached out to Ecology, offering to file an *amicus* brief (friend of the court document) to support the State's biosolids program. Ecology accepted the offer. Since this was a new path for the association, a volunteer team formed to help gather resources and create a strategic alliance to bring the voice of biosolids to the table. Sixteen organizations joined the team, including national assistance from the National Association of Clean Water Agencies. This partnership prepared two briefs – one representing biosolids generators and the other speaking for biosolids customers. Northwest Biosolids worked directly with the law firm of Beveridge & Diamond in preparing the briefs and with local legal representation in reviewing and submitting these documents. The key arguments in these briefs were:

- Upholding the Wahkiakum County ban will revert Washington state to a patchwork of biosolids regulations, replacing the harmonious state biosolids rule that has worked well.
- This ban is not a fact-based challenge, leaving room for additional similar bans.

- Biosolids regulations are formulated around sound science; they present the most restrictive requirements that can also be applied to any other amendment available (e.g. septage).
- This ban will be far-reaching; it will not only impact Washington, but states, counties, and small towns across the country.

Ecology, Wahkiakum County, Lewis County, and Northwest Biosolids presented oral arguments to the Court of Appeals on July 1, 2014. Biosolids representatives were present to show support. The judges asked challenging questions of all parties around critical areas: ordinances, the meaning of the RCW 70.95, and whether this was an outright ban. At one point, Wahkiakum's lawyer was scolded for not answering a judge's question. The *Daily News* wrote "The Justices asked pointed questions for each side, and at one point gave Bigelow a mild rebuke when he ignored or misunderstood a question from Justice Linda C.J. Lee." The Court was tasked with determining whether the County's Class B biosolids and septage ban irreconcilably conflicted with the state law given the arguments presented by the County, Ecology, and their *amici* – Lewis County in support of Wahkiakum, and, in support of Ecology: Natural Selection Farms, Inc. and Boulder Park, Inc. (farm *amici*) and Northwest Biosolids Management Association, National Association of Clean Water Agencies, Washington Association of Sewer & Water Districts, and the town of Cathlamet, the largest town in Wahkiakum County (public *amici*).

In July 2014, the Court of Appeals released their decision in *Washington State Department of Ecology vs. Wahkiakum County*, unanimously reversing the County biosolids ban and reinforcing the legislature's original intent to support biosolids recycling in the State.

The Court's published opinion outlines their comprehensive review of both sides, with resounding support to uphold the duty assigned to Ecology by the legislature to safely manage and regulate biosolids in the State. The Court emphasized that the County's list of arguments were "unpersuasive" and even "misstated" the legislature's intent to "implement a biosolids management program "to the maximum extent possible" that ensures biosolids are "reused as a beneficial commodity."

The farm *amici* brief was even cited, underlining the effectiveness of the permitting process for land application of biosolids. "Farmers have come to rely on the well-established and uniform state regulation of land application for biosolids for planning and investment."

The Court also highlighted State Supreme Court cases that supported their decision to uphold the current biosolids regulatory framework.

Ecology and the *amici* supporting them were adamant about the influence one ban could have in other counties across the state and nation. The Court agreed, stating that "if local governments have the power to ban land application of biosolids, land application of biosolids could be banned throughout the state, clearly thwarting the legislature's purpose of recycling biosolids through land application rather than landfill disposal or incineration. The County's ordinance thwarts the express purpose of the legislature and, thus, is irreconcilable with state law and unconstitutional."

Wahkiakum County petitioned the State Supreme Court to reverse the Court of Appeal's decision. That five panel court denied the petition, supporting the Court of Appeals ruling that, "the County's ordinance is unconstitutional because it irreconcilably conflicts with state law. Accordingly, we reverse the superior court's order granting summary judgment in favor of the County and remand for entry of summary judgment in favor of Ecology."

This court case exemplifies the key role regional associations serve in the biosolids field. Without the coordination of Northwest Biosolids and support from our members, partnering associations and legal team, this biosolids ban that threatened biosolids near and far may have not been overturned. This court ruling has become part of the case law supporting biosolids recycling, signifying a true beacon of support for responsible biosolids management regulations and the programs that operate within them – a unified biosolids rule supported by a unified court decision. This was a landmark decision and a historic chapter for biosolids recycling, enabled, in large part, by an effective regional group.

Learn more at www.nwbiosolids.org.

North East Biosolids and Residuals Association (NEBRA)
NEBRA was created in 1997 because of pressures and trends similar to those that led to the creation of Northwest Biosolids ten year earlier: public upset about biosolids recycling projects, changing and inconsistent state and local regulations, and lack of a unified response to concerns.

NEBRA formed as a 501(c)(3) charitable non-profit registered in New Hampshire and with the U. S. Internal Revenue Service (IRS). Stakeholders hotly debated its mission resulting in carefully-chosen wording: "to *promote*" the environmentally sound recycling" of biosolids and other residuals in New England (ultimately, the scope of the organization reached into eastern Canada – Quebec and the Maritimes). Founding members of NEBRA chose to emphasize *promoting* biosolids and other residuals, directing a proactive approach.

An interim Board of Directors raised initial funds for NEBRA in early 1998, and, in February, hired Ned Beecher as Coordinator (subsequently to become Executive Director). 2017 is NEBRA's 20th year.

Like other non-profits, NEBRA is continually challenged in securing adequate funding. While there are dozens of organizations and individuals who have steadfastly and continuously supported NEBRA with their memberships year after year, new memberships come slowly. NEBRA's most significant offering is not a tangible benefit you can touch. Yes, members get conference and workshop discounts and such, but the biggest benefit of NEBRA and other regional groups is just this: *being there*. Without voices speaking for biosolids and other organic residuals, who would support them? Who would provide the myriad technical details – the scientific basis and the experience of thousands of wastewater treatment programs? If every biosolids program had to compile and explain all the information there is about biosolids, it would be expensive and inefficient. Having one thorough repository of information and experience in each region makes sense. And, for this, instead of spending tens of thousands of dollars, each public wastewater agency, engineering firm, biosolids management company, and biosolids user pays just a few hundred dollars a year. The cost efficiency of a collaborative effort is extraordinary.

Like similar groups, NEBRA has become a knowledgeable resource and advocate for our members. It is like an insurance policy – something members pay for that is ready and able to help them at a moment's notice. As one leading biosolids management company president said, "In six months or a year from now, when the next public concern about the safety of biosolids is raised in the press or on the internet, I want NEBRA to be there to provide accurate and scientifically based information." That's what "being there" is all about. But this benefit is challenging to describe.

If Northwest Biosolids had not existed, biosolids bans might now be spreading through counties in Washington State. In New England, when NEBRA formed, there were about 40 towns passing biosolids bans. By the early 2000s, that trend had stopped. In New Hampshire, NEBRA collaborated with farming and other groups to oppose such local ordinances and negative legislation. NEBRA led the effort for the NH Legislature to proactively adopt the term "biosolids" in state law. When public concerns flared up in Nova Scotia around the new Halifax biosolids program, NEBRA supported local stakeholders with information, workshops, and media outreach. For the past decade, there have been few significant public upsets about biosolids use in the NEBRA region.

NEBRA is not the sole cause of this turnaround; other stakeholders and common sense have made biosolids recycling a normal activity. But NEBRA has served a critical role in helping biosolids/residuals management proponents work together by providing a central repository of information and coordinating efforts.

Now, in the past two years, NEBRA can point to two key successes in advancing biosolids recycling in the region's most influential state: Massachusetts. These efforts, led by NEBRA, are helping keep open and expand markets and end uses for biosolids through regulatory reform. What is the dollar benefit to members? It's hard to put a number on it. But keeping regulations from unnecessarily restricting biosolids use means lower costs for local agencies seeking outlets for biosolids.

Since the 1980s, Massachusetts (MA) "sludge" regulations included ceiling limits on trace elements (heavy metals) that were different from – and preceded – those in the federal Part 503 regulations. The MA limit for molybdenum in biosolids was set at 25 mg/kg generally and 10 mg/kg if the biosolids were applied to forage crops. This conservative, cautionary limit had no basis in science, but, despite recommendations that the MA Department of Environmental Protection (MA DEP) change it, it remained in place until 2016. The unnecessary Mo limit resulted in very little of Boston's heat-dried biosolids being used in MA, despite strong interest by farmers and others.

To address the concern, NEBRA coordinated with stakeholders and held a free workshop attended by MA DEP staff and others at which leading scientists (e.g. Dr. Rufus Chaney, USDA; Dr. George O'Connor, Univ. of FL; and Rick Stevens, U. S. EPA), some of whom had done the research on biosolids Mo and soils, presented their findings and recommendations. Continued exchange of information with MA DEP staff – and gentle urging for action by NEBRA – led to a regulation change that became effective in 2016, with a new Mo standard of 40 mg/kg, the number identified by the risk assessment presented at the workshop. Now, most Boston biosolids can be applied in state. The monetary value of lowered transport costs alone is substantial.

Similarly, in the past two years, NEBRA has led scrutiny of new regulations created rather suddenly by the MA Department of Agricultural Resources (MDAR). These rules severely restrict applications of phosphorus (P) on any soils for which tests indicate adequate or excessive total P. While NEBRA and its members strongly agree with the environmental importance of restricting P applications to protect water quality, the particular MDAR regulations are poorly written, broad, and blunt. They stymy the use of organic residuals (biosolids, food scraps, composts, digestates, etc.), which the state is actively diverting from landfills. It's a classic example of one well-intentioned regulatory program conflicting with another, because of lack of understanding of the nuances.

In response, NEBRA proposed a workshop on P in organic residuals, including biosolids, to advance regional understanding of the subtleties of P dynamics. The idea blossomed so that UMass Extension took it on, producing the workshop on November 2, 2016; 135 people attended from MA and surrounding states (see https://www.nebiosolids.org/managing-p-in-organic-residuals-applied-to-soils). Key stakeholders' understanding was advanced.

Following on that success, NEBRA continues to advocate for a more thoughtful, nuanced policy toward applications of P in organic residuals in MA, arguing that the benefits of their use are significant and that excess P applied in them is often not as environmentally important because of the binding effects that come with biosolids (Fe, Al, Ca, etc.). NEBRA is developing an industry best practice document to guide its members and other stakeholders in managing P in organic residuals. This document could be a template for regulatory guidance, helping ensure UMass Extension and MDAR also rely on P site indices that include P source coefficients and tests for P saturation and water extractable P to determine application rates of composts, digestates, and biosolids.

If NEBRA had not existed, these two regulatory matters would likely have continued without scrutiny, effectively shutting down local markets for local biosolids products.

Learn more at www.nebiosolids.org.

Mid-Atlantic Biosolids Association (MABA)

The Mid-Atlantic Biosolids Association (MABA) also formed in 1997. It is an advocacy organization for biosolids resource recovery, with a mission to communicate the benefits of biosolids resources within the biosolids community and the communities they serve. MABA works to coordinate a regional voice for communicating philosophy and practices to enhance biosolids management; provide continuing education and information-sharing opportunities; effectively advocate for biosolids with regulators, public officials, key professionals, and the public; equip members with tools for development and implementation of laws, regulations, and guidelines; and support biosolids research.

MABA currently contracts with Bill Toffey, Effluential Synergies, to provide executive services. Cheryl Paulin serves as MABA's membership and committee communication lead, and Charlie Alcorn is the online and news services coordinator.

One of the primary services MABA has offered recently is the dissemination of information through its blog, providing technical commentary on a variety of topics of interest to the biosolids community and showcasing positive articles about the recovery of resource value from biosolids. MABA also coordinates an Annual Meeting and Symposium presenting technical information and networking opportunities, as well as other symposia and biosolids sessions at state water environment association conferences.

Learn more at www.mabiosolids.org.

California Association of Sanitation Agencies – Biosolids Program

The California Association of Sanitation Agencies (CASA) is a non-profit trade association which was formed in 1956 and now represents more than 90% of the sewered population of California. CASA advocates on behalf of its members on all issues related to wastewater treatment to regulators, legislators, and all interested stakeholders. CASA has 7 full time staff led by Executive Director Roberta Larson, plus contracts with both a federal and state lobbyist. CASA added a Biosolids Program Manager in 2001 and hired Greg Kester to fill that role in 2006. The position evolved into the Director of Renewable Resource Programs in 2013 and now encompasses biosolids, renewable energy, recycled water, and all recyclable products of wastewater treatment.

A key focus for CASA is increasing awareness and understanding among Congress, the state legislature, and regulatory agencies regarding how the wastewater sector helps achieve many of the recently enacted legislative mandates intended to mitigate climate change.

State regulators have provided an exemption from solid waste regulations for POTWs that receive anaerobically digestible material (i.e., FOG, food waste, etc.) for co-digestion as long as they comply with permit conditions imposed by the State Water Boards. CASA worked closely with multiple state regulatory agencies to secure this exemption to ensure that only one state agency regulates wastewater plants. Such co-digestion is an ever-increasing practice in California, since it helps mitigate climate change by diverting raw organics from landfills and increases renewable energy.

CASA is currently working with the California Department of Food and Agriculture on the Governor's 2015 Healthy Soils Initiative (HSI). Biosolids help achieve every one of the HSI objectives, and CASA is ensuring those benefits are understood by all.

CASA has assembled a research team and developed a proposal to demonstrate and quantify the benefits of using biosolids to reclaim fire-ravaged land. The project has been accepted as a Targeted Collaborative Research project by WE&RF and funding is actively being sought to execute the work. Severe fires have been occurring at an increasing rate across the country and the globe, and biosolids restore organic matter and soil health, reduce erosion, and increase vegetative growth. CASA has also worked closely with regulators and university researchers to advance the use of biosolids to reclaim superfund and other mine sites, brownfields, and wetland restoration.

In the judicial arena, CASA was a co-plaintiff with several of our member agencies in long-term litigation challenging a Kern County ordinance prohibiting the land application of biosolids. The lawsuit, led by the City of Los Angeles, yielded a highly favorable trial court ruling in November 2016 that the ordinance violates state law and exceeds the counties police power. There has been no word at the time of this writing on whether Kern County will appeal the decision.

Learn more at http://casaweb.org/renewable-resources/biosolids/# .

Virginia Biosolids Council (VBC)

The Virginia Biosolids Council (VBC) formed as an organization to support recycling of biosolids in Virginia through information and education on the beneficial use and safety of biosolids. The VBC membership includes municipal wastewater treatment facilities and companies that work to keep Virginia waters clean and offer quality biosolids products for land application. Members abide by a Code of Good Practice that was set forth to offer voluntary guidance to Council members for their responsibility in maintaining high performing biosolids beneficial use programs. This Code has helped to reduce biosolids complaints from 203 in 2008 to 29 in 2016 and regulatory warnings from 11 in 2008 to 0 in 2016.

Just recently, at the end of 2016, in a landmark case, the Richmond, Virginia, Circuit Court upheld the new Department of Environmental Quality's (DEQ) Biosolids Regulations against a Potomac/Shenandoah Riverkeeper legal challenge. The facilitation done by the VBC played a pivotal role in this decision. It began many years ago with the service and participation of VBC among the various committees created to first support the transition of the biosolids program from the Virginia Health Department to Virginia's DEQ, and then through participation in the various technical committees that supported the development of the comprehensive regulations.

Like the other regional biosolids groups, VBC has fought the local battles and prevailed for biosolids recycling.

Learn more at www.virginiabiosolids.com/.

STRENGTHENING THE BIOSOLIDS NETWORK IN NORTH AMERICA

Solids management is one of the largest costs incurred by WRRFs. All available options for end use or disposal are at least occasionally challenged here or there. Application to soils can be environmentally sound and cost efficient, but it will always be particularly subject to public concerns and requires continual outreach and education. It's complex.

The goal that many in biosolids management share is the long-term sustainable recovery and utilization of the resources in biosolids. To achieve this goal, not only must WRRFs conduct high-quality biosolids management operations, but they must continue to provide information and support research and science and outreach and education. Staffed regional biosolids groups have proven to be the most cost-effective structures for this work.

But commitment to the proactive information compilation and outreach work of regional groups is not as strong as it might be. Within the regions where biosolids groups operate – and including at the national level in the U. S.– there are less than 6 FTEs working in independent, objective non-profit roles to advance biosolids management and policy.

Organization	Pop. for region	FTE 2016	million / FTE	Annual budget/ Program cost
CASA	39,250,017	1.0	39.25	
MABA	59,366,577	0.6	59.37	$35,000
NEBRA	24,916,525	0.8	31.15	$95,000
NW Biosolids	22,810,999	1.0	22.81	$213,000[1]
VBC	8,411,808	0.5	16.82	
NACWA (U.S. reach)	100,000,000	0.25		
WEF (U.S. reach)	300,000,000	1.5		
U. S.	**323,148,587**	**5.7**	**57.19**	

Meanwhile, the number of FTEs working on biosolids at U. S. EPA and state regulatory agencies (other objective, independent organizations) has dwindled substantially over the past decade. This decline is on top of the decline noted in the 2007 national biosolids survey (NEBRA et al., 2007).

It appears that a reasonable level of effectiveness can be achieved when regional biosolids groups have dedicated staffing of 0.1 – 0.6 FTE per million population. That is not a large commitment and does not require that much funding. But that minimum level of commitment is key to the ongoing defense and success of resource recovery from biosolids.

LEARNING FROM OTHER PRODUCT TRADE ASSOCIATIONS

Most product categories have trade associations, at least at the international and national level. For example, for fertilizer, there are at least two large national groups: AAPFCO and The Fertilizer Institute. The latter "is the voice of the fertilizer industry, representing the public policy, communication, stewardship and sustainability and market intelligence needs of fertilizer producers, wholesalers and retailers as well as the businesses that support them with goods and services." Replace the word "fertilizer" with "biosolids" and you get a sense of what the regional biosolids groups provide.

[1] This includes $96,800 of funding for biosolids research at local universities.

There are also state groups, such as the Organic Fertilizer Association of California and the Florida Fertilizer and Agrichemical Association (FFAA). The focus of FFAA echoes that of regional biosolids groups: "the Florida Fertilizer & Agrichemical Association has branched out to represent its fertilizer, limestone, crop protection, and ag biotech member companies professionally in the legislative, regulatory and public opinion arenas. FFAA's main objectives today - To promote the responsible use of plant protection and nutrient products in Florida" (Florida FAA, 2017).

We can learn from such trade associations in other product categories. For example, the Semiconductor Industry Association (SIA) is huge, of course: it's central to the modern economy and generates huge profits in comparison to the biosolids industry. But their vision – or that of other trade associations – can serve as a template for the vision of biosolids associations:

The Semiconductor Industry Association (SIA) is the voice of the U.S. semiconductor industry.... Semiconductors – microchips that control all modern electronics – enable the systems and products we use to work, communicate, travel, entertain, harness energy, treat illness, and make new scientific discoveries. The semiconductor industry directly employs nearly a quarter of a million people in the U.S. In 2015, U.S. semiconductor company sales totaled $166 billion, and semiconductors make the global trillion dollar electronics industry possible. SIA seeks to strengthen U.S. leadership of semiconductor manufacturing, design, and research by working with Congress, the Administration and other key industry stakeholders to encourage policies and regulations that fuel innovation, propel business and drive international competition

Here's a rewrite for biosolids (*wording changes shown in italics*):

The *Biosolids Association* is the voice of the *biosolids* industry *in the _______ region.... Biosolids – soil amendments and fertilizers manufactured from recycled municipal wastewaters* – enable systems and products *that protect water quality, recycle nutrients and carbon, and create sustainable, ecological communities*. The *biosolids* industry directly employs _____ [we don't know how many – that's another piece of work to be done!] in the *region U.S. / Canada / etc.* In 2015, U.S. *biosolids* sales totaled $_____ [we don't know how much – that's another piece of work to be done!], and *biosolids help* make the *national billion?*-dollar *farming* industry possible. *The Biosolids Association* seeks to strengthen leadership of *biosolids* manufacturing, design, and research by working with *states, provinces,* Congress, the Administration and other key industry stakeholders to encourage policies and regulations that fuel innovation, propel business, and drive competition.

Trade associations, including the regional biosolids groups, are formed to take control of an industry's future. With EPA and states continuing to reduce staffing for biosolids, where will the knowledge and experience reside? As of 2002, "each of the 10 EPA Regions have between 0.2 and 2 full-time employees (FTEs), and a total nationwide of 8.8 FTEs, working in all areas of biosolids management. The EPA ORD has 2 FTEs devoted to the program, and EPA headquarters has 4.8 FTEs" (National Research Council, 2002). Since 2002, that number has diminished considerably, likely down by at least half. State regulatory agencies have also shifted priority away from biosolids. Add to that the retirement of leading, highly-experienced EPA, USDA, and state staff, and the result is not only fewer FTEs, but less experience in the remaining FTEs devoted to biosolids management and policy. Who is making up for that shortfall? We cannot rely on government to be strong voices for biosolids when needed; this has been experienced over and over again in local conflicts. It's not their role. The burden is on us, on the regional biosolids groups, on WEF, on NACWA - on our profession.

WHAT MORE NEEDS TO BE DONE?

You might ask, what more needs to be done in the area of biosolids management and policy? After all, it seems biosolids recycling is here to stay, right? Not necessarily. And certainly not everywhere. There are ongoing threats to its viability in British Columbia, western New York, Maryland, and other local areas in the mid-Atlantic. Because of historic institutional bias, Connecticut has no formal program to allow its utilities to apply biosolids to soils in the state. There are bans on biosolids use in California counties, New York, and New Hampshire towns, as well as curbs on use in many other locales around the U. S. and Canada.

And, while some biosolids products sell for more than $50 per yard, most are given away. This wide discrepancy indicates an immature, under-developed market. Biosolids regional groups help further develop markets by promoting quality products, fending off unnecessary restrictions, and advancing public understanding.

There is plenty of work to be done. Other examples:

- Interact more with agronomy and environmental groups, to advance their understanding of the value of biosolids recycling to local economies and soils.
- Interact more with other trade associations in related fields (e.g. solid waste or soil or mulch groups). (We do have interactions with a few key compost and biogas groups, but that's limited.)
- Compile data and track progress – We still know little about how we are doing on a continent-wide scale. The last robust data for the U. S. is from 2004 (NEBRA et al., 2007). There is no robust data for the management of Canadian biosolids. And related data would be helpful, such as data on the critical success factors noted in *Enabling the Future* (see below).
- Adapt programs to address issues that are uncovered in tracking data; for example, more training may be needed as biosolids professional skills and knowledge are lost to retirements.

Critical Success Factors (from WEF, 2013)

To manifest the developing paradigm of maximizing biosolids resource recovery, biosolids management professionals will need to pay attention to the following critical success factors:

- The skills and knowledge of biosolids management professionals;
- The age distribution of biosolids management professionals;
- The strength and capacity of biosolids-focused organizations;
- The status (number, tone, and complexity) of federal and state policies, laws, and regulations pertaining to biosolids;
- The quality of biosolids products (trace elements, trace chemicals, pathogens, nutrient balance, odors);
- The quality of biosolids management, including the constant of continual improvement;
- The level of federal and state policy support for biosolids resource recovery;
- The level of agricultural, conservation, and environmental group support;
- The level of public support (trust);
- The levels of public and private funding for biosolids management infrastructure, training, and operations; and
- The trend in biosolids resource recovery and the rate of biosolids recycling.

In addition, the *Enabling the Future* report outlined "five steps toward maximum resource recovery" from biosolids:

1. • Increase the capacity and skills of the biosolids profession by providing more biosolids management information, training, and other support.
2. • Advance policies, legislation, and regulations that allow for – and create incentives for – resource recovery.
3. • Continue to improve the quality of biosolids and biosolids management programs of all sizes and in all regions.
4. • Increase communications and outreach with environmental and agricultural groups and the general public about the beneficial uses of biosolids and the importance of maximizing resource recovery from these materials.
5. • Advocate for increased funding for these initiatives, as well as for infrastructure needed to maximize resource recovery.

Biosolids groups are the best – and possibly only – organizations that will ensure this work gets done. If we don't take this on, if we don't strongly advocate for quality biosolids management programs, who will? Do we want to control our own future?

CONCLUSION

The current regional biosolids groups, along with the efforts of NACWA and WEF and the WEF state biosolids committees around the continent, are effective – and cost-effective. Where they have focused their efforts – California, Virginia, New England, the Northwest – biosolids recycling is advancing and public upsets are few and far between. In those areas, significant challenges to biosolids use have been rebuffed. In contrast, some areas without the focus of such organizations have, in recent years, been embroiled in biosolids controversy: Texas is still feeling the impacts of years of public conflict that led to significantly more restrictive state regulations.

Our biosolids profession has to do more to take our future in our own hands. Throughout all industries, this is done through trade and professional associations. Such associations are only truly effective when they have paid staff, when it is someone's primary responsibility to advance sustainable, quality biosolids management and products. The late 1990s was likely the high point in numbers of FTEs at U. S. EPA, state agencies, and in biosolids associations. There has been a steady decline ever since. When U. S. EPA funding trailed off in 2011, WEF saw a net loss of biosolids staffing of at least one FTE. And, from then through 2016, not one FTE has been added to any of these organizations. Now, in 2017, WEF has added a full-time biosolids position.

More is needed. The documented success of the current regional biosolids groups – as described above – can and should be replicated elsewhere. Such organizations deserve strong support. The existing organizations hope to see – and will always help in – the formation of other regional groups.

REFERENCES

AAPFCO, http://www.aapfco.org/
Florida Fertilizer and Agrichemical Association, http://www.ffaa.org/, accessed 1/12/2017.
Illinois Fertilizer & Chemical Association, https://ifca.com/
Moss et al., 2013. *Enabling the Future: Advancing Resource Recovery from Biosolids.* Alexandria, VA: Water Environment Federation.
National Research Council. 2002. *Biosolids Applied to Land: Advancing Standards and Practices.* National Academy of Sciences.
NEBRA et al. 2007. *A National Survey of Biosolids Regulation, Quality, End Use, and Disposal.* https://www.nebiosolids.org/about-biosolids/
Mulch & Soil Council, http://www.mulchandsoilcouncil.org/
Organic Fertilizer Association of California, http://www.organicfertilizerassociation.org/
Sustainable Food Trade Association, http://www.sustainablefoodtrade.org/

Not All Dryer Products are Created Equal

Lisa Boudeman[1]

[1]Material Matters

Keywords: biosolids, thermal drying, product quality

EXECUTIVE SUMMARY

Thermal drying is an accepted wastewater industry process for generating a biosolids product that meets the United States Environmental Protection Agency (USEPA) 40 Code of Federal Regulations (CFR) Part 503 Class A standards for both pathogen reduction (PR) and vector attraction reduction (VAR). Utilities and engineers alike often consider biosolids drying as a processing technology that will generate a high quality, marketable product. While the wastewater treatment industry typically groups all biosolids drying technologies together into one category for PR and VAR compliance, not all biosolids drying technologies and dried products are created equal. Each dryer technology operates under distinct conditions (temperature, heating method, screening, recycle flows, etc.), which greatly influences the characteristics of the dried biosolids product.

Unfortunately, many thermally dried biosolids products are not desirable or marketable in beneficial use markets. Experience from numerous utilities has shown that a Class A dried product can be just as costly to manage as Class B or unclassified biosolids products. Therefore, it is critical for utilities to understand that installing a biosolids dryer is not an end all solution to developing a sustainable biosolids management program. The utility must fully understand the dryer technology that will be installed, as well as the corresponding product that will be generated based on their plant processes. Furthermore, a utility must recognize that in some cases, biosolids drying, regardless of the dryer processing type, may not be a suitable processing technology to meet their economic and beneficial use goals.

INTRODUCTION

Thermal drying is an industry standard biosolids processing method in which dewatered wastewater solids are dried by direct or indirect contact with hot gases to meet Class A Exceptional Quality (EQ) Standards (assuming pollutant concentrations are below EPA 503 Table 3 concentrations). Thermal drying meets EPA 40 CFR Part 503 regulatory requirements for Pathogen Reduction through Class A - Alternative 5, process to further reduce pathogens (PFRP) heat drying, where biosolids are heated to 80°C and, in turn, dried to greater than or equal to 90% total solids (TS). Additionally, thermal drying typically meets EPA Part 503 regulatory requirements for vector attraction reduction (VAR) via Option 8, drying solids to greater than or equal to 90% TS.

Utilities select thermal drying to meet their individual goals, which are influenced by a combination of economic, social, and environmental pressures. Major drivers for installing dryer technology include reduction of beneficial use (or disposal) costs, generation of a Class A EQ product, flexibility (i.e. ability to enter into non-agricultural markets), anaerobic digestion gas utilization, reduced storage needs, and abatement of nuisance odors. In a majority of cases, utilities choose thermal drying, at least in part, for economic reasons. Thermal drying reduces the total volume of product generated, which in turn, reduces costs associated with transportation, disposal, land application and/or other beneficial use outlets. In other cases, utilities select thermal drying to meet Class A EQ requirements, which allows for greater flexibility in the use and management of the product. When used in the agriculture market, Class A EQ products typically have fewer setback requirements and fewer notification requirements than Class B products. Alternatively, utilities may feel pressure to move out of the agricultural market due to the increasing amount of regulatory pressures, which include phosphorus-related restrictions, winter land application bans, and setback requirements; or population growth and expansion, which can reduce the size and availability of local farms. Non-agricultural markets include lower value markets such as soil blending and alternative fuel source, and higher value markets that include fertilizer-blending, golf courses, and sod production.

While certain thermally dried biosolids products have successfully entered into non-agricultural markets, many do not have the physical characteristics (i.e. low dust, hardness, no inorganic materials) or chemical characteristics (low odor, nutrient profile) necessary to advance into these markets. Thermally dried products vary considerably depending on many factors, including dryer type, type and degree of solids digestion, and post processing. Figure 1 depicts the wide range of physical characteristics found in thermally dried wastewater treatment solids.

Figure 1. Biosolids with a wide range of physical and chemical characteristics, dependent on upfront solids processing, dryer type, and post-processing.

BIOSOLIDS QUALITY FACTORS

Dryer Type

There are three major types of biosolids dryers that generate Class A biosolids product: direct and indirect. Direct dryers dry wastewater solids by sending hot gases into direct contact with biosolids to evaporate moisture. The most common direct dryers utilized today are rotary drum dryers and belt dryers. Direct dryers commonly screen the large and small particles out of the dried product, which generates a uniform and consistently sized final product. The screened product is recycled to the front of the dryer to act as a seed to be blended with additional cake solids. Product recycling helps to reduce the "sticky" phase (45-60% TS) during which biosolids can be very difficult to manage, and increases the hardness of the final product.

In indirect dryers, the wastewater solids are separated from direct contact with a hot medium (typically steam or oil) by a metal surface. The biosolids are dried by contact with a series of heated metal plates or paddles connected by a centralized turning shaft. The movement of the shaft moves the solids along the hot surface to enhance moisture evaporation. Common indirect dryers utilized today include tray and paddle dryers. Indirect dryers may or may not have product screening following the dryer. Products that are not screened will have fines to make the product dusty, and will contain large granules that can be difficult to manage. Figure 2 shows product processed through a Kruger Biocon belt dryer, an Andritz DDS drum dryer, and an Andritz Gouda indirect dryer.

Figure 2. A comparison of direct and indirect dryer products. From left to right is are products from a Kruger Biocon belt dryer, an Andritz DDS drum dryer, and an Andritz Gouda indirect dryer.

Type and Degree of Digestion
The type and degree of digestion will influence the odor profile, nutrient content, hardness, and dustiness of a thermally dried biosolids product. Digestion, and its influence on dried product quality, can be divided into three categories: no digestion, aerobic digestion or extended aeration, and anaerobic digestion. The fibers found in undigested solids makes it very difficult for particles to compact, which results in a dried product that is soft and dusty, with low bulk density. Undigested solids also have the greatest amount of volatile solids (>75% TS), the organic compounds that are used as food for microbes or potential vectors, relative to other products. Therefore, these products will typically have a more offensive and more intense odor relative to other products, especially when rewetted. Because undigested solids do not experience a high degree of nitrogen destruction, undigested solids will typically have the greatest amount of nitrogen relative to digested products.

Aerobically digested wastewater solids typically have some level of volatile solids reduction, resulting in a cake product with lower volatile solids content (~65-70% TVS). Dried aerobically processed biosolids will have an odor profile that is less offensive and less intense than undigested, thermally dried products. Aerobic digestion breaks down the fibrous particles found in primary wastewater solids, which allows the particles to form more tightly together, resulting in a harder pellet with a higher bulk density.

Anaerobic digestion typically enhances both the physical and chemical characteristics of thermally dried biosolids products. Anaerobic digestion conventionally reduces volatile solids by at least 38%, which generates a product with a volatile solids content that is typically well below 65%. As such, thermally dried, anaerobically digested biosolids products will have an odor profile that is typically non-offensive, with a low to moderate intensity, and characterized as a coffee or chocolate powder smell. The additional volatile solids (and fiber) destruction characterized by anaerobic digestion, allow the solids particles to bind closer together, to form a harder, denser granule. Figure 3 shows two products with different digestion processes processed through the same dryer; the product on the left is undigested, whereas the product on the right is processed through biological aerated filtration.

Figure 3. A depiction of how digestion can impact biosolids quality. The product on the left is undigested, whereas the product on the right is processed through biological aerated filtration.

Post Processing

"Post processing" refers to optional processing steps following the drying process to improve product quality that includes screening, crushing, pelletizing, and the addition of dedusting oil. Screening is a process in which biosolids are sifted through a series of plates with holes of a certain diameter to remove both the coarse and fine particles. In some dryers, screenings are recycled back to the head of the dryer and blended with the cake product to avoid the "sticky" phase of the biosolids product. In other cases, the screened particles are disposed in a landfill.

Pelletizing is a process in which the biosolids produced from the dryer goes through a pellet mill to compress the particles (i.e. increase the bulk density of the product), and to make a more uniform product. Figure 4 shows the process of a product transformed from a low bulk density (~20 lb./cu. ft.) noodle, to a high bulk density pellet (>40 lb./cu. ft.). Dedusting oil is mineral-based oil sprayed onto the finished biosolids product to suppress dust.

Figure 4. Dried biosolids dried on a Kruger Biocon belt dryer (left) are crushed (middle) and pelletized (right) to increase the density and reduce the volume of biosolids product.

CASE STUDIES

Ocean County Utilities Authority. Ocean County, New Jersey

The Ocean County Utilities Municipal Authority, located in Ocean County, New Jersey, operates three wastewater treatment plants: the South Plant (average daily flow (ADF) 7.2 mgd), the Central Plant (ADF 21.2 mgd), and the North Plant (ADF 21.7 mgd). Each treatment plant operates very effective mesophilic anaerobic digesters, which typically operate between 60 and 65% volatile solids destruction. The solids from each are ultimately transported to the Central Plant where the biosolids are dried in an Andritz drum dryer system (DDS) 40 rotary dryer.

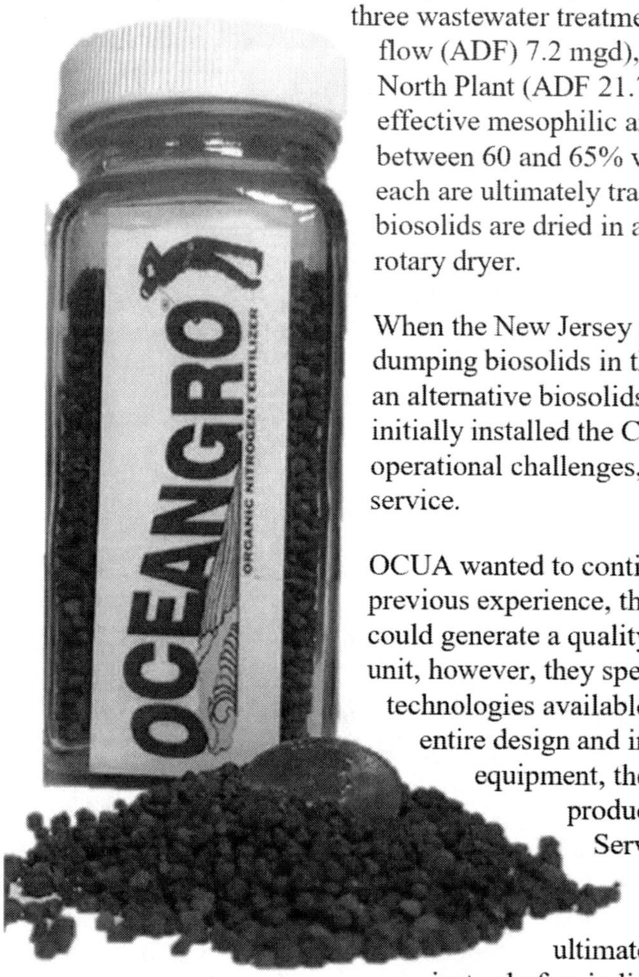

When the New Jersey regulations prohibited landfilling and ocean dumping biosolids in the 1980s and 1990's, OCUA needed to find an alternative biosolids management method. The OCUA initially installed the Carver-Greenfield process, which, due to operational challenges, was removed after only a few years in service.

OCUA wanted to continue with drying technology, and with their previous experience, they knew they wanted to install a dryer that could generate a quality product. Before installing another drying unit, however, they spent time learning about each of the drying technologies available in the market at the time. Throughout the entire design and installation of the dryer and ancillary equipment, the chief focus was on the quality of the final product. Mr. David Ertel, the director of Central Services at the time, understood that the "raw material (i.e. the material entering the dryer) is what makes the product". They ultimately chose to install a three pass direct dryer instead of an indirect dryer because the direct dryer generated a harder, more uniform pellet that can be distributed into high value markets (e.g. fertilizer blending).

In order to prevent inorganics from entering the final product, the solids are screened multiple times throughout the treatment process. Inorganics are first removed with a coarse three-inch screen at the headworks, and again via a fine rotary screen immediately preceding gravity belt thickener. The second screen avoids rag build up in the digester, and prevents inorganics from contaminating the final product. The digested product is then dewatered via an Ashbrook 2 M belt filter press to generate a cake that is 16-17% TS. The product is then transferred by a shaftless screw conveyor to the Andritz rotary dryer. The dried product is screened in a rotary screener to a size guide number (SGN) of 150 and transported to an adjacent silo. The final product is sprayed with a light oil to seal the dust as it leaves the product silo and enters into the truck. By waiting to spray Armoz mineral dedusting oil on the product just prior to distribution, this avoids the oil from being absorbed into the product.

OCUA routinely conducts quality analysis to maintain product quality. The facility regularly tests for granular sizing and hardness with a rototap screening machine, which measures the distribution of pellet sizes and tests for generation of fine particles when the product is subject to force. They also test for product density by using a fertilizer density scale.

Initially OCUA distributed (i.e. sold) the product into the bulk agriculture market. Due to the consistent product quality and ongoing marketing efforts, they have sold an increasing volume of product into high value markets including fertilizer blending and golf courses. Over time, they have also adjusted the screen size of the dried product to generate a smaller pellet that was demanded by their customers.

Camden County Utilities Authority. Camden County, New Jersey
The Camden County Municipal Utilities Authority, located in Camden County, New Jersey, operates the Delaware One Water Pollution Control Facility (WPCF). The Delaware One WPCF has an average daily flow of 54 mgd, and generates roughly 15,000 dry metric tons of solids each year. In the mid-2000s, the CCMUA elected to install a dryer to divert their solids from landfill to a beneficial end use. CCMUA installed three indirect Komline-Sanderson paddle dryers at the Delaware One WPCF in 2012, with the goal to divert solids as an alternative fuel source for a cement plant.

After starting up the dryers, the CCMUA ran into a number of unforeseen challenges. The undigested solids contained a large volume of fibrous material, which clogged various parts of the dryer. This dryer manufacturer did not encounter this challenge in previous installations because this was the first installation in which primary solids were fed directly to the dryer. To address this challenge, the utility installed a strain press for fiber removal following the primary clarifiers, and pumped to a second holding tank. While the strain press helped ameliorate the problem, fibers continue to be a challenge at the plant.

Additionally, the product was very fine and lightweight, which made it very difficult to convey. To address this, the dryer operator needed to revise the dryer operation to add a "dense phase" within the dryer to try to increase the bulk density. The fine particle size caused bridging in the pneumatic conveyance system at the cement plant, in which the product would be stuck coming out of the transportation trucks, and/or clog the plant's fuel feeding system. Therefore, although undigested biosolids would hypothetically be the ideal alternative fuel source because the energy has not been removed through digestion, this product did not meet the cement kiln's needs and is no longer accepted there.

Because the CCMUA's biosolids are not digested, the product has a relatively intense, unfavorable odor. If managed in appropriate areas, the product may be suited for certain agricultural operations.

CCMUA plans to install anaerobic digesters at the Delaware One WPCF in 2018. Based on Komline-Sanderson's experience with other wastewater treatment plants, it anticipated that the addition of anaerobic digestion would substantially improve the product's quality by reducing a majority of the fibrous material, which will prevent clogging and will allow the product to compact better. Additionally, the volatile solids destruction will improve the product's odor profile, and allow the product to be readily distributed into other beneficial use markets.

SUMMARY AND CONCLUSIONS

Although biosolids drying is grouped into one Class A PR Alternative and one VAR option, biosolids drying technologies vary greatly in their operating parameters and the characteristics of the final products generated. Every step of the wastewater treatment process – from influent characteristics, to digestion, to post-dryer processing – affects biosolids quality, and, in turn, the beneficial use outlets that can receive the product. Many utilities have experienced that installing a dryer is not an end-all solution for finding a beneficial use outlet and/or reducing biosolids management costs; in some cases, "go to" outlets (e.g. alternative fuel, agriculture, or landfill) may not be viable for some dried products.

To minimize unanticipated challenges associated with installing a dryer, a utility must carefully establish its goals for installing a dryer and managing the thermally dried biosolids. The utility must diligently evaluate the products generated from each dryer technology under consideration; if possible, conduct a pilot with a sample of its product, or at a minimum, obtain products with similar up-front processing. If the product(s) do not have the physical and/or chemical characteristics suitable for local outlets, then the utility should consider alternative biosolids processing methods that may be better suited to meet the stated economic and beneficial use goals.

Merging Innovation and Resource Recovery into Daily Operations

By: Manon Fisher[1], Natalie Sierra[2], Ben Jordan[3], and Karri Ving[1]

[1]San Francisco Public Utilities Commission
[2]Brown and Caldwell
[3]Independent Consultant

Key Words
Resource recovery, innovation, biosolids, sustainability

Abstract
The San Francisco Public Utilities Commission (SFPUC) Wastewater Enterprise (WWE) has recently rebranded itself as a resource recovery utility. Within this new identity comes a mission to identify technologies and/or program areas that can be pursued to further goals associated with recovering resources from wastewater, biogas, biosolids, and nutrients – all previously thought of simply as wastes. Nationally, and in California in particular, this idea of resource recovery has led to the flowering of a number of innovative programs and technologies. The WWE has identified a need to assess these opportunities and identify those that best support the WWE and the core mission and values of the SFPUC.

To this end, WWE has developed screening tools to assess resource recovery concepts and potential projects. These tools have been used by the recently formed Resource Recovery Roundtable (R2 Roundtable), a cross-disciplinary group established to evaluate and prioritize projects that merit further consideration by WWE management. The intent is to have tools readily available for WWE staff to identify and communicate the information that is essential to perform an initial evaluation. When completed by the R2 Roundtable, each evaluation provides a written record and transparent demonstration of how the agency would fulfill its resource recovery objectives while making responsible use of limited staff and resources. This paper discusses the tools developed, and how the process has evolved since initial implementation.

Introduction
The San Francisco Public Utilities Commission (SFPUC) provides water, power, and sewer services to the residents of the City and County of San Francisco. Within the SFPUC, the Wastewater Enterprise (WWE) is responsible for the operations and maintenance of the sewer system, and has recently rebranded itself as a resource recovery utility. Within this new identity is a mission to identify technologies and program areas that further recover resources from wastewater, biogas, biosolids and nutrients - all previously thought of simply as wastes.

WWE staff has identified a need to assess these opportunities in a comprehensive manner and identify those opportunities that best support the utility's core mission and values of sustaining the resources entrusted to our care.

To accomplish this task, WWE Management established a Resource Recovery Roundtable (R2 Roundtable) in early 2014 as a means of facilitating collaborative discussion and dialogue across resource categories, focusing on the potential for multi-benefit strategic initiatives. In 2015, resource recovery staff began the process of developing the framework and screening tools to assess incoming resource recovery concepts and potential projects across the four key resource categories. These tools were intended for use by WWE staff and the R2 Roundtable to evaluate projects that merit further consideration by WWE management. In addition, the framework helps formalize and streamline the innovation process while ensuring transparency and access to all WWE staff.

Internally to WWE, the process of using the screening tools has been called the "Resource Recovery Innovation Platform," because it describes how the WWE reviews, evaluates and decides upon innovative processes, ideas, and technologies within its core business. In some cases, these may be concepts that are new to WWE, but are established elsewhere. In other cases, topics may be presented at conferences or presentations that are more theoretical in nature. Ultimately, concepts that are vetted through the Innovation Platform are documented and memorialized in an online SharePoint site and made available to the entire organization. These topics may ultimately be implemented within the WWE or through partnership with external organizations, but regardless of the implementation the process is made accessible to everyone.

Developing the Innovation Platform

For most wastewater treatment plants innovation takes place regularly, even if it is not outwardly called "innovation". The intention behind developing a resource recovery framework and innovation platform was to demystify this process and open it up to all WWE staff. Doing so also presented an opportunity to formalize and streamline the process currently in place, and through the Resource Recovery Roundtable, ensure cross functional representation throughout the innovation process. Ultimately, developing a clear framework has helped foster strategic acumen among WWE staff and ensures that strategic innovation is integrated across functional groups in key resource recovery areas.

Drivers and Supporting Internal Initiatives

The review process and associated prioritization tools developed were designed to align with existing WWE goals and policies, understanding that any new resource recovery project should provide a logical link back to one or more goals contained within other WWE or SFPUC planning documents. The six overriding source documents used in establishing the Innovation Platform were the SFPUC Strategic Plan, WWE's Strategic Business Plan, WWE's Sewer System Improvement Plan, the SFPUC Sustainability Plan, SFPUC's Technology Policy, and SFPUC's Community Benefits Policy. Key points taken from each are summarized in Table 1.

Table 1. Supporting Documents for Innovation Platform Development

Guidance Document	Relevant Goals and Objectives
SFPUC Strategic Plan	Objectives within the Environmental Stewardship goal area include: • Sustainably manage natural resources and physical systems to protect the people, water, land, and ecosystems that we affect • Develop, coordinate, and communicate a comprehensive and consistent approach to mitigate and adapt to climate change • Be resource efficient in all business operations • Investigate the feasibility of implementing an environmental management system
WWE Strategic Business Plan	Objectives associated with environmental stewardship include: • Anticipate, advocate, and influence regulations and policies that impact WWE services • Advance resource recovery and pollution prevention solutions to maximize ratepayer value and minimize WWE's environmental footprint • Ensure climate change adaptation and mitigation planning is integrated into WWE planning process when appropriate • Develop long-term plans, policies, programs, and standards using an integrated watershed planning approach to optimize system performance
Sewer System Improvement Plan	Establishes levels of service a capital project must meet: • Provide a compliant, reliable, resilient, and flexible system that can respond to catastrophic events • Integrate Green and Grey Infrastructure to Manage Storm water and Minimize Flooding • Provide Benefits to Impacted Communities • Modify the System to Adapt to Climate Change • Achieve Economic and Environmental Sustainability ○ Beneficially reuse 100% of biosolids ○ Beneficially reuse 100% of biogas ○ Use non-potable water sources to meet facility non-potable water demands • Maintain Ratepayer Affordability
SFPUC Strategic Sustainability Plan	Provides five durable sustainability goals for the agency: • Provide high quality services • Engage our public and invest in our communities • Promote a green and sustainable city • Invest in our people • Plan for the future
SFPUC Technology Policy	The 2012 SFPUC Technology Policy establishes guidelines for the research, testing, and deployment of both new and established technologies. Ratepayer benefit and consistency with triple bottom line (TBL) principles must be established prior to full scale deployment of a given technology.
SFPUC Community Benefits Policy	The 2011 Community Benefits Policy outlines strategies for the SFPUC to achieve benefits to the surrounding communities in the course of its day-to-day activities. Beneficial outcomes include workforce development, economic development, and beneficial land development.

Process and Tool Development

The Innovation Platform process, including all supporting tools, was developed by a small team of WWE staff and consultants through a series of workshops. Tools were developed so that every WWE staff person could articulate why a topic has merit, why additional resources are needed, and why something should move forward. PowerPoint templates, and guidelines for crafting a clear argument have proved very valuable in helping staff, who may not typically perform these job functions, articulate why their idea has merit and should be attributed additional resources.

Interim and final work products were presented to the R2 Roundtable at monthly meetings; only approved processes and products are described in this Technical Memorandum. The process as developed has been summarized in a flow chart, presented here as Figure 1, and is intended to be adaptive.

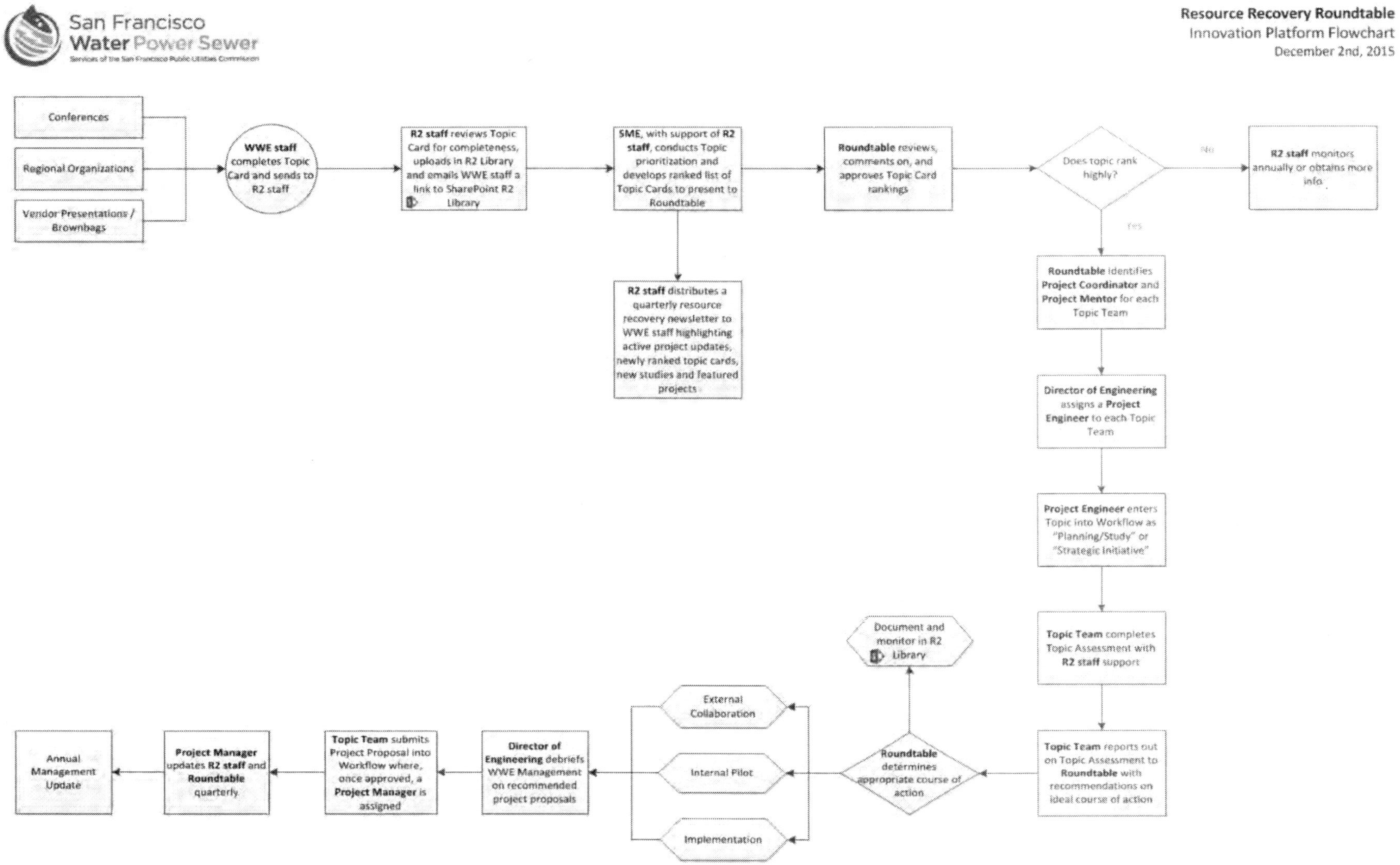

Figure 1. Innovation Platform Flowchart (Source: Wastewater Enterprise)

Implementing the Innovation Platform

With the tools and platform established, the R2 staff began the process of implementing the platform, with the intent of revisiting and modifying the process and tools, as needed, after the first year. Implementation required identification of staff roles and responsibilities and internal communication regarding the Innovation Platform. The identified roles and tools within the Innovation Platform are discussed in this section; modifications made during implementation are addressed in the Discussion section.

Roles and Responsibilities

At the center of the Innovation Platform is the engagement of the WWE staff across multiple disciplines in incorporating innovation into the WWE's day-to-day work. The ability, for example, to further investigate an innovative concept brought forward from a conference such as WEFTEC was one of the drivers behind creation of this process. Moreover, while information might be obtained from vendors or consultants, the majority of the information gathering and topic development is ultimately performed by WWE staff. It is anticipated that in most cases, consultant support will not be needed until a topic reaches the point of implementation and a specific pathway is selected.

Role	Responsibility
WWE STAFF	• Submit Topic Cards on innovation of interest
R2 STAFF	• Support WWE staff in the production of Topic Cards, Prioritization Memos and Topic Assessments • Organize and maintain the Topic Card Library and ensure information is accurate and up to date • Host R2 Roundtable meetings • Track progress on how R2 activities integrate with and achieve the utility's larger strategic objectives
R2 ROUNDTABLE	• Hand-picked by the WWE Assistant General Manager • Represent every division within WWE • Governing board responsible for weighing in on the merits of Topic Cards brought forward by WWE staff
SUBJECT MATTER TEAMS	• Hand-picked by the WWE Division Manager • Teams organized by resource category (biosolids, biogas, nutrients, water) and include representation from experts in Operations, Maintenance, Engineering, Planning and Regulatory, and the Collection System • Responsible for assessing Topic Cards within their resource category of expertise
TOPIC TEAMS	• Assigned by the R2 Roundtable and Director of Engineering once a topic is determined to be of enough importance to assign resources • Consists of a project engineer, project coordinator and a project mentor

Project Concept Introduction and Topic Card

As a first step, any WWE staff may bring forth an idea for evaluation. Typically, such ideas are brought forward following attendance at professional conferences, but may also come from reading journal articles, interactions with other agencies, or technology vendors. The proposer then fills out a WWE Topic Card (Attachment A) with as much information as is available. The purpose of the Topic Card is two-fold: 1) to capture incoming ideas in a standardized fashion that can then be used to populate a Topic Card library and 2) to gather enough information for the in-house subject matter teams to assess the idea.

Topic cards were designed to be simple enough for any WWE staff to complete, and completing the form formalizes the beginning of the data gathering process. This new program enables staff at all levels to put forth ideas in a more formal fashion. Moreover, WWE Management now requires staff that attend conferences to seek out novel practices and technologies and submit a Topic Card alongside their requests for reimbursement. All submitted Topic Cards are uploaded to the Topic Card Library, where they may be seen by any and all interested parties.

The Topic Card library allows for a searchable database of ideas, accessible to all WWE staff via a SharePoint site. R2 staff then review the Topic Card for completeness such that the Subject Matter Team can evaluate the topic and determine whether it merits review by the R2 Roundtable. The review process ensures that all submitted Topic Cards at least undergo an initial review and annual revisit.

Quarterly, R2 staff distributes a newsletter to WWE staff highlighting new topic cards. The newsletter also includes information regarding topic card prioritization, active resource recovery project updates, and new studies.

Screening Process

A significant part of the screening process involves information gathering such that R2 staff and ultimately, the R2 Roundtable, have enough information to understand where new resource recovery concepts can potentially integrate with WWE's day-to-day business, or where practices can be modified to improve resource recovery. In some cases, this exercise may end with data gathering and summation, or it may go on to either pilot testing or full-scale implementation. The screening process was developed such that each step of the process requires more information development. While introduction of a Topic Card is meant to be a simple and accessible step for any WWE staff, topics that advance through the screening process ultimately result in a report to WWE Management requiring input from the subject matter team.

The first step in the overall process is gathering sufficient information so as to assess compatibility of the concept with stated WWE and SFPUC goals and objectives, as well as WWE's business practices. While the Topic Card provides the initial information gathering, the next step in the screening process allows for a comparative ranking of concepts (Topic Prioritization).

Topic Prioritization is a process conducted by the Subject Matter Team, who have the expertise necessary to evaluate the merits of a Topic Card within a given subject area. If the Subject Matter Team agrees that a concept addresses a core WWE concern, and merits further analysis, they perform a simple evaluation to determine whether enough information is currently available (either directly from the Topic Card as completed by the concept proposer, or by working with the concept proposer to obtain more information). Three possible outcomes of this process have been identified:

1) If insufficient information is available, for example, because a technology is simply too nascent to evaluate further, R2 staff will log the Topic Card in the Topic Card library, where it can be referenced in the future if and when the concept evolves further. The Topic Card library is searchable such that relevant information on any given topic can easily be accessed. If R2 staff determines that there is insufficient information available, they log all the available information in the Topic Card library, but do not move the topic onward for further review by the R2 Roundtable.

2) If sufficient information is available for a preliminary evaluation, the Topic Card is evaluated by the Subject Matter Team, to ensure it is consistent with WWE's current business practices. If it is not, the topic, along with the reason for the current incompatibility, is logged in the R2 library.

3) If the Topic Card has sufficient information and is deemed consistent with WWE's business practices, R2 staff convenes an evaluation meeting with the relevant Subject Matter Team to evaluate the Topic Card. R2 staff, supported by the Subject Matter Team complete a Topic Prioritization.

The intent of the Topic Prioritization is to provide both the R2 Roundtable and, ultimately, WWE Management with the information they need to make a decision as to whether to advance a topic for incorporation into WWE practices. The Topic Prioritization began as an Excel based tool that quantitatively ranked the topic using evaluation criteria developed by the R2 staff. Criteria included relevancy to existing internal planning documents and potential risk associated with inaction. While this tool allowed for a clear, quantitative means of determining projects relevance and fit within the organization, in practice, WWE staff found it to not provide sufficient information on the merits of a topic, and found that quantifying the value of a project was too subjective. It ultimately evolved into a more qualitative prioritization tool (Prioritization Memorandum) that allowed for more objective and insightful input.

Prioritization Memorandum

The intent of the Resource Recovery Prioritization memo is to evaluate the topic with respect to stated SFPUC and WWE goals and objectives, and capture staff recommendations for the best course of action. As described above, R2 staff convenes a meeting with the relevant SME(s), along with any staff or managers who might be interested or impacted by the topic. During the meeting, the topic and its relevance are presented, and those present must develop staff recommendations. The person with the most appropriate experience, or an assigned delegate, is assigned to develop the Prioritization Memo capturing the meeting discussion and suggested next steps. The prioritization is then presented at the R2 Roundtable for review and approval of staff recommendations.

Roundtable Evaluation and Review of Topic

Topic cards and their associated staff recommendations from the prioritization memo are then brought forward to the R2 Roundtable for review and evaluation. Designated WWE staff then makes recommendations to the R2 Roundtable with respect to the promise of the concept, and the R2 Roundtable provides oversight and feedback on the topic. The R2 Roundtable has two options when reviewing the Topic Card and associated prioritization. The Roundtable may decide that the concept is insufficiently relevant to WWE's business practices, or involves too much uncertainty or risk to develop further. In this case, the topic is logged in the R2 library, with the associated documents and rationale for not pursuing the topic further and the R2 staff will initiate a reevaluation for the topic on an annual basis.

The R2 Roundtable may also decide that more information is needed to proceed further; in this case, R2 staff will work with the project proposer, designated WWE staff or a team of staff to develop that information. If, however, the topic has been evaluated to the Roundtable's satisfaction, then the Roundtable works to develop a presentation to WWE Management. The presentation is developed by a member of the Subject Matter Team or R2 staff as appropriate.

As mentioned previously, the ultimate decisions of the WWE staff and the R2 Roundtable are summarized in a quarterly resource recovery newsletter for WWE staff. The newsletter includes information on newly introduced and prioritized topic cards, and the status of existing topic cards. Information is also provided on active projects, new studies, and external collaborations to help create WWE wide awareness of resource recovery activities.

Topic Presentation

The primary use of the topic presentation is to assist staff in organizing the available information on the topic under consideration for presentation to Management. The information presented can help decision makers to fully assess a new project in order to determine the best course of action (e.g. monitor, pilot, or implement). Information developed during the topic presentation can ultimately be used later in the process if the topic is advanced as a project proposal by management. The Topic Presentation tool is a PowerPoint presentation template designed to guide staff in developing the information necessary to fully evaluate the topic with respect to benefits to the WWE, capital costs, potential for cost recovery, and other resources needed. The questions within the tool are meant to guide the user through information that can help describe the proposal to those outside the process, as well as to capture any information that may have been obtained from meetings with vendors, consultants, or academics.

The first set of questions within the tool is designed to frame the project and may have already been answered by the project proposer in the Topic Card. These put the project within the larger context of the WWE's and SFPUC's goals. Subsequent questions address other expressed values of the WWE – responses will help guide work for projects that move past the screening step, such as a business case evaluation or risk analysis. Where information cannot be completed, the team may be directed to address the reasons why. The comments section of the Power Point template provides guidance for answering the questions presented.

Completion of the questions within the tool requires back and forth dialogue between the Subject Matter Team, R2 staff, and the R2 Roundtable. Ideally, a completed screening tool should contain all the information necessary for WWE senior management to assess the proposal, should the R2 Roundtable elect to promote it for full implementation.

The key areas for information development have been parsed into sections within the tool as follows:

Ratepayer Value – Value is not directly related to cost, rather, it points to how a technology or process may help the WWE fulfill its stated mission with respect to regulatory compliance, operational efficiencies, and use of resources.

Financial Costs and Benefits – This section allows the team to describe capital and recurring costs associated with the technology or process, as well as cost recovery mechanisms including tip fees or grants and low interest loans.

Environmental and Regulatory Impacts – This section allows the team to describe any positive or negative impacts that the technology or process may have on WWE's environmental performance. This may range from the generation of side streams to greenhouse gas emissions to offsets of resource uses, such as potable water or electricity.

Risk Mitigation – New processes or technologies can provide ways to mitigate risk in a number of critical areas including response to emergencies, climate change, or resource depletion (e.g. potable water during a drought). This section allows the team to articulate any ways in which the proposed topic can address these areas of risk. The risk associated with implementing the new technology can also be identified and mitigation approaches offered.

Community Benefits – The SFPUC is the first public utility in the nation to adopt a Community Benefits policy, and as such strives to be a good neighbor to all whose lives or neighborhoods are directly affected by SFPUC operations. This section provides the opportunity to the team to describe how the proposed topic may or may not augment community benefits. They may also indicate if community benefits are best articulated later in the process, once the implementation decision has been made.

Highest and Best Use – At the heart of resource recovery is the concept of finding new and better uses for existing resources, many of which were formerly disposed, such as biosolids or grease. Topics that provide a different or better use for WWE resources can help WWE execute its core business in more efficient ways. These benefits can be articulated by the team in this section.

Implementation Potential – The more novel or lab—scale a proposal is, the less likely it is that it would be implemented full-scale at WWE facilities. Information presented in this section directly aids the R2 Roundtable in deciding the scale of implementation – pilot or full-scale- for a viable topic.

Next Steps – In this section the team provides information on what their vision is for the next steps, based on the information collected. The four choices identified are: document and monitor, external collaboration, internal pilot, or implementation. Each of the four courses of action results in the topic being included in an annual management review and five-year topic updates.

> ***Document and Monitor*** **-** In this case, the Roundtable makes a determination that the project concept is not appropriate for WWE to pursue at this time, such as the existence of certain constraints that render the topic infeasible. The team would provide written information on the constraints, as well as what criteria, policies, or funding mechanisms would need to change in order for the topic to be revisited. Information contained within the Topic Assessment tool is documented by R2 staff in the R2 library for future reference. Interested WWE staff may continue to investigate the project concept to allow the project to be considered at a future date, at which time it may shift into one of the other pathways listed below.
>
> ***External Collaboration*** **-** The recommendation for an external collaboration would likely come about if the technology or process is relatively unproven and either a) external partnerships have been identified or b) the costs of implementation, even at the pilot level, are prohibitive. External

collaboration can take the form of working with university researchers (either individually or as part of a consortium), mitigating cost and/or risk by siting the pilot or demonstration work at a utility that is somehow better suited to host, and concept development through one of the WWE membership agencies such as WERF. The latter option can prolong concept development but can have other advantages with respect to bringing more resources to bear on the concept evaluation.

Internal Pilot - Internal pilots are an appropriate recommendation when the topic is relatively unproven and the feasibility of the topic is highly dependent on some unique feature of the SFPUC as a utility. It is possible for a topic to move from the external collaboration phase into the internal pilot phase if specific questions of concern were not addressed during the external collaboration phase. Unique features of the SFPUC include but are not limited to its suite of services, extensive land holdings, the urban setting of its treatment plants, the particular characteristics of its wastewater, biosolids, and biogas, and its combined sewer system. Traditionally, internal pilots have been conducted by the WWE's Process Engineering and Research Group with the cooperation of Operations, Maintenance, and Laboratory staff.

Implementation - There may be concepts that are so well developed and suggest great advantages for the WWE such that the Roundtable and WWE Management elect to move toward implementation. This option would be selected if each section of the Topic Assessment demonstrated clear advantages and few potential pitfalls to implementation. Budget and staff capacity need to be identified but selection of this option requires WWE Management approval regardless of whether there are costs associated with it.

Management Approval

Once a decision is made on a topic, the decision must be presented to Management. As WWE Engineering plays a critical role in this process, the WWE Director of Engineering takes the lead on presenting the final R2 Roundtable decisions to WWE Management. Semiannually, at the existing WWE Senior Management meeting, Management review and comment on the topic cards, topic prioritization, and R2 Roundtable recommendations. This allows Management oversight to ensure that resource recovery efforts are in line with other management goals and objectives. Management may also choose to allocate resources to further develop a topic, as deemed appropriate. R2 staff compiles a report in advance of the semiannual management meeting. This report is updated with Management decisions following completion of the semiannual meeting.

When approved by management, project proposals are then entered into the Workflow Process where a project manager is assigned to direct execution of the work. In addition to the project manager, a project team is created – consisting of a project coordinator, project engineer and project mentor. Such roles are assigned in coordination and cooperation with the relevant supervisory staff. The project team then coordinates with the R2 staff and R2 Roundtable on a quarterly basis, to ensure communication on projects developed from the Innovation Platform. This also allows R2 staff to easily update Management when needed.

Monitoring and Annual Review

While staff is encouraged to submit Topic Cards year-round, it is not likely that a large percentage of these topics will move beyond the documentation step due to limited staff and resource constraints. None the less, a key component of the Innovation Platform is that every topic is uploaded to the SharePoint Topic Card Library where it can be monitored and updated as new information becomes available. Annually, the Subject Matter Team will review previously submitted Topic Cards and determine whether or not enough has changed to warrant further analysis or implementation of a topic.

Results and Discussion

The Innovation Platform and its associated screening tools were presented to the R2 Roundtable for review and approval in January, 2016. Subsequently, the process was opened up to submittal of new topics and concepts, with the intention of revisiting the process at the end of the year to discuss the successes that should continue, or the aspects of the process that should change. The process was developed with the intention of being iterative so as to gauge what worked best for WWE staff, and will continue to be reassessed annually to ensure the process remains as innovative as the technologies and projects being reviewed.

Since January, 2016 R2 staff have received 44 Topic Cards across four resource categories; 14 biosolids, five biogas, eight water, seven nutrients as well as 10 under the broader category of optimization. In addition, the R2 Roundtable has convened for 12 meetings and provided tremendous insight and staff recommendations for how to prioritize innovation and overall programmatic initiatives.

When the R2 team first developed the Platform it was a response to an informal decision making process that involved minimal documentation and was not reflective of the great work being done to advance the organization. In an attempt to shift the culture towards one of documentation and process, the R2 team developed the Innovation Platform. After one year of implementation, R2 staff could see that the process in action was a bit too formal, with many meetings and more formality than WWE staff was used to. It was evident that the process developed was a bit of an overcorrection, and as such opened up the risk of having individuals bypass the system for an easier path.

To remedy this issue, the R2 team took a step back in late 2016, after one year of implementation, to determine what steps of the process absolutely must be documented. The objective was to determine the path of least resistance for staff where essentials are documented, and all views are represented, but the process does not feel too overbearing or rigid. Below is a copy of the first process developed, with the revised process beneath to show where streamlining occurred.

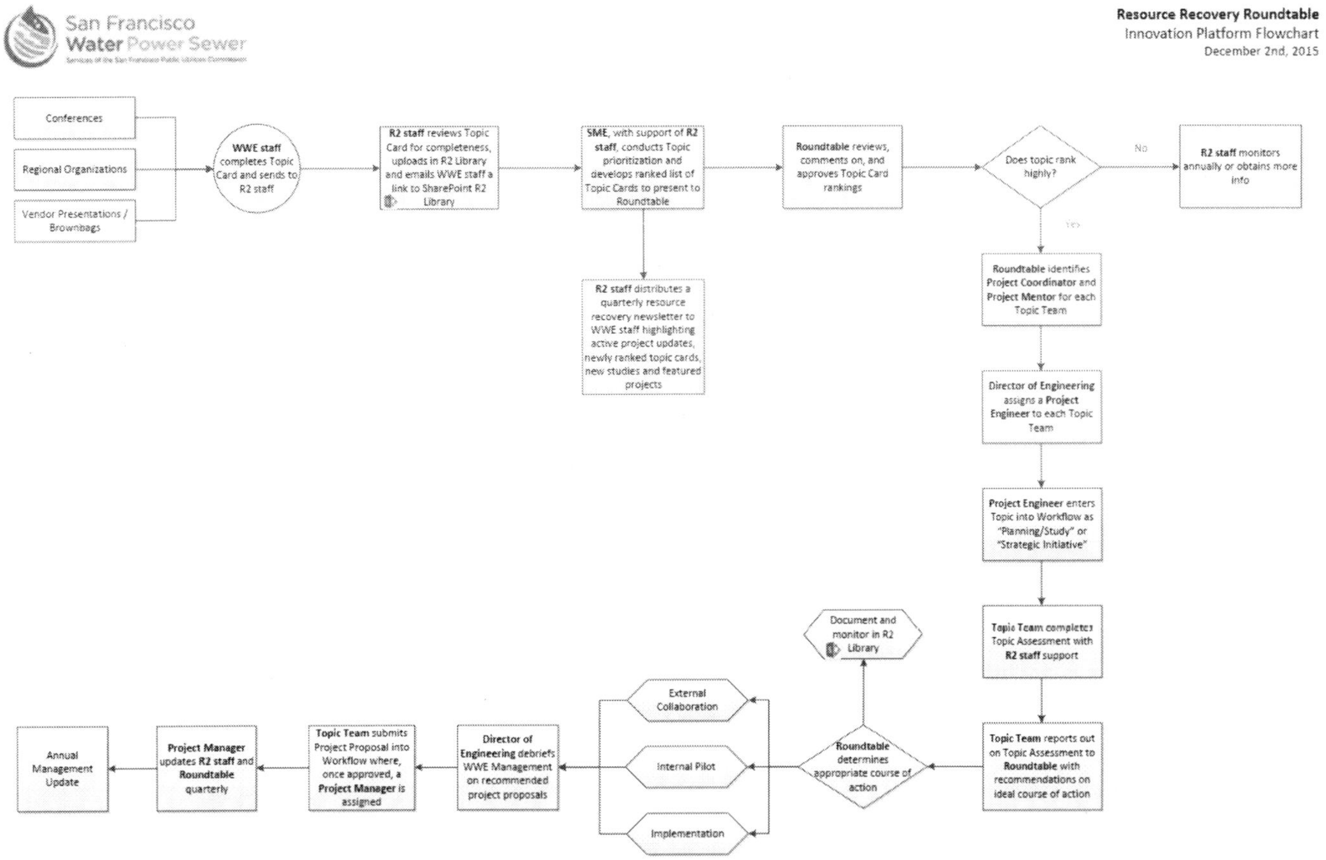

Figure 2. 2015 Innovation Platform Flowchart (Source: Wastewater Enterprise)

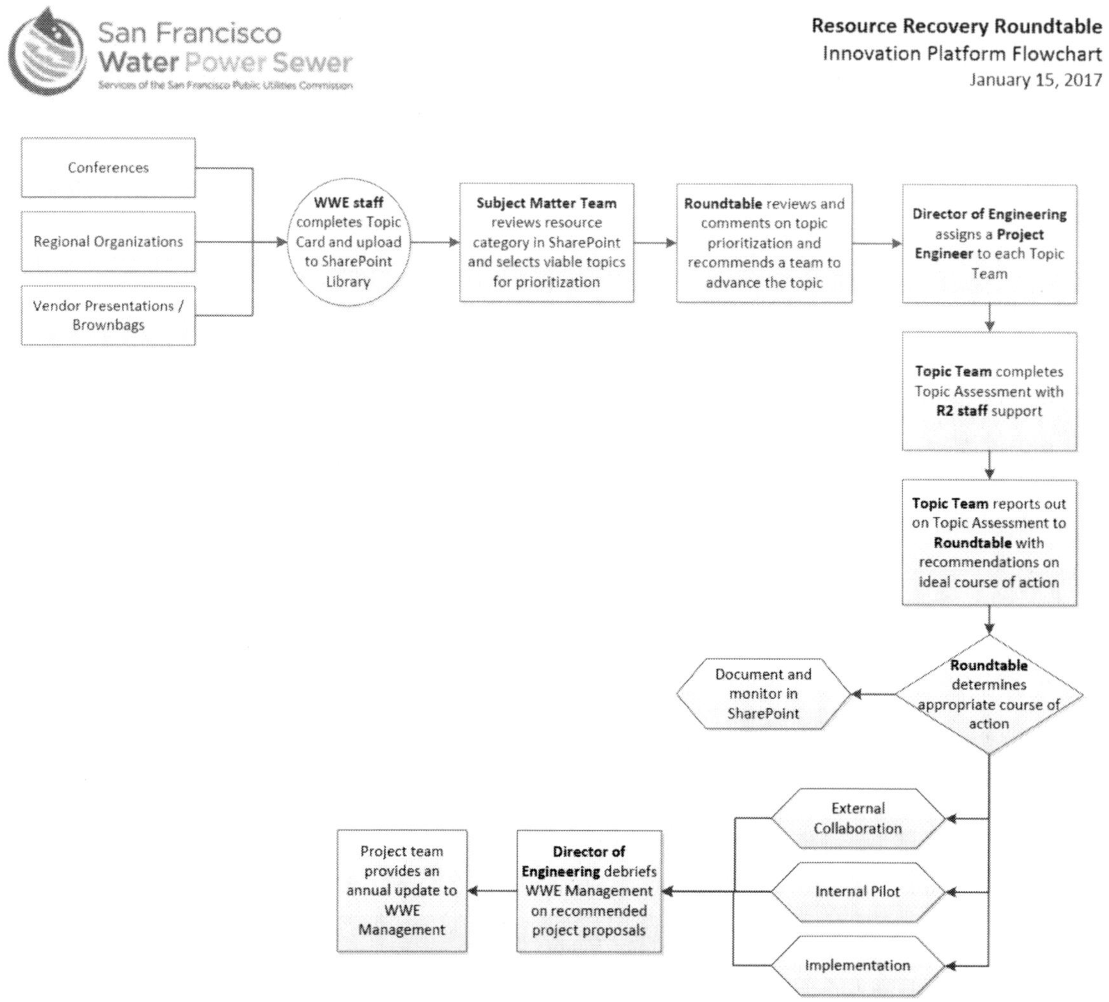

Figure 3. 2017 Innovation Platform Flowchart (Source: Wastewater Enterprise)

Despite the somewhat cumbersome process, preliminary feedback from the R2 Roundtable was very positive, and primarily focused on the value of documenting ideas and decisions. Doing so has reduced the likelihood of siloed decision making and lost documentation or information, while simultaneously increasing the defensibility of decisions to the public and increasing stakeholder representation in the decision making process.

It was determined by the R2 Roundtable that Topic Cards and the Prioritization Memo were the most valuable aspects of the Platform. With these two documents in hand, WWE staff can easily enter topic information into the existing WWE Project Proposal Workflow process if it is determined that a topic has enough merit to assign resources to it. Over time, these tools have evolved to better communicate to upper management and laterally to the Subject Matter Team. As articulated earlier, the intent of these tools is to equip WWE staff with the resources necessary to advocate for and make defensible arguments in favor of why an innovative technology, process or partnership should be pursued.

In addition to the value of documentation, the R2 Roundtable identified the Subject Matter Teams as a key component of the Innovation Platform. As mentioned above, the Subject Matter Team plays a key role in assessing the suitability and merit of a topic card. Wastewater Utilities carry the expertise of regulatory and policy staff, engineers, scientists, and operators within the same organization. To leverage this expertise and ensure appropriate representation from each group, the Subject Matter Teams were carefully selected by resource category to have representation from each of WWE's five relevant divisions – Operations, Maintenance, Engineering, Planning and Regulatory Compliance, and the Collection System. This cross disciplinary expertise benefits innovation because a topic can be assessed holistically rather than just for its technical merits or regulatory implications. Ensuring that each topic receives input from operators, engineers and regulatory specialists helps reduce the chance of a fatal flaw and continues the transition away from siloed decision making.

Table 1. Subject Matter Teams

	Water	**Biogas**	**Biosolids**	**Nutrients**
Operations	Andy Clark	Andy Clark	Johnny Johnson	Andy Clark
Maintenance	Joel Prather	Joel Prather	Joel Prather	Joel Prather
Engineering	Joe Wong	Meei-Lih Ahmad Nohemy Revilla	Ravi Krishnaiah Alex Miot	Domenec Jolis
Regulatory	Laura Pagano	Meei-Lih Ahmad	Adrian Santiago	Amy Chastain
Collection System	Manon Fisher	N/A	Karri Ving	N/A

The role of the Subject Matter Team has proved to play a much larger role than originally expected. Primarily, there are far more ideas than there are resources to fund and implement them. As such, it is critical that an expert group of individuals representing all of the needs of the organization can properly weigh in on the most appropriate process or technology to solve the problem at hand. In addition, and perhaps more importantly, the potential of the Subject Matter Teams to act as a professional development incubator was unexpected but will be a valuable asset moving forward. Subject Matter Experts were handpicked by management, however each Expert is also advised to include a practitioner who can be developed alongside them and eventually become an expert themselves. This alleviates some of the burden placed on the Subject Matter Expert and also helps ensure that WWE continues to invest in emerging leaders in their respective fields.

Another key insight since the implementation of the Innovation Platform is the need for a strong Project Team. As mentioned earlier, a Project Team enters the process once a topic has been deemed worthy by WWE Management to assign additional staff and resources. Through experience with successful and unsuccessful projects, it has been determined that a successful project team should consist of the following three roles:

Project Engineer: The person who performs engineering analysis and determines how a give project fits into WWE's operations.

Project Coordinator: The person who assists in project development and conducts research and outreach to relevant stakeholders.

Project Mentor: A senior level WWE staff person capable of guiding a project with experience in budgeting, working with a range of WWE staff and management.

R2 Staff recognized that all successful resource recovery, and wastewater-wide project were lead by a project team with the above roles. While it is not necessary for each of these roles to be filled by a unique individual, there must be at least two people on the team who can fulfil the above roles. This team is still highly supported by R2 staff and the R2 Roundtable, however without a strong team the likelihood of a project not moving forward substantially increases.

Moving Forward

Over the next year, the R2 Roundtable will transition from hands on development of the Innovation Platform to an active governing board that weighs in on the implementation of resource recovery initiatives. The ongoing implementation of the Innovation Platform will take similar shape to the original process, with a few adaptations to reflect the above results. On an ongoing basis, WWE staff and Management will bring forward ideas in the form of Topic Cards and submit them to R2 staff for review and filing in the R2 Topic Card library.

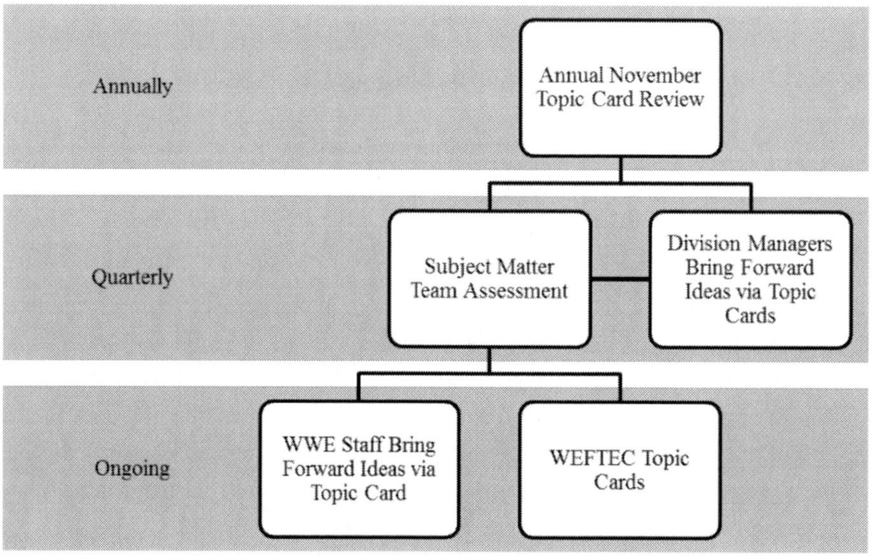

Figure 4. Topic Card Review and Prioritization Process

Conclusions

The Innovation Platform was expected to be adaptive, and change over time as it is increasingly adopted by WWE staff. The Platform is supported as a means of providing a transparent methodology for WWE staff to introduce, analyze and develop concepts that can help the WWE become a better utility. In addition, it provides the means of communicating these concepts throughout the WWE, helping to bring universal awareness of innovation and strategic thinking to all interested staff.

References

2016-2018 Wastewater Enterprise Strategic Business Plan, Wastewater Enterprise, August 2016.

Strategic Sustainability Plan. San Francisco Public Utilities Commission. March 2011.

SFPUC Technology Policy, San Francisco Public Utilities Commission. September 2012.

SFPUC Community Benefits Policy, SFPUC Resolution 11—0008. San Francisco Public Utilities Commission. January 2011.

Anaerobic codigester feeding pattern drives long-chain fatty acid bioconversion kinetics and syntrophic community structure

Ryan M. Ziels[a], David A.C. Beck[b], H. David Stensel[a]

[a] Department of Civil and Environmental Engineering, University of Washington, Seattle, WA

[b] eScience Institute, University of Washington, Seattle, WA

Abstract:

Experimental anaerobic codigesters were fed with dairy cattle manure and oleate ($C_{18:1}$) with either a pulse-fed or continuous-fed regime up to a maximum loading of 64% of the influent COD over 204 days. Maximum substrate conversion rates of oleate ($q_{max,oleate}$) and acetate ($q_{max,acetate}$) reached significantly higher values in the pulse-fed codigester compared to the continuous-fed codigester. The effluent acetate concentration in the continuous-fed codigester exceeded 3,000 mg/L at the highest oleate loading rate, whereas the pulse-fed codigester maintained low effluent acetate levels. LCFA-degrading *Syntrophomonas* bacteria were significantly enriched in both pulse-fed and continuous-fed LCFA codigesters, but higher relative and absolute concentrations of *Syntrophomonas* 16S rRNA genes were observed in the pulse-fed codigester. $q_{max,oleate}$ and $q_{max,acetate}$ both correlated with the concentrations of *Syntrophomonas* and *Methanosaeta* 16S rRNA genes, respectively. Improved quantitative predictions of bioconversion kinetics were achieved with multiple-linear regression models based on the absolute abundance of high-resolution *Syntrophomonas* and *Methanosaeta* OTUs. Pulse-feeding LCFA during anaerobic codigestion thus selected for higher microbial bioconversion kinetics and functional stability, which were related to changes in the fine-scale syntrophic community structure.

Keywords: Codigestion, LCFA, biomethane, biogas, syntrophy, methanogen, microbial

1. Introduction

Anaerobic digestion of agricultural livestock manure can be a cost-effective solution to minimize greenhouse gas emissions while treating the organic waste through the recovery of biomethane. Nonetheless, the low biomethane potential of animal manures (Labatut et al., 2011) can inhibit the economic viability of agricultural digesters (Bishop and Shumway, 2009). Fats, oils, and greases (FOG) are appealing substrates for codigestion because they have a methane yield potential that is approximately 2.7-times greater than livestock manures on a per mass-basis (Labatut et al., 2011). When added to the digestion process, fats and oils are rapidly hydrolyzed into glycerol and long-chain fatty acids (LCFA) (Cirne et al., 2007; Hanaki et al., 1981), with over 90% of the methane potential of lipids maintained in the released LCFA (Sousa et al., 2009).

The conversion of LCFA into methane is considered the rate-limiting step for degrading FOG in anaerobic digesters (Cirne et al., 2007; Hanaki et al., 1981). LCFA-degradation under methanogenic conditions involves an obligatory syntrophic partnership between acetogenic β-oxidizing bacteria and acetate- and hydrogen-utilizing methanogenic archaea (Schink, 1997; Sousa et al., 2009). The bacterial species known to β-oxidize LCFA in syntrophy with a hydrogen- or formate- consuming partner have so far been identified within only two families, *Syntrophomonadaceae* and *Syntrophaceae* (Sousa et al., 2009). The slower biodegradation of LCFA relative to lipid hydrolysis in anaerobic digesters can be problematic for process stability, as LCFA are known inhibitors of methanogenesis at high concentrations (Angelidaki and Ahring, 1992; Hanaki et al., 1981; Lalman and Bagley, 2002; Rinzema et al., 1994). LCFA accumulation during anaerobic digestion can limit the kinetics of their own biodegradation by inhibiting syntrophic β-oxidizing bacteria (Hanaki et al., 1981) as well as hydrogenotrophic and aceticlastic methanogens (Hwu and Lettinga, 1997; Lalman and Bagley, 2002). Increasing microbial resistance and LFCFA bioconversion kinetics is thus critical for maximizing biomethane recovery and improving process economics of anaerobic digesters through FOG codigestion.

The relationship between microbial community structure and LCFA bioconversion kinetics in anaerobic digestion processes has not been well characterized. Previous efforts have predicted LCFA degradation kinetics in anaerobic digesters with modeling-based approaches (Palatsi et al., 2010; Zonta et al., 2013). However, those biokinetic models were not calibrated with actual measurements of LCFA-degrading biomass, and thus relied upon speculations about the community composition. Qualitative shifts in microbial community structure during LCFA degradation have been observed using molecular fingerprinting techniques (Baserba et al., 2012; Hatamoto et al., 2007; Shigematsu et al., 2006; Sousa et al., 2007), but the typically low concentration of syntrophic β-oxidizing bacteria in digester biomass has limited their accurate detection and quantification with such methods (Menes and Travers, 2006). By monitoring syntrophic populations with quantitative PCR (qPCR) during FOG codigestion with municipal wastewater sludge, Ziels et al. (2016) showed that effluent LCFA concentrations were positively correlated to the FOG loading rate normalized to the concentration of LCFA-degrading *Syntrophomonas* bacteria. This finding implicated that the syntrophic community within the FOG codigester had a specific rate of LCFA conversion. Thus, we hypothesized that different species of LCFA-degrading bacteria and/or methanogenic partners could have different bioconversion kinetics, and their abundances could be differentially selected by the environmental conditions of a digester.

Biological selective pressures could potentially be incorporated into digester operating strategies to improve process stability by favorably managing the biomass composition, as the functional stability of anaerobic digesters has been linked to microbial biodiversity (Briones and Raskin, 2003; Werner et al., 2011). Since FOG degradation relies on the activity of the metabolically-specific syntrophic LCFA β-oxidizing bacteria and methanogenic archaea (Schink, 1997), improving the functional stability of those populations could potentially lead to more robust FOG codigester performance. Digester microbial community structures can be influenced by substrate type (Sundberg et al., 2013), temperature (De Vrieze et al., 2015; Westerholm et al., 2015), solids retention time (Hao et al., 2016), free ammonia (Westerholm et al., 2015), and feeding pattern (Conklin et al., 2006; De Vrieze et al., 2013; Mulat et al., 2016). Of these parameters, the digester feeding pattern may be more easily controlled from an operational standpoint. Yet, the impacts of feeding pattern on the structure and functional stability of LCFA-degrading syntrophic communities during codigestion of LCFA-rich wastes remain unknown.

The objective of this research was to investigate the effects of anaerobic codigester feeding pattern on: (1) LCFA bioconversion kinetics, (2) digester functional stability and performance, and (3) microbial community structure and diversity. Anaerobic codigesters fed with dairy manure and oleate were operated with either a pulse-fed or continuous-fed regime to compare their respective substrate conversion kinetics and microbial community structures. Quantitative models were developed to predict LCFA-conversion kinetics based on high-resolution syntrophic community structure and absolute abundance.

2. Materials and Methods

2.1 Anaerobic Digester Set Up and Operation:

Three identical glass 5.5 L anaerobic semi-continuous stirred tank reactors were inoculated with effluent sludge from a mesophilic plug-flow anaerobic digester treating dairy manure at a nearby farm. The digesters were operated with a 4.0 L working volume and a 20-day hydraulic retention time (HRT), which was also equal to the solids retention time (SRT). The digesters were intermittently mixed for 2 minutes every 10 minutes with an axial impeller mixer (5VB, Cleveland Mixer), and temperature was controlled at 35°C in an environmental chamber. Daily biogas production was measured with wet tip gas meters (Wet Tip Gas Meter Co., Nashville), and the methane composition of the biogas was measured at least twice weekly at the end of the daily feeding cycle.

Fresh cow dairy manure feedstock was collected from the influent mixing tank of the same dairy farm digester that supplied the inoculum sludge. The manure feedstock was blended for 1 minute before being stored in ~1L aliquots at -20°C for preservation. Single manure aliquots were thawed at 4°C daily before use. Three batches of manure were collected throughout the entire experimental period. Manure was manually fed into the reactors once daily after withdrawing 200 mL of sludge.

Digester operation was segregated into three experimental phases: (1) start-up period with manure-only feeding (days -50 to 0); (2) Phase I of oleate codigestion with manure (days 1 to 84); and (3) Phase II of oleate codigestion with manure (days 85 to 204). Manure was fed to the digesters at a loading rate of 1.3 ± 0.2 g VS/L-d during all experimental phases.

During codigestion Phase I, two of the digesters received an aqueous solution of sodium oleate (>97%, Tokyo Chemical Industry Co.) at 10% of the daily influent feed volume, while the third digester was fed only manure as the control. Oleate was used as a surrogate for FOG. The organic loading rate (OLR) to the codigesters was adjusted over time via the concentration of the oleate feed, while the total feed volume was fixed in order to maintain a constant SRT. The two codigesters were operated with an identical OLR on a daily average basis, but their oleate feeding frequencies differed. One codigester received the oleate feed in semi-continuous additions occurring every 6-hrs; this codigester feed regime is hereby referred to as *continuous-fed*. The other codigester received the oleate feed in a pulse every 48-hrs, with its average daily cumulative feed volume equal to 10% (v/v); this codigester feed regime is hereby referred to as *pulse-fed*. The OLR of oleate was increased stepwise approximately every SRT (20-days) during Phase I up to 27% of the influent feed COD (w/w) over the course of 80 days. On day 85, the vessel of the pulse-fed codigester was incidentally cracked, and thus its operation was terminated.

During Phase II, two oleate codigesters were operated in parallel: (1) the control digester from Phase I was subjected to oleate codigestion under the pulse-feeding regime, hereby termed *pulse-fed codigester v.2*; and (2) the operation of the continuous-fed codigester from Phase I was maintained. The OLR to the pulse-fed codigester v.2 in Phase II was increased about every 10 days until day 150. Stepwise increases in OLR were conducted until oleate reached 64% of the influent COD (w/w) to both codigesters on day 150. Due to acetate accumulation in the continuous-fed codigester at an oleate loading of 64% influent COD, the oleate loading was reduced to 62% influent COD on day 184.

2.2 Bioconversion Kinetics of Oleate

The biomass developed in the study digesters was used as inoculum for a series of batch assays conducted on days 0, 41, and 84 during Phase I, and on days 129, 170, and 204 for both Phase II codigesters as well as on day 148 for the pulse-fed codigester v.2. Assays were conducted in 30-mL glass serum bottles that were initially purged with N_2:CO_2 (80:20) before adding 15 mL digester biomass, and were subsequently sealed with butyl rubber septa and purged again with N_2:CO_2 (80:20). An aqueous solution of sodium oleate (100 g/L) was added to the vials via syringe to reach initial oleate concentrations of: 0 (control), 100, 175, 350, 750, 1800, 4500 mg/L. Triplicate vials were established for every condition tested. Gas production was measured in each vial using a digital manometer (Series 490, Dwyer Instruments, Michigan City). The methane concentration in the headspace was sampled with each pressure measurement. The background methane production from the control was subtracted from each substrate-amended treatment. Methane production rates were adjusted to standard temperature and pressure (STP), and were normalized to the amount of VS added to the vials initially.

2.3 Oleate Inhibition of Aceticlastic Methanogenic Activity

To test inhibitory effects of oleate on methanogenic activity, aceticlastic maximum methane production rates were measured in batch assays with oleate added at various initial concentrations. The aceticlastic maximum methane production rate ($q_{max,acetate}$) was measured according to the procedure described by Conklin et al. (2006). Vial preparation and digester biomass sampling was conducted as described above in Section 2.2. A series of triplicate vials received 50 mM sodium acetate, while one set of triplicate vials was not amended with any substrate to serve as the control. For the acetate-amended vials, sodium oleate was subsequently added at initial concentrations of: 0, 1.5, 4, 8, 12, and 16 g/L. Methane production was monitored in all vials over a 6-hr period based on headspace pressure and composition. The $q_{max,acetate}$ was calculated based on linear regression of methane production curves as described by Conklin et al. (2006).

2.4 Analytical Methods

Total COD (TCOD), total solids (TS) and volatile solids (VS) of the manure feed and digester effluents were measured 3-times weekly. TCOD was measured with High-Range COD digestion vials (Hach Co., Loveland, CO) according to Hach Method 8000. TS and VS were measured according to Standard Methods (Eaton and Franson, 2005). Ammonia nitrogen (NH3-N) was measured twice weekly in filtered (0.45 μm) digester sludge and manure feed using the High-Range Ammonia Nitrogen kit (TNT AmVer (Salicylate), Hach Co.). Prior to daily wasting and feeding of the digesters, pH was measured (EcoSens pH100A, YSI, Yellow Springs, Ohio) immediately after withdrawing samples.

Effluent sludge samples for volatile fatty acid (VFA) analysis were initially centrifuged at 14,000 *g* for 10 minutes. The supernatant was serial filtered with 0.45 and 0.22 μm polyethersulfone (PES) syringe filters, and 100 μL of 10% formic acid was then added to 1 mL of filtrate to acidify the sample. VFA were measured with a GC-FID (Shimadzu GC-2010) equipped with a BP-21 column (30 m×0.32 mm I.D., 0.25 μm film thickness, SGE). LCFA were extracted from digester effluent according to the protocol described by Ziels et al. (2015), and were analyzed by GC-FID with a BP-21 column (30 m×0.32 mm I.D., 0.25 μm film thickness; SGE). Biogas samples (~0.5-1 mL) were collected with a syringe equipped with a Teflon Mininert valve, and the methane composition was analyzed using a SRI 8610C GC-FID with a Supelco Alumina Sulfate Plot column (50 m, 0.53 mm i.d.).

2.5 DNA Extraction and Quantification

DNA was extracted in triplicate from digester biomass samples collected on the same days as the batch oleate conversion kinetic assays. Briefly, ~15 mL of digester sludge was collected via catheter syringe and transferred to pre-purged, O_2-free centrifuge tubes on ice. The digester sludge was aliquoted into 2 mL microcentrifuge tubes on ice in an anaerobic glove chamber. The 2 mL samples were then centrifuged at 14,000 *g* for 10 min at 4°C, immediately decanted, and stored at -80°C. DNA was extracted using the PowerMicrobiome™ kit (MO BIO Laboratories Inc., Carlsbad, CA) using phenol:chloroform:isoamyl during bead-beating, according to the manufacturer's instructions. Extracted DNA was quantified with the Quant-iT HS DNA kit using a Qubit 2.0 fluorometer (Invitrogen, Carlsbad, CA).

2.6 Quantitative PCR

Quantitative PCR (qPCR) was conducted in accordance to Ziels et al. (2015) targeting 16S rRNA gene fragments of the syntrophic β-oxidizing bacterial genera of *Syntrophomonas*, as well as the domains *Bacteria* and *Archaea* (Yu et al., 2005). More detailed descriptions of the reaction conditions, primers and probes, and calibration standards are given in the Supporting Information. qPCR was conducted on triplicate DNA extractions from each sampling point, and gene copy numbers were normalized per g VS of digester biomass.

2.7 16S rRNA Gene Amplicon Sequencing

Amplicon sequencing of 16S rRNA genes was conducted on triplicate DNA extractions from each sampling point. Briefly, 16S rRNA gene fragments from all samples were amplified with barcoded and indexed V4-V5 primers 515F-Y (5′-GTGYCAGCMGCCGCGGTAA) and 926R (5′-CCGYCAATTYMTTTRAGTTT) (Parada et al., 2015), and the products were pooled and sequenced on a 2x300bp Illumina MiSeq run by the United States Department of Energy Joint Genome Institute. Raw sequence data is available through the JGI Portal (http://jgi.doe.gov) under project ID number 1105525.

2.8 Bioinformatics

To obtain high resolution into community composition without generating spurious operational taxonomic units (OTUs), amplicon reads were denoised into their exact sequences using the DADA2 pipeline (Callahan et al., 2016). Reads were first quality filtered by trimming the first 10 bp from all sequences, truncating forward reads at 265 bp, truncating reverse reads at 180 bp, and filtering all reads based on a maximum expected error of 2 using DADA2. Parameters for the DADA2 error model were initially calculated on dereplicated reads of a random subset of 20 samples, and the estimated error parameters were subsequently used to denoise all dereplicated sample reads into their exact sequences. Forward and reverse sequences were then merged with DADA2 using a minimum overlap of 20 bp and zero allowed mismatches. Merged denoised sequences were truncated to 390 bp and clustered into 99.5% OTUs following chimera removal with UPARSE (Edgar, 2013), resulting in 5014 non-singleton OTUs. Representative sequences of the 99.5% OTUs were classified against the SILVA SSU Ref NR dataset (v.123) using the RDP classifier (Wang et al., 2007).

Relative abundance values were calculated based on normalized sequence counts obtained with DESeq2 (Love et al., 2014), which accounted for the different sequencing depths among samples. Diversity estimates were based on Hill diversity numbers (^{q}D) of orders $q = 0$, 1, and 2 (Jost, 2006) using DESeq2 normalized sequence counts.

2.9 Regression Analysis

A stepwise regression approach was used to identify a subset of microbial predictor variables (i.e. 99.5% OTUs) that significantly explained maximum methane production rates from oleate and acetate ($p < 0.05$). Predictor variables for the $q_{max,oleate}$ regression model included the inferred absolute abundances of 13 *Syntrophomonas* OTUs; and the predictor variables for the $q_{max,acetate}$ regression model consisted of inferred absolute abundances of three *Methanosaeta* OTUs. Inferred absolute abundances were calculated by multiplying OTU relative abundances by the target group 16S rRNA gene copies per g VS obtained by qPCR. The triplicate amplicon sequence profiles for each sampling point were included individually in the model development and validation. Prior to model fitting, multicollinearity between OTU abundances was examined using Pearson's correlation, and OTUs with correlation coefficients greater than 0.7 were removed from the dataset. Stepwise feature selection was conducted with the remaining OTU abundance datasets, and model selection was cross-validated based on 10 sub-samplings repeated 10 times using the caret package in R (Kuhn, 2008).

3. Results

3.1 LCFA feeding pattern impacted codigester performance and bioconversion kinetics

The impacts of codigester LCFA feeding pattern on process performance and bioconversion kinetics were investigated in two operational phases over 204 days. The increase in influent LCFA had no significant affect on the codigester effluent VS concentrations, indicating that LCFA addition with both feeding strategies did not impair the manure VS destruction efficiency and that most or all of the LCFA fed was degraded. The pH of the codigesters was on average 7.2 and 7.5 by the end of Phase I and II, respectively.

During Phase I, the average daily methane production rate was similar for the two codigesters, which increased with each incremental increase in the feed LCFA. By the end of Phase I, the methane production rate for the codigesters was on average 70% higher than that for the manure-fed control (average of 1700 mL/d versus 1000 mL/d). Similar increases in methane production during Phase II were observed with each incremental increase in LCFA loading to the codigesters. At the highest LCFA loading of 64% of the influent COD during Phase II, the daily methane production from both codigesters was on average 440% higher than that for the Phase I manure-fed control (average of 5500 mL/d versus 1000 mL/d). The average specific methane yield was 8% higher for the pulse-fed digester relative to the continuous-fed codigester at the highest LCFA loading during Phase II, but the difference was not statistically significant due to the large daily variation in gas production with the pulse-feeding regime.

Acetate was the only VFA that accumulated during LCFA codigestion. The effluent acetate concentration for both digesters was always below 40 mg/L during Phase I, indicating that the rates of aceticlastic activity matched that of acetogenesis. When the LCFA was increased to 64% of the influent COD during Phase II, the effluent acetate concentration in the pulse-fed codigester v.2 was elevated to 500-1000 mg/L after 24 hrs of LCFA feeding, but again reached low levels by the end of the 48-hour LCFA feed cycle time. In contrast, the effluent acetate concentration in the continuous-fed digester began to accumulate after day 150 when the feed LCFA COD fraction was increased to 64%, and did not reach a final lower acetate concentration as was observed with the pulse-fed codigester. The feed LCFA loading was reduced from 64% to 62% of the influent COD on day 184 to abate acetate accumulation within the continuous-fed codigester; yet the effluent acetate concentration steadily increased to over 3,000 mg/L, suggesting that the high LCFA loading led to digester imbalance in the continuous-fed codigester.

Batch kinetic assays were conducted to examine LCFA inhibition of the maximum acetate utilization rate ($q_{max,acetate}$). Values of $q_{max,acetate}$ were typically inhibited at oleate concentrations above 4 g/L, with the exception of day 204 when the $q_{max,acetate}$ of both Phase II codigesters was inhibited by over 60% at an oleate concentration of 4 g/L. The maximum initial LCFA concentration that occurred within any digester after feeding would have been 2.7 g/L in the pulse-fed codigester starting on day 150, due to the 48-hr LCFA pulse feed cycle (based on an OLR of 6.0 g COD/L-d, 64% of influent COD as oleate, and 2.9 g COD/g oleate). Effluent

LCFA concentrations remained below 150 mg/L throughout Phases I and II, indicating that the codigester oleate levels likely remained below inhibitory threshold concentrations for aceticlastic methanogenesis during operation. The $q_{max,acetate}$ values of both codigesters were similar at the end of Phase I, but in Phase II the pulse-fed codigester v.2 had a $q_{max,acetate}$ that was 35% higher than the continuous-fed codigester by day 204 (Table 1). The $q_{max,acetate}$ of the pulse-fed codigester v.2 biomass surpassed that of the continuous-fed codigester on day 170 (Table 1), when the LCFA loading to both digesters was equal. Notably, the $q_{max,acetate}$ of both codigesters decreased by over 50% between days 170 and 204 at the highest LCFA loadings tested; yet the acetate utilization kinetics in the pulse-fed codigester v.2 still remained higher (Table 1). Pulse-feeding oleate during codigestion therefore selected for higher acetate utilization kinetics as well as an increased aceticlastic resistance to oleate.

The higher $q_{max,acetate}$ values reached in the pulse-fed codigester relative to the continuous-fed codigester also coincided with higher maximum oleate conversion rates ($q_{max,oleate}$) in the pulse-fed systems by the end of Phase I and II (Table 1). By the end of Phase I, the $q_{max,oleate}$ of the pulse-fed codigester was 16% higher than the continuous-fed codigester. When the LCFA loading was equal between the Phase II codigesters on days 170 and 204, the $q_{max,oleate}$ was 16% and 20% higher in the pulse-fed codigester v.2 than the continuous-fed codigester, respectively. Biokinetic modeling of the oleate conversion rates also showed that the predicted $q_{max,oleate}$ for the pulse-fed codigester was 65% and 27% higher than the continuous-fed codigester at the end of Phase I and II, respectively. During Phase I, the maximum rate of oleate conversion in the batch assays was observed at initial oleate concentrations typically between 175-350 mg/L, above which the consumption rate was inhibited. During Phase II, the inhibitory concentration of oleate for both codigesters sludges increased toward ~1800 mg/L. These inhibitory levels were below the observed level of toxicity for aceticlastic methanogenesis, indicating that substrate inhibition of oleate conversion was likely attributed to decreased activity of syntrophic bacteria and/or the hydrogen utilizing methanogenic partners. Overall, these results show that the LCFA feeding regime during codigestion impacted the specific kinetics of syntrophic LCFA conversion, with pulse-fed systems selecting for higher conversion rates.

Table 1: Maximum conversion rates for oleate ($q_{max,oleate}$) and acetate ($q_{max,acetate}$) observed in the batch kinetic assays with digester sludges collected throughout Phases I and II. Values in parentheses indicate one standard deviation.

Reactor	Day	Oleate q_{max} (mL CH_4/g VS-d)	Acetate q_{max} (mL CH_4/g VS-d)
Phase I Digesters			
Control	0	24 (2)	45 (4)
	41	11 (0)	35 (4)
	84	15 (1)	39 (1)
Continuous-Fed Co-digester	0	27 (2)	47 (5)
	41	34 (1)	40 (3)
	84	56 (4)	64 (6)
Pulse-Fed Co-digester	0	25 (4)	37 (3)
	41	23 (1)	32 (2)
	84	65 (2)	61 (8)
Phase II Digesters			
Pulse-Fed Co-digester	129	49 (1)	88 (8)
	148	124 (9)	104 (19)
	170	168 (4)	142 (9)
	204	108 (1)	70 (8)
Continuous-Fed Co-digester	129	167 (0)	137 (11)
	170	144 (5)	111 (13)
	204	81 (2)	52 (5)

3.2 Microbial community structure and diversity was altered by the codigester feeding pattern

By the end of Phase I, the *Bacteria* and *Archaea* community profiles were similar between the pulse-fed codigester and the continuous-fed codigester (Figs. 1A and 2A). Correspondence analysis on *Bacteria* and *Archaea* OTU read abundances showed that all digester community profiles were closely clustered on day 84. A notable change in the *Bacteria* community structure during Phase I involved an increase of β-oxidizing *Syntrophomonas* sequences to 9% ± 0.2 and 4% ± 0.1 of bacterial reads in the pulse-fed and continuous-fed codigesters by day 84, respectively, which were both significantly higher (p<1e-3) than the control at 0.5% ± 0.04 (Figure 2A). While the fraction of *Syntrophomonas* in the pulse-fed codigester was almost double that of the continuous-fed codigester by day 84 (p<0.001; Figure 2A), the continuous-fed codigester had significantly higher absolute concentrations of total *Bacteria* 16S rRNA genes at the end of Phase I (p<0.01, respectively; Figures 1B and 2B). Consequentially, the absolute concentration of *Syntrophomonas* 16S rRNA genes measured by qPCR was similar in the pulse-fed codigester and continuous-fed codigester on day 84 (Figure 3B).

Both the microbial community abundance and composition was altered by the LCFA feeding strategy during Phase II. Correspondence analysis revealed that the *Bacteria* and *Archaea* OTU read abundance profiles diverged between the pulse-fed codigester v.2 and continuous-fed codigester beginning on day 129 onward. The *Archaea* population in the pulse-fed codigester v.2 had significantly elevated relative fractions of *Methanosaeta, Methanoculleus,* and *Methanospirillum* sequences in comparison to the continuous-fed codigester by day 170 (p<0.001; Figure 1A). In contrast, the continuous-fed codigester *Archaea* population had a significantly higher fraction of *Methanobacterium* than the pulse-fed codigester v.2 by day 170 (p<0.001; Figure 1A). By day 204, the total *Archaea* 16S rRNA gene concentration was similar in both codigesters (Fig. 2C). The *Bacteria* community of the pulse-fed codigester v.2 had a significantly higher fraction of β-oxidizing *Syntrophomonas* (p<0.001) by day 170 in Phase II relative to the continuous-fed codigester (Fig. 2A). The relative fraction of *Syntrophomonas* in the *Bacteria* sequence library increased to 43% ± 2 by day 204 in the pulse-fed codigester v.2, which was almost double that of the continuous-fed codigester at 22% ± 0.4 (Fig. 2A). The total *Bacteria* 16S rRNA gene concentration reached similar levels in both codigesters by day 204 (Fig. 2C), indicating that the absolute abundance of *Syntrophomonas* was higher in the pulse-fed codigester due to its higher community fraction. This result was confirmed with qPCR on total *Syntrophomonas* 16S rRNA genes, which showed a higher concentration in the pulse-fed codigester v.2 relative to the continuous-fed codigester on day 204 (p<0.001; Fig. 3C). The increased abundance and prevalence of *Syntrophomonas* in both codigesters suggests that it played an important role in β-oxidizing the feed LCFA, while its growth was impacted by the LCFA feeding strategy.

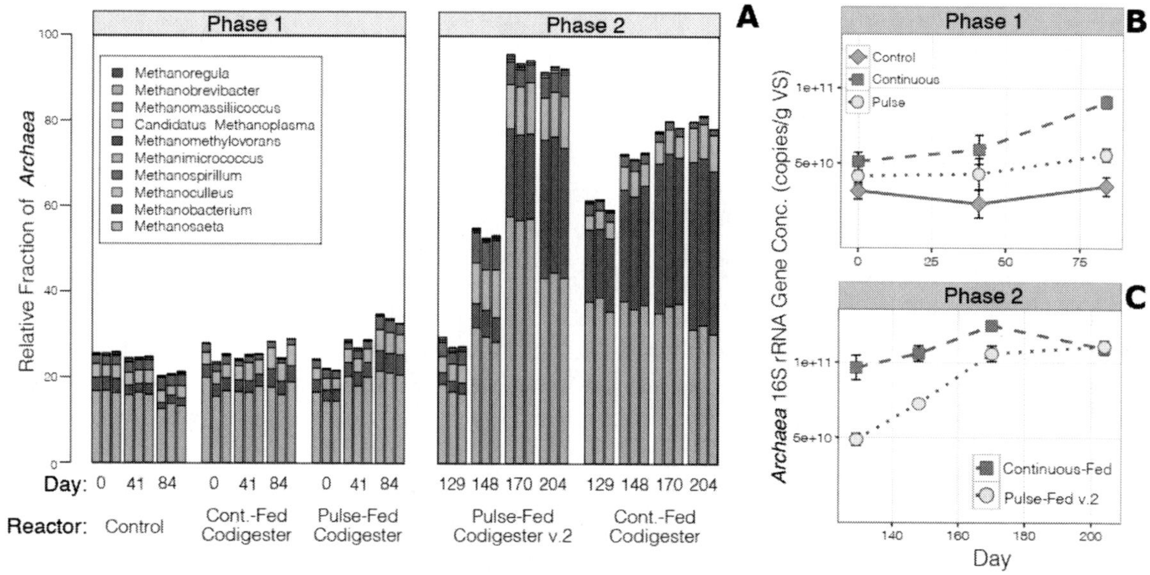

Figure 1: (A) Relative fractions of the 12 most abundant genera in the *Archaea* 16S rRNA amplicon sequence libraries of the digesters during Phases I and II. Profiles from triplicate DNA extractions are shown at each time point. (B) Concentration of *Archaea* 16S rRNA genes in the digesters during Phase I measured by qPCR. (C) Concentration of *Archaea* 16S rRNA genes in the digesters during Phase II measured by qPCR. Error bars represent one standard deviation based on triplicate DNA extractions.

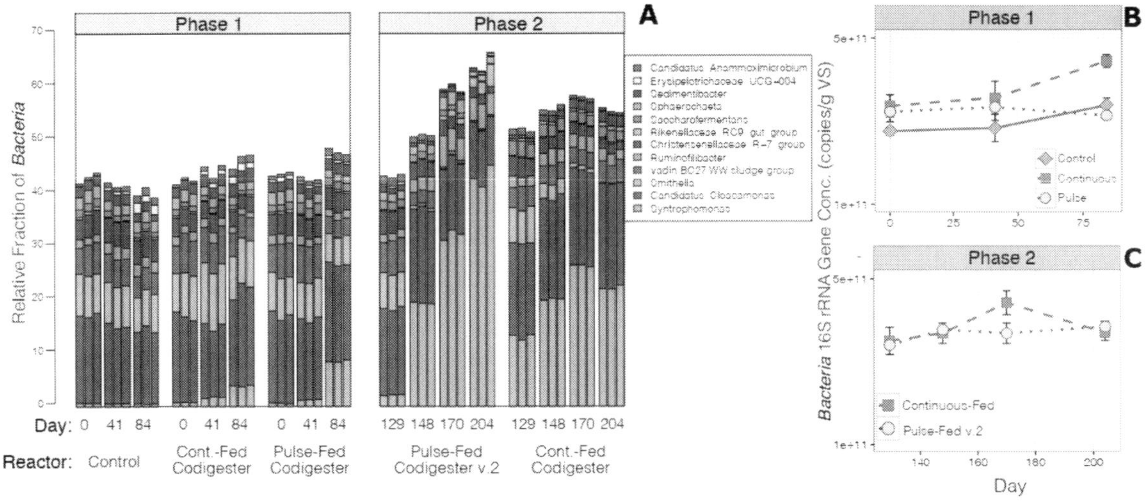

Figure 2: (A) Relative fractions of the 12 most abundant genera in the *Bacteria* 16S rRNA amplicon sequence libraries of the digesters during Phases I and II. Profiles from triplicate DNA extractions are shown at each time point. (B) Concentration of *Bacteria* 16S rRNA genes in the digesters during Phase I measured by qPCR. (C) Concentration of *Bacteria* 16S rRNA genes in the digesters during Phase II measured by qPCR. Error bars represent one standard deviation based on triplicate DNA extractions.

Denoising the amplicon reads into their exact sequences with DADA2 (Callahan et al., 2016) prior to clustering at 99.5% similarity allowed for a high-resolution analysis of community composition without generating spurious OTUs. Such high-resolution analysis showed that at least 85% of *Syntrophomonas* classified sequences were distributed among 13 OTUs (Fig. 3A). Time-series profiles of the dominant *Syntrophomonas* OTUs showed divergence based on codigester LCFA feeding patterns (Fig. 3A). These results indicate that the fine-scale community structure of LCFA β-oxidizing bacteria in the codigesters was impacted by the LCFA feeding pattern. In contrast, over 95% of the sequences classified within the aceticlastic genus of *Methanosaeta* were comprised by 3 OTUs, which had similar relative abundance profiles between the codigester feed regimes. The hydrogen-utilizing genus *Methanobacterium* was represented at over 90% by a single OTU. The codigester LCFA feed pattern therefore impacted the overall *Archaea* community composition and abundance to a greater extent than the fine-scale *Archaea* community composition.

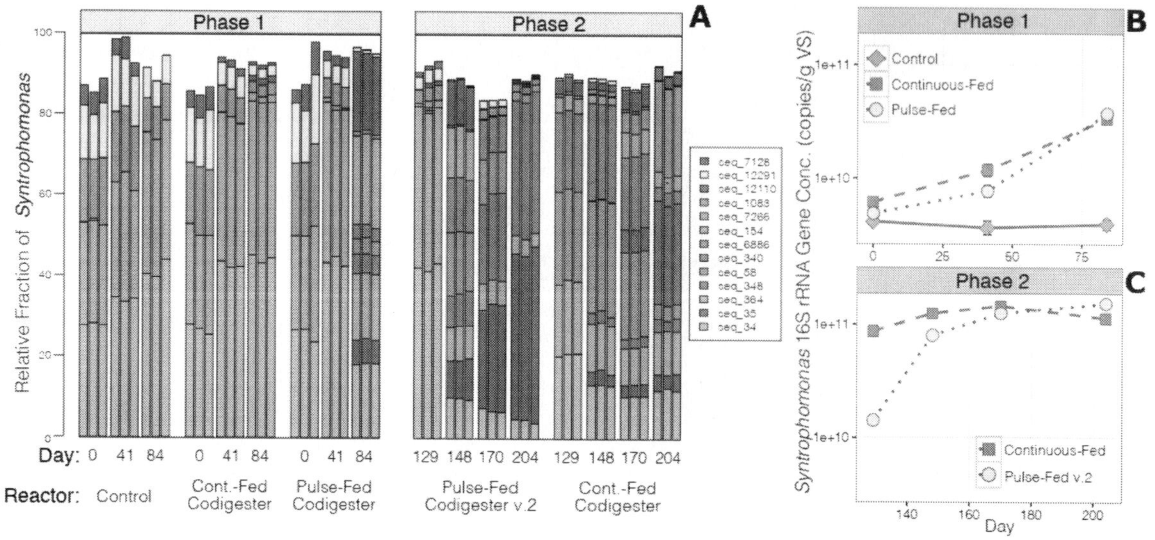

Figure 3: (A) Relative fractions of the 13 most abundant OTUs classified within the *Syntrophomonas* genus of the digesters during Phases I and II based on 16S rRNA amplicon sequencing. Profiles from triplicate DNA extractions are shown at each time point. (B) Concentration of *Syntrophomonas* 16S rRNA genes in the digesters during Phase I measured by qPCR. (C) Concentration of *Syntrophomonas* 16S rRNA genes in the digesters during Phase II measured by qPCR. Error bars represent one standard deviation based on triplicate DNA extractions.

3.3 Methane production kinetics were predicted by syntrophic community abundance and composition

A linear regression model was established to examine the relationship between maximum oleate conversion rates ($q_{max,oleate}$) measured during batch kinetic assays with all digester sludges (16 tests total) and the 16S rRNA gene concentration of total *Syntrophomonas* in the digester sludges. The total *Syntrophomonas* 16S rRNA gene concentration was a significant predictor of $q_{max,oleate}$ values (p<0.001). The specific slope of the linear regression was 8.6 x 10^{-10} mL CH_4/16S rRNA gene-day, and the overall model fit had an R^2 of 0.73.

To examine the importance of syntrophic bacteria community structure, a multiple linear regression model was developed to predict $q_{max,oleate}$ values based on the absolute concentrations of a selected set of *Syntrophomonas* 99.5% OTUs. Feature selection and cross-validation showed that the linear model provided the greatest R^2 by incorporating the absolute concentrations of six *Syntrophomonas* OTUs. The resulting multiple linear regression model had an improved R^2 of 0.92 in comparison to the linear regression model that was dependent only on the total *Syntrophomonas* 16S rRNA gene concentration.

A similar approach showed that the absolute concentration of total *Methanosaeta* 16S rRNA genes was a significant predictor of the aceticlastic $q_{max,acetate}$ values (p=0.001), although the R^2 of the correlation was only 0.51. A multiple-regression model was developed to predict $q_{max,acetate}$ values based on the absolute 16S rRNA gene concentrations of three *Methanosaeta* OTUs, which resulted in an R^2 of 0.74. Thus, oleate and acetate conversion kinetics of the codigester biomasses were best predicted based on fine-scale community structure.

4. Conclusions

This study assessed the impacts of codigester LCFA feeding pattern on microbial community structure and bioconversion kinetics. The following can be concluded from the results of this study:

- Overall *Bacteria* and *Archaea* community profiles diverged based on the codigester LCFA feeding pattern.
- *Syntrophomonas* bacteria were significantly enriched in both pulse-fed and continuous-fed LCFA codigesters. Higher relative and absolute concentrations of *Syntrophomonas* 16S rRNA genes were observed in the pulse-fed codigester v.2. The pulse-fed LCFA codigester v.2 reached higher relative fractions of aceticlastic *Methanosaeta* by the end of Phase II operation at high LCFA loadings. The codigester LCFA feeding strategy thus impacted the composition of the syntrophic community.
- Higher $q_{max,oleate}$ values were reached in the pulse-fed LCFA codigesters in Phase I and II operation. The $q_{max,acetate}$ values were also higher in the pulse-fed codigester by the end of Phase II operation. These results indicate that the anaerobic codigester LCFA feeding pattern influenced microbial bioconversion kinetics.
- $q_{max,oleate}$ and $q_{max,acetate}$ both correlated significantly with the concentrations of total *Syntrophomonas* and *Methanosaeta* 16S rRNA genes, respectively. Improved model accuracy was obtained through multiple-linear regression based on the absolute concentration of fine-scale OTUs. This method could be used to monitor key microbial OTUs that significantly contribute to substrate conversion kinetics in anaerobic digesters.

References:

Angelidaki, I., Ahring, B.K., 1992. Effects of free long-chain fatty acids on thermophilic anaerobic digestion. Appl. Microbiol. Biotechnol. 37, 808–812. doi:10.1007/BF00174850

Baserba, M.G., Angelidaki, I., Karakashev, D., 2012. Effect of continuous oleate addition on microbial communities involved in anaerobic digestion process. Bioresour. Technol. 106, 74–81. doi:10.1016/j.biortech.2011.12.020

Bishop, C.P., Shumway, C.R., 2009. The Economics of Dairy Anaerobic Digestion with Coproduct Marketing. Appl. Econ. Perspect. Policy 31, 394–410. doi:10.1111/j.1467-9353.2009.01445.x

Briones, A., Raskin, L., 2003. Diversity and dynamics of microbial communities in engineered environments and their implications for process stability. Curr. Opin. Biotechnol. 14, 270–276. doi:10.1016/S0958-1669(03)00065-X

Callahan, B.J., McMurdie, P.J., Rosen, M.J., Han, A.W., Johnson, A.J.A., Holmes, S.P., 2016. DADA2: High-resolution sample inference from Illumina amplicon data. Nat. Methods 13, 581–583. doi:10.1038/nmeth.3869

Cirne, D.G., Paloumet, X., Björnsson, L., Alves, M.M., Mattiasson, B., 2007. Anaerobic digestion of lipid-rich waste—Effects of lipid concentration. Renew. Energy 32, 965–975. doi:10.1016/j.renene.2006.04.003

Conklin, A., Stensel, H.D., Ferguson, J., 2006. Growth Kinetics and Competition Between Methanosarcina and Methanosaeta in Mesophilic Anaerobic Digestion. Water Environ. Res. 78, 486–496. doi:10.2175/106143006X95393

De Vrieze, J., Saunders, A.M., He, Y., Fang, J., Nielsen, P.H., Verstraete, W., Boon, N., 2015. Ammonia and temperature determine potential clustering in the anaerobic digestion microbiome. Water Res. 75, 312–323. doi:10.1016/j.watres.2015.02.025

De Vrieze, J., Verstraete, W., Boon, N., 2013. Repeated pulse feeding induces functional stability in anaerobic digestion. Microb. Biotechnol. 6, 414–424. doi:10.1111/1751-7915.12025

Eaton, A.D., Franson, M.A.H., 2005. Standard Methods for Examination of Water & Wastewater. Amer Public Health Assn.

Edgar, R.C., 2013. UPARSE: highly accurate OTU sequences from microbial amplicon reads. Nat. Methods 10, 996–998. doi:10.1038/nmeth.2604

Hanaki, K., Matsuo, T., Nagase, M., 1981. Mechanism of inhibition caused by long-chain fatty acids in anaerobic digestion process. Biotechnol. Bioeng. 23, 1591–1610. doi:10.1002/bit.260230717

Hao, L., Bize, A., Conteau, D., Chapleur, O., Courtois, S., Kroff, P., Desmond-Le Quéméner, E., Bouchez, T., Mazéas, L., 2016. New insights into the key microbial phylotypes of anaerobic sludge digesters under different operational conditions. Water Res. 102, 158–169. doi:10.1016/j.watres.2016.06.014

Hatamoto, M., Imachi, H., Yashiro, Y., Ohashi, A., Harada, H., 2007. Diversity of Anaerobic Microorganisms Involved in Long-Chain Fatty Acid Degradation in Methanogenic Sludges as Revealed by RNA-Based Stable Isotope Probing. Appl. Environ. Microbiol. 73, 4119–4127. doi:10.1128/AEM.00362-07

Hwu, C.-S., Lettinga, G., 1997. Acute toxicity of oleate to acetate-utilizing methanogens in mesophilic and thermophilic anaerobic sludges. Enzyme Microb. Technol. 21, 297–301. doi:10.1016/S0141-0229(97)00050-1

Jost, L., 2006. Entropy and diversity. Oikos 113, 363–375. doi:10.1111/j.2006.0030-1299.14714.x

Kuhn, M., 2008. Caret package. J. Stat. Softw. 28.

Labatut, R.A., Angenent, L.T., Scott, N.R., 2011. Biochemical methane potential and biodegradability of complex organic substrates. Bioresour. Technol. 102, 2255–2264. doi:10.1016/j.biortech.2010.10.035

Lalman, J., Bagley, D.M., 2002. Effects of C18 long chain fatty acids on glucose, butyrate and hydrogen degradation. Water Res. 36, 3307–3313. doi:10.1016/S0043-1354(02)00014-3

Love, M.I., Huber, W., Anders, S., 2014. Moderated estimation of fold change and dispersion for RNA-seq data with DESeq2. Genome Biol. 15, 1.

Menes, R.J., Travers, D., 2006. Detection of fatty acid beta-oxidizing syntrophic bacteria by fluorescence in situ hybridization. Water Sci. Technol. 54, 33. doi:10.2166/wst.2006.483

Mulat, D.G., Jacobi, H.F., Feilberg, A., Adamsen, A.P.S., Richnow, H.-H., Nikolausz, M., 2016. Changing Feeding Regimes To Demonstrate Flexible Biogas Production: Effects on Process Performance, Microbial Community Structure, and Methanogenesis Pathways. Appl. Environ. Microbiol. 82, 438–449. doi:10.1128/AEM.02320-15

Palatsi, J., Illa, J., Prenafeta-Boldú, F.X., Laureni, M., Fernandez, B., Angelidaki, I., Flotats, X., 2010. Long-chain fatty acids inhibition and adaptation process in anaerobic thermophilic digestion: Batch tests, microbial community structure and mathematical modelling. Bioresour. Technol. 101, 2243–2251. doi:10.1016/j.biortech.2009.11.069

Parada, A.E., Needham, D.M., Fuhrman, J.A., 2016. Every base matters: assessing small subunit rRNA primers for marine microbiomes with mock communities, time series and global field samples. Environ. Microbiol. 18, 1403–1414. doi:10.1111/1462-2920.13023

Rinzema, A., Boone, M., Knippenberg, K. van, Lettinga, G., 1994. Bactericidal Effect of Long Chain Fatty Acids in Anaerobic Digestion. Water Environ. Res. 66, 40–49.

Schink, B., 1997. Energetics of syntrophic cooperation in methanogenic degradation. Microbiol. Mol. Biol. Rev. 61, 262–280.

Shigematsu, T., Tang, Y., Mizuno, Y., Kawaguchi, H., Morimura, S., Kida, K., 2006. Microbial diversity of mesophilic methanogenic consortium that can degrade long-chain fatty acids in chemostat cultivation. J. Biosci. Bioeng. 102, 535–544. doi:10.1263/jbb.102.535

Sousa, D.Z., Pereira, M.A., Smidt, H., Stams, A.J.M., Alves, M.M., 2007. Molecular assessment of complex microbial communities degrading long chain fatty acids in methanogenic bioreactors. FEMS Microbiol. Ecol. 60, 252–265. doi:10.1111/j.1574-6941.2007.00291.x

Sousa, D.Z., Smidt, H., Alves, M.M., Stams, A.J.M., 2009. Ecophysiology of syntrophic communities that degrade saturated and unsaturated long-chain fatty acids. FEMS Microbiol. Ecol. 68, 257–272. doi:10.1111/j.1574-6941.2009.00680.x

Sundberg, C., Al-Soud, W.A., Larsson, M., Alm, E., Yekta, S.S., Svensson, B.H., Sørensen, S.J., Karlsson, A., 2013. 454 pyrosequencing analyses of bacterial and archaeal richness in 21 full-scale biogas digesters. FEMS Microbiol. Ecol. 85, 612–626. doi:10.1111/1574-6941.12148

Wang, Q., Garrity, G.M., Tiedje, J.M., Cole, J.R., 2007. Naïve Bayesian Classifier for Rapid Assignment of rRNA Sequences into the New Bacterial Taxonomy. Appl. Environ. Microbiol. 73, 5261–5267. doi:10.1128/AEM.00062-07

Werner, J.J., Knights, D., Garcia, M.L., Scalfone, N.B., Smith, S., Yarasheski, K., Cummings, T.A., Beers, A.R., Knight, R., Angenent, L.T., 2011. Bacterial community structures are

unique and resilient in full-scale bioenergy systems. Proc. Natl. Acad. Sci. 108, 4158–4163. doi:10.1073/pnas.1015676108
Westerholm, M., Müller, B., Isaksson, S., Schnürer, A., 2015. Trace element and temperature effects on microbial communities and links to biogas digester performance at high ammonia levels. Biotechnol. Biofuels 8, 154. doi:10.1186/s13068-015-0328-6
Yu, Y., Lee, C., Kim, J., Hwang, S., 2005. Group-specific primer and probe sets to detect methanogenic communities using quantitative real-time polymerase chain reaction. Biotechnol. Bioeng. 89, 670–679. doi:10.1002/bit.20347
Ziels, R.M., Beck, D.A.C., Martí, M., Gough, H.L., Stensel, H.D., Svensson, B.H., 2015. Monitoring the dynamics of syntrophic β-oxidizing bacteria during anaerobic degradation of oleic acid by quantitative PCR. FEMS Microbiol. Ecol. 91, fiv028. doi:10.1093/femsec/fiv028
Zonta, Ž., Alves, M.M., Flotats, X., Palatsi, J., 2013. Modelling inhibitory effects of long chain fatty acids in the anaerobic digestion process. Water Res. 47, 1369–1380. doi:10.1016/j.watres.2012.12.007

Resource Recovery Cooperatives for the Security of Consistent, Quality Codigestion Feedstock

David McNeil, City of Tempe/Tempe Grease Cooperative

Abstract:

Codigesting wastewater utilities typically have regulatory control of FOG at the food service establishment (FSE), and ownership of FOG at the wastewater plant after a tipping transaction, but lose control within the shoulders of the supply chain. This loss of control between the initial regulatory transaction and the WWTP tipping transaction creates uncertainties related to codigestion feedstock security, feedstock quality and contamination, operational logistics and costs at the plant, and related economics. Tempe Arizona's use of a retailers' cooperative model for FOG collection and recovery gives the utility an opportunity, through contract, for total control over the feedstock throughout the supply-chain, and for better long-term feedstock security over the operational life of FOG receiving and codigestion facilities.

Background

Wastewater utilities with codigestion facilities typically share the complementary goals of reducing the negative impacts of fats, oils, and grease (FOG) on municipal sewer infrastructure, and recovering FOG for renewable energy production at wastewater treatment plant (WWTP) digesters. A codigesting utility's ideal logistical lifecycle for a molecule of FOG is to effectively capture it at commercial food service establishments (FSEs) before it enters sewers, to transport it to digesters at a wastewater treatment plant (WWTP), and to fully recover and codigest it to generate renewable biogas. Current industry practices attempt to realize this lifecycle through a sequence of regulatory and retail transactions throughout the collection-transport-delivery supply chain:

1) Sewer utilities engage in individual regulatory transactions with FSEs to ensure compliance;
2) To meet their obligations under #1, restaurants engage in individual retail transactions with waste haulers for grease trap cleaning and maintenance at, and waste removal from, the premise; and
3) Utilities engage in individual retail transactions with grease haulers responsible for #2 to obtain ownership at the WWTP, where it is then codigested.

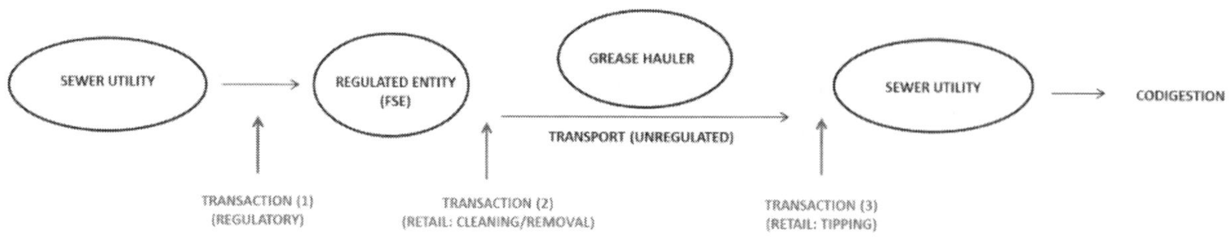

FIGURE 1: TRANSACTIONAL LIFECYCLE OF FOG REMOVAL AND RECOVERY

Problem

This transactional structure fails to efficiently and effectively achieve utility FOG goals at several levels, which contributes to reported FSE compliance rates with municipal FOG-related ordinances of <30% in major U.S. cities and uncollected grease rates between 800 and 1,700 lbs/restaurant/year (EPA 833-F-07-007; EPA-833-F-12-003), and underachievement of FOG recovery and codigestion at our nation's WWTPs (Shen et al, 2015; Fillmore, et al, 2016).

First, the sewer industry's regulatory transactional structure often fails to achieve compliance at the FSE because it holds restaurants accountable for the service practices of third-party grease haulers. Accountability for cleaning practices is passed sequentially from the sewer authority to the hauler through the restaurant, requiring restaurateurs to enforce the utility's service quality requirements. This delegation of enforcement from the regulatory to the retail transaction fails because restaurateurs typically do not have the knowledge or skill sets to effectively enforce requirements through oversight of this very specialized plumbing service. The sewer agency is the sole expert of what constitutes effective and compliant cleaning practices, yet there is no direct accountability link between the waste hauler and the sewer authority (see Figure 2).

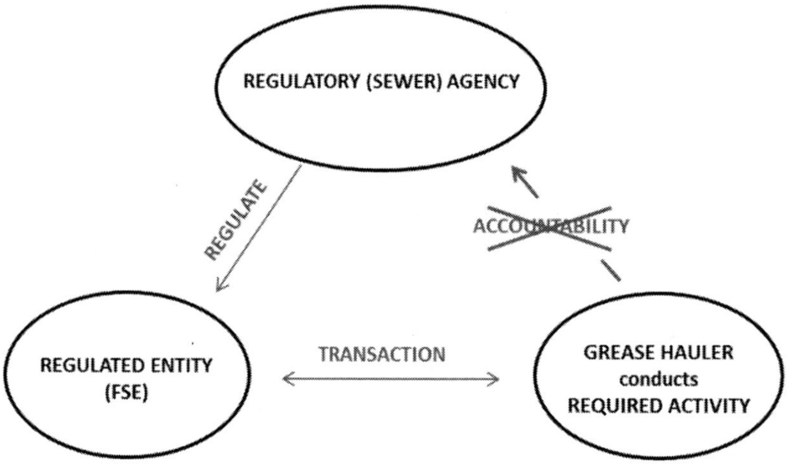

FIGURE 2: FOG ACCOUNTABILITY GAP

Additionally, due to the fact that enforcement responsibility is delegated to the restaurant premise manager and regulatory criteria is limited to the cleaning practices at the premise only, regulatory control stops at the property boundary, beyond which both the restaurant and utility have no authority. Unregulated transport between FSEs and the final disposition site, a WWTP or other, creates obstacles to achieving both sewer management and feedstock security goals.

Unscrupulous haulers can discharge illegally to cut disposal costs, sometimes back to the same sewers that their service was intended to protect. Uncontrolled transport can also result in feedstock contamination and digester upset, as WWTPs receiving FOG loads have no control or assurance that hauler routes excluded collection of materials that could upset digester operations. Operating a WWTP as a commercial tipping facility is not familiar industry for municipal utilities, and uncertainty regarding both consistent quantity and quality of feedstock translates to potential operational challenges for digesters, elaborate monitoring, handling, treatment and decontamination processes at receiving plants, and uncertainty about gross energy production (Parry, 2014).

Lastly, receiving FOG through a commercial waste disposal transaction creates uncertainty regarding that transaction's contribution to cost-recovery for codigesting WWTPs. Market manipulations and competively fluctuating landfill costs add uncertainty to a market that is already vulnerable to fluctuating energy prices (Parry, 2014). Despite these shortcomings, most codigestion facilities continue to recover feedstock through commercial tipping operations.

The traditional FOG collection and recovery structure (Figure 1) seats the utility at both the front end and the back end of the transactional sequence (i.e., the utility initiates the first regulatory transaction that compels waste removal by the restaurant, and the sewer utility is on the receiving end of the retail transaction that secures the feedstock from haulers), but suffers from the fact that the utility is absent from transactions, and transactional control of FOG, within the supply chain. If the sewer utility could gain quality control over cleaning and removal service at the FSE, and retain both ownership and control of FOG throughout transport between the FSE and WWTP, the quality control failures inherent in the traditional FOG structure could be resolved.

Tempe Grease Cooperative

The City of Tempe Arizona's Grease Cooperative (TGC), a municipal FOG collection and recovery cooperative run by the city's wastewater utility in partnership with community restaurants and partner grease hauling vendors, may be just the answer to retaining service and feedstock quality control throughout the FOG supply chain. Tempe found this solution by searching for a business model that would secure utility ownership of FOG waste/feedstock, and found such a model that to Tempe's knowledge had never been implemented by a local government: *retailers' cooperatives.*

Retailers' cooperatives have long been used by groups of independent retailers to leverage their collective purchasing power to obtain competitive advantages in the marketplace by acquiring discounts from suppliers and sharing marketing costs through a central buying organization (see Figure 3).

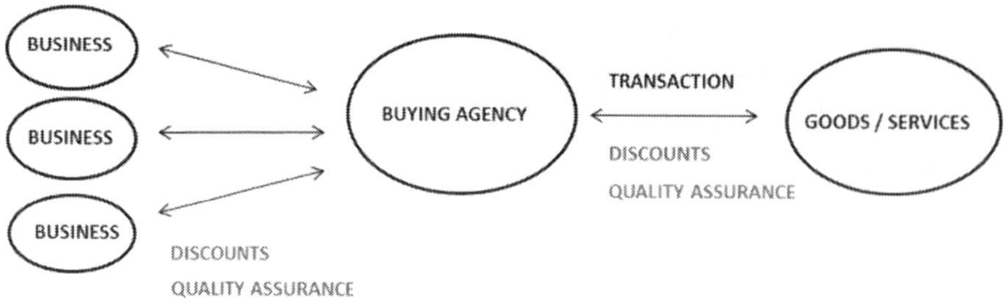

FIGURE 3: RETAILERS' COOPERATIVES

Through the TGC, the City of Tempe serves as that buying agent for compliance requirements on behalf of local restaurants. The TGC is the first program in the world in which a regulatory sewer authority brokers both pricing and service quality for grease hauling services on behalf of regulated food service establishments (FSEs) that voluntarily enroll in this program as an alternative to being held responsible for the practices of their hired vendors (see Figure 4). As the central buying organization, the utility provides member restaurants with both discounts and compliance assurance under collective contracts that are established and administered by the utility. As a result, partners in the TGC collectively establish more sustainable plumbing and sewer infrastructure for the region, and create a more livable city by reducing odors and sewer overflows at member facilities and community-wide, all at a lower cost for both the utility and its regulated users.

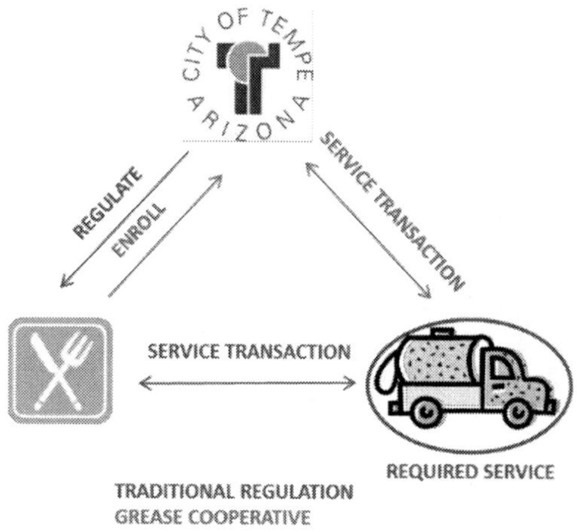

FIGURE 4: TEMPE GREASE COOPERATIVE

Unlike traditional FOG regulation, transaction management allows the City to manage services beyond the property boundary of the regulated business. In addition to GPS and other tracking requirements to ensure complete transparency in the transport and disposition stages of service, contracts with haulers establish city ownership of hauled waste, securing its availability and manageability as codigestion feedstock.

The FOG collection and recovery cooperative model gives the utility an opportunity, through contract, for total control over the feedstock from grease trap to codigester. Regulatory authority at the FSE allows the use of kitchen best management practices (BMPs) to manage feedstock quality at the source. Grease trap sizing criteria and pumping requirements can be designed with maximum FOG recovery in mind. Collection routes for vendors can be established, tracked, and enforced to eliminate the risk of contamination from septage or other non-FOG sources, and can even be designed to establish the optimal mix of sources for each load based on FSE-specific FOG information and digester requirements. FSE pump-out schedules can be established to best equalize delivery timing and reduce onsite storage needs. If needed, member FSEs could also be solicited to contribute pre-consumer food waste to further optimize the codigestion recipe. Cradle-to-grave ownership puts the utility in the driver's seat for the entire lifecycle of removal and recovery.

At the WWTP , the utility no longer has to staff and operate a regulated commercial waste disposal facility, but instead opens the plant gates only for Coop deliveries at pre-established times based on contractual schedules. Along with reductions in costs of pretreatment and decontamination resulting from better and more consistent feedstock quality, service and delivery schedules should result in lower staffing requirements and reduced costs.

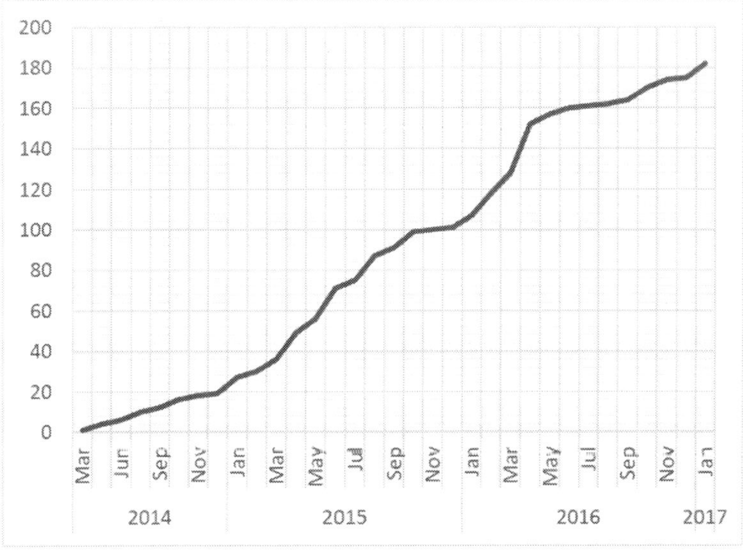

FIGURE 5: TGC MEMBER GROWTH

After three years of implementation, 182 food service establishments, or over one in six regulated food businesses in Tempe, has enrolled in the Tempe Grease Cooperative, understanding that TGC membership is the brand for restaurants that care about their business, their community, their environment, and the disposition of their waste. Based on current enrollment, the Tempe Grease Cooperative is securing almost 600,000 gallons of total grease trap waste annually.

Having made the collection program a reality, Tempe is currently developing a 5-year strategic plan for the TGC, both to resolve administrative challenges identified in the first several years and to complete its core goal of recovering the city's FOG for codigestion. This latter goal is a challenge for Tempe because the city does not own or operate a WWTP within its jurisdiction, but instead has ownership stake in the regional 91st Avenue WWTP operated by the City of Phoenix. The TGC's strategic initiatives over the next 5 years include efforts to regionalize the cooperative to benefit other cities within the Phoenix region and potentially throughout the state, to explore regional partnerships for the delivery of FOG to WWTPs, to engage restaurant industry organizations in the effort, to develop software to facilitate administration of the TGC, and to partner with other utilities and research organizations to study the economic benefits of innovative supply-chain solutions to some of the obstacles to FOG codigestion.

FIGURE 6: TGC RESOURCE RECOVERY TRENDS

Conclusion

For utilities that are already codigesting at independently owned and operated WWTPs, Tempe's resource recovery model provides an opportunity to better manage FOG at the source, to secure ownership, quality, and disposition control throughout the FOG supply chain, to receive a consistently known quantity and quality feedstock, and to establish a more facility-friendly and lower risk method of securing feedstock.

The development and implementation of the Tempe Grease Cooperative not only capitalizes on existing and future waste-to-energy opportunities, but also demonstrates the value of innovative approaches to engaging private enterprise in public policies and executing innovative business delivery frameworks to create more sustainable communities. The TGC establishes a blueprint for using group discounts to promote organics recovery and other sustainable practices community-wide using the purchasing power of municipal governments. The TGC embodies the vision of a City that embraces innovation and partnership to best serve its businesses, citizens, and visitors, and a restaurant community that believes in doing the right things for both business and its environment.

REFERENCES

Fillmore, L., Deines, A., Case for Codigestion at Water Resource Recovery Facilities, BioCycle, September 2016

Parry, D.L., Co-Digestion of Organic Waste Products with Wastewater Solids, Water Environmental Research Foundation, OWSO5R07, 2014

Shen, Y., Linville, J.L, Urgun-Demirtas, M., Mintz, M.M., Snyder, S.W., An Overview of Biogas and Utilization at Full-Scale Wastewater Treatment Plants (WWTPs) in the United States: Challenges and Opportunities Towards Energy-Neutral WWTPs, Renewable and Sustainable Energy Reviews, May 2015

U.S. Environmental Protection Agency, Office of Water, National Pretreatment Program, Fact Sheet: Controlling Fats, Oils, and Grease Discharges from Food Service Establishments, EPA-833-F-07-007; EPA-833-F-12-003, July 2007, updated September, 2012

Codigestion Study Helps Turn Industrial Pre-Treatment Problem to Renewable Energy Benefit

Scott Hardy[1], David Ornelas[2], Edward Keenan[1], Brandt Miller[1], Michelle Brown[1], Chamindra Dassanayake[1], Spyros Pavlostathis[3], Mohammad Abu-Orf[1]

[1] Hazen and Sawyer
[2] El Paso Water Utility
[3] Georgia Institute of Technology

ABSTRACT

El Paso Water (EPW) noticed a spike in influent biochemical oxygen demand (BOD) at one of their treatment plants that was causing process issues in the Spring of 2015. EPW, which treats on average 60 mgd of wastewater at four treatment facilities, determined that the spike in BOD coincided with the start of a new biofuels facility discharging a high strength waste (HSW) into their collection system. EPW and Hazen and Sawyer (Hazen) investigated and quantified the process impacts of this new waste stream on the treatment plant. Treatability studies performed by Hazen identified that the HSW was non-toxic to the activated sludge process but required EPW to run an additional blower which increased their operational cost. In excess of 100,000 mg/ L COD, the HSW seemed to be an excellent candidate for co-digestion which if viable could reduce EPW aeration costs and increase electricity generation with their existing combined heat and power (CHP) system. Characterization and ultimate digestibility testing was performed by Georgia Institute of Technology (Georgia Tech). The results from Georgia Tech were combined with the digester capacity analysis to estimate the digester gas production for codigesting with the specific industrial waste. Degradability of the HSW doubled when codigested along with primary and waste activated sludge (WAS) compared to digesting it by itself. Irregularities in HSW volatile solids testing is also discussed. EPW is moving forward with a full scale pilot. The power savings from fully utilizing the existing 1 MW CHP system provides a favorable payback for constructing the HSW receiving station.

KEY WORDS: Codigestion, high strength waste, anaerobic digestion, digester gas, digestibility

INTRODUCTION

El Paso Water (EPW) noticed a spike in influent biochemical oxygen demand (BOD) at one of their treatment plants that was causing process issues in the Spring of 2015. EPW, which treats on average combined wastewater flow of 60 mgd at four treatment facilities, determined that the spike in BOD coincided with the start of a new biofuels facility discharging high strength waste (HSW) into their collection system. EPW contracted with Hazen and Sawyer (Hazen) to investigate and quantify process impacts of this new waste stream on their treatment plant. Treatability studies performed by Hazen identified that the HSW was non-toxic to the activated sludge process but required EPW to run an additional blower which increased their operational cost. In excess of 100,000 mg/ L COD, the HSW seemed to be an excellent candidate for co-digestion which if viable could reduce EPW aeration costs and increase electricity generation with their existing combined heat and power (CHP) system. With excess anaerobic digester capacity and needing additional digester gas to maximize the beneficial use of their CHP system, EPW initiated a co-digestion study with Hazen to explore co-digesting the HSW at their 27.7

mgd Haskell Street Wastewater Treatment Plant (HSWWTP). In addition to the HSW, EPW wanted Hazen to investigate the co-digestability of a sugar waste from another industry.

ANAEROBIC DIGESTER CAPACITY ANALYSIS

The HSWWTP has four anaerobic digesters, each with a diameter of approximately 80 feet. Digester No. 1 is the oldest digester and was constructed in 1942. Digesters No. 2 and No. 3 were added twelve years late in 1957, and Digester No. 4 is the newest digester which was constructed in 1977. **Table 1** summarizes additional digester design criteria for each of the anaerobic digesters.

The anaerobic digesters receive both primary and thickened waste activated sludge (TWAS) which is pumped to the digesters from the mixed sludge blend tank. The mixed sludge pump station has a total four progressive cavity pumps, each dedicated to a single digester.

Table 1: Anaerobic Digester Design Criteria

Digester No.	Original Construction Date	Existing Cover Type	Diameter (ft.)	Side Water Depth (ft)	Digester Volume (MG)	Total Digester Volume (MG)
1	1942	Floating	80	30	1.11	4.46
2	1957	Floating	80	30	1.09	
3	1957	Floating	80	30	1.09	
4	1977	Floating	80	31.5	1.17	

Three operational scenarios were considered for determining the volumetric capacity, or the amount of HSW that could be co-digested in the anaerobic digesters.

- **Scenario 1** – Assumes three digesters are in service, which is the current operating scenario with one out of service for improvements. The total HSW and digester feed (PS + TWAS) would be equally distributed between three digesters.
- **Scenario 2** - Assumes all four digesters are in service, which is a likely future operating scenario. The total HSW and digester feed (PS + TWAS) would be equally distributed between four digesters.
- **Scenario 3** – Assumes a single digester is setup for receiving HSW. This scenario may apply if the plant decides to accept HSW on an immediate needs basis to alleviate costs associated with hauling the GAF waste. For this analysis Digester No. 4 was considered as the digester that would receive the HSW with PS and TWAS being split equally among all digesters in service.

The total volumetric loading to the digesters under max month flow (MMF) conditions was 163,628 gpd, which was obtained from historical data (January 2013 through October 2015). The digester capacity was evaluated to maintain a minimum solids retention time (SRT) of 15 days and a volatile solids loading rate of less than 0.16 lbs. of volatile solids (VS) per day per cubic foot of digester volume.

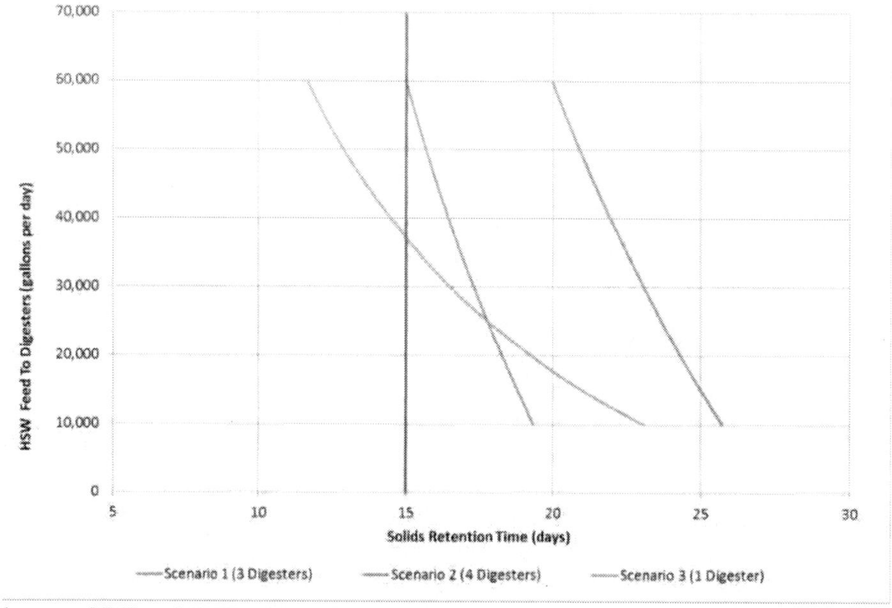

Figure 1: Digester SRT at MMF with HSW Addition

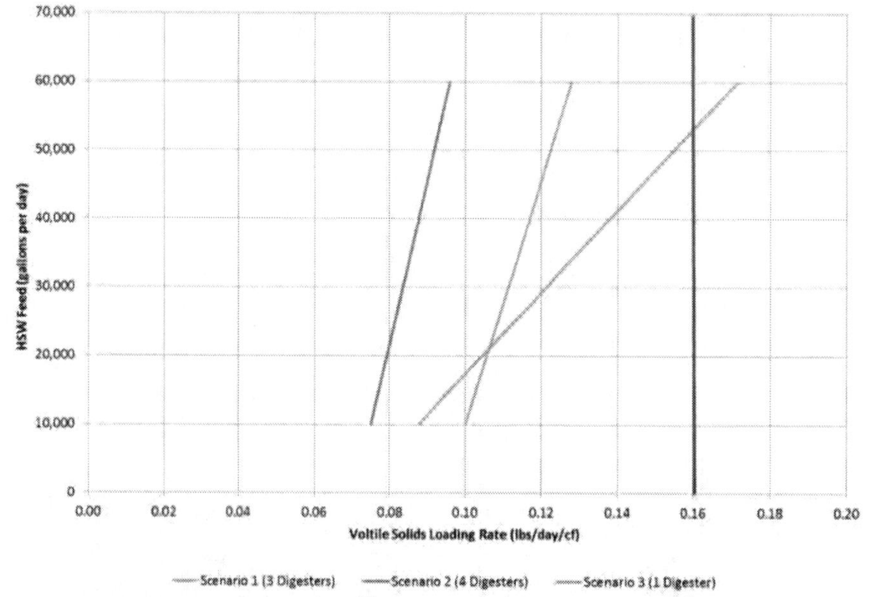

Figure 2: Digester Volatile Solids Loading Rate at MMF with HSW Addition

Under their current operating condition with three digesters in service, the digesters are volume limited and can accept up to 60,000 gallons per day of HSW under MMF conditions to maintain a digester SRT of 15 days. Under this condition, the digesters are not volatile solids limited.

HSW CHARACTERIZATION

The HSW was sampled 8 times in a period of two months and characterized by EPW. One of the samples was also characterized by GA Tech and the results are presented in **Table 2**.

Table 2: HSW Characterization

Parameter	Average	Range	GA Tech
pH	8	7.8 - 9.2	8.2
TSS (mg/L)	2,695	1,580 – 3,840	
VSS (mg/L)	2,623	1,530 – 3,750	
VS (%)	0.72	0.32 – 2.3	3.0
TS (%)	1.0	0.6 – 2.6	3.7
VS/TS	0.72	0.54 – 0.88	0.81
COD (mg/L)	215,250	121,000 – 307,000	184,600
sCOD (mg/L)	175,000	107,000 – 249,000	182,456
Total COD:VS	32	8.9 - 58	6.17
NH_4-N (mg-N/L)	550	2 – 2,700	17
TKN (mg/L)	4.0	3.2 – 5.6	

The HSW is high in soluble COD and low in TSS, which makes it an excellent feedstock for codigestion because it should increase digester gas production without increasing solids production. The pH is around 8, which should help maintain stable digester pH. The ammonia is highly variable and ranges from 2 to 2,700 mg/L NH_4-N; however, the average ammonia loading based on available volume of the HSW is minimal compared to the influent ammonia loading of the plant.

Volatile Solids Measurement Considerations

Sample characterization indicated that the HSW has a COD:VS ratio of greater than 6. A COD:VS ratio greater than 4 is unrealistic. In order to further investigate this high COD/VS fraction, GA Tech extracted the HSW using isooctane, while another sample of equal volume was first dried at 105° C for 20 hours (similar condition to that used to gravimetrically determine the TS content), the dried residue was then re-dissolved in deionized (DI) water and then extracted using isooctane. The two isooctane extracts were then injected in a gas chromatography unit equipped with a mass spectrometric detector (GS-MSD) and the total ion chromatograms (TIC) were evaluated. The MS spectra of significant peaks were compared to those in the National Institute of Standards and Technology (NIST) library for identification. It is noteworthy that sample drying affected the composition of the HSW sample and led to the oxidation of several compounds identified in the raw HSW sample. Based on the total area count of identified and non-identified compounds, sample drying resulted in a loss of about 26% of volatile mass. Therefore, the conditions used for the standard gravimetric analysis of the HSW sample for the measurement of the TS (and thus VS) concentration must have contributed to the loss of volatile mass, resulting in an unrealistically high total COD/VS ratio. If digester feed rates are based on

VS rather than COD, the loss of volatile mass would result in underestimating the VS loading and thus overfeeding the digester.

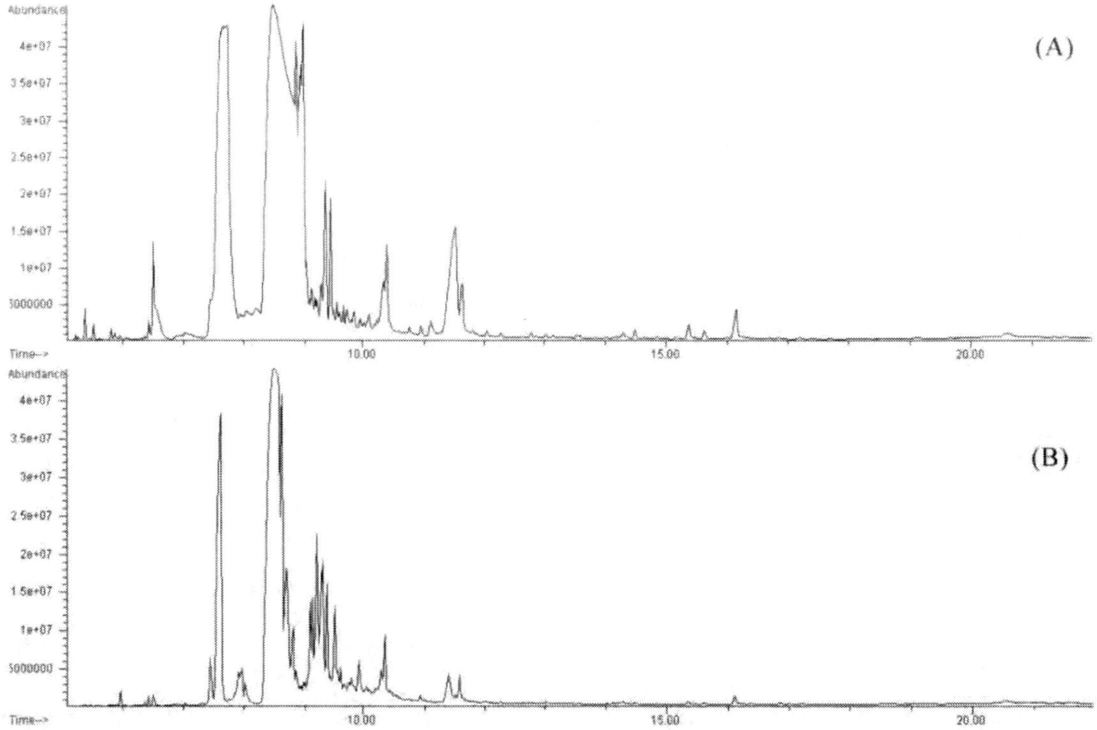

Figure 1: Total Ion Chromatogram (TIC) of the HSW Sample Before (A) and after drying at 105°C (B).

BENCH-SCALE ULTIMATE DIGESTIBILITY TESTING

The primary sludge (PS), waste activated sludge (WAS), HSW and sugar waste (SugW) were tested individually and together to determine the ultimate digestibility. The detailed procedure for the bench-scale ultimate digestibility testing is presented by Tandukar and Pavlostathis (2015). The results are shown in **Figure 2.**

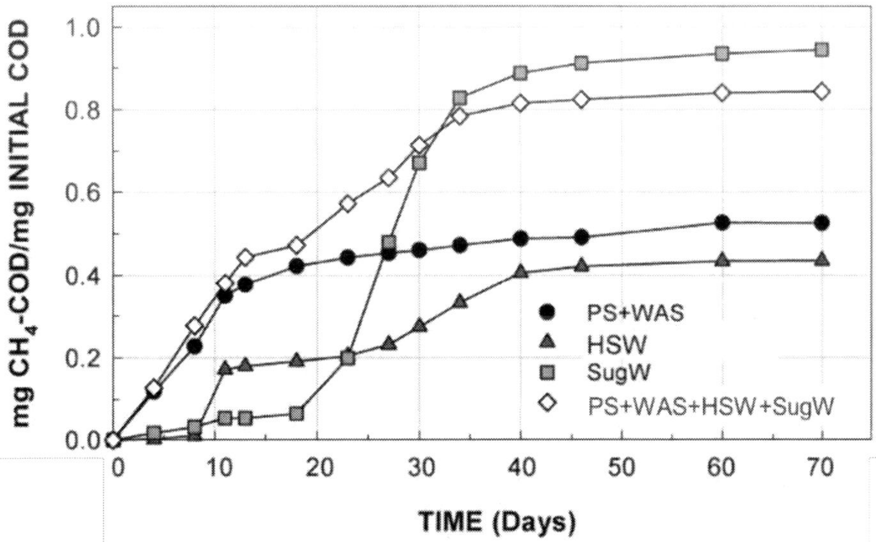

Figure 2 – Ultimate Digestibility of HSW

When each waste stream is digested separately with a seed from the HSWWTP digesters, the sugar waste had the highest degradability of > 95%, while the HSW was only 42%. The combined primary and waste activated sludge showed a degradability 50%. However, when all of the wastes are combined, the overall digestibility was 85%. Previous studies have shown that additional degradation of the PS and WAS is achieved during co-digestion. Assuming a 10% increase in digestibility of the PS and WAS, the degradability of the HSW is estimated to be 83%. The above results indicate that co-digestion lead to synergistic impact on co-digestion, resulting in improved overall digestion.

Digester Gas Projection

Using the ultimate digestibility data, the additional digester gas production from the co-digestion of only the HSW was estimated using the availability of HSW and impacts to digester solids residence time, as shown in **Table 3** assuming 83% digestibility of the HSW. EPW is considering the sugar waste to codigest at one of their other wastewater treatment facilities. The HSW is projected to triple the current digester gas production.

Table 3: Project Digester Gas Production with Co-Digestion of HSW

HSW Flow (gpd)	Digester SRT (Days)	COD Load (kg/day)	Estimated Gas Production from HSW Addition (scfm)	Current Annual Average Gas Production (scfm)	Total Gas Production (scfm)
10,000	22.4	7,000	88	163	251
20,000	21.0	14,001	177		340
30,000	19.7	21,001	265		428
40,000	18.6	28,002	354		517
50,000	17.7	35,002	442		605
60,000	16.7	42,000	530		693

HSW RECEIVING SYSTEM

The HSW receiving facility was designed to accept a total of 60,000 gallons per day which is the maximum expected volume if the facilities accept loads of the HSW. Since the HSW has been processed at the production facility, it is not expected that there will be debris, rocks, flatware or other large debris in the waste stream. Because the waste stream is expected to be generally clean, the receiving station can be simplified. See **Figure 3** for proposed layout of the HSW receiving system.

Figure 3 – Proposed HSW Receiving System Layout

HSW Receiving System

Given that the HSW receiving station will initially only consider accepting waste from one HSW supplier and tipping fees are not expected to be imposed, a keypad station for tracking hauler deliveries is not necessary. A receiving station for the HSW stream will consist of a quick connect, motorized or manual valve and optional flow meter for metering volumes of deliveries.

After the receiving station, the HSW will pass through a basket screen which will protect downstream equipment, such as pumps, in the event any unexpected debris in the HSW. Because the material is generally clean, a grinder/macerator is not provided. The trucks connecting to the receiving station will also need to be positioned on a spill containment pad to contain any spills.

HSW Storage Tanks

Three 20,000 gallon reinforced fiberglass tanks are being proposed for storage of the HSW prior to it being fed to the anaerobic digesters. Since the HSW being considered does not have a high fat, oil or grease content, heating of the tanks is not necessary. Each tank will be equipped with ultrasonic level sensors mounted on the top of the tank to confirm the level /volume of HSW within each tank. A containment area around the tanks will be required to contain any leaks or spills from the tanks. The tanks considered for this evaluation are 14 feet in diameter and 24 feet tall.

Unloading and Mixing Pumps

Two unloading pumps with the ability to pump to any of the three HSW storage tanks are being proposed. The unloading pumps will be located downstream of the basket screen and will be sized to unload a truck in less than fifteen minutes. The pumps proposed for this evaluation are rotary lobe type pumps, although other pump types can be considered during detailed design. In some designs the unloading pump can serve as a mixing pump as well, but this requires the addition of three way valves so the pump can pull from either the receiving system for unloading or the storage tank for mixing. To simplify operation of the system, dedicated unloading and mixing pumps are being considered. Each unloading pump will have a capacity of 300 gpm, and an approximate motor size of approximately 10 hp. Each mixing pump will be sized for a capacity of 50 gpm and will have an approximate motor size of 3 hp. The mixing pump is sized not to keep tank contents completely mixed, but to provide tank turnover and keep the waste from stratifying.

HSW Feed Pumps

Each storage tank will have dedicated rotary lobe feed pump for transferring the HSW to the anaerobic digesters. The feed pumps will be sized to pump the total volume of a given tank to the digesters within a 24-hour period. Equalizing HSW feed to the digesters is important for steady operation of the digesters. The feed pumps will discharge to a common discharge header where a magnetic flow meter will measure the HSW being pumped to the digesters. It is proposed that the HSW will be pumped to the Mixed Sludge Pump Station where it can be mixed and distributed to the digesters using the existing digester feed pumps. Each feed pump will be sized at 50 gpm and will have an approximate motor size of 3 hp.

Odor Control

Odor control is a critical component of the HSW receiving station design. Initial investigation into the odor control system indicates that carbon media will provide the best means to control potential odors in the HSW storage tank. A more in depth evaluation of alternative media can be performed during design. Only four air changes per hour is recommended when the tank is empty. It is important to note that the tanks will not be empty for a majority of the time so the actual air changes will be higher. For example, if the tanks are half full, the air changes will increase to 8 air changes per hour. Based on four air changes per hour, the proposed system will have an airflow rate of approximately 1,000 cfm per tank or a total of 3,000 cfm for all three tanks. For purposes of this evaluation, a Purafil deep bed scrubber was considered.

COST BENEFIT ANALSYIS

Energy savings options were evaluated for the HSWWTP that included alternatives with and without utilizing the additional biogas from the codigestion of the HSW. **Table 4** summarizes the options evaluated.

HSWWTP electrical service is Time-of-Use Municipal Pumping Service Rate and has On-Peak, Shoulder-Peak and Off-Peak rate structure. Based on historical power usage and the current electrical rate structure, the annual electrical cost baseline was 12,500 MW with an annual cost of $880,000 in 2016 dollars. This does not include operation of the CHP system. Natural gas cost for HSWWTP was $0.61 per 100 cubic feet of natural gas (CCF).

Table 4: Energy Saving Options

Scenario	Annual Energy Cost Offset	Percent of Digester Gas Flared	Capital Cost Implications
No HSW and Excess Digester Gas:			
Use natural gas in 1MW generator during shoulder-peak and on-peak periods only	$73,000 / yr.	100%	Low
Use digester in boilers for digester heating	$88,000 / yr.	~ 70% (~ 115 SCFM)	Low
Combination of the two alternatives above	$161,000 / yr.	70%	Low
Utilize Excess Digester Gas from HSW Codigestion			
Use digester gas in 1 MW CHP at full load and recovery heat for digester heating only[1]	$615,000 / yr.	45% (~ 220 SCFM)	Low
Use digester gas in 1 MW CHP at full load and recovery heat for digester heating only; and use excess digester gas for current natural gas uses throughout the plant[2]	$721,000 / yr.	< 5%	High

Footnotes:
1. The CHP system is capable of meeting the digester heating demands year round based on seasonal digester heating demand calculations. The excess heat would be wasted in this alternative.
2. Other natural gas demands include administration building boilers and natural gas driven aeration blowers. Excess digester gas typically meets all plant natural gas demands between April and October based on the monthly average total plant demand from FY08/09 to FY15/16. Purchasing natural gas for building heating boilers may be required between November and March depending on the weather conditions.

EPW decided to move forward with codigesting the HSW and using the digester gas in the 1 MW CHP, as an initial step. Utilizing the excess digester gas for other natural gas uses will be considered at a later date.

The estimated capital cost for construction of the HSW receiving station specific for the biodiesel waste, and digester gas infrastructure improvements to handle the increased gas flow is presented in **Table 5**. Based on the project annual energy savings, the simple payback period is 4.5 years.

Table 5: Simple Payback Period

System / Component	Conceptual Level Opinion of Probable Construction Cost
HSW Receiving Station[1]	$1,900,000
Digester Gas Infrastructure Improvements[1]	$850,000
Total Construction Cost[1]	$2,750,000
Annual Energy Savings	$615,000
Simple Payback Period	4.5 years

Footnotes:
1. Cost does not include engineering and administration.

CONCLUSION

Bench scale results showed synergistic effects on co-digestion of HSW and sugar waste with primary and secondary sludge in this study. El Paso Water opted to consider co-digesting the HSW at this time. A conceptual level design for a receiving station provided El Paso Water with easy to operate and maintain method of receiving and delivering the HSW to its digesters.

Results showed favorable cost benefit analysis for this practice. The plant has injected the HSW into one of their digesters and noticed a significant increase in digester gas production with stable digester operation. Design on the HSW receiving facility and digester gas improvements is scheduled to start in Spring 2017.

REFERENCES

Tandukar, Madan; Pavlostathis, Spryos G. (2015) *Co-digestion of municipal sludge and external organic wastes for enhanced biogas production under realistic plant constraints*; Water Research, Vol. 87, pp. 432-445.

Effects of Biosolids Addition and Alkalinity Sources on High-Solids Anaerobic co-Digestion (HS-AcD) of Food Waste and Green Waste

P. Dixon[1], P. Bittencourt[2], E. Lee[1], M. Wang[1], E. Jimenez[1], Q. Zhang[1], and S. J. Ergas[1]

1. Department of Civil & Environmental Engineering, University of South Florida, 4202 E Fowler Ave, ENB 118, Tampa, FL 33620, (813) 974-2957, phillipdixon@mail.usf.edu
2. Department of Mechanical Engineering, University of South Florida, 4202 E Fowler Ave, ENC 2300, Tampa, FL 33620, 813-974-3539, paulacampos@mail.usf.edu

ABSTRACT

High-Solids Anaerobic co-Digestion (HS-AcD) of biosolids with the organic fraction of municipal solid waste (OFMSW) has potential to increase biogas production and improve the economic sustainability of both solid waste and biosolids management. In this study, biochemical methane potential (BMP) assays and life cycle cost assessments (LCCA) were used to investigate effects of biosolids and alkalinity source addition on bioenergy production and economic sustainability during HS-AcD. Co-digestion of biosolids with food waste (FW) and green waste (GW, also known as yard waste) resulted in decreased methane (CH_4) yield compared with digestion of FW and GW alone. Both crushed oyster shells and limestone improved methane yields compared to BMPs carried out with no added alkalinity source. A life cycle cost analysis (LCCA) was carried out for Hillsborough County in Florida that showed that co-digestion of biosolids with FW and GW improved life cycle costs due to high revenues for electricity and heat as well as avoided cost of OFMSW and biosolids management. Among all digestion options tested in this study, HS-AcD using FW, GW, and biosolids with addition of oyster shells was the most economically feasible.

KEYWORDS: Bioenergy production, biosolids, high solids anaerobic co-digestion, life cycle cost analysis, nutrient recovery, solid waste

INTRODUCTION

High-Solids Anaerobic co-Digestion (HS-AcD) of biosolids with the organic fraction of municipal solid waste (OFMSW) has a number of benefits over liquid anaerobic digestion (L-AD). High-Solids Anaerobic Digestion (HS-AD) systems generally operate between 15-40% total solids (TS) and have been shown to have the following advantages over L-AD: reduced parasitic energy losses, less water usage, and smaller reactor size. In addition, HS-AcD systems have little or no production of liquid side-streams which require further treatment. Furthermore, they allow for the recovery of nutrients as a stabilized compost (or digestate) that can be used as a soil amendment (Hinds *et al.* 2016). HS-AcD of OFMSW and biosolids is particularly attractive to Florida wastewater facilities because biosolids management is increasingly expensive due to limited L-AD capacity for biosolids, more stringent recent regulations on land application of biosolids, and the high cost of biosolids disposal in landfills (Robert and Forbes 2011).

Advantages of co-Digestion of OFMSW with biosolids include the potential to increase the economic sustainability of HS-AD of OFMSW by improving waste characteristics, increase overall biogas production, improve methane (CH_4) yields, and divert OFMSW and biosolids from landfills and/or incinerators (Chen *et al.* 2014). OFMSW is a heterogeneous material that includes food waste (FW), green waste (GW, also referred to as yard waste), and paper products (Fernández Rodríguez *et al.* 2012). Only about 40% of OFMSW has been found to be easily biodegradable (Sosnowski *et al.* 2003), with paper and GW being the least biodegradable components (Hamzawi *et al.* 1998). Incorporation of biosolids as a co-substrate has the potential to provide a more readily bioavailable substrate mixture with an improved alkalinity content and carbon to nitrogen (C/N) ratio.

HS-AD of FW can result in the rapid production of VFAs as sugars and fatty acids are broken down (Chen *et al.* 2014). When the rate of VFA production exceeds the methanogens' utilization rate, the pH in the reactor will decrease, the methanogens will become inhibited, and the reactor's microbial activity will cease. The total VFA concentrations should be maintained below 10,000 mg/L to prevent inhibition of the methanogens (Khanal 2011). This can be remediated by adding a solid phase alkalinity source to the reactor's contents, such as limestone (Zhang *et al.* 2008). Prior research by our group has shown that crushed oyster shells can provide a slow release source of alkalinity (Sengupta *et al.* 2007). On average crushed oyster shells are composed of 95% calcium carbonate ($CaCO_3$) and are also a low-cost waste product of the oyster industry (Hamester *et al.* 2012).

Although HS-AcD has many advantages as mentioned above, the economic viability is a primary consideration in decision-making. Previous economic studies focus on liquid anaerobic co-digestion of FW and biosolids (Kim *et al.* 2011) and HS-AD of FW (Rogoff and Clark 2014). The assessment of the economic sustainability of the HS-AcD is still lacking. Life cycle cost analysis (LCCA) is a useful tool to assess the economic sustainability, which is able to quantify the total cost of the assets by taking all the cost factors into consideration over the full life cycle, including transportation, construction, operation and maintenance, and disposal. Thus, the LCCA helps to select the most economic option with the lowest complete cost of ownership.

The overall goal and objectives of this research are to improve the economic sustainability of organic waste management for public utilities. Specific objectives include: 1) investigate bioenergy production and nutrient recovery from HS-AcD of OFMSW with and without biosolids addition, 2) compare limestone and crushed oyster shells as alkalinity sources for HS-AcD, and 3) evaluate the economic benefits of HS-AcD.

MATERIALS & METHODS

Biochemical Methane Potential (BMP) Assays: BMP reactors were set up as described in Hinds *et al.* (2016). As shown in **Table 1**, different mixtures (substrates) were used to test the effects of biosolids and alkalinity sources (crushed oyster shells and limestone) addition on the performance of HS-AcD of FW and GW. For each experimental condition, seven replicate reactors were run in 250-ml serum bottles at a constant mesophilic temperature of 35°C. Blanks containing only seed sludge and controls containing the experimental mixtures FW+YW and FW+YW+B with seed sludge and without any alkalinity sources were run in 100-ml serum bottles in triplicate. Blank BMPs were used to track the CH_4 production from the inoculum (determine an inoculum baseline for the experiment), and the controls were digested to determine baseline for the experiment with no alkalinity.

The study inoculum consisted of dewatered anaerobically digested sludge from the Northeast Clearwater Treatment Facility in Clearwater, Florida. FW waste was prepared as described by

Ariunbaatar *et al.* (2014) and consisted of: apples, banana peels, oranges, mixed greens, carrots, egg shells, hard boiled eggs, bread, potatoes, and cooked chicken. GW was based on the typical composition of GW in Florida and contained: oak tree leaves, pine needles, grass, and shrubbery cuttings. Biosolids consisted of dewatered (via screw press) waste activated sludge (WAS) from the Falkenburg Advanced Wastewater Treatment Plant in Tampa, FL. The equivalent amount of crushed oyster shells and limestone were added to the reactors provided an alkalinity of 3,000 mg/L (as $CaCO_3$).

Analytical Methods: Biogas was initially collected on a daily basis from the headspace of each digestion bottle, but was collected less frequently as biogas production rates decreased. Biogas volume was determined using a 50-ml frictionless syringe. The CH_4 content of the biogas was measured using the sodium hydroxide (NaOH) displacement method as described by the American Society for Testing and Materials, ASTM (2002). BMPs were sacrificed periodically and the following chemical parameters of the leachate were analyzed using *Standard Methods* (APHA 2012): pH, alkalinity, TS, and volatile solids (VS). Leachate ammonium (NH_4^+) was analyzed using a Timberline Model TL-2800 Ammonia Analyzer (Boulder, CO, USA). VFAs were measured using Hach test kit 10240 (Loveland, CO, USA); VFA values are reported as the equivalent amount of acetic acid. The methane yield was calculated by subtracting the methane produced by the blank (or inoculum) from the total cumulative methane production from the BMP and then dividing that number by the substrate g VS from the BMP (see Eq. 1 below). CH_4 yield values were adjusted to standard temperature and pressure (STP, 273.2 K and 101.3 kPa).

$$Methane\ Yield = \frac{BMP\ mL\ CH_4 - \left[Blank\ mL\ CH_4 \times \left(\frac{SS\ BMP\ g\ VS}{SS\ Blank\ g\ VS}\right)\right]}{Total\ BMP\ g\ VS - SS\ BMP\ g\ VS} \quad \text{(Eq. 1)}$$

In this equation, *BMP mL* CH_4 is the total volume of CH_4 produced by the BMP, *Blank mL* CH_4 is the total volume of CH_4 produced by the blank, *SS BMP g VS* is the volatile solids mass of the BMP's seed sludge, *SS Blank g VS* is the volatile solids mass of the Blank's seed sludge, and *Total BMP g VS* is the volatile solids mass of the entire BMP reactor.

Economic analysis: LCCA for full-scale HS-AcD scenarios with varying biosolids addition and alkalinity sources were conducted using the present value (PV) method. The scenarios for this study were as follows: 1) FW and GW with oyster shells (FW+GW w/OS), 2) FW, GW, and biosolids (B) with no additional alkalinity source (FW+GW+B), 3) FW, GW, and biosolids with oyster shells (FW+GW+B w/OS), and 4) FW, GW, and biosolids with limestone (FW+GW+B w/L). The LCCA included infrastructure, operation and maintenance (O&M), collection and transportation (C&T) costs, and revenues from beneficial products including electricity, heat, and digestate. The life cycle cost (LCC) was computed as follows:

$$LCC = C_I + C_{O\&M} \times UPV^* + C_{C\&T} \times UPV - (C_{R,t\&b\&h} \times UPV + C_{R,e} \times UPV^*) \quad \text{(Eq. 2)}$$

where C_I is the initial cost, $C_{O\&M}$ is the O&M cost, $C_{C\&T}$ is the C&T cost, $C_{R,t\&b\&h}$ are the revenues from tipping fee saving and digestate and heat sales, $C_{R,e}$ is the revenue from electricity sale, UPV is a uniform present value factor, and UPV* is a non-uniform present value factor. The discount or interest rate and the escalation rate used to calculate UPV and UPV* were assumed to be 1.9% (the average rate for 10 years) and 0.65%, respectively (EERC 2017, USIR 2016).

In this analysis, the available amount of the wastes was estimated based on the waste production from Hillsborough County in Florida. **Figure 1** shows a production and management flow diagram for FW, GW, and biosolids in 2015 for Hillsborough County. In 2015, 100% of the FW and 39% of the GW was used in the county's resource recovery facility (incineration Waste to Energy) to generate electricity, while 56%, 1.4%, and 3.3% of the GW were used in mulch/organic soil production, composting, and landfill cover, respectively. Wastewater treatment facilities in the county produced 128,000 tons of biosolids and 81% were disposed in landfills in 2015. To estimate the total amount of organic wastes available for HS-AcD, it was assumed that 40% of the produced commercial FW (37,700 tons/yr) was diverted from the current waste flow to the HS-AcD resulting in a mixture ratio of FW, GW, and biosolids of 1.0:1.0:1.2 by TS. The residential FW was not considered in this analysis since there is no separate collection system for residential FW in Hillsborough County. The total amount of organic wastes for this analysis was approximately 113,000 tons/yr. For HS-AcD using FW and GW, the same amount of organic wastes (a mixture ratio of FW and GW as 1:1 by TS) was used, which accounted for 60% of the produced commercial FW (56,600 tons/yr).

The HS-AcD system was assumed to have the same configuration as a BIOFerm Dry Fermentation system, which is a mesophilic, batch, and single-stage technology (BIOFerm 2014). The system is comprised of garage style fermenters, a percolation tank, biogas storage tank, biofilter, and combined heat and power (CHP) unit. It was assumed that the operating conditions for the HS-AcD system were the same as the experimental conditions with a 28-day retention time. The initial cost was estimated based on the data obtained from current installations of BIOFerm and existing literature (BIOFerm 2017, Self-Reliance 2010, Strimbu 2016, Vavrin *et al.* 2014). **Figure 2** shows the capital costs as a function of the operating capacity for existing BIOFerm systems in the U.S. The HS-AcD capital cost was estimated based on a regression model shown in **Figure 2**. The O&M cost covers the costs for all O&M activities, including processing feedstock, labor, and chemical use. **Table 2** shows the existing data for O&M costs in U.S. ranging from $34 to $72 per ton. To be conservative, an O&M cost of $72/ton (Vavrin *et al.* 2014) was used in this study. The C&T cost for the wastes was $0.10/mile/ton, which was based on Faucette *et al.* (2002). Transportation vehicles were assumed to have a haul loading of 30-tons, with an average travel distance of 50-miles round trip (Faucette *et al.* 2002). The cost of the oyster shells was assumed to be zero because they were considered as wastes from local processing industries, while the cost of limestone was $1.3/kg. Since, small amounts of oyster shells and limestone were used in the HS-AcD, transportation costs of these materials were not considered in this analysis.

A sensitivity analysis was conducted to determine how changes in input parameters impact the LCC results. For this study, the following input parameters were modified by ±10% and the LCC values were recalculated: amount of organic wastes, travel distance, tipping fee, and electricity, heat, and digestate sale prices. The sensitivity index (I) was calculated as follows:

$$I = \frac{(LCC_{i+1} - LCC_{i-1})/LCC_i}{(F_{i+1} - F_{i-1})/F_i} \qquad \text{(Eq. 3)}$$

where LCC_{i+1}, LCC_i, and LCC_{i-1} are the LCC associated with ±10% input parameter changes, and F_{i+1}, F_i, and F_{i-1} are input parameters for i+1 (10% increasing parameters), i (base parameters), and i-1 (10% decreasing parameters). The parameters used for the LCCA are shown in **Table 3**.

RESULTS AND DISCUSSION

BMP Assays: A comparison of CH_4 yields for FW+GW with and without biosolids is shown in **Figure 3**. CH_4 yields were significantly lower when biosolids were added to FW and GW. This may have been due to the recalcitrance of the WAS under mesophilic digestion conditions or differences in substrate to inoculum ratios (S/I) used in BMPs with and without biosolids (**Table 1**). WAS has been found to very recalcitrant in its nature. Specifically, the organic carbon in biosolids consists of two parts that differ in their degree of biodegradability: a labile fraction (53-71%) that can be quickly mineralized and a recalcitrant fraction (29-45%) that is not available or resistant to microorganisms (Torri *et al.* 2014, Walton *et al.* 2001). In addition, S/I greatly affects BMP performance. It was found in this study that BMPs with higher S/I ratios required longer retention times to produce as much methane as reactors with lower S/I ratios. Throughout the literature it is also noted that high S/I ratios result in longer hydraulic retention times (HRT) to degrade organic matter (González-Fernández and García-Encina 2009). Therefore, at ideal S/I ratios, reactor imbalances due to VFA accumulation and long retention times to degrade organic matter can be avoided. The result of no imbalances in a reactor is greater CH_4 yields. Additional experiments with the same S/I ratio are currently underway to eliminate the differences in methane production caused by varying S/I ratios. CH_4 yield results observed in this study **(Table 5)** are comparable to those achieved in prior studies for HS-AcD of FW+GW (Chen *et al.* (2014)) and OFMSW and biosolids (Zhang *et al.* (2008)).

As expected, NH_4^+ concentrations increased and VS concentrations decreased over time as organic matter degraded **(Table 4)**. NH_4^+ concentrations were not found to be in the toxic range (1,500-1,700 mg/L) for AD in any of the digestion sets (Gerardi 2003). NH_4^+ release was higher in digestion sets with biosolids addition due to the higher nitrogen content of the WAS compared with FW+GW. This results in a compost product in the solid phase with higher total nitrogen content, potentially making it more valuable as a fertilizer. In addition, if the biosolids were treated in an L-AD system the resulting sidestream would need to be treated in a sidestream process or recycled back to the headworks resulting in high energy and chemical costs and potentially disrupting the mainstream treatment process. VS concentrations were reduced by an average of 3.6% in all HS-AcD digestion sets, with the greatest VS reduction in FW+GW+B with oyster shells (4.7%) (Hinds *et al.* 2016).

VFA concentrations were found to be similar between digestion sets **(Table 5)**. It was found that all sets maintained VFA concentrations much below levels considered inhibitory to methanogenisis (>10,000 mg/L) (Khanal 2011). pH and alkalinity in all HS-AcD digestion sets were also maintained at levels appropriate for methanogenesis (pH>6.5 and alkalinity>1,000 mg/L as $CaCO_3$) (Tchobanoglous and Burton 1991). Alkalinity was found to be higher for the sets with biosolids addition **(Table 5)**. CH_4 yields were significantly higher when an alkalinity source was added, with no significant differences between lime and crushed oyster shells (**Figure 4**). This may have been because of VFA production and localized alkalinity imbalances within micro-niches in the reactor due to incomplete mixing (Veeken and Hamelers 2000), but requires more research to determine. The main advantages of adding biosolids in HS-AcD are increased overall bioenergy production, recovery of nutrients and diversion of biosolids from land application or landfilling, which are discussed further in the LCCA results.

Economic analysis: Table 6 shows LCCA results for four different digestion options over 25 years. The largest revenue was generated from electricity sale for all the digestion options, followed by heat and digestate sale. All options would save tipping costs for landfilling of GW and biosolids. The tipping fee for 3% of the GW (5,000 tons/yr) could be saved for the HS-AcD using FW+GW w/OS, while the other digestion options were able to save the tipping fee associated with 3% of the GW (5,000 tons/yr) as well as 35% of the biosolids (45,300 tons/yr). The C&T costs were not significant in the LCC results (< 5%), while the O&M costs were the largest contributor to the LCCs for all digestion options. For HS-AcD using FW+GW+B w/L, the O&M cost was the highest among others due to the use of limestone (109 kg/ton organic wastes). The cost of limestone ($1.3/kg) was a major contributor for increasing the O&M cost. Considering HS-AcD using FW+GW w/OS and FW+GW+B w/OS, addition of biosolids was able to achieve higher revenues due to the increased CH_4 production and digestate quantities, as well as the avoided tipping cost for biosolids landfilling. For the three mixtures, including FW+GW w/OS, FW+GW+B, and FW+GW+B w/L, the sum of the initial and O&M costs greatly exceeded the annual revenues, making the systems not economically feasible. The most economical option was the HS-AcD using FW+GW+B w/OS due to its high CH_4 production.

Overall, annual O&M costs were significant when determining economic feasibility of the systems (shown in **Figure 5**). For instance, all digestion options, except FW+GW+B w/L, were economically feasible when an annual O&M cost with $35/ton was applied. Considering the annual O&M cost of $72/ton, however, only the FW+GW+B w/OS option was economically feasible. Thus, the LCC results can be sensitive to the annual O&M cost. The result for the sensitivity analysis is shown in **Table 7**. The result shows that the FW+GW+B w/OS HS-AcD option is more sensitive to changes in all tested input parameters among the others. The travel distance had a minimal impact on the LCCs. The LCCs are more sensitive to electricity cost and amount of organic waste; the electricity cost is directly related to revenue for electricity sale, while the amount of organic wastes is closely related to the initial and O&M costs as well as revenues for electricity and heat sales.

CONCLUSIONS

The overall goal of this research is to improve the economic sustainability of organic waste management for public utilities. The following conclusions can be drawn from this work:

- Addition of biosolids decreased CH_4 yields but increased the economic sustainability of HS-AcD by increasing overall CH_4 production, and revenues for electricity, heat, and digestate sales, as well as the avoided tipping cost for landfilling.
- Although pH, alkalinity and VFA concentrations were in the healthy range for methanogenisis in all digestion sets, the addition of limestone and crushed oyster shells increased CH_4 yields and revenues for electricity and heat sales.
- Although there were no significant differences between CH_4 yields with limestone and oyster shells, oyster shells were a low-cost waste product resulting in decreased LCC for this option.
- Addition of both biosolids and oyster shells in HS-AcD was the most economically sustainable option, making that system the most economically feasible due to increased revenues from electricity and heat sales.
- By diverting OFMSW from landfills the landfill leachate quality is also improved. By avoiding onsite L-AD of biosolids, the production of sidestreams requiring further

treatment to remove nutrients is avoided. Thus, additional consideration of HS-AcD of OFMSW and biosolids for both energy and nutrient recovery is warranted.

ACKNOWLEDGEMENTS

This material is based upon work supported by the National Science Foundation under Grants No. 1243510 and 1156905, and the Hinkley Center for Solid and Hazardous Waste Management. Any opinions, findings, and conclusions or recommendations expressed in this material are those of the authors and do not necessarily reflect the views of the funding agencies.

Table 1. Experimental Set-Up Based on VS

Mixture	Alkalinity Source	FW (g VS)	GW (g VS)	Biosolids (g VS)	Inoculum (I) (g VS)	S/I Ratio
FW+GW	-	1.2	1.0	0.0	1.2	1.8
FW+GW+B	-	1.2	1.0	1.0	1.2	2.7
FW+GW	Oyster Shells	3.5	3.0	0.0	3.5	1.8
FW+GW+B	Oyster Shells	3.5	3.0	3.1	3.5	2.7
FW+GW	Limestone	3.5	3.0	0.0	3.5	1.8
FW+GW+B	Limestone	3.5	3.0	3.1	3.5	2.7
Seed Sludge (Blank)	-	0.0	0.0	0.0	1.6	N

Notes:
B = Biosolids (Dewatered WAS)
FW = Food Waste
GW = Green Waste
I = Inoculum (L-AD Effluent)
N/A = Not Applicable
S = Substrate
VS = Volatile Solids

Table 2. Annual Operation and Maintenance Costs for HS-AcD in the U.S.

$/ton of Feedstock	Inflation-Adjusted to 2017 Dollars ($/ton)	References
37.00	40.72	Authority 2010
40-55	44.55-61.25	Self-Reliance 2010
34.00	37.42	Self-Reliance 2010
72.00	72.99	Vavrin, McNamara et al. 2014

Table 3. Input Parameters for the LCCA analysis

Input	**Value**	**References**
Discount or Interest Rate (%)	1.9	USIR 2016
Escalation Rate (%)	0.65	EERC 2017
Average Hauling Distance (miles)	50	Assumed
Collection and Transportation Rate ($/mile/ton)	0.1	Faucette, Governo et al. 2002
Tipping Fee ($/ton)	20	County 2016
Limestone ($/kg)	1.3	Survey 2017
Limestone Consumption (kg/ton organic wastes)	109	Obtained from our experiments
Oyster Shells ($/kg)	0	Assumed
Oyster Shells Consumption (kg/ton organic wastes)	82	Obtained from our experiments
Heating Value (kWh/m^3)	9.94	Passos and Ferrer 2015
Combined Heat and Power Efficiency: Heat (%) Electricity (%)	 49.5 37.3	BIOFerm 2017
Electricity Rate ($/kWh)	0.08	EIA 2016
Heat Rate ($/kWh)	0.01	Moriarty 2013
Stabilized Biosolids Price ($/ton)	11.2	Schwarzenegger 2010
Life cycle Cost Analysis Period (yr)	25	Assumed

Table 4. TS, VS, and NH_4^+-N of the Biochemical Methane Potential (BMP) Assays

Mixture	**TS (%)**		**VS (%)**		**NH_4^+-N (mg/L)**	
	D_0	**D_{43}**	**D_0**	**D_{43}**	**D_0^1**	**D_{43}**
FW+GW	12.5 (±0.0088)	NS	10.3 (±0.011)	NS	110	NS
FW+GW+B	13.2 (±0.0076)	NS	11.0 (±0.0080)	NS	120	NS
OS FW+GW	12.3 (±0.0051)	10.8 (±0.014)	10.1 (±0.0066)	7.71 (±0.0067)	110	490 (±0.12)
OS FW+GW+B	13.5 (±0.0037)	9.40 (±0.0068)	11.2 (±0.0035)	6.47 (±0.0021)	130	570 (±18)
L FW+GW	13.0 (±0.016)	9.17 (±0.00024)	10.5 (±0.018)	6.60 (±0.00015)	110	440 (±3.5)
L FW+GW+B	13.3 (±0.0057)	10.3 (±0.0018)	10.8 (±0.0037)	7.62 (±0.00042)	120	540 (±1.5)

Notes:

1 = No Replicates Sampled
D = Day
L = Limestone
NH_4^+-N = Ammonium Nitrogen
NS = Not Sampled
OS = Oyster Shells
TS = Total Solids
VS = Volatile Solids

Table 5. VFA, pH, and Alkalinity of the Biochemical Methane Potential (BMP) Assays.

Mixture	VFA (mg CH_3COOH/L)		pH		Alkalinity (mg $CaCO_3/L$)		CH_4 (ml CH_4/g VS fed)	
	D_0	D_{44}	D_0	D_{44}	D_0	D_{44}	D_{31}	D_{44}
FW+GW	160 (±2.8)	NS	7.4 (±0.099)	NS	510 (±49)	NS	256 (±22)	266 (±15)
FW+GW+B	110 (±2.8)	NS	7.3 (±0.064)	NS	660 (±85)	NS	173 (±9.0)	184 (±6.0)
OS FW+GW	180 (±11)	100 (±4.2)	7.4 (±0.15)	8.3 (±0.0)	770 (±350)	2,100 (±16)	265 (±7.0)	271 (±8.0)
OS FW+GW+B	180 (±1.4)	110 (±4.9)	7.2 (±0.028)	8.4 (±0.014)	870 (±330)	2,400 (±18)	203 (±4.0)	216 (±4.8)
L FW+GW	140 (±2.8)	90 (±0.14)	7.4 (±0.17)	8.3 (±0.0)	700 (±190)	1,900 (±7.1)	233 (±15)	241 (±15)
L FW+GW+B	160 (±71)	110 (±1.4)	7.4 (±.14)	8.4 (±0.0)	580 (±88)	2,200 (±0.0)	211 (±4.1)	225 (±4.5)

Notes:
$CaCO_3$ = Calcium Carbonate
CH_4 = Methane
VFA = Volatile Fatty Acid

Table 6. Life Cycle Cost for Different HS-AcD Digestion Options

Item	FW+GW w/OS ($)	FW+GW+ B ($)	FW+GW+B w/OS ($)	FW+GW+B w/L ($)
Initial Cost	38,410,000	38,410,000	38,410,000	38,410,000
Operation and Maintenance cost	174,526,000	174,526,000	174,526,000	491,508,000
Collection and transportation cost	373,000	373,000	373,000	373,000
Tipping fee saving	1,978,000	19,896,000	19,896,000	19,896,000
Electricity sale	145,430,000	142,118,000	157,261,000	173,139,000
Heat sale	19,638,000	19,190,000	21,235,000	23,379,000
Digestate sale	21,925,000	21,925,000	22,376,000	22,226,000
Life cycle cost	24,339,000	10,180,000	-7,460,000	291,652,000

Table 7. Results of Sensitivity Analysis for Different HS-AcD Options

Items	FW+GW w/OS	FW+GW+B	FW+GW+B w/OS	FW+GW+B w/L
Amount of Organic Waste	7.65	16.65	-22.66	1.67
Travel Distance	0.02	0.04	-0.05	0.001
Electricity Sale Cost	-5.98	-13.96	21.08	-0.59
Heat Sale Cost	-0.81	-1.89	2.85	-0.08
Digestate Sale Cost	-0.90	-2.15	3.00	-0.08
Tipping Cost	-0.08	-1.95	2.67	-0.07

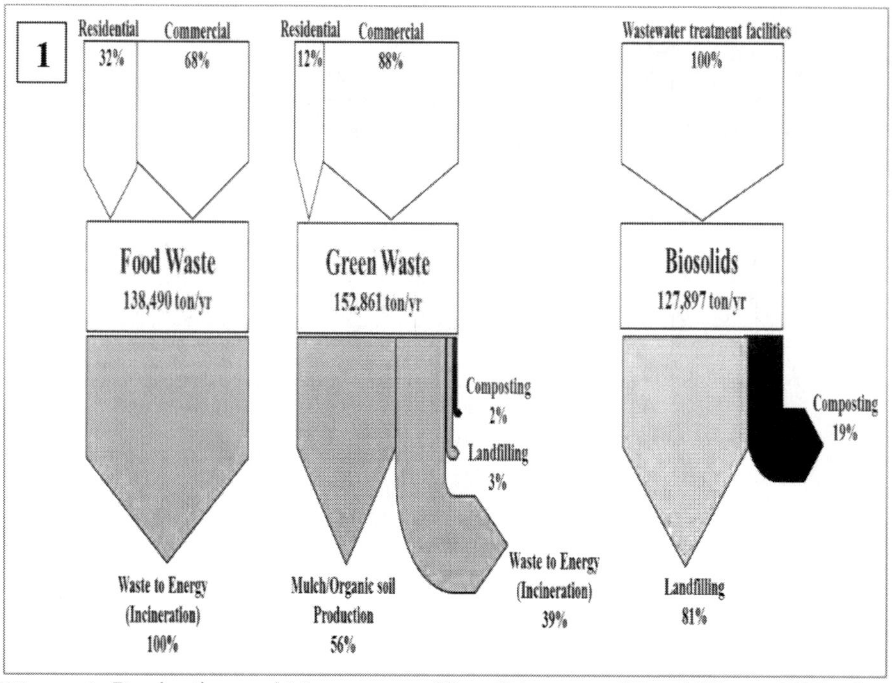

Figure 1. Production and Management Flow Diagram of Food Waste (FW), Green Waste (GW), and Biosolids

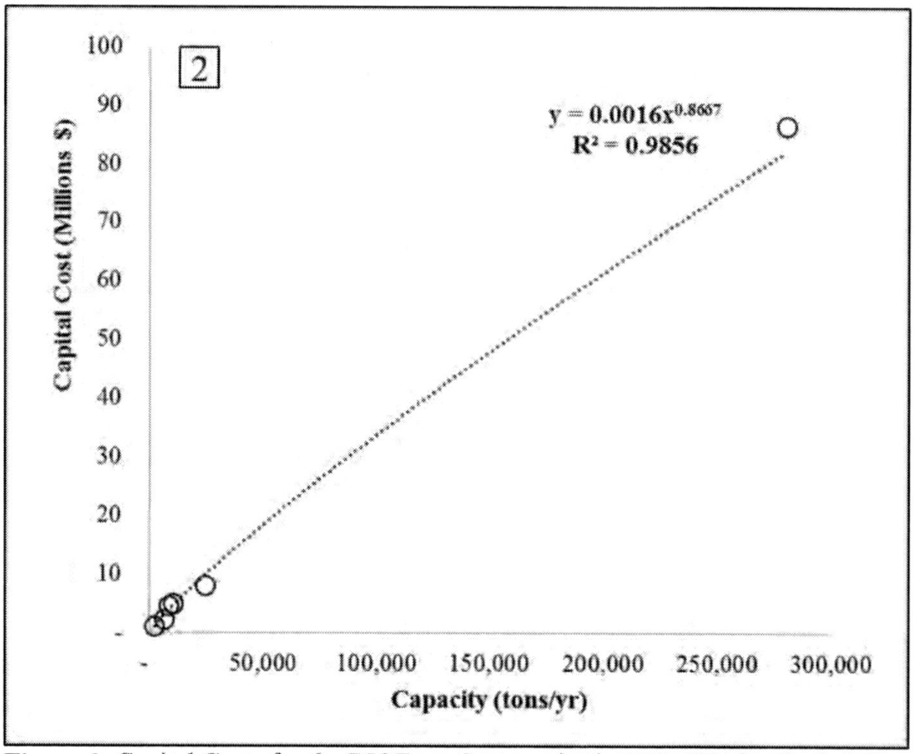

Figure 2. Capital Costs for the BIOFerm Systems in the U.S. (Circle: Capital Costs Obtained from Literature; Dashed Line: A Regression Model Curve)

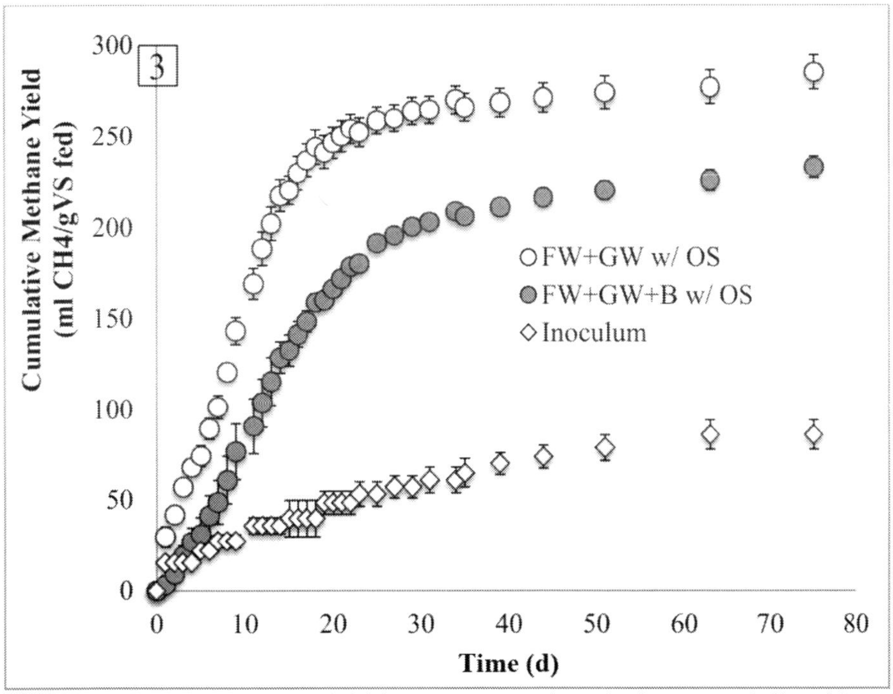

Figure 3. Cumulative CH_4 Yield for FW+GW w/Crushed Oyster Shells with and without Biosolids

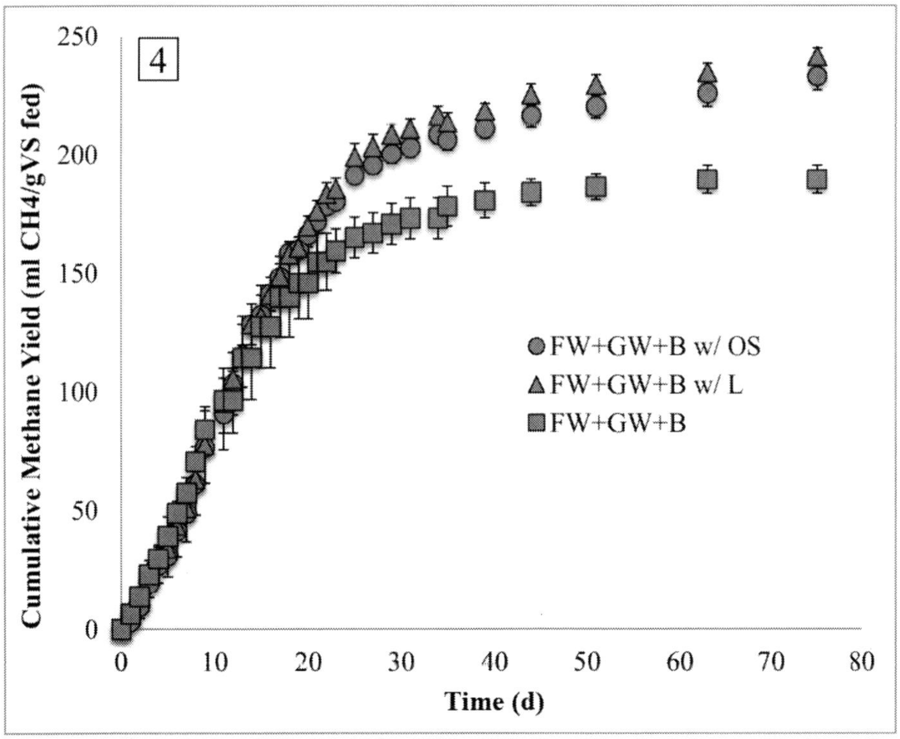

Figure 4. Cumulative CH_4 Yield for FW+GW+B w/Crushed Oyster Shells and Limestone as Alkalinity Sources

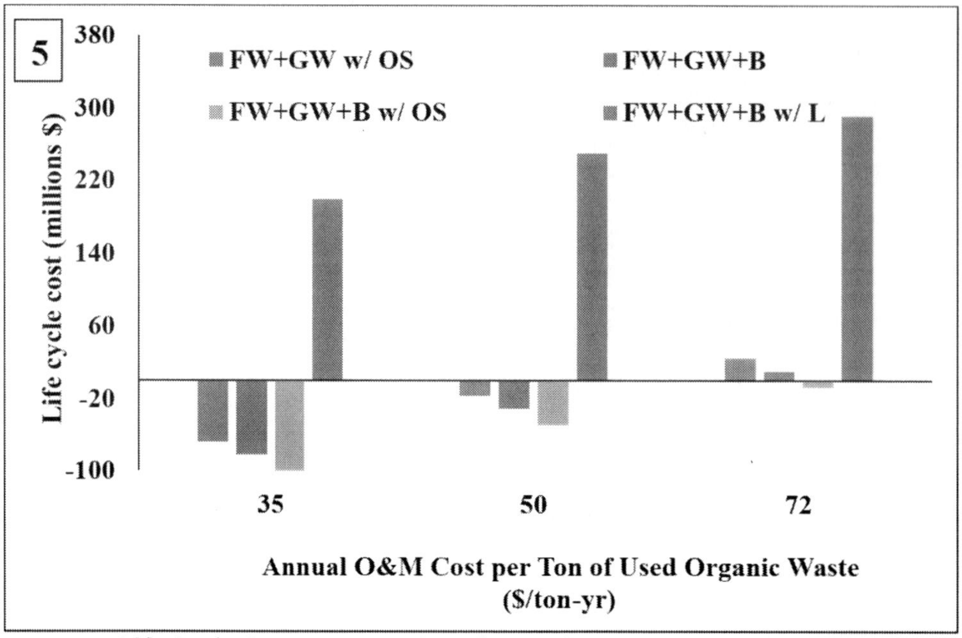

Figure 5. Life Cycle Costs as a Function of Annual O&M Cost per Ton of Organic Waste Used in the High-Solid Anaerobic co-Digestion (HS-AcD) with the Four Considered Digestion Options

REFERENCES

APHA, A., WEF (2012) *Standard Methods for the Examination of Water and Wastewater*, https://www.standardmethods.org/.

Ariunbaatar, J.; Panico, A.; Frunzo, L.; Esposito, G.; Lens, P. N. L.Pirozzi, F. (2014) *Enhanced anaerobic digestion of food waste by thermal and ozonation pretreatment methods*; Journal of Environmental Management, *146*, 142-149.

ASTM, Standard Test Method for Gas Content (Nonacidic) of Insulating Liquids by Displacement with Carbon Dioxide. In *ASTM D1827-92*, ASTM International: West Conshohocken, PA, 2002, www.astm.org.

Authority, H. W. M. Food Waste Diversion and Utilization in Humboldt County. http://www.hwma.net/sites/default/files/humboldt_regional_food_waste_digester_feasibility_study.pdf (accessed February 12, 2017).

BIOFerm Dry Fermentation. Anaerobic Digestion Systems for Solid Input Materials. http://www.biofermenergy.com (accessed February 12, 2017).

BIOFerm FAQ (n.d.). http://www.biofermenergy.com/anaerobic-digestion-2/faq/ (accessed February 12).

Chen, X.; Yan, W.; Sheng, K.Sanati, M. (2014) *Comparison of High-Solids to Liquid Anaerobic co-Digestion of Food Waste and Green Waste*; Bioresource technology, *154*, 215-221.

County, H., Solid Waste Rates, Fees, Charges and Assessment for Collection Services, Resource Recovery, and Disposal Facilities. 2016, http://www.hillsboroughcounty.org/library/hillsborough/media-center/documents/solid-waste/trash-and-recycle/scalehousefees.pdf?la=en.

EERC Applied Economics Office, National Institute of Standards and Technology. http://energy.gov/eere/femp/eerc-download.

EIA Electricity Data. https://www.eia.gov/electricity/monthly/epm_table_grapher.cfm?t=epmt_5_6_a (accessed February 12, 2017).

Faucette, L.; Governo, J.Graffagnini, B. (2002) *Commercial foodwaste composting feasibility study for south metro Atlanta*; Engineering Outreach Service, Faculty of Engineering, University of Georgia. Athens, GA.

Fernández Rodríguez, J.; Pérez, M.Romero, L. I. (2012) *Mesophilic anaerobic digestion of the organic fraction of municipal solid waste: Optimisation of the semicontinuous process*; Chemical Engineering Journal, *193–194*, 10-15.

Gerardi, M. H., *The microbiology of anaerobic digesters*. John Wiley & Sons: 2003.

González-Fernández, C.García-Encina, P. A. (2009) *Impact of substrate to inoculum ratio in anaerobic digestion of swine slurry*; Biomass and Bioenergy, *33*, (8), 1065-1069.

Hamester, M. R. R.; Balzer, P. S.Becker, D. (2012) *Characterization of calcium carbonate obtained from oyster and mussel shells and incorporation in polypropylene*; Materials Research, *15*, (2), 204-208.

Hamzawi, N.; Kennedy, K.McLean, D. (1998) *Anaerobic digestion of co-mingled municipal solid waste and sewage sludge*; Water Science and Technology, *38*, (2), 127-132.

Hinds, G. R.; Mussoline, W.; Casimir, L.; Dick, G.; Yeh, D. H.Ergas, S. J. (2016) *Enhanced methane production from yard waste in high-solids anaerobic digestion through inoculation with pulp and paper mill anaerobic sludge,* ; Environmental Engineering Science, in press.

Khanal, S. K., *Anaerobic biotechnology for bioenergy production: principles and applications*. John Wiley & Sons: 2011.

Kim, M.-H.; Song, Y.-E.; Song, H.-B.; Kim, J.-W.Hwang, S.-J. (2011) *Evaluation of food waste disposal options by LCC analysis from the perspective of global warming: Jungnang case, South Korea*; Waste management, *31*, (9), 2112-2120.

Moriarty, K. (2013) *Feasibility study of anaerobic digestion of food waste in St. Bernard, Louisiana*; NREL.

Passos, F.Ferrer, I. (2015) *Influence of hydrothermal pretreatment on microalgal biomass anaerobic digestion and bioenergy production*; water research, *68*, 364-373.

Robert, H.Forbes, J. (2011) *The changing landscape of biosolids management in Florida: the 21st century's first decade & predictions for the next one*; Florida Water Resources J, (6), 54-60.

Rogoff, M. J.Clark, B. J. (2014). *Anaerobic Digestion – What Are the Economics? Waste Advantage Magazine* July 2014. pp 22-25. https://wasteadvantagemag.com/anaerobic-digestion-economics/

Schwarzenegger, A. (2010) *ANAEROBIC PHASED SOLIDS DIGESTER PILOT DEMONSTRATION PROJECT.*

Self-Reliance, I. f. L. Update on Anaerobic Digester Projects Using Food Wastes in North America. https://ilsr.org/update-on-anaerobic-digester-projects-using-food-wastes-in-north-america/ (accessed February 12, 2017).

Sengupta, S.; Ergas, S. J.Lopez-Luna, E. (2007) *Investigation of solid-phase buffers for sulfur-oxidizing autotrophic denitrification*; Proceedings of the Water Environment Federation, *2007*, (2), 1139-1159.

Sosnowski, P.; Wieczorek, A.Ledakowicz, S. (2003) *Anaerobic co-digestion of sewage sludge and organic fraction of municipal solid wastes*; Advances in Environmental Research, *7*, (3), 609-616.

Strimbu, J. (2016) *Feasibility Study of Biogas Generated Electricity for the Austin E. Knowlton Center for Equine Science at Otterbein University.*

Survey, U. S. G. Limestone. https://www.usgs.gov/ (accessed February 12, 2017).

Tchobanoglous, G.Burton, F. L. (1991) *Wastewater engineering*; Management, *7*, 1-4.

Torri, S. I.; Corrêa, R. S.Renella, G. (2014) *Soil carbon sequestration resulting from biosolids application*; Applied and Environmental Soil Science, *2014*.

USIR United States Inflation Rate 1914-2017. http://www.tradingeconomics.com/united-states/inflation-cp (accessed February 12, 2017).

Vavrin, J. L.; McNamara, I.States, U. *Base-Wide Biodigester for Basecamps in an Operational Environment*; DTIC Document: Engineer Research and Development Center (U.S.) & Construction Engineering Research Laboratory, 2014.

Veeken, A.Hamelers, B. (2000) *Effect of substrate-seed mixing and leachate recirculation on solid state digestion of biowaste*; Water Science and Technology, *41*, (3), 255-262.

Walton, M.; Herrick, J. E.; Gibbens, R. P.Remmenga, M. D. (2001) *Persistence of municipal biosolids in a Chihuahuan desert rangeland 18 years after application*; Arid Land Research and Management, *15*, (3), 223-232.

Zhang, P.; Zeng, G.; Zhang, G.; Li, Y.; Zhang, B.Fan, M. (2008) *Anaerobic co-digestion of biosolids and organic fraction of municipal solid waste by sequencing batch process*; Fuel Processing Technology, *89*, (4), 485-489.

Lifting the Fog from FOG Receiving

Rashi Gupta[1*], Daniel L. Meacham[1], Phil Parkins[2], Travis A. Peacock[3], Anthony L. Montoya[3]

[1] Carollo Engineers, Costa Mesa, California
[2] Carollo Engineers, Phoenix, Arizona
[3] Albuquerque Bernalillo County Water Utility Authority, New Mexico
* Email: RGupta@carollo.com

ABSTRACT

Fats, oils, and grease (FOG) often causes problems within collection systems, but represents a potentially high-energy feedstock for anaerobic digestion that can increase onsite energy production. Numerous facilities have implemented FOG receiving systems to reduce collection system overflows and increase digester gas production. However, as the number of such facilities has grown, so have reports of operational difficulties. Ten facilities were surveyed to determine a number of parameters, including operational practices, system configuration and equipment used, operation and maintenance (O&M) requirements, and benefits realized after implementation of FOG receiving. Digestion of FOG did increase digester gas production at most facilities, allowing several to meet or exceed most of their energy demands. However, FOG handling and system maintenance can be cumbersome and the delivered product can be highly variable in quantity, quality, and characteristics. Harnessing the energy inherent in FOG can significantly offset energy demand, but careful consideration of the practical aspects of FOG receiving is necessary for sustainable and successful operations.

KEYWORDS: FOG Receiving, Operation and Maintenance, System Equipment, Materials of Construction, FOG Quality, Digester Gas Production, Lessons Learned

INTRODUCTION

A number of wastewater utilities have constructed receiving stations for fats, oils, and grease (FOG) where FOG haulers can discharge grease trap waste. Where FOG receiving has been practiced, the intent of such stations has been to remove the problematic FOG stream from the sewers and concentrate it in receiving tanks from which it can be metered into the anaerobic digestion system. Anaerobic digestion of this material, in addition to typical municipal sludge and other external feedstocks, can increase production of digester gas that can then be harnessed to produce power and/or heat for the utility. Some of these facilities have reportedly achieved net zero energy status by producing energy sufficient to meet or exceed their plants' demands. In addition to reducing power purchases from utilities by increasing the production of renewable energy, these facilities have also sought to reduce clogging of sewer piping and meet the demands of FOG haulers seeking to reduce hauling distances.

However, as the number of such facilities has grown, so have reports of operational difficulties, inconsistent FOG availability, variable gas production, and higher than expected maintenance requirements. Preliminary feasibility studies have typically focused on FOG availability, tipping fees, and projected digester gas/energy production. Few studies have focused on the actual operation and maintenance (O&M) associated with these systems. Hence, the O&M expectations of plant staff are based on limited information and the actual O&M for FOG receiving typically becomes apparent only after start-up of the systems.

METHODOLOGY

For the reasons noted in the introduction, the Albuquerque Bernalillo County Water Utility Authority (the Water Authority) in New Mexico, while interested in the potential benefits of receiving FOG at its Southside Wastewater Reclamation Plant (SWRP), decided to study the operating histories of such systems at currently operating plants before constructing a FOG receiving station. To develop an up-to-date understanding of FOG receiving, the project team surveyed several FOG facilities in the United States that were being implemented or already in operation in 2015. The survey generally focused on the parameters listed in Table 1. This information was intended to aid the Water Authority in determining the feasibility of a FOG receiving station at the SWRP.

Table 1. Information Sought During Surveys.

Category	Information
Facility Overview	Wastewater treatment facility capacity Size/capacity of FOG Receiving Station Project drivers Years in operation
FOG Characteristics and Quality Control	FOG sources Quantity received FOG quality Sampling parameters Quality control measures
FOG Receiving System	Process schematic Piping and valves Equipment Changes in system since installation

Table 1. Information Sought During Surveys.

Category	Information
Construction Costs and Schedule	Costs Schedule
Operations and Maintenance (O&M)	Operating procedures Digester feed control parameters Reported changes in digester gas production O&M issues Staffing requirements Tipping fees/FOG acceptance charges

Of the 11 facilities shown in the map in Figure 1, 10 were successfully contacted and participated in the survey. Medium and large-capacity treatment plants were prioritized in the survey to better match the size of the SWRP. The survey was primarily conducted through telephone calls and electronic communication with operations staff, but some facilities were also visited in person. Due to the preference of some surveyed facilities, results are summarized in this paper as general conclusions and findings or provided without identification of specific facilities.

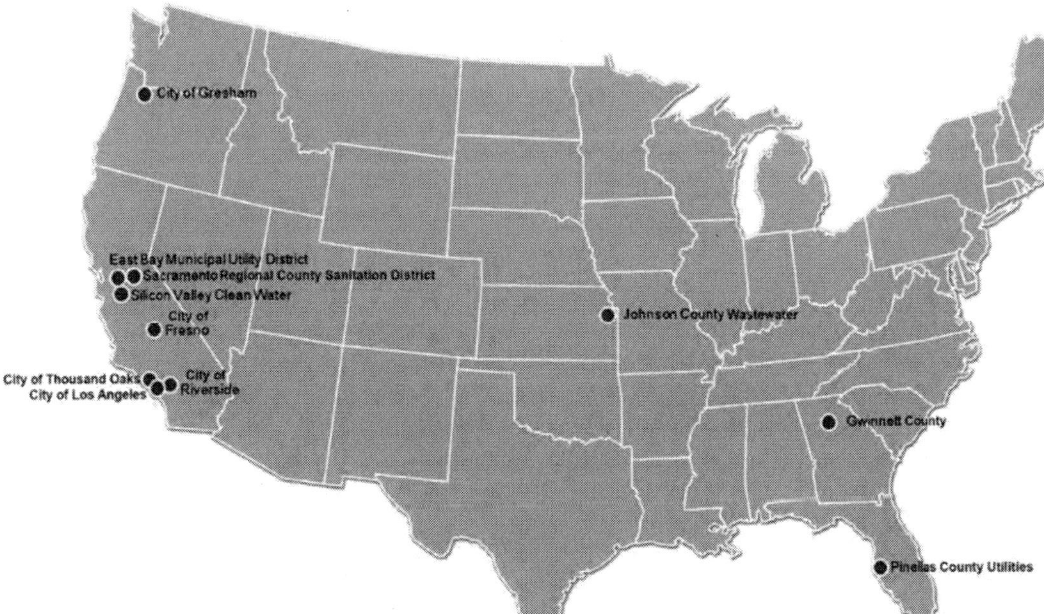

Figure 1. Surveyed Facilities with FOG Receiving Stations.

SURVEY RESULTS

Facility Overview

Surveyed wastewater treatment facilities have implemented FOG Receiving Stations for a variety of goals, including:

- Reduced utility costs through increased digester gas production and subsequent power/heat generation.
- Reduced clogging and impacts on collections systems.
- Improved sustainability and reduced greenhouse gas emissions by digesting rather than disposing of FOG at landfills.
- Revenue generation.

Most of these facilities have been in operation from 3 to 11 years, but one system was awaiting start-up at the time of the survey. The total average dry weather flow (ADWF) treatment capacity, FOG receiving station capacity, and FOG storage capacity for the surveyed facilities are shown in Table 2.

Table 2. Design Capacity of Surveyed Treatment Facilities and FOG Systems.

Parameter	Range	Median Value
Plant ADWF Capacity (mgd)[a]	12.5 - 450	34
FOG Receiving Station Throughput Capacity (gpd)[b]	12,000 - 175,000	30,000
FOG Receiving Station Storage Capacity (gal)[c]	5,000 - 60,000	28,500

[a] mgd = million gallons per day
[b] gpd = gallons per day
[c] gal = gallons

FOG Characteristics and Quality Control

The FOG product is typically pumped out of restaurant grease traps and delivered to the receiving stations by private haulers. The material is considered "brown grease" rather than "yellow grease" and characteristics vary significantly from one source to another. Seven of the facilities regularly tested certain parameters to assess FOG quality or determine if haulers were in compliance with their permit requirements. However, there were also facilities that did not regularly monitor the characteristics of incoming FOG. Table 3 summarizes the testing intervals and reported values for regularly monitored parameters. Monitoring of material quality during delivery is typically limited to visual inspection and sometimes, pH measurements. Sampling is typically done with grab samples, either from each truck load or from the material stored in the receiving tanks. While pH can be quickly tested in the field with litmus strips, other analyses are conducted at facility laboratories after the material has already been delivered.

Table 3. Reported FOG Characteristics, Monitored Parameters, and Monitoring Frequency.

Parameter	Number of Facilities Regularly Monitoring	Typical Frequency of Measurement[a]	Range of Reported Values[b], (median)
Quantity of FOG Received	9	With delivery or daily total	1,500 - 105,000 gpd, (18,500 gpd)
pH	5	With delivery or daily composite	2 - 6, (4)
Total Solids (TS)	5	With delivery or daily composite	0.5% - 20%, (8.6%)
Volatile Solids (VS)	5	With delivery or daily composite	Less than 20% to 100%, (79%)
Chemical Oxygen Demand (COD)	2	With delivery or daily composite	5,000 - 202,000 mg/L, (50,000 mg/L)

[a] Frequency of testing is based only on facilities that reported testing intervals and regularly monitored parameters.
[b] Some facilities indicated that they test for these parameters, but did not report measured values.

While most facilities prohibit the use of septage trucks for delivered FOG, it was reported that some haulers use the same trucks for both septage and FOG. This results in some contaminants typically associated with septage being delivered with the FOG. Many facilities found FOG to be abrasive, containing significant quantities of grit and other objects such as rocks, concrete, metal debris, and materials found in septage or restaurant waste. The FOG tends to be acidic, and facilities have reported that the low pH may cause equipment damage.

Three of the surveyed facilities contract with single, local FOG haulers who are responsible for delivering a consistent, screened product with certain specifications for product pH, solids content, and volatile solids concentration. While this practice limits the number of haulers that can utilize the receiving station, it does allow for better control of FOG quality and consistency. Even with this level of control, the material can still be grit-laden and abrasive.

Seven of the surveyed facilities require FOG haulers to apply for a permit before they can offload FOG at the receiving stations. This allows multiple haulers to offload and maximizes the utility of the receiving station for haulers within the area. However, it was also reported that this practice makes it difficult to monitor or control the quality of the FOG received. Permitted haulers that are found to deliver excessively contaminated product or product that causes process upsets due to its characteristics are typically warned and their permits can be revoked upon multiple infractions. In general, these facilities also reported wide variation in daily FOG deliveries. Several trucks might deliver on one day followed by one or two days of very few loads. One facility with a design capacity of 30,000 gpd received an average of 1,500 gpd over approximately one year, with many days without any delivery and a few days exceeding

50,000 gallons. These fluctuations impact system economics, digester gas production, and subsequent gas utilization and cogeneration systems. Most surveyed facilities use their FOG receiving tanks, low-pressure digester gas storage, and/or digester feed control procedures to minimize fluctuations in gas production.

FOG Receiving System

Each surveyed facility had common process needs, but varied in specific equipment and material selection. Figure 2 illustrates a general process flow diagram that includes major process equipment used at various facilities and their frequency of use.

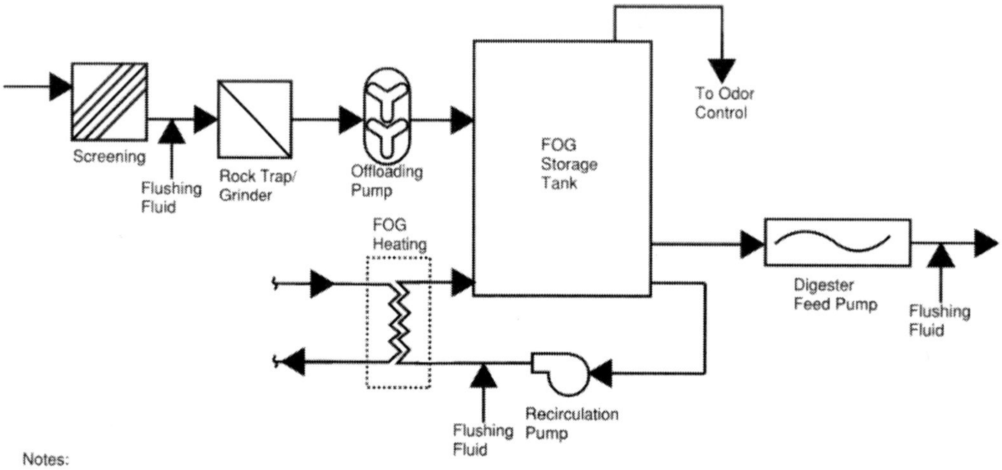

Notes:

1. 3 of 10 facilities utilized a manual bar screen or basket strainer to screen FOG upstream of equipment.
2. 7 of 10 facilities used a combination rock trap/macerator or inline grinder.
3. 7 of 10 facilities used an offloading pump. Offloading pumps were used for systems servicing haulers without on-board air-assist or pumping systems. Facilities with below-grade receiving tanks or air-assisted hauling trucks did not have offloading pumps.
4. 8 of 10 facilities use pump recirculation to mix FOG, 5 of these facilities utilized their offloading pumps to perform this task.
5. Every facility utilized a tank to store FOG. A single facility also allowed rocks to settle in the bottom of their storage tank to remove them from the system.
6. 6 of 10 facilities utilized some system of FOG heating, including heat exchangers and/or heat tracing.
7. Every facility utilized feed pumps to transport FOG from storage to digesters.
8. Sludge, hot water, or plant water are used for flushing FOG lines.

Figure 2. General FOG Receiving Station diagram, showing major components used by surveyed facilities.

Several types of equipment and various materials of construction are used for system components at the surveyed facilities. Table 4 lists the equipment types used for different components of the FOG receiving system and the major materials of construction as reported. Some facilities modified their FOG receiving stations after gaining experience with the delivered FOG product. Many of these changes were intended to increase equipment service life or reduce maintenance needs, and were driven by the contaminants, abrasion, and/or corrosive nature of the FOG material. Systems that did not originally have FOG screening were refitted with some type of screening or straining device, although this was noted as a particularly maintenance-intensive part of the system. Different types of elastomers were tried for valves and pumps, with

Buna-N being the elastomer most often used. However, all reported elastomers were susceptible to some degree to damage from the FOG stream. Steel materials were generally replaced with stainless steel components to better handle corrosion. Some systems that had been installed without the ability to heat FOG initially were retrofitted with heat tracing or sludge recirculation systems to minimize clogs. No mechanical grit removal systems were reported at any of the surveyed facilities. It was found that grit settles in the system piping and FOG receiving tanks, and is periodically removed manually. Some facilities purposely use the tanks to settle grit and minimize grit carryover into the digesters.

Table 4. Equipment Types and Materials of Construction Used by Surveyed Facilities.

System Component	Equipment and Materials of Construction Used at Different Facilities	Number of Facilities
Bar Screens/ Strainers	Manual bar screen before inlet to storage tanks	2
	Steel basket strainer with stainless steel internals	1
Rock Trap/ Macerators or Grinders	Combination rock trap/macerator, hot dipped galvanized and hardened steel	3
	Combination rock trap/macerator, hot dipped galvanized and hardened steel housing with stainless steel blades and cutting screen	2
	Inline grinder, stainless steel internals or materials not reported	2
	Use of FOG storage tank as rock trap	1
Offloading Pumps	Rotary lobe, NBR/Buna-N lobe casing with steel or stainless steel wear plates	3
	Rotary lobe, NBR/Buna-N lobe casing with tungsten carbide wear plates and housing	1
	Rotary lobe, stainless steel lobes with rubber tips	1
	Peristaltic hose, materials not reported	1
	Centrifugal chopper pump, NBR/Buna-N elastomers with ductile iron casing and hardened cast steel impeller	1

Table 4. Equipment Types and Materials of Construction Used by Surveyed Facilities.

System Component	Equipment and Materials of Construction Used at Different Facilities	Number of Facilities
Above-Grade FOG Receiving Tanks	Steel tank in rectangular, cylindrical-flat bottom, or cylindrical-conical bottom shape	2
	Stainless steel tank in cylindrical-flat bottom or cylindrical-conical bottom shape	1
	Cylindrical FRP tank	3
	Cylindrical polyethylene	2
Below-Grade FOG Receiving Tanks	Rectangular concrete	2
Recirculation Pumps	Rotary lobe, NBR/Buna-N lobe casing with steel or stainless steel wear plates (offloading pump used for recirculation)	2
	Rotary lobe, NBR/Buna-N lobe casing with tungsten carbide wear plates and housing (offloading pump used for recirculation)	1
	Rotary lobe, stainless steel lobes with rubber tips (offloading pump used for recirculation)	1
	Centrifugal chopper pumps with ductile iron casing and hardened steel impeller	3
	Centrifugal chopper pumps with all stainless steel internal components and casing	2
	Submersible centrifugal chopper pumps, with hardened steel internals, epoxy coating and lining	1
Heating System	Tube-in-Tube heat exchanger with FOG recirculation and hot water as heat source	3
	Heat tracing and insulation of piping and/or FOG tank	3
	FOG delivered heated	1
	Steam Injection (installed, but not used)	1

Table 4. Equipment Types and Materials of Construction Used by Surveyed Facilities.

System Component	Equipment and Materials of Construction Used at Different Facilities	Number of Facilities
Digester Feed Pumps	Progressing cavity with NBR/Buna-N elastomers	5
	Progressing cavity with custom silicone stators	1
	Rotary lobe with NBR/Buna-N elastomers	2
	Rotary lobe, NBR/Buna-N encasement with tungsten carbide-coated housing	1
Odor Control	Carbon canister for tank vent	4
	Chemical or biological odor scrubber with fans	1
Piping	NBR/Buna-N hose	1
	Glass-lined ductile iron	6
	Unlined stainless steel	1
	PVC	2
Isolation/Control Valves	Plug values with NBR/Buna-N plug facing	6
	316 SST full-port ball valves with RTFE seats	1
	PVC ball valves with RTFE seats and FKM/Viton O-rings	2

Construction Costs and Schedule

Where reported, construction costs for the FOG receiving stations ranged from $250,000 to $4 million in the year of system construction. Costs for the seven facilities that reported a value are presented in Table 5. Construction durations were reported between 6 to 18 months. Construction schedule and costs were related to system size, redundancy, complexity, construction method, and intended life of the receiving station. Some systems were either constructed as "temporary" systems by plant staff or as part of larger expansion projects. Costs specifically for the FOG receiving station were estimated by survey respondents when constructed as part of a larger project.

Some facilities received project funding through the American Recovery and Reinvestment Act and a few other agencies also took advantage of available state funding to construct their FOG receiving stations. Several of the "temporary" systems included in the survey were in operation for a number of years after being initially constructed by plant staff. These systems were often simpler than larger, permanent facilities and provided with less redundancy and lower cost materials of construction.

Table 5. Reported Construction Costs for FOG Receiving Stations.

Facility[a]	FOG Storage Tank Capacity, gallons	Reported Construction Cost[b]	Year Constructed
A	60,000	\$3 million	2015
B	20,000	\$2 million	2010
C	27,000	\$2.1 million	2013
D	45,000	\$2.9 million	2011
E	5,000	\$0.25 million	2008
F	30,000	\$2.3 million	2013
G	43,500	\$3 million to \$4 million	2011

[a] Facilities are not specifically identified due to preference of some survey respondents. Facilities identified alphabetically in this table may differ from those identified similarly in other tables.
[b] Reported costs are presented for the year in which the facility was constructed.

Operation and Maintenance
Operating Procedures. Operational requirements and procedures differ based on type and size of FOG receiving system. Some facilities allow FOG deliveries only during staffed hours so that operations personnel are onsite to monitor deliveries, collect samples, and operate the FOG system. Operators at these facilities can track haulers and inspect delivered product so they are better able to control FOG quality than those facilities that allow material delivery during off-hours. Some receiving stations that do not have staff onsite during delivery are fitted with key card systems that track hauler data and provide haulers access to the automated system based on unique access cards. However, gathering information such as FOG source and sampling necessary to capture FOG characteristics typically requires the presence of staff during deliveries. One surveyed facility has operated with deliveries during staffed and non-staffed hours. This facility found that FOG quality tended to be more consistent and the product less contaminated when deliveries were allowed only when plant operators were present.

When plant operators are present at FOG delivery, they typically collect samples from each load for subsequent analysis. Without staff presence at each delivery, samples are typically collected from the FOG tank or piping system to provide a general picture of the FOG delivered throughout the day. Most facilities test for pH, total solids (TS) and volatile solids (VS) content. Very few facilities test for chemical oxygen demand (COD). Sampling results for these parameters help identify potential sources of problematic FOG and avoid digester upsets.

Operations for FOG stations at the surveyed facilities vary based on system design. However, typical operations for a station with an above-grade receiving tank and components similar to that presented in Figure 2 include the following steps:

1. FOG Delivery: The FOG hauler checks in with an access card, a guard at the facility gate, or plant staff. The hauler connects the truck hose to a connection point at the receiving station. The offloading is started either by the hauler through an onboard air-assist system or at the receiving system's offloading pump. As the pump starts, the system rock trap/macerator also starts. The contents of the hauler's truck are transferred through the screen/strainer and rock trap/macerator into the FOG receiving tank. Valves within the system can be manual or automated. Manual valves are positioned by plant staff before system start and automated valves direct FOG flow per their control logic. If present, staff members collect samples for subsequent analysis.
2. FOG Recirculation: After the truck is emptied, valving is switched to allow a dedicated recirculation pump or the offloading pump to recirculate FOG within the tank. Some systems include a heat exchanger through which the FOG is pumped. Heat is typically supplied by hot water from the digester heating system. Systems with heat tracing typically turn the heat tracing system on at a setpoint low temperature.
3. Debris Removal: After the truck is emptied, the system screen/strainer and rock trap are cleared of debris and any spilled FOG is washed away to maintain a reasonably clean working area. Clearing debris from the screen/strainer is typically a manual and time consuming task that is completed by plant staff. Some strainer baskets become too heavy to lift without a hoist mechanism. Debris is typically transferred into covered trash bins that are periodically emptied for landfill disposal.
4. Digester Feed: FOG is typically blended into a sludge stream that is pumped into the digester. This sludge stream can be digester feed sludge (thickened primary or secondary sludge) or it can be the heated recirculation sludge that maintains digester operating temperatures. Separate FOG feed into poorly mixed areas of the digester can cause the FOG to float and develop into a mat rather than being properly entrained within the digester contents. Specific operational practices for how to control FOG feed to the digester are described below.

Digester Feed Control. As noted earlier, most of the surveyed facilities reported wide variation in daily FOG deliveries. If carried into the digester feed cycle, these fluctuations can impact digester performance, digester gas production, and subsequent gas utilization and cogeneration systems. Most surveyed facilities equalize the feed within their FOG receiving tanks, meter FOG with variable speed digester feed pumps, and utilize digester feed control procedures to maintain acceptable operational parameters. However, based on survey responses, there was no "standard" way of controlling FOG feed to the digesters. Each of the various methods below was reportedly used by different facilities to control FOG feed to their digesters:

- FOG tank is emptied into digesters over an operator adjustable timeframe.
- FOG is fed into digesters based on the level in the FOG tank, with no digester-related controls.

- A small quantity of FOG is fed to digesters and gradually increased over time, as digesters acclimate to the new food source. This was done after foaming issues, which were believed to be due to excess FOG loading into the digesters, causing an upset in the biomass.
- FOG is fed directly into digesters, maintaining an adjustable FOG:sludge ratio by volume within the digester. The reported volumetric ratio was approximately 30 to 40% FOG and 60 to 70% sludge.
- FOG is fed to the digesters while maintaining total solids loading into the digesters by reducing sludge feed.
- FOG is first mixed with an equal volume of primary sludge, then fed to digesters.
- FOG is fed solely based on maintaining the same volumetric loading of FOG in the digesters each day.
- FOG feed is determined by internally developed control system designed to optimize gas production.

Among facilities with more than three digesters, two blended FOG with digester feed sludge in a tank or piping and pumped the blend to all digesters in the system. Three facilities fed FOG only to specific digesters within the facility. Facilities that had three or fewer digesters typically fed FOG to all digesters within the system.

Reported Changes in Digester Gas Production. Of the facilities surveyed, half were able to quantify increases in digester gas production after the addition of FOG. The remaining facilities were not able to quantify the increase, but most reported that gas production did increase after FOG digestion was implemented. Two facilities even reported that digestion of FOG has played a key role in them achieving energy neutrality status. A third facility noted that its combination of FOG and food waste digestion, cogeneration, and solar power production has offset almost 90-percent of the plant's energy demands. On the other hand, one facility reported that FOG volume received at their station was too low to determine any impact on digester gas production and another plant indicated that they had expected a larger increase than what they actually experienced.

As presented in Table 6, reported digester gas production increases were between 30 to 100%, depending on FOG quantity and quality. Some facilities accept external feedstock beyond just FOG, so some increases in digester gas production may be related to these other feedstocks. One facility that has separately accepted FOG and food waste indicated that digestion of FOG increased gas production significantly more than the food waste they received but did not provide quantitative data on the difference.

Table 6. Reported Increases in Digester Gas Production.

Facility[a]	FOG Storage Tank Capacity, gallons	Average Quantity of FOG Received, gal/day	Reported Percent Increase in Digester Gas Production[b]	Notes
A	10,000 to 17,000	900 to 105,000[c]	Approx. 100%	Note e
B	20,000	4,000 to 9,000	Approx. 30%	—
C	45,000	10,000 to 50,000	Approx. 57%	Note e
D	NR[d]	Approx. 25,000	Approx. 57% to 97%	—
E	43,500	18,000 to 24,000	Approx. 100%	Note e

[a] Facilities are not specifically identified due to preference of some survey respondents. Facilities identified alphabetically in this table may differ from those identified similarly in other tables.
[b] Reported percent increase relative to gas production before implementation of FOG receiving.
[c] Range based on wide variety of FOG vessels used to deliver product to facility and acceptance of material during each of 3 shifts. Downstream blend tank used to further equalize FOG and produce an acceptable FOG:sludge ratio for combined digester feed.
[d] NR = Not Reported
[e] Facility accepts FOG and other feedstocks.

O&M Issues. All facilities agreed that FOG is a difficult, sometimes odorous, material to handle that requires considerable attention and maintenance. In most cases, maintenance needs for these FOG receiving stations were higher than originally anticipated by O&M staff. Many of the previously noted changes in FOG system materials and configuration were made to mitigate high maintenance requirements. At the time of the survey, two of the facilities reported that they were considering some system changes to reduce system maintenance and increase equipment life. Even though maintenance needs have been high, most facilities indicated that the net benefits in terms of increased digester gas production were worth this effort. Staff members at these facilities had accepted the maintenance intensity and had accounted for the labor effort and materials replacement costs within their operational plans. Table 5 summarizes some of the reported maintenance issues, tasks performed, and task frequency.

Table 7. Reported O&M Issues Experienced with FOG Receiving.

Reported Issue	Task	Reported Frequency
Screens/strainers clog with grease, rags, and debris	Manual cleaning of screens/strainers	Daily or after each load
Rock trap filled with debris	Manual draining and cleaning of rock trap	After each load to weekly

Table 7. Reported O&M Issues Experienced with FOG Receiving.

Reported Issue	Task	Reported Frequency
Damage/wear of macerator blades	Replace damaged cutting blades	Once to twice per year
Wear of macerator screen	Replace screen	Once per year
Damage/wear of elastomers in system pumps	Replace elastomers or elastomer-cased pump components (lobes, lobe tips, stators, etc.)	3 to 6 months
Damage/wear of pump housings	Replace pump housing, possibly with harder or more corrosion-resistant material	6 to 36 months
Damage/wear of pump wear parts	Replace wear parts, possibly with harder material	As needed
Standard pump upkeep	Inspection, pump component adjustment, and lubrication	Weekly to annually, depending on task
Damage/wear of elastomers in system valves	Inspect and replace valves or elastomer-cased valve components (plugs, flapper disks, etc.), possibly with different valve types and/or materials	6 to 36 months
Pipe clogs or FOG mat formation in receiving tanks, especially during cold weather	Pipe flushing and manual destruction of mats	During cold weather and as part of normal operations
FOG tank upkeep	Draining, inspection, and removal of grit, debris, and rocks	Semi-annually

Staffing Requirements. The reported O&M requirements of FOG receiving stations require significant staffing. Specific staffing requirements are dependent on system size and complexity, level of automation, FOG acceptance protocols, quality control measures, and facility philosophy on FOG system operation. Reported staffing requirements for facilities that were able to provide this information are presented in Table 8. As shown, reported staffing requirements ranged from one full-time employee (FTE) at quarter-time to two FTEs. Depending on each system, staffing is required for the following tasks:

- Permitting FOG haulers and monitoring permit compliance
- Inspecting FOG loads, sample collection, and analysis

- Regular inspection and operation of FOG station valves and pumping, maceration, heating, mixing, digester feed, and odor control systems
- Removing captured materials from screens, strainers, rock traps, and tanks
- General housekeeping of FOG receiving area
- Regular, preventative maintenance of system components, including odor control
- Repair and replacement of damaged system components
- Collecting and monitoring FOG data for delivered material quantity, characteristics, impacts on digester feed, and impacts on digester gas production
- Monitoring of digester performance and troubleshooting as necessary
- Determination and collection of tipping fees
- Management of contracts, if any, with FOG haulers

As noted previously, some facilities accept FOG only during manned hours under staff supervision so they dedicate staff members to the FOG system to inspect the FOG load and operate the FOG system. These facilities typically also require sampling and analysis of each FOG load. Other facilities allow FOG delivery during unmanned hours so staff time is not required at every delivery. Accurate projections of staffing requirements must consider each facility's staffing philosophy during FOG delivery and subsequent sampling and system operation.

Table 8. Reported Staffing Requirements.

Facility[a]	FOG Storage Tank Capacity, gallons	Reported Staffing Requirements[b]
A	10,000 to 17,000	1 FTE
B	20,000	1 FTE
C	27,000	1 to 2 FTE
D	45,000	2 FTE
E	43,500	0.25 FTE

[a] Facilities are not specifically identified due to preference of some survey respondents. Facilities identified alphabetically in this table may differ from those identified similarly in other tables.
[b] FTE = Full-time employee equivalent

Tipping Fees/FOG Acceptance Charges. In some cases, agencies were able to charge a tipping fee for haulers to deliver FOG to the treatment plant. As shown in Table 9, these fees ranged from $0.01 to $0.09 per gallon. Facilities located in areas with competing FOG receiving stations or disposal alternatives reported downward pressure on tipping fees. One facility that originally charged $0.11 per gallon has reduced the tipping fee to $0.08 per gallon. Tipping fees were also reported to impact the quantity of FOG delivered to receiving stations. If the fees are considered too high by the haulers, they typically find lower cost options and do not deliver the product to the receiving station. This has a significant impact on system economics, especially if a consistent revenue stream was expected during a feasibility assessment of the FOG system. As noted previously, one facility with a design capacity of 30,000 gpd received an average of 1,500 gpd over approximately one year, with many days without any delivery and a few days exceeding 50,000 gallons. The expected tipping fees from this system have not been realized and system economics have been dramatically impacted by the limited FOG deliveries.

Table 9. Reported Tipping Fees.

Facility[a]	Reported Tipping Fees	Reported Annual Tipping Fee Revenue	Notes
A	$0.08/gal	-	Notes b and c
B	$0.09/gal	-	Note d
C	$0.03/gal	-	Note b
D	-	$300,000/year	Note b
E	$0.01/gal	-	—
F	$0.08/gal	-	—
G	-	$300,000/year	Note b

[a] Facilities are not specifically identified due to preference of some survey respondents. Facilities identified alphabetically in this table may differ from those identified similarly in other tables.
[b] Facility accepts FOG and other feedstocks.
[c] Reported fee is specifically for FOG.
[d] Facility reported that delivered FOG volumes were significantly lower than expected.

DISCUSSION OF RESULTS

Wastewater treatment facilities have implemented FOG receiving stations for various reasons. Beyond sustainability goals and reducing collection system maintenance, the prospect of increasing digester gas production and offsetting energy demand is a major driver for FOG receiving. Digestion of FOG has played a part in allowing two of the surveyed facilities to produce enough energy onsite to become energy-neutral (or energy-exporting) plants. A combination of FOG receiving, additional external feedstock, anaerobic digestion, and cogeneration comprise the energy portfolio for those energy-neutral facilities. Many of the other plants with FOG receiving stations also reported increases in digester gas production that partially offset total energy costs. While these facilities reported some increase in digester gas production, the increases were sometimes lower than expected and gas production varied with the actual quantity and quality of FOG received.

In addition to the actual digester gas and energy production that is realized after implementation, tipping fees significantly impact system economics. Tipping fees are highly variable and can change due to market conditions. Receiving stations that are in areas with competing facilities may not receive expected volumes of FOG or associated revenue. Similarly, plants that charge relatively high tipping fees compared to other options for the local haulers may drive the market elsewhere. However, those facilities that are located in areas without major competition can successfully charge market-rate tipping fees and realize additional revenue to improve their overall bottom line. Appropriate tipping fees must balance market conditions with the operating costs and revenue goals of the wastewater treatment plant. System economics are heavily dependent on both the cost savings from energy production and revenue generated from tipping fees. Hence, conservative planning based on a potential range of FOG quantities, tipping fees, and projected gas production can help facilities make well-informed decisions.

System economics are also impacted by initial capital costs and the ongoing cost for O&M. Economic analysis should consider not only the FOG receiving station but also the facility's digestion system. Implementation of FOG receiving at facilities with excess digestion capacity is more likely to be economically feasible than for facilities that require construction of new digesters. Adding new digestion volume would significantly impact capital costs. When analyzing the capacity of the digestion system to handle FOG, the ancillary systems associated with the digesters should also be considered in terms of capacity and suitability. The digester heating, gas, and mixing systems would all be impacted by the addition of FOG. Digester gas piping, utilization, and flaring systems should have sufficient capacity to handle projected gas production. Heat demand will increase with any additional digester feed, and the mixing system should be capable of entraining FOG into the digester contents rather than allowing it to float and form a mat within the tank. If any of these systems require modifications to accommodate FOG, the associated capital and O&M costs should be included in the system economics.

Receiving stations can vary in size, complexity, redundancy, level of automation, equipment, and materials of construction. All of these factors affect both the capital cost and staffing requirements. Construction costs for most of the FOG receiving stations included in the survey exceeded $2 million and staffing requirements ranged from one FTE at quarter-time to two FTEs full-time. When considering costs associated with staffing, all of the tasks required for FOG receiving should be accounted for, from initial permitting and contract development to system operations, sampling, monitoring, and maintenance. The overall cost of the system can then be compared to potential cost savings and revenue generation to determine system economics. If the initial economic analysis does not reflect the potential O&M costs for these types of systems, proper staffing may not be available, and system operation may suffer after the station is built and started up.

The quality and characteristics of FOG can vary greatly, and the material is often delivered with numerous contaminants such as rocks, grit, metal, and concrete debris. The abrasiveness, low pH, and chemical composition of this material have been reported as contributing factors in the damage of FOG system components. As a result, FOG receiving stations can be O&M-intensive and may require additional staffing to sustain successful operations. The mechanical components, the methods of handling FOG, and the operation of FOG receiving stations differ from one facility to another. In general, equipment life for FOG system components is significantly shorter than equipment life for the same components in sludge or similar applications. Though some materials may work well with one aspect of FOG, such as low pH, the same material may not be suitable for other characteristics like abrasion. The most suitable materials balance performance in abrasive and corrosive environments. However, various materials and equipment types all experience some level of wear and eventually require replacement. Regular inspections and preventative maintenance are necessary to identify maintenance needs early and sustain successful operations.

The chemical characteristics of FOG also impact system performance and digester gas production. If material is delivered with high water content relative to volatile organics content, it may simply add to the hydraulic load on the digester without providing a high-strength feedstock that increases digester gas production. Facilities that are able to monitor and control FOG quality are more likely to have consistent and stable digester performance. One way to control FOG quality is to contract with a single FOG hauler and require that hauler to provide material within specifications. This method can reduce variability in the FOG quality and quantity received, but it also limits the use of the receiving station. Facilities that prefer to maximize use of their receiving stations can permit multiple local haulers and revoke permits if delivered product causes operational problems. Having staff present during FOG delivery allows for initial material inspection and sample collection that can also improve the quality and consistency of received FOG.

While there was no "standard" way of controlling FOG feed to the digesters, all of the facilities monitored digester performance to avoid process upsets. Some FOG may have surfactants or other characteristics that exacerbate foaming within the digester. Slowly metering the material into the digester feed or sludge recirculation lines is a common practice that may minimize sudden impacts on the digester, reduce foaming, and allow for better FOG entrainment. Three of the surveyed facilities currently control FOG feed by maintaining a specific FOG:sludge ratio by volume (approximately 30:70 to 40:60). Some facilities also track FOG as a percent of total solids or volatile solids feed to the digestion system. However, a feed regime based on COD may ultimately be the best control parameter to optimize digester performance and track digester gas production. For better tracking of FOG quality and feed control, FOG samples should be regularly analyzed for pH, TS, VS, and COD. Better tracking of digester gas production with and without FOG would provide more comprehensive data that can be used to project impacts of FOG digestion on gas production.

Some facilities accept both FOG and other high-strength feedstocks like food processing wastes (FPW). In general, FOG has higher energy value than most FPW but these other waste products tend to be less contaminated and more consistent than FOG. If a facility is weighing FOG receiving against co-digestion of FPW, the potential digester gas production and associated O&M requirements for both products should be considered relative to facility goals. Food processing wastes also differ from FOG in carbon-to-nitrogen ratios. While FOG primarily consists of carbon-rich lipids, certain FPW contain high levels of nitrogen that can impact the liquid treatment train through ammonia in the returned dewatering filtrate stream. The relative impacts of FOG or FPW digestion on overall solids production and dewatering processes may also differ. These types of process impacts should be considered if a facility is considering digestion of FOG, FPW, or other organic feedstocks.

CONCLUSIONS

While FOG receiving and digestion can be maintenance intensive, it also represents a means to increase onsite production of renewable energy, reduce collection system overflows, and provide local disposal options for FOG haulers. Many lessons have been learned from operating facilities about how to plan, design, operate, and maintain a sustainable FOG receiving system. The necessary amount of O&M effort is dependent on system design and FOG characteristics, and the required staff, labor, and investments in equipment and materials must be seriously considered during feasibility assessment. The associated costs in time and material must be studied and weighed against the potential benefits in terms of digester gas production, tipping fees, and improvements for the collections system. Implementation of these lessons learned and improvements to current operating procedures will yield program success as more FOG receiving systems are considered and brought online.

Making a well-informed decision on FOG receiving requires:

- Careful definition of the facility's economic, environmental/sustainability, and operational goals. It is important for all stakeholders to achieve consensus on these goals and their relative weights/priorities.
- Determination of the quantity and quality of FOG potentially available in the local area.
- Understanding of the local FOG market, current FOG disposal options, and associated costs to FOG haulers. This is necessary to determine tipping fees that are competitive and sustainable.
- Conservative planning based on a range of FOG quantity, tipping fees, and projected gas production.
- Determination of existing FOG hauler quality control and what quality control measures the haulers would be willing to implement if required for acceptance at the FOG receiving station.
- Acceptance of the O&M intensity of FOG receiving systems and provision of required staffing.
- Incorporation of lessons learned and successful design features into FOG receiving system designs.

The Water Authority is currently surveying the local FOG haulers to determine FOG availability, quality, and current market conditions. The final decision regarding implementation of a FOG receiving station will be made collectively by all stakeholders, including the O&M personnel that will be required to sustain the system.

ACKNOWLEDGEMENTS

The project team thanks all of the facilities that participated in the survey.

Synergistic Effects of Codigesting Preprocessed Food Waste Slurry

Michael Keleman[1*], M. Kim[2], M.M.I. Chowdhury[3], George Nakhla[2,3]

[1] InSinkErator, Racine, Wisconsin.
[2] Chemical and Biochemical Engineering, University of Western Ontario, London, Ontario.
[3] Civil and Environmental Engineering, University of Western Ontario, London, Ontario.
[*] Email: michael.keleman@emerson.com.

ABSTRACT

Anaerobic digestion is the most widely used technology to produce biogas such as methane and hydrogen from the decomposition of organic compounds. The effectiveness of the process depends on the stability of the consecutive reactions i.e. hydrolysis, acidification, acetogenesis, and methanogenesis. The process is widely used in sludge treatment for stabilization and production of methane gas. Due to the recent demand of renewable energy, anaerobic digestion has also been used for treating biodegradable wastes; for instance, the organic fraction of municipal solids wastes, wastewater treatment biosolids, and various food and beverage wastes (Arsova, 2010). Particularly, anaerobic digestion of food wastes (FW) is also considered as one of the effective methods of waste management. However, digestion stability can be hampered when FW is used as single substrate due to the high variability of its composition depending on its source (Lacovidou et al., 2012). Thus, the use of FW as co-substrate for municipal sludge digestion has emerged to enhance sludge digestibility, and increase energy generation to facilitate the achievability of energy-neutral wastewater treatment.

KEYWORDS: Food wastes, Co-digestion, Anaerobic digestion, Primary sludge, Thickened waste activated sludge, COD/N ratio.

INTRODUCTION

As wastewater treatment plants endeavor to become water resource recovery facilities - Utilities of the Future, codigestion of sewage sludge with food waste to enhance methane production in anaerobic digesters is gaining interest. Centralized preprocessing of the organic fraction of municipal waste, especially commercial food waste, is one approach utilized to prepare slurry for delivery in bulk directly to anaerobic digesters. This avoids transportation through sewers which requires primary clarification to separate the solids for digestion and is therefore more efficient. Grind2Energy™ is a recent innovation that utilizes decentralized source separation and preprocessing of commercial food waste with a 10 HP grinder and minimal water to produce a slurry. However, Grind2Energy™ slurry is stored onsite within a large storage tank for up to 7 to

14 days before delivery to water resource recovery facilities. This preprocessing and storage facilitates biological activity where hydrolysis, fermentation and acidification creates an energy rich, readily biodegradable and virtually contaminant free slurry to enhance methane production. There is some research on quantifying optimal loading of codigestion feedstock to avoid process upset. However, to better understand the impacts of codigesting preprocessed Grind2Energy™ slurry, researchers at Western University conducted an experimental analysis to evaluate and model various loading rates on anaerobic digesters and quantify the resulting methane and solids production.

The beneficial effects of FW as co-substrate for sludge anaerobic digestion include improvement of methane yield and acceleration of methane production (Koch et al., 2016; Lacovidou et al., 2012). For instance, according to a study by Sosnowski et al. (2008), methane production per kg volatile solid destroyed was 0.32 L/kgVS for sewage sludge only, 0.23 L/kgVS for FW only, and 0.44 L/kgVS for a mixture of sewage sludge (75% dry solids) and FW (25% dry solids). The positive effect of FW on sludge digestion performance is related to the increased carbon to nitrogen (C/N) ratio and enhanced kinetics due to the addition of food wastes. In fact, the optimal C/N ratio for anaerobic digestion is 15:1-30:1, much greater than the 6:1 to 9:1 of wastewater sludges (Koch et al., 2016; Lacovidou et al., 2012). The average C/N ratio of 19 food waste samples reported in different studies is 16:1 (Kim et al., 2015), and thus co-treatment of FW and wastewater sludges could be beneficial to energy production and solids destruction. Moreover, the addition of easily degradable FW helps to overcome the limitation of slow hydrolysis, which is the rate limiting step in sludge digestion. A recent study by Koch et al. (2016) reported that methane production in a full scale WWTP digester was enhanced from 0.31 L/kgVS with raw sludge (C/N ratio of 8.8) to 0.39 L/kgVS with addition of FW with a C/N ratio of 17.7 at 10% (W/W), resulting in a C/N ratio of 9.7. Although the 9.7 was still lower than the aforementioned optimal C/N ratio (15:1-30:1), the substantially high hydrolysis rate of food wastes triggered enhancement of acidogenesis and methanogenesis (Koch et al., 2016).

Research was recently completed under the direction of Dr. George Nakhla of the University of Western Ontario to explore the codigestion of Grind2Energy food waste slurry (G2EFWS) and wastewater biosolids. G2EFWS samples used for this experiment were collected from systems in WI where food waste from supermarkets is ground with minimal water and retained in a large tank for up to two weeks prior to delivery to digesters. The objective of this study was to explore the long-term effect of food wastes on co-digestion with primary sludge (PS) and thickened waste activated sludge (TWAS). This study elucidates the co-digestion performance and explores its synergistic effects.

EXPERIMENTAL PLAN

Five reactors with a total volume of 2.4 L and a working volume of 1.8 L each were used as semi-continuous flow anaerobic digesters (Figure 1). Solids retention time (SRT) was controlled at 20 days through feeding and wasting at 90 mL daily. Reactors were completely mixed using a mixer (Stir-Pak, model SSM20). Initially, feeding and wasting for the system operation was done on continuous flow using peristaltic pumps (Masterflex L/S Tubing Pumps, Model 77200-62); however, due to unstable feeding and wasting associated with clogging, the feed pattern was switched to semi-continuous flow through manually feeding and wasting using a syringe at the same time each day with wastage performed prior to feeding.

RESULTS

Codigestion performance of five different digesters fed with different concentrations of food wastes achieved COD removal efficiencies of 47% (sludge only, COD/N ratio of 21), 52% (sludge only, COD/N ratio of 27), 61% (10% FW, COD/N ratio of 23), 64% (20% FW, COD/N ratio of 25), and 69% (40% FW, COD/N ratio of 27). Increasing the COD/N ratio positively affected digestion performance reflected by COD removals from reactors treating sludge only. In addition, co-digestion also enhanced digester performance compared with the performance of sludge only (52%, COD/N ratio 27) possibly due to the positive effect of biodegradability of food wastes on anaerobic digestion. The first-order COD removal rates in the FW digesters ranged from 0.075 d^{-1} to 0.11 d^{-1}, more than double the control of 0.042 d^{-1}. Methane production per g influent COD were 1.3-1.8 fold higher in co-digestion than digestion with sludge alone, reflecting the synergistic effect of food waste digestion on energy generation. Additional methane production through synergistic effects accounted for a minimum of 18%-20% of the overall methane production in the co-digestion reactors. Food wastes COD removal efficiencies were estimated as 85%-96% while FW VSS destruction efficiencies ranged from 55% to 70%.

SUMMARY AND CONCLUSIONS

The experimental data showed that the digested sludges TSS concentrations in the FW digesters were not different than the control, implying that sludge disposal costs will not be significantly impacted. The most significant finding of this study demonstrated that the increased loadings of G2EFWS do not deteriorate digester performance, but on the contrary improve the digestibility of primary and WAS solids. This clearly dispels a common myth in the industry that the addition of FW does in fact decrease the available capacity for sludges. Applying the analysis results to a typical digester system provide important implications of the findings of this study. For a wastewater treatment plant utilizing anaerobic digestion with a population equivalent of

100,000 emphasizes the beneficial impact of FW co-digestion on methane production and digestion kinetics.

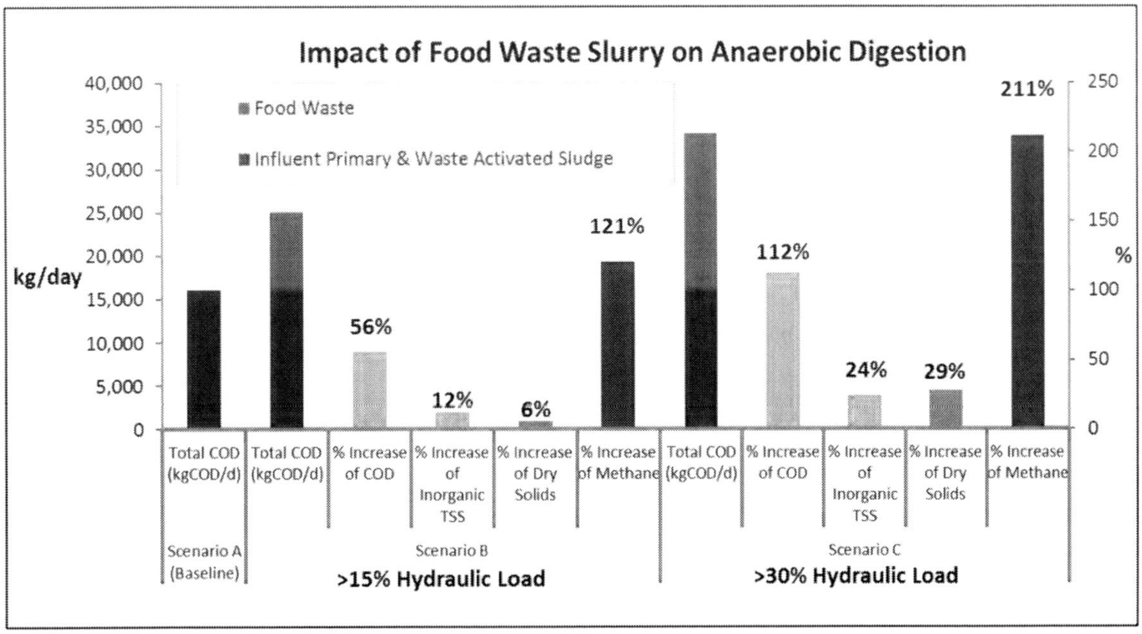

ACKNOWLEDGEMENTS

The authors gratefully acknowledge Emerson Electric Corp. for funding this project and the University of Western University for conducting the research.

REFERENCES

Kim, M., et al. (2016) Synergism of co-digestion of food wastes with municipal wastewater treatment biosolids. *Waste Management.*

CLEVELAND EXPERIENCE
THREE NEW FLUID BED MUNICIPAL SLUDGE INCINERATORS AND A STEAM TURBINE TO GENERATE POWER

D. Hancock, CEM, Process Manager, Southerly Wastewater Treatment Center
T. Shively, PG, Construction Manager, Southerly Wastewater Treatment Center
T. Vasel, P.E., Plant Construction Manager, Southerly Wastewater Treatment Center
K. Dangtran, Ph.D., Vice President, Biosolids Group, SUEZ
*L. Takmaz, Ph.D., Product Manager, Incineration, SUEZ

ABSTRACT

Southerly Wastewater Treatment Center is located at 6000 Canal Road, Cuyahoga Heights, Ohio and is owned and operated by the North East Ohio Regional Sewer District (NEORSD). The plant serves more than 600,000 people living in the area and has an average wastewater treatment capacity of 473 million liter per day (125 million gallon per day, mgd). Maximum wastewater treatment capacity is 1.5 billion liter per day (400 mgd). The plant has another 1.3 billion liter per day (335 mgd) additional capacity for the primary treatment of storm water.

Southerly plant has three new fluid bed incinerators (one stand by) with three heat recovery systems producing superheated steam to generate power through a steam turbine, which is designed to handle two incinerators due to one incinerator being on stand-by during normal operation. Turbine is designed to generate 2.4 MW when two incinerators are in operation with 11,975 kg/hr (26,400 lb/hr) steam flow. Maximum power generation rate for the turbine is 3.54 MW, when three incinerators are in operation having 18,144 kg/hr (40,000 lb/hr) steam flow. All three fluid bed units are of type hot wind-box with refractory arch air distributor and refractory lined windbox. Each fluid bed unit has the capacity to incinerate 90.72 metric dry ton per day (100 US dry ton per day) (DTPD) total solids.

Unit No. 2 passed the performance testing on Sept. 3, 2014. Unit No. 1 passed the performance testing on Sept. 23, 2014. Unit No. 3 passed the performance testing on Oct. 7th, 2014. All three fluid bed units have satisfied the emission requirements listed in the air permit and project specifications. All three units have been retrofitted in 2015 with the mercury removal system to be in compliance with US EPA MACT Quad M emission limits. Stack emission tests in 2015 and 2016 indicated that all three retrofitted units were in compliance with MACT limits for an existing fluid bed incinerator.

Each fluid bed unit is equipped with a waste-heat recovery system employing a primary heat exchanger and a waste heat boiler. Primary heat exchanger (shell and tube design) is used to preheat the fluidization air to approximately 584 C (1,083 F) minimizing the auxiliary fuel usage during the steady state operation. Each unit was designed to be autogenous at 68% volatile, 28% total solids, and 5,833 kcal/kg (10,500 BTU/lb) sludge heat value based on volatile part of dry solids including skimmings (processed grease and fats) introduced into the sludge cake. Flue gas from the fluid bed reactor is passed through a primary heat exchanger and then discharged into a waste heat boiler to produce 6,895 kg/hr (15,200 lb/hr) superheated steam at 371 C (700 F), 44 barg – 4,400 kpa (635 psig). From waste heat boiler, flue gas is sent to a plume suppression heat exchanger and then discharged into a wet scrubber to remove particulate and acid gases (SO_2, HCl). Wet scrubber is not equipped with a caustic injection system. Clean flue gas discharged from the wet scrubber flows through the mercury removal system and an ID Fan before being discharged into the plume suppression heat exchanger (shell and tube design). ID Fan is installed between the mercury removal system and plume suppression heat exchanger to maintain vacuum conditions inside the waste heat boiler to prevent any potential flue gas/ash leak into the building. Plume suppression heat exchanger is used to suppress the visible plume in stack by increasing the clean flue gas temperature to 121C (250F).

The text provides the detailed information on the process, the scope of equipment installed and the information regarding the performance test results.

Southerly fluid bed units have been successfully started up, commissioned, and passed the performance testing in 2014, 2015, and 2016, respectively. All three units have met the emission requirements listed in the air permit, project specifications and MACT limits. The successful operation at Southerly plant has shown that the improved thermal oxidizer design incorporating a power generation system can reduce the operational expenditures significantly and is also an economical, environmentally friendly, and cost effective solution for sludge disposal.

KEYWORDS

Fluidized bed incineration, sludge disposal, wet scrubber, power generation, MACT, mercury

* Corresponding Author

BACKGROUND

Southerly Wastewater Treatment Center (SWWTC) is located outside Cleveland (OH) and serving the greater Cleveland area. It was during the 1970s, the plant decided on installing sludge thermal conditioning (Zimpro) system requiring large amount of steam. Plant had four multiple hearth incinerators each equipped with a waste heat recovery boiler to supply the steam required by the Zimpro system. The NEORSD later decided to replace the four multiple hearth sewage sludge incinerators with three new fluid bed units each having 100 DTPD capacity. In addition, it was decided to decommission the existing Zimpro system and to install a new chemical conditioning system (Reference 2) and nine new high solids dewatering centrifuges. It was estimated that having a fluid-bed incineration system with waste-heat recovery would result in savings of $1 million dollar per year compared to the existing system by reducing the natural gas consumption by 98%. Initial construction cost was estimated to be in the range of $90 to $120 million. It was also estimated that selling electricity at $0.03/kwh would have a pay back of 20 to 25 years. In 2008, a validation workshop took place at SWWTC from August 11 to 15th. The validation workgroup recommended that NEORSD should proceed with the three new fluid bed incinerators with waste heat recovery. It was projected that two fluid-bed units would generate about 2.6 - 3 MW of electricity. Including the steam system with power-generation capability added $22 million to the project. The payback period was estimated to be 11 years when using the generated power on site, instead of selling it to a utility. A $0.07/kWh savings was projected based on using the power generated on-site through a steam turbine, instead of purchasing the power from a utility.

SYSTEM DESCRIPTION

Figure 1 shows the process flow diagram for each unit, which comprises the NEORSD's Renewable Energy Facility (REF). Each REF unit has a reactor with a hot windbox, refractory-arch dome, and a teardrop shaped freeboard (Figure 2). Reactor diameter and height are 9.86 m (32.33 ft) and 12.20 m (40 ft), respectively. Four sludge ports are used to feed the dewatered sludge into the sand bed. Reactor is designed to have at least 7 seconds residency time to have a complete combustion resulting in lower emissions. Primary heat exchanger (shell and tube design) is installed downstream of the reactor to preheat the fluidization air up to 584C (1,083F) to have autogenous combustion. Further waste heat recovery is accomplished by installing a water tube waste heat boiler to generate superheated steam at 371 C (700 F) and 44 barg (635 psig). Steam generation rate per train is 6,895 kg/hr (15,200 lb/hr). Following the waste heat boiler, a plume suppression heat exchanger (shell and tube design) is installed to eliminate visible plume in stack. After the plume suppression heat exchanger, a wet scrubber and tray cooler is installed to remove particulate and acid gas from the flue gas. Mercury removal modules are installed after the scrubber to remove mercury from the flue gas. ID fan is installed after the mercury removal modules to maintain negative pressure inside the waste heat boiler. Steam from two trains is combined into a single stream and sent to a steam turbine (single stage, single valve, condensing type) to generate power (Figure 3). Boiler ash is removed through a dry ash system and sent to an ash sump (Figure 4 and 5). Scrubber wet ash slurry from each train is also sent to the same ash sump, where ash slurry from all three trains is mixed together and sent to an ash lagoon through the ash pumps. Ash sump has its own mixing system to ensure ash particles are not settling at the bottom of the sump resulting in non-uniform consistency in the sump and plugging of the ash lines.

Steam system is designed to recycle the condensate discharged from the steam turbine. After passing through the main condenser, condensate is sent back to the condensate tank. Cooling water for the main condenser is supplied by the plate and frame heat exchanger. On the hot side of plate and frame heat exchanger (Figure 6), condensate from main condenser is used as hot medium. On the cold side, scrubber water is used as cold medium. The steam recycling system also utilizes an auxiliary steam condenser.

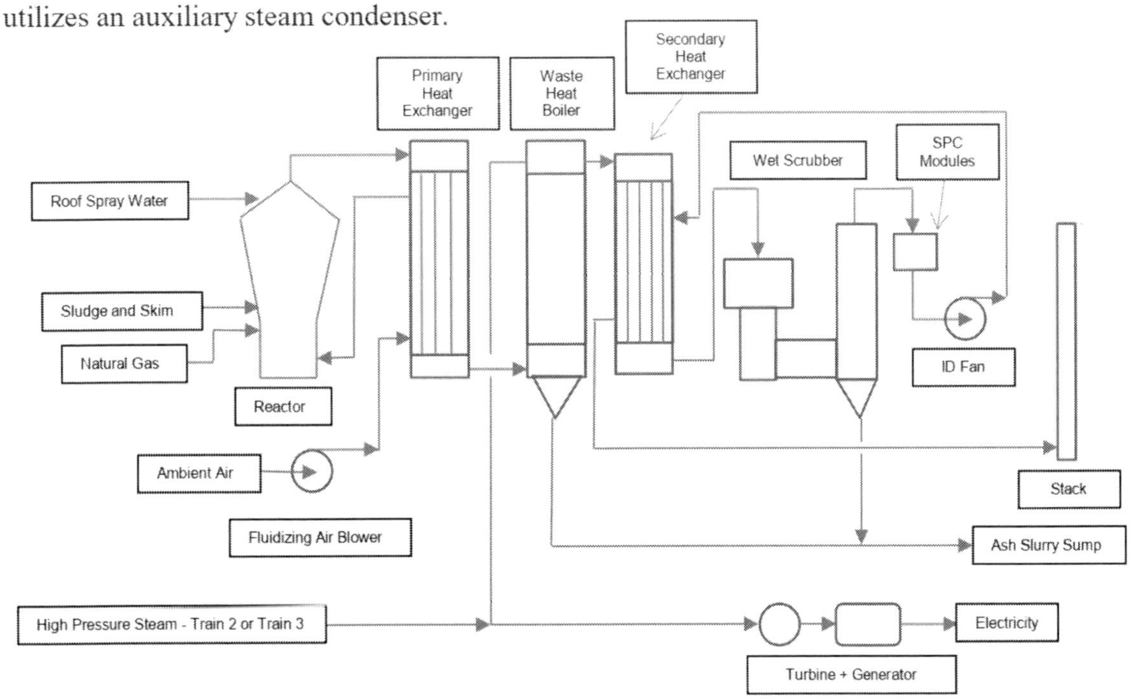

Figure 1: REF Process Flow Diagram.

Steam turbine is coupled with a generator to produce electricity. When the steam turbine is down or tripped, steam from waste heat boiler is sent to the auxiliary condenser through a desuperheater and then recycled through the steam system before being sent back to the waste heat boilers as feed water. In addition, the steam turbine has a bypass line to be used during turbine downtime. The turbine is designed to generate 2.4 MW when two incinerators are in operation with 11,975 kg/hr (26,400 lb/hr) steam flow. The maximum power-generation rate for the turbine is 3.54 MW, when three incinerators are in operation having 18,144 kg/hr (40,000 lb/hr) steam flow.

COMMISSIONING AND PERFORMANCE TESTING

The REF fluid-bed incinerators were started-up, commissioned, and underwent performance testing in 2014. Unit No. 2 passed the performance testing on Sept. 3, 2014. Unit No. 1 passed the performance testing on Sept. 23, 2014. Unit No. 3 passed the performance testing on Oct. 7, 2014. All three fluid-bed units have satisfied the emission requirements listed in the air permit and project specifications.

Figure 2: REF fluid bed reactor.

Figure 3: REF steam turbine.

Figure 4: REF waste heat boiler dry ash removal system.

Figure 5: REF waste heat boiler dry ash transporter to ash sump.

Figure 6: REF steam system plate and frame heat exchanger.

During the performance testing, sludge, ash, and water samples were collected and analyzed. Air Compliance Testing, Inc. based in Cleveland, Ohio was contracted to perform the stack testing on all three units. Table 1 shows the stack emission-test results on Unit No. 2 showing compliance with project specifications and air permit. Results of stack testing on Unit No. 1 and No. 3 are presented in Table 2 and 3 respectively.

Table 1: Unit No. 2 stack emission test results on Sept. 3, 2014.

Parameter	Limit	Test Results (Unit No. 2)
PM TOTAL	0.18 kg (0.4 lb)/dry ton solids burned	0.08 kg (0.17 lb)/dry ton solids burned
PM 10	0.18 kg (0.4 lb)/dry ton solids burned	0.06 kg (0.14 lb)/dry ton solids burned
CO	0.73 kg (1.6 lb)/dry ton solids burned	0.13 kg (0.28 lb)/dry ton solids burned
NOx	1.7 kg (3.75 lb)/dry ton solids burned	0.44 kg (0.97 lb)/dry ton solids burned
VOC	0.32 kg (0.7 lb)/dry ton solids burned	0.009 kg (0.02 lb)/dry ton solids burned
SO2	0.73 kg (1.6 lb)/dry ton solids burned	0.44 kg (0.973 lb)/dry ton solids burned
Mercury	1.5 kg (3.3 lb) per day	0.028 kg (0.0624 lb) per day
Beryllium	0.003 kg (0.007 lb) per day	0.00007 kg (0.00015 lb) per day
Lead	0.002 kg (0.005 lb) per day	0.00045 kg (0.000984 lb) per day
Visible Emission	< 20 % Opacity	none

Table 2: Unit No. 1 stack emission test results on Sept. 23, 2014.

Parameter	Limit	Test Results (Unit No. 1)
PM TOTAL	0.18 kg (0.4 lb)/dry ton solids burned	0.05 kg (0.11 lb)/dry ton solids burned
PM 10	0.18 kg (0.4 lb)/dry ton solids burned	0.04 kg (0.09 lb)/dry ton solids burned
CO	0.73 kg (1.6 lb)/dry ton solids burned	0.01 kg (0.023 lb)/dry ton solids burned
NO_X	1.7 kg (3.75 lb)/dry ton solids burned	0.7 kg (1.54 lb)/dry ton solids burned
VOC	0.32 kg (0.7 lb)/dry ton solids burned	0.01 kg (0.02 lb)/dry ton solids burned
SO2	0.73 kg (1.6 lb)/dry ton solids burned	0.35 kg (0.769 lb)/dry ton solids burned
Mercury	1.5 kg (3.3 lb) per day	0.025 kg (0.0552 lb) per day
Beryllium	0.003 kg (0.007 lb) per day	0.00005 kg (0.0001 lb) per day
Lead	0.002 kg (0.005 lb) per day	0.00068 kg (0.0015 lb) per day
Visible Emission	< 20 % Opacity	none

Table 3: Unit No. 3 stack emission test results on Oct. 7, 2014.

Parameter	Limit	Test Results (Unit No. 3)
PM TOTAL	0.18 kg (0.4 lb)/dry ton solids burned	0.054 kg (0.12 lb)/dry ton solids burned
PM 10	0.18 kg (0.4 lb)/dry ton solids burned	0.042 kg (0.093 lb)/dry ton solids burned
CO	0.73 kg (1.6 lb)/dry ton solids burned	0.017 kg (0.037 lb)/dry ton solids burned
NOx	1.7 kg (3.75 lb)/dry ton solids burned	0.463 kg (1.02 lb)/dry ton solids burned
VOC	0.32 kg (0.7 lb)/dry ton solids burned	0.032 kg (0.07 lb)/dry ton solids burned
SO2	0.73 kg (1.6 lb)/dry ton solids burned	0.207 kg (0.456 lb)/dry ton solids burned
Mercury	1.5 kg (3.3 lb) per day	0.0207 kg (0.0456 lb) per day
Beryllium	0.003 kg (0.007 lb) per day	0.00008 kg (0.00018 lb) per day
Lead	0.002 kg (0.005 lb) per day	0.000337 kg (0.000744 lb) per day
Visible Emission	< 20 % Opacity	none

MACT RETROFIT & COMPLIANCE

After passing through the performance testing, all three units have been retrofitted with the mercury removal system supplied by EnviroCare International in 2015. Figure 7 shows the mercury removal system installed on one of the units. The mercury removal system includes patented Gore modules installed in a single stainless steel vessel. Table 4 shows the results of stack testing on all three units. Unit No. 2 was the first incinerator tested for MACT on Dec. 1, 2015. Units. 1 and 2 were tested on Jan. 20 and Jan. 21, 2016, respectively. All three units have been in compliance with the MACT emission limits for an existing incineration system.

Table 4: Stack test results on all three incinerators.

		MACT	No. 1	No. 2	No. 3
Cd	mg/dscm	0.0016	<0.00017	<0.00025	<0.00030
CDD/CDF, TMB	ng/dscm	1.2	-	-	-
CDD/CDF, TEQ	ng/dscm	0.1	0.0043	N/A	0.0177
CO	ppmvd	64	<1.3	<2.26	<1.2
HCl	ppmvd	0.51	<0.1	<0.18	<0.091
Hg	mg/dscm	0.037	<0.014	<0.0028	<0.0065
NOx	ppmvd	150	80.2	102.4	60.6
Opacity	%	0	0	0	0
Pb	mg/dscm	0.0074	<0.00029	0.00076	<0.00033
PM	mg/dscm	18	1.104	<1.03	0.83
SO2	ppmvd	15	8.1	<5.92	8.5

Figure 7: Mercury removal system installed on one of the REF incinerators.

MERCURY REMOVAL TECHNOLOGY

SPC Gore technology, marketed by EnviroCare International, was developed for use by the power industry for mercury removal as part of the air-pollution control system. This patented technology has been used as part of the air-pollution control system installed on municipal sewage-sludge fluid-bed incinerators very recently. It is a new, patented technology and is a very simple cost effective way of reducing the amount of mercury in flue gas to meet the MACT emission limit (MMMM). SPC Vessels can be installed inside or outside the building. Currently, SUEZ has seven incineration systems in the US using the SPC Gore modules to remove mercury from flue gas. All seven incinerators have passed the stack emission testing by meeting the MACT mercury limit in stack (0.037 mg/dscm).

CONCLUSION

Overall, the NEORSD's REF fluid bed incinerators have been operating and generating power since the commissioning in 2014. All three units have met the stack emission limits at the design burning capacity. It is a renewable energy facility and carbon-neutral. All three incinerators have been successfully retrofitted with the mercury removal system in 2015 supplied by EnviroCare International based in California. Stack test results in 2015 and 2016 showed that all three units have been in compliance with the MACT emission limits for an existing fluid-bed incineration plant. After the retrofit, the REF incinerators are currently serving the community in greater Cleveland area.

REFERENCES

NEORSD Environmental News (2008), *"Savings on all side: Plan for new incinerators to burn wastewater Biosolids is environmentally friendly, economically viable, and now approved by State and experts"*, (www.neorsd.org/environotes).

Robert P. Dominak (2009), *"How the NEORSD's Biosolids Program Got Green"*, (http://www.wefnet.org/nbp/January%202010/NBP_Bob%20Dominak_012710.pdf)

Detailed Market Analysis Helps Focus Capital Decisions for Orange County Sanitation District

Natalie Sierra (Brown and Caldwell), Ron Alexander (RAA), Steve Wilson (Brown and Caldwell), Deirdre Bingman (OCSD), Tom Meregillano (OCSD), Sharon Yin (OCSD), Jeff Mohr (OCSD), and Daniel Buhrmaster (Black and Veatch)

The Orange County Sanitation District (OCSD) is in the process of conducting a Biosolids Master Plan (BMP), which will result in capital and programmatic decisions to shape OCSD's biosolids program in the decades to come. A unique aspect of this master planning effort was the fact that the planning process began with an analysis of biosolids markets. This analysis included an extensive market research to evaluated the market potential for biosolids products generated by OCSD, either at its wastewater treatment plants (WWTPs) or off-site at a merchant facility, thus supporting OCSD's biosolids management program over the long term. The market-based work helped select biosolids products with market potential to support OCSD's Biosolids Policy, most notably a commitment to a diverse portfolio of off-site management options that support the recycling of biosolids. Thus, capital improvements made would result in a sustainable biosolids end use program.

The market analysis considered base products generated at OCSD's facilities (biosolids cake products), as well as further processing steps used to generate new products. Such products included soil blends, compost, and thermally dried product. The range of products was then paired with feasible markets, which were then evaluated individually. OCSD's depth of experience in beneficial biosolids use placed a special emphasis on analyzing whether any product generated had a true, reliable market within the larger region encompassing Southern California and Western Arizona.

Market research on the regional horticultural (lawn/garden), agricultural and fertilizer industries was completed to better define the available markets for the potential biosolids products and processing options, and to determine the products' value and marketability.. Demographic data indicate that significant regional capacity exists within these markets. A review of emerging markets for environmentally beneficial applications (e.g., dewatered cake for land reclamation, compost for erosion control, etc.) was also conducted, although these markets are considered to be less developed in the region with respect to biosolids products.

Generally speaking, experience with biosolids products in the agricultural and horticultural markets is strong in Southern California. California agriculture, a dominant industry in the region, can absorb large volumes of fertilizer-type products, whether thermally dried biosolids or compost, and use of biosolids in this sector is expected to continue. Interest in use of biosolids for land reclamation can further expand available acreage. Continued marketing and outreach by either OCSD or its contractors will help support ongoing regional use of biosolids products.

The market analysis was then fed into a comprehensive evaluation of both products and markets. Two sets of evaluation criteria were developed—one market-based set for evaluation of the product-market pairs and a second set for evaluation of the products (and associated processing technologies). Each criterion was assigned a weighted value by the team. The product-market pairs were evaluated based on the market research conducted and the joint consultant-OCSD team evaluated the products and their associated processes. The resulting scores were presented and refined in a workshop held with OCSD. Table 1 below presents the final set of products and markets selected at this workshop.

This market-based evaluation resulted in recommendations for products and markets to be carried forward into an analysis of on-site and off-site technologies that were screened to select the on-site solids handling process. The alternatives recommended from the market analysis were paired with specific thickening and stabilization technologies for the creation of end-to-end alternatives. End-to-end alternatives were then screened using net present value analysis and triple-bottom line criteria for the creation of a recommended program. On and off-site management options will be summarized in a biosolids management plan, to be completed in February 2017, which will provide a roadmap to get from today's program to the selected program for the future.

Table 1. Final Products and Markets Screened for OCSD's Biosolids Master Plan

Product	Highest Scoring Markets
Class A Compost	Soil Blending
	Bulk Agriculture, CA
	Bulk Agriculture, AZ
	Bulk Horticulture/Landscaping
	Distribution as Bagged Product
	Golf Course and Other Specialty
Class B Cake	Bulk Agriculture, AZ
Class A Cake	Bulk Agriculture, CA
	Bulk Agriculture, AZ
	Land Reclamation
Class A Thermal Hydrolysis Product (THP) Soil Blend	Bulk Horticulture/Landscaping
Class A THP Cake	Bulk Agriculture, CA
	Soil Blending
	Land Reclamation
Class A Soil Blend	Bulk Horticulture/Landscaping
Class A High Quality Granule	Bulk Agriculture, CA
	Bulk Agriculture, AZ
	Bulk Horticulture/Landscaping
	Distribution as a Bagged Product
	Fertilizer Blending
Partially Dried Class B Cake	Bulk Agriculture, AZ
Class A Partially Dried Cake	Bulk Agriculture, AZ
	Bulk Agriculture, CA
Class A THP Partially Dried Product	Bulk Agriculture, CA
	Bulk Agriculture, AZ
	Soil Blending
	Bulk Horticulture/Landscaping
	Land Reclamation

AUTHOR INDEX

AUTHOR INDEX

AUTHOR INDEX

Water Environment Federation (WEF)
601 Wythe Street
Alexandria, Virginia 22314
USA

ISBN 978-1-5108-6297-5

CURRAN ASSOCIATES INC.
proceedings
.com